# ECOLOGY AND FIELD BIOLOGY

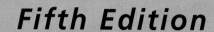

*Fifth Edition*

# ECOLOGY AND FIELD BIOLOGY

Robert Leo Smith
*West Virginia University*

HarperCollinsCollegePublishers

Acquisitions Editor: Liz Covello
Developmental Editor: Vicki Cohen
Project Editor: Brigitte Pelner
Design Supervisor: Mary Archondes
Text and Cover Designer: Mary Archondes
Cover Illustration/Photo: Akaka Falls, Hawaii. G. Frad Lewis. © tony stone worldwide
Art Studio: Robert L. Smith, Jr.
Photo Researcher: Michelle Ryan
Electronic Production Manager: Valerie A. Sawyer
Manufacturing Manager: Helene G. Landers
Electronic Page Makeup: Sarah Johnson
Printer and Binder: RR Donnelley & Sons Company
Cover Printer: The Lehigh Press, Inc.

For permission to use copyrighted material, grateful acknowledgment is made to the copyright holders on pp. A-1 to A-3, which are hereby made part of this copyright page.

Ecology and Field Biology, Fifth Edition

Library of Congress Cataloging-in-Publication Data

Smith, Robert Leo
    Ecology and field biology / Robert Leo Smith. – 5th ed.
        p.    cm.
    Includes bibliographical references and index.
    ISBN 0-06-500976-2
    1. Ecology.   2. Biology–Fieldwork.   I. Title.
QH541.S6   1995                                         95-34933
574.5–dc20                                              CIP


96 97 98 9 8 7 6 5 4 3 2

To the $F_2$ Generation II

Kevin, Meghan, and Jonathan

# Brief Contents

# Detailed Contents

# Preface

Prefaces are written to be read, although I suspect that more readers flip past them than read them. Personally, I read the preface first, and perhaps several times throughout the course of using the text. In some books the preface is the most absorbing part, which may say a lot for the preface or very little for the book. In technical and text books, the preface is the only place where authors give us any real reason for writing the book, how they think it should be used, and what they hope readers will gain from it.

The first edition of *Ecology and Field Biology* appeared in 1966. The objectives of this fifth edition remain the same as those of the first and subsequent editions, which are to present a balanced introduction to ecology—plant and animal, theoretical and applied, physiological and behavioral, and population and ecosystem. A text in general ecology should provide an overview of the field and present the basic ideas and concepts with which students should be familiar. Previous editions of *Ecology and Field Biology* have demonstrated this, and the fifth edition continues this necessary vehicle for learning the fundamentals of ecology.

Another objective is to provide a readable, flexible text around which instructors can develop their own courses according to their time available. Since many curricula restrict ecology to a one-semester or even a quarter course, it is difficult, if not impossible, to adequately cover general ecology in that amount of time. Although the chapters logically build upon one another, instructors can cover chapters out of sequence and their students will still have a thorough understanding of ecology.

## REVISIONS IN THE TEXT

Since the first edition appeared over 30 years ago, ecology has experienced major growth and development. Revisions of the text have reflected those changes and with each thorough update, I have retained the most important and fundamental elements of ecology and incorporated new developments and approaches.

The revision has resulted in the following shifts in the presentation of topics:
- Chapter 2 on experimental ecology has been extensively revised.
- The two previously separate chapters on soil have been incorporated into one.
- Comparative ecosystem ecology now follows Part 2 on ecosystem ecology, providing easier access for students and instructors to relate general discussions on ecosystem function, such as nutrient cycling, to specific ecosystem and ecophysiological adaptations.
- Integration of various topics such as acid deposition and global change into appropriate chapters rather than segregation into separate peripheral chapters. For example, effects of air pollution and acid rain are included in the discussion in Chapter 12 on biogeochemical cycling. Topics relative to conservation biology such as on genetics of small populations and viable populations are covered in Chapter 20; extinction is discussed in Chapter 22; human exploitation of natural populations is found in Chapter 25; and habitat fragmentation and edge conditions are covered in Chapter 28.
- The text points out the relevance of ecological concepts to ecological and environmental issues and those areas in which research is most active or needed.

## ORGANIZATION

The text is divided into six parts. Part 1 is the introduction. Chapter 1 provides a brief overview of ecology—what it is, how it developed, and its importance and relevance to current environmental issues. Chapter 2 presents ecology as an experimental science. Although students may have some grasp of the scientific method

when they take their first ecology course, it is from the context of laboratory experimentation and not a field approach. This chapter introduces students to hypothesis testing, the use and place of predictive models in ecology, their advantages and weaknesses, and the difficulty ecologists face when asked to provide highly predictive or firm answers to current environmental issues.

Part 2 emphasizes the physical environment and adaptations of organisms to it. A major emphasis in this section is physiological ecology. Chapter 3 introduces the concepts of adaptation and homeostasis which are basic to an appreciation of both physiological ecology and natural selection. Chapter 4 discusses climate, which greatly influences the environmental conditions under which organisms live. The chapter ends with an overview of how climate influences the distribution of vegetation. Chapters 5, 6, and 7 cover the three major abiotic influences on organisms: moisture, temperature, and light, mostly from an ecophysiological point of view. Chapter 8 considers the importance of nutrients to plants and animals, the sources, and small-scale cycling of those nutrients. Chapter 9 explores soil, its development, and its role as a major storehouse of both nutrients and moisture.

Part 3 deals with basic ecosystem ecology. Chapter 10 introduces the concept of the ecosystem and discusses the two major processes in ecosystem function: photosynthesis and decomposition. Chapter 11 explores energy flow in ecosystems including primary production and its estimation, secondary production, and food chains and food webs. Chapter 12 is devoted to biogeochemical cycles. The effects of excessive nutrient input or pollution, including acid deposition, are incorporated in this chapter.

Part 4 on comparative ecosystem ecology has been a hallmark of this text. Chapters 13–16 provide a comparative study of the major ecosystems relative to the structure and function, including differences in structure, energy flow, nutrient cycling, and adaptations of organisms to those environments. The common thread through these chapters is the nitrogen cycle, which serves to emphasize the functional differences among the various major ecosystems.

The material in Part 4 not only introduces concepts unique to certain ecosystems, such as nutrient spiraling in stream and microbial loops in marine ecosystems, but also provides students with the background information they need to understand the major problems receiving so much attention today in ecology and environmental conservation—the destruction of old growth forests, tropical deforestation, wetland values and losses, habitat fragmentation, and acidification.

All of these chapters contain a good deal of ecophysiology and evolutionary adaptations as well as ecosystem structure and function. This coverage provides the instructor with a source of additional material that can be incorporated with topics in other chapters. Chapter 13 provides an overview of grasslands, tropical savannas, shrublands, deserts, and tundra. Chapter 14 is devoted to various forest ecosystems from the taiga to the tropics. Chapter 15 provides a survey of lentic and lotic ecosystems, with an emphasis on the differences in the pathways of energy flow, and energy flow between the two types of ecosystems and nutrient cycling in flowing water. Also discussed are the delineation of wetlands and the effects of dams on lotic systems. Chapter 16 deals with the marine environment from the intertidal zones to the open sea. It includes discussions of hydrothermal vents, role of planktonic microbes in marine food webs, mangrove tidal swamps, and nutrient cycling in coral reef ecosystems.

Part 5 deals with population ecology and is one of the most heavily revised sections of the text. Chapter 17 covers demography and considers both unitary and modular populations. Chapter 18 considers population growth and density-dependent and density-independent influences, with examples from both plant and animal populations. Chapters 19 and 20 deal with behavioral ecology. Chapter 19 expands on intraspecific competition introduced in Chapter 18. It covers dispersal and the part of social behavior involving social dominance and territoriality. Chapter 20 is concerned with mating systems, sexual selection, reproductive effort, and parental care. Chapter 21 considers population genetics with an emphasis on the genetics of small and fragmented populations, effective population size, and viable populations, topics associated with conservation biology.

Interspecific relations make up Chapters 22–27. Chapter 22 looks at interspecific competition. Chapter 23 is an overview of predation theory, including optimal foraging theory, which leads to discussions of predator-prey systems in Chapters 24 and 25. Chapter 24 discusses plant-herbivore interactions, including plant defenses and herbivore responses. Chapter 25 looks at carnivorous predators and their prey, including intraspecific predation, or cannibalism, and intraguild predation. This chapter includes a discussion on predator-prey cycles and human exploitation of prey populations. Chapter 26 explores the world of parasitism and disease, as well as social parasitism, including their population dynamics and evolutionary responses. Chapter

27 ends Part 5 with a consideration of the coevolution of the various facets of mutualism, its possible origins, and effects on interacting populations.

Part 6 on the community covers much of what is considered community ecology, although no distinct line can be drawn between community ecology and population ecology. Chapter 27 examines community structure, both physical and biological, island biogeography, and the roles of predation, competition, parasites and diseases, and mutualism on community structure. Special emphasis is given to horizontal heterogeneity, edges, ecotones, and habitat fragmentation. These topics are basic to landscape ecology and relate back to the material covered in Chapter 21 on the effects of habitat fragmentation and isolation on the genetics of small populations.

Chapters 29 and 30 consider community change. Chapter 29 examines the scale and frequency of disturbance and its role in community dynamics, structure, and development. Chapter 30 covers succession, strongly influenced by disturbance. This chapter includes a detailed discussion of the current theories of succession and discusses mechanistic models of the phenomenon. The chapter concludes with a discussion of past communities from the Pleistocene era on.

## TOPICS NEW TO THIS EDITION

New topics in this edition include:
- Importance of statistical power tests in ecological studies
- Effects of impoundments on river ecology
- Mechanistic models of succession
- Delineation of wetlands
- Global change
- Nitrogen overfertilization and forest decline
- Microbial plankton in marine food webs
- Sunspot influence on snowshoe hare cycles
- Intraguild predation
- Parasitism and sexual selection

## PEDAGOGY

The fifth edition features the following pedagogy:
- Chapter-opening list of concepts
- Topical outline of each chapter
- End-of-chapter summary
- Review questions that include several critical-thinking questions
- Key terms boldfaced upon first introduction
- Glossary of key terms at the back of the text
- Cross references to related topics found in other chapters

## ILLUSTRATION PROGRAM

Back in 1966, the first edition featured the use of spot drawing of organisms in graphs. This feature has been adopted in other ecology texts. The fifth edition of *Ecology and Field Biology* introduces a new advance in ecology texts—the use of full color throughout. The illustration program, a mixture of the old and new, adds an exciting visual appeal to the text. We have retained the superb pen and ink work from the first edition by the late Ned Smith and have avoided detracting from it by adding color. All of the remaining illustrations have been redesigned, redrawn, and generated on the computer by Robert Leo Smith, Jr. All of the photographs, with a few necessary exceptions, are new and in color. They have been carefully selected to supplement and enhance the text, and don't simply add color to it. Many new illustrations have also been added. The old end maps have been replaced by two new ones. The inside front cover features the distribution of world vegetation based on Holdridge classification. The inside back cover features the distribution of world vegetation based on ecoregions.

# APPENDIX

In the previous edition, the Appendix was published as a separate supplement called the Student Resource Manual. This arrangement was less than satisfactory because the manual did not always travel with the text. Due to an overwhelming number of requests, the Student Resource Manual has been reincorporated in this edition, and can be found as an Appendix. In this edition, I have retained what I and some reviewers consider the most popular and useful sections and eliminated the less useful sections on estimating primary and secondary production, and environmental measurements. I have also eliminated references to field identification manuals and annotated selected references. The literature explosion has made the inclusion of a list of the latter more and more idiosyncratic. However, I have retained the list of Journals of Interest to Ecologists, because it is good for students and instructors to know the array of over 170 journals that in one way or another relate to ecology.

This Appendix provides a manual of methodology for the following:

- Sampling and terrestrial vegetation
- Sampling aquatic vegetation
- Dendrochronology
- Palynology
- Estimating animal population size
- Determining population dispersion and interspecific association
- Community similarity
- Community ordination
- Species diversity
- Population structure
- Life and fecundity tables
- Reproductive values
- Rate of increase

For those instructors who wish to incorporate computer programs into their courses, I highly recommend the following two books: *Ecological Methods* by Krebs (HarperCollins 1990) and *Statistical Ecology* by Ludwig and Reynolds (Wiley 1989). I use computer programs in my courses, with some reservations. Some students may be likely to toss data into a computer program without understanding what the program involves. To overcome this, I have the students work out diversities, ordinations, and population problems without computers first. Only then do they understand what is involved in their calculations.

The most difficult and time-consuming part of the text to prepare has been the bibliography, which consists of several thousand titles. New entries were checked against text citations and old titles were deleted. Citations of older references have been retained where they are appropriate. Truth does not change over the years. However, a great number of new titles have been cited, necessary to thoroughly update the text.

This edition has received input from a number of reviewers listed in the acknowledgments who pointed out inaccuracies and suggested changes. Some errors, however, may slip through; I would appreciate if you would call them to my attention.

# ANCILLARIES

A complete set of supplementary materials is available to support the use of *Ecology and Field Biology*, Fifth Edition, by both students and instructors:

- Instructor's Manual
- Student Workbook
- Test Bank
- TestMaster
- Overhead Transparencies

# ACKNOWLEDGMENTS

No textbook is the product of the author alone. Although the author writes the text, the material of which it is composed represents the work of hundreds of ecological researchers who have spent lifetimes in the field and

laboratory. Their published works on experimental results, observations, and conceptual thinking provide the raw material out of which a textbook is fashioned.

Revisions of a textbook depend heavily on the input of users. The fifth edition has received input from a number of instructors and students who pointed out inconsistencies, inaccuracies, suggested changes, and the incorporation of new material. I took their suggestions seriously, and incorporated many of them; the book is much improved from their input. I am deeply grateful to the following reviewers for their helpful comments and suggestions on how to improve this edition: Donald L. Beaver, Michigan State University; Edmund E. Bedecarrax, City College of San Francisco; William D. Bowman, University of Colorado, Boulder; Mark Brinson, East Carolina University; Stanley H. Faeth, Arizona State University; Howard E. Hunt, Louisiana Technical University; Thomas C. Kane, University of Cincinnati; Robert J. Kinucan, Sul Ross State University; Kenneth M. Klemow, Wilkes University; G. Wayne Minshall, Idaho State University; John G. Morris, Florida Institute of Technology; Robert K. Neely, Eastern Michigan University; Arthur K. Neuburger, Kansas Wesleyan University; Erik T. Nilsen, Virginia Polytechnic Institute and State University; H. Larimore Ragsdale, Emory University; Howard I. Russock, Western Connecticut State University; Thomas Sasek, Northeast Louisiana University; Frederick A. Servello, University of Maine; Merrill H. Sweet, Texas A&M University; Kristiina Vogt, Yale University; Daniel C. Wilhoft, Rutgers University; J.C. Zieman, University of Virginia.

Tad Day, University of Arizona, provided help and advice on ecological effects of ultraviolet radiation on plants. Jim McGraw, Biology Department, West Virginia, gave helpful advice on phenotypic plasticity in plants and other aspects of plant population biology. John Sencindiver of the Soil Science Department, Division of Plant Science, West Virginia University, clarified some questions raised by reviewers. Hillar Klandorf, Division of Animal Science, West Virginia University, helped clarify my discussions of homeostasis and photoperiodism in animals. Mike Huston, Oak Ridge National Laboratory, provided an advanced copy of several chapters from his book *Biological Diversity* that proved helpful in revising the chapters on community ecology. Ralph Bailey, U. S. Forest Service, provided information on and the map of Ecoregions of the World (inside back cover).

Two persons contributed greatly to this edition. Barbara Conover, the initial developmental editor, efficiently summarized the comments of the reviewers, which aided considerably in revising the manuscript. In addition, she skillfully performed the onerous task of editing the large bibliography, which included eliminating titles no longer cited, picking up missing citations, and pointing out inconsistencies between text and bibliographic citations. I am especially grateful to Dr. Kenneth Klemow of Wilkes University, who provided a detailed chapter-by-chapter analysis of the entire revised text. His input improved the clarity and focus of the revision. He was especially helpful in pointing out potential problem areas for students and being certain that I emphasize that ecology is a dynamic and not a dogmatic science.

As in the past, this text is sort of a family industry, but even more so with this edition. My first son Robert Leo, Jr., a graphic artist, rendered all of the color graphics under considerable time pressure. His familiarity with the text, artistic ability, and skill at computer graphics allowed a close collaboration in the development of illustrations, not possible otherwise. My second son Thomas Michael, Environmental Science Department, University of Virginia, contributed Chapter 2 on ecological experimentation and models and provided input to the chapters on succession and community structure, as well as the Holdridge Life Zone map (inside front cover).

The book could not have arrived at its present stage without the patience, help, and encouragement of the staff at HarperCollins. Glyn Davies, former editor-in-chief, gave the project his much appreciated strong support and encouragement. Michelle Ryan, photo researcher at Scott, Foresman, did an excellent job of selecting many new photos needed for color illustrations. It was a real chore choosing photos to match my specific requirements for each subject, and she did so with much patience and care. Ecologically knowledgeable Kristin Zimet, copy editor for the fourth edition, did an outstanding job on fine-tuning this fifth edition. Vicki Cohen, developmental editor for the last stages of the book, steered the manuscript through the final stages, making sure captions, figures, and other parts of the manuscript fell into place. New production methods and the conversion of the text to full color added a new dimension to production, at least for the author. The task of pulling the manuscript and art through the final stages into a book fell to the project editor, Brigitte Pelner, who did a great job of keeping the proofs and author on track, and design supervisor, Mary Archondes, who took great care in keeping the art program organized.

Through it all, my wife Alice took care (and still does) of all the problems of living, while I spent my evenings and weekends, with few to spare, working on this book. She has patiently endured book widowhood for years. To her I am very grateful.

*Robert Leo Smith*

Part *1*

# Introduction

# Ecology: Its Meaning and Scope

*Concepts*

1. Ecology is the study of the structure and function of nature.
2. Ecology developed from many roots, but its beginnings trace back to natural history and plant geography.
3. Ecology has branched into many subdivisions, many of them specialized.
4. The principles of ecology form the scientific basis for the solution to many environmental problems.

# ECOLOGY DEFINED

What is ecology? Ask nonecologists and they will probably answer that ecology has something to do with the environment or with saving it. Before the 1960s, few of them could have given you any answer. If you had asked a biologist in the same time period, you would probably have gotten some vague answer implying that ecology was "quantified natural history." Ecology only became a household word in the 1960s, through the environmental movement, and then its popular meaning became confused with environmentalism.

The origin of the word *ecology* is the Greek *oikos,* meaning "household," "home," "place to live." Clearly, ecology deals with the organism and its place to live, its environment. Ecologists continue to work toward a deeper definition. Here is a sampling of their attempts:

"the study of structure and function of nature." (Odum 1971:3)

"the scientific study of the distribution and abundance of animals." (Andrewartha 1961:10)

"the scientific study of the interactions that determine the distribution and abundance of organisms." (Krebs 1985:4)

"the scientific study of the relationships between organisms and their environments." (McNaughton and Wolfe 1979:1)

"the study of the relationships between organisms and the totality of the physical and biological factors affecting them or influenced by them." (Pianka 1988:4)

"the study of the adaptation of organisms to their environment." (Emlen 1973:1)

"the study of the principles which govern temporal and spatial patterns for assemblages of organisms." (Fenchel 1987:12)

"the study of the patterns of nature and how those patterns came to be, and how they change in space and time." (Kingsland 1985:1)

"the study of organisms and their environment—and the interrelationships between the two." (Putman and Wratten 1984:13)

"the study of the relationship between organisms and their physical and biological environments." (Ehrlich and Roughgarden 1987:3)

The author of the term, Ernst Haeckel, defined ecology as "the body of knowledge concerning the economy of nature—the investigation of the total relationships of the animal both to its inorganic and its organic environment; including, above all, its friendly and inimical relations with those animals and plants with which it comes directly or indirectly into contact—in a word ecology is the study of all those complex interrelations referred to by Darwin as the conditions for the struggle for existence."

None of these definitions really is satisfactory. They are either too restrictive or too vague, and the original definition relates only to animal ecology. Most definitions apply to population ecology and overlook ecosystem function. The subject has outgrown them.

For now, let us use a wider working definition. *Ecology* is the study of the structure and function of nature. Structure includes the distribution and abundance of organisms as influenced by the biotic and abiotic elements of the environment; and function includes all aspects of the growth and interaction of populations, including competition, predation, parasitism, mutualism, and transfers of nutrients and energy among them.

The term *ecology* is derived from the same root word as *economics,* "management of the household." Ecology, then, could be considered as the economics of nature. Wells, Huxley, and Wells (1939) commented that "Ecology is really an extension of economics to the whole world of life." Some economic concepts, such as resource allocation, cost-benefit ratios, and optimization theory, have found a place in ecology.

# THE DEVELOPMENT OF ECOLOGY

Just as there is no consensus on the definition of ecology, so there is no agreement on its beginnings. It is more like a multistemmed bush than a tree with a single trunk. Some historians trace the beginnings of ecology to Darwin, Thoreau, and Haeckel; others to the Greek scholar Theophrastus, a friend and associate of Aristotle, who wrote about the interrelationships between organisms and the environment.

## Plant Ecology

The modern impetus to ecology came from the plant geographers. They discovered that, although plants differed in various parts of the world, certain similarities and differences demanded explanation. They looked to climate as a possible answer, because similar climates supported similar vegetation. Carl Ludwig Willdenow (1765–1812), one of the early influential plant geographers, championed this explanation. His ideas caught the attention of a wealthy young Prussian naturalist, Friedrich Heinrich Alexander von Humboldt (1769–1859) (Figure 1.1a). He sailed in 1799 with the French botanist Aimé Bonpland for a five-year expedition through tropical Spanish America. They traveled though Mexico, Cuba, Venezuela, and Peru, and explored the Orinoco and Amazon Rivers. Humboldt described these travels and tropical America's plant and animal life in a 30-volume work, *Voyage to the Equatorial Regions.* In these books Humboldt described vegetation in terms of physiognomy, correlated vegetation types with environmental characteristics, and coined the term *association.*

(a)                                          (b)                                          (c)

**Figure 1.1** Plant geographers. (a) Friedrich Heinrich Alexander von Humboldt. (b) Johannes Warming. (c) F. E. Clements.

Travel to the tropics continued to stimulate plant ecology. A Danish botanist, Johannes Warming (1841–1924), spent three years studying tropical vegetation in Brazil (Figure 1.1b). Thirty years later he described the vegetation in a book in which he introduced such concepts as dominants and subdominants and noted the effects of fire and time. His major contribution was the book *Plantesamfund: Grundtrak af den okologiske Plantegeografi,* published in 1895. In this book he unified plant morphology, physiology, taxonomy, and biogeography and emphasized the importance of moisture, temperature, and soil for vegetation. The book fixed the modern concept of ecology and greatly influenced its development. Following Warming, another tropical traveler, the German botanist Andreas Schimper (1856–1901) published *Plant Geography on a Physiological Basis,* a book heavy on plant morphology and light on plant physiology. Schimper explained regional differences in vegetation as a function of moisture and temperature.

Other plant geographers stayed closer to home. Nevertheless they made major advances in plant ecology, especially the emerging concept of plant succession—vegetation change over time. Anton Kerner (1831–1898) was one of these geographers. Commissioned to survey the vegetation of eastern Hungary and Transylvania, he described plant succession in *Plant Life of the Danube Basin.* He pioneered the use of experimental transplant gardens at various elevations in the Tyrolean Alps to study the growth and behavior of plants taken from alpine and lowland sites. Later, the Polish botanist Jozef Paczoski (1864–1941) described how plants modify their environment by creating microenvironments and introduced such concepts as shade tolerance,

competition, plant succession, and the role of fire. Because his book was published in Slavic, it was belatedly discovered by ecologists outside the Slavic world. He is now recognized as the father of plant sociology or *phytosociology.*

The study of plant communities developed along separate paths in Europe and America. In Europe plant ecologists concentrated on describing the plant community. Christen Raunkiaer (1860–1938) of Denmark contributed a scheme of life form classification and quantitative methods of sampling vegetation, the data from which could be treated statistically. Later Josias Braun-Blanquet developed methods of community sampling, data reduction, and the classification and nomenclature of plant communities. A. E. Tansley, however, urged a more experimental approach to plant ecology. His views on ecology and research anticipated by years the type of ecological studies that emerged in the 1970s.

In America plant ecologists were far more interested in how plant communities develop. Settlement was destroying the original forests and grasslands, creating profound changes in vegetation. Interest in these disturbances led to pioneer studies of the dynamics of vegetation, especially plant succession. A doctoral study on the succession of plant life on the Indiana sand dunes by H. E. Cowles (1897) established plant succession as one of the central concepts of modern ecology.

F. E. Clements (1916) further developed the concept of succession by investigating its causes and consequences. Dogmatic and convincing, Clements quickly became the major theorist of plant ecology in the United States (Figure 1.1c). He gave ecology a hierarchical framework, introduced innumerable terms (no longer used) and the idea of

environmental indicators, and developed an organismal theory of plant ecology that still colors ecology today.

## Animal Ecology

Animal ecology developed later than plant ecology and along lines divorced from it. The beginnings of animal ecology can be traced to two Europeans, R. Hesse of Germany and Charles Elton of England. Elton's *Animal Ecology* (1927) and Hesse's *Tiergeographie auf logischer grundlage* (1924), translated into English as *Ecological Animal Geography,* strongly influenced the development of animal ecology in the United States. Charles Adams and Victor Shelford were two pioneering animal ecologists there. Adams published the first textbook on animal ecology, *A Guide to the Study of Animal Ecology* (1913). Shelford wrote *Animal Communities in Temperate America* (1913).

Shelford gave a new direction to ecology by stressing the interrelationship of plants and animals. Ecology became a science of communities. Some earlier European ecologists, particularly the marine biologist Karl Mobius, had developed the general concept of the community. In his essay "An Oyster Bank Is a Biocenose" (1877), Mobius explained that the oyster bank, although dominated by one animal, was really a complex community of many interdependent organisms. He proposed the word *biocenose* for such a community. The word comes from the Greek meaning "life having something in common."

The appearance in 1949 of the encyclopedic *Principles of Animal Ecology* by five second generation ecologists from the University of Chicago—W. C. Allee, A. E. Emerson, Thomas Park, Orlando Park, and K. P. Schmidt—pointed the direction modern ecology was to take. It emphasized trophic structure and energy budgets, population dynamics, and natural selection and evolution.

Still another area of biology, animal behavior, grafted its branch onto ecology. Although Darwin, Wallace, and others described activities of animals, the formal study of animal behavior began with George John Romanes (1848–1894), who introduced the comparative method of studying nonhuman animals to gain insights into human behavior. His approach depended largely on inferences, but C. Lloyd Morgan (1852–1936), an English behaviorist, emphasized the use of direct observation and experiment.

After the early 1900s, animal behavior study developed along four major lines. One was the study of behavioral mechanisms, perceptual and physiological. It became known as *behaviorism.* A second, more relevant to ecology, was the study of the function and evolution of behavior, including comparative physiology. This study became known as *ethology.* The three major founders of ethology were Konrad Lorenz, noted for his studies of genetically programmed behavior; Niko Tinbergen, who developed the scheme of four areas of inquiry (causation, development, evolution, and function); and Karl von Frisch, who pioneered studies of bee communication and behavior. After World War II a third field of animal behavior, wedded to ecology, appeared. It was *behavioral ecology,* which investigates the way animals interact with their living and nonliving environments, with a special emphasis on how that behavior is influenced by natural selection. Behavioral ecology in 1975 begot a controversial offspring, *sociobiology,* pioneered by E. O. Wilson in *Sociobiology: The New Synthesis.* Sociobiology, concentrating on field observations of social groups of animals, applies the principles of evolutionary biology to the study of social behavior in animals. It became controversial when some writers attempted to apply it to humans.

## Physiological Ecology

*Physiological ecology,* or *ecophysiology,* is concerned with the responses of individual organisms to temperature, moisture, light, nutrients, and other factors of the environment. Early plant physiologists studied photosynthesis and plant growth, including the influence of environment on growth. Justus Leibig (1840) investigated the role of limited supplies of nutrients on the growth and development of plants and came up with the "law of the minimum." F. F. Blackman (1905) extended this idea of limiting factors to include a maximum—a plant could get too much of a good thing. Soon both plant physiologists and plant ecologists began to work out the physiological relationships among plants, climate, atmosphere, and soil. As plant physiologists deciphered the mechanisms of photosynthesis and water relations in plants, ecophysiologists related these functions to plant distributions and adaptations. After World War II, the field grew rapidly. New instrumentation, experimental techniques, and rapidly advancing knowledge allowed ecophysiologists to study the interactions of plant physiology and enviromental responses in the field and in laboratory growth chambers.

Animal ecophysiology developed out of animal physiology, concerned at first with the functioning of the human body. V. E. Shelford (1911) stimulated the study of animal ecophysiology when he applied the concept of limiting factors to animals in a "law of tolerance." This law linked the physiology of an organism to its environment. He suggested that organisms have both a negative and an optimal response to environmental conditions, and that these responses influence their distribution. This idea stimulated investigations into how such physiological responses as thermoregulation, energy metabolism, and water balance relate to the environmental conditions in which animals live.

Observers noted that certain plants and animals use chemical substances for defense, and that some plant exudates inhibit the growth of associated species. They began to investigate chemical substances in the natural world. There were studies of the role of chemicals in species recognition, courtship, and defense as well as studies of the chemicals

themselves. Such work has grown into the specialized field of *chemical ecology.*

## Population Ecology

As plant ecology was arising out of plant geography, other developments were under way. One was the voyage of Charles Darwin on the *Beagle,* during which he collected numerous biological specimens, made detailed notes, and mentally framed his view of life on Earth (Figure 1.2). Darwin (1809–1882) observed the relationships between organisms and environment. He attributed the similarities and dissimilarities of organisms within continental land masses and among continents to geographical barriers separating the inhabitants. He noted from his collection of fossils how successive groups of plants and animals, distinct yet obviously related, replaced one another over geological time.

In developing his theory, Darwin was influenced by the writings of Thomas Malthus (1766–1834). An economist, Malthus (1798) advanced the principle that populations grew in geometric fashion, doubling after some period of time. Experiencing such rapid growth, a population would outstrip its food supply. Ultimately the population would be restrained by a "strong, constantly operating force— among plants and animals the waste of seeds, sickness, and premature death. Among mankind, misery and vice."

**Figure 1.2** Charles Darwin.

From this concept Darwin developed the idea of "the survival of the fittest" as a mechanism of natural selection and evolution.

Meanwhile, unknown to Darwin, an Austrian monk, Gregor Mendel (1822–1884), was studying in his garden the transmission of inheritable characters from one generation of pea plants to another. The work of Mendel would have answered a number of Darwin's questions on the mechanisms of inheritance and provided for his theory of natural selection the firm base it needed. Belatedly, Darwin's theory of evolution and Mendelian genetics were combined to form the study of evolution and adaptation, two central themes in ecology. The theoretical basis of the role of inheritance in evolution was advanced by Sewell Wright (1931), R. S. Fisher (1930), and J. Haldane (1932, 1954), who developed the field of *population genetics.*

The Malthusian concept of population growth and limitations stimulated the study of population dynamics. P. F. Verhulst (1838) of Italy formulated the mathematical basis for population growth under limiting conditions. Verhulst's work, expanded by R. Pearl and L. J. Reed (1929), was the basis for the contributions of A. Lotka and V. Volterra (1926) to the study of population growth, predation, and interspecific competition. Their work established the foundations of *population ecology,* concerned with population growth, regulation, and intraspecific and interspecific competition. The mathematical models of Lotka and Volterra were tested experimentally in the Soviet Union by G. F. Gause (1934) with laboratory populations of protozoans and in the United States by Thomas Park (1954) with flour beetles. Many of the concepts of population genetics have been combined with ideas from population ecology to make up the field of *evolutionary ecology,* concerned with the interactions of population dynamics, genetics, natural selection, and evolution.

Ecologists now have a body of theory relating to competition, population growth, life-history strategies, resource utilization, niche, coevolution, community structure, food webs, and the like. Theoretical ecologists take theories and equations developed in pure mathematics, physics, and even economics and apply them to ecological questions. They attempt to provide a substantial mathematical foundation for ecological concepts, upon which predictions can be based. Theoretical ecologists have stimulated new insights into relationships among species, utilization of resources, and life-history patterns. Critics of theoretical ecology argue that it suffers from too many hypotheses that are untested or untestable in the field.

## Ecosystem Ecology

Early plant ecologists were concerned mostly with terrestrial vegetation; but in Europe a group of biologists was interested in the natural history of fresh waters. Prominent among these biologists were A. Thienemann (1931) and

F. A. Forel (1901). Thienemann developed an ecological approach to freshwater biology. He introduced the ideas of organic nutrient cycling and trophic feeding levels, using the terms *producers* and *consumers.* Forel was more interested in the physical parameters of freshwater habitats, particularly lakes. He described thermal stratification and internal seiches within lakes. In his monograph on Lake Leman, he introduced the term *limnology* for the study of freshwater life. In a way limnology became the forerunner of ecosystem ecology.

Early concepts were further developed by S. A. Forbes, an entomologist at the University of Illinois and the Illinois State Laboratory of Natural History (Illinois Natural History Survey). He wrote a classic of ecology, "The Lake as a Microcosm," about the interrelations of life in a lake, particularly through food chains, and the role of natural selection in the regulation of numbers of predators and prey.

Unrelated to limnology, but destined to have an important influence on its future and that of ecology, was the work of Edgar Transeau in an Illinois cornfield. Transeau was not an ecologist, much less a limnologist. He was interested in improving farm production by understanding the photosynthetic efficiency of the corn plant. His landmark paper "The Accumulation of Energy in Plants" (1926) marked the beginning of the study of primary production and energy budgets.

Thienemann's and Transeau's work stimulated the study of lakes by E. A. Burge and by C. Juday of the Wisconsin Natural History Survey. In a classic paper, "The Annual Energy Budget of an Inland Lake," Juday summarized not only the accumulation of energy by aquatic plants over a year but also its movement through various feeding groups, including the decomposers.

The work of Juday and Birge influenced a young limnologist at the University of Minnesota, R. A. Lindeman. Lindeman was interested in exploring plant succession in terms of energy. He turned his attention to Cedar Bog Lake in Minnesota. In a 1942 paper, "The Trophic-Dynamic Aspect of Ecology," Lindeman described succession in terms of energy flow through the lake ecosystem. He showed how short-term processes of feeding or trophic relationships affected the long-term changes in the lake. This paper, a significant advance in ecology, marked the beginning of *ecosystem ecology.*

Preceding Lindeman's contribution was the theoretical work of a physical chemist, A. J. Lotka. In his book *Elements of Physical Biology* he introduced thermodynamic principles of energy transformations in biology along the lines of physical chemistry. He considered food webs and the cycles of carbon dioxide, phosphorus, nitrogen, and water; and he viewed Earth as a single energy-transforming system. Most ecologists overlooked Lotka's contribution. However, his ideas and Lindeman's study stimulated further pioneering work on energy flow and nutrient budgets by G. E. Hutchinson (1957, 1969) and H. T. and E.P. Odum in the

1950s. J. Ovington (1962) in England and Rodin and Bazilevic (1967) in the Soviet Union investigated nutrient cycling in forests. The increased ability to measure energy flows and nutrient cycling by means of radioactive tracers and to analyze large amounts of data with computers permitted the development of *systems ecology,* the application of general systems theory and methods to ecology.

## Cooperative Studies

Ecology has developed from so many roots and has grown so many branches that it probably will always remain, as Robert McIntosh (1980) calls it, "a polymorphic discipline." Attempts to reduce ecology to a set of basic principles have not been successful. Ecology ranges over many diverse areas—marine, freshwater, and terrestrial. It involves all taxonomic groups, from bacteria and protozoa to mammals and forest trees, at all levels—individuals, populations, ecosystems. Any of these levels and groups may be studied from various points of view—behavioral, physiological, mathematical, chemical. As a result ecology, by necessity, involves isolated groups of specialists.

Pulling some of these groups together in the 1960s was the International Biological Program, known as IBP. A growing concern over environmental problems facing the world prompted the program. In the United States the IBP, initiated in 1967, focused on a cooperative study and analysis of ecosystems, including the tundra, the coniferous forest, the eastern deciduous forest, the desert, and Mediterranean types. The goals, as summarized by McIntosh (1976) included: (1) understanding the interactions of the many components of complex ecological systems; (2) exploiting this understanding to increase biological productivity; (3) increasing the capacity to predict the effects of environmental impacts; (4) enhancing the capacity to manage natural resources; and (5) advancing the knowledge of human genetic, physiological, and behavioral adaptations.

IBP's greatest contribution was to increase our understanding of processes in ecosystems, particularly photosynthesis and productivity, water and mineral cycling, decomposition, and the role of detritus. IBP did not provide any definitive theoretical foundations for ecology. It lacked strong organization and coordinated direction. However, it did advance modern ecosystem ecology. Summaries of IBP research, still being published in numerous volumes, provide a base for future ecological research.

## TENSIONS WITHIN ECOLOGY

The complexity of ecology's past has led some scientists into opposing camps. Often these deep controversies have been, or can be made, productive.

## Plant Versus Animal Ecology

The first major split in ecology was the failure of plant ecology and animal ecology to meet on common ground. In England plant ecology was influenced strongly by A. E. Tansley and animal ecology by Charles Elton. At that time one journal, *The Journal of Ecology,* sponsored by the British Ecological Society, covered the field of ecology. In a few years Elton started *The Journal of Animal Ecology.* The two, plant ecology and animal ecology, went their separate ways.

In the United States, the split was less amicable. Early on, a controversy developed over the term *ecology.* Botanists decided at the Madison (Wisconsin) Botanical Congress in 1893 to drop the *o* from *oecology* and adopt an anglicized spelling. Zoologists refused to recognize the term at all. The entomologist William Morton Wheeler complained that botanists had usurped the word, and had distorted the science. He urged zoologists to drop the term and adopt the word *ethology.*

The schism was widened by a more fundamental difference in approach. Plant ecologists ignored any interaction between plants and animals. In effect, they viewed plants as growing in a world without parasitic insects and grazing herbivores. For years plant and animal ecologists went their separate ways. F. E. Clements and V. E. Shelford began to bring the two sides together with *Bio-Ecology* (1939), in which they suggested that plants and animals be considered as interacting components of broad biotic communities or biomes.

## Organismal Versus Individualistic Ecology

Although the division between plant and animal ecology narrowed, a new division was to plague ecology. It had its roots in the ideas of Clements, who strongly influenced philosophical ideas in ecology. Clements viewed the plant community as an organism. Like an individual organism, vegetation moved through several stages of development, from youthful colonization of bare ground to a mature, self-reproducing climate in balance with its climate determined environment. The climax was the end or goal toward which all vegetation progressed. If disturbed, vegetation responded by retracing its developmental stages to the climax again.

Clements' organismal approach was not lost on animal ecologists. In the United States the zoologist and animal behaviorist William Morton Wheeler, an international authority on ants and termites, advanced the idea that ant colonies behave as organisms. They carry out such functions as food gathering, nutrition, self-defense, and reproduction. Basing his ideas on those of C. Lloyd Morgan, a biological philosopher, Wheeler applied **emergence theory** to ecology. He proposed that natural associations have certain emergent properties as aggregations of organisms—predators and prey, parasites and hosts—that arose from lower levels of organization. All levels occurred together in an ecological community or *biocenosis.* The biocenosis modified its component species through behavioral changes and new levels of integration. Everything in the biocenosis was related to everything else. His view of a tight but orderly nature contrasted with the chaotic effect imposed on nature by humans.

This organismic, levels-of-hierarchy view of nature advanced by Clements and Wheeler captured the thinking of that influential group of ecologists at the University of Chicago, the authors of *The Principles of Animal Ecology,* Allee, Park, Park, Emerson, and Schmidt. In that book they stated that the organismic concept of ecology was "one of the fruitful ideas contributed by biological science to modern civilization."

Although the organismal concept dominated ecology until the early 1960s, many ecologists refused to accept it. Clement's organismic concept had its critics, notably H. A. Gleason and A. E. Tansley. In 1926 Gleason published "The Individualistic Concept of the Plant Association." In it he argued that the plant association was hardly an organism capable of self-reproduction. Instead, he argued, each community is unique. It arises randomly through environmental selection of seeds, spores, and other reproductive parts of plants that enter a particular area. The English ecologist A. E. Tansley, once enamored with the organismic concept, ultimately rejected it too. Vegetation, he allowed, might be called a quasi-organism, but certainly not an organism or a complex organism. In fact, Tansley rejected the whole idea of a biotic community as anthropomorphic. No social relationship exists among plants or between plants and animals as the term connotes, he argued. In its place Tansley substituted the term **ecosystem.** He viewed plants and animals as components of a system that also included physical factors.

## Holism Versus Reductionism

By the mid-1960s the individualistic concept of Gleason had supplanted the organismic concept—almost. Many of its philosophical and functional attributes lived on in the "new ecology" of the 1960s. The new ecology, as defined by E. P. Odum (1964, 1971), is a "systems ecology," an "integrative discipline that deals with supraindividual levels of organization."

According to this concept ecosystems develop from youth to maturity. Each stage of development exhibits some of its own unique characteristics. Interactions among populations and between plants and animals result in a hierarchical organization. This organization involves interacting components that produce large functional wholes. The outcome is the emergence of new system properties that are not evident at the level below (Odum 1971, 1982). These emergent properties account for most of the changes in species and growth that take place over time. The approach is **holistic** (studying the total behavior or attributes of a complex system) because systems are considered too complex to

study in bits. Because the whole is greater than the sum of its parts, ecosystems can be studied only as functional units.

This holistic approach has critics who take a **reductionist** approach. They consider that the ecosystem is in fact the sum of its parts. By understanding how each part—the species, their numbers, and characteristics—functions, we can discover how the whole system operates. Rather than guiding the evolution of species, the nature of ecosystems results from the evolution of species.

Fenchel (1987:17) puts the reductionist's point of view well: "I find the entire argument as nonsensical as stating that an alarm clock is qualitatively different from its constituent wheel, bolts, and springs. A holist approach to an alarm clock . . . is to observe that when wound it will run. To arrive at a real understanding of the device one must take it apart in order to see how it works . . . to take a reductionist's approach."

The holist would counter that studying the wheels, bolts, and springs tells nothing about the way the whole system functions, what the clock really does. You could study a few separate components, but they are outside the context of the whole clock. Only when all parts of the system are functioning as a unit can the clock function. Then its emergent property, telling time, becomes apparent.

Is the sum of the parts of the clock greater than the whole or not? Allen and Starr (1982) in their book *Hierarchy* argue that the whole problem of emergent properties is a matter of scale and assert that some properties of the whole are emergent and cannot be derived from the behavior of the parts alone. They also point out that ecosystem models of holists are simply large-scale reductionism. Ecosystem ecologists cannot possibly study a model of an entire ecosystem. They can only study pieces of it. The only major difference between a reductionist and a holist is that the holist studies larger pieces, made up of assembled parts studied by the reductionists.

What keeps ecosystem ecologists (holists) and population and evolutionary ecologists (reductionists) apart is their approach to ecology. Population ecologists focus on species' interactions with their environment in the broadest sense. They are interested in the historical or ultimate reasons why natural selection favored different adaptive responses among species over evolutionary time. Ecosystem ecologists are more interested in the how of current or proximate outcomes of the functional interactions at the population, community, and ecosystem levels. These differences may not be as great as they appear.

What can bring the two groups together? Population ecologists could approach population growth and population interactions such as mutualism, parasitism, predation, and competition as interacting systems (Berry 1981) and as components of a hierarchy of systems. Systems ecologists could integrate some evolutionary theory into system models, particularly in the area of ecosystem development and organization (Loehle and Peckmann 1988). Food web theory, for

example, crosses the line into both evolutionary and systems ecology, involving both species interactions and the transfer of energy and nutrients through a hierarchy. Ecosystem functioning ultimately depends upon species adaptations, which are the outcomes of evolution. For example, efficiency of water use by certain ecosystems such as grasslands and deserts results from the water use efficiency of the individual plants. The natural assemblage of plants and animals that comprise the living component of an ecosystem is not a random collection of species but rather one that has been determined by the competitive abilities and other attributes of the component species (H. Odum 1983).

## Theoretical Versus Applied Ecology

Ecological theories and models help us understand the human impact on environments. They provide a basis for ecosystem and natural resource management, preservation, and restoration. All of these activities make up *applied ecology*. For years theoretical and academic ecologists viewed applied ecology as an intellectual lightweight. Applied ecologists, for their part, often ignored theory, even when it could be of practical use. Fortunately ecologists of both persuasions now recognize that solutions applied to environmental problems must be based on sound theory developed through research.

Applied ecology began to take shape in the 1930s. In 1932 Herbert Stoddard pointed out the role of fire in the control of plant succession in his book *The Bobwhite Quail*. This topic was ignored by academic plant ecologists. Aldo Leopold pioneered the application of ecological principles to the management of wildlife in his classic *Game Management* (1933). In *Forest Soils* (1954) H. L. Lutz and R. F. Chandler discussed nutrient cycles and their role in the forest ecosystem. J. Kittredge pointed out the impact of forests on the environment in *Forest Influences* (1948).

We have begun to apply ecosystem and theoretical ecology more intensively to resource management in the past decade, even though economics often takes precedence over protection. Forestry, once concerned with the raising of trees for harvest, now emphasizes biomass accumulation, nutrient cycling, the effects of timber harvesting on nutrient budgets, and the role of fire in forest ecosystems. Specialists in range management are interested in the functioning of grassland ecosystems, the effects of grazing intensities on aboveground and belowground production by plants, and the structure of grassland communities. Wildlife managers, who once emphasized only game species, now consider the entire wildlife spectrum, including species not hunted. The range of interest covers both population ecology of wildlife and the maintenance and management of plant communities as wildlife habitat. Wildlife managers have developed an interest in population genetics, especially as it relates to the effects of hunting on game species and to the restoration of endangered species.

A related developing field comprising both applied and theoretical ecology is *conservation biology*. It has been defined as "the science of scarcity and diversity" (Soule 1986). Conservation biology addresses the problems of gross habitat destruction and a great reduction in population size of species.

A second developing field tied to applied ecology is **landscape ecology.** It is concerned with spatial patterns in landscape and how they develop, with emphasis on the role of disturbance, including human impacts (Forman and Godron 1986).

A third new field is **restoration ecology.** It applies experimental research to the restoration of ecosystems on highly disturbed lands (Jordan et al. 1987).

Although applied ecology has been around since the early 1930s, it did not gain visibility until the 1970s, when ecology became involved in social, political, and economic issues. This involvement grew out of public awareness of the problems of pollution, toxic wastes, overpopulation, and a degraded environment. Although the public treated these issues as if they were new, ecologists had grappled with environmental problems for years. The ecological movement had its roots in Europe, especially Germany. An early founder of political ecology was Ernst Haeckel, who coined the term *ecology*. From Germany it moved to northern Europe, Great Britain, and the United States. In England the animal ecologist Charles Elton helped found the Nature Conservancy. The plant ecologist A. E. Tansley founded the British Ecological Society and was active in the conservation movement.

In the United States George Perkin Marsh in 1885 called attention to the effects of poor land use on the human environment in his dramatic book *Man and Nature.* In the 1930s F. E. Clements urged that the Great Plains be managed as grazing land and not be broken by the plow. The plant ecologist Paul Sears wrote *Deserts on the March* (1935) in response to the Great Plains Dust Bowl of the 1930s. William Vogt's *Road to Survival* (1948) and Fairfield Osborn's *Our Plundered Planet* called attention to the growing population-resource problem. Aldo Leopold's *A Sand County Almanac* (1949), which called for an ecological land ethic, was read largely by those interested in wildlife management until the 1970s, when it became the bible of the environmental movement.

Rachel Carson did more than anyone else to bring environmental problems to the attention of the public (Figure 1.3). Since the publication of her book *Silent Spring* (1962), people have become more aware that chemical poisons and other pollutants are recycled through the environment. Once castigated as more fiction than fact, Carson's predictions came only too true as carnivorous birds fell victim to toxic chemicals. With a ban on DDT in the United States, some eagles, hawks, and osprey began a gradual comeback. Carson made people quick to recognize other continuing chemical dangers, such as dioxin and PCBs.

**Figure 1.3** Rachel Carson.

In the past quarter century since people became concerned about growing environmental degradation, how has the situation changed? We started off well enough with environmental legislation: the National Environmental Policy Act (1969), designed to protect the environment from overzealous development and to mitigate losses, the Endangered Species Act (1975, amended 1982 and 1994), and the Clean Water Act (1977, amended 1981, 1987, 1994), among others. The early enthusiasm for a quality environment, however, has dampened; government is less sensitive about environmental issues, and funding for environmental protection and research, especially at the federal level, has shrunk. During the 1980s there was even an environmental backlash at the federal level, as the administration attempted to undo all the environmental progress made during the previous two decades.

There has, of course, been progress. Water quality has improved considerably, and the air above some of our cities is cleaner. However, we have discovered that our environmental problems are not only more difficult to solve than once believed; many are growing worse. Toxic wastes pollute groundwater and land. Air is becoming more polluted worldwide. Haze has cut visibility in the eastern United States by more than 50 percent in the past 40 years. Acid rain affects lakes and streams. Increased concentrations of carbon dioxide and ozone threaten climatic stability. Roads

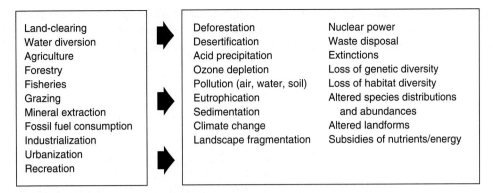

**Figure 1.4** Human activities lessen the sustainability of the biosphere.

cut into open country and suburban expansions eat away at the hinterlands and farmlands. Continued deforestation in both temperate and tropical regions is fragmenting wildlife habitat, increasing the rate of extinction. A rapidly growing urban and suburban population with increasing interest in outdoor recreation is placing intolerable pressures on state and national parks that threaten their ecological integrity. Even the oceans have not escaped, as human debris and chemicals have been deadening the seas and destroying marine life. In spite of surplus agricultural production, wetlands are still being drained for more cropland at an alarming rate, threatening the very existence of already dangerously declining wetland wildlife. All of these activities have impacted regional and global ecological processes and have lessened Earth's ability to support a diversity of life, including humans (Figure 1.4). With the human population growing at the rate of 1.8 percent annually, its pressures on Earth's resources will accelerate.

Among the many environmental problems facing humanity, three broad areas are critical: global change, biological diversity, and sustainability. We have made worldwide changes in climate, in land and water use, and in landscape patterns. We are causing dramatic change in the distribution, abundance, and number of species. The loss of diversity can affect the stability of communities and populations upon which our economy depends, as exemplified by the loss of commercially important fish species. This rapid diminution of Earth's resources affects our ability to sustain both natural and managed ecosystems and human life itself.

The basis and solution of our environmental problems are ecological in nature. To this end The Ecological Society of America has developed a three-pronged Sustainable Biosphere Initiative, involving research, education, and environmental decision-making (Lubchenco et al. 1991). Research priorities focus on the critical areas of global change, biological diversity, and sustainable ecological systems. Researchers seek answers to such problems as the responses of ecological systems to stress, development and application of ecological theory to the management of ecological systems, and an ecological understanding of the effects of introduced species, pests, and pathogens.

Unfortunately, attempts to apply sound ecological principles to environmental problems often run headlong into economic, political, and social opposition, as witnessed by the debates over old-growth forest, regulation of fishing, land zoning, and wetland preservation. Successful application requires a citizenry that understands ecology and its importance. We need ecological education at all levels. Ecological principles need to be clearly understood by economists, engineers, lawyers, businesspeople, and politicians, all of them decision makers who can hurt or improve the environment. Most decision-makers are unaware of the facts, do not understand basic ecological concepts, or are even hostile to environmental considerations for political, economic, or special reasons. An educated public can make decision-makers more responsible.

The future of human life on Earth depends on far more ecological knowledge than we now possess, even though we are not applying all we know. For the first time in the history of Earth, *Homo sapiens* has become the completely dominant organism, changing Earth and its diversity of life at will with little regard for the consequences. It is little wonder, then, that some of the most intellectually challenging problems in ecology lie in that transition zone between theoretical and applied ecology.

## SUMMARY

Ecology, difficult to define precisely, is the study of the interrelations of organisms with their total environment, physical and biological. Its origins are diverse, but a main root goes back to early natural history and plant geography. They evolved into the study of plant communities, ecosystems, trophic levels, and energy budgets. A major branch that developed out of the natural history of animals was the study of natural selection and evolution, beginning with the major contributions of Darwin and Wallace. It further branched into evolutionary ecology, population genetics, population ecology, and theoretical ecology. Another branch gave rise to behavioral ecology, concerned with the way animals in-

teract with their living and nonliving environment as influenced by natural selection, and to physiological ecology, concerned with the physiological responses of individual organisms to environmental factors. Studies of chemical reactions of organisms to their environment stimulated the development of chemical ecology. It concerns the uses of chemicals by plants and animals as attractants, repellents, and defensive mechanisms, their evolution and chemical structure. Ecology has so many roots and branches that we call it polymorphic.

As its various disciplines expand, ecology is becoming fragmented into specialties with a growing lack of communication among them. One tension developed between organismic and individualistic ecology. Another persists between holistic ecosystem ecology and reductionist evolutionary and population ecology. The differences, however, are not insuperable.

Applied ecology is concerned with the application of ecological principles to major environmental and resource management problems. Traditionally, applied ecology meant forest, range, wildlife and fishery management. Recently applied ecology has spawned the new fields of conservation biology, restoration ecology, and landscape ecology. The future quality of human life in all its aspects and the sustainability of Earth depend upon our ability to recognize and apply ecological principles to its management.

## REVIEW QUESTIONS

1. Define ecology.
2. Why was plant geography a stimulus for the development of modern ecology?
3. What differences separate the organismal concept of ecology from the individualistic concept?
4. How do the two concepts in question 3 relate to holism and reductionism in ecology?
5. What is applied ecology, and how does it relate to theoretical and ecosystem ecology?
6. Refer to *The Sustainable Biosphere Initiative* (Lubchenco et al. 1991), select one of the research topics, and discuss how the results of such research would relate to our environmental problems.
7. Why is ecology not taken as seriously as it should be by the public and decision-makers?

# Experimentation and Models

## *Concepts*

1. A hypothesis is a statement of causality that can be tested experimentally.
2. Two approaches to developing and testing hypotheses are the inductive method and the deductive method.
3. Testing hypotheses involves the analysis of data collected by observation and experimentation.
4. Useful in ecological studies are models, explicit sets of hypotheses relating pattern and process. Models may be analytical or simulation.
5. To be useful, models must be validated with experimental and field data.

The general goal of science is to understand the variation in pattern and function of the natural world and to predict future events. The specific goal of ecology is to understand the patterns and processes related to life on Earth: how plants and animals have adapted to various environments and how they interact among themselves and with their abiotic environment. To gain insight ecologists, like all scientists, develop possible explanations about the causes of observed phenomena. These possible explanations are called hypotheses. A **hypothesis** is a statement about an observation that can be tested experimentally.

# INDUCTIVE AND DEDUCTIVE APPROACHES

There are two approaches to the development and testing of hypotheses, inductive and deductive. In the **inductive method,** the scientist gathers empirical data and from it arrives at a generalization. The inductive method proceeds from specific observations to a general conclusion.

Consider ecologists monitoring grassland productivity by clipping all the grass from a series of plots for five consecutive summers. They observe that grassland productivity appears to be higher in years with abundant rainfall during the growing season than in years of drought. From this observation they develop the hypothesis that grass productivity is related to soil moisture. They have gone from the specific observation of grassland production and rainfall at a given site to a general hypothesis relating grassland productivity to soil moisture.

Now suppose they position rain shields over certain plots, reducing the rainfall reaching the ground under them. By adjusting these shields over the course of several growing seasons, they test the general hypothesis that grass productivity is linearly related to annual precipitation or some measure of soil moisture. Given positive results, they may extend the experiment to areas having different soils and regions receiving different annual rainfall. This method is called the **deductive method.** The experimenters have gone from a general idea about plant response to water and patterns of rainfall to a specific prediction relating grassland productivity to soil moisture at a given site.

Although the distinction between these two approaches may at times seem subtle, they are very different basic approaches to developing an understanding of our world. The inductive approach is the formation of general principles from specific observations. The deductive approach is the prediction of specific events from general principles. Each method has its advantages, and ecologists employ both methods.

# COLLECTING DATA

Testing hypotheses entails first the collection of data. There are two ways to get data: by direct observation or by experimentation. Some ecologists prefer direct observation,

or "natural experiments" (Diamond 1986). For example, they are studing changes over time in Yellowstone National Park since the dramatic fires. Another natural experiment might study the relationship between the abundance of standing dead trees (snags) and the population density of woodpeckers. Because woodpeckers use snags as nest sites, the abundance of snags might be a limiting resource to population size. By counting the number of woodpeckers and the number of dead trees in various forest stands, ecologists could compare the two. If the woodpecker density is higher on sites with more snags, then the observations support the hypothesis.

The observed patterns, however, only show a correlation; they do not settle the question of cause and effect. There may be other factors responsible for the variation in woodpecker population density, which may also be related to the density of snags. If the density of large trees on which to feed limits the population, you might expect a relationship between woodpecker population density and stand age. In this case, the factor causing the variation in population density among stands is the availability of suitable foraging sites and not nesting sites. It just so happens that the abundance of nesting sites (standing dead trees) is correlated to the causal variable of food abundance (stand age). To help determine cause and effect an ecologist may employ an experimental approach in which one or more variables are directly manipulated. This rigorous establishment of causation is what separates science from guess work.

The experimental approach, unlike the collection of observations from unmanipulated systems, directly determines the response of one variable (the **dependent variable**) to changes in some other variable (the **independent variable**). The investigators control the independent variable. They manipulate the variable in a predetermined way (*treatments*) and monitor the response of the dependent variable. For example, in a laboratory experiment to test a plant's growth response to increased $CO_2$, the concentration of $CO_2$ is the independent variable and plant growth is the dependent variable.

One important constraint in experiments is the need to control the variation in other independent variables that may influence the response of the dependent variable. In the experiment on plant growth response to elevated $CO_2$, the investigator must be sure that all other factors, such as moisture, temperature, and light, that may influence plant growth are the same for all treatments. Otherwise it would be difficult to interpret the plant response as a direct function of $CO_2$ alone.

Even though the investigator tries to reduce all possible differences among individuals or units that are being used for the experiment and maintain a rigid control over environmental conditions, there will always be some amount of variation among individual plants. One source of variation difficult to control is genetic variation among individuals. Because such factors are often impossible to control, the in-

(a)

(b)

**Figure 2.1** Although ecologists carry on experimental studies in the laboratory, much of their work is done in the field. (a) A grid established in an oak forest to study the effects of thinning and insect defoliation on the growth and spread of the fungus *Armillaria mellea*. *Armillaria mellea* (honey mushroom) is the cause of root rot in oaks, resulting in the death of trees, especially those of low vigor. (b) Collection of fungal samples from soil blocks removed along the grids.

vestigator must examine the response of a number of individuals rather than relying on a single observation. This approach is known as *replication,* and the individuals receiving the treatment are the *replicates.* The investigator must determine the number of replicates required for each experiment to account for this uncontrolled variation. The decision will relate to the degree of variation within the population being examined.

Because the purpose of the experiment is to examine response to treatment, we have to know what would happen without it. A group of individuals must be used as a control. The control group does not receive the treatment, but otherwise is handled exactly like the treated individuals. The control forms the basis for comparison with the individuals receiving the treatment. In the experiment on $CO_2$, the investigator would grow the control plants under a current normal (ambient) concentration of $CO_2$, providing an estimate of expected growth under baseline conditions.

Ecologists conduct experiments in the laboratory and in the field (Figure 2.1). Ecological experiments carried out in the laboratory usually involve synthetic communities of one or a few species. On the basis of these simplified systems, such experiments can reveal a wide range of possible outcomes or relationships that the investigator can evaluate in the field under more natural and complex conditions.

Field experiments involve the manipulation of one or more independent variables in the natural system. The investigator accomplishes this manipulation by adding or removing a species, by erecting exclosures to prevent access to a space or resource, or by adding or withholding resources. Unlike the laboratory experimenter, the field investigator often finds it impossible to control the many independent variables that may influence the response of the dependent variable. This circumstance limits the investiga-

tor's ability to assign a causal relationship between the response of the dependent variable and the treatments. In this case the investigator must include a set of experimental units or plots as controls. The investigator uses this set of field controls for comparison with the set of units to which the treatments are applied. This comparison, discussed later, will form the basis of testing the hypothesis of the influence of the independent variable on the dependent variable.

Field experiments are generally on a larger scale than laboratory experiments, and in some ways are more "realistic" in that the objects under study have not been isolated and controlled in a laboratory environment. However, there are limitations to field studies. One such limitation is the problem of adequately choosing the sample or replicates of the objects under study.

An essential assumption for hypothesis testing is that the replicates be independent. By *independent* we mean that the units are interspersed in time and space so that the response of any one unit has no influence on the response of any other unit. If we were to examine the response of plant growth to soil nutrition, we might plant individuals in separate pots and apply varying levels of fertilizer. If we planted several individuals in a single pot, the individuals in that pot could not be used as replicates. The response of each would be directly influenced not only by the treatment, but by the others. The individual plants would not represent independent observations. The observable unit now would become the population of plants in each pot. The error of using nonindependent replicates in an experiment is called *pseudoreplication* (Hurlbert 1984).

Pseudoreplication is a particular problem in field experiments. Suppose investigators want to examine the effects of clearcutting a forest on nitrate leaching from the watershed. They establish a number of sample stations at

various points along a number of streams in the watershed. As a control, they set up a similar number of stations in an adjacent watershed that has not been clearcut. They collect water samples at each station once a month for three months over the summer and determine the nitrate concentrations in the samples. The investigators then analyze the data, using each observation from each station as a replicate for comparison between cut and uncut watersheds.

Although it may appear at first as an adequate experimental design, this design suffers from pseudoreplication. First the investigators must determine whether the streams within each of the two watersheds are independent. If water is exchanged between streams (for example, one is a tributary of another), then the observations are not independent in space; the state of one sample unit is influenced by the state of the other. Second, the observations at any station are not independent in time. The state of the stream during the first month of observation will have a direct influence on the state of the stream during the second month. In fact, each watershed may represent only a single independent observation. If so, the data are inadequate for statistical analysis, because of a lack of replication.

## TESTING HYPOTHESES

Assume we have developed a hypothesis and collected the data necessary to test it. Our next step is to test the hypothesis and reach a conclusion. In most cases the hypothesis is a statement concerning the effects of the independent variable on the dependent variable. In the case of plant growth response to elevated $CO_2$, the dependent variable is plant growth and the independent variable is the concentration of $CO_2$. Our experimental data would consist of growth measures (such as biomass gain over a period of time) for plants grown under ambient levels of $CO_2$ (our control samples) and elevated levels (treatment samples). Given our understanding of photosynthesis, we would hypothesize that elevated levels of $CO_2$ result in increased plant growth.

We test the hypothesis statistically by creating a null hypothesis ($H_0$) and accepting an alternative hypothesis ($H_1$). The **null hypothesis** is a statement of no difference between control and treatment units; we say, the independent variable has no significant effect on the dependent variable. The *alternative hypothesis* is a statement of significant difference between control and treatment units. In our experiment the null hypothesis would state that the elevated levels of $CO_2$ have no effect on growth. The reason for the null hypothesis becomes apparent when the nature of statistical tests used for hypothesis testing is examined.

Suppose that in an effort to test the hypothesis that elevated $CO_2$ causes increased plant growth we obtain weight data (shown in Figure 2.2) for 25 individual plants of the same species grown under ambient $CO_2$ (320 ppm) and dou-

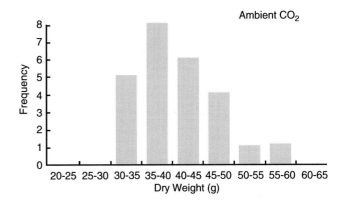

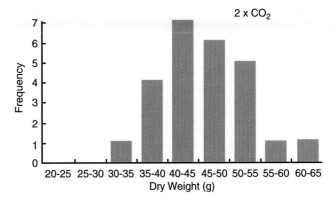

**Figure 2.2** Frequency distribution of dry weights for 25 plants grown under ambient and two times ambient concentrations of $CO_2$.

bled $CO_2$ (720 ppm) concentrations. We grew the plants from seed under controlled laboratory conditions so we could interpret any differences in growth as a function of treatment ($2X\ CO_2$). The plants grown under ambient $CO_2$ levels provide a control. Our null hypothesis ($H_0$) is no difference in the growth of individual plants between treatment and control; our alternative hypothesis ($H_1$) is that a difference does exist. To test this null hypothesis, we must develop some measure of expectation based on it.

Ideally we might expect that each of the 25 plants in the control would grow identically over the period of the experiment because we grew them under the same conditions. However, as the data show in Figure 2.2, the plants vary in their growth response (values of biomass accumulated). This variation among individuals in the control may be caused by a variety of reasons, but the major reason is genetic variation. The important point is that some degree of inherent variation will exist independent of treatments. We must develop an approach for evaluating this variation. We want to know if the estimates of plant growth under the two levels of $CO_2$ are indeed different, or if they simply represent the natural variation among individuals within the population sampled.

We develop our measure of expectation under the null hypothesis by examining the possible outcomes of sampling from a single population. Let us assume that the natural variation in the growth of individuals of the species we are

studying approximates a normal (bell-shaped) distribution. The normal distribution (Figure 2.3) is a special class of statistical distributions defined by two parameters, the *mean* ($\mu$) and the **variance** ($\sigma$). The population mean is a measure of central tendency, which we estimated from the replicates as $\mu = \Sigma\, x_i/n$ where $x_i$ are the values of the dependent variable for the replicates and $n$ is the number of replicates (observations). Sigma squared ($\sigma^2$) is the population variance, a measure of dispersion about the mean. It is estimated from the sample as $\sigma^2 = \Sigma\,(x_i - x)^2/n - 1$. The square root of the variance is the **standard deviation** (s.d.). The normal distribution is symmetrical about the mean (Figure 2.3). Also 68, 95, and 99 percent of all observations fall between the mean and one, two, and three standard deviations. Although many populations of observations do not perfectly fit the normal distribution, the normal distribution provides a convenient approximation (Figure 2.4). (See Sokal and Rohlf 1981 for detailed discussion of the normal distribution and parametric statistics.)

Let us now assume, first, that we have complete knowledge of all individuals of the plant species we are studying when grown under ambient levels of $CO_2$, and second, that the distribution of values approximates the normal distribution. We can characterize the population by its mean.

Suppose we now choose at random two samples of 25 plants each from this population. If we plot the distribution of these values for each sample (Figure 2.5) we would expect differences between the distributions based on chance alone. We can calculate a mean value for each of these samples as we did for the population as a whole. These sample means are estimates of the population mean, but we do not always expect them to be identical to the population mean because they are based only on a sample of the population.

We can now represent the difference between the two samples by the difference between the mean values of growth for each sample of 25 plants. If we return the individuals to the population and repeat this procedure of sampling over and over again, we can generate a large number of possible differences between the mean values of growth for the two samples of 25 observations drawn from the same population. In some cases the mean values for the samples will be identical or very close; for other samples the differences between the means will be quite large. If we know the mathematical function representing the actual population distribution (as with the normal distribution), we can calculate the probability of having a given difference in the calculated means between the two samples from the same population. That is, we can define the probability that we could observe a given difference in the mean value of growth determined for our treatment and control plants if they were both samples from the same population.

## Types of Error

In our plant experiment, it appears as if the observed differences in growth are the result of random variation in the population rather than the treatment (elevated $CO_2$). We can

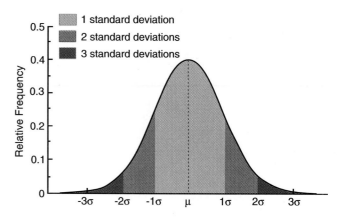

**Figure 2.3** Areas under the normal probability density function and the cumulative normal distribution functions. (Adapted from Sokal and Rohlf 1981).

now make a judgment whether we should reject the null hypothesis. However, we have merely calculated the probability that the two samples have come from the same population. The decision to reject the null hypothesis depends upon how willing we are to reject the null hypothesis even though it is true. This type of error is called *Type I error,* and the probability is called *alpha* ($\alpha$). Alpha, therefore, is the

**Figure 2.4** (a) Populations with a normal distribution can have different means ($\mu$) but the same standard deviations ($\sigma$). (b) Some populations have the same means but different standard deviations.

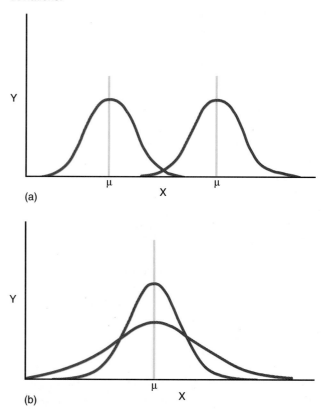

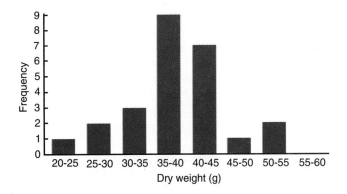

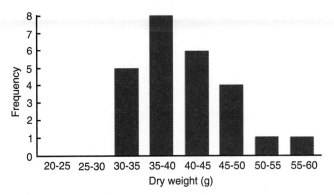

**Figure 2.5** Frequency distribution of dry weights for two samples of 25 plants taken from the same population ($\mu = 40$, $\sigma = 7$).

chance we could observe a difference between two sample means equal to or larger than the difference calculated from our samples, even though there is no significant effect due to treatment. For a given normal distribution and sample size, alpha becomes smaller and smaller as the difference between the two mean values increases. In general, scientists use an alpha of 0.05 or 5 percent, often referred to as the 95 percent level of significance. This level means that there is only a 5 percent chance of having a difference equal to or larger than that observed between the two samples if the null hypothesis is true.

The converse of alpha, the probability of failing to reject a false null hypothesis, is called *beta* ($\beta$). This probability occurs when we assume that the difference in the observed sample means is within the accepted range of possible values (defined by alpha) under the null hypothesis and therefore the samples are from the same population, when in fact the observed difference between the sample means is due to the treatment. We call this error a *Type II error*.

In a way a scientist testing a hypothesis is like a judge attempting to determine the guilt or innocence of a person charged with a crime. Not having witnessed all of the pertinent events of the crime, the judge does not really know whether the person is guilty. Instead, based on the evidence presented, the judge convicts or acquits. We hope that the guilty person is convicted, and the innocent person is acquitted. In the United States, our judicial system has put the burden of proof on the prosecution; the accused is assumed in-

nocent until proven guilty. We minimize the probability of convicting the innocent even though this probability may often mean acquitting the guilty.

Likewise, scientists do not know whether a hypothesis is true or false. They want to accept a true hypothesis and reject a false one. The value of alpha is essentially the chance of being wrong by rejecting a true null hypothesis. If scientists choose a very small value of alpha, they will not reject many true null hypotheses (Type I error); however, they run a high risk of failing to reject those that are false (Type II error). Scientists are like the judge who places a greater importance on not convicting the innocent (Type I error), accepting the risk of acquitting the guilty.

## Statistical Power Test

It is easy for us to fall into the trap of assuming that the null hypothesis has to be distinctly true or false. In reality, hypotheses are statistical, based on probability. If the probability of Type I error is sufficiently small, that is if $p < 0.05$ (or some acceptable level of alpha), the null hypothesis is rejected and the alternative is accepted. What is the probability that our tests will result in the rejection of the null hypothesis (no treatment effect or no difference between mean values for the two samples) at a particular alpha level, when the null hypothesis is really false—that there is an effect—and the alternative is true? That probability is given by a statistical power test, where power equals $1 - \beta$ (Table 2.1). When $(1 - \beta)$ is very small, the test is considered to be very high power; the probability is low that we will not make a Type II error often. But if $(1 - \beta)$ is high, then the power is low and we can have little confidence in conclusions based on the failure to reject the null hypothesis (Toft and Shea 1983).

Power varies as a function of alpha, sample size, and effect size (Rotenberry and Wiens 1985). The effect size is the degree to which the phenomenon under consideration is present in the population, or how far it departs from the null hypothesis. As the size of the effect increases, the statistical ability (power) to detect that effect increases. The smaller the effect, the more difficult it is to detect. If we expect a large effect and fail to detect it (do not reject the null hypothesis), we can be reasonably certain (small $\beta$) that it is not present. Therefore, when a statistical test estimates that

**Table 2.1  Decision Table for Hypothesis Testing and Associated Probabilities**

|  | DECISION | |
| --- | --- | --- |
| *"Truth"* | *Accept $H_0$* | *Reject $H_0$* |
| $H_0$ True | No Error $1 - \alpha$ | Type I Error $\alpha$ |
| $H_0$ False | Type II Error $\beta$ | No Error $1 - \beta$ |

*Source:* From Toft and Shea 1983.

Type I error is sufficiently small ($p < 0.05$), we reject the null hypothesis and go with the alternative hypothesis. If the probability is not sufficiently small, we do not reject the null hypothesis.

Of what importance is the ability to detect Type I or Type II errors, beyond the investigator's interest? Finding no effect when in fact there is one is risky in applied ecology (as well as in toxicology and health sciences), where research guides the management of natural resources and other aspects of the environment. Studies of effects of acid rain on forest growth, pesticides on nontarget species, and increased $CO_2$ levels on crops have economic, political, and social implications. When ecologists fail to reject a null hypothesis, decision-makers conclude that there is no effect, and take no protective action (Toft and Shea 1983, Peterman 1990). On the other hand, rejection of a null hypothesis of an effect when there is none, sends a different message. Decision-makers may decide to play it safe, being overprotective.

To make an informed decision on environmental policy, decision-makers should consider β, the probability of making a Type II error, and the size of the detectable difference (Toft and Shea 1983, Gerrodette 1987, Peterman 1990). Only researchers can present this information. Without it, policy-makers may go wrong. Making a Type I error, incorrectly concluding that an effect exists when it really does not, they could impose unnecessary and costly controls and regulations. Making a Type II error, they could overlook a problem, resulting in economic losses, sickness, and environmental degradation (Peterman 1990).

# MODELS AND PREDICTIONS

A model is an abstract representation of a real system. It lets us predict the response of a dependent variable(s), based on an explicit assumption or set of assumptions. The goal of the scientific process is to understand the patterns of variation in nature, and models allow us to do that.

We can view hypotheses as models. The hypothesis that plant growth will increase under elevated $CO_2$ is an explicit assumption based on the physiological understanding of photosynthesis. We can test the model experimentally and either accept or reject it. Further, we can test the generality of this model with other species, or examine the interaction with other independent variables, such as temperature. This model is qualitative, because it predicts only a direction of variation (increased growth) in the dependent variable rather than the amount of variation. In most cases, once a qualitative relationship has been established, a quantitative model is sought. The new model can be statistical or nonstatistical.

## Statistical Models

Statistical models predict the value of the dependent variable based on mathematical functions. An example of statistical models is the simple linear regression model: $y = a + bx$, where $y$ is the dependent variable, $x$ is the independent variable, $a$ is the parameter describing the $y$ intercept and $b$ is the parameter describing the slope of the line. The $y$ intercept is the value of $y$ when $x$ is equal to zero. The slope of the line quantitatively relates the change in $y$ per unit change in $x$. The parameters, $a$ and $b$, are solved mathematically based on the paired observations of $x$ and $y$.

For example, a long-standing theory is that a close relationship exists between the density of small trees and shrubs and the density of snowshoe hares (*Lepus americanus*). Litvaitis, Sherburne, and Bissonette (1985) established two 49 ha study sites in regenerating spruce-fir and hardwood stands. They determined snowshoe hare densities in spring and fall by mark-recapture methods, quantified habitat use with fecal pellet counts, measured intensity of twig-clipping by the hares, and sampled the density of understory vegetation. The results of their study, summarized in Figure 2.6, showed a linear relationship between hare density and stem density. The hares increased as understory stem density increased. The regression equation ($y = 0.000046x - 1.06$) is a statistical model for predicting the density of hares ($y$) given a value of stem density for the plot ($x$).

One important constraint of such statistical models is that we may not extrapolate beyond the bounds of observations. The model can be used legitimately to predict hare density for plots that have a stem density between 20 and $60 \times 10^3$ stems per hectare. Since no observations were made beyond these bounds, we have no information as to how hare density might vary with still higher density of stems. With increasing density of stems, hare density may continue to increase, level off because of some other factor, or even decline because the stem density impedes mobility. To predict the effect of stem density beyond the bounds of these observations, additional plots with stem densities above 60 and below 20 would have to be included in the analysis. Another way of looking at it is that the model is a description of the correlation between the two variables, not a statement of causation.

## Nonstatistical Models

Many nonstatistical mathematical models go beyond a description of the relationship between the dependent and independent variables and assign a mechanism to the parameters. For example, ecologists have used data on grassland productivity from arid and semiarid grasslands to develop a statistical model relating grassland productivity to annual rainfall (Figure 2.7). Although the model provides a good estimate of expected productivity (dependent variable) within the range of rainfall (independent variable) examined, it does not address the mechanism by which this relationship comes to exist.

To accomplish this goal, we need a much more complex model of plant processes. Coughenour and associates (1985) developed such a model for the grasslands of the Serengeti in East Africa. This model simulates the processes of photosynthesis and respiration as well as the cycling of nutrients

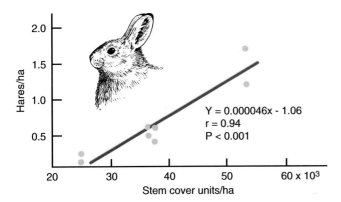

**Figure 2.6** Relationship between stem density and estimated snowshoe hare density in Maine, 1981–1983. (From Litvaitis et al. 1985.)

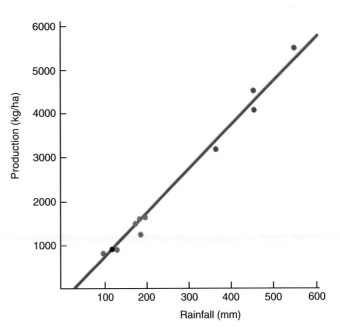

**Figure 2.7** Production of the grassland of Namibia in relation to annual rainfall. (From H. Walter 1973.)

through the ecosystem (Figure 2.8). The model is made up of a series of mathematical equations that describe $CO_2$ uptake, respiration, loss of water through the stomata, and other plant functions. These equations describe physical and biochemical processes. They are not statistical descriptions of observed patterns based on sampled observations as in Figure 2.5. The model can predict patterns of net primary productivity, evapotranspiration, and nutrient cycling through grassland ecosystems as a function of soil, rainfall, and herbivory. Although the model addresses the specific mechanisms responsible for the relationship between rainfall and grassland productivity, it requires much data to define the parameters relating each of these processes to plant growth. Although this approach is a valuable contribution to understanding the grassland ecosystem, the statistical model may, within the bounds of observation, prove to be a better predictor for the purposes of management. Both types of models are essential in the development and application of science.

**Analytical Models** A major dichotomy within the class of nonstatistical models is that of analytical and simulation models. Two approaches currently used by ecologists to model density-dependent population dynamics provide an example of this dichotomy. One type of model used is the logistic population model:

$$dN/dt = rN(K - N)/K$$

where $N$ is some measure of the population size such as the number of individuals or biomass, $t$ is time, $r$ is the instantaneous growth rate expressed in per capita units of $N$, and $K$ is the carrying capacity or the maximum sustainable value of $N$ for the given environment. In contrast to statistical models, the parameters are assigned a biological meaning with respect to the process being simulated. Figure 2.9 illustrates the general form of the logistic model.

The logistic model is an analytical model because the equation can be solved. For example, consider a hypothetical population of an initial size of 30 with the parameters of $r = 0.186$ and $K = 50$. If we insert these parameters into the logistic model and solve, we arrive at a predicted population size of 38 in six years and 49 in twenty years. Thus the behavior of the model can be analyzed with respect to the parameters. Also, the logistic model is deterministic, because only one solution exists for a given set of parameters ($r$ and $K$) and initial population size ($N$).

**Simulation Models** Simulation models cannot be solved analytically. They require the use of a computer. An example is the group of models known as individual-based population models. In contrast to the logistic population growth model, individual-based population models simulate each individual organism within the population rather than an aggregated parameter describing the population as a whole. For example, the parameter $r$ in the logistic equation is a population estimate of the average net fecundity rate per individual and $K$ represents the carrying capacity of the habitat in terms of maximum sustainable population size. In contrast, individual-based models simulate the establishment, growth, reproduction and mortality of each individual in the population, with these processes responding directly to availability of resources like water, light, and nutrients. Fecundity rates may vary among individuals based on their size and age, and the carrying capacity of the habitat may vary as a function of resource availability. The population response is viewed as the composite of individuals. One advantage of this approach is that it is not necessary to make

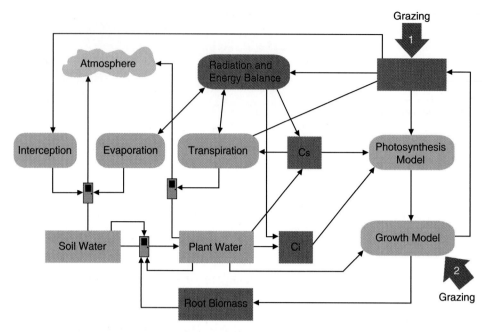

**Figure 2.8** Nonstatistical model of the Serengeti grassland ecosystem. The diagram depicts the components of the system—soil water, plant water, leaf area, and root biomass (designated by rectangles)—and the processes, flows of water and nutrient transfers among the compartments. All components are affected by grazing, which reduces the leaf area of plants and removes the growing points at the tips of roots and stems. The squares $C_s$ and $C_i$ are stomatal and internal conductances of water. Doors indicate flow regulation. (From Coughenour et al. 1985.)

two assumptions associated with the logistic model: (1) all individuals within the population are identical—the unique features of each are unimportant; (2) the population is perfectly mixed—there are no local spatial interactions of any important magnitude (Huston et al. 1988). In some cases these assumptions are appropriate; a plantation of trees might be uniform. They are inappropriate, for example, where competitive ability is size-dependent (as in a mixed-age stand of trees) or where reproduction is dependent on the establishment and maintenance of territory (as among territorial animals).

An example of individual-based models is the forest gap model (Botkin et al. 1971; Shugart and West 1977). Forest gap models simulate the establishment, growth, reproduc-

tion, and mortality of individual trees on a forest stand. Each individual responds to the environment of the forest stand (temperature, light, water, and nutrients) and in return influences the environment of the stand (for instances, tall trees shade smaller ones). These models are much more complex than the population-level models, in that each individual in the population must be simulated through time. The models require a great amount of information on the life history of species to relate the processes of growth, reproduction, and mortality to environmental conditions. In fact, the models must simulate (1) the environmental conditions of the forest stand; (2) the response of plants to that environment; (3) how plants modify the environment of the plot, influencing the other individuals in the stand.

Both modeling approaches have their advantages and disadvantages. The more appropriate approach depends upon the objectives of the investigator and the question being addressed.

## Validation

Once a model is developed, the investigator must validate it. Does the model really agree with the behavior of the real-life system it is to mimic? To validate the model, the investigator must collect an independent set of data in the field, incorporate it into the model, and determine whether the predictions of the model based on the new

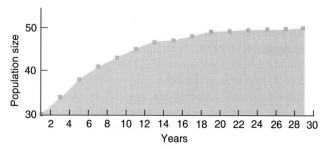

**Figure 2.9** Logistic growth curve involving two parameters: $r = 0.186$ (rate of growth) and $K = 50$ (carrying capacity).

set of data compare with data used to frame the model (Jeffers 1988).

Validation is the objective test of how much confidence we can place in the model. It measures quantitatively the extent to which the output of the model agrees with the behavior of the real-life system (Jeffers 1988). In other words, is the model capable of producing empirically correct predictions? The process compares the model's results with the data upon which it is based as well as with new sets of observations that are independent of the data used to frame the model. Validation indicates (1) how well the model predicts both the data used in its construction and other independent data, and (2) how consistently the model predicts other occurrences. This is a most important step. Ecology has many models, but few are well validated.

Fahrig and Merriam (1985) used a simulation model approach to examine the consequences of habitat patchiness on the population dynamics of white-footed mice (*Peromyscus leucopus*) inhabiting forest patches in an agricultural landscape. The question that the ecologists addressed was whether population dynamics within a patch were influenced by the degree to which it is isolated from other patches.

The experimenters developed a patch dynamics model that was used to simulate changing sizes of resident mice populations in a series of interconnected habitat patches. The model was structured to simulate a series of habitat patches (forest woodlots) located in a landscape of otherwise unsuitable habitat (agricultural fields). These habitat patches were either isolated or connected to other patches (Figure 2.10). Connected patches allow for movement of individuals between patches, whereas isolated patches do not. The model had two components of population dynamics: (1) within-patch dynamics, and (2) between-patch dynamics. Within-patch dynamics included recruitment into the population from reproduction, movement of individuals from one age class to the next, and age-specific mortality. Between-patch dynamics included as rates of immigration and emigration, which were dependent on the state of the population in each patch and its connection to or isolation from other patches.

After conducting model simulations, the investigators collected field data on population dynamics of white-footed mice inhabiting seven woodlots to compare real-life situations with model simulations. Some of these woodlots were surrounded by agricultural fields; other woodlots were connected by fencerows, which functioned as corridors between forest patches.

The model predicted that population growth is lower in isolated than in connected woodlots but that population growth rates within a patch type did not differ. The field study of the populations in the seven woodlots showed similar results. Thus investigators were able to verify their model for their particular locale. Such verification, however, does not mean that the model is applicable to other areas without further testing.

Although usually regarded as predictors, models, even unvalidated ones, have other uses. They can suggest differ-

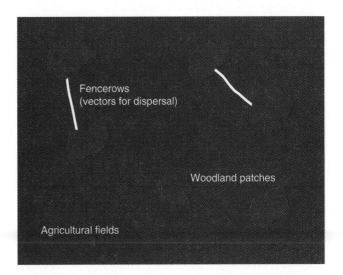

**Figure 2.10** Diagram for a simulation model of isolated and interconnected woodland islands in an agricultural landscape.

ent solutions to a problem and yield insights that might otherwise go undiscovered. They can also be used together with empirical methods to generate new hypotheses and to design more powerful research programs.

## SUMMARY

Ecology has moved from a descriptive to an empirical and experimental approach emphasizing hypothesis testing, statistical analysis, and development of models to provide new insights and develop new hypotheses. A hypothesis is a statement about causative agents that can be tested experimentally.

Two approaches to testing hypotheses are the inductive method, which goes from the specific to the general, and the deductive method, which goes from the general to the specific. The inductive approach is useful for investigating correlations between classes of facts. In the deductive method, the investigator develops a research hypothesis, collects data to support or refute it, develops a mathematical model, attempts to fit the model to the data, and then, if necessary, modifies the model.

Testing hypotheses entails the collection of data by direct observation or by experimentation. The latter involves simplification by manipulating one or a few variables while holding others constant. Both laboratory and field experiments involve manipulations of one or a few independent variables in natural communities.

By experimentation we determine the response of one variable, the dependent, to variations in another independent variable or variables manipulated by treatments. We must replicate treatments to account for uncontrolled variations among experimental units. Controls that do not receive the treatment form the basis for comparison. The results allow the investigator to reject or accept the hypothesis. Rejection

or failure to reject a hypothesis may involve a Type I or a Type II error. Rejection of a hypothesis even though it is true is a Type I error; failure to reject a false hypothesis is a Type II error.

A model is an abstraction and simplification of natural phenomena developed to predict a new phenomenon or to provide insights into existing ones. A verbal or graphic model may serve as the basis of a more formal mathematical model. A mathematical model may be statistical or nonstatistical. Nonstatistical models may be analytical or simulation. Analytical models are mathematical formulations that can be solved directly. Simulation models take on a variety of forms, including differential equations. Because they cannot be solved analytically, simulation models require the use of a computer to arrive at a solution.

Once developed, the model should be validated. Validation is the explicit and objective test of the basic hypothesis. It measures quantitatively the extent to which the output of the model agrees with the behavior of the real-life system.

## REVIEW QUESTIONS

1. What is a hypothesis?
2. How do the inductive and deductive approaches to testing hypotheses differ? What are advantages and weakness of each?
3. What distinguishes a dependent variable from an independent variable?
4. What is the importance of replicates and controls in ecological studies? What is the problem of pseudoreplication?
5. What is a model? Why are models useful in ecological research?
6. What is the difference between statistical and nonstatistical models? How do analytical models differ from simulation models?
7. Comment on this statement by C. J. Krebs: "Hypotheses without data are not very useful, and data without hypotheses are wasted." What does Krebs mean?
8. Examine some papers in such ecological journals as *Ecology* and *Journal of Animal Ecology,* both in early and recent years. How do they differ in experimental approach? Are the hypotheses clearly stated? Are the data collected adequate for testing the hypotheses?

## CROSS-REFERENCES

Photosynthesis, 154–159; grassland function, 233–236; population growth, 392–396.

*Part 2*

# The Organism and Its Environment

# Adaptation

## Concepts

1. Adaptation is any heritable behavioral, morphological, or physiological trait that maintains or increases the fitness of an organism to live under a given set of environmental conditions.
2. Adaptation, measured as fitness of the organism, is the result of natural selection.
3. An organism is adapted to function between some upper and lower limits of environmental conditions. This adaptiveness is expressed as the law of tolerance.
4. Organisms must maintain a rather constant internal environment compared to external environments. They maintain this conformity through homeostatic mechanisms.

Why does the range of the balsam fir (*Abies balsamea*), a boreal tree, stop in the northern United States, except for a few outliers to the south? Although individual trees survive as specimen trees farther south, it does not reproduce successfully there. Why has the exotic Royal Paulownia (*Paulownia tomentosa*) tree, introduced into the United States from China as an ornamental, been able to escape from cultivation and thrive in vacant land and along roadsides in the southern United States? Of some 30 species of exotic birds introduced into the United States from Europe, why have only a few, especially the starling (*Sturnus vulgaris*), the house sparrow (*Passer domesticus*), and the rock dove (*Columba livia*), successfully colonized North America? Answers to such questions relate to the adaptive ability of these species.

## THE MEANING OF ADAPTATION

**Adaptation** is one of those terms in ecology that has been burdened with different meanings. The most common one is any behavioral, morphological, or physiological trait that is assumed to be the result of natural selection. A second meaning is any physiological or morphological feature or form of behavior used to explain the ability of an organism to live where it does. A third definition is a change in physical, physiological, or behavioral traits that results from some current environmental pressure, such as "adapting" to a change in temperature.

To understand adaptation we must turn to some basics. Consider a local population of any organism living in a given environment. It consists of a number of individuals of different ages that possess heritable variations. Some of these individuals die early in life; others survive to maturity but fail to reproduce; and still others reproduce and leave behind various numbers of offspring. The fate of individuals hinges on genetically determined characteristics that enable them to cope with the physical and biological environment.

Among the reproducing individuals, some will leave more offspring than others. These individuals are considered more fit than the others because they contribute the most to the population gene pool. Organisms that leave few or no offspring contribute little or nothing to the gene pool and so are considered less fit.

Fitness, however, is more than just a numbers game. Assume that individual A leaves four offspring and individual B leaves two offspring. By our general definition of fitness, we would say that A is more fit than B. However, suppose that of the four individuals produced by A only one reproduces, while both offspring of B reproduce. Although A left more offspring, B actually contributes proportionately more to the next generation, making B the more fit. It is not necessarily the number of offspring an individual leaves behind that measures fitness; rather it is the number of its descendants that influence the heritable characteristics of the population. The **fitness** of an individual is measured by the contribution it makes to future generations.

The differential reproductive success or fitness of individual organisms comes about through the process of **natural selection.** Under a given set of environmental conditions, those individuals most able to cope with the situation are selected for, and those unable to do so are selected against. Thus natural selection favors any heritable structural or behavioral characteristic that increases fitness.

The heritable characteristics an organism possesses, it owes to past generations. Its ancestors, in effect, experienced a selective screening of various combinations of genetic characteristics that produced the present set. The possession of these characteristics enables the organism to match the features of its environment. As long as environmental conditions are similar to those of the past, the organism is adapted to the environment. If conditions change significantly, then the fitness and even the survival of individuals will be in jeopardy. Adaptation, then, is any heritable behavioral, morphological, or physiological trait that maintains or increases the fitness of an organism under a given set of environmental conditions.

The key to adaptability is the diversity of genetic characteristics (genotypes) found among the local populations of a species. Each local population exists under a particular set of environmental conditions, and its members have evolved genetic adaptations to that environment. Such populations adapted to local conditions are **ecotypes.** Ecotypes are most prevalent among plants and sessile organisms. Whereas mobile organisms can move to more favorable environmental conditions as the situation demands, plants and sessile organism are fixed in place. They cannot escape the environmental conditions under which they grow. The organisms either adapt to the environment or perish.

The genetic responses of individuals in a local population living under a particular set of conditions are expressed in phenotypes. The **phenotype** is the physical expression of the genotype and its interaction with the environment. Through its effects on the phenotype, the genotype is responsible for the manner in which the individual interacts with its environment. These phenotypic responses may be morphological (physical characteristics), physiological (functional characteristics), or phenological (the timing of growth, flowering, and other life history changes stimulated by seasonal changes in the environment).

Within the local habitats, responses to variations in microenvironments result in a phenotype adapted to these microconditions. The part of phenotypic variation among individuals in a local population that can be attributed to their microenvironments is **phenotypic plasticity** (Scheiner 1993).

Consider the perennial Arctic herbaceous plant *Dryas actopetela,* whose local populations occupy two habitats: the

dry ridges and stony slopes known as fellfields, and snow-beds where plants are protected overwinter by snow and the soil is moist. J. McGraw and J. Antonovics (McGraw and Antonovics 1983; McGraw 1985a, b) studied the physiology and population dynamics of the populations occupying the two habitats. Some of the differences between the two are genetically fixed. The snowbed *Dryas* has longer petioles (the stalk that attaches the leaf to the stem) than the fellfield *Dryas,* larger leaves with more marginal teeth, higher specific leaf weight, and a flowering period up to two months later.

In their study, the two plant ecologists transplanted fellfield *Dryas* to snowbeds and snowbed *Dryas* to the fellfields, along with controls transplanted within their "home" sites. Many of the characteristics of both *Dryas* remained distinct, even though less pronounced. However the petiole length, leaf area, marginal teeth on the leaf and specific leaf weight of the transplanted snowbed *Dryas* more nearly matched those of their fellfield neighbors. The fellfield plants did not exhibit such morphological plasticity.

More pronounced was the variation in survival during the different stages of the life cycle. When the two *Dryas* were planted together in the habitats, the fellfield type germinated equally well in both habitats, but the snowbed type showed poorer germination in the fellfield. The fellfield types had poorer survival than the snowbed plants after the first winter and at the end of the first year in the snowbed, but there was no initial difference in adult survival. The snowbed ecotype, however, survived high adult mortality in the fellfield environment.

On the basis of experimental field manipulations, the two ecologists found that light and nutrients were the most significant environmental features involved in differential survival. The fellfield *Dryas* were less shade-tolerant and grew poorly when covered by shade cloth or overtopped by associated grasses; the plasticity in leaf size of the snowbed *Dryas* reduced self-shading where the plants were crowded. The addition of nitrogen and phosphorus stimulated the growth of snowbed plants, but not the fellfield plants, crowded by growth of overtopping grasses. Thus both physiologically and morphologically, the snowbed *Dryas* exhibited much greater phenotypic plasticity than the fellfield *Dryas*.

Although some adaptive features of an organism may be the result of natural selection, others may have different origins (Gould 1982, Harper 1982). Some current adaptive traits may have shifted into current use from a different previous use. For example, certain mating displays among birds have apparently derived from grooming behavior (Tinbergen 1951, Johnsgard 1965). A structure or feature currently in use may have derived from one that had no previous function. Not every feature of an organism is necessarily functional. If it is neutral in its effects, or if it is an accidental by-product of other selection changes, a feature

may be retained because it is not subject to any selection pressures. As such, it becomes a source of raw material for future evolution.

## TOLERANCE

The adaptiveness of an organism to its environment is exhibited by its ability to function between upper and lower limits within a range of environmental conditions. The role of environmental limitations in the response of organisms to their environment was recognized as early as 1840 by the German organic chemist Justus von Liebig. In his book *Organic Chemistry and Its Application to Agriculture and Physiology,* Liebig described plant chemistry and growth and set forth his conclusions in a simple statement, revolutionary for his day: "The crops of a field diminish or increase in exact proportion to the diminution or increase of the mineral substance conveyed to it in manure."

What Liebig was saying is that each plant requires certain kinds and quantities of nutrients. If one of these mineral substances is absent, the plant fails to grow or dies. If the substance is present in minimal quantities only, the growth of the plant will be minimal. This statement became known as the *law of the minimum.*

The law of the minimum, strictly applied, carries with it some restrictions. It applies only under steady-state conditions, with all other resources being in excess of needs. If the quantity of the limiting resource increases, so that the rate of growth of the plant increases, then that substance is no longer limiting, but another might become so, as the increased growth places more demand on other resources. Further, the concentration or availability of another nutrient may modify or influence the availability or use of the minimum one. For example, if availability of phosphorus is low, molybdenum, another essential plant nutrient needed in very small amounts, may be sufficient for the plant's needs. If the availability of phosphorus increases to a point that it stimulates additional plant growth, molybdenum will become the limiting element. This concept applies not only to nutrients but also to other factors of the environment.

Too little of a nutrient or other resource, such as light, may be harmful; but too much of a good thing can be just as bad, and thus limiting. F. F. Blackman (1905) advanced this concept. He pointed out that a maximum quantity of a resource tolerated by an organism would limit response as well. Blackman's observations became known as the *law of limiting factors.*

The law of the minimum and the law of limiting factors operate within environmental conditions. The reproduction, growth and functioning of individual organisms depend upon the amount of the essential environmental requirements presented to them in minimal quantities during the most critical season of the year. Thus, for many birds and

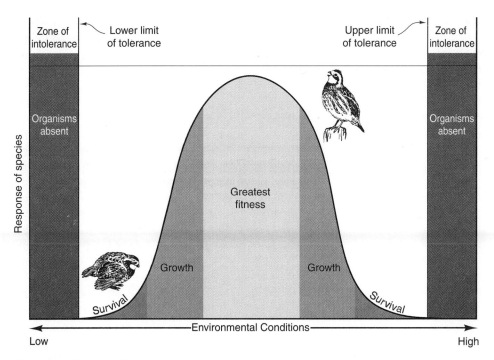

**Figure 3.1** The law of tolerance.

mammals, the availability of food is critical not during the winter but between the end of winter and early spring when food resources have been exhausted. Organisms, then, live within a range of too much and too little, the limits of tolerance. This concept of an organism's response to environmental conditions limiting its reproductive success, growth, and ability to survive was incorporated by V. E. Shelford in 1913 into the *law of tolerance.*

The law of tolerance can be illustrated as a bell-shaped tolerance curve (Figure 3.1). The ordinate or *x* axis represents the gradient or range of a particular environmental factor. The abscissa or *y* axis represents the response of the species or individuals of a species on that gradient. The upper or middle part of the curve embraces the optimal state for reproduction or fitness. Descending parts of the graph represent conditions under which individuals grow but possess lower fitness, survive but not reproduce, and fail to survive.

The tolerance curves are broad for many organisms. They are able to exist within a wide range of values for a particular environmental factor, such as salinity, temperature, or humidity. Other organisms, even life stages within the same species, have a narrow range of tolerance, concentrated at either end of the total tolerance curve. For example, the adult blue crab (*Callinectes sapidus*) of the American East and Gulf Coasts can live in salinities that range from seawater (about 34 ppt) to nearly fresh water. Eggs and larvae of the blue crab can survive only in salinities above 23 ppt.

The range of tolerance is not fixed. As seasons and conditions change, individuals may acclimate to them and shift the tolerance curves to the right or to the left (Figure 3.2) Consider, for example, fish inhabiting a pond in which the

water temperature changes from spring through winter. As the water warms in spring, the tolerance of fish for warmer temperatures gradually increases; at the same time their tolerance for lower temperatures decreases. Similarly, as the water cools in fall and winter, the tolerance for low temperatures increases, while the tolerance for high temperatures decreases. Thus a temperature that would be lethal for a fish in winter, if the fish were suddenly exposed to it, can be tolerated in summer. Although the tolerance ranges shift with the season, all the shifting takes place within the adaptive physiological limits of the organism. This plastic, relatively short-term response of an individual to exposure to changing natural environments is **acclimatization.**

The law of tolerance explains much about the local and geographical distribution of species. Most are restricted by a condition or conditions beyond the tolerable limits of the most sensitive stage of their life cycle. An organism may have a wide range of tolerance for some substance or condition, but a narrow range for another, and will thus be limited

**Figure 3.2** Seasonal shifts in tolerance ranges.

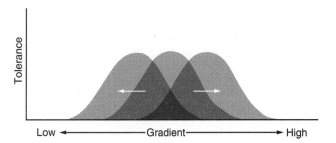

by the condition for which it has a narrow range. A mature tree, for example, may survive and grow under environmental conditions outside its natural range, but cannot reproduce because its seedlings are outside their tolerance range. In some cases, a less than optimal state in one condition may lower the limits of tolerance for another; or the organism may achieve its best fitness at some intermediate point between interacting factors.

# HOMEOSTASIS

Daily and seasonally, most organisms are confronted with a changing external environment. In spite of it, they have to maintain a fairly constant internal environment. They need some means of regulating internal conditions such as body temperature, water balance, pH, and amount of salts in fluids and tissues, relative to external ones. They take in substances from the environment and use them in cellular chemical reactions; but they also have to release to the environment both excessive intake of substances and metabolic waste products to maintain a fairly constant internal state.

Animals that allow some aspects of their internal environment, such as temperature, more or less to follow external conditions are called *poikilotherms*. Reptiles are examples. Others, notably birds and mammals, maintain constant internal environments through metabolic activity; they are called **homeotherms.** Poikilotherms may regulate some of their internal environment, such as salt concentrations; and some exceptional homeotherms, such as bats and hummingbirds, may allow their body temperatures to drop to conform to the environment. The maintenance of a relatively constant internal environment in a varying external environment is called **homeostasis.**

Homeostasis involves the flow of external environmental information into a system with a biological homeostatic device or mechanism that responds to changes. We are surrounded by all sorts of mechanical homeostatic devices. Consider the thermostat that controls the furnace. If we wish the temperature of the room to be 20°C (68°F), we establish that set point on the thermostat. When the temperature of the air falls below that point, a temperature-sensitive device within the thermostat trips the switch that turns on the furnace. When the temperature reaches the set point, the temperature-sensitive device responds by shutting off the furnace heating element. This type of response that halts or reverses the movement away from and returns it to the set point is called **negative feedback.** The state of the system and its inputs are inversely related.

If the thermostat fails to function properly and does not shut the furnace off, then the furnace continues to burn, the temperature continues to rise, and eventually the furnace overheats. Assume the excessive heat from the furnace sets fire to surrounding walls. That fire spreads through the building, and then to those surrounding it. The fire feeds upon itself, expanding and growing more intense. This continued movement away from the set point is called **positive feedback.** The measure of this feedback is directly related to input. Examples of positive feedback are compound interest, the geometric growth of populations, and even the growth of cancerous cells.

All organisms maintain some sort of homeostasis. To appreciate how homeostasis operates in living systems, consider how your body responds to a sharp drop in temperature (Figure 3.3). The normal temperature or set point for humans is 37° C (98.6° F). When the environmental temperature drops, sensory mechanisms in the skin detect it and send a message to the hypothalamus in the brain. The hypothalamus (which behaves much like a thermostat) sends messages to higher brain centers that result in voluntary acts, such as putting on a coat or moving closer to heat. It also sends a message to the autonomic nervous system that triggers shivering, an involuntary muscular exercise producing more heat. The hypothalamus stimulates the adrenal gland to release the hormone epinephrine into the bloodstream. This hormone speeds up the conversion of liver glycogen to glucose. At the same time the hypothalamus stimulates the pituitary gland to release a thyroid-stimulating hormone, causing the thyroid to increase the release of its hormone, thyroxine. Both hormones stimulate body cells to increase their respiratory activity, producing body heat. As the body temperature increases, the hypothalamus detects the change and slows down heat production. Functioning by negative feedback, the hypothalamus maintains an optimal body temperature.

**Figure 3.3** Homeostatic control of your body temperature.

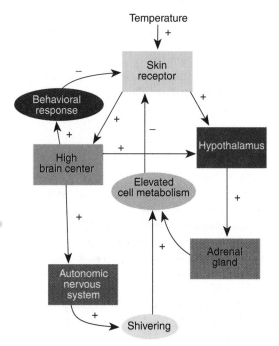

If you experience extreme environmental temperatures, the homeostatic system may break down. If the environmental temperature drops too low, the homeostatic system is unable to generate enough heat to sustain normal internal body temperature. The body cools, and metabolic processes slow down, further decreasing body temperature. Unless cooling is halted, you would die by freezing. If the environmental temperature becomes too hot, the body cannot lose heat fast enough to hold the temperature about the set point. In an attempt to control the situation, body metabolism speeds up, further increasing body temperature, eventually ending in heatstroke or death. In these situations the feedback is positive, reinforcing change and driving the system to lower and lower or higher and higher values.

The big difference between living and mechanical systems is that in living systems the set point is not firmly fixed as it is in mechanical systems. Instead, organisms have a limited range of tolerances, called *homeostatic plateaus*. The homeostatic system works within the maximum and minimum values by using negative feedback to regulate activity about the set point. If the system deviates from the set point, a negative feedback response—a control mechanism—inhibits any strong movement away from the set point. If it fails to do so, positive feedback drives the response away from the set point.

Among the homeotherms the maintenance of homeostasis is largely physiological, aided by behavioral and morphological mechanisms. To avoid summer heat, birds and mammals may seek shade; to keep warm in winter, they add an insulating layer of fur, feathers, or fat through physiological responses to changing daylengths. Poikilotherms maintain temperature homeostasis by gaining heat from the environment. Their control of homeostasis is largely behavioral. With responses involving negative feedback loops that send messages back to the brain, poikilotherms seek shade or warmth to maintain body temperatures. Many of their other homeostatic responses, however, are purely physiological, such as those of fish maintaining their internal osmotic pressure in water. The role of homeostasis will become much more evident in the chapters that follow.

# SUMMARY

To survive, function, and leave reproducing offspring, organisms must be adapted to their environment. Although the term *adaptation* has several meanings, ecologically it refers to any heritable behavioral, morphological, or physiological trait that maintains or increases the fitness of an organism under a given set of environmental conditions. Organisms acquire their adaptations through the selective forces of the environment. The most adapted or most fit contribute to the next generation; the less fit do not. The key to adaptation is the genetic variability of local populations

that exist under particular environments and have evolved genetic adaptations to them. Genetic diversity and phenotypic plasticity, the physical expression of the interaction between genotype and the environment, within a population enable individuals to respond to short-term or long-term changes in the environment.

The adaptiveness of organisms involves their ability to function within some upper and lower limits within the range of environmental conditions. The response of organisms to limiting environmental conditions is expressed in the law of tolerance as measured by fitness. The ranges of tolerance are not fixed but vary seasonally within the total physiological tolerance of the organism. The ranges of tolerance influence distribution of organisms. Those organisms possessing a wide range of tolerances will be the most widely distributed.

Despite daily and seasonal environmental changes, organisms must maintain some equilibrium in their internal environment. The maintenance of a relatively constant internal environment in a varying external environment is called homeostasis. Homeostasis involves negative feedback responses. An organism responds physiologically and behaviorally to maintain an optimal internal environment.

# REVIEW QUESTIONS

1. Define adaptation.
2. How do organisms become adapted to their environment? How does adaptation relate to natural selection?
3. Contrast the law of the minimum with the law of tolerance.
4. What are ecotypes?
5. Define homeostasis. How does homeostasis function through negative feedback?
6. P. T. Boag and P. R. Grant (1986) found that during very dry years when the abundance of small seeds declined faster than large seeds, larger individuals of the medium ground finch *Geospiza fortis* on the island of Daphne Major experienced higher survival than smaller individuals. High survival related to the ability of the larger birds to feed on large, hard seeds. The population suffered high mortality, with most of the survivers being the larger individuals. Discuss this phenomenon relative to natural selection, adaptation, and the role of genetic variability in a local populations.

# CROSS-REFERENCES

Plant responses to moisture, 70–75; animal responses to moisture, 75–77; plant responses to temperature, 83–85; animal responses to temperature, 85–95; plant adaptations to light intensity, 102–104; photoperiodism, 106–110; seasonality, 110–112; nutrients and plants, 112–124, tundra, 250–261.

# Climate

## Outline

## Concepts

1. Climate is a product of weather over time.
2. Solar radiation is the major determinant of climate.
3. Temperature change in rising and descending air masses is an adiabatic process.
4. All circulation systems within the atmosphere and oceans are driven by solar radiation.
5. A persistent physical effect of Earth's rotation is the Coriolis effect.
6. Climate has a pronounced influence on the global and regional distribution of plants and animals.
7. Microclimates have a more pervasive influence on the local distribution of plants and animals organisms than the general climate of a region.
8. Major vegetational units reflecting climatic patterns are biomes.

Today's temperature, the TV weather reporter informs us, is so many degrees above or below normal; and the rainfall for the month is below normal. The reporter refers to the means or averages of temperature and precipitation over a certain period of time. These averages provide the norm about which meteorologists can describe daily weather fluctuations. A combination of temperature, moisture, precipitation, and winds for a given place expressed as means or averages describes its *climate*. Because of the daily, monthly, and yearly fluctuations, descriptions of climate never exactly match from one period to another, nor do they give us an exact picture. What might have been typical climate for a region 50 years ago may not be the climate there today.

Climate is a product of weather, which is air in motion, driven by unequal heating. Heating of the atmosphere involves an exchange of heat between the air and the surface area over which it flows. Loss of heat by an air mass over one region is balanced by the gain of heat elsewhere, but only over a long period of time. The short-term imbalance of heat gains and losses is responsible for most disturbances in the atmosphere. Masses of cold air and masses of warm air often clash along a moving front (the boundary separating two air masses with different properties) to produce major storms.

Major air masses differ in several ways, including where they originate, how they move, and where their frontal zones come into conflict. This variability provides the basis of our classification of climate. World climates fall into three major groups. Group I includes tropical and equatorial air masses. Group III includes the cold climates dominated by polar air masses. Squeezed between and subject to all of the vagaries of tropical and polar air masses is Group II, the temperate climates of the middle latitudes. Polar air masses from the north and tropical air masses from the south sweep over the middle latitudes. Where the two masses meet becomes a zone of intense conflict, producing highly variable and fluctuating weather conditions. For that reason ecologists look not to meterologic records but to plants, highly sensitive indicators of long-term climatic conditions, to produce a better generalization of climate.

# SOLAR RADIATION: THE KEY TO CLIMATE

Looking at a photograph of Earth taken from space, you can easily get the impression that all parts of the planet equally share the solar radiation reaching it. They do not. The polar regions receive much less solar radiation than the tropical and temperate regions. The sun seems to shine equally everywhere on the globe, but it does not; there are great inequalities. Variations in Earth's surface, water, rock, sand, soil, snow cover, and vegetation, along with Earth's daily rotation and yearly revolution about the sun, influence how the heat of the sun is absorbed by Earth and distributed over the globe. These variations and inequalities create different patterns of heating and cooling that influence Earth's climates.

## Fate of Solar Radiation

Solar radiation travels more or less unimpeded toward Earth until it contacts Earth's atmosphere. The amount of solar radiation that reaches that point (a height of 83 km) as measured on a surface held perpendicular to the sun's rays is 1.98 ly/min, a value known as the **solar constant.** (A langley is equal to one gram calorie of heat received or emitted by one square centimeter of surface.) Although called a constant, it actually fluctuates because of variations in ultraviolet outputs of the sun and solar flares.

However bright the sun may seem, only 50 percent of the solar energy traveling to Earth makes it through the atmosphere to Earth's surface. The surface receives approximately 1 ly/min in the form of short-wave radiation (electromagnetic wave lengths under 3 microns). What happens to all the incoming energy? If you consider the amount of solar radiation that reaches the atmosphere as 100 percent, 25 percent is reflected from clouds and atmosphere back to space along with another 5 percent by Earth's surface (Figure 4.1). Another 25 percent is absorbed by dust, water vapor, and carbon dioxide in the atmosphere. Thus reflection and absorption remove 55 percent of the solar radiation as it arrives. Earth's surface absorbs the remaining 45 percent as short-wave radiation. This energy is radiated back as long-wave radiation (electromagnetic wavelengths longer than 3 microns) and is absorbed by the atmosphere. Of this energy, 12 units escape to outer space from the atmosphere, and 33 units are absorbed by water vapor and carbon dioxide in the atmosphere that act much like a blanket over Earth. This heat is radiated back to Earth, giving some stability to surface temperatures.

In passing through the atmosphere, the energy of certain wavelengths is absorbed, so a limited spectrum of energy reaches Earth's surface (Figure 4.2). The atmosphere removes nearly all of the ultraviolet radiation. Atmospheric gases scatter shorter wavelengths, giving a bluish color to the sky and causing Earth to shine out in space, as evidenced by photographs taken from the moon. Water vapor scatters radiation of all wavelengths, so an atmosphere with much water vapor is whitish—thus the grayish appearance of a cloudy day. Dust scatters long wavelengths to produce reds and yellows in the atmosphere. Because of the scattering of solar radiation by dust and water vapor, part of it reaches Earth as diffuse light from the sky, called *skylight*. This skylight enables us to see in shaded areas and in twilight. Infrared radiation that reaches Earth and is sensed as heat (sensible heat) is absorbed and a portion is reradiated back as far infrared (4 to 100 µ). What we see as light is visible ra-

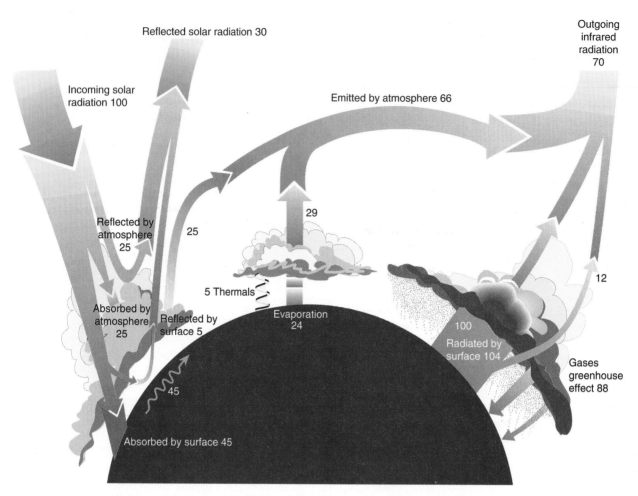

**Figure 4.1** Earth's radiational heat balance. The distribution of incoming short-wave solar radiation is indicated as percentages of 100 units reaching the top of the atmosphere. Reradiation creates the greenhouse effect, which tends to stabilize Earth's temperature. (After Schneider 1987.)

diation that can be separated down into a spectrum that ranges from violet to red. This part of solar radiation has little to do with Earth's heat budget, but it is of inestimable importance in energy fixation through photosynthesis.

## Albedo

Earth does not absorb all the solar radiation impinging on it. Earth's surface reflects back a percentage of the solar radiation, called **albedo.** The amount reflected back determines how fast and to what degree the surface is heated. Water's surface has a low albedo for direct rays, approximately 2 percent, and a very high albedo for low angle rays, so there is glare on the water in late afternoons of summer. The albedo for snow and ice is high, 45 to 90 percent; hence the problem of snow blindness on sunny days in a snowy landscape. For forests and grassland, the albedo ranges from 5 to 30 percent, and for clouds overhead it is about 90 percent.

Because of the reflectance of the surface and the angles of the sun's rays, Earth's albedo varies from region to region

throughout the year. Albedos for land masses do not vary much from one to another during the summer. From tundra to desert in North America, summer albedos are about 16 percent. Greens and browns of the summer landscape absorb far more solar radiation than they reflect. During winter, the story is different. The treeless, snow-covered tundra has a winter albedo of about 85 percent, compared to 46 to 50 percent for forest and grassland at 45° to 55° N latitude, and 18 to 19 percent for western desert and shrubland and eastern croplands and woodlands at 25 to 35° N latitude.

Global albedos measured from outer space beyond the atmosphere range from high values of 50 to 60 percent at the polar regions to lows from 20 to 30 percent in tropical and equatorial latitudes. Although the Southern Hemisphere has less land mass and more ocean than the Northern Hemisphere, their albedos are similar. This similarity suggests that global albedos are influenced more by cloud cover than by surfaces of land and sea. However, the albedo of the Northern Hemisphere is more irregular than that of the Southern Hemisphere.

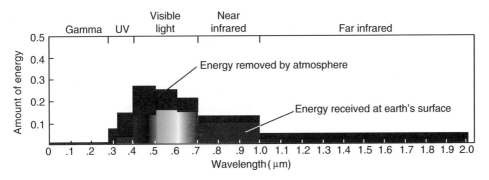

**Figure 4.2** Energy in the solar spectrum before and after depletion by the atmosphere from a solar altitude of 30°. (From Reifsnyder and Lull 1965.)

*Air warmer near ground ⇒ greater atm. pressure*

## Adiabatic Processes

Earth is a giant heat machine driven by the heating and cooling of an atmosphere that obeys the law of gases. A gas allowed to expand becomes cooler; the same volume of gas compressed into a smaller space becomes warmer. Expanded into a larger space, the molecules are more widely separated and collide less frequently, resulting in a drop in sensible heat. When the same volume of gas is compressed, the molecules are closer together and collide more frequently, raising sensible heat. Such a process, in which heat is neither lost to nor brought in from the outside, is termed an **adiabatic process.** The temperature change is called the *dry adiabatic temperature change.*

This adiabatic process occurs when air rises and sinks in the atmosphere. Atmospheric pressure near the ground is greater than at high altitudes, and therefore the air is warmer near the ground. If a parcel of air is warmer than its environment, it becomes positively buoyant and ascends. As this parcel of dry air rises, it moves into an area of lower atmospheric pressure and cools at the rate of 10° C per km (1° C per 100 m). When an air mass sinks to a lower elevation, the air compresses and warms at the same rate. This rate of temperature change with change in elevation is called the *dry adiabatic lapse rate.*

As a mass of air moves up or down, the moisture it contains condenses as it cools, because cool air holds less moisture than warm. As the moisture condenses, the air gains some heat of condensation (heat released when water cools), generating sensible heat that counteracts the cooling. The rate of cooling now takes on a reduced value known as the *moist adiabatic lapse rate.* The moist adiabatic lapse rate has no single value, because it varies with the temperature of and the moisture in the air; but the rate averages about 6° C per km.

Under ideal conditions air rises and cools and sinks and heats at adiabatic rates; but differential heating of land and water areas and lower regions of the atmosphere creates instability in the atmosphere. Earth by day is heated by short-wave solar radiation, which is absorbed in different amounts over the land depending on vegetation, slope, soil, season, and so on. Lower layers of the atmosphere, heated by the Earth's surface, rise in small volumes; colder air falls.

Consider a warm air mass near the surface lifted by turbulence to a higher elevation and cooling adiabatically. As it rises it is surrounded by cooler and thus heavier air. If its lapse rate is greater than 1° C/100 m, the air mass is still buoyant, and it continues to rise and cool at a rate faster than adiabatic. As it rises it pulls in more air behind it, creating eddies of turbulent adiabatic air. This turbulence creates an *unstable* air mass. Thus a lapse rate greater than dry adiabatic favors vertical motion and instability (Figure 4.3).

As air rises and adiabatic cooling continues, the air temperature approaches **dew point** (the temperature at which the air will be saturated with as much water as it can hold). The dew point temperature decreases at the rate of 0.2° C per 100 m. Dew point and air temperature approach each other at the rate of about 0.9° C per 100 m. As the air mass reaches dew point, the water it contains condenses to form clouds of fog. This condensation of moisture in rising turbulent air masses gives substance to unstable air masses you see as developing thunderheads on a hot summer day.

Unstable air conditions result when rising air masses cool at a lapse rate faster than adiabatic. If a rising air mass cools at a lapse rate less than 1° C/100 m, it resists upward movement and may sink lower. Such an air mass is not subject to eddies and turbulence. When such conditions prevail, an air mass is *stable* (Figure 4.3). Thus a volume of air whose temperature lapse rate is less than the dry adiabatic rate is stable. If the lapse rate is the same as dry adiabatic, then the air mass is *neutral,* neither resisting nor favoring an upward push into the atmosphere.

This behavior of air masses has a significant influence on climate, from the development of local thundershowers in the mountains to massive air movements on a global scale. It provides the thermals used by soaring hawks and migrating birds. It affects formation of fog and the local and regional concentration and dispersion of atmospheric pollutants.

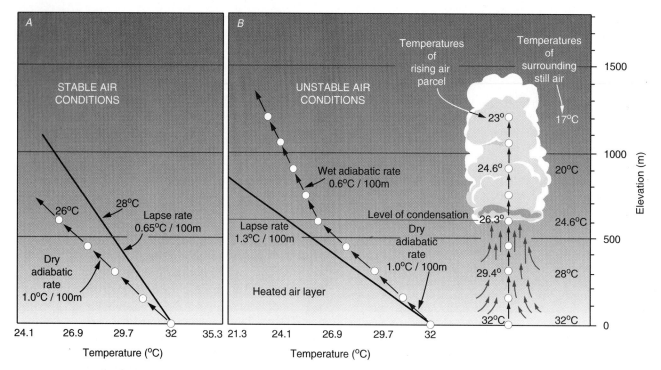

**Figure 4.3** Adiabatic processes plotted for hypothetical stable and unstable air masses. Note that in stable air conditions, the value of the lapse rate is greater than the normal dry adiabatic lapse rate of still air. In unstable conditions the lapse rate is less than normal dry adiabatic rate. As the parcel of air rises, it is adiabatically cooled at the dry rate. When condensation begins, the wet adiabatic rate takes over and liberated heat is added to the rising air, still keeping it warmer than the surrounding air. These temperature differences are illustrated in the diagram of the developing thunderhead.

# EFFECTS OF ROTATION

## The Coriolis Effect

Added to the behavior of warming and cooling air masses is the effect of a rotating Earth on air masses. Earth spins on its axis, turning from west to east. An object or parcel of air moving over Earth (or any other rotating body) tends to move in a straight line and will travel with Earth at the same speed. (For this reason we do not experience the tremendous speed of Earth's rotation.) At the equator, the surface of Earth, far from its axis, moves most rapidly at 425 m/sec (39,000 km or 24,000 mi a day). At 60° latitude the surface of Earth moves at 232 m/sec. By the law of motion a moving object will maintain its momentum unless acted upon by some outside force such as friction. If an object or air mass moves north from the equator, where it is travelling most rapidly, it will keep its same momentum as it moves north into a region of slower rotation. This momentum will cause the object or air mass to veer in the direction of Earth's rotation; its path will be deflected to the right (east). If the object moves southward from the equator, it will also curve to the right, but the deflection will appear to be to the left or west relative to the Northern Hemisphere. This deflection is explained as an imaginary force, the **Coriolis force** (Figure 4.4). For this reason we get a clockwise motion of air masses

and oceans in the Northern Hemisphere and a counterclockwise motion in the Southern Hemisphere.

## Movement of Air Masses

Sunlight does not strike Earth uniformly (Figure 4.5a). Because of Earth's shape, the sun's rays strike more directly on the equator than on the polar regions, so lower latitudes get more heat. Further increasing the inequity of heat distribution is latitude, the inclination of Earth's axis at an angle of 23½°, and the revolution of Earth about the sun. These factors combine to determine the amount of solar radiation reaching any point of Earth at any time.

Earth's surface at all times lies half in the sun's rays and half in shadow, marked by a dividing line, the circle of illumination. The circle of illumination is bisected by the equator and always lies at right angles to the sun's rays. At the equator Earth experiences roughly 12 hours of daylight and 12 hours of night throughout the year, whereas the polar regions experience 24 hours of night in winter and 24 hours of daylight in summer (Figure 4.5b). Only two times a year, at the vernal and autumnal equinoxes (March 20 or 21, December 22 or 23), does the circle of illumination pass through the poles. At the summer and winter and solstices, the circle of illumination is tangent to the Antarctic and Arctic Circles. Thus at the time of the winter solstice, the

*Factors that determine solar radiation reach Earth*

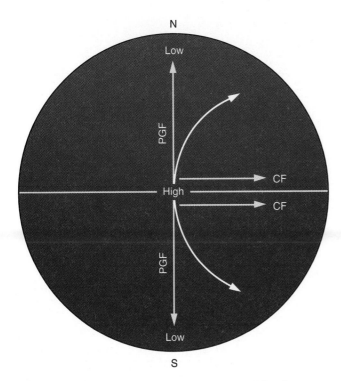

**Figure 4.4** Coriolis deflection in the Northern and Southern Hemispheres. PGF is the pressure gradient force from high pressure near the equator to low pressure near the polar regions. CF is the Coriolis force.

sun remains below the horizon at the North Pole and above the horizon at the South Pole. During the summer solstice, the reverse situation is true. Therefore the lower latitudes, especially the equatorial regions, have more and constant heat, and are always warmer than the polar regions.

Air heated close to the surface at the equatorial regions rises until it reaches the stratosphere, where temperature no longer decreases with altitude. There the air masses that now possess the same or lower temperature than the stratosphere are blocked from any further upward movement. With more air rising, air masses are forced to spread north and south to-

ward the poles. As the air masses approach the poles, they cool, become heavier, and sink. This heavier cold air then is pulled southward and northward by the void created as bottom air heated at the equator is forced to rise (Figure 4.6).

If Earth were stationary and without any irregular land masses and oceans, the atmosphere would flow in this unmodified circulatory pattern. However, Earth's surface is irregular; there is differential heating; and the Earth, as we know, spins, creating the Coriolis effect. On the rotating Earth, air currents in the Northern Hemisphere move north, but not for long. The Coriolis effect deflects the winds to the right, changing their direction to the northeast or east. Frictional drag slows the northward movement and causes the air masses to pile up at about 30° N latitude, where they lose heat by radiation (Figure 4.6). This combination of piling and heat loss forces some of the air to descend, producing a cell of semipermanent high pressure at about 60° N latitude. Air that has descended flows both northward toward the pole and southward toward the equator. The northward-flowing air currents become the prevailing westerlies, and the southward-flowing air, also deflected to the right, becomes the northeast trade winds of the lower latitudes. (These terms refer to the originating direction of the wind, not the direction in which they are blowing.)

The air that stayed aloft gradually moves northward, continues to lose heat, descends at the polar region, and loses additional heat at a surface already cooled by insufficient solar radiation and the albedo of the ice packs. This dense mass of cold air flows southward toward the equator. On its way it is deflected to the right and becomes the polar easterlies.

Similar flows take place in the Southern Hemisphere, but the air flow is deflected to the left by the rotating Earth. Thus the cold winds of winter in the Southern Hemisphere are south winds, and the warm winds of spring flow down from the north.

This pattern of rising and descending air forms tubes about the Earth. The direct circulation cells near the equator are called Hadley cells, and the indirect mid-latitude cells

**Figure 4.5** Altitude of the sun (a) at the equinoxes and (b) at the solstices. Latitude, the inclination of Earth's axis at an angle of 23°, and the revolution of Earth about the sun determine the amount of solar radiation reaching any point on Earth at any time.

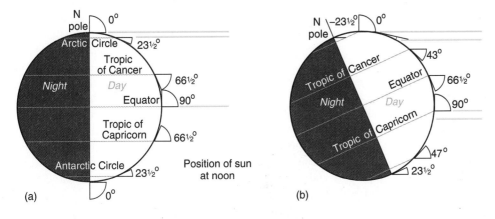

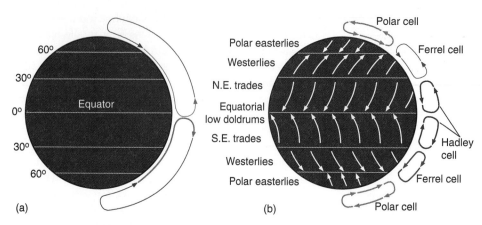

**Figure 4.6** Circulation of air cells and prevailing winds (a) on an imaginary, nonrotating Earth and (b) on a rotating Earth.

are called Ferrel cells, after the meteorologists who described them.

The region where the northeast and southeast trade winds meet is known as the *intertropical convergence* (Figure 4.7). A region of stable air, it tends to be dry. Although tropical regions about the equator always are exposed to warm temperatures, the sun is directly over the equator only two times a year, at the spring and fall equinoxes. At the summer solstice, the sun is directly over the Tropic of Cancer; at the winter solstice, it is directly over the Tropic of Capricorn, summer in the Southern Hemisphere. At these times the Earth receives the most intense solar radiation, heating land, water, and air masses. The rising tropical air leads to instability and much rain. As the junction of rising and falling tropical air masses shifts north and south, it brings on the wet and dry seasons in the tropics. Because air and land heat slowly, a time lag of about one month develops between the change in the vertical orientation of the sun and the shift in the intertropical convergence. For this reason the intertropical convergence does not move as far north and south as does the sun (Figure 4.7). Such a

lag is also the reason why in the Northern Hemisphere, July is the hottest month of summer even though the sun is directly overhead in June.

The interaction of wind and heating produces more or less permanent high-pressure cells known as subtropical highs in the Atlantic and Pacific Oceans; winds and cooling produce low pressure cells such as the Aleutian and Icelandic lows. The highs are more pronounced during the summer months, the lows during the winter months. Also produced are monsoon winds, dry winds that blow from continental interiors to the oceans in summer, and winds heavy with moisture that blow from the oceans to the interior in winter, bringing with them heavy rains.

Last, there are moving air masses with their cyclonic and anticyclonic frontal systems. These major air circulations are responsible for the changing swirls or cloud patterns seen over Earth from space.

In the Northern Hemisphere, winds in high pressure cells, called **anticyclones,** flow clockwise and move outward and downward from the center of the system. Because air moves down from high altitudes, highs are characterized

**Figure 4.7** Shifts of the intertropical convergence, producing rainy seasons and dry seasons. Note that as the distance from the equator increases, the dry season is longer and the rainfall is less. These oscillations result from changes in the altitude of the sun between the equinoxes and the solstices as diagramed in Figure 4.5. (From H. Walter 1977.)

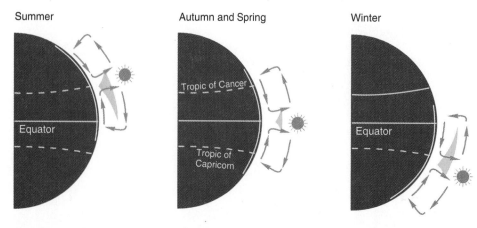

by minimum cloudiness and little precipitation. A high-pressure area is surrounded on all sides by low pressure. Low-pressure cells, called *cyclones,* flow counterclockwise. The air moves inward and upward, resulting in cooling, increased relative humidity, and with adequate moisture, precipitation. In the Southern Hemisphere winds move counterclockwise around a high-pressure system and clockwise around a low-pressure system.

## Ocean Currents

The upper or surface waters of the ocean are constantly in motion. This motion, reflected in waves and currents, is caused largely by winds blowing across the surface. These wind-driven ocean currents, modified by Coriolis forces, transport enormous quantities of water across vast distances. In doing so, ocean currents also become a mechanism for the transport of sensible heat from equatorial regions to Arctic regions.

In the absence of any land masses, ocean currents could circulate unimpeded around the globe, as does the flow of ocean waters around Antarctica. Continental land masses, however, divide the ocean into two main bodies, the Atlantic and the Pacific. Both oceans are unbroken from high latitudes north and south to the equator; and both are bordered by land masses on either side that deflect ocean currents (Figure 4.8).

Each ocean is dominated by two great circular water motions or **gyres,** each centered on a subtropical high-pressure area north and south of the equator. Within each gyre the current, in response to the Coriolis force, moves clockwise in the Northern Hemisphere and counterclockwise in the Southern Hemisphere. The movements of the currents are also influenced by the prevailing winds, the trades or tropical easterlies on the equator side and the prevailing westerlies on the pole side. The eastward-flowing equatorial countercurrent separates the two gyres, north and south. This current results from the return of lighter (less dense) surface water piled up on the western side of the ocean basin by the equatorial currents.

As currents flow westward they become narrower and increase their speed. Deflected by the continental basin and pushed by the Coriolis force, they turn poleward, carrying warm water with them. The two major currents in the Northern Hemisphere are the Gulf Stream in the Atlantic and the Kuroshio (or Japanese) Current in the Pacific. The Gulf Stream flows north from the Caribbean, presses close to Florida, swings along the southeast Atlantic Coast of North America and divides (Figure 4.8). One part becomes the Norway Current, carrying warm water past Scotland, warming and dampening the climate of Great Britain (which lies in about the same latitiude as central Canada). The other part of the current swings south as the Canary Current, completing the gyre. The Kiroshio Current gives rise to the south-flowing warm California Current, which bathes the Pacific Northwest in rain and fog, responsible for the growth of the magnificent Pacific Northwest coniferous rain forest. Counterparts in the Southern Hemisphere are the Australian Current in the Pacific and the Brazil Current in the Atlantic. The counterpart of the Canary Current in the South Atlantic is the Benguela Current, which flows along the African Coast; in the Pacific it is the Humboldt Current.

As the California and Humboldt Currents swing westward to complete the gyres, they drag along with them water

**Figure 4.8** Ocean currents of the world: (1) Antarctic West Wind Drift; (2) Peru Current (Humboldt); (3) South Equatorial Current; (4) Counter-Equatorial Current; (5) North Equatorial Current; (6) Kuroshio Current; (7) California Current; (8) Brazil Current; (9) Benguela Current; (10) South Equatorial Current; (11) Guinea Current; (12) North Equatorial Current; (13) Gulf Stream; (14) Norwegian Current; (15) North Atlantic Current; (16) Canaries Drift; (17) Sargasso Sea; (18) Monsoon Drift (summer east; winter west); (19) Mozambique Current; (20) West Australian Current; (21) East Australian Current. Dashed arrows represent cold water. (From Coker 1947.)

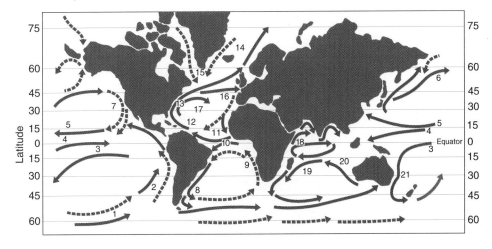

from the coast. To replace the displaced water, deep cold water rises to the surface, bringing with it a load of nutrients from the bottom. The surfacing of this cold, nutrient-rich water into the sunlit area of the ocean is termed **upwelling**. There heavy phytoplankton production supports an abundance of marine life, especially fishes and birds. Because these waters are colder than the adjacent land, the currents supply little moisture to it, resulting in a string of deserts along the coast of Mexico and South America.

The combination of sea surface temperature, ocean currents, and the atmosphere can have profound effects of climate. Changes in east-west zonal circulations of the atmosphere, as well as north-south movements, influence easterly and westerly flows of winds. These movements, coupled with displacements of major ocean currents, can have pronounced effects on climate over the short term.

An example is warming of the surface waters of the ocean off the South American coast around Christmas time, a phenomenon called El Niño, meaning "Little Boy" (referring to the Christ child). El Niño represents one extreme of the Southern oscillation, an irregular interannual fluctuation in the Pacific involving major changes in rainfall patterns; its opposite is La Niña (Philander 1989). In normal years—the years between the two extreme events—strong southeast trade winds drive warm equatorial waters moving southward along the Peruvian coast westward into the Pacific, away from the coast. As these waters move westward, cold water from the deep comes to the surface.

The westward-driven waters concentrate in the Pacific west of the international dateline. At the La Niña, temperatures of the sea surface in the central and eastern Pacific are unusually low and the trade winds are very strong. This phenomenon may contribute to summer droughts in North America. As the warm waters driven westward by the trade winds increase in the western Pacific, high pressure develops near Darwin, Australia, a strong tropical low-pressure area builds in the western Pacific near Tahiti (Rasmussen 1985), and the trade winds collapse. The low-pressure area and warm ocean waters move eastward toward South America. The eastern tropical Pacific experiences an increase in rainfall; west of the international dateline rainfall decreases, and the western Pacific experiences a drought.

When this eastward flow reaches its other extreme, the El Niño, about every five years, it reverses the westward flow of the Equatorial Current and blocks the upwelling of cold water along the South American coast. Surface waters of the ocean may warm up as much as 5° C, no nutrients replenish the surface water, and phytoplankton production, the base of the marine food web, drops with catastrophic effects on marine life. Schooling fish and the marine birds—the pelicans, boobies, and cormorants—that depend on them disappear; less familiar warmwater fish may invade the area; seabird reproduction declines or may fail altogether. El Niño also affects ecosystems from central Alaska to Chile, brings drought to agricultural areas of South Africa, the United States, and Canada, brings torrential rains to usually dry areas, and dislocates the rainy season in the tropics.

Periodically, El Niño events become unusually intense. One such major event occurred in 1982–1983. Not only did it disrupt fisheries, but it caused catastrophic losses among Galapagos fur seals (*Arctocephalus galapagoensis*), flooded the nests of sea birds nesting on Christmas Island and elsewhere in the Pacific, caused reproductive failure of guano birds along the South American coast, and destroyed much of the California kelp, largely because of high swells (Glynn 1988).

## REGIONAL CLIMATES

The massive circulation patterns of coupled atmosphere and ocean currents affect the broad global climatic patterns. The climate of any given region or locality, however, results from a combination of patterns of temperature and moisture. These patterns are influenced not only by latitude, but also by the location of the region within the continental land masses. Nearby bodies of water and geographical features such as mountains further influence weather and climate.

### Topographical Influences

Mountains influence regional climates in two ways: by modifying the patterns of precipitation and by creating climatic differences with altitude. As an air mass is intercepted by a mountain range, it ascends and cools at a dry adiabatic rate. When the moisture reaches its condensation point, it cools at the moist adiabatic rate and drops its moisture on the windward side. The air, no longer saturated, descends on the leeward slope, warms at the dry adiabatic rate, and picks up moisture from the land (Figure 4.9). As a result, the windward side of a mountain range supports more moisture-loving vegetation than the leeward side, where drier conditions exist in the rain shadow. Thus in North America the westerly winds that flow over the Sierra Nevada drop their moisture on the west-facing slopes, which support excellent forest growth, whereas the leeward sides are the hot deserts of the Southwest and the cool deserts of sagebrush in the Northern Great Basin. In the Appalachians the windward side supports northern hardwood forests and the leeward side oak-pine forests.

In mountainous country climate changes on an altitudinal gradient. The change in climate going upslope mimics the broad climatic changes experienced by going to higher latitudes. As a rule of thumb, temperatures drop about 1.5 to 3° C for every 300 m rise in elevation, so climates of higher altitudes bear little resemblance to those of the lowlands about their bases. Above the base region is the montane level, which has declining temperatures and increasing relative humidity as the altitude increases. At the highest levels the climate is very cold, and on top of the highest mountains is a land of perpetual ice and snow. This broad rule holds for mountains from tropics to boreal regions, although individual mountains possess their own individual weather conditions.

**Figure 4.9** Formation of a rain shadow.

We can obtain a picture of regional climates in terms of an annual progression of temperature and precipitation from a climograph. A **climograph** is a plot of the mean monthly temperatures against mean monthly relative humidities or precipitation (Figure 4.10). Connecting the points for each month forms an irregular polygon, which we can compare with polygons for other areas.

Climographs give us a broad picture of regional climates, but provide little information on the ecological and biotic interactions with temperature and precipitation. The geogra-

pher C. W. Thornthwaite (1948) developed a technique of associating climatic data with seasonal surpluses and deficits of water in the soil (Figure 4.11). The availability of soil water is determined by the amount of water drawn from the soil by plant transpiration and by direct evaporation, collectively called **evapotranspiration,** relative to the amount of water replaced in the soil by precipitation. Evapotranspiration is difficult to measure directly, but it can be estimated by potential evapotranspiration, which increases by a factor of 2 for each 10° C rise in temperature. *Potential evaporation is*

**Figure 4.10** Temperature-moisture climographs. (a) The hot, dry desert climate differs graphically from the cool, temperate, moist climate of the East. Data for the graph are mean temperature and precipitation from 1941 to 1950 for Yuma, Arizona, and Albany, New York. (b) Conditions on the rain shadow side and the high rainfall side in the Appalachian Mountains in West Virginia. Numbers in parentheses indicate month.

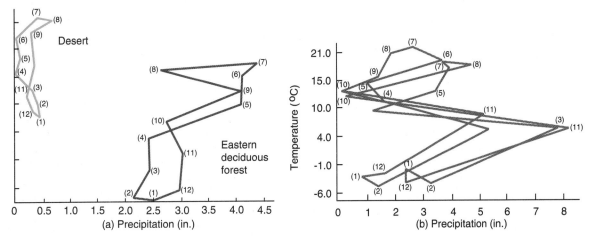

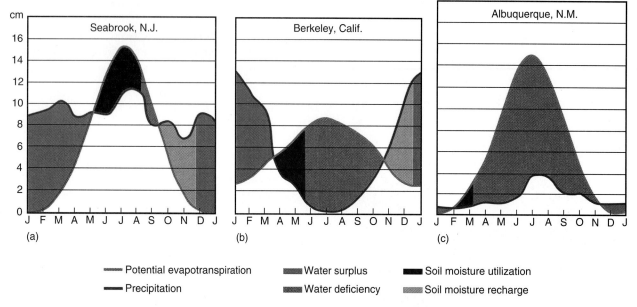

**Figure 4.11** Relationship between precipitation and potential evapotranspiration (evaporation from soil plus transpiration from plants) for three regions of North America: (a) deciduous forest, (b) semiarid shrubland, and (c) desert. Note the seasonal differences in water deficiencies in the three areas. (From Thornthwaite 1948.)

the amount of water that could be drawn from the soil if the availability of water were unlimited. Potential evaporation thus can be computed as a function of air temperature through the year. A curve of the calculated potential evapotranspiration over time can be contrasted with curves of seasonal precipitation to predict seasonal soil water surpluses and deficits (Figure 4.11). If precipitation input to the soil exceeds potential evapotranspiration through all seasons, the soil remains saturated through the year. If in any season the precipitation curve falls below the potential evaporation curve, it suggests that the soil will dry out and plant and animal life will face seasonal drought conditions.

## Humidity

When weather reporters tell us that the relative humidity is 50 percent, they are referring to water vapor content in the air. Water vapor gets into the air by evaporation from moist surfaces and bodies of water and from transpiration by plants. When water evaporates, more molecules leave the surface of the water (or any wet object) than enter it. When the number of molecules leaving the water equals the number returning, the air over the water is *saturated.*

In the air, water vapor acts as an independent gas that, like air, has weight and exerts pressure. The amount of pressure water vapor exerts independent of dry air is **vapor pressure.** The pressure that water vapor exerts when the air is saturated is *saturation vapor pressure.*

For any given temperature there is a fixed quantity of water that air can hold in a given space. Warm air can hold

more water than cool air and therefore has greater vapor pressure than cool air, which becomes saturated with less water vapor per unit volume (Figure 4.12) Because saturation vapor pressure is a function of temperature, the difference between the saturation vapor pressure and the actual vapor pressure at any given temperature represents the *vapor pressure deficit.* Water vapor in the atmosphere moves in response to differences in vapor pressure between any two levels, such as the surface of a pond and the air above. The rate of movement is proportional to vapor pressure differences, or the vapor pressure gradient.

The amount of water vapor a given volume or weight of air holds is expressed in terms of humidity. The weight of water vapor per unit volume of air is *absolute humidity* and is measured in grams per cubic meter. *Relative humidity,* with which we are most familiar, is the amount of water vapor actually in the air relative to what it could hold if saturated. Thus if the actual amount of water vapor in the air is only 20 percent of the amount it could hold at a given temperature, the relative humidity is 20 percent. If the air holds all the water vapor it can, it is saturated and the relative humidity is 100 percent. Relative humidity generally is higher at night and early morning when the air temperatures are lower; it is lower by day when temperatures increase.

In any one area relative humidity varies widely from one place to another, depending upon terrain. Variations in humidity are most pronounced in mountainous country. Low elevations warm up and dry out earlier in the spring than high elevations, and soil moisture becomes depleted in summer. Because of daytime heating, relative humidity is lower

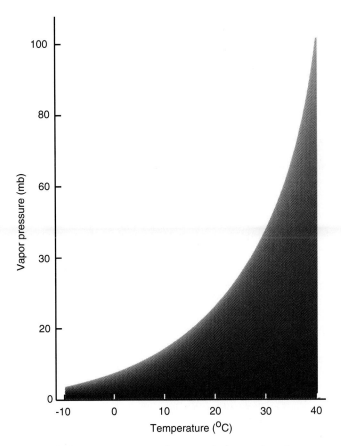

**Figure 4.12** Vapor pressure of saturated air at various temperatures.

in the afternoon in the valley and on slopes directly exposed to the sun; it is highest on tops of mountains. At night the valley bottoms are most humid. However, the daily range of humidity is greatest in the valleys and the lowest at higher elevations.

## Inversions

Earth's surface heats up by day. By night gradual radiation cools the surface air above it (Figure 4.13). This layer of air gradually deepens as the night progresses and forms a nighttime surface temperature inversion in which the temperature increases with height. This night inversion is gradually eliminated by morning surface heating (Figure 4.14).

Such inversions are particularly pronounced in hilly and mountainous country in summer when the air mass is stable and the weather is calm and clear (Figure 4.15). At night air in the valley cools next to the ground, forming a weak surface inversion. At the same time cold dense air flows down slopes from the hill or mountaintop. Together they cause the inversion to become deeper and stronger, and the cold dense air is trapped beneath a layer of warm air. In mountainous areas the top of the night inversion is usually below the main ridge. If air is sufficiently cool and moist, fog may form in the valley. Smoke from industry and other heated pollutants released in such an inversion will rise only until their temperature equals that of the surrounding air, a point called the thermal belt or "warm slope zone." Then smoke flattens out and spreads horizontally just below the thermal belt. These inversions break up when surface air warms during the day to create vertical convections and turbulence or when a new air mass moves in.

**Figure 4.13** Radiant heating of Earth. (a) During the day heat gains exceed heat loss. (b) At night there is a net cooling of the surface.

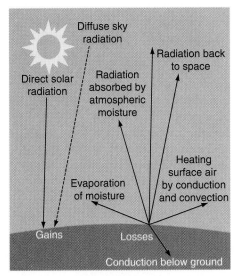

(a) Daytime surface
heat exchange

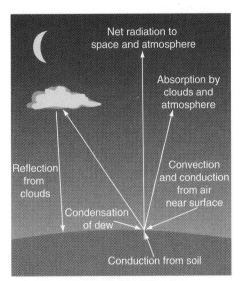

(b) Nighttime surface
heat exchange

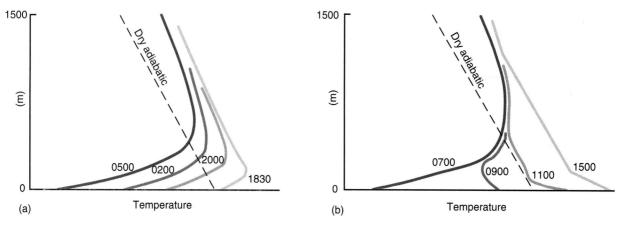

**Figure 4.14** Formation and elimination of a nighttime surface inversion typical on clear, cool nights. (a) As the ground cools rapidly after sundown, a shallow surface inversion forms (1830). As cooling continues during the night, the inversion deepens from the surface upward, reaching its maximum depth just before dawn (0500). (b) After sunrise, the surface begins to warm and the night surface inversion (0700) is gradually eliminated during the forenoon of a clear summer day. A surface superadiabatic layer and a dry-adiabatic layer above deepen until they reach their maximum depth about midafternoon (1500). (After Schroeder and Buck 1970.)

Similar but more widespread inversions occur when a high-pressure area stagnates over a region. In a high-pressure area the air flow is clockwise and spreads outward. The air flowing away from the high must be replaced, and the only source for replacement air is from above. Thus surface high-pressure areas are regions of sinking air movements from aloft, call *subsidence.*

When high-level winds slow down, heavy cold air at high levels in the atmosphere tends to sink. As the parcel of air sinks, it is compressed, heats, and becomes drier. A layer of warm air then develops at a higher level in the atmosphere (Figure 4.16) with no chance to descend. It hangs several hundred to several thousand feet above the earth, forming a **subsidence inversion.** Such inversions tend to prolong the period of stagnation and increase the intensity of air pollution. Subsidence inversions that bring about our highest concentrations of pollution are often accompanied by lower-level radiation inversions.

Along the west coast of the United States, and occasionally along the east coast, the warm seasons often produce a

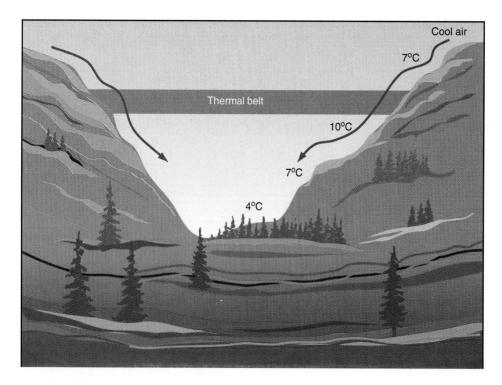

**Figure 4.15** Topography plays an important role in the formation and intensity of nighttime inversions. (Adapted from Schroeder and Buck 1970.)

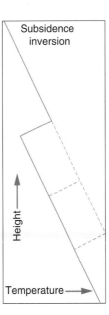

**Figure 4.16** Descent of a subsidence inversion. The movement of the inversion is traced by successive temperature measurements, indicated by the dashed lines. The nearly horizontal dashed lines indicate the descending base of the inversion. The solid line indicates temperature. The temperature lapse rate in the descending layer is nearly dry adiabatic. The bottom surface is marked by a temperature inversion. Two features, temperature inversion and a marked decrease in moisture, identify the base of the subsiding layer. Below the inversion is an abrupt rise in the moisture content of the air. (After Schroeder and Buck 1970.)

coastal or *marine inversion* (Figure 4.17). In this case cool, moist air from the ocean spreads over low land. This layer of cool air, which may vary in depth from a hundred to several thousand meters, is topped by warmer, drier air, which also traps pollutants in the lower layers.

# MICROCLIMATES

When the weather report states that the temperature is 24° C and the sky is clear, the information may reflect the general weather conditions for the day; but on hill and mountain slopes, on cliff tops, in valleys, in crannies and pockets, on the surface of the ground, and in and beneath vegetation, the climate is different. Heat, moisture, air movements, and light all vary greatly from one local site to another to create a range of microclimates.

## North-Facing and South-Facing Slopes

The greatest microclimatic differences exist between north-facing and south-facing slopes (Figure 4.18). South-facing slopes in the Northern Hemisphere receive the most solar energy. North-facing slopes receive the least energy.

At latitude 41° N (about central New Jersey and southern Pennsylvania) midday insolation on a 20° slope is, on the average, 40 percent greater on the south-facing slopes than on the north-facing slopes during all seasons. This differ-

**Figure 4.17** During a marine inversion, cool air from the ocean moves in beneath the heated layer above. (After Schroeder and Buck 1970.)

xeric → dry       mesic → moist

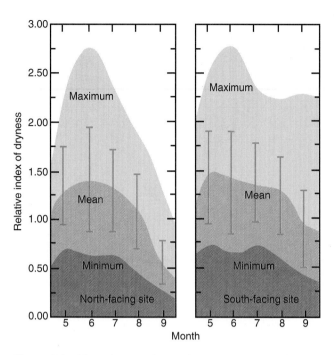

**Figure 4.18** Microclimates of north-facing and south-facing slopes as described by the seasonal course of the radiative index of dryness during the 1948–1982 period on the Little Laurel Run watershed in the Appalachian Mountains of northern West Virginia (39°41′ N, 79°45′ W). The radiative index of dryness is the ratio of seasonal sums of net radiation to those of latent heat of precipitation. When over a period of time net radiation is greater than the latent heat of precipitation, there is an excess of energy and the site experiences a dry period. On both the north-facing and south-facing slopes, the excess of energy supply over water supply follows a seasonal course. In June, for example, net radiation may exceed the latent heat of evaporation by a factor of 2.75. Note, however, that the excess of energy supply over water supply on the south-facing slope lasts longer than that on the north-facing slope. Therefore north-facing slopes tend to be more mesic and cooler than south-facing slopes. (After Tajchman, Harris, and Townsend 1988).

ence has a marked effect on the moisture and heat budget of the two sites. High temperatures and associated low vapor pressures induce evapotranspiration of moisture from soil and plants. The evaporation rate often is 50 percent higher, the average temperature higher, the soil moisture lower, and the extremes of all of these are more variable on south-facing slopes. Thus the microclimate ranges from warm and dry (xeric) conditions with wide extremes on south-facing slopes to cool and moist (mesic), less variable conditions on north-facing slopes. Xeric conditions are most highly developed on the top of south-facing slopes, where air movement and soil drainage are greatest, whereas the most mesic conditions are at the bottom of the north-facing slopes.

The whole north-facing and south-facing slope complex is the result of a long chain of interactions: Solar radiation influences moisture regimes; the moisture regime influences the species of trees and other plants occupying the slopes

(Figure 4.19); the species of trees influence mineral recycling, which is reflected in the nature and chemistry of the surface soil and the nature of the herbaceous ground cover.

Being mobile, few if any animals are typical of only north-facing or south-facing slopes, as far as we now know. However, their movements may be limited to some extent by the differences in conditions and food supplies on the slopes. Mule deer (*Odocoileus hemionus*) tend to use south-facing slopes more heavily in winter and early spring and north-facing slopes in summer (Taber and Dasmann 1958). In the central Appalachians the red-backed vole (*Clethrionomys gapperi*), normally an inhabitant of cool fir (*Abies*), spruce (*Picea*), aspen (*Populus*), and northern hardwood forests throughout its range, is restricted in its local distribution to forested, mesic, north-facing slopes. Those species of soil invertebrates intolerant of humidity, such as some mites, can exist only in a dry habitat and therefore are confined to the south-facing slope.

## Climate Near the Ground

On a summer afternoon the temperature under a calm, clear sky may be 28° C at 1.83 m (6 ft), the standard level of temperature recording. On or near the ground—at the 5 cm level—the temperature may be 5° C higher; and at sunrise, when the temperature for the 24-hour period is the lowest, the temperature may be 3° C lower at ground level (Biel 1961). Thus in an open field in the mideastern United States, the afternoon temperature near the ground may be equivalent to the temperature at 1.83 m in Florida, 700 miles to the south; and at sunrise the temperature may be equivalent to the 1.83 m temperature in southern Canada. Even greater extremes occur above and below the ground surface. In New Jersey, March temperatures about the stolons of clover plants 1.5 cm above the surface of the ground may be 21° C, while 7.5 cm below the surface the temperature about the roots is − 1° C (Biel 1961). The temperature range for a vertical distance of 9 cm is 22° C. Under such climatic extremes most organisms exist.

The chief reason for the great differences between temperature at ground level and at 1.83 m is solar radiation. During the day in an open field, the soil, the *active surface* (where most solar energy is absorbed), absorbs solar radiation, which comes in short waves, and radiates it back as long waves to heat a thin layer of air above it (Figure 4.20). Because air flow at ground level is almost nonexistent, the heat radiated from the surface remains close to the ground. Temperatures decrease sharply in the air above this layer and in the soil below. The heat absorbed by the ground during the day is reradiated by the ground at night. This heat is partly absorbed by the water vapor in the air above. The drier the air, the greater is the outgoing heat and the stronger is the cooling of the surface of the ground and the vegetation. Eventually the ground and the vegetation are cooled to the dew point, and water vapor in the air may condense as

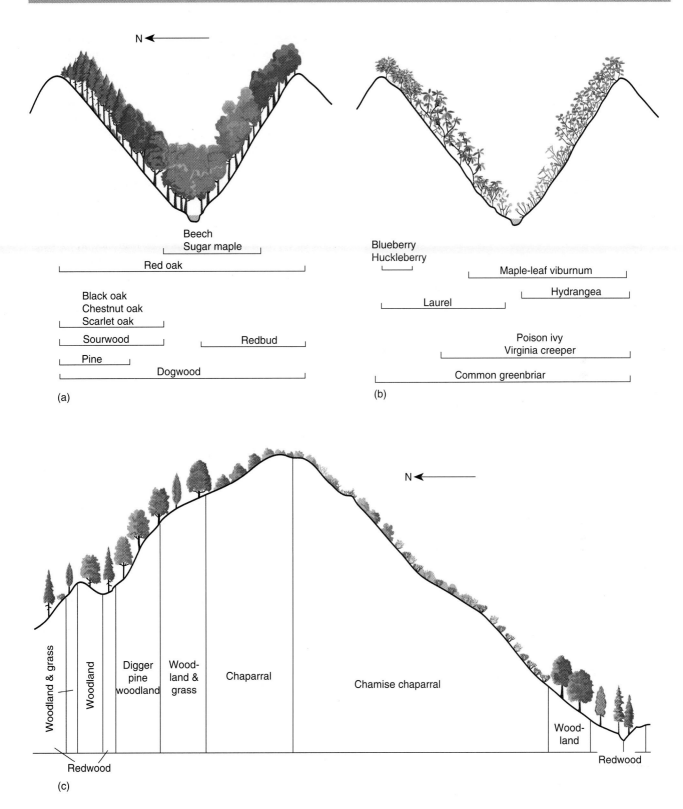

**Figure 4.19** Influence of microclimate on the type and distribution of vegetation on north-facing and south-facing slopes. (a) and (b) The hill country of southwestern West Virginia. (c) The Point Sur area in California. In each case note the similarity of vegetation on the lower parts of north-facing and south-facing slopes.

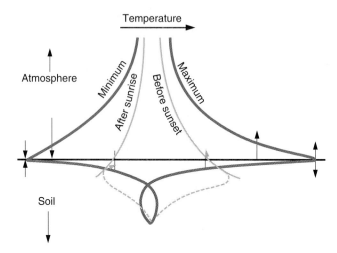

**Figure 4.20** Idealized vertical temperature profiles near the active surface at times of maximum, minimum and intermediate surface temperatures. (Adapted from Gates 1962.)

dew on the vegetation and ground. After a heavy dew a thin layer of chilled air lies over the surface, the result of rapid absorption of heat in the evaporation of dew.

## Influences of Vegetation and Soil

By altering wind movement, evaporation, moisture, and soil temperatures, vegetation influences the microclimate of an area, especially near the ground. Temperatures at ground level in the shade are lower than those in places exposed to the sun and wind. Average maximum soil temperatures at 2.5 cm below the surface in a closed-canopy northern hardwoods forest and an aspen-birch forest in New York State were 15.5° C from mid-May to late June and 20° C (absolute maximum 25° C) from early July to mid-August (Spaeth and Diebold 1938). Maximum soil temperatures at 2.5 cm in open-canopied chestnut-oak forests were 19° C from mid-May to late June and 24° C (absolute maximum, 34° C) from July to mid-August.

Vegetation also reduces the steepness of the temperature gradient and influences the height of the active surface. With absent or very thin vegetation, temperature increases sharply near the soil; but as plant cover increases in height and density, the leaves of the plants intercept more solar radiation (Figure 4.21). Plant crowns then become the active surface, or more precisely the active layer. As a result temperatures are highest just above the dense crown surface and lowest at the surface of the ground. Maximum absorption of solar radiation in tall grass occurs just below the upper surface of the vegetation, whereas in short grass maximum temperatures are at ground level (Waterhouse 1955). (Among broad-leafed herbaceous plants daily maximums occur on the upper leaf surfaces.) At night minimum temperatures are some distance above the ground because the air is cooled above the tops of plants and the dense stalks prevent the chilled air from settling to the ground.

Within dense vegetation such as a forest, air movements are reduced (Figure 4.22). Complete calm exists at ground level. This calm is an outstanding feature because it influences both temperature and humidity and creates a favorable environment for insects and other animals.

Vegetation deflects wind flow up and over its top. If the vegetation is narrow, such as a windbreak or a hedgerow, the microclimate on the leeward side may be greatly affected. Deflection of wind produces an area of eddies immediately behind the vegetation, in which the wind speed is low and small particles such as seeds are deposited (Figure 4.23). Beyond is an area of turbulence, in which the climate tends to be colder and drier than normal. If some wind passes through the barrier and some goes over it, no turbulence develops, but the mean temperature behind the barrier is high in the morning and lower in the afternoon.

Humidity changes greatly from the ground up. Because evaporation takes place at the surface of the soil or at the active surface of plant cover, the vapor content (absolute humidity) decreases rapidly from a maximum at the bottom to atmospheric equilibrium above. During the night little dif-

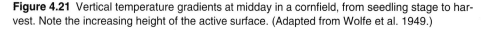

**Figure 4.21** Vertical temperature gradients at midday in a cornfield, from seedling stage to harvest. Note the increasing height of the active surface. (Adapted from Wolfe et al. 1949.)

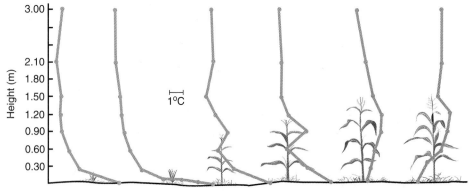

Increasing temperature (°C)

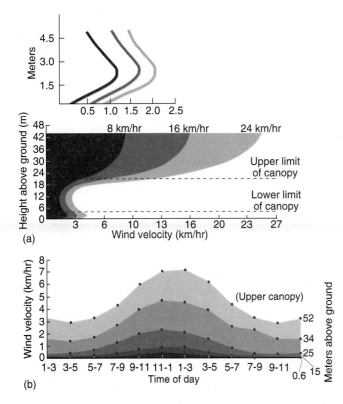

**Figure 4.22** (a) Distribution of wind velocities with height as affected by the timber canopy of coniferous forests for wind velocities of 8, 16, 24 km/hr at 43 m above the ground. (b) Average wind velocity during a June day inside a coniferous forest with a cedar understory in northern Idaho. Note the decrease in velocity near the ground. The inset is a detail of wind movement from ground level to 4.5 m. (From Gisborne 1941.)

ference exists above and on the ground. Within growing vegetation, however, relative humidity is much higher within and below plant cover than above it. In fact, near-saturation conditions may exist in part because of transpiration.

Soil properties also influence microclimates. Dark-colored soils are better absorbers of heat than light-colored soils, because light-colored soils reflect heat. Dry soils are poorer conductors of heat than moist soils. Poor conductivity and reflectivity are two reasons why temperatures are cool at relatively shallow depths in dry sandy soils. In a soil that conducts heat well, considerable heat energy will be transferred into the soil, from which it radiates to the surface at night (see Figure 4.20).

## Microclimate and Habitat

Microclimates on a larger scale, such as slope position and aspect (position relative to the sun), are of minimal influence to most plants and animals. What does affect them is the microclimate on an even smaller scale—what conditions are like on the lee side of shrubs and trees, in tree cavities and logs, beneath logs, stones, and leaf litter, and in caves. These are the places where most animals spend their days, moving about to locate the most favorable conditions.

Seeking thermal cover, a place to escape from the extremes of weather, is a major activity of animals from insects to large mammals. Thermal cover influences their movements, feeding, nesting, and resting sites. Large mammals, such as deer and elk (*Cervus elaphus*), seek thickets of conifers and evergreen shrubs, where wind flow is reduced and where it is warmer in winter and cooler in summer (Ozoga and Gysel 1972, Edgerton and McConnell 1976). Cavity-nesting and cavity-roosting birds seek hollows of trees, where wood is an excellent insulator against heat and

**Figure 4.23** The effect of a hedgerow or windbreak on wind flow. Note the area of relative calm close to the screen and the degree of turbulence some distance beyond.

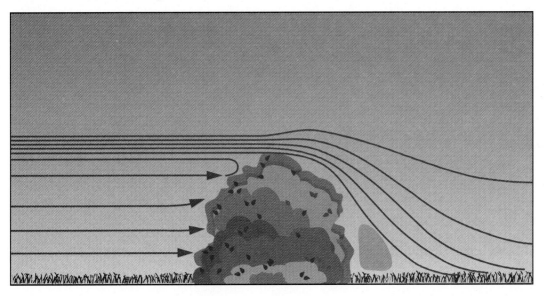

cold (Kendeigh 1961). Salamanders seek leaf litter and decayed logs in summer, where the microclimate is cool and humid. Tropical salamanders inhabit bromeliads and banana plants, whose leaves afford cooler and more stable temperatures than their surroundings (Feder 1982). Tropical log-dwelling salamanders move to the warmer, sunlit side of the log during the cooler times of day and to the cooler, more humid portion of the log during the heat of the day.

Vertebrates are not alone in responding to microclimatic conditions. The microdistribution of many insects is influenced by microclimatic conditions. Some seek cool, moist places; others seek dry, warm conditions. Slugs (Limacidae) stay hidden beneath moist litter and rotting logs by day and move about only during the humid nights. Bark beetles (Scolytidae) inhabiting fallen logs avoid the driest, warmest parts of the log and seek the dampest section of the log near the ground to lay their eggs. Nectarivorous hummingbird flower mites (Gamasidae: Ascidae) inhabiting *Heliconia wagneriana* flowers seek the most shaded locations within the blooms and avoid poorly shaded parts and flowers (Dobkin 1985).

Plants, too, respond to microclimatic conditions. Unable to move, they depend on colonization of most favorable microclimatic sites. Herbaceous growth in a dense forest, for example, may be limited to microsites that are flooded with large flecks of sunlight. Some of the most pronounced microclimatic effects on plants occur in areas of convex slopes and low concave surfaces. These places have much lower temperatures at night, especially in winter, much higher temperatures during the day, especially in summer, and a higher relative humidity. The concave surfaces radiate heat rapidly on still, cold nights, and cold air flows in from surrounding higher levels. On such sites the air temperature may be 8° C lower than surrounding terrain, causing a temperature inversion on a small scale. Because low ground temperatures in these areas tend to result in late spring frosts, early fall frosts, and a subsequent short growing season, these depressions are called **frost pockets** (Geiger 1965).

The pockets need not be deep. Spurr (1957) found that the minimum temperatures in small depressions only 1 to 1.2 m deep were equivalent to those of a nearby valley 60 m below the general level of the land. Such variations in temperature due to local microrelief can strongly influence the distribution and growth of plants. Tree growth is inhibited; and because the low surfaces more often than not accumulate water as well as cold air, such sites may contain plants of a more northern distribution. Frost pockets may also develop in small forest clearings. The surface of the tree crowns channels cold air into the clearings as terrestrial radiation cools the layer of air just above.

## Urban Microclimate

The building of cities has altered not only the slope, soil, and vegetation of the land they occupy, but also the atmosphere, creating a distinctive urban microclimate.

The urban microclimate is a product of the morphology of the city and the density and activity of its occupants. In the urban complex, stone, asphalt, and concrete pavement and buildings with a high capacity for absorbing and reradiating heat replace natural vegetation with low conductivity of heat. Rainfall on impervious surfaces is drained away as fast as possible, reducing evaporation. Metabolic heat from masses of people and waste heat from buildings, industrial combustion, and vehicles raise the temperature of the surrounding air. Industrial activities, power production, and vehicles pour water vapor, gases, and particulate matter into the atmosphere in great quantities. The effect of this storage and reradiation of heat is the formation of a heat island about cities (Figure 4.24) in which the temperature may be 6° to 8° C higher than the surrounding countryside (Landsberg 1970; SMIC 1971).

Heat islands are characterized by high temperature gradients about the city. Highest temperatures are associated with areas of highest density and activity; temperatures decline markedly toward the periphery of the city (Figure 4.25). Although detectable throughout the year, heat islands are most pronounced during summer and early winter and are most noticeable at night when heat stored by pavements and buildings is reradiated to the air. The magnitude of the heat island is influenced strongly by local climatic conditions such as wind and cloud cover. If the wind speed, for example, is above some varying critical value, a heat island cannot be detected.

During the summer the buildings and pavement of the inner city absorb and store considerably more heat than does the vegetation of the countryside. In cities with narrow streets and tall buildings the walls radiate heat toward each other instead of toward the sky. At night these structures slowly give off heat stored during the day. Although daytime differences in temperature between the city and the country may not differ noticeably, nighttime differences become pronounced shortly after sunset and persist through the night. The nighttime heating of the air from below coun-

**Figure 4.24** Idealized scheme of nighttime circulation above a city in clear, calm weather. A heat island develops over the city. At the same time a surface inversion develops in the country. As a result cool air flows toward the city, producing a country breeze. Lines are temperature isotherms; arrows represent wind. (Adapted from H. Landsberg 1970.)

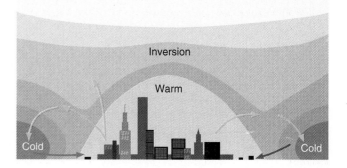

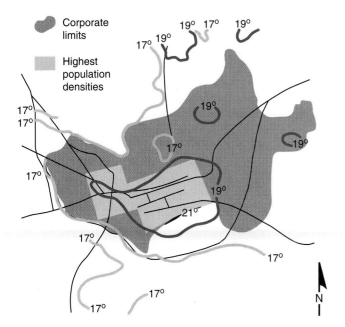

**Figure 4.25** Thermal pattern of night air in a small city, Chapel Hill, North Carolina. The highest temperatures are inside the corporate limits, where the population and activity are the greatest. (From Kopec 1970.)

**Table 4.1   Climate of the City Compared to the Country**

| Element | Comparison |
| --- | --- |
| Condensation nuclei and particles | 10 times more |
| Gaseous admixtures | 5–25 times more |
| Cloud cover | 5–10 percent more |
| Winter fog | 100 percent more |
| Summer fog | 30 percent more |
| Total precipitation | 5–10 percent more |
| Relative humidity, winter | 2 percent less |
| Relative humidity, summer | 8 percent less |
| Radiation, global | 15–20 percent less |
| Duration of sunshine | 5–15 percent less |
| Annual mean temperature | 0.5°–1.0° C more |
| Annual mean wind speed | 20–30 percent less |
| Calms | 5–20 percent more |

*Source:* Adapted from H. E. Landsberg 1970.

teracts radiative cooling and produces a positive temperature lapse rate while an inversion is forming over the countryside. This, along with the surface temperature gradient, sets the air in motion, producing "country breezes" to flow into the city.

In winter solar radiation is considerably less because of the low angle of the sun, but heat accumulates from human and animal metabolism and from home heating, power generation, industry, and transportation. In fact, heat contributed from these sources is 2.5 times that contributed by solar radiation. This energy reaches and warms the atmosphere directly or indirectly, producing more moderate winters in the city than in the country.

Urban centers influence the flow of wind. Buildings act as obstacles, reducing the velocity of the wind up to 20 percent of that of the surrounding countryside, increasing its turbulence, robbing the urban area of the ventilation it needs, and inhibiting the movement of cool air in from the outside. Strong regional winds, however, can produce thermal and pollution plumes, transporting both heat and particulate matter out of the city and modifying the rural radiation balance a few miles downwind (Clarke 1969; Oke and East 1971).

Throughout the year urban areas are blanketed with particulate matter, carbon dioxide, and water vapor. The haze reduces solar radiation reaching the city by as much as 10 to 20 percent. At the same time, the blanket of haze absorbs part of the heat radiating upward and reflects it back, warming both the air and the ground. The higher the concentration of pollutants, the more intense is the heat island.

The particulate matter has other microclimatic effects. Because of the low evaporation rate and the lack of vegetation, relative humidity is lower in the city than in surrounding rural areas, but the particulate matter acts as condensation nuclei for water vapor in the air, producing fog and haze. Fogs are much more frequent in urban areas than in the country, especially in winter (Table 4.1).

Another consequence of the heat island is increased convection over the city. Updrafts, together with particulate matter and large amounts of water vapor from combustion processes and steam power, lead to increased cloudiness over cities and increased local rainfall both over cities and over regions downwind. An evidence of weather modification by pollution is the increase in precipitation and stormy weather about La Porte, Indiana, downwind from the heavily polluted areas of Chicago, Illinois, and Gary, Indiana, and close to moisture-laden air over Lake Michigan. Since 1925 there has been a 31 percent increase in precipitation, a 35 percent increase in thunderstorms, and a 240 percent increase in the occurrence of hail (Changnon 1968).

## CLIMATE AND VEGETATION

Life, however, responds to climatic variations. Plants, in particular, are sensitive indicators of climate. Over their long evolutionary history, they have adapted to the limits of temperature and moisture by assuming certain physical forms, dimensions, and physiological attributes that enable them to cope with the climate. At one extreme, where temperatures are hot and precipitation is high, tropical forests developed; at the other extreme of cold temperatures and low precipitation, Arctic tundra resulted.

## Vegetation Zonation

These climate-induced patterns of vegetation are influenced not only by latitude, but also by the location of regions within a continental land mass, particularly in mountain areas. As mountains extend upward they reach into colder air, and the vegetation reflects the change. Approximately every 300 m (1000 ft) rise in elevation is equivalent to 160 km (100 mi) in latitude. Thus altitudinal changes in patterns of vegetation reflect latitudinal changes. These vegetation patterns show up as continental and altitudinal zones of vegetation.

Continental zonation of vegetation is pronounced in North America (Figure 4.26). In Northern North America two broad belts of vegetation, the tundra and boreal forest, stretching east and west, are influenced by temperature. Below these two belts the vegetation patterns are controlled mostly by precipitation and evaporation, the latter influenced considerably by temperature (Thornthwaite 1927). Because available moisture becomes less from east to west, vegetation follows a similar pattern, with belts running north and south. Humid regions along the coast support forest vegetation. West of the eastern forest region is a subhumid

**Figure 4.26** Vegetational zones of North America. (Courtesy U.S. Department of the Interior, Fish and Wildlife Service.)

Kilometers
200  0  200    600    1000
         400     800

☐ 1  Arctic
■ 2  Open boreal
■ 3  Closed boreal
■ 4  Northern hardwood
■ 5  Aspen parkland
■ 6  Montane woodland
■ 7  Pacific rain forest
■ 8  Eastern deciduous forest
■ 9  Grasslands

■ 10  Oak-savanna — Transition grassland
■ 11  Northern desert — cool
■ 12  Southern desert — hot
■ 13  Mesquite-grassland
■ 14  Piñon-juniper-oak
■ 15  Chaparral-oak woodland
■ 16  Southeast evergreen
■ 17  Mexican pine and pine-oak
■ 18  Tropical areas combined

zone where precipitation is less than evaporation. Here the ratio of precipitation to evaporation is about 60 to 80 and the land supports a tallgrass prairie. Further west is semiarid country, where the precipitation to evaporation ratio is 20 to 40. It supports a shortgrass prairie. Beyond the shortgrass prairie are desert grassland and the southwestern desert. Western North America is dominated by mountains and montane vegetation.

In Eurasia the zonation of natural vegetation, highly modified in places by centuries of human disturbance, more nearly follows that of temperature, resulting in broad belts of vegetation running east and west. In Africa zonation also consists of broad east-west belts of mediterranean shrubland, desert, tropical grassland and savanna, tropical forest (western Africa), and tropical shrub forest.

In mountainous country on all continents, vegetation zones reflect climatic changes on an altitudinal gradient (Figure 4.27). The vegetation in the lowlands is characteristic of the region. Above the base region is the montane level, where the vegetation changes from shrub, woodland, chaparral, or deciduous forest (depending upon the geographical location of the mountain range) to montane coniferous forests, alpine tundra, and at the summit of the very highest mountains, perpetual ice and snow. There are, of course, many variations among the continents, but this basic pattern persists.

Biomes European ecologists call the broad vegetation patterns of the world **plant formations.** F. E. Clements and V. E. Shelford (1939) classified world plant formations. They accepted broad plant formations as the biotic units and then associated animals with the plants. This approach works fairly well, because animal life does depend upon a plant base. Clements and Shelford called these biotic units **biomes.** Each biome consists of a distinctive combination of plants and animals, and each is characterized by a uniform life form of vegetation (see map inside back cover), such as grassland or coniferous forest. Each biome includes atypical areas associated with sites of stress or recovery from disturbance.

Depending on how finely you want to classify them, there are at least nine major terrestrial biomes or biome-types: tundra, taiga, temperate forest, temperate rain forest, tropical rain forest, savanna, temperate grasslands, shrubland, and desert. When Whittaker (1975) plotted these various biome-types on gradients of mean annual temperature and mean annual precipitation, they formed a distinctive climatic pattern (Figure 4.28). The graph is a reasonable approximation of global vegetation patterns. As the dashed lines indicate, many types intergrade with one another. Soil and exposure to fire can influence which one of the several biome-types occupies an area. The structure of biome-types is further influenced by whether the climate is marine or continental. The same amount of

precipitation, for example, can support grassland, shrubland, or even forest.

The various biome-types rarely have sharp boundaries. One type grades into another. The combination of species varies as individual species respond to environmental gradients. Such gradients are called **ecoclines.**

If we were to sample vegetation along a transect cutting across a number of biome-types, such as up a high mountain slope or north to south across a continent, we would discover that plant communities change gradually. If we were to run this transect across mid-continental North America beginning in the moist, species-rich forests of the Appalachians and ending in the desert, the transect would follow a gradient of vegetation types on a climatic moisture gradient. It would run from mesophytic forests through oak-hickory forests, oak woodlands with a grassy understory, tallgrass prairies (now corn land), mixed prairies (wheatland), shortgrass plains, desert grasslands, and finally desert shrublands (Figure 4.29a). Likewise, a transect from southern Mexico to the Arctic would move along a climatic temperature gradient from tropical forest through subtropical forest, temperate deciduous forest, temperate mixed forest, boreal coniferous forest, and tundra (Figure 4.29b).

Life Zones Another useful approach to classification and study of ecosystems is the Holdridge life zone system (Holdridge 1964, 1967; Holdridge et al. 1971). It is based on the assumptions that (1) mature, stable vegetational types represent discrete vegetation units recognizable throughout the world and (2) geographical boundaries of vegetation correspond closely to boundaries between climatic zones (Holdridge et al. 1971). Vegetation is determined largely by an interaction of temperature and rainfall.

The Holdridge system divides the world into **life zones** arranged according to latitudinal regions, altitudinal belts, and humidity provinces (Figure 4.30). The boundaries of each zone are defined by mean annual precipitation and mean annual biotemperature. Biotemperature is defined as the sum of mean monthly temperatures of those months with a mean temperature above $0°$ C divided by 12. Altitudinal belts and latitudinal regions are defined in terms of mean biotemperatures and not in the usual terms of latitude or elevation.

The Holdridge system is based on three levels of classification: (1) climatically defined life zones; (2) associations, which are subdivisions of life zones based on local environmental conditions; and (3) further local subdivisions based on actual cover or land use. The term **association,** as used by Holdridge, means a unique ecosystem, a distinctive habitat or physical environment and its naturally evolved community of plants and animals.

The subdivision into zones or associations is based on an interaction between environment and vegetation as defined by biotemperature, precipitation, moisture availabil-

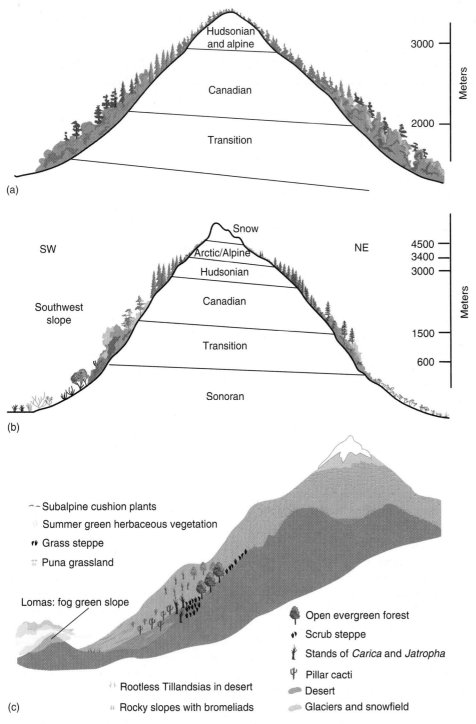

**Figure 4.27** Altitudinal zonation of vegetation in mountains. (a) Mount Marcy in the Adirondack Mountains of New York State. (b) The Rocky Mountains (generalized). (c) Complex cross-section of the Andes at the latitude of Lima, Peru, from the coast to the high mountains. (After H. Walter 1979.)

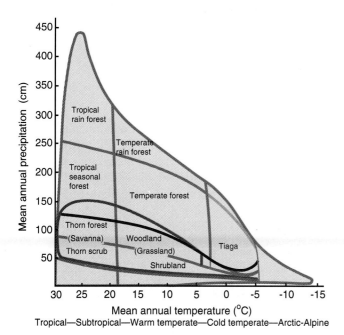

**Figure 4.28** Pattern of world plant formations or biomes in relation to temperature and moisture. Where climate varies, soil can shift the balance between such types as woodland, shrubland, and grass. The dashed line encloses a wide range of environments in which either grassland or one of the types dominated by woody plants prevail in different areas. (After Whittaker 1970a:167.) (See also map inside back cover.)

ity, potential evapotranspiration, and potential evapotranspiration ratio.

Precipitation as considered by Holdridge is the total amount of precipitation for a region rather than seasonal distribution. Long-term average annual precipitation tends to be partially correlated with its seasonal distribution (in terms of soil moisture availability) and life zone boundaries as reflected in vegetation. This fact leads to a classification at the association level.

Moisture availability is the supply of moisture available for plant growth. A function of the relationship between precipitation and potential evapotranspiration, it determines the biologically significant condition of humidity. For all major components of vegetation, soil rather than direct precipitation supplies water necessary for physiological activity. In any given climatic situation moisture available to plants varies with those soil characteristics that influence filtration, storage, and so on. These variations determine association classifications.

Potential evapotranspiration from land vegetation is essentially a function of biologically positive heat balance or biotemperature. Potential evapotranspiration from land vegetation is a hypothetical figure, not a directly measurable figure against which other moisture values may be compared. It is the amount of water that would be transpired under constantly optimal conditions of soil moisture and plant cover. It is calculated by multiplying mean annual temperature by an experimentally derived constant, 59.93.

Potential evapotranspiration ratio is the ratio of the mean annual potential evapotranspiration to average total annual precipitation. It provides an index of biological humidity conditions. A ratio of 1.00 indicates precipitated moisture is exactly equal to potential evapotranspiration for a long-term

**Figure 4.29** Gradients of vegetation in North America from east to west and north to south. (a) The east-west gradient reflects precipitation. The transect does not cut across the Rocky Mountains. (b) The north-south gradient reflects temperatures.

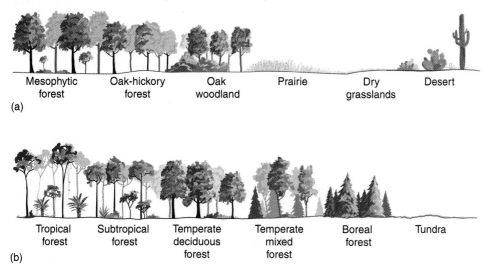

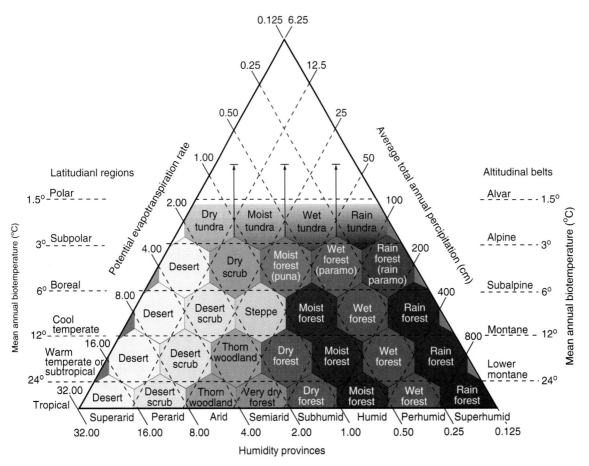

**Figure 4.30** The Holdridge life zone system for classifying plant formations. The life zones are determined by a gradient of mean annual biotemperatures by latitude and altitude, the ratio of potential evapotranspiration to annual precipitation, and total annual precipitation. (After Holdridge 1967. See also map inside front cover.)

average year. As the ratio increases, precipitation is progressively less than water lost and climates are increasingly arid. As the ratio decreases, the climates become progressively more humid.

## Ecoregions

A different attempt at the classification of the biotic world is the ecoregion approach, developed largely by R. G. Bailey (1976) of the U.S. Forest Service. It is based on the ecoregion concept of J. M. Crowley (1967). The classification scheme involves a synthesis of climate types, vegetation associations, and soil types into a single geographic, hierarchical classification, which reflects both ecological properties and spatial patterns (Table 4.2).

An *ecoregion* is a continuous geographical area across which the interaction of climate, soil, and topography are sufficiently uniform to permit the development of similar types of vegetation (Figure 4.31). Boundaries between ecoregions, however, are arbitrary, drawn where the associations of the two regions cover approximately the same area.

Within an ecoregion certain ecological communities may be similar to those of other regions.

The classification involves a hierarchy. The largest category is the *domain,* a subcontinental area of broad climatic similarity. Domains are divided into *divisions,* determined by isolating areas of differing vegetation and regional climates. Divisions are divided into *provinces,* which correspond to broad vegetational regions having a uniform regional climate and the same type or types of zonal soils. Provinces are subdivided into sections and smaller categories useful in local areas.

The ecoregion classification relates more to management of forest, range, and wildlife resources than to ecology. Its purpose is to provide a foundation for ecological management of resources. Because each region has its own distinctive flora, fauna, climate, soil, and landform, each requires its own approach to management. The ecoregion approach provides a means of studying the problems of resource management on a regional basis. It provides a framework for the

**Table 4.2** Characteristics of Some First-Order and Second-Order Ecoregions

| Domain | Division | Temperature | Rainfall | Vegetation | Soil |
|--------|----------|-------------|----------|------------|------|
| Polar | Tundra | Mean temperature of warmest month <10°C | Water deficient during the cold season | Moss, grasses, and small shrubs | Tundra soils (entisols, inceptisols, and associated histosols) |
| | Subarctic | Mean temperature of summer is 10° C; of winter, −3° C | Rain even throughout the year | Forest, parklands | Podzols (spodosols and associated histosols) |
| Humid temperate | Warm continental | Coldest month below 0° C, warmest month <22° C | Adequate throughout the year | Seasonal forests, mixed coniferous-deciduous forests | Gray-brown podzolic (spodosols, alfisols) |
| | Hot continental | Coldest month below 0° C, warmest month >22° C | Summer maximum | Deciduous forests | Gray-brown podzolic (alfisols) |
| | Subtropical | Coldest month between 18° C and −3° C, warmest month >22° C | Adequate throughout the year | Coniferous and mixed coniferous-deciduous forests | Red and yellow podzolic (ultisols) |
| | Marine | Coldest month between 18° C and −3° C, warmest month <22°C | Maximum in winter | Coniferous forests | Brown forest and gray-brown podzolic (alfisols) |
| | Prairie | Variable | Adequate all year, excepting dry years; maximum in summer | Tallgrass, parklands | Prairie soils, chernozems (mollisols) |
| | Mediterranean | Coldest month between 18° C and −3° C, warmest month >22° C | Dry summers, rainy winters | Evergreen woodlands and shrubs | Mostly immature soils |
| Dry | Steppe | Variable, winters cold | Rain <50 cm/yr | Shortgrass, shrubs | Chestnut, brown soils, and sierozems (mollisols, aridisols) |
| | Desert | High summer temperature, mild winters | Very dry in all seasons | Shrubs or sparse grasses | Desert soils (aridisols) |
| Humid tropical | Savanna | Coldest month >18° C, annual variation <12° C | Dry season with <6 cm/yr | Open grassland, scattered trees | Latosols (oxisols) |
| | Rain forest | Coldest month >18° C, annual variation <3° C | Heavy rain, minimum 6 cm/month | Dense forest, heavy undergrowth | Latosols (oxisols) |

*Note:* Names in parentheses in the soil column are soil taxonomy orders (USDA Soil Survey Staff 1975); see Table 9.2.
*Source:* Bailey 1978, map.

organization and retrieval of data in a resource inventory. The concept has been widely accepted in the area of applied ecology.

# SUMMARY

Weather is air in motion, driven by unequal heating, whereas climate is the summation of weather over time. Solar radiation, the major source of thermal energy for Earth, is the major determinant of climate. The prevailing winds, ocean currents, and rainfall patterns over the planet are all affected by variations in heat budgets. The heat budget in turn is influenced by the position of incoming radiation on a spherical Earth, the distribution of land and water masses, the albedo (the percentage of solar radiation reflected back toward space), and Earth's daily rotation. The climate of any given region is a combination of patterns of temperature and moisture, which are influenced by latitude and location in the continental land mass. Mountain ranges influence regional climates by intercepting air flow, causing

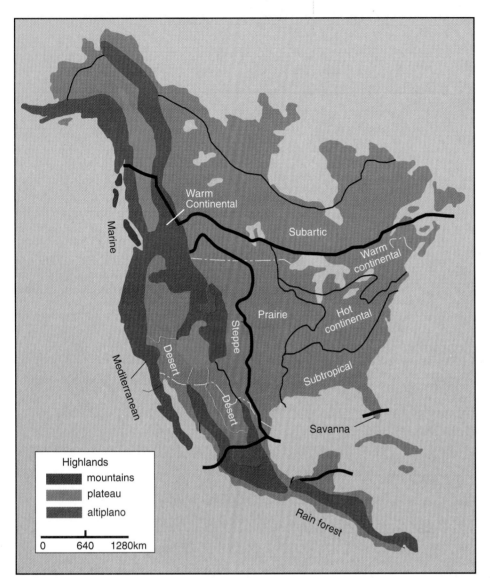

**Figure 4.31** Ecoregions of North America. The heavy lines outline the domains: polar, dry, humid temperate, and humid tropical. The thin lines demark divisions (see Table 4.2). (From Bailey 1978.)

its moisture to drop out on the windward side and creating an area of dry air on the leeward side.

Daily heating and cooling cause air masses to rise and sink. Under certain conditions, the temperature of air masses increases with height rather than decreases. Such an air mass is stable, creating a temperature inversion that can trap atmospheric pollutants close to the ground. Inversions break up when air close to the ground heats, causing it to circulate and rise through the inversion, or when a new air mass moves into the area.

Although regional climates determine general conditions, the specific climatic conditions under which organisms live vary considerably within an area. These variations, or microclimates, are influenced by topography, aspect,

height above the ground, and other factors. Most pronounced are environmental differences between the ground level and the upper strata of vegetation and between north-facing and south-facing slopes.

Climate has a pronounced effect on the distribution of vegetation. We have used the broad differences in climatically-induced vegetation patterns as basis for classification. The most familiar classification is the biome-type concept, which groups plants and animals of the world into distinctive life forms such as tropical forest and tundra. Boundaries of biome-types, or life zones, as they are known in Europe, coincide with the major plant formations of the world. Climatic variations with altitude produce changes in vegetation that mimic those of latitude. More

refined, and somewhat more complex, is the Holdridge life zone system, which divides the world into zones defined by mean annual precipitation and mean annual biotemperature. A third classification system is the ecoregion scheme. It refines the biome-types into hierarchies based on large geographical areas in which soils, temperature, and precipitation are sufficiently uniform to permit the development of similar types of vegetation. This classification provides an ecological foundation for the management of regional resources.

## REVIEW QUESTIONS

1. What is the major determinant of climate?
2. Contrast short-wave and long-wave radiation relative to Earth.
3. What effect does the greater heating of lower latitudes have on the air circulation on Earth?
4. What is the Coriolis force? How does it influence global air circulation and ocean currents?
5. What is an inversion? How do nightime valley inversions and subsidence inversions occur?
6. What effect would cloudiness and wind speed have on the degree and depth of an air temperature inversion near the active surface at night?
7. What is the significance of microclimates to local plant and animal distribution?
8. How does the pattern of temperature and moisture influence global vegetation? Mountain vegetation? How is this influence reflected in the vegetation of North America? Europe? Africa?
9. Why in many humid middle-latitude regions is forest growth greater on north-facing slopes than on south-facing slopes?
10. Contrast the biome, ecoregion, and life zone concepts of classification of major vegetation types.

## CROSS-REFERENCES

Plant responses to moisture, 70–75; animal responses to moisure, 75–77; thermal energy exchange, 80–82; plant responses to temperature, 80–83; animal responses to temperature, 85–95; seasonality, 110; nature of energy, 168–169; greenhouse effect, 205–206; grasslands, 226–236; savannas, 236–239; shrublands, 239–246; deserts, 246–250; forests, 264–288; tundra, 250–261.

*Chapter* **5**

# Water Balance

## Outline

## Concepts

1. Water has unique properties that make it essential for life.
2. The water on Earth cycles among land, oceans, and atmosphere.
3. Plants gain water from the soil through a water potential gradient maintained by transpiration.
4. Plants have evolved means of coping with too little or too much water.
5. The means by which animals maintain a water balance vary with the type of environment in which they live.

Leonardo da Vinci wrote, "Water is the driver of nature." Although most ecologists would now argue that solar radiation is really the driver of nature, da Vinci did not overstate the importance of water to life on Earth. Without the cycling of water, nutrients such as calcium and phosphorus could not make their endless odyssey through the ecosystem, ecoysystems could not function, and life could not be maintained. Strategies for obtaining and conserving water have shaped the nature of terrestrial organisms and communities. Means of living within it have been the pervading influence on life in the water. The distribution of plants and animals is influenced by when, where, and how much precipitation, runoff, and evaporation occur. Thus water is essential to life on Earth.

## THE STRUCTURE OF WATER

Water is unique. It can vanish as a gas in the atmosphere, reappear as dew, fog, clouds, rain, and snow, and in the cold change into a solid. Water can change into all these forms because of the physical arrangement of its hydrogen atoms and hydrogen bonds.

A molecule of water consists of one atom of oxygen (atomic number = 8) and two atoms of hydrogen (atomic number = 1), bonded together covalently (Figure 5.1a). The outer orbitals of the oxygen atom have six electrons, but are able to accommodate eight. The orbital of hydrogen has one electron, but is able to accommodate two. By sharing, the two hydrogen atoms become covalently bonded to the oxygen atom to form a molecule of water. The remaining four, or two pairs, of unshared electrons on the oxygen's outer shell tend to concentrate in a direction away from the O–H bond. If you draw a tetrahedron to enclose the water molecule, the two hydrogen nuclei occupy two corners of the tetrahedron and the unshared pairs of electrons occupy the other two corners. Because oxygen's outer electrons form this tetrahedron, the two hydrogen bonds are separated by a 105° angle. As a result the water molecule is not symmetrical, but V-shaped. The oxygen nucleus of the water molecule attracts the shared electrons more strongly than do the two smaller hydrogen nuclei. This unequal distribution of charges results in a slightly negative charge on the oxygen side of the molecule and a slightly postive charge on the hydrogen side, so the water molecule is polarized.

Because of this polarity, the positively charged region (hydrogen) of one water molecule is attracted to the negatively charged region (oxygen) of another (Figure 5.1b). This electrostatic attraction forms a hydrogen bond that acts as a bridge between water molecules. The 105° angle of association between the hydrogen atoms encourages an open tetrahedral-like arrangement of water molecules, with hydrogen atoms acting as connecting links between them. This association accounts for the latticelike arrangement of clusters of water molecules.

The lattice is at its most highly organized state when water is frozen. Each oxygen atom is hydrogen-bonded to two hydrogen atoms, and each hydrogen is covalently bonded to one oxygen atom. Ultimately each water molecule is bound to four other water molecules by means of hydrogen atoms. One such unit built upon another gives rise to a lattice with large open spaces (Figure 5.1c). Water molecules so structured occupy more space than they would in the liquid form. As a result water expands upon freezing and ice floats.

When heat is applied to ice, the molecules have a higher kinetic energy (energy of motion) and the weakened bonds are broken, causing the lattice to partially collapse. Although water molecules now move more freely and form a liquid, they can pack more tightly together. The volume occupied by the water molecules decreases and the density increases, until water achieves its greatest density at 3.98° C. At this temperature contraction of the molecules brought about by the partial collapse of the lattice balances the thermal expansion of the warming molecules. As water is heated above 3.98° C, more hydrogen bonds break and the distance between the molecules increases, resulting in a lower density. Given enough kinetic energy, individual water molecules can leave ice or liquid water and enter the atmosphere as gaseous water vapor. The vaporization of ice is called *sublimation* and the vaporization of a liquid is *evaporation*. Both processes occur more rapidly as the temperature increases. However, evaporation occurs most frequently when water is heated above its boiling point, 100° C at sea level.

Evaporation involves considerable amounts of heat; 536 cal are needed to overcome the attraction between water molecules and to convert 1 g of water at 100° C into vapor. Heat lost at the point of evaporation returns at the point of condensation (the conversion from vapor to liquid). Such phenomena play a major role in meteorological cycles and the evaporative cooling of organisms.

Seawater behaves somewhat differently. Seawater is defined as water with a minimum salinity of 24.7 o/oo (o/oo = parts per thousand). Its density, or rather its specific gravity relative to that of an equal volume of pure water (sp. gra. = 1) at atmospheric pressure, is correlated with salinity. At 0° C the density of seawater with a salinity of 35 o/oo is 1.028. The density of seawater increases at a lower temperature and decreases at higher temperatures. No definite freezing point exists for seawater. Ice crystals begin to form at a temperature that varies with salinity. As pure water freezes out, the remaining unfrozen water becomes more salty, which further lowers its freezing point. If the temperature decreases enough, a solid block of ice crystals and salt ultimately forms.

## PHYSICAL PROPERTIES

Water can store tremendous quantities of heat with a relatively small rise in temperature. It can do so because of the large number of hydrogen bonds. Energy applied to water

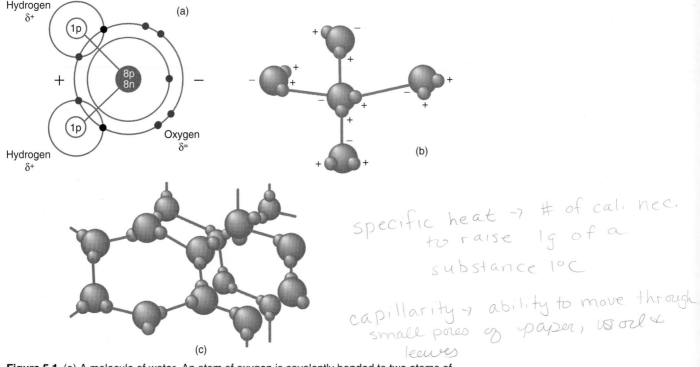

Figure 5.1 (a) A molecule of water. An atom of oxygen is covalently bonded to two atoms of hydrogen. The angle between the two hydrogen bonds is 105°. (b) Hydrogen bonds between water molecules. The partial positive charge of the hydrogen atoms of one molecule attracts a negatively charged oxygen atom of another molecule to form a hydrogen bond between them. (c) In ice, below 0°C, hydrogen bonds hold water molecules rigidly in place to form a crystalline lattice.

goes into breaking the bonds and not into raising the kinetic energy (temperature) of the molecules. Thus water has a high *specific heat,* the number of calories necessary to raise 1 g of a substance 1° C. The specific heat of water is by definition unity or 1. The specific heat of water is exceeded by only a few other liquids, such as ammonia.

Not only does water have a high specific heat; it also possesses the highest heat of fusion and heat of evaporation of all known substances that are liquid between temperatures of 0 to 100° C. Water must lose considerable quantities of heat before it can change from a liquid to a solid form, ice; and conversely, ice must absorb considerable heat before it can convert to a liquid. Approximately 80 calories of heat are needed to convert 1 g of ice at 0° C to a liquid state at 0° C. This amount of heat is equivalent to that needed to raise the same quantity of water from 0° to 80° C.

Hydrogen bonding and polarity give water two other unique properties, cohesion and adhesion. *Cohesion* is the holding together of like substances; the hydrogen bonds account for the cohesiveness of water. *Adhesion* is the holding together of unlike substances. The polarity of water molecules allows them to form hydrogen bonds with other polar molecules and ions. These properties account for the *capillarity* of water, its ability to moves through the pores of paper, the fine pores of soil, and the conducting ducts of leaves. This capillarity is due in part to the adhesion of water to an electrosta-

tic surface and the cohesion of water molecules. Water molecules are repelled by nonpolar molecules, such as oil, which explains why oil and water do not mix.

Molecular cohesion of water accounts for *surface tension.* Water molecules below the surface are symmetrically surrounded by other molecules of water. The forces of attraction are the same on all sides. At the water's surface, the molecules exist under a different set of conditions. Below is a hemisphere of strongly attractive water molecules; above is a much smaller attractive force of the air. The molecules on the surface therefore tend to contract and become taut.

In aquatic habitats surface tension is a barrier to some organisms and support for others. It is a force that enables water to creep upward into fine tubes such as the pores of soil and the conducting network of plants. Aquatic insects and plants have evolved structural adaptations that prevent the penetration of water into the tracheal systems of the former and the stomata and internal air spaces of the latter.

Hydrogen bonding gives water another property, **viscosity.** Compared to other liquids, the viscosity of water is high. Observe a liquid flowing through a clear glass tube. The liquid moving through the tube behaves as if it were parallel concentric layers flowing one over the other. The rate of flow is greatest at the center; because of friction between the layers, the flow decreases toward the sides of the tube. This type of resistance between layers is called *lateral*

or *laminar viscosity.* You can observe this phenomenon along the side of any stream or river with uniform banks. The water at streamside is nearly still, whereas the current in the center may be swift.

Viscosity of flowing water is complicated by another type of resistance, *eddy viscosity.* Water masses pass from one layer to another, creating turbulence both horizontally and vertically. Eddy viscosity, which is many times greater than laminar viscosity, is biologically important, especially for planktonic organisms. It prevents them from drifting down to the deep and brings nutrients to them.

Viscosity is the source of frictional resistance to objects moving through water. Because this resistance is 100 times that of air, animals must expend considerable muscular energy to move through the water. This resistance has been an important selection pressure in the evolution of a stream-lined fusiform body shape in fish and such marine mammals as whales and dolphins and of rod-like shapes, protuberances, and other features of phytoplankton and zooplankton that modify their sinking rates.

# THE WATER CYCLE

## Distribution of Water

Although water in a stream or a spring rain appears separate, it is part of a worldwide whole distributed in land, sea, and atmosphere and unified by the water cycle. It is influenced by solar energy, by the currents of the air and oceans, by heat budgets, and by water balances of land and sea. Through historical time the amount of free water has remained relatively stable, although the balance between land and sea has fluctuated. According to the Russian hydrologist Shinitnikov (Kalinin and Bykov 1969), we are passing from a humid period in Earth's history to a dry one, in which land areas are losing water at the rate of 430 k³ (105 mi³) a year; the oceans are gaining that amount, rising on the average of 1.2 mm (0.05 in.) per year.

Water occurs in six general forms in the biosphere: vapor, fresh surface water (streams and lakes), groundwater, snow and ice, salt water (oceans), and water inside the bodies of organisms. Oceans cover 71 percent of Earth's surface (Figure 5.2). With a mean depth of 3.8 km (2.36 mi), they hold 93 to 97 percent of all the earth's waters (depending on the estimate used). Fresh water represents only 3 percent of the planet's water supply. Of the total fresh water on Earth, 75 percent is locked up in glaciers and ice sheets, enough to maintain all the rivers of the world at their present rate of flow for the next 900 years. Thus about 2 percent of the world's water is in ice. This leaves less than 1 percent of the world's water available as fresh liquid water. Freshwater lakes contain 0.3 percent of the freshwater supply, and at

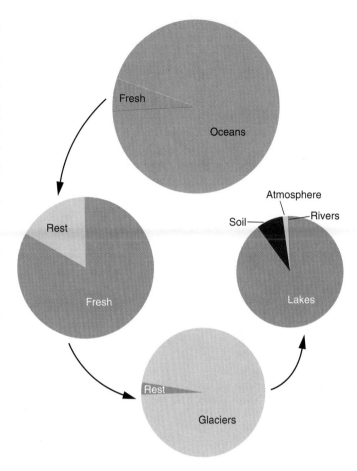

**Figure 5.2** The distribution of Earth's water supply.

any one time rivers and streams contain only 0.005 percent of it. Soil moisture accounts for approximately 0.3 percent. Another small portion of the earth's water is tied up in living material.

Groundwater accounts for 25 percent of our fresh water. Groundwater fills the pores and hollows within the earth just as water fills pockets and depressions on the surface. Estimates, necessarily rough and inaccurate, place renewable and cyclic groundwater at $7 \times 10^6$ km³ (Nace 1969), or approximately 11 percent of the freshwater supply. Some of this groundwater is "inherited," as in aquifers in desert regions, where the water is thousands of years old. Because inherited water is not rechargeable, heavy use of these aquifers for irrigation and other purposes is depleting the supply. Loss is occurring in the aquifers of the midwestern United States and Saudia Arabia. In the foreseeable future the supply could be exhausted. A portion of the groundwater, approximately 14 percent, lies below 1000 m. Known as fossil water, it is often saline, and does not participate in the hydrological cycle.

The atmosphere, for all its clouds and obvious close association with the water cycle, contains only 0.035 percent

fresh water. If this atmospheric water were to condense on Earth, it would form a layer of water no more than one inch thick, yet it is the atmosphere, and its relation to land and ocean, that keeps the water circulating over Earth.

## The Local Water Cycle

Outside it is raining. The rain strikes the house and runs down the windows and walls into the ground. It disappears into the grass, drips from the leaves of trees and shrubs, and trickles down the trunks. When the rain stops, the windows and walls dry. The entire episode—the spring shower, the infiltration into the ground, the throughfall in the trees and bushes, the runoff from the walks, the evaporation—epitomizes the water cycle on a local scale (Figure 5.3).

*Precipitation* is the driving force of the water cycle. Whatever its form, precipitation begins as water vapor in the atmosphere. When air rises it is cooled adiabatically, and when it rises beyond the temperature level at which condensation takes place, clouds form. The condensing moisture coalesces into droplets 1 to 100 μ in diameter and then into rain droplets with a diameter of approximately 1000 μ (1 mm). Where temperatures are cold enough, ice crystals may form instead. Particulate matter smaller than 10 μm in the atmosphere act as nuclei on which water vapor condenses. At some point the droplets or ice crystals fall as some form of precipitation. As the precipitation reaches the earth, some of the water reaches the ground directly, and some is intercepted by vegetation, litter on the ground, and urban structures and streets. It may be stored in lakes and ponds, run off, or in time infiltrate the soil.

Because of *interception,* various amounts of water evaporate into the atmosphere without ever reaching the soil surface. Grass in the Great Plains may intercept 5 to 13 percent of the annual precipitation (average loss, 8 percent), and litter beneath a stand of grass may intercept 3 to 8 percent of annual precipitation (average loss, 4 percent) (Corbett and Crouse 1968). Precipitation striking forest trees or forest canopy must penetrate the crowns before it reaches the ground. A forest in full leaf in the summer can intercept a significant portion of a light summer rain.

In a deciduous forest in summer a greater proportion of rainfall is intercepted during a light shower than during a heavy rain. During a light shower the rain does not exceed the storage capacity of the canopy, and the water held by the leaves subsequently evaporates. Water exceeding the storage capacity of the canopy either drips off the leaves as **throughfall** or runs down the stem, twigs, and trunk as **stemflow.** Water then enters the soil in a narrow band around the base of the tree. In winter deciduous trees intercept very little precipitation.

The amount of rainfall intercepted depends upon the type and the age of the forest. In general, conifers intercept more

**Figure 5.3** The water cycle, showing major pathways through the ecosystem.

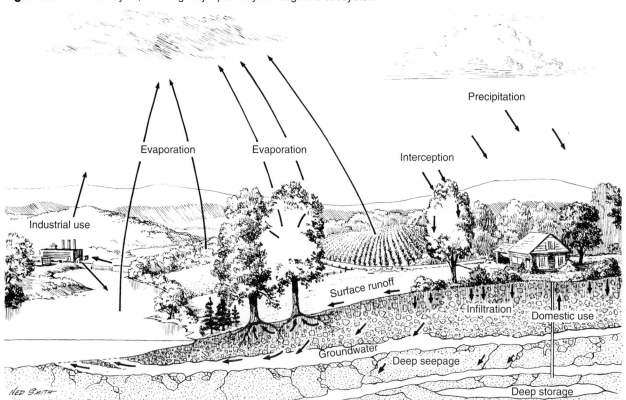

rainfall on an annual basis than hardwoods. Mature pine stands, for example, will intercept 20 percent of a summer rainfall and 14 to 18 percent of the annual precipitation, whereas an oak forest will intercept 24 percent of a summer rain, but only 11 percent of the total annual precipitation (Lull 1967).

In urban areas a great portion of rain falls on roofs, sidewalks, and streets, which are impervious to water. The water runs down gutters and drains to be hurried off to rivers. The percentage of rainfall striking impervious surfaces varies with the nature of the urban area. Nearly 100 percent of the surface in commercial areas may be impervious to rainfall. In residential areas 25 to 80 percent may be impervious, depending on how closely the buildings are spaced; in suburban areas, only 8 percent may be impervious (shopping malls excluded) (Antoine 1964; Felton and Lull 1963). Interception in urban areas is about 16 percent of total annual precipitation, approximately the same amount intercepted by deciduous forests in summer. Residential areas intercept about 13 percent (Lull and Sopper 1969).

The precipitation that reaches the soil moves into the ground by **infiltration.** The rate of infiltration is governed by soil, type of vegetation, slope, and the characteristics of the precipitation itself. In general, the more intense the rains, the greater the rate of infiltration, until the infiltration capacity of the soil, determined by soil porosity, is reached. Because water moves through the soil by the action of two forces, capillary attraction and gravity, soils with considerable capillary pore space have initial rapid infiltration rates, but the pores rapidly fill with water. Water will infiltrate longer into soils with a high proportion of noncapillary pore space. Vegetation that tends to roughen the surface retards surface flow and allows the water to move into the soil. Slope, impeding layers of stone or frozen soil, and other conditions influence the rate at which water moves into the soil.

Long wet spells and heavy storms may saturate the soil, and intense rainfall or rapid melting of snow can exceed the infiltration capacity of the soil. When the soil can no longer absorb the precipitation, the water becomes *overland flow.* Sheet flow over the surface is changed to channelized flow as the water becomes concentrated in depressions and rills. This process can be observed even on city streets as water moves in sheets over the pavement and becomes concentrated in streetside gutters. Again the amount of runoff and associated erosion of soil depends upon slope, texture of the soil, soil moisture conditions, and the type and condition of vegetation.

In the undisturbed forest infiltration rates usually are greater than intensity of rainfall, and surface runoff does not occur. In urban areas infiltration rates range from zero to a value exceeding the intensity of rainfall where soil surface is open and uncompacted. Because of low infiltration, surface runoff from urban areas might be as much as 85 percent of the precipitation (Lull and Sopper 1969). Because they are

so compacted by frequent tramping and mowing, lawns have a low infiltration rate. Felton and Lull (1963) have demonstrated that water infiltrates lawns at an average rate of 0.25 mm/min, compared to 0.142 mm/min in forest soil.

Water entering the soil will **percolate,** or seep down to an impervious layer of clay or rock to collect as groundwater. From here the water finds its way into springs, streams, and eventually to rivers and seas. A great portion of this water is utilized by humans for domestic and industrial purposes, after which it reenters the water cycle by discharge into streams or into the atmosphere, as gravitational water.

A part of the water is retained in the soil. This portion, held by capillary forces between the soil particles, is called **capillary water.** The maximum amount of water that a soil can hold at one-third atmosphere of pressure after gravitational water is drained away is **field capacity.** Highly porous, sandy soils have a low field capacity, whereas fine-textured and humic soils have a high field capacity. Humus (material that develops as a result of decomposition of organic matter) may retain 100 to 200 percent of its own weight in water. For each 2.5 cm of humus about 2 cm of water is stored for as long as two days after a rain and is slowly discharged into streams (Lull 1967). Storage in urban areas obviously is considerably less, almost nothing for paved areas.

Water remaining on the surface of the ground, in the upper layers of the soil, and collected on the surface of vegetation, and water in the surface layers of streams, lakes, and oceans returns to the atmosphere by *evaporation.* Evaporation is the movement of water molecules from the surface into the atmosphere at a rate governed by how much moisture the air contains (vapor-pressure deficit).

As the surface layers of soil dry out, a barrier through which little soil water moves develops and evaporation ceases. Further water losses from the soil take place through plants. Plants take in water from the soil through the roots and lose it through the leaves and other organs in a process called **transpiration.** Transpiration is the evaporation of water from internal surfaces of leaves, stems and other living parts and its diffusion from the plant. In the presence of sufficient moisture leaves remain turgid with the stomata fully open, which permits an easy inflow of carbon dioxide and a loss of water vapor. This loss continues as long as moisture is available for roots in the soil, as long as the roots are capable of removing water from the soil, and as long as the amount of energy striking the leaf is enough to supply the necessary latent heat of evaporation. Thus plants can continue to remove water from the soil until the capillary water is exhausted. The total flux of evaporating water—from the surfaces of the ground and vegetation—is called *evapotranspiration.*

The temperate deciduous forest and the urbanized areas represent two environmental extremes in water cycling. In comparison to the forest, urbanized areas are characterized by reduced interception, less infiltration, much less soil

moisture storage, less evapotranspiration, and reduced water quality. Urban areas also exhibit increased overland or surface flow, increased runoff, and increased peak flows of streams and rivers.

## The Global Water Cycle

Once evaporated into the atmosphere or carried away by surface runoff, the water involved in the local hydrological cycle enters the global water cycle. The molecules of water that fell in the spring shower might well have been a part of the Gulf Stream a few weeks before and perhaps spent some time in the Amazon tropical rain forest before that. The local storm is simply a part of the mass movement and circulation of water about Earth, a movement we see in the changing cloud patterns over the face of Earth in satellite pictures on the evening weather programs. The atmosphere, oceans, and land masses form a single gigantic water system that is driven by solar energy and gravity.

The atmosphere is the one key element in the world's water system. At any one time the atmosphere holds no more than a 10 to 11-day supply of rainfall in the form of vapor, clouds, and ice crystals. The turnover rate of water molecules is rapid. The average resident time for a water molecule in the atmosphere is approximately ten days. Detention of precipitation is maximum in the polar regions of the Northern and Southern Hemispheres. In tropical regions the period of maximum detention is the early part of the rainy season.

Atmospheric water comes from evaporation from land and sea; therefore the amount of moisture in the atmosphere at any given point around the globe depends upon global variations in the rate of evaporation. Evaporation is considerably greater at lower latitudes than at higher latitudes because of more direct solar radiation. The atmosphere over the oceans holds more global moisture as well as more moisture per average volume of air than over land. Oceans provide 84 percent of total annual evaporation, considerably more than they receive in return from precipitation. Land areas contribute 16 percent, receiving more water than evaporates from them.

Moisture in the atmosphere moves with the general circulation of the air. Air currents, hundreds of kilometers wide, are like giant unseen rivers moving above Earth. Only part of this moisture falls as precipitation in any one place. For example, each year the United States receives on the average an unequally distributed 6000 km³ of precipitation, but the liquid equivalent of water vapor passing over the country is approximately ten times that much.

Precipitation on land in excess of evaporation is eventually transported to the sea by rivers. Rivers are the prime movers of water over the globe and carry many more times the amount of water than their channels hold. By returning water to the sea, rivers tend to balance the evaporative deficit of the oceans. Evaporation, precipitation, detention, and transportation maintain a stable water balance on Earth. Consider the amount of water that falls on Earth in terms of 100 units (Figure 5.4). On the average 84 units are lost from the ocean by evaporation, while 77 units are gained from precipitation. Land areas lose 16 units by evaporation and gain 23 units by precipitation. Runoff from land to oceans makes up 7 units, which balances the evaporative deficit of the ocean. The remaining 7 units circulate as atmospheric moisture.

Examined from a global point of view, the water cycle connects every place on Earth. "Local" water problems—shortages, floods, and pollution—are actually global; and local water management schemes affect the planet as a whole. Problems result not because an inadequate amount of water reaches Earth, but because it is unevenly distributed, especially to human population centers, and misused. Because humanity has strongly interjected itself into the

**Figure 5.4** Global water budget. The mean annual global precipitation of 83.6 cm has been converted into 100 units.

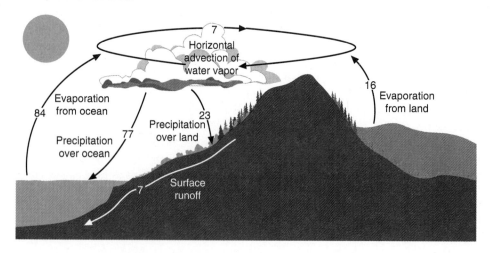

water cycle, the natural usable water resources have decreased, and water quality has declined. The natural water cycle has not been able to compensate for destructive human actions.

# PLANT RESPONSES TO MOISTURE

It is paradoxical that in order to gain water from the soil plants must lose water to the atmosphere. Plants take in water from the soil through the roots and lose it through the leaves and other organs in a process called transpiration. Transpiration is the evaporation of water from internal surfaces of leaves, stems, and other living parts and its diffusion from the plant into the atmosphere. To appreciate how water moves from the soil and plant to the atmosphere, we must understand some other aspects of water.

For water to move or diffuse from soil solution into the roots it must pass through the cell membranes of the root. Some membranes are permeable; they do not impede the movement of substances through them. Others are much more selective; they allow some substances to pass through them but not others. Such membranes are semipermeable. Semipermeable membranes are relatively permeable to water and are permeable to other substances.

The general tendency of molecules is to move from a region of high concentration to a low concentration. This movement or diffusion accounts for the spread of a solute (dissolved substance) throughout a solvent (the medium that dissolves the solute). For example, consider a solution of salt enclosed in a semipermeable membrane lowered into a container of pure water. The membrane allows the molecules of water to flow through it, but not the solutes, the ions of sodium and chlorine. The concentration of water molecules on the outside of the membrane is high; on the inside of the membrane it is low. The water will move across the membrane into the solvent until the concentration of water molecules is equalized on both sides of the membrane. This process is called **osmosis.**

As water molecules (and other substances) move across a semipermeable membrane, the diffusion generates a certain amount of pressure that slows the movement of water across it. The amount of pressure needed to stop its movement across a membrane is called *osmotic pressure.* Osmotic pressure accounts for the turgor achieved by plant cells when the water supply is adequate. As plants face internal water stress, the concentration of water molecules in the cells decreases, and water, when available, moves from the soil solution into the plant. The tendency of a solution to attract water molecules from areas of high to low concentration is called its **osmotic potential.** The osmotic potential of a solution depends on its concentration. The higher the concentration of solutes (and thus the lower the concentration of water molecules) the lower is the solution's osmotic potential and the greater is its tendency to gain water.

## The Soil-Plant-Atmosphere Continuum

Water in the soil moves along an energy gradient into the roots with their higher solute concentration and lower water potential. What prevents the roots from achieving equilibrium with the energy of soil water and pulls the water up through the plant is a water potential gradient from the soil to the atmosphere (Figure 5.5). Plants pull water from the soil, where the water potential is the highest, to the atmosphere, where water potential is the lowest.

When the relative humidity of the atmosphere drops below 100 percent, the water potential of the atmosphere rapidly becomes negative, increasing the capacity of the air to evaporate water and drive transpiration. Attracted by the low water potential of the atmosphere, moisture from the surface of and between the mesophyll cells within the leaf evaporates and escapes through the stomata. To replace the water lost, more water is pulled from the xylem (hollow conducting tubes throughout the plant) in the leaf veins. Transpiration and replacement of water at the cell's surface make the surface water potential more negative. Negative water potential reaches down to the fine rootlets in contact with soil particles and pores. This tension pulls water from the root and up through the stem to the leaf. The root's water potential declines so that more water moves from the soil into the root.

The loss of water through transpiration continues as long as the amount of energy striking the leaf is enough to supply the necessary latent heat of evaporation, moisture is available for roots in the soil, and the roots are capable of removing water from the soil. Unless soil moisture is replenished by precipitation, eventually only a only a thin film of water remains around the soil particle. This is *hydroscopic water,* unavailable to plants.

During the growing season, plants continuously remove water from the soil at variable rates. These rates depend on daily environmental conditions that influence water potential and upon the plant's resistance to water loss, which is highest in the leaves. Resistance from the stomata is probably the most important regulator of water loss through the plant. The stomatal openings are also the entrance sites for $CO_2$ moving into the plant. As long as the stomata are open to take in carbon dioxide, they are also allowing water to escape. For the plant this is no problem as long as the soil is moist; but during a period of a short or protracted drought, the plant must either close the stomata, shutting off the supply of $CO^2$ and reducing photosynthesis, or keep them open

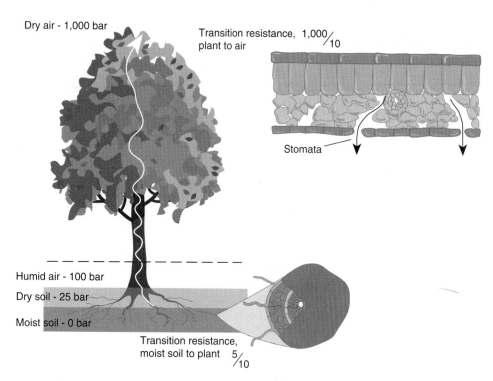

Dry air - 1,000 bar

Transition resistance, 1,000/10
plant to air

Stomata

Humid air - 100 bar

Dry soil - 25 bar

Moist soil - 0 bar

Transition resistance,
moist soil to plant    5/10

**Figure 5.5** Water potential gradient and resistance between soil, plant, and atmosphere. The steepest part of the water potential gradient occurs between the leaf surface and the dry air. The greatest resistance to water loss is in the leaf. The water potential gradient is measured in bars. (1 bar = 0.987 atmospheres = $10^6$ ergs/g = 0.1 MPa. MPa is a Megapascal, a pressure unit directly relatable to energy per unit mass. Pure water has 0 MPa). The insets show the uptake at the root and pathway of water loss.

and continue to lose water. The less water a plant loses during the fixation of one molecule of $CO_2$, the higher is its **water-use efficiency.**

Lack of water is a major selective force in the evolution of the plant's ability to cope with moisture stress. One large group of terrestrial plants, the mosses, and their nonplant associates, algae, fungi, and lichens, has no protective mechanism against water loss. The internal water status of this group tends to match atmospheric moisture conditions. As water becomes less available nonvascular plants—the mosses and liverworts—as well as lichens and terrestrial algae dry out. Their cells shrink without disturbing the fine protoplasmic structures within them and their vital processes gradually become suppressed. When moisture conditions improve, they imbibe water and the cells fill and resume normal functioning (Larcher 1980). Such organisms (called **poikilohydric**) restrict their growth to moist periods, and their biomass is always small.

Other plants (termed **homeohydric**), chiefly ferns and seed plants, are able to maintain, within limits, a stable water balance mostly independent of fluctuations of atmospheric moisture conditions. That ability is made possible by vacuoles that store water within the cell, a protective cuticle that slows down evaporation, stomata that regulate transpi-

ration, a combination of osmotic pressure and turgor pressure of water within the cells, and an extensive root system.

## Response to Deficits

Leaves reflect drought stress in plants. Some, such as evergreen rhododendrons, show drought stress by an inward curling of the leaves. Others show it in a wilted appearance caused by a lack of turgor in the leaves. Wilting is an adaptation that allows leaves to reduce water loss by reducing the amount of solar radiation hitting them. A less obvious, but more significant response is the closure of the stomata. This closure reduces transpirational loss of water, but it also raises the internal temperature of the leaf, causing heat stress that can affect protein synthesis (Walbot and Cullis 1985). Stomatal closure reduces $CO_2$ diffusion into the leaf. Reduced diffusion ultimately results in reduced plant growth, and smaller plants, leaves, buds and other parts (Kramer 1983). Prolonged drought inhibits the production of chlorophyll, causing the leaves to turn yellow or later in the summer to exhibit premature autumn coloration. As conditions worsen, deciduous trees may prematurely shed their

leaves, the oldest ones dying first. Such premature shedding can result in dieback of twigs and branches. Deciduous trees of the temperate regions drop their leaves in fall, avoiding winter drought, and the deciduous trees of the African savanna do likewise at the onset of the dry season.

Other physiological changes take place during drought. Some water-stressed plants reduce their osmotic potential by accumulating in the leaves such inorganic ions as calcium, magnesium, potassium, and sodium (Mattson and Haack 1987), amino acids, sugars and sugar alcohols (Kramer 1983). A decrease in soil water, increase in soil temperature, and reduced root growth alter mineral uptake from the soil. These changes may make drought-stressed plants more vulnerable to insect attack (Mattson and Haack 1987).

Conifers and other evergreens experience winter drought. When winter temperatures are warm enough to thaw the ice in the water ducts of woody plants, the trees lose this water by transpiration through the leaves. Because the soil is frozen, making water unavailable, the plants cannot replace the water lost by transpiration. The result is browning of needles and dieback of twigs from dehydration.

## Adaptations to Drought

Plants in semiarid and desert regions have evolved adaptive mechanisms to survive and reproduce in a dry environment. One method is reducing the leaf area of the canopy by dropping the leaves during periods of extreme drought and developing new leaves at the onset of rain. Although such a response usually results in decreased photosynthetic production, some plants maintain a bit of their photosynthetic capacity by increasing photosynthetic activity in green stems (Nobel 1985).

The angle of leaves to the sun affects the heat load experienced by the plant. When water is limiting, many species of plants reorient the angle of their leaves so that their surfaces are parallel to the sun's rays (Ehleringer and Forseth 1980, Schultz et al 1987). This shift in position reduces leaf temperatures and transpiration rates. With adequate water, these plants orient their leaves perpendicular to the sun's rays.

Another means of conserving water is succulence, the storage and retention of water within the plant's tissues. Notable are the cacti (Figure 5.6a). Because their root systems are shallow, cacti can draw up considerable amounts of water during periods of rain. The plants swell rapidly as they store water in enlarged vacuoles of the cells. During periods of drought stress the cacti draw on this store of water. Desert plants that employ the CAM type of photosynthesis close their stomata during the day; they open them at night and take in and fix $CO_2$ for use during the day. For this reason these plants experience a minimal loss of moisture through transpiration.

Plants meet moisture stress by means of various leaf adaptations. In some the leaves are small, the cell walls thickened, the stomata smaller but more numerous per unit of surface area, the vascular system denser, and the palisade tissue—columnar photosynthetic cells within the leaf—more highly developed than the mesophyll. These structural features increase the ratio of the internal exposed cellular surface of the leaf to the external surface. Many have leaves covered with hairs that scatter incoming solar radiation and increase the depth of the boundary layer—a thin layer of still air close to the leaf surface (Ehleringer 1983). Others have leaves coated with waxes and resins that reflect light and reduce their absorptivity (Mulroy and Rundel 1977).

Root adaptations provide still other means of maintaining water balance. Some woody plants, such as cottonwood (*Populus deltoides*), willows (*Salix* spp.) and mesquite (*Prosopis* spp.), have deep roots that reach an underground water supply. Growing in arid and semiarid regions, these plants, known as **phreatophytes,** have roots in constant contact with a fringe of capillary water above the water table. Other plants, such as the desert creosote bush (*Larrea tridentata*), have spreading root systems covering a large shallow surface area. Such root systems enable them to secure the maximum amount of water when it is available.

Some plants adopt an ephemeral life cycle in which the population survives the dry period as dormant seeds ready to sprout quickly when the rains come (Ehleringer 1985). They bloom and set seed within a short period before moisture is completely gone, and then return to seed dormancy (Figure 5.6b). The cycle may be based on annual wet and dry seasons. The plant seeds may remain dormant for a number of years, waiting for adequate moisture.

Aridity and drought, even seasonal in nature, influence growth and survival of plants. Seeds arriving on dry sites are subject to dehydration. To germinate successfully, they need to absorb quantities of water. If they do germinate, the seedlings often die after their early stages of development. Trees that become established on perennially dry sites rarely grow tall. They grow slowly, with just enough photosynthetic production to achieve microscopic annual growth. By producing only small amounts of living tissue, they reduce drought stress and avoid senescence and death. Such trees tend to be very long-lived. Extreme examples are the bristlecone pines of the Rocky Mountains and the Great Basin of the United States, *Pinus aristata* and *P. aristata* var. *longaeva,* which live 6,000 to 10,000 years.

Plants of the salt marsh grow in a physiologically dry substrate because salinity limits the amount of water they can absorb; and what they do take in contains a high level of solutes. These plants, known as *halophytes,* compensate for an increase in sodium and chlorine uptake by diluting internal solutions with water stored in the tissues. Some plants exhibit high internal osmotic pressure many times that of freshwater and terrestrial plants, possess salt-secreting glands, secrete heavy cutin on their leaves, and are succulent. Others mechanically remove salts at the root mem-

(a)

(b)

**Figure 5.6** Plants adapted to drought. (a) Barrel cacti, like this red barrel cactus (*Ferocactus* spp.) of the southwestern United States desert, store available water in their tissues. (b) The brief desert rainy season brings on a profusion of flowering annual plants.

branes. Plants of saline deserts encounter similar problems. Not only do they exist in a saline soil; they also must endure dry conditions.

## Responses to Flooding

Too much water can place as much stress on plants as too little water. Symptoms of flooding are similar to those of drought, including stomatal closure, yellowing and premature loss of leaves, wilting, and rapid reduction in photosynthesis (Jackson and Drew 1984, Kozlowski and Pallardy 1984). The causes, however, are different.

Excess water around the roots stimulates the plant to resist the movement of water through the roots to the shoots by wilting. Flooding also brings about morphological changes in the plant. Deprived of oxygen because water displaces air in the soil pores, adventitious roots grow horizontally along the oxygenated zone of the soil surface, where anaerobically generated toxins are absent. These roots emerge from the submerged part of the stem when original roots succumb, to replace them. Perennially high water tables force some plants found in poorly drained soils—for example, red

maple (*Acer rubrum*) and white pine (*Pinus strobus*)—to develop shallow, horizontal root systems, which make them highly susceptible to drought and windthrow.

Prolonged flooding often results in dieback or death of plants, particularly woody species (Figure 5.7a). The response of trees to flooding depends upon its duration, the movement of water, and the species of tree. Trees are injured more by standing in stagnant water than in oxygenated flowing water. If a tree's roots are flooded for more than half the growing season, death results (Kozlowski 1984). Look for standing dead trees about beaver dams and along highways where road construction has impeded soil drainage.

## Adaptations to Flooding

Growing plants need both sufficient water and a rapid gas exchange with their environment. Much of this exchange takes place between the roots and soil atmosphere. When soil pores are filled with water, no gas exchange can take place. The plants experiencing depressed $O_2$ levels, in effect, are asphyxiated; they drown. Roots depend upon

(a)

(b)

**Figure 5.7** (a) Trees flooded by a beaver dam will eventually die. (b) "Knees" or pneumatophores rise from the later roots of swamp bald cypress. These knees may be beneficial as aeration organs and help anchor the trees.

gaseous $O_2$ to carry on aerobic respiration. Without sufficient oxygen, the roots cannot respire aerobically and are forced to shift into anaerobic metabolism. In addition, anaerobic conditions in the soil inhibit uptake and transport of ions, and depress the concentration of nitrogen, phosphorus, and potassium in the shoot. These conditions also cause the plant to accumulate ethylene. Ethylene gas, a growth hormone, especially under flooded conditions (Jackson 1985), is highly insoluble in water and is produced normally in small amounts in the roots. Under flooded conditions ethylene diffusion from the roots is slowed and oxygen diffusion into the roots is inhibited. Ethylene, therefore, accumulates to high levels in the root. This stimulates adjacent cortical cells in the cortex of the root to lyse and separate to form interconnected gas-filled chambers (Drew et al. 1979, Jackson 1982) called **aerenchyma.** These chambers, typical of hydrophyllic plants, allow some exchange of gases between submerged and better aerated roots.

Hydrophyllic plants have evolved certain adaptations that allow them to cope with a waterlogged environment. Most have gas-filled chambers through which oxygen diffuses from the shoots to the roots. These interconnected gas spaces throughout the leaves and roots occupy nearly 50 percent of the plant tissue. Some plants of cold, wet, alpine tundras have similar spaces in leaf, stem, and root to carry oxygen to the roots (Keeley et al. 1984). In contrast to nonwetland species, wetland plants lose less oxygen from the roots, thus maintaining oxygen concentrations (Jackson and Drew 1984).

A few woody species can grow on permanently flooded sites. Among them are bald cypress (*Taxodium distichum*), mangroves, willows, and water tupelo (*Nyssa aquatica*). Cypresses experiencing fluctuating water tables develop knees or **pneumatophores,** specialized growths of the root system, which help them survive (Figure 5.7b). Pneumatophores on mangroves provide oxygen to roots during tidal cycles. These plants may also modify their metabolism to adapt to short-term hypoxia and reduce ethanol toxicity (Hook 1984).

# ANIMAL RESPONSES TO MOISTURE

The responses of animals to moisture changes are more complex than those of plants. Animals possess fewer mechanisms to rid the body of excess water and solutes or to conserve them. These mechanisms range from the contractile vacuoles of protozoans to the complex kidney and urinary systems of birds and mammals. Further, animals possess a protective outer covering that protects against passive water loss, and they can move to avoid unfavorable situations or to find new water.

## Adaptations for Environments

If an organism is not to dehydrate, its input of water must equal losses. Moisture is a constant problem in two environments— salt water and the desert. Many organisms of these two environments use the same strategy to overcome the problem.

In the marine environment invertebrates, such as clams and sponges, have no problem with osmotic imbalance because they have the same osmotic pressure as seawater. However, other animals of marine and brackish environments have cells that are more dilute than seawater. To survive in this situation, these animals must inhibit the loss of water by osmosis through the body wall and prevent an accumulation of salts in the system. Marine teleost (bony) fish absorb salty water into the gut. They secrete magnesium and calcium through the kidneys and pass these ions off through urinary ducts as a partially crystalline paste. The fish excrete sodium and chlorine by pumping the ions across membranes of special cells in the gills. This pumping process is called **active transport.** It involves the movement of salts against a concentration gradient at the cost of metabolic energy. Sharks and rays retain urea to maintain a slightly higher concentration of salt in the body than in surrounding seawater. Birds of the open sea can utilize seawater because they possess a special salt-secreting gland located on the surface of the cranium (Schmidt-Nielsen 1960). Gulls (Laridae), petrels (Hydrobatidae), and other sea birds excrete from these glands fluids in excess of 5 percent salt. Petrels and other tube-nosed swimmers forcibly eject the fluid through the nostrils; others drip the fluid out of the internal or external nares.

Among the reptiles the diamondbacked terrapin (*Malaclemys terrapin*), an inhabitant of the salt marsh, lives in a variably saline environment. The terrapin maintains a relatively high osmotic pressure in its cells when the water is dilute; yet it possesses the ability to accumulate substantial amounts of urea in the blood through the functioning of the salt glands located near the eyes when it finds itself in water more concentrated than 50 percent seawater.

The kidney, the main avenue for the elimination of salt among marine mammals, achieves the same importance for water conservation in mammals of arid regions. The mammalian kidney, an efficient mechanism for the maintenance of water balance, involves a countercurrent multiplier system, one of the two countercurrent systems in the animal body. (The other is the countercurrent heat exchanger; see Chapter 6.) In both systems fluids move in opposite directions in closely aligned vessels.

The countercurrent multiplier system within the mammalian kidney maintains water and solute balance in the body. The kidneys possess an active tubular transfer system, which involves a hairpin turn as a part of each tubule, known as the loop of Henle. This loop enables filtrate to flow in opposite directions in the descending and ascending parts, creating a countercurrent flow across a steep osmotic gradient. This system serves to concentrate the urine and return water to the body. The final concentration of urine depends upon the length of the loop; the longer the loop, the more concentrated the urine. This efficient water recovery is essential for all animals living in an environment where water is limiting.

Marine mammals that have to drink seawater possess large kidneys with long loops. Porpoises have highly developed the ability to eliminate salts rapidly. The urine of marine mammals is hyperosmotic, possessing a greater osmotic pressure than blood and seawater.

The problem facing terrestrial mammals living in an arid environment is to conserve water, which they can accomplish through a highly efficient kidney. The kangaroo rat of the southwestern North American desert and its ecological counterparts, the jerboas (*Jerboa* spp.) and gerbils (*Gerbillus* spp.) of Africa and the Middle East, and the marsupial kultarr (*Antechinomys laniger*) of Australia, feed on dry seeds and dry plant material even when succulent green plants are available. They rarely drink. Instead, these mammals obtain water from their own metabolic processes. To

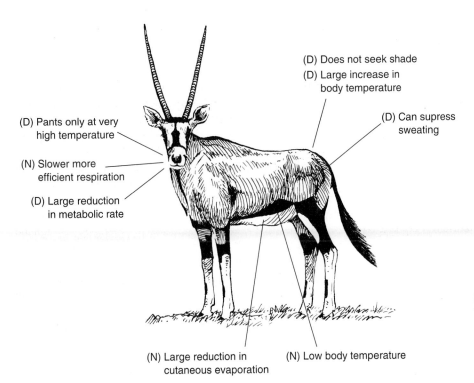

(D) Does not seek shade
(D) Large increase in body temperature

(D) Can supress sweating

(D) Pants only at very high temperature

(N) Slower more efficient respiration

(D) Large reduction in metabolic rate

(N) Large reduction in cutaneous evaporation

(N) Low body temperature

**Figure 5.8** The physiological adaptations to aridity and heat of the African ungulate, the oryx. D = day; N = night. (Adapted from M. S. Gordon 1972, based on data from C. R. Taylor 1969.)

aid in its retention, these desert rodents, like marine mammals, also have exceptionally long loops of Henle that produce highly concentrated urine (24 percent urea and salt, twice the concentration of seawater). To conserve water further these rodents remain by day in sealed burrows. They possess no sweat glands, and their feces are dry. In addition, some desert mammals can tolerate a certain degree of dehydration. Desert rabbits may withstand water losses up to 50 percent of their body weight; camels can tolerate a 27 percent loss.

Extreme adaptations to aridity exist among some of the African antelopes. Outstanding is the oryx (*Oryx leucoryx*) (Figure 5.8). Many African ungulates migrate to escape the heat and dryness, but the oryx remains. During the day it stores heat in its body, causing **hyperthermia,** a substantial rise in body temperature. That rise reduces evaporative losses, because evaporative loss is due to difference between the body and air temperature. The oryx further reduces daytime evaporative losses by suppressing sweating and panting only at very high temperatures. It reduces its metabolic rate, lowering the rate of internal production of calories and thus the amount of needed evaporative cooling. By night the oryx reduces its nonsweating evaporation across the skin by reducing its metabolic rate below that of daytime. The oryx's respiratory rates are proportional to its respiratory efficiency and inversely proportional to body temperature. A cool animal breathes more slowly than a warm one, using a greater proportion of inspired oxygen. With a lowered nighttime body temperature, the saturation level for water vapor in the exhaled air is lower. The oryx normally does not drink water.

It depends upon metabolic water and grasses and succulent shrubs. In fact, the oryx can obtain all the water it needs by eating food containing an average of 30 percent water.

Other animals in arid environments possess adaptations suggestive of those of arid land plants. One is to become dormant during periods of environmental stress. The animal may adopt an annual cycle of active and dormant stages. For example, Couch's spadefoot toad (*Scaphiopus couchi*) of the southwestern desert of the United States spends the dry period—lasting for eight to nine months—in an underground burrow lined with a gelatinous substance that reduces evaporative losses through the skin. It emerges when rainfall saturates the ground, moves to the nearest puddle, mates, and lays eggs. Young tadpoles hatch in a day or two, mature rapidly, and metamorphose into functioning adults. They dig their own retreats in which to estivate until the next rainy period.

Some desert birds, like marine birds, utilize a salt gland to maintain a water balance. Otherwise marine, desert, and salt marsh birds have the same basic adaptations for the conservation of water and the elimination of sodium and chlorine ions. Because birds normally secrete semisolid uric acid as an adaptation related to weight reduction for flying, they are well-fitted for water conservation in arid environments.

## Responses to Drought and Flooding

Animals move to more favorable environments when moisture conditions become critical. Of course, under extreme conditions, animals can die from dehydration or drown.

Drying of marshes, for example, causes nesting losses among waterfowl and muskrats and makes them highly vulnerable to predation. Flooding and heavy rains can drown the young of ground-nesting birds and mammals. Waterlogged soils force earthworms to the surface, where many of them die.

Drought, however, can have more pervasive effects on animal abundance and distribution. During the wet season on the African savanna, the African buffalo (*Syncerus caffer*) finds food abundant; but during the dry season, the quality of food declines as the grasses dry. Buffalo become more selective, seeking green leaves, moving to the moist riverine habitat, breaking into smaller units, and utilizing different areas. As the dry season progresses, buffalo become less selective, consuming dry stems and leaves they otherwise would have rejected. As food quantity and quality decline, food availability for the buffalo declines (Sinclair 1977). The more buffalo present, the less food there is available for each individual.

Eventually, available protein drops below the levels needed by the body, and the animals use up their fat reserves. Undernourished and lacking the protein intake necessary to maintain the immune system, old animals become most vulnerable to the diseases and parasites they normally harbor. The number that die depends upon the rapidity with which the adults use up their energy reserves before the coming of the rainy season. If the next season sees more rainfall and that rainfall extends sporadically into the dry season, the mortality of adults the following year is reduced. Rainfall influences the quantity and quality of dry season forage, which in turn influences mortality of adult buffalo and thus may regulate the density of buffalo populations (Sinclair 1977).

Drought stress promotes outbreaks of leaf-eating insects by influencing thermal and nutritional conditions that favor insect growth (Mattson and Haack 1987). High temperatures raise the rate and efficiency of enzymatic reactions in insects. They enhance the insects' detoxification systems, allowing them to override plant defenses. High temperatures may also promote genetic changes favoring insect population growth. The increased contents of nitrogen, minerals, and sugars in the leaves of drought-stressed plants provide a rich food for the leaf-eating insects.

Moisture influences the speed of development and fecundity of some insects. If the air is too dry, the eggs of some locusts and other insects become quiescent. There is an optimum humidity at which the nymphs of some insects develop the fastest. Others, such as the nondiapause migratory grasshopper *Melanoplus sanguinipes,* require a minimum soil moisture before eggs will develop (Mukerji and Gage 1978). Excessive moisture may be directly lethal to some insects, but more important, it encourages the development and spread of pathogenic microorganisms, especially fungi, among insect populations (Ferro 1987, Martinat 1987).

## SUMMARY

Water is a unique substance, related to molecular structure. Depending upon its temperature, it can take the form of a gas, liquid, or solid. It has the ability to absorb considerable quantities of heat with a small rise in temperature and gives up heat just as slowly. Water has high viscosity, which affects its flow, and exhibits a high surface tension, caused by a stronger attraction of water molecules for each other than for the air above the surface. These properties are important ecologically and biologically.

Most of Earth's water is in the oceans. Less than 1 percent is available as free water. This water moves through the water cycle, involving precipitation, interception, surface flow, and evaporation. The key to the water cycle is the atmosphere. Evaporation into and precipitation from the atmosphere, detention in oceans and land, and atmospheric transport maintain the global water balance. Oceans lose more water by evaporation and gain less from precipitation than land areas. The loss is made up by runoff from land.

Atmospheric moisture is expressed in terms of relative humidity, the measure of the amount of water vapor actually in the air relative to the amount of water vapor the air could hold if saturated. The higher the temperature, the more water vapor the air can hold. Relative humidity is higher by night than by day and is greater at high elevations than at low. It varies widely with topography, vegetation, wind, and temperature. Another useful measure of air moisture is dew point, the temperature at which water vapor in the atmosphere condenses as water droplets.

The stress of too much or too little water is a major selective force in the evolution of plants and animals. Ferns, mosses, lichens, and algae have no means of regulating internal moisture and therefore follow environmental moisture conditions. Most plants maintain a stable water balance independent of environmental fluctuations, by affecting osmotic potential and turgor pressure, collectively known as water potential. Under normal moisture conditions water potential decreases from soil to the leaf, pulling up water.

Plants respond to moisture deficits first by closing the stomata, reducing transpiration. Severe drought decreases photosynthesis, causes leaves to turn yellow and shed, and even kills.

Some plants avoid drought by becoming ephemeral. Some reduce transpiration by closing their stomata, increasing leaf thickness, possessing a waxy cuticle, storing water in cells (succulence), or shedding leaves during the dry season. Other plants extend roots into areas of soil where the water has not been tapped or reach deep to a permanent water supply. Plants in saline environments may accumulate and excrete salt through the leaves or store water in them.

Too much water also creates stress. In extreme conditions anaerobic respiration in the soil asphyxiates plants. Flooding accumulates toxic material in a plant, depresses

the concentration of ions and oxygen, and may stimulate the growth of adventitious roots along the oxygenated soil surface. External symptoms suggest drought stress. Prolonged flooding can result in death.

Wetland plants have adaptations to flooding. Notable are aerenchyma tissues, which form interconnected gas spaces through the leaves to the roots, carrying oxygen to the waterlogged roots.

Moisture or the lack of it has a major influence on the distribution of plants, both geographically and locally. Some are restricted to moist or mesic sites, others to dry or xeric sites.

Animals' responses to moisture stresses are more complex. Most animals maintain a water balance by the excretory system. This system is most complex in mammals, whose countercurrent multiplier system in the nephron of the kidney reabsorbs water from urine against an osmotic gradient. Many animals inhabiting saline environments have salt-excreting glands. Animals of arid regions may reduce water loss by becoming nocturnal, producing highly concentrated urine, using only metabolic water, and tolerating a certain degree of dehydration. Drought stress can lower food quality and cause death. Drought also promotes outbreaks of leaf-eating insects by influencing thermal and nutritional conditions in plants that favor insect growth.

## REVIEW QUESTIONS

1. What are the unique structural properties of water?
2. What are specific heat, latent heat, viscosity, and surface tension of water? What is the ecological importance of these properties?
3. Compare the precipitation–evaporation ratios of land and ocean in the global water cycle.
4. Follow the fate of a spring shower from cloud to groundwater and identify the various processes involved.
5. Explain the relationship among water potential, transpiration, and a plant's extraction of water from the soil.
6. How do plants cope with drought conditions?
7. What is the difference between drought resistance and drought tolerance?
8. What major adaptations enable arid land plants to live in their dry environment?
9. How do the mammalian kidney and related structures in other organisms enable them to maintain a water and solute balance with their environment?
10. If marine mammals and desert rodents can survive in a physiologically arid environment without fresh water, why can't humans?

## CROSS-REFERENCES

Humidity, 45; microclimates, 50–53; climate and vegetation, 54–60; thermal energy exchange, 80–82; soil moisture, 68, 133; photosynthesis, 154–159; $C_3$, $C_4$, and CAM plants, 154–157; lakes and ponds, 290–301; streams, 301–313; freshwater wetlands, 313–323; salt marsh, 348–357; desert, 246–250.

# Thermal Balance

## Outline

## Concepts

1. Organisms maintain a thermal balance with their environment by means of radiation, convection, conduction, and evaporation.
2. The body temperatures of most organisms vary with the environment; the body temperatures of birds and mammals remain fairly constant.
3. Ectothermic organisms depend upon the heat of the environment to maintain their body temperatures; endothermic ones maintain their body temperature by metabolical heat.
4. Organisms have evolved means of adapting to the stresses of heat and cold.
5. Optimum temperatures for life processes influence the distribution of organisms.

All organisms live in a thermal environment characterized by heat and temperature. There is a difference between the two. *Heat* is a form of energy possessed by all substances that results from the random motion of molecules within the substance. The quantity of heat possessed by a substance depends upon the kinetic energy of its molecules and its size. A large substance holds more heat than a small one. *Temperature* is the immediate direct measure of the average kinetic energy possessed by individual molecules of a substance. Temperature expresses the substance's intensity of hotness and measures its tendency to give up heat.

All organisms depend upon a certain temperature or range of temperatures to carry on their metabolic processes. Low temperatures slow metabolic processes because of their effect on rates of chemical reactions. Very high temperatures denature enzymes and thus slow reactions. Somewhere in between these extremes organisms find their optimal temperatures for metabolic and other activities. However, temperatures vary greatly on Earth's surface both in space and time.

The environmental temperatures experienced by most organisms result, directly or indirectly, from solar radiation. The amount of solar radiation reaching any point on Earth at any time varies with the time of year, slope, aspect, cloud cover, time of day and other factors. Seasonal fluctuations can be extreme from one point to another. In North Dakota, for example, where the annual mean temperature is between 3° and 9° C, temperatures fluctuate from a low of 43° C in winter to 49° C in summer. In West Virginia, where the mean annual temperature is 12° C, temperatures range from −37° to 44° C. Environmental temperatures in any area differ between sunlight and shade and between daylight and dark. Surface temperatures of soil may be 30° higher in the sunlight than in the shade. Daytime temperatures are often 17° C higher than nighttime temperatures; on deserts this spread may be as high as 40° C. Temperatures on tidal flats may rise to 38° C when exposed to direct sunlight and sink to 10° C within a few hours when the flats are covered by water.

The ability to withstand extremes in temperatures varies widely among living organisms, but there are temperatures above and below which no life can exist. A temperature of 52° C is about as high as any animal and protozoan can still grow and multiply. Observed thermal limit for the survival of metabolically active vascular plants ranges from about +60° C to −60° C in different species. Some hot-spring algae can live in water as warm as 73° C under favorable conditions (Brock 1967); and some Arctic algae can complete their life cycles in places where temperatures barely rise above 0° C. Nonphotosynthetic bacteria inhabiting hot springs can actively grow at temperatures greater than 90° C (Bott and Brock 1969, Brock 1979).

## THERMAL ENERGY EXCHANGE

Living organisms are intimately associated with an energy environment. The effect that it has depends upon the organism's ability to cope with a thermal budget and thermal

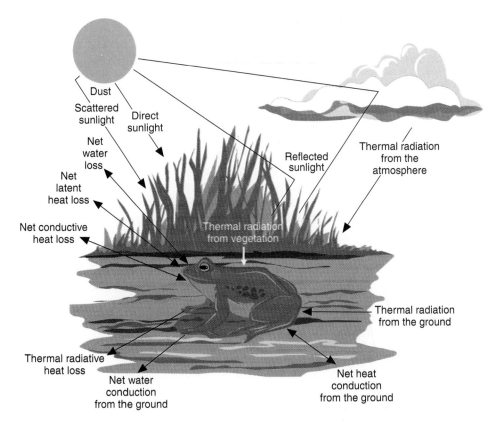

Dust
Scattered sunlight
Direct sunlight
Net water loss
Net latent heat loss
Net conductive heat loss
Reflected sunlight
Thermal radiation from the atmosphere
Thermal radiation from vegetation
Thermal radiation from the ground
Thermal radiative heat loss
Net water conduction from the ground
Net heat conduction from the ground

**Figure 6.1** Exchange of energy between a frog and its environment. The frog receives short-wave radiation directly from the sun or scattered by clouds and objects; it receives long-wave radiation from the substrate, vegetation, and atmosphere. The frog loses energy by emitting long-wave infrared radiation; by exchanging energy with ambient air; and by convection, evaporation, and condensation. It will also lose or gain energy by conduction of heat to and from its body core and by conduction and thermal radiation to the ground. (After Tracy 1976:295.)

stress. The relationship involves the exchange of energy between the organism and the environment and the maintenance of a balance between heat energy gained and heat energy lost (Figure 6.1). In effect, thermal energy absorbed from the environment plus metabolic energy produced must equal thermal energy lost from the body and energy stored.

A major source of heat transfer is *radiation*. This radiation may be direct sunlight, which can be intense. At noon in summer on a clear day direct solar radiation can amount to 976.9 w/m²/min (1.4 cal/cm°/ min). Another source is skylight or diffuse radiation, sunlight scattered by moisture and dust in the atmosphere. This heat may amount to 139.5 w/m²/min (0.2 cal/cm²/min). A third source is reflected sunlight bounced off objects in the environment. It can range from 69.78 to 209.3 w/m²/min (0.1 to 0.3 cal/cm²/min), depending upon the type of surface reflecting the sunlight. In addition, thermal radiation, long-wave infrared radiation, is emitted from the surfaces of soil, rocks, organisms and all other objects in the environment at their ambient temperatures (Figure 6.2). The rate at which a surface emits radiation relates to its surface temperature and its ability to give off thermal radiation. Still another source is the heat given off as a product of metabolism. In metabolic oxidation one mole of glucose gives off 686 kcal of energy, only 40 kcal of which is trapped as ATP; the rest goes off as heat. Just as the organism gains heat by *radiation* from the environment, so it loses heat to the environment by infrared or long-wave radiation.

Another source of heat transfer is **conduction,** the direct transfer of heat from one substance to another. The amount of heat lost (or gained) by conduction varies with the surface area exposed, the separation and temperature difference of the two surfaces, the thickness of insulation, such as fur, and conductivity. Air, for example, has low conductivity; water has high conductivity.

**Convection** is the transfer of heat by the circulation of fluid (liquid or gas). Convection may occur naturally in the fluid surrounding an object, such as swirling air above a hot stove; or it may be forced, with pressures from fluids passing by or over an object, such as air flow generated by a moving fan. The amount of sensible heat (heat that can be felt) transferred by convection depends upon the shape and area of the organism, the velocity of the fluid, and the physical properties of the fluid. Forced convection can speed up another process, *evaporation,* discussed in the previous chapter. Evaporation cools an organism because in changing from a liquid to a vaporous state, water must absorb a substantial amount of heat. Evaporation is influenced by the vapor pressure gradient between air and the object.

Influencing all forms of heat gain and loss by radiation, conduction, convection, and evaporation is the boundary layer. Every organism, in fact every object, has a boundary layer of still air adhering to the surface (see Gates 1972).

**Figure 6.2** Energy exchange—absorption and emission—in a meadow. $Q$ = net radiation; $V$ = evaporation; $L$ = sensible heat convection; $B$ = soil heat flux; figures are cal/cm². (a) By day the active layer lies between 35 and 55 cm; it absorbs 45 percent of net radiation. The lowermost active layer absorbs 28 percent. During input 80 percent of the radiant energy is used for the evaporation of water, 15 percent for sensible heat, 5 percent to raise soil temperature. (b) At night radiation as well as heat exchange is reversed. (After Cernusa 1976.)

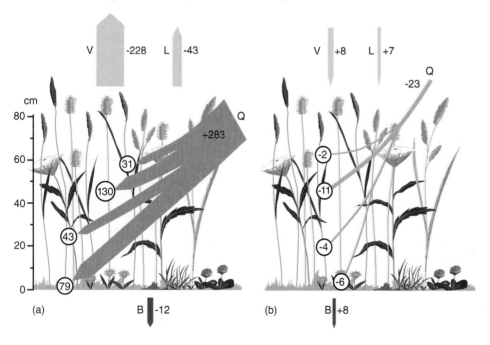

The resistance this layer offers to absorption and loss of heat is influenced by the size, shape, texture, and orientation of the organism, wind speed, and temperature differences between the surface of the organism and the air.

The relationship of all these components is net energy exchange. Net heat gain by solar radiation, infrared radiation, conduction, convection, and metabolism plus gains in energy storage must equal the total heat lost by radiation, conduction, convection, evaporation, and losses in energy storage. More simply, we can say heat gained equals heat lost.

One aspect is important: Heat produced continuously by organisms is lost passively to the environment. Such loss can take place only when the ambient or surrounding temperature is lower than the core body temperature. When ambient temperature equals core temperature, the route for passing heat off to the environment is lost. When ambient temperature exceeds body temperature, the flow is reversed and heat moves from the environment to the organism.

# TEMPERATURE AND METABOLISM

Temperature affects metabolic processes of organisms by influencing the kinetics of chemical reactions and the effectiveness of enzymes. All organisms can be divided roughly into two major groups: those whose body temperature is regulated by physiological means, represented by birds and mammals, and those whose body temperatures vary with the environmental conditions. This last group includes all other organisms, from protists to plants to amphibians and reptiles. The thermal effects on metabolism differ between the two groups. For birds and mammals, body temperature remains fairly constant; the organisms respond to changing environmental temperatures by increasing metabolic activity. For all others, body temperatures and thus tissue temperatures change with environmental temperatures. Among such organisms with variable tissue temperatures the rate of metabolism at rest rises nearly exponentially with temperature increase. This relationship between a rise in environmental temperature and the rate of metabolism is described by van't Hoff's reactive rate/temperature rule: for every 10° C rise in temperature the rate of oxygen consumption and therefore metabolic rate doubles. This factorial increase in metabolism over each 10° C increment of temperature is called the *temperature coefficient,* or $Q_{10}$:

$$Q_{10} = R_T / R_{(T-10)}$$

where $R$ is the rate at any given body temperature $T$, and $R_{(T-10)}$ is the rate at body temperature $T - 10°$ C.

For example, suppose that an organism has a metabolic rate of 1.8 cal/h at a body temperature of 20° C and a rate of 1 cal/h at 10°. $Q_{10}$ would be 1.8. For every 10° rise in temperature, the metabolic rate would increase by a factor of 1.8. The $Q_{10}$ resting metabolism for most organisms whose body temperatures follow environmental temperatures is around 2. However, metabolic rates are not truly exponential, so $Q_{10}$ is not a constant over all temperatures. The coefficient can vary within a particular range of temperatures. For example, between 10° and 20° C $Q_{10}$ might be 2.0, but between 20° and 30° it might be only 1.4. Thus $Q_{10}$ reflects the thermal sensitivity of an organism at various temperatures. What happens when the thermal balance changes suddenly? Consider an early spring plant stimulated into early flowering and growth by a few weeks of warm weather. Suddenly a cold spell exposes the plant to freezing temperatures. Also consider a fish living in warm water of 20° C. It is suddenly immersed in water of 4° C. Both the plant and the fish experience an abrupt temperature change. The flowers and new growth freeze; the fish suffers thermal shock. In neither case is the organism able to respond physiologically to such big temperature changes over such a short time. If these organisms had experienced a gradual directional change toward those temperatures for days or weeks, they would have physiologically acclimated and would have tolerated the cold.

Now suppose a fish living in water of 20° C is placed in water of 10° C. The fish's metabolism drops rapidly but not fatally. If the fish is now held at 10° C for four to five weeks, its metabolic rate will return to or approach its previous level. Such a response tends to reduce the effect of temperature change on metabolism.

Plants respond in a similar manner. Billings et al. (1971) raised alpine sorrel (*Oxyria digyna*) from seed at a uniform temperature in a greenhouse for four months. They subdivided the plants into three groups and grew them for five to six months at three different temperature regimes: warm (32/21° C day/night), medium (21/10° C), and cold (12/4° C). Then they measured replicates for net photosynthesis at temperatures ranging from 10° to 43° C. The optimum temperatures for photosynthesis differed for the groups. Optimum temperature for photosynthesis for plants acclimated to the warm temperature regime was 28° C; for those held at medium temperature, 21.5° C; and for those held at the cold temperature, 17° C. The plants, like the fish, showed differences in physiological states after long-term exposure to a particular environmental condition.

Such responses are called **acclimation.** In general, acclimation is an adjustment to artificially imposed conditions in a laboratory. Acclimation is not restricted to temperature alone. Organisms may acclimate to humidity, oxygen and carbon dioxide supplies, food availability, and other conditions. Understanding acclimation responses is useful at two levels: in studying how organisms adjust their physiological state to a particular environmental condition and in transplanting organisms to new environments. For example, it is important that hatchery-reared fish are acclimated to the temperature of the water in which they are to be released.

Under natural conditions, however, the environment never differs in just one or two parameters. Many other

ones, such as humidity for plants, and food availability for fish, are simultaneously involved. The difference in the physiological states that appear after long-term exposure to different natural environmental conditions is **acclimatization.** Plants acclimatize to the colder temperatures of fall and winter, and fish acclimatize to the seasonal changes in water temperatures and other conditions. As they do, their tolerance levels to sudden exposure to high or low temperatures also shift seasonally. A temperature a plant or fish could easily tolerate in late fall or winter would be intolerable in late spring and summer.

# PLANT RESPONSES TO TEMPERATURE

As we saw in Chapter 4, the thermal environment in which a plant lives at any one time can vary immensely from one part of its body to another. Roots are generally buffered from temperature extremes by the soil, while the aboveground structures experience widely varying temperatures. To complicate the thermal environment of a plant, some leaves, twigs, and buds are exposed to sun while others are shaded. Trees, cacti, and other plants with large stem diameters experience higher temperatures on the sunny side than the shaded side of the stem (Nobel 1978). Smooth-barked trees, particularly beech (*Fagus grandifolia*) and bigtooth aspen (*Populus grandidentata*), exposed to intense solar radiation are subject to sunscald.

Plants regulate their temperature by losing part of the energy they absorb. They maintain their heat balance by reradiation, convection, and transpiration. Reradiation accounts for one-half of the energy plants absorb (Gates 1968). Convection across the boundary layer transfers heat from the leaf to cooler air. Transpiration accounts for additional heat loss when energy is expended in converting water to water vapor.

Plants do not follow ambient temperatures exactly. During warm summer days and beneath the desert sun, plant tissue temperatures may be above ambient. For example, desert succulent cacti have surface temperatures 10° to 15° C above ambient (Nobel 1978); winter annuals may be warmer than the surrounding air (Regehr and Bazzaz 1976); and some desert ephemerals may have temperatures below ambient (Mulroy and Rundel 1977).

Although plants have both miminal and maximal temperatures for survival, they can grow and reproduce only with a range of optimal temperatures. Further, in each part of its life cycle a plant may have a different set of optimal temperatures. For example, the temperature necessary to stimulate germination may be lower than the temperature that favors flower development. All of these optimal temperatures vary among species, among ecotypes within a species, and among individuals in a population.

Plants living in a permanently high or low temperature regime have evolved physiological properties enabling them to endure their environment, but plants living in moderate environments must keep adjusting. Their survival depends upon their ability to avoid or tolerate thermal stress. Seasonally, avoidance usually involves adaptation, such as dormant bulbs and tubers, leaf fall and associated dormancy during winter or dry seasons, and completion of a life cycle during periods of favorable temperatures. Some plants possess certain plastic morphological adaptations. For example, sun leaves of oaks are smaller and have deeper lobes than shade leaves, increasing the surface area exposed to the air for cooling. Desert plants may have very small leaves or no leaves at all, and carry on photosynthesis through the stems. Short-term tolerance involves a response to heat or cold at cellular or subcellular levels or by some behavioral means. Under heat stress some plants hang their leaves parallel rather than horizontal to the sun's rays, reducing the surface area exposed to solar radiation (Grime 1966, Mooney et al. 1977, Ehleringer 1980, Ehleringer and Forseth 1980).

## Heat Stress

Heat affects the physiological processes of plants. On a hot summer day a sunlit leaf can become too warm for photosynthesis (Figure 6.3). Following an early morning burst of activity, net photosynthesis drops and respiration becomes dominant. This decline in net photosynthesis comes about because of a rapid increase in respiration and photosynthetic inactivation at high temperatures (Jones 1992).

How different species or even races within species respond photosynthetically to high temperatures appears to be genetically determined in part. For example, Pearcy (1976, 1977) studied the populations of the perennial big saltbush (*Atriplex lentiformis*), that grows both in southern coastal California and in Death Valley. Cloned plants from separate populations showed similar growth rates in a controlled environment (Figure 6.4). Leaves of the two clones had similar photosynthetic responses when grown at moderate temperatures (23° C day/18° C night), but in a hot temperature regime (43°/30° C) the desert clones grew better than the coastal clones. The desert population shifted its temperature responses so that photosynthesis improved in the heat. The coastal clone, however, was adversely affected by high temperatures; the rate of photosynthesis fell. The coastal clone obviously lacked the ability to acclimate to high growth temperatures.

When exposed to a rapid rise in heat, plants shut down normal protein synthesis and replace it with a set of heat shock proteins that aid in short-term survival. If the heat persists, it disrupts the protein structure of the plant. Most plants die at temperatures between 44° and 50° C.

Many species of plants, especially cacti, can acclimate to high temperatures. Such plants generally have high levels of bound water and high cytoplasmic viscosity. They are able to carry on protein synthesis at a sufficiently high rate in the face of rising temperatures to equal the protein breakdown

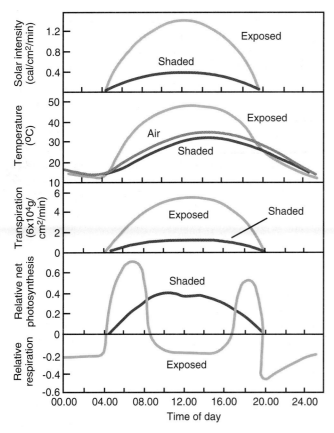

**Figure 6.3** The effect of temperature on the net photosynthesis of an exposed and a shaded leaf. Graphs top to bottom show: incident solar radiation, air and leaf temperatures, transpiration rates, and metabolic rates. The exposed leaf is inhibited photosynthetically when its temperature is high. (From Gates 1968b:6.)

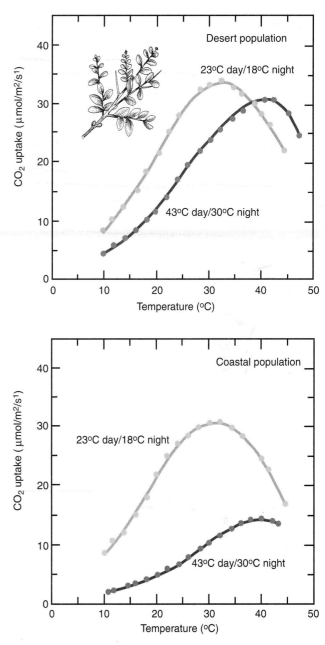

**Figure 6.4** The temperature dependence of photosynthetic uptake by leaves of cloned plants of big saltbush (*Atriplex lentiformis*) from a coastal and a desert population. Both clones were grown under high light (1700–18000 mol quanta $m^2$/sec) and normal concentrations of $CO_2$ and $O_2$. (After Pearcy 1977:485).

rate and thus avoid ammonia poisoning (Steponkus 1981). Heat tolerance of plants varies seasonally and with developmental stages and aging. Germinating and young plants and growing organs are more sensitive to heat than adult organs.

## Cold Stress

The ability of plants to tolerate cold temperatures has a genetic component. Plants of tropical and subtropical regions and the leaves of evergreen shrubs and trees in warm coastal regions have little resistance to low temperature and may suffer lethal damage at temperatures just above freezing (Larcher and Bauer 1981). Plants in seasonally cold climates build up a seasonal tolerance to frost. When temperatures drop below the minimum for growth, these plants become dormant, even though respiration and photosynthesis may continue slowly.

h plants, however, are subject to damage from
and frost damage, the degree of which depends
magnitude of the drop in temperature, and how
old persists. An outburst of cold air from an in-

vading polar front, especially if accompanied by rapid radiational cooling at night, can cause chilling or frost damage. Plants sensitive to chilling experience damage to cell membranes, and lose electrolytes rapidly. Frost damage is more severe. If freezing occurs slowly enough, ice crystals form outside the cells. This intercellular ice draws additional water out of the cells, dehydrating them. The degree of dehydration depends upon the thermodynamic balance between inside and outside the cell and the con-

centration of solutes within the cell. If the temperature falls too rapidly for this dehydration to take place, ice crystals form within the cell and damage the cell structure by expanding the cell wall. When the plant tissues thaw, the cellular contents spill out, producing the watery appearance characteristic of frozen plants.

If the onset of cold is slow, many plants can acclimate to the cold. During the growing season some plants avoid chilling and frost damage by increasing their sugars and sugar alcohols to lower the freezing point of cell fluids. This results in the **supercooling** of cell sap for short periods of time. Cell sap lowers to a temperature somewhat below freezing without freezing immediately.

Plants in seasonally cold climates build up a tolerance to frost coupled with winter dormancy. These plants develop frost tolerance or hardening in a stepwise process in the fall of the year. The first level of hardening develops at +5° to 0° C. At this stage the plants can survive moderate frost. If frost is progressive and persistent, plants reach their limit of potential freezing tolerance. Generally, hardening occurs when the cells produce organic compounds, such as sugars, amino acids, and nontoxic substances, which reduce the freezing point and act as antifreeze. Plants retain their tolerance to cold until growth starts in spring. Then a sudden drop in temperature may kill tissues that were able to survive far lower temperatures during the winter.

Tolerance to freezing is not uniformly distributed throughout the plant. Roots, bulbs, and rhizomes are most sensitive to freezing, succumbing to temperatures between −10° and −30° C. Often cold weather will not kill these structures because they are protected underground. Buds of woody plants, too, vary in their tolerance. Terminal buds of trees are less resistant to cold than lateral buds; most resistant are the basal reserve buds on the twigs. These buds are the ones that will replace new spring growth killed by a late frost. Woody stems are more cold resistant than leaves and buds. In fully frost-hardened twigs and trunks, the cambium is most resistant.

Further resistance to chilling and frost damage is obtained by insulation. Some species of Arctic and alpine and early spring flowers of temperate regions possess hairs that act as heat traps and prevent cold injury. The interior temperature of cushion-type and rosette plants may be 20° C higher than the surrounding air. Thus some plants can carry on photosynthesis at low temperatures (Regehr and Bazzaz 1976). Low temperatures and frozen soil, however, can reduce the uptake of water. This loss of uptake coupled with the drying effects of winter winds can produce winter drought.

## ANIMAL RESPONSES TO TEMPERATURE

Thermal relations between an animal and its environment can be appreciated best by considering a thermal model of the animal body (Figure 6.5). The core of the animal body,

its deep interior, is assumed to be at a uniform temperature $T_b$. The ambient temperature of the environment is $T_e$. The temperature of the body's surface, $T_s$, because of the boundary layer and other factors, will differ from both $T_b$ and $T_e$. Separating the body core from the body surface are layers of muscle tissue and fat, across which temperature gradually changes from $T_b$ to $T_e$.

To maintain core body temperature, the animal must balance losses or gains by changes in metabolic heat production. Thus thermal balance in the core of the animal is influenced by heat produced by metabolism, heat stored, heat flow to skin as affected by the thickness and conductivity of fat and fur, hair or feathers, heat flow to the ground, and heat lost by evaporation. A general formula for heat balance in animals is:

$$M = K_o(T_b - T_e)$$

where $M$ is effective net metabolic heat production, $K_o$ is overall thermal conductance of the organism; $T_b$ is body core temperature, and $T_e$ is environmental temperature. $K_o$ is a constant for a given set of conditions, variable among species. It is a function affected by surface area, conduction, convection, and radiation. The larger the value of $K_o$, the greater is the loss of heat.

How animals confront thermal stress is influenced heavily by their physiology and environment. Because air has a lower specific heat than water and absorbs less solar radiation, terrestrial animals are subject to more radical and potentially dangerous changes in their thermal environment than

**Figure 6.5** Schematic representation of an animal body. Body core temperature is at $T_b$, the environmental temperature is $T_e$, the surface temperature is $T_s$, and $L$ is the thickness of the outer layer of the body.

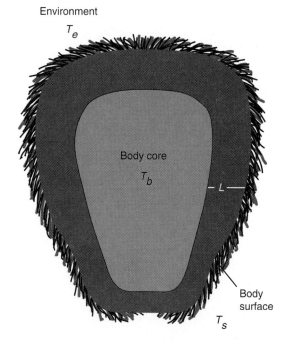

Environment

$T_e$

Body core

$T_b$

L

Body surface

$T_s$

aquatic organisms. Incoming solar radiation can produce lethally high temperatures, and radiational loss of heat to the air, especially at night, can result in lethally low temperatures. Aquatic animals live in a more stable energy environment, but they have a lower tolerance of temperature change.

Physiologically animals can be divided into three groups: (1) those that maintain a fairly constant internal temperature regardless of external temperature, such as birds and mammals; (2) those that allow their body temperature to vary with ambient, such as invertebrates, fish, amphibians, and reptiles; and (3) those that sometimes regulate their body temperature and sometimes do not, such as bees and bats. Animals belonging to the first group are **homeotherms** (*homeo,* "the same"), to the second group **poikilotherms** (*poikilos,* "manifold" or "variegated"), and to the third group **heterotherms** (*hetero,* "different").

Homeotherms maintain a rather constant body temperature in the face of changing environmental temperatures by means of their own oxidative metabolic production of heat, called **endothermy.** Poikilotherms maintain their body temperature by using sources of heat energy such as solar radiation and reradiation rather than metabolism. This mechanism of maintaining body temperature is called **ectothermy.** Heterotherms utilize both endothermy and ectothermy, depending upon environmental situations and metabolic needs. The terms *homeotherm* and *endotherm,* and *poikilotherm* and *ectotherm,* are often used synonymously, but there is a difference. The terms *ectotherm* and *endotherm* emphasize the mechanisms by which body temperatures are determined; the other terms emphasize the nature of the variations in body temperature.

## Poikilotherms

Poikilotherms, such as amphibians and insects, have a high thermal conductance between the body and the environment and a low metabolic rate. They gain body heat easily from the environment and lose it just as fast. Within the range of temperatures that poikilothermic animals can tolerate, the rate of metabolism, and therefore oxygen consumption, increases (Figure 6.6). Their $Q_{10}$ or thermal sensitivity is approximately 2. Rising temperatures increase the rate of enzymatic activity, which controls metabolism and oxidation of carbohydrates. Lacking any homeostatic devices, terrestrial poikilotherms must depend upon some behavioral control over body temperatures. Lizards may vary their body temperatures no more than 4° to 5° C when active and amphibians 10° C. The range of body temperatures over which ectotherms carry out their daily activities is called the *active temperature range* (ACT). By limiting their ACT, poikilotherms can adjust their physiological and developmental processes to a limited range at a low metabolic cost. Within this ACT, many poikilotherms have a low thermal sensitivity to a certain range of acute temperature changes. For example, barnacles of the intertidal zone may have a body

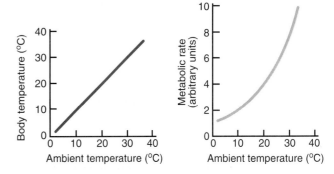

**Figure 6.6** Relationship among body temperature, resting metabolic rate, and ambient temperatures in poikilotherms. (a) Body temperature is a function of ambient temperature. (b) Resting metabolism is a function of body temperature. (After Hill and Wyse 1989:83.)

temperature of 30° C at low tide when exposed to sun and air. Immersed in ocean water of 14° C at high tide, their body temperature drops to 16° C, yet their resting rates of metabolism show little thermal sensitivity within that temperature range (Newell and Northcroft 1965). The evolution of such low thermal sensitivity adds some stability to their metabolic processes in an environment of recurring sharp changes in temperature.

Being ectothermic has advantages and disadvantages. Prisoners of environmental temperatures, poikilotherms of temperate regions, such as snakes, can become highly active only when the temperature is sufficiently warm. Because metabolic activity declines with temperature, these animals become sluggish in the cool of morning and evening. Similarly, they have to restrict their active life to the late spring, summer, and early fall. During periods of intense physical activity, when energy consumption is high, poikilotherms depend upon the anaerobic breakdown of glycogen for energy. This breakdown results in an accumulation of lactic acid in the tissues that can be oxidized only after activity ceases. Anaerobic metabolism severely limits bursts of poikilothermic activity to a few minutes because of physical exhaustion. This tendency to exhaustion is one reason why so many predatory terrestrial poikilotherms such as snakes and alligators secure prey by ambush rather than by chase.

Poikilotherms allocate more energy to biomass production than to metabolism. Because they do not depend upon internally generated body heat, they can reduce metabolic activity during periods of temperature extremes and of food or water shortage. Low energy demands enable poikilotherms to colonize areas of limited food and water, such as deserts. Because they do not have the problem of metabolic heat loss, poikilotherms are not limited to any minimum size or definite shape. Many are single-celled, such as paramecia and amoebas; others, such as earthworms, millipedes, and snakes, have cylindrical bodies. Such characteristics enable poikilotherms to exploit resources and habitats unavailable to homeotherms. On the other hand, the same metabolic re-

strictions impose an upper size limit: ectotherms would not be able to absorb enough heat to warm a very large body. For this reason some paleontologists argue that the large dinosaurs had to be endothermic. A counterargument is that large ectotherms could develop and maintain body temperatures above air temperatures in a tropical environment because their low surface-to-volume ratio (see page 91) would limit cooling.

### Adaptations to High Temperature

Although poikilotherms as a group do not have a lower limit to body size, nevertheless size is an important aspect to their life (Stevenson 1985). The rate of heating and cooling in a poikilotherm decreases as size increases. That fact determines the behavioral options open to a poikilotherm for controlling body core temperature ($T_b$). Consider a beetle about 10 mg in weight. Because of its small weight, it heats and cools quickly, but rarely can it raise its body temperature above ambient. However, it does have the ability to control its $T_b$ by moving in and out of the sun. If the temperature of a rock or a log on which it rests becomes too hot, it can easily find a shaded crevice. This ability allows very small poikilotherms to make maximum use of favorable environmental temperatures.

Associated with the beetle on a rock may be a small lizard of about 0.5kg. It, too, is small enough to heat up fast to an active temperature that allows it to make maximum use of favorable thermal conditions during the day. If the temperature becomes too hot, the lizard can quickly retreat to a cooler site. By changing locations, the lizard can maintain some level of stability in its body temperature.

For large terrestrial poikilotherms, those over 10 kg, such as large lizards and tortoises, life is considerably different. Because they heat slowly, large poikilotherms need a much higher operative temperature than smaller ones to reach the same active $T_b$. For this reason they do not have much range of $T_b$ in which to operate. Further, because large poikilotherms have difficulty locating shelter from environmental extremes, they are more or less restricted to environments with small seasonal fluctuations in temperature. Large poikilotherms, however, have the advantage of moving freely about in space and time during the day because their large body mass protects them from environmental extremes.

Most aquatic poikilotherms (fish and aquatic invertebrates) generally encounter temperature fluctuations of lesser magnitude and usually have more poorly developed behavioral and physiological thermoregulatory capabilities than terrestrial forms. Immersed in a watery environment, most of these animals do not maintain any appreciable difference between their body temperature and that of the environment. Any heat produced in the muscles is transferred to blood flowing through them, and carried to the gills and skin, where it is lost to the water by convection and conduction. Thus fish are ideal poikilotherms. Because of the close relationship between body temperature and environmental temperature, fish are readily victimized by any rapid change in environmental temperatures.

Aquatic poikilotherms adjust to their thermal environment largely through acclimation, because there is little they can do to change their surroundings. Terrestrial forms can adapt by behavioral responses. We can view the total range of the thermal environment of an aquatic poikilotherm as a group of zones (Figure 6.7). A central zone of **thermal tolerance** is the range of temperature within which an aquatic poikilotherm, particularly fish, have the highest survival. Within this range fish may seek certain preferred temperatures. The zone of thermal tolerance is bounded by an upper and lower zone of *thermal resistance*. This zone includes the temperature ranges within which the organism can survive for an indefinite period. Outside of the upper and lower bounds of thermal resistance are the zones of upper and lower **incipient lethal temperature.** At these temperatures a stated fraction of a fish population (generally 50 percent) will die with prolonged exposure when brought to it rapidly from a different temperature.

The incipient lethal temperature is not fixed. Its value is affected by the previous thermal history of the organism. Within limits, poikilotherms can adjust or acclimatize to higher or lower temperatures. If the organism lives at the

**Figure 6.7** Thermal tolerance of a hypothetical fish in relation to thermal acclimatization. (From Coutant 1970:350.)

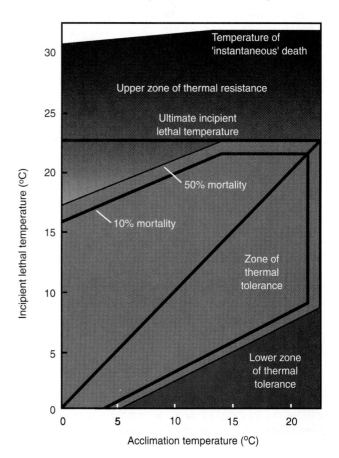

higher end of the tolerance range, it acclimatizes so that both the lethal temperatures and the lower limits are higher than if it were living within the cooler end of the range. Aquatic poikilotherms can adjust slowly to seasonal temperatures.

Once acclimatized to a given temperature, fish adjust more readily to an increase than to a decrease in temperature. Acclimatization both increases the length of time an organism can survive to an elevated or lowered temperature and raises the maximal or lowers the minimal temperature it can survive for a given length of time. However, acclimatization has upper and lower limits, and ultimately a temperature will be reached that will be lethal. This temperature is the ultimate incipient lethal temperature (Coutant 1970).

Amphibians present a somewhat different situation. Permanently aquatic forms, such as many salamanders, maintain body temperature in the same manner as fish: they seek preferred temperatures within their habitat. For semiterrestrial and terrestrial frogs and salamanders, adjusting to temperature change is more complex. These salamanders exhibit seasonal variations in body temperature and possess little in the way of behavioral thermoregulation. Generally they are restricted to moist, shaded environments that are thermally homogeneous in a given season. Semiterrestrial frogs, such as bullfrogs and green frogs, are able to exert considerable control over their body temperature, which does not, as is often assumed, simply follow air temperature. By basking in the sun (**heliothermism**), frogs can raise their body temperature as much as 10° C above ambient temperature. Because of associated evaporative water losses, such amphibians must either be near water or partially submerged (Figure 6.8) Forms that live near water also use evaporative cooling through the skin to reduce body heat loads. By changing position or location or by seeking a warmer or cooler substrate, amphibians can maintain body temperatures within a narrow range of variation (Lillywhite 1970).

Reptiles have their own response to temperature. Most are terrestrial and lack the buffering effects of water. Exposed to widely fluctuating temperatures of the terrestrial environment, reptiles must possess more refined means of temperature regulation.

Reptiles exhibit little relationship between their core body temperature and ambient temperature, even though they are poikilothermic. Evaporative cooling by panting and by water loss through the skin keeps the body temperature from reaching **critical thermal maximum** (CTM), the temperature at which the animal's capacity to move is so reduced that it cannot escape from thermal conditions that will lead to its death (Cowles and Bogert 1944). Thus reptiles possess some of the basic physiological mechanisms so highly developed in endotherms.

The simplest way for a reptile to regulate body temperature is heliothermism. Basking in the sun raises the core body temperature. When it reaches a preferred level, the animal moves to the shade and remains there until the body temperature drops below the preferred range. The reptile then returns to the sun.

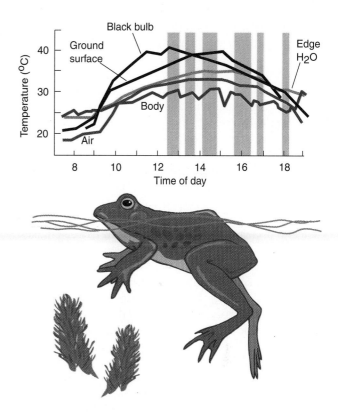

**Figure 6.8** The body temperature of a bullfrog measured telemetrically. Dips in the black bulb temperature (maximum possible emittance, as measured by a true black body) indicate the effects of cloud cover, convection, or both. Water temperature around the pond's edge varies from one location to another by as much as 2° to 3°C. While in shallow water a frog may show a higher body temperature than that recorded for the edge water. Note how uniform a temperature the bullfrog maintains by moving in and out of the water. (From Lillywhite 1970:164.)

More elaborate behavior common to many lizards is *proportional control.* If ambient temperature is lower than preferred, the lizard can spread its ribs, flatten its body, and orient itself so that its body is at right angles to the sun to gain the maximum amount of heat. If the temperature is too high, the lizard can pull its ribs together and orient its body parallel to the sun, decreasing the surface area exposed. Other behavioral means include burrowing into the soil and possibly changing color. Thus reptiles have at their disposal a variety of behavioral mechanisms useful for temperature regulation (Figure 6.9).

Some reptiles possess physiological mechanisms allowing them some control over maintenance of a preferred body temperature. At least four families of lizards (Iguanidae, Gekkonidae, Varanidae, and Agamidae) can control the rate of change in body temperature by changing the rate of heartbeat and varying the rate of metabolism.

Adaptations to Low Temperatures  Many poikilothermic animals of temperate and Arctic regions must endure long periods of below-freezing temperatures in winter.

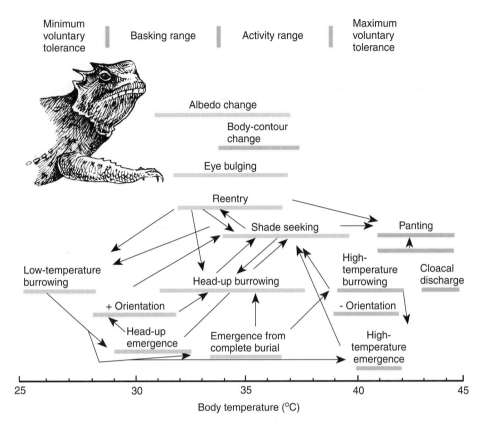

Minimum voluntary tolerance  Basking range  Activity range  Maximum voluntary tolerance

Albedo change

Body-contour change

Eye bulging

Reentry

Shade seeking  Panting

Low-temperature burrowing

Head-up burrowing

High-temperature burrowing  Cloacal discharge

+ Orientation  - Orientation

Head-up emergence  Emergence from complete burial  High-temperature emergence

25    30    35    40    45

Body temperature (°C)

**Figure 6.9** Behavioral mechanisms in the regulation of body temperature by the horned lizard (*Phrynosoma coronatum*). (From Heath 1965.)

They escape the cold through supercooling and resistance to freezing.

**Supercooling** takes place when the body temperature falls below freezing without freezing body fluids. The amount of supercooling that can take place is influenced by the presence of certain solutes, such as glycerol, sorbitol, and mannitol, all polyhydric alcohols, in the body. Some Arctic fish, certain insects of temperate and cold climates, and reptiles exposed to occasional cold nights employ supercooling.

Some intertidal invertebrates of high latitudes and certain aquatic insects survive the cold by actually freezing and then thawing when the temperature moderates. In some, more than 90 percent of the body fluids may become frozen, and the remaining fluids contain highly concentrated solutes, including polypeptides and glycopeptides. Ice forms outside shrunken cells, and muscles and organs are distorted. After thawing, they quickly resume normal shape. Other animals, particularly Arctic and Antarctic fish and many insects, resist freezing because of the presence of glycerol in body fluids. Glycerol protects against freezing damage and lowers the freezing point, increasing the degree of supercooling. Wood frogs (*Rana sylvatica*), spring peepers (*Hyla crucifer*), and gray tree frogs (*H. versicolor*) can overwinter just beneath the leaf litter because they accumulate glycerol in their body fluids, enabling them to resist freezing (Schmid 1982). Dormant poikilotherms also experience such physiological changes as lower blood sugar, increased liver glycogen, high antifreeze and increased carbon dioxide levels in the blood, and darkened skin.

In addition to supercooling, many insects exhibiting frost hardiness enter a resting stage called **diapause,** characterized by a cessation of feeding, growth, mobility, and reproduction. Among many insects diapause is a genetically determined, obligatory resting stage before development can proceed. It is timed mostly through photoperiod and is associated with falling temperatures. Diapause prevents the appearance of a sensitive stage of development at a time when low temperatures would kill. Diapause ends with the lengthening of photoperiod and the return of warm temperatures.

## Homeotherms

Birds and mammals escape the thermal constraints of the environment by being endothermic. They maintain their thermal metabolic optimum by oxidizing glucose and other energy-rich molecules. They regulate the gradient between body and air temperatures by seasonal changes in insulation (the type and thickness of fur, structure of feathers, and layer of fat), which poikilotherms do not possess, by evaporative cooling, and by increasing or decreasing metabolic heat production. Homeothermy allows these animals to remain active regardless of environmental temperatures, although at high energy costs.

High body temperatures may have evolved from an inability of large animals to dissipate rapidly the heat produced during periods of high activity (Heinrich 1976, 1979). That situation would favor enzyme systems that function at a high

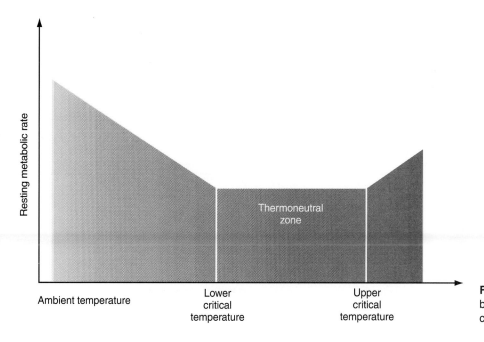

**Figure 6.10** General resting metabolic response of homeotherms to changes in ambient temperature.

temperature with a set point around 40° C. Otherwise the animal would expend an enormous amount of water for evaporative cooling to ambient temperatures. The ability to operate at high temperatures provides homeotherms with great endurance and the means of remaining active at low temperatures.

However, homeothermy has placed a size restraint on animals possessing it. A close relationship exists between body size and basal metabolic rate of a resting, fasting ani-

mal. For each species there is a range of environmental temperatures within which the metabolic rates are minimal, the **thermoneutral zone** (Figure 6.10). Outside this zone metabolism increases. The basal metabolic rate, as measured by oxygen consumption, is proportional to body mass raised to the 3/4 (0.75) power. As body weight increases, the weight-specific metabolic rate decreases. Conversely, as body mass decreases, basal metabolism increases (exponentially with very small body size) (Figure 6.11).

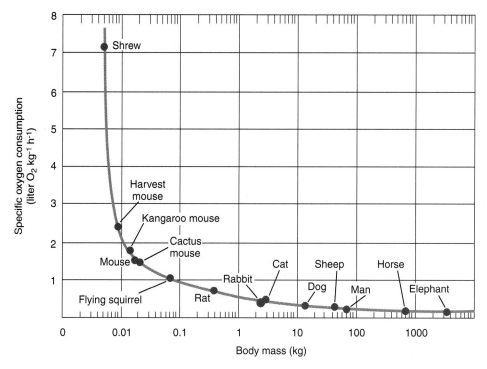

**Figure 6.11** Observed rates of specific oxygen consumption per unit body mass by various mammals. Oxygen consumption increases rapidly with decreasing body mass. Note that the abscissa has a logarithmic scale and the ordinate an arithmetic scale. (After Schmidt-Nielson 1979:1885.)

The relationship is represented by the equation:

$$V_{O_2} = 0.676 \, M_b^{0.75}$$

where $M_b$ = body weight in kilograms and $V_{O_2}$ = oxygen consumption in liters of $O_2$. Birds have a somewhat higher metabolic rate:

$$V_{O_2} = 0.679 \, M_b^{0.723}$$

This relationship of oxygen consumption to body size, a straight line with a slope of approximately 0.75, holds for a wide range of organisms, from unicellular ones to reptiles, and even some forest trees.

Within any taxonomic group of endotherms, small animals have a higher metabolic rate per unit of body weight than large ones. Part of the reason lies in the ratio of surface area to body mass. An animal loses heat to the environment in proportion to the surface area exposed. Given the same environmental conditions, a large animal loses proportionately less heat to the environment than a small one. To maintain a constant internal temperature, small endothermic animals have to burn energy rapidly. In fact, weight-specific rates of small endotherms rise so rapidly that below a certain size they could not meet energy demands. Five grams is about as small as an endotherm can be and still maintain a metabolic heat balance. Few endotherms ever get that small. Two who do are the world's smallest mammals: the pygmy white-toothed shrew (*Suncus etruscus*) of Africa, which weighs 2 gm as an adult, and Kitti's hog-nosed bat (*Craseonycteus thonglongyai*) of Thailand.

Because of the conflicting metabolic demands of maintenance of body temperature and growth, most young birds and mammals are born in altricial state—blind, naked, and helpless—and begin life as ectotherms. Depending upon the body heat of parents allows the young to allocate most of their energy to growth.

To regulate heat exchange between themselves and the environment, homeotherms utilize a number of physiological and morphological mechanisms. One is a countercurrent heat exchanger (Figure 6.12). Some mammals, particularly those of the Arctic, such as porpoises and whales, have extensive areas in the extremities where the veins are closely juxtaposed to the arteries. Much of the heat lost by outgoing arterial blood is picked up or exchanged to the returning venous blood. Thus the venous blood is warmed on its return and reenters the body core only slightly cooler than the outgoing arterial blood. Such vascular arrangements are common in the legs of mammals and birds, and in the tails of rodents, especially the beaver (*Castor canadensis*).

The arteries and veins of many endotherms are divided into a large number of small, parallel, intermingling vessels that form a discrete vascular bundle or net known as a **rete.** Within the rete the blood flows in two directions and a heat exchange takes place. Such a rete in the head functions to cool the highly heat-sensitive brain of the oryx, an African antelope exposed to high daytime temperatures (Figure

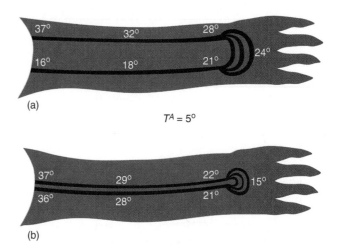

(a)

$T^A = 5°$

(b)

**Figure 6.12** A model of countercurrent flow in the limb of a mammal, showing hypothetical temperature changes in the blood (a) in the absence and (b) in the presence of countercurrent heat exchange.

6.13). Among some Arctic mammals, such as the Arctic fox (*Alopex lagopus*), countercurrent circulation prevents excessive heat loss through the extremities. The footpads of this fox as well as wolves and sled dogs routinely cool to near 0° C (Irving 1972, Hill 1976).

Physiologically it is more difficult for homeotherms to adapt to high temperatures than to cold. To maintain $T_b$ they must lose heat to the environment. If the ambient temperature is the same or exceeds the core body temperature, then heat will flow from the environment into the animal and in-

**Figure 6.13** The desert gazelle can keep a cool head in spite of a high body core temperature. Arterial blood passes in small arteries through a pool of venous blood cooled by an evaporative process as it drains from the nasal region. (After Taylor and Lyman 1972.)

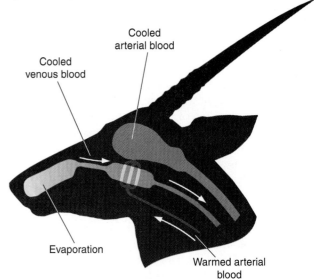

Cooled
arterial blood

Cooled
venous blood

Evaporation

Warmed arterial
blood

duce heat stress. To counteract this stress many animals employ evaporative cooling, largely through sweating and panting, but at the risk of reducing their water balance.

Birds do not sweat, and an insulating covering of feathers inhibits water loss through the skin. Birds lose body heat largely through radiation, conduction, and convection. When conditions demand it, birds can decrease their heat load by evaporative cooling through panting. However, panting requires work, and work only adds more metabolic heat. Some groups of birds, particularly the goatsuckers (Caprimulgidae), owls (Strigidae), pelicans (Pelecanidae), boobies (Sulidae), doves (Columbidae), and gallinaceous (chickenlike) birds, avoid this dilemma by *gular fluttering,* movements of parts of the gullet. Evaporative cooling by gular fluttering uses less energy than panting.

Another response to excessive heat, especially in birds, is hyperthermia. A rise in the body temperature adjusts the difference between the body and the environment, and thermal homeostasis is established at higher temperatures. (Hyperthermia among mammals was considered in Chapter 5.)

Parts of the body may serve as thermal windows in the radiation of heat. The large ears of such desert mammals as the kit fox (*Vulpes macrotis*) and the antelope jackrabbit (*Lepus alleni*) function as efficient radiators to the cooler desert sky (Schmidt-Nielsen 1964). By seeking shade, where the ground temperatures are low and solar radiation is screened out, or by sitting in depressions, where radiation from the hot ground surface is obstructed, the jackrabbit can radiate 5 kcal/day through its two large ears (400 cm²). This loss is equal to one-third of the metabolic heat produced in a 3-kg rabbit. Such a radiation loss alone reduces enough heat without much loss of water.

Some animals of arid regions simply avoid heat by adopting nocturnal habits and remaining underground or in the shade during the day. Some desert rodents that are active by day periodically seek burrows and passively lose heat through conduction by pressing their bodies against burrow walls.

Within a range of temperature birds and mammals can maintain their core body temperatures by a change in the insulating thickness of hair, fur, feathers, and fat. Prior to winter many mammals acquire a heavier coat of hair that thins out with the coming of warm weather. Birds can add to their thermal stability in cold of winter by fluffing their feathers. During sudden or prolonged cold spells there is a point at which insulation is no longer effective and the animals must maintain body heat by increased metabolism (Figure 6.14). This point is the *critical temperature* (Figure 6.10). It varies greatly among tropical and arctic animals (Scholander, Hock et al. 1950; Scholander, Walters et al. 1950). Tropical birds and mammals exposed to temperatures below 23.5° to 29° C increase their heat production. If air temperature is lowered to 10° C the tropical animal must triple its heat production; and if lowered to freezing, the animal is no longer

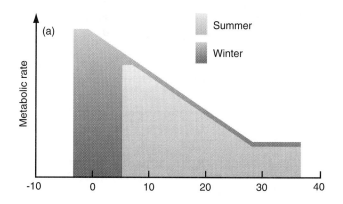

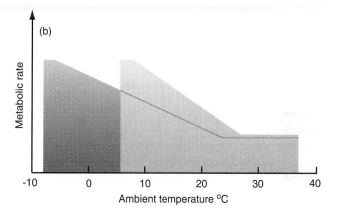

**Figure 6.14** (a) Simple metabolic acclimatization. Note that in winter the metabolic rate increases as the ambient temperature declines. The plateau on the left indicates maximal metabolic rates. (b) Simple insulatory acclimatization. Note how insulation reduces the metabolic rate in winter and permits tolerance of much lower temperatures.

able to produce heat as rapidly as it is being lost. Arctic small mammals, on the other hand, do not increase their heat production until the air temperature has fallen to −29° C. Large Arctic mammals, like the muskox (*Ovibos moschatus*), can sustain the coldest weather without heat beyond that produced by normal basal metabolism. Eskimo dogs and Arctic foxes can sleep outdoors at temperatures of −40° C without stress. This ability is not due to any difference in metabolism itself but to effective insulation and cold acclimatization.

When body temperatures fall below the critical level, endotherms increase metabolism by shivering. *Shivering,* the uncoordinated, involuntary, high-frequency contraction of skeletal muscles, converts chemical energy to thermal energy just as voluntary muscular contractions do. Among birds exposed to cold, shivering and voluntary muscular activity are primary sources of extra heat.

Mammals acclimated to cold temperatures can increase their heat production by *nonshivering thermogenesis* (Hill and Wyse 1988). Nonshivering thermogenesis is the genera-

tion of heat from the metabolism of brown fat, a highly vascular brown adipose tissue around the head, neck, thorax, and major blood vessels. Brown fat occurs in the young of most species and in mammals that hibernate; it increases in mass when animals are chronically exposed to low temperatures (Smith and Horwitz 1969, Chaffee and Roberts 1971). Heat generated from the metabolism of this fat is transported to the heart and brain. Nonshivering thermogenesis allows mammals to be active at lower temperatures than if they had to depend on shivering alone. This form of heat production and exercise are additive; the two together can be important in maintaining body temperatures.

## Heterotherms

Among both poikilotherms and homeotherms are species that sometimes regulate their body temperature and sometimes do not. Such animals are called *heterotherms*. These animals at different stages of their daily and seasonal cycle or under certain environmental situations take on the characteristics of both endotherms and ectotherms. They can undergo rapid, drastic, repeated changes in body temperature.

Insects are ectothermic and poikilothermic; yet in the adult stage most species of flying insects are heterothermic. When flying, they have high rates of metabolism with heat production as great as or greater than homeotherms, and show the same general relationship between body mass and energy metabolism (Figure 6.15). They reach this high metabolic state in a simpler fashion than homeotherms because they are not constrained by pulmonary and cardiovascular pumping systems. Insects take in oxygen by demand through openings or spiracles on the body wall and transport it throughout the body by a tracheal system. Nor do they have to maintain equal temperatures through the body.

Temperature is critical to the flight of insects. Most cannot fly if the temperature of the thoracic muscles is below 30° C; nor can they fly if the muscle temperature is over 44° C. This constraint means the insect has to warm up before it takes off, and it has to get rid of excess heat in flight. With wings beating up to 200 times per second, flying insects can produce a prodigious amount of heat.

Some insects, such as butterflies and dragonflies, can warm up by orienting their bodies and spreading their wings to the sun. Most warm up by shivering the flight muscles in the thorax. Moths and butterflies may vibrate their wings to raise thoracic temperatures above ambient (Pivnick and McNeil 1986, Rawlins 1980). Bumblebees do the same through abdominal pumping without any external wing movements. They use their flight muscles uncoupled from the wings (Heinrich 1976, 1979). They must maintain a constant thoracic temperature during flight and not lose too much heat through the abdomen. To accomplish both, the bee must maximize its rate of heat loss at a high ambient temperature and minimize its heat loss at a low one.

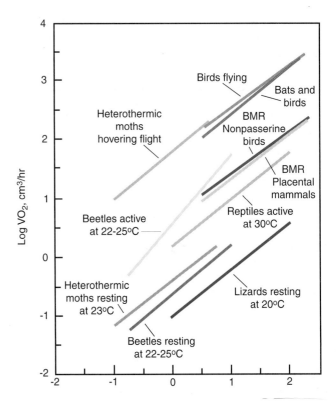

**Figure 6.15** The regression of energy metabolism on body mass of heterothermic insects and some selected terrestrial vertebrates at rest, during terrestrial activity, and during flight. (From Bartholomew 1981:60.)

One heat retention mechanism is insulation. In the bumblebee, the sphinx moth and other moths, and butterflies, the thorax and dorsal surface of the abdomen is insulated with pile (which appears as fine hair). The ventral surface of the abdomen, free of pile, acts as a thermal window. The bumblebee has abdominal air sacs that act as another insulating device to retard heat from the thorax to the abdomen. The blood flow between the cooler abdomen and the warm thorax is heated and cooled by a countercurrent mechanism. Cool blood pumped forward by the heart is warmed in the aorta by passing through the flight muscles and, if necessary, can be cooled on the way back to the abdomen (Figure 6.16). Without such heat retention mechanisms cool blood circulating into the thorax would cool the muscles, interfering with their function. In flight on a hot summer day, the insect may need to reduce the thoracic temperatures to prevent overheating. By increasing bloodflow to the central abdomen, bumblebees in particular lose excess heat to the air. In very warm situations, bumblebees can resort to evaporative cooling by regurgitating fluids, wetting the proboscis, and moving it about in the air.

When crawling about flowers securing nectar, flying insects except the bumblebee suddenly become ectothermic, and thoracic temperatures may drop to ambient. Bumblebees maintain a high thoracic temperature when crawling

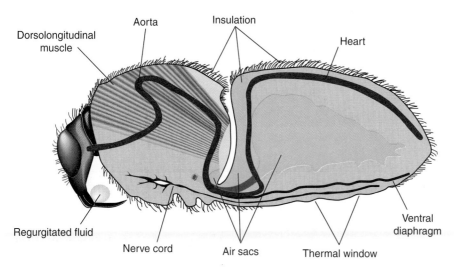

Dorsolongitudinal muscle
Aorta
Insulation
Heart
Regurgitated fluid
Nerve cord
Air sacs
Thermal window
Ventral diaphragm

**Figure 6.16** Temperature regulation in the bumblebee involves insulation on the thorax and the dorsal side of the abdomen. The heart pumps cool blood (hemolymph) to the thorax, where it can be warmed or cooled. (After Heinrich 1976:564.)

about flowers, permitting immediate flight without another warmup.

Certain vertebrate poikilotherms that exhibit some degree of thermoregulation employ countercurrent heat exchangers. The swift, highly predaceous tuna and mackerel sharks possess a rete in the band of dark muscle tissue that aids a sustained swimming effort. Metabolic heat produced in the muscle warms up the venous blood, which gives up heat to the adjoining, newly oxygenated blood returning from the gills. Such a countercurrent heat exchange increases the power of the muscles, because warm muscles are able to contract and relax more rapidly.

To escape extreme temperatures, a few homeotherms become heterothermic, allowing their core body temperatures to fall to near ambient temperature. The simplest response is daily *torpor,* experienced by a number of birds, such as hummingbirds (Trochilidae) and poorwills (*Phalaenoptilus nuttallii*), and by small mammals such as bats, pocket mice (*Perognathus* spp.), kangaroo mice (*Microdipodops pallidus*), and even white-footed mice. Such daily torpor is not necessarily associated with any scarcity of food or water. Rather it seems to have evolved as a means of reducing energy demands over that part of the day or night during which the animals are inactive. Arousal returns body temperature rapidly to normal as the animal renews its metabolic heat.

A deeper state of torpor is **hibernation,** characterized by a cessation of coordinated locomotory movements; a reduction of heart rate, respiration, and total metabolism; and a body temperature below 10° C. The animal, however, retains the ability to warm up spontaneously and to emerge periodically from the hibernating state using only endogenously generated heat (A. R. French 1988).

Entrance into hibernation is a controlled physiological process akin to sleep (J. M. Walker et al. 1979), difficult to

explain and difficult to generalize from one species to another. Some hibernators, such as the woodchuck (*Marmota monax*), feed heavily in late summer to build up large fat reserves, from which they will draw energy during hibernation. Others, like the chipmunk, lay up a store of food in underground chambers instead. All hibernators, however, have to acquire a metabolic regulatory mechanism different from that of the active state. Tissues of deeply hibernating mammals must be able to function adequately at both of the temperatures ranges at which they exist during the year.

During hibernation heartbeat is lowered or even intermittent, respiration is reduced, and carbon dioxide levels in the blood are increased, building up respiratory acidosis (Lyman et al. 1982). Acidosis affects cellular processes, inhibits glycolysis, lowers the threshold for shivering, and reduces the metabolic rate. No hibernator remains torpid continuously. All awaken spontaneously from time to time and then drop back into torpor. The chipmunk (*Tamias striatus*) with its large store of seeds spends much less time in torpor than the larger fat-storing hibernators. Many chipmunks may not enter torpor at all, even though they remain underground for several months.

As the animal arouses from hibernation it hyperventilates without any change in breathing intervals, $CO_2$ in the body decreases, and blood pH rises. These changes are followed by shivering in the muscles, resulting in high lactic acid production and a rise in metabolism (Figure 6.17).

Hibernation in the black bear (*Ursus americanus*) and grizzly bear (*Ursus arctos*) is unique, if bears can be considered true hibernators or even heterotherms. Bears do not undergo profound hypothermia. It would take much too long to warm up a large body. Their large size allows them to build sufficient fat reserves to last them through the period of winter dormancy even though metabolism is near normal. Bears move into hibernating state gradually in the fall when acorns

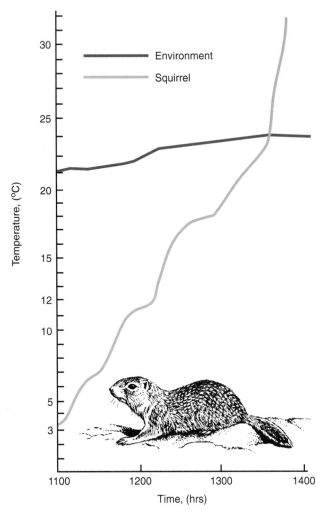

**Figure 6.17** Rise in the body temperature of the Arctic ground squirrel upon arousal from hibernation, stimulated by shivering. With a second rapid rise, at 24° C, the animal opens its eyes and sits up. (After Mayer 1960.)

degraded into amino acids that are reincorporated into plasma proteins (R. A. Nelson et al. 1983, R. A. Nelson and Beck 1984).

For the female bear hibernation is also the time of giving birth to two to four young. The blind, almost naked cubs are only eight inches long and weigh 220 to 280 g. By the time the female comes out of hibernation, the young suckling on the sleeping sow will weigh 1.8 to 2.5 kg.

A dormancy similar to that of hibernation is employed by some desert mammals and birds during the hottest and driest parts of the year. Such summer dormancy is called **estivation.** Because little evidence exists to show that estivation is triggered by either heat or drought and the physiological processes are similar, J. W. Hudson (1973) considers it synonymous with hibernation. L. R. Walker et al. (1979) suggest that estivation be termed *shallow torpor* instead.

## TEMPERATURE AND DISTRIBUTION

Understanding the reasons for the distributional ranges of species is one of the goals of ecology. One critical influence on distribution is temperature. The optimal temperature range for completing a species' life cycle, especially reproduction, imposes the greatest restriction on its distribution. For example, the adult northern lobster (*Homarus americanus*) of the North Atlantic coastal waters can live at temperatures ranging from 17° C to 0° C, but it can breed only in waters warmer than 11° C.

The vampire bat (*Demodus rotundus*) ranges from central Mexico to northern Argentina. Both the northern and southern limit of its range parallel the 10° C minimal isotherm of January (McNab 1973) (Figure 6.18). Its distributional limits are associated with the bat's poor capacity for temperature regulation at ambient temperatures below 10° C.

Temperature is a major influence on the natural distribution of plants on a continental, regional, and local scale. Their ranges often coincide with a certain minimal or maximal isotherm; but some plants, such as red spruce and balsam fir, occupy local pockets of habitat outside their major range. The northward distributions of certain plants possessing the $C_4$ photosynthetic pathway (see Chapter 10), especially the grasses, appear to be limited by cool temperatures (Stowe and Teeri 1978). The role of temperature is more evident in the distribution of $C_4$ grasses and grasses with the $C_3$ photosynthetic pathway on an elevational gradient along tropical mountain slopes on Hawaiian volcanoes (Rundel 1980) and Mt. Kenya (Tieszen et al. 1979). On the Hawaiian volcanoes $C_4$ species dominate the low, hot elevations and $C_3$ species dominate the cool, high elevations (Figure 6.19). The vegetational change with elevation and thus temperature suggests that $C_3$ plants outcompete $C_4$ plants at cool temperatures and vice versa. The mean maximum daily temperature at the midpoint of replacement both in Hawaii and Africa is about 22° C.

and berries are abundant. They gorge themselves, consuming up to 20,000 calories a day. They seek a den in a cave, beneath an overhanging rock, or in a large hollow tree. Their temperature drops only slightly from 37° C to 35° C, within normal limits for an active bear. Their heartbeat decreases from a summer sleeping rate of 40 to 50 beats per minute to 8 to 10 beats per minute. Their metabolism drops to 50 to 60 percent of normal, requiring the burning of some 4000 calories of energy a day. This high metabolic rate ultimately results in a 20 to 25 percent loss in body weight. During winter dormancy bears do not drink, eat, defecate, or urinate, although urine forms and enters the urinary bladder (Nelson 1980). Even after bears emerge from hibernation five to six months later, they may not eat or drink for several weeks.

Bears can endure a long period of food deprivation and maintain a high rate of metabolism because they can control urea metabolism, so uremia and dehydration do not occur. Urea, water, and other constituents in urine are absorbed by the bladder wall and reenter the bloodstream. There urea is

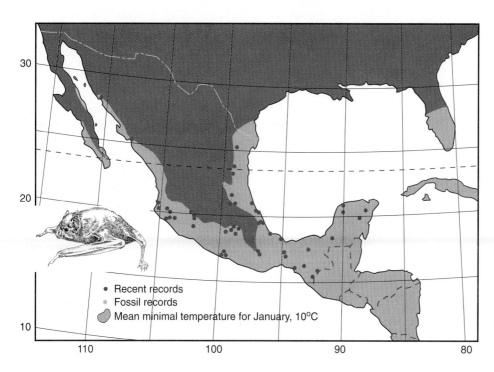

**Figure 6.18** The northward distribution of the vampire bat (*Desmodus rotundus*) parallels closely the 10° C minimal isotherm for January. Its southward distribution in Argentina and Chile is also limited by the same isotherm. (After McNab 1973:140.)

On a continental scale, the ranges of forest trees also reflect temperature. In southern Australia the distribution of snow gum (*Eucalyptus pauciflora*) differs from that of scribbly gum (*E. rossi*) on the basis of temperature alone (Austin et al. 1984) (Figure 6.20). Snow gum can occupy a much wider range of temperature than scribbly gum because its seeds do not ger-

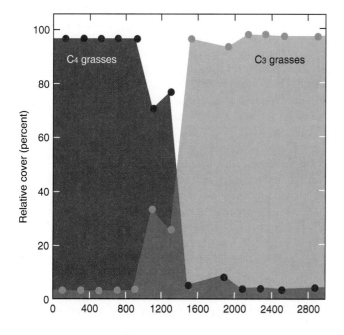

**Figure 6.19** Temperature influences plant distribution. Distribution of $C_3$ and $C_4$ grasses, as measured by relative ground cover, along an elevational gradient in the Hawaii Volcanoes National Park. (After Rundel 1980:355.)

minate until they are least likely to be affected by snow, drought, or high temperature (Abrecht 1985), and it has the ability to maintain high photosynthetic activity over a range of temperature optimums, closely tracking the long-term mean monthly temperatures (Slatyer 1977). The northernmost limit of the evergreen live oak (*Quercus virginiana*) is defined by average annual minimum temperature between −3.9° C and −6.7° C. Paper birch (*Betula papyrifera*), a cold climate species, is found as far north as the 12° C July isotherm and seldom grows naturally where the average July temperature exceeds 21° C. The distribution of black spruce (*Picea mariana*) follows a similar pattern.

Some plants can grow successfully as individuals outside their natural range, but will not reproduce. Blueberries (*Vaccinium* spp.), for example, will not flower or fruit successfully unless chilled. Other plants fail to grow to normal size because the twigs freeze back in winter or are killed by late spring frosts. Such plants are restricted within their natural ranges to areas where temperatures are favorable for growth and reproduction.

It is often difficult to separate the influence of temperature from that of moisture, because a close interaction exists between the two in terrestrial environments. In fact, the two determine in large measure the climate of a region and the distribution of animals and plants.

The role of temperature and its relationship with moisture in the distribution and abundance of species is the subject of intensive research relating to global climate change (for example, see MacDonald and Sertario 1990; Kareiva, Kindsolver, and Huey 1993). How will organisms respond to increasing temperature? What effect will global climate change have on flora of the tundra and

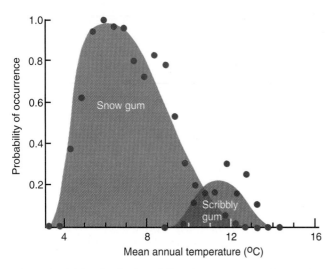

**Figure 6.20** The distribution of *Eucalyptus* trees as influenced by maximum and minimum mean annual temperatures.

taiga? Will ranges of temperate plants shift northward? What plants and animals will succumb or be supplanted by competitors better adapted to warmer temperatures? What effect will such changes in plant communities have on associated animal and microbial life? How will increasing temperature affect agricultural plants? These and hundreds of other questions about the effects of global climate change beg for answers.

## SUMMARY

Organisms live in a thermal environment characterized by heat and temperature. Heat is a form of energy that results from the random motion of molecules within a substance. Temperature is a measure of a substance's tendency to give up heat. Organisms maintain a balance between heat energy gained from and lost to the environment. Heat gains come from direct and reflected sunlight, diffuse radiation, long-range infrared radiation, and metabolism. Heat is lost to the environment by infrared radiation, conduction, convection, and evaporation. Heat loss to the environment can occur only when the ambient temperature is less than body core temperature.

Response to the thermal energy environment differs among organisms. Most organisms—from protists, plants, and invertebrates to reptiles—are poikilothermic. They gain their heat energy from the environment. Their rate of metabolism, as measured by oxygen consumption per unit of body weight, approximately doubles for every 10° C rise in temperature in accordance with van't Hoff's rule. Birds and mammals maintain their body temperatures by internally generated body heat. They are known as homeotherms.

Physiological processes such as photosynthesis and the various phases of life cycle of plants have an optimum range of temperature. Plant mechanisms to resist extremes of heat and cold include reflectivity of leaves and bark, leaf size and shape, orientation of leaves toward the sun, and frost hardening, the synthesis of protective antifreeze substances in cells.

Animal responses are more diverse. Animals depending upon the environment as a source of heat are ectothermic. Their body temperatures tend to follow ambient temperatures. Because of their variable body temperature these animals are termed poikilotherms. They regulate their body temperatures behaviorally by moving in and out of warmer and cooler areas. Animals that depend upon internally produced heat to maintain body temperatures are endothermic. They maintain a rather constant body temperature independent of the environment. Many animals are heterothermic. Depending upon the environmental and physiological conditions, they function either as endotherms or ectotherms. They regulate body temperature metabolically or allow it to drop to ambient temperature, increasing or conserving energy.

Within limits poikilotherms acclimatize to higher or lower temperatures by avoiding temperature extremes, by adjusting tolerances to given temperature ranges, by heliothermism, and in some instances by evaporative cooling. Some reptiles change their body temperatures by changing the rate of heartbeat and metabolism. Some cold-tolerant poikilotherms utilize supercooling, the synthesis of glycerol in body fluids to resist freezing in winter. Homeotherms and heterotherms acclimatize by seasonal changes in body insulation, vascularization, evaporative heat loss, and nonshivering thermogenesis. Most homeotherms and heterotherms employ a well-developed countercurrent circulation, the exchange of body heat between arterial and venous blood and their equivalent in insects. This exchange retains body heat by reducing heat loss through body parts, or cools blood flowing to such vital organs as the brain.

Some animals enter a state of dormancy during environmental extremes to reduce the high energy costs of staying warm or cool. They slow their metabolism, including heartbeat, respiration, and body temperature. Birds such as hummingbirds and mammals such as bats undergo daily torpor, the equivalent of deep sleep, without the extensive metabolic changes of seasonal torpor. Hibernation involves a whole rearrangement of metabolic activity to run at a very low level.

Temperature regimes set upper and lower limits within which organisms can grow and reproduce, influencing the spatial distribution of plants and animals. Global warming could affect the distribution of organisms.

## REVIEW QUESTIONS

1. Describe radiation, conduction, convection, and evaporation. What is the role of each in energy exchange between the organism and its environment?

2. What is the relationship between temperature and metabolism? What is the temperature coefficient, and how does it relate to van't Hoff's rule?

3. How do plants respond to heat and cold?

4. Contrast poikilothermy, homeothermy, and heterothermy. How do they relate to ectothermy and endothermy?

5. Why are small endotherms so rare?

6. Why should the camel, a desert mammal, have such a heavy coat of hair?

7. Bats are small hibernating mammals. They do not store food, nor do they put on a heavy layer of fat. How do bats survive the long winter hibernation?

8. Compare the oxygen consumption per kilogram of body weight of three mammals, one 10 kg, the second 100 kg, and the third 200 kg. Use the formulas on page 82.

9. Consider a population of fish acclimatized to the warm water outflow into a river from a power generating station. Discuss the physiological stresses experienced by the fish and consequences if the power plant shut down for one week.

10. In what way does temperature limit the distribution of organisms? How might global warming affect the distribution of organisms?

## CROSS-REFERENCES

Adaptation, 30–31; evaporation, 44; humidity, 45; microclimate, 50–53; climate and vegetation, 58–60; drought resistance, 71–72; hyperthermia, 76; photosynthesis 154–159; $C_3$ and $C_4$ plants 154–157; deserts, 246–250.

# Light and Biological Cycles

## Concepts

1. Visible light, with wavelengths between 400 and 740 nm, is photosynthetically active radiation. Ultraviolet and infrared fall beyond it.
2. Light has spectral quality, intensity, duration, and directionality. It may be reflected, absorbed, or transmitted. All these properties are of great ecological importance.
3. Shade tolerance is a measure of a plant's ability to grow at low light intensities.
4. Light influences the daily and seasonal activities of organisms.
5. Organisms possess a circadian rhythm of about 24 hours that becomes synchronized to a 24-hour environmental rhythm of light and dark.
6. An organism's biological clock is set by changes in daylength.
7. Seasonality, the yearly cycle of plant and animal activity, is mediated by periodicity.

Day and night, sunlight and shade, cloudy days, moonlit nights—the play of light has pronounced effects on life. It influences the distribution of plants and animals on land and in water, and it affects their daily activities and seasonal changes. This chapter focuses on the ecological importance of light and considers some of the physiological mechanisms that underlie the large-scale patterns in nature.

# THE NATURE OF LIGHT

Light is solar radiation in the visible range to humans, embracing wavelengths of 400 to 740 nanometers (nm); a nanometer is one-billionth of a meter (Figure 7.1). Collectively, these wavelengths are known as **photosynthetically active radiation (PAR)** because they contain the wavelengths used in photosynthesis. Light with shorter wavelengths than the visible range is ultraviolet light—UV-A with wavelengths of 315 to 380 and UV-B with wavelengths of 280 to 315. Light with wavelengths longer than the visible range is near infrared, with wavelengths of 740 to 4000 nm, and far infrared or thermal radiation, from 4000 to 100,000 nm.

In addition to its spectral quality, light has three other components, intensity, duration, and directionality. Light is transported in discrete bundles called *photons*. A mole of photons ($6.02 \times 10^{23}$ photons) equals an Einstein (E). One E $= 1.7 \times 10^5$ Joule (J); a Joule is a unit of energy. The intensity of PAR is measured as the proton flux density or the total number of protons within the visible light spectrum striking a given area. It is expressed in microeinsteins per square centimeter per unit time ($\mu e/cm^2/sec$). At midday on a clear day the flux of PAR is approximately 2000 $\mu e/m^2/sec$.

The intensity of light varies daily and seasonally. It is influenced by its angle of incidence, the angle at which it strikes a surface. The angle of incidence of solar radiation depends upon the altitude of the sun—its height above the horizon. During a clear day the highest intensity of light occurs at

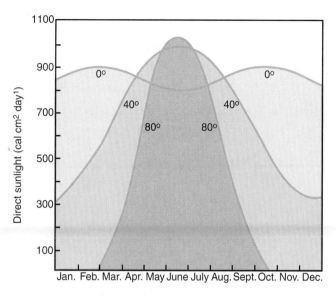

**Figure 7.2** Amount of direct sunlight incident upon a horizontal surface at the outer extremity of Earth's atmosphere as a function of time of year for north latitudes of 0°, 40°, and 80°. (From Gates 1972:28.)

noon when the sun is overhead and the lowest at dawn and twilight when the angle of incidence has a long path through the atmosphere. The seasonal intensity of light changes with latitude (Figure 7.2). In the Northern Hemisphere sunlight is most intense in June at 80° N latitude (near the North Pole) and at 40° N latitude (New York City); in winter at 80° N latitude the intensity is zero. At low latitudes (the Tropics) there is little seasonal fluctuation in direct light intensity, but note that direct sunlight is greatest at 0° latitude during the time of winter in the Northern Hemisphere.

Duration of light is also seasonal. It, too, is a function of latitude and season. On the first day of summer in the Northern Hemisphere, June 21, daylength at 10° N latitude is nearly 24 hours; at 40° N latitude, 17 hours; and at 0° latitude about 12 hours. On the first day of winter at 40° the length of day is 9 hours. At the spring and fall equinoxes in

**Figure 7.1** A portion of the electromagnetic spectrum, separated into solar and terrestrial radiation. (After Halverson and Smith 1979:6.)

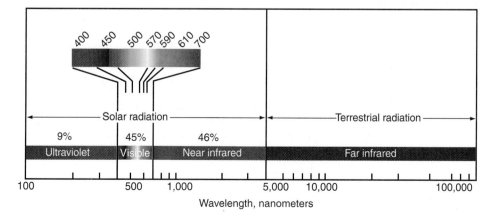

March and September, the length of day is 12 hours everywhere in the world.

The directionality of light shifts daily and seasonally. The sun is nearly overhead in early summer; but as winter approaches, the altitude of the sun drops toward the horizon. In mountainous country south-facing and west-facing slopes receive the most light; deep narrow valleys may experience shorter days because of shading by opposite slopes. On a microscale many organisms find themselves always shaded on one side. Plants growing on the north side of a house or in a wood are overshaded.

Light impinging on an object may be absorbed, reflected, or transmitted through it. For example, when light hits a leaf, plants reflect about 70 percent of infrared wavelengths striking the leaves perpendicularly. They reflect only 6 to 12 percent in the visible range. The degree of reflection varies with the nature of the leaf. Leaves with whitish hairs and cuticles reflect more light than deep green leaves. Plants reflect green light more strongly (10 to 20 percent) than·red and orange light (3 to 10 percent). Because they reflect green light most strongly, leaves appear green. Only about 3 percent of UV light is reflected. Much of the visible radiation is absorbed. Chloroplasts in the plant absorb 70 percent of PAR. Chlorophyll absorbs most of the red light and uses it in photosynthesis. A remaining fraction of light is transmitted through the leaf, the amount depending upon its thickness and structure. Thin leaves transmit more light than thick ones. Transmission is greatest in those wavelengths that are also reflected. Thus plants transmit mostly green and far infrared wavelengths.

In both grasslands and forest the leafy canopy intercepts most of the sunlight that floods open spaces (Figure 7.3). As light penetrates the canopy, different wavelengths filter out—the light becomes attenuated. The degree of attenuation is influenced by the density of the canopy, the optical properties of the leaves, and the number and sizes of gaps in the canopy. Thus light conditions within the forest vary from unfiltered light coming through the gaps, creating changing sunflecks on the ground, to filtered light in which the spectrum is altered. Blue and red are the most highly attenuated wavelengths, whereas green and far infrared wavelengths pass through relatively unaltered. Far infrared wavelengths are less strongly attenuated than visible light. These changes in spectral quality vary among forests. For example, beneath a canopy of a deciduous forest on a clear day far infrared radiation dominates with lesser amounts of blue, green, and red (Vezina and Boulter 1966), giving an observer the impression of standing in reddish-green shade. In the depths of the forest only far infrared wavelengths prevail. In the more open pine forest, little difference exists in the transmission of various wavelengths. On cloudy days selective absorption of spectral wavelengths is lower, and transmission is higher in both forests.

Only about 1 to 5 percent of the light striking the canopy of a typical temperate hardwood forest in summer reaches

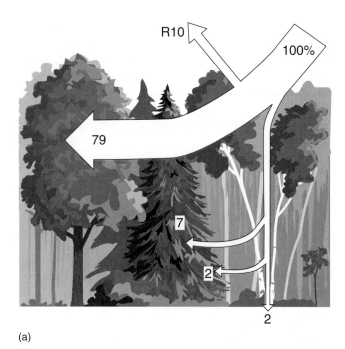

(a)

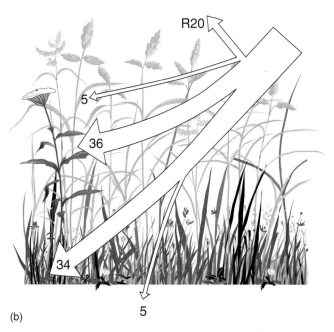

(b)

**Figure 7.3** Attenuation of radiation. (a) In a boreal mixed forest 10 percent of the incident photosynthetically active radiation (PAR) is reflected in the upper crown, and the greatest absorption occurs in the crown. (b) In a meadow 20 percent of the photosynthetically active radiation is reflected from the upper surface, whereas the greatest absorption occurs in the middle and lower regions. (Adapted from Larcher 1980:7.)

the forest floor. In a tropical rain forest only 0.25 to 2 percent gets through. More light travels through pine stands—about 10 to 15 percent—but densely crowned spruce allows only 2.5 percent of open sunlight to reach the forest floor. However, even in dense forests sunlight penetrates the canopy during the day as sunflecks. In fact, sunflecks can account for 70 to 80 percent of the solar energy reaching the forest floor. They enable many plants of the forest floor to endure the shaded conditions on the ground (Chazdon and Pearcy 1991). Woodlands with trees possessing relatively open crowns, such as oaks and birches, allow light to filter through. There light dims gradually, as it does in grasslands. In grasslands most of the light is intercepted by the middle and lower layers of vegetation.

Light that strikes the surface of water is also attenuated (Figure 7.4). Depending upon the angle at which light strikes the water, a greater or lesser amount is reflected from the surface. For example, only 6 percent of the light striking the surface of water at an angle of incidence of 60° from the perpendicular to the surface of the water is reflected; at 80°, 35 percent is reflected. The higher the sun is from the horizon, the more light will enter the water.

Light that enters the water is absorbed rapidly. Only about 40 percent reaches a meter deep in clear lake water. Moreover, water absorbs some wavelengths more than others. First to go is visible red light and infrared light in wavelengths greater than 750 nm. This absorption reduces solar energy by one-half. In clear water yellow goes next, followed by green and violet, leaving only blue wavelengths to penetrate deeper water. A fraction of blue light is lost with increasing depth. In the clearest of seawater, only about 10 percent of the blue wavelengths reaches more than 100 m.

Of course, not all water is clear. Freshwater lakes, streams, and ponds, and coastal waters have a certain degree of turbidity, which greatly affects light penetration. Because of the high level of yellow substances, such as clay colloids and fine detrital material washed into the water from terrestrial ecosystems, the blue waveband is most strongly attenuated and is removed at very shallow levels. Green penetrates most deeply, and where concentrations of yellow materials are high, red wavelengths may penetrate as far as the green. In very yellow water red is the last wavelength to be extinguished. Waters with organic stains, such as bog lakes and ponds, appear dark, and nutrient-poor lakes and ponds, low in phytoplankton, are clear and bluish. Waters supporting a dense growth of phytoplankton take a decidedly greenish appearance. Heavy growth of phytoplankton and cyanobacteria can shut out light to deeper waters just as effectively, if not more so, than the dense canopy of a deciduous or tropical rain forest.

As light is attenuated through a canopy of vegetation or water, the intensity of light diminishes because of absorption. This decrease in intensity is measured by the **extinction coefficient,** the ratio between the intensity of light at the surface and the intensity at a given depth of water or of the vegetative canopy. In water the extinction coefficient is influenced by such factors as turbidity and density of phytoplankton growth. In forests and grasslands the extinction coefficient is influenced by **leaf area index (LAI).** The leaf area index is the total surface area of leaves above a given area of ground:

$$LAI = total\ leaf\ area/ground\ area$$

A leaf area index of 3, for example means that a given area of ground would be covered by three times that area of leaves. The greater the leaf area index, the greater is the attenuation of radiation. Grasslands with low LAI because of upright and angled leaves have an extinction coefficient of less that 0.5; in plant communities with horizontal leaves, such as herbaceous plants, the extinction coefficient may be 0.7 or greater. In forests the extinction coefficient will be very high in the canopy.

Related to the extinction coefficient is **compensation intensity,** the intensity at which plants can no longer carry on sufficient photosynthesis to overcome respiratory energy. Few plants have a compensation intensity of less than 1 percent of sunlight. Phytoplankton can carry on photosynthesis only to the depth at which photosynthetically active radiation falls to 1 percent of that just below the surface.

## PLANT ADAPTATIONS TO LIGHT INTENSITY

The major role of light in the life of plants is as an energy source for photosynthesis. Some plants, both aquatic and terrestrial, can thrive in less light than others. They are able

**Figure 7.4** The spectral distribution of solar energy at Earth's surface and after it has been modified by passage through varying depths, measured in meters, of pure water. Note how rapidly red wavelengths are attenuated. At approximately 10 m red light is depleted; but at 100 m, blue wavelengths still retain nearly one-half their relative intensity. (From G. Clarke, 1939.)

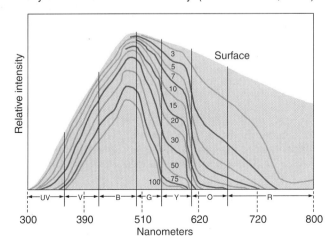

to maintain a better balance between and photosynthesis and respiration than others. The relative abilities of plants and of leaves of plants to carry on photosynthesis at low light intensities is termed their **shade tolerance.**

## PAR Radiation

How a plant responds to high and low light intensities depends in part on its tolerance to shade. Shade avoiders or sun-tolerant plants grow best in open sites receiving full sunlight. They establish themselves rapidly on disturbed sites, tolerate extremes of dryness and wetness, and achieve fast growth in the open. They carry on photosynthesis more efficiently in full sunlight and have a high rate of photosynthesis and a high rate of respiration. They rapidly convert photosynthate into growth.

By contrast, shade-tolerant plants are not competitive with sun-tolerant plants in full sun. They have a lower rate of photosynthesis and, more important, a lower rate of respiration. For those reasons they grow more slowly in all environments. Many shade-tolerant plants have the ability to carry on photosynthesis at low light intensities. They can open their stomata rapidly in dim light, taking advantage of brief sunflecks (Chazdon 1988, Chazdon and Pearcy 1991). These plants may live suppressed in the shaded understudy of a forest or shrubland for many years, but when released from overhead shade, they respond with rapid growth (Grime 1966).

When forced to grow in the shade, sun plants may respond by growing rapidly in height in an attempt to emerge from the shade. (Observe the growth of tomato plants or full-sun annual flowers planted in the shade.) Their leaves are widely spaced, reducing shade on individual leaves. Rapid growth and thin leaves make them highly susceptible to drought. Thin cell walls of their stems reduce supporting tissue and subject them to fungal infections (Grime 1966).

Shade tolerance or intolerance, however, may involve more than light. Moisture, too, plays a role. Plants of the understory may experience a shortage of soil moisture resulting from competition from overhead vegetation. Intolerant plants might be able to fix energy at a rate higher than required to balance respiration, but the rate is not enough to allow roots to grow into deeper soil where moisture is available.

Because leaves are the principal photosynthetic organs of plants, their shape and size reflect the plants' reaction to shade. Plants that show little plasticity in leaf shape in either sun or shade seem to function only under one set of environmental conditions. Other plants, especially forest trees, are more flexible. The leaves of the upper canopy of the tree are exposed to intense light and the leaves low in the canopy are shaded. These trees change leaf shape and size in response to light conditions (Figure 7.5). Sun leaves are smaller, thicker, and more deeply lobed than shade leaves, with well-developed support and conduction systems (veins). Shade

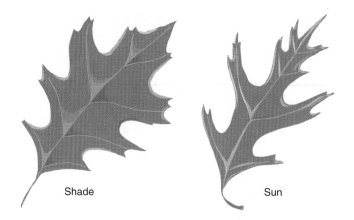

**Figure 7.5** Morphological differences in sun and shade leaves of red oak. The shade leaves are much larger with shallow lobes.

leaves are thinner, weakly lobed, or not lobed; they have a large surface area per unit of weight, fewer stomata, and less support and conduction tissue. In part their structure reflects less heat, moisture, and physical stress. When shade leaves or shade plants are suddenly exposed to full light, they lose an excessive amount of moisture and experience light damage to the chloroplasts. The shock kills them. The branch or plant may survive by growing new leaves better adapted to high light. When shaded bark on trunks of trees is exposed to direct sunlight, the light stimulates adventitious buds in the bark to develop into epicormic branches (Figure 7.6).

Typical among shade-tolerant tree species are sugar maple (*Acer saccharum*), white cedar (*Chamaecyparis thyoides*), and Eastern hemlock (*Tsuga canadensis*) (Table 7.1). These species exist successfully under a dense forest canopy at low light intensities but reach maximum growth rates only when the canopy is opened. Shade-tolerant understory plants include ferns and various herbaceous plants, among them violets (*Viola* spp.). Many forest herbaceous plants grow and flower early in the spring before the forest canopy is fully developed and in the fall when the canopy thins, allowing light to reach the forest floor. Thus sun and shade plants have adapted to colonize only certain habitats. These differential abilities influence community composition and succession.

Because of the rapid attenuation of light with depth, most aquatic plants live in the equivalent of a shaded environment, but with a major difference. Far infrared wavelengths are strongly absorbed by water, but not by terrestrial foliage. The red-to-far red ratio in water is between 3.6 and 4.7 to 1, compared to 1.15 for the forest interior.

Aquatic plants can so modify the light environment for plants in the lower water column that they greatly reduce PAR. Because phytoplankton species can become photoinhibited at high light intensities at the surface, especially in sunny weather, they carry on high photosynthetic rates deeper down. Some species of phytoplankton move up and

**Figure 7.6** Epicormic branches develop along the trunks of forest trees suddenly exposed to full sunlight.

**Table 7.1  Relative Shade Tolerance of Some North American Trees**

| | *Very Intolerant* | |
|---|---|---|
| Jack pine | Whitebark pine | Aspens |
| Virginia pine | Pin cherry | Paper birch |
| Longleaf pine | Tamarack | Black locust |
| Lodgepole pine | Western larch | Cottonwoods |

| | *Intolerant* | |
|---|---|---|
| Red pine | Junipers | Black cherry |
| Shortleaf pine | Sycamore | Sweet gum |
| Loblolly pine | Sassafras | Black walnut |
| Ponderosa pine | Scarlet oak | Red alder |
| Eastern red cedar | Yellow-poplar | Madrone |

| | *Intermediate* | |
|---|---|---|
| Eastern white pine | Douglas-fir | |
| Western white pine | Silver maple | Oaks |
| Sugar pine | Hickories | White ash |
| Slash pine | Yellow birch | Elms |

| | *Tolerant* | |
|---|---|---|
| Spruces | Grand fir | Basswood |
| Redwood | White fir | Red maple |
| Western red cedar | Alpine fir | |

| | *Very tolerant* | |
|---|---|---|
| Eastern hemlock | Balsam fir | Sugar maple |
| Western hemlock | Pacific silver fir | American beech |
| Mountain hemlock | Pacific yew | Flowering dogwood |
| Red spruce | White cedar | Hop hornbeam |

down through the water column to escape inhibitory effects of high light and to reach the depth at which light intensity is most favorable. Photosynthetic ability is further influenced by seasonal variations in water temperature and light, and by the adaptations of phytoplankton and macrophytes for different wavelengths of visible light. Most macrophytes grow at the depth at which light intensity is most favorable.

In marine environments red, green, and brown algae typically are found in different proportions with increasing depth. That pattern has been explained by the different absorptive spectra of their pigments, affecting their photosynthetic ability. Red algae have been considered typical of the deepest water because they contain phycobilin pigments that absorb green wavelengths common in deep water. Green algae with chlorophyll *a* and *b,* similar to terrestrial vegetation, have been considered mainly inhabitants of shallow water. Brown algae with chlorophyll *a* and *c* and a special carotenoid pigment, fucoxanthin, have been considered the seaweeds of intermediate depths.

Many recent studies, however, point out that zonation of seaweeds is not that simple, and that no significant difference exists in either the minimum or the maximum depths inhabited by red, green, and brown algae. Red algae do not have an exclusive claim to the deepest waters; and the green algae, supposedly restricted to shallow water, may grow in greater abundance at depths of 90 to 100 m than red algae (Figure 7.7). In Hawaiian waters, green algae may penetrate as deep as red algae and more deeply than the brown. Seaweeds appear to respond physiologically more to light intensity than to spectral quality of light (Ramus 1983). We still do not know what role these accessory pigments play.

## Ultraviolet Radiation

Of considerable ecological significance is solar ultraviolet radiation reaching Earth. Most of the ultraviolet radiation is absorbed by the stratospheric ozone layer. Solar UV-B (280–320 nm) radiation decreases on a latitudinal gradient from the tropics, where the ozone layer is the thinnest, to the

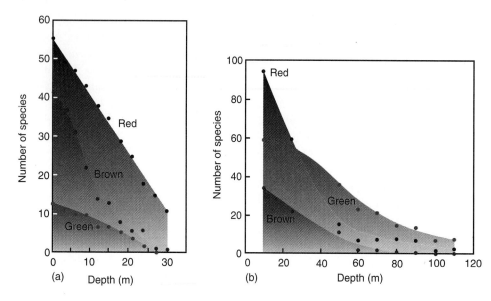

**Figure 7.7** Variations in the distribution of red, green, and brown algae with depth. (a) In New England waters red algae show a gradual decline in the number of species with depth. Brown algae are more abundant at shallow depths. The less abundant green algae have more species adapted to light conditions at deeper waters. (From Kirk 1983:294). (b) In the clear tropical waters of the Caribbean, all three types of marine benthic algae reach much greater depths. Below 50 m, however, green algae dominate at all depths. (From Kirk 1983:296.)

poles, where the layer is the thickest. Solar UV-B radiation increases about 14 to 18 percent per 1000 m elevation.

In recent years this stratospheric ozone layer is being diminished by increasing concentrations of human-made ozone-destroying chemicals such as chlorofluorocarbons. This depletion is most pronounced over the poles and the tropics (Kerr 1992). This depletion allows an increase of ultraviolet radiation, especially UV-B radiation, specifically the wavelengths between 290 and 315 nm. (Caldwell et al. 1989). Laboratory and greenhouse experiments show that UV-B radiation can damage DNA, partially inhibit photosynthesis, alter the growth form of plants, and reduce yield (Caldwell et al. 1989, Tevini and Teramura 1989, Teramura 1990). These damaging effects, however, have not been clearly demonstrated in field-grown plants.

Plants have evolved defenses against UV-B radiation reaching the interior of leaves. One line of defense is leaf reflectance. The major barriers to UV-B penetration are the epidermal cells, containing anthocyanins, colorless flavenoids, and other phenols that absorb UV-B radiation yet transmit PAR to the interior of the leaf.

Plants exhibit a wide range in their ability to screen ultraviolet radiation. Tropical and alpine plants, naturally exposed to high levels of ultraviolet radiation, more effectively block UV-B radiation than temperate species. However, even within these plants, some attenuate UV-B better than others. Day et al. (1992) compared the screening ability of a group of Rocky Mountain plants growing at high elevations. The leaf epidermis of herbaceous dicotyledonous plants was

the least effective. Eighteen to 41 percent of UV-B radiation reached the interior of the leaf. The needles of conifers were the most effective, attenuating essentially all incident UV-B radiation. This effectiveness is due in part to the fact that UV-B pigments are found not only in cell vacuoles but in epidermal cell walls. Between the two extremes are grasses and woody dicots. These researchers attributed the differences to the cost effectiveness of such defenses. Because their longer-lived needles are exposed to larger lifetime doses of UV-B radiation, conifers may have evolved highly effective screening mechanisms. On the other hand, grasses and herbaceous and woody dicots with short-lived seasonal leaves would not have evolved such screening properties.

Increased UV-B radiation may have more impact on animals than on most plants. Especially vulnerable are light-pigmented organisms, especially humans, who are susceptible to ultraviolet light-induced skin cancer.

Faced with a potential increase of UV-B radiation, we know little about its effects. Research concentrated on crop plants indicates that about one-half of the three hundred species are sensitive to UV-B radiation. We know little about its effects on forest tree species and other native plants. We do not have a clear understanding of plant defenses against UV-B radiation. Can plants repair damage to the interior of the leaf? Is resistance to UV-B radiation inherited? Will radiation cause inheritable genetic changes? Will UV-B radiation give a selective advantage to resistant species and thus alter plant diversity? Scores of important questions need to be answered.

# PHOTOPERIODISM

No one can ignore the rhythms of Earth—its recurring daily and seasonal changes. Life of the daylight hours retreats before the night, and life of the night retreats before the coming of daylight. As the year progresses, the seasons change. The progression of lengthening and shortening days is marked by the flush of new growth and the senescence of the old, by the arrival and departure of migrating birds. These rhythms are driven by the daily rotation of Earth on its axis and its 365-day revolution about the sun. Through evolutionary time life became attuned to and synchronized with the daily and seasonal changes in the environment.

## Circadian Rhythms

At dusk in the forests of North America, a small squirrel with silky fur and large black eyes emerges from a tree hole. With a leap the squirrel sails downward in a long sloping glide, maintaining itself in flight with broad membranes stretched between its outspread legs. Using its tail as a rudder and brake, it makes a short, graceful upward swoop that lands it on the trunk of another tree. This is *Glaucomys volans,* the flying squirrel, perhaps the most common of all our tree squirrels. Because of its nocturnal habits, this mammal is seldom seen. Unless disturbed, it does not come out by day. It emerges into the forest world with the arrival of darkness; it retires to its nest before the first light of dawn.

The squirrel's day-to-day activities form a 24-hour cycle. The correlation of the onset of activity of the flying squirrel with the time of sunset suggests that light has some regulatory effect on the activity of the squirrel. If the flying squirrel is brought indoors and confined under artificial conditions of night and day, it will restrict its periods of activity to darkness, and its periods of inactivity to light. Whether the conditions under which the squirrel lives are 12 hours of darkness and 12 hours of light or 8 hours of darkness and 16 hours of light, the onset of activity always begins shortly after dark.

Such behavior in itself does not mean that the squirrel has any special timekeeping mechanisms. It could easily be responding behaviorally to nightfall and daybreak. However, if the same squirrel is kept in constant darkness, it still maintains its pattern of activity and inactivity from day to day in the absence of all time cues. Under these conditions the squirrel's activity rhythm deviates from the 24-hour periodicity (Figure 7.8). Its daily cycle of activity and inactivity under the conditions of constant darkness (when it normally is active) varies from 22 hours and 58 minutes to 24 hours and 21 minutes, the average being less than 24 hours (DeCoursey 1961). Because the cycle length deviates from 24 hours, the squirrel gradually drifts out of phase with the periods of daylight and night of the external world.

This innate rhythm of activity and inactivity covering approximately 24 hours is characteristic of all living organisms except bacteria. Because these rhythms approximate, but seldom match, the periods of Earth's rotation, they are called **circadian rhythms** (from the Latin *circa,* "about," and *dies,* "day"). The period of the circadian rhythm, the number of hours from the beginning of a period of activity one day to the beginning of activity on the next, is called the **free-running cycle.** In other words, the rhythm of activity exhibits a self-sustained oscillation under constant conditions of dark or light.

Circadian rhythms have a strong genetic component and are transmitted from one generation to another. They are affected little by temperature changes, are insensitive to a great variety of chemical inhibitors, and are not learned from or imprinted upon the organism by the environment. They influence not only the times of physical activity and inactivity but also physiological processes and metabolic rates. They do not have any special adaptation to specific local or regional environmental conditions. What circadian rhythms do is provide a mechanism by which organisms can maintain synchrony with their environment.

Thus plants and animals are influenced by two daily periodicities, the external rhythm of 24 hours and the internal circadian rhythm of approximately 24 hours. If the two rhythms are to be in phase, some external environmental "timesetter" must adjust the endogenous rhythm to the exogenous. The most obvious cues are temperature and light. Of the two the master timesetter is light. It brings the circadian rhythm of organisms into phase with their external environment.

## The Biological Clock

The circadian rhythms and their sensitivity to light are the mechanisms underlying the biological clock, the timekeeper of physical and physiological activity in living things. In one-celled organisms and plants, the clock appears to be located in individual cells. In multicellular animals the clock is associated with the brain.

Skillful surgical procedures have allowed circadian physiologists to discover the location of the physiological clock in some mammals, birds, and insects. In most insects studied, the clock, including the photoreceptors, is located either in the optic lobes or in the tissue between the optic lobes and the brain. In the cockroach and cricket the receptors for the entrainment of circadian rhythms of locomotion and stridulation are located in the compound eye, but the controlling clock is in the brain (Beck 1980, Saunders 1982). In birds the clock evidently is located in the pineal gland, in the lower central part of the brain (Farner and Lewis 1971, Gwinner 1978). In mammals the clock appears to be located in a part of the hypothalamus just above the optic chiasm, the place where the optic nerves from the eyes intersect, although the pineal gland also plays a part.

To function as a timekeeper, the clock has to have an internal mechanism with a natural rhythm of approximately

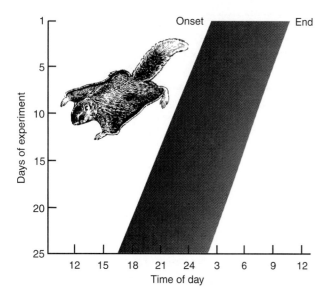

**Figure 7.8** Drift in the phases of activity of a flying squirrel held in continuous darkness at 20° C for 25 days. Note that the onset of activity becomes gradually later each day. (After DeCoursey 1960:51.)

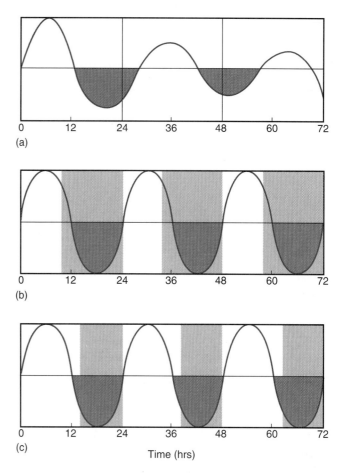

**Figure 7.9** The Bunning model of the entrainment of circadian rhythms to daylength. (a) Oscillations of the clock cause an alternation of half-cycles with quantitatively different sensitivities to light (white vs. solid). The free-running clock in continuous light or continuous darkness tends to drift out of phase with the 24-hour photoperiod. (b) Short-day conditions allow the dark to fall into the white half-cycle. (c) In the long day the light falls into the solid half-cycle. (From Bunning 1960:253.)

24 hours, which can be reset by recurring environmental signals, such as the changes in the time of dawn and dusk. The clock has to be able to run continuously in the absence of any environmental timesetter; and it has to be able to run the same at all temperatures. Cold temperatures must not slow it down, nor warm temperatures speed it up.

Two basic models of biological clocks have been proposed. One is an oscillating circadian rhythm sensitive to light (Bunning 1964) (Figure 7.9). The cycle or time measuring process begins with the onset of light or dawn. The first half is light-sensitive, and the second half requires darkness. Short-day effects are produced when light does not extend into the dark period. Long-day effects are produced when light does extend into the dark period. Because this simple model does not explain all photoperiodic responses, a variation is a two-oscillator model in which one oscillation is regulated by dawn and the other by dusk. There are more complex models of the clock, but this basic model underlies most of them.

Is there a population of clocks? Does one master clock drive all other clocks? It appears that the biological clock is organized as a hierarchy of clocks. The various overt rhythms influencing physiological and behavioral phenomena are controlled by "slave" or subservient clocks coupled to a master clock. These clocks or pacemakers may be integrated groups of cells within organs such as the central nervous system, where they have specific timekeeping functions. When the master clock is reset by a light signal, it in turn resets the other clocks. One slave clock may take longer to resynchronize than another with a different period. This resetting of the master clock results in transient disturbances of the phase relationships among the subservient clocks.

This desynchronization of the overt physiological rhythms of the body causes, for example, the discomforts of jet lag.

How biological clocks function is the domain of the physiologist. Ecologists are more interested in their adaptive value. One adaptive value is that the biological clock lets organisms anticipate environmental cues.

Certain circadian activities, for example, relate to aspects of the environment that are ecologically more important than light or dark per se. The transition from night to day is accompanied by such environmental changes as a rise in humidity and a drop in temperature. Woodlice, centipedes, and millipedes, which lose water rapidly in dry air, spend the day in the dark and damp under stones, logs, and leaves. They emerge at dusk when the humidity of the air is more favorable. These animals show an increased tendency to escape from light as the length of time they spend in darkness increases. On the other hand, their intensity of response to

low humidity decreases with darkness. Thus these invertebrates come out at night into places too dry for them during the day; and they quickly retreat to their dark hiding places as light comes (Cloudsley-Thompson 1956, 1960).

The circadian rhythms of many organisms relate more to the biotic than the physical aspects of their environment. Predators, such as insectivorous bats, must relate their feeding activity to the activity rhythm of their prey. Moths and bees must visit flowers when they are open to obtain nectar. The flowers must have a rhythm of opening and closing that coincides with the time when insects that pollinate them are flying. The clock is the most energy-efficient way to adapt to the periodicities of the environment.

## Critical Daylengths

In the northern and southern latitudes the daily periods of light and dark lengthen and shorten with the seasons. The activities of plants and animals are geared to the changing seasonal rhythms of night and day. The flying squirrel, for example, starts its daily activity with nightfall, regardless of the season. As the short days of winter turn to the longer days of spring, the squirrel begins its activity a little later each day (Figure 7.10). Most animals and plants of temperate regions have reproductive periods that closely follow changing daylengths of the seasons. For most birds the height of the breeding season is lengthening days of spring; for deer the mating season is the shortening days of fall. Trilliums (*Trillium* spp.) and violets bloom in the lengthening days of spring before the forest leaves are out and an abundance of sunlight reaches the forest floor. Asters (*Aster* spp.) and goldenrods (*Solidago* spp.) flower in the shortening days of fall.

The signal for these responses is **critical daylength.** When the duration of light (or dark) reaches a certain portion of the 24-hour day, it inhibits or promotes a photoperiodic response of an organism to changing daylength (Figure 7.11). Critical daylength varies among organisms, but it usually falls somewhere between 10 and 14 hours. Through the year plants and animals compare that time scale with the actual length of day or night. As soon as the actual daylength or nightlength is greater or lesser than the critical daylength, the organism responds appropriately. Some organisms can be classed as day-neutral, not affected by daylength, but rather controlled by some other influence such as rainfall or temperature. Others are short-day or long-day organisms. **Short-day organisms** are those whose reproductive or other seasonal activity is stimulated by day-lengths shorter than their critical daylength. **Long-day organisms** are those whose seasonal responses such as flowering and reproduction are stimulated by daylengths longer than the critical daylength. Short-day responses, particularly in plants, can be inhibited by a flash of light in the dark period, their subjective night in the circadian rhythm (Figure 7.11). The

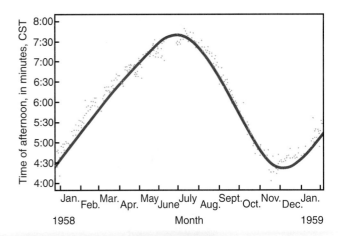

**Figure 7.10** Onset of running wheel activity for one flying squirrel in natural light conditions throughout the year. The graph is the time of local sunset through the year. (From DeCoursey 1960:50.)

plants then respond as if they were experiencing a long day. Horticulturists exploit these short-day and long-day responses of plants to force plants to come into bloom through the year.

Many organisms possess both long-day and short-day responses. Because the same duration of dark and light occurs two times a year, as the days lengthen in spring and shorten in fall, the organisms could get their signals mixed. For them the distinguishing cue is the direction from which the critical daylength is approached. One critical daylength is

**Figure 7.11** The influence of photoperiod on the time of flowering in long-day and short-day plants. If exposure to light is experimentally controlled, short-day plants are stimulated to flower under short-day conditions, are inhibited from flowering under long-day conditions, and respond to an interruption of a long dark period as though they had been exposed to long-day conditions. Long-day plants do not flower under short-day conditions but only under long-day conditions and under interrupted short-day conditions. Animals show a similar response, particularly in reproductive cycles.

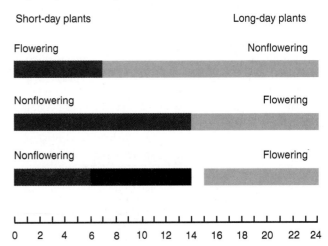

reached as long days move into short and another as short days move into long.

For example, *diapause,* a stage of arrested growth in insects over winter in the temperate regions (Chapter 6), is controlled by photoperiod. The time measurement in such insects is precise, usually falling in a light phase somewhere between 12 and 13 hours. A quarter-hour difference in the light period can determine whether an insect goes into diapause or not (Saunders 1982). Adkisson (1966) found that the cotton bollworm (*Heliothis zea*) failed to enter diapause if the larvae were exposed to a light period of 13.25 hours; its critical day length is 13 hours. Experimentally, the bollworm terminated diapause most rapidly under photoperiods of 14 hours. Thus to the cotton bollworm the shortening days of late summer and fall forecast the coming of winter and call for diapause; and the lengthening days of late winter and early spring are the signals for the insect to resume development, pupate, emerge as an adult, and reproduce.

Gonadal development and spring migratory behavior in birds increase with daylength. This was experimentally demonstrated over 60 years ago when Rowan (1925) forced juncos (*Junco hyemalis*) into the reproductive stage out-of-season by artificial increases in daylength. Since then numerous experimental studies involving many species of birds have demonstrated that the reproductive cycle in birds is controlled by physiological response to daylength (Farner 1959, 1964a, 1964b; Wolfsson 1959). After the breeding season, the gonads of birds regress spontaneously. During

this time light cannot induce gonadal activity. The short days of early fall hasten the termination of this period. Progressively shorter days of winter begin to stimulate the birds again. The lengthening days of early spring then bring the birds into reproductive stage.

In mammals photoperiod influences both activity such as food storage (Muul 1969) and reproduction. Consider, for example, such seasonal breeders as sheep and deer, whose reproductive cycle is initiated in the shortening days of fall. Their pineal gland (associated with the biological clock) produces a hormone, melatonin, which controls the activation of the reproductive cycle. Because more melatonin is produced during the dark cycle of the circadian rhythm, these animals receive a high concentration of melatonin for a longer period of the day as the days lengthen. This increase in melatonin reduces the sensitivity of the hypothalamus to the negative feedback of steroids from the ovaries and testes, allowing the anterior pituitary to release pulses of luteinizing hormone (LH), increasing follicular growth in the ovaries and sperm production in the testes.

In the male deer development of antlers as well as the reproductive cycle relate to photoperiod. The lengthening days of spring stimulate an increase in growth hormones and prolactin. (Figure 7.12). These hormones stimulate the growth of antlers in the spring and early summer. During the shortening days of late summer, growth hormones and prolactin decrease. Under the influence of melatonin, testosterone and sexual resurgence increase. In the presence of

**Figure 7.12** The seasonal course of hormonal levels during the annual cycle of the white-tailed deer and its relationship to antler growth. Note the response of growth hormones and prolactin to lengthening days and the decline in prolactin and the increase in the production of testosterone during the shortening days of fall. (After Goss et al. 1974.)

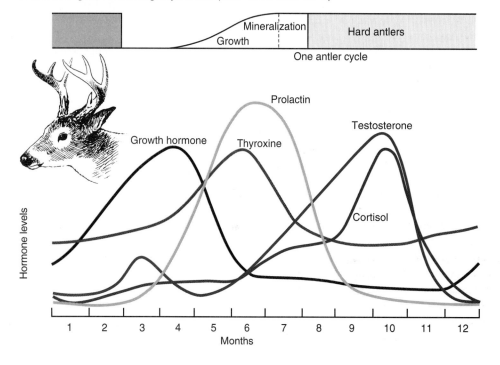

testosterone antler growth ceases, velvet sheds, and antlers harden by the onset of the season of rut. The decline of testosterone in winter results in the shedding of antlers. Normally the deer is in velvet about one-third of the year. When the duration of that year is changed artificially by altering the frequency of daylength, deer replace antlers as often as two, three, or four times a year, or only once every other year, depending on how much the light cycle has been altered. Thus growth, development, and loss of antlers and the reproductive cycle are controlled by the hormones, testosterone and prolactin, whose production is mediated by the photoperiodic responses of the hypothalamus.

# TIDAL AND LUNAR CYCLES

For some organisms tidal and lunar rhythms are of greater ecological importance than light-dark cycles. Animals that inhabit the intertidal zones of the sea show rhythms in their behavior that coincide with cycles of high and low tides. Internal timing processes entrained to tidal cycles have been demonstrated in a number of intertidal animals such as the European shore crab (*Carcinus maenas*) and fiddler crabs (*Uca minax* and *U. crenulata*), held in the laboratory under constant environmental conditions. These persistent tidal rhythms are comparable to circadian rhythms in animals of other environments. They mimic the ebb and flow of tides every 12.4 hours, one-half the lunar day of 24.8 hours (the interval between successive moonrises). (For reviews see Palmer 1990 and 1976, Enright 1975.)

Reproduction in some marine organisms is restricted to a period that bears some relationship to tides. These rhythmic phenomena occur every lunar cycle of 28 days, or in some instances every semilunar cycle of 14 to 15 days. Among these species is the grunion (*Leusethes tenuis*), a small California fish that swarms in from the sea to lay eggs on sandy beaches, and the intertidal midge (*Clunio marinus*). These periodicities are so exact that activities of these animals can be predicted ahead of time. Laboratory studies confirm the entrainment of activity cycles to moonlight (Enright 1975).

# SEASONALITY

Who is not aware of the seasonal changes in plants and animals—the unfolding of leaves in spring and the dropping of leaves in fall, the blooming of flowers and the ripening of seeds, the migration of birds? Biological events recurring with the passage of seasons and influenced by the interaction of light, temperature, and moisture are *seasonality*. The study of the causes of the timing of these events, of biotic and abiotic forces affecting them, and the interrelation among phases of the same or different species is called **phenology** (Leith 1974).

Seasonality in temperate and arctic regions results largely from changes in light and temperature. Seasonality in tropical regions is keyed to rainfall. In a broad way seasonal changes in temperature and light regimes result in alternate warm and cold periods. However, the progression is gradual, and in temperate zones seasons can be identified as early or late spring, early or late fall, and so on. In the tropics the seasons are alternately wet and dry, and their onset is abrupt. The beginning of the rainy season is a dependable environmental cue by which plants and animals become synchronized to seasonal changes. In the tropics the division between the dry season and the wet season is marked by 100 mm in rainfall. The dry season in tropical rain forest has less than 100 mm rainfall per month. In extreme conditions, of course, no rain falls. The wet season has more than 100 mm rainfall per month. The onset of the rainy season, which may last up to six months, varies with the movement of the intertropical convergence. (See Figure 4.7). For this reason the wet season and the dry season are predictable.

Phenological responses reflect altitudinal and latitudinal changes in light, temperature, and moisture. The advance of spring in the temperate regions is marked by progressively later flowering of the same species of trees and herbs across a region, depending upon elevation. Although these progressive changes are pronounced across broad geographical areas (Figure 7.13), distinct variations exist within a given region. These variations reflect local microclimates, which act as selection pressures on local populations, resulting in ecotypic variations in environmental response. In tropical regions flowering and fruiting and leafy growth of plants reflect the alternation of wet and dry seasons (Figure 7.14). The coming of the rainy season is marked by a flush of vegetative growth just as warming spring temperatures trigger leafy growth in temperate regions. Over much of the seasonal tropics flowering and fruiting coincide with the dry season (Janzen 1967). Some species flower at the end of the rainy season, when soil moisture is still high. Other species flower at the end of the dry season.

Seasonal activities of animals center about reproduction and changes in food resources. Appearance and disappearance of many species of insects is usually associated with termination or initiation of diapause. In temperate regions diapause is usually controlled by photoperiod. Insects appear when food is abundant and go into diapause during unfavorable periods of the year. Growth after diapause is influenced by temperature and moisture.

In a similar manner photoperiod influences the seasonality of birds in temperate regions (Figure 7.15). Seasonal phases such as migration, reproduction, and molt usually do not overlap. In winter birds channel energy demands into survival and thermoregulation. Within this pattern some seasonal environmental mechanisms may modify bird activity. Migration, triggered by photoperiod and physiological processes, is highly correlated with weather patterns. Spring migration occurs with the onset of a warm front with an air flow to the north or northeast. Fall migrations to the south appear to be timed to start after the passage of a cold front with its flow of continental polar air.

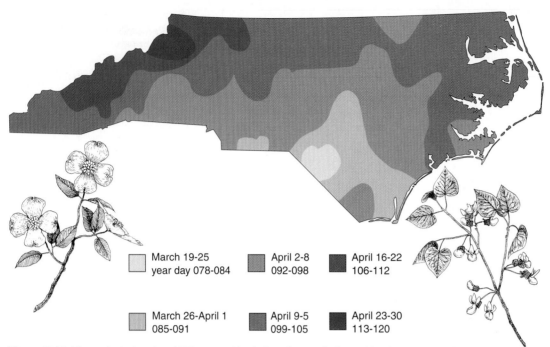

**Figure 7.13** The arrival of spring 1970 across North Carolina, as indicated by the opening of the flowers of dogwood (*Cornus florida*) and redbud (*Cercis canadensis*). (From Leith 1974.)

| | | |
|---|---|---|
| March 19-25 year day 078-084 | April 2-8 092-098 | April 16-22 106-112 |
| March 26-April 1 085-091 | April 9-5 099-105 | April 23-30 113-120 |

Among mammals photoperiod influences reproductive and related behavior and physiological responses (see page 109). The reproductive cycle of the white-tailed deer (Figure 7.16), for example, is initiated in the fall, and the young are born in spring when the highest quality food for lactating mother and young is available. In tropical Central America, the home of numerous species of fruit-eating (frugivorous) bats, the reproductive periods track the seasonal production of food. The birth periods of frugivorous bats coincide with fruiting peaks. Thus young are born when both females and

young will have adequate food. Insects and other arthropods reach their greatest biomass early in the rainy season in the Costa Rican forests (Buskirk and Buskirk 1976). At this time the insectivorous bats give birth to their young.

# SUMMARY

Light has a profound effect on the ecology of almost all individuals, species, and communities. Visible light, the electromagnetic spectrum between the wavelengths of 400 and 740 nm, is known as photosynthetically active radiation (PAR). Short wavelengths between 280 and 380 nm are ultraviolet light; wavelengths longer than 740 nm are infrared. In addition

**Figure 7.14** Synchronization of flowering and fruiting in rain forest tree species in Golfito, Costa Rica, with mean monthly rainfall. Note that flowering and fruiting reach their highest during the dry season, the months of January, February, and March. (Adapted from D. H. Janzen 1967.)

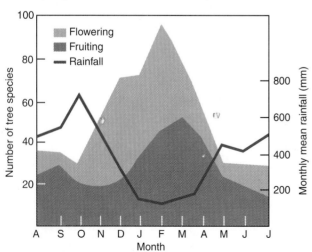

**Figure 7.15** Events in the annual life cycle of a migratory songbird. The strong seasonal periodicity of the events relates to daylength. (From Weise 1974:140.)

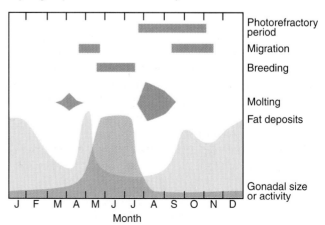

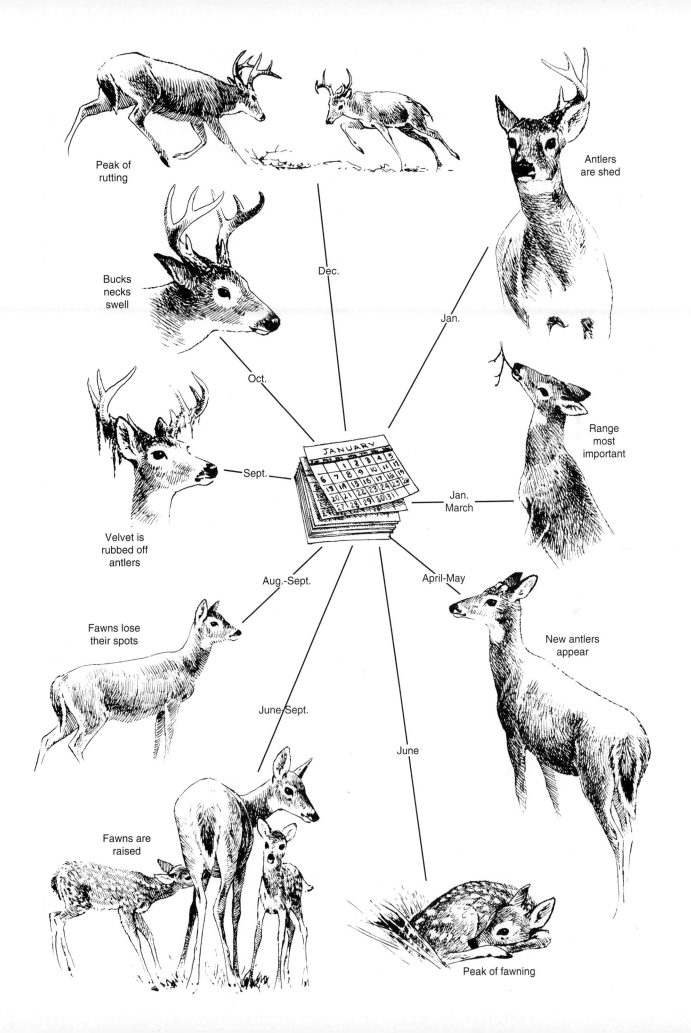

Peak of
rutting

Antlers
are shed

Dec.

Bucks
necks
swell

Jan.

Oct.

Range
most
important

Sept.

Velvet is
rubbed off
antlers

Jan.
March

Aug.-Sept.

April-May

Fawns lose
their spots

New antlers
appear

June-Sept.

June

Fawns are
raised

Peak of fawning

to its spectral qualities, light also possesses intensity, duration, and directionality, all of which vary diurnally and seasonally. Light impinging on an object may be reflected, absorbed, or transmitted through it. Plants reflect green light most strongly and absorb red wavelengths used in photosynthesis. Plants transmit the same wavelengths they reflect and absorb.

Light passing through a canopy of vegetation or through water becomes attenuated. Certain wavelengths drop out before others. On a clear day in a forest green and far red wavelengths pass through relatively unaltered. In pure water red and infrared light are absorbed first, followed by yellow, green, and violet; blue penetrates the deepest. Natural water is rarely clear; attenuation of wavelengths is strongly affected by turbidity, organic stains, and phytoplankton. The measure of the decrease in intensity between the surface and a given depth is the extinction coefficient. That intensity of light at which plants are able to balance photosynthesis with respiration is the light compensation point.

Plants that function best under different light intensities may be classified as sun plants (shade-intolerant) or shade plants (shade-tolerant). Each group is adapted to certain light regimes. Shade-adapted plants and leaves have low photosynthetic, respiratory, metabolic, and growth rates and are resistant to fungal infections. Sun plants and leaves have a high rate of respiration, are adapted to high light intensity, rarely reach light saturation, and are highly susceptible to fungal infections under low light.

Depletion of the ozone layer in the stratosphere by human-introduced pollutants is allowing increased penetration of ultraviolet light in the wavelengths of 250 to 315 nm, known as UV-B. This narrow band of wavelengths can adversely affect the photosynthetic abilities and growth of plants as well as cause skin cancer in animals. Many plants, especially those of the tropics and alpine regions and conifers, possess defenses against ultraviolet light. Flavonoids and phenolics in the epidermal cells of leaves absorb UV-B wavelengths before they reach the interior of the leaf.

Daily periodicities of plants and animals are under the influence of day-night cycles. Daily activities follow circadian rhythms, free-running under constant conditions with an oscillation that has its own inherent frequency. For most organisms the inherent clock deviates slightly from 24 hours. These circadian rhythms are synchronized with the 24-hour cycles of light and dark by a biological clock set by the major external timesetter, light. The onset and cessation of activities of most organisms are usually synchronized with dusk and dawn, the response depending upon whether the organisms are diurnal (light-active) or nocturnal (dark-active).

The biological clock is useful not only to synchronize the daily activities of plants and animals with night and day, but also to time the activities with the seasons of the year. The possession of a self-sustained rhythm with approximately the same frequency as that of environmental rhythms enables organisms to "predict" such advance situations as the coming of spring. It brings plants and animals into a reproductive state at a time of year when the probability for survival of offspring is the highest; it synchronizes within a population such activities as mating and migration, dormancy and flowering. Other periodicities, particularly among marine organisms, are strongly influenced by lunar cycles and by tidal cycles associated with them.

Seasonal changes in light along with accompanying changes in temperature and precipitation bring about seasonal periodicities in recurring biological events, the study of which is phenology. In temperate regions seasonality is influenced by changes in light and temperature and in tropical regions by rainfall. Such periodicities are reflected in leaf growth, flowering, and fruiting of plants and in the breeding cycles, feeding activities, and migration of animals.

## REVIEW QUESTIONS

1. Contrast a shade-tolerant plant with a shade-intolerant plant.
2. Would you expect shade tolerance to be constant for a given species? Why or why not?
3. What is a circadian rhythm, and how does it relate to the 24-hour environmental cycle of light and dark?
4. What is a biological clock, and how is it related to the circadian rhythm?
5. What is the adaptive value of photoperiodism?
6. If UV-B radiation increases, what might the effects be on tropical plants? Temperate plants? Alpine and Arctic plants?
7. Prepare a graph of the waking times of selected local birds, such as robins and cardinals, marked by the first morning songs, against time of local sunrise. (For the dedicated only; this project requires early rising, especially as the spring wears on.) What conclusions can you draw from the graph? How do your daily observations relate to seasonal periodicity?
8. Relate circadian rhythms and biological clocks to jet lag. Compare the change in activities (sleep, meals, and so on) a flight from New York to Frankfurt, from New York to Tokyo, and reverse for each. What happens to timesetting the biological clock?

## CROSS-REFERENCES

Adaptation, 30–31; solar radiation, 36–38; intertropical convergence, 41; thermal energy exchange, 80–82; photosynthesis, 154–159; tides, 329–330.

**Figure 7.16** (opposite) Seasonal cycle of the white-tailed deer. The annual cycle is attuned to the decreasing daylength of fall, during which the breeding season begins, and to the lengthening days of spring, when antler growth begins.

# Nutrients

## Concepts

1. Certain nutrients are essential for the growth and reproduction of all organisms.
2. The ultimate sources of nutrients for all organisms are mineral soil, organic matter, and wet and dry atmospheric deposition.
3. Plant species have evolved different abilities to use available nutrients.
4. Nutrients are retained and recycled within an ecosystem through short-term and long-term storage in organisms and detritus.
5. Nutrient availability influences the nature, abundance, and distribution of plants and animals.
6. The nutritive status of plant consumers depends upon the composition of plants they eat.

Daily we are reminded about nutrient needs. Advertisements ask us: are we getting enough calcium or iron? Is our salt intake too high? Are we giving our pets and domestic animals an adequate diet? We pay lawn-care specialists to assure that our lawns and gardens are adequately fertilized. What about the plants and animals in the wild, whose only source is the environment in which they live? What nutrients do plants and animals need? Where do they get them?

## ESSENTIAL NUTRIENTS

Living organisms require some 30 to 40 chemical elements for growth, development and metabolism (Table 8.1). Some of these elements are needed in large amounts. Known as **macronutrients,** they include, among others, oxygen, hydrogen, nitrogen, calcium, carbon, phosphorus, potassium, magnesium, sulfur, sodium, and chlorine. Others, needed in much lesser quantities, are **micronutrients** or trace elements. They include, among others, copper, zinc, boron, manganese, molybdenum, cobalt, iodine, selenium, and iron. Some are essential to all organisms; others appear to be essential for some, but not for all organisms. If micronutrients are lacking, plants and animals fail as completely as if they lack nitrogen, calcium, or any other major element.

## NUTRIENT SOURCES AND CYCLING

Plants absorb nutrients from soil and water, and in some cases, from the atmosphere. All consumer organisms, from carnivores to the decomposer bacteria and fungi, directly or indirectly depend upon plants.

What is the original source of the nutrients that support life? The basic sources are the weathering of mineral soil, decomposition of organic matter, nitrogen fixation, atmospheric gases, deposition of atmospheric particles, and ocean salt spray. Being stationary, plants must obtain their nutrients from a highly localized area of soil, water, and air. Because most consumers are mobile, they can be more selective, but they still depend upon the success of plants in sequestering nutrients in their tissues. Plants depend upon local nutrient cycles, part of the larger biogeochemical cycles discussed in Chapter 12.

The availability of essential nutrients for terrestrial plants depends heavily on the nature of the soil. Many soil materials are deficient in nutrients upon which plants depend, affecting plants and herbivores. Supplementing nutrients in the soil are nutrients carried by rain, snow, air currents, and animals. Precipitation brings appreciable quantities of nutrients, called **wetfall** (Eaton et. al. 1973, Patterson 1975, Likens et al. 1985). Some of these nutrients, such as tiny dust particles of calcium and sea salt, form the nuclei of raindrops; other are solutions of trace gases. Additional nutrients, such as sulfates and nitrates, wash out of the atmosphere as the rain falls. For some nutrients the amount brought in by airborne particles and aerosols, collectively called *dryfall,* may exceed that carried in by precipitation (Table 8.2). Dust and vapors were found to supply well over half of the input of calcium, potassium, nitrates, and sulfate to the Walker Branch experimental oak forest at Oak Ridge, Tennessee (Lindberg et al. 1986). Seventy to 90 percent of rainfall striking the forest canopy reaches the forest floor. As it drips through the canopy (throughfall) and runs down the stems (stemflow), rainwater picks up and carries with it nutrients deposited as dust on leaves and stems together with nutrients leached from them. Therefore rainfall reaching the forest floor is richer in calcium, sodium, potassium, and other nutrients than rain falling in the open at the same time (Tamm 1951, Madgwick and Ovington 1959, Eaton, Likens, and Bormann 1973, Patterson 1975). Throughfall in an English oak woodland accounted for 17 percent of the nitrogen, 37 percent of the phosphorus, 72 percent of the potassium, and 97 percent of the sodium added by the canopy to the soil; the remainder was added by fallen leaves (Carlisle et al. 1966).

Although stemflow contributes only 5 percent of the total rainfall reaching the forest floor, it is so concentrated about the trunk that for some species the moisture it supplies is five to ten times as great as the rainfall nearby. The amount of stemflow reaching the ground varies with the species. Smooth-barked beech has considerably more stemflow than oaks, whose bark absorbs water (Patterson 1975). In North American forests of balsam fir (*Abies balsamea*), epiphytic lichens remove nitrogen and add calcium and magnesium to stemflow (Lang et al. 1976). Although throughfall provides more nutrients because of its large volume of flow, stemflow provides a more concentrated nutrient solution. Again stemflow concentration is species-specific. Beech and hickories return considerably more calcium and potassium than oaks; and pines return smaller amounts of calcium, magnesium, potassium, and manganese than hardwoods (Patterson 1975).

The nutrients carried to the forest floor by throughfall, stemflow, and other sources are taken up in time by tree roots and translocated to the canopy. How such internal cycling works was demonstrated in a pioneering study by Witherspoon et al. (1962, 1964), using radioisotopes of elements to quantify nutrient cycling in ecosystems. The object was to follow the pathway of a radiolabeled trace element through a forest ecosystem. Cesium behaves like potassium. It is highly mobile, cycles rapidly in an ionic form, and is easily leached from plant surfaces by rainfall. Moreover, because a known quantity of the element could be traced, the amounts of the element apportioned to wood, twigs, and leaves could be determined.

Witherspoon inoculated the trunks (boles) of 12 white oak trees with 20 microcuries of $^{134}Cs$. He followed gains, losses, and transfers of this isotope in the trees and soil. About 40 percent of the $^{134}Cs$ inoculated into the oaks in April moved into the leaves in early June (Figure 8.1). Leaching of radiocesium from the leaves began when the first rains fell after inoculation. By September this loss amounted to 15 percent of the maximum concentration in

**Table 8.1** Some Essential Elements

## MACRONUTRIENTS

| Element | Role |
|---|---|
| Carbon (C) Oxygen (O) Hydrogen (H) | Basic constituents of all organic matter. |
| Nitrogen (N) | Utilized only in fixed form: nitrates, nitrites, ammonium. Component of chlorophyll and enzymes; building block of proteins. |
| Calcium (Ca) | In animals needed for acid–base balance, clotting of blood, and contraction and relaxation of heart muscles; controls movement of fluid through cells; gives rigidity to skeletons of vertebrates; forms shells of mollusks, arthropods, and one-celled Foraminifera. In plants combines with pectin to give rigidity to cell walls; essential to root growth. |
| Phosphorus (P) | Necessary for energy transfer in living organisms; major component of nuclear material of cells. Animals require a proper ratio of Ca:P, usually 2:1 in the presence of vitamin D, wrong ratio in vertebrates causes rickets. Deficiency in plants arrests growth, stunts roots, and delays maturity. |
| Magnesium (Mg) | In living organisms essential for maximum rates of enzymatic reactions involving transfer of phosphates from ATP to ADP. Integral part of chlorophyll and middle lamella of plants; involved in protein synthesis in plants. In animals activates more than 100 enzymes. Deficiency in ruminants causes a serious disease, grass tetany. |
| Potassium (K) | In plants, involved in the formation of sugars and starches. In animals, involved in synthesis of protein, growth, and carbohydrate metabolism. |
| Sulfur (S) | Basic constituent of protein. Plants utilize as much sulfur as they do phosphorus. Excessive sulfur is toxic to plants. |
| Sodium (Na) | Maintenance of acid–base balance, osmotic homeostasis, formation and flow of gastric and intestinal secretions; nerve transmission; lactation; growth and maintenance of body weight. Toxic to plants along roadsides when used to treat highways. |
| Chlorine (Cl) | Enhances electron transfer from $H_2O$ to chlorophyll in plants. Role in animals similar to that of sodium, with which it is associated in salt (NaCl). |

## MICRONUTRIENTS

| Element | Role |
|---|---|
| Iron (Fe) | In plants involved in the production of chlorophyll; part of complex protein compounds that serve as activators and carriers of oxygen and as transporters of electrons in mitochrondia and chloroplasts. In iron-rich respiratory pigment hemoglobin in blood of vertebrates and hemolymph of insects; thus essential for the functioning of every organ and tissue of the animal body. Synthesis into hemoglobin and hemolymph occurs throughout life. Deficiency results in anemia. |
| Manganese (Mn) | In plants enhances electron transfer from $H_2O$ to chlorophyll and activates enzymes in fatty acid synthesis. In animals necessary for reproduction and growth. |
| Boron (B) | Fifteen functions ascribed to boron in plants, including cell division, pollen germination, carbohydrate metabolism, water metabolism, maintenance of conductive tissue, translocation of sugar. Deficiency in plants causes stunted growth in leaves and roots and yellowing leaves. |
| Cobalt (Co) | Required by ruminants for the synthesis of vitamin $B_{12}$ by bacteria in the rumen. Deficiency results in anemia. |
| Copper (Cu) | In plants concentrated in chloroplasts; influences photosynthetic rates; involved in oxidation-reduction reactions; enzyme activor. Excess interferes with phosphorus uptake, depresses iron concentration in leaves, reduces growth. Deficiency in vertebrates causes poor utilization of iron, resulting in anemia and decreased calcification of bones. |
| Molybdenum (Mo) | In free-living nitrogen-fixing bacteria and blue-green algae acts as catalyst for conversion of gaseous nitrogen into usable form. High concentration in ruminants causes "teart" disease characterized by diarrhea, debilitation, and permanent fading of hair color. |
| Zinc (Zn) | In plants needed in the formation of growth substances (auxins); associated with water relations; component of several enzyme systems. In animals functions in several enzyme systems, especially the respiratory enzyme carbonic anhydrase in red blood cells. Deficiency in animals causes a dermatitis, parakeratosis. |
| Iodine (I) | Involved in thyroid metabolism. Deficiency results in goiter, hairlessness, and poor reproduction. |
| Selenium (Se) | In function closely related to vitamin E; prevents white muscle disease in ruminants. Borderline between requirement level and toxicity is narrow. Excess results in loss of hair, sloughing of hooves, liver injury, and death. |

**Table 8.2** Annual Atmospheric Deposition of Major Ions into an Oak (*Quercus*) Forest, Walker Branch, Oak Ridge, Tennessee

| | ATMOSPHERIC DEPOSITION (mEq/m²/yr) | | | | | |
| Process | $SO_4^{2-}$ | $NO_3^-$ | $H^+$ | $NH_4^+$ | $Ca^{2+}$ | $K^+$ |
|---|---|---|---|---|---|---|
| Precipitation | 70 ± 5 | 20 ± 2 | 69 ± 5 | 12 ± 1 | 12 ± 2 | 0.9 ± 0.1 |
| Dry deposition | | | | | | |
|   Fine particles | 7 ± 2 | 0.1 ± 0.02 | 2.0 ± 0.9 | 3.6 ± 1.3 | 1.0 ± 0.2 | 0.1 ± 0.05 |
|   Coarse particles | 19 ± 2 | 8.3 ± 0.8 | 0.5 ± 0.2 | 0.8 ± 0.3 | 30 ± 3 | 1.2 ± 0.2 |
|   Gas | 62 ± 7 | 26 ± 4 | 85 ± 8 | 1.3 | 0 | 0 |
| Total deposition | 160 ± 9 | 54 ± 4 | 160 ± 9 | 18 ± 2 | 43 ± 4 | 2.2 ± 0.3 |

*Source:* From Lindberg et al. 1986.

the leaves. Seventy percent of this rainwater loss reached the mineral soil; the remaining 30 percent found its way into the litter and understory. When the leaves fell in autumn, they carried with them twice as much radiocesium as had leached from the crown. Over the winter, half was leached to the mineral soil. Of the radiocesium in the soil, 92 percent still remained in the upper 10 cm nearly two years after inoculation. Eight percent of the cesium was confined to an area within the crown perimeter, and 19 percent was located in a small area about the trunk. In spring cesium retained over winter in the wood and minimal transfers from the soil and litter moved back into the leaves. This quantified study provided early insights and a general model of internal cycling and retention of elements in forest trees.

A fraction of the nutrient pool in the forest ecosystem is involved in short-term cycling. Nutrients taken up by trees are returned to the forest floor by litterfall, throughfall, and stemflow (Figure 8.2). However, a portion of the nutrient uptake is stored in tree limbs, trunk, bark, and roots as accumulated biomass. This portion is effectively removed from short-term cycling. Some of the nutrients accumulate in the litter and in the living biomass of consumer organisms, including decomposers of the forest floor, from which they are recycled at various rates. Organic matter has a key role in recycling nutrients because it prevents rapid losses from the system. Large quantities of nutrients are bound tightly in organic matter structure; they are not readily available until released by activities of decomposers. However, some nutri-

**Figure 8.1** The cycle of ¹³⁴Cs in white oak, an example of the pathways of nutrients through plants. The figures are for an average of 12 trees at the end of the 1960 growing season. (After Witherspoon et al. 1962; courtesy Oak Ridge National Laboratory.)

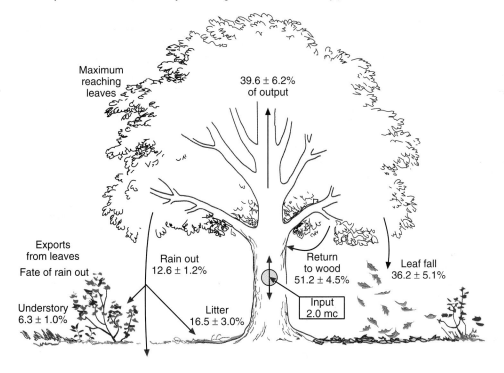

Maximum reaching leaves

39.6 ± 6.2% of output

Exports from leaves

Fate of rain out

Rain out 12.6 ± 1.2%

Return to wood 51.2 ± 4.5%

Leaf fall 36.2 ± 5.1%

Understory 6.3 ± 1.0%

Litter 16.5 ± 3.0%

Input 2.0 mc

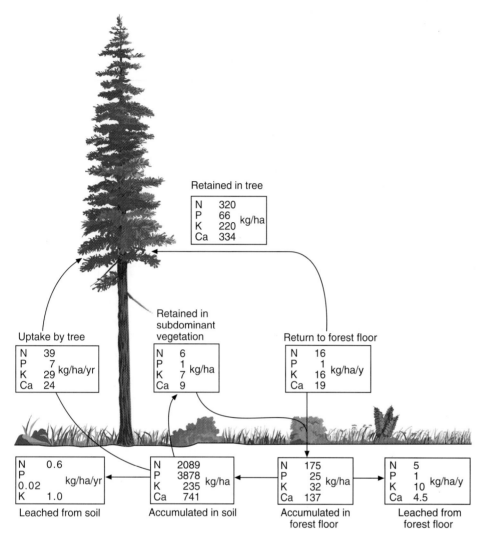

Retained in tree

| N | 320 | |
|---|---|---|
| P | 66 | kg/ha |
| K | 220 | |
| Ca | 334 | |

Uptake by tree

| N | 39 | |
|---|---|---|
| P | 7 | kg/ha/yr |
| K | 29 | |
| Ca | 24 | |

Retained in subdominant vegetation

| N | 6 | |
|---|---|---|
| P | 1 | kg/ha |
| K | 7 | |
| Ca | 9 | |

Return to forest floor

| N | 16 | |
|---|---|---|
| P | 1 | kg/ha/y |
| K | 16 | |
| Ca | 19 | |

Leached from soil

| N | 0.6 | |
|---|---|---|
| P | 0.02 | kg/ha/yr |
| K | 1.0 | |

Accumulated in soil

| N | 2089 | |
|---|---|---|
| P | 3878 | kg/ha |
| K | 235 | |
| Ca | 741 | |

Accumulated in forest floor

| N | 175 | |
|---|---|---|
| P | 25 | kg/ha |
| K | 32 | |
| Ca | 137 | |

Leached from forest floor

| N | 5 | |
|---|---|---|
| P | 1 | kg/ha/y |
| K | 10 | |
| Ca | 4.5 | |

**Figure 8.2** Nutrient cycling of nitrogen, phosphorus, potassium, and calcium in a second-growth Douglas-fir forest. (Based on data from Cole, Gessel, and Dice 1967.)

ents are leached from the soil and carried out of the ecosystem by underground waterflow to streams. These losses may be balanced by the withdrawal of nutrients from deep soil reserves and by the weathering of parent rock material.

Considerable quantities of nutrients are withdrawn permanently from ecosystems by harvesting (Table 8.3), especially in farming and logging. In such ecosystems these losses must be replaced by artificial fertilization; otherwise the ecosystem becomes impoverished.

The major sources of nutrients for aquatic life are inputs from the surrounding land in the form of drainage water, detritus, and sediment, and from precipitation. Availability of

**Table 8.3** Comparison of the Average Annual Yield and Nutrient Removal by a 16-Year-Old Loblolly Pine Plantation with That of Agricultural Crops

| | | REMOVAL | | | |
|---|---|---|---|---|---|
| Crop | Yield | N | P | K | Ca |
| | Tons/ha | Kilograms/ha | | | |
| Loblolly pine, whole tree | 14.5 | 17.5 | 2.4 | 12.6 | 12.8 |
| Loblolly pine, pulpwood | 8.75 | 6.5 | 0.9 | 5.1 | 6.4 |
| Corn (grain) | 11.75 | 130.5 | 29.7 | 37.3 | — |
| Soybeans (beans) | 3.0 | 145.0 | 14.85 | 46.7 | — |
| Alfalfa (forage) | 10.0 | 212.5 | 23.25 | 185.6 | 75.4 |

*Source:* Data from Jorgensen and Wells 1986

nutrients in the open water of lakes and ponds is influenced by the turnover of nutrients in phytoplankton and zooplankton. These aquatic systems lack the long-term biological retention of nutrients typical of forested systems. Major long-term storage takes place in deep bottom sediments, where nutrients may be unavailable for a long time. Nutrients in shallow sediments are most efficiently utilized by rooted aquatics. Flowing water aquatic systems are highly dependent on a steady input of detrital material from the watersheds through which they flow. Retention of nutrients in flowing water ecosystems is difficult, but it is aided by logs and rocks, which hold detritus in place, and by algal uptake of nutrients.

The exchange of nutrients between the abiotic environment and living organisms is depicted in Figure 8.3. Essential to nutrient flow are plants or primary producers and decomposers that draw upon and add to the available nutrient pool and pass nutrients to the consumers. The supply of available nutrients is governed by: (1) inputs of nutrients into the detrital pool by plants (producers) and consumers; (2) the rate at which the decomposers break down the detritus and release the nutrients it contains; (3) the amounts of detritus and available nutrients that go into long-term storage in soil, humus, and sediment; and (4) the release of nutrients from the long-term reserve. The manner and rate at which producers and consumers use nutrients depend upon the size of the abiotic nutrient reserves, the proportion of the total supply of nutrients sequestered in living biomass, and the speed of nutrients' passage from the detrital pool to an available form.

Nutrient cycling in an ecosystem is influenced by the nature of its organisms. Organisms such as phytoplankton and zooplankton in aquatic systems are short-lived, grow rapidly, absorb nutrients quickly, and just as quickly return them to the available nutrient pool. Other organisms, such as forest trees, are large, grow slowly, and store large quantities of nutrients in their biomass for much longer periods.

Each species contributes differently to the overall nutrient cycling within an ecosystem. Trees, for example, sequester varying amounts of elements in short-term and long-term nutrient pools in their structural components—wood, bark, twigs, roots, and leaves. Each species differs in its concentrations of nutrients and its ability to recycle them. These differences were noted by Day and McGinty (1975) in a study of nutrient cycling by several species of trees on the Coweeta Watershed at Franklin, North Carolina (Table 8.4). One was a canopy tree, chestnut oak (*Quercus prinus*); the second, an understory species, flowering dogwood (*Cornus florida*); and the third an understory evergreen, rhododendron (*Rhododendron maximum*).

Chestnut oak had the largest standing crop of nutrients, as we would expect of a large tree. Within the tree, leaves held the most potassium and magnesium, and bark had the largest concentration of calcium. Among the three species the leaves of chestnut oak had the highest standing crop of nitrogen.

The evergreen rhododendron had the largest standing crop of leaf biomass of the three species. Its thick, long-lived leaves held more of calcium and magnesium than its other parts. Because the leaves are evergreen, the nutrients are recycled over a period of seven years instead of the one year typical of the deciduous chestnut oak and flowering dogwood.

Dogwood, a small tree, had the smallest biomass, but it had a high leaf-to-wood ratio and the distinction of possessing the highest concentration of calcium in its leaves. Dogwood concentrated over three times as much calcium per unit leaf biomass as did chestnut oak, and one and a half times as much as rhododendron. The small dogwood recycled 66 percent as much calcium as the chestnut oak and 150 percent more than rhododendron. Thus differences in the nutrient content of various species in an ecosystem, the size of the short-term and long-term nutrient pools, and the rates at which the plants recycle them have a pronounced influence on nutrient cycling.

## NUTRIENTS AND PLANTS

Plants require some 16 essential elements: carbon (C), hydrogen (H), oxygen (O), nitrogen (N), phosphorus (P), sulfur (S), potassium (K), calcium (Ca), magnesium (Mg), iron (Fe), manganese (Mn), zinc (Zn), copper (Cu), molybdenum (Mo), boron (B), and chlorine (Cl). Not all plants need them in the same quantities or in the same ratios, but all do require specific amounts.

Each plant species has a specific ability, not duplicated by other species, to exploit a nutrient supply. Plants growing in the same environment draw on slightly different nutrient sources. Shallow-rooted plants, for example, use the nutrient supply in the upper surface soil, whereas plants with taproots draw on deeper supplies of nutrients. Some species growing on soils poor in nutrients are adapted to low nutrient levels and show little response to increased availability, whereas species growing on highly fertile soils have become adapted to higher levels of nutrition. The effects of nutrient availability, especially nitrogen and phosphorus, on plant

**Figure 8.3** Model of nutrient flow in an ecosystem. Arrows indicate the direction of flow of nutrients between compartments. (Adapted from Webster, Waide, and Patten 1975:13.)

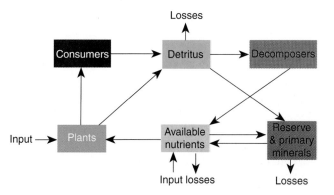

*cations → positive ions*

**Table 8.4** **Mean Nutrient Concentrations in Three Forest Trees, Coweeta Hydrologic Laboratory, Franklin, NC\***

| Species | Bark | Wood | Twigs | Leaves |
|---|---|---|---|---|
| | Potassium, % dry weight | | | |
| Chestnut oak | 0.13 ± 0.04 | 0.18 ± 0.04 | 0.39 ± 0.03 | 1.26 ± 0.06 |
| Flowering Dogwood | 0.34 ± 0.06 | 0.18 ± 0.02 | 0.72 ± 0.04 | 1.44 ± 0.11 |
| Rhododendron | 0.13 ± 0.06 | 0.24 ± 0.04 | 1.55 ± 0.77 | 0.52 ± 0.26 |
| | Calcium, % dry weight | | | |
| Chestnut oak | 1.25 ± 0.17 | 0.09 ± 0.01 | 0.68 ± 0.06 | 0.58 ± 0.07 |
| Flowering Dogwood | 2.36 ± 0.26 | 0.11 ± 0.01 | 0.80 ± 0.06 | 1.85 ± 0.11 |
| Rhododendron | 0.30 ± 0.10 | 0.07 ± 0.31 | 0.99 ± 0.24 | 1.20 ± 0.29 |
| | Magnesium, % dry weight | | | |
| Chestnut Oak | 0.07 ± 0.01 | 0.01 ± 0.001 | 0.08 ± 0.02 | 0.14 ± 0.01 |
| Flowering Dogwood | 0.18 ± 0.03 | 0.04 ± 0.01 | 0.13 ± 0.01 | 0.42 ± 0.03 |
| Rhododendron | 0.02 ± 0.02 | 0.01 ± 0.01 | 0.15 ± 0.05 | 0.19 ± 0.06 |

| Species | Twigs | Leaves | Twigs | Leaves |
|---|---|---|---|---|
| | Nitrogen, % dry weight | | Phosphorus, % dry weight | |
| Chestnut oak | 0.43 ± 0.04 | 2.30 ± 0.06 | 0.15 ± 0.01 | 0.16 ± 0.004 |
| Flowering dogwood | 0.22 ± 0.02 | 2.11 ± 0.06 | 0.10 ± 0.002 | 0.16 ± 0.01 |
| Rhododendron | 1.01 ± 0.14 | 1.25 ± 1.47 | 0.21 ± 0.04 | 0.13 ± 0.02 |

\*Error terms are ± 1 standard error.
*Source:* From Day and McGinty 1975:741.

abundance and distribution are reflected in the competitive abilities of plants.

Nowhere has this competitive ability been better demonstrated than on the long-term grassland fertilizer trial plots at Rothamstead Experimental Farm in England. Among the plots, established in 1856, were some containing natural vegetation that probably resembled the original. The unfertilized plots supported some 60 species of higher plants, including all of the species of plants found on the fertilized plots and species restricted to the natural plots (Thurston 1969). Species diversity on the natural plots was high, and no species was clearly dominant. The vegetation was short and the yield of hay low. On plots that received applications of phosphorus, potassium, sodium, and magnesium, but no nitrogen, legumes such as clover became dominant at the expense of other species. The addition of nitrogen discouraged legumes, reduced their growth, and encouraged grasses. Nutrient addition to grassland enhances the growth of a few highly competitive and lush-growing grasses at the expense of creeping and rosette-type species (Willis 1963). Thus species composition changes under different nutrient regimes (McGraw and Chapin 1989).

## Calcicoles and Calcifuges

Soil acidity affects the nutrient uptake of plants and indirectly affects their distribution. A close relationship exists between available calcium and the degree of soil acidity as measured by pH. pH is based on the logarithms of the concentration of hydrogen and hydroxyl ions in solution. The logarithmic pH scale goes from 0 to 14, with a pH of 7 denoting a neutral solution. A pH greater than 7 denotes an alkaline solution, and a pH of less than 7 an acidic solution. A pH of 5 has 10 times the hydrogen concentration of a solution with a pH of 6; a solution with a pH of 4 has 10 times more hydrogen ions than one of pH 5, and 100 times more than one of pH 6. In the soil hydrogen ions exist in solution and adsorbed to the surface of solid particles of clay and soil organic matter. Basic or positive ions (called cations) of calcium replace the negative hydrogen ions in soil solution and on the soil particles. Therefore soils high in calcium have a pH of around 7, whereas acidic soils have a pH of 6.5 and less.

Many plants grow well in mildly acidic or acid soils; other grow best in high calcium (often called high lime) soils. Plants have been broadly classified as **calcicole** (lime-loving), **calcifuge,** (lime-hating) and **neutrophilus,** tolerant of either condition (Larcher 1980). This relationship, however, is more than a simple one between plants and calcium. Calcium availability is only one of a number of pH-related soil conditions, including the toxicity of iron, aluminum, and other heavy metal ions, and deficiencies of nitrogen, phosphorus, and magnesium. Calcium deficiency and low pH are closely associated only because calcium is the most abundant cation in the soil.

(a)

(b)

**Figure 8.4** (a) Prairie blazing star *(Liatris pycnostachya),* a calcicole plant, grows in the calcium rich soils of the prairie region. (b) Mountain laurel (*Kalmia latifolia*), one of the heaths, is a typical calcifuge.

True calcifuges, such as rhododendrons and azaleas (Figure 8.4), have a low lime requirement and can live in soils with a pH of 4.0 and less. Highly acidic soils associated with calcifuges invariably have a high concentration of aluminum and iron ions, toxic to most plants. Calcifuge plants are tolerant of aluminum and have a high demand for iron. They either possess specific sites within the cytoplasm where aluminum may accumulate harmlessly, chelate aluminum, or precipitate it at the cell surface. Calcifuge plants are especially sensitive to calcium. If grown on calcareous soils, they suffer from lime chlorosis, a disease in which roots and leaves become stunted and the leaves yellowed. This sensitivity may be due in part to the plant's mechanism for chelating aluminum, which has an affinity for iron at higher pH. This deficiency can be corrected by the foliar application of ferrous iron salts. It is the ability to grow in the presence of toxic ions, especially aluminum, that sets calcifuge plants apart from calcicoles.

This difference can create a problem for gardeners where calcifuge plantings of azaleas and rhododendron grow adjacent to lawns of calcicole grasses such as fescue and bluegrass (*Poa pratensis*). Keeping the lawn grasses healthy requires the application of lime, but maintaining the evergreen shrubs, requires an acid fertilizer.

True calcicolous plants, such as alfalfa (*Medicago sativa*), blazing star (*Chamaelirium luteum*), and southern redcedar (*Juniperus silicicola*) are restricted to soils of high pH, not because they have any particular demand for calcium, but because they are susceptible to aluminum toxicity, acidity, and other factors influenced by calcium. Aluminum toxicity begins to appear at a pH of 4.5. Free aluminum accumulates on the surface of the root and in the root cortex. It interacts with phosphorus to form highly insoluble compounds. Calcareous soils tend to be porous and well-drained, with lower moisture and higher temperatures than noncalcareous soils. These conditions restrict certain plants, less competitive on more mesic sites, to calcium-rich soils.

## Plants of Serpentine and Toxic Soils

Heavy metals, such as iron, nickel, chromium, cesium, zinc, and cobalt, are toxic to plants, causing chlorosis and stunted growth. They interfere with the uptake of other nutrients and inhibit root growth and penetration. For these reasons high concentrations of heavy metals can be a powerful force of natural selection, resulting in flora specialized to endure their toxic effects. This effect most evident on mine slag heaps and the chemically distinct serpentine soils. **Serpentine soils** are derived from ultrabasic magnesium silicate rocks, usually greenish in color. These soils are very high in magnesium and iron, high in nickel, chromium, and cobalt, and low in calcium, phosphorus, sodium, and aluminum. Serpentine soils possess a distinctive flora tolerant of those conditions. Plant life consists of rare and endemic species and ecotypes of nonserpentine species that have developed a special tolerance to serpentine soils (Kruckeberg 1954, Franklin and Dyrness 1973). **Endemics** are species restricted to certain specialized habitats; **ecotypes** are ecological races that are well adapted to a local set of conditions. These endemics and ecotypes appear to substitute

magnesium for other bases and to tolerate nickel and chromium in their tissues at levels highly toxic to other plants (Walker 1954).

The transition in vegetation between serpentine and adjacent soils is often sharp, marked by plants stunted in appearance. In Oregon, for example, open, stunted stands of pine replace Douglas-fir; in California chaparral vegetation replaces oak woodland; and in New Zealand tussock grassland replaces southern beech (*Nothofagus*) (Whittaker 1954).

The evolution of tolerance to heavy metals has produced certain species that are able to minimize toxic effects (Figure 8.5). Their roots' uptake mechanism may exclude heavy metals from the plant; or the plant may carry the metals unchanged in its tissues. A few species are site-specific for heavy metals (Antonovics et al. 1971), indicating the presence of those metals in the soil. For example, in England and Europe the pansy (*Viola calaminaria*) and the pennycress (*Thlaspi calaminare*) grow only on zinc carbonate and silicate calamine soils. (Their names derive from the soils on which they grow.). In central Africa a group of copper-tolerant plants, especially *Becium homblei,* are used as indicator plants for the possible location of commercial ore deposits (Howard-Williams 1970).

The industrial age has produced large areas of soils contaminated with heavy metals, including mine tailings, coal spoils, strip-mined lands, and landfills. These areas have favored the rapid evolution within 30 to 100 years of ecotypes

of certain plants, particularly grasses, tolerant of heavy metals (Law et al. 1977, McNeilly 1987). The evolution of such species depends upon three factors: (1) the selection of tolerant seedlings from a normal surrounding population; (2) continued natural selection for metal tolerance and against susceptible genotypes, despite gene flow from the surrounding population; and (3) selection for the ability to survive in physically harsh, largely xeric, nutrient-poor environments. Such selection usually favors a tolerance for a specific heavy metal such as zinc or nickel associated with the site, and not a broad range of them. These ecotypes generally are poor competitors in adjacent normal habitats. Races of tolerant grasses are now being used to revegetate toxic soils (R. A. M. Smith and Bradshaw 1979, Gemmel and Goodman 1980).

## Plants of Saline Habitats

Soils high in sodium chloride (such as solonchaks or alkaline soils of semiarid regions in the western United States and Mexico, Australia, the Middle East, Central Asia, and northern Africa, tidal marshes, coastal dunes, and mangrove swamps) support their own distinctive vegetation of *halophytes* or salt plants. Halophytes grow in soil with more than a 0.2 percent salt content. They must obtain water from soil with a higher osmotic pressure than normal water. Some species have a high concentration of salt in their root cells

**Figure 8.5** Serpentine soil is characterized by a greenish color and a paucity of vegetation. This particular area is in Utah.

and are able to absorb water by osmosis. Some species are succulent; they store water in the cells of leaves and stem and thus dilute the ions they take in. Others excrete excess salt through salt glands in the leaves; the salt is washed away by rain. Still others accumulate ions in tissues away from metabolic sites and shed the leaves and their accumulated salts (Greenways and Murrs 1980).

Few halophytes are obligates, requiring a saline habitat (Barbour 1970). Halophytes are adapted to saline habitats through their ability to handle excess salts. Most halophytes of coastal marshes and swamps, such as saltmarsh grass and mangroves, grown best at low salinity; their growth decreases as salinity increases. Others, such as glassworts (*Salicornia*) (Figure 8.6) grow best at moderate salinity; their growth declines at low and high salinity.

Halophytes are salt-resistant. Within limits, they can maintain homeostasis and carry on metabolic functions in the presence of excess salt. Some nonhalophytes growing along seashores, such as salt-spray rose (*Rosa rugosa*) and beach plum (*Prunus maritima*), also exhibit salt resistance. Salt resistance is an important consideration in the selection of trees and crop plants for growing in saline soils of semi-arid regions. Some species of pines, Douglas-fir, balsam fir, and black cherry (*Prunus serotina*) are particularly vulnerable to direct salt damage, whereas larch (*Larix*), paper birch (*Betula papyrifera*), and aspen are relatively salt-resistant.

## Atmospheric Pollution Injury

Atmospheric pollutants, particularly sulfur dioxide and oxides of nitrogen, have a pronounced adverse effect on the growth of plants and the composition of plant communities. Pollutants are carried by winds and air currents to regions distant from their sources, a problem discussed in more detail in Chapter 12. Under low dosage the vegetation becomes a sink for the contaminants, which may be supplemental nutrients. At intermediate doses tree species or individuals suffer impaired metabolism and nutrient stress. This problem reduces the plants' competitive ability and subjects them to insect attack and disease. High dosages may kill one or many species of plants (MacKenzie and El-Ashry 1989, W. H. Smith 1990)

The San Bernardino National Forest northeast of Los Angeles has been in the path of air masses heavily contaminated with photochemical oxidants (ozone and PAN) for 30 years. These pollutants have caused the decline and death of ponderosa pine (*Pinus ponderosa*), largely through bark beetle infestation of pollution-stressed trees. As the pines die, the forest composition is shifting to more pollution-tolerant white fir and incense cedar (see W. H. Smith 1990).

Sulfur dioxide in high doses can destroy vegetation. There are several well-known examples, but the best documented is the destruction of a mixed boreal forest in the vicinity of an iron smelter in Wana, northern Ontario. In that area all woody vegetation with the exception of some

damaged elder within 8 km of the smelter was destroyed, and no continuous plant cover remained. The forest canopy was eliminated within a zone extending 16 to 19 km away from the smelter. Shrubs interspersed with some scattered white spruce (*Picea glauca*) and paper birch dominated the zone from 19 to 27 km. Beyond 27 km the forest was intact, but many trees exhibited stress (Gordon and Gorham 1963).

Increased atmospheric deposition of ammonium and nitrates, which in limited quantities can have a beneficial effect on nitrogen-limited ecosystems, such as northern and high altitude forest, may be overfertilizing the system. Both trees and microorganisms seem to be receiving more nitrogen than they can handle (Aber et al. 1989). This increase results in a continuing growth of foliage into the summer. The new growth does not have time to become frost-hardened and so is killed back in winter. Increased growth depletes other nutrients, especially phosphorus, causing phosphorus deficiency. No longer stimulated to seek new nutrient sources, roots slow their growth and root biomass decreases, impairing the ability of trees to absorb water during dry periods. Ultimately the trees die.

## NUTRIENTS AND CONSUMERS

Consumers, notably the higher animals and arthropods, require mineral elements and 20 amino acids, of which 14 are essential. Amino acids make up proteins and vitamins. These needs differ little among vertebrates and invertebrates. Insects, for example, have the same dietary requirements as vertebrates, although they need more potassium, phosphorus, and magnesium than vertebrates and less calcium, sodium, and chlorine (Dodd 1973). Directly or indirectly, the source of these essential nutrients is plants. For this reason, the quantity and quality of plants affect the nutrition of consumers. When food is of limited quantity, consumers may suffer from acute malnutrition, leave the area, or starve. When food is of low quality, it affects reproduction, health, and longevity.

The problem facing consumers is the conversion of plant tissue to consumer tissue. Plants and their consumers have different chemical compositions. Animals are high in fat and proteins, which they use as structural building blocks. Plants are low in proteins and high in carbohydrates, much of it in the form of cellulose and lignins in cell walls. Nitrogen is major constituent of protein. In plants the carbon to nitrogen ratio is 40 to 1. In mammals the ratio is about 14 to 1. The plant feeders or herbivores face the task of converting cellulose and a limited supply of plant protein into animal tissue. Few herbivores have the enzymes necessary to break down cellulose and lignin; most of them require help from symbiotic bacterial flora to digest plant tissue.

The highest quality plant food for herbivores, vertebrate and invertebrate, is proteinous nitrogen (Mattson 1980,

**Figure 8.6** *Salicornia,* a succulent annual or perennial, depending upon the species, is a true halophyte, able to exist only in saline situations.

Strong et al. 1984). As the nitrogen content of their food increases, assimilation of plant material improves, increasing growth, reproductive success, and survival. Nitrogen is concentrated in the growing tips, new leaves, and buds of plants. Its content declines as leaves and twigs mature and become senescent. Herbivores have adapted to this period of new growth. Herbivorous insect larvae are most abundant early in the growing season and complete their growth before the leaves mature. Vertebrate herbivores give birth to their young at the start of the growing season, when the most protein-rich plant foods will be available for their growing young.

Although food selection is strongly influenced by availability and season, herbivores do show some preference for the most nitrogen-rich plants, which they probably detect by taste and odor (Verme and Ullrey 1984). Chemoreceptors in the nose and mouth of deer encourage or discourage consumption of certain foods. During drought, nitrogenous compounds are concentrated in certain plants, making them more attractive and vulnerable to herbivorous insects (Mattson and Haack 1987).

The need for quality foods differs among herbivores. Ruminant animals, for example, can subsist on rougher or lower quality forage because bacteria in the rumen can synthesize such requirements as vitamin $B_1$ and certain amino acids from simple nitrogenous compounds. Therefore the caloric content and the nutrients in a certain food might not reflect its real nutritive value for the ruminant. Nonruminant herbivores require more complex proteins and may carry bacterial symbionts in the cecum. Seed-eating herbivores exploit the concentration of nutrients in the seeds. Such animals are not likely to have dietary problems.

Among the carnivores, quantity is more important than quality. Carnivores rarely have a dietary problem because they consume animals that have resynthesized and stored protein and other nutrients from plants in their tissues. Carnivores are simply converting animal tissue with the same chemical composition as their own into carnivore flesh.

Mineral availability and deficiencies appear to influence the abundance and relative fitness of some animals. One essential nutrient that has received attention is sodium, the most variable nutrient in forest and arctic ecosystems (Jordan et al. 1972). In areas of sodium deficiency in the soil, herbivorous animals face an inadequate supply of sodium in their diets. The problem has been noted in Australian herbivores (Blair-West et al. 1968), in African elephants (*Loxodonta africana*) (Weir 1972), in rodents (Aumann and Emlen 1965, Weeks and Kirkpatrick 1978), in white-tailed deer (Weeks and Kirkpatrick 1976), and in moose (*Alces alces*)(Belovsky 1981b, Belovsky and Jordan 1981).

Sodium deficiency can influence the distribution, behavior, and physiology of mammals, especially the herbivores. The spatial distribution of elephants on the Wankie National Park in central Africa appears to be closely correlated with the sodium content of drinking water, with the greatest number of elephants found at waterholes with the highest sodium content. Three herbivorous mammals, the European rabbit (*Oryctolagus cuniculus*), the moose, and the white-tailed deer in parts of their range, experience sodium deficiencies. In sodium-deficient areas in southwestern Australia, the European rabbit builds up reserves of sodium in its tissues during the nonbreeding season. These reserves appear to be exhausted near the end of the breeding season, which ends abruptly. During the breeding season, the rabbits

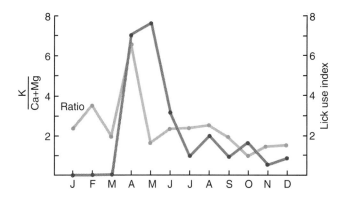

**Figure 8.7** Monthly ratio of K to Ca + Mg in the diet compared to an index of the intensity of use of mineral licks by white-tailed deer in Indiana. Use of mineral licks by deer was most intense when the ratio of K to Ca + Mg in the diet was the lowest in the spring and the mineral needs the greatest. (From Jones and Weeks 1985:132.)

selectively graze on sodium-rich plants to the point of depletion, exhausting sodium availability.

Ruminants face severe problems of mineral deficiencies in spring. Attracted by the flush of new growth, deer, bighorn sheep (*Ovis canadensis*), mountain goats (*Oreamnos americanus*), elk (*Cervus elaphus*), and domestic cattle and sheep feed on new succulent grass, but with high physiological costs. Spring vegetation is much higher in potassium relative to calcium and magnesium than during the rest of the year (Figure 8.7). This high intake of potassium stimulates the adrenal gland to increase the excretion of aldosterone (Weeks and Kirkpatrick 1976). Aldosterone is the principle mineral corticoid that promotes retention of sodium by the kidney. While aldosterone stimulates the re-

tention of sodium, it also facilitates the excretion of potassium and magnesium. Because concentrations of magnesium in soft tissues and skeletal stores are low in herbivores, these animals experience magnesium deficiency. This deficiency results in a rapid onset of diarrhea and often a neuromuscular derangement or tetany. The deficiency comes late in gestation for females and at the beginning of antler growth for male deer and elk, a time when mineral demands are high.

To counteract this mineral imbalance in the spring, the large herbivores seek mineral licks, places in the landscape where animals concentrate to satisfy their mineral needs by eating mineral-rich soil (Figure 8.8). Although sodium chloride is associated with mineral licks, animal physiologists hypothesize that it is not sodium the animals seek but magnesium, and in the case of bighorn sheep, mountain goats, and elk, calcium as well (R. L. Jones and Hanson 1985).

Moose of the northern forest ecosystem face a particular problem obtaining sufficient sodium (Belovsky and Jordan 1981). Northern terrestrial vegetation contains only 3 to 28 ppm of sodium, far below the nutritional requirements of ruminants. To circumvent sodium stress, moose feed in summer on submerged aquatic plants, which contain between 2 and 9400 ppm. Moose may consume from 63 to 95 percent of the production of submerged aquatics, particularly pond lilies. Such heavy consumption appears to trigger declines in aquatic vegetation. The resulting decreased availability of sodium may be one reason for the periodic decline of moose populations in parts of their range (Belovsky 1981b).

The size of deer, their antler development, and their reproductive success all relate to nutrition (Moen and Severinghaus 1983). Other factors being equal, only deer obtaining high quality foods grow large antlers. Deer on diets low in calcium, phosphorus, and protein show stunted

**Figure 8.8** A mineral lick used by white-tailed deer.

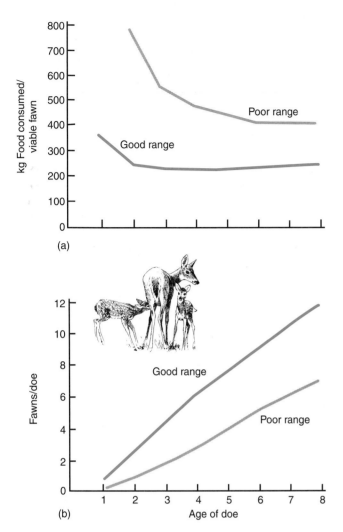

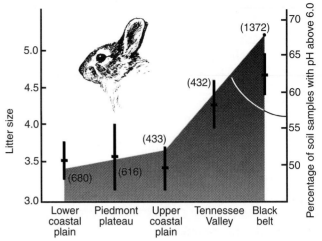

**Figure 8.10** The mean size of second cottontail litters from five Alabama soil regions plotted against the percentage of soil samples above 6.0 pH, indicative of high fertility soils. Vertical lines give the range of two standard deviations. The numbers in parentheses are the numbers of soil samples analyzed for each soil region. Note the strong relationship between soil fertility and litter size. (From E. P. Hill 1972:1201.)

**Figure 8.9** Differences in the reproductive success of female white-tailed deer on good and poor ranges in New York State. (a) Food consumed per viable fawn on poor range (Adirondack Mountains) was much greater than food consumed on good range (Western New York). (b) Reproductive success was considerably greater on good range than poor range. (Data from Cheatum and Severinghaus 1950.)

growth, and bucks develop only thin spiky antlers (French et al. 1955). Reproductive success of does is highest where food is abundant and nutritious. On the best range in New York State, 1.71 fawns on the average were born for each reproductive doe (Figure 8.9). On a poor range, however, average fawn production per doe was only 1.06 (Cheatum and Severinghaus 1950).

Other studies show a more general relationship between the level of soil fertility and the health of mammals. Studies of the relationship between soil and cottontail rabbits in Missouri showed no significant difference in body weight of rabbits collected from areas with soils of contrasting fertility (Williams 1965), but there was a positive correlation between soil fertility and fecundity both in Missouri (Williams and Cashey 1965) and in Alabama (Hill 1972) (Figure 8.10).

# SUMMARY

All living organisms require certain nutritive elements essential to reproduction, growth, survival, and distribution. Organisms vary in their requirements and tolerances for different elements. The macronutrients, including nitrogen, calcium, phosphorus, potassium, and sodium, are required in relatively large quantities by all living organisms. Others, the trace elements, such as copper, zinc, and iron, are needed in lesser and often minute quantities; yet without them organisms will die or become impaired as if they lacked one of the major nutrients.

The basic nutrient sources for life are the mineral soil, organic matter, and wet and dry atmospheric depositions. Nutrients are retained and recycled through the ecosystem through the processes of deposition, leaching, uptake by plants, consumption, decomposition, and short-term and long-term storage in living biomass, dead organic matter, and soil.

Nutrients are constantly added, stored, and removed from ecosystems. A measure of this inflow and outflow makes up a nutrient budget. The rate at which nutrients are added, stored, and recycled over short-term and long-term periods influences the nature of the ecosystem.

Each species has specific abilities or requirements to exploit a nutrient supply in its own manner. This fact is often reflected in the competitive abilities of different plants. Some plants grow well on acid soil, are tolerant of aluminum, and are sensitive to calcium. These plants are known as calcifuges. Other plants are susceptible to aluminum toxicity

and do best in soils high in calcium. They are calcicoles. A few plants have evolved tolerances to toxic heavy metals such as copper, nickel, and zinc. A special group of plants, the halophytes, is tolerant of high salinities. Having adverse effects on plants are atmospheric pollutants such as sulfur dioxide.

The source of nutrients for animals, directly or indirectly, is plants. Low concentrations of nutrients in plants can have adverse effects on the growth, development, and fitness of herbivores. Three essential nutrients that influence the distribution, behavior, and fitness of grazing herbivores are sodium, calcium, and magnesium. Among herbivores the quality of food, especially its protein content and digestibility, is of critical importance. Among carnivores quantity is more important than quality. They consume animals that have already converted plant nutrients to animal flesh.

## REVIEW QUESTIONS

1. Distinguish between a macronutrient and a micronutrient.
2. What is the nutritional importance of the following nutrients for plants and animals: calcium, magnesium, sodium, iron, and boron?
3. What are the basic sources of nutrients for living organisms? How do they become available to plants? To animals?
4. What is dryfall? Wetfall? Throughfall? Stemflow? From a nutrient standpoint, what is the significance of each to plants?
5. Describe the short-term cycling of nutrients in a forest.
6. In what ways are nutrients retained and cycled from long-term and short-term reserves within an ecosystem?
7. Consider the nutrient reserves in the forest illustrated in Figure 8.2. What percentage of nutrients would be removed if the tree were cut and removed? What might be the fate of the nutrients accumulated in the forest floor? Why?
8. Distinguish between calcicoles and calcifuges. What are the outstanding characteristics of each?
9. What is serpentine soil? What are the effects of heavy metals on plant tolerance and distribution?
10. What are halophytes? Where would you find them?
11. How does the nutrient level in plants affect the well-being of animals?
12. During the late 1930s and early 1940s, when the United States was trying to overcome the problems of the dust bowl and severe soil erosion in the South, soil conservationists came up with the slogan, "Poor soils make poor people." They were referring to the quality and quantity of crops grown on poor soil. Relate this slogan to human nutrition and food problems facing developing countries. How do the concepts in this chapter apply to the growing interest in organic farming?

## CROSS-REFERENCES

Adaptation, 30–31; soil chemistry, 133–134; decomposition, 159–165; carbon cycle, 199, 206–210; nitrogen cycle, 210–212; sulfur cycle, 210–212; phosphorus cycle, 212–213; acid deposition, 214–216; Ecosystem functions: grassland, 233–236; shrubland, 245–246; desert, 248–250; forest, 268–272, 277–280, 285–287; tundra, 258–261; lakes and ponds, 297–298; wetlands, 321–328, 354–355; flowing water, 305–309; marine, 334–336; mycorrhizae, 528, 584.

# Soils

## Outline

## Concepts

1. Soil is the thin mantle of weathered mineral and organic matter that supports terrestrial life.
2. Chemical and physical weathering produce a profile with distinctive soil horizons.
3. Soil has physical properties, including texture and moisture, and chemical properties that influence the life that soil supports.
4. Soil development is strongly influenced by the nature of the organic matter deposited on its surface and by the soil organisms it supports.
5. Major processes in soil formation are podzolization, laterization, calcification, salinization, and gleyization.
6. Each combination of climate, vegetation, soil material, and time results in unique soils that can be broadly classified into 11 major soil orders.

Soil is the foundation of terrestrial communities. It is the major site of decomposition of organic matter and the return of mineral elements to the nutrient cycle (see Chapter 10). From it plants obtain the minerals and water they need for photosynthesis. Vegetation, in turn, influences the development of soil, its chemical and physical properties, and its organic content. Controlling much of the decomposition of organic matter and soil-forming processes are an abundance and diversity of soil organisms, including bacteria, fungi, and soil invertebrates. Thus soil acts as a pathway between the organic and mineral worlds.

# DEFINITIONS OF SOIL

As familiar as it is, soil is difficult to define. Indeed, one eminent soil scientist, a pioneer of modern soil studies, Hans Jenny (1980), will not give an exact definition of soil. In his book *The Soil Resource* (1980:364), he writes:

> Popularly, soil is the stratum below the vegetation and above hard rock, but questions come quickly to mind. Many soils are bare of plants, temporarily or permanently; or they may be at the bottom of a pond growing cattails. Soil may be shallow or deep, but how deep? Soil may be stony, but surveyors (soil) exclude the larger stones. Most analyses pertain to fine earth only. Some pretend that soil in a flower pot is not soil, but soil material. It is embarrassing not to be able to agree on what soil is. In this pedologists are not alone. Biologists cannot agree on the definition of life and philosophers on philosophy.

However, soils have been variously defined. One definition is that soil is a natural product formed from weathered rock by the action of climate and living organisms. Another common definition states that soil is a collection of natural bodies of Earth that is composed of mineral and organic matter and is capable of supporting plant growth. The term "natural bodies" emphasizes that soils are different. Further, if soils are considered natural bodies, the concept becomes the basis for soil classification.

In a more extensive definition, soil is the weathered outer layer of Earth's crust, which ranges from a thin film to thick layers, composed of weathered rock materials and organic matter interspersed with pores filled with air and water, and which supports plant life. This layer of soil differs from the material beneath it in color, structure, other physical characteristics, and in chemical composition. Soil, moreover, is not just an abiotic environment supporting plants. It is teeming with life—billions of minute animals, bacteria, and fungi. The interactions between the abiotic and biotic make soil a living system.

Soil is also recognized as a three-dimensional unit or body, possessing length, width, and depth. A three-dimensional soil body large enough that we can study all its physical and chemical properties and its horizons is called a **pedon.** Pedons, the basic unit in the study of soils, range in area from one to 100 square meters.

# THE SOIL PROFILE

Soil, Hans Jenny (1980:6) wrote, "is a body of nature that has its own internal organization and history of genesis." Let us begin with its internal organization. A fresh cut along a roadbank or an excavation tells something about a soil. A close-up and even cursory look reveals bands and blotches of color from the surface downward. Closer examination, even handling the material, reveals changes in texture and structure. Any vertical cut through a body of soil or pedon is the **soil profile.** The apparent layers are called the **horizons.** Each horizon has a characteristic set of features, particularly color, that distinguishes it from other horizons. Each horizon has its own thickness, texture, structure, consistency, porosity, chemistry, and composition.

In general soils have five major horizons: *O,* an organic layer, and *A, E, B,* and *C,* the mineral layers. Below the four may lie the *R* or nonsoil horizon (Figure 9.1). In some soils the horizons are quite distinct. In other soils the horizons form a continuum with no clear-cut distinction between one horizon and another.

The *O horizon* is the surface layer, formed or forming above the mineral layer and composed of fresh or partially decomposed organic material that has not been mixed into mineral soil. It is usually absent in cultivated soils. This layer and the upper part of the next horizon, *A,* constitute the zone of maximum biological activity. Both are subject to greatest changes in soil temperatures and moisture conditions, contain the most organic carbon, and are the sites where most or all decomposition takes place. The *O* horizon is further subdivided into an *Oi* or litter layer and *Oa* or humus layer. The *Oi* layer fluctuates seasonally. In temperate regions it is thickest in the fall, when new litter is added, and thinnest in the summer after decomposition has taken place.

The *A* **horizon** (once called $A_1$) is the upper layer of mineral soil with a high content of organic matter. It characterized by an accumulation of organic matter and by the loss of some clay, inorganic minerals, and soluble matter. The *E horizon* (once labeled $A_2$) is the zone of maximum leaching (eluviation, thus the label *E*). Its chemistry and structure have been altered by the downward movement of suspended and dissolved material through weathering and leaching.

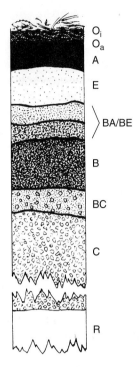

**Figure 9.1** A generalized profile of the soil. Rarely does any one soil possess all of the horizons shown. *Oi*, loose leaves and organic debris. *Oa*, organic debris partially decomposed (fermented) or matted. *A*, dark-colored horizon with high content of organic matter mixed with mineral matter. *E*, horizon of maximum leaching. Prominent in spodosols, it may be faintly developed in other soils. *B*, zone of maximum accumulation of clay minerals or of iron and organic matter. Subhorizons include *BA* or *BE*, transitional to *B*, but more like *B* than *A*. *BC* or *CB*, transitional to *C*. *C*, the weathered material, either like or unlike the material from which the soil presumably formed. A gley layer may occur, as well as layers of calcium carbonate, especially in grassland. *R*, consolidated bedrock.

The *E* horizon is further characterized by the development of granular, platelike, or crumblike structure. Some soils possess both an *A* and an *E* horizon. Other soils lack one or the other, depending on the nature of weathering and the incorporation of organic matter with mineral soil.

The **B horizon** is the zone of illuviation, collection of leached material. It accumulates silicates, clay, iron, aluminum, and humus, from the *E* horizon. It develops a characteristic physical structure involving blocky, columnar, or prismatic shapes (see page 133). Below the *B* horizon in some soils may be compact, slowly permeable layers, clay pans and fragipans. A *clay pan* possesses much more clay than the horizon above it. It is very hard when dry and stiff when wet. A **fragipan** is a brittle, seemingly cemented, subsurface horizon low in organic matter and clay but high in silt or very fine sand. When dry a fragipan is very hard, and

when wet it tends to rupture suddenly if pressure is applied. Both clay pans and fragipans interfere with root and water penetration in the soil.

The **C horizon** contains weathered material, either like or unlike the material from which the soil is presumed to have developed. Some active weathering takes place in this horizon, but it is little affected by soil formation. Below the *C* horizon is unweathered material, the *R horizon*.

## PROPERTIES OF SOIL

### Physical Properties

Soils are distinguished by differences in their physical and chemical properties. Physical properties include color, texture, structure, depth, and moisture. All are highly variable from one soil to another.

Color    Color has little direct influence on the function of a soil, but considered with other properties, it can tell a good deal about the soil. Color is one of the most useful and important characteristics for the identification of soil (see page 145). In temperate regions, brownish-black and dark brown colors, especially in the *A* horizon, generally indicate considerable organic matter. The *B* horizon of well-drained soils may range anywhere from very pale brown to reddish and yellowish color; dark brown and blackish colors, especially in the lower horizons, indicate poor drainage. However, it does not always follow that dark-colored soils are high in organic matter. Soils of volcanic origin, for example, are dark in color acquired from their parent material of basaltic rocks. In warm temperate and tropical regions, dark clays may have less than 3 percent organic matter.

Red and yellow soils derive their colors from the presence of iron oxides, the bright colors indicating good drainage and good aeration. Red and yellow colors increase from cool regions to the equator. Other red soils obtain their color from parent material such as red lava rock and not from soil-forming processes. Well-drained yellowish sands are white sands containing a small amount of organic matter and such coloring material as iron oxide. Quartz, kaolin, carbonates of calcium and magnesium, gypsum, and various compounds of ferrous iron give whitish and grayish colors to the soils. Grayish colors indicate permanently saturated soils in which iron is in the ferrous form. Imperfectly and poorly drained soils are mottled with blotches of various shades of yellow-brown and gray. The colors of soils are determined by the use of standardized color charts, notably the Munsell color charts.

**Texture** The texture of a soil is determined by the proportion of different-sized soil particles (Figure 9.2). Texture is partly inherited from parent material and partly a result of the soil-forming process.

Particles are classified on the basis of size into gravel, sand, silt, and clay. Gravel consists of particles larger than 2.0 mm and is not considered part of the fine fraction of soil. Sand ranges from 0.05 to 2.0 mm, is easily seen, and feels gritty. Silt consists of particles from 0.002 to 0.05 mm in diameter, which can scarcely be seen by the naked eye and feel and look like flour. Clay particles, less than 0.002 mm, are too small to be seen under an ordinary microscope. Clay controls the most important properties of soils, including plasticity—the ability to change shape when pressure is applied and retain that shape when the pressure is removed—and the exchange of ions between soil particles and soil solution. The nature of a soil's texture is the percentage (by weight) of sand, silt, and clay. Based on these proportions soils are divided into textural classes (Figure 9.2).

Texture relates to pore space in the soil and therefore plays a major role in the movement of air and water in the soil, penetration by roots, and water storage capacity. In an ideal soil, particles make up 50 percent of the volume; the other 50 percent is pore space. Pore space includes spaces within and between soil particles, as well as old root channels and animal burrows. Coarse-textured soils possess large pore spaces that favor rapid water infiltration and rapid drainage. Up to a point, the finer the texture, the smaller the pores, and the greater the available active surface for water adherence and chemical activity. Very fine-textured soils, such as heavy clays, easily become compacted if plowed, stirred, or walked upon. They are poorly aerated and difficult for roots to penetrate. The condition can be alleviated by the addition of organic matter to break up clods and give a more loamy texture, the best for plant growth.

Soil particles are held together in clusters or shapes of various sizes, called *aggregates* or **peds.** The arrangement of these aggregates is called **soil structure.** As with texture, there are many types of soil structure. Soil aggregates are classified as granular, crumblike, platelike, blocky, subangular, prismatic, and columnar (Figure 9.3). Structure is influenced by texture, plants growing on the soil, other soil organisms, and the soil's chemical status.

**Depth** The layer of soil upon which life depends is not spread evenly across the landscape. Its depth varies from

**Figure 9.2** A soil texture chart showing the percentages of clay (below 0.002 mm), silt (0.002 to 0.05 mm), and sand (0.05 to 2.0 mm) in the basic soil textural classes. For example, a soil with 60 percent sand, 30 percent silt, and 10 percent clay would be considered a sandy loam.

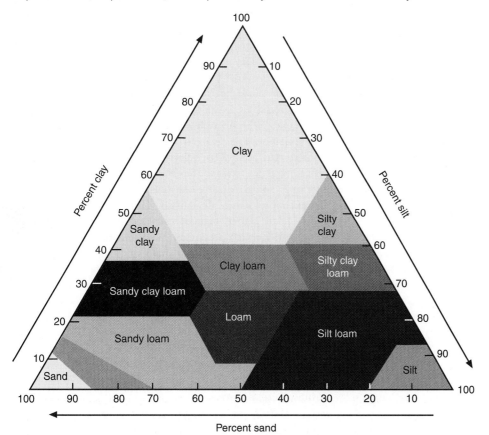

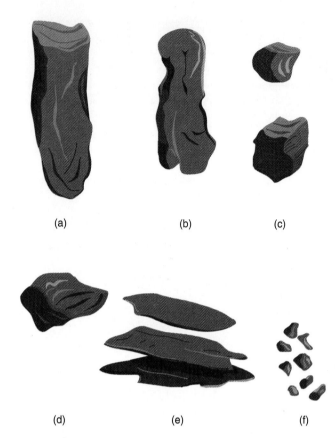

**Figure 9.3** Some types of soil structure. (a) Prismatic, (b) columnar, (c) angular blocky, (d) subangular blocky, (e) platelike, (f) granular.

place to place depending on slope, weathering, parent material, and vegetation. Soils developed under native grassland tend to be several meters deep, whereas soils developed under forests are relatively shallow with an *A* horizon of about 15 centimeters and a *B* horizon of about 60 centimeters. Soils at the bottom of slopes, on level ground, and on alluvial plains tend to be deep, whereas soils on ridgetops and steep slopes tend to be shallow, with bedrock close to the surface. Natural fertility and water-holding capacity of such soils are low.

Moisture  Dig into the surface layer of a soil about a day after a soaking rain and note the depth of water penetration. Unless the soil is a heavy clay, you should discover that the transition between wet surface soil and dry soil is sharp. The water that fell on the ground infiltrated the soil, filling the pore spaces and draining into the dry soil below. Depending upon the amount of water, the downward flow or percolation halts within two to three days and the water hangs in the soil capillaries (Chapter 5).

The maximum amount of water the soil will hold following the drainage of gravitational water is called **field capacity (FC).** It represents the moisture conditions in each individual soil after the large pore spaces have drained fully. A

more precise definition is the amount of water retained under negative (suction) pressure of approximately 1 MPa to 3.5 MPa for most soils. Field capacity is influenced heavily by soil texture, and to a lesser extent by clay minerals, stoniness, and soil structure. Sand has 30 to 40 percent of its volume in pore space, and clays and loams 40 to 60 percent.

Field capacity, however, does not represent the water available to plants. That is **available water capacity (AWC),** the water retained between field capacity and the **permanent wilting point (PWP).** It represents the soil's renewable storage volume. The permanent wilting point occurs when the soil dries out to a point at which plants are unable to absorb enough water from the soil to replace that lost by transpiration.

The ability of a soil to retain moisture against drainage determines soil moisture regimes. Texture plays an important role. Heavy clays and clay loams retain more moisture than light sandy soils. AWC is lowest in coarse-textured soils and maximum in medium-textured soils. In fine-textured soil, films of water are held on soil particles at increasing high tensions and become less available to plants. The amount of organic matter in the soil also adds to a soil's water holding capacity, as does stoniness.

The topographic position of a soil affects the movement of water both on and in the soil. Water tends to drain downslope, leaving soils on higher slopes and ridgetops relatively dry, and creating a moisture gradient from ridgetops to streams. However, after a dry period more moisture may be stored from rain on the upper slope than on the more moist lower slope.

In all there are five drainage classes. (1) *Well-drained* soils are those in which plant roots can grow to a depth of 90 cm without restriction due to excess water. (2) On *moderately well-drained* soils plant roots can grow to a depth of 50 cm without restriction. (3) *Somewhat poorly drained* soils restrict the growth of plant roots beyond a depth of 36 cm. (4) *Poorly drained* soils are wet most of the time and are usually characterized by alders, willows, and sedges. (5) On *very poorly drained* soils water stands on or near the surface most of the year.

The delineation of drainage classes of soils is important ecologically and economically. Somewhat poorly drained soils mark the beginning of wetland ecosystems. Historically, poorly drained soils were (and many still are) "improved" for agricultural and other uses by ditching and installing drainage tiles. Such drainage has been slowed by (often controversial) wetland laws. Drainage classes are also important considerations for the selection of construction and highway sites.

## Chemical Properties

Soils also have different chemical properties. These properties are influenced by parent material, vegetation, moisture, clay organic matter content, and other factors. Chemical

elements in the soil are dissolved in solution, are a constituent of organic matter, and are adsorbed on particles. These ions move from soil to plant, from plant to animals, and into the biogeochemical cycle (see Chapter 12). In soils ions are limited in their mobility because they are closely held to particles of clay and humus.

Because of its unique structure, clay controls many of the important properties of soil. The basic clay mineral consists of three elements: aluminum ($Al^{3+}$), silica ($Si^{4+}$), and oxygen ($O^{2-}$). The base of the clay mineral is silica, which attracts four oxygens to form a tetrahedron (Figure 9.4). These tetrahedrons are bound together by oxygen to form unit cells, $(Si_2O_5)_n$. These cells repeat in an orderly fashion to form lengthy sheets. The other element, aluminum, reacts with oxygen in a similar manner to form sheets of octahedrons. The silica sheets bond to the aluminum sheets in layers to form platelike particles known in colloidal chemistry as **micelles** (Figure 9.5). Some clay particles consist of one sheet of silica and one sheet of aluminum held together by hydrogen bonds to form 1 : 1 clays (kandites). An example of such a clay is kaolinite, $Al_4Si_4O_{10}(OH)_8$. (Figure 9.5a). Others bond two silica layers to one aluminum layer to form 2:1 clays (smectites), an example of which is the common clay mineral, montmorillonite (Figure 9.5b). A general formula for smectites is $Al_4Si_8O_{20}(OH)_4nH_2O$. A third type of clay mineral has two layers of silica and one layer of aluminum

**Figure 9.4** (a) Structure of clay colloids or micelles. Basic building units of aluminosilicate crystals are silica, $Si^{4+}$, a tetrahedron, and aluminum, $Al^{3+}$, an octrahedron. (b) Sheets of aluminum and silica held together with shared bonds.

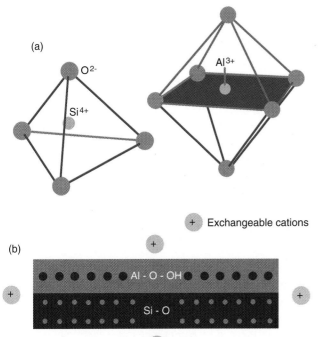

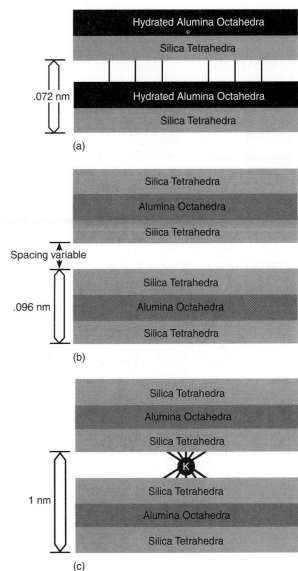

**Figure 9.5** Units of clay. (a) Kaolinite, a 1:1 clay, is composed of pairs of silica and alumina sheets held together by hydrogen bonds. (b) A unit of montmorillonite, a 2:1 clay, consists of a silica sheet on each side of an alumina sheet. The interlattice space varies with the amount of water present. (c) A unit of illite consists of silica sheet on either side of an alumina sheet. Adjacent micelles are held together by potassium (K) bridges.

with magnesium ($Mg^+$) or potassium ($K^+$) holding the layers together (Figure 9.5c).

There are differences among all these clays. In 1:1 clays the associated hydroxyl atoms balance the structure electrostatically. In kaolinite the oxygen atoms on the surface of the silicate layer face the hydroxyl atoms of the aluminum layer. The oxygen and hydroxyl atoms form hydrogen bonds that hold the layers tightly together, so water and cations cannot enter the interlayer spaces. This situation does not exist in the 2:1 clays. They have twice as many Si atoms as Al atoms in a unit cell, twice as many oxygen atoms, half the

number of hydroxyls, and a variable number of water molecules. Because oxygen atoms at the surface of unit cells face each other, the bonding is weak, and water and its associated cations can enter the clay particles. For this reason the units' cells vary in size.

An important feature of the micelles, especially of the 2:1 clays, is that one element can substitute for another without changing the structure. For example, Al may substitute for Si in the tetrahedral sheet. Since the Si atom has a charge of 4+ and the Al atom only 3+, the unit cell would have a net negative charge of −1. Similarly, iron ($Fe^{2+}$ and $Fe^{3+}$) or magnesium ($Mg^{2+}$) may substitute for $Al^{3+}$ in the octahedral sheet, changing the electrical charge. Substituting one atom for another without changing the arrangement of the atom or the morphology of the mineral is called *isomorphous substitution.*

One of the most important features of clay minerals, this substitution results in net negative charges that must be balanced by positively charged cations. *Exchangeable cations* are loosely held on the surface of the micelles and can be replaced easily by others. The edges and sides, negatively charged, act as highly charged anions. They attract cations, water molecules, and organic substances. The total number of negatively charged exchange sites on clay and humus particles that attract positively charged cations is called the **cation exchange capacity (CEC)** (Figure 9.6). The cation exchange capacity represents the net negative charges possessed by the soil. These negative charges enable a soil to prevent the leaching of its positively charged nutrient cations.

Exchange sites are occupied by such ions as calcium ($Ca^{2+}$), magnesium ($Mg^{2+}$), potassium ($K^+$), sodium ($Na^+$), and hydrogen ($H^+$). Some of these ions, especially $Al^{3+}$ and $H^+$, cling more tenaciously to micelles than do others. Less tenacious, in descending order, are $Ca^{2+}$, $Mg^{2+}$, $K^+$, $NH^+$, and $Na^+$. These latter ions are more easily displaced from the exchange site, and their place may be taken by aluminum or hydrogen ions. The percentage of sites occupied by ions other than hydrogen is called the *percent base saturation.* Acidic soils have a low percent base saturation because they have a high number of exchangeable hydrogen ions. Soils with a high cation exchange capacity are potentially fertile. Soils high in both CEC and base saturation are fertile unless they are saline or contain toxic heavy metals.

Positively charged cations and negatively charged anions are dissolved in soil solution and occupy the exchange sites on the clay and humus particles. Cations occupying the exchange sites are in a state of dynamic equilibrium with similar cations in solution. Cations in soil solution are continuously being replaced by or exchanged with cations on the clay and humus particles in response to changes in the concentration in the soil solution. For example, the removal of certain cations from soil solution by the roots of plants reduces the concentration of those cations and enhances the release of cations from the micelles.

Hydrogen ions added by rainwater, by acids from organic matter, and by metabolic acids from roots and microorganisms increase the concentration of hydrogen ions in the soil solution and displace other cations, such as $Ca^{2+}$, on the micelles. As more and more hydrogen ions replace other cations, the soil becomes increasingly acidic. Acidity is one of the most familiar of all chemical conditions in the soil. Typically soils range from a pH of 3 (strongly acid) to a pH of 9 (strongly alkaline). Soils just

**Figure 9.6** An example of cation exchange reactions when $Ca(OH)_2$ (lime) is added to an acidic soil. The aluminum ions in acid soil have been replaced by calcium ions, and the replaced aluminum ions react with soil water to form insoluble hydroxides and oxides. They will remain so as long as the soil stays neutral.

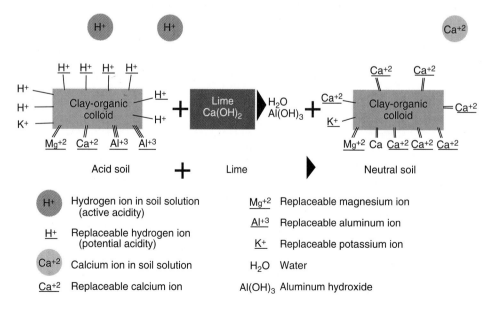

over a pH of 7 (neutral) are considered basic, and those of 5.6 or below are acid. As soil acidity increases, the proportion of exchangeable $Al^{3+}$ increases and $Ca^{2+}$, $Na^+$, and other cations decrease. Such changes bring about not only nutrient deprivation in plants and microorganisms but also aluminum toxicity.

## THE LIVING SOIL

Soil possesses several outstanding characteristics as a medium for life. It is relatively stable structurally and chemically. The underground climate is far less variable than above-surface conditions. The atmosphere remains saturated or nearly so, until soil moisture drops below a critical point. Soil affords a refuge from high and low extremes in temperature, wind, evaporation, light, and dryness. These conditions allow soil fauna to make easy adjustments to unfavorable conditions. On the other hand, soil hampers movement. Except to such channeling species as earthworms, pore space is important. It determines living space, humidity, and gases.

Only a part of the upper soil layer is available to most soil animals as living space. Spaces within the surface litter, cavities walled off by soil aggregates, pore spaces between individual soil particles, root channels, and fissures—all are potential habitats. Most soil animals are limited to pores and cavities larger than themselves.

Water in the pore spaces is essential: the majority of soil life is active only in water. Soil water is usually present as a thin film lining the surfaces of soil particles. This film contains, among other things, bacteria, unicellular algae, protozoa, rotifers, and nematodes. Most of the organisms are restricted in their movements by the thickness and shape of the water film in which they live. Some soil animals, such as millipedes and centipedes, are highly susceptible to desiccation and avoid it by burrowing deeper.

When excessive water floods pore spaces, typical after heavy rains, conditions are disastrous for some soil inhabitants. If earthworms cannot evade flooding by digging deeper, they come to the surface, where they often die from ultraviolet radiation. Many small species and immature stages of larger centipedes and millipedes are immobilized by a film of water and are unable to overcome the surface tension imprisoning them.

A wide diversity of life is found in the soil (Figure 9.7). The number of species of bacteria, fungi, protists, and representatives of nearly every invertebrate phylum found in the soil is enormous. Over 250 species of Protozoa live in

**Figure 9.7** Life in the soil. This drawing shows only a small fraction of the kinds of organisms that inhabit the soil and litter. Note the fruiting bodies of fungi, which are consumed by animals, vertebrate and invertebrate.

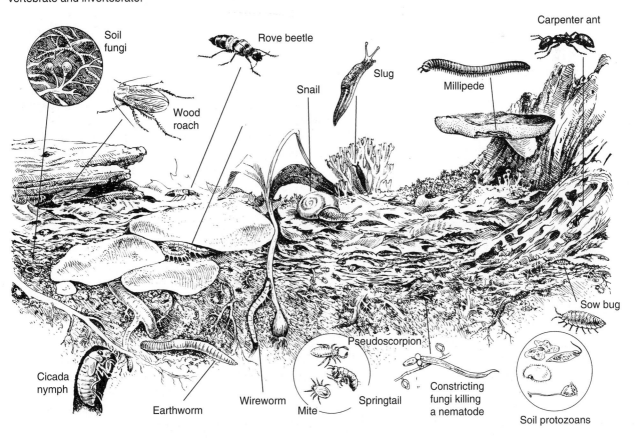

English soils (Sandon 1927). In the soil of beech woods in Austria live at least 110 species of beetles, 229 species of mites, and 46 species of snails and slugs (Franz 1950).

Dominant among the soil organisms are bacteria, fungi, protozoans, and nematodes. Flagellated protozoans range from 100,000 to 1,000,000 per gram of soil, amoebas from 50,000 to 500,000, and ciliates up to 1000 (Waksman 1952). Nematodes occur in the millions per hectare of soil (Overgaard 1949). These organisms obtain their nourishment from the roots of living plants and from organic matter. Some protozoans and free-living nematodes feed selectively on bacteria and fungi.

Living within the pore spaces of the soil are the most abundant and widely distributed of all forest soil animals, the mites (Acarina) and springtails (Collembola), which make up over 80 percent of the total number of animals in the soil (Salt et al. 1948). Flattened dorsoventrally, they are able to wiggle, squeeze, and even digest their way through tiny caverns in the soil. They feed on fungi or search for prey in the dark interstices and pores of the organic mass.

The more numerous of the two, both in species and numbers, are the mites, tiny eight-legged arthropods from 0.1 to 2.0 mm in size. The most common mites in the soil and litter are the Orbatei. They live mostly on fungal hyphae that attack dead vegetation as well as on the sugars digested by the microflora of evergreen needles.

The Collembolae are the most generally distributed of all insects. Their common name, springtail, is descriptive of the remarkable springing organ at the posterior end, which enables them to leap comparatively great distances. The springtails are small, from 0.3 to 1 mm in size. They consume decomposing plant materials, largely for the fungal hyphae they contain.

Prominent among the larger soil fauna are the earthworms (Lumbricidae). Earthworm activity consists of burrowing through the soil. Burrowing involves ingestion of soil, the ingestion and partial digestion of fresh litter, and the subsequent egestion of both mixed with intestinal secretions. Egested matter is defecated as aggregated castings on or near the surface of the soil or as a semiliquid in intersoil spaces along the burrow. These aggregates produce a more open structure in heavy soil and bind light soil together. In this manner earthworms improve the soil environment for other soil organisms by creating larger pore spaces and by mixing organic matter with the mineral soil.

Feeding on the surface litter are millipedes. They eat leaves, particularly those in which some fungal decomposition has taken place. Lacking the enzymes necessary for the breakdown of cellulose, millipedes live on the fungi contained within the litter. The millipedes' chief contribution is the mechanical breakdown of litter, making it more vulnerable to microbial attack, especially by saprophytic fungi.

Accompanying the millipedes are snails and slugs. Among the soil invertebrates they possess the widest range of enzymes to hydrolyze cellulose and other plant polysaccharides, possibly even the highly indigestible lignins.

Not to be ignored are termites (Isoptera), white wingless, social insects. Termites, together with some dipteran and beetle larvae, are the only larger soil inhabitants that can break down the cellulose of wood. They do so with the aid of symbiotic protozoans living in their gut. Termites dominate the tropical soil fauna. In the tropics termites are responsible for the rapid removal of wood, dry grass, and other materials from the soil surface. In constructing their huge and complex mounds, termites move considerable amounts of soil.

The detrital-feeding organisms support the predaceous soil organisms. Small arthropods are the principal prey of spiders, beetles, pseudoscorpions, predaceous mites, and centipedes. Protozoans, rotifers, myxobacteria, and nematodes feed on bacteria and algae. Various predaceous fungi live on bacteria-feeders and algal-feeders.

## SOIL GENESIS

Soil formation or genesis is the creation of soil from a non-soil parent material. This process involves the breakdown of parent material into smaller particles, rearranging and changing their mineral structure, adding organic matter, producing clays, and creating horizons.

### Soil-Forming Factors

Hans Jenny (1941) proposed that any individual soil or soil property results from the interaction of five independent soil-forming factors: parent material, climate, biota, topography, and time.

Parent Material Soil parent material, the unconsolidated mass from which soils are formed, is parent rock or transported material (Table 9.1). Rocks are residual or in-place parent material. They may be igneous, sedimentary, or metamorphic. *Igneous rocks* form when a hot mixture of elements, called magma, cools. If the magma cools at Earth's surface, it forms extrusive igneous rocks; if it cools below the surface, it forms intrusive igneous rocks. *Sedimentary rocks* form from material deposited in lakes and oceans. Over time and under the pressure of overlying materials, layers of these accumulated sediments consolidate into rocks. Over geological time, these rocks have been exposed by uplift and wearing away of mountains. *Metamorphic rocks* are formed when either igneous rocks or sedimentary rocks are heated or subjected to intense pressure at considerable depths within Earth. The original minerals melt and form new minerals. The composition of all these rocks largely determines the chemical composition of the soil.

Other parent materials are transported by wind, water, glaciers, and gravity. Because of the diversity of materials, transported soils are commonly more fertile than soils derived from in-place parent materials.

**Table 9.1  Types of Parent Material**

| Parent Material | Characteristics |
| --- | --- |
| Residual | In place. |
| Sedimentary rock | Deposition and compaction; shale, limestone. |
| Igneous rock | Cooling of magma; granite and basalt. |
| Metamorphic rock | Changed igneous and sedimentary; slate and marble. |
| Transported | Carried by wind, water, ice, gravity. |
| Glacial ice | |
| Till | Unsorted loose material of particles, from clay to boulders, carried in, ahead of, or under glaciers. |
| Outwash | Water-deposited material from glacial melt. |
| Water Deposition | |
| Alluvium | Eroded soil material carried and deposited by rivers and streams. |
| Wind Deposition | |
| Aeolian | Shifting sand. |
| Loess | Nonstratified silt mixed with scattered particles of sand and clay. |
| Gravity | |
| Colluvium | Erosional deposits on hillsides. |
| Organic soils | Derived from plant material; peat. |

**Climate**  Temperature and rainfall, which vary with elevation and latitude, govern the rate of weathering of rocks, decomposition of minerals and organic matter, and leaching and movement of weathered materials. In this manner climate directly influences the accumulation of soil parent materials and the development of soil horizons. Furthermore, climate influences the nature of plant and animal life in a region, both of which are important in soil development.

**Biotic Factors**  Vegetation, animals, bacteria, and fungi all contribute to the formation of soil. Vegetation is largely responsible for the organic matter in the soil and the color of the surface layer, and it influences the nutrient content of the soil. The role of the biota in soil development is discussed later.

**Topography**  Topography, or the lay of the land, influences the amount of water that enters the soil. More water runs off and less enters the soil on steep slopes than on level land. Therefore soil is less well-developed on steep slopes and horizons are often indistinct and shallow. On low and flat land, extra water enters the soil, and the subsoil may be wet and grayish (see page 142). Topography also influences the rate of erosion and downhill transport of soil material.

**Time**  The weathering of rock material, the accumulation, decomposition, and mineralization of organic material, the loss of minerals from the upper surface, gains in minerals and clay in the lower horizons, and horizon differentiation—all require considerable time. Well-developed soils in equilibrium with weathering, erosion, and biotic influences may require 2,000 to 20,000 years for their formation; but soil differentiation from parent material may take place in as short a time as 30 years (Crocker and Major 1955, Thurman and Sencindiver 1986). Certain acid soils in humid regions develop in 2000 years because the leaching process is speeded by acidic materials. Parent materials heavy in texture require a much longer time to develop into "climax" soils, because of an impeded downward flow of water. Soils develop more slowly in dry regions than in humid ones. Soils on steep slopes often remain young regardless of geological age because rapid erosion removes soil nearly as fast as it is formed. Floodplain soils age little through time, because of the continuous accumulation of new materials. Young soils are not as deeply weathered as and are more fertile than old soils, because they have not been exposed to the leaching process long. Old soils tend to be infertile because of long-time leaching of nutrients without replacement from fresh material.

## Weathering

The formation of soil begins with the weathering of rocks and their minerals. Primary minerals such as the feldspars (aluminosilicates with calcium, potassium, or sodium) transform into secondary minerals. These secondary minerals include a variety of clay minerals that may weather into still other kinds of clay minerals.

The parent material is exposed to mechanical weathering. Under the combined action of water, wind, and temperature, rock surfaces flake and peel away. Water seeps into crevices, freezes, expands, and cracks the rock into smaller pieces.

Accompanying this mechanical weathering and continuing long afterwards is chemical weathering, decomposing primary minerals. These minerals, particularly the aluminosilicates, are converted to secondary minerals, particularly clays such as kaolinite and montmorillonite. As iron is especially reactive with water and oxygen, iron-bearing minerals are prone to rapid decomposition. Iron remains oxidized in the red ferric state or is reduced to the gray ferrous state. The clay particles produced are shifted and rearranged within the mass by percolating water and on the surface by runoff, wind, or ice.

Because of variations in slope, climate, and native vegetation, many different soils can develop on the same parent materials. The thickness of parent material, the kind of rock, and the degree of weathering affect fertility and water relations of the soil.

## Biotic Influences

Plants and animals have a pronounced influence on soil development. In time plants colonize the weathered material. Plant roots penetrate and further break down the parent ma-

terial. They pump nutrients up from its depths and add them to the surface. In doing so plants recapture minerals carried deep into the soil by weathering processes. Through photosynthesis, plants capture the sun's energy and add a portion of it in the form of organic carbon to the soil. Plant debris enables bacteria, fungi, earthworms, and other soil organisms to colonize the area.

Decomposition ultimately converts organic compounds to inorganic ones. Higher organisms in the soil—millipedes, centipedes, earthworms, mites, springtails, grasshoppers, and others—consume fresh material and leave partially decomposed products in their excreta. Microorganisms further reduce this material into carbohydrates, proteins, lignins, fats, waxes, resins, and ash. These compounds break down into inorganic products.

The fraction of organic matter that remains is called **humus,** dark-colored, noncellular, chemically complex organic material whose characteristic constituents are humin, a group of unchanged plant chemicals, and other organic compounds such as fulvic acid and humic acid. Decomposition proceeds so slowly that the amount of organic material changes little each year. The formation of new humus balances annual losses by decomposition. The ratio of the formation of new humus and the destruction of old determines the amount of humus in the soil.

Soil organisms strongly influence the development of the O and A horizons of the soil, especially in temperate forest regions. Their role was pointed out by a Danish forester, P. E. Muller, in 1879 and 1884. Two types of humus formation exist in temperate forests, mull and mor. They result from an interaction of physical, chemical, and especially biological mechanisms (Figure 9.8).

Mor Characteristic of dry or moist acid habitats, especially heathland and coniferous forest, **mor** has a well-defined, unincorporated, and matted or compacted organic deposit resting on mineral soil (Figure 9.9a). It results from an accumulation of litter that is slowly mineralized and remains unmixed with mineral soil. Therefore a sharp break exists between the O and A horizons.

Slow though decomposition may be, it is the manner in which the process proceeds that distinguishes mor from other humus types. The main decomposing agents are fungi, both free-living and mycorrhizal, which tend to depress soil animal activity and produce acids; nitrifying bacteria may be absent. The fungi consume mostly the vascular cells of leaves and leave behind a residue of mesophyll tissue. Proteins within the leaf litter are stabilized by protein-precipitating material, making them, in some cases, resistant to decomposition. Because of limited volume, pore space, acidity, type of litter involved, and the nature of its breakdown, mor is inhabited by a small biomass of soil animals. They have little mechanical influence on the mineral soil. Instead, these organisms live in an environment of organic material cut off from mineral soil beneath.

Mull **Mull,** on the other hand, results from a different process. Characteristic of mixed and deciduous woods on fresh and moist soils with a reasonable supply of calcium, mull possesses only a thin scattering of litter on the surface, and the mineral soil is high in organic matter (Figure 9.9c). All organic materials convert to true humic substances. Because of animal activity, they are inseparably bound to the mineral fraction, which absorbs them like a dye. There is no sharp break between the O and A horizons. Because of less acidity and a more

**Figure 9.8** The sequence of humus types and related processes. Note the inverse relationship between bacteria and fungi as the humus sequence goes from mor to mull, as well as the pronounced changes in invertebrate life. (From J. A. Wallwork 1973:53.)

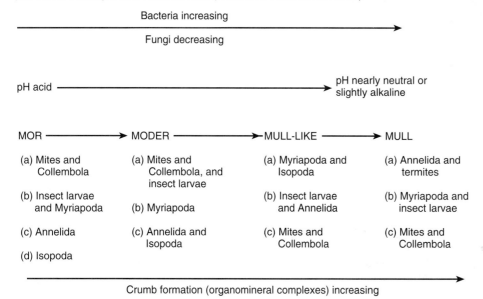

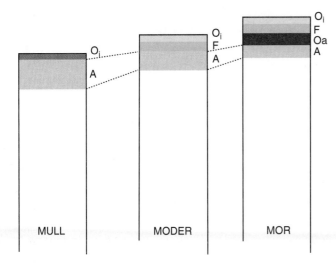

**Figure 9.9** Comparison of layers in mull, moder, and mor humus. *F* is the fermentation layer.

equitable base status, bacteria tend to replace fungi as the chief decomposers, and conversion of ammonium produced in the early breakdown of litter to nitrates by soil bacteria is rapid. Soil animals are more diverse and possess a greater biomass, reflecting a more equitable distribution of living space, oxygen, food, and moisture, and a smaller fungal component. This faunal diversity is one of mull's greatest assets, because humification flows through a wide variety of organisms with differing metabolisms. Not only do these soil animals fragment plant debris and mix it with mineral particles, enhancing microbial and fungal activity, but they also incorporate humified material with mineral soil. This constant interchange of material takes place from the surface to soil and back again. Plants extract nutrients from the soil and deposit them on the surface. Then the soil flora and fauna reverse the process.

**Moder** On the continuum from mull to mor lies **moder** (Figure 9.9b). In this humus type, plant residues are transformed into the droppings of small arthropods, particularly Collembola and mites. Residues not consumed by the fauna are reduced to small fragments, little humified and still showing cell structure. The droppings, plant fragments, and mineral particles all form a loose, netlike structure held together by chains of small droppings. In acid mor the washing action of rainwater destroys the shape of the droppings. Under heavy precipitation humus leached from the droppings acts as a binding substance to form a dense, matted litter approaching a mor. On the continuum between moder and mull, we often see the droppings of large arthropods, which take in considerable quantities of mineral matter with food. However, moder differs from mull in its higher organic content, restricted nitrification, and more or less mechanical mixture of the organic components with the mineral. The two are held together by humic substances, yet separable. In other words, the organic crumbs are deficient in mineral matter, in contrast to mull, in which mineral and organic parts are inseparably bound.

Finally, there is **skeletal humus,** or raw humus, appearing much like mor. It has not developed into any of the other three types.

# SOIL DEVELOPMENT

Soil continues to develop and differentiate over time (Figure 9.10). Major processes in soil development include podzolization, laterization, calcification, salinization, and gleyization.

## Podzolization

**Podzolization** (from Russian *podzol,* "ash") is the depletion of bases and the movement of clays from the surface layers to the *B* horizon (Figure 9.10a). Insoluble iron and aluminum oxides of the surface soil combine in complex ways with organic acids and humus. These oxides and clays are carried downward by rainwater into a *B* horizon, characterized by an accumulation (illuviation) of organic matter and oxides of iron and aluminum. Left behind under the most intense podzolization process is a light or ash-colored *A* horizon from which the process gains its name. The organic horizon is a mor with a layer of fermented litter on top of a layer of humus unmixed with mineral soil. Such soils are called **Spodosols.** Highly acid, they develop under coniferous forests in cool humid regions.

The podzolic process extends to other soils, notably Alfisols and Ultisols (see Table 9.2). Alfisols are associated with humid temperate deciduous-coniferous and deciduous forest regions. Like the Spodosols, these soils have accumulations of clay in the *B* horizon; but unlike the Spodosols, they have little accumulation of organic matter on the surface. Instead organic matter tends to be well-mixed by soil animals to form a darkened, organically enriched *A* horizon (melanization). In some soils the division between the *A* and *B* horizons may be hard to distinguish, and the *E* horizon may be missing.

The related *Ultisols* have a lower base saturation than Alfisols. Characteristic of warmer climates, Ultisols are more intensely weathered and have redder color because of the release of iron from the silicates. Such reddish soils, characteristic of the southeastern United States and tropical regions, also are in part a product of another soil-forming process, laterization.

## Laterization

In humid subtropical and tropical forested regions of the world, where rainfall is heavy and temperatures high, the soil development processes are much more intense. Because temperatures are uniformly high, weathering in these regions is almost entirely chemical, brought about by water and its dissolved substances. The residues from this weathering—bases, silica, aluminum, hydrated aluminosilicates, and iron oxides—become soluble (Figure 9.10b). Because

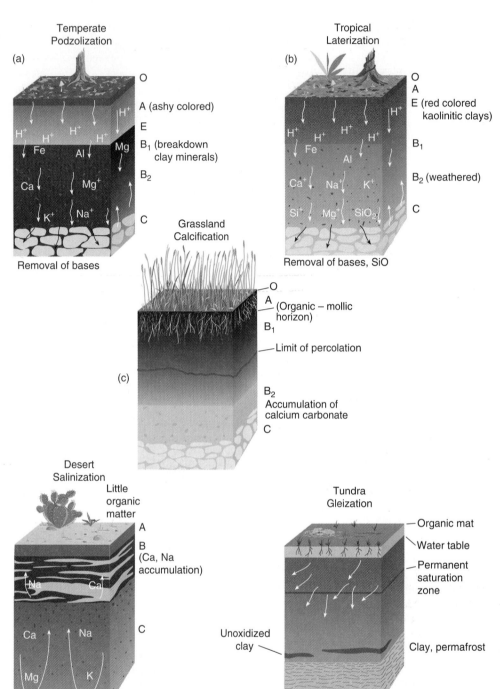

**Figure 9.10** Soil-development processes. (a) Podzolization, characteristic of cool, humid climates. (b) Laterization, a process of warm wet climates. (c) Calcification, common to semiarid climates. (d) Salinization in dry to very dry climates. (e) Gleization, characteristic of cold, wet climates, as well as of other climates when water is perched in the *B* or *C* horizon.

precipitation usually exceeds evaporation, the water movement is almost continuously downward. With only a small quantity of electrolytes present in the soil water because of continual leaching, silica and aluminosilicates are carried downward, while sesquioxides (clays lacking a silica base) of aluminum and iron remain behind. The sesquioxides are relatively insoluble in pure rainwater, but the silicates tend to precipitate as a gel in solutions containing humic substances and electrolytes. If humic substances are present, they act as protective colloids about iron and aluminum oxides and prevent their precipitation by electrolytes. The end product of such a process is a soil composed of silicate and

hydrous oxides, clays, and residual quartz, deficient in bases, low in plant nutrients, and intensely weathered to great depths.

The large amount of residual iron and aluminum left after the depletion of silica and bases becomes enriched as hydrous oxides, forming a variety of often brilliant reddish colors in the upper part or *E* horizon, which may be especially deep. For this reason these soils may lack distinct horizons. In some tropical soils the subsoil hardens irreversibly into plinthite when exposed to sun and air (Richter and Babbar 1991). This soil-forming process is called **laterization.** True lateritic soils are called **Oxisols.**

## Calcification

The subhumid-to-arid and temperate-to-tropical regions of the world—the plains and prairies of North America, the steppes of Russia, the veldts and savannas of Africa, and the pampas of South America—support grassland vegetation. Dense root systems extend many feet below the surface. Each year nearly all of the vegetative material above ground and a part of the root system are turned back to the soil as organic matter. Although the material decomposes rapidly the following spring, it is not completely gone before the next cycle of death and decay begins. Soil inhabitants mix the humus with mineral soil, developing a soil high in organic matter.

Because the amount of rainfall in grassland regions generally is insufficient to remove calcium and magnesium carbonates, they are carried down only to the average depth that percolating waters reach (Figure 9.10c). Grass maintains a high calcium content in the surface soil by absorbing large quantities from the lower horizons and redepositing them on the surface. Little clay is lost from the surface. This process of soil development is called **calcification.** The soils so formed are called **Mollisols** (from Latin *mollis,* "soft").

Soils developed by calcification have a distinct, thick *A* horizon and an indistinct *B* horizon, characterized by an accumulation of calcium carbonate. The *A* horizon is high in organic matter and in nitrogen, even in tropical and subtropical regions.

## Salinization

Arid and semiarid regions have sparse vegetation. Because plant growth is limited, little organic matter and nitrogen accumulate in the soil. Scant precipitation results in slightly weathered and slightly leached soils high in plant nutrients. The

horizons are usually faint and thin. Within these regions are areas where soils contain excessive amounts of soluble salts, either from the parent material or from the evaporation of water draining in from adjoining land. Infrequent rains penetrate the soil and carry the material downward to a certain depth. The water evaporates, leaving a more or less cemented horizon of sodium, calcium, or magnesium salts below the surface. This horizon is called **caliche.** This soil-forming process is called **salinization** (Figure 9.10d). Soils of arid and semiarid regions are **Aridosols,** soils low in humus, high in base content, and often containing gypsum, clay, or carbonate horizons.

## Gleyization

Calcification, laterization, and podzolization take place on well-drained soil. Under poorer drainage conditions a different soil development process is at work. The amount of water that passes through or remains in the soil determines the degree of oxidation and breakdown of soil minerals. Iron in soils where water stays near or at the surface of the ground most of the time is reduced to ferrous compounds. These compounds give a dull gray or bluish color to the horizons. This process, called *gleyization,* may result in compact, structureless horizons (Figure 9.10e).

**Gley soils** are high in organic matter because more organic matter is produced by vegetation than can be broken down by humification. An absence of soil microorganisms and anaerobic conditions greatly reduce humification. Gleyization is the characteristic soil-forming process in cold wet situations, especially in the tundra; but it is also common in other regions where the water table remains perched above the *B* and *C* horizons.

An example of the gleyization process in a temperate Alfisol is depicted in Figure 9.11. On gentle to moderate

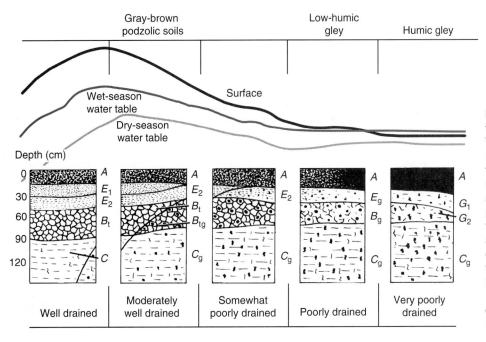

**Figure 9.11** Effect of drainage on the development of an alfisol. Wetness increases from left to right. The diagram represents the topographic position the profiles might occupy. Note that the strongest soil development takes place on well-drained sites where weathering is maximum. The least amount of weathering takes place on the very poorly drained soils where the wet season water table lies above the surface of the soil. *G* or *g* indicates mottling; *t* indicates translocated silicate clays. (Adapted in part from Knox 1952.)

slopes, where drainage conditions are improved, gleyization is reduced and takes place deeper in the profile. As a result the subsoil shows varying degrees of mottling of grays and browns. On hilltops, ridges, and steep slopes, where the water table is deep and the soil well-drained, the subsoil is reddish to yellowish-brown from oxidized iron compounds.

## Other Processes

Some soils have developed in other ways. One group is the **Vertisols.** These soils are very high in clays that swell in response to wetting and shrink while drying. Such wetting and drying causes frequent deep cracking.

**Histosols** include all organic soils. They form by an accumulation of organic matter because of high water tables, poor drainage, or conditions of extremely high precipitation and low evaporation. Histosols are found both in the tropics and in the Arctic tundra.

Entisols and Inceptisols have large *O* horizons and poorly developed *E* horizons. **Entisols** lack a *B* horizon, whereas **Inceptisols** have a developing *B* horizon. Both soils are found wherever soil-forming processes have been interrupted by flooding, erosion, and glaciation or slowed by cold temperatures. They are generally considered to be young soils. **Andosols** derive from volcanic ash deposits and are high in organic matter. These soils, usually highly fertile, are important in Central America.

## CLASSIFICATION AND MAPPING

To apply what we know about soils, we must classify and map them. The mostly widely used system is the United States Comprehensive Soil Classification System (1982) developed by the Department of Agriculture Soil Conservation Service to standardize soil nomenclature throughout the world. There are other similar approaches to classification, including the FAO-UNESCO system (see Landon 1984), which is somewhat easier to use in the field. It replaces the old system of Russian origin.

The United States system consists of 11 main orders (Table 9.2). Their worldwide distribution is determined largely by climate and vegetation (Figure 9.12). These orders are further subdivided into suborders, great groups, families, and series. The names of the lower classification units always end on the formative syllable of their respective order preceded by other syllables connotative of various

---

**Table 9.2** The Eleven Major Soil Orders

| Order | Derivation and Meaning | Description | Location |
|---|---|---|---|
| Entisol | Coined from *recent* | Dominance of mineral soil materials; absence of distinct horizons. | All climates, any vegetation; young soils; floodplains, glaciated areas, mountains. |
| Inceptisol | L. *inceptum,* "beginning" | Texture finer than loamy; little translocation of clay; often shallow; little development of horizons. | All climates and vegetation; developing soil. |
| Vertisol | L. *verto,* "invert" | Dark clay soils; wide, deep cracks when dry. | Areas with clay-textured parent material; climate can dry and crack soil. |
| Aridosol | L. *aridus,* "arid" | Dry for extended periods; low in humus; high in base content; may have clay, carbonate, gypsum horizons. | Desert and semiarid regions. |
| Mollisol | L. *mollis,* "soft" | Surface horizons dark brown to black with soft consistency; rich in bases. | Semihumid regions; native grassland vegetation. |
| Spodosol | Gr. *spodos,* "ashy" | Light gray, whitish *A* horizon on top of a black and reddish *B* horizon; high in extractable iron and aluminum. | Cool humid regions; coniferous forests. |
| Alfisol | Coined from *Al* and *Fe* | Shallow penetration of humus; translocation of clay; well-developed horizons; not strongly leached. | Humid, temperate regions; deciduous and coniferous-deciduous forests. |
| Ultisol | L. *ultimus,* "last" | Intensely leached; strong clay translocation; low base content; red or yellow soils. | Warm humid to tropical climate; forest vegetation. |
| Oxisol | Fr. *oxydé,* "oxidized" | Highly weathered soils; red, yellow, or gray; rich in kaolinite, iron oxides, and often humus. | Tropical and subtropical climates and vegetation. |
| Histosol | Gr. *histos,* "organic" | High content of organic matter. | Bogs, marshes, muck. |
| Andosol | Jpn. *ando,* "black" and *do,* "soil" | Developed from volcanic ejecta; not highly weathered; upper layers dark-colored. | Volcanic regions, especially tropical. |

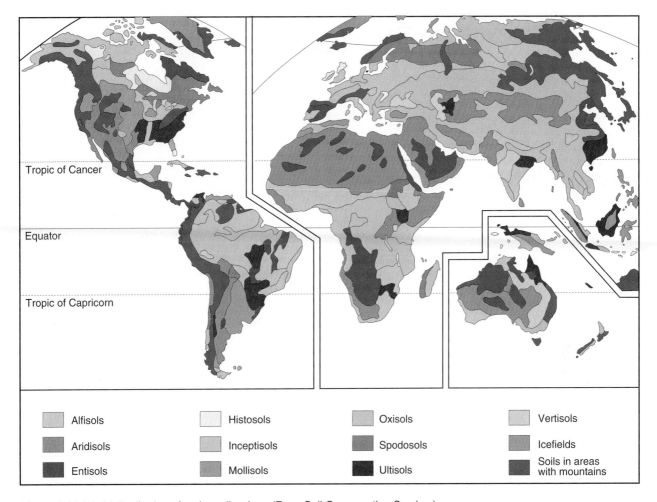

**Figure 9.12** World distribution of major soil orders. (From Soil Conservation Service.)

soil properties. For example, a suborder of Mollisol is Aquoll with *aqu* meaning "water" and *oll* referring to M*oll*isol.

## Grouping Soils

All soils formed from a particular type of parent material that are similar in all characteristics and arrangement of the soil profile, except for the texture of the *A* horizon, make up a **soil series.** The soil series is the lowest taxonomic level in the soil classification scheme.

In the United States system soil series are named after the locality in which they were first described. For example, the Ovid Series in New York was named after the town of Ovid and the Miami after the Miami River in western Ohio (not to be confused with Florida's Miami). Like species among plants and animals, soil series are defined in terms of the largest number of differentiating characteristics and occur in fairly limited areas. Higher categories of taxonomy combine series into larger groupings, distinguished by fewer differentiating properties. At the highest level of classification, classes correspond roughly to broad climatic zones.

Locally every soil series has its neighboring soil series with unlike properties (Figure 9.13), reflecting changes in slope, drainage patterns, and soil materials. These soil series may have abrupt boundaries or form a continuum. If several related soils found side by side have developed from the same parent material but differ mainly in natural drainage and slope, they form a **catena,** from the Latin for "chain." In Figure 9.13, the groups of soils from C through G, all derived from loess, make up a catena.

Although the major soil development process defines broad soil patterns, local patterns are controlled by topography. As the topography of an area develops over time through the action of physical and chemical weathering, a group of soils develops along with it. This group of soils, which may include several catenas, is known as a **toposequence.** In Figure 9.13 the groups of soils from A through G represent a toposequence, ranging from well-drained ridgetop to very poorly drained soils and alluvial deposits. If all the soil-forming factors remain constant except for vegetation, then vegetation leaves its imprint on soil formation. Soils so developed form a *biosequence.* In Figure 9.13 soils D (Seaton) formed under forest and E (Joy) formed under

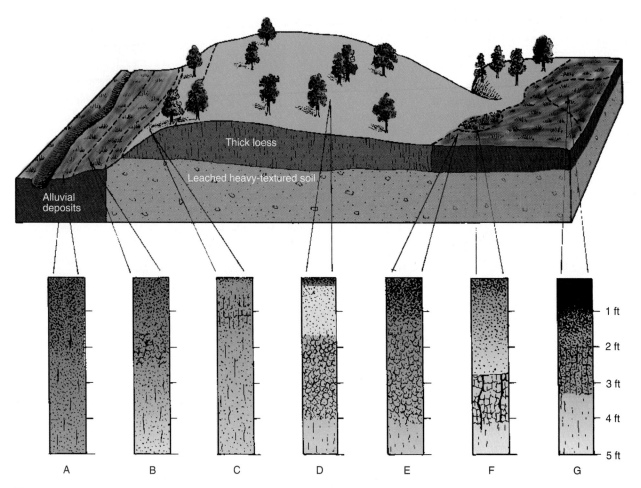

**Figure 9.13** Topography and vegetation acting together produce a toposequence of soil series. This diagram shows the normal sequence of seven soil series from the Mississippi to the uplands in Illinois. It also illustrates how bodies of soil types fit together in the landscape. Boundaries between adjacent bodies are gradations or continuums, rather than sharp lines.

The lower part of the diagram pictures the profiles of the seven soils, showing the color and thickness of the surface horizon and the structure of the subsoil. Note how the natural vegetation has influenced surface color and how topographic position and distance from the bluff have influenced subsoil development.

Profile A (Sawmill) is an Entisol, a bottomland soil formed from recent sediments and not subject to much weathering. Profile B (Worthen) at the foot of the slope is also an Entisol. Developed from recent alluvial material, it shows little structure. Profile C (Hooper) on the slope developed from a thick loess on top of leached till, while the soil on the bottom of the slope developed directly from the till. Profile D (Seaton) is an upland soil formerly covered with timber. It possesses a light surface color and lacks structure, the result of rapid deposition of loess during early soil formation, holding soil weathering to a minimum. These two soils probably should be considered Entisols because they lack any well-formed horizons, which are the product of strong soil weathering. Profile E (Joy) represents upland soil developed under grass. Note the dark surface and the lack of structure, again the result of a rapid deposition of loess. Profile F (Edgington) is a depressional wet spot. Extra water flowing in from adjacent fields increased the rate of weathering, resulting in a light grayish surface and subsurface and a blocky structure to the subsoil. This indicates strongly developed gley soil. The depth of subsoil suggests that considerable sediment has been washed in from the surrounding area. Profile G (Sable) represents a depressional upland prairie soil. The deep, dark surface and the coarse blocky structure mark it as a Mollisol. Abundant grass growth produced the dark color. (After Veale and Wascher 1956.)

grass make up a biosequence. Groups of soils of different ages derived from the same parent materials but subject to soil-forming processes over different lengths of time make up **chronosequences.** Examples occur in areas of retreating glaciers, volcanic flows, and sand dunes.

When soils are mapped, unlike soils may be grouped together into **associations** for reasons of scale or practicality. When occurring in inseparable patterns, soils are mapped as *complexes.* In detailed mapping, a soil series can be subdivided into types, phases, and variants.

## Mapping Soils

The distribution of specific kinds of soils across the landscape is delineated on a soil map (Figure 9.14). The kinds of soil mapped may cover from a few acres to several hundred acres. These mapped units may include small areas of other soils too small to map separately.

Soil maps are developed through soil surveys, which date back to 1900. Modern soil mapping is done by the Soil Conservation Service in cooperation with other governmental agencies and soil scientists at Agricultural Experiment Stations of land grant universities.

As their first step in mapping soils, soil scientists collect all available information relevant to local soils: geology, climate, vegetation, topographic maps, and aerial photographs. They begin mapping by broadly delineating boundaries of soil types on aerial photographs. Then they visit the various sites, sampling the soil by spade and soil auger to determine the characteristics of the soil profile. They identify the soil in the field, refine the plotting of the soil boundaries, and place the identifying symbols on the map. Finally, they describe the soils, interpret the survey, and publish the the results in a Soil Survey Report.

The Soil Survey Report describes the soil map units and the typical pedons for each soil series. It assesses the management of the soil for crops and pasture, its susceptibility to erosion, and its usefulness for crop production, woodland management, and wildlife habitat. It also provides information on the suitability of various soils for road construction, housing development, building sites, sanitary facilities, water management, and recreation. Unfortunately, many site development engineers and real estate developers do not make use of this information. Their failure to do so results in damage and loss from soil instability.

## MISMANAGED SOIL

Soil descriptions relate to undisturbed soils. Over much of the landscape, soils have been highly disturbed. Soils have

**Figure 9.14** Soil map of a section of Jefferson County, West Virginia, indicating soil associations and soil series. The symbol H indicates the Hagerstown series and C the Chilhowie series. These soils are deep, well-drained, and on fairly level land. The symbols indicate the characteristics of the soils. For example, HbB is Hagerstown silt loam, 2 to 6 percent slope; HgB is very rocky silt loam, 2 to 6 percent loam, HgC is very rocky silt loam, 6 to 12 percent slope; and CdB is Chilhowie silt clay, 2 to 6 percent slope. (Courtesy of Soil Conservation Service.)

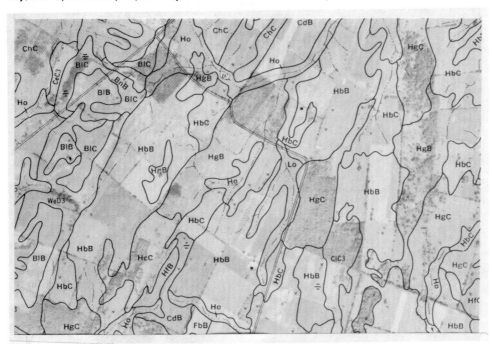

been buried under fill, overturned and moved about by excavations, surface mining, and road construction, and exposed to erosion by wind and water. Upper horizons have been mixed by agricultural plowing and tillage and compacted by heavy machinery and trampling.

Soil compaction occurs when any weight pushes the soil particles together and reduces the size of the pores. Greatest compaction occurs under wet and moist conditions. Moist soil particles easily slide over one another. Heavy machinery from large tractors to construction equipment, trampling on lawns and playing fields, concentrated use of pathways and hiking and riding trails, and off-road use of all-terrain-vehicles all compact the soil. Compacted soil cannot absorb water, so the water flows across the surface, resulting in *soil erosion*—the carrying away of soil particles by wind and water.

Disturbance of topsoil by agriculture, construction, and road building exposes it to erosion. Stripped of its protective vegetation, soil is removed by wind and water faster than it can be formed. Loss of the upper layers of humus-charged, granular, highly absorptive topsoil exposes the humus-deficient, less stable, less absorptive, and highly erodable layers beneath. If the subsoil is clay, it absorbs water so slowly that heavy rains produce a highly abrasive and rapid runoff. Soil of the uplands ends up in muddied rivers, where it affects wildlife, settles out in dams, shortening their useful life, and builds up river deltas.

The intensity of water erosion is influenced by slope, the kind and condition of soil, and rainfall. The least conspicuous type of water erosion is *sheet erosion,* a more or less even removal of soil over a field. When runoff tends to concentrate in streamlets instead of moving evenly over a slope, the cutting force is increased and *rill erosion* produces small channels downslope. Where concentrated water cuts the

same rill long enough, or where runoff in sufficient volume cuts deeply into the soil, highly destructive gullies result (Figure 9.15). *Gully erosion* often begins in wheel ruts made by off-road vehicles in fields and forest, logging roads and skid trails, and livestock and hiking trails.

Bare soil, finely divided, loose, and dry, as it often is after tillage, is ripe for wind erosion. Very fine particles of dust are picked up by the wind and carried as dust clouds. Often dust particles are lifted high in the atmosphere and carried for hundreds and even thousands of miles. This problem occurred in the Great Plains during the droughty Dust Bowl days of the 1930s, and it is happening on an increasing scale worldwide today.

## ■ SUMMARY

Soil is the foundation of terrestrial ecosystems. It is the site of decomposition of organic matter and of the return of mineral elements to the nutrient cycle. It is the habitat of animal life, the anchoring medium for plants, and their source of water and nutrients. Soil begins with the weathering of rock and minerals, which involves the leaching and carrying away of mineral matter. Its development is influenced by climate, vegetation, parent material, and topography. Plants rooted in weathering material further break down the substratum, pump up nutrients from its depths, and add all-important organic material.

Through decomposition and mineralization this dead organic matter is converted into humus, an unstable product that is continuously being formed and destroyed by mineralization. Humus is grouped into three types: mor, characteristic of acid habitats, whose chief decomposing agents are

**Figure 9.16** Concentrated flow of water across the soil surface can result in massive gully erosion that feeds on itself to tear away at the land.

fungi; mull, characteristic of deciduous and mixed wood-lands, whose chief decomposing agents are bacteria; and moder, characteristic of deciduous forests, which is highly modified by the action of soil animals.

Strongly influencing soil development are organisms in the soil, from bacteria and fungi to earthworms. Larger organisms live in pore spaces and channels in the soil. Bacteria, protozoans, and other microorganisms live in the thin film of water surrounding soil particles. Most microorganisms obtain their nourishment from organic matter. Larger soil inhabitants feed on fresh litter and other detrital material, and on bacteria and fungi. These detrital feeders support an array of predacious soil inhabitants, from mites to spiders.

As a result of the weathering process, accumulation and breakdown of organic matter by soil organisms, and the leaching of mineral matter, horizons or layers form in the soil. Five horizons are commonly recognized, although all are not necessarily present in any one soil: the *O,* or organic, layer; the *A* horizon, characterized by accumulation of organic matter; *E,* the zone of leaching of clay and mineral matter; *B,* in which mineral matter accumulates; and *C,* the underlying material, exclusive of bedrock. These horizons divide into subhorizons.

Soil profile development is influenced over large areas by vegetation and climate. In grassland regions the chief soil-forming process is calcification, in which calcium accumulates at the average depth reached by percolating water. In arid regions salt accumulates closer to the surface, so calcification proceeds to salinization. In forest regions, podzolization—the leaching of calcium, magnesium, iron, and aluminum from the upper horizon, and the retention of silica—takes place. In tropical regions laterization, in which silica is leached and iron and aluminum oxides are retained in the upper horizon, is the major soil-forming process. Gleyization takes place in poorly drained soils. Organic matter decomposes slowly, and iron is reduced to the ferrous state and is leached away.

Differences between soils and between horizons within soils are reflected by variations in texture, structure, and color. Texture of the soil is determined by the proportion of soil particles of different sizes: sand, silt, and clay. Texture is important in the movement and retention of water in the soil. Soil particles, particularly the clay-humus complex, are the key to nutrient availability and the cation exchange capacity of the soil—the number of negatively charged sites on soil particles that can attract positively charged ions. Cation exchange capacity relates to percent base saturation, the percentage of sites occupied by ions other than hydro-gen. Soils with a high cation exchange capacity are potentially fertile.

Each combination of climate, vegetation, soil material, topography, and time results in a unique soil, of which the smallest repetitive unit is the pedon. Individual soils with the same profile characteristics and derived from the same parent material make up a soil series. Soil series are further classified into families, great groups, suborders, and orders. Soil mapping helps us use land appropriately.

# REVIEW QUESTIONS

1. What is soil?
2. How do soil profiles develop?
3. What are the roles in soil development of physical and chemical weathering, plant life, soil organisms, and climate?
4. What is the basic structure of clay? How do clays differ in structure?
5. What is cation exchange capacity? How does it affect nutrient availability and soil acidity?
6. Explain the differences among mull, moder, and mor organic layers.
7. Compare the following forms of soil development: podzolization, calcification, salinization, laterization, and gleyization.
8. What are the eleven major soil orders?
9. From your local or state Soil Conservation Service office obtain a soil survey for your area. What are the local soil series? Relate local soils to agricultural development, urban development, soil erosion problems, and forest distribution.
10. Soil profiles have been highly disturbed by agriculture and other activities. How does tillage affect the soil profile? The organic horizon and soil organic matter content? What effect might the application of soil pesticides have on soil organisms and soil development processes? (For some answers see Tate 1987.)

# CROSS-REFERENCES

Local water cycle, 67; soil-plant-atmosphere continuum, 70–71; nutrient sources and cycling, 116–120; calcicoles and calcifuges, 121–122; nutrients and consumers, 124–127; decomposition, 159–165.

*Part 3*

# The Ecosystem

# Concept of the Ecosystem

## Concepts

1. An ecosystem is an interacting system of living organisms and their physical environment.
2. Ecosystems consist of three basic components or subsystems—producers, consumers, (including decomposers), and inorganic and organic substances—which exchange energy and nutrients.
3. The two essential processes in energy flow and nutrient cycling are photosynthesis and decomposition.
4. Plants employ three types of photosynthesis based on the methods of carbon fixation: $C_3$, $C_4$, and CAM.
5. Decomposition is a sequential process involving leaching, fragmentation, catabolism, anabolism, immobilization, and mineralization.
6. Decomposition involves all consumers; the true decomposers are bacteria, fungi, and a diversity of detritivores and microbivores.

Imagine yourself in a spacecraft far from Earth. Through the window, the planet appears as a bluish ball suspended in the black void of space. Kaleidoscopic patterns of whites, blues, reds, and greens are changing across its surface as it turns and as clouds swirl across it. It hangs alone, self-contained, dependent on the outer reaches of space only for the energy of sunlight. It is close enough to the sun to be warmed by radiant energy; yet it does not become overheated or overcooled. It is protected from damaging radiation by an atmosphere unlike that of any of its sister planets.

As the spacecraft approaches Earth, the changing patterns sharpen into broad outlines of mountains, deserts, plains, and seas. In the vast area from outer space to the core of Earth, it is only at the narrow interface of land, air, and water that life exists. This thin blanket of life surrounding Earth is called the **biosphere** (Figure 10.1).

Supporting the biosphere is the **lithosphere,** the rocky material of Earth's outer shell from the surface to about 100 km deep. The thin, uppermost layer of the crust, at the most a few thousand meters deep, is the substrate of life—the foundation on which it rests, the material of which it is made, and its primary source of nutrients.

The body of liquid on or near the surface of Earth is the **hydrosphere.** The hydrosphere includes all Earth's liquid waters (oceans, lakes, and underground water or aquifers), frozen water (polar ice caps and glaciers), and the small amount of water vapor in the atmosphere.

As the spaceship comes even closer to Earth, grasslands, forests, croplands, rivers, lakes, and oceans come into focus. Each is physically and biologically different, and each is inhabited by different organisms that are well adapted to the environment in which they are found. In each energy is fixed by plants and transferred to animal components. Nutrients are withdrawn from the substrate, deposited in the tissues of plants, cycled from one feeding group to another, released by decomposition to the soil, water, and air, and then recycled. Each of these regions constitutes an ecosystem. These ecosystems are not independent. Energy and nutrients in one find their way to another. Ultimately all parts of Earth are interrelated.

## COMPONENTS OF ECOLOGICAL SYSTEMS

In 1935 the English ecologist A. E. Tansley coined the term *ecosystem.* The concept is based on the idea of a physical system:

> The more fundamental conception is . . . the whole system (in the sense of physics) including not only the organism-complex, but also the whole complex of physical factors forming what we call the environment. . . . We cannot separate (the organisms) from their special environment with which they form one physical system. . . . It is the system so formed which (pro-

**Figure 10.1** Structural features of the outer part of Earth. The atmosphere, the biosphere, the lithosphere, and the hydrosphere are often collectively called the ecosphere.

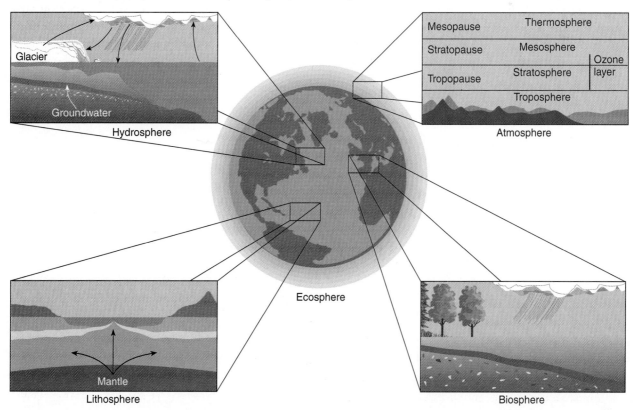

vides) the basic units of nature on the face of the earth. . . . These *ecosystems,* as we may call them, are of the most various kinds and sizes.

An **ecosystem** is basically an energy-processing and nutrient-regenerating system (Figure 10.2). It has two major parts, the biotic and the abiotic. The biotic part consists of all interacting organisms living in the area, the **community.** The abiotic part embraces the physical environment with which the organisms of the community interact. The biotic and abiotic exchange energy and materials. Populations are the subsystems through which the system functions.

For ecosystems like streams, lakes, ponds, and forest stands surrounded by open land, the boundaries appear to be distinct; for most ecosystems they do not. Apparent boundaries, however, are misleading, because no ecosystem stands alone. Energy and materials flow from one ecosystem to another. A stream ecosystem, for example, is strongly influenced by the terrestrial ecosystems through which it flows.

Inputs into the system are both biotic and abiotic. The *abiotic inputs* are energy, inorganic substances ($CO_2$, C, N, $O_2$, and so on), and organic compounds (proteins, carbohydrates, humic acids, inorganic matter). Radiant energy, both heat and light, imposes constraints on the system by influencing temperature and moisture regimes, seasonality, and photosynthetic activity. Temperature and moisture determine, in part, what organisms can live in a system and affect its productive capability. The *biotic inputs* include other organisms that move into the ecosystem as well as influences imposed by other ecosystems in the landscape.

In simplest terms, all ecosystems, aquatic and terrestrial, consist of three basic components: the producers, the consumers (including the decomposers), and inorganic and dead organic matter. The **producers** or *autotrophs,* the energy-capturing base of the system, are largely green plants and algae. They fix the energy of the sun and manufacture food from simple inorganic and organic substances. Autotrophic metabolism is greatest in the upper layers of the ecosystem—the canopy of the forest and the surface water of lakes and oceans.

The **consumers** or *heterotrophs* utilize the food stored by the autotrophs, rearrange it, and finally decompose the complex materials into simple, inorganic substances. In this role they influence the rate of energy flow and nutrient cycling.

The heterotrophic component is often subdivided into two subsystems, consumers and decomposers. The consumers feed largely on living tissue, whereas the decomposers break down dead matter or **detritus** into inorganic substances. No matter how they are classified, all heterotrophic organisms are consumers, and all in some way act directly or indirectly as decomposers. Heterotrophic activity in the ecosystem is most intense where organic matter accumulates—in the upper layer of the soil, in the litter of terrestrial ecosystems, and in the sediments of aquatic ecosystems.

The third component consists of dead organic matter, dissolved organic and inorganic substances in aquatic systems, and the soil matrix. Dead organic matter comes from plant and consumer remains; decomposers act on it. Such matter is the basis of the internal cycling of nutrients in the ecosystem.

Traditionally, $CO_2$, $O_2$, and mineral nutrients derived from the weathering of materials and from precipitation are considered as abiotic components of the ecosystem. Here they are considered as inputs (Figure 10.2).

The driving force of the system is the energy of the sun, which causes all other inputs to circulate through the system. Outflows from one subsystem become inflows to another. While energy is used and dissipated as heat of respiration, the chemical elements from the environment are being recycled. Consumers regulate the speed at which nutrients turn over in the system.

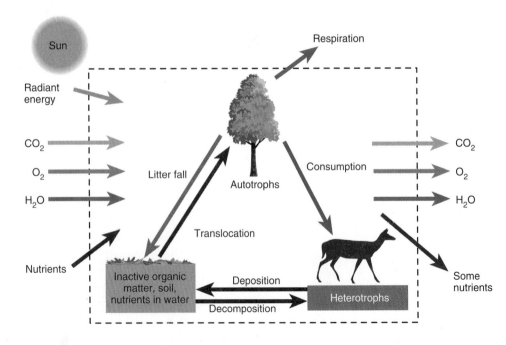

**Figure 10.2** Schematic diagram of an ecosystem. The dashed line represents the boundary of the system. (Adapted from O'Neill 1976.)

# ESSENTIAL PROCESSES

The movement of energy and nutrients through the ecosystem follows a route starting with photosynthesis and ending with decomposition. These are the two essential processes. Photosynthesis fixes light energy and incorporates nutrients into active plant tissue. Decomposition dissipates energy and reduces organic matter to inorganic substances.

## Photosynthesis

Almost all ecosystems are driven by the energy that is fixed and organic compounds that are produced by the photosynthetic activity of autotrophs like green plants and algae. Energy enters the ecosystem as visible light and is stored as reduced carbon compounds (mostly sugars) by plants through photosynthesis. Those reduced compounds are then oxidized through a series of biochemical reactions to form molecules containing less chemical energy. The reactions liberate energy, some of which is used to drive energy-requiring processes like active transport, mitosis, and protein synthesis. By-products of metabolism include water, carbon dioxide, and nitrogenous compounds, which are recycled through the system, and respiratory heat, which is dissipated.

Photosynthesis occurs in all green parts of vascular plants, bryophytes, and algae. The formula for photosynthesis is:

$$6CO_2 + 6H_2O \rightarrow 6CO_2 + 12H + 6O \rightarrow (CH_2O)_6 + 6O_2$$

The reduction of six moles of $CO_2$ to one mole of sugar $(CH_2O)$ results in the removal of six moles of $H_2O$ and the production of six moles of $O_2$. In essence, photosynthesis involves the transfer of electrons from water to carbon dioxide. Further syntheses produce free amino acids, proteins, fatty acids and fats, vitamins, nucleic acids, pigments, coenzymes, and secondary compounds. The synthesis of various products may take place in different parts of the plant. Mature leaves of certain species of plants produce only simple sugars, while young shoots and rapidly developing leaves produce fats, proteins, and other constituents.

Chlorophyll-bearing vascular plants and algae, both terrestrial and aquatic, account for most of global photosynthesis. Photosynthetic bacteria that use hydrogen, hydrogen sulfide, and various organic compounds instead of water as electron donors make minor contributions.

Photosynthetic Pathways Plant physiologists have learned that photosynthesis involves a complex sequence of metabolic reactions, which we will outline briefly. When light energy strikes a leaf, a portion of that energy is absorbed by the chlorophyll pigments in the chloroplast and transferred to an electron of a specific chlorophyll molecule. That electron is said to be excited and jumps to a nearby receptor molecule. Without the input of energy, the electron would stay with the chlorophyll. Almost immediately after receiving the electron, the receptor molecule donates it to a series of electron-carrier molecules, also in the chloroplast. The passage of the electron results in the synthesis of adenosine triphosphate (ATP) from adenosine diphosphate (ADP). Ultimately the electron is passed to oxidized nicotinamide adenine dinucleotide phosphate (NADP+), yielding the strong reductant NADPH. Meanwhile a molecule of water is split to form an electron, a proton, and an oxygen atom. The electron replaces the one lost to the chlorophyll molecule.

A second set of reactions generates carbohydrates from the fixation of $CO_2$ followed by the reduction of NADPH and the input of energy from ATP. In most plants the first step of this process links $CO_2$ from the atmosphere with a five-carbon sugar called ribulose bisphosphate (RuBP) (Figure 10.3). This reaction is catalyzed by a single enzyme, *ribulose bisphosphate carboxylase/oxygenase (Rubisco)*, which has been called the most important enzyme in all nature. Almost immediately two molecules of a 3-carbon compound, 3-phosphoglycolate (PGA) are produced. The plant converts some PGA into 6-carbon sugars, such as glucose and fructose; but most of the PGA is recombined to reform RuBP molecules that accept more $CO_2$. This photosynthetic pathway involving the 3-carbon PGAs is called the *Calvin-Benson cycle* or $C_3$ cycle (Figure 10.3a), and plants employing it are known as **$C_3$ plants** (Figure 10.4a). It is the dominant photosynthetic pathway of plants of temperate regions.

The $C_3$ pathway has one drawback. Rubisco not only has an affinity for $CO_2$; it also will catalyze reactions with oxygen. This reaction forms a 2-carbon molecule, phosphoglycolate, and one PGA, in contrast to the 2 PGAs formed in the reaction with $CO_2$. Further metabolism of phosphoglycoate results in the release of $CO_2$. The oxygenation of RuBP and the eventual release of $CO_2$ is called **photorespiration.** It uses more energy that it saves and reduces the efficiency of $C_3$ photosynthesis (see Ogren and Challot 1982). Photorespiration poses a severe problem in arid climates, where plants must close their stomata to prevent water loss. When stomata are closed, $CO_2$ within the leaf is depleted and $O_2$ accumulates.

Some plants found in warm to hot and semiarid areas have another pathway, called $C_4$ photosynthesis, facilitated by an internal leaf anatomy different from that of $C_3$ plants (Figure 10.5). $C_3$ plants have vascular bundles surrounded by a layer or sheath of large colorless cells; cells containing chlorophyll are irregularly distributed throughout the mesophyll of the leaf. Plants with the $C_4$ cycle have vascular bundles surrounded by a sheath or wreath of cells rich in chlorophyll, mitochondria, and starch. Mesophyll cells are arranged laterally around the bundle sheaths and have few chloroplasts.

**$C_4$ plants** have acceptor molecules of phosphoenolpyruvate (PEP), a three-carbon compound, within the chloroplasts of the mesophyll cells. $CO_2$ entering the leaf reacts initially with PEP carboxylase in the mesophyll cells to form oxaloacetate, which is soon converted to malic acid or aspartic acid, each having four-carbon molecules (Figure

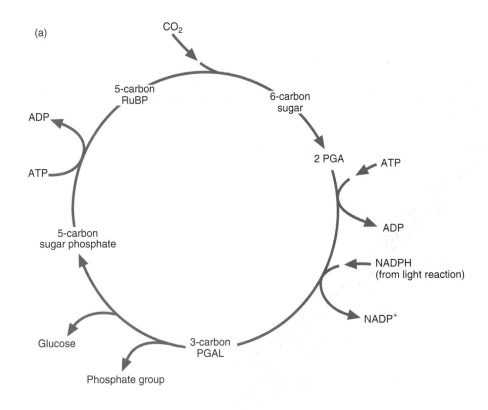

(a)

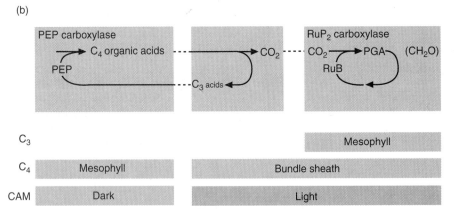

(b)

**Figure 10.3** (a) The Calvin-Benson or $C_3$ cycle. (b) Basic features of $C_3$, $C_4$, and CAM photosynthetic pathways compared. Carboxylations in the $C_4$ take place in different cells; in CAM plants, they take place in the same cell but at different time of day. (Adapted from Jones 1992:169.)

10.3b). These molecules are transported internally to the specialized bundle sheath cells. There the plant enzymatically breaks down malic and aspartic acids to release the fixed $CO_2$. Then the $CO_2$ is fixed again, using the normal $C_3$ enzymes to form three-carbon molecules as in the Calvin cycle. The remaining 3-carbon fragment is pyruvic acid, which diffuses back to the mesophyll cells and forms more PEP. $C_4$ plants do not replace the Calvin cycle. Rather PEP serves as a $CO_2$ pump to concentrate the $CO_2$ for the Calvin cycle within the bundle sheath cells (Figure 10.3b). As a result the concentration of $CO_2$ within the bundle sheath cells is significantly higher than in either the mesophyll cells or the surrounding atmosphere.

The extra step in the fixation of $CO_2$ gives $C_4$ plants a certain advantage. PEP carboxylase has a much higher affinity for $CO_2$ than does Rubisco, which has a tendency to catalyze $O_2$ when the relative concentration of $CO_2$ within the leaf is low and the concentration of $O_2$ is high. The PEP carboxylase in the mesophyll cells of $C_4$ plants captures $CO_2$ in the outer leaf tissue and releases $CO_2$ to the bundle sheath cells in the inner tissues. This process draws down the $CO_2$ in the mesophyll to near zero and increases the $CO_2$ pressure in the chloroplasts of the bundle sheaths, allowing $CO_2$ to compete effectively with oxygen for binding with RuBP. This ability to fix $CO_2$ at very low levels allows $C_4$ plants to carry on photosynthesis when the temperature is high or moisture is low, conditions that close down stomatal openings in the leaf to reduce loss of moisture. By creating a $CO_2$ sink in the mesophyll of the leaf, $C_4$ plants increase the steepness of the gradient of $CO_2$ concentration from the leaf

(a)

(b)

(c)

**Figure 10.4** (a) Most temperate herbaceous plants, like May-apple (*Podophyllum peltalum*), possess the $C_3$ photosynthetic cycle. (b) Switch grass (*Panicum vergatum*) is a warm season $C_4$ plant (c) Barrel cactus possesses the CAM photosynthetic cycle.

to the atmosphere. Carbon dioxide can then diffuse into the leaf when stomata are closed.

The $C_4$ pathway appears to be an adaptation to environments with high light intensity, high water stress, and high leaf temperatures. Such an environment places an extra premium on efficient $CO_2$ fixation while reducing water loss through the stomata. $C_4$ plants have a high water use efficiency—the number of grams of dry weight gained per kilogram of water transpired during the growing season. Compared to $C_4$ plants, $C_3$ plants have a low light saturation threshold (Figure 10.6). They also have a lower optimum temperature range (16° to 25° C) for photosynthesis than $C_4$ plants (30° to 45° C) (Figure 10.7).

The $C_4$ method is not found in algae, byrophytes, ferns, gymnosperms, or the more primitive angiosperms. $C_4$ species are mostly grasses (Figure 10.4b) native to tropical and subtropical regions, and some shrubs and dicot herbaceous plants of arid and saline regions. These are environments where high water use efficiency is important. Grasses make up about half of the known $C_4$ species. Most North American species of $C_4$ grasses have a subtropical distribution (Teeri and Stowe 1976, Teeri 1979). No known $C_4$ grasses grow on the tundra. The $C_4$ pathway occurs in unrelated taxa, so it evolved several times (Ehleringer and Monson 1993).

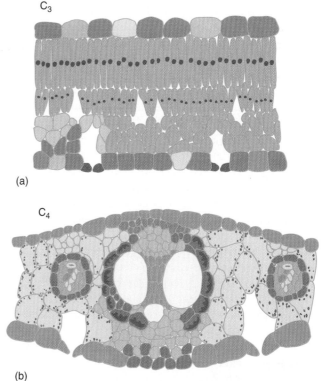

$C_3$

(a)

$C_4$

(b)

**Figure 10.5** Cross section of a leaf (a) of a $C_3$ plant and (b) of a $C_4$ plant. Note the bundle sheath cells surrounding the vascular bundle.

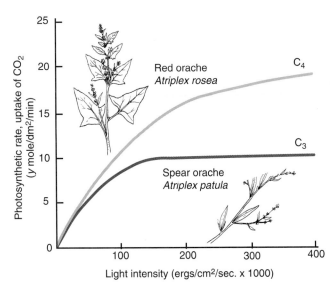

**Figure 10.6** Effect of changes in light intensity on the photosynthetic rates of $C_3$ and $C_4$ plants grown under identical conditions of a 16-hour day, 25° C at day, 20° C at night with ample water and nutrients. The $C_3$ species spear orache (*Atriplex patula*) exhibits a decline in the rate of photosynthesis, as measured by $CO_2$ uptake, as light intensity increases. The $C_4$ species red orache (*Atriplex rosea*) shows no such inhibition. (After Bjorkman 1973:53.)

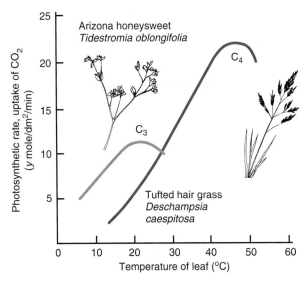

**Figure 10.7** Effect of changes in leaf temperature on the photosynthetic rates of $C_3$ and $C_4$ plants. $C_3$ plants, represented by tufted hair grass (*Deschampsia caespitosa*) of the Arctic tundra, exhibit a decline in the rate of photosynthesis as the temperature of the leaf increases. They also reach maximum photosynthetic output at a lower temperature than $C_4$ species, represented by Arizona honeysweet (*Tidestromia oblongifolia*). This desert species increases its rate of photosynthesis as the temperature of the leaf increases up to about 50° C. (After Bjorkman and Berry 1973.)

The ability of $C_4$ plants to carry on photosynthesis in full sun, at high temperatures, and in low water regimes gives $C_4$ plants a competitive edge over $C_3$ plants in those situations only. $C_3$ plants, in spite of their photorespiratory limitations, are more productive than $C_4$ plants in low light and cool temperatures. For this reason $C_4$ plants do not dominate vegetation. They grow poorly in shade, and their advance into temperate regions is restricted by low temperatures.

Some agricultural crops, such as sugar cane and corn, are $C_4$ plants; others, such as potatoes, are $C_3$ species with a low photosynthetic rate (see Bjorkmam and Berry 1973, Bassham 1977). Agricultural plants escape competition only because humans eliminate weeds, many of which are $C_4$ plants, including crabgrass (*Digitaria*), pigweed (*Setaria*), and barnyard grass (*Echinochloa*) (Teeri 1979).

Since $C_3$ plants, unable to concentrate $CO_2$, are carbon-limited and $C_4$ plants are not, how would each respond to the projected rise in atmospheric $CO_2$? Under current atmospheric conditions the ratio of $CO_2$ to oxygen is low, and the concentration of $CO_2$ in the chloroplasts of $C_3$ leaves is 1000 times less than $O_2$ (Ehleringer et al. 1991). This conspicuous presence of oxygen allows photorespiration to take place in $C_3$ plants. If atmospheric $CO_2$ rises, the concentration of $CO_2$ in the leaves of $C_3$ plants will increase and photorespiration will decrease. For example, if the concentration of $CO_2$ increased by 50 percent, $C_3$ plants would experience a 50 percent reduction in photorespiration. $C_4$ plants, however, would not show a response to an increase in $CO_2$, because these plants are already carbon-saturated at current $CO_2$ levels of 350 ppm. $C_3$ plants would have an advantage in growth over $C_4$ plants under conditions of rising atmospheric $CO_2$ (Ehleringer and Monson 1993). However, if rising $CO_2$ causes an increase in temperature and a decrease in rainfall, $C_4$ plants might benefit.

Yet another photosynthetic process, known as the **Crassulacean Acid Metabolism (CAM)**, is found among a number of succulent semidesert plants in some 15 families, including Cactaceae (Figure 10.4c), Euphorbiaceae, and Crassulaceae from which the method of carbon fixation received its name. CAM plants do not have specialized bundle sheath cells, and they open their stomata during the cool nights rather than by day to minimize water loss. During the night PEP carboxylase fixes $CO_2$ and accumulates large quantities of malic acid in the vacuoles of the mesophyll cells (Figure 10.3b). During the day the plants close their stomata to retain water. Malic acid enzymatically gives up the $CO_2$, which along with respiratory $CO_2$ is fixed by RuBP and incorporated into the $C_3$ cycle. $CO_2$ concentration remains high in the leaf during the day because it cannot diffuse through the closed stomata. As a result Rubisco operates in a high $CO_2$ environment during the day, eliminating photorespiration.

Although appearing to be similar in function to $C_4$ plants, CAM plants differ; they separate the initial fixation of $CO_2$

from the Calvin cycle at different times, rather than in different locations as in $C_4$ plants (Osmond 1978). The CAM pathway is slow and inefficient, but well adapted to a rigorous arid environment. Therefore CAM plants are native to deserts.

**Leaf Area Index** Photosynthesis is carried on by individual leaves, but the net photosynthetic production of a plant or community of plants depends upon the contribution of all leaves. That contribution depends upon their position in the vegetative profile and the degree to which the leaves above deplete the light resource for the leaves below.

The intensity of light reaching a plant is influenced by the local light regime, the modification of that light by the structure of vegetation, and the position of the leaf in the vegetation profile (Figure 10.8). Leaf-to-light interception can be described by the *leaf area index (LAI)* (Figure 10.9). The LAI is the ratio of leaf area per unit of ground area. The LAI is usually low at the beginning of the growing season and increases with full leaf and plant maturity. A low LAI indicates less leaf area than ground area and represents wasted sunlight on the ground; a high LAI indicates energy wasted by respiration of shaded leaves. Most plants have an optimal LAI, the point at which there is minimal shading of one leaf by another and net photosynthesis is maximum.

The optimal LAI of a plant or plant community depends upon the intensity of light radiation, the shape and arrangement of leaves in the canopy, and the angle of the sun. All of these change seasonally, daily, and even hourly. The optimal LAI increases with leaf area up to a point, as light intensity increases and as leaves are more perpendicular to the ground. Leaves that are perpendicular to incoming light (horizontal leaves) intercept the most light, but a number of layers of such leaves reduces the amount of light reaching

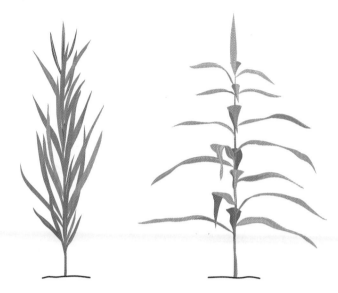

**Figure 10.8** Although horizontal leaves capture the most sunlight, the upper layers shade the lower, reducing the overall interception of light. An upright orientation of leaves is more efficient in capturing the sun's energy. That arrangement is typical of communities in which plants are growing closely, as among the grasses.

the lower ones. Leaves growing at an angle to the ground require a much higher LAI to intercept the same quantity of light as horizontal leaves; but as their LAI increases above a certain value, angled leaves carry on photosynthesis at a faster rate. Such an adaptation in leaf position, resulting in maximum photosynthesis rate despite a high LAI, is characteristic of corn, grasses, beets, turnips, and other row crops (Loomis et al. 1967).

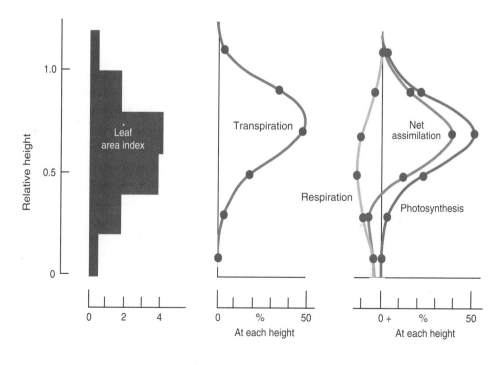

**Figure 10.9** The greatest amount of transpiration and photosynthesis takes place where the LAI is the highest. (After Baumgartner 1968.)

Efficiency of Photosynthesis The photosynthetic efficiency of converting energy of the sun to organic matter can be assessed in two ways: by the amount of energy required for the evolution of a molecule of oxygen or by the ratio of calories per unit area of harvested vegetation to solar radiation intercepted.

In terms of energy input, photosynthesis is an efficient process. To release 1 mole of oxygen and to fix 1 mole of carbon dioxide a $C_3$ plant needs an estimated 320 kcal of light energy. For each mole of oxygen evolved, approximately 120 kcal of energy are fixed. This corresponds to a gross efficiency of approximately 38 percent. Efficiencies calculated for isolated chloroplasts and for some algae are 21 to 33 percent (Bassam 1965, Kok 1965, Wassink 1968).

From the standpoint of calories stored in relation to light energy received, the efficiency is considerably less. The usable spectrum, 0.4 to $0.7\mu$ wavelengths, is only about one-half the total energy incident upon vegetation. Highest short-term net efficiency measured over periods of weeks of active growth may be 12 to 19 percent (Wassink 1968). In most instances, however, photosynthetic efficiency is computed either for the year or for the growing season. Efficiency in converting radiation to biomass for coniferous forests of the northern temperate region ranges from 0.1 to 3 percent, for deciduous forests from 0.5 to 1 percent, and for deserts from 0.01 to 0.2 percent (Webb et al. 1983). A Puerto Rican tropical rain forest has an estimated net photosynthetic efficiency of 7 percent (H. T. Odum 1970). Croplands range from 3 to 10 percent; corn, a $C_4$, has an efficiency of about 9 percent; and sugar cane, also a $C_4$ plant, about 8 percent (Cooper 1975). The net photosynthetic efficiency of land areas is about 0.3 percent, and of the ocean about 0.13 percent. Total yields of solar energy on Earth amount to about 0.15 to 0.18 percent (Wassink 1968).

## Decomposition

Decomposition is the breakdown of energy-rich organic matter by consumers to $CO_2$, $H_2O$, and inorganic nutrients such as nitrates, phosphates, and sulfates. Whereas photosynthesis involves incorporation of solar energy, carbon dioxide, water, and inorganic nutrients into organic biomass, decomposition involves the release of energy and the conversion of organic nutrients into inorganic ones. Decomposition is a complex of many processes, including fragmentation, change in physical structure, ingestion, egestion, and concentration. These processes are accomplished by a diversity of organisms linked in highly tangled food webs. All consumers, herbivores and carnivores, function as decomposers. As they digest food, they break down organic matter, change it, or release it partially decomposed to other feeding groups. True decomposition, however, is accomplished largely by the true decomposers—bacteria, fungi, and detritivores that feed on dead matter of all kinds.

The Decomposers The innumerable organisms involved in decomposition fall into several major functional groups. Organisms most commonly associated with decomposition are the **microflora,** comprising the bacteria and fungi. Bacteria may be aerobic, requiring oxygen for metabolism; or they may be anaerobic, able to carry on their metabolic functions without oxygen by using inorganic compounds, such as sulfates, as the oxidant. This type of respiration by anaerobic bacteria, commonly found in aquatic muds and sediments and in the rumen of ungulate herbivores, is fermentation. **Fermentation,** which converts sugars to organic acids and alcohols, inefficiently breaks down organic matter, lowers the pH of the substrate, and favors fungal activity. Many decomposer bacteria are *facultative anaerobes.* They use oxygen when it is present, but in its absence they can use inorganic compounds as their energy source. Others are **obligate anaerobes,** meaning they cannot survive with oxygen.

Bacteria are the dominant microfloral decomposers of animal matter. Fungi are the major decomposers of plant material. They extend their hyphae into the organic material to withdraw nutrients. Fungi range from "sugar fungi," which feed on highly soluble, easily available organic compounds such as glucose, to more complex hyphal fungi that invade tissues (Figure 10.10a).

Bacteria and fungi secrete enzymes into plant and animal material to break down the complex organic compounds. Some of the products are absorbed as food. Once one group has exploited the material to its ability, another group of bacteria and fungi able to utilize the remaining material more resistant to decomposition, such as cellulose and lignin, moves in. Thus a succession of microflora occurs in the decomposing organic matter until the material is finally reduced to inorganic nutrients.

Decomposition is aided by the fragmentation of leaves, twigs, and other dead organic matter, collectively called *detritus.* Litter-feeding invertebrates, the **detritivores,** do this work. They fall into four major groups: (1) microfauna, represented by protozoans; (2) mesofauna, whose body length falls between 100 μm and 2 mm, represented by mites (Figure 10.10b), springtails, and potworms; (3) macrofauna, between 2mm and 20mm, represented by nematodes, caddisfly larvae, and mayfly and stonefly nymphs; and (4) megafauna over 20 mm, represented by snails, earthworms, and millipedes (Figure 10.10c), and in aquatic systems by mollusks and crabs. These detritivores feed on plant and animal remains and on fecal material.

Energy and nutrients incorporated in bacterial and fungal biomass do not go unexploited in the decomposer world. Feeding on bacteria and fungi are the **microbivores.** Making up this group are protozoans such as amoebas, springtails (Collembola), nematodes, larval forms of beetles

**Figure 10.10** Examples of decomposers. (a) Fungi are major decomposers of leaf litter in the forest. (b) Mites are among the most abundant small detritivores. Some species are predacious. (c) Earthworms and millipedes are large detritivores important in the fragmentation of leaf litter.

(Coleoptera), flies (Diptera), and mites (Acarina). Smaller forms feed only on bacteria and fungal hyphae. Because larger forms may feed on both microflora and detritus, members of this group are often difficult to separate from detritivores.

Microbivores act as regulators of decomposition in two important ways: by controlling the abundance and distribution of fungi and by stimulating microbial activity (Lussenhop 1992, Seastedt 1984). Microbivores are important dispersers of fungal spores through the soil. By selectively grazing certain fungal populations, microarthropods—for example, Collembola—influence fungal distribution. They may also favor efficient or inefficient decomposer fungi by reducing competition among the hyphae of various fungal species and between bacteria and fungi. Grazing soil arthropods can so reduce microbial populations that they delay ordinary decomposition. Moderate grazing promotes microbial activity by preventing the microflora from overpopulation and by maintaining it at a level of maximum productivity or rate of division. Thus microbivores may prevent senescence of bacterial and fungal populations. In addition microarthropods stimulate fungal and bacterial growth by the direct return of mineral nutrients through their feces.

**Stages of Decomposition** Decomposition of dead plant and animal matter moves through several stages, from deposition to the final breakdown into inorganic nutrients (Figure 10.11). Early stages of decomposition involve **leaching,** the loss of soluble sugars and other compounds carried away by water. An abiotic process, it results in weight loss and changes in chemical composition. Early stages also involve **fragmentation,** the reduction of leaves and other organic matter into smaller particles. It may be accomplished physically by wind and trampling or chemically by digestion. Detritivores oxidate organic compounds, releasing energy, and degrade them into smaller and simpler products, processes collectively called **catabolism.** Conversion of materials from organic to inorganic form is **mineralization.** It results in the gradual disintegration of dead organic matter into nutrients available to primary producers and microbes.

The same decomposer organisms resynthesize these compounds into decomposer tissue in processes called **anabolism** (Figure 10.11). For a time microbes tie up these nutrients in their biomass. As long as the microbes live, these nutrients are unavailable for recycling. This incorporation of nutrients into living microbial biomass is known as *nutrient immobilization* (Figure 10.12). The amount of mineral matter that detritivores can process and incorporate in biomass varies greatly among them, especially seasonally (Figure 10.13). Immobilization of nutrients—particularly potassium, calcium, and nitrogen—can affect primary production. The amount of nutrients available for primary producers depends in part on the magnitude of uptake and release through death by microbial decomposers.

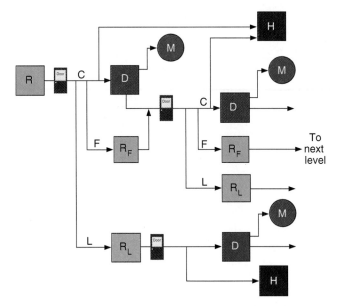

**Figure 10.11** A model of pathways in the decomposition of dead plant and animal matter (R). Initial decomposition involves fragmentation (F), of chemically unchanged litter, leaching (L) of soluble materials in an unchanged chemical form to another site, and catabolism (C), breakdown of complex molecules into smaller ones. Part of the catabolized compounds enters humus (H), and part is resynthesized into decomposer tissue (D). In time part of the decomposer tissue is mineralized (M); part becomes new decomposer tissue. Similar pathways continue at each stage of decomposition until the original material is completely mineralized. (Adapted from Swift, Heal and Anderson 1979:51, 53, 54.)

**Decomposition in Action** Microbial decomposition of plant leaves can begin while the leaves are still on the plant. Living plant leaves produce varying quantities of exudates that support an abundance of surface microflora (Ruinen 1962). These organisms feed on the exudates and on any cellular material that sloughs off. The same exudates account for the nutrients leached from the leaves during a rain. In tropical rain forests leaves are heavily colonized by bacteria, actinomycetes, and fungi.

Meanwhile, other organisms are using organic material from the roots of living plants. The soil region immediately surrounding the roots, known as the **rhizosphere,** and the root surface itself (the **rhizoplane**) support a host of microbial feeders on root litter and root exudates. Root exudates may consist of simple sugars, fatty acids, and amino acids. In fact some 10 sugars, 21 amino acids, 10 vitamins, 11 organic acids, 4 nucleotides, and 11 miscellaneous compounds have been identified in the rhizosphere. Obviously, not all such exudates occur in the rhizosphere of all plants. The absence of certain exudates can influence the quantitative and qualitative differences in the microflora of rhizospheres of different plant species (Clark 1969, Lussenhop 1992).

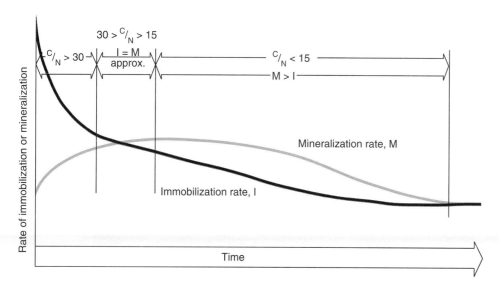

**Figure 10.12** The relationship between immobilization rate and mineralization rate during the decomposition of plant material. When the ratio of carbon to a nutrient (C/X), in this case nitrogen (C/N) is high, the immobilization rate is high and mineralization rate is low.

**Figure 10.13** Seasonal immobilization of nitrogen in forest litter by mesofauna and microbes. Microbial activity is greatest during the warmer months of the year. Microbes immobilize much greater quantities of nitrogen than mesofaunal arthropods, earthworms, and nematodes. Mesofaunal immobilization is much more constant throughout the year. Earthworm immobilization is the lowest during the summer, when litter is minimal. (After Ausmus, Edwards, and Witkamp 1976:411.)

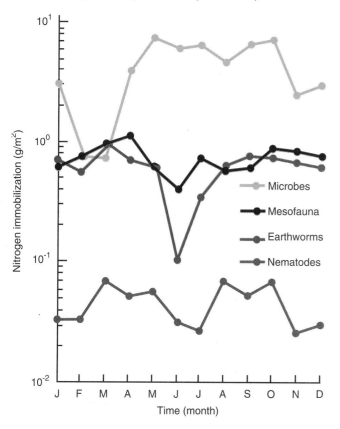

When the plant body becomes senescent, decomposition accelerates. The plant is invaded by both bacteria and fungi. If moisture is sufficient, fungi colonize dead stems of living grass plants. A favorite point of invasion is the point where the leaf is attached to the stem (the node). The species of fungal flora involved are influenced by the length of the internodes (the distance along the grass stem between leaves), the height of leaves from the ground (the closer to the ground, the more humid the habitat), and the species of grass (Hudson and Webster 1958; J. Webster 1956–1957). Fungi infect pine needles five to six months before the needles fall (Burges 1963). Destruction of the palisade layers of deciduous leaves by leaf miners opens up the affected leaves to microbial attack while they are still hanging on the tree. Still, most decomposition does not take place until the dead vegetation comes in contact with the soil.

Once on the ground, plant litter is subject to attack by other bacteria and fungi, mostly the latter. Among the first to invade the material are the sugar-consuming fungi, such as *Penicillium,* which use the readily decomposable organic compounds. When the glucose is gone, the debris is invaded by other bacteria that feed on more complex carbohydrates, hemicellulose, and cellulose (McClaugherty et al. 1985). The rate at which these organisms feed on the debris depends upon moisture and temperature. Higher temperature favors more rapid decomposition, and continuous moisture is more favorable than alternate wetting and drying.

The accessibility of detritus to the microflora is aided by detritivores that open holes and break the litter into smaller parts. The action of such litter feeders as millipedes and earthworms can increase exposed leaf area up to 15 times (Ghilarov 1970). Because the net assimilation of plant detritus by litter-feeders is on the average less than 10 percent—

they utilize only the easily digested proteins and carbohydrates—a great deal of material with concentrated mineral matter passes through the gut of these organisms. This fecal material is readily attacked by other microbes. Some litter-feeders, particularly the earthworms, enrich the soil with vitamin $B_{12}$. In addition, they bury surface litter and mix organic matter with the soil, bringing the material in contact with other microbes (Hartenstein 1986). James (1991) found that in a Kansas tallgrass prairie, earthworms annually processed 4 to 10 percent of the A horizon. Earthworms processed approximately 10 percent of the total soil organic matter in the top 15 cm, equivalent to 100 to 300 percent of plant annual belowground production each year. Although the mineral pool contained in the detritivores' biomass and their contribution to energy flow is relatively small (about 4 to 8 percent), the detritivores still make a major indirect contribution to decomposition (van der Drift 1971, Luxton 1982).

The importance of detritivores has been demonstrated in many experiments involving the use of nylon litter bags to exclude mesofauna and macrofauna from the litter sample or the use of naphthalene, which drives away arthropods but does not inhibit the activity of microflora. Witkamp and Olsen (1963) placed white oak leaves, some confined in litter bags and some unconfined, in pine, oak, and maple stands in Tennessee in November. By the following June both the unconfined and confined leaves showed a similar loss in weight. However, after June the unconfined leaves showed a sudden increase in microbial activity. Before June both types of leaves lost weight by the breakdown of easily decomposable substances through the action of microflora and microfauna. Because the unconfined leaves were broken into fragments by the detritivores as well as by birds, mice, wind, and rain, they were more available to microorganisms for further decay. Experiments by Edwards and Heath (1963), Witkamp and Crossley (1966) and others show that the suppression of activities of detritivores results in a marked slowdown in the rate of microbial decomposition. In the absence of these invertebrates, the decomposition of wood is slowed by half. Not only do the detritivores physically fragment the substrate, they also inoculate it with fungi and bacteria (Ghilarov 1970). Without the activity of the detritivores, nutrient elements could stagnate in the litter, bound energy in the ecosystem would increase, and the growth of plants and animals would decrease.

Decomposition of animal matter is more straightforward than decomposition of plant matter. The chemical breakdown of flesh does not require all of the specialized enzymes needed to digest plant matter. Decomposition is accomplished largely by bacteria rather than fungi and by certain arthropods such as blow flies (Calliphoridae)—if scavengers and large detritivores such as crows, vultures, and foxes leave any carcass behind for decomposers.

In summer and fall, when temperatures are high, microbial activity and colonization of dead flesh by blow flies (Calliphoridae) is intense. Blow fly maggots emerging from eggs laid in the carcass can consume a small mammal carcass in seven to eight days (Putman 1978a, b). Between bacteria and maggots, 70 percent of the organic material of a small mammal carcass is released, leaving only about 30 percent of the carcass behind as hair and bones. Because of low temperatures, decomposition in winter and spring is restricted to the relatively slow activity of microorganisms. The carcass, nearly 85 percent remaining, gradually becomes mummified through reduced pH and chemical changes in the tissues and is eventually fragmented and scattered.

Unlike the bodies of dead animals, most fecal material represents an already highly decomposed substrate. However, the dung of large grazing herbivores still contains an abundance of partially digested organic matter that provides a rich resource for specialized detritivores, in addition to bacteria, fungi, and earthworms. Among them are many species of flies that lay their eggs in dung, upon which the larvae will feed. The most notable are the dung beetles (Figure 10.14) (Scarabaeinae, Aphodiinae, and Geotrupinae). Some species of the Scarabaeinid dung beetles live beneath the dung upon which their larvae feed. Others, the tumblebugs, form a mass of dung into a ball in which they lay their eggs, roll it a distance, dig a hole, and bury it as a food supply for the larvae. Aphodiinae dung beetles tunnel and form a dung ball underground. The earth-boring dung beetles, Geotrupinae, spend most of their lives in deep burrows, usually beneath carrion or dung. The female lays her eggs in a plug of dung at the end of the burrow.

**Influences on Decomposition** The rate of decomposition is influenced by a number of abiotic and biotic variables. Both temperature and moisture greatly influence microbial activity. Alternate wetting and drying and continuous dry spells tend to reduce both the activity and popula-

**Figure 10.14** Dung beetle.

tions of microflora (Swift et al. 1979). Slope exposure, especially as it relates to temperature and moisture, and type of vegetation can increase or decrease decomposition. Witkamp (1963) found that bacterial counts on north-facing slopes were nine times higher in hardwoods than in coniferous stands, but counts from hardwood and coniferous stands on south-facing slopes did not differ. This finding undoubtedly was due to drier conditions on south-facing slopes.

The species composition of leaves in the litter had the greatest influence. Easily decomposed and highly palatable leaves from such species as redbud, mulberry, and aspen support higher populations of decomposers than litter from oak and pine, which is high in lignin. Earthworms have a pronounced preference for such species as aspen, white ash, and basswood, do not entirely consume sugar maple and red maple leaves, and do not eat red oak leaves at all (J. Johnston 1936). In a European study (Lindquist 1942) earthworms preferred the dead leaves of elm, ash, and birch, ate sparingly of oak and beech, and did not touch pine or spruce needles. Millipedes likewise show a species preference (van der Drift 1951). Thus decomposition of litter from certain species proceeds more slowly than litter from others. On easily decomposable material initially high populations and diversity of microflora decline with time as energy is depleted. On more resistant oak and pine litter, initially low population densities increase as decomposition proceeds (Witkamp 1963).

Decomposer organisms, like all other living things, have energy and nutrient requirements, which they meet by digesting energy-rich dead organic matter. The availability of any particular nutrient depends upon the ratio of energy supply to nutrient supply, expressed in terms of a carbon-nutrient ratio, C:X. For microorganisms, the nutrient most limiting is nitrogen, so the ratio is usually expressed C:N. The C:N ratio for plants is 40–80:1; for bacteria and fungi the C:N ratio is 10–15:1. When the C:X ratio is high, then the supply of that nutrient relative to the energy source is low. If the nutrients they require are limited, decomposers will incorporate these nutrients into their own biomass, immobilizing them (Figure 10.12). For example, if the C:N ratio is

greater than 30:1, any ammonium produced by the metabolic activities of the microbial organisms is reassimilated into microbial biomass as rapidly as it is formed. As decomposers use up the immediately available N, further decomposition becomes inhibited unless decomposers can draw nitrogen from another source, such as the soil. In this case the decomposers may deplete soil nitrogen to a point where plants experience nitrogen deficiency.

Thus as long as the supply of energy or carbon is high and the demand for a particular nutrient is greater than that available, the limited nutrient is conserved (immobilized) in the biomass of the decomposers. As carbon is used and reduced to $CO_2$, the ratio of C to X declines. When the ratio of the detrital material drops to the ratio of the organisms themselves, usually 20:1 or below, the nutrient is no longer in demand (Table 10.1), and mineralization takes place (Figure 10.12).

**Decomposition in Aquatic Ecosystems** Decomposition in aquatic ecosystems follows a pattern similar to that in terrestrial ecosystems, but with some major differences, influenced by the water. Decomposition still involves leaching, fragmentation, colonization of detrital particles by bacteria and fungi, and consumption by detritivores and microbivores.

In flowing water ecosystems, leaves, twigs, and other particulate matter are colonized by aquatic fungi. One group of aquatic arthropods, called shredders, fragment the organic particles and in the process eat bacteria and fungi on the surface; downstream collectors filter from the water fine particles and fecal material from the shredders. Grazers and scrapers feed on algae, bacteria, fungi, and organic matter collected on rocks and large debris (all discussed in Chapter 15). Algae take up nutrients and dissolved organic matter from the water.

In still, open water of ponds, lakes, and sea, the dead organisms and other organic material, called particulate organic matter (POM), drifts toward the bottom. On its way this POM is constantly ingested, digested, and mineralized until much of the organic matter settles on the bot-

**Table 10.1 Carbon–Nutrient Ratios for Resource and Decomposer Organisms, and Concentration During Uptake (carbon contents assumed for each level are wood, 47 percent; fungus, 45 percent; insect, 46 percent)**

|  | C:N | C:P | C:K | C:Ca | C:Mg |
|---|---|---|---|---|---|
| Resource (wood) | 157 | 1424 | 224 | 147 | 1022 |
| Fungal decomposer | 26 | 94 | 136 | 11 | 346 |
| Concentration | 6 | 15 | 2 | 13 | 3 |
| Animal decomposer | 6 | 51 | 66 | 157 | 235 |
| Concentration (to wood) | 26 | 28 | 3 | 1 | 4 |
| (to fungus) | 4 | 2 | 2 | xs | 2 |

tom in form of humic compounds. How much settles depends in part on the depth of the water through which the particulate matter falls. In shallow water much of it may arrive in relatively large packages, to be further fragmented and digested by bottom-dwelling detritrivores such as crabs, snails, and mollusks (Newell 1965). Bacteria too work on the bottom or benthic organic matter. Bacteria living on the surface can carry on aerobic respiration, but within a few centimeters deep the oxygen supply is exhausted. Under this anoxic condition, a variety of bacteria capable of anaerobic respiration take over decomposition. With oxygen depleted, these bacteria employ other inorganic electron acceptors, in particular $NO_3$ near the top of the bottom mud, resulting in denitrification; $Fe^{3+}$ and $SO_4$ in the middle layers, resulting in sulfate and iron reduction; and $HCO^{3-}$ in the deep muds, resulting in methane production.

Dissolved organic matter (DOM) in the water column is another story. A major source of this DOM is the macroalgae, phytoplankton, and zooplankton inhabiting the open water. Phytoplankton and other algae excrete quantities of organic matter at certain stages of their life cycle, particularly during rapid growth and reproduction. During photosynthesis the marine alga *Fucus vesiculosus* produces as exudate on the average 42 mg C/100 g dry weight of algae/hr. Total exudate accounts for nearly 40 percent of the net carbon fixed (Sieburth and Jensen 1970). Twenty-five to 75 percent of the regeneration of nitrogen and phosphorus takes place in the presence of microorganisms by cellular breakdown (*autolysis*) of phytoplankton and zooplankton and the exuded matter going into solution, rather than by bacterial decomposition (Johannes 1968). In fact, 30 percent of the nitrogen contained in the bodies of zooplankton is lost by autolysis within 15 to 30 minutes after death, too rapidly for any bacterial action to occur.

Although phytoplankton is able to take up organic and inorganic compounds, only bacteria in the water are able to assimilate the many kinds of dissolved organic molecules that occur in low concentrations in the water, convert them into bacterial biomass, and make them available for other feeding groups, notably zooplankton. Dissolved organic matter then becomes a substrate for the growth of bacteria. Both dissolved and colloidal matter condense on the surface of air bubbles in the water, forming organic particles on which bacteria flourish (Riley 1963, R. T. Wright 1970, Cole 1982). In effect, bacteria concentrate these nutrients by incorporating them into their own biomass and make them available for other feeding groups, notably zooplankton. As in terrestrial ecosystems the utilization of these organic nutrients by bacteria both increases and immobilizes nutrients. Bacterial consumption can reduce the supply of available nutrients to phytoplankton, thus reducing algal blooms.

Ciliates and zooplankton eat bacteria and in turn excrete nutrients in the form of exudates and fecal pellets in the wa-

ter. Zooplankton, too, in the presence of an abundance of food, consumes more than it needs and will excrete half or more of the ingested material as fecal pellets. These pellets make up a significant fraction of the suspended material. Bacteria attack these pellets to obtain the nutrients and growth substances they contain. Thus the cycle starts again.

# SUMMARY

The various biotic units that make up the biosphere, the thin layer of life on Earth, are known as ecosystems. They may be as large as vast, unbroken tracts of forest and grasslands or smaller than a pond. The ecosystem is an energy-processing system, receiving abiotic and biotic inputs. The driving force is the energy of the sun. Abiotic inputs include oxygen, carbon dioxide, and nutrients from weathering of Earth's crust and from precipitation. Biotic inputs include organic materials from surrounding ecosystems.

The ecosystem itself consists of three components: (1) the autotrophs, producers that fix energy of the sun; (2) the heterotrophs, consumers and decomposers that utilize the energy and nutrients fixed by the producers and return nutrients to the system; and (3) dead organic material and inorganic substrate that act as short-term nutrient pools and maintain cycling of nutrients within the system.

The most basic processes in the functioning of the ecosystem are photosynthesis and decomposition. Photosynthesis is the process by which green plants utilize the energy of the sun to convert carbon dioxide and water into carbohydrates. Most plants utilize the Calvin cycle alone, which involves the formation of a three-carbon phosphoglyceric acid used in subsequent reactions. Other plants utilize a four-carbon process, in which carbon dioxide taken into the leaf reacts to form malic or aspartic acid, stored in mesophyll cells. The $CO_2$ fixed in these compounds is then released to the Calvin cycle in the bundle sheath cells. $C_3$ and $C_4$ plants possess structural and physiological differences that are important ecologically. $C_4$ plants can carry on photosynthesis at higher leaf temperatures, at higher light intensities, and at lower $CO_2$ concentrations than $C_3$ plants. Succulent plants of the semiarid deserts utilize the Crassulacean Acid Metabolism (CAM). These plants open their stomata and fix $CO_2$ as malic acid by night. By day they close their stomata and utilize both fixed and respiratory photosynthesis in the Calvin cycle.

Involved in the return of nutrients to the ecosystem and the final dissipation of energy are the decomposer organisms. True decomposer organisms are bacteria, fungi, and other associated organisms. The detritivores, an array of organisms from microfauna such as protozoans to earthworms and caddisflies, fragment and digest large detrital material into smaller pieces more easily attacked by

microflora. Feeding on the microflora are the microbivores that act as regulators of microbial populations, preventing senescence and stimulating bacterial and fungal growth. Decomposers delay the loss of and regulate the movements of nutrients by immobilizing them in decomposer tissue and by releasing nutrients through mineralization. Initial decomposition rates are high because of the leaching of soluble compounds and consumption of highly palatable tissue by microbes. Later decomposition slows because less decomposable material remains. In terrestrial ecosystems bacteria and fungi take the major role in decomposition; in aquatic ecosystems bacteria and fungi act more as converters, whereas phytoplankton and zooplankton take a major part in the cycling of nutrients.

## REVIEW QUESTIONS

1. What is an ecosystem?
2. What are the three major components of an ecosystem? How are they related? What is the driving force of an ecosystem?
3. What two functional processes are absolutely essential, and why?
4. Compare the photosynthetic process in $C_3$, $C_4$, and CAM plants and related structural and functional differences.
5. What is the significance of leaf area index (LAI)?
6. What are the functions of the major groups of decomposers?
7. What are the sequential stages of decomposition?

## CROSS-REFERENCES

Characteristics of water, 64; soil humus, 139–140; primary production, 169–175; secondary production, 178–181; detrital food chain, 185–187; energy flow, 189–191; trophic levels, 191–192; food webs, 192–193, 336, 621–623; biogeochemical cycles, 196–221.

# Ecosystem Energetics

## *Concepts*

1. Energy, potential and kinetic, is the ability to do work.
2. Energy flow is governed by the laws of thermodynamics.
3. Light energy, fixed by autotrophs as primary production, is transferred to and stored by heterotrophs as secondary production.
4. The major pathways of energy through ecosystems are the detrital and grazing food chains.
5. Energy is recycled through ecosystems via complex food webs involving both pathways before its final dissipation by decomposition.
6. Efficiencies of energy transfer vary widely among autotrophs and poikilothermic and homeothermic heterotrophs.

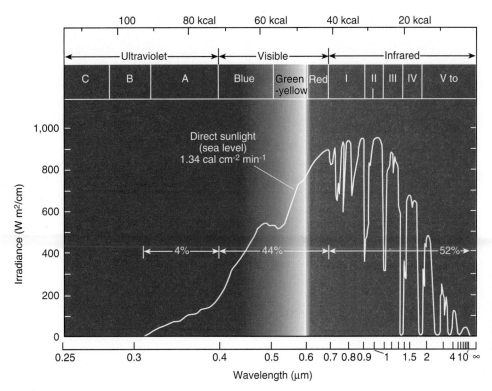

**Figure 11.1** Spectral distribution of solar radiation, showing the division of the spectrum involved in plant processes. The quantum content, expressed as kcal, is indicated for each frequency of incident radiation. Compare Figures 6.1 and 7.1. (Adapted from Gates 1965:9.)

The sunlight that floods Earth is a source of two forms of energy that keep the planet functioning (Figure 11.1). It arrives as light in little packets of energy known as photons (Chapter 7). When these photons reach land and water, they are transformed into another form of energy, **heat,** which warms Earth, heats the atmosphere (Chapter 6), drives the water cycle (Chapter 5), and causes currents of air and water (Chapter 4). Those photons that reach plants, especially leaves and other green parts, are transformed into **photochemical energy** used in photosynthesis (Chapter 10). That photochemical energy, fixed in carbohydrates and other compounds, becomes the source of energy for other living organisms.

## THE NATURE OF ENERGY

### Energy Defined

**Energy** is defined as the ability to do work. Work is what happens when a force acts through a distance, expressed as force × distance. Energy is either potential or kinetic. **Potential energy** is energy at rest. It is capable of and available for work. **Kinetic energy** is energy in motion, doing work. Work that results from the expenditure of energy can either store or concentrate energy (as potential energy) or arrange or order matter without storing energy.

Energy is measured in several units. The SI (Systeme Internationale) measure of a unit of energy is the *joule;* a joule equals 4.168 one-gram calories. A *calorie* is the amount of heat necessary to raise 1 gram (= 1 ml) of water 1° C at 15° C. When large quantities of energy are involved, a *kilogram calorie* (kcal) or Calorie is more appropriate. A kilocalorie is the amount of heat required to raise 1 kilogram (or 1 liter) of water 1° C at 15° C. Another often-used unit of measure is the *British thermal unit* (Btu), the amount of heat necessary to raise 1 pound of water 1° F. One Btu equals 252 g cal.

The transfer of energy involves the movement or flow of units of energy from one point to another, known as **flux** or power, measured in **watts per square meter per second** (= $4.1868 \times 10^3$ or 4186.8 kcal /sec). This flux of energy requires an **energy source,** the point from which energy flows, and an **energy sink** or receiver, the point to which it flows. Without a sink for heat energy, the sun could not be an energy source. Earth receives energy from the sun, absorbs a part of it, and gives up energy as heat to another sink, outer space.

Energy content and flow depend upon the actions of individual molecules. Characteristically, thermal energy is distributed rapidly among all molecules in a system without necessarily causing a chemical reaction. The effect of thermal energy is to set the molecules into a state of random motion and vibration. The hotter the object, the more the molecules are moving, vibrating, and rotating. These motions spread from a hot body to a cooler one, transferring energy in molecular terms from one to the other.

The energy of light waves, on the other hand, causes electronic transitions within atoms and molecules, called **excitations.** These excitations can lead to photochemical reactions.

## Laws of Thermodynamics

Two laws describe the expenditure and storage of energy. The **first law of thermodynamics** regards the conservation of energy. It states that energy is neither created nor destroyed. It may change forms, pass from one place to another, or act upon matter, transforming it to energy in various ways. However, regardless of what transfers and transformations take place, no gain or loss in total energy of a system occurs. All of it can be accounted for. Energy is simply transformed from one form or place to another.

For example, when wood is burned to ash, the potential energy present in the wood equals the kinetic energy released. Heat is evolved to the surroundings, so we call this type of reaction **exothermic.** On the other hand, energy from the surroundings may be drawn into a reaction. Here, too, the first law holds true. In photosynthesis, for example, the molecules of the products (glucose) store more energy than the reactants (carbon dioxide and water). The extra energy comes from sunlight. Again there is no gain or loss in total energy. When energy from the outside flows into a system to raise it to a higher energy state, the reaction is **endothermic.**

Although the total amount of energy in any reaction, such as burning wood, does not increase or decrease, much of the potential energy is degraded in quality and becomes unable to perform further work. This energy ends up as heat, serving to disorganize or randomly disperse the molecules. The measure of this relative disorder is called entropy.

This degradation of energy is the subject of the **second law of thermodynamics,** which makes an important generalization about energy transfer. It states that when energy is transferred or transformed, part of the energy is lost as waste; it assumes a form that cannot be passed on any further. When coal is burned in a boiler to produce steam, some of the energy creates steam that performs work, but part of the energy is dispersed as heat to the surrounding environment. The same thing happens to energy in an ecosystem. As energy is transferred from one organism to another in the form of food, a large part of the energy is degraded as heat through metabolic activity, with a net increase in entropy. The remainder is stored as living tissue. However, biological systems do not seem to conform to the second law, for the tendency of life is to produce order out of disorder, to decrease rather than increase entropy.

The second law, theoretically, applies to the isolated, closed system, in which there is no exchange of energy or matter between the system and its surroundings. Closed systems are largely physical laboratory artifacts. Closed systems approach thermodynamic equilibrium, a point at which all energy has assumed a form that cannot do work. A closed system tends toward a state of minimum free energy (energy available to do work) and maximum entropy, whereas an open system maintains a state of higher free energy and lower entropy. In other words, the closed system tends to run down; the open one does not. As long as there is a constant input of free energy and matter to the system and a constant outflow of entropy (in the form of heat and waste), the system maintains a steady state. Life is an open system maintained in a steady state.

Energy entering the biosphere as visible light is stored during photosynthesis in energetic covalent bonds found mainly in glucose. From that point energy and materials move through the ecosystem. As they do, biochemical changes reduce matter into compounds of less potential chemical energy. The chemical changes are accompanied by the production of metabolic heat, which eventually goes into the energy sink. This loss of heat is accompanied by a loss of carbon dioxide, water, and nitrogenous compounds that are cycled through the biosphere. Although some energy is irrevocably lost from the biosphere into outer space, some of it is stored in the system.

## PRIMARY PRODUCTION

The flow of energy through the ecosystem starts with its fixation by plants and algae through photosynthesis (Chapter 10). During that process solar energy hits chloroplasts within the cells, setting off reactions that produce carbohydrates and other organic molecules from carbon dioxide and water. The bonds of these organic molecules are energy-rich, containing energy from the sun. Thus all organic compounds of plants—cellulose, starch, proteins, nucleic acids, and secondary compounds—contain high amounts of chemical bond energy. Indeed, the entire plant body, with the exception of water and mineral nutrients like calcium and phosphorus, contains the energy-rich products of photosynthesis. Energy accumulated by plants is called production, or more specifically **primary production** because it is the first and most basic form of energy storage in an ecosystem. The rate at which energy is stored by photosynthetic activity is known as **primary productivity.** All of the sun's energy that is assimilated, total photosynthesis, is **gross primary production (GPP).** Like other organisms, plants require energy for reproduction and maintenance. The energy for these needs is provided by a reverse of the photosynthetic process, **respiration (R).** Energy remaining after respiration and stored as organic matter is **net primary production (NPP).** This flow of energy can be expressed by the formula:

$$\text{gross primary production (GPP)} = \text{net primary production (NPP)} + \text{respiration (R)}$$

Production, both gross and net, is usually measured as the rate at which energy or biomass is produced per unit area per unit time, or productivity. This rate is expressed in such terms as kilocalories per square meter per year (kcal/m$^2$/yr), a measure of energy, or grams per square meter per year (g dry wt/m$^2$/yr), a measure of biomass. If you know the caloric value of the plant material, you can convert dry weight per unit area to kilocalories.

The accumulated organic matter found on a given area at a given time is the **standing crop biomass.** Biomass is usually expressed as grams dry weight of organic matter per unit area (g dry wt/m²). Biomass differs from productivity, which is the rate at which organic matter is created by photosynthesis. Biomass present at any given time is not the same as total production. It does not include any plant tissue eaten by herbivores or lost through death prior to the period of measurement or take into account future growth. High standing crop biomass does not necessarily imply high productivity. The size of a standing crop represents accumulated biomass, not a rate of production. Small organisms turn over rapidly, being eaten nearly as fast as they are produced, and therefore accumulate minimal biomass. The size of their standing crop has little relationship to their productivity.

## Energy Allocation

Net primary production represents the storage of organic matter in plant tissue in excess of respiration. Plants budget this fixed energy or net income for different uses. A portion is allocated to **growth,** the buildup of components such as stems and leaves that promote the further acquisition of energy and nutrients (Chapin et al. 1990). A portion goes to **storage,** which is photosynthate built up in the plant for future for growth and other functions (Chapin et al. 1990). This storage involves accumulation, reserve formation, and recycling. **Accumulation** is the increase of compounds that do not directly support growth. These include carbon compounds such as such as starch and fructose, nitrogen as specialized storage proteins, and mineral ions. Accumulation occurs when the resource supply exceeds demands of growth and maintenance. It is most prevalent in species with inherently slow growth rates, such as trees. **Reserve formation** involves the synthesis of storage compounds from resources that otherwise would be allocated directly to promote growth. By **recycling** material from aging tissue to new growth, the plant is able to retain compounds that otherwise would be lost to litter. How plants budget their energy resources tells much about their life history and how they respond to environmental conditions and stresses.

With a limited energy budget and changing needs through its life history, a plant has to divert resources from one use to another during the year. Failure to maintain a balanced energy budget kills the plant. Early in its life history a plant spends net production on growth—new leaves and stems—that will add to its earning power of increased photosynthesis. Once a plant has built up capital in vegetative growth, it diverts energy from and often at the expense of growth to a reserve budget from which it can draw energy when needed, such as the replacement of leaves lost to insects. Later in the life cycle the plant spends a considerable amount of its photosynthate on flowers and then fruits. The plant withdraws resources from its reserve budget and vegetative capital to spend on reproduction, for its fitness depends on such allocations.

This withdrawal is evident in maturing annual and perennial plants. Because the plant has not only diverted resources away from the large lower leaves on its stem, but has actually withdrawn material from them, the lower leaves become brown or shriveled. The plant supports only a sufficient number of small upper leaves to meet maintenance needs. All of the remaining energy goes into reproduction and storage.

As you might expect, budget allocations differ among the many types and species of plants. It will even vary among closely related plants or members of the same species growing under different environmental conditions.

The energy budget of a plant growing in a favorable environment looks much different from one of the same species in a less favorable situation. A plant growing in a resource-impoverished environment expends more of its photosynthate on the growth of roots to reach scarce nutrients or moisture. It has less energy available for growth of leaves and stem, and ultimately for reproduction. The same species living on a nutrient-rich environment expends less on roots and more on vegetative growth. Some plants use more nutrients than they need, a strategy called luxury consumption. In the presence of an abundance of nitrogen, many plants increase their uptake of that element, decrease their allocation to roots, increase their investments in leaves and stems, accumulate aboveground biomass, and even fail to allocate sufficient energy to reproduction (see Fitter 1986).

Resource allocations differ considerably among annuals, perennial herbs, and woody plants. An annual plant (Figure 11.2) gets its initial start from the minimal amount of energy contained in the seed. Once the first seedling leaves begin photosynthesis, the plant invests up to 60 percent of its production to growth of aboveground vegetative biomass. When the plant has sufficient growth in leaf, stem, and root, the annual plant diverts production to flowers; only about 10 to 20 percent now goes to leaves. As the fertilized flowers develop into seeds, the plant withdraws energy from leaves and roots and sends 90 percent of its current photosynthetic production into growth and ripening of seeds, its only means of ensuring survival to the next year (Larcher 1980).

Perennials (Figure 11.3) have a different lifestyle, and their energy budget reflects it. Many, such as asters, goldenrods, and daisies begin growth in the spring from energy reserves stored in roots and in some species, such as saxifrage (*Saxifraga*) and roundleaf ragwort (*Senecio obovatus*), the rosettes of basal leaves. Once growth begins, the plant puts all of its production into vegetative structure, laying down these reserves as capital to draw upon for flowering. Once the plant is ready to flower, it diverts its production away from storage to production of flowers and seeds, even to the point of sacrificing some of the lower leaves on the stalk. Near the end of the season, perennial plants translocate energy from their aging leaves to the roots, where the en-

**Figure 11.2** Annual plants, such as foxtail (*Setaria*), allocate most of their production to growth and ripening of seeds. Seeds are their only means of overwinter survival.

**Figure 11.3** Perennial plants, like this ox-eye daisy (*Chrysanthemum leucanthemum*), start growth in spring from energy stored in roots, to which they return energy in the fall. During the flowering period the plant diverts most of its energy to reproduction.

ergy will be available for initiating next season's growth (Larcher 1980).

Other herbaceous plants, such as white trillium (*Trillium grandiflorum*) (Figure 11.4) and spring beauty (*Claytonia* spp.), take a different approach. They draw on energy reserves stored in the root to support early flowers or other reproductive structures and early leaves in the spring. In late spring and early summer they produce more leaves that generate photosynthate that is sent back into the roots to build up capital for next year's blooms. When the plant has accomplished that, it ceases production and aboveground biomass dies.

Woody plants must invest their energy into more items, with proportionately more going to woody tissue and roots. In spring the tree draws upon and uses up at least one-third of the reserves to get its new leaves started. Once photosynthesis begins, the deciduous trees budget their energy production for growth of leaves, then flowers, fruit, new cambium, new buds, and final deposits of starch in roots and bark, roughly in that order. A tree's most expensive drain on its annual energy budget is the production of fruit. A pine tree expends 5 to 15 percent of its annual photosynthetic production on cones; beech and oaks may spend more than 20 percent of their net energy income on nuts and acorns; and apple trees can use up 35 percent on apples (Larcher 1980). In fact, fruit production is so expensive that most trees can afford the luxury of a good crop of fruit only once every three to seven years. When they do, trees draw down their carbohydrate reserves and limit the growth of other

**Figure 11.4** White trillium (*Trillium grandiflorum*), an early spring flower, draws on energy reserves in the roots to support early blooms and early vegetative growth. Then they send their production back into the roots to build up reserves for the following spring.

components. At the end of the growing season, deciduous trees withdraw carbohydrates and nutrients from the leaves and send them back to the roots.

On an ecosystem-wide basis, however, involving individuals of a number of species with different life histories, and annual variations in reproductive activity, energy allocation over a season will average out somewhat differently. For example, based on annual biomass accumulation, trees of a young oak-pine forest on Long Island, New York, allocated 25 percent of net primary production to stem wood and bark, 40 percent to roots, 33 percent to twigs and leaves, and 2 percent to flowers and seeds. Shrubs budgeted 54 percent of net primary production to roots, 21 percent to stems, 23 percent to leaves, and 2 percent to flowers and fruit (Whittaker and Woodwell 1969).

Overall, the proportionate allocation of net primary production to belowground and aboveground biomass tells much. A high root-to-shoot ratio (R/S) indicates that most of the production goes into the supportive function of plants and that most of their active biomass is belowground. A large root biomass enables plants to reach water and nutrients. Thus plants with a large root biomass are more effective competitors for water and nutrients and can survive more successfully in infertile and harsh environments. Plants with a low R/S ratio have most of their biomass above ground and assimilate more light energy, resulting in higher productivity.

Differences in root-to-shoot ratios characterize various ecosystems. Sedge and grass meadows of the tundra, plants characteristic of an environment with a long, cold winter and a short growing season, have R/S ratios ranging from 5 to 11. Tundra shrubs may range from 4 to 10. Further south, midwest prairie grasses have an R/S ratio of about 3, indicative of cold winters and low moisture supply. Forest ecosystems, with their high aboveground biomass, have a low R/S ratio. For the Hubbard Brook forest in New Hampshire the R/S ratio (based on data of Gosz et al. 1976) for trees is 0.213, for shrubs 0.5, and for herbs 1.0. As we could predict, the R/S ratio increased from the trees down to the herbaceous layer.

Changes in R/S ratios reflect the plants' response to stress. Grasses may respond to grazing stress by concentrating more of their net production in roots (Andrews et al. 1975). In heavily grazed plots on shortgrass prairie, grasses allocated 69 percent of net primary production to roots, 19 percent to shoots, and 12 percent to basal crowns. In contrast, lightly grazed prairie allocated 60 percent of net primary production to roots, 22 percent to shoots, and 18 percent to crowns.

## Biomass Distribution

Aboveground biomass is distributed vertically in the ecosystem. The vertical distribution of leaf biomass in terrestrial systems, and of floating-leaf vegetation and the concentration of phytoplankton in the surface waters of aquatic systems, influences the penetration of light. The degree of light penetration, in turn, influences the distribution of production in the ecosystem (Figure 11.5).

The region of maximum productivity in aquatic ecosystems is not the upper strongly sunlit surface (strong sunlight inhibits photosynthesis) but some depth below, depending upon the clarity of the water and the density of plankton growth. As depth increases, light intensity decreases until it reaches a point at which the light received by the phytoplankton is just sufficient to meet respiratory needs and production equals respiration. This point is known as **compensation intensity.**

In the forest ecosystem a similar situation exists. The greatest amount of photosynthetic biomass as well as the highest net photosynthesis is not at the top of the canopy, but at some point below maximum light intensity. In spite of wide differences in plant species and types of ecosystems, the vertical profiles of biomass are similar.

## Ecosystem Productivity

The productivity of ecosystems varies widely over the globe (Figure 11.6). It is strongly influenced by temperature and rainfall (Figure 11.7). Net productivity for a variety of ecosystem types is summarized in Table 11.1. The most productive terrestrial ecosystems are tropical forests with high rainfall and warm temperatures; their net productivity ranges between 1000 and 3500 g/m²/yr. Temperate forests, where rainfall and temperature are lower, range between 600 and 2500 g/m²/yr (Whittaker and Likens 1975). Shrublands such as heath balds and tallgrass prairie have net productions of 700 to 1500 g/m²/yr (Whittaker 1963; Kucera et al. 1967). Desert grasslands produce about 200 to 300 g/m²/yr, whereas deserts and tundra range between 100 and 250 g/m²/yr (Rodin and Bazilevic 1967). These differences in net primary productivity from tropic to arctic regions are reflected in litter production, which in tropical forests ranges between 900 and 1500 g/m²/yr, in temperate forests 200 and 600 g/m²/yr, and in arctic and alpine regions 0 and 200 g/m²/yr (Bray and Gorham 1964).

Net productivity of the open sea is generally quite low (Figure 11.8). The productivity of the North Sea is about 170 g/m²/yr, the Sargasso Sea 180 g/m²/yr. However, in some areas of upwelling, such as the Peru Currents, net productivity can reach 1000 g/m²/yr.

Consistently high productivity usually results from an energy subsidy to the system. This subsidy may be a warmer temperature, greater rainfall, circulating or moving water that carries in food or additional nutrients, or, in the case of agricultural crops, the use of fossil fuel for cultivation and irrigation, the application of fertilizer, and the control of pests. Swamps and marshes, ecosystems at the interface of land and water, have net productivity ranging from 900 to 3300 g/m²/yr. Estuaries, because of input of nutrients from rivers and tides, and coral reefs, because of input from

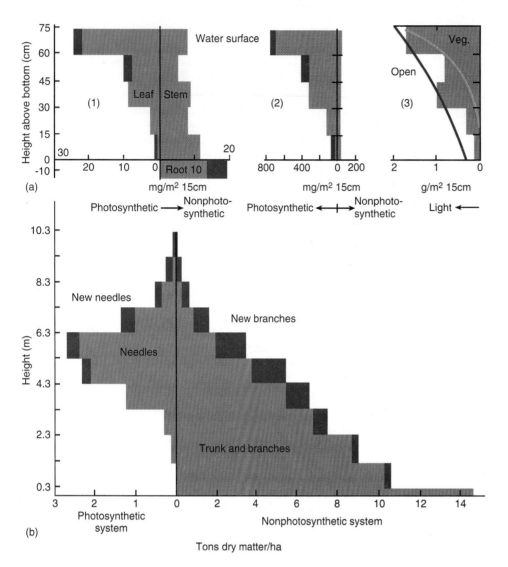

**Figure 11.5** Vertical distribution of production and biomass in aquatic and terrestrial communities. (a) Three graphs for the pondweed (*Potamogeton*) community: (1) division of biomass into leaf, stem, and root (solid areas represent winter buds); (2) concentration of chlorophyll in the plant community; (3) leaf area and light profile. The orange line is light in the community; the blue line represents light in open water. Note that less light reaches greater depths in the community because of vegetation.(After Ikusima 1965.) (b) Structure and productive system (foliage) of a pine-spruce-fir forest in Japan. (After Monsi 1968.)

changing tides, have a net productivity between 1000 and 2500 g/m²/yr. Among agricultural ecosystems sugar cane has a net productivity of 1700 to 1800 g/m²/yr, hybrid corn 1000 g/m²/yr, and some tropical crops 3000 g/m²/yr.

In any ecosystem, annual net production (accumulation of biomass) changes with age. For example, a Scots pine plantation achieved maximum production of $22 \times 10^3$ kg/ha at the age of 20; it then declined to $12 \times 10^3$ kg/ha at 30 years of age (Ovington 1961). Woodlands apparently achieve their maximum annual production in the pole stage (20 to 30 cm), when the dominance of the trees is the great-

est and the understory is at a minimum (Figure 11.9). The understory makes its greatest contribution during the juvenile and mature stages of the forest. As age increases, more and more of the gross production is needed for maintenance, and less remains for net production.

Net production also declines from a young ecosystem such as a weedy field or an agricultural crop to a mature plant community such as a forest. As plant communities approach a stable or steady-state condition (see Chapter 29), more of the gross production is used for the maintenance of biomass and less goes into newly added organic matter.

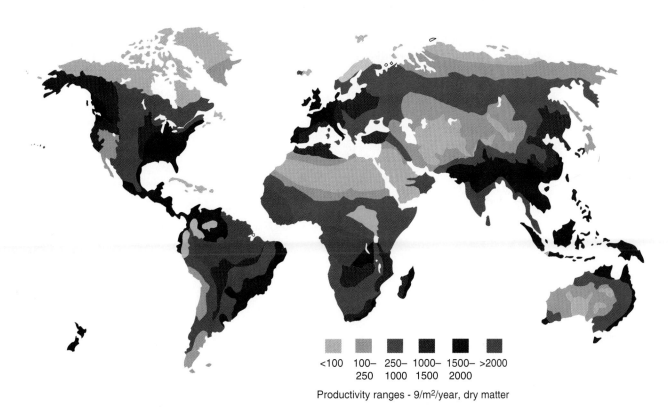

**Figure 11.6** A map of world terrestrial primary production. Note the high productivity of tropical regions. (Based on Golley and Leigh 1972.)

Productivity ranges - 9/m²/year, dry matter

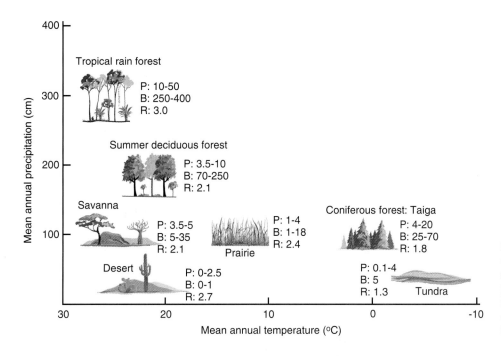

**Figure 11.7** Climatic distribution of primary production, biomass, and radiation input. P = primary production (tn/ha); B = biomass (tn/ha); R = solar radiant input (kcal/m²/yr 0.3–3.0 µ). (From Etherington 1975:355.)

**Table 11.1** Net Primary Productivity and Plant Biomass of World Ecosystems

| Ecosystems (in Order of Productivity) | Area (10⁶ km²) | Mean Net Primary Productivity per Unit Area (g/m²/yr) | World Net Primary Productivity (10⁹ mtr/yr) | Mean Biomass per Unit Area (kg/m²) |
|---|---|---|---|---|
| CONTINENTAL | | | | |
| Tropical rain forest | 17.0 | 2000.0 | 34.00 | 44.00 |
| Tropical seasonal forest | 7.5 | 1500.0 | 11.30 | 36.00 |
| Temperate evergreen forest | 5.0 | 1300.0 | 6.40 | 36.00 |
| Temperate deciduous forest | 7.0 | 1200.0 | 8.40 | 30.00 |
| Boreal forest | 12.0 | 800.0 | 9.50 | 20.00 |
| Savanna | 15.0 | 700.0 | 10.40 | 4.00 |
| Cultivated land | 14.0 | 644.0 | 9.10 | 1.10 |
| Woodland and shrubland | 8.0 | 600.0 | 4.90 | 6.80 |
| Temperate grassland | 9.0 | 500.0 | 4.40 | 1.60 |
| Tundra and alpine meadow | 8.0 | 144.0 | 1.10 | 0.67 |
| Desert shrub | 18.0 | 71.0 | 1.30 | 0.67 |
| Rock, ice, sand | 24.0 | 3.3 | 0.09 | 0.02 |
| Swamp and marsh | 2.0 | 2500.0 | 4.90 | 15.00 |
| Lake and stream | 2.5 | 500.0 | 1.30 | 0.02 |
| Total continental | 149.0 | 720.0 | 107.09 | 12.30 |
| MARINE | | | | |
| Algal beds and reefs | 0.6 | 2000.0 | 1.10 | 2.00 |
| Estuaries | 1.4 | 1800.0 | 2.40 | 1.00 |
| Upwelling zones | 0.4 | 500.0 | 0.22 | 0.02 |
| Continental shelf | 26.6 | 360.0 | 9.60 | 0.01 |
| Open ocean | 332.0 | 127.0 | 42.00 | 1.00 |
| Total marine | 361.0 | 153.0 | 55.32 | 0.01 |
| World total | 510.0 | 320.0 | 162.41 | 3.62 |

*Source:* Adapted from Whittaker and Likens 1973.

Thus the ratio of gross production to biomass declines through time (Figure 11.10).

Productivity varies considerably not only among different types of ecosystems, but also among similar systems and within one system from year to year. Productivity is influenced by such factors as nutrient availability, moisture, especially precipitation, temperature, length of the growing season, animal utilization, and fire. For example, the herbage production of a grassland may vary by a factor of eight between wet and dry years (Weaver and Albertson 1956). Overgrazing of grasslands by cattle and sheep or defoliation of forests by such insects as the saddled prominent and gypsy moth can seriously reduce net production. Fire in grasslands may increase productivity if moisture is normal, but reduce it if precipitation is low (Kucera et al. 1967). An insufficient supply of nutrients, especially nitrogen and phosphorus, can limit net productivity, as can the mechanical injury of plants, atmospheric pollution, and the like.

Although the size of the standing crop is not synonymous with high productivity, the size of the standing crop does influence the capacity to produce. A pond with too few fish or a forest with too few trees does not have the capacity to utilize the energy available. On the other hand, too many fish or too many trees means less energy available to each individual. Crowding lowers the efficiency of use and influences the storage and transfer of energy through the ecosystem.

## Estimating Primary Productivity

Estimating primary productivity in an ecosystem is difficult. It is hard to measure the photosynthetic rate of photosynthetic tissue and the respiration of nonphotosynthetic tissue. We are left with techniques that measure accumulation and loss of plant biomass through time. Plant biomass (dry weight) representing net production is the best integrator of the difference between gross primary production and respiration. However, net production (biomass) can be low because initial $CO_2$ fixation or gross production is low or because $CO_2$ loss through respiration is high. These limitations must be appreciated in any determinations of productivity. Some techniques estimate productivity indirectly by relating the amounts of biomass accumulation to net primary production, or to the quantities of oxygen or carbon dioxide released or used in production.

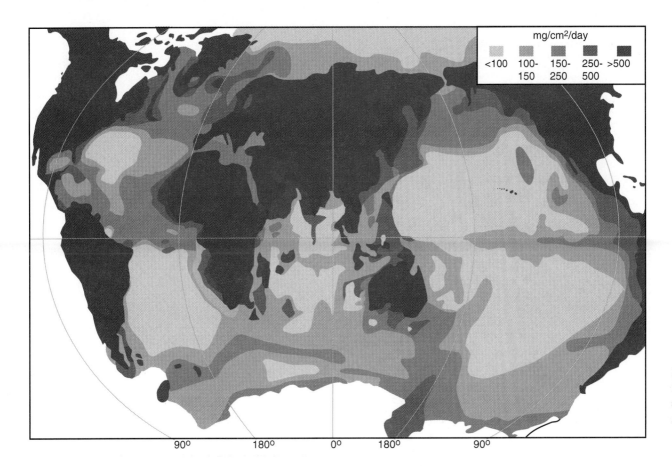

**Figure 11.8** Geographical variations in the primary productivity of the ocean. Note that the highest productivity is near the coastal zones; the lowest productivity is in the open sea.

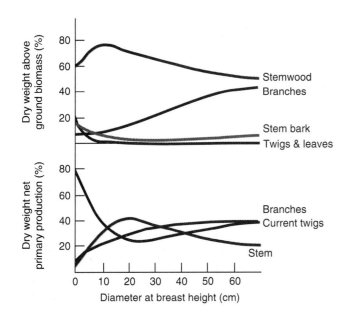

**Figure 11.9** Relation of aboveground biomass and production to size of tree for 63 sample trees of three major species common to the deciduous forests of eastern North America—sugar maple, yellow birch, and beech. Note that as trees increase in size the ratio of branches to stems increases. In larger trees branches account for a greater proportion of net primary production than stems. In smaller trees current leaves and twigs account for the greater percentage of primary production. This percentage declines rather rapidly as trees approach pole stage (20 to 30 cm diameter breast height), then increases as trees mature. (From Whittaker et al. 1974:239.)

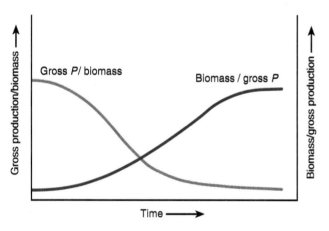

**Figure 11.10** Change through time in ratios between gross community photosynthesis and biomass or production efficiency, and between biomass and gross community photosynthesis, or maintenance efficiency. Note the early high production efficiency and the later accumulation of biomass as a result of production. (After G. D. Cooke 1967:71.)

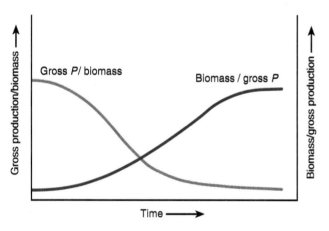

**Light and Dark Bottle Method** Aquatic ecologists commonly use the **light and dark bottle method.** It is based on the premise that the amount of oxygen produced is proportional to gross production, because one molecule of oxygen is produced for each atom of carbon fixed. They suspend two bottles containing a given concentration of phytoplankton at a level from which they obtained the samples. One bottle is black to exclude the light; the other is clear to admit light. In the light bottle a quantity of oxygen proportional to the total organic matter fixed (gross production) is produced by photosynthesis. At the same time the phytoplankton is consuming oxygen for respiration. Thus the amount of oxygen remaining in the bottle is proportional to the amount of fixed organic matter remaining after respiration of net production. In the dark bottle phytoplankton is using but not producing oxygen. Thus the quantity of oxygen used, obtained by subtracting the amount of oxygen left in the bottle at the end of the run (usually 24 hours) from the quantity at the start, gives a measure of respiration. The amount of oxygen in the light bottle added to the amount used in the dark provides an estimate of total photosynthesis or gross production.

This method has its problems. Some of the respiration attributed to phytoplankton may be bacterial. The phytoplankton population may increase in the light bottle during the experimental run but not in the dark bottle. Also, the procedure is based on the assumption that respiration in the dark is the same as that in the light.

**Radiocarbon Tracer Method** One of the most useful and sensitive methods aquatic ecologists employ to determine productivity, especially in lakes and oceans of low productivity, is the measurement of $^{14}C\text{-}CO_2$ uptake (Peterson 1980). Basically the method involves the addition of a quan-

tity of radioactive carbon as a carbonate to a sample of water in clear and opaque bottles containing the natural phytoplankton population. After a period of time, usually 24 hours, the investigators strain the plankton material from the water, wash it, and dry it. Then they take counts of radioactivity and from them calculate the amount of carbon dioxide fixed in photosynthesis. This estimate is based on the assumption that the ratio of activity of $^{14}C\text{-}CO_2$ added to the activity of phytoplankton is proportionate to the ratio of the total carbon available to phytoplankton to that assimilated.

This technique, too, has its deficiencies. Like the light and dark bottle method, it does not discriminate between respiration by phytoplankton and bacteria, the uptake and release of $^{14}C$ by bacteria and zooplankton, and the abilities of different species of phytoplankton to use available light.

**Chlorophyll Concentration** A third method aquatic ecologists use is the estimation of production from chlorophyll concentration and light intensity through the water column. This technique evolved from the discovery by plant physiologists that a close relationship exists between chlorophyll concentration and photosynthesis at any given light intensity (Ryther and Yentsch 1957). The relationship remains constant for different species of phytoplankton.

This method involves the determination of chlorophyll content of phytoplankton in a given volume of water. Because all plants need chlorophlyll to carry on photosynthesis, the amount of chlorophyll in a given amount of water is a direct measure of the total biomass or standing crop of phytoplankton it contains. The technique is simple. Aquatic ecologists chemically extract the chlorophyll from the water samples and measure the amount of color in a colormeter. The deeper the color, the greater is the concentration of chlorophyll and therefore phytoplankton biomass.

The problem with this technique is that chlorophyll concentration varies with phytoplankton species and even within the cells of species. Further, the extraction technique may alter the chlorophyll. Nevertheless, the technique does provide an estimate. It is well-suited for survey work in aquatic ecosystems.

**Carbon Dioxide Flux** One of the most useful methods for estimating primary productivity in terrestrial ecosystems is the measurement of the uptake of carbon dioxide in photosynthesis and its release in respiration (Woodwell and Botkin 1970, Hillbert and Oechel 1987). In this method a sample of the community, which may be a twig and its leaves, a segment of a tree stem, the ground cover and soil surface, or even a portion of the total community, such as an on-site sample of grassland or arctic tundra, is enclosed in a clear plastic tent. Air is drawn through the enclosure at a known rate, and the carbon dioxide concentration of the incoming and outgoing air is measured with an infrared gas analyzer. The assumption is that any carbon dioxide removed from the incoming air during the day has been incorporated into organic matter. Therefore the quantity of carbon dioxide in the light bag is equivalent to photosynthesis mi-

nus respiration. A similar sample may be enclosed in a dark bag. The amount of carbon dioxide produced in the dark bag is a measure of respiration when photosynthesis shuts down. The quantity of $CO_2$ in the light and dark bags added together estimates gross production.

Because the enclosure of a segment of a community necessarily alters the environment of the sample, the aerodynamic method can be used instead (Lemon 1960, 1983; Monteith and Szeicz 1960; Saugier and Ripley 1974). It is based on the assumption that the reduction of carbon dioxide in a canopy of vegetation is equivalent to net photosynthesis. The method involves the periodic measurement of the vertical gradient of carbon dioxide concentration by means of sensors arranged in a series from the ground to a point above the vegetation. The difference in the flux or flow between layers represents the amount of carbon dioxide used by the foliage of each layer. The rate of diffusion or flux within the canopy is proportional to the gradient of carbon dioxide and to wind speed and turbulence within the canopy. One shortcoming of the technique is its inability to distinguish between soil and plant contributions to the carbon dioxide flux and to determine the uptake of carbon dioxide by roots and its transport to the chloroplasts.

**Harvest and Dimension Analysis** Widely used to estimate production in terrestrial ecosystems is the harvest method. It is most useful for estimating the production of cultivated land, range, and communities of annual plants where production starts from zero at seeding or planting time, becomes maximum at harvest, and is subject to minimal use by consumers.

Briefly, the technique involves clipping or removing vegetation at periodic intervals and drying the samples to a constant weight. To obtain an accurate estimate the production of plant biomass must be sampled throughout the growing season and the contribution of each species must be determined. Different species of plants reach their peak production at different times during the growing season. The difference in standing crop biomass between harvest periods expressed as grams per square meter per unit time provides an estimate of net primary productivity. Caloric value of the material can be determined through use of a calorimeter, and the biomass can be converted to calories. Net primary productivity is then expressed as kilocalories per square meter per year.

Harvesting aboveground material estimates only aboveground productivity. Estimation of belowground productivity requires the sampling of root biomass, which is difficult at best. Although the roots of some annual and crop plants may be removed from the soil, the task becomes more difficult with grass and herbaceous species, and even more so with forest trees. With the exception of annual plants, investigators face the almost impossible task of separating new roots from old. They have the added problem of estimating the turnover of short-lived small roots and root material and

the variability of the sample. One technique to estimate small root production is direct observation of root growth in glass-sided root pits (Neubould 1968). Another, harvest-type technique, is the extraction of fine roots from soil core samples (Singh 1984). A new indirect technique showing considerable promise involves intense, frequent sampling of fine roots, as just described, together with the measurement of nitrogen fluxes between vegetation, soil organic matter, mineralization, inputs from precipitation and mineralization, and nitrogen availability (Nadelhoffer et al. 1985). These measurements account for all nitrogen fluxes except for belowground litter or turnover of fine roots. The transfer of nitrogen from vegetation to soil organic matter by the way of belowground litter can be calculated indirectly as the difference between total nitrogen flux and the measurements of all other fluxes.

Because plants of different age, size, and species make up the forest community, ecologists use a modified harvest technique known as **dimension analysis** (Whittaker and Woodwell 1968, Whittaker and Marks 1975). Dimension analysis involves the measurement of height, diameter at breast height (dbh), and diameter growth rate of trees in a sample plot. Investigators cut, weigh, and measure a set of sample trees and determine their current growth and age by growth ring analysis. They also determine total weight, both fresh and dry, of the leaves and branches as well as the weight of trunk and limbs. Often they excavate and weigh root biomass and determine biomass of ground vegetation and litterfall. By various calculations, ecologists obtain the net annual production of wood, bark, leaves, twigs, roots, and flowers. From this information they estimate biomass and production of trees and other vegetation in the sample unit and then interpolate it for the whole forest. Because energy utilized by plants and plant material consumed by animals and microorganisms are not accounted for, both the harvest method and dimension analysis estimate net community rather than net primary production.

Data obtained from such intensive sampling provides information to develop regression equations for particular forest types, needed to estimate standing crop and productivity from less intensive sampling. After dimension analysis ecologists and foresters can predict standing crop by using such data as diameter$^2$ × height, and annual productivity by measuring current wood growth.

# SECONDARY PRODUCTION

Net production is the energy available to the heterotrophic components of the ecosystem. Theoretically, at least, all of it is available to the herbivores or even to the decomposers, but rarely is it all utilized in this manner. The net production of any given ecosystem may be dispersed to another food chain outside of the ecosystem by humans, wind, or water. Much

of the living material is physically unavailable to the grazers—they cannot reach many plants or plant parts. The organic matter of live organisms is unavailable to decomposers and detritus-feeders, and that of dead plants may not be relished by grazers. The amount of net production available to herbivores may vary from year to year and from place to place. The quantity consumed varies with the type of herbivore and the density of the population. Once consumed, a considerable portion of the plant material, again depending upon the kind of plant and the digestive efficiency of the herbivore, may pass through the animal's body undigested. A grasshopper assimilates only about 30 percent of the grass it consumes, leaving 70 percent available to the detrital food chain (Smalley 1960). Mice, on the other hand, assimilate about 85 to 90 percent of what they consume (Golley 1960; R. L. Smith 1962).

Energy, once consumed, either is diverted to maintenance, growth, and reproduction or is passed from the body as feces, urine, and fermentation gases (Figure 11.11). The energy content of feces is transferred to the detritus food chain. Of the energy left after these losses, part is utilized as **heat increment,** which is heat required for metabolism above basal or resting metabolism. The remainder of the energy is *net energy,* available for maintenance, production, and reproduction. It includes energy involved in capturing or harvesting food, muscular work expended in the animal's daily routine, and energy needed to keep up with the wear

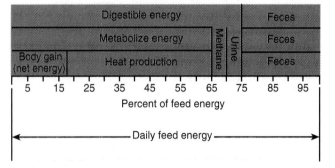

**Figure 11.11** Relative values of the end products of energy metabolism in white-tailed deer. Note the small amount of net energy gained (body weight) in relation to that lost as heat, gas, urine, and feces. The deer is a herbivore, a first-level consumer. (After Cowan 1962:5.)

and tear on the animal's body. The energy used for maintenance is lost as heat.

Maintenance costs, highest in small, active, warm-blooded animals, are fixed or irreducible. In small invertebrates energy costs vary with temperature, and a positive energy balance exists only within a fairly narrow range of temperatures. Below 5° C spiders become sluggish, cease feeding, and have to utilize stored energy to meet metabolic needs. At approximately 5° C energy absorbed into living cells (assimilated) approaches energy lost through respiration. From 5° to 20.5° C spiders assimilate more energy than they respire. Above 25° C, their ability to maintain a positive energy balance declines rapidly (Van Hooke 1971).

Energy left over from maintenance and respiration goes into the production of new tissue, fat, growth, and new individuals. This net energy of production is **secondary production** or **consumer production.** Within secondary production there is no portion known as gross production. What is analogous to gross production is actually assimilation. Secondary production is greatest when the birthrate of the population and the growth rates of the individuals are the highest. This peak usually coincides, for evolutionary reasons, with the time when net primary production is also the highest.

This scheme is summarized in Figure 11.12. It is applicable to any consumer organism, herbivore or carnivore, poikilotherm or homeotherm. Herbivores are the energy source for carnivores, and when they are eaten, not all of the energy in their bodies is utilized. Part of it goes unconsumed, and again metabolic losses must be accounted for. At each transfer considerably less energy is available for the next consumer level.

The energy budget of a consumer population can be summarized by

$$C = A + F + U$$

where C is the energy ingested or consumed, A is the energy assimilated, and F and U are the energy lost through feces and nitrogenous wastes.

The term A can be refined further:

$$A = P + R$$

where P is secondary production and R is energy lost through respiration. U, representing nitrogenous wastes, should be included as part of A (A = P + R + U), because they are involved in the homeostasis of organisms. Because of the difficulty of separating nitrogenous from fecal wastes, U typically is not included in A. Thus

$$C = P + R + F + U$$

or secondary production is

$$P = C - R - F - U$$

Just as net primary production is limited by a number of variables, so is secondary production. The quantity, quality

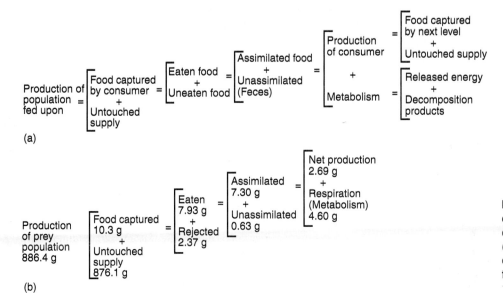

(a)

(b)

**Figure 11.12** (a) General model of components of energy metabolism in secondary production. (b) A field example for a ground-dwelling spider population. (Data from Moulder and Reichle 1972:496.)

(including nutrient status and digestibility), and availability of net production are three limitations. So, too, is the degree to which primary and available secondary production are used.

Secondary production can be examined from the viewpoint of three different ratios (Table 11.2). One is the ratio of assimilation to consumption or ingestion, A/I, the measure of the efficiency of the consumer at extracting energy from the food it consumes. It relates to food quality and effectiveness of digestion. The second is the ratio of production to assimilation, P/A. It is a measure of the efficiency of a consumer in incorporating assimilated energy into new tissue (growth), or secondary production. These two ratios underlie the magnitude of a third ratio of production to consumption, P/I. It indicates how much energy consumed by the animal is available to the next group of consumers.

The ability of the consumer population to use the energy it ingests varies with the type of consumer and the species (Table 11.3). Homeotherms use about 98 percent of their assimilated energy in metabolism and only about 2 percent in secondary production. Poikilotherms, on the average, convert about 44 percent of their assimilated energy in secondary production. They turn a greater proportion of their assimilated energy (A) into biomass (P). However, there is a major difference in assimilation efficiency between poikilotherms and homeotherms. Poikilotherms have an efficiency of about 42 percent, whereas homeotherms have an average efficiency of over 70 percent. Therefore the

**Table 11.2  Definitions of Selected Energy Efficiencies ($t$ = time)**

| Between Trophic Levels | | |
|---|---|---|
| Energy intake (Lindeman) efficiency | $\dfrac{P}{L}$ | $\dfrac{\text{gross production}}{\text{light absorbed}}$ |
| Assimilation efficiency | $\dfrac{A_t}{A_t - 1}$ | $\dfrac{\text{assimilation } t}{\text{assimilation } t - 1}$ |
| Production efficiency | $\dfrac{P_t}{P_t - 1}$ | $\dfrac{\text{production of biomass } t}{\text{production of biomass } t - 1}$ |
| **Within Trophic Levels** | | |
| Production efficiency | $\dfrac{P_t}{A_t}$ | $\dfrac{\text{production at } t}{\text{assimilation at } t}$ |
| Growth efficiency | $\dfrac{P_t}{I_t}$ | $\dfrac{\text{production at } t}{\text{energy ingested at } t}$ |
| Assimilation efficiency | $\dfrac{A_t}{I_t}$ | $\dfrac{\text{assimilation at } t}{\text{energy intake at } t}$ |

**Table 11.3** Assimilation Efficiency and Production Efficiency for Homeotherms and Poikilotherms

| Efficiency | All Homeotherms | Grazing Arthropods | Sap-feeding Herbivores | Lepidoptera | All Poikilotherms |
|---|---|---|---|---|---|
| Assimilation | | | | | |
| A/C | 77.5 ± 6.4 | 37.7 ± 3.5 | 48.9 ± 4.5 | 46.2 ± 4.0 | 41.9 ± 2.3 |
| Production | | | | | |
| P/C | 2.0 ± 0.46 | 16.6 ± 1.2 | 13.5 ± 1.8 | 22.8 ± 1.4 | 17.7 ± 1 |
| P/A | 2.46 ± 0.46 | 45.0 ± 1.9 | 29.2 ± 4.8 | 50.0 ± 3.9 | 44.6 ± 2.1 |

*Source:* Based on data from Andrzejewska and Gyllenberg 1980.

poikilotherm has to consume more calories than a homeotherm to obtain sufficient energy for maintenance, growth, and reproduction.

A wide range of conversion efficiencies exists among various feeding groups (Table 11.4). Production efficiency in plants (net production/light absorbed) is low, ranging from 0.34 percent in some phytoplankton to 0.8 to 0.9 percent in grassland vegetation. Herbivores use plant production with varying degrees of efficiency, depending upon whether they are poikilotherms or homeotherms. Because they eat foods already converted to animal tissue, carnivores, both poikilothermic and homeothermic, have high assimilation efficiencies. On the North American midwestern grasslands, average herbivore production efficiency, involving mostly poikilotherms, ranges from 5 to 16 percent. Carnivores have production efficiencies ranging from 13 to 24 percent.

# FOOD CHAINS

Net production is the base upon which the rest of life on Earth depends. This energy stored by plants is passed along through the ecosystem in a series of steps of eating and being eaten known as a **food chain.**

Food chains are descriptive. When worked out diagrammatically, they consist of a series of arrows, each pointing from one species to another, for which it is a source of food.

In Figure 11.13, for example, the marsh vegetation is eaten by the grasshopper, the grasshopper is consumed by the shrew, the shrew by the marsh hawk or the owl.

No one organism lives wholly on another; the resources are shared, especially at the beginning of the chain. The marsh plants are eaten by a variety of invertebrates, birds, mammals, and fish; and some of the animals are consumed by several predators. Thus food chains become interlinked to form a **food web,** the complexity of which varies within and between ecosystems.

## Components

Herbivores Feeding on plant tissues is a whole host of plant consumers, the herbivores. Herbivores are capable of living on diets high in cellulose. In doing so, they face several dietary problems. Their diets are rich in carbon, but low in protein. Most of the carbohydrates are locked in indigestible cellulose and the proteins in chemical compounds. Lacking the enzymes needed to digest cellulose, herbivores from mollusks, insects, and other invertebrates to grazing mammals depend upon specialized bacteria and protozoa to digest cellulose and proteins and to synthesize fatty acids, amino acids, proteins, and vitamins. For most vertebrates bacteria and protozoans are concentrated in the foregut or hindgut, where they carry on anaerobic fermentation. (A few cellulose-consuming wood beetles and wasps depend upon fungi. These insects carry fungal spores with them externally when they invade new wood.)

**Table 11.4** Consumer Efficiency (Secondary Production/Secondary Consumption)

| Habitat | Growing Season, Days | Producers | | Herbivores | | Carnivores | |
|---|---|---|---|---|---|---|---|
| | | Production | % Efficiency | Production | % Efficiency | Production | % Efficiency |
| Shortgrass plains | 206 | 3.767 | 0.8 | 53 | 11.9 | 6 | 13.2 |
| Midgrass prairie | 200 | 3.591 | 0.9 | 127 | 16.5 | 37 | 23.7 |
| Tallgrass prairie | 275 | 5.022 | 0.9 | 162 | 5.3 | 15 | 13.9 |

*Source:* Based on data from Andrzejewska and Gyllenberg 1980.

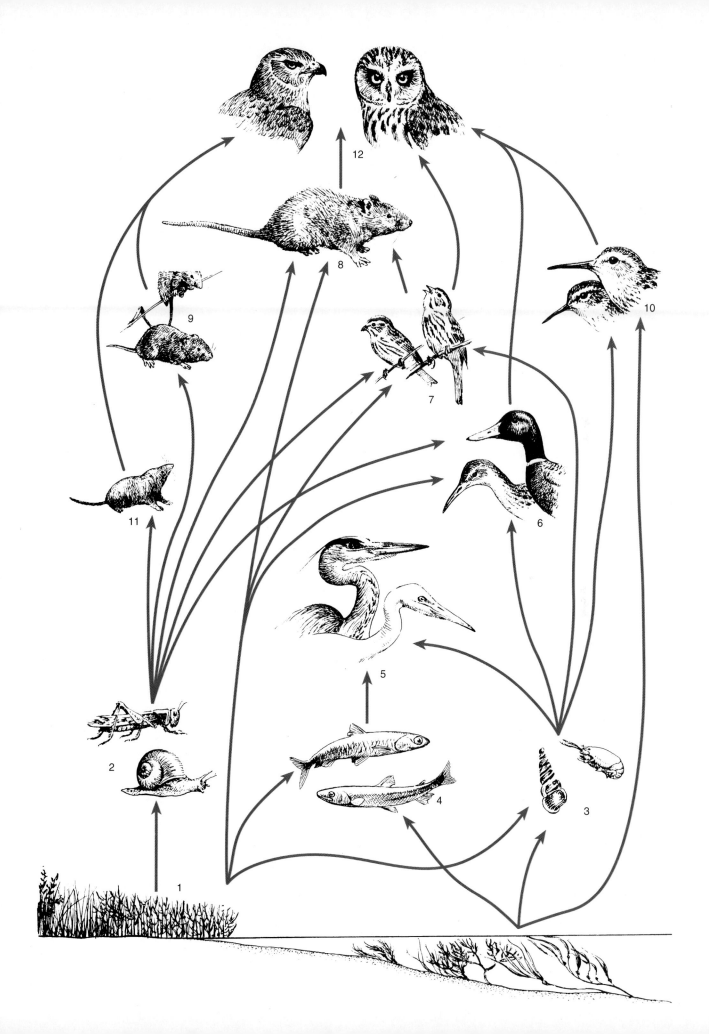

Ruminants, such as cattle and deer, are outstanding examples of herbivores anatomically specialized for the digestion of cellulose. They possess a highly complex digestive system, consisting of a four-compartment stomach—the rumen, reticulum, omasum, and abomasum or true stomach (Figure 11.14)—and a long intestine. The rumen and reticulum, inhabited by anaerobic bacteria and protozoans, function as fermentation vats. The ruminant hosts provide the microbes with the cellulose substrate on which to act. The hosts also regulate the pH and chemistry of the stomach environment by adding bicarbonates and phosphates from their highly developed salivary glands (Hungate 1975, Hofmann 1989).

As ruminants graze they chew their food hurriedly. The consumed material descends to the rumen and reticulum, where it is softened to a pulp by the addition of water, kneaded by muscular action, and fermented by bacteria. At leisure the animals regurgitate the food, chew it more thoroughly to reduce plant particle size, making it more accessible to microbes, and swallow it again. The mass again enters the rumen. Finer material (about 5 µl) is pulled into the reticulum, and from there it is forced by contraction into the third compartment, or omasum. The opening into the omasum dictates the size and amount of material entering it. In the omasum the material is further digested and finally forced into the abomasum, or true glandular stomach.

The digestive process carried on by the microorganisms produces short-chain volatile fatty acids. These acids are rapidly absorbed through the wall of the rumen into the bloodstream and are oxidized to form the ruminant's chief source of energy. Part of the material in the rumen is converted to methane and lost to the animal, and part remains as fermentation products. Many of the microbial cells involved are digested in the abomasum to recapture still more of the energy and nutrients. In addition to fermentation, the bacteria also synthesize B-complex vitamins and essential amino acids.

Variations exist in the digestive processes of ruminants and the kinds of forage they use. In general, three feeding types have evolved among ruminants, none of which is restricted to any one taxonomic group (Hofmann 1989). Grass and roughage eaters, such as cattle, sheep, and water buffalo (*Bubalus bubalus*), which make up about 25 percent of the 150 ruminants, including six domestic species, are adapted to eat fibrous food, forage rich in cellulose. They graze for

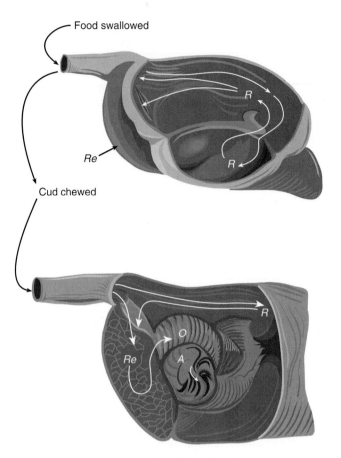

**Figure 11.14** Anatomy and function of the ruminant stomach. The four-compartment stomach consists of the rumen (R), reticulum (Re), omasum (O), and abomasum (A). As indicated by the arrows, the consumed food enters the rumen. The ruminant regurgitates boluses of the fermented material (cud) and rechews it. Finer material enters the reticulum, then the omasum and abomasum. The coarser material reenters the rumen for further fermentation.

long periods of time, then spend a nearly equal period resting and ruminating. About 40 percent of the ruminant species are concentrate selectors. These include the giraffe (*Giraffa* spp.), greater kudu (*Tragelaphus strepsiceros*), dikdiks (*Madoqua* spp.), and roe deer (*Capreolus capreolus*). Their digestive systems are far less suited to digest plant fibers. They seek high quality, easily digested forage

**Figure 11.13** (Opposite) A midwinter food web in a *Salicornia* salt marsh (San Francisco Bay area). Producer organisms are marsh and aquatic plants. Marsh plants (1) are consumed by terrestrial herbivorous invertebrates, represented by grasshopper and snail (2), and by herbivorous marine and intertidal invertebrates (3). Fish, represented by smelt and anchovy (4), feed on vegetable matter from both ecosystems. The fish in turn are eaten by first-level carnivores, represented by the great blue heron and common egret (5). Continuing through the web are the following omnivores: clapper rail and mallard duck (6), savanna and song sparrows (7), Norway rat (8), California vole and salt marsh harvest mouse (9), and the least and western sandpipers (10). The vagrant shrew (11) is a first-level carnivore. Top carnivores are the marsh hawk and short-eared owl (12). (Adapted from R. F. Johnson 1956:99.)

rich in soluble cell contents, such as soft, juicy dicot plants. They graze for short periods of time, followed by short periods of rumination. The plant material stays for only a short time in the relatively small rumen. About 35 percent of the ruminants are intermediates between the other two types. These ruminants include red deer (*Cervus elaphus*), reindeer (*Rangifer tarandus*), impala (*Aepyceros melampus*), and gazelles (*Gazella* spp.). They eat a mixed diet, but forage selectively. They avoid fiber as long as possible, and are influenced by seasonal fluctuations in forage quality. The white-tailed deer (*Odocoileus virginianus*), for example, eats low-fiber forages such as dicot herbs and leaves, young woody growth, and acorns, as long as they are available. During the winter season their only available food is largely fibrous woody plant material. Like concentrate feeders, these intermediates cannot digest fibrous forage efficiently.

Grazing marsupials, such as the kangaroos, also possess a ruminant type of digestive system, which has arisen by parallel evolution. Although the stomach is not divided into four compartments, it has regions analogous to rumen, reticulum, abomasum, and omasum.

Most of the digestion in ruminants takes place in the foregut. Among nonruminants such as rabbits and horses, digestion takes place less efficiently in the hindgut. Nonruminants have simple stomachs, long intestinal tracts that slow the passage of food through the gut, and a well-developed caecum, a blind pouch attached to the colon of the intestine, where fermentation takes place. Among such herbivores fermentation follows rather than preceeds the small intestine. Such hindgut fermentation results in incomplete digestion. Some herbivores, however, are able to process large quantities of forage rapidly. Some nonruminants and some invertebrate herbivores such as isopods and millipedes resort to **coprophagy,** the ingestion of fecal material for further extraction of nutrients. In lagomorphs—rabbits, hares, and pikas—part of ingested plant material enters the caecum and part enters the intestine to form dry pellets. In the caecum the ingested material is processed by microorganisms and is expelled into the large intestine as soft, moist pellets surrounded by a proteinaceous membrane. The soft pellets, much higher in protein than and lower in crude fiber than the hard pellets, are reingested. The amount of feces recycled by coprophagy ranges from 50 to 80 percent. The reingestion is important because it provides bacterially synthesized B vitamins and ensures a more complete digestion of dry material and a better utilization of protein (McBee 1971).

**Carnivores** Herbivores are the energy source for carnivores, the flesh-eaters. Organisms that feed directly upon the herbivores are termed first-level carnivores or second-level consumers. First-level carnivores represent an energy source for second-level carnivores. Still higher categories of carnivorous animals feeding on secondary carnivores exist in some communities. As the feeding level of carnivores increases, their numbers decrease and their fierceness, agility, and size

increase. Finally the energy stored at the top carnivore level is used by the decomposers.

Unlike herbivores, carnivores are not faced with digesting cellulose or finding high quality food. Because little difference exists in the chemical composition of the flesh of prey and the flesh of predators, there is no problem in digestion and assimilation of nutrients from their prey. Carnivores regurgitate indigestible bones, hair, feathers, and chitin. Their major problem is obtaining a sufficient quantity of food.

Not needing to digest complex celluloses, carnivores have short intestines and simple stomachs. The stomach is little more than an expanded hollow tube with muscular walls. It stores and mixes foods, and its glands add mucus, enzymes, and hydrochloric acid to speed digestion.

**Omnivores** Not all consumers confine their feeding to one level. The omnivores consume both plants and animals. The red fox (*Vulpes vulpes*) feeds on berries, small rodents, and even dead animals. Thus it is herbivore, carnivore, and scavenger. Some fish feed on both plant and animal matter. The basically herbivorous white-footed mouse (*Peromyscus leucopus*) feeds also on insects, small birds, and bird eggs. The food habits of many animals vary with the seasons, with stages in the life cycle, and with their size and growth (Figure 11.15). Such animals are considered omnivores, and their feeding habits are called omnivory.

**Decomposers** Conventionally, decomposers make up the final consumer group. Their basic function is to release the nutrients in dead organic matter back into the nutrient cycle. Although frequently considered unclassifiable in the general scheme of food chains, decomposers function as herbivores or carnivores, depending upon their food. Decomposition was discussed in Chapters 9 and 10.

There are two groups of decomposer organisms, macroorganisms and microorganisms. The macroorganisms or detritivores include such small detritus-feeding animals as mites, earthworms, millipedes, and snails in terrestrial ecosystems and crabs, mollusks, and worms in aquatic ecosystems. Acting as reducer-decomposers, they ingest organic matter, break it down into smaller pieces, mix it with the substrate, excrete it, spread fungal spores, break down microbial antagonisms, and even add substances to the material that stimulate microbial growth. They consume bacteria and fungi associated with detritus, as well as protozoans and small invertebrates that cling to the material. True decomposers are the microorganisms—bacteria and fungi. A succession of these organisms absorbs a portion of the detrital material as food, leaving unconsumed material as food for still other microorganisms.

**Other Feeding Groups** Several other feeding groups are also involved in energy transfer. **Parasites** spend a considerable part of their life cycle living on or in and drawing their nourishment from their hosts. Most do not kill their hosts, but some insect larvae, the **parasitoids,** slowly consume their host and then transform into another stage of

*Micropogon undulatus*, Atlantic Croaker

Food categories

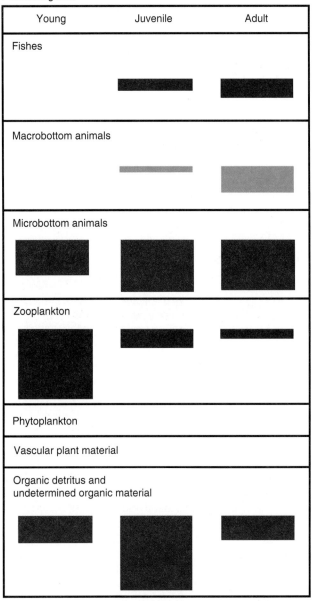

**Figure 11.15** Trophic spectra for young, juvenile, and adult stages of the Atlantic croaker from Lake Pontchartrain, Louisiana. The young fish subsist largely on zooplankton, the juveniles on organic matter, the adults on bottom animals and fish. (After Darnell 1961:566.)

their life cycle. Functionally, parasites are specialized carnivores and herbivores.

**Scavengers** are animals that eat dead plant and animal material. Among them are termites and various beetles that feed in dead and decaying wood, and crabs and other marine invertebrates that feed on plant particles in water. Botflies, dermestid beetles, vultures, and gulls are among the many animals that feed on animal remains. Scavengers may be either herbivores or carnivores.

**Saprophytes** are plant counterparts of scavengers. They draw their nourishment from dead plant and animal material, chiefly the former. Because they do not require sunlight as an energy source, they can live in deep shade or dark caves. Examples of saprophytes are Indian pipe (*Monotropa uniflora*) and beechdrops (*Epifagus virginiana*). The majority are herbivores, but others feed on animal matter.

## Major Food Chains

Within any ecosystem there are two major food chains, the grazing food chain and the detrital food chain (Figure 11.16). In most terrestrial and shallow-water ecosystems, with their high standing crop and relatively low harvest of primary production, the detrital food chain is dominant. In deep-water aquatic ecosystems, with their low biomass, rapid turnover of organisms, and high rate of harvest, the grazing food chain may be dominant.

The amount of energy shunted down the two routes varies among communities. In an intertidal salt marsh less than 10 percent of living plant material is consumed by herbivores and 90 percent goes the way of the detritus-feeders and decomposers (Teal 1962). In fact, most of the organisms of the intertidal salt marsh obtain the bulk of their energy from dead plant material. In a Scots pine plantation 50 percent of the energy fixed annually is utilized by decomposers (Figure 11.17). The remainder is removed as yield or is stored in tree trunks (Ovington 1961). In some communities, particularly those that are under-grazed, unconsumed organic matter may accumulate and remain out of circulation for some time, especially when conditions are not favorable for microbial action. The decomposer or detritus food chain receives additional materials from the waste products and dead bodies of both the herbivores and carnivores.

Detrital Food Chains   The detrital food chain is common to all ecosystems, but in terrestrial and littoral ecosystems it is the major pathway of energy flow. In a yellow-poplar forest (*Liriodendron*), 50 percent of gross primary production goes into maintenance and respiration, 13 percent is accumulated as new tissue, 2 percent is consumed by herbivores, and 35 percent goes to the detrital food chain (Edwards, unpublished, cited by O'Neill 1976). Two-thirds to three-fourths of the energy stored in a grassland ecosystem that is ungrazed by cattle is returned to the soil as dead plant material, and less than one-fourth is consumed by herbivores (Hyder 1969). Of

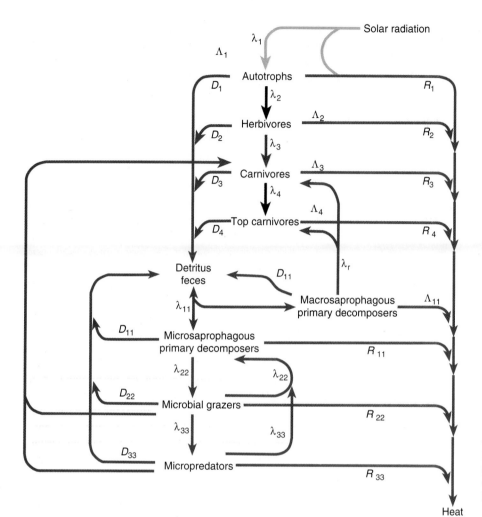

Figure 11.16 Model of a detrital and grazing food chain. Two pathways lead from the autotrophs, one to the grazing herbivores and the other to detrital feeders. Note the connections between the two food chains. (After Paris 1968.)

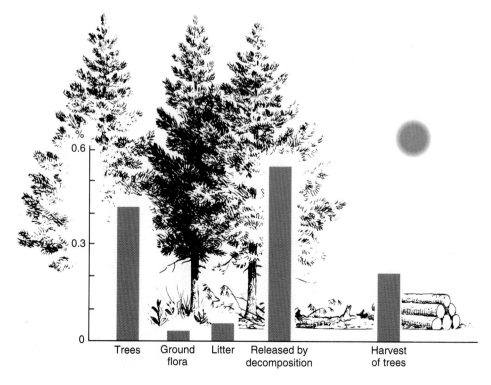

Figure 11.17 The fate of the 1.3 percent of solar energy assimilated as net production by a 23-year-old Scots pine plantation. (After J. D. Ovington 1961).

the quantity consumed by herbivores, about one-half is returned to the soil as feces. In the salt marsh ecosystem the dominant grazing herbivore, the grasshopper, consumes just 2 percent of the net production available to it (Smalley 1960).

Energy flow through a detrital food chain is difficult to measure, although the use of radioactive tracers gives some idea of energy flow through selected food webs. Andrews et al. (1975), basing their estimates on a number of studies, determined that microbial activity accounted for 99 percent of total saprophytic assimilation in a shortgrass prairie ecosystem. Saprophagic grazers such as nematodes accounted for the remaining 1 percent.

Gist and Crossley's (1975) study of a selected invertebrate population living in forest litter provides an example of a detrital food chain (Figure 11.18). Although no data for energy flow were collected, the flux of radioactive calcium and phosphorus permitted the construction of a food web. The quantity of elements provided some idea of the energy flow. The litter is consumed by five "herbivorous" groups: millipedes (Diploda), orbatid mites (Cryptostigmata), springtails (Collembola), cave crickets (Orthoptera), and pulmonate snails (Pulmonata). The mites and springtails are the most important of these litter feeders. These herbivores are preyed upon by small spiders (Araneidae) and predatory mites (Mesostigmata). The predatory mites feed on annelids, mollusks, insects, and other arthropods. Spiders feed on predatory mites; springtails, pulmonate snails, small spiders, and cave crickets are preyed upon by carabid beetles; whereas medium-sized spiders feed on cave crickets and other insects. The medium-sized spiders, in turn, become additional items in the diet of the beetles. Beetles, spiders, and snails are eaten by birds and small mammals, members of the grazing food chain not shown on the original web.

Through predation detrital food webs feed into grazing food chains at higher consumer levels. Detrital and grazing food webs are separate compartments in an ecosystem only at the detritus and primary consumer levels.

Food chains involving saprophages may take two directions (as shown in Figure 11.16), toward the carnivores or toward microorganisms. The role of these feeding groups in the final dissipation of energy has already been mentioned; but they also are food to numerous other animals. Slugs eat the larvae of certain flies and beetles, which live in the heads of fungi and feed on the soft material. Mammals, particularly the red squirrel and chipmunks, eat woodland fungi. Dead plant remains are food sources for springtails and mites, which in turn are eaten by carnivorous insects and spiders. They in turn are energy sources for insectivorous birds and small mammals. Blowflies lay their eggs in dead animals, and within 24 hours the maggot larvae hatch. Unable to eat solid tissue, they reduce the flesh to a fetid mass by enzymatic action in which they feed on the proteinaceous material. These insects are food for other organisms.

Grazing Food Chains The grazing food chain is the one most obvious to us. Cattle grazing on pastureland, deer browsing in the forest, rabbits feeding in old fields, and insect pests feeding on garden crops represent the basic consumer groups of the grazing food chain. In spite of its conspicuousness, the grazing food chain, as pointed out earlier, is not the major food chain in terrestrial and many aquatic ecosystems. Only in some aquatic ecosystems do the grazing herbivores play a dominant role in energy flow.

Although voluminous data exist on phytoplankton productivity, filtration rates by grazing zooplankton, and production efficiencies of zooplankton (for summary, see Wetzel

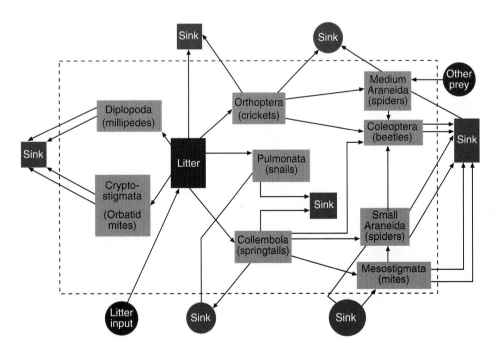

**Figure 11.18** A detrital food web involving litter-dwelling invertebrates of a hardwood forest in the southeastern United States. The dashed line represents the boundaries of the system. Note that the detritus food chain involves a herbivorous component—millipedes and mites to the left and snails, crickets, and springtails in the center. The herbivores support a carnivorous component. The detrital food web, like the grazing food web, can become complex. (After Gist and Crossley, Jr. 1975:86.)

1975), few data are available on the flow of energy, rate of grazing, biomass turnover rates for phytoplankton, and turnover of zooplankton biomass within the same aquatic system. Hillbricht-Ilkowska (1974) studied the relationship between primary production of phytoplankton and consumer production in freshwater lakes. He found that the direct utilization of primary production by filter-feeding zooplankton was intense and its efficiency of energy transfer was high when the size and structure of phytoplankton were favorable for filter feeding. In lakes where the net form of phytoplankton was dominant, direct utilization was low and energy transfer inefficient. As an example, in lakes in which the production of phytoplankton ranged from 200 to 1200 kcal/m²/yr, production of filter-feeding zooplankton amounted to 10 to 250 kcal/m²/yr with a transfer efficiency of 2 to 30 percent. In lakes in which filter-feeding zooplankton production ranged from 50 to 150 kcal/m²/yr, the production of predacious zooplankton ranged from 2 to 50 kcal/m²/yr with an energy transfer efficiency of 5 to 40 percent.

In terrestrial systems a small proportion of primary production goes by way of the grazing food chain. Over a three-year period only 2.6 percent of net primary production of a yellow-poplar forest was utilized by grazing herbivores, although the holes made in the growing leaves resulted in a loss of 7.2 percent of photosynthetic surface (Reichle et al. 1973). In a study of energy flow through a shortgrass prairie ecosystem, involving ungrazed, lightly grazed, and heavily grazed plots, Andrews et al. (1974) found that on the heavily grazed prairie cattle consumed 30 to 50 percent of aboveground net primary production. In both lightly and heavily grazed plots about 40 to 50 percent of energy consumed by the cattle is returned to the ecosystem and the detrital food chain as feces (Dean et al. 1975). In the heavily-grazed Serengeti plains of East Africa 33 to 66 percent of annual aboveground production is consumed. In some instances consumption may reach 99 percent (McNaughton 1985, McNaughton and Georgiadis 1988).

Although the aboveground herbivores are the conspicuous grazers, belowground herbivores can have a pronounced impact on primary production and the grazing food chain. Andrews et al. (1975) found that belowground herbivores consisting mainly of nematodes (Nematoda), scarab beetles (Scarabaeidae), and adult ground beetles (Carabidae) accounted for 81.7 percent of total herbivore assimilation on the ungrazed shortgrass prairie, 49.5 percent on the lightly grazed prairie, and 29.1 percent on the heavily grazed prairie. Ninety percent of the invertebrate herbivore consumption took place below ground, and 50 percent of the total energy was processed by nematodes. On the lightly grazed plots prairie cattle consumed 46 kcal/m² during the grazing season and the belowground invertebrates consumed 43 kcal/m². When a nematicide was added to a midgrass prairie, aboveground net production increased 30 to 60 percent. Thus belowground herbivores can impose a greater stress on the ecosystem than aboveground herbivores.

Size of both food items and consumers has considerable influence on the direction a food chain takes, because there are upper and lower limits to the size of food an animal can capture. Some animals are large enough to defend themselves successfully. Some foods are too small to be collected economically; it takes too long to secure enough to meet the animal's metabolic needs. Thus the upper limit to the size of an animal's food is determined by its ability to handle and process, and the lower limit by its ability to secure enough.

There are exceptions, of course. By injecting poisons, spiders and snakes kill prey much larger than themselves; wolves hunting in packs can kill an elk or a caribou. The idea that food chains involve animals of progressively larger sizes is true only in a general way. Where parasites are involved, the opposite situation exists. The larger animals are at the base, and as the number of links increases, the size of the parasites become smaller.

A grazing food chain has been carefully worked out (Golley 1960) for old field vegetation, meadow mice, and weasels (Figure 11.19). The mice are almost exclusively herbivorous, and the weasels live mainly on mice. The vegetation converts abut 1 percent of the solar energy into net production, or plant tissue. The mice consume about 2 percent of the plant food available to them, and the weasels about 31 percent of the mice. Of the energy assimilated, the plants lose about 15 percent through respiration, the mice 68 percent, and the weasels 93 percent. The weasels use so much of their assimilated energy in maintenance that a carnivore preying on weasels could not exist.

In a very general way, energy transformed through the ecosystem by way of the grazing chain is reduced by a magnitude of 10 from one level to another. Thus if an average of 1000 kcal of plant energy is consumed by herbivores, about 100 kcal is converted to herbivore tissue, 10 kcal to first-level carnivore production, and 1 kcal to second-level carnivores. The amount of energy available to second- and third-level carnivores is so small that few organisms could be supported if they depended on that source alone. For all practical purposes, each food chain has from three to four links, rarely five (Pimm 1982). The fifth link is distinctly a luxury item in the ecosystem.

Supplementary Food Chains Other feeding groups, such as the parasites and scavengers, form supplementary food chains in the community. Parasitic food chains are highly complicated because of the life cycle of the parasites. Some parasites are passed from one host to another by predators in the food chain. External parasites (ectoparasites) may move from one host to another. Other parasites are transmitted by insects from one host to another through the bloodstream or plant fluids.

Food chains also exist among parasites themselves. Fleas that parasitize mammals and birds are in turn parasitized by a protozoan, *Leptomonas*. Chalcid wasps lay eggs in the ichneumon or tachinid fly grub, which in turn is parasitic on other insect larvae.

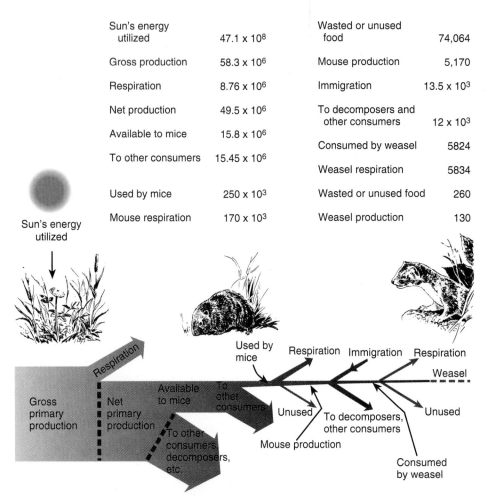

| Sun's energy utilized | $47.1 \times 10^8$ |
| Gross production | $58.3 \times 10^6$ |
| Respiration | $8.76 \times 10^6$ |
| Net production | $49.5 \times 10^6$ |
| Available to mice | $15.8 \times 10^6$ |
| To other consumers | $15.45 \times 10^6$ |
| Used by mice | $250 \times 10^3$ |
| Mouse respiration | $170 \times 10^3$ |

| Wasted or unused food | 74,064 |
| Mouse production | 5,170 |
| Immigration | $13.5 \times 10^3$ |
| To decomposers and other consumers | $12 \times 10^3$ |
| Consumed by weasel | 5824 |
| Weasel respiration | 5834 |
| Wasted or unused food | 260 |
| Weasel production | 130 |

**Figure 11.19** Energy flow through a food chain in an old field community in southern Michigan. The relative size of the blocks suggests the quantity of energy flowing through each channel. Values are cal/ha²/yr. (Based on data from Golley 1960.)

# MODELS OF ENERGY FLOW

The concept of energy flow in ecological systems is one of the cornerstones of ecology. The model was first developed by Raymond Lindeman in 1942 in a study of the trophic dynamic structure of Lake Mendota in Wisconsin. The trophic dynamic concept is based on the assumptions that the laws of thermodynamics hold for plants and animals, that plants and animals can be arranged into feeding groups or trophic levels, that at least three trophic levels—producer, herbivore, and carnivore—exist, that the net energy content of one trophic level is passed on to the next trophic level above, and that the system is in equilibrium.

Whether at the producer or consumer level, energy flow through the ecosystem is mediated by the individual (Figure 11.20a). A quantity of energy in food is consumed. Part of it is assimilated and part is lost as feces, urine, and gas. Part of the assimilated energy is used for respiration, and part is stored as new tissue which can be used to some extent as an energy reserve. Some is used for growth and the production of new individuals.

A model of energy flow through a population (Figure 11.20b) shows some additions. Growth in individual biomass

becomes changes in size of standing crop. Part of the biomass goes to predators and parasites, and there are gains and losses of energy and biomass from and to other ecosystems.

The population boxes can be fitted into several trophic levels and linked to form a model of energy flow through the ecosystem (Figure 11.20c). In this model there is no attempt to separate out the decomposers and detritus feeders, because they must fall into one of the several trophic levels when the food web is collapsed.

To provide for the grazing and detritus paths of energy transfer, Wiegert and Owens (1971) propose a somewhat different model of energy flow, with transfers through two channels beyond the autotrophs (Figure 11.21). They define all organisms utilizing living material as **biophages,** and all organisms utilizing nonliving matter as **saprophages.** First-order biophages, who eat living plants, are the traditional herbivores; whereas first-order saprophages feed on dead plant material as well as organic material egested by first-order biophages. First-order biophages in turn are utilized at death by second-order saprophages, which are functional carnivores. In turn, first-order saprophages may be preyed upon by second-order biophages. Although this model, like others, cannot separate out organisms that occupy several

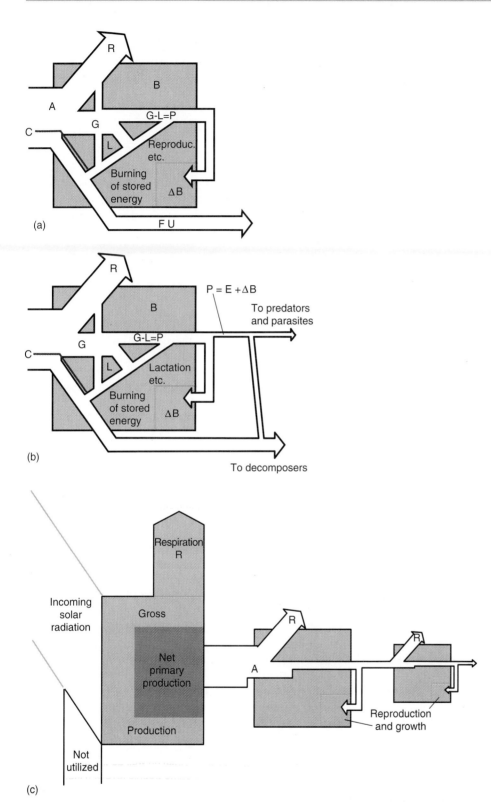

**Figure 11.20** Models of energy flow: (a) through the individual organism, (b) through the population, and (c) through the ecosystem. Note the losses and the portion of energy accumulated as growth in the organism. The ecosystem model considers the decomposers as occupying one of several trophic levels rather than as on a separate trophic pathway. A = assimilation, B = biomass, ΔB = changes in biomass, C = consumption, E = material removed from population to ecosystem, FU = rejecta, G = total biomass growth, L = decrement of biomass through weight loss and death, R = respiration, P = production.

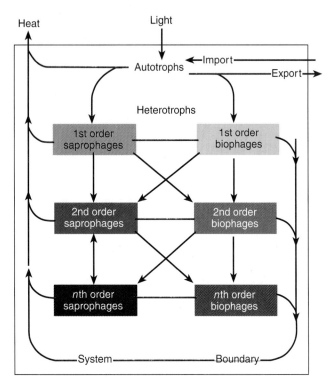

**Figure 11.21** Model of two-channel energy flow. The model separates energy flow into two pathways, one utilized by organisms that feed on living material, the biophages, and the other utilized by organisms that feed on nonliving material, the saprophages. (After Weigert and Owen 1971:71.)

trophic levels, it does modify the Lindeman model so that the decomposers and detritivores are assigned to functional components.

## Trophic Levels and Ecological Pyramids

Feeding relationships in an ecosystem may be considered in terms of trophic or consumer levels. From a functional rather than a species point of view, all organisms that obtain their energy in the same number of steps from the producers belong to the same trophic level in the ecosystem. The first trophic level belongs to the producers, the second level to the herbivores, and the higher levels to the carnivores. Some consumers occupy a single trophic level, but many others, as we have seen, occupy more than one trophic level.

The concept of consumer layers was advanced by Charles Elton (1927) in his classic book *Animal Ecology*. He pointed out the great difference in the numbers of organisms involved in each step of the food chain. Consumers at the lower end of the chain are the most abundant. In successive links of consumption, carnivores decrease rapidly in number and increase in size until there are very few carnivores at the top. This concept is known as the **pyramid of numbers** (Figure 11.22a). However, consumer levels based on numbers is misleading. Although consumers at the base of the pyramid

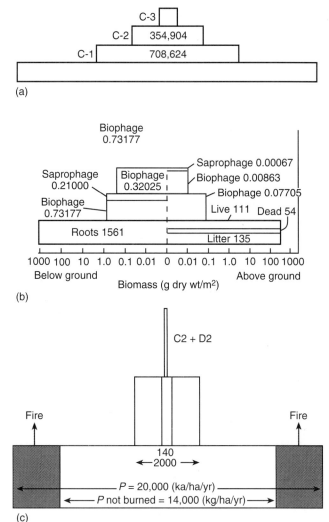

**Figure 11.22** Three kinds of ecological pyramids. (a) A composite pyramid of numbers based on vegetation in an old field in Michigan (Evans and Cain 1952) and an animal census of two pastures and a meadow in New York (Wolcott 1937). (After E. Odum 1983:152.) (b) A pyramid of biomass for a northern shortgrass prairie for July. The base of the pyramid represents biomass (g dry weight/m²) of producers; the second (middle) level, primary consumers; and the top, secondary consumers. Aboveground biomass (right) and belowground biomass (left) are separated by a dashed vertical line. The trophic level magnitudes are plotted on a horizontal logarithmic scale. The compartments are divided on a vertical linear scale according to live, standing dead, and litter biomass or biophagic and saprophagic consumer biomass. Unlike the conventional pyramids, this one recognizes the detrital as well as the grazing food chain components on the same trophic levels. (After French, Steinhorst, and Swift 1979.) (c) A pyramid of energy for the Lamto Savanna, Ivory Coast. P is primary production; C1, primary consumers; C2, secondary consumers; D1, decomposers of vegetable matter; D2, decomposers of animal matter. Again, the detrital and grazing food chains have been collapsed into the same trophic levels. (After La Motte 1975:216.)

may be most abundant, there is great deal of variation in their size and numbers. Plants are eaten by herbivores ranging in size from very small arthropods to elephants.

More realistic is a **pyramid of biomass** (Figure 11.22b). If the numbers of consumers at each feeding level are multiplied by their weight, we arrive at a pyramid of biomass. The pyramid of biomass indicates by weight or other measurement of living material the total bulk of organisms of fixed energy present at any one time—the **standing crop.** Because some energy or material is lost through respiration at each link, the total mass supported at each level is limited by the energy stored or the rate of energy storage in the level below. In general, the biomass of producers is greater than the biomass of the herbivores they support, and the biomass of the herbivores is greater than that of the carnivores. This relationship results in a gradually sloping pyramid, particularly for terrestrial and shallow-water communities where the producers are large, accumulation of organic matter is great, life cycles are long, and the rate of harvesting is low.

However, for some ecosystems, the pyramid of biomass is inverted. Primary production in aquatic ecosystems, such as lakes and open seas, is concentrated in phytoplankton. Planktonic algae have a short life cycle, multiply rapidly, accumulate little organic matter, and are heavily exploited by herbivorous zooplankton. At any point the standing crop is low. As a result the base of the pyramid is much smaller than the structure it supports.

Pyramids of numbers and biomass tell us something about structure, but little about function. When production, amount of biomass accumulated, is considered in terms of energy, the pyramid suggests the amount of energy flow from one level to another. The basis of the **pyramid of energy** (Figure 11.22c) is the quantity of energy fixed, stored, and passed on to the next trophic level. Such an approach eliminates the biases of numbers and size and emphasizes the role organisms play in the transfer of energy. For example, some organisms have a small biomass, but they assimilate and pass on considerably more energy than some organisms with larger biomass. On a pyramid of biomass these smaller organisms would appear much less important in the ecosystem than they really are.

The concept of trophic levels has weaknesses. The typical trophic structure is based on plants or primary production and the transfer of energy from primary production to higher levels. It discounts detrital material and decomposers and saprophages. In other words, it basically is a description of the grazing food chain, although there have been attempts to incorporate biophages and saprophages into both pyramids of biomass and energy (Figure 11.22b). Consumers, especially those above the herbivore level, often occupy more than one trophic level, and their contribution to the biomass and energy flow at each level must be apportioned. Moreover, the trophic level concept does not take into account the availability of energy. All of the energy at any level is not available to consumers. Among green plants, for example, there is a difference in energy available to consumers from various plant parts (Odum and Biever 1984). Certain

structures, because of chemical or morphological characteristics, are not eaten. The concept also gives the false impression that energy does not cycle through ecosystems. Actually, a unit of energy reaching the decomposer bacteria and fungi is not necessarily dissipated, as the concept implies; some of it becomes part of the decomposer biomass to be recycled up through the food web again.

**Food Webs** Trophic relationships in nature are not simple, straight-line food chains. Numerous food chains mesh into a complex food web with links leading from producers through an array of primary and secondary consumers. Producers support numerous primary consumers, which support many higher consumers.

It is hard to apply the concept of trophic levels to food webs, because some organisms feed on more than one trophic level. Just what is the position of the sparrows in Figure 11.13? They feed both on primary producers (seeds), putting them on level 2, and on herbivorous insects, placing them on level 3. Further, how do we measure the amount of energy that a species contributes to each of the trophic levels it occupies? Some species have different food habits during different stages or seasons of their life cycle (see Figure 11.15). How do such species figure in determining the maximum number of trophic levels? How does physical stress, such as turbulence in aquatic ecosystems, affect the basal trophic level, or the consumers? Could environmental conditions eliminate a trophic level? Recent interest in the nature and role of trophic structure and food webs in community structure and energy flow makes such questions important.

Ecologists working with food web theory have arrived at a new view of trophic levels (Cohen and Newman 1988, Yodzis 1988, Cohen and Briand 1984, Pimm 1982). They consider species in a food web as **trophospecies,** ones that are trophically similar to the same prey and same predators, and classify them as basal, intermediate, and top species. Basal species, which include both primary producers and detritus, are at the bottom of the food web and feed on no other species. The top species, occupying the apex of the food web, are ones on which no other species feed; and intermediate species, neither basal nor top, may feed on more than one trophic level (omnivores). Predators are species that feed on other species in the web, and prey are species that are fed upon by some other species.

Current interest in food webs has arisen out of two long-standing generalizations about food webs. The first is that the maximum number of trophic levels in a food web is limited by the second law of thermodynamics, the inefficiency of energy transfer from one trophic level to another. The second is that complex food webs are more stable than simple ones. In the 1970s these two generalizations were questioned (May 1973, Pimm and Lawton 1977, Pimm 1982). Scientists began to ask new questions: What determines the size of food webs and the number of trophic levels? How are food webs organized and structured? How are food webs affected by the invasion of new species and the loss of current

ones? Are complex food webs more stable than simple ones? What processes, including population dynamics and energy flow, are involved in the pattern of food webs?

Food chains, in general, tend to be short, rarely exceeding four links (Pimm 1982). No matter how productive an ecosystem, the number of links remains essentially the same. Increased productivity of an ecosystem does not support longer food chains, as often assumed. It may support more species and thus more complexity, but as the number of species in the web increases, the connectance decreases.

In any food web there are four possible connectances: (1) top and intermediate; (2) different intermediates; (3) intermediate and basal; (4) top and basal. The proportion in each is about the same in every habitat: 35 percent in the first category, 30 percent in the second, 27 percent in the third, and 8 percent in the fourth. Rarely does the top link with the basal. The proportion of species within each web that are basal, intermediate, and top appears to be independent of species richness (Briand and Cohen 1984).

What happens if a species is removed from or reduced in a food web? Theoretically, there is little effect if the species removed is a generalized prey species or a generalist predator species, or a member of a simple, straight-line food chain (Pimm 1980). If the species removed is a top predator, then the effect cascades down through the intermediate to the basal. Some of the best examples come from experiments in fisheries and lake productivity (Carpenter, Kitchell, and Hodgson 1985). Consider a top piscivore, a predatory fish such as a bass. The bass reduces the biomass of vertebrate planktivores, allowing invertebrate planktivore biomass and large zooplankton herbivores to increase, and causing small herbivore zooplankton and primary production to decrease. Conversely, if the top predator is reduced, vertebrate planktivores increase, invertebrate planktivores and large herbivorous zooplankton decrease, and small herbivore zooplankton and primary production increase. Thus the structure of a food often is affected adversely when a new or exotic species, such as a carnivore or grazing herbivore, is introduced into an ecosystem, or when a key species is removed from the food web. For this reason interest in food web structure extends from theoretical ecology to applied ecology.

# SUMMARY

A basic characteristic of ecosystems is the flow of energy. The source of that energy is sunlight reaching Earth. Light energy is fixed by the autotropic component of ecosystems and made available to the heterotrophic component.

Energy is the ability to do work. Energy is either potential energy, capable of and available for work, or kinetic energy, energy in motion or doing work. The expenditure and storage of energy is described by two laws of thermodynamics. The first law, the conservation of energy, states that en-

ergy is neither created or destroyed. The second law states that at each transfer or transformation of energy, part of that energy assumes a form that cannot be passed on any further because it is randomly dispersed, often as heat. Randomly dispersed energy is called entropy. Ultimately closed systems run down. Open systems, such as ecosystems, however, with their constant input and outflow, maintain a steady state in spite of the second law of thermodynamics.

Total energy fixed as organic molecules by autotrophs is gross primary production. Energy remaining after respiratory costs accumulates as fixed energy or net production. Plants allocate this energy to aboveground and belowground vegetative growth—roots, leaves, stems, flowers, fruit, and seed.

Ecosystem productivity, the rate of production per unit of time, is strongly influenced by temperature and rainfall. Most productive are tropical rain forests, certain agricultural crops, estuaries, and coral reefs. Least productive are deserts, tundra, and open ocean.

Methods of estimating primary production include the light and dark bottle technique, which measures the consumption and production of oxygen under light and dark conditions; uptake of a radioactive tracer, $^{14}C$; uptake and release of $CO_2$ during photosynthesis by plants or segments of plants in enclosed chambers; chlorophyll concentration; and harvest and dimension analysis.

Primary production flows through the heterotrophic component of the ecosystem to be fixed as secondary production. Energy consumed and assimilated goes to maintenance, growth, and reproduction. Among homeotherms most of energy assimilated goes to maintenance; among poikilotherms maintenance costs are much less, so most of the assimilated energy goes to growth. However, homeotherms assimilate much more of the energy they consume than do poikilotherms.

Herbivores are the primary consumers. The herbivores, in turn, are a source of energy for the carnivores. At each step or transfer of energy in the food chain a considerable amount of potential energy is lost in respiration, until the available energy is so small that few organisms can be supported. Organisms above the herbivore level often utilize several sources of energy, including plants, and thus become omnivores. All food chains eventually end with the decomposers, mainly bacteria and fungi, which reduce the remains of plants and animals into simple substances.

Energy flow in ecosystems may take two routes. One goes through the grazing food chain; the other goes through the detritus food chain, in which the bulk of the production is utilized as dead organic matter by decomposers and saprovores. Much of the energy flowing into the detrital food chain is recycled through upper levels of both the detrital and grazing chains before its final dissipation.

The loss of energy at each transfer limits the number of trophic levels in the food chain to three to five. Biomass declines at each level. A plot of the weight of individuals at each successive trophic level results in a sloping pyramid. In

certain aquatic situations, where there is a rapid turnover of small aquatic consumers, the pyramid of biomass may be inverted. Energy, however, always decreases from one trophic level to another.

Energy flow involves complex trophic relationships known as food webs, about which food web theory has grown. This theory considers the size, organization, and structure of food webs, as influenced by environment, numbers of, invasion by, and loss of species, and the relationship of one trophic level to another.

The ratio of energy flow from one trophic level of an ecosystem to the next or between individual organisms is ecological efficiency. Because efficiencies are dimensionless, several different ratios can be determined. Some of the most useful are assimilation efficiencies, growth efficiencies, and utilization efficiencies.

Although knowledge of energy flow in ecosystems is fragmentary and difficult to come by, the concept of energy flow is a valuable basis for understanding both ecosystem functioning and the relationship of humans to their environment.

# REVIEW QUESTIONS

1. What conditions are necessary for energy flow?
2. What are primary production, primary productivity, gross primary production, net primary production, standing crop, and respiration?
3. What is the significance of the root/shoot ratio to production in terrestrial ecosystems?
4. In what manner do plants allocate energy during the year?
5. What is secondary production?
6. What is the difference in energy allocation and efficiency between homeotherms and poikilotherms?
7. What are the two major food chains, and how do they interrelate?
8. Consider your diet for one week and then work out a food web of which you were a part. On how may levels did you function? What was the original source of the food? How do food webs involving modern urban humans differ from a natural food web?
9. Report on what happens to food webs and biodiversity when a new species, especially a predator, is introduced into an ecosystem. For example, consider the introduction of the peacock bass (*Cichla ocellaris*) into Gatum Lake, Panama Canal, of the largemouth bass (*Micropterus salmonides*) into Lake Atitlan, Guatemala, or of bait fish into North American rivers and lakes.

# CROSS-REFERENCES

Biotic influences on soil, 138–140; the living soil, 136–137; process of photosynthesis, 154–159; photosynthetic efficiency, 159; process of decomposition, 161–165; plant-herbivore interactions, 522–524; plant defenses, 524–530; predator-prey relations, 530–541; intraguild predation, 542–543.

# Biogeochemical Cycles

*Concepts*

1. Cycles through the biosphere involve biological, geological, and chemical compartments.
2. The two general types of biogeochemical cycles are gaseous and sedimentary.
3. The carbon cycle is involved in major energy transfers in the ecosystem.
4. The nitrogen cycle is driven by microbial processes.
5. Human impacts on the oxygen, carbon, nitrogen, sulfur, and phosphorus cycles result in major ecological disturbances.
6. Heavy metals, chlorinated hydrocarbons, and radionuclides follow biogeochemical cycles.

The living world depends upon the flow of energy (Chapter 11) and the circulation of materials through ecosystems. Both influence the abundance of organisms, the rate of their metabolism, and the complexity and structure of the ecosystem. Energy and materials flow through the ecosystem together as organic matter; one cannot be separated from the other. Certain elements are essential for energy fixation and flow; mineral cycling cannot function without the input of energy (Figure 12.1). One feeds upon the other in an apparent positive feedback loop: the more nutrients available, the more energy flows; increased energy flow produces greater cycling of nutrients. However, negative feedback is also at work: limited nutrients mean reduced energy flow; less energy allows less cycling of nutrients. In essence two forces are at work. One is the tendency within ecosystems for living organisms to store resources in the form of new tissue. The other is the tendency for organic matter to degrade into inorganic compounds.

Nutrients move through the ecosystem in **biogeochemical cycles.** These cycles involve chemical exchanges of elements among atmosphere, rocks of Earth's crust, water, and living things. "Bio" in the term refers to living organisms; "geo" refers to the rocks, air, and water; and "chemical" refers to the chemical interactions involved. Biogeochemical cycles are of two general types, the gaseous and the sedimentary. The main pools of nutrients possessing a gaseous cycle are the atmosphere and the ocean, and for that reason such nutrients have pronouncedly global circulation patterns. The main reservoirs of sedimentary cycles are the soil and the rocks of Earth's crust. Both types of cycles involve biological and nonbiological agents, both are driven by the flow of energy, and both are tied to the water cycle.

## GASEOUS CYCLES

**Gaseous cycles** circulate many elements basic to biological life—oxygen, carbon dioxide, nitrogen, and hydrogen. A few cycles, especially sulfur, and to a certain extent nitrogen, possess both a gaseous component and a sedimentary component.

### The Oxygen Cycle

Oxygen, the by-product of photosynthesis, is very active chemically. It combines with a wide range of chemicals in

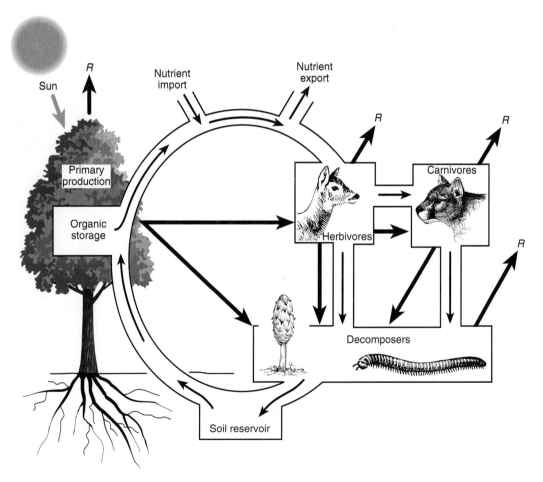

**Figure 12.1** The interrelationship between nutrient cycling and energy flow in the ecosystem. (Smith 1976:16.)

Earth's crust, and it reacts spontaneously with organic compounds and reduced substances. It is involved in the oxidation of carbohydrates, releasing energy, carbon dioxide, and water. Its primary role in biological oxidation is that of an electron acceptor, due to a high electron negativity. The breakdown and decomposition of organic molecules proceed primarily by dehydrogenation. Hydrogen is removed by enzymatic action from organic molecules in a series of reactions and is finally accepted by oxygen, forming water.

Oxygen, although necessary for life, can be toxic to certain organisms, like anaerobic bacteria. Higher organisms have evolved means to protect themselves from the toxic effects of molecular oxygen. At the same time these organisms have also evolved elaborate mechanisms to ensure a supply of oxygen. Plants have stomata to regulate the movement of gases into and out of the leaf. Some animals obtain oxygen from air or water by diffusion through the skin; others have gills or lungs. Organisms have an array of catalysts such as iron-containing molecules to mediate the transfer of hydrogen to oxygen molecules.

The major supply of free oxygen supporting life is in the atmosphere. It comes from two significant sources. One is the photodissocation of water vapor, in which most of the hydrogen released escapes into outer space. The other source is photosynthesis, active only since life began on Earth. Because photosynthesis and respiration are comple-

mentary, involving the release and capture of oxygen, one would seem to balance the other, so no significant quantity of oxygen would accumulate in the atmosphere. At some time in Earth's history the amount of oxygen introduced into the atmosphere had to exceed the amount involved in the decay of organic matter and oxidation of sedimentary rocks. Part of the atmospheric oxygen has to represent that which would have been used to decompose organic matter buried as the unoxidized reserves of photosynthesis—coal, oil, gas, and organic carbon in sedimentary rocks (Kasting 1993, Knoll 1986).

The major nonliving oxygen pool consists of molecular oxygen, water, and carbon dioxide, all intimately linked in photosynthesis and other oxidation-reduction reactions. Oxygen is biologically exchangeable in such compounds as nitrates and sulfates, utilized by organisms that reduce them to ammonia and hydrogen sulfide.

Because oxygen is so reactive, its cycling is complex. It reacts rapidly with reduced organic matter in the soil and reduced mineral constituents of Earth's crust. Buried organic matter that escapes oxidation conserves atmospheric oxygen by contributing its portion to replace the oxygen consumed by weathering of rocks, thus stabilizing the atmospheric reservoir (Figure 12.2). The amount of oxygen remaining in the atmosphere after oxidation of organic and inorganic matter is on the order of 1 part per 10,000, but the amount in the

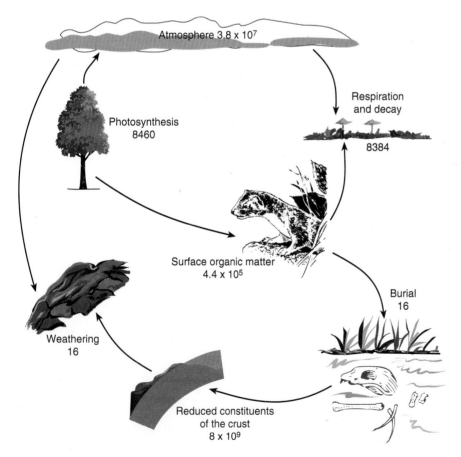

**Figure 12.2** The oxygen cycle. Bold letters denote reservoirs of oxygen or reduced matter with which oxygen can react. The units are $10^{12}$ moles of either oxygen or the capacity of the reservoir to combine with oxygen. Processes are indicated by lightface. The rates are expressed in $10^{12}$ moles per year of oxygen or equivalent capacity to combine with oxygen. (Adapted from Walker 1980.)

Atmosphere $3.8 \times 10^7$

Photosynthesis 8460

Respiration and decay 8384

Surface organic matter $4.4 \times 10^5$

Burial 16

Weathering 16

Reduced constituents of the crust $8 \times 10^9$

atmosphere remains rather stable and resistant to short-term changes of 100 to 1000 years. The residence time of atmospheric oxygen before its removal by respiration and decay and its replacement by photosynthesis is on the order of 4500 years (Walker 1984).

As a constituent of carbon dioxide, oxygen circulates freely throughout the biosphere. Some carbon dioxide combines with calcium to form carbonates. Oxygen combines with nitrogen to form nitrates, with iron to form ferric oxides, and with many other minerals to form various other oxides. Part of the atmospheric oxygen that reaches the higher levels of the troposphere reduces to ozone ($O_3$) in the presence of high energy ultraviolet radiation.

**The Ozone Layer**   Most of the ozone resides in the stratosphere, 10 to 50 km above Earth (Figure 12.3). The stratosphere is characterized by increasing temperature in contrast to the colder temperatures below because the upper relatively thin layer of ozone absorbs most of the ultraviolet radiation and radiates infrared wavelengths to outer space. A downward intrusion of stratospheric air introduces sufficient ozone to the troposphere to initiate photochemical processes in the lower atmosphere. Most importantly, it shields Earth against biologically harmful solar ultraviolet radiation.

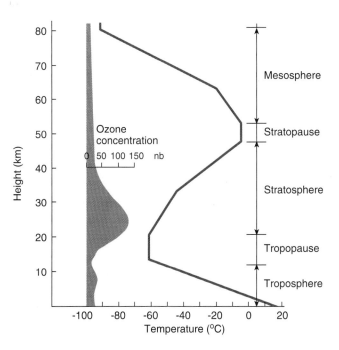

**Figure 12.3** The ozone layer in the stratosphere. Absorption of longer ultraviolet rays in the atmosphere acts upon oxygen molecules ($O_2$) to produce ozone ($O_3$), most of which is concentrated in a layer in the stratosphere at approximately 20 to 35 km above Earth. The layer shows both seasonal changes in height (highest in summer and lowest in winter) and latitudinal changes in height (higher in low latitudes). The tropopause and stratopause are areas in Earth's atmosphere where temperature is constant with height. (After Gates 1993:190.)

Ozone in the outer atmosphere is maintained by a cyclic photolytic reaction. Production of ozone requires the breakage of the O—O bond in $O_2$. Once freed, the O atoms rapidly combine with $O_2$ to form $O_3$:

$$O + O_2 + M \rightarrow O_3 + M$$

where M is any third-body molecule such as $N_2$ or $O_2$ that aids in the reaction but is not changed by it. In the stratosphere the $O_2$ bond is broken (dissociated) by solar radiation:

$$O_2 + hv \rightarrow O + O$$

where $h$ is Planck's constant (cal/sec) and $v$ is frequency of the light. This stratospheric reaction consumes a large amount of solar energy.

At the same time a reverse reaction consumes $O_3$:

$$O_3 + hv \rightarrow O + O_2$$
$$O + O_3 \rightarrow 2O_2$$
Net reaction: $2O_3 + hv \rightarrow 3O_2$

A number of other catalytic reactions tend to destroy $O_3$ and maintain a balance between $O_2$ and $O_3$, especially the gaseous oxides of nitrogen and hydrogen. For example, NO consumes $O_3$ in the following catalytic cycle:

$$NO + O_3 \rightarrow NO_2 + O_2$$
$$O_3 + hv \rightarrow O + O_2$$
$$NO_2 + O \rightarrow NO + O_2$$
Net reaction: $2O_3 + hv \rightarrow 3O_2$

The NO consumed in the first reaction is replaced by another molecule available for further destruction of ozone in the final reaction, and $O_3$ is consumed.

Under natural conditions in the atmosphere a balance exists between the rate of ozone formation and its rate of destruction. This balance exists largely in the stratosphere. However, in recent times a number of anthropogenic and even biological catalysts injected into the stratosphere are reactive enough to cause a decrease in stratospheric ozone (Lal and Holt 1991, McElroy and Salawith 1989, Cicerone 1987). Among them are chlorofluorocarbons ($CCl_2F_2$ and $CCl_3F$), methane ($CH_4$), both natural and anthropogenic, and nitrous oxide ($NO_2$) from denitrification and synthetic nitrogen fertilizer. Of particular concern is chlorine monoxide (ClO) derived from chlorofluorocarbons in aerosol spray propellants (banned in the United States and becoming banned worldwide), refrigerants, solvents, and other sources. Its chlorine can be in harmless chlorine nitrates ($ClNO_3$) or it can be involved in the catalytic destruction of ozone (Figure 12.4):

$$Cl + O_3 \rightarrow ClO + O_2$$
$$O_3 + hv \rightarrow O + O_2$$
$$ClO + O \rightarrow Cl + O_2$$
Net reaction: $2O_3 + hv \rightarrow 3O_2$

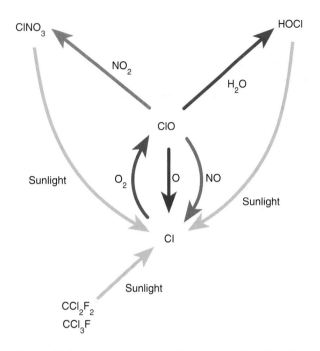

**Figure 12.4** The cycle of ozone destruction induced by chlorine monoxide (ClO) derived from human use of chlorofluorocarbons. The chlorine can either be incorporated in harmless chlorine nitrates or be involved in the photolytic dissociation of ozone. (After Kerr 1987:1182.)

In 1975 atmospheric scientists first discovered a pronounced springtime thinning in the ozone over the South Pole. Since then the thinning has developed into an ozone hole, where the ozone has been depleted 50 to 95 percent. The detection of chlorine monoxide in the hole suggested that ClO was the problem. Tiny ice particles in the cold lower stratosphere catalyzed chemical reactions that converted the chlorine in CFCs into ozone-destroying ClO (Jones 1989). In 1990 other atmospheric scientists noted a similar thinning in the ozone layers over the Arctic (Hofmann and Deshler 1991). This ozone loss was attributed not only to CFCs but also to natural aerosols of sulfuric acid that immobilize nitrogen oxides. Nitrogen oxides protect the ozone by tying up chlorine into harmless forms. With nitrogen oxides immobilized, chlorine forms ClO. More surprising was the discovery in 1991 of sheets of ClO as far south as Cuba (Kerr 1992). The presence of ClO in the north temperate latitudes could be attributed to its escape from the Arctic air mass. Some atmospheric scientists, however, believe that small ice particles associated with the strong updrafts of tropical air masses catalyze the production ClO and account for summertime as well as wintertime ozone depletion. Whatever the causes, the ozone layer in temperate latitudes is eroding at the rate of 3 percent per decade over latitude of New York, and 5 percent over Buenos Aires and Sydney. Even if world nations abide by the Montreal agreement to eliminate CFC production by the year 2000, the amount of stratospheric ClO globally will increase to 4.1 ppb from the current levels of 3.4 ppb.

Scientists are concerned over the reduced amount of ozone, because it allows disproportionate amounts of ultraviolet radiation to penetrate Earth's atmosphere. For example, under the conditions of an overhead sun and average amounts of $O_3$, a 1 percent decrease in ozone would result in a 20 percent increase in ultraviolet penetration at wavelengths of 250 nm, a 250 percent increase at 290 nm, and a 500 percent increase at 287 nm (Walker 1977). Although solar ultraviolet wavelengths of less than 240 nm are absorbed by atmospheric $O_2$ and $O_3$, only $O_3$ is effective for wavelengths between 240 and 320 nm (Cicerone 1987). Such reductions in the ozone layer can have adverse ecological effects such as altering DNA, causing severe sunburn and increasing skin cancer in humans, and inhibiting photosynthesis in some terrestrial plants and phytoplankton.

## The Carbon Cycle

Because it is a basic constituent of all organic compounds and a major element in the fixation of energy by photosynthesis, carbon is so closely tied to energy flow that the two are inseparable. In fact, the measurement of productivity is commonly expressed in terms of grams of carbon fixed per square meter per year. The source of all fixed carbon, both in living organisms and fossil deposits, is carbon dioxide, $CO_2$, found in the atmosphere and dissolved in the waters of Earth. To trace its cycling through the ecosystem is to redescribe photosynthesis and energy flow (Figure 12.5).

The cycling of carbon dioxide involves assimilation by plants and conversion to glucose. From glucose plants synthesize polysaccharides, proteins, and fat and store them in the form of plant tissue. When digesting plant tissue, herbivores synthesize these compounds into other carbon compounds. Meat-eating animals feed on herbivores and the carbon compounds are redigested and resynthesized into other forms. Some of the carbon is returned directly by both plants and animals in the form of $CO_2$ as a by-product of respiration. The remainder for a time becomes incorporated in the living biomass.

Carbon contained in animal wastes and in the protoplasm of plants and animals is released eventually by assorted decomposer organisms. The rate of release depends upon environmental conditions such as soil moisture, temperature, and precipitation. In tropical forests most of the carbon in plant detritus is quickly recycled, for there is little accumulation in the soil. In drier regions, such as grasslands, considerable quantities of carbon are stored as humus. In swamps and marshes, where dead material falls into the water, organic carbon is not completely mineralized; it is stored as raw humus or peat and is only slowly broken down.

Similar cycling takes place in the freshwater and marine environments. Phytoplankton utilizes the carbon dioxide that has diffused into the upper layers of water or is present as carbonates and converts it into carbohydrates. The carbohydrates so produced pass through the aquatic food chains.

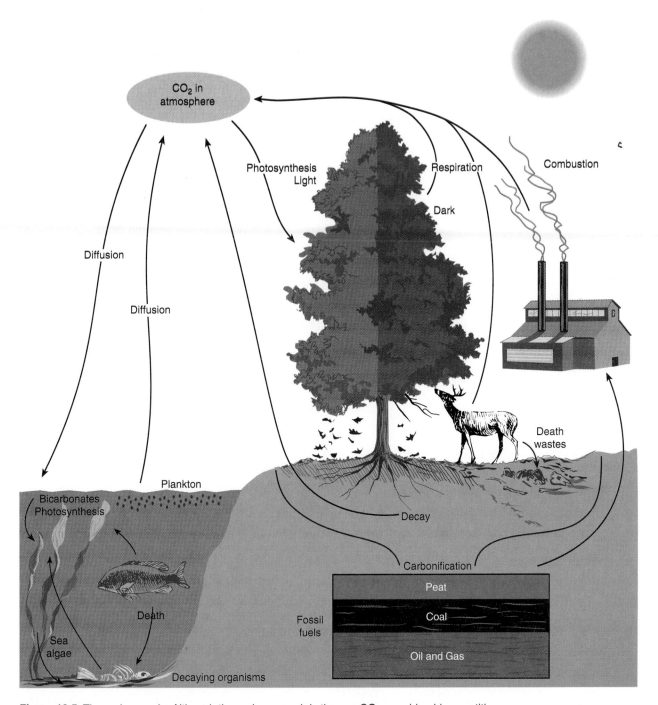

**Figure 12.5** The carbon cycle. Although the main reservoir is the gas $CO_2$, considerable quantities are tied up in organic and inorganic compounds of carbon in the biosphere.

The carbon dioxide produced by respiration is reutilized by the phytoplankton in the production of more carbohydrates. Under proper conditions a portion is reintroduced into the atmosphere. Significant portions of carbon, bound as carbonates in the shells of mollusks and foraminifers, become buried in the bottom mud at varying depths when the organisms die. Isolated from biotic activity, that carbon is removed from cycling and becomes incorporated into bottom sediments, which through geological time may appear on the surface as limestone rocks or as coral reefs. Other organic carbon is slowly deposited as gas, petroleum, and coal at an estimated global rate of 10 to 13 g/m2/yr.

Local Patterns The concentration of carbon dioxide in the atmosphere around plant life fluctuates throughout the day (Figure 12.6a). At daylight, when photosynthesis begins, plants start to withdraw carbon dioxide from the air and the concentration declines sharply by about 25 percent.

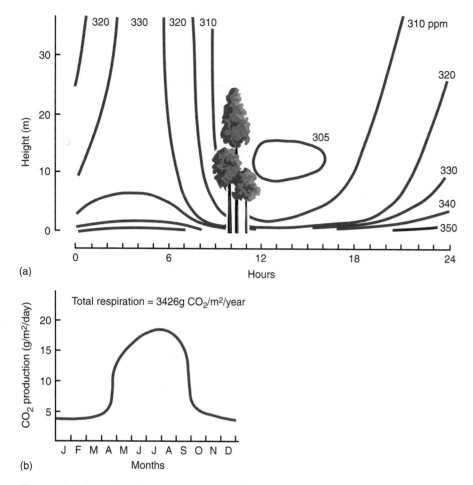

**Figure 12.6** Diurnal and seasonal patterns of carbon cycling in a forest. (a) Daily flux of $CO_2$. Note the consistently high level of $CO_2$ on the forest floor, the site of microbial respiration. Atmospheric $CO_2$ in the forest is lowest from midmorning to late afternoon. $CO_2$ levels are highest during the night, when photosynthesis is shut down and respiration is pumping $CO_2$ into the atmosphere. (From Baumgartner 1968, after Miller and Rusch 1960.) (b) Changes in $CO_2$ production over the course of a year in a pine forest, based on respiration rates during inversions. Respiration rates as measured by $CO_2$ production are highest during the summer, when photosynthesis and decomposition are the greatest. (Woodwell and Dykeman 1966.)

By afternoon, when the temperature is increasing and the humidity is decreasing, the respiration rate of plants increases, and the assimilation rate of carbon dioxide increases. By sunset, when the light phase of photosynthesis ceases and the light-independent phase (Calvin cycle) shortly thereafter. Carbon dioxide is no longer being withdrawn from the atmosphere, and its concentration in the atmosphere increases sharply. A similar diurnal fluctuation takes place in aquatic ecosystems.

There is also an annual course in the production and utilization of carbon dioxide (Figure 12.6b). This seasonal change relates both to temperature and to growing seasons. In the spring, when land is greening and phytoplankton is actively growing, the daily production of carbon dioxide is high. Measured by nocturnal accumulation in spring and summer, the rate of carbon dioxide production by respiration may be two to three times higher than winter rates at the

same temperature. The rate increases dramatically about the time of the opening of buds and falls off just as rapidly about the time the leaves of deciduous trees start to drop in the fall. This increased production of $CO_2$ does not accumulate in the atmosphere, however, because of the high rate of $CO_2$ fixation by photosynthesis.

**The Global Pattern** The carbon budget of the earth is closely linked to the atmosphere, land, and oceans and to the mass movements of air around the planet (Figure 12.7). The carbon pool involved in the global carbon cycle amounts to an estimated 55,000 Gt (Gt is a gigaton, equal to 1 billion or $10^9$ metric tons). Fossil fuels account for an estimated 10,000 Gt. The oceans contain 93 percent of the active carbon pool, that is, about 35,000 Gt, mostly as bicarbonate ions ($HCO_3^-$) and carbonate ions ($CO_3^-$). Dead organic matter in the oceans accounts for 1650 Gt of carbon, and living

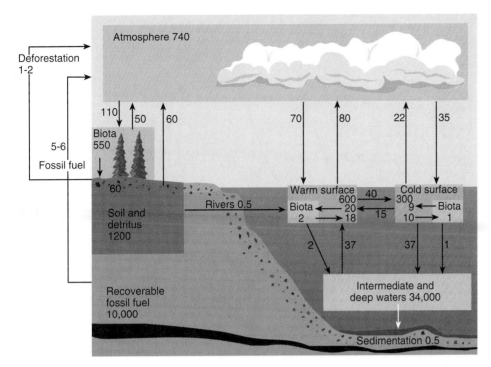

**Figure 12.7** Global compartments of $CO_2$. The diagram shows the interconnected subsystems. Arrows indicate the major fluxes (in Gt/yr) and pool sizes (in Gt) of carbon. (Edmonds 1992:14.)

matter, mostly phytoplankton, 3 Gt. The terrestrial biosphere contains an estimated 1200 Gt as dead organic matter and 550 Gt as living matter. The atmosphere, the major coupling mechanism in the cycling of $CO_2$, holds about 740 Gt of carbon (Baes et al. 1977).

In the sea the surface water acts as the site of main exchange of carbon between atmosphere and ocean. The ability of the surface waters to take up $CO_2$ is governed largely by the reaction of $CO_2$ with the carbonate ion to form bicarbonates:

$$CO_2 + CO_3^- + H_2O \Leftrightarrow 2HCO_3^-$$

In the surface water carbon circulates physically by means of currents and biologically through assimilation by phytoplankton and movement through the food chain. Only about 10 percent of the carbon that comes up from the deep in upwellings is used in organic production; the other 90 percent goes back to the deep (Broecker 1973).

Most of the carbon in the land mass is in slowly decomposing (exchanging) dead organic matter. A small amount is in rapidly exchanging material. Exchanges of $CO_2$ between the land mass and the atmosphere (uptake in photosynthesis and release by decomposition) are nearly in equilibrium (Figure 12.7).

Of considerable importance in the terrestrial carbon cycle is the proportion of detrital carbon (contained in dead organic matter on the soil's surface and in the underlying mineral soil) to the living organic pool. Estimates place the amount of detrital carbon at 1200 Gt, compared to 550 Gt for the total world terrestrial biomass. Much of this detrital

carbon is in the lower soil profile, where it exceeds carbon in the surface soil by a factor of 25. The lower soil layers of tropical forests and both tropical and temperate grasslands contain a large percentage of soil carbon, whereas soil carbon in temperate forests and arctic regions is confined largely to the surface soil layers.

The average amount of carbon per unit of soil profile increases from the tropical regions poleward to the boreal forest and tundra (Table 12.1). Low values for the tropical forest reflect high rates of decomposition, which compensate for high productivity and litterfall. Frozen tundra soil and waterlogged soils of swamps and marshes have the greatest accumulation of detritus because decay is inhibited by moisture, low temperature, or both. Detritus greatly exceeds biomass in boreal forests and tundra. Detritus likewise exceeds biomass in temperate grasslands, but in tropical savannas the reverse situation exists, probably because of recurring fires. World output of carbon from soil respiration in terrestrial ecosystems amounts to approximately 60 Gt per year.

The global cycle of atmospheric carbon dioxide, like local cycles, exhibits a marked annual variation, particularly in the land-dominated Northern Hemisphere (Figure 12.8). Carbon dioxide content of the atmosphere up to the lower stratosphere begins to decline markedly during early spring north of the 30° N latitude, when photosynthetic activity and increasing biomass storage withdraw more carbon dioxide from the atmosphere than is replaced by respiration. The concentration of carbon dioxide reaches a low in August. By October the level increases again. The decline of $CO_2$ during the arctic summer suggests a rapid and heavy removal of car-

**Table 12.1 Distribution of Detritus and Biomass by Ecosystem Types**

| Ecosystem Type | Mean Total Profile Detritus (kg C/m²) | CV* (%) | World Area (10⁹ ha) | Total World Detritus (10⁹ mtn C) | Total World Biomass (10⁹ mtn C) |
|---|---|---|---|---|---|
| Woodland and shrubland | 6.9 | 59 | 8.5 | 59 | 22.0 |
| Tropical savanna | 3.7 | 87 | 15.0 | 56 | 27.0 |
| Tropical forest | 10.4 | 44 | 24.5 | 255 | 460.0 |
| Temperate forest | 11.8 | 35 | 12.0 | 142 | 175.0 |
| Boreal forest | 14.9 | 53 | 12.0 | 179 | 108.0 |
| Temperate grassland | 19.2 | 25 | 9.0 | 173 | 6.3 |
| Tundra and alpine | 21.6 | 68 | 8.0 | 173 | 2.4 |
| Desert scrub | 5.6 | 38 | 18.0 | 101 | 5.4 |
| Extreme desert | 0.1 | — | 24.0 | 3 | 0.2 |
| Cultivated land | 12.7 | — | 14.0 | 178 | 7.0 |
| Swamp and marsh | 68.6 | 63 | 2.0 | 137 | 13.6 |
| Total | | | 147.0 | 1456 | 826.9 |

*CV = coefficient of variation = standard deviation/mean × 100.
*Source:* Adapted from W. H. Schlesinger 1977.

bon dioxide by the photosynthetic activity of arctic plant life. This scrubbing effect is maximum in August, when arctic ecosystems are most active and polar ice is at a minimum.

**Human Impact** The $CO_2$ flux among land, sea, and atmosphere has been disturbed by a rapid injection of carbon dioxide in the atmosphere from the burning of fossil fuels (Figure 12.9) and from the clearing of forests. Clearing increases the input of $CO_2$ from burning trees and decomposing organic matter. Further, it reduces the forest sink for the storage of $CO_2$ scrubbed from the atmosphere by photosynthesis and biomass accumulation.

Adding to the problem of increasing carbon dioxide are increases in other atmospheric gases, especially methane ($CH_4$). Its major sources are ruminant animals, microbial decomposition in swamps, marshes, and tundra, and industrial gases released to the atmosphere. Over time (resident time

of 3.2 years) methane oxidizes to $H_2O$, especially in the stratosphere. Atmospheric methane has approximately doubled over the past 200 years (Figure 12.10). This increase is linked to human population growth and to increased cattle ranching and rice paddy production.

$CO_2$ concentrations in the atmosphere have varied over the past 160,000 years. Analysis of the $CO_2$ trapped in air bubbles in a 2083 m deep ice core recovered by the Soviet Antarctic Expeditions at Vostok reveals the changes. Levels were low during glacial periods and higher during interglacial periods (Figure 12.11). During the Last Glacial Maximum levels of $CO_2$ were low, about 200 ppm by volume (ppmv). $CO_2$ increased during the glacial-Holocene transition. At the beginning of the Industrial Revolution atmospheric $CO_2$ was about 260 ppmv. Since then, input of $CO_2$ has been increasing at an exponential rate (Figure 12.12) to about 353 ppmv in 1992. Up to 1950, two-thirds of

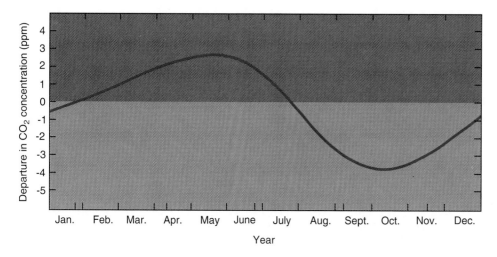

**Figure 12.8** Seasonal variations in atmospheric concentrations in $CO_2$ during a typical year at Barrow, Alaska. (After Pearman and Hyson 1981.)

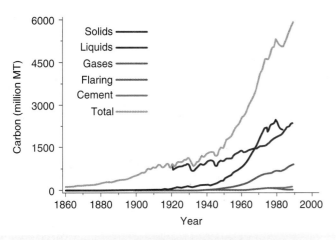

**Figure 12.9** Global $CO_2$ emissions from fossil fuel burning, gas flaring from oil fields and refineries, and cement production. Note the exponential increase in emissions since 1960.

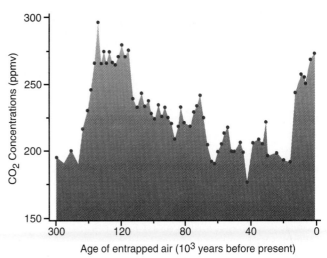

Age of entrapped air ($10^3$ years before present)

**Figure 12.11** Fluctuations in atmospheric $CO_2$ as derived from air samples taken from the 2038 m deep ice core at the Vostoc Station, Antarctica. (Barnola et al. 1991:5.)

the carbon dioxide injected into the atmosphere came from biospheric sources and one-third from anthropogenic sources. Since 1950, most $CO_2$ has come from the burning of fossil fuels (Broecker et al. 1979), the destruction of the world's forests, and change in land use (Houghton et al. 1983). An increase of about 12 percent in the mean annual concentration in 32 years has been corroborated by data collected from worldwide monitoring stations.

Only about 40 to 50 percent of the total $CO_2$ injected into the atmosphere remains there. What happens to the rest? This question is the object of considerable research. Such research is critical because the answer relates directly to global warming (see below). Current assessments of the global carbon budget assume a balance between Earth and atmosphere (Figure 12.5). The global budget, however, is

not balanced (Table 12.2) (Edmonds 1992). A significant fraction, about 22 percent, of anthropogenic carbon cannot be accounted for.

The ocean is a major carbon sink (Tans et al. 1990, Quay et al. 1992). The uptake of anthropogenic $CO_2$ by oceans depends heavily upon eddy current circulation between the surface and deep water. Oceans account for about 28 percent of the total $CO_2$ disposition.

The other potential major sink is the terrestrial biota. Its ability to absorb a significant fraction is controversial (Hobbie et al. 1984). Some ecologists (Botkin 1977, Houghton et al. 1983) argue that terrestrial vegetation is a net source, not a sink. Terrestrial vegetation, especially

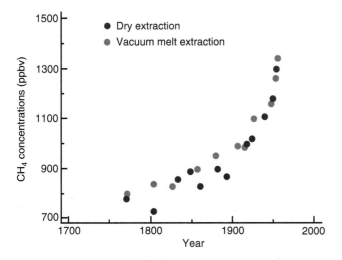

**Figure 12.10** Concentration of atmospheric methane over the past 200 years as indicated by air samples taken from ice cores at Siple Station, Antarctica. (Stauffer et al. 1991:221.)

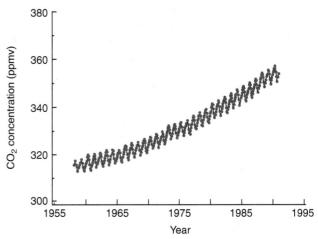

**Figure 12.12** Rise in monthly atmospheric $CO_2$ concentrations as obtained from Mauna Loa in Hawaii. The record shows a 12 percent increase in the mean annual concentration in 32 years. (Keeling and Wharf 1991.)

**Table 12.2** Average Global Carbon Fluxes
for the 1980s

| Sources and Sinks | PgC/yr |
|---|---|
| Emissions from fossil fuels into the atmosphere | $5.4 \pm 0.5$ |
| Emissions from net deforestation and land-use change | $1.6 \pm 1.0$ |
| **Total Net Emissions** | **$7.0 \pm 1.5$** |
| Accumulation in the atmosphere | $3.4 \pm 0.2$ |
| Uptake by the ocean | $2.0 \pm 0.8$ |
| **Total Disposition** | **$5.4 \pm 1.0$** |
| *Net Imbalance (the "missing carbon" sink)* | *$1.6 \pm 1.4$* |

*Source:* IPCC (1990: 1.14).

forests, does store carbon in new growth and detrital material, but whether its storage is greater than its contribution to atmosperic carbon is unknown. Are other sinks such as coral reefs and coastal wetlands being overlooked? Current assessments of the carbon cycle assume that exchanges between land masses and the atmosphere are in equilibrium over the long term. That assumption is being challenged (Edmonds 1992, Lugo 1992). Scientists are now questioning whether the atmosphere and the biota were ever in equilibrium. We do not yet adequately understand the carbon cycle.

There is no disagreement that atmospheric $CO_2$ has increased by 25 percent over the past 100 years or so. If atmospheric $CO_2$ continues to increase at its present rate of 0.39 percent per year, its concentration in the atmosphere will reach 400 ppm by about the year 2030. In view of our increasing dependence on fossil carbon, our need to understand the fate of excess carbon and the nature of carbon sinks is pressing.

**The Greenhouse Effect** Continuing increase of atmospheric $CO_2$ and $CH_4$ is intensifying the **"greenhouse effect."** Acting as a shield over Earth, the greenhouse gases, $CO_2$ and $CH_4$, allow incoming shortwave radiation to penetrate the atmosphere but absorb outgoing longwave radiation and redirect some of it toward Earth (Figure 12.13). These natural greenhouse gases are necessary to sustain life on Earth. They keep Earth's temperature about 33° C higher than it would be without them. However, as the concentration of greenhouse gases increases, it traps more heat. Each doubling of $CO_2$ concentration results in a doubling of the average temperature of the troposphere, with a greater increase at the higher latitudes and a smaller increase at the lower latitudes. Molecule for molecule, methane is more effective than $CO_2$ in contributing to the greenhouse effect, because its infrared absorptions fall into wavelength regions not strongly absorbed by existing concentrations of $H_2O$, $CO_2$, and $O_3$, and because it contributes to an increase in heat-absorbing stratospheric water vapor (see Blake and Rowland 1988).

Intensive studies and modeling of atmospheric changes in $CO_2$, (DOE 1986, Kasting and Ackerman 1986, COHMAP 1988) seek to predict potential outcomes if global warming should occur. Current models predict a global warming of 1° C relative to the base year 1850 by the year 2000. Such a rise would lengthen the growing season in the northern United States and southern Canada by 10 days. It could lessen precipitation in the central plains, bring an earlier summer, and increase evaporation from the soil (Manabe and Wetherald 1986, Parry, 1992). At the current rate of increase in atmospheric $CO_2$, temperatures could rise 2° to 3° C by the year 2100.

Such rises in temperature would have pronounced ecological effects, let alone social and economic impacts. One effect would be a short-term rise in plant productivity and related phenomena. $C_3$ plants would have the most to gain (Lemon 1983, Pary 1992). They would increase their net productivity because $CO_2$ would no longer be limiting; they would increase their water-use efficiency because the stomatas would not need to open fully to gain sufficient $CO_2$, reducing transpiration; and photorespiration would cease to be a problem. Provided moisture did not become limiting or the temperatures become too warm, $C_3$ plants would become competitive with $C_4$ plants, which would not respond to increased $CO_2$. However, increased net photosynthesis carries with it certain demands. With more carbohydrates available for growth, plants would need more water and more nutrients. Without that, increased $CO_2$ would be to no avail.

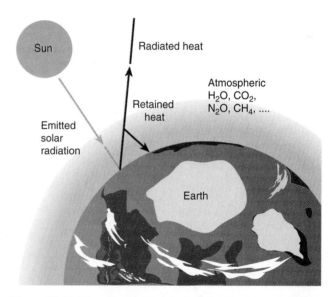

**Figure 12.13** The greenhouse effect. Solar shortwave radiation easily penetrates Earth's atmosphere, although some radiation is blocked by dust and pollutants. Carbon dioxide and other atmospheric gases (ozone, methane) inhibit the escape of longwave radiation back to outer space, heating Earth. If the retention of more longwave radiation than normal is not balanced by some blockage of incoming shortwave radiation, Earth will continue to warm.

Increases in plant productivity could have other outcomes. Increased growth could result in changes in allocation of photosynthate to various parts of the plant and lower in changes in leaf area and leaf structure. Increased branching in herbaceous and woody plants and lower root-to-shoot ratios could change the competitive position of plants, influencing ecosystem composition.

A general warming undoubtedly would result in a northward shift of ecosystems toward the poles and up mountains. Temperature changes (and associated hydrological balances) have had a profound influence on changes in vegetational distribution through Earth's history (see Woodward 1987). Simulation models of global warming (Solomon 1986, Woodward 1987, COHMAN 1988, Smith et al. 1992) involving changes in temperature and precipitation predict major vegetational changes. Tundra, boreal, and deciduous forests would shift northward, and deciduous forests would expand as coniferous forests decrease or replace tundra. Verification of these models by comparing paleoecological data with model simulations shows them to be highly accurate in their predictive powers.

Other associated outcomes have been hypothesized for global warming (Peters and Darling 1985). Melting of permafrost and decomposition of tundra peat would release methane and carbon dioxide, increasing the input of these two gases to the atmosphere. Southern animals would extend their ranges northward and be confronted with new competitors and a loss of habitat; Arctic animals would face extinction. Small relict populations, lacking genetic diversity to adapt to changing climates and habitats, and left behind in limited refuges in a changing environment, would perish. Along continental coastlines rising sea levels would gradually submerge estuaries, tidal marshes, and human settlements.

This scenario has its critics, and in fact suggests some advantages, such as longer growing seasons and fewer droughts in some areas, and warmer northern regions. The fact that climate has been warming at less than the predicted rates gives critics an argument that the effect of a buildup of greenhouse gases is overblown. The slower rise may be attributed to the presence of pollution haze in the atmosphere, which reflects enough solar energy back to outer space to slow down global warming. However, the aerosols are more short-lived than greenhouse gases, so at constant emission levels, the greenhouse gases eventually will win.

## The Nitrogen Cycle

Nitrogen is an essential constituent of protein, a building block of all living material. It is also a major constituent, about 79 percent, of the atmosphere. The paradox is that in its gaseous state, nitrogen is unavailable to most life. It must first be converted to some chemically usable form. Getting it into that form comprises a major part of the nitrogen cycle.

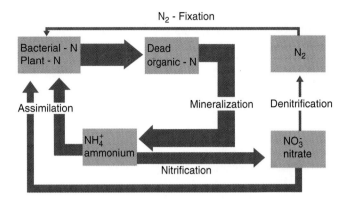

**Figure 12.14** The bacterial processes involved in nitrogen cycling. The width of each arrow is an approximation of the process rate. (Blackburn 1983:64.)

**Processes** The nitrogen cycle consists of four processes (Figure 12.14). These are fixation, the conversion of nitrogen in its gaseous state to a usable form; mineralization or ammonification, the conversion of amino acids in organic matter to ammonia; nitrification, the oxidation of ammonia to nitrites and nitrates; and denitrification, the reduction of nitrates to gaseous nitrogen. Most of the nitrogen cycle is driven by microbes.

**Fixation** converts gaseous nitrogen, $N_2$, to ammonia and nitrates. Ammonia, $NH_4$, is the product of biological fixation. Nitrates are the product of high-energy fixation by lightning; occasionally cosmic radiation and oxygen in the atmosphere combine into nitrates, which come to Earth in rainwater as nitric acid, $H_2NO_3$. Estimates suggest that less than 8.9 kg N/ha/yr arrives on Earth by high-energy fixation.

Biological fixation, the more important method, makes available 100 to 200 kg N/ha, roughly 90 percent of the fixed nitrogen contributed to Earth each year. In biological fixation molecular (or gaseous) nitrogen, $N_2$, is split into two atoms:

$$N_2 \rightarrow 2N$$

This step is energy-expensive, because the two nitrogen atoms are connected by a triple bond. It requires an input of 160 kcal for each mole (28 g) of nitrogen. The free N atoms can combine with hydrogen to form ammonia, with the release of about 13 kcal of energy:

$$2N + 3H_2 \rightarrow 2NH_3$$

This fixation is accomplished by mutualistic bacteria living in association with leguminous and root-noduled nonleguminous plants, by free-living bacteria, and by cyanobacteria (blue-green algae). In agricultural ecosystems approximately 200 species of nodulated legumes are the preeminent nitrogen fixers. In nonagricultural systems some 12,000 species of organisms, from bacteria and blue-green

algae to nodule-bearing plants, are responsible for nitrogen fixation.

Legumes, the most conspicuous of the nitrogen-fixing plants, have a mutualistic relationship with members of the bacterial genus *Rhizobium*. Rhizobia are aerobic, non-spore-forming, rod-shaped bacteria (Figure 12.15). They live in the immediate surroundings of plant roots, called the **rhizosphere.** Stimulated by secretions and enzymes from the legumes, swarming rhizobia enter the root hairs, where they multiply and increase in size. This invasion and growth results in swollen, infected root hair cells that make up the central tissues of the root nodules. Inside the nodules the bacteria change from rod-shaped to a nonmobile form that carries on nitrogen fixation. Although the process is well-known, the mechanisms are still the subject of intensive study.

A large number of nonleguminous nodule-bearing plants, most of them early pioneering species, grow on sites where soil is low in nitrogen. These plants make significant contributions of nitrogen to wildlands. Among such plants are alder (*Alnus*), New Jersey tea (*Ceanothus*), and Russian olive (*Elaeagnus*).

Also contributing to the fixation of nitrogen are free-living soil bacteria. The most prominent of the 15 known genera are the aerobic *Azotobacter* and the anaerobic *Clostridium*. *Azotobacter* prefers soils with a pH of 6 to 7 that are rich in mineral salts and low in nitrogen. *Clostridium* is ubiquitous, found in nearly all soils. Both genera produce ammonia as the first stable end product. Free-living and symbiotic bacteria both require molybdenum as activators and are inhibited by an accumulation of nitrates and ammonia in the soil.

Cyanobacteria are another important group of largely nonsymbiotic nitrogen-fixers. Of some 40 known species the most common are in the genera *Nostoc* and *Calothrix*, found both in soil and aquatic habitats. Cyanobacteria are often pioneers on bare mineral soil. Especially successful in water-logged soils, they are nitrogen-fixers in the rice paddies of Asia. Cyanobacteria are perhaps the only fixers of nitrogen over a wide range of temperatures in aquatic habitats from Arctic to Antarctic seas to freshwater ponds and hot springs. Like bacteria, cyanobacteria require molybdenum for nitrogen fixation.

Other plants may be involved in nitrogen fixation. In humid tropical forests epiphytes growing on tree branches and bacteria and algae growing on leaves may fix appreciable amounts of nitrogen. Certain lichens have been implicated in nitrogen fixation (Henriksson and Simu 1971). Lichens with nitrogen-fixing ability possess nitrogen-fixing cyanobacteria species as their algal component.

**Mineralization** or **ammonification** of organic nitrogen is the major step in the nitrogen cycle. In this process, proteins and nucleic acids in dead plant and animal material are broken down by bacteria and fungi to amino acids. The amino acids are oxidized to carbon dioxide, water, and ammonia, with a yield of energy; for example:

$$CH_2NH_2COOH + \tfrac{1}{2}O_2 \rightarrow 2CO_2 + H_2 + NH_3 + 178\ kcal$$

Ammonium, or the ammonia ion, is absorbed directly by plant roots, incorporated into amino acids, and passed through the food chain. Some of the ammonia is dissolved in water, part is trapped in the soil, and some is fixed in clay minerals.

**Nitrification** is a biological process in which ammonia is oxidized to nitrate and nitrite, yielding energy. Two groups of organisms are involved. *Nitrosomonas* bacteria use the ammonia in the soil as their sole source of energy. They can promote its oxidation first to nitrous acid and water and then to nitrite:

$$NH_3 + 1\tfrac{1}{2}O_2 \rightarrow HNO_2 + H_2 + 165\ \text{kcal}$$
$$HNO_2 \rightarrow H^+ + NO_2$$

Energy left in the nitrite ion is exploited by another group of bacteria, the *Nitrobacter*, which oxidizes the nitrite ion to nitrate with a release of a small amount of energy:

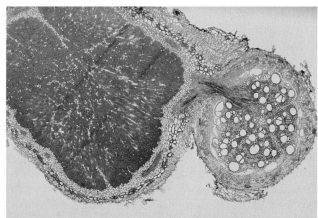

**Figure 12.15** *Rhizobium* bacteria and root nodules.

$$NO_2 + \tfrac{1}{2}O_2 \rightarrow NO_3$$

In nitrification, *Nitrosomonas* oxidizes 35 moles of nitrogen for each mole of $CO_2$ assimilated; *Nitrobacter* oxidizes 100 moles.

Nitrification is generally beneficial, but not always. Nitrification involves the conversion of slowly leached forms of nitrogen into readily leached nitrates. If quantities of nitrates are large enough and sufficient water percolates through the soil, nitrates can be removed faster than they can be taken up by plant roots. This situation results in the pollution of water. An abundance of nitrates also leads to increased losses of gaseous nitrogen.

Nitrates are a necessary substrate for **denitrification,** in which the nitrates are reduced to gaseous nitrogen by certain organisms to obtain oxygen. The denitrifiers, represented by fungi and the bacteria *Pseudomonas,* are facultative anaerobes. They prefer an oxygenated environment, but if oxygen is limited, they can use $NO_3^-$ instead of $O_2$ as the hydrogen acceptor. In doing so, they release $N_2$ in the gaseous state as a by-product:

$$C_6H_{12}O_6 + 4NO_3 \rightarrow 6CO_2 + H_2O + 2N_2$$

**The Global Nitrogen Cycle** The global nitrogen cycle follows the pathway of the local nitrogen cycle, only on a grander scale. The global cycle with estimated pool sizes is diagrammed in Figure 12.16. The sources of fixed nitrogen under global conditions are biological and high-energy fixation of atmospheric nitrogen, additions of inorganic nitrogen in rain from such sources as fixed "juvenile" nitrogen from volcanic activity, absorption of ammonia from the atmosphere by plants and soil, and nitrogen accretion from windblown aerosols, which contain both organic and inorganic forms of nitrogen. In terrestrial ecosystems, ammonia and nitrates are taken up by plants and converted to organic nitrogen (amino acids) and moved through the food chain. Dead organic matter is broken down into ammonia, then into nitrates and nitrites. Nitrates may be taken up directly by plants, immobilized by microbes, stored in decomposing humus, or leached into streams, lakes, and eventually seas.

In aquatic ecosystems, nitrogen is cycled in a similar manner, except that the large reserves, such as those contained in soil humus, are lacking. Life in the water contributes organic matter and dead organisms that undergo decomposition and subsequent release of ammonia and nitrates.

Tracer studies with $^{15}N$, a short-lived, nonradioactive isotope, show that in marine ecosystems, ammonia is recycled rapidly by phytoplankton (Dugdale and Goering 1967). As a result little ammonia exists in natural waters, and nitrate is used only in the virtual absence of ammonia. In addition to biological cycling, there are small but steady losses from the biosphere to the deep sediments of the ocean and to sedimentary rocks. In return there is a small addition of new nitrogen from the weathering of igneous rocks and juvenile nitrogen from volcanic activity.

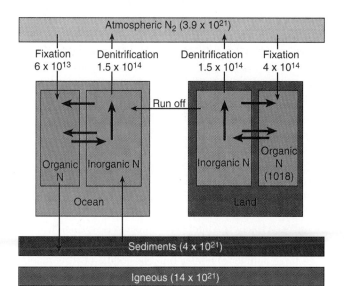

**Figure 12.16** The global nitrogen cycle. Pool sizes, in parentheses, are given as gN. The quantitatively most important transfer rates (gN per year) are indicated by thick arrows. (From Blackburn 1983:64, based on data from Fenchel and Blackburn 1979.)

Under natural conditions, nitrogen loss by denitrification, volatilization, leaching, erosion, wind-blown aerosols, and transportation out of the system is balanced by biological fixation and other sources. Both chemically and biologically, terrestrial and aquatic ecosystems constitute a dynamic equilibrium system in which a change in one phase affects the other.

**Human Impact** Human intrusion into the nitrogen cycle upsets its natural equilibrium either by reducing nitrogen availability or by overloading the system. Conversion of forests and grasslands to cropland results in a steady decline in nitrogen content of the soil. Mixing and breaking up the soil exposes more organic matter to rapid decomposition. Removal of nitrogen through harvested crops and grazing causes additional losses. Harvest of timber results in a heavy outflow of nitrogen from the forest ecosystem, not only in the timber removed, but also in the short-term nitrate losses from the soil.

On the other hand, excessive amounts of nitrogen may be added to the system, creating various problems. Heavy applications of commercial fertilizer disturb the natural balance between fixation and denitrification by reducing the former and increasing the latter. A considerable portion of the added nitrogen may be leached to the groundwater as nitrates. Animal wastes, especially from livestock concentrated in large feedlots, are another source of nitrates in groundwater. Excessive quantities of nitrates in groundwater become health-endangering pollutants.

A third source of nitrate pollution is human waste, particularly sewage, treated or otherwise, released into rivers, lakes,

and estuaries. In spite of the magnitude of water pollution from agricultural sources, human effluents contribute even heavier loads, especially from municipal sewage treatment.

Automobiles and power plants are the major sources of nitrogenous pollutants in the atmosphere, particularly nitrogen oxides. The primary nitrogenous air pollutant is nitrogen dioxide, $NO_2$, a pungent gas that produces a brownish haze. In the atmosphere nitrogen dioxide is reduced by ultraviolet light to nitrogen monoxide and atomic oxygen:

$$NO_2 \rightarrow NO + O$$

Atomic oxygen reacts with molecular oxygen to form ozone (see p. 198):

$$O_2 + O \rightarrow O_3$$

Ozone, in a never-ending cycle, reacts with nitrogen monoxide to form nitrogen dioxide and oxygen:

$$NO + O_3 \rightarrow NO_2 + O_2$$

This cycle illustrates only a few of the reactions that nitrogen oxides undergo or trigger. In the presence of sunlight atomic oxygen from nitrogen dioxide also reacts with a number of reactive hydrocarbons to form radicals. These radicals then take part in a series of reactions to form still more radicals that combine with oxygen, hydrocarbons, and nitrogen dioxide. As a result, nitrogen dioxide is regener-

ated, nitrogen monoxide disappears, ozone accumulates, and a number of secondary pollutants form, including formaldehydes, aldehydes, and peroxyacytnitrates, known as PAN. All of these pollutants, especially PAN and ozone, are very toxic and injure many forms of plant life exposed to them.

One outcome of nitrogenous atmospheric pollution is an increased deposition of nitrogen, which benefits ecosystems that are traditionally nitrogen-limited, especially northern and high altitude forests. Such ecosystems, however, can suffer from too much of a good thing. Because nitrogen is limiting, these forests are efficient at retaining and recycling nitrogen from precipitation and organic matter. Only a few normally lose any significant amount of nitrates to streams. Many of these forests are receiving more nitrogen in the form of ammonium and nitrates than the trees and their associated microbial populations can handle and accumulate.

The first response to increased availability of nitrogen in a nitrogen-limited ecosystem is increased growth of both canopy and stems. Evidence suggests, however, that increased levels of nitrogen can lead to the decline and dieback of coniferous forests at high elevations (Figure 12.17) (Aber et al. 1989). If the increased growth in foliage continues into summer, the late new growth may not have time to become frost-hardened and is killed during the winter. Overstimulated by nitrogen, tree growth exceeds the availability of other necessary nutrients in the soil, particu-

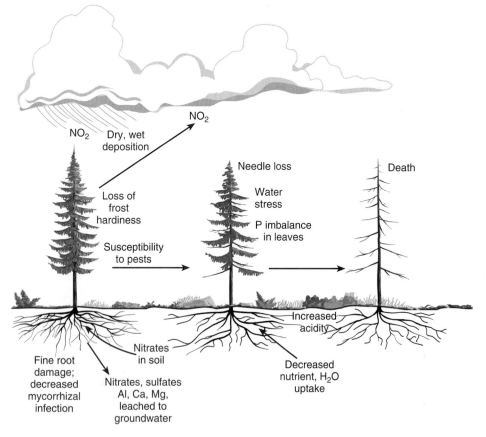

**Figure 12.17** Excess inputs of nitrogen in forest ecosystems, according to the nutrient limitation hypothesis, affect many processes that result in the decline and death of forest trees and convert the ecosystem from a nitrogen sink to a nitrogen source.

larly phosphorus, and the tree begins to experience nutrient deficiencies. Experimental evidence suggests that the production of fine roots and ectomycorrhizae, which take up nutrients from the soil, is lower on sites rich in nutrients, especially nitrogen, than in nutrient-poor soils, and root turnover is higher. Trees on nutrient-poor soils have a longer-lived and higher density root system and a lower turnover of root biomass. These conditions help the trees to scavenge nutrients from poor soils. As nitrogen levels increase, root biomass decreases because sufficient nitrogen is nearby. A smaller root biomass then inhibits the uptake of nutrients other than nitrogen and impairs the ability of trees to pull water from the soil during periods of drought.

As nitrogen deposition increases, ammonium levels in the soil increase. Excess ammonium in the soil stimulates nitrification, denitrification, and mobility of nitrates in the soil, even at a low pH. As excess nitrates are leached from the soil, they increase anion movement through the soil, releasing aluminum ($Al^{3+}$), as has been observed in the spruce stands of the Great Smoky Mountains. This aluminum finds its way to aquatic ecosystems, increasing their acidification. Increased nitrification and denitrification can also increase the emission of nitrous oxides to the atmosphere, adding to the greenhouse effect. Eventually nitrate losses approach nitrogen deposition.

# SEDIMENTARY CYCLES

Mineral elements required by living organisms, such as sulfur, phosphorus and calcium, come initially from inorganic sources. The available forms occur as salts dissolved in soil water or in streams, lakes, and seas. Like the other elements, mineral elements cycle between living organisms and their abiotic environment.

**Sedimentary cycles** vary from one element to another, but essentially each has two abiotic phases: the salt solution phase and the rock phase. Mineral elements found in rocks come directly from Earth's crust and are released slowly by weathering. Then as soluble salts they enter the water cycle. Unless they are absorbed by plants, they move through the soil to streams and lakes and eventually reach the seas, where they remain indefinitely. Other salts return to Earth's crust through sedimentation. They become incorporated into salt beds, silts, and limestone; after weathering they again enter the cycle.

Plants and many animals fulfill both their micro- and macromineral requirements directly from mineral solutions in the environment. Other animals acquire the bulk of their minerals from plants and animals they consume. After death the minerals return to the soil and water through decay.

Most elements enter the sedimentary cycle. Sulfur and phosphorus will serve as examples. The sulfur cycle has gaseous and sedimentary components. Phosphorus, on the other hand, is wholly sedimentary.

# The Sulfur Cycle

Sulfur has a long-term sedimentary phase, in which it is tied up in organic material (coal, oil, and peat) and inorganic form (pyritic rocks and sulfur deposits). It is released by the weathering of rocks, erosional runoff, decomposition of organic matter, and industrial production and is carried to terrestrial and aquatic ecosystems in a salt solution. However, the bulk of sulfur first appears in the gaseous phase as a volatile gas, hydrogen sulfide ($H_2S$), in the atmosphere. It comes from several sources: the combustion of fossil fuels, volcanic eruptions, the surface of oceans, and gases released in terrestrial and aquatic decomposition. Hydrogen sulfide quickly oxidizes into another volatile form, sulfur dioxide ($SO_2$). Atmospheric sulfur dioxide, soluble in water, is carried back to Earth in rainwater as weak sulfuric acid, $H_2SO_4$.

Whatever the source, sulfur in soluble form is taken up by plants. Starting with photosynthesis, it is incorporated through a series of metabolic processes into such sulfur-bearing amino acids as cystine. From the producers the sulfur in amino acids is transferred to consumers (Figure 12.18).

Excretions and death carry sulfur in living material back to the soil and to the bottoms of ponds, lakes, and seas, where sulfate-reducing bacteria release it as hydrogen sulfide or as a sulfate. Other microorganisms in the forest soil convert inorganic sulfate to organic forms and incorporate it into forest ecosystems (Swank et al. 1983). Colorless sulfur bacteria both reduce hydrogen sulfide to elemental sulfur and oxidize it to sulfuric acid. Photosynthetic green and purple bacteria utilize hydrogen sulfide in the presence of light as an oxygen acceptor in the reduction of carbon dioxide. Best known are the purple bacteria found in salt marshes and in mud flats of estuaries. These organisms are able to carry the oxidation of hydrogen sulfide as far as sulfate, which may be recirculated and taken up by the producers or may be used by sulfate-reducing bacteria. The green sulfur bacteria can carry the reduction of hydrogen sulfide to elemental sulfur.

Sulfur, in the presence of iron and under anaerobic conditions, will precipitate as ferrous sulfide, $FeS_2$. This compound is highly insoluble under neutral and alkaline conditions and is firmly held in mud and wet soil. Sedimentary rocks containing ferrous sulfide (called pyritic rocks) may overlie coal deposits. Exposed to the air in deep and surface mining, the ferrous sulfide oxidizes and in the presence of water produces ferrous sulfate ($FeSO_4$) and sulfuric acid:

$$2FeS_2 + 7O_2 + 2H_2O \rightarrow 2FeSO_4 + H_2SO_4$$

In other reactions ferric sulfate ($Fe_2SO_4$) and ferrous hydroxide ($FeOH_3$) are produced:

$$12FeSO_4 + 3O_2 + 6H_2O \rightarrow 4Fe_2(SO_4)_3 + 4Fe(OH_3) \downarrow$$

In this manner sulfur in pyritic rocks, suddenly exposed to weathering by human activities, discharges heavy slugs of sulfuric acid, ferric sulfate, and ferrous hydroxide into aquatic

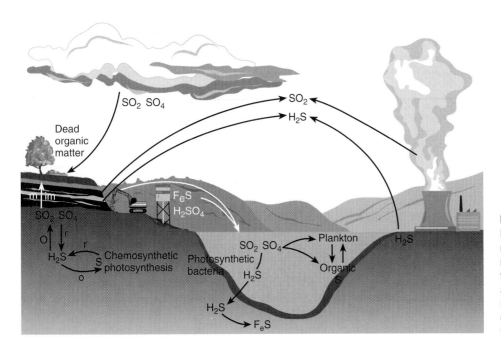

**Figure 12.18** The sulfur cycle. Note the two components, sedimentary and gaseous. Major sources from human activity are the burning of fossil fuel and acidic drainage from coal mines. Major natural sources are volcanic eruptions and decomposition of organic matter.

ecosystems. These compounds destroy aquatic life. Mining has converted hundreds of miles of streams and rivers in the eastern United States to highly acidic water (Figure 12.19).

**The Global Sulfur Cycle** The gaseous phase of the sulfur cycle permits circulation on a global scale. The atmosphere contains not only sulfur dioxide and hydrogen sulfide but sulfate particles as well. The sulfate particles become part of dry deposition; the gaseous forms combine with moisture to be recirculated by precipitation. The concentration of sulfur dioxide in rainwater falling over land, accord-

ing to estimates, is about 0.6 mg/liter and over oceans is 0.2 mg/liter, excluding sea spray.

It is almost impossible to estimate the biological turnover of sulfur dioxide, because of the complicated cycling within the biosphere. Erikisson (1963) estimates that the net annual assimilation of sulfur by marine plants is on the order of 130 million tons. Adding the anaerobic oxidation of organic matter brings the total to an estimated 200 million tons. Both industrially emitted sulfur and fertilizer sulfur are eventually carried to the sea; these two sources probably account for the 50-million-ton annual increase of sulfur in the ocean.

**Figure 12.19** Streams polluted by acid mine drainage are easily identified by the "yellow boy" or ferric sulfates. Such acidification has been the fate of most streams in coal-mining regions.

**Human Impact** Annually we pour into the atmosphere some 147 million tons of sulfur dioxide. Seventy percent of it comes from the burning of coal. Once in the atmosphere gaseous sulfur dioxide reacts with moisture to form sulfuric acid.

Sulfuric acid in the atmosphere has a number of effects. It is irritating to the respiratory tract in concentrations of a few parts per million. In a fine mist or absorbed in small particles, it can be carried deep into the lungs to attack sensitive tissue. High concentrations of sulfur dioxide (over 1000 micromilligrams/m³) have been implicated as a prime cause in many air pollution disasters characterized by higher than expected death rates and increased incidence of bronchial asthma.

Plants exposed to atmospheric sulfur are injured or killed outright. Injury to plants is caused largely by acidic aerosols during periods of foggy weather, during light rains, or during periods of high relative humidity and moderate temperatures. Pines, more susceptible than broadleaf trees, react by partial defoliation and reduced growth. Exposure of plants to sulfur dioxide with as low a concentration as 0.3 ppm for 8 hours can produce both acute and chronic injury.

## The Phosphorus Cycle

Phosphorus, unlike sulfur, is unknown in the atmosphere, and none of its known compounds have appreciable vapor pressure. Therefore the phosphorus cycle can follow the hydrological cycle only partway, from land to sea (Figure 12.20).

Under undisturbed natural conditions the source of phosphorus in the soil is the mineral apatite, a phosphate of calcium, $Ca_5(PO_4)_3(F, Cl, OH)$. Phosphorus also occurs in secondary forms as compounds of calcium and iron, and in organic combination. Under the best of conditions, the amount is small. Phosphates, such as calcium triphosphate, $Ca_3(PO_4)_2$, are freely soluble only in acid solutions and under reducing conditions. Its natural limitation in aquatic ecosystems is emphasized by the explosive growth of algae in water receiving heavy discharges of phosphorus-rich wastes (Stevenson and Stoermer 1982).

The main reservoirs of phosphorus in the biosphere are rock and natural phosphate deposits, from which the element is released by weathering, by leaching, by erosion, and by mining for agricultural use. Some of it passes through terrestrial and aquatic ecosystems as organic phosphorus by the way of plants, grazers, predators, and parasites; and it is returned to the ecosystem by excretion, death, and decay. In terrestrial ecosystems organic phosphates are reduced by bacteria to inorganic phosphates. Some are recycled to plants, some become unavailable in chemical compounds, and some are immobilized in microorganisms. Some of the phosphorus of terrestrial ecosystems escapes to lakes and seas, both as organic phosphates and as particulate organic matter.

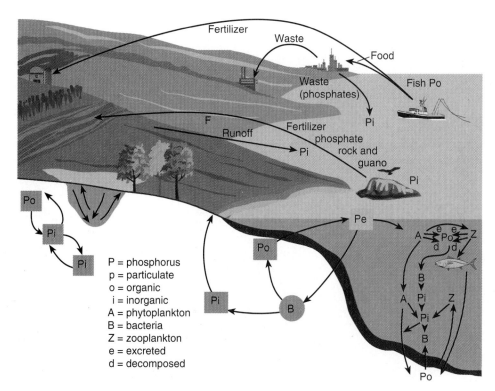

**Figure 12.20** The phosphorus cycle in terrestrial and aquatic ecosystems.

**Aquatic Ecosystems** In marine and freshwater systems the phosphorus cycle involves four fractions. (1) Particulate phosphorus is the largest reservoir. It includes both dead particulate matter and phytoplankton. (2) Inorganic phosphates, mostly soluble orthophosphate ($PO_4^{3-}$), come from various sources, both aquatic and terrestrial. Because this phosphorus is taken up rapidly by phytoplankton, it has a very short turnover time. (3) An organic phosphorus compound with low molecular weight (about 250) is excreted by organisms, especially zooplankton. (4) A soluble macromolecular colloidal phosphorus is derived from organic phosphorus. The major exchange takes place between the first two fractions, particulate phosphorous and inorganic phosphates. The organic compound fraction converts to the colloidal compound fraction. Both of these, especially, the colloidal form, release phosphate to the soluble inorganic fraction. This phosphate is rapidly recycled through the plankton (Lean 1973a, 1973b)

Phosphorus in phytoplankton may be ingested by zooplankton or detritus-feeding organisms. Zooplankton, in turn, may excrete as much phosphorus daily as is stored in its biomass (Pomeroy et al. 1963). By excreting phosphorus zooplankton is instrumental in keeping the aquatic cycle going. More than half of the phosphorus zooplankton excretes is inorganic phosphate, which is taken up by phytoplankton. In some instances 80 percent of this excreted phosphorus is sufficient to meet the needs of phytoplankton. The remainder of the phosphorus in aquatic ecosystems is in organic forms. These are utilized by bacteria, which fail to regenerate much dissolved inorganic phosphate themselves. Instead the bacteria are eaten by microbial grazers, who then excrete the phosphate they ingest (Johannes 1968).

Part of the phosphorus in aquatic ecosystems is deposited in deep and shallow sediments. Precipitated largely as calcium compounds, much of it becomes immobilized for long periods of time in bottom sediments. In marine ecosystems seasonal overturns and upwellings return some of the phosphorus to photosynthetic zones, where it is available to phytoplankton. The phytoplankton then maintains the phosphorus in the lighted zone during the period of active growth. During this time much of the phosphorus is tied up in organic matter, and only by a rapid turnover in its populations can phytoplankton meet its phosphorus requirements.

The role of organisms in the cycling of phosphorus in aquatic ecosystems is further illustrated by cycling in the tidal marshes. In the intertidal marshes of the southern United States, the marsh grass *Spartina alternifolia* withdraws phosphorus from subsurface sediments. Half of the phosphorus withdrawn is fixed in plant tissue. The other half is leached from leaves by rain and tides and carried out by tidal exchange (Reimold 1972). When the grass dies, abundant animal life in the marsh and adjacent waters and tidal creeks use the detritus as food.

**Human Impact** Human activities have altered the phosphorus cycle. To maintain phosphorus levels in the soil, agriculturists must apply phosphate fertilizer. The source of that fertilizer is phosphate rock. Because of the abundance of calcium, iron, and ammonium in the soil, most of the phosphate applied as fertilizer becomes immobilized as insoluble salts, and little escapes in runoff.

Part of the phosphorus used as fertilizer is removed in crops when harvested. Transported far from the point of fixation, this phosphorus eventually is released as waste in the processing and consumption of food. Concentration of phosphorus in wastes of food-processing plants and of feedlots adds a quantity of phosphates to natural waters. Even greater amounts come from urban areas, where phosphates are concentrated in sewage effluents.

Most of the phosphorus enrichment of aquatic ecosystems comes from sewage disposal plants. Primary sewage treatment removes only 10 percent of the total phosphorus. Secondary treatment removes only 30 percent at best. Feedlots contribute runoff, but phosphorus has such a strong affinity for the soil particle that the problem is not as severe. Sewage contributes nearly all of the phosphorus reaching rivers and lakes.

Phosphorus is more intimately involved in the overenrichment or **eutrophication** of freshwater ecosystems than nitrogen. Of the three nutrients required for aquatic plant growth, potassium is usually present in excess, nitrogen is supplemented by fixation, and phosphorus tends to be precipitated in the sediments and cannot be supplemented naturally. Therefore phosphorus is usually limiting, and in the presence of a fuller supply, algae respond with luxurious growth.

## Heavy Metals: Lead

The accumulation of toxic materials in the biosphere has brought biogeochemical cycles forcibly to the attention of the public. Heavy metals such as lead, mercury, and cadmium, have always cycled through natural ecosystems in trace amounts. Human activities, however, have markedly increased their concentration in the environment. These metals then, pass through the food chain in such a way that they become increasingly concentrated in organisms occupying the upper trophic levels.

Lead is an excellent example of a heavy metal injected into the ecosystem by human activity. Like sulfur and nitrogen, it moves great distances through the atmosphere from its point of origin. Automobiles burning leaded gasoline poured most of the lead into the air, until the use of unleaded gas was banned in the United States, Japan, Brazil, and European Common Market countries. Mining, smelting, and refining of lead, lead-consuming industries, coal combustion, burning of refuse and sewage sludge, and the burning and decay of lead-painted surfaces add additional quantities to the atmosphere.

Because lead is emitted into the air as very small particles, less than 0.5 μm in diameter, it is widely distributed to

all parts of Earth. Concentrations are highest near point sources of pollution. When automobiles used leaded gasoline, roadsides received heavier depositions than remote areas. Urban areas, close to industrial sources, may have a flux rate greater than 3000 g/ha/yr, whereas remote areas may experience a flux rate of less than 20 g/ha/yr.

Lead particles settle on the surface of soil and on vegetation. Forest canopies are particularly efficient at collecting lead from the atmosphere as dry deposition (W. Smith 1990). Lead accumulates in the canopies during the summer and is carried to the ground by rain as throughfall and stemflow and by leaf fall in autumn. Between 1975 and 1984, the Hubbard Brook Experimental Forest in New Hampshire, remote from any major sources of lead contamination, experienced an input of 190 g/ha/yr. It had an output of only 6 g/ha/yr, leaving the rest behind in the soil. Forests in the White Mountains of New Hampshire, exposed to air masses from industrial and urban areas, accumulated 200 g/ha/yr. A closed-canopy spruce forest in Germany experienced a lead flux of 756 g/ha/yr in precipitation collected beneath the canopy compared to 405/g/ha/yr in adjacent open fields.

Small amounts of lead occur naturally in some soil. In uncontaminated areas this background lead amounts to 10 to 20 μg per gram of dry soil. In the forest soil, lead becomes bound to organic matter in the litter layer and reacts with sulfate, phosphate, and carbonate anions in the soil. In such an insoluble form, lead moves slowly, if at all, into the lower horizons. Its residence time in the upper soil layer is around 5000 years.

Once in the soil and on plants, lead enters the food chain. Plant roots take up lead from the soil. Leaves pick it up from contaminated air or from particulate matter resting on the leaf. Lead is then taken up by herbivorous insects and grazing mammals, who pass it on to higher consumers. This uptake through the food chain is most pronounced along highway roadsides where residual quantities persist. Microbial systems also pick up lead and immobilize substantial quantities of it.

The long-term increase in concentrations of atmospheric lead in the industrialized areas of Earth has resulted in significant increases of lead in humans. The average body burden of lead among adults and children in the United States is 100 times greater than the natural burden, and existing rates of lead absorption are 30 times the level in preindustrial society. An intake of lead can cause mental retardation, partial paralysis, loss of hearing, and death.

# ACID DEPOSITION

Ever since the Industrial Revolution the inputs of anthropogenic sulfur dioxide and nitrogen oxides have been increasing, until such inputs equal the amount injected naturally, about 75 to 100 million tons annually. Most of the inputs are over the Northern Hemisphere. The major sources are power plants, industrial complexes, motor vehicles and internal combustion engines. The latter are the major sources of nitrogen oxides ($NO_x$).

Once in the atmosphere, sulfur dioxide and nitrogen oxides become mixed. Some of the pollutants return to Earth soon as particulate matter and airborne gases, known as **dry deposition;** a major portion is transported far away. The pathway it takes is strongly influenced by the general atmospheric circulation (National Research Council 1983). During their atmospheric transport $SO_2$ and $NO_x$ and their oxidative products become involved in complex chemical reactions involving hydrogen chloride and other compounds, oxygen, and water vapor. These reactions dilute solutions of strong acids, notably nitric and sulfuric acids (Figure 12.21). Ultimately they come to Earth in acid rain, snow, and fog, known as **wet deposition.**

The acidity of precipitation downwind from major industrial centers in eastern North America and northern and central Europe has increased at least 2 to 16 times over that of precipitation in geographical areas remote from industrial pollution (Galloway et al. 1984). The annual pH of rain and snow in those continental regions averages between 4 and 4.5 (Figure 12.22), and it may range from 2.3 to 4.6, compared to a more natural pH of 5.6 (Likens and Bormann 1974). Added to precipitation is acidic fog whose pH may range from 2.2 to 4.0 (Waldman et al. 1982).

## Aquatic Ecosystems

Acid deposition, wet or dry, can have a pronounced effect on aquatic and terrestrial ecosystems. Those that are not well buffered are highly sensitive to acid inputs. Acid precipitation has been implicated in the acidification of mountain streams and lakes in the northern United States, especially the Adirondacks, eastern Canada, and Scandinavia (National Research Council of Canada 1981, Swedish Ministry of Agriculture 1982).

Acidic inputs into aquatic ecosystems come from rainfall, snowmelt, and groundwater leaching from adjacent watersheds. Of particular importance may be snowfall that accumulates over winter and melts quickly in the spring, discharging a huge dose of acidic water into streams and lakes (Hornbeck et al. 1976). This sudden, heavy input can drop the pH levels of aquatic ecosystems quickly. More important, it releases aluminum ions, which at the levels of 0.1 to 0.3 mg/l retard growth, gonadal development, and egg production of fish and increase fish mortality. The aluminum, rather than acidity, which some species of fish can tolerate, may be the most significant pollutant (Schofield and Trojnar 1980), although recent studies suggest that acidity alone is sufficient to cause mortality in developing brook trout (Hunn et al. 1987).

Acidification of lakes reduces bacterial activity in sediments (Rao et al. 1984), inhibiting decomposition and nutrient regeneration. Reduced nutrient regeneration affects

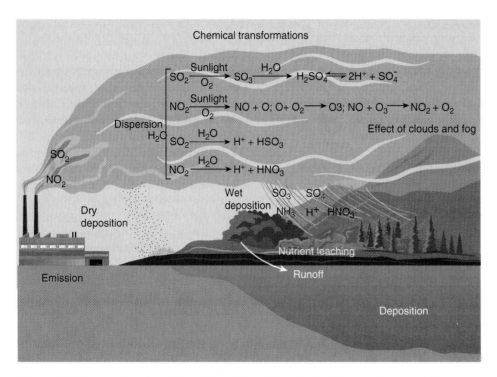

**Figure 12.21** Formation of acid depositions. Excessive sulfur and nitrogen in several forms are being poured into the atmosphere. They are converted to sulfates, sulfides, and sulfuric and nitric acids and carried to Earth.

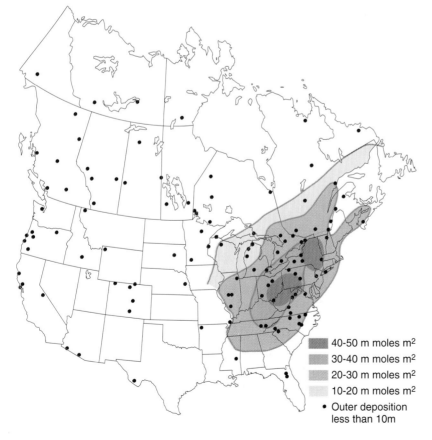

**Figure 12.22** Zones of mean annual wet deposition of hydrogen ions (m moles /m²) weighted by precipitation amount in the United States and Canada in 1980. (From Impact Assessment, work group 1, United States-Canada Memorandum of Intent on Transboundary Air Pollution, final report, January 1983.)

aquatic food webs by reducing phytoplankton and invertebrate populations upon which fish depend for food. Loss of fish reduces survival and reproductive success of such fish-eating birds as herons and loons.

Acidic runoff from snowmelt that fills small basins in the soil creates temporary breeding pools for amphibians. Even in regions not experiencing acidification of lakes and streams, it inhibits the reproduction of frogs and salamanders (Pierce 1985). Most vulnerable to acidity is the fertilization stage, because sperm apparently disintegrate at low pHs (Schlichter 1981). This effect may be a cause of the rapid decline of amphibians. Tolerance to acidity among amphibians tends to increase through the larval to adult stages.

## Terrestrial Ecosystems

The effects of acid precipitation on terrestrial ecosystems are more difficult to document or demonstrate, except in very localized vicinities of smelters or acid mine drainage. Except for the strongly podzolic ones, soils throughout much of eastern North America are sufficiently buffered to neutralize acidic precipitation. In some cases acidic precipitation can improve soil fertility at first by addition of sulfur (Krug and Frink 1983); but in time the deposition may exceed the soil's capacity to utilize it.

However, over a longer period of time acid precipitation can have adverse effects. Substantial leaching of nutrients on acidic soils can reduce nutrient availability and increase the solubility of aluminum ions, as they are replaced by hydrogen ions on the soil particles. Free aluminum affects the structure and function of fine roots, reducing their ability to take up nutrients and moisture. Further, acid rainfall can inhibit the activity of fungi and bacteria in the soil, reducing the rate of humus production, mineralization, and the fixation of nutrients. All of these interactions result in nutrient-deficient soils.

Little evidence exists to show that acid rain has any direct effect on most terrestrial plants. Acid rain, intercepted by vegetation, leaches nutrients, particularly cations like calcium, magnesium, and potassium from the leaves and needles. This leaching, a normal process in nutrient cycling, has little effect on the trees' health, provided the trees can replace the nutrients lost by uptake from the soil. However, forests at high elevations are frequently enveloped by mists and fog. Cloud droplets are more acidic and hold higher concentrations of other pollutants than rainfall. When immersed in fog, needle-leafed conifers effectively comb moisture out of the air. Their wet surfaces permit the uptake of the pollutants the moisture contains. As the water evaporates, it leaves behind high concentrations of pollutants. Some of these pollutants may be washed off during the next rain and deposited on the soil. Added to this wet deposition is a dusting of dry deposition that leaches nutrients from the leaves (Smith 1990).

Forests, especially coniferous ones, are declining in North America and Europe. Over Central Europe Norway spruce, pine, beech, and oak suffer from chlorosis and defoliation. Fourteen percent of all Swiss forests, 22 percent of Austria's forests, and 29 percent of Holland's forests show symptoms of decline. In North America similar symptoms have developed in the spruce and fir forests of the high Appalachians and in the pine forests of southern California.

Forest decline is not a new phenomenon. During the past 100 to 200 years our forests have experienced several declines with different species affected. What sets the current decline apart from all others is differences among the symptoms in the past and the similarity of symptoms among species today. Past declines could be attributed to natural stresses, such as drought and disease. What causes forest decline and dieback today is not established, but the widespread similarity of symptoms suggests a common cause, air pollution. All of the affected forests are in the path of pollutants from industrial and urban sources.

Forests close to the point of origin of pollutants experience the most direct effects of air pollution. Their decline and death can be directly attributed to air pollution. Little evidence exists that acid precipitation alone, is the cause of forest decline and death at more distant points. Acid deposition, however, in all its aspects can so weaken the trees that they succumb to other stresses such as drought and insect attack. The stressed stands of Fraser fir (*Abies fraseri*) in the Great Smoky Mountains are succumbing to the attacks of the introduced balsam woolly adelgid (*Adelges piceae*). The once deep, fragrant stands of Fraser fir, especially on the windward side and peaks of the Great Smoky Mountains, are now stands of skeleton trees (Figure 21.23).

Besides the economic and aesthetic loss of trees, air pollution and acidic rain alter succession by changing the species composition of the affected forests. Just as the chestnut blight shifted dominance in the central hardwoods forest from chestnut to oaks, so air pollution is shifting dominance from pines and other conifers to deciduous trees more tolerant of air pollution.

## CHLORINATED HYDROCARBONS

Of all human intrusions into biogeochemical cycles, none has done more to call attention to materials cycling than the widespread application of DDT (dichlorodiphenyltrichloroethane). Its use is now banned in the United States. During World War II this newly developed pesticide was used in huge quantities to control disease-carrying and crop-destroying insects, particularly mosquitoes and boll weevils. As early as 1946 Clarence Cottam of the U. S. Fish and Wildlife Service called attention to DDT's damaging effects on ecosystems and nontarget species. The impact of pesticides on ecosystems, however, remained obscure until Rachel Carson (1962) wrote *Silent Spring,* a book that ex-

**Figure 12.23** Forest decline in the Great Smokies. Air pollution makes trees vulnerable to the attacks of the balsam woolly adelgid.

posed the dangers of hydrocarbons. The detection of DDT in tissues of animals in the Antarctic (Risebrough et al. 1976), far removed from any applied source of the insecticide, emphasized the fact that DDT does indeed enter the global biogeochemical cycle and become dispersed around Earth.

In the late 1960s scientists discovered that another chlorinated hydrocarbon, PCB, was also accumulating in foodstuffs, many species of fish and birds, and in humans. PCB, a major toxic waste, is a generic name for a number of synthetic organic compounds characterized by biphenyl molecules containing chlorine atoms. They are widely used in the electronic industry and in plastics, solvents, and printing inks. PCBs are discharged into the environment through sewage outfalls and industrial discharges.

DDT, PCBs, and other chlorinated hydrocarbons have certain characteristics that enable them to enter global circulation. Because they are highly soluble in lipids or fats and not very soluble in water, they tend to accumulate in the lipids of plants and animals. They are persistent and stable and have a half-life of approximately 20 years. They have a vapor pressure high enough to ensure direct losses from plants. They can become adsorbed by particles or remain as a vapor. In either state they can be transported by atmospheric circulation and then return to land and sea with rainwater.

DDT, like most insecticides, is applied on a large scale by aerial spraying (Figure 12.24). Half or more of a toxicant applied in this manner is dispersed to the atmosphere and never reaches the ground (Tarrant 1971). On the ground or on the water's surface the pesticide is subject to further dispersion. Pesticides reaching the soil are lost through volatilization, chemical degradation, bacterial decomposition, runoff, and the harvest of crops.

In flowing water, chlorinated hydrocarbons are subject to further distribution and dilution as they move downstream. Insecticides released in oil solution penetrate to the bottom and cause mortality of fish and aquatic invertebrates (see reviews in Pimentel 1971a, Cope 1971). Trapped in the bottom rubble and mud, the insecticide may continue to circulate locally and kill for some days.

In lakes and ponds emulsifiable forms of DDT tend to disperse through the water, but not necessarily in a uniform way. DDT in oil solutions tends to float on the the surface and move about in response to the wind. PCBs, however, accumulate in bottom sediments, adsorbed in silt and fine particles, which hold them in the aquatic environment.

Eventually these chlorinated hydrocarbons reach the ocean and associate with surface slicks, where their concentration may be 10,000 times greater than in lower waters. These slicks, which attract plankton, are carried by ocean currents. In the oceans part of the residues may circulate in the mixed layer, some may be transferred below the thermocline to the abyssal waters, and more may be lost through sedimentation of organic matter. Although considerable amounts of DDT and other chlorinated hydrocarbons are transported by water, these amounts are insignificant from a viewpoint of global circulation. The major transport of chlorinated hydrocarbon residues takes place in the atmosphere.

Not only does the atmosphere receive the bulk of pesticidal sprays (well over 50 percent of that applied), but it also picks up that fraction volatilized from soils, vegetation, and water. The adsorption of DDT residues to airborne particulate matter increases the capacity of the atmosphere to hold DDT. Thus the atmosphere becomes a large circulating

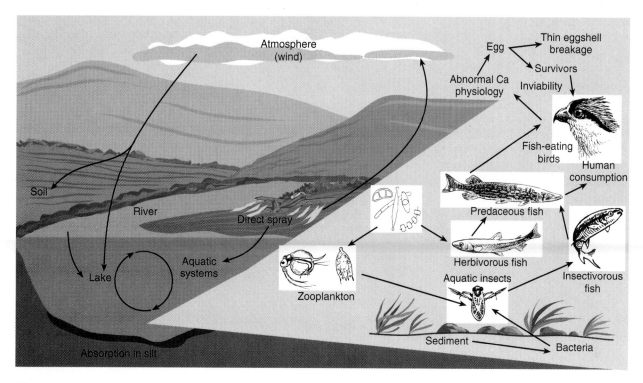

**Figure 12.24** The movement of chlorinated hydrocarbons in terrestrial and aquatic ecosystems. The initial input comes from spraying on vegetation. A large portion fails to reach the ground and is carried on water droplets and particulate matter through the atmosphere.

reservoir of DDT and other chlorinated hydrocarbons (SCEP 1970). This fact is especially important because DDT is still widely used in other parts of the world.

Although the quantity of residues of some chlorinated hydrocarbons, especially the banned DDT, may be small, the residual concentrations are still sufficient to have a deleterious effect on marine, terrestrial, and freshwater ecosystems (Beyer and Gish 1980, Matthiessen 1985). DDT, its degradation product DDE, and PCBs tend to concentrate in the fatty tissues of living organisms, where they undergo little degradation (see Menzie 1969, Bitman 1970, Peakall 1970).

In the animal body, these chlorinated hydrocarbons interfere with calcium metabolism. They block ion transport by inhibiting the enzyme ATPase, which makes energy available to the cell. The reduced transport of ionic calcium across membranes can cause death. They also inhibit the enzyme carbonic anhydrase, which in birds is essential for the deposition of calcium carbonate in the eggshell and for the maintenance of a pH gradient across the membranes of the shell gland.

The high solubility of chlorinated hydrocarbons in lipids magnifies their concentration. Most of the chlorinated hydrocarbons contained in the food is retained in the fatty tissue of the consumers. Because they break down slowly, chlorinated hydrocarbons accumulate to high and even toxic levels. They are passed on to consumers in the next trophic level, where again they are retained and accumulated. The

carnivores in the top level of the food chain receive massive amounts of chlorinated hydrocarbons.

As long as the rate of accumulation of DDT is not too great, birds tend to eliminate some of it, which reduces the body burden. Fish, however, continuously exposed to PCBs in bottom sediments, concentrate them in their tissues (Stein et al. 1987), where the chemical residue is often higher than that of DDT. In some polluted waters, such as the Hudson River in New York State and parts of the Great Lakes, fish have such high levels they are not fit for human consumption.

Since the prohibition of DDT in the United States in 1972, birds of prey are recovering from low reproductive rates induced by DDT. The recovery of the bald eagle, osprey, brown pelican, and other birds occupying high trophic levels attest to the highly detrimental effects of chlorinated hydrocarbons have had on the reproduction of birds.

Despite strict laws regulating the use of DDT, the ecological threat of pesticides has not lessened. Chlorinated hydrocarbons, including DDT, are used extensively in other parts of the world, notably Central and South America and Asia. There, the problems associated with these insecticides still persist. They affect not only the native fauna, but also migratory birds from the Northern Hemisphere that come in contact with the insecticides on their wintering grounds. Quantities of pesticides are sent back north to the United States, Canada, and Europe on fruits and vegetables grown for the winter market.

More than 500 million kilograms of a wide range of chemical pest control materials (collectively called pesticides) are used annually in the United States. Herbicides make up 60 percent of this total, insecticides 24 percent, and fungicides, 16 percent. Of these pesticides 341 million kilograms are used on agricultural crops and pastures, 55 million by government and industry, 4 million on forests, and a surprising 55 million in and around urban and suburban homes. The most concentrated use is about the home grounds, not in agricultural fields. The dosage of pesticides about homes is 14 kg/ha compared to 3 kg on agricultural crops. As much as one-third of these household pesticides is never used and is thrown into the trash and ultimately into the environment. These excess pesticides, together with losses from croplands and roadsides, expose pests to widespread selective pressures, increasing their resistance to pesticides.

# RADIONUCLIDES

Ever since the atomic bomb ushered in the atomic age, the impact of nuclear radiation on life on earth has been a major concern. Involved are high-energy, short wavelength radiations, known as **ionizing radiations.** They are so called because they are able to remove electrons from some atoms and attract them to other atoms, producing positive and negative ion pairs. Of greatest interest is ionizing electromagnetic or gamma radiation, which has a short wavelength, travels a great distance, and penetrates matter easily.

Sources of gamma radiation are atomic blasts from weapons testing, nuclear reactors, and radioactive wastes. By-products of both weapons testing and nuclear reactors such as zinc-65 ($^{65}$Zn), strontium-90 ($^{90}$Sr), cesium-137 ($^{137}$Cs), iodine-131 ($^{131}$I), and phosphorus-32 ($^{32}$P) are radioactive. When the uranium atom is split or fissioned into smaller parts, it produces, in addition to tremendous quantities of energy, a number of new elements or fission products, including strontium, cesium, barium, and iodine. Some of these fission products last only a few seconds; others can remain active for several thousand years. These radioactive elements enter the food chain and become incorporated in living organisms, in which they can cause cancer and genetic defects.

In the same atomic reaction some particles with no electrical charges, called neutrons, get in the way of high-energy particles. Nonfission products are the result. They include the radioisotopes of such biologically important elements as carbon, zinc, iron, and phosphorus, which are useful in tracer studies.

Both fission and nonfission products, together called **radionuclides,** are released to the atmosphere by nuclear testing and by wastes from nuclear reactors, hospitals, and research laboratories unless carefully handled. Later they return to earth along with rain, dust, and other material as radioactive fallout. Once the isotopes reach the earth, they enter the food chain and become concentrated in organisms in amounts that exceed by many times the quantities in the surrounding environment. In effect, local radiation fields develop in the tissues of plants and animals.

Of particular concern is the radioactive output of nuclear power plants, especially since the Three Mile Island and Chernobyl accidents. Pressurized water reactors, commonly used in nuclear power plants, release low levels of radioactivity to the air and condenser water. When water passes through the intense neutron flux of the reactor, it is contaminated by radioactivity. Trace elements in the water are activated, producing radioisotopes. Added to them are radioactive corrosive products from the surface of metal cooling tubes. Except for tritium, most of the radioisotopes are removed in a radioactive waste removal process, which creates additional problems of nuclear waste disposal. Those left have a short half-life and rapidly decay below detection levels.

## Terrestrial Ecosystems

Radionuclides disperse in terrestrial ecosystems by gaseous, particulate, and aerosol deposition and in liquid and solid wastes. Plants intercept particulate radionuclide contaminants, absorb them through the foliage, and take them up from soil and litter. From the plants radionuclides move through the ecosystem along the food chain.

Strontium-90 and cesium-137 are two of the most destructive radioactive materials released into the biogeochemical cycle (Figure 12.25). Ecologically they behave like calcium and follow it in the cycling of nutrients. Both easily enter the grazing food chain, especially in regions with high rainfall or abundant soil moisture and with low levels of calcium and other mineral nutrients in the soil. One such region, the arctic tundra, has been subject to heavy nuclear fallout from weapons testing in the past. It received another input as fallout from the Chernobyl nuclear power plant explosion in 1986.

Lichens, the dominant plants of the tundra, absorb virtually 100 percent of the radioactive particles and gases drifting onto them. From lichens the contaminants $^{90}$Sr and $^{137}$Cs travel up the food chain, from caribou and reindeer to wild carnivores and humans.

Early studies of the effects of nuclear fallout on human food chains involved Eskimos and northern Alaskan Indians (Palmer et al. 1963, W. C. Hanson 1971). Caribou are a major food source for northern Alaskan natives, who kill the animals during the northward migration in spring and stockpile the meat for late spring and early summer food. Because the caribou feed all winter on lichens, their flesh in spring during the study contained three to six times as much $^{137}$Cs as it did in the fall. In spring the Indians showed a corresponding rise in $^{137}$Cs level, often 50 percent (Figure 12.26).

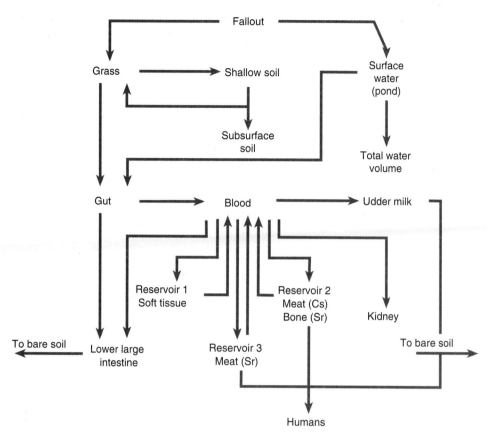

**Figure 12.25** Radionuclide cycling through the food chain. Strontium and cesium are transferred through fallout to the grazing food chain and on to humans through meat and milk from reindeer, cattle, and sheep. This cycling occurred during the early days of nuclear weapon testing and more recently following the nuclear plant explosion at Chernobyl. (Courtesy Oak Ridge National Laboratory.)

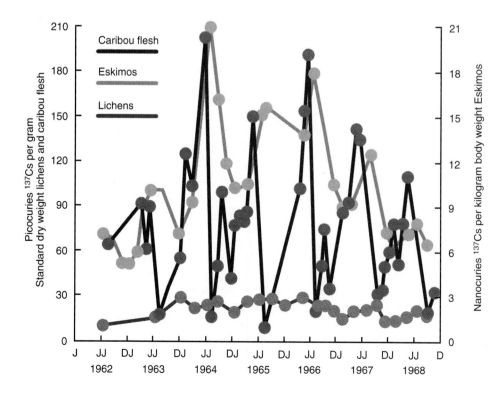

**Figure 12.26** Cesium-137 concentrations in lichens, caribou flesh, and Eskimos at Anaktuvuk Pass, Alaska, during the period 1962–1968. Note the relationship between the concentration of cesium-137 in caribou flesh and the amount in humans. As the concentration in caribou declined seasonally, so did the concentration in humans. (From W. C. Hanson 1971.)

The level decreased when the people changed to a diet of fish in the summer.

The accident at Chernobyl produced a nuclear cloud that drifted northward over the Arctic, depositing much of its contamination over arctic Eurasia (Davidson et al. 1987) and causing a repeat of the story of the 1960s. Radioactive-contaminated lichens passed strontium and cesium on to the reindeer, making both meat and milk unsuitable for human consumption, severely impacting the Laplanders' economy and their future.

In vertebrate food chains strontium and cesium usually accumulate throughout the body at higher trophic levels, but cobalt, ruthenium, iodine, and some other radionuclides do not. Some, such as iodine, concentrate in certain tissues. In arthropod food chains potassium, sodium, and phosphorus accumulate, whereas strontium and cesium do not.

## Aquatic Ecosystems

Radionuclides contaminate aquatic ecosystems largely through waste from nuclear power plants and from the nuclear processing industry. Radioactive materials that enter the water become incorporated in bottom sediments and circulate between mud and water. Some become absorbed by bottom-dwelling insects and fish downstream from the source. In fact, they may be exposed to chronic low-level radiation. Under such conditions a sort of equilibrium is established between retention in the organisms, the bottom sediments, daily input, and decay.

The concentration of radionuclides does not necessarily increase consistently through the food chain. In many situations the concentration decreases at higher trophic levels, because radionuclides tend to concentrate in the bones of fish and the shells of mollusks and do not move up into higher trophic levels.

In spite of considerable study, we still know little about the uptake, assimilation, distribution in tissues, turnover rates, and equilibrium levels of radionuclides in various ecosystems. As nuclear power plants and radioactive medical wastes increase, our knowledge of the behavior of radionuclides must become more sophisticated. Knowledge of the hazards must extend not only to humans, but also to the biota upon which we depend.

## SUMMARY

Nutrients flow from the living to the nonliving components of the ecosystem and back in a perpetual cycle. By means of these cycles plants and animals obtain nutrients necessary for their survival and growth.

There are two basic types of biogeochemical cycles: the gaseous, represented by oxygen, carbon, and nitrogen, whose major pools are in the atmosphere; and the sedimentary, represented by the sulfur and phosphorus cycles, whose major pools are in Earth's crust.

Because it is so reactive chemically, oxygen, the main life support on Earth, has a complex cycle involving oxidative reactions with organic matter and reduced mineral components of Earth's crust. An important constituent of the atmospheric reservoir of oxygen is ozone ($O_3$). The ozone layer blocks out much of the ultraviolet portion of solar radiation. Injection of chlorofluorocarbons and nitrogen and other oxides into the atmosphere is causing destruction of stratospheric ozone, with serious ecological implications.

The carbon cycle is so closely tied to energy flow that the two are inseparable. It involves the assimilation of and respiration of carbon dioxide by plants, its fixation in the form of carbohydrates in plant and animal tissue, its release through respiration, the mineralization of detritus, soil respiration, accumulation of carbon in standing crop biomass, and withdrawal into long-term reserves. The carbon dioxide cycle exhibits both annual and diurnal fluctuations. The equilibrium of carbon dioxide exchange among land, sea, and air has been disturbed by rapid injection of $CO_2$ into the atmosphere by burning fossil fuels. Over one-half of the input is removed from the atmosphere by oceans and terrestrial and marine vegetation. Increased $CO_2$ and methane ($CH_4$) concentrations in the atmosphere have the potential of raising the average temperature of Earth by several degrees, with serious ecological implications. The effects are unknown and are the topic of active research.

The nitrogen cycle is characterized by the fixation of atmospheric nitrogen by mutualistic nitrogen-fixing bacteria associated with roots of many plants, largely legumes, and cyanobacteria. Other processes are ammonification, the breakdown of amino acids by decomposer organisms to produce ammonia; nitrification, the bacterial oxidation of ammonia to nitrate and nitrates; and denitrification, the reduction of nitrates to gaseous nitrogen.

Humans pour nitrogen dioxide into the atmosphere and nitrates into aquatic ecosystems. The major sources of nitrogen oxides are automobiles and power plants. Nitrogen dioxide is reduced by ultraviolet light to nitrogen monoxide and atomic oxygen. These substances react with hydrocarbons in the atmosphere to produce a number of pollutants, including ozone and PAN. Both PAN and ozone make up photochemical smog, a pollutant harmful to plants and animals. Excessive quantities of nitrates are added to aquatic ecosystems by improper use of nitrogen fertilizer on agricultural crops, by animal wastes, and by sewage effluents. Nitrogen oxides have also been implicated in forest decline.

The sedimentary cycles involve two phases, salt solution and rock. Minerals become available through the weathering of Earth's crust, enter the water cycle as a salt solution, take diverse pathways through the ecosystem, and ultimately return to Earth's crust through sedimentation.

The phosphorus cycle is wholly sedimentary, with reserves coming largely from phosphate rock. Much of the phosphate

used as fertilizer becomes immobilized in the soil, but great quantities are lost in wastes carried by sewage effluents.

The sulfur cycle is a combination of the gaseous and sedimentary cycles, because it has reservoirs in Earth's crust and in the atmosphere. It involves a long-term sedimentary phase in which sulfur is tied up in organic and inorganic deposits, is released by weathering and decomposition, and carried to terrestrial and aquatic ecosystems in salt solution. A considerable portion of sulfur is cycled in the gaseous state, circulating on a global scale. Sulfur enters the atmosphere from volcanic eruptions, the surface of the ocean, gases released by decomposition, and the combustion of fossil fuel. Entering the gaseous cycle initially as hydrogen sulfide, sulfur quickly oxidizes to sulfur dioxide. Sulfur dioxide, soluble in water, is carried to Earth as weak sulfuric acid. Whatever the source, sulfur is taken up by plants and incorporated into sulfur-bearing amino acids, later to be released by decomposition. Injected into the atmosphere by industrial consumption of fossil fuels, sulfur dioxide has become a major pollutant, killing plants, causing respiratory afflictions in humans and animals, and contributing to wet and dry acid deposition.

Industrial use of such heavy metals as lead and mercury, always present at low levels in the biosphere, has significantly increased their occurrence. Both pose potential and actual health problems as they enter the food chain.

Acid deposition, by cloud droplets, precipitation, and dry particles is implicated in the acidification of lakes and streams in northeastern North America and Scandinavia where the soils are poorly buffered and in the decline of spruce forests in eastern North America and Europe.

Of still more consequence globally are the chlorinated hydrocarbons. Used in insect control and for industrial purposes, these hydrocarbons have contaminated global ecosystems and entered food chains. Because they become concentrated at higher trophic levels, chlorinated hydrocarbons harm predacious animals most, often interfering with their reproductive capabilities.

Radionuclides from nuclear weapons testing, from nuclear power plants, and from hospital wastes can enter and become concentrated in food chains, particularly grazing ones, and transfer to higher trophic levels.

Further intensive study of the cycling of nutrients and toxic materials is necessary to protect ecosystems and to develop safer disposal of toxic wastes.

## REVIEW QUESTIONS

1. What are the distinguishing characteristics of gaseous and sedimentary biogeochemical cycles?
2. What pools of $CO_2$ are involved in the global carbon cycle?
3. What are the possible consequences of increasing $CO_2$ and $CH_4$ inputs into the atmosphere?
4. Describe the following processes in the nitrogen cycle: fixation, ammonification, nitrification, denitrification.
5. Through what three compartments does the phosphorus cycle move in aquatic ecosystems?
6. What is the impact of chlorinated hydrocarbons on the ecosystem?
7. Why do ecologists have difficulty accepting the hypothesis that acid deposition is the direct cause of forest decline?
8. Why should we be concerned about greenhouse gases and global warming?

## CROSS-REFERENCES

Solar radiation, 36–37, 100–102; atmospheric circulation, 39–42; ocean gyres, 42; oxygen molecule, 64–65; water cycle, 67–70; nutrients, 117; micronutrients and macronutrients, 116; nutrient flows, 116–120; soil profile, 130–131; soil chemistry, 133–136; soil organic matter, 139–140; decomposition, 159–165; mineralization, 162; C:N ratios, 162, 164; seasonal overturn in lakes, 290–293; nutrient spiraling in streams, 309–310; peatlands, 317. Nitrogen cycling: grassland, 236; savanna, 245–246; desert, 249; tundra, 260–261; forest, 270–272, 278–280, 286; mutualism, 582–596.

Part *4*

# Comparative Ecosystem Ecology

# Grassland to Tundra

*Concepts*

1. Native grasslands occupy regions subject to periodic drought with rainfall between 25 and 75 cm/yr and characterized by an accumulation of soil organic matter.
2. Primary productivity of grasslands is directly related to precipitation and is heavily influenced by grazing.
3. Savannas, characterized by a continuous well-developed layer of grass and an open discontinuous layer of shrubs or trees, occur in regions with wet and dry seasons and rainfall of 25 to 200 cm/yr.
4. Shrublands, diverse in nature, range from mediteranean types in arid and semiarid regions to heathlands of cool-to-temperate climates and successional shrublands.
5. Deserts occur where rainfall ranges between 7 cm and 40 cm, potential evaporation exceeds rainfall, and productivity is low.
6. Tundras are cold-dominated ecosystems of high lattitudes and high altitudes. Low temperatures, short growing seasons, and low availability of nutrients dictate the nature of tundra life.

# GRASSLAND

When the European explorers looked out across the prairies for the first time, they witnessed a scene they had never before experienced. Nowhere in all western Europe had they seen anything similar. Lacking any other name, the explorers named these grasslands "prairie," from the French for "meadows."

This land was the North American prairie and plains, the climax grassland that occupied the midcontinent. It was one of several great grassland regions in the world, including the steppes of Russia, the pusztas of Hungary, the South African veldt, and the South American pampas (see inside cover). In fact, at one time grasslands covered about 42 percent of the land surface of the world, but today much of that area is under cultivation.

All grasslands have in common a climate characterized by high rates of evaporation, periodic severe droughts, a rolling-to-flat terrain, and animal life that is dominated by grazing and burrowing species. They occur largely where rainfall is between 25 and 75 cm/yr, too light to support a heavy forest growth and too heavy to encourage a desert. Grasslands, however, are not exclusively a climatic formation. Most of them require periodic fires for maintenance, renewal, and elimination of incoming woody growth.

## Grassland Vegetation

Grasslands are dominated by grasses, members of the Family Poaceae (Gramineae). Grasses are distinctive because their stems, called culms, produce narrow leaves that grow from their bases. This growth form allows grasses to be grazed or mowed with minimal mortality.

Grasses are either sod-formers or bunchgrasses. As the names imply, the former develop a solid mat of grass over the ground and the latter grow in bunches (Figure 13.1). The space between bunchgrasses is often occupied by other plants, usually herbs. Orchard grass (*Dactylis glomerata*), broomsedge (*Andropogon virginicus*), crested wheatgrass (*Agropyron desertorum*), and little bluestem (*Andropogon scoparius*) are typical bunchgrasses that form clumps by the erect growth of all the culms and spread at the base by shoots that develop from underground stems (tillering). Sod-forming grasses, which include such species as Kentucky bluegrass (*Poa pratensis*) and western wheatgrass (*Agropyron smithii*), spread by underground stems. Some grasses may be either sod or bunch, depending upon the local environment. Big bluestem (*Andropogon gerardii*) will develop a sod on a rich, moist soil and form bunches on a dry soil. Grasses are also categorized as cool season or warm season grasses. Cool season grasses, like bluegrass, put on growth in the spring and become senescent in summer. Warm season grasses, like switch grass (*Panicum virgatum*) put on most growth and flower in the summer.

**Figure 13.1** Growth forms and root penetration (maximum depth of about 2.5 m) of a sod grass (right) and a bunchgrass (left).

Associated with grasses are a variety of forbs, especially legumes. Cultivated haylands and pastures usually are planted to a mixture of grasses and such legumes as alfalfa (*Medicago sativa*) and red clover (*Trifolium pratense*). With them may grow unwanted plants such as mustard (*Brassica* spp.), dandelion (*Taraxacum officinale*), and daisy (*Aster* spp.). Successional grasslands often consist of native grasses and introduced species such as timothy (*Phleum pratense*) and bluegrass, and an assortment of herbaceous plants, including cinquefoil (*Potentilla* spp.), dewberry (*Rubus* spp.), and goldenrods (*Solidago* spp.). On the prairie legumes and asters are important components of climax grasslands (Weaver 1954).

## Grassland Types

Grassland is the largest of the four major vegetational formations, accounting for about 24 percent of Earth's vegetation. Much of the world's natural grasslands have been converted to cropland, supporting such cultivated grains (grasses) as corn, wheat, oats, and barley, or to cultivated grasslands. In North America grasslands dominate much of the midcontinent with a gradation from mesic grasslands beginning at the eastern forest grassland ecotone to the semiarid grasslands of the west (Figure 13.2). For convenience ecologists have grouped the North American grasslands associations into three major types: tallgrass, mixed-grass, and shortgrass prairies. Considerable diversity exists within each group, on both an east-west and north-south gradient.

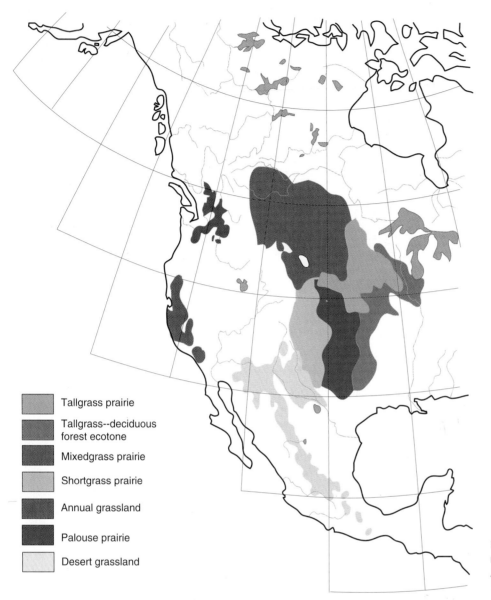

Tallgrass prairie

Tallgrass--deciduous forest ecotone

Mixedgrass prairie

Shortgrass prairie

Annual grassland

Palouse prairie

Desert grassland

**Figure 13.2** Distribution of the major grasslands of North America. (Adapted from Sims 1988.)

**Cultivated and Successional Grasslands** Grasslands in normally forested regions are either cultivated (Figure 13.3a) or successional (Figure 13.3b). In highly developed agricultural areas, such as eastern and central North America, Great Britain, and Europe, cultivated grasslands are the major representatives of their class, although a few natural types do exist. By clearing the forests, humans developed grasslands principally as a source of food for livestock and secondarily for landscaping. In some agricultural regions, notably New England and the Great Lakes states in North America, grasslands abandoned by agriculture have reverted to forest. In other regions, especially Britain, some grasslands have persisted for centuries, becoming a sort of climax community supporting its own distinctive vegetation (see Duffey 1974).

Cultivated grasslands can be classified as permanent, in grass over seven years and managed for hay, pasture, lawns, and golf courses; temporary or rotational, plowed every three to five years for crop production; and rough, marginal, unimproved, semiwild lands used principally for grazing.

Ecologically, permanent hayfields and grazing lands differ from rotational hayfields. Permanent haylands, more common in Britain than in North America, and permanent grazing lands consist of species that are adapted to periodic defoliation by cutting and grazing and produce their maximum growth in spring.

Rotational or temporary hayfields are dominated by two or three cultivated species, usually two grasses and a legume. Such hayfields support denser and heavier growth than permanent and successional grasslands. Management of such grasslands involves fertilization, mowing for hay, and at regular intervals, plowing to grow other crops, such as corn or small grains in the crop rotation. Such hayfields can provide an excellent habitat for grassland wildlife, but early

(a)

(b)

**Figure 13.3** (a) Pastureland dominated by perennial ryegrass (*Lolium perenne*) in England. (b) Successional grassland in New York state with invading herbaceous plants and scattered shrubs.

mowing destroys nesting cover at the beginning or height of the nesting season.

**Tallgrass Prairie** The tallgrass prairie extends from the forest-grassland ecotone in Wisconsin, Indiana, and Illinois where fire controlled the encroachment of oak forests into the grassland to Minnesota, eastern South Dakota, Nebraska, Kansas, and Oklahoma. Except for scattered remnants, such as the Sand Hills of Nebraska and the Osage and Flint Hills of Oklahoma and Kansas, the tallgrass prairie has been converted to cornfields and other cropland (Figure 13.4a). Big bluestem is the dominant grass of the tallgrass prairie, particularly in moist lowlands. Associates include switch grass and Indian grass (*Sorghastrum nutans*) and a diversity of forbs, especially legumes and composites (Compositae). Drier uplands in the tallgrass prairie are dominated by bunch-forming needlegrass (*Stipa* spp.), side oats grama (*Bouteloua curtipendula*), and the prairie dropseed (*Sporobolus heterolepis*). Much of this grassland is interspersed with trees and shrubs, especially along streams and lower slopes of rolling hills.

**Mixed-Grass Prairie.** West of the tallgrass prairie region extending from the southern parts of the central Canadian Provinces through the Dakotas, western Kansas, and Texas is the mixed-grass prairie (Figure 13.4b) (Sims 1988). It is an ecotone between the tallgrass prairie and the shortgrass prairie or plains. Mid-height grasses occupy lowlands and shortgrass the higher elevations. The mixed prairie embraces largely the needlegrass-grama grass associations with needlegrass-wheatgrass associations dominating lower topographic positions on medium-textured soils (Coupland 1950). Because the mixed prairie experiences great variability in precipitation, its aspect varies widely from year to year. In moist years mid-height grasses are prevalent, whereas in dry years shortgrasses and forbs are dominant. The grasses are largely cool season bunch species.

**Shortgrass Prairie or Plain** South and west of the mixed prairie and grading into the desert is the shortgrass plains or steppe, a country too dry for most midgrasses. It extends from the Nebraska panhandle, southeastern Wyoming, and eastern Colorado southward to New Mexico and Texas (Figure 13.5a). The shortgrass plain reflects a climate where the rainfall is slight and infrequent (25 to 43 cm in the west, 51 cm in the east), the humidity low, the winds high, and the evaporation rapid. Shallow-rooted, the shortgrasses use moisture in the upper soil layers, beneath which is a permanent dry zone into which the roots do not penetrate. Sod-forming blue grama (*Bouteloua gracilis*) and buffalo grass (*Buchloe dactyloides*) dominate the shortgrass plains. On wet bottomlands, switch grass, Canada wild rye (*Elymus canadensis*), and western wheatgrass (*Agropyron smithii*) replace blue grama and buffalo grass. Because of the dense sod, fewer forbs grow on the plains.

Just as the tallgrass prairie was destroyed by the plow, so has much of the shortgrass plains been ruined by overgrazing and by plowing for wheat. Because of the low available moisture, the region could not support wheat farming. Drought, lack of a tight sod cover, and winds turned much of the shortgrass plains into the the Dust Bowl of the 1930s, recovery from which has taken years.

**Desert Grassland** From southeastern Texas to southern Arizona into Mexico lies the desert grassland. It is similiar in many respects to the shortgrass plains except that triple awn grass replaces buffalo grass (Humphrey 1958). Composed largely of bunchgrasses, the desert grasslands are widely interspersed with other vegetation types, such as mesquite (*Prosopis*) woodland. The climate is hot and dry.

(a)

(b)

**Figure 13.4** (a) A remnant tallgrass prairie in Iowa. (b) The mixed-grass prairie has been called "daisy land" because of the diversity of its wildflowers.

Rain falls only during summer (July and August) and winter (December to February) in amounts that vary from 30 to 41 cm in the western parts to 51 cm in the east. Evaporation is rapid, up to 203 cm a year. Annual grasses germinate and grow only during the summer rainy season, whereas annual forbs grow mostly in the cool winter and spring months. Overgrazing by domestic livestock, fire suppression, and erosion of the thin topsoil have allowed the invasion of mesquite and larrea.

**Palouse Prairie** The intermountain region between the Rocky Mountains and the Cascade Range, extending from Southwest Canada, eastern Washington, and Oregon supports another type of grassland, the Palouse prairie. Because the region is dominated by cool-season short bunchgrasses and sagebrush (*Artemisia* spp.), it is often called shrub steppe. Originally the dominant grasses were bluebunch wheatgrass (*Agropyron spicatum*) and bluebunch fescue (*Festuca idahoensis*). Highly sensitive to grazing, these grasses were

**Figure 13.5** Shortgrass prairie in western Wyoming.

largely replaced by cheatgrass (*Bromus tectorum*), Kentucky bluegrass (*Poa pratensis*), and such forbs as sunflower (*Helianthus* spp.) and prickly lettuce (*Lactuca* spp.).Thus, like other grasslands, the Palouse prairie has been greatly altered by cultivation, grazing, and plant introductions.

**Annual Grassland** Confined largely to the Central Valley of California is annual grassland. It is associated with mediterrranean-type climate (see mediterranean-type shrublands), characterized by winter precipitation and hot, dry summers. Growth occurs during early spring and most plants are dormant in summer, turning the hills a dry tan color accented by the deep green foliage of scattered California oaks (*Quercus* spp.). The original vegetation was perennial bunchgrasses dominated by purple needlegrass (*Stipa pulchra*) and an array of other perennial and annual grasses and forbs. Unable to withstand heavy grazing and competition from exotic annual species, the original California grasslands have been replaced by vigorous annual grasses well-adapted to a mediterranean-type climate. Dominant species are wild oats (*Avena fatua*) and slender oat grass (*Avena barbata*).

**Other Grasslands** At one time the great grasslands of the Eurasian continent extended from eastern Europe to western Siberia south to Kazakhstan. The Eurasian steppes, treeless except for ribbons and patches of forest, are divided into four latitudinal belts from mesic meadow steppes in the north to semiarid grasslands in the south. The meadow steppes, extending south from the boreal forest, occupy a region of chernozem soil in which the rainfall is 50 to 60 cm. Once outstandingly beautiful in spring and summer, little remains of the meadow steppes, turned under the plow for cereal grains. Further south where rainfall is 40 to 50 cm, grasslands are dominated by bunchgrasses, mostly *Stipa*

species. In the central Asian steppes with their cold, dry spring grasses are replaced by woody and herbaceous species of *Artemisia*. About the Black Sea and in Kazakhstan, where the humidity is higher, steppe vegetation is dominated by feather grasses (*Stipa* spp.) and fescues.

In the southern hemisphere the major grasslands are in southern Africa and in South America. The Orinoco River basin in Venezuela is a vast grassy plain called the **llano.** Dominated by a variety of tropical grasses, it supports an extensive cattle industry (Figure 13.6). Further south the **pampas** extend westward in a large semicircle from Buenos Aires to cover about 15 percent of Argentina. In the eastern part of the pampas, dominated by tallgrasses, rainfall exceeds 90 cm well distributed throughout the year. South and west, where rainfall is about 45 cm, semidesert vegetation becomes prominent. South into Patagonia, where precipitation averages about 25 cm, the pampas change to open steppe grasses dominated by *Stipa* and *Festuca*. These pampas have been modified by the introduction of European forage grasses and alfalfa. The eastern tallgrass pampas have been converted to wheat and corn.

The grasslands of southern Africa (not to be confused with savannas) occupy the eastern part of a plateau 1500 to 2000 m high in the Transvaal and the Orange Free State. Most of the rainfall comes in the summer, brought in by moist air masses from the Indian Ocean. The heaviest rainfall is in the east; the lowest is in the west where grasslands grade into semiarid shrubland known as the Karoo. Once inhabited by great herds of antelopes, these grasslands are now agricultural lands.

Australia has four types of grassland. Arid bunchgrass occupies the northern part of the continent, where the rainfall averages between 20 and 50 cm, mostly in the summer. Areas with less than 20 cm rainfall support arid hummock

**Figure 13.6** (a) Humid tropical grassland in Brazil with gallery forest in the background. (b) Dry tropical grassland in Ethiopia.

(a)

(b)

grasslands. Coastal grasslands, dominated by *Sporobolus,* occupy the tropical summer rainfall region. Coastal areas where the precipitation is between 50 and 100 cm support subhumid grasslands. Most of these grasslands have been changed by fertilization, introduced grasses and legumes, and sheep grazing.

## Structure

Vegetation  Grasslands possess three strata—roots, ground layer, and herbaceous layer (Figure 13.7). The root layer is more pronounced in grasslands than in any other major ecosystem. Half or more of the plants' biomass is hidden in the soil. Most of the root biomass rather uniformly occupies the upper 16 cm of the soil profile and decreases with depth. The depth to which roots of grasses extend is considerable. Little bluestem reaches 1.3 to 1.7 m and forms a dense mat to 0.8 m (J. Weaver 1954). Roots of blue grama and buffalo grass penetrate vertically to 1 m. In addition, many grasses possess underground stems, or rhizomes, that serve both to propagate the plants and to store food. On the end of the rhizome, which has both nodes and scalelike leaves, is a terminal bud that develops into aerial stems or new rhizomes. Rhizomes of most species grow at shallow depths, not over 10 to 14 cm deep. Forbs such as goldenrods and asters possess large woody rhizomes and fibrous roots that add to the root mat in the soil. Some, such as snakeroot (*Eupatorium*), have extensive taproots 5 m long. Among

hayland plants, alfalfa possesses a taproot that grows to considerable depth.

The roots of all grassland plants are not confined to the same general area of the soil but develop in three or more zones. Shallow-rooted species seldom extend much below 60 cm. Others grow well below the shallow-rooted species, but seldom deeper than 1.5 m. Deep-rooted plants extend even further into the soil and absorb relatively little moisture from the surface soils. Thus plant roots absorb nutrients from different depths in the soil, depending upon moisture (J. Weaver 1954, Wieland and Bazzaz 1975).

The ground layer is dominated by mosses and plants with a rosette-type growth form, such as dandelions. It is characterized by low light intensity during the growing season, temperatures cooler than ambient on the ground, and reduced wind flow. Even though the grass tops may move like waves of water, air close to the ground is calm. Conditions on grazed lands are different. Because the grass cover is closely cropped, the ground layer is subject to higher temperatures and greater wind velocity.

The seasonally variable herbaceous layer consists of three or more strata, more or less variable in height, according to grassland type. Ground layer plants, such as wild strawberry, dandelion, cinquefoil, and violets grow upward to make the first stratum. As the growing season progresses, these plants become hidden beneath the middle and upper layer. The middle layer consists of shorter grasses and such herbs as wild mustard, coneflower (*Ratibida columnaris*),

**Figure 13.7** (a) Profile of a grassland showing physical stratification during the summer, energy flow, structure, and stratification of the physical environment. (b) Leaf area indexes, as indicated by bar graphs, at different levels in a grassland for both green and dead plant structures.

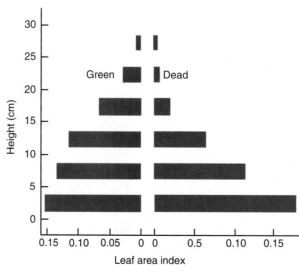

(a)                                                                 (b)

and asters. The upper layer consists of the flowering heads of tall grasses and forbs, such as goldenrods, most conspicuous in late summer.

Mulch Grasslands accumulate a layer of mulch on the ground. The oldest layer consists of decayed and fragmented remains of fresh mulch. Leafy fresh mulch consists of dead, often standing, residual herbage (thatch to lawn owners). Three or four years must pass before natural grassland herbage will decompose completely. The amount of accumulated mulch often is enormous. On a tallgrass prairie it may be two to three times the amount of annual production (Knapp and Seastedt 1986).

Grazing reduces mulch, as do fire and mowing (see Table 13.2). Light grazing tends to increase the weight of humic mulch at the expense of fresh (Dix 1960); moderate grazing results in increased compaction of mulch by grazing animals, which favors increase in microbial activity and a subsequent reduction in both fresh and humic mulch. Burning reduces both fresh and humic mulch, but the mulch structure returns two or three years after a fire on lightly grazed and ungrazed lands (Tester and Marshall 1961, Hadley and Kieckhefer 1963). Mowing greatly reduces fresh mulch and in a matter of time humic mulch as well. An unmowed prairie accumulated 10 metric tn of humic/ha; a similar prairie, mowed, had less than 4 metric tn (Dhysterhuis and Schmutz 1947).

An accumulation of mulch has both positive and negative effects. It conserves soil moisture by reducing evaporation, but it also retards infiltration of precipitation. It alters the microclimate and physiology of emerging shoots, so $CO_2$ uptake is reduced. By insulating the soil surface from solar radiation, mulch reduces root productivity and inhibits the activity of soil microbes and invertebrates. It provides nesting cover for grassland birds and offers cover for small mammals.

Animal Life All types of grasslands support similar forms of animal life. Much of it exists within the strata of vegetation: roots, ground layer, and herb cover. Invertebrates, particularly insects, occupy all strata at some time during the year. In winter insect life is confined largely to the soil, litter, and grass crowns as pupae and eggs. In spring, soil occupants are mostly earthworms and ants. Ground and litter layers harbor scavenger carabid beetles and predaceous spiders. Because this layer provides only limited supports for webs, there are more hunters than web builders among spiders.

Life in the herbaceous layer varies as the strata become more pronounced from summer to fall. Here invertebrate life is most diverse and abundant. Homoptera, Coleoptera, Orthoptera, Diptera, Hymenoptera, and Hemiptera are all represented. Insect life reaches two highs during the year, a major peak in summer and a less well-defined one in fall.

Mammals are the most conspicuous vertebrates of the grasslands, and the majority of them are herbivores. A large and rich ungulate fauna evolved on the grasslands. The bison (*Bison bison*) and pronghorn antelope (*Anti-* *locarpa americana*) of North America were surpassed only by the richer and more diverse ungulate fauna of the East African plains. In Australia, marsupial kangaroos (Figure 13.8) are the grazing equivalent of the placental ungulates. South America lacks an ungulate fauna. Their closest equivalents are the large grazing rodents that live in herds, the capybara (*Hydrochaeris hydrochaeris*). Today herds of cattle and flocks of sheep have replaced most of the native grazing herbivores.

Grassland animals share some outstanding traits. Hopping or leaping is a common method of locomotion. Strong hind legs enable a variety of animals, such as grasshoppers, jumping mice, jackrabbits, and gazelles, to rise above the grass where visibility is unimpeded. Speed, too, is well developed. Some of the world's fastest mammals, such as antelopes and the cheetah, live in grasslands. Many of the rodents are adapted to digging and burrowing. Because of dense grass and lack of trees for song perches, some grassland birds have conspicuous flight songs that advertise territory and attract mates.

Animal life in cultivated and successional grasslands depends upon human management for the maintenance of habitat. Mowing for hay, a major management tool, destroys habitat at a critical time of year. Nests of rabbits, mice, and birds are exposed at the height of the nesting season. Losses from both mechanical injury and predation are often heavy, although most species remain to complete or reattempt nesting activity. Pasturelands more often than not are so badly overgrazed that they support little vertebrate life.

Successional grasslands, usually lacking a heavy cover of grass, do not usually support as wide a variety of life as

**Figure 13.8** The eastern gray kangaroo (*Macropus giganteus*) is one of the dominant grazing marsupials of Australian grasslands.

hayfields. Such grasslands are inhabited by grasshopper sparrows (*Ammodramus savannarum*), vesper sparrows (*Pooecetes gramineus*), and meadow mice (*Microtus* spp.); but deep-grass species such as meadowlarks (*Sturnella* spp.) and bobolinks (*Dolichonyx oryzivorus*) are few, if not absent.

## Function

Grassland types, species composition, productivity, and other attributes are directly related to rainfall and temperature. Primary production is lowest where precipitation is lowest and temperatures are high. It is highest where mean annual precipitation is greater than 80 cm and mean annual temperature is above 15° C (see data in Coupland 1979; Breymeyer and Van Dyne 1980). Production, however, is most directly related to precipitation (Figure 13.9 and Table 13.1). The greater the precipitation, the higher is production. This relationship reflects in part the reduction of water stress and the enhancement of nutrient uptake and in part differences in water use efficiency.

Grasses of arid regions use water inefficiently, although most of the water is evaporated from the soil. Adaptations among desert grasses are not for the efficient use of water but for survival and reproduction under adverse conditions. In semiarid country where water is limiting, plants have evolved adaptations to make the maximum use of water. In humid, tallgrass country the vegetative canopy is dense and intercepts most of the solar radiation, so adaptations of plants are for efficient capture of light at the expense of efficient use of water.

Production is mirrored in biomass accumulation (Table 13.1). Aboveground green biomass among the array of grasslands studied in the International Biological Program (IBP) ranges from 50 g/m² in arid grasslands to 827 g/m² in subhumid grasslands. Mean underground biomass ranges from 45 to 4707 g/m². In most tropical grasslands, that biomass is usually less than 1000 g/m² (median 200 g/m²). Except for a short seasonal period of maximum aboveground biomass, belowground biomass is two to three times that aboveground.

The relationship of belowground to aboveground biomass is best expressed in terms of root-to-shoot ratios. Among grasslands the ratios range from 0.2 to 10.3. The ratio is smallest in tropical grasslands, a mean of 0.8. The ratio for most temperate grasslands is between 4 and 5; for the tall-grass and mixed grass prairie the ratio may be 6 or 7.

Primary productivity ranges from 82 g/m²/yr in semiarid regions to 3396 g/m²/yr in subhumid tropical grasslands. In natural and seminatural temperate grasslands, production ranges from 98 to 2430 g/m²/yr. Much of the net production of grassland does not appear as aboveground biomass. In ungrazed grasslands 75 to 85 percent of the photosynthate is translocated to the roots for storage below ground. In grazed grasslands reduced allocation of carbon below ground and increased allocation to aboveground production results in decreased root biomass.

Grasslands have the highest index of productivity per unit of standing crop (20 to 55 percent) among all terrestrial ecosystems. The tundra during its short growing season has an index of 10 to 20 percent. The index for the boreal forest is only 2 to 5 percent and for the temperate deciduous forest, 8 to 10 percent. Of course, a great deal of energy in woody ecosystems goes into the maintenance of high standing crops, which grasslands do not support. Efficiency of energy capture by grasslands is 0.75 percent for net primary production. Light to moderate grazing increases that efficiency to 0.87 percent.

Grasslands evolved under grazing pressures of ungulates since the Cenozoic. Grasses have adapted to grazing by compensatory growth for tissue damaged or removed by grazing (Sims and Singh 1971, McNaughton 1979, Risser 1985; for extensive discussions see Coupland 1979, Breymeyer and Van Dyne 1980). Such compensation involves increased photosynthetic rates in residual tissue; reallocation of nutrients and photosynthates from one part of the plant to another, especially from roots to stems; removal of photosynthetically inefficient mature tissue to make room for young tissue and subsequent increase in light intensities to stimulate new growth; reduction of leaf senescence, pro-

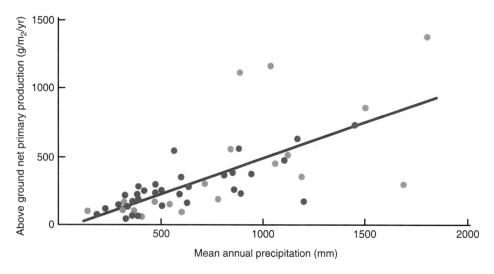

**Figure 13.9** Relationship between aboveground primary production and mean annual precipitation for 52 grassland sites around the world. North American grasslands are indicated by dark green dots. (From Lauenroth 1979:10.)

**Table 13.1** Mean Standing Crop of Selected Grasslands, Growing Season (g/m², oven dry)

| Standing Crop | Desert Ungrazed | Desert Grazed | Shortgrass Plains Ungrazed | Shortgrass Plains Grazed | Mixed Prairie Ungrazed | Mixed Prairie Grazed | True Prairie Ungrazed | True Prairie Grazed |
|---|---|---|---|---|---|---|---|---|
| Above ground | 81.2 | 49.5 | 69.0 | 42.9 | 154.2 | 94.6 | 256.1 | 220.8 |
| Cool season grass | 0 | 0 | 6.6 | 2.6 | 101.6 | 9.5 | 4.6 | 24.2 |
| Warm season grass | 51.3 | 9.4 | 42.6 | 27.1 | 45.8 | 82.1 | 227.1 | 152.6 |
| Old dead | 0.9 | 1.1 | 25.9 | 18.3 | 80.2 | 58.5 | 204.0 | 44.0 |
| Mulch | 68.7 | 52.3 | 97.5 | 89.3 | 448.3 | 239.2 | 111.2 | 206.5 |
| Below ground | 197.0 | 185.0 | 1600.0 | 1800.0 | 1213.0 | 2086.0 | 893.0 | 781.0 |
| Rainfall (cm) | 23 | 30 | 38 | 94 | | | | |

*Source:* Adapted from Lewis 1971.

longing the active photosynthetic period of remaining tissues; hormonal control of meristems, stimulating tillering and leaf growth; conservation of soil moisture by reducing transpiration surface; and recycling of nutrients through dung and urine (McNaughton 1979).

Numerous experimental studies of herbivore-grass interactions by plant ecologists and range managers have demonstrated that moderate grazing stimulates primary production of grasslands (Dyer et al. 1993). McNaughton (1985) points out that moderate grazing on the Serengeti grasslands stimulated production up to twice the levels in ungrazed control plots. IBP grassland studies examined grazed and ungrazed systems from a functional point of view. In general, total net primary production was greater on grazed than on ungrazed sites. However, grazed grasslands channeled more carbon belowground than did ungrazed grasslands (Table 13.2). Grazed grasslands allocated about 63 percent of total net primary production to belowground net production and about 37 percent to aboveground net production. Ungrazed systems sent about 52 percent of total net primary production belowground and about 48 percent above. In contrast, ungrazed treatments had an aboveground production efficiency of 0.26 percent and the grazed 0.33 percent (Sims and Singh 1971).

Another response of grasslands to grazing is a change in species composition (Sims 1988). Some grasses and forbs are sensitive to intensive grazing pressure and tend to disappear while other species increase (Table 13.3). On desert grasslands of North America black grama (*Bouteloua eriopoda*) is replaced by desert shrubs; on shortgrass plains, which are the most stable under grazing pressure, blue grama and prickly pear increase; on mixed-grass prairies midgrasses decrease and shortgrasses and sedges increase. On tallgrass sites tallgrasses disappear and little bluestem and tall dropseed increase; if grazing pressure is heavy, the site may be invaded by the weedy Japanese brome or chess (*Bromus japonicus*) (Lewis 1971).

Although large herbivores are the most conspicuous consumers, invertebrates are the most important grazers. The aboveground biomass of invertebrates—including plant consumers, saprovores, and predators—ranges from 1 to 50 g/m², while grazing mammals amount to about 2 to 5 g/m². Belowground invertebrates exceed 135 g/m². The major aboveground consumers are grasshoppers, and the major belowground consumers are nematodes. Nematodes account for 90, 95, and 93 percent of all belowground herbivory, carnivory, and saprophagous activity, respectively. They ac-

**Table 13.2** Net Primary Production and Efficiency of Energy Capture by Ungrazed and Grazed Grasslands Within the Growing Season

| Type | Above Ground Net Production (kcal/m²) | Above Ground Efficiency (%) | Below Ground Net Production (kcal/m²) | Below Ground Efficiency (%) | Total Net Production (kcal/m²) | Total Efficiency (%) |
|---|---|---|---|---|---|---|
| Grazed | | | | | | |
| Desert | 456 | .06 | 625 | .09 | 1081 | .13 |
| Shortgrass plains | 476 | .10 | 3285 | .70 | 3761 | .80 |
| Mixed prairie | 444 | .10 | 1810 | .41 | 2254 | .51 |
| True prairie | 2048 | .42 | 1701 | .35 | 3749 | .77 |
| Ungrazed | | | | | | |
| Desert | 688 | .09 | 489 | .07 | 1177 | .16 |
| Shortgrass plain | 568 | .12 | 2153 | .45 | 2721 | .57 |
| Mixed prairie | 788 | .18 | 1264 | .29 | 2052 | .47 |
| True prairie | 1348 | .27 | 872 | .17 | 2220 | .44 |

*Source:* Adapted from Sims and Singh 1971.

**Table 13.3** Response of Grassland Species to Grazing Disturbance, Grassland Biome, 1970

| Grassland | Species | % Community Standing Crop | |
| --- | --- | --- | --- |
| | | Ungrazed | Grazed |
| Desert | Black grama (*Bouteloua eriopoda*) | 53 | 8 |
| | Broom snakeweed (*Xanthocephalum sarothrae*) | 15 | 26 |
| | Russian thistle (*Salsola kali*) | 3 | 25 |
| Shortgrass plains | Blue grama (*Bouteloua gracilis*) | 63 | 72 |
| | Prickly pear (*Opuntia polyacantha*) | 5 | 12 |
| Mixed prairie | Wheatgrass (*Agropyron smithii*) | 43 | 9 |
| | Blue grama (*Bouteloua gracilis*) | 9 | 37 |
| | Buffalo grass (*Buchloe dactyloides*) | 28 | 44 |
| True prairie | Little bluestem (*Andropogon scoparius*) | 69 | 23 |
| | Indian grass (*Sorghastrum nutans*) | 7 | t |
| | Japanese chess (*Bromus japonicus*) | 0 | 22 |
| | Tall dropseed (*Sporobolus asper*) | 2 | 37 |

*Source:* Lewis 1971.

count for 46 to 67 percent of root and crown consumption, 23 to 85 percent of fungal consumption, and 43 to 88 percent of belowground predation. Aboveground herbivores consume 2 to 7 percent of primary production, while belowground herbivores consume 13 to 46 percent.

Energy flow through small mammal populations in grasslands may be high relative to their biomass but low relative to the standing crop biomass (Risser et al. 1981). Small mammals consume only about 4 percent of the herbage available. Energy flow through breeding bird populations may be even smaller. Birds probably exert little influence on energy flow and storage of energy and nutrients in grassland ecosystems (Wiens 1973).

Not only is a large proportion of primary production consumed below ground, but a greater proportion is used at each trophic level (Figure 13.10). Invertebrates convert about 9 to 25 percent of ingested energy to animal tissue, whereas vertebrate grazers, such as sheep and cattle, convert 3 to 15 percent to animal tissue. However, most of the primary production, aboveground especially, goes to the decomposers, dominated by fungi, whose biomass is two to seven times that of bacteria. Overall, the decomposer biomass exceeds that of invertebrates.

Green plant consumers are important in the cycling of nutrients in grassland ecosystems. Invertebrate consumers are highly inefficient in assimilating ingested material. Much of their intake is deposited as feces or frass. Because the nutrients in the frass are in a highly soluble form, they are returned rapidly to the system. Large grazing herbivores return a portion of their intake as dung, which becomes a major pathway of nutrient cycling (Figure 13.11). They harvest nutrients over a large area and concentrate them into a small area (McNaughton 1985). A well-developed coprophagous fauna speeds the decay of manure and accelerates the activity of bacteria in feces.

**Figure 13.10** Grassland production (in boxes) and consumption (between boxes), representing means and standard errors for desert, tallgrass, mixed, and shortgrass sites in North American grasslands. Production on one trophic level is represented as a proportion of production at the next trophic level and efficiency as the ratio of production to consumption. (From French et al. 1979:185.)

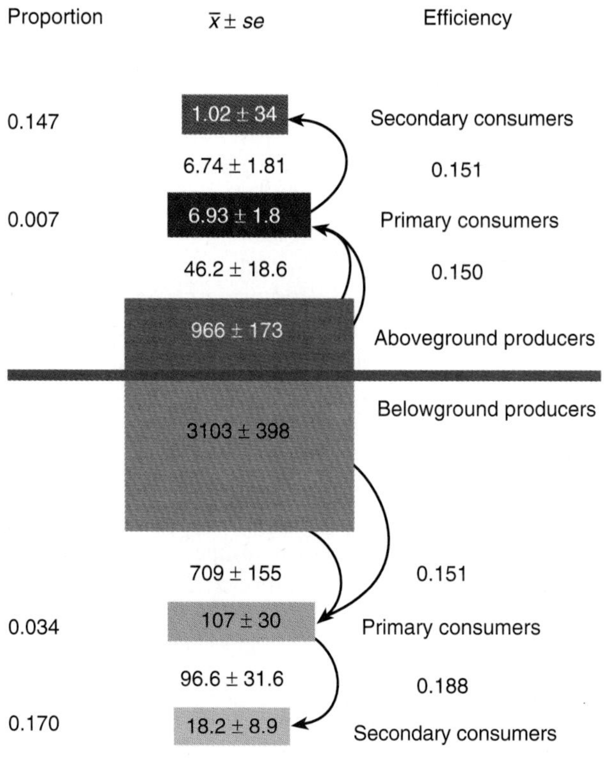

| Proportion | $\bar{x} \pm se$ | Efficiency |
| --- | --- | --- |
| 0.147 | 1.02 ± 34 | Secondary consumers |
| | 6.74 ± 1.81 | 0.151 |
| 0.007 | 6.93 ± 1.8 | Primary consumers |
| | 46.2 ± 18.6 | 0.150 |
| | 966 ± 173 | Aboveground producers |
| | | Belowground producers |
| | 3103 ± 398 | |
| | 709 ± 155 | 0.151 |
| 0.034 | 107 ± 30 | Primary consumers |
| | 96.6 ± 31.6 | 0.188 |
| 0.170 | 18.2 ± 8.9 | Secondary consumers |

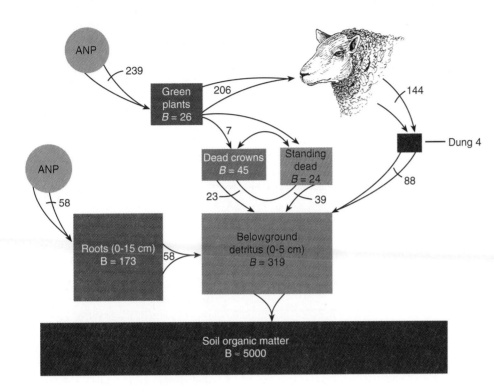

**Figure 13.11** Carbon (energy) flow in a grazed pasture, indicating the role of grazing herbivores in the functioning of grassland ecosystems. Boxes represent the retention of carbon (g/m²). Arrows represent the flow of carbon (g/m²/yr). ANP = annual net productivity; B = biomass. Note the high proportion of energy that moves through the grazing herbivore. (From Breymeyer 1980:803.)

Nitrogen provides an example of nutrient cycling in grasslands (Figure 13.12). About 90 percent of nitrogen in grasslands is tied up in soil organic matter, 2 percent in litter, 5 percent in live and dead plant cover, 1 percent in dead shoots, and about 0.8 percent in soil microflora, and a very small amount in aboveground and belowground invertebrates. When nitrogen is limiting, most of it is shunted to green herbage, where it remains during the growing season. Some of this is consumed by herbivores. Part is translocated below ground to roots. Much of this nitrogen is moved above ground to new growth the following season. Some is retained in standing dead leaves and returned to the litter and soil surface, where it is acted upon by fungi and bacteria. Part becomes immobilized in microbial biomass but is quickly mineralized upon the death of microbes and reenters the nitrogen cycle. Still another fraction is tied up in humus and resists decomposition. Turnover, however, is fairly rapid, and most of the nitrogen that enters a green plant one growing season will reenter another the following year.

Central to nutrient cycling in grasslands is the accumulation of mulch or detritus. A large standing crop of mulch can have a detrimental effect on nitrogen cycling, particularly in tallgrass ecosystems (Knapp and Seastedt 1986). Mulch intercepts rainfall from which microbes can assimilate inorganic nitrogen directly before it reaches plant roots. The mulch inhibits nitrogen fixation by blue green and free-living nitrogen-fixing microbes. Periodic grassland fires clear away the mulch layer and release other nutrients in detritus to the soil, but nitrogen, equal to about two years of nitrogen inputs to the system through rainfall, is lost to the atmosphere. Periodic fires, however, stimulate the growth of nitro-

gen-fixing leguminous forbs and improve conditions for earthworms.

Because of their great economic importance as grazing land, grasslands have been the subject of intensive research. This research investigates grassland ecosystem functioning, the evolutionary adaptations of grasses to grazing, and the application of ecological principles to range and grassland management. In spite of years of study, our knowledge of grassland ecology is still limited. We know little about the relationship between aboveground and belowground consumers and grassland productivity. We lack good techniques to measure belowground production. Still debated is the relationship of overgrazing to grassland productivty. There are other questions. What is the effect of grazing, especially in semiarid and arid grasslands, on fossorial animals and on key species of plants? Does their loss impair ecosystem functioning? What is the relationship between periodic droughts and overgrazing in converting arid grasslands to desert? These and other questions await answers in grasslands.

## TROPICAL SAVANNAS

The one ecosystem that defies general description is the tropical **savanna.** The problem is an old one, involving even its name (Hills 1965, Sarmiento and Monasterrio 1975, Bourliere and Hadley 1983). The word in its several origins, largely Spanish, referred to grasslands or plains. Over time the word was applied to an array of vegetation types representing a continuum of increasing cover of woody vegeta-

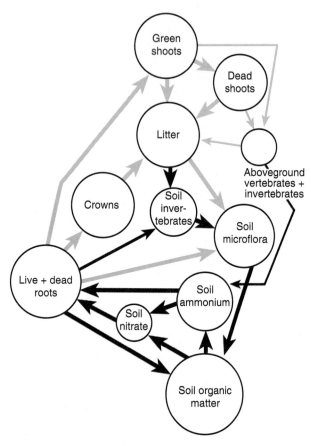

**Figure 13.12** The nitrogen cycle in a shortgrass prairie. Circle size is proportional to the logarithm of the average standing crop in each compartment in grams of nitrogen per square meter. The width of the arrows is proportional to the average annual flows between compartments. The black arrows are rough estimates; we lack knowlege of belowground process. (Adapted from *Frontiers,* Academy of Natural Sciences of Philadelphia, 1978.)

faces of little relief, often old alluvial plains. Because the soils are low in nutrients and moisture, they cannot support a forest. Precipitation exhibits extreme seasonal fluctuations; in South American savannas the soil water regime may fluctuate from excessively wet to extremely dry, often below the permanent wilting point. Savannas are subject to recurrent fires, and the dominant vegetation is fire-adapted. Grass cover with or without woody vegetation is always present. When present, the woody component is short-lived, with individuals seldom surviving for more than several decades (except for the African baobab trees). Detrital-processing termites are a conspicuous component of savanna animal life.

## Structure

The major and most essential stratum of the savanna ecosystem is grass, mostly bunch, with no vertical structure. The woody component adds one or two more vertical layers, ranging from about 50 to 80 cm when small woody shrubs are present to about 8 m in the tree savannas. Highly developed root systems make up the larger part of the living herbaceous biomass. The root system is concentrated in the upper 10 cm but extends down to about 30 cm. Savanna trees have extensive horizontal roots that go below the layer of grass roots. They have a high root-to-shoot ratio (R/S) in sharp contrast to typical forest trees, which have a low R/S ratio. Competition may exist between grass and woody vegetation for soil moisture, but more intense competition takes place between trees, accounting for the spacing patterns of woody vegetation (Gutierrez and Fuentes 1979, T. M. Smith and Walker 1983, T. M. Smith and Grant 1986).

In contrast to the poorly developed vertical structure is a well-developed, although often unapparent, horizontal structure. The bunch grasses form an array of clumps set in a matrix of open ground, creating patches of low vegetation with frequent changes in microclimatic conditions. The addition of woody growth, widely spaced shrubs and trees, lends horizontal structure to the soil. Trees add some organic matter and nutrients to the soil beneath them, reduce evapotranspiration, resulting in increased herbaceous and woody shrub growth, and provide patches of shade. On the African savanna large grazing herbivores rest in the shade during the heat of the day and concentrate nutrients from dung and urine beneath the trees (Figure 13.15.)

Savannas support or are capable of supporting a large and often varied assemblage of herbivores, both grazing and browsing, particularly in Africa. This element of large grazing herbivores is missing from the South American savannas, where it is replaced in part by a number of deer, tapirs, and capybara, the largest living rodent (Ojasti 1983). In spite of their visual dominance, large ungulates consume only about 10 percent of primary production. Dominant herbivores are the invertebrates, including acrid mites, acridid grasshoppers, seed-eating ants, and detrital-feeding dung beetles and termites (Lamotte 1975, Gillon 1983).

tion, from open grassland to widely spaced shrubs or trees to closed woodland with a grass understory (Figure 13.13).

Savannas cover much of central and southern Africa, western India, northern Australia, large areas of northern and east-central South America, and some of Malaysia. Some savannas are natural. Others are seminatural or anthropogenic, brought about and still maintained by centuries of human interference.

Moisture appears to be the major determinant of savannas (Figure 13.14), a function of both rainfall (amount and distribution) and soil texture (Sarmiento and Monasterrio 1975; Tinley 1982). Fire, soil nutrients, and herbivores are important modifiers of the basic vegetation that the water regime promotes (Huntley and Walker 1982).

## Characteristics

Savannas, in spite of their vegetational differences, exhibit some common characteristics. Savannas occur on land sur-

**Figure 13.13** Savanna ecosystems in southern Africa: (a) grass savanna; (b) shrub savanna; (c) tree savanna. (d) A tree savanna in Venezuela.

## Function

Primary production in savannas is not well documented. Because of the wide diversity of savanna types, it is difficult to make any strong generalizations. Probably a wide range of production exists between grass savanna on one end of the gradient and tree savanna and woodlands on the other (Lamotte 1975, Rutherford 1978, Lamotte and Bourliere 1983). Primary production is initiated at the beginning of the wet season when moisture releases nutrients from materials accumulated in the dry season and stimulates nutrient translocation from the roots to new growth. A quick flush of growth into grass and woody plants follows. Large root systems efficiently transfer nutrients from the soil to above-ground biomass.

The nutrient pool, especially nitrogen, is low. Its circulation in a tree savanna (Bernhard-Reversat 1982, Sarmiento 1984) provides some insight into the function of savanna ecosystems. Nitrogen content and organic matter of the soil are low. Both are concentrated largely in the upper 10 cm of soil, with the highest concentration beneath the scattered trees. The first rains of the wet season trigger mineralization. Its rate, 5 to 8 percent of the total N per year, is fairly high. It is the function of the number of days the soil was wet for a given temperature and N content. Bernard-Reversat (1981) found that the nitrogen flux between soil and vegetation was higher under the trees than in the open, because of reduced evaporation, (which kept the soil moist), more organic matter accumulation, and more herbaceous growth. Savanna trees, especially the African acacias, exhibited tight internal

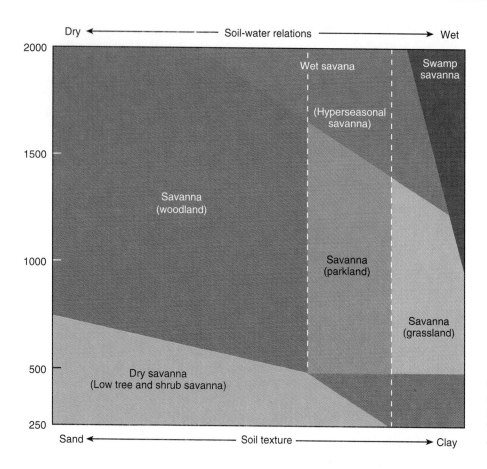

Dry ← ———— Soil-water relations ————→ Wet

**Figure 13.14** Classification of savannas of the world, based on annual rainfall and soil moisture as influenced by soil texture. (After R. W. Johnson and Tothill 1985.)

**Figure 13.15** An impala seeks the afternoon shade provided by a clump of acacia trees in a South African savanna.

cycling. Nitrogen concentration in the leaves decreased as the dry season approached, with maximum withdrawal before leaf fall. The trees transferred some of the nitrogen into new woody growth, but much of it went to the root reserve, where N accumulated for the flush of new season growth.

A similar tight circulation exists in neotropical savannas (Sarmiento 1984). An important portion of nitrogen, 66 percent of it, is translocated to the perennial roots, while another portion remains in the dry standing biomass (Figure 13.16). Most of the nitrogen in the dry aboveground biomass is lost to the atmosphere by volatilization if fire sweeps the savanna; otherwise a fraction will be transferred to the soil through leaching effects of rainwater.

## SHRUBLANDS

Covering large portions of the arid and semiarid world is climax shrubby vegetation (see inside cover). In addition climax shrubland exists in parts of temperate regions, because historical disturbances of landscapes have seriously affected their potential to support forest vegetation (Eyre 1963). Among such shrub-dominated plagioclimaxes—brought about by ecosystems and maintained by human disturbances—are the moors of Scotland and the macchia of South America. Outside these regions, shrublands represent

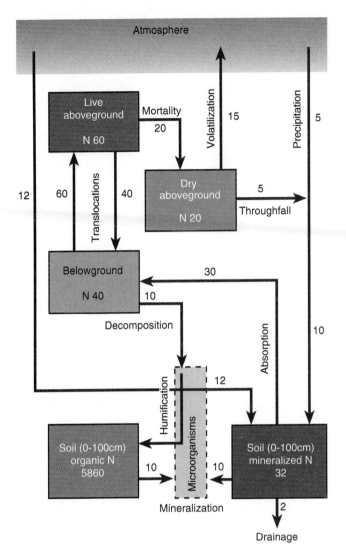

**Figure 13.16** The annual nitrogen cycle in the seasonal savanna at Barinas, Venezuela. The aboveground biomass at maximum density accumulates 60 kg/ha/N. As the aboveground growth dries, about 66 percent of the nitrogen translocates to the roots; the remainder remains in dry standing crop. Of this N an estimated 25 percent is incorporated into the mineral nitrogen pool of the soil by throughfall. Some of the nitrogen in the dry aboveground material is lost through volatilization when burned. In the roots 25 percent of the biomass is recycled through decomposition. In this situation about 10 kg/ha/yr passes through soil microorganisms to the organic nitrogen pool. The value of nitrogen accumulated in the pools are in tm/ha, and the values of the flows are in tm/ha/yr. (After Sarmiento 1984:191.)

a stage in the successional sequence leading back to forest. There they exist as second-class citizens of the plant world (McGinnes 1972), given little attention by botanists, who tend to emphasize dominant plants. Too little work has been done on successional shrub communities.

## Characteristics of Shrubs

Shrubs are difficult to characterize. They have, as McGinnes (1972) points out, a "problem in establishing their identity."

They constitute neither a taxonomic nor an evolutionary category (Stebbins 1972). One definition says that a **shrub** is a plant with woody persistent stems, no central trunk, and a height of 4.5 to 6 m. Size does not set shrubs apart, though, because under severe environmental conditions many trees will not exceed that size. Some trees, particularly coppice stands, are multiple-stemmed, and some shrubs have large single stems. Shrubs may have evolved either from trees or herbs (for detailed discussion on evolution of shrubs, see Stebbins 1972).

The success of shrubs depends largely on their abilities to compete for nutrients, energy, and space (West and Tueller 1972). In certain environments shrubs have many advantages. They have less energetic and nutrient investment in aboveground parts than trees. Their stemmy growth affects light interception, heat dissipation, and evaporative losses. The multistemmed forms influence interception and stemflow of moisture, increasing or decreasing infiltration into the soil (Mooney and Dunn 1970b). Because most shrubs quickly form extensive root systems, they can utilize soil moisture deep in the profile. This feature gives them a competitive advantage over trees and grasses in regions where the soil moisture recharge comes during the nongrowing season. Because they have a high root-to-shoot ratio, shrubs draw less nutrient input into aboveground biomass and more into roots. Their perennial nature allows immobilization of limiting nutrients and slows nutrient recycling, favoring further shrub invasion of grasslands.

Subject to strong competition from herbs, some climax shrubs, such as chamise (*Adenostoma fasciculatum*), may inhibit the growth of herbs by allelopathy, the secretion of substances toxic to other plants (McPherson and Muller 1969). Only when fire destroys mature shrubs and degrades the toxins do herbs appear in great numbers. As the shrubs recover, herbs decline. Seeds of herb species affected apparently have evolved the ability to lie dormant in the soil until released from suppression by fire.

## Types of Shrublands

Mediterranean-Type Shrublands In five regions of the world lying for the most part between 32° and 40° north and south of the equator are areas characterized by a mediterranean climate: the semiarid regions of western North America, the regions bordering the Mediterranean, central Chile, the Cape region of South Africa, and southwestern and southern Australia. The mediterranean climate is characterized by hot, dry summers with at least one month of protracted drought and cool, moist winters. At least 65 percent of the annual precipitation falls during the winter months and for at least one month the temperature remains below 15° C (Aschmann 1973).

All five areas (see inside covers) support physiognomically similar communities of xeric broadleaf evergreen shrubs and dwarf trees, known as broad sclerophyll vegetation, with herbaceous understory (for detailed descriptions

see McKell et al. 1972, di Castri and Mooney 1973; di Castri, Goodall, and Specht 1981). Although vegetation in all mediterranean-type ecosystems exhibits strong convergence (see Mooney 1977), each has evolved its own distinctive flora and fauna. Mediterranean vegetation evolved from tropical floras and developed in dry summer climates that did not exist until the Pleistocene (Raven 1973).

In addition to similar forms, vegetation in each of the mediterranean systems also shows similar adaptations to fire (see Mooney and Conrad 1977) and to low nutrient levels in the soil. In the mediterranean systems of the Northern Hemisphere annuals make up 50 percent of the species and 10 percent of the plant genera; 40 percent of the species are endemics.

There are variations of the basic mediterranean-type ecosystems (see di Castri 1981). In the Mediterranean region shrub vegetation often results from forest degradation and falls into three major types. The **garrigue,** resulting from degradation of pine forests, is low, open shrubland on well-drained to dry calcareous soil. The **maqui,** replacing cork oak forests, is higher thick shrubland in areas of more rainfall. The **mattoral,** further subdivided into high, middle, low, and scattered (Tamales 1981), appears to be equivalent to the North American chaparral (Soriano 1972, Tomaselle 1981a). The Chilean mediterranean system, also called mattoral, varies across topographic positions from the coast to slopes of coastal ranges and the foothills of the Andean cordillera (Rundel 1981). Much of the South African mediterranean shrubland is typically heathlands known as **fynbos,** discussed later. The more typical mediterranean-type shrubland, dominated by broad-sclerophyll woody shrubs, goes by the names of **strandveld** and **renosterveld.**

In southwest Australia the mediterranean shrub country, known as **mallee,** is dominated by low-growing *Eucalyptus,* 5 to 8 m high, with broad sclerophyllous leaves. There are six types of mallee ecosystems, which intergrade. Of these, three fall into mediterrean-type ecosystems. with a typically grassy and herbaceous understory (Figure 13.17). The other three types occur on oligotrophic soils and fall under the category of heathland shrubs. Razed by fire at irregular intervals, the mallee ecosystem differs markedly from typical mediterranean-type ecosystems in its summer growth rhythm. Mediterranean-type shrublands typically initiate new growth during the spring; but the mallee retains its summer growth rhythm from the subtropical Tertiary, out of phase with the mediterranean climate of the area. Growth takes place during the summer, the driest part of the year, drawing on water conserved in the soil during wet winter and spring showers (Specht 1981).

In North America the sclerophyllic shrub community is known as **chaparral,** a word of Spanish origin meaning a thicket of shrubby evergreen oaks. California chaparral (Figure 13.18) is dominated by scrub oak (*Quercus dumosa*) and chamise (*Adenostoma fasciculatum*). Another shrub type, also designated as chaparral, is associated with the Rocky Mountain foothills in Arizona, New Mexico, and Nevada, and elsewhere. It differs from California chaparral in two ways. It is dominated by Gambel oak (*Quercus gambelii*) and other species and lacks chamise; and it is summer-active and winter-deciduous, whereas California chapparal is evergreen, winter-active, and summer-dormant (see Hanes 1981).

For the most part mediterranean-type shrublands lack understory and ground litter, are highly inflammable, and are heavy seeders. Many species require the heat and scarring action of fire to induce germination. Others sprout vigorously after a fire.

For centuries periodic fires have roared through mediterranean-type vegetation, clearing away the old growth, mak-

**Figure 13.17** Tall shrub mallee in Victoria, Australia is an example of a mediterranean-type shrubland dominated by *Eucalyptus.* Note the canopy structure and the open understory of grass at the beginning of the spring rains. This type of vegetation supports a rich diversity of bird life.

**Figure 13.18** Southern California mixed chaparral is highly prone to fire.

ing way for the new, and recycling nutrients through the ecosystem. When humans intruded into this type of vegetation, they changed the fire regime, either by attempting to exclude fire completely or by overburning (see Trabaud 1981). In the absence of fire, chaparral grows tall and dense and yearly adds more leaves and twigs to those already on the ground. During the dry season the shrubs, even though alive, nearly explode when ignited. Once set on fire by lightning or humans, an inferno follows, as victims of the November 1993 fires in California would attest.

After fire the land returns either to lush green sprouts coming up from buried root crowns or to grass, if a seed source is nearby. Grass and vigorous young sprouts are excellent food for deer, sheep, and cattle. As the sprout growth matures, chaparral becomes dense, the canopy closes, the litter accumulates, and the stage is set for another fire.

Because of the rough terrain characteristic of the mediterranean-type ecosystem some areas have remained relatively undisturbed, especially in California and South Africa. But in Australia and the Mediterranean basin, human activity, especially livestock grazing and fruit and vegetable farming, have degraded the broad-leaf sclerophyllous vegetation (see Trabaud 1981).

**Northern Desert Shrub**  In the Great Basin of North America, the northern, cool, arid region lying west of the Rocky Mountains, is the **northern desert shrub.** The climate is continental, with warm summers and prolonged cold winters. Although this region is perhaps more appropriately considered a desert, it is one of the most important shrublands in North America (Figure 13.19). Its physiognomy differs greatly from the southern hot desert, and the dominant vegetation is shrub. The vegetation falls into two main associations: one is sagebrush, dominated by big sagebrush (*Artemisia tridentata*), which often forms pure stands; the other is shadscale, (*Atriplex confertifolia*), a C4 species, and other chenopods (see Caldwell 1985), halophytes tolerant of saline soils. Inhabiting this shrubland are pocket and kangaroo mice (Heteromyidae), lizards, sage grouse (*Centrocerus urophasianus*), sage thrasher (*Oreoscoptes montanus*), sage sparrow (*Amphispiza belli*), and Brewer's sparrow (*Spizella breweri*), four birds that depend upon sagebrush for their existence.

A similar type of shrubland exists in the semiarid inland of southwestern Australia. Numerous chenopod species, particularly the saltbushes of the genera *Atriplex* and *Maireana,* form extensive low shrublands on low riverine plains (Figure 13.20).

**Heathlands**  Typically **heathlands** have been associated with cool to cold temperate climatic regions of northwestern Europe. It was probably coincidental that the original name came to refer to land dominated by Ericaceae. The word **heath** comes from the German *Heide* meaning "an uncultivated stretch of land," regardless of the vegetation. It just so happened that in parts of Germany uncultivated or waste land was dominated by Ericaceae. In reality heathlands are found in all parts of the world, from the tropics to polar regions and from lowland to subalpine and alpine altitudes.

Heathland vegetation is an assemblage of dense to middense growth of ancient or primitive genera of angiosperms: evergreen sclerophyllous shrubs and subshrubs such as Daphne heath *Brachyloma daphnoides,*  all adapted to fire (Figure 13.21). Heathland shrubs have leaves with thick cuticles, sunken stomata, thick-walled cells, and hard and waxy upper surfaces. Many species have leaves with small

**Figure 13.19** Northern desert shrubland in Wyoming is dominated by sagebrush. Although classified as cold desert (see page 247), sagebrush forms one of the most important shrub types in North America.

surface area—less than 25mm and termed leptophyll—and others roll their edges in toward the midrib. Although mostly associated with heathlands, many heathland shrubs such as laurel and blueberries (*vaccinium*) are present, at least locally, in other ecosystems such as the deciduous forest, as shrubby understory.

Heathlands invariably occur on oligotrophic soils especially deficient in phosphorus and nitrogen. Heathlands are most extensive in the arctic regions. Other extensive areas occur in the mediterranean-type regions of South Africa, where they are known as fynbos (Kruger 1979), and in

southeastern and western Australia (see George et al. 1979). In subtropical to tropical climates true heathlands are confined to alpine areas and to lowland, oligotrophic soils subject to seasonal waterlogging. Some heathlands, such as the heather-dominated moors of Scotland, are human-induced and can be maintained only by periodic fires (Gill and Groves 1981, Gimingham 1981).

There are two distinct heathland ecosystems: dry heathlands and wet heathlands. Dry heathlands are on well-drained soils and subject to seasonal drought; wet heathlands are subject to seasonal waterlogging. In wet

**Figure 13.20** Saltbush shrubland in Victoria, Australia is dominated by *Atriplex*. It is an ecological equivalent to the shrublands of the Great Basin in North America.

**Figure 13.21** Heathlands, dominated by ericaceous shrubs, have a similar physiognomy around the world. Typical is this heathland along the southern coast of Australia.

heathlands graminoids (grasses and sedges) may become codominant with heathland shrubs; and in extreme wet heathlands the grass component is suppressed by *Sphagnum* moss. Both height and foliage cover of heathlands vary considerably with the habitat. They are divided according to height of the uppermost stratum: shrubs >2m, scrub; shrubs 1–2 m, tall heathland; shrubs 25–100 cm, heathland; and shrubs <25 cm, dwarf heathland.

**Successional Shrublands** On drier uplands shrubs rarely exert complete dominance over herbs and grass. Instead the plants are scattered or clumped in grassy fields. The open areas between are filled with the seedlings of forest trees, which in the sapling stage occupy the same ecological position as tall shrubs (Figure 13.22). Typical are the hazelnut (*Corylus* spp.), forming thickets in places, sumacs (*Rhus* spp.), chokecherry (*Prunus virginiana*), and shrub dogwoods (*Cornus* spp.).

On wet ground the plant community often is dominated by tall shrubs and contains an understory intermediate between that of a meadow and a forest (Curtis 1959). In northern regions the common tall shrub community found along streams and around lakes is the alder thicket, composed of alder or alder and a mixture of other species such as willow and red osier dogwood. Alder thickets persist for some time before being replaced by the forest. Out of the alder country, the shrub or carr community (*carr* is an English name for wet-ground shrub communities) occupies the low places. Dogwoods are some of the most important species in the

carr. Growing with them are a number of willows, which as a group usually dominate the community.

Shrub thickets are valued as food and cover for game; and many shrubs, such as hawthorn, blackberry, sweetbrier, and dogwoods, rank high as game food. However, the overall value of different types of shrub cover, its composition, quality, and the minimum amounts needed, have never been assessed. There is some evidence (Egler 1953, Niering and Egler 1955) that even where the forest is the normal end of succession, shrubs can form a stable community that will persist for many years. If incoming tree growth is removed either by selective spraying or by cutting, shrubs eventually form a closed community resistant to further invasion by trees. This fact could have wide application to the management of power line rights-of-way.

## Structure

Shrub ecosystems, successional or mature, are characterized by woody structure, increased stratification over grasslands, dense branching on a fine scale, and low height, under 8 m. Typically there are three layers: a broken upper canopy, an irregular low shrub canopy, and a grass/herbaceous layer. This structure, of course, may vary. Dense shrubland may have only a canopy layer; and stratification often decreases as the shrubs reach maximum height, particularly in seral shrublands. Patchy distribution of shrubs varies with the vegetation type. Heathlands may exhibit minimal patchiness. Mediterranean-type shrublands, notably the mattoral and the

**Figure 13.22** In eastern North America and in northern and western Europe shrublands are usually successional communities. This old field in southern New York state is in the early stages of shrub invasion.

mallee, may have pronounced horizontal stratification. Greatest patchiness probably occurs in successional shrublands with scattered clumps of invading shrubs and trees.

Shrub communities have their own distinctive animal life. Successional shrub communities support not only species common to shrubby edges of forests and shrubby borders of fields but a number of species dependent upon them, such as bobwhite quail (*Colinus virginianus*), cottontail rabbit (*Sylvilagus floridanus*), prairie warbler (*Dendroica discolor*), and yellow-breasted chat (*Ictera virens*). In Great Britain some shrub communities, especially hedgerows, have been stable for centuries and many forms of animal life, invertebrate and vertebrate, have become adapted to or dependent upon them. Among these species are the whitethroat (*Sylvia communis*), linnet (*Acanthis cannabina*), blackbird (*Turdus merula*), and hedgehog (*Erinaceus europaeus*).

Climax shrub communities have a complex of animal life that varies with the region (Bigalke 1979, Dwyer et al. 1979, Blondel 1981, Cody 1973, Schoddee 1981). Within the mediterranean-type shrublands and heathlands similarity in habitat structure and in the nature and number of niches has resulted in pronounced parallel and convergent evolution among bird species (Blondel 1981, Cody 1973).

## Function

The functional aspects of shrubland ecosystems are poorly studied, but data from a few mediterranean-type systems provide some interesting insights into nutrient cycling.

In the California chaparral precipitation falls mostly in the cool winter months. Most of the plant growth and flowering is concentrated in spring; 75 percent of the flowering plants, half of them annuals, bloom in May at the end of the rainy season. However, the evergreen dominant can fix $CO_2$ throughout the year. The greatest daily carbon gain is in the spring; the lowest during summer drought; and in winter with its short photoperiod, carbon fixation is minimal (Mooney 1981, Mooney and Miller 1985). The amount fixed in summer depends upon the amount of rainfall received during the previous winter. Yearly accumulation of aboveground biomass for a sclerophyllous plant of intermediate age is about 1000 kg/ha (Mooney and Parsons 1973).

The soils of mediterranean-type ecosystems are low in nutrients and are especially deficient in nitrogen and phosphorus. The cycling of these nutrients appears to be tight, conservative, and seasonal (Mooney and Miller 1985), reflecting rainfall (or the lack of it) and associated microbial activity, high during the wet season. A flush of microbial activity, which involves decomposition of humus and mineralization of nitrogen and carbon, follows wetting of dry soil. Nitrate accumulation depends upon a progressive drying period after rains, when the topsoil gradually dries out to an increasing depth. Improved soil aeration and increasing soil temperature favor rapid bacterial nitrification and the retention of nitrate ions in the soil. As the dry cycle continues, nitrates accumulate and remain fixed in the topsoil along with other nutrients (Schaefer 1973). When the rains arrive and wet the soil, the concentration of nutrients stimulates a flush of growth. If heavy rains suddenly enter dry topsoil, quantities of nutrients may be lost by leaching. In the California chaparral much of the nitrogen returned to the soil is lost through erosion (Mooney and Parsons 1973).

Some plants of the mediterranean systems exhibit some nutrient conservation mechanisms. *Ceanothus,* an early successional species in the California chaparral, is a nonleguminous nitrogen fixer (Mooney and Parsons 1973). In the Australian mallee *Atriplex vesticana,* a dominant plant, lowers the nitrogen content of the surrounding soil and concentrates nitrogen through litterfall in the soil directly beneath the plant. More nitrogen, however, is withdrawn from the soil than is returned by litterfall, which represents about 10 percent of the total plant nitrogen. Litter has lower nitrogen and phosphorus content than fresh leaves, which suggests that the plant withdraws nitrogen and phosphorus from the leaf into the stem before the leaf falls.

Australian sclerophyllous shrubs also exhibit phosphorus conservation, involving three mechanisms. First, in aging leaves phosphorus, like nitrogen, is recirculated through the plant with minimal losses to litter. Second, a fine mat of mycorrhizal roots penetrates the decomposing litter and takes up phosphorus. Third, polyphosphate forming and hydrolyzing enzymes are present in the roots of sclerophyllous plants. As orthophosphate is released from decomposing litter in spring, it is stored in roots as long-chain polyphosphate. When growth begins, the polyphosphate seems to be hydrolyzed back to orthophosphate and transported to the growing shoots (Specht 1973).

The best available data for mineral cycling in a mediterranean-type ecosystem are for a 17-year-old *Quercus coccifera* garrigue in southern France (Figure 13.23). The total aboveground mineral mass is 773 kg/ha; 629 kg is in wood, and 144 kg in leaves. Calcium is the most important element, amounting to 485 kg/ha, followed by nitrogen, 159 kg/ha, and potassium, 85 kg/ha. The mean annual incorporation of nutrients into perennial organs amounts to 37 kg/ha, of which two-thirds is calcium. Annual return of elements by the way a litter amounts to 75.2 kg/ha. Net primary productivity is on the order of 3.4 tn/ha/yr with a mean annual production of litter of 2.3 tn/ha (Lossaint, 1973, Rapp and Lossaint 1981).

The ecology of mediterranean shrublands has been studied intensively over the past quarter century. Nevertheless a number of questions remain, especially relating to management. What is the effect on native plants and animals of seeding grass on chaparral hills after fire? What is the relationship of these unnatural grasslands to the fire ecology of the region? Does the present unnatural fire frequency of once every two or three decades alter nutrient pools and eliminate certain species? What is the effect of absence of fire on long-term community structure, productivity, and species diversity? Answers to such questions are important in the maintenance of mediterranean ecosystems and the welfare of the human populations occupying them.

## DESERTS

Geographers define **deserts** as land where evaporation exceeds rainfall. No specific level of rainfall serves as a criterion; deserts range from extremely arid ones to those with sufficient moisture to support a variety of life. Deserts have been classified according to rainfall into *semideserts,* with precipitation between 15 and 40 cm per year; *true deserts,* with precipitation below 12 cm per year; and *extreme deserts,* areas with rainfall below 7 cm per year. Deserts, which occupy about 26 percent of the continental area, occur in two distinct belts between 15° and 35° latitude in both the Northern and Southern hemispheres—around the Tropic of Cancer and the Tropic of Capricorn.

Deserts are the result of several forces. One force that leads to the formation of deserts and the broad climatic regions of Earth is the movement of air masses over Earth. High-pressure areas alter the course of rain. The high-pressure cell off the coast of California and Mexico deflects rainstorms moving south from Alaska to the east and prevents moisture from reaching the southwest. In winter high-pressure areas move southward, allowing winter rains to reach

**Figure 13.23** Annual turnover of macronutrients in the *Quercus coccifera* garrigue at Le Puech de Juge near Montpellier in Southern France. (After Lossaint 1973.)

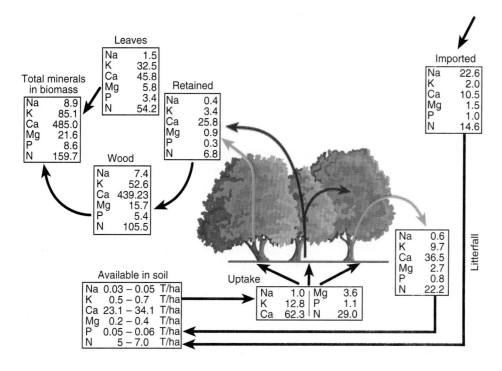

southern California and parts of the North American desert. Winds blowing over cold waters become cold also; they carry little moisture and produce little rain. Thus the west coast of California and Baja California, the Namib desert on coastal southwest Africa, and the coastal edge of the Atacama in Chile may be shrouded in mist, yet remain extremely dry.

Mountain ranges also play a role in desert formation by causing a rain shadow on their lee side. The High Sierras and the Cascade Mountains intercept rain from the Pacific and help maintain the arid conditions of the North American desert. The low eastern highlands of Australia block the southeast trade winds from the interior. Other deserts, such as the Gobi and the interior of the Sahara, are so remote from the source of oceanic moisture that all the water has been wrung from the winds by the time they reach those regions.

## Characteristics

All deserts have in common low rainfall, high evaporation (7 to 50 times as much as precipitation), and a wide daily range in temperature, from hot by day to cool by night. Low humidity allows up to 90 percent of solar insolation to penetrate the atmosphere and heat the ground. At night the desert yields the accumulated heat of the day back to the atmosphere. Rain, when it falls, is often heavy, and unable to soak into the dry earth, it rushes off in torrents to basins below.

The topography of the desert, unobscured by vegetation, is stark and, paradoxically, partially shaped by water. Unprotected, the soil erodes easily during violent storms and is further cut away by the wind. Alluvial fans stretch away from eroded, angular peaks of more resistant rocks. They join to form deep expanses of debris, the *bajadas.* Eventually the slopes level off to low basins, or *playas,* which receive waters that rush down from the hills and water-cut canyons, or *arroyos.* These basins hold temporary lakes after the rains, but water soon evaporates and leaves behind a dry bed of glistening salt.

The aridity of the desert may seem inimical to life; yet in the deserts life does exist, surprisingly abundant, varied, and well adapted to withstand or circumvent the scarcity of water.

Deserts are not the same everywhere. Differences in moisture, temperature, soil drainage, topography, alkalinity, and salinity create variations in vegetation cover, dominant plants, and groups of associated species. There are warm deserts and cool deserts, extreme deserts and semideserts, ones with sufficient moisture to verge on being grasslands or shrublands, and gradations between those extremes within continental deserts. However, there is a certain degree of similarity among warm deserts and cool deserts of the world. The cool deserts, including the Great Basin of North America, the Gobi, Takla Makan, and Turkestan deserts of Asia, and high elevations of warm deserts are dominated by *Artemisia* and chenopod shrubs, and may be considered shrub steppes. The northern part of the North American Great Basin is dominated by nearly pure stands of big sagebrush (see Figure 13.19) and the southern part by shade scale (*Atriplex confertifolia*) and bud sage (*Artemisia*). The warm deserts range from no or scattered vegetation to some combination of chenopods, dwarf shrubs, and succulents. (For a detailed review of world deserts see Evanari, Noy-Meir, and Goodall 1985, 1986). The deserts of southwestern North America—the Mojave, the Sonoran, and the Chihuahuan—are dominated by creosote bush (*Larrea divaricata*) and bursage (*Franseria* spp.). Areas of favorable moisture support tall growths of *Acacia,* saguaro (*Cereus giganteus*), paloverde (*Cercidium* spp.), and ocotillo (*Fouquieria* spp.).

## Structure

Woody-stemmed and soft brittle-stemmed shrubs are characteristic desert plants (Figure 13.24). In a matrix of shrubs grows a wide assortment of other plants, the yucca, cacti, small trees, and ephemerals. In the Sonoran, Peru-Chilean, South African Karoo, and southern Namib deserts, large succulents rise above the shrub level and change the aspect of the desert far out of proportion to their numbers (Figure 13.25). The giant saguaro, the most massive of all cacti, grows on the bajadas of the Sonoran desert. Ironwood, smoketree, and paloverde grow best along the banks of intermittent streams.

Both plants and animals are adapted to the scarcity of water either by drought evasion or by drought resistance. Drought-evading plants flower only in the presence of moisture. Known as ephemerals, they persist as seeds during drought periods, ready to sprout, flower, and produce seeds when moisture and temperature are favorable. There are two periods of flowering in the North American deserts, after winter rains come in from the Pacific Northwest and after summer rains move up from the southwest out of the Gulf of Mexico. Some species flower only after summer rains, and a few bloom during both seasons. If rains fail, these ephemeral species do not bloom. Drought-evading animals, like their plant counterparts, adopt an annual life or go into estivation or some other stage of dormancy during the dry season. If extreme drought develops during the breeding season, birds fail to nest and lizards do not reproduce.

Belowground biomass in the desert can be as patchy as aboveground biomass. Some drought-evaders are deep-rooted woody shrubs, such as mesquite (*Prosopis* spp.) and tamarisk (*Tamarix*), whose taproots reach the water table, rendering them independent of water supplied by rainfall. Others, such as *Larrea* and *Atriplex,* are deep-rooted perennials with laterals that extend as far as 15 to 30 m from the stems. Still other perennials are drought-resisters, such as the various species of cactus, with shallow roots often no more than a few centimeters below the surface. Ephemerals have shallow and poorly branched roots reaching a depth of about 30 cm, where they pick up moisture quickly from light rains (see Drew 1979).

The desert floor is stark, a raw mineral substrate devoid of a continuous litter layer. Dead leaves, bud scales, and

(a)

**Figure 13.24** Two hot deserts dominated by woody, brittle-stemmed shrubs. (a) Mojave Desert in Nevada. (b) Desert in the Northern Territory of Australia. The sparseness and spacing of the shrubs reflects the extreme aridity of these areas.

(b)

dead twigs accumulate in wind-protected areas beneath the plants and in depressions in the soil.

## Function

The desert ecosystem differs from other ecosystems in at least one important way: the input of precipitation is highly discontinuous. It comes in pulses as clusters of rainy days 3 to 15 times a year, of which only 1 to 6 may be large enough to stimulate biologic activity. Thus the desert ecosystem experiences periods of inactive steady states broken by periods of production and reproduction. Both processes are stimulated by rain and continue through short periods of adequate moisture; when water is scarce again, both production and biomass return to some low steady state.

Primary production in the desert depends upon the proportion of available water used and the efficiency of its use. Data from various deserts in the world (for a summary, see Noy-

Meir 1973, 1974) suggest that annual net primary production of aboveground vegetation varies from 30 to 200 g/m².

The amount of biomass that accumulates and rate of turnover (ratio of production to biomass) depend upon the dominant type of vegetation. In deserts such as the Sonoran, where trees, shrubs, and cacti dominate, annual productivity is about 10 to 20 percent of the aboveground standing crop biomass of 300 to 1000 g/m². In deserts with perennial-type vegetation, annual production is 20 to 40 percent of biomass of 150 to 600 g/m². Annual or ephemeral communities have a 100 percent turnover of both roots and aboveground foliage; annual production is the same as peak biomass. These turnover rates are higher than those of forests and tundra. The ratio of belowground biomass to aboveground for perennial grasses and forbs is between 1 and 20, and for shrubs between 1 and 3. In general, warm desert plants do not have a high root-to-shoot ratio. The root biomass is relatively small. Cool desert plants have a higher root-to-shoot ratio.

**Figure 13.25** Saguaro dominates the aspect of this portion of the Sonoran Desert in the southwestern United States.

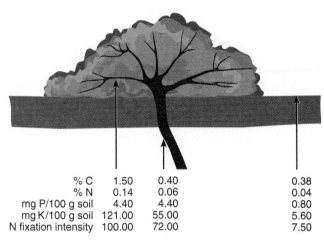

| | | | |
|---|---|---|---|
| % C | 1.50 | 0.40 | 0.38 |
| % N | 0.14 | 0.06 | 0.04 |
| mg P/100 g soil | 4.40 | 4.40 | 0.80 |
| mg K/100 g soil | 121.00 | 55.00 | 5.60 |
| N fixation intensity | 100.00 | 72.00 | 7.50 |

**Figure 13.26** Organic content and carbon content of desert soil beneath and away from the base of an Egyptian desert shrub, *Anabasis aretioides.* A value of 100 is attributed to the soil surface, under the plant canopy. (From Binet 1981:327.)

Nutrient cycling in arid ecosystems is tight, and two major nutrients, phosphorus and nitrogen, are in short supply. Much of the nutrient supply is tied up in plant biomass, living and dead. The tissues of desert plants have higher concentrations of nutrients than plants of mesic environments, and they tend to retain certain elements before shedding any parts. For example, the nitrogen and phosphorus of fallen phyllodes (the enlarged and commonly flattened leaf stalks that function as leaves) of *Acacia aneura* in the northern Australian desert decrease markedly at the time of major phyllode fall, but the contents of potassium, calcium, and magnesium show little change. The desert plant *Artemesia* translocates phosphorus and potassium back to the twig before shedding its leaves. The nutrients retained in the shed parts collect beneath the plants, where microclimate conditions created by the shrubs favor biological activity. The soil is further enriched by animals attracted to the shade (Binet 1981, West 1979, 1981; West and Skuiins 1978). The plants, in effect, create islands of fertility beneath themselves.

This fact is illustrated in Figure 13.26, showing the organic carbon and nutrient levels beneath and away from the base of an Egyptian desert shrub, *Anabasis aretioides.* The percentages of carbon and nitrogen are over three times those of the open soil, and the levels of phosphorus and potassium five and 21 times as great. Adding to primary production in the desert are lichens, green algae, and

cyanobacteria (blue-green algae), abundant as soil crusts. Their biomass ranges up to 240 kg/ha, and they have an unusually high rate of nitrogen fixation. As pointed out in Figure 13.26, the intensity of fixation is greatest beneath the desert shrub, over ten times greater than in the open.

In spite of their aridity, desert ecosystems support a surprising diversity of animal life, notably herbivorous species (Figure 13.27). Grazing herbivores of the desert tend to be generalists and opportunists in their mode of feeding. They consume a wide range of species, plant types, and parts. Desert sheep feed on succulents and ephemerals when available and then switch to woody browse during the dry period. As a last resort herbivores consume dead litter and lichens. Small herbivores—the desert rodents, particularly the family Heteromyidae and ants—tend to be **granivores,** feeding largely on seeds.

These granivores are important in the dynamics of desert ecosystems. One of the notable small herbivores of the Sonoran Desert is the harvester ant (*Pogonomyrmex occidentalis*), which lives on seeds stored in underground granaries. During periods of drought these ants gather mainly the seeds of several species of perennials. During winter rains, when annual plants flower and seed, ants gather the seeds of these plants. By using different plants, the ant obtains a constant food source throughout the desert year.

Herbivores can have a pronounced effect on primary producers, especially if they are more abundant than the land's capacity to support them. Once grazers have utilized annual production, they consume plant reserves, especially in long dry periods. Overgrazing and overbrowsing can so weaken the plant that the vegetation is destroyed or irreparably damaged. Areas protected from grazing, especially by cattle and goats, have a higher biomass and a greater percentage of palatable species than grazed areas.

Herbivores in a shrubby desert, under most conditions, consume only a small part of the aboveground primary pro-

**Figure 13.27** Large herbivores inhabit the less environmentally extreme regions of the Kalahari Desert of southern Africa.

duction, but seed-eating herbivores can consume most of the seed production. Chew and Chew (1970) found that small grazing herbivores (the jackrabbit and kangaroo rat) used only about 2 percent of the aboveground net primary production, but ate 87 percent of the seed production, a rate of consumption that could have a pronounced effect on the composition of plant populations. How pronounced has been demonstrated by J. R. Brown and Heske (1990). For 12 years they removed three species of kangaroo rats (*Dipodomys* spp.) from plots of a Chihuahuan Desert shrub habitat near a zone of a natural transition from desert to grassland. These rodents feed heavily on larger seeds and burrow in the soil. Their removal resulted in a threefold increase in density of tall perennials and annual grasses, a sharp decrease in desert shrubs, and invasion of rodent species typical of arid grasslands.

Desert carnivores, like the herbivores, are opportunistic feeders, with few specialists. Most desert carnivores, such as foxes and coyotes, have mixed diets that include leaves and fruits, and even insectivorous birds and rodents eat plant foods. Omnivory rather than carnivory and complex food webs seem to be the rule in desert ecosystems.

The detrital food chain seems to be less conspicuous in the desert than in other ecosystems, but nevethless it is important. Most functional and taxonomic groups of soil microorganisms exist in the desert, from bacteria, fungi, and actinomycetes to nematodes, mites, and termites. In some deserts considerable amounts of nutrients may be tied up in termite structures, to be released when structures are destroyed. In the North American warm deserts rates of decomposition of both surface and buried litter seem to be independent of rainfall, in part because soil fauna in the desert are active in the absence of moisture (Whitford et al. 1981, Santos et. al 1984). Bacterial and fungal decomposition is

important. Its rate appears to be influenced by free-living soil nematodes. Their populations in turn are influenced by predatory mites (Santos et al. 1984). Although short-term litter decomposition does not seem to be related to yearly fluctuations in rainfall, in North American warm deserts it is highly correlated with long-term rainfall patterns, because it shapes the structure of the desert soil community. We know too little about the structure and the functioning of desert ecosystems, and the relationship of desert life to physical factors of the environment in the various desert types.

## TUNDRAS

Encircling the top of the world is a region of cold, often desertike conditions, the Arctic **tundra.** The word comes from the Finnish *tunturia,* meaning "a treeless plain" (Figure 13.28). At lower latitudes similar landscapes, the alpine tundra, occur at the peaks of tall mountains. In the Antarctic a well-developed tundra is lacking because of a small land area and deep ice sheets. Arctic or alpine, the tundra is characterized by low temperatures, low precipitation (cold air can carry very little water vapor), and a short growing season ranging from 50 to 60 days in the high Arctic tundra to 180 days in low altitude alpine tundra.

The arctic tundra is a land dotted with lakes and transected by streams. Its vegetation ranges from tall shrubs (2–5 m high), to dwarf shrub heath (5–20 cm high), graminoids, and mosses (Bliss 1988). Where the ground is low and moist, extensive bogs exist. On higher, drier areas and places exposed to the wind, vegetation is scant and scattered, and the ground is bare and rock-covered. These regions are the **fellfields,** an anglicization of the Danish *fjoeldmark,* or rock deserts. Bliss (1981) divides the arctic tundra

**Figure 13.28** The wide expanse of the arctic tundra.

into the polar desert, with less than 5 percent cover, polar semidesert, and tundra with 100 percent cover and wet to moist soil (Figure 13.29).

## Characteristics

Frost molds the tundra landscape. Alternate freezing and thawing and the presence of a permanent frozen layer in the ground, the *permafrost,* create conditions unique to the arctic tundra. The sublayer of soil is subject to annual thawing in spring and summer and freezing in fall and winter. The depth of thaw may vary from a few centimeters in some places to half a meter in others. Below the thaw depth the ground is always frozen solid and is impenetrable to both water and roots. Because the water cannot drain away, flatlands of the Arctic are wet and covered with shallow lakes and bogs. This

**Figure 13.29** Major subdivisions of the North American tundra, including the High Arctic, which holds polar deserts and polar semi deserts, and the Low Arctic. (After Bliss 1981.)

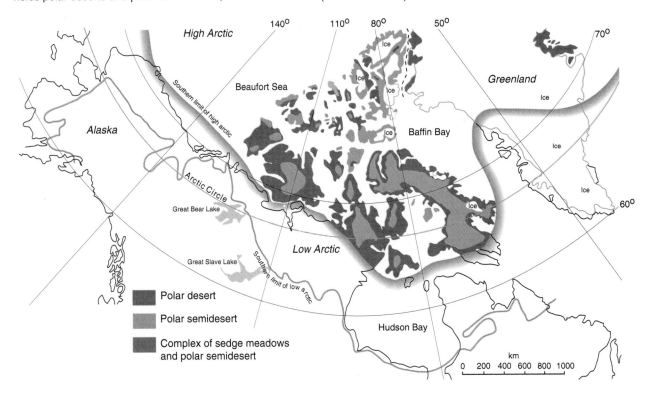

reservoir of water lying on top of the permafrost enables plants to exist in the driest parts of the Arctic.

The symmetrically patterned landforms so typical of the tundra result from frost. The fine soil materials and clays, which hold more moisture than the coarser materials, expand while freezing and then contract upon thawing. This action tends to push the larger material upward and outward from the mass to form the patterned surface.

Typical nonsorted patterns associated with seasonally high water tables are frost hummocks, frost boils, and earth stripes (Figure 13.30). Frost hummocks are small earthen mounds up to 1.5 m in diameter and 1.3 m high, which may or may not contain peat. Frost boils are formed when the surface freezes across the top, trapping the still unfrozen muck beneath. As this bulge chills and expands, the mud is forced up through the crust. Raised earth stripes, found on moderate slopes, appear as lines or small ridges flowing downhill. They apparently are produced by a downward creep or flow of wet soil across the surface of the permafrost.

Sorted patterns are characteristic of better-drained sites. The best known are the stone polygons, the size of which is related to frost intensity and the size of the material (Johnson and Billings 1962). The larger stones are forced out to a peripheral position, and the smaller and finer mater-

ial, either small stones or soil, occupies the center. The polygon shape may result from an accumulation of rocks in desiccation cracks formed during drier periods. These cracks appear as the surface of the soil dries out, in much the same way as cracks appear in bare, dry, compacted clay surfaces in temperate regions. On the slopes, creep, frost-thrusting, and downward flow of soil change polygons into sorted stripes running downhill. Mass movement of supersaturated soil over the permafrost forms solifluction terraces, or "flowing soil." This gradual downward creep of soils and rocks eventually rounds off ridges and other irregularities in topography. This molding of the landscape by frost action, called *cryoplanation,* is far more important than erosion in wearing down the arctic landscape.

In the alpine tundra permafrost exists only at very high elevations and in the far north, but frost-induced patterns— small solifluction terraces and stone polygons—are still present. The lack of permafrost results in drier soils; only in alpine wet meadows and bogs do soil moisture conditions compare with the Arctic. Precipitation, especially snowfall and humidity, is higher in the alpine than in the arctic tundra, but the steep topography results in a rapid runoff of water.

The arctic and alpine regions share many features that characterize the tundra biome. The vegetation of the tundra is structurally simple (Figure 13.31). The number of species tends to be small, growth is low, and most of the biomass and functional activity are confined to a few groups. Most of the vegetation is perennial and reproduces vegetatively rather than by seed. Although it appears homogeneous, the pattern of vegetation is diverse. Small variations in microtopography result in a steep gradient of moisture. The combination of microrelief, snowmelt, frost heaving, and aspect, among other conditions, produces an almost endless change in plant associations from place to place (Polunin 1955).

Although the environment of the arctic and alpine regions is somewhat similar, the vegetation of the two differs in species composition and adaptation to light. Mooney and Billings (1961) conducted a series of studies to compare physiological and growth responses in arctic and alpine populations of alpine sorrel (*Oxyria digyna*) (Figure 13.32) They found that the plant exhibits increased production of flowers and decreased production of rhizomes in the southern portion of its range. Northern populations of the plant have a higher photosynthetic rate at lower temperatures and attain a maximum rate at lower temperatures. Alpine plants reach the light saturation point at higher intensities than the arctic plants, which are adapted to lower light intensities. Arctic plants require longer periods of daylight for photosynthetic activity and growth. The further north the geographic origin of the plant, the more slowly the plant grows under short photoperiod.

South of the arctic tundra and below the alpine tundra is a transition zone between forest and tundra, the edge of which marks the tree line. The tree line is a zone of gnarled trees, stunted by the desiccating force of the wind, an area known as the **Krummholz** or "crooked wood" (Figure

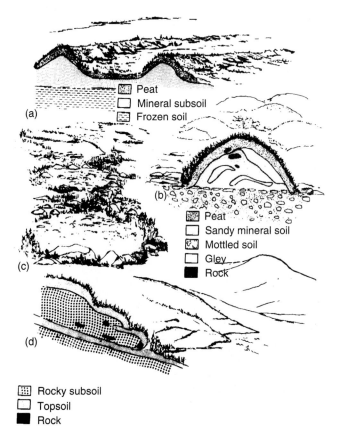

**Figure 13.30** Patterned land forms of the tundra region: (a) unsorted earth stripes; (b) frost hummocks; (c) sorted stone nets and polygons; (d) solifluction terrace. (Diagrams adapted from Johnson and Billings 1963.)

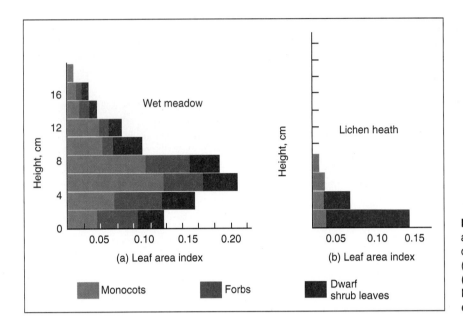

**Figure 13.31** Structure of vegetation in arctic tundra. (a) Maximum leaf area indices (m²/m²) at different plant heights (single surface) for a wet meadow. (b) Leaf area indices for a lichen heath. Note the structural simplicity. (After Berg et al. 1975).

13.33a). The Krummholz in the North American alpine region is best developed in the Appalachian and Adirondack mountains. On the high ridges, trees begin to show signs of stunting far below the timber line. The stunting increases with altitude until spruces and birches, deformed and semiprostrate, form carpets 0.6 to 1 m high, impossible to walk through but often dense enough to walk upon. Where strong winds come in from a constant direction, the trees are sheared until the tops resemble close-cropped heads. However, the trees on the lee side of the clumps grow taller than those on the windward side.

In the Rocky Mountains the Krummholz is much less marked, for there the timber line ends almost abruptly with little lessening of height (Figure 13.33b). Most of the trees are flagged; that is, the branches remain only on the lee side. In the alpine regions of Europe, the Krummholz is characterized by dwarf mountain pine (*Pinus mugo*), which forms dense thickets 1 to 2 m high on calcareous soils. On acid soil dwarf mountain pine is replaced by dwarf juniper (*Juniperus communis nana*). In the Australian Brindabella Range the Krummholz is marked by pockets of low, twisted snow gum (*Eucalyptus pauciflora*) (Figure 13.34).

**Figure 13.32** Alpine sorrel (*Oxyria digyna*) growing on the arctic tundra.

(a)                                    (b)

**Figure 13.33**  The Krummholz. (a) In the Rocky Mountains the tree line is sharply defined. (b) Dwarf spruce and fir grow in narrow pockets that hold snow in winter.

**Figure 13.34**  (a) The tree line in the Australian Alps in the Brindabella Range. (b) Low twisted growth of snow gum (*Eucalyptus pauciflora*) growing in protected pockets gives way to low-growing heaths.

(a)                                                                                         (b)

Wind, cold, and winter desiccation are regarded as the cause of the dwarfed and misshapen condition of the trees. The ability of some tree species, such as mountain (mugo) pine, to show a Krummholz effect may be genetically determined. Eventually conditions become too severe even for the prostrate forms, and trees drop out completely except for those that have taken root behind the protection of high rocks. These causes are only speculative. Much needs to be learned about the interaction of growth form, tolerance to cold, and spatial distribution of limited nutrients and water near the tree line (Stevens and Fox 1991). All of these factors could be involved.

## Structure

### Arctic Tundra
In spite of its distinctive climate and many endemic species, the tundra does not possess a vegetation type unique to itself. Structurally the tundra is a grassland and mixed shrubland.

In the Arctic only those species able to withstand constant disturbance of the soil, buffeting by the wind, and abrasion from wind-carried particles of soil and ice can survive. Although such an environment produces a vegetation that appears homogeneous, it is not. A combination of microrelief, snowmelt, frost heaving, and aspect, among other conditions, produces an endless change in plant associations from spot to spot. In the arctic tundra low ground is covered with cottongrass (*Eriophorum virginicum*) and other sedges, dwarf heath, and a sphagnum moss complex. Well-drained sites support heath shrubs, dwarf willows, and birches, dryland sedges and rushes, herbs, mosses, and lichens. The driest and most exposed sites—the flat-topped domes, rolling hills, and low-lying terraces, all usually covered with coarse, rocky material and subject to extreme action by frost—support only sparse vegetation. Plant cover consists of scattered heaths and mats of mountain avens, as well as crustose and foliose lichens.

Topographic location and snow cover delimit a number of arctic plant communities. Steep, south-facing slopes and river bottoms support the most luxurious and tallest shrubs, grasses, and legumes. Cottongrass dominates the gentle north-facing and south-facing slopes, reflecting higher air and soil temperatures and greater snow depth. Pockets of heavy snow cover create two types of plant habitats, the snow patch and the snowbed. Snow patch communities occur where wind-driven snow collects in shallow depressions and protects the plants beneath. Snowbeds, typical of both arctic and alpine situations, are found where larger masses of snow accumulate because of topographic peculiarities.

The conditions unique to the Arctic result in part from three interacting forces: permafrost, vegetation, and the transfer of heat. Permafrost is sensitive to temperature changes. Any natural or human disturbance, however slight, can cause the permafrost to melt. Because the permafrost itself is impervious to water, it forces all the water to move above it. Thus surface water becomes conspicuous even

though precipitation is low (see R. J. E. Brown and Johnson 1964, R. J. E. Brown 1970). Vegetation protects the permafrost by shading, which reduces the heating of the soil. It retards warming and thawing of the soil in summer and increases the average temperature in winter. If the vegetation is removed, the depth of thaw is 1.5 to 3 times that of the area still retaining the vegetation. Accumulated organic matter and dead vegetation further retard the warming of the soil in summer, even more than a vegetative cover. Thus vegetation and organic matter, acting as positive feedback, impede the thawing of permafrost and act to conserve it (Pruitt 1970, Bliss et al. 1973).

In turn permafrost, acting as negative feedback, chills the soil, retarding the general growth of both aboveground and belowground parts of plants and the activity of soil microorganisms. It also impoverishes the aeration and nutrient content of the soil (Tyrtikov 1959). The effect is more pronounced the closer the permafrost comes to the formation of shallow root systems.

The tundra holds some fascinating animals, even though the diversity of species is low. Animals of the Arctic are mostly circumpolar in distribution. They include the caribou or reindeer (*Rangifera tarandus*), muskox (*Ovibos moschatus*), arctic hare (*Lepus arcticus*), and arctic ground squirrel (*Alopex lagopus*). Some 75 percent of the species of birds of the North American tundra are common to the European tundra (Udvardy 1958).

Muskox and caribou are the dominant large herbivores. Muskox (Figure 13.35) are intensive grazers with low herd movements. In summer they feed on grasses, sedges, and dwarf willow in the valleys and plains; in winter they move up to the windswept ridges where snow cover is scant. Caribou are extensive grazers, spreading over the tundra in summer to feed on sedges and grasses, and migrating southward to the taiga in winter to feed on lichens.

Intermediate-sized herbivores include arctic hares, which feed on willows. The smallest and dominant herbivore over much of the arctic tundra are the cyclic lemmings (*Lemmus* spp.), which feed on fresh green sedges and grasses. Species of herbivorous birds are few, dominated by ptarmigan and migratory geese.

The major arctic carnivore is the wolf (*Canis lupis*). (The polar bear is a marine predator.) The wolf preys on muskox, caribou, and lemmings when they are abundant. Medium-sized predators include the Arctic fox, which feeds on the arctic hare. The smallest mammalian predators, the least weasel (*Mustela rixosa*) and the short-tailed weasel (*Mustela erminea*), feed principally on lemmings and the nestlings and eggs of birds. Major avian predators, the snowy owl (*Nyctea scandiaca*) and the hawk like jaeger (*Stercorarius*), prey heavily on lemmings. Except for the wolf, whose population remains relatively stable when free from hunting by humans, the fortunes of most arctic carnivores rise and fall with the flood and ebb of lemming life.

The arctic tundra, with its expanse of ponds and boggy ground, is the haunt of a myriad of waterfowl, sandpipers,

**Figure 13.35** The gregarious musk ox grazes on sedges and grasses in summer and browses on willows and other shrubs in winter.

and plovers, which arrive when the ice is out and return south before winter sets in.

Invertebrate life is scarce, as are amphibians and reptiles, for whom the short summers and cold are barriers. Some snails are found in the arctic tundra about Hudson Bay. Insects, reduced to a few genera, are nevertheless abundant, especially in mid-July. The insect horde consists of blackflies (Simuliidae), deer flies (Tabinidae), and mosquitoes (Culicidae).

**Alpine Tundra** The alpine tundra is a land of rock-strewn slopes, bogs, alpine meadows dominated by hair grass (*Deschampsia*), and shrubby thickets (Figure 13.36a). Cushion and mat-forming plants, rare in the Arctic, are important in the alpine tundra. Low and ground-hugging, they are able to withstand the buffeting of the wind. The cushionlike blanket traps heat, and the interior of the cushion may be 20° C warmer than the surrounding air, a microclimate that is used by insects.

In spite of similar conditions, however, only about 20 percent of the plant species of the arctic and alpine tundra are the same, and they are of different ecotypes. Lacking in the Rocky Mountain alpine tundra are heaths and a heavy growth of lichens and mosses between other plants; lichens are confined mostly to rocks, and the ground is bare between the plants. The alpine tundras of the Appalachian Mountains of North America and the Alps of Europe, however, are dominated by heaths and sedge meadows, and mosses are common. The Australian alpine region supports a growth of heaths on rocky sites, and wet areas are covered with sphagnum bogs, cushion heaths, and sod tussock grasslands (Figure 13.36b).

Alpine tundras also occur above the treeline of the high mountains in tropical regions: Central America, South America, Africa, Borneo, New Guinea, Java, Sumatra, and Hawaii. Tropical alpine regions undergo great seasonal variation in rainfall and cloud cover but experience little variation in mean daily temperature. Seasonal evenness in temperature is balanced by a strong daily fluctuation in temperature from below freezing conditions at night to hot summerlike temperatures during the day. Such a diurnal freeze-thaw cycle is unique to tropical alpine regions.

Tropical alpine tundras support tussock (bunch) grasses, small-leafed shrubs, and heaths; but the one feature that sets tropical alpine vegetation apart from the temperate alpine is the presence of unbranched to little-branched, giant, treelike rosette plants (Figure 13.37) (A. P. Smith and Young 1987). Although the genera and species differ among tropical alpine areas (for example, *Senecio* species in Africa and *Espeletia* species in the Andes), their growth forms and physiology are strikingly similar, suggesting convergent evolution. These species are the antithesis of the usual low-growing tundra species. The higher the elevation, the taller these plants grow, reaching up to 6 m. Many of these giant rosettes have well-developed water-storing pith in the center of the stem. They retain dead rosette leaves about the stem, which apparently act as insulation against the cold, and secrete a mucilaginous fluid about the bases of leaves that seems to function as a heat storage device. The leaves also possess dense pubescent hairs that reduce convective loss of heat (Meinzer and Goldstein 1984).

The alpine tundra, extending upward like islands in the mountain ranges of the world, is small in area and holds few characteristic animal species. The alpine regions of western North America are inhabited by pikas (*Ochotona* spp.), marmots (*Marmota* spp.), mountain goat (*Oreamnos americanus*) (not a goat at all but related to the South American

(a)

(b)

**Figure 13.36** Alpine tundra (a) in the Rocky Mountains and (b) in the Australian Alps. Although the species of vegetation are different, the growth forms are convergent. The physiognomy is similar.

**Figure 13.37** Tropical alpine tundra on Mt. Kenya, East Africa. Such alpine plants are able to resist freezing temperatures. Plant height tends to increase with elevation.

chamois *Rupicapra rupicapra*), mountain sheep (*Ovis* spp.), and elk. Sheep and elk spend their summers in the high alpine meadows and winter on the lower slopes. The marmot, a mountain woodchuck, hibernates over winter, whereas the pika cuts and stores grass in piles to dry for winter. Some rodents, such as voles and pocket gophers (*Thomomys*), remain under the snow and ground during winter.

The alpine regions contain a fair representation of insect life. Flies and mosquitoes are scarce, but springtails, beetles, grasshoppers, and butterflies are common. Because of ever-present winds, butterflies fly close to the ground; other insects have short wings or no wings at all. Insect development is slow; some butterflies take more than two years to mature, and grasshoppers three.

## Function

**Primary Production** Net annual primary production varies markedly across the tundra, depending upon the plant community considered (Bliss 1975, Wielgolaski 1975a). Because of the microenvironmental conditions influenced by soil, slope, aspect, exposure to wind, snow depth, and drainage conditions, plant communities change rapidly over short distances.

Different components of the plant community make different contributions to net productivity, depending upon the site. In the hummocky sedge-moss meadow of the Devon Island IBP site, Canada, aboveground net primary productivity of vascular plants amounted to 44.7 g/m$^2$/yr, whereas mosses contributed 33 g/m$^2$/yr and lichens contributed nothing. In the raised beach community dominated by lichens and cushion plants, vascular plants contributed 17.8 g/m$^2$/yr, mosses 2 g/m$^2$/yr, and lichens 25 g/m$^2$/yr. In an alpine wet meadow, vascular plants contributed 254 g/m$^2$/yr and mosses 173 g/m$^2$/yr.

Much of the primary production of vascular plants is below ground rather than above. The data emphasize this important functional aspect of the tundra. The net annual aboveground primary production of vascular plants in a wet sedge-moss meadow at Devon Island was 45 g/m$^2$ above ground and 129 g below; a wet meadow on an alpine tundra at the Hardangervidda IBP site in Norway had a biomass of 254 g/m$^2$ and 1316 g/m$^2$ below ground. (For summaries see Bliss et al. 1973, Wielgolaski et al. 1981.)

Although total production of the tundra is low because of a short growing season, daily primary production rates of 0.9 to 1.9 g/m$^2$ in the arctic tundra and 2.2 g/m$^2$ in some alpine tundras are comparable to those of some temperate grasslands. The efficiency of primary production of the Devon Island tundra (Table 13.4) ranges from 0.03 percent for the polar desert to 1 percent for the hummocky sedge-moss meadow (Bliss 1975). Efficiency of a Norwegian tundra ranges from 0.6 percent for a dwarf heath community to 2.4 percent for a willow thicket (Wielgolaski and Kjelvik 1975). By comparison, efficiency of temperate grassland ecosystems ranges from 0.33 to 3.8 percent.

**Biomass** Biomass reflects the pattern of net production. At Devon Island, for example, biomass for the polar desert amounts to 270 g/m$^2$ and for the hummocky sedge-moss meadow 3208 g/m$^2$. More live biomass exists below ground than above in most tundra communities. How much depends upon the nature of the tundra community. On Devon Island the hummocky sedge-moss community has a shoot-to-root ratio of 1:12, the cushion plant-lichen community 1:0.06; and the wet sedge communities of both the arctic and alpine tundra, 1:8.9.

Dead biomass on the tundra also accumulates below rather than above ground. The entire monocot portion of the vegetation grows and dies each season. Most of the herbaceous and woody plants accumulate little above ground and rapidly turn over energy in the living portion of the system. The belowground portion is more persistent, with roots lasting for two to ten years, reflecting low mortality and slow decomposition. Because decomposition is slowest on wet

**Table 13.4** Net Annual Production and Efficiency for Various Tundra Plant Communities and Components (Growing Season of 50 to 60 Days)

| Component | Net Production (kJ/m$^2$) | Total Radiation (kJ/m$^2$) | Efficiency (%) Total Radiation | PAR* |
|---|---|---|---|---|
| Polar desert (plateau) | 138 | $11.72 \times 10^5$ | 0.01 | 0.03 |
| Polar semidesert (raised beach) | 711 | $11.13 \times 10^5$ | 0.06 | 0.16 |
| Sedge-moss meadows (all meadows) | 2845 | $9.04 \times 10^5$ | 0.31 | 0.79 |
| Hummocky sedge-moss meadow | 3506 | $9.04 \times 10^5$ | 0.39 | 1.03 |
| Total lowland | 1916 | $9.62 \times 10^5$ | 0.20 | 0.50 |

*PAR (photosynthetically active radiation) = 40 percent of total radiation; efficiency was calculated on the basis of radiation received during the growing season for each component part and the contribution (%) of that component to the total.
*Source:* Bliss 1975.

sites, the difference between aboveground and belowground biomass is greatest there. Decomposition is faster on well-drained and on nutrient-rich sites. The greatest accumulation of belowground biomass paradoxically is found in those sites where net productivity is the lowest.

**Decomposers** Most of the production of the tundra enters the detrital food chain (Figure 13.38). At Devon Island, for example, grazing by lemmings accounts for only about 3 to 4 percent of the aboveground standing crop, except during cyclic highs. On the hummocky sedge-moss meadow 2 percent of the plant production is used by herbivores; 98 percent of the primary production is channeled to microbivores and saprovores. The latter are a wide variety of soil organisms, dominated by protozoans. Among important soil invertebrates are pot worms (Enchytraeidae), nematodes, fly larvae, and various crustaceans (McLean 1980). Annual production and consumption among these organisms are greater than in the herbivore system, even in a lemming high (Bunnel et al. 1975).

However, the major detrital consumers are bacteria and fungi (Bunnell et al. 1980; Flanagan and Bunnell 1980). Tundra soils contain a diversity of soil bacteria. They occur in about the same abundance as in temperate soils. Fungi mycelia in tundra soils are as abundant or more so than in temperate mull and mor soils. As in other terrestrial systems, fungi appear to be more important than bacteria. Like primary producers, decomposers are restricted in their activities by the cold (Bliss et al. 1973, Bliss 1975, Bunnell et al. 1975, Rosswall 1975).

**Consumers** Major herbivores of the tundra include waterfowl, ptarmigan, lemmings, hares, muskox, and caribou (reindeer). In some parts of the arctic tundra lemmings are the dominant herbivores. During cyclic highs, these rodents may eat over 25 percent of the aboveground primary production. Of the food they eat lemmings return about 70 percent as feces, which accumulate in winter and release quantities of nutrients during the spring thaw (Bunnell et al. 1975). During cyclic highs lemmings can hold primary production to 3 to 48g/m². By reducing the litter layer, these animals can also reduce the insulation of the soil and increase the depth of thawing. In doing so, lemmings can influence the composition and nature of the tundra plant community.

In parts of the arctic tundra lemmings are not significant grazing herbivores. Their place is taken by muskox and caribou. These two ungulates have an average standing crop of 0.17 kg/km² in the Canadian and Alaskan tundra, low compared to the ungulate biomass of 140 kg/km² of the African savanna. Muskox remove less than 1 percent of potential primary production. However, muskox are selective grazers. On restricted areas grazed on Devon Island, the animals removed 80 to 85 percent of the herbage available. Decomposition of their dung is slow, requiring 5 to 10 years (Bliss 1975). Preying on the herbivores are a number of carnivores. Arctic foxes consume 23 to 44 kcal/kg/body weight, mostly lemmings, in winter. An efficient assimilator, the fox rarely passes as feces more than 5 percent of the total energy ingested. Extremely effective as predators, weasels may eat up to 20 percent of the lemming population. The snowy owl, another lemming predator, has a winter food requirement of

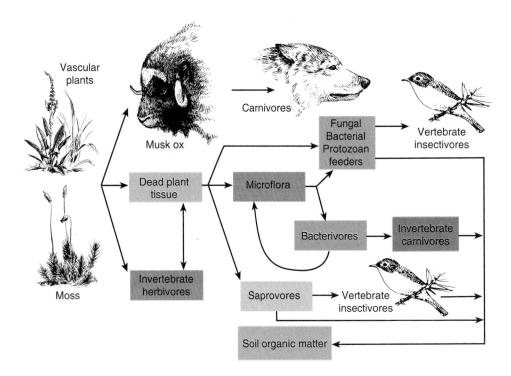

**Figure 13.38** Energy flow diagram for a sedge-moss meadow system. (From Bliss 1975.)

four to seven lemmings a day. Not only can the predators act as a force to drive the lemming population to a low, but the predators themselves are affected by a scarcity of lemmings. When lemming populations are low, arctic fox experience reproductive failures and snowy owls cannot exist on the tundra. When forced southward, the owls face an uncertain fate.

Nutrients   Arctic and alpine tundras are low in nutrients because of a short growing season, cold temperatures, low precipitation, and restricted decomposition. Of all terrestrial ecosystems, the arctic tundra has the smallest proportion of its nutrient capital in live biomass. Dead organic material functions as a nutrient pool. Most of this nutrient capital is not directly available to plants. Pools of soluble soil nutrients, especially nitrogen and phosphorus, are small relative to exchangeable pools, which in turn are small relative to nonexchangeable pools (Figure 13.39) (Bunnell et al. 1975) As a result nutrient cycling in the tundra is conservative. Vascular plants retain and reincorporate nutrients, especially nitrogen, phosphorus, potassium, and calcium, in their tissues.

Leaching of nutrients is minimal, occurring mostly in the spring at the beginning of the growing season. As the snow melts, runoff increases, carrying with it animal debris that accumulates over winter. Spring rains leach senescent plant material produced the previous growing season. Rising temperatures stimulate decomposition. Because 60 percent of the active roots are in the upper 5 cm of the soil, the root mass once thawed takes up most of the nutrients. However, early in the season, plants have to compete with microbes, whose uptake of nutrients can exceed those of plant roots (Chapin et al. 1980). In summer vascular plants leak nutrients, especially phosphorus and potassium, from the cuticle of the leaves to the surface, where they are washed off by summer rains. These nutrients are often taken up by bryophytes, which absorb or adsorb nutrients before they reach the ground. Nutrients incorporated into mosses are slowly released. Thus byrophytes function as a temporary nutrient sink.

In some tundra ecosystems lemmings become involved in nutrient cycling. Foraging in meadows and ridges, but building their nests and defecating in troughs, lemmings transport nutrients from one place to another. Higher levels of soil nutrients, especially phosphorus, in polygon troughs, and higher rates of decomposition in these microsites may reflect lemming activity (Bunnell et al. 1975).

Two nutrients, nitrogen and phosphorus, are most limiting. Two major sources are precipitation and biological fixa-

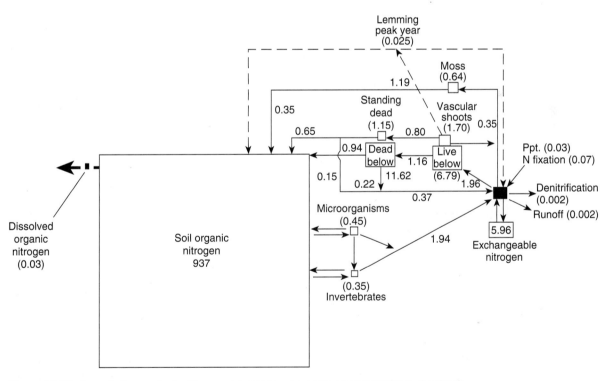

**Figure 13.39** Annual nitrogen budget in a moist meadow on arctic coastal tundra to the depth of 20 cm. The area of each box is proportional to its N in g/m². Values next to the arrows indicate annual fluxes in g/m²/yr. Solid lines represent years of average lemming populations. Years of high lemming populations are indicated by dashed lines. Amount of leaching from vascular plants is unknown; and mosses are assumed to get all of their nutrients from the soil (which they probably do not). Note the high proportion of N in the soil organic matter. (After Chapin et al. 1980:471.)

tion. Nitrogen fixation is accomplished by both anaerobic and aerobic free-living bacteria; by cyanobacteria in soil, in water, and especially in mosses where they live epiphytically; and by lichens, particularly in the Fennoscandian tundra (Granhall and Lid-Torsvik 1975, Kallio and Kallio 1975). The input from precipitation is usually lower than the input from biological fixation, but at some places the two are equal.

Nitrogen accumulates in tissues in greater quantities than other nutrients, making much of it unavailable for recycling. At Barrow research station in Alaska about 65 percent of the gross input, 59 mg/m$^2$, is stored in the system as living and dead organic matter (Bunnell et al.1975). Circulation in the system is restricted by decomposers, whereas flux rates are more or less controlled by nitrogen-fixing microorganisms. Small quantities of nitrogen are lost through leaching.

Phosphorus can be very limiting in the tundra ecosystem. Phosphorus apparently controls the rate of production of new leaves by controlling the rate at which nutrients are removed from older leaves and translocated to new ones (Bunnell et al. 1975). Although the accumulations of nitrogen, potassium, and calcium are similar from site to site across the tundra, the accumulation of phosphorus is highest at the most productive sites.

Because available pools of both nitrogen and phosphorus are small, a constant turnover between exchangable and soluble pools is necessary to replenish the quantity absorbed by plants. Bunnell et al. (1975) estimate that at the Barrow site soluble and exchangable nitrogen must turn over 11 times during a growing season to meet plant needs, and phosphorus must be replenished 200 times during the growing season or 3 times during the day. Plants depend on nutrients released by decomposition, the uptake of which is aided by mycorrhizae.

## SUMMARY

Natural grasslands occupy regions where rainfall is between 25 cm and 70 cm a year. Many persist through the intervention of fire and human activity. Once covering extensive areas of the globe, native grasslands have shrunk to a fraction of their original size because of conversion to cropland and grazing lands. Disappearing along with the native grasslands were the native grazing herbivores, replaced in part by domestic livestock. Conversion of forests into agricultural lands, the planting of hay and pasture fields, and development of successional grasslands on disturbed sites have extended the range of some grassland animals into once forested regions. Successional and climax grasslands are dominated by sod-formers, bunchgrasses, or both. Depending upon their fire history and degree of grazing, grasslands accumulate a layer of mulch that retains moisture, influences the character and composition of plant life, and provides shelter and nesting sites for some animals.

Productivity varies considerably, influenced by precipitation. It ranges from 82 g/m$^2$/yr in semiarid grasslands to 40 times that much in subhumid, tame, and cultivated grasslands. The bulk of primary production goes underground to the roots. To a point, grazing stimulates primary production. Although the most conspicuous grazers are the large herbivores, the major consumers are invertebrates. The heaviest consumption takes place below ground, where the dominant herbivores are nematodes. Most of the primary production goes to decomposers. Nutrients are recycled rapidly. A significant quantity goes to the roots, to be moved above ground to next year's growth.

Savannas are grasslands with woody vegetation. They are characteristic of regions with alternating wet and dry seasons. Difficult to characterize precisely, savannas range from grass with an occasional tree to shrub and tree savannas. The latter grade into woodland and thornbush with an understory of grass. Much of the nutrient pool is tied up in plant and animal biomass, but nutrient turnover is high with little accumulation of organic matter.

Shrublands, which go by different names in various parts of the world, dominate regions with a mediterranean-type climate in which winters are mild and wet and summers are long, hot, and dry. Successional shrublands occupy land in transition from grassland to forest. Such shrubland may remain stable for years.

Shrublands characteristically have a densely branched woody structure and low height. The success of shrubs depends upon their ability to compete for nutrients, energy, and space. In semiarid situations, shrubs have numerous competitive advantages, including structural modifications that affect light interception, heat losses, and evaporative losses. Growth in mediterranean-type shrublands is concentrated at the end of the wet season, when nutrients in solution and a relative abundance of moisture produce a flush of vegetation. Nutrient cycling, especially of nitrogen and phosphorus, is tight. Many plants translocate nutrients from leaves to stem and roots before leaf fall; others concentrate nitrogen in litterfall, which the plants take up again quickly in the wet season.

Deserts occupy about one-quarter of Earth's land surface and are largely confined to two worldwide belts, around the Tropic of Cancer and the Tropic of Capricorn. Deserts result largely from the movement of air masses, rain-blocking mountain ranges, and remoteness from sources of oceanic moisture. Two types of deserts exist: cool deserts, exemplified by the Great Basin of North America, and hot deserts.

The desert is a harsh environment in which plants and animals have evolved to become drought-evaders or drought-resisters. Functionally deserts are characterized by low net production, by opportunistic feeding patterns for herbivores and carnivores, and by a detrital food chain that is less important than in other ecosystems. Crustlike growths of blue-green algae on the desert floor fix quantities of nitrogen, but most of it is lost to the atmosphere.

The alpine tundra of the high mountain ranges in lower latitudes and the arctic tundra that extends beyond the tree

line of the far north are at once similar and dissimilar. Both have low temperatures, low precipitation, and a short growing season. Both possess a frost-molded landscape and plant species whose growth rates are slow. The arctic tundra has a permafrost layer; rarely does the alpine tundra. Arctic plants require longer periods of daylight than alpine plants and reproduce vegetatively, whereas alpine plants propagate themselves by seed. Over much of the Arctic, the dominant vegetation is cotton grass, sedge, and dwarf heaths. In the alpine tundra cushion and mat-forming plants, able to withstand buffeting by the wind, dominate exposed sites, while cotton grass and other tundra plants are confined to protected sites. Net primary production is low because of a short growing season, although daily primary production rates are comparable to those of temperate grasslands. Biomass accumulates below rather than above ground. In spite of an assemblage of grazing ungulates and rodents, most production goes to decomposers. Major detrital consumers are bacteria and fungi. Nutrient levels are low, and nutrients tend to accumulate and become stored in living and dead plant material unavailable for recycling. Circulation is restricted by limited activity of decomposers, while the flux rates are controlled by nitrogen-fixing organisms. Because pools of both nitrogen and phosphorus are small, constant turnover between exchangeable and soluble pools is necessary to replenish the quantity absorbed and retained by plants.

## REVIEW QUESTIONS

1. Why is the root system so important in the grassland ecosystem?
2. What is the role of mulch in grassland ecosystems?
3. How have grasses adapted to grazing? (Refer also to Chapter 24.)
4. Contrast the production and function of the above-

ground and belowground components of grassland ecosystems. Why does so much energy flow and nutrient cycling take place belowground?
5. What distinguishes savannas from grasslands both in structure and function? Under what climatic conditions have they developed?
6. Describe mediterranean-type shrublands.
7. Contrast the benefits of evergreen and deciduous leaves in the mediterranean-type shrublands.
8. What climatic forces lead to deserts? What human influences?
9. In what ways are plants and animals of the desert adapted to aridity? (See also Chapter 5.)
10. Compare and contrast nitrogen cycling in grassland, savanna, mediterranean-type shrubland, desert, and tundra ecosystems.
11. Argue that successional shrublands are not wastelands.
12. Compare and contrast the arctic tundra with the alpine tundra with respect to environmental conditions and physiological responses of tundra life.
13. What are the major differences between the alpine tundras of the temperate and tropical regions?
14. Where does most of the net primary production of the tundra accumulate and why?

## CROSS-REFERENCES

Climate, 36–48; microclimate, 48–53; climate and vegetation, 54–60; water use efficiency, 70–71; plant adaptations to aridity, 71–73; animal adaptations to aridity, 75–76; adaptations to cold, 82, 84–85, 88–89, 92–95; nutrient cycling, 116–120; C$_4$ plants, 154–155; decomposition, 159–165; herbivores, 181, 183–184; nitrogen cycle, 206–210; population cycles, 403–406, 543–545; plant defenses, 524–529; grazing disturbance, 645–646; succession, 656–677.

# Forests

## Concepts

1. Coniferous, deciduous, and tropical forests consist of many types determined by broad environmental conditions.
2. Vertical stratification in forests influences environmental conditions and diversity of life.
3. Nutrient cycling in coniferous forests is characterized by short-term cycling apparently dependent upon mycorrhizal fungi on the roots, and accumulation of nutrients in biomass, particularly the foliage and bole.
4. Nutrient cycling in deciduous forests is marked by the internal cycling of nutrients between leaves and roots, the release of nutrients from the litter to the soil by microbial decomposition, and uptake by roots.
5. Nutrient cycling in tropical forests involves rapid decomposition of litterfall and rapid uptake by roots together with a retranslocation of phosphorus and other elements.

Of all the vegetation types of the world, none is more widespread or more diverse than the forest (see inside cover). A map of world vegetation shows forest growth in distinct bands around the Northern Hemisphere. Progressing southward from the tundra are consecutive belts of coniferous, temperate deciduous, and near the equator, tropical forests. The latter are most extensive in the Southern Hemisphere. Within each band of global forest is a diversity of forest types.

Most forests with which we are familiar are relatively young second, third, or even fourth-growth (secondary) forests. They differ from those 150 to 450+ years old, known as **old-growth** (formerly called virgin). Old-growth forests, nearly gone and the subject of a bitter controversy in the Pacific Northwest, consist of trees of various sizes and ages. Old-growth stands are dominated by long-lived individuals and contain significant amounts of woody debris and downed logs. They possess more diversity in structure and function than secondary forests, and are characterized by highly evolved complex relationships between plants and animals. These relationships are destroyed when they are cut.

Regardless of type, all forests possess large aboveground biomass. This biomass creates several layers of vegetation that influence environmental conditions, including light, moisture, temperature, wind, and carbon dioxide.

# CONIFEROUS FORESTS

## Types

Boreal Forest or Taiga    The **taiga** (Russian for "coniferous forest") forms a circumpolar belt of coniferous forest throughout the Northern Hemisphere. Its northern limit is roughly along the July 13° C isotherm, the southern extent of the Arctic front in summer. Its southern limit, much less abrupt, is more or less marked by the winter position of the Arctic front, the July 18° C isotherm, just north of the 58° N latitude. In Eurasia it begins in Scandinavia and extends across the continent to northern Japan. In Europe the forest is dominated by Norway spruce (*Picea abies*); in Siberia by Siberian spruce (*P. obovata*), Siberian stone pine (*Pinus sibirica*), and larch (*Larix sibirica*); and in the Far East by Yeddo spruce (*P. jaezoensis*). In North America the taiga extends from Labrador across Canada and through Alaska to the Brooks Range. It is dominated by four genera of conifers, *Picea, Abies, Pinus,* and *Larix,* and two genera of deciduous trees, *Populus* and *Betula.* Dominant trees include black spruce (*P. mariana*) and jack pine (*Pinus banksiana*). Although the taiga throughout may have the same general appearance, its vegetation exhibits important regional differences.

The boreal forest has four major vegetation zones: (1) the forest–tundra ecotone, characterized by open stands of stunted spruce, lichens, and moss; (2) the open boreal woodland, characterized by lichen–black spruce woodland (Figure 14.1a); (3) the main boreal forest with continuous stands of coniferous trees and a moss and low shrub understory broken up by poplar and birch on disturbed areas (Figure 14.1b); and (4) the boreal–mixed forest ecotone where the boreal forest grades into the mixed hardwood-conifer forest of southern Canada and the northern United States (Oechel and Lawrence 1985).

Throughout much of its extent the boreal forest experiences great seasonal fluctuations of temperature. These

**Figure 14.1** The boreal forest extends from the taiga (a) dominated by black spruce at its northern limits to the spruce-birch forest (b) at its southern limits.

(a)

(b)

freeze-thaw cycles affect shallow-rooted trees, nutrient-poor podzolic soils, and depending upon the region, the presence or absence of permafrost. Occupying, for the most part, glaciated land, the taiga is also a region of cold lakes, bogs, rivers, and alder thickets.

**Temperate Rain Forests** South of Alaska along the Pacific Coast the coniferous forest differs from the northern boreal forest, both floristically and ecologically. The reasons for the change are both climatic and topographic. Moisture-laden winds move in from the Pacific, meet the barrier of the Coast Range, and rise abruptly. Suddenly cooled by this upward thrust into the atmosphere, the moisture in the air is released as rain and snow in amounts that can exceed 600 cm/yr. During the summer, when winds shift to the northwest, the air is cooled over chilly northern seas. Although rainfall is low, cool air brings in heavy fog, which collects on the forest foliage and drips to the ground to add 127–130 cm or more of moisture. This land of superabundant moisture, high humidity, and warm temperatures supports the temperate rain forest. Its luxuriant vegetation is dominated by a variety of conifers well adapted to wet, mild winters, dry warm summers, and nutrient-poor soils (see Franklin and Dyrness 1973, Waring and Franklin 1979, Franklin and Waring 1979, Lassoi, Hinckley, and Grier 1985). These forests (Figure 14.2) are dominated by western hemlock (*Tsuga heterophylla*), mountain hemlock (*T. mertensiana*), Pacific silver fir (*Abies amabilis*), and Douglas-fir, all trees with both high foliage and stem biomass: Further south, where precipitation remains high, grows the redwood (*Sequoia sempervirens*) forest, occupying a strip of land about 724 km long.

**Montane Coniferous Forests** The air masses that drop their moisture on the western slopes of the Coast Range descend the eastern slopes, heat, and absorb moisture, creating the conditions that produce the Great Basin desert (see Chapter 13). The same air rises up the western slopes of the Rockies, cools, and drops moisture again, although far less than on the Coast Range. Here in the Rocky, Wasatch, Sierra Nevada, and Cascade Mountains several coniferous forest associations develop (see Franklin and Dyrness 1973, Pfister et al. 1977). In the southwestern United States these coniferous forests occur between 2500 and 4200 m elevation, and in the northern United States and Canada between 1700 and 3500 m elevation. At high elevations in the Rocky Mountains, where winters are long and snowfall is heavy, grows the subalpine forest, dominated by Engelmann spruce (*Picea engelmannii*) and subalpine fir (*Abies lasiocarpa*) (Figure 14.3a). Middle elevations have stands of Douglas-fir and lower elevations are dominated by open stands of ponderosa pine (*Pinus ponderosa*) and open to thick stands of the early successional pioneering conifer, lodgepole pine (*P. contorta*) (Figure 14.3b).

Similar forests grow in the Sierras and Cascades. There high elevation forests consist largely of mountain hemlock,

**Figure 14.2** The temperate rain forest embraces the redwood and sequoia forests of California and the Douglas-fir–western hemlock–Sitka spruce forests of the northwest Pacific Coast. Typical is this redwood stand in northern California with typical undergrowth of sorrel.

red fir (*Abies magnifica*) and lodgepole pine. Sugar pine (*Pinus lambertiana*), incense cedar (*Libocedrus decurrens*), and the largest tree of all, the giant sequoia (*Sequoiadendron giganteum*), which grows only in scattered groves on the west slopes of the California Sierras.

A deciduous successional, occasionally permanent species common both to the montane and the boreal forest is trembling aspen (*Populus tremuloides*), the most widespread tree of North America (Figure 14.4).

Montane coniferous forests dominated by pine also occur in Mexico and Guatemala. In Europe forests of Scots pine (*Pinus sylvestris*) (Figure 14.5a) and Norway spruce (*Picea abies*) (Figure 14.5b) extend across the mountain ranges to Spanish Sierra Nevada. Both species are planted as plantation trees and as ornamentals in the United States.

**Woodlands** In parts of western North America where the climate is too dry for montane coniferous forests, we find piñon-juniper woodlands in which *Pinus* and *Juniperus* are

(a)

**Figure 14.3** (a) Subalpine montain forest in the Rocky Mountains is dominated by subalpine fir (*Abies lasiocarpa*). (b) The drier lower slopes support ponderosa pine.

(b)

always associated (Figure 14.6). These ecosystems are characterized by open growth small trees with a well-developed understory of grass and shrubs. The tree crowns rarely touch and the root systems extend two to three times the diameters of the crowns. Piñon-juniper woodlands are found from the front range of the Rocky Mountains to the eastern slopes of the Sierra Nevada foothills. These woodlands have been greatly changed by livestock grazing, fuel harvesting, and exclusion of fire. Such disturbances increase the density of trees and reduce the cover of grass (West 1988).

Southern Pine Forests  The pine forests of the coastal plains of the South Atlantic and Gulf States (Figure 14.7) are usually considered part of the temperate deciduous forest because they represent a successional rather than final stage. These pines maintain their presence by a competitive advantage over hardwoods on nutrient-poor, dry sandy soil and by adaptation to fire. At the northern end of the coastal pine forest in New Jersey, pitch pine (*Pinus rigida*) is the dominant species. There a scrub ecotype of pitch pine and a dwarf form of blackjack oak (*Quercus marilandica*), rarely exceeding 3 meters in height, form extensive stands known as the pine barrens. Further south, loblolly (*P. taeda*), longleaf (*P. autralis*), slash (*P. elliotii*), and sand (*P. clausa*) pine are most abundant. Where fires are allowed to burn or controlled burning is practiced, the understory is open and dominated by wiregrass (*Aristida*). In the deep south saw palmetto

**Figure 14.4** Quaking aspen (*Populus tremuloides*) is the dominant deciduous tree in the western montane forest.

(a)

(b)

**Figure 14.5** (a) A Scots pine forest in Scandinavia. (b) A Norway spruce forest in the Carpathian mountains.

**Figure 14.6** A piñon-juniper woodland in Utah.

**Figure 14.7** Homogenous stands of pine cover much of the lowland coastal plain and Piedmont of the southern United States. Many of these stands are dominated by loblolly pine.

(*Serenoa repens*) and scrub palmetto (*Sabal etonia*) are common in the understory.

## Structure

Coniferous forests can be divided into three broad classes according to growth form and general growth behavior: (1) pines with straight, cylindrical trunks, whorled spreading branches, and a crown density that varies from the dense crowns of red and white pine to the open thin crowns of Virginia, jack, Scots, and lodgepole pine; (2) spire-shaped evergreens, including spruce, fir, Douglas-fir, and (with some exceptions) the cedars, with more or less tall pyramidal crowns, gradually tapering trunks, and whorled, horizontal branches; and (3) deciduous conifers such as larch and baldcypress with pyramidal, open crowns that shed their needles annually. Growth form and behavior influence animal life and other aspects of various coniferous ecosystems.

Vertical stratification in coniferous forests is not well-developed. Because of a high crown density and deep shade, the lower strata are poorly developed in spruce and fir

forests. The ground layer consists largely of ferns and mosses with few herbs. The maximum canopy development in spire-shaped conifers is about one-third down from the open crown. This feature gives such forests a profile different from that of pines. Pine forests with a well-developed high canopy lack lower strata. However, old stand, open-crowned pines may have three strata, an upper canopy, a shrub layer, and a thin herbaceous layer. The litter layer in coniferous forests is usually deep, poorly decomposed, and on top of instead of mixed in with the mineral soil (mor; see Chapter 9). It inhibits the germination and development of woody understory and herbaceous plants.

Animal life in the coniferous forest varies widely, depending upon the nature of the stand. Soil invertebrate litter fauna is dominated by mites. Earthworm species are few and their numbers low. Insect populations, although not diverse, are high in numbers and, encouraged by the homogeneity of the stands, often destructive. Spruce budworm (*Choristonema fumiferana*), found throughout the boreal forest, attacks balsam fir and spruces. Related species attack jack pine and red pine. Sawflies (*Neodiprion*) attack a wide variety of pines including pitch, Virginia, shortleaf, and loblolly.

A number of bird species are closely associated with coniferous forests. In North America they include such species as chickadees (Paridae), kinglets (Sylviidae), pine siskins (*Carduelis pinus*), crossbills (*Loxia* spp.), purple finch (*Carpodacus purpureus*), and hermit thrush (*Hylocichla guttata*) (Figure 14.8). Related species, the tits and grosbeaks, are common to European coniferous forests.

Species diversity varies. In general, coniferous forests of northeastern and southeastern North America and the Sierra Nevada mountains support the richest avifaunas (Wiens 1975). Bird densities are highest in the Pacific Northwest and lowest in immature northeastern coniferous forests. In the latter more than 50 percent of the individuals are migratory neotropical warblers, whereas in western coniferous forests less than 10 percent are warblers. Foliage-gleaning insectivorous birds are dominant in all types of coniferous forests.

Except for strictly boreal species, such as the pine marten (*Martes americana*) and lynx (*Felis lynx*), mammals have much less affinity for coniferous forests. Most are associated with both coniferous and deciduous forest. The white-tailed deer, moose, black bear, and mountain lion (*Felis concolor*) are examples. Their north-south distribution seems to be limited more by climate, especially temperature, than by vegetation. The red squirrel (*Tamiasciurus hudsonicus*), associated with coniferous forests, is quite common in deciduous woodlands in the southern part of its range.

## Function

Function, particularly nutrient budgets and nutrient cycling, has been studied more extensively in forests than in any other ecosystems (see Reichle 1981). One type studied in

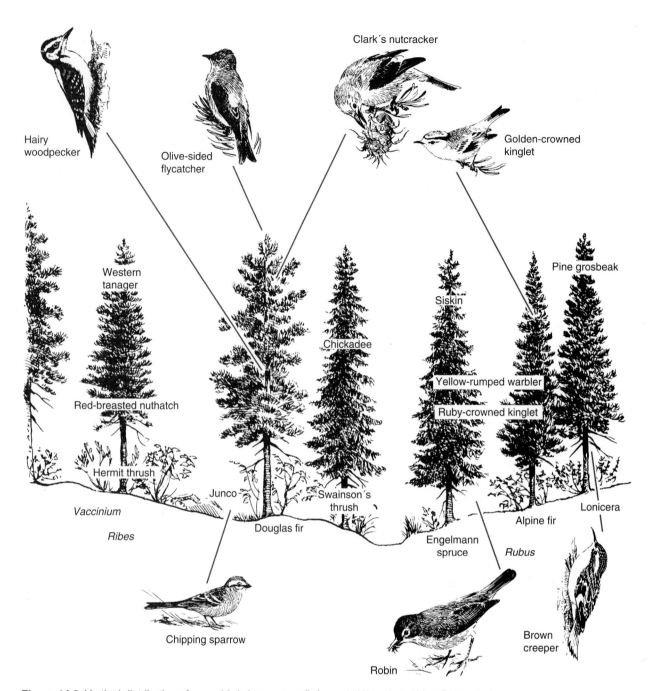

**Figure 14.8** Vertical distribution of some birds in a spruce-fir forest in Wyoming. (After Salt 1967.)

detail is the Douglas-fir coniferous forest, including young and old-growth stands, of the Pacific Northwest (Cole et al. 1969, Cole and Rapp 1981, Johnson et al. 1982). Contrasts between a 36-year-old and a 450-year-old Douglas-fir stand are summarized in Table 14.1 and Table 14.2. The distribution of organic matter and nitrogen differs between the two stands. The old stand has a considerably larger total biomass, both living and dead, than the younger stand. Although the two have similar foliar biomass, the young stand has 4 percent of its living biomass in foliage, higher than the old-growth stand with 2 percent. Most of the living

biomass in both age classes is in branches and bole. This biomass amounts to about 80 percent in the young stand and 86 percent in the 450-year-old stand. Sixteen percent of the living biomass of the young stand is in roots, compared to 12 percent of the old stand. Litter accounts for 18 percent of the detrital organic matter and the soil 82 percent. In sharp contrast, litter in the old stand accounts for 55 percent of detrital organic matter and the soil 45 percent. Litter organic matter exceeds soil organic matter because of the accumulation and slow decomposition of needles and the long-term decomposition of large fallen trunks and limbs.

**Table 14.1** Organic Matter Distribution, kg/ha (% in parentheses) for Young and Old-Growth Douglas-Fir and Young Loblolly Pine Ecosystems

| Component | Douglas-Fir 36-Yr-Old | Douglas-Fir 450-Yr-Old | Loblolly Pine 16-Yr-Old |
|---|---|---|---|
| Overstory | | | |
| Foliage | 9,097 (4) | 8,906 (2) | 9,700 (4) |
| Branches | 22,031 (11) | 48,543 (8) | 27,900 (12) |
| Stemwood | 121,687 (60) | 472,593 (78) | 132,180 (57) |
| Bark | 18,728 (9) | * | 18,460 (8) |
| Roots | 32,986 (16) | 74,328 (12) | 44,550 (19) |
| Total | 204,529 | 604,370 | 232,490 |
| Subordinate vegetation | 1,010 | 9,864 | |
| Forest floor | | | |
| Wood | 6,345 (1) | 55,200 (1) | |
| Litter & humus | 16,427 (17) | 43,350 (54) | |
| Total | 22,772 | 98,550 | |
| Soil | 111,552 (82) | 79,250 (45) | |
| Total ecosystem | 339,863 | 792,034 | |

*Source:* Data for Douglas-fir from Johnson et al. 1982; for loblolly pine from Jorgensen and Wells 1986.
*Value for bark included in stemwood.

The nitrogen budget emphasizes the importance of internal cycling and storage. The young stand has most of its N in foliage and bole, whereas the old stand has most of its N in bole and root (Table 14.2). The 450-year-old stand accumulates considerably more nitrogen, but has a lower percentage of it in the soil. The young stand has 98 percent of its detrital nitrogen accumulated in the soil. Input of nitrogen into both systems is low and approximately the same, about 2 kg/ha /yr (Table 14.3). The old stand returns somewhat more nitrogen to the forest floor than the young stand, mostly because of greater stemflow and throughfall. The young stand has a considerable greater N requirement and a larger uptake than the old stand, 25.1 versus 14.8 kg/ha/yr. This uptake amounts to about 55 percent of the young stand's requirement and 44 percent of that for the old stand. In both stands the deficiency of uptake is compensated by recycling elements within the biomass. This internal cycling is also typical of other nutrients except for calcium and magnesium. Uptake of those nutrients far exceeds requirement, so cycling within the tree is not necessary.

**Table 14.2** Nitrogen Distribution (% in parentheses), kg/ha, for Young and Old-Growth Douglas-Fir and Young Loblolly Pine Ecosystems

| Component | Douglas-Fir 42-Yr-Old | Douglas-Fir 450-Yr-Old | Loblolly Pine 16-Yr-Old |
|---|---|---|---|
| Overstory | | | |
| Foliage | 102 (32) | 75 (14) | 81 (26) |
| Branches | 61 (19) | 49 (9) | 60 (19) |
| Stemwood | 77 (24) | 189 (36) | 78 (24) |
| Bark | 48 (15) | 162 (30) | 36 (11) |
| Roots | 32 (10) | 58 (11) | 64 (20) |
| Total | 320 | 533 | 319 |
| Subordinate vegetation | 6 | | |
| Forest floor | | | |
| Wood | 14 | 132 | |
| Litter & humus | 161 | 434 | |
| Total | 175 | 566 | 307 |
| Soil | 2809 | 4300 | 1750 |
| Total ecosystem | 3310 | 5399 | 2376 |

*Source:* Data for Douglas-fir from Johnson et al. 1982; for loblolly pine from Jorgensen and Wells 1986.

**Table 14.3** Nitrogen Transfers (kg/ha/yr) in Young and Old-Growth Douglas-Fir and Young Loblolly Pine Ecosystems

| Component | Douglas-Fir 42-Yr-Old | Douglas-Fir 450-Yr-Old | Loblolly Pine 16-Yr-Old |
|---|---|---|---|
| Input | 1.67 | 2.0 | 5.4 |
| Return to forest floor | | | |
|   Throughfall | 0.53 | 3.4 | 4.1 |
|   Litterfall | 25.4 | 25.6 | 33.4 |
| Total | 25.93 | 29.0 | 37.5 |
| Within vegetation | | | |
|   Requirement | 45.8 | 33.3 | 66.0 |
|   Redistribution | 20.7 | 18.5 | 11.6 |
|   Uptake | 25.1 | 14.8 | 54.4 |

*Source:* Data for Douglas-fir from Johnson et al. 1982; for loblolly pine from Jorgensen and Wells 1986.

Contrasting with the Douglas-fir stands of the Pacific Northwest is loblolly pine of the coastal plains of the southeastern United States. Loblolly pine grows on a wide range of soils, from poorly drained lowland sites, where it does best, to dry, poor upland sites. This pine grows rapidly, achieving a diameter of 20 cm and a height of 18 m by age 20. In this short period of time loblolly pine accumulates more organic matter than and as much nitrogen as 36-year Douglas-fir. The biomass distribution in young Douglas-fir and loblolly pine is approximately the same among all components. However, loblolly pine accumulates twice as much nitrogen in its roots as does the Douglas-fir, whereas Douglas-fir sequesters more nitrogen in its foliage. Of 2376 kg of N per ha in the loblolly pine stand, 13 percent is tied up in plant biomass and an equal amount in the forest floor. A 36-year old Douglas-fir stand ties up 11 percent of the N in plant biomass and old-stand Douglas-fir, 8 percent. Thus at an early age loblolly pine ties up a large portion of the site's nutrients (Table 14.3) (a trait that gives conifers the reputation of being accumulator plants), with maximum accumulation by age 30 (Figure 14.9).

Some coniferous ecosystems scavenge nutrients directly from rainfall. This scavenging is accomplished by microcommunities of algae and lichens occupying canopy leaves in balsam fir forests (Lang, Reiners, and Heier 1976) and western coniferous forests (D. W. Johnson et al. 1982, Carroll 1979). Old-growth Douglas-fir supports in its canopy a complex biological community, including primary producers, consumers, and decomposers (Figure 14.10). Cyanophycophyllous lichens fix atmospheric nitrogen. Organic nitrogen lost through leaching from lichens combines with canopy moisture to form a dilute organic solution that in turn is taken up by microorganisms and other canopy epiphytes. Part of this microbial production is consumed by canopy arthropods. These little nutrient cycles in the canopy tend to influence and even restrict the amount of nutrients that reach the forest floor by throughfall and stemflow.

Ground moss is an important but overlooked component of some coniferous forests. Binkley and Graham (1981) found that in an old-growth Douglas-fir stand, ground-layer mosses, which account for only 0.13 percent of the aboveground biomass, add 5 percent to the estimated aboveground

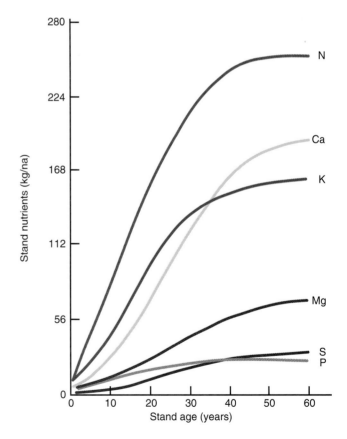

**Figure 14.9** Accumulation of macronutrients in whole trees in loblolly pine stands on fertile sites. Note the rapid accumulation of nitrogen and calcium. (Adapted from Jorgensen and Wells 1986:9.)

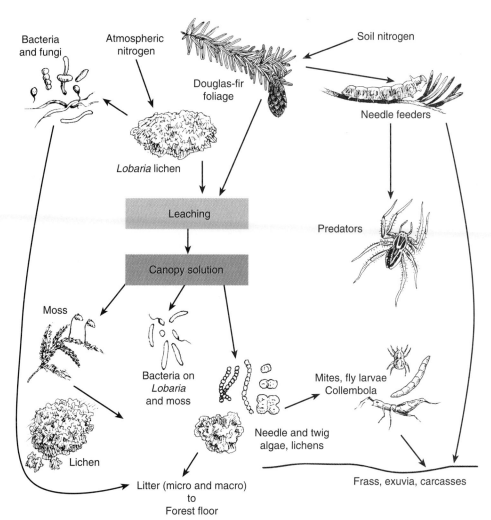

**Figure 14.10** Nitrogen cycle in the canopy of old-growth Douglas-fir. Such microecosystems probably exist in other forests, including tropical rain forest. Communities of primary producers (lichens) and biophage and saprophage consumers conserve and recycle nutrients such as nitrogen and influence nutrient return to the forest floor by leaching and throughfall. (After Johnson et al. 1982:193.)

primary production, 5 percent to the uptake of calcium and potassium, and 10 percent to the uptake of nitrogen and phosphorus.

Litter-decomposing fungi have a role in nutrient cycling in coniferous stands. Needles make up most of the litterfall, and they are infected first by fungi that remove most of the nitrogen and phosphorus, leaving mostly cellulose behind. It becomes incorporated in the fermentation zone of litter, to be acted upon by other fungi.

Stimulating fungal activity in the fermentation zone is the rain of pollen in spring. This pollen rain supplies, among other nutrients, N, K, P, and Ca. Although the level of nutrient input is inconsequential to tree growth, it can be the main nutrient source for fungi in the fermentation zone in summer. Nitrogen and phosphorus apparently stimulate the fungi to complete litter decomposition and release other elements to tree growth.

Fungal hyphae or rhizomorphs in turn concentrate nutrients, especially nitrogen, phosphorus, sodium, and several trace elements, in their tissues and fruiting bodies. Thus fungi act as a living sink of nutrients. Resistant to leaching,

the fungal rhizomorphs hold these biologically important elements in the litter (Stark 1973, 1974).

One component of forest ecosystems too often overlooked is dead wood. The mass of dead wood waxes and wanes with tree mortality and disturbances (Lang 1985) and never achieves equilibrium within a stand (Lang 1985, Long 1982). Woody litterfall usually increases over time (Long 1982) and becomes most conspicuous in old temperate forests, both deciduous and coniferous, where logs and large limbs decay slowly. Fallen trees and large limbs made up over 71 percent of the forest floor mass in a 250-year-old oak forest in New Jersey, (Lang and Forman 1975) and over 60 percent in old growth Douglas-fir stands (Grier and Logan 1977, Sollins et al. 1980).

Dead wood in the form of large standing dead trees or snags (Figure 14.11) or downed trunks and limbs make up a unique and critical component of the forest ecosystem (Maser and Trappe 1984, Maser et al. 1979). Standing dead trees provide essential nesting and den sites for cavity-nesting birds and mammals, and food and foraging areas. Downed dead trees, which may make up 10 to 20 percent of the ground surface in forests, provide food, protected path-

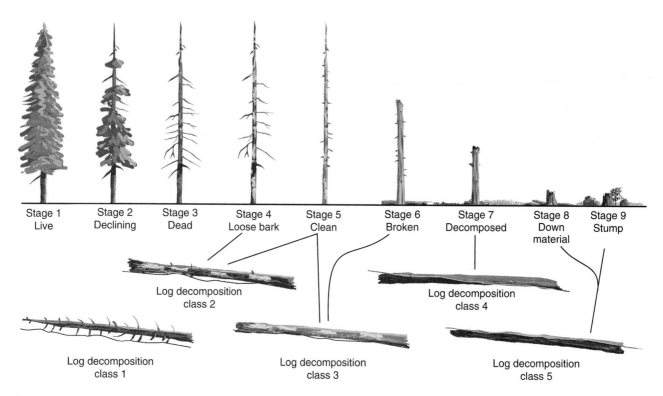

| Stage 1 Live | Stage 2 Declining | Stage 3 Dead | Stage 4 Loose bark | Stage 5 Clean | Stage 6 Broken | Stage 7 Decomposed | Stage 8 Down material | Stage 9 Stump |

Log decomposition class 2

Log decomposition class 4

Log decomposition class 1

Log decomposition class 3

Log decomposition class 5

**Figure 14.11** An important component of all forest ecosystems is dead wood, both standing and down. Snags or dead trees go through a process of decay, each stage of which supports its own group of animal life. When dead trees fall, they enter one of the first four decomposition classes. (After Thomas et al. 1979:64.)

ways for small mammals and reproductive sites for certain woody plants. The elimination of standing and fallen dead trees can greatly impoverish animal life in the forest.

# TEMPERATE BROADLEAF FORESTS

Temperate forests, in spite of their name, do not exist in a temperate environment. They face extreme fluctuations in daily and seasonal temperatures that stress the physiological activity of plants and animals. Deciduous forests are leafless during the winter and in northern regions remain leafless for most of the year. They are exposed to droughts and in places, flooding. In spite of their intemperate environment, temperate forest ecosystems are able to maintain high productivity.

Temperate broadleaf forests can be classified into temperate deciduous forests, temperate woodlands, and temperate evergreen forests. Like grasslands, the temperate broadleaf forests have been highly disturbed and modified by humans. Nearly all of the original forests have been cut for lumber or cleared agriculture, mining, and urban development. Many of the forests we see today are second and third growth

stands, or are new stands that have reclaimed abandoned agricultural lands through successional processes. Some of these forests are unmanaged and natural processes predominate. Others are managed, typically by thinning, and by the removal of undesirable species and poorly formed trees.

## Types

**Temperate Deciduous Forests** The temperate deciduous forest once covered large areas of Europe and China, parts of South America and the middle American highlands, and eastern North America. The deciduous forests of Europe and Asia have largely disappeared, cleared for agriculture and settlements. The dominant trees include European beech (*Fagus sylvatica*), pedunculate oak (*Quercus robor*), ashes (*Fraxinus* spp.), birches (*Betula* spp.), and elms (*Ulmus* spp.) (Figure 14.12). Because of glacial history, the diversity of tree species in European deciduous forests does not compare with that in North America or China.

In eastern North America the temperate deciduous forest consists of a number of forest types that intergrade. The northern segment of the deciduous forest complex is the hemlock-white pine-northern hardwoods forest, which occupies southern Canada and extends southward through the northern United States and along the high Appalachians into

**Figure 14.12** A European deciduous forest dominated by oak and ash.

North Carolina and Tennessee. Beech, sugar maple, black cherry, red oak, and white pine are the chief components. White pine was once an outstanding tree of the forest, but because most of it was cut before the turn of the century, it now grows only as a successional tree on abandoned land and as scattered trees through the forest.

On relatively flat glaciated country with its deep, rich soil grow two somewhat similar forests, the beech-sugar maple forest, found mostly from southern Indiana north to central Minnesota and east to central New York; and the sugar maple-basswood forest, found from Wisconsin to Minnesota south to northern Missouri.

Occupying much of eastern central North America is the extensive central hardwoods forest. The central hardwoods can be divided into three major types. (1) The **cove,** or mixed mesophytic, forest consists of a large number of species dominated by yellow-poplar (Figure 14.13a). This forest, which reaches its best development on the northern slopes and deep coves of the southern Appalachians, is one of the most magnificent in the world. Much of its original grandeur has been destroyed by fire and high-grading—removing the best trees and leaving the poorest trees and unwanted species. However, even in second- and third-growth stands, its richness is apparent. (2) On drier sites and south-facing slopes and drier mountains grows the oak-chestnut forest. The chestnut (*Castanea dentata*) was eliminated as a timber tree by the introduced chestnut blight. Its place has been taken over by the oaks, so that distinctive forest type is now largely oak forest. (3) Oak-hickory forests (Figure 14.13b) dominate the western edge of the central hardwoods in the Ozarks and the forests along the prairie river systems.

**Temperate Woodlands** In the southwestern United States, Mexico, and California open oak and oak and pine woodlands are transitional from coniferous forests to grassland and desert. Woodlands in southern Arizona, New Mexico, and northern Mexico are dominated by oaks, especially Emory oak (*Quercus emoryi*), junipers, and pine, or pure oak woodlands. In the Rocky Mountains, particularly Utah, occur oak-sagebrush woodlands. Forming a ring about the Great Valley of California is still another type—evergreen-oak woodlands, dominated by blue oak (*Quercus douglasii*) with a grassy undergrowth (Figure 14.14).

**Temperate Evergreen Forests** In several subtropical areas of the world are extensive mixed forests of both broadleaf evergreen and coniferous trees. Such forests include the eucalyptus in Australia (Figure 14.15a), anacardia gallery forests of South America and New Caledonia, and false beech (*Nothofagus* spp.) forests in Patagonia and Australia (Figure 14.15b). Temperate evergreen forests also occur in the Caribbean and on the North American continent along the Gulf Coast, in the hummocks of the Florida Everglades, and in the Florida Keys. Depending upon location, these forests are characterized by oaks, magnolias, gumbo-limbo (*Bursera simaruba*), and royal and cabbage palms.

(a)

(b)

**Figure 14.13** (a) A cove hardwood forest in Virginia. Redbud and dogwoods blooming in the understory add color to this forest dominated by yellow-poplar. (b) An oak-hickory forest in the Ozarks.

**Figure 14.14** A blue oak woodland in California.

(a)

(b)

**Figure 14.15** (a) An Australian mixed eucalyptus forest in the Grampian Range in western Victoria, Australia. (b) False beech (*Nothofagus*) forest in New Zealand.

## Structure

Highly developed, uneven-aged deciduous forests usually consist of four strata (Figures 14.16, 14.17). The upper canopy consists of dominant and codominant trees, below which is the lower tree canopy of saplings and understory trees, and the shrub layer. The ground layer consists of herbs, ferns, and mosses.

Even-aged stands, the results of fire, clear-cut logging, and other large-scale disturbances (see Chapter 30), often have poorly developed strata beneath the canopy because of dense shade. The low tree and shrub strata are thin, and the ground layer also is poorly developed, except in small gaps.

The diversity of animal life is associated with stratification and the growth forms of plants (see Chapter 28). Some animals, particularly forest arthropods, spend the major part of their life in a single stratum; others range over two or more strata. The greatest concentration and diversity of life in the forest occurs on and just below the ground layer. Many animals, particularly the soil and litter invertebrates, remain in the upper levels of the soil. Others, such as mice, shrews, ground squirrels, and forest salamanders, burrow into the soil or litter for shelter and food. Larger mammals, like deer, live on the ground layer and feed on herbs, shrubs, and low trees. Birds move rather freely among several strata, but favor one layer over another. Ruffed grouse (*Bonasa um-*

*bellus*) and ovenbird (*Seiurus aurocapillus*) occupy the ground layer but move into the upper strata to feed, roost, or advertise territory.

Other species occupy the upper strata—shrub, low tree, and canopy layers. The red-eyed vireo (*Vireo olivaceus*) inhabits the lower tree stratum and the wood pewee (*Contopus*

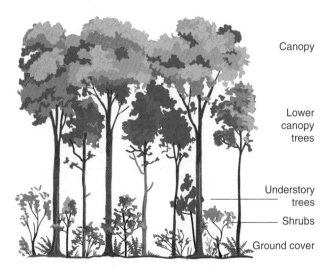

Canopy

Lower canopy trees

Understory trees

Shrubs

Ground cover

**Figure 14.16** Stratification in an eastern deciduous forest.

**Figure 14.17** Leafless, this deciduous forest in late fall reveals its vertical structure.

*virens*) the lower canopy. The black-throated green warbler (*Dendroica virens*) and scarlet tanager (*Piranga olivacea*) live in the upper canopy. Squirrels are mammalian inhabitants of the canopy, and woodpeckers, nuthatches, and creepers live in the open space of tree trunks between shrubs and the canopy.

The physical stratification of the forest influences its microclimates. The highest temperatures are in the upper canopy, because this stratum intercepts solar radiation. Temperatures tend to decrease through the lower strata. The most rapid decline takes place from the leaf litter down through the soil. The temperature profile changes through the 24-hour period. At night the temperatures are more or less uniform from the canopy to the floor.

Temperature stratification varies seasonally. In fall when the leaves drop and the canopy thins, temperatures fluctuate more widely at the various levels. Maximum temperatures decrease from the canopy downward, but rise again at the litter surface. The soil, no longer shaded by an overhead canopy, absorbs and radiates more heat than in summer. This late winter and early spring warming of the forest soil, in part, stimulates the blooming of early spring woodland flowers.

Humidity in the forest interior is high in summer because of plant transpiration and poor air circulation. During the day, when the air warms and its water-holding capacity increases, relative humidity is lowest. At night, when temperature and moisture-holding capacities are low, relative humidity rises. The lowest humidity in the forest is a few feet above the canopy, where air circulation is best. The highest humidity is near the forest floor, the result of evaporation of moisture from the ground and settling of cold air from the strata above.

Bathed in full sunlight, the uppermost layer of the canopy is the brightest part of the forest. Down through the forest strata light intensity dims to only a fraction of full sunlight. In an oak forest only about 6 percent of the total midday sunlight reaches the forest floor; brightness of light at the forest floor is about 0.4 percent of that of the upper canopy.

Light intensity and its spectral qualities within the forest vary seasonally. The forest floor receives its maximum illumination during early spring before the leaves appear. A second lower peak of maximum illumination during the growing season occurs in the fall. The darkest period is midsummer. Light intensity during summer is highly variable from point to point and time to time as sun shines through gaps in the canopy, creating sun flecks. These tiny patches of bright light influence the distribution of herbaceous vegetation on the forest floor and contribute disproportionately to their daily energy budget (Chazdon and Perry 1991).

## Function

Energy flow and nutrient cycling have been assessed for several deciduous forest stands, three of which are the Hubbard Brook northern hardwoods forest in New Hampshire, Walker Branch mesic hardwoods forest at Oak Ridge, Tennessee, and an oak-pine forest at Brookhaven, Long Island.

Energy flow is measured most conveniently in terms of carbon pools and carbon fluxes. Carbon budgets for the mesic yellow-poplar (*Liriodendron tulipifera*)–oak (*Quercus* spp.) forest at Oak Ridge (Reichle et al. 1973) and a young oak–pitch pine (*Pinus rigida*) forest at Brookhaven (Whittaker and Woodwell 1969, Woodwell and Botkin

**Table 14.4** Metabolism of Two Deciduous Forests

| | Parameter | Liriodendron Forest | Quercus-Pinus Forest |
|---|---|---|---|
| Total standing crop | TSC | 8.76 | 5.96 |
| Net primary production | NPP | 0.73 | 0.60 |
| Relative production | NPP/TSC | 8.3% | 10% |
| Autotroph respiration | $R_A$ | 1.44 | 0.68 |
| | $R_A$/TSC | 0.16 | 0.11 |
| Heterotroph respiration | $R_H$ | 0.67 | 0.29 |
| Ecosystem respiration | $R_E = R_A + R_H$ | 2.11 | 1.01 |
| Net ecosystem production | $NEP = NPP - R_H$ | 0.06 | 0.28 |
| Annual decay | | 0.70 | 0.36 |

*Note:* Units of measure are kg C/m$^2$ and kg C/m$^2$/yr.
*Source: Liriodendron* data after Reichle et al. 1973; *Quercus-Pinus* data from Woodwell and Botkin 1970 in National Academy of Science 1974.

1970) are summarized in Table 14.4. In some respects the oak-pine stand and the yellow-poplar stand are similar, even though they have different standing crops, 5.96 and 8.76 kg C/m$^2$ respectively (Table 14.4). The two forests have comparable net primary production and autotrophic respiration. Both lose approximately the same amount of photosynthetic surface to insects, about 3 percent of net annual production.

However, the two stands differ considerably at the heterotrophic level. Heterotrophic respiration for yellow-poplar is more than twice that of the oak-pine forest This large heterotrophic respiration plus the large autotrophic respiration result in total ecosystem respiration for yellow-poplar over two times that of the oak-pine forest. The main difference in heterotrophic respiration is annual decomposition of 700 g C/m$^2$ for yellow-poplar compared to 360 g C/m$^2$ for the oak-pine. Much of this difference comes from the resistance of oak leaves and pine needles to rapid decomposition. The ratio of net ecosystem production (NEP) to total standing crop (0.05 for oak-pine and 0.007 for yellow-poplar) indicates that the young oak-pine forest is accumulating carbon seven times as fast as the older yellow-poplar stand.

Mineral cycling has been studied extensively in several temperate forest ecosystems, including an oak-pine forest (Whittaker and Woodwell 1969), a northern hardwoods forest (Likens et al. 1971, Likens 1976, Bormann and Likens 1979), a European oak forest (Duvigneaud and Denaeyer-DeSmet 1970), and many others (see Cole and Rapp 1981).

Nutrient cycling may be considered as a balance between inputs to the biological system and outputs or losses from the system through streamflow through the watershed. The difference represents the amount of recharge to the system from the soil pool. In the oak-pine forest at Brookhaven, Long Island, New York, total input from precipitation for the four cations potassium, calcium, magnesium, and sodium amounts to 2.5 g/m$^2$, while losses to the water table range from 3.7 to 5.2 g/m$^2$ (Woodwell and Whittaker 1968). Output from the Hubbard Brook forest in New Hampshire, a northern hardwoods system, summarized in Table 14.5, exceeds the input of calcium, magnesium, and sodium, while potassium shows a gain through input (Likens et al. 1967).

These and other budgets suggest several characteristics of mineral cycling in temperate deciduous forests. One is that the uptake of nutrients (annual increment of elements associated with bole and branch wood plus annual loss through litterfall, leaf wash, and stemflow) does not meet requirements (annual increment of elements associated with bole and branch wood plus current foliage production) (Table 14.6). For example, in the Walker Branch yellow-poplar–oak stand, the uptake of nitrogen amounts to 58.1 kg ha/yr, only 66 percent of requirement. Deficiencies have to be met by cycling nutrients within the tree biomass. Nutrient accumulations in tree biomass (Table 14.7) form a considerable pool of nutrients, most of which is unavailable for short-term recycling. Forty-nine percent of N, 35 percent of

**Table 14.5** Inputs and Losses of Nutrients of the Hubbard Brook Forest (kg/ha)

| | Ca | Mg | Na | K |
|---|---|---|---|---|
| Input | 3.0 ± 0 | 0.7 ± 0 | 1.0 ± 0 | 2.5 ± 0 |
| Output | 8.0 ± 0.5 | 2.6 ± 0.06 | 5.9 ± 0.3 | 1.8 ± 0.1 |
| Loss/Gain | −5.0 ± 0.5 | −1.9 ± 0.06 | −4.9 ± 0.3 | +0.7 ± 0.1 |

*Source:* Data from Likens et al. 1967.

**Table 14.6** Annual Element Balance of a 30- to 80-Year-Old Yellow-Poplar–Oak Forest, Oak Ridge, Tennessee (kg/ha/yr)

| | N | P | K | Ca | Mg |
|---|---|---|---|---|---|
| Requirement | 87.9 | 6.3 | 47.5 | 82.6 | 21.7 |
| Uptake | 58.1 | 3.4 | 40.0 | 87.8 | 12.4 |
| Internal recycling | 29.8 | 2.9 | 7.5 | −5.2 | 9.3 |

*Source:* Data from Cole and Rapp 1981:359.

**Table 14.7** Nutrient Balance Sheet for 30- to 80-Year-Old Yellow-Poplar–Oak Forest, Oak Ridge, Tennessee (kg/ha)

|  | N | P | K | Ca | Mg |
|---|---|---|---|---|---|
| *Input/output* |  |  |  |  |  |
| Atmosphere | 8.7 | 0.54 | 1.0 | 9.1 | 1.1 |
| Leaching to watershed | 1.8 | 0.02 | 6.8 | 147.5 | 77.1 |
| Loss/gain | +6.9 | +0.52 | −5.8 | −138.4 | −76.0 |
| *Internal cycling* |  |  |  |  |  |
| Litterfall | 36.2 | 2.7 | 19.1 | 58.3 | 8.3 |
| Throughfall | 12.0 | 0.4 | 18.4 | 21.9 | 3.4 |
| Total | 48.2 | 3.1 | 37.5 | 80.2 | 11.7 |
| *Live accumulation* |  |  |  |  |  |
| Woody biomass | 189 | 15 | 127 | 462 | 38 |
| Roots | 122 | 22 | 132 | 211 | 19 |
| Foliage | 78 | 6 | 45 | 75 | 21 |
| Total | 389 | 43 | 304 | 748 | 78 |
| *Detrital accumulation* |  |  |  |  |  |
| Forest litter | 187 | 11 | 14 | 294 | 22 |
| Soil rooting zone | 7,300 | 1,400 | 36,000 | 6,300 | 8,700 |
| Total | 7,487 | 1,411 | 36,014 | 6,594 | 8,722 |

*Source:* Data from Cole and Rapp 1981:394.

P, 42 percent of K, 62 percent of Ca, and 49 percent of Mg are incorporated into woody biomass; 31, 51, 43, 28 and 24 percent respectively in root biomass; and 20, 14, 15, 10, and 27 percent respectively in foliage.

Considering the whole Walker Branch yellow-poplar–oak forest ecosystem, 7 percent of N, 3 percent of P, 12 percent of Ca, and 1 percent each of K and Mg are stored in vegetation. The litter layer is the most important nutrient pool, because it is quickly decomposed (average turnover time of four years) and nutrients recycled, although the bulk of the nutrient pool is in mineral soil. Nutrients stored in living biomass, especially in roots, are translocated and recycled through the living biomass, particularly the foliage. The foliage, in turn, translocates a considerable portion of its nutrients back to roots before leaf fall. However, mineral cycling can be maintained only if nutrients are pumped from soil reserves or are released through the weathering of parent materials.

The role of the various components in nutrient cycling is illustrated by long-term studies of the nitrogen cycle in the deciduous forests at Hubbard Brook, New Hampshire (Bormann et al. 1977), Walker Branch, Oak Ridge, Tennessee (Henderson et al. 1973), and Coweeta North Carolina (Swank and Crossley 1988). Both studies point out that natural forest ecosystems tend to accumulate and cycle large amounts of nitrogen. At both sites most of the nitrogen (87 to 90 percent) is incorporated in the mineral soil horizons. The remaining nitrogen is in vegetation and the forest floor (Figure 14.18).

Deciduous forests appear to differ functionally from coniferous forests in the magnitude and nature of nutrient cycling. The differences are graphed in Figure 14.19. It compares nutrient cycling in a mixed oak forest and a spruce

**Figure 14.18** The nitrogen cycle in a mixed mesophytic forested watershed at Coweeta, North Carolina. Nitrogen pools in ecosystem components are given in Kg/ha. (After Monk and Day 1988.)

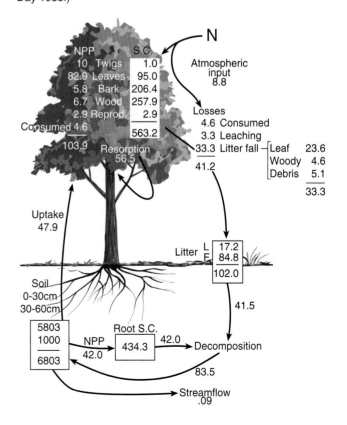

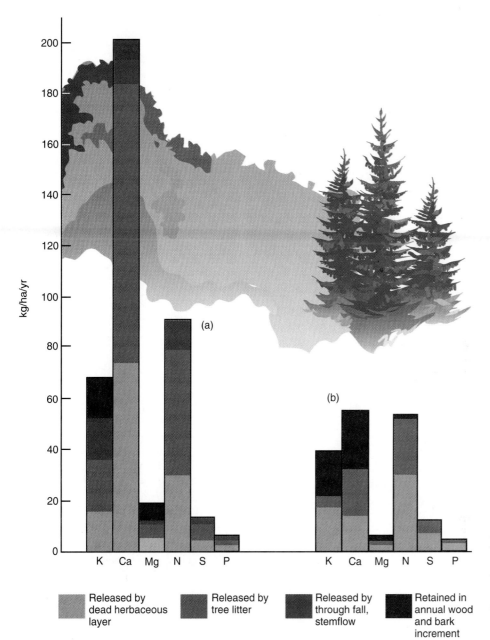

kg/ha/yr

(a)

(b)

K Ca Mg N S P

K Ca Mg N S P

Released by dead herbaceous layer

Released by tree litter

Released by through fall, stemflow

Retained in annual wood and bark increment

**Figure 14.19** Comparison of the biological cycle of nutrients (a) in a deciduous forest, mixed oak forest (*Quercus*) at Virelles, Belgium and (b) in a coniferous forest, a spruce forest (*Picea*) at Mitwart, Belgium. Productivity for both forests is 14.6 tn/ha/yr. (After Duvungneaud and Denaeyer-De Smet 1975.)

forest, both having an annual production of 14.6 tn/ha/yr. From these data it is evident that considerably more nutrients are cycled through the deciduous forest than through the coniferous forest and that the spruce retains more nutrients in its biomass than the oak. Spruce retains more nitrogen, sulfur, and phosphorus than it returns through litterfall, throughfall, stemflow, and dead parts of the herbaceous layer. The oak forest, on the other hand, returns more of all elements than it retains. The deciduous forest is much more efficient at recycling calcium. The oak forest recycles calcium largely through litterfall, whereas the spruce forest returns considerable quantities through stemflow and throughfall. Although the deciduous forest takes up more potassium than the spruce forest, it retains less. The spruce forest retains nearly half its uptake of potassium. This retention of

nutrients by conifers relates to their evergreen leaves. Conifers store nutrients and carbohydrates in their evergreen canopies rather than in stem and roots. For this reason they cannot mobilize these resources to replace foliage lost to herbivores or other causes, as deciduous forests do. This physiological difference explains in part why conifers appear to be nutrient accumulators.

## TROPICAL FORESTS

We are accustomed to think that tropical forests are only lush rain forest covering equatorial regions. Rain forests are but a part of a highly diverse tropical forest complex including cloud forests, rain forests, seasonal forests, and dry

forests. Tropical forests cover two times as many life zones (Holdridge) as the relatively homogeneous temperate deciduous forests and seven times as many as the even more homogeneous boreal forests (Lugo and Brown 1991). Such a wide diversity makes it difficult to compare tropical forests to other forest types, or even to make general statements about them. This difficulty has resulted in many misconceptions about tropical forest structure and function.

## Types

Tropical Rain Forests The tropical **rain forest** (Figure 14.20), so named in 1898 by the German botanist A. F. W. Schimper (see Chapter 1), comes in at least 30 to 40 types, which include the monsoon forest, the evergreen savanna forest, the evergreen mountain forest, tropical evergreen alluvial forest, as well as the true equatorial lowland tropical rain forest (see H. T. Odum 1970, Richards 1972, Whitmore 1984, Mabberley 1983). They once formed a worldwide belt about the equator. The largest continuous rain forest is found in the Amazon basin of South America. West and central Africa and the Indo-Malaysian regions are other major locations of tropical rain forest. Smaller types of rain forest, now virtually gone, occur on the eastern coast of Australia, the windward side of the Hawaiian Islands, and the east coast of Madagascar. In all of these areas rain forests grade into temperate and subtemperate rain forest.

Tropical rain forests grow where seasonal changes are minimal. The mean annual temperature is about 26° C, the mean minimum rarely goes below 25° C, and the difference

(a)

(b)

(c)

(d)

**Figure 14.20** (a) Amazonian tropical rain forest. Note the canopy emergents. (b) Lowland tropical rain forest in peninsular Malaysia. The prominent canopy trees are *Shorea,* representative of the dipterocarp family. (c) Eucalyptus rain forest in Australia. This rain forest is largely confined to scattered stands in Queensland. (d) Rain forest in Hawaii.

in temperature through the year is less than 4° C. Heavy rainfall occurs throughout the year, not less than 100 mm in any month for two out of every three years. What constitutes heavy rainfall varies with the region. The tropical rain forest regions in Latin America receive about 4000 cm rain annually and those in Africa about 1500 cm. Under such perpetual midsummer conditions plant activity continues uninterrupted, resulting in luxurious growth.

Tree species number in the thousands. A 10-square kilometer area of tropical rain forest may contain 1500 species of flowering plants and up to 750 species of trees. The richest is the lowland tropical forest of peninsular Malaysia, which contains some 7900 species. There, one of the major families, Dipterocarpaceae, contains 9 genera and 155 species, of which 27 are endemic. (The Asian dipertocarps have 12 genera and 470 species). The few rain forest communities with single dominants are limited to areas of particular combinations of soils and topography.

The interior of a tropical rainforest is impressive. The tree trunks are straight, smooth, and slender, often buttressed with planklike extensions of roots (Figure 14.21), and reach 25 to 30 m before expanding into crowns with large, leathery, simple leaves (for the complex architecture of tropical tree see Hallee, Oldeman, and Tomlinson 1978). Climbing plants, the lianas, long, thick, and woody, hang from trees like cables, and epiphytes grow on trunks and limbs. Undergrowth of the dark interior is sparse, consisting of shrubs, herbs, and ferns. Litter decays so rapidly that the clay soil, more often than not, is bare. The tangled vegetation popularly known as *jungle* is secondary (second-growth) forest that develops where primary forest has been disturbed. Tropical rain forests account for millions of species of flora and fauna, and one-half of all known plant and animal species. Rain forests support 5 to 10 million species of arthropods, with more to be discovered.

**Tropical Seasonal Forests**  Tropical rain forests grade into semi-evergreen and semideciduous seasonal forests. They are characterized by less rainfall, more variable temperatures, and a dry season during which about 30 percent of the upper canopy tree species lose their leaves. In the Indo-Malaysian forests, new leaves emerge about a month after the coming of the monsoon rains. In other regions leaves emerge about a month before the rainy season. Such forests are most common in southeastern Asia, India, South America, and Africa, and along the Pacific side of Mexico and Central America.

**Tropical Dry Forests**  Dry forests make up approximately 42 percent of all tropical forests (Figure 14.22). We overlook them and their importance in our concern about tropical rain forests (Murphy and Lugo 1986). Africa and tropical islands hold the largest proportion of dry tropical forest. They comprise about 22 percent of South American and 55 percent of Central American forested areas. Most of the original dry forests are gone, especially in Central America and India. Many of them have been converted to agricultural and grazing land, or they have regressed through disturbance to thorn woodland, savanna, and grassland.

Tropical dry forests experience a dry period, the length of which varies with latitude. The more distant the forest is from the equator, the longer is the dry season, up to eight months. During the dry period, trees and shrubs drop their leaves. Before the start of the rainy season, which may be wetter than the wettest time in the rain forest, the trees begin to leaf. During the rainy season the landscape becomes uniformly green.

**Figure 14.21** Planklike buttresses help support tall rain forest trees.

**Figure 14.22** A tropical dry forest in Africa. The plants in the middle distance are aloes, succulent herbaceous plants of the lily family.

## Structure

The tropical rain forest can be divided into five general layers, but stratification is often poorly defined because many tree species have the same growth plan but differ in size (Hallee, Oldeman, and Tomlinson 1978, Tomlinson 1983, Brunig 1983). Stratification is most apparent in the undisturbed forest. The uppermost or emergent layer consists of trees over 60 to 80 m high, whose crowns rise above the rest of the forest to form a discontinuous canopy (Figure 14.23). The second layer, consisting of trees about 50 m high, forms another lower discontinuous canopy. Not clearly separated from one another, these two layers form an almost complete canopy. The third layer, the lowest tree stratum, is continuous, often the deepest layer, and is well defined. The fourth layer, usually poorly developed in deep shade, consists of shrubs, young trees, tall herbs, and ferns. The fifth stratum is the ground layer of tree seedlings and low herbaceous plants.

A conspicuous part of the rain forest is plant life dependent on trees for support. Such plants include epiphytes, climbers, and stranglers. Epiphytes, such as orchids, bromeliads, and aroids (members of the arum family), attach themselves to a host tree and obtain their nutrients from air, rainwater, and organic debris trapped by their aerial roots. Some epiphytes are important in recycling minerals leached from the canopy. Climbers are vinelike plants that reach the tops of trees and expand into the form and size of a tree crown. Climbers grow prolifically in openings, giving rise to the image of the impenetrable jungle. Stranglers start life as epiphytes. They send roots to the ground and increase in number and girth until they eventually encompass the host tree and claim the crown limbs as support for their own leafy growth.

The mature tropical forest, like the mature temperate forest, is a mosaic of continually changing vegetation. Death of tall trees, brought about by senescence, lightning, wind storms, hurricanes, defoliation by caterpillers, and other causes creates gaps (see Chapter 30), which shade-intolerant pioneer species quickly fill (Poore 1968, Hartshorn 1978, Doyle 1981). These trees are replaced eventually by shade-tolerant late successional species; but continuous random disturbances across the forest insures persistence of the species in the mature forest. A high frequency of treefall—Poore (1968) estimated that in his 12 ha Malaysian study area, one-half had experienced gap formation—may account for the low density of large trees (1 m+ dbh) in mature rain forests.

Most tropical rain forest trees complete height growth when they have achieved only about one-third to one-half of their final bole diameter. Thus stratification or layering results when a group of species of similar mature height dominate a stand (Whitmore 1984). Layering is also influenced by crown shape, which in turn is correlated with tree growth. Young trees still growing in height have a single stem and a tall narrow crown; they are *monopodial*. Mature trees have a number of large limbs diverging from the upper stem or trunk; they are *sympodial*. This change happens when the bud of the main stem axis ceases to grow and the lateral buds take over their role. This process repeats itself, adding to crown growth and producing a pattern that suggests the spokes of an umbrella. Looking up into the canopy, the observer gains the impression that the crowns of the tree fit together like a jigsaw puzzle with the pieces about a meter apart (Figure 14.24). This growth pattern is called *crown shyness*.

Layering of vegetation influences the internal microclimate of the forest. The crowns of emergent trees experience

**Figure 14.23** Stratification of vegetation and animal life in a Malaysian rain forest.

conditions similar to open land. Through the canopy the level of $CO_2$ and the amount of humidity increase and temperature and evaporation decrease. From the ground to 1 m the levels of $CO_2$ are high and humidity stands at 90 percent; temperature on the average is 6° C cooler than outside forest cover, and experiences a strong nocturnal inversion (Bourgeron 1983). Light decreases rapidly down through the canopy. The amount of light that reaches the floor of a Malaysian rain forest is about 2 to 3 percent of incident radiation; half of that comes from sun flecks, about 6 percent from breaks in the canopy, and 44 percent from reflected and transmitted light (Mabberley 1983).

Stratification of animal life in the tropical rain forest, however, is pronounced (Figure 14.23). Harrison (1962) recognized six distinct feeding strata. (1) A group feeding above the canopy consists largely of insectivorous and some carnivorous birds and bats. (2) A top of the canopy group— a large variety of birds, fruit bats, and other species of mammals—feeds on leaves, fruit, and nectar. A few are insectivorous and mixed feeders. (3) Below the canopy, a zone of tree trunks, is a world of flying animals—birds and insectivorous bats. (4) Also in the middle canopy are scansorial mammals which range up and down the trunks, entering the canopy and the ground zone to feed on the fruits of epiphytes growing on tree trunks, on insects, and on other animals. (5) Large ground animals make up the fifth feeding group. This includes large mammals and a few birds, living on the ground and lacking climbing ability, that are able to reach up into the canopy or cover a large area of forest. They include the large herbivores and their attendant carnivores. (6) The final feeding stratum includes the small ground and undergrowth animals, birds and small mammals capable of some climbing, that search the ground litter and lower parts of tree trunks for food. This stratum includes insectivorous, herbivorous, carnivorous, and mixed feeders.

Animal life in the tropical rain forest is largely hidden, either by dense foliage of the upper strata or by the cover of night. Birds are largely arboreal, and although brightly colored, remain hidden in the dense foliage. Ground birds are small and dark-colored, difficult to see. Mammals appear scarcer than they really are, for they are largely nocturnal or arboreal. Ground-dwelling mammals are small and secretive. Tree frogs and insects are most conspicuous at evening, when their tremendous choruses are at full volume. Insects are most diverse at forest openings, along streams, and at forest margins, where light is more intense, temperatures fluctuate, and air circulates freely. Highly colored butterflies, beetles, and bees are common. Among the unseen invertebrates, hidden in loose bark and in axils of leaves, are snails, worms, millipedes, centipedes, scorpions, spiders, and land planarians. Termites are abundant in the rain forest and play a vital role in the decomposition of woody plant material. Together with ants, they are the dominant insect

**Figure 14.24** A view up through the canopy of a tropical rain forest reveals crown shyness. The upper canopy has sympodial crown structure.

life. Ants are found everywhere in the rain forest, from the upper canopy to the forest floor, although in common with other rain forest life, the majority tend to be arboreal.

Many specialized interactions among plants and animals exist in tropical rain forests. Plants, often widely dispersed in the forest, depend for pollination upon birds, bats, and insects, especially beetles, bees, moths, and butterflies (see Procter and Yeo 1973, Baker et al. 1983, Howe and Westley 1988) and for seed dispersal on fruit-eating birds, bats, rodents, and primates. Heavy predation on seeds by insects may result in wide dispersal of tree species (Janzen 1971). Other interactions involve repellent toxins to discourage predation by herbivores and even insect-plant mutualisms in which insects such as ants live in hollow stems and prevent other insects from gaining entrance or feeding on the plant (Janzen 1967) (see Chapter 24).

## Function

The most intensely studied tropical forest is the one at the Puerto Rico Nuclear Center. H. T. Odum (1970) and his associates worked up an energy budget for that forest. Incoming solar radiation amounted to 3830 kcal/m$^2$/day. Gross production amounted to 131 kcal/m$^2$day, of which 116 kcal was used in respiration, leaving a net production of 15.2 kcal/m$^2$/day as determined by gas analysis. Roots were responsible for 60 percent of the respiration, leaves for 33 percent, and trunks, branches, and fruit for the remainder. Net productivity as measured by biomass accumulation was 16.31 kcal/m$^2$/day. Cumulative biomass addition through wood growth was 0.72 kcal/m$^2$/day or 3.8 percent of net production. The remainder of net production passed through the grazing and detrital food chains.

Odum's energy budget is for a specific rain forest. Because site, soil, and other conditions of different tropical forests vary widely, productivity also varies widely (Figure 14.25). However, a few generalizations can be made about their energy budgets. Tropical forests use 70 to 80 percent of their energy intake in maintenance and 20 to 30 percent for net production. Average gross primary production is about 67 mtn/ha/yr or $28 \times 10^3$ kcal/m$^2$/yr (Golley 1972). Mean annual net production is about 21.6 T/ha/yr. This amount exceeds temperate forests, averaging about 13 T/ha/yr, by a factor of 1.7 and boreal forests, averaging 8 T/ha/yr, by a factor of 2.7 (Golley and Farnsworth 1973). However, tropical and temperate forests differ somewhat in their efficiency of production. Efficiency in this case is defined as the sum of energy stored in wood, leaves, fruit, and litter divided by total solar energy available to the community. C. F. Jordan (1971, 1983) found that the high overall productivity of tropical forests relates more to foliage than to wood production. The rate of wood production is not any greater than in the temperate hardwoods forest, but the rate of leaf and litter production is higher in the tropics.

A large standing biomass is typical of tropical ecosystems. The tropical rainforest averages about 300 T/ha, compared to 150 T/ha for a temperate forest. Tropical rain forests tend to concentrate proportionately more calcium, silica, sulfur, iron, and sodium and less potassium and phosphorus than temperate forests (Figure 14.26). Mineral concentrations vary widely among tropical forests, influenced by site, soil, and climate (Table 14.8) (Vitousek and Sanford 1986). Concentrations of all major nutrients are higher on more fertile soils, exemplified by the Panamanian forests, than on the infertile oxisol/ultisol soils that are especially low in phosphorus, potassium, and calcium. It follows that

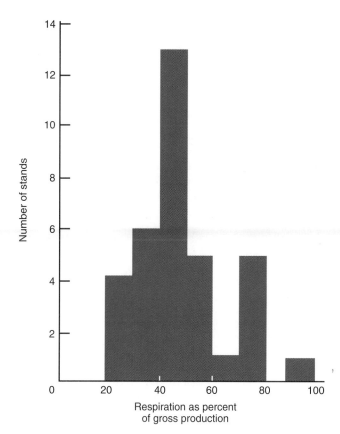

**Figure 14.25** Variation in production in tropical rain forests, shown by autotrophic respiration (R) as a percent of gross primary production (GPP). All types of vegetation are represented in this figure. Variation among stands makes it difficult to generalize about primary production in tropical forests. (From Golley 1972.)

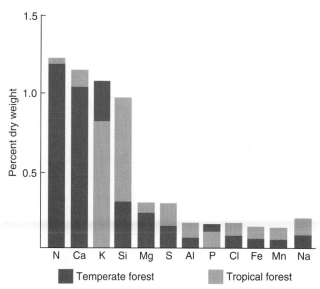

**Figure 14.26** Comparison of concentration of elements in temperate and tropical forests. Tropical forests have a higher concentration of nitrogen in their biomass. Temperate hardwood forests have higher concentrations of phosphorus and potassium. (From Golley 1975.)

tropical forests on more fertile soils return more litter with higher concentrations of nutrients to the forest floor than do forests on other soils (Table 14.9). Phosphorus, always of low availability because of long-term weathering, rapid adsorption of the element by sesquioxide clays, and immobilization by decomposers, is most limiting on infertile soils. Forests on infertile soils cycle phosphorus more efficiently than forests on moderately fertile sites, by retranslocating phosphorus from leaves prior to litterfall. Montane tropical forests fit no one pattern because of the variability of their mountainous sites. Throughfall is the major pathway for cycling of potassium in all forests.

Except for nitrogen, nutrient budgets, like those of the temperate forest, can be assessed by comparing inputs from rainfall with outputs by stream discharge or by using soil lysimeters. The difference between the two represents inputs from the weathering of the soil and soil materials. Data for a tropical moist forest in Costa Rica indicate that inputs of phosphorus and potassium by precipitation exceed outputs, but considerably more calcium and magnesium are lost from the system than are gained by rainfall (Table 14.10). This finding and others suggest considerable input of these two nutrients from the soil reservoir and conservation of phos-

phorus and potassium by rapid internal cycling. Internal cycling is aided by the rapid return of nutrients leached by throughfall, retention of nutrients by fungal rhizomorphs, and uptake by mycorrhizal fungi.

In the temperate forest, bacteria and fungi, the main agents of decay, release nutrients directly into the mineral soil, where they are subject to leaching. The fine roots of these trees are woven into the matrix of mineral soil. Fine roots of tropical trees, however, are concentrated in the well-aerated upper 2 to 15 cm of humus, and only a few roots penetrate the upper layer of mineral soil (Cornforth 1970b, C. F. Jordan 1979, 1982a,b).

Symbiotically associated with the roots are mycorrhizal fungi. The roots of tropical forests support an abundance of mycorrhizal fungi that attach fine roots to dead organic matter by hyphae and rhizomorph tissue. The mycorrhizae transfer nutrients directly from dead organic matter to the living roots with a minimum of leakage to the soil, so minerals remain tied up in living and dead organic matter (Went

**Table 14.8  Nutrient Content of Aboveground Biomass in Moist Tropical Forest Ecosystems**

| Site | Biomass (T/ha) | Nutrients (kg/ha) | | | | |
|------|----------------|------|------|------|------|------|
| | | N | P | K | Ca | Mg |
| Panama | 316 | | 158 | 3020 | 3900 | 403 |
| Brazil | 406 | 2430 | 59 | 435 | 432 | 201 |
| Venezuela | 335 | 1084 | 40 | 302 | 260 | 69 |

*Source:* Data from Vitousek and Sanford 1986.

**Table 14.9  Dry Mass and Nutrient Content of Litterfall**

| Site | Litterfall T/ha | Nutrients (kg/ha) | | | | |
|---|---|---|---|---|---|---|
| | | N | P | K | Ca | Mg |
| *Moderately Fertile Soil* | | | | | | |
| Panama | 11.1 | 195 | 15 | 47 | 216 | 26 |
| Costa Rica | 8.1 | 135 | 6 | 20 | 59 | 16 |
| Sarawak | 11.5 | 110 | 4 | 26 | 240 | 20 |
| *Infertile Oxisols/Ultisols* | | | | | | |
| Ivory Coast | 11.9 | 170 | 8 | 28 | 61 | 51 |
| Brazil | 7.3 | 106 | 2 | 13 | 18 | 14 |
| Malaysia | 8.9 | 100 | 2 | 32 | 70 | 18 |
| *Montane Forest* | | | | | | |
| Papua New Guinea | 7.6 | 90 | 5 | 28 | 95 | 19 |
| Sarawak | 11 | 86 | 3 | 13 | 21 | 16 |
| Hawaii | 5.2 | 37 | 2 | 12 | 84 | 10 |

*Source:* Adapted from Vitousek and Sanford 1986:153.

and Stark 1968). This concentration and rapid recycling of nutrients in the uppermost layers of the soil explains why many tropical soils cleared of their forest vegetation are nutrient-poor and cannot support agricultural crops for more than a few years, in contrast to temperate forests.

## Fate of Tropical Forests

In the 1800s North America, especially the United States, experienced the most rapid and massive deforestation of a continent by humans in Earth's history. Well over 80 percent of the rich hardwoods were destroyed, and the great pine forests of the Great Lakes region experienced a devastation from which the land has never recovered. Added to this was the elimination of all but a minuscule fraction of the prairie ecosystem. This destruction caused the extinction of many species, including the passenger pigeon (*Ectopistes migratorius*) and the ivory-billed woodpecker (*Campephilus principalis*). The extent of this deforestation is masked today by return of forest to previously cut-over and abandoned agricultural land, although the replacement forests lack the diversity and grandeur of the original.

Today, the latter part of the 1900s, the tropical forest regions of the planet are experiencing an even faster rate of

**Table 14.10  Input–Output Balances for Moderately Fertile Tropical Moist Forest, Costa Rica (kg/ha/yr)**

| Element | Precipitation Input | Hydrologic Output |
|---|---|---|
| P | 0.17 | 0 |
| K | 5.4 | 3.6 |
| Ca | 3.1 | 5.7 |
| Mg | 2.6 | 8.5 |

massive deforestation (Figure 14.27). By the end of this century at the present rate of destruction most of the rain forests of southeast Asia will be gone, together with the patches in Australia, West Africa, and Madagascar; and within 25 years little will remain of the Amazonian rain forest. What makes the deforestation of the tropical areas so critical, compared to the deforestation of North America, is that much of the rain forest is poorly adapted to agriculture; and once cleared, burned, planted, and abandoned, it is slow to return to forest, if ever.

Lost in the destruction of the tropical rain forest is its high diversity of plant and animal life. Although occupying only 7 percent of Earth's surface, the tropical rain forest regions hold 50 to 80 percent of the world's plant species. Most of the animal species are endemic, resident, and nonmigratory. Once the habitat is destroyed, the species go extinct. But not only is the fauna endangered; for the first time in human history, hundreds of plant species also face extinction. They do so not only because of habitat destruction but also because of the loss of fauna upon which they depend for pollination and seed dispersal. The great diverse tropical regions are a source of food, medicinal plants little studied and utilized, genetic

**Figure 14.27** The tropical rain forests of the world face destruction by land-clearing schemes, the fate of this piece of Amazonia rain forest.

diversity, and economic opportunity poorly managed for the future (see Myers 1983, E. O. Wilson 1988, Collins 1990).

# SUMMARY

Coniferous, deciduous, and tropical forests are the three major types of forest ecosystems. The coniferous forest, which forms a vast belt encircling the northern part of the Northern Hemisphere, is typical of regions where the summers are short and winters are long and cold. The deciduous forest, richly developed in North America, Europe, and eastern Asia, grows in a region of moderate precipitation and mild temperatures during the growing season. A diversity of tropical forests occur in equatorial regions under environmental conditions that range from high humidity, heavy rainfall, minimal seasonal changes, and an annual mean temperature of about 28° C to seasonal dryness and annual mean temperatures of about 17° C. All three types are more or less stratified into layers of vegetation, including a canopy, subcanopy, and ground layer. Accompanying this vegetative stratification is stratification of light, temperature, and moisture. The canopy intercepts light and precipitation; the forest floor is shaded through the year in most coniferous and tropical rain and moist forest, and in late spring and summer in the deciduous forest.

Coniferous and deciduous forests hold different species of animal life, but seasonal adaptations are similar. The greatest concentration and diversity of life are on and just below the ground layer. Other animals live in the various strata from low shrubs to the canopy. The tropical rain forest has pronounced feeding strata from above the canopy to the forest floor, and many of its animals are strictly arboreal. Whatever the forest, the trees that compose it create different environments that ultimately dictate the kinds of plants and animals that can live within it.

Although gross primary productivity of forest ecosystems is high, so much of the GPP is allocated to the maintenance of forest structure that effective production is low: deciduous forest, 33 percent; coniferous forest, 45 percent; and tropical rain forest, 12 percent. Mineral cycling is tight. Nutrients accumulate in woody biomass to form a pool unavailable for short-term cycling. Although the bulk of the nutrients is in mineral soil, the most important pools in mineral cycling are root mortality, litterfall, and foliar leaching. Internal cycling of some nutrients, especially nitrogen and phosphorus, is important in nutrient conservation. In most forest systems, only a small fraction of nutrients is lost from the system through streamflow.

Coniferous forests exhibit short-term cycling between litterfall and uptake by trees. At the same time conifers appear to be accumulators, removing elements from the soil in quantities large enough to upset nutritional balances in ecosystems. Nutrient cycling in conifers appears to depend on mycorrhizal fungi, fungal biomass, and root activity. Fungi act as nutrient sinks, but a symbiotic relationship exists between mycorrhizae and roots in the uptake of nutrients.

Tropical forests tend to concentrate nutrients, especially phosphorus, in leaves, retranslocate nutrients within the plant, and rapidly take up and recycle nutrients from quickly decomposing litterfall. Roots in rain forests and moist forests are concentrated in the upper soil layer in close contact with the litter. There a symbiotic relationship between roots and mycorrhizae facilitates nutrient uptake.

The tropical forest and coniferous forest contrast with the deciduous forest, in which microbial decomposition releases nutrients to the mineral soil, in turn providing the pathway through the roots for nutrient uptake.

# REVIEW QUESTIONS

1. Compare and contrast coniferous, deciduous, and tropical forests in structure and nutrient cycling.
2. What is the importance of dead wood in a forest stand? What are the ecological effects of removing dead wood for firewood and other uses, as is usual in European forests?
3. Name and characterize the several broad types of tropical forests.
4. If tropical forests soils are so nutrient-poor, how can they support such a high plant biomass and diversity?
5. Why is it difficult to compare tropical forests with temperate forests?
6. Why has the tropical dry forest been most heavily impacted by humans? (Review Murphy and Lugo 1986.)
7. An ecological hypothesis (often considered a dogma) holds that net primary production (NPP) of a forest decreases as stand age increases because much of the gross primary production (GPP) goes into maintenance—the increased respiration of woody tissue. Ryan and Waring (1992) found little significant difference between the maintenance respiration of a 40 year-old and a 245-year-old stand of logdepole pine. If these results hold, how can you account for the decline in NPP with age in a forest stand? Suggest several hypotheses.
8. Lugo and Brown (1991) point out that comparisons of ecosystem functions between tropical and temperate forests are often poor and inaccurate. Why should this be so?

# CROSS-REFERENCES

Relative humidity, 45; Holdridge life zones, 56, 58–59; light, 100; internal nutrient cycling, 116–120; soil chemistry, 133–136; mor humus, 139; tropical soils, 140–141; soil groups, 144–146; decomposition, 159–165; primary production, 169–175; nitrogen cycle, 210–212; phosphorus cycle, 212–213; mycorrhizae, 528, 584–585; mutualism, 582–587.

# Freshwater Ecosystems

## Concepts

1. Freshwater ecosystems fall into two broad groups, lentic ecosystems contained in a basin and lotic or flowing water.
2. Nutrient cycling and energy flow in lentic ecosystems are largely contained within the basin, but are heavily influenced by inputs from the surrounding landscape.
3. The structure of lotic ecosystems is shaped by flowing water.
4. Nutrient cycles in lotic systems involve spiraling, the retention and uptake of nutrients moving downstream.
5. Wetlands are best defined by considering vegetation, hydrological conditions, and soil.
6. The nature of a wetland is influenced by its hydroperiod.

Global aquatic ecosystems fall into two broad classes defined by salinity—freshwater ecosystems and saltwater ecosystems. The latter include inland brackish and salt water, as well as marine and estuarine habitats. Freshwater ecosystems, the study of which is known as **limnology,** are conveniently divided into two groups: **lentic** or standing water habitats and **lotic** or running water habitats.

# LENTIC ECOSYSTEMS

Lakes and ponds are inland depressions containing standing water. They vary in size from small ponds of less than one hectare to large seas covering thousands of square kilometers. They range in depth from one meter to over 2000 meters.

Lakes and ponds arise in many ways. Some North American lakes were formed by glacial erosion and deposition. Glacial abrasion of slopes in high mountain valleys carved basins, which filled with water from rain and melting snow to produce tarns (Figure 15.1a, b). Retreating valley glaciers left behind crescent-shaped ridges of rock that dammed up water behind them. Numerous shallow kettle lakes and potholes formed in glaciated areas in northeastern North America and northwestern Europe (Figure 15.1b).

Lakes are also formed by the deposition of silt, driftwood, and other debris in the beds of slow-flowing streams. Loops of streams that meander over flat valleys and flood plains often become cut off, forming crescent-shaped oxbow lakes (Figure 15.1c).

Shifts in Earth's crust, either by uplifting of mountains or the breaking and displacement of rock strata, causing part of a valley to sink, develop depressions that fill with water. Craters of extinct volcanoes may fill with water (Figure 15.1d), or landslides can block streams and valleys to form new lakes and ponds. In a given area all natural lakes and ponds have the same geological origin and the same general characteristics; but because of varying depths at the time of origin, they may represent several stages of development.

Many lakes and ponds are formed by nongeological activity. Beavers dam up streams to make shallow but often extensive ponds. Humans intentionally create artificial lakes by damming rivers and streams for power, irrigation, and water storage, or by constructing small ponds and marshes for water, fishing, and wildlife. Quarries and strip mines fill with water to form other ponds.

## Physical Characteristics

Temperature The high specific heat of water gives the aquatic environment a more stable temperature regime than the terrestrial environment. It is not subject to sharp daily fluctuations in temperature, and seasonal changes come on gradually. Slow warming and cooling of the surface water result in seasonal stratification of temperature and mixing of water in many lakes and ponds, and to a certain degree in seas (Figure 15.2).

In summer the intensity of the sun heats the surface water. The higher the temperature of the surface water, the lighter it becomes, so it floats on the colder, denser water beneath. The warm upper layer is not easily mixed with the denser below, establishing a mixing barrier between them. The freely circulating surface water with a small but variable temperature gradient is the **epilimnion** (Greek for "upper lake") (Figure 15.2). Below the thermal barrier is a middle mass of water, the **metalimnion** ("middle lake"). This middle mass is characterized by a steep and rapid decline in temperature of about 1° C for each meter of depth. This temperature gradient is known as the **thermocline.** Unfortunately, the two terms are used interchangeably to define the middle layer. Below these two layers is the **hypolimnion** ("lower lake"), a deep cold layer cut off from the air above.

With the coming of autumn, the air temperature falls. The surface water loses heat to the atmosphere through convection, conduction, and evaporation. The temperature of the surface water drops, and the metalimnion sinks. The epilimnion increases until it includes the entire lake. The temperature is now uniform from top to bottom, the lake waters circulate, and oxygen and nutrients are recharged throughout the lake. This seasonal mixing is called the **overturn.** The fall overturn, stirred by the wind, may last until ice forms.

As the surface water cools below 4° C, it becomes lighter and remains on the surface. If the climate is cold enough, the surface water freezes; otherwise it remains close to 0° C. A slight inverse stratification may develop, in which the water becomes warmer up to 4° C with depth. The water immediately beneath the ice may be warmed by solar radiation through the ice. Because this warming increases its density, this water subsequently drops to the bottom, where it mixes with water warmed by heat conducted from bottom mud. The result is a higher temperature at the bottom, although the overall stability of the water is undisturbed. As the ice melts in spring, the surface water again reaches 4° C and streams downward, aiding the wind in mixing the water. After the entire water mass is 4° C, even the slightest winds can cause a complete circulation of water between the surface and the bottom. During this spring overturn, nutrients sequestered on the bottom, the oxygen in the surface waters, and the plankton within are mixed. Later in the season the surface water warms, and summer stratification develops.

This general picture of seasonal changes in temperature stratification does not apply to every lake and pond. In shallow lakes and ponds temporary stratification of short duration may occur; in others stratification may exist without a thermocline. In some very deep lakes the metalimnion simply descends during periods of overturn, but does not disappear. In such lakes the bottom water never becomes mixed with the top layers.

(a)

**Figure 15.1** Some types of lakes. (a) A cirque, a steep-sided, semi-circular, glacial lake. (b) A rock basin glacial lake or tarn in the Rocky Mountains. (c) Potholes dot the glacial landscape of the north central United States. (d) an oxbow lake formed when a bend in a river was cut off from the main channel. (e) Lake formed by filling of a volcanic crater.

(b)

(c)

(d)

(e)

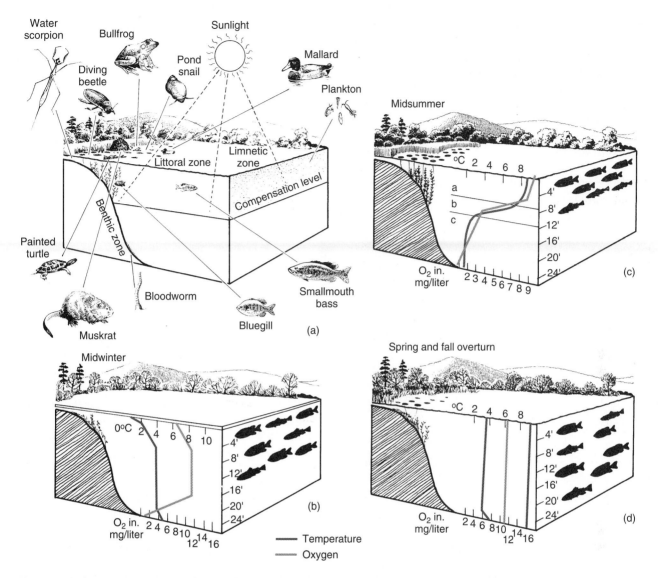

**Figure 15.2** Seasonal variations in the stratification of oxygen and temperature and the distribution of aquatic life in lake ecosystems. (a) This generalized picture of a lake in midsummer shows the major zones—littoral, limnetic, profundal, and benthic. The compensation level is the depth at which light is too low for photosynthesis. Surrounding the lake is a variety of organisms typical of a lake community. (b, c) The distribution of oxygen and temperature during the different seasons affects the distribution of fish life. The narrow fish silhouettes represent trout, or cold-water species. The wider silhouettes are bass, or warm-water species. Note the pronounced horizontal stratification in midsummer. (d) Oxygen and temperature curves are nearly vertical during the spring and fall over-

**Oxygen** Rarely is oxygen limiting in terrestrial environments; but even at saturation levels, the availability of oxygen in aquatic environments is meager and problematic. Oxygen enters the water by absorption from the atmosphere and by photosynthesis. The amount of oxygen and other gases water can hold depends upon pressure, temperature, and salinity. The solubility of a gas in water decreases as the temperature rises. For this reason cold water holds more oxygen than warm water. Solubility decreases as salinity increases and increases as pressure increases.

Oxygen absorbed by surface water is mixed with deeper water by turbulence and internal currents. In shallow, rapidly flowing water and in wind-driven sprays, oxygen may reach and maintain saturation and even supersaturated levels, because of the increase of absorptive surfaces at the air-water interface. Water loses its oxygen through increased temperatures, increased respiration of aquatic life, and aerobic decomposition.

During the summer, oxygen, like temperature, may become stratified in lakes and ponds. The amount of oxygen

usually is greatest near the surface, where an interchange between water and atmosphere, further stimulated by the stirring action of the wind, takes place. The quantity of oxygen decreases with depth because of decomposition in the bottom sediments. In some lakes oxygen varies little from top to bottom; every layer is saturated for its temperature and pressure. Water in some lakes is so clear that light penetrates below the depth of the thermocline and encourages the growth of phytoplankton. Because of photosynthesis, the oxygen content may be greater in deep water than on the surface.

During spring and fall overturn, when water recirculates through the lake, oxygen becomes replenished in deep water. In winter the reduction of oxygen in unfrozen water is slight, because bacterial decomposition is reduced by the cold and water at low temperatures holds the maximum amount of oxygen. Under ice, however, oxygen depletion may be serious, causing a heavy winterkill of fish.

### Carbon Dioxide, Alkalinity, and pH

Carbon dioxide, another atmospheric gas in water, behaves much differently than oxygen. Unlike oxygen, carbon dioxide combines chemically with water, so it occurs both in free and bound states. Like oxygen, free carbon dioxide in fast-flowing water is in equilibrium with the atmosphere, according to the laws of the behavior of gases. Because the concentration of $CO_2$ in the atmosphere is low, the concentration of $CO_2$ is also low in water, about 0.5 cc/1 at 10° C and 0.2 cc/1 at 24° C. Considerably more $CO_2$ is held as carbonate and bicarbonate ions.

$CO_2$ dissolved in water combines with water to form carbonic acid:

$$CO_2 + H_2O \rightleftharpoons H_2CO_3$$
$$H_2CO_3 \rightleftharpoons HCO_3^- + H^+$$
$$HCO_3^- \rightleftharpoons H^+ + CO_3^=$$

Carbon dioxide in simple solution and as $H_2CO_3$ is free carbon dioxide. Carbon dioxide in bicarbonate ions $HCO_3^-$ and in the carbonate ion $CO_3^=$ is called combined or bound. In the presence of an acid, combined carbon dioxide is converted to free $CO_2$.

The amount of acid needed to free $CO_2$ is the measure of the water's **alkalinity,** the amount of anions of weak acid in water and of the cations (other than protons) balanced against them. $H_2CO_3$ dissociates, releasing the hydrogen ion and thus affecting the **pH,** the measure of the degree to which water is acid. At neutral (pH 7) most of the $CO_2$ is present as $HCO_3^-$ (Figure 15.3). At a high pH more $CO_2$ is present as $CO_3^=$ than at a low pH, where more carbon dioxide is present in the free condition. Addition or removal of $CO_2$ affects pH, and a change in pH affects $CO_2$.

Bicarbonate and carbonate ions (and other anions of weak acids) form a **buffer system,** which resists changes in pH. The buffer capacity of solutions is determined by

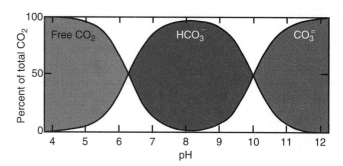

**Figure 15.3** Theoretical percentages of $CO_2$ in each of its three forms in water in relation to pH.

the abundance of their anions, which is directly related to alkalinity.

Seawater and hard fresh water (having much calcium and a high pH) are highly buffered. The pH of fresh water varies from 3 to 10. Rainwater has a pH of 5 to 6. The usual range in streams and lakes is 6.5 to 8.5, but many lakes and streams have a pH as low as 3.5 seasonally, depending upon the watershed. Waters draining from watersheds dominated geologically by limestone will have a much higher pH and be well buffered compared to waters from watersheds dominated by acid sandstone and granite. The pH of soft waters (without much calcium) can fluctuate widely, especially when receiving acid rain.

In soft waters carbon dioxide may be reduced by photosynthesis of aquatic plants. If photosynthesis is high, corresponding respiration of organic matter, absorption from the atmosphere, and absorption from groundwater may not be enough to balance the losses. This fact can affect the physiology of aquatic invertebrates, especially in the oxygen affinity and alkalinity of the blood and the development of the exoskeleton.

A close relationship exists between the formation and dissolution of calcium carbonate and the photosynthetic and respiratory activity of aquatic plants:

$$(C_aHCO_3)_2 \rightleftharpoons C_aCO_3 + H_2O + CO_2 \overset{assimilated*}{\underset{precipitated*}{}}$$

In hard water (high pH), plant activity removes carbon dioxide and causes the precipitation of calcium carbonate. In soft water, $Ca^{++}$ and $CO^=$ tend to remain in solution. The limited supplies of $Ca^{++}$ and $CO_3^=$ in soft water restrict the distribution of some aquatic invertebrates.

### Light

The depth to which light penetrates is limited by the turbidity of the water and the absorption of light rays. On this basis lakes and ponds can be divided into two basic layers, the **trophogenic zone,** roughly corresponding to the epilimnion; and the lower **tropholytic zone,** where decomposition is most active. The latter zone is about the same as the hypolimnion. The boundary between the two zones is

the **compensation depth,** where photosynthesis balances respiration (Figure 15.2).

## Structure

Unlike most terrestrial ecosystems, lentic ecosystems have well-defined boundaries—the shoreline, the sides of the basin, the surface of the water, and the bottom sediment. Within these boundaries gradations of light, oxygen, and temperature profoundly influence life in the lake, its distribution, and its adaptations.

Lentic ecosystems can be subdivided into vertical strata and horizontal zones based on photosynthetic activity. The **littoral zone,** or shallow water zone, is the one in which light penetrates to the bottom (Figure 15.4). The area is occupied by rooted plants, such as water lilies, rushes, and sedges. Plants and animals found here vary with water depth, and a distinct zonation of life exists from deeper water to shore. Floating aquatics such as pond lilies and pondweeds colonize the deeper water. In shallow water beyond the zone of floating plants grow the **emergents,** plants whose roots and lower stems are immersed in water and whose upper stems and leaves stand above water. Among these emergents are plants with narrow tubular or linear leaves such as bulrushes and cattails (Figure 15.5a). Associated with them are such broadleaf emergents as pickerelweed and arrowheads (Figure15.5b). The distribution and diversity of plants vary with water depth and its fluctua-

tions. Within the sheltering beds of emergent plants animal life is abundant.

The **limnetic zone,** or open water zone, extends to the compensation level. It is inhabited by suspended organisms, the **plankton,** and free-swimming organisms, or **nekton,** such as fish.

Dominating the limnetic zone is **phytoplankton** (Figure 15.6a), including diatoms, desmids, and filamentous algae. Because these tiny plants carry on photosynthesis in open water, they are the base upon which limnetic life depends. Suspended with the phytoplankton are small animals, or **zooplankton** (Figure 15.6b), which graze on the minute plants. These animals form an important link in energy flow in the limnetic zone.

Phytoplankton, responding to temperature, oxygen, and light, is distributed vertically in the water column. Light, of course, sets the lower depth at which phytoplankton can exist. By its own growth, phytoplankton limits light penetration and thus reduces the depth at which it can live. Within these limits the depth at which various species of phytoplankton live is influenced by optimum conditions for their development. Some phytoplankton species live just beneath the water's surface; others are more abundant a few feet beneath, whereas those requiring colder temperatures live deeper still.

Because animal plankton feeds on phytoplankton, most of it, too, is concentrated in the trophogenic zone. Like phytoplankton, animal plankton is often stratified; but its strati-

**Figure 15.4** Zonation of emergent, floating, and submerged vegetation at the edge of a lake or pond. Such zonation does not necessarily reflect successional stages but rather response to water depth.

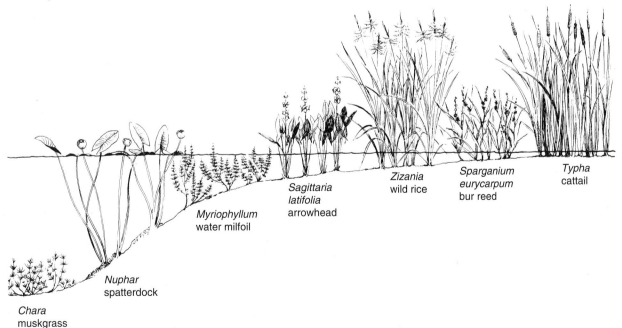

*Zizania* wild rice

*Sparganium eurycarpum* bur reed

*Typha* cattail

*Sagittaria latifolia* arrowhead

*Myriophyllum* water milfoil

*Nuphar* spatterdock

*Chara* muskgrass

(a)

(b)

**Figure 15.5** (a) Cattails dominate the littoral zone of many ponds to the exclusion of other emergent species. (b) A stand of arrow arum (*Peltandra virginica*) marks the edge of the littoral zone and the beginning of deep water.

fication changes seasonally, because most zooplankton is capable of independent movement. In winter some planktonic forms are distributed evenly to considerable depths; in summer they concentrate in those layers most favorable to them and their stage of development. At that season zooplankton undertakes a vertical migration during some part of the 24-hour period. Depending upon the species, these minute animals spend the day or night in the deep water or on the bottom and move up to the surface during the alternate period to feed on phytoplankton.

During the spring and fall overturns, plankton is carried down. At the same time nutrients, released by decomposition

**Figure 15.6** (a) Phytoplankton. (b) Zooplankton.

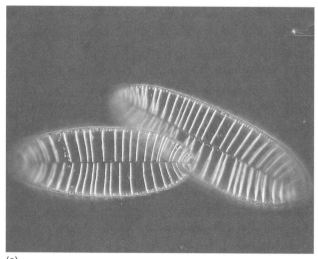

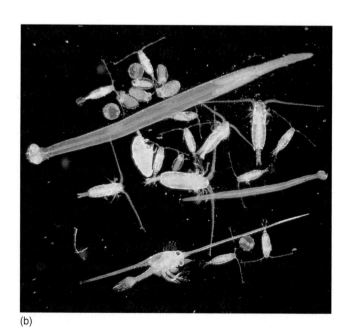

(a)

(b)

in the tropholytic zone, are carried upward into the impoverished upper layers. In spring, when surface waters warm and stratification again develops, phytoplankton has access to both nutrients and light. A spring bloom develops, followed by a rapid depletion of nutrients and a reduction in planktonic populations, especially in shallow water.

Feeding on both phytoplankton and zooplankton are nekton organisms. In the limnetic zone fish make up the bulk of the biomass. Their distribution is influenced mostly by temperature, oxygen, and food supply. For example, during the summer, bass and pike inhabit the warmer eplimnion water where food is abundant. In winter they retreat to deeper and warmer water. Lake trout, on the other hand,

move to cooler depths as summer advances. During the spring and fall overturns, when oxygen and temperature are fairly uniform throughout, both cold-water and warm-water species occupy all levels.

Common to both the trophogenic and tropholytic zones is the **benthic zone,** or bottom zone. The benthic organisms collectively are known as the **benthos.** The bottom ooze is region of great biological activity, so great that oxygen curves for.lakes and ponds show a sharp drop in the water just above the bottom. Because organic muck lacks oxygen, the dominant organisms there are anaerobic bacteria. Under anaerobic conditions, decomposition cannot proceed to inorganic end products. When amounts of organic matter

**Figure 15.7** Model for nutrient cycling and energy flow in a lake ecosystem. Meteorological, geological, and biological inputs enter from the watershed. Nutrients and energy move through a number of pathways. Part accumulates in bottom sediments. (Based on Likens and Bormann 1974 and Rich and Wetzel 1978.)

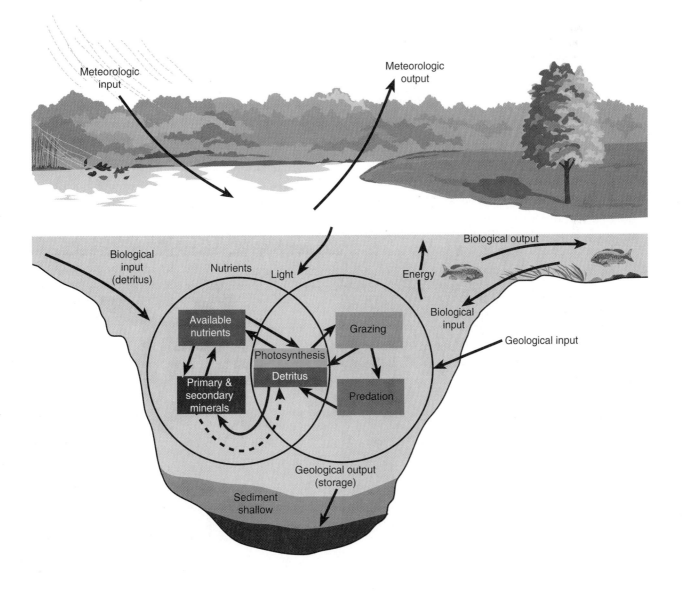

reaching the bottom are greater than can be used by bottom fauna, odoriferous muck rich in hydrogen sulfide and methane results. Therefore, lakes and ponds with highly productive limnetic and littoral zones often have an impoverished fauna on and near the bottom. Life in the bottom ooze is most abundant in lakes with a deep hypolimnion, in which oxygen is still available.

## Function

In many ways a lake might be considered a self-contained ecosystem, but in fact, lentic ecosystems are strongly influenced by inputs of nutrients from sources outside the basin (Figure 15.7). Nutrients and other substances move across the boundaries of the lentic system along biological, geological, and meteorological pathways (Likens and Bormann 1975).

Wind-borne particulate matter, dissolved substances in rain and snow, and atmospheric gases represent meteorological inputs to the system. Meteorological outputs are small, mainly spray aerosols and gases such as carbon dioxide and methane. Geological inputs include nutrients dissolved in groundwater and inflowing streams and particulate matter washed into the basin from the surrounding terrestrial watershed. Geological outputs include dissolved and particulate matter carried out of the basin by outflowing waters and nutrients incorporated in deep sediments, which may be removed from circulation for a long period of time. Biological inputs and outputs, relatively small, include animals such as fish that move into and out of the basin. Energy input is largely sunlight, and energy output is heat of respiration (R). The lentic ecosystem receives its hydrological input from precipitation and the drainage of surface waters. Outputs involve seepage through walls of the lake basin, subsurface flows, evaporation, and evapotranspiration. Within the lentic ecosystem nutrients move among three compartments, dissolved organic matter, particulate organic matter, and primary and secondary minerals. Nutrients and energy move through the system by the grazing and detrital food chains.

Although studies of lake metabolism have emphasized the phytoplankton-zooplankton grazing food chain, in reality lakes, like terrestrial communities, are dominated by the detrital food chain. The reason for an emphasis on the grazing food chain in lakes has been a preoccupation with the open-water zone and a disregard of the littoral zone, which adds significantly to lake productivity and supplies substantial quantities of detritus to the system. Detritus is all dead organic carbon. It includes particulate and dissolved organic carbon (POC and DOC) from external sources that enter and cycle within the system, and organic matter lost to a particular trophic level by such nonpredatory losses as egestion, excretion, and secretion (Rich and Wetzel 1978; Wetzel 1975).

Lake ecosystems function mostly within a framework of organic carbon transfer. The central pool, which comes from both internal and external sources, represents the major flow through the system. Particulate organic carbon comes from three sources: (1) imports, such as leaf fragments, into the system from the outside; (2) the littoral zone; (3) the limnetic zone. Most of the detrital metabolism takes place in the benthic zone, where particulate matter is decomposed, and in the limnetic zone during sedimentation.

Primary production is carried out in the limnetic zone by phytoplankton and in the littoral zone by macrophytes. The ratio of these contributions varies among lentic systems. Phytoplankton production is influenced by nutrient availability in the water column. If nutrients are not limiting and the only losses are respiratory, the rate of net photosynthesis and biomass accumulation is high. In fact, a linear relationship exists between phytoplankton production and phytoplankton biomass (Brylinsky 1980). However, as phytoplankton biomass increases, shading and respiration per unit surface increases, and net photosynthesis and thus production declines. When nutrients are low, respiration and mortality increase, reducing net photosynthesis and thus biomass. However, if zooplankton grazing and bacterial decomposition are high, nutrients are recycled rapidly, resulting in a high rate of net photosynthesis even though the concentration of nutrients and biomass accumulation are low.

Rooted aquatic plants, the macrophytes, also contribute heavily to lake production. Their maximum biomass is close to annual cumulative net production. The ratio of macrophytic production to microphytic production is influenced by the fertility of the lake. Highly fertile lakes support a heavy growth of phytoplankton that shades out macrophytes and reduces their contribution. In less fertile lakes where phytoplankton production is low, light penetrates the water and rooted aquatics grow. Macrophytes are little affected by nutrient exchange in the water column. Rooted aquatics draw on nutrients from the sediments rather than from open water.

Nutrient transfers within lentic ecosystems take place largely between the water column and sediments, and involve uptake by phytoplankton, zooplankton, bacteria, and other consumers as well as sedimentation in both the water column and benthic muds (Figure 15.8). In spring when phytoplankton bloom is at its height, nitrogen and phosphorus become depleted in the trophogenic zone, in part because of the high rate of photosynthesis, the high rate of sinking of dead phytoplankton, and the high rate of sedimentation.

In summer conditions change. Because of a decline in phytoplankton in the trophogenic zone and a slower sinking rate, as much N and P enters solution as is taken up by phytoplankton in photosynthesis. In the tropholytic zone, N and P increase in the dissolved and particulate pools and in bottom sediments. Phosphorus, in particular, becomes trapped in the hypolimnion and remains there, unavailable to phytoplankton until the fall overturn.

Macrophytes, however, can influence this transfer of phosphorus from sediments to the water column to

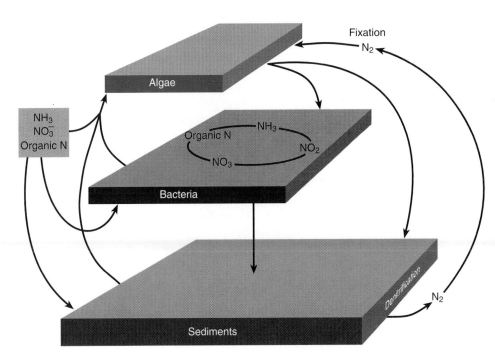

**Figure 15.8** A model of nitrogen cycling in a lentic ecosystem showing the relationship between water column and sediments. Sediments are both a storehouse and a source of nutrients in the lentic system. (Based on Golterman and Kouwe 1980:138.)

phytoplankton, and at the same time increase the accretion of bottom sediments. Rooted macrophytes draw phosphorus from the sediment. S. Carpenter (1981) in a study in Lake Wingra, Wisconsin, found that macrophytes doubled the amount of sedimentary phosphorus made available to phytoplankton that it otherwise would have had to obtain from direct release from the sediment. Macrophytes obtained 73 percent of their shoot phosphorus from sediments, eventually making much of it available to phytoplankton. This uptake stimulates both the production of macrophytic biomass and phytoplankton biomass, adding to sediment accumulation on the bottom. Sediment accumulation provides new areas for colonization and additional phosphorus for phytoplankton. Macrophytes enhance the recycling of phosphorus by mobilizing it from the sediments. Such mobilization and recycling accelerate the enrichment of lakes.

Although the productivity of a lake is related to the nutrient richness of its waters, other internal forces influence it. Any two lakes with a similar nutrient load may differ in productivity, a difference that relates to the allocation of chlorophyll *a* among the major primary producers, the phytoplankters, which vary in their rates of metabolic activity and of nutrient recycling, both size-dependent (Carpenter and Kitchell 1984). Small phytoplankters, for example, have higher maximum growth rates overall, higher maximum growth at low levels of nutrients, and a lower sinking rate than large phytoplankton species.

Grazing on the phytoplankters are zooplankton, essential to the recycling of nutrients, particularly N and P. Depending on the size relationship of the dominant zooplankters to

phytoplankters, these herbivorous zooplankters can influence the species composition and size structure of the phytoplankton community. These zooplankers, in turn, are preyed upon by insects, crustaceans, and small fish; and they in turn become prey for larger fish. Interactions among these feeding groups flow down through the food web, influencing productivity at each trophic level.

## Nutrient Status

A close relationship exists between land and water ecosystems. Primarily through the hydrological cycle one feeds upon the other. The water that falls on land runs from the surface or moves through the soil to enter streams, springs, and eventually lakes. The water carries with it silt, clay, organic matter, and nutrients in solution, all of which enrich aquatic ecosystems. Human activities, including road building, logging, mining, construction, and agriculture, add an additional heavy load of silt, organic matter, and nutrients, especially nitrogen and phosphorus. The outcome is nutrient enrichment of aquatic systems, or **eutrophication.**

The term *eutrophy* (from the Greek, "well nourished") means a condition of being nutrient-rich. The opposite of eutrophy is **oligotrophy,** the condition of being nutrient-poor. The terms were introduced by the German limnologist C. A. Weber in 1907, for the development of peat bogs. E. Naumann later associated the terms with phytoplankton production in lakes: Eutrophic lakes found in fertile lowland regions hold high populations of phytoplankton; oligotrophic lakes, typical of rocky upland regions, contain little

(a)

(b)

**Figure 15.9**  (a) An oligotrophic lake in Maine. (b) A eutrophic lake. Note the floating algal mats on the water.

plankton. However, this concept of oligotrophy and eutrophy ignores the input of highly productive litoral zones.

Oligotrophic Systems  Oligotrophic lakes (Figure 15.9a) are characterized by a low surface-to-volume ratio, water that is clear and appears blue to blue-green in the sunlight, bottom sediments that are largely inorganic, and a high oxygen concentration through the hypolimnion (Figure 15.10). The nutrient content of the water is low; although nitrogen may be abundant, phosphorus is highly limiting. Low nutrient availability results from a low input of nutrients from external sources. This factor in turn causes a low production of organic matter, particularly phytoplankton. Low organic production results in low rate of decomposition and high oxygen concentration in the hypolimnion. These oxidizing conditions are responsible for low nutrient release from the sediments. The lack of decomposable organic substances results in low bacterial populations and slow rates of microbial metabolism. Although the number of organisms in oligotrophic lakes may be low, the diversity of species is often high. Fish life is dominated by members of the salmon family.

When nutrients in moderate amounts are added to oligotrophic lakes, they are rapidly taken up and circulated. If nitrogen and carbon are in excess and phosphorus is limiting, the addition of phosphorus would stimulate growth; if nitrogen is limiting, the addition of that element would do the same. In most oligotrophic lakes phosphorus rather than nitrogen is limiting (Vallentyne 1974; Wetzel 1975). As increasing quantities of nutrients are added to the lake, it begins to change from oligotrophic to **mesotrophic** (having a moderate amount of nutrients) to eutrophic. This change has been happening at an increasing rate to clear oligotrophic lakes around the world.

Eutrophic Systems  A typical eutrophic lake (Figure 15.10) has a high surface-to-volume ratio; that is, the surface area is large relative to depth. It has an abundance of nutrients, especially nitrogen and phosphorus, that stimulate a heavy growth of algae and other aquatic plants (Figure 15.9b). Increased photosynthetic production leads to increased regeneration of nutrients and organic compounds, stimulating further growth. Phytoplankton becomes concentrated in the upper layer of the water, giving it a murky green cast. The turbidity reduces light penetration and restricts biological productivity to a narrow zone of surface water. Algae, inflowing organic debris and sediment, and the re-

**Figure 15.10**  Comparison of oligotrophic and eutrophic lakes.

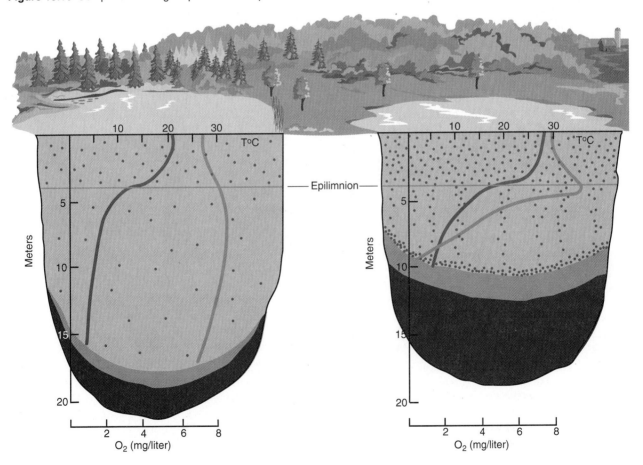

mains of rooted plants drift to the bottom, adding to the highly organic sediments. On the bottom bacteria partially convert dead matter into inorganic substances. The activities of these decomposers deplete the oxygen supply of the bottom sediments and deep water to a point where the deeper parts of the lake are unable to support aerobic forms of life, and aerobic decomposition ceases. The amount of oxygen needed for oxidative decomposition is greater than the amount of oxygen available. The needed amount is called **biochemical oxygen demand (BOD).**

As the basin continues to fill, the volume decreases, and shallowness speeds the cycling of available nutrients and further increases plant production. Positive feedback carries the system to eventual extinction—the filling in of the basin and the development of a marsh, swamp, and ultimately a terrestrial community.

In fact, "galloping eutrophication" has been changing naturally eutrophic lakes into **hypertrophic** ones. An excessive nutrient content results from a heavy influx of wastes, raw sewage, drainage from agricultural lands, river basin development, runoff from urban areas, and burning of fossil fuels. This accelerated enrichment, which results in chemical and environmental changes in the system and causes major shifts in plant and animal life, has been called **cultural eutrophication** (Hasler 1969).

Dystrophic Systems   Lakes that receive large amounts of organic matter from surrounding watersheds, particularly in the form of humic materials that stain the water brown, are called **dystrophic.** Although the productivity of dystrophic lakes is considered low, this refers only to planktonic production. Dystrophic lakes generally have highly productive littoral zones, particularly those that develop bog flora. This littoral vegetation dominates the metabolism of the lake ecosystem, providing a source of both dissolved and particulate organic matter (see Wetzel and Allen 1970, Wetzel 1975).

Marl Systems   A fourth type of system, the **marl lake,** contains extremely hard water due to inputs of calcium over a long period of time. A hard-water lake is relatively unproductive and remains so because of the reduced availability of nutrients, even though carbonates remain high. Under certain conditions in these lakes, phosphorus, iron, magnesium, and other nutrients form insoluble compounds and are lost to the system. Sodium and potassium are low, but nitrogenous compounds are high. Because of low photosynthetic productivity, nitrogen remains unutilized. Calcium is often supersaturated, and carbonates, especially calcium carbonate, and humic dissolved organic matter precipitate to form marl deposits on the bottom. If carbonate inputs from drainage are reduced or depleted or if sediments build up high enough to support littoral vegetation and the growth of *Sphagnum,* the marl lake gradually develops into a bog (for details, see Wetzel 1972, 1975.)

# LOTIC ECOSYSTEMS

Lotic ecosystems, streams and rivers, are characterized by continuously moving water. Current cuts the channel, molds the character of the stream, and influences the lives of organisms inhabiting flowing waters.

## Physical Characteristics

Streams may begin as outlets of ponds or lakes, or they may arise from springs and seepage areas (Figure 15.11a). Added in varying quantities is surface runoff, especially after heavy or prolonged rains and rapid snowmelt. Because precipitation, the source of all runoff and subsurface water, varies seasonally, the rate and volume of streamflow may fluctuate widely over months, weeks, or even days from flood conditions to dry channels.

As water drains away from its source, it flows in a direction dictated by the lay of the land and the underlying rock formations. Its course may be determined by the original slope and its regularities; or the water, seeking the least resistant route to lower land, may follow the joints and fissures in bedrock near the surface and shallow depressions in the ground. Whatever its direction, water is concentrated into rills that erode small furrows, which soon grow into gullies.

Water, moving downstream, especially where the gradient is steep, carries with it a load of debris that cuts the channel wider and deeper and that sooner or later is deposited within or along the stream. At the same time erosion continues at the head of the gully, cutting backward into the slope and increasing its drainage area.

Just below its source, the stream may be small, straight, and swift, with waterfalls and rapids. Further downstream, where the gradient is less and the velocity decreases, meanders become common. They are formed when the current, deflected by some obstacle on the floor of the channel, by projecting rocks and debris, or by the entrance of swifter currents, strikes the opposite bank. As the water moves downstream, it is thrown back to the other side again. These abrasive forces create a curve in the stream, on the inside of which the velocity is slowed and the water drops its load. Such cutting and deposition often cause valley streams to change course and to cut off the meanders to form oxbow lakes. When the water reaches level land, its velocity is greatly reduced, and the load it carries is deposited as silt, sand, or mud.

At flood time the material carried by the stream is dropped on the level lands over which the water spreads to form floodplain deposits. These floodplains, which humans have settled so extensively, are a part of the channel that the river takes back at flood time, a fact that few people recognize. The current at flood time is swiftest in the normal channel of the stream and slowest on the floodplain. Along the margin of the channel and the floodplain, where the

(a)

(b)

**Figure 15.11** (a) This small lake marks the beginning of a stream. (b) A satellite view of the delta of the Mississippi River, which marks the end of a river system that had its beginning source as a small stream.

rapid water meets the slow, the current is checked and all but the fine sediments are dropped on the edges of the channel. Thus the deposits on the floodplain, which become alluvial soils, are higher on the immediate border and slope off gradually toward the valley side.

When a stream or river flows into a lake or sea, the velocity of the water is suddenly checked and the load of sediment is deposited in a fan-shaped area at the inlet point to form a delta (Figure 15.11b). Here the course of the water is broken into a number of channels, which are blocked or opened with subsequent deposits. As a result, the delta is characterized by small lakes and swampy or marshy islands. Material not deposited at the mouth is carried further out to open wa-

ter, where it settles on the bottom. Eventually the sediments build up above the water to form new land surface.

Velocity of the current molds the character of a stream. This velocity varies from stream to stream and within the stream itself. It is influenced by the size, shape, and steepness of the stream channel, the roughness of the bottom, the depth, and the rainfall.

The velocity of flow influences the degree of silt deposition and the nature of the bottom. The current in the riffles is too fast to allow siltation, but coarser silt particles drop out in the smooth or quiet sections of the stream. High water increases the velocity; it moves bottom stones, scours the streambed, and cuts new banks and channels. In very steep

streambeds, the current may remove all but very large rocks and leave a boulder-strewn stream.

Flowing water also transports nutrients to and carries waste products away from many aquatic organisms and may even sweep them away. Balancing this depletion of bottom fauna, the current continuously reintroduces bottom fauna from areas upstream. Similarly, as nutrients are washed downstream, more are carried in from above. For this reason, the productivity of primary producers in streams is 6 to 30 times that of those in standing water (Nelson and Scott 1962). The transport and removal action of flowing water benefits such continuous processes as photosynthesis.

## Structure

**Fast Water** Fast or swiftly flowing streams are, roughly, all those whose velocity of flow is 50 cm/sec or higher (A. Nielsen 1950). At this velocity the current will remove all particles less than 5 mm in diameter and will leave behind a stony bottom. The fast stream is often a series of two different but related habitats, the turbulent riffle and the quiet pool (Figure 15.12). The waters of the pool are influenced by processes occurring in the rapids above, and the waters of the rapids are influenced by events in the pool.

Riffles are the sites of primary production in the stream (see Nelson and Scott 1962). Here the aufwuchs or periphyton assume dominance and occupy a position of the same importance as the phytoplankton of lakes and ponds. The **aufwuchs** consist chiefly of diatoms, blue-green and green algae, and water moss. Extensive stands of algae grow over rocks and rubble on the streambed and form a slippery covering. Growth during favorable periods may be so rapid that the stream bottom is covered in ten days or less (Blum 1960). Many small algal species are epiphytes growing on top of or in among other algae.

The outstanding feature of much of this algal growth is its ephemeral nature. Scouring action of water and the debris it carries tears away larger growth, epiphytes and all, and sends the algae downstream. As a result there is a constant contribution from upstream to the downstream sequence.

Above and below the riffles are the pools. Here the environment differs in chemistry, intensity of current, and depth. Just as the riffles are the sites of organic production, so the pools are the sites of decomposition. They are the catch basins of organic materials, for here the velocity of the current is reduced enough to allow a part of the load to settle out. Pools are the major sites for free carbon dioxide production during the summer and fall.

Overall production in a stream is influenced in part by the nature of the bottom. Pools with sandy bottoms are the least productive, because they offer little substrate for either aufwuchs or animals. Bedrock, although a solid substrate, is so exposed to currents that only the most tenacious organisms can maintain themselves. Gravel and rubble bottoms support the most abundant life because they have the great-

**Figure 15.12** Two different but related habitats in a stream: riffles (foreground) and a pool (background).

est surface area for the aufwuchs, provide many crannies and protected places for insect larvae, and are the most stable (Figure 15.13a). Food production decreases as the particles become larger or smaller than rubble. Insect larvae, on the other hand, differ in abundance on the several substrates. Mayfly nymphs are most abundant on rubble, caddisfly larvae on bedrock, and Diptera larvae on bedrock and gravel (Pennak and Van Gerpen 1947).

The width of the stream also influences overall production. Bottom production in streams 6 m wide decreases by one-half from the sides to the center; in streams 30 m wide it decreases by one-third (Pate 1933). Streams 2 m or less in width are four times as rich in bottom organisms as those 6 to 7 m wide. This is one reason why headwater streams make such excellent trout nurseries.

**Slow Water** As the current slows, a noticeable change takes place in streams (Figure 15.14). Silt and decaying organic matter accumulate on the bottom, and fine detritus

(a)

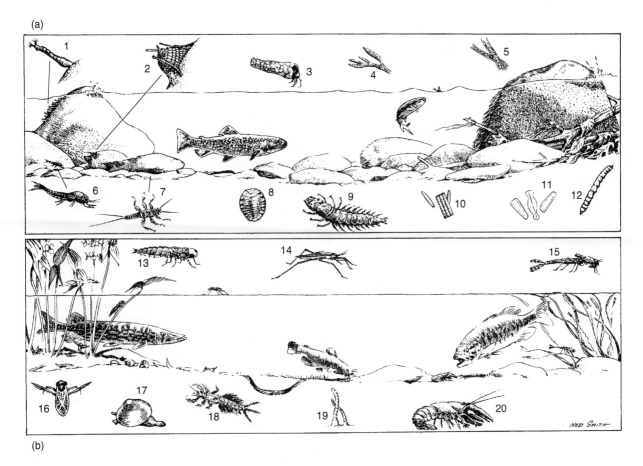

(b)

**Figure 15.13** Comparison of life in a fast streams (a) and a slow stream (b). (1) Blackfly larvae (Simuliidae); (2) net-spinning caddisfly (*Hydropsyche* spp.); (3) stone case of caddisfly; (4) water moss (*Fontinalis*); (5) algae (*Ulothrix*); (6) mayfly nymph (*Isonychia*); (7) stonefly nymph (*Perla* spp.); (8) water penny (*Psephenus*); (9) hellgrammite (dobsonfly larva, *Corydalis cornuta*); (10) diatoms (*Diatoma*); (11) diatoms (*Gomphonema*); (12) cranefly larva (Tipulidae); (13) dragonfly nymph (Odonata, Anisoptera); (14) water strider (*Gerris*); (15) damselfly larva (Odonata, Zygoptera); (16) water boatman (Corixidae); (17) fingernail clam (*Sphaerium*); (18) burrowing mayfly nymph (*Hexegenia*); (19) bloodworm (Oligochaeta, *Tubifex* spp.); (20) crayfish (*Cambarus* spp.). The fish in the fast stream is a brook trout (*Salvelinus fontinalis*). The fish in the slow stream are, from left to right: northern pike (*Esox lucius*), bullhead (*Ameiurus melas*), and smallmouth bass (*Micropterus dolommieu*).

from upstream is the main source of energy. Faunal organisms are able to move about to obtain their food, and a plankton population develops. The composition and configuration of the stream community approaches that of standing water.

With increasing temperatures, decreasing current, and accumulating bottom silt, organisms of the fast water are replaced by organisms adapted to these conditions (Figure 15.13b). Brook trout and sculpin give way to smallmouth bass and rock bass, the dace to shiners and darters. With current at a minimum, many resident fish lack the strong lateral muscles typical of the trout and have compressed bodies that permit them to move with ease through masses of aquatic plants. Mollusks, particularly *Sphaerium* and *Pisidium,* and pulmonate snails, crustaceans, and burrowing mayflies re-

place the rubble-dwelling insect larvae. Only in occasional stretches of fast water in the center of the stream do remnants of headwater-stream organisms still live.

As the volume of water increases, as the current becomes even slower, and as the silt deposits become heavier, detritus-feeders, particularly mollusks, increase. Rooted aquatics appear. Emergent vegetation grows along the riverbanks, and duckweeds float on the surface. Indeed, the whole aspect approaches that of lakes and ponds, even to zonation along the river margin.

Abundant decaying matter promotes the growth of protozoan and other plankton populations. Scarce in fast water, plankton increases in numbers and species in slow water. Rivers have no typical plankton of their own. Those found there originate mainly from backwaters and lakes. In gen-

**Figure 15.14** A slow stream edged meanders through a growth of willows.

eral, plankton populations in rivers are not nearly as dense as those in lakes. Time is too short for much multiplication of plankton because little time is needed for a given quantity of water to flow from its source to the sea. Also occasional river rapids, often long, kill many plankton organisms by vi-

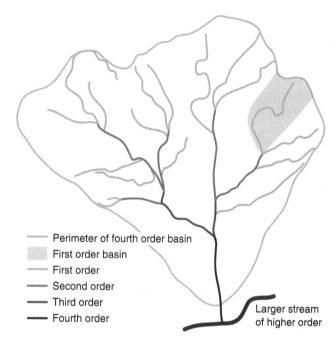

Perimeter of fourth order basin
First order basin
First order
Second order
Third order
Fourth order
Larger stream of higher order

**Figure 15.15** Stream orders and basins.

olent impact against suspended particles and the bottom. Aquatic vegetation filters out this minute life as the current sweeps it along.

Stream Order Because streams become larger on their course to a river and are joined along the way by many others, streams at any one point in the landscape can be classified according to order (Figure 15.15). A small headwater stream without any tributaries is a first order stream. When two streams of the same order join, the stream becomes one of higher order. If two first-order streams unite, the resulting stream becomes a second-order one; and when two second-order streams unite, the stream becomes a third-order one. A stream can increase in order only when a stream of the same order joins it. Order cannot be increased with the entry of a lower order stream. In general, headwater streams are orders 1 to 3; medium-sized streams, 4 to 6; and rivers, greater than 6.

Streams receive runoff from the surrounding land as overland flow. The land area contributing to the flow for any one stream is its **basin** (Figure 15.15). A first-order stream collects overland flow from the slopes of a first-order basin. A second-order stream receives flow from first order-streams, and thus first-order basins, as well as its own direct overland flow. The same pattern repeats for third and higher order streams. These basins are also called watersheds. A **watershed** is a body of land bounded above by a ridge or water divide and below by the level at which the water drains from the basin. The size of a watershed tends to increase geometrically with the mean basin areas of successive stream orders.

## Function

The lotic or flowing water system is open and largely heterotrophic, especially in headwater streams. A major source of energy and nutrients is detrital material from the outside (Figure 15.16). Much of this organic matter comes in the form of leaves and woody debris dropped from streamside vegetation, collectively called coarse particulate organic matter (CPOM) (particles larger than 1 mm) (Figure 15.17). Another type of organic input is fine particulate organic matter (FPOM), material less than 1 mm, including leaf fragments, invertebrate feces, and precipitated dissolved organic matter. A third input is dissolved organic matter (DOM), material less than 0.5 micron in solution. One source of DOM is rainwater dropping through overhanging leaves, dissolving the nutrient-rich exudates on them. Another is subsurface seepage, which brings nutrients leached from adjoining forest, agricultural, and residential lands. Many streams receive inputs from dumping of industrial and residential effluents. Supplementing this detrital input is autotrophic production in streams by diatomaceous algae growing on rocks and by rooted aquatics such as water moss (*Fontinalis*).

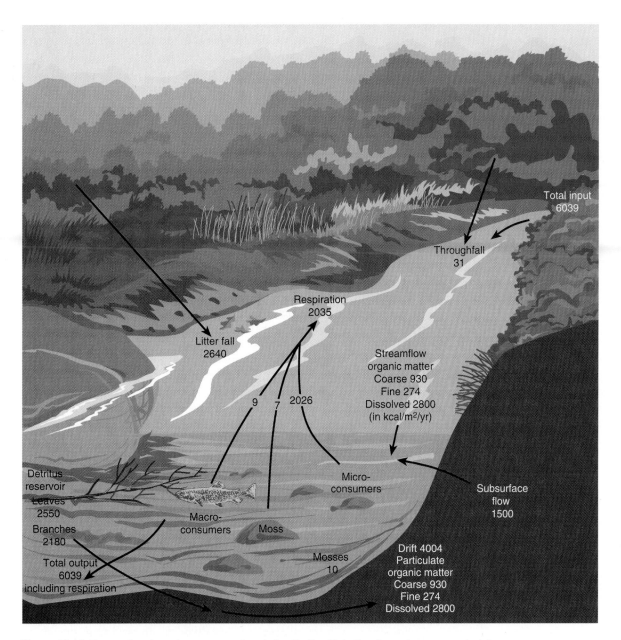

**Figure 15.16** Energy flow in a stream ecosystem in kcal/m²/yr. Note the dependence on materials from terrestrial sources and inflow from upstream and the roles of coarse particulate organic matter (CPOM), fine particulate organic matter (FPOM), and dissolved organic matter (DOM). Primary production contributes little to energy flow. Energy values are based on Bear Brook, Hubbard Forest, New Hampshire. (Data from Fisher and Likens 1973.)

The processing of this organic matter involves both physical and biological mechanisms. In fall, leaves drift down from overhanging trees, settle on the water, float downstream, and lodge against banks, debris, and stones. Soaked with water, the leaves sink to the bottom, where they quickly lose 5 to 50 percent of their dry matter as water leaches soluble organic matter from their tissues. Much of this DOM is either incorporated onto detrital particles or precipitated to become part of FPOM. Another part is incorporated into microbial biomass. Once softened, leaves and other debris are processed by a number of species of invertebrates, which can be placed into several functional groups (Figure 15.18).

Functional Groups  Within a week or two, depending upon the temperature and the plant species involved, the surface of the leaves is colonized by bacteria and fungi, largely aquatic hyphomycetes. Fungi are more important on CPOM because large particles offer more surface for mycelial development (Cummins and Klug 1970). Bacteria are associated more with FPOM. Microorganisms degrade cellulose

**Figure 15.17** Much of the detrital input in this first-order stream comes with leaf fall in autumn. Leaves cover the surface of the pool.

**Figure 15.18** Model of structure and function in a system, showing the processing of leaves and other particulate matter and dissolved organic matter. (Adapted from Cummins 1974:663.)

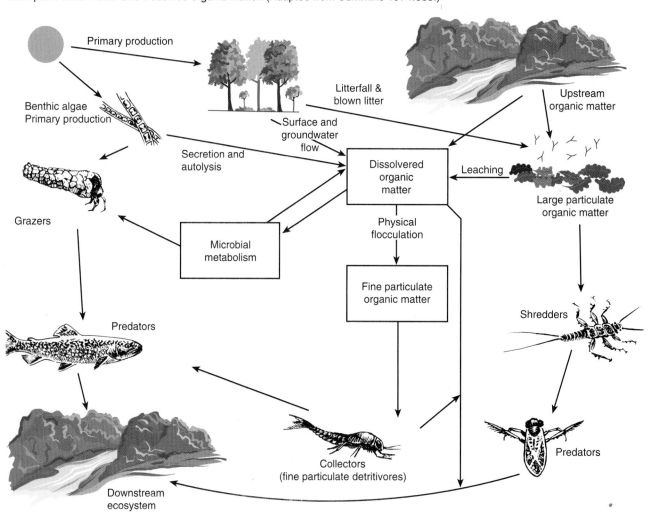

and metabolize lignin. Their populations form a layer on the surface of leaves and detrital particles that is much richer nutritionally than the detrital particles themselves (Anderson and Cummins 1979). Leaves and other detrital particles are attacked by a major feeding group, the *shredders,* invertebrates that feed on leaves and other large organic particles. Among the shredders are the larvae of craneflies (Tipulidae) and caddisflies (Trichoptera), the nymphs of stone-flies (Plecoptera), and crayfish. They break down the CPOM, feeding on the material not so much for the energy it contains but for the bacteria and fungi growing on it (Cummins 1974, Cummins and Klug 1979). In doing so they skeletonize the leaf by feeding on the softer portions. Shredders assimilate about 40 percent of the material they ingest and pass off 60 percent as feces.

Broken up by shredders and partially decomposed by microbes, the leaf material along with fecal material becomes part of the FPOM, which also includes some precipitated DOM. Drifting downstream and settling on the stream bottom, FPOM is picked up by another feeding group of stream invertebrates, the *filtering* and *gathering collectors.* The filtering collectors include, among others, the larvae of blackflies (Simuliidae) with filtering fans, and net-spinning caddisflies, including *Hydropsyche.* Gathering collectors, such as the larvae of midges, pick up particles from stream bottom sediments. Collectors obtain much of their nutrition from bacteria associated with fine detrital particles.

Whereas shredders and collectors feed on detrital material, another group feeds on algal coating of stones and rubble. They are the *scrapers,* which include the beetle larvae, popularly known as the water penny (*Psephenus* spp.), and mobile caddisfly larvae. Much of the material they scrape loose enters the drift as FPOM. Behaviorally and morphologically, scrapers are adapted to maintain their position in the current either by flattening to avoid the main force of flow or by weighting down with heavy mineral cases.

Feeding on mosses and filamentous algae are the *piercers,* consisting largely of microcaddisflies. Another group, associated with woody debris, are the *gougers,* invertebrates that burrow into waterlogged limbs and trunks of fallen trees.

Feeding on the detrital feeders and scrapers are predaceous insect larvae such as large stoneflies, the powerful dobsonfly larvae (*Corydalus cornutus*), stream salamanders, and fish such as sculpin (*Cottus* spp.) and trout. Invertebrate predators employ either ambush or searching strategies. Some engulf the prey whole or in pieces; others pierce the body of prey and suck out all or part of the contents. Predaceous aquatic insects detect prey largely by mechanical cues (Peckarsky 1982), and are size-selective opportunists. Prey species in turn have evolved morphological and behavioral defenses, including flattened body shapes, protective cases (caddisflies), and cryptic coloration. Predators, especially those in headwater streams, do not depend solely on aquatic insects; they also feed heavily on terrestrial invertebrates that fall or are washed into the stream.

Because of current, quantities of CPOM and FPOM and invertebrates tend to move downstream to form a traveling benthos called drift. This process is normal in streams, even in the absence of high water and abnormal currents (for reviews see Hynes 1970, Waters 1972). Drift is so characteristic of streams that a mean rate of drift can serve as an index of the production rate (Pearson and Kramer 1972). This drift is essential to the production processes of downstream systems (Wallace et al. 1982).

**Energy Flow and Nutrient Cycling** Energy flow and nutrient cycling in a lotic ecosystem have been documented for the well-studied, small, forested Bear Brook in Hubbard Forest of northern New Hampshire (Fisher and Likens 1973). That budget is summarized in Figure 15.16. Over 90 percent of the energy input comes from the surrounding forested watershed or from upstream. Primary production by mosses accounts for less than 1 percent of the total energy supply. Algae are absent from the brook. Inputs from litter and throughfall account for 44 percent of the energy supply, and geological inputs account for 56 percent. Energy is introduced in three forms: CPOM, represented by leaves and other debris; FPOM, represented by drift and small particles; and DOM. In Bear Brook 83 percent of the geologic input from sediments and particulate mineral matter and 47 percent of the total energy input is in the form of DOM. Sixty-six percent of the organic input is exported downstream, leaving 34 percent to be utilized locally.

Nutrient cycling is more difficult to assess, because of the very open nature of the lotic system. Triska et al. (1984) estimated the nitrogen budget of a small stream draining a 10.1 ha watershed in the old-growth Douglas-fir H. J. Andrews Experimental Forest in Oregon (Table 15.1). The major annual input of dissolved nitrogen, amounting to 15.3 g/N/m$^2$, came from two major sources: hydrological and biological. Hydrological inputs, largely subsurface flow in the form of seeps, accounted for most of the input, 11.1 g/m$^2$. Biological inputs accounted for 4.2 g/m$^2$. Particulate organic nitrogen contributed 11.9 g, of which CPOM made up 7.2 g and FPOM 4.7 g. Total output of nitrogen was 11.3 g/m$^2$. Of this amount 2.5 g was particulate organic matter dominated by CPOM (1.9 g), and 8.8 g was dissolved N, less than the input because of biological uptake. The stream intercepted and transported nearly all of the nitrogen lost from the 10 ha watershed, yet effectively retained and processed a considerable portion of it. Nitrogen transported downstream would be utilized there for microbial consumption.

A major problem for flowing water ecosystems is retention of nutrients upstream. Although nutrient cycling is downhill in all ecosystems, the problem in lotic systems is the constant movement of the substrate—water—away from the system. Nutrients in terrestrial and lentic systems are recycled more or less in place. An ion of a nutrient passes from

**Table 15.1** Nitrogen Budget (g/m$^2$) for a Small Coniferous Forest Stream (Watershed 10, H. J. Andrews Experimental Forest, Oregon)

| Inputs | | Outputs | |
|---|---|---|---|
| Dissolved organic N pool | 15.25 | Total nitrogen output | 11.36 |
| Hydrological | 11.06 | Dissolved N | 8.81 |
| DON | 10.56 | DON | 8.38 |
| NO$^3$ – N | 0.50 | NO$^3$ – N | 0.43 |
| Biological | 4.19 | Particulate organic N | 2.53 |
| N-fixation | 0.76 | FPOM | 1.66 |
| Thoughfall | 0.30 | CPOM | 0.87 |
| Litterfall | 1.35 | Insect emergence | 0.02 |
| Blown leaves | 1.78 | | |
| Particulate organic N pool | 11.93 | | |
| FPOM | 4.77 | | |
| CPOM | 7.16 | | |

*Source:* Data from Triska et al. 1984.

soil or water column to plants and consumers back to soil or water in the form of detrital material or exudates. It is then recycled within the same segment of the system, although losses do occur. Cycling is essentially temporal.

Lotic systems have an added spatial cycle. Nutrients in the form of DOM and POM move continually downstream. How rapidly they are carried depends upon water flow and physical and biological retention in place. The greater the flow, the more rapid the loss. Physical retention involves storage in wood detritus such as logs and snags in the stream, accumulation of debris in pools formed behind boulders, leaf sediments, and beds of macrophytes. Recycling is biological, controlled by uptake and storage in animal and plant tissue.

The process of nutrient uptake, transformation, and release with downstream transport may be pictured as a longitudinal *spiral* (Figure 15.19). The combined processes of nutrient cycling and downstream transport are termed **spiraling.** One cycle in the spiral is the distance required for the uptake of an ion of a nutrient by consumer organisms, its transformation into particulate form, and its return to water, where it is available for reutilization. The distance an ion travels in dissolved form before being taken up is its *uptake length.* The distance it travels in particulate form before being released to DOM is its *turnover length.* The uptake length plus the turnover length is the *spiral length.* The longer the spiral length, the less efficient is the stream in retaining nutrients; the shorter the length, the more efficient.

Tight spiraling of nutrients in flowing water ecosystems depends upon the retention of leafy detritus in place long enough to allow the biological component of the stream, especially the shredders and microbes, to process organic matter. This delay is especially important in rapidly flowing headwater streams (Bilby 1989). Logs, snags, and other woody debris in the channel and along the banks, plus large rocks, act as dams, intercepting leafy detritus and forming

pools that collect sediments. These debris dams create a diversity of physical habitats within the stream and influence the nature of the invertebrate community (Figure 15.20). In small streams (orders 0 to 2) wood and debris and sediments stored behind them may make up 50 percent of the stream area, and in larger streams (orders 3 to 4) 25 percent of the stream area (Triska et al. 1982).

What are the measured lengths of spirals? Webster et al. (1991) measured the short-term spiral lengths of marked phosphorus in three small heavily shaded woodland streams at the Coweeta Hydrologic Laboratory in the Appalachian Mountains of North Carolina. Uptake lengths ranged from 3 m to 81 m. They were influenced by velocity and by temperature of the water. Nutrient uptake lengths are much shorter in the warmer water of summer than in the colder water of winter (D'Angelo et al. 1991).

## The River Continuum

Lotic ecosystems involve a continuum of physical and biological conditions from the headwaters to the mouth (Vannote et al. 1980, Minshall et al. 1983). The upper reaches of the headwaters (orders 1 to 3) are usually swift, cold, and in forested regions shaded. Riparian vegetation reduces light, inhibiting autotrophic production (Hawkins et al. 1982), and contributes more than 90 percent of organic input into streams as terrestrial detritus. Even when headwater streams are exposed to sunlight and autotrophic production exceeds heterotrophic inputs, organic matter produced invariably enters detrital food chains (Minshall 1978). Dominant organisms are shredders, processing large litter and feeding on CPOM, and collectors, processors of FPOM. Populations of grazers are minimal, reflecting the small amount of autotrophic production, and predators are mostly small fish—sculpins, darters, and trout. Headwater streams are accumulators, processors, and transporters of particulate

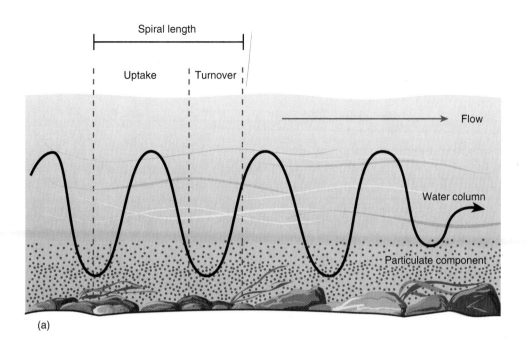

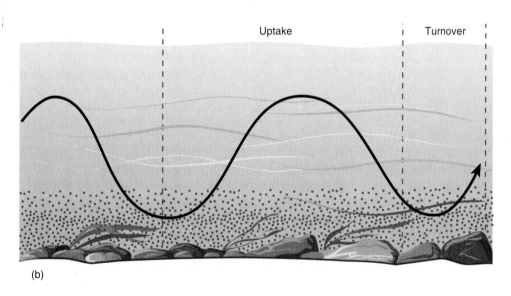

**Figure 15.19** Nutrient spiraling in a lotic ecosystem between particulate organic matter and the water column. Uptake and turnover take place as nutrients flow downstream. (a) represents tight spiraling, and (b) more open spiraling. The tighter the spiraling, the more efficient. (Adapted from Newbold et al. 1982:630.)

organic matter from terrestrial systems. As a result the ratio of gross primary production to community respiration is less than 1. Consumer organisms utilize long-chain and short-chain organic compounds for transport downstream. Organisms of headwater streams are adapted to a narrow temperature range, to a reduced nutrient regime, and to maintenance of their position in the current.

As streams increase in width to medium-sized creeks and rivers (orders 4 to 6), the importance of riparian vegetation

and its detrital input decreases. The lack of shading results in higher temperature. As the gradient declines, the current slows and becomes more variable. The diversity of microenvironments supports a greater diversity of organisms. An increase in light and temperature and a decrease in terrestrial input encourage a shift from heterotrophy to autotrophy, relying on algal and rooted plant production. Gross primary production now exceeds community respiration. Because of the lack of CPOM, shredders disappear, and collectors,

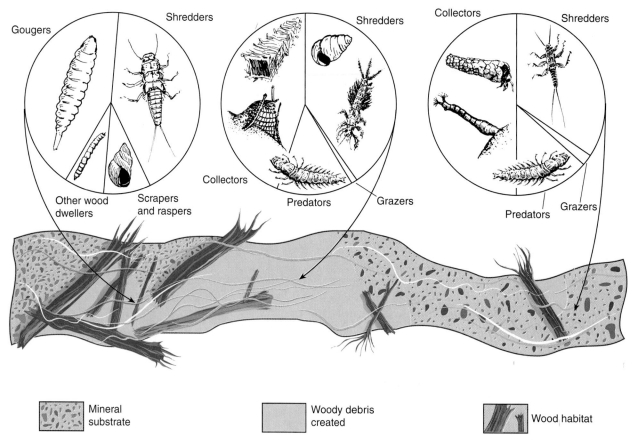

**Figure 15.20** Large woody debris modifies the stream habitat, slowing the flow, retaining organic debris for processing by invertebrates, and adding to the diversity of biota. Note the addition of another functional group, the gougers, where woody debris is prominent. Woody debris habitat favors collectors and shredders, especially on small-order streams.

feeding on FPOM transported downstream, and grazers, feeding on autotrophic production, become the dominant consumers. Predators show little increase in biomass but shift from cold-water species to warm-water species.

As the stream order increases from 6 through 10 and higher, riverine conditions develop. The channel is wider and deeper. The volume of flow increases, and the current becomes slower. Sediments accumulate on the bottom. Both riparian and autotrophic production decrease, with a gradual shift back to heterotrophy. A basic energy source is FPOM, utilized by bottom-dwelling collectors, now the dominant consumers. However, slow, deep water and dissolved organic matter (DOM) support a minimal phytoplankton and associated zooplankton population.

Throughout the downstream continuum, the lotic community capitalizes on upstream feeding inefficiency. Downstream adjustments in production and the physical environment are reflected in changes in consumer groups (Figure 15.21). Through the continuum the lotic ecosystem achieves some balance between the forces of stability, such as natural obstructions in flow that aid in retention of nutri-ents upstream, and the forces of instability, such as drought, flooding, and temperature fluctuations.

## Regulated Rivers and Streams

Few of the world's rivers are free-flowing. Most have been dammed, many repeatedly, channelized, or otherwise altered, modifying the river continuum (Cummins 1988). Such rivers, whose water levels and water flows are controlled, whose channels have been straightened, and whose waters have been isolated from their flood plains by levees, are termed **regulated rivers** (Figure 15.22).

Regulated rivers differ significantly from free-flowing ones. Under normal conditions, free-flowing streams and rivers experience seasonal fluctuations in flow. Snowmelt and early spring rains bring scouring high water; summer brings low water levels that expose some of the streambed and speed decomposition of organic matter along the edges. Life of the lotic environment has adapted its life cycle to these seasonal changes. Damming a river or stream interrupts both nutrient spiraling from upstream and the lotic

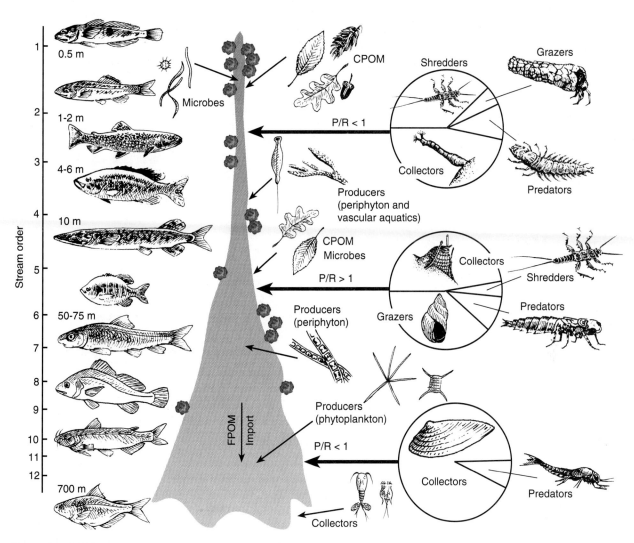

**Figure 15.21** Changes in consumer groups along the river continuum. The headwater stream is strongly heterotrophic, dependent on terrestrial input of detritus, and the dominant consumers are shredders and collectors. As stream size increases, the input of organic matter shifts from particulate organic matter to primary production from algae and rooted vascular plants. The major consumer groups are now collectors and grazers. The zone at which the shift occurs depends upon the degree of shading. As the stream increases to river, the lotic system shifts back to heterotrophy, supported by inputs of fine particulate organic matter and dissolved organic matter. A phytoplankton population may develop. The bottom consumers are collectors, mostly bottom-dwelling organisms.

continuum. Downstream flow is greatly reduced as a pool of water fills behind the dam. This pool develops characteristics similar to those of a natural lake, yet retains some features of the lotic system, such as a constant inflow of water. Heavily fertilized by decaying detrital material on the newly flooded land, the lake develops a heavy bloom of phytoplankton and in tropical regions dense growths of floating plants. Species of fish, often exotics adapted to lake-like conditions and introduced to provide a commercial or sport fishery, replace fish of flowing water.

The type of pool allowed to develop behind the dam depends upon the purpose of the dam and has a strong effect on downstream conditions. Some are single-purpose dams for flood control or water storage; others are multiple-purpose dams, providing hydroelectric power, irrigation water, and recreation, among other uses. Flood control dams have a minimum pool; the dam fills only during a flood, at which time inflow exceeds outflow. Engineers release the water slowly from the dam so that outflow exceeds inflow, but minimizes downstream flooding. In time water in the dam slowly recedes to original pool depth. During flood and post-flood periods the river below carries a strong flow for some time, scouring the river bed. During normal times flow below the dam is stabilized. If the dam is for water storage,

**Figure 15.22** Dammed river.

the reservoir holds its maximum pool; but during periods of water shortage and drought, drawdown of the pool can be considerable, exposing large expanses of shoreline for a long time and stressing or killing littoral life. Only a minimal quantity of water is released downstream, usually an amount required by law, if such exists. Hydroelectric and multiple-purpose dams hold a variable amount of water, determined by demand for power. Pools move up and down except during periods of power production, which involves pulsed releases strong enough to wipe out or dislodge benthic life downstream.

Dams with a large pool of water become stratified with a well-developed epilimnion, metalimnion, and hypolimnion. If water is discharged from the upper stratified layer of the reservoir, the effect of the flow downstream is similar to that of a natural lake. Warm, nutrient-rich, well oxygenated water creates highly favorable conditions for some species of fish below the spillway and on downstream. If the discharge is from the cold hypolimnion, downstream receives cold, oxygen-poor water carrying an accumulation of iron and other minerals and a concentration of soluble organic materials. Such conditions, inimical to stream life, may persist for hundreds of kilometers downstream before the river reaches anything near normal conditions. The use of gated selective withdrawal structures or induced artificial circulation to increase oxygen concentration reduces such problems at some dams.

Impacts of dams on lotic systems are compounded when a number of multipurpose dams are built on a river. The amount of water released and moving downstream becomes less with each dam until eventually all available water is consumed and the river simply dries up. That is the situation on the Colorado River, the most regulated river in the world. The river is nearly dry by the time it reaches Mexico.

Effects of dams go beyond simply changing the nature of the lotic system. Large dams on such rivers as the Columbia in North America interfere with the migratory patterns of anadromous fish, such as salmon, which come in from the sea and ascend rivers to spawn. Dams obstruct the upstream movement of fish. Although fish ladders provide some assistance, many local populations have been excluded from their traditional spawning streams. Even if spawning is successful, unnatural timing of high and low flows in the rivers may induce premature seaward migration of the young or lengthen their downstream passage, exposing them to high temperatures. Nearly 90 percent of juveniles perish on the journey from their home stream to the estuary.

## WETLANDS

What is a wetland? The answer to this question is critical, for upon it rests the fate of many wetlands, among the most endangered ecosystems. Viewed as wastelands to be put to more productive uses, wetlands have been drained for agriculture, urban expansion, industrial sites, home sites, and dumps. Since settlement the United States has lost over 53 percent of its wetlands. Although wetlands make up only 5 percent of the land surface of the lower 48 states and 12 percent if Alaska and Hawaii are included, strong pressure to drain much of the remaining wetlands persists. A number of states and the federal government have instituted regulations to protect remaining wetlands for the values they provide. This protective legislation centers about the definition of a wetland.

Defining a **wetland** would seem to be simple enough. Wetlands do have certain characteristics: (1) the presence of water at or near the surface; (2) soils that differ from those of adjacent uplands, and (3) vegetation adapted to wet conditions (Mitsch and Gosselink 1993). Such an answer, however, is not enough. The problem is that wetlands are a halfway world between aquatic and terrestrial ecosystems and exhibit some characteristics of each. Combining the various characteristics and using them to draw a line between wetlands and uplands to obtain a precise definition is difficult. More often than not, attempts to define wetlands involve less science and more politics and economics. A piece of land classified as a wetland becomes off limits to development.

Some wetlands are easy to distinguish. A water area supporting submerged plants such as pondweed, floating plants such as pond lily, and emergents such as cattails and sedges is unquestionably a wetland. But what about a piece of ground where the soil is more or less permanently wet and supports some ferns and such trees as red maple (*Acer rubrum*), which also grow on the uplands? Where do you draw the line between wetlands and uplands on this gradient of soil wetness? On the aquatic side, at what point does a wetland end and a true aquatic habitat with open water begin?

A number of wetland definitions exist (for a review see Mitsch and Gossleink 1993). A widely accepted scientific definition is presented in *Classification of Wetlands and Deepwater Habitats of the United States* (Cowardin et al. 1979):

> Wetlands are transitional between terrestrial and aquatic systems where the water table is usually at or near the surface or the land is covered by shallow water. . . . Wetlands must have one or more of the following three attributes: (1) at least periodically, the land supports predominately hydrophytes; (2) the substrate is predominately undrained hydric soil; and (3) the substrate is nonsoil and is saturated with water or covered by shallow water at some time during the growing season of each year.

A regulatory definition is used by the U. S. Army Corps of Engineers (1984):

> The term "wetlands" means those areas that are inundated or saturated by surface or ground water at a frequency and duration sufficient to support, and under normal circumstances do support, a prevalence of vegetation typically adapted for life in saturated soil conditions. Wetlands generally include swamps, marshes, bogs, and similar areas.

The latter definition emphasizes vegetation cover as the main criterion. It is obvious that vegetation alone should not define a wetland. Vegetation serves as an indicator, within limitations, but also involved are soils and the hydrological conditions that created the wetland in the first place. Wetlands range along a gradient of permanent to periodically flooded conditions to permanently to periodically saturated soil at some time during the growing season and support hydrophytic (water-loving) vegetation (Figure 15.23) (Tiner 1991, Lyon 1993). Wetland soils are hydric; that is, they are saturated, flooded, or ponded long enough during the growing season to develop anaerobic conditions in the upper part (Soil Conservation Service 1987). These soils are either mineral or organic. Mineral soils have less than 25 to 35 percent organic matter on a dry weight basis and have a developed soil profile with a gray to blue-gray color brought about by gleization. Organic soils have an organic content greater than 25 to 35 percent, are composed primarily of plant materials in various stages of decomposition, are dark in color, and have higher water-holding capacities than mineral soils.

Hydrophytic vegetation consists of those plants adapted to grow in water or on soil or a substrate that is periodically deficient in oxygen because of excess water. Hydrophytic plants include several groups: (1) obligate wetland plants, such as the submerged pondweeds, floating pond lily, and emergent cattails and bulrushes, and trees such as bald cypress; (2) facultative wetland plants, or amphibious plants, such as certain sedges and alder that can grow in standing water or saturated soil and rarely grow in other sit-

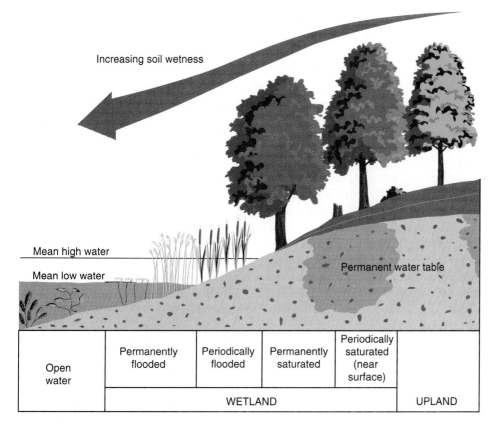

Increasing soil wetness

Mean high water

Mean low water

Permanent water table

| Open water | Permanently flooded | Periodically flooded | Permanently saturated | Periodically saturated (near surface) | |
|---|---|---|---|---|---|
| | WETLAND | | | | UPLAND |

**Figure 15.23** Location of wetlands along a soil moisture gradient. (After Tiner 1991.)

## Table 15.2  Types of Wetlands

| Type | Site Characteristics | Plant and Animal Populations |
| --- | --- | --- |
| *Inland Fresh Areas* | | |
| Seasonally flooded basins or flats | Soil covered with water or waterlogged during variable periods, but well drained during much of the growing season; in upland depressions and bottomlands | Bottomland hardwoods to herbaceous growth |
| Fresh meadows | Without standing water during growing season; waterlogged to within a few inches of surface | Grasses, sedges, rushes, broadleaf plants |
| Shallow fresh marshes | Soil waterlogged during growing season; often covered with 15 cm or more of water | Grasses, bulrushes, spike rushes, cattails, arrowhead, smartweed, pickerelweed; a major waterfowl production area |
| Deep fresh marshes | Soil covered with 15 cm to 1 m of water | Cattails, reeds, bulrushes, spike rushes, wild rice; principal duck-breeding area |
| Open fresh water | Water less than 3 m deep | Bordered by emergent vegetation such as pondweed, naiads, wild celery, water lily; brooding, feeding, nesting area for ducks |
| Shrub swamps | Soil waterlogged; often covered with 15 cm or more of water | Alder, willow, buttonbush, dogwoods; nesting and feeding area for ducks to limited extent |
| Wooded swamps | Soil waterlogged; often covered with 0.3 m of water; along sluggish streams, flat uplands, shallow lake basins | North: tamarack, arborvitae, spruce, red maple, silver maple; South: water oak, overcup oak, tupelo, swamp black gum, cypress |
| Bogs | Soil waterlogged; spongy covering of mosses | Heath shrubs, *Sphagnum,* sedges |
| *Coastal fresh areas* | | |
| Shallow fresh marsh | Soil waterlogged during growing season; at high tide as much as 15 cm of water; on landward side, deep marshes along tidal rivers, sounds, deltas | Grasses and sedges; important waterfowl areas |
| Deep fresh marshes | At high tide covered with 15 cm to 1 m of water; along tidal rivers and bays | Cattails, wild rice, giant cutgrass |
| Open fresh water | Shallow portions of open water along fresh tidal rivers and sounds | Vegetation scarce or absent; important waterfowl areas |
| *Inland Saline Areas* | | |
| Saline flats | Flooded after periods of heavy precipitation; waterlogged within few inches of surface during the growing season | Seablite, salt grass, saltbush; fall waterfowl-feeding areas |
| Saline marshes | Soil waterlogged during growing season; often covered with 0.61 to 1 m of water; shallow lake basins | Alkali hard-stemmed bulrush, wigeon grass, sago pondweed; valuable waterfowl areas |
| Open saline water | Permanent areas of shallow saline water; depth variable | Sago pondweed, muskgrasses; important waterfowl-feeding areas |
| *Coastal Saline Areas* | | |
| Salt flats | Soil waterlogged during growing season; sites occasionally to fairly regularly covered by high tide; landward sides or islands within salt meadows and marshes | Salt grass, seablite, saltwort |
| Salt meadows | Soil waterlogged during growing season; rarely covered with tide water; landward side of salt marshes | Cord grass, salt grass, black rush, waterfowl-feeding areas |
| Irregularly flooded salt marshes | Covered by wind tides at irregular intervals during the growing season; along shores of nearly enclosed bays, sounds, etc. | Needlerush, waterfowl cover area |
| Regularly flooded salt marshes | Covered at average high tide with 15 cm or more of water; along open ocean and along sounds | Atlantic: salt marsh cord grass; Pacific: alkali bulrush, glassworts; feeding area for ducks and geese, |
| Sounds and bays | Portions of saltwater sounds and bays shallow enough to be diked and filled; all water landward from average low-tide line | Wintering areas for waterfowl |
| Mangrove swamps | Soil covered at average high tide with 15 cm to 1 m of water; along coast of southern Florida | Red and black mangroves |

*Source:* Adapted from Shaw and Fredine 1956.

uations; (3) facultative species, such as red maple, that have about a 50:50 probability of growing in either wetland or nonwetland situations; and (4) facultative upland species, such as beech and Kentucky bluegrass, that have a 1 to 30 percent probability of growing in a wetland. It is these last plants that are critical in determining the upper limit of a wetland on the soil moisture gradient—a situation where species designation alone is insufficient.

For example, some species of trees usually associated with uplands adapt and grow quite well in wetland environments. One from eastern North America is red maple. It thrives in the drier uplands, but is also a conspicuous species in forested wetlands. In the uplands, red maple has a deep taproot; in wetland situations the tree has a shallow root system that enables it to avoid anaerobic stress. The red maple is a facultative species that evolved ecotypes adapted to different soil moisture conditions. Black gum (*Nyssa sylvatica*), too, grows in both upland and wetland situations. Pitch pine (*Pinus rigida*), associated with the drier ridgetops of the southern Appalachians, has ecotypes that grow in poorly drained soils and muck of swamps. In fact, pitch pine is a dominant species in the wetlands of the extensive New Jersey pine barrens. Hemlock, a shallow-rooted species, and white pine are also at home in wetland situations.

The point is that individual species of vegetation alone do not define a wetland. Although important indicators, especially in the wettest situations, ecotypes of upland species confuse the situation and make it difficult to draw a line on the gradient between wetland and upland. It is essential to

**Figure 15.24** A glacial prairie marsh supporting a heavy stand of emergent vegetation.

consider the hydrologic conditions and soil properties along with the vegetation to identify and delineate a wetland.

## Types of Wetlands

A wide variety of wetlands from fresh water to salt exists, and classifying them for management and conservation has presented problems. An older, shorter, but still generally useful classification appears in Table 15.2. A much more comprehensive hierarchical classification is presented by Cowardin et al. (1979).

Wetlands most commonly occur in three topographic and hydrologic situations. Many developed in shallow basins ranging from upland topographic depressions to filled-in lakes and ponds. These are **basin wetlands.** Other wetlands developed along shallow and periodically flooded banks of rivers and streams. These are **riverine wetlands.** A third type occurs along the coastal areas of larger lakes and oceans. They are known as **fringe wetlands.** Some of the best developed fringe wetlands are mangrove communities associated with tropical marine environments and considered in Chapter 16.

What separates the three types is the direction of water flow. Water flow in the basin wetlands is vertical, involving precipitation and capillary flow. In riverine wetlands water flow is unidirectional. In fringe wetlands flow goes in both directions because it involves rising lake levels or tidal action. The flows may bring in and carry away nutrients, and they may physically stress systems by exporting or importing excessive amounts of nutrients and sediments.

Wetlands may have deep or shallow water, or occupy soils that are water-saturated to various degrees. Wetlands dominated by emergent herbaceous vegetation are **marshes** (Figure 15.24). Supporting reeds, sedges, grasses, and cattails, marshes are essentially wet grasslands. Wetlands dominated by woody vegetation, or forested wetlands, are commonly called **swamps.** Deep-water swamps are dominated by cypress, tupelo, and swamp oaks (Figure 15.25a). Shrub swamps dominated by alder and willows are known as carrs or scrub-shrub swamps (Figure 15.25b). Along large river systems may occur extensive tracts of **riparian woodlands** (Figure 15.26) that are occasionally or seasonally flooded by river waters but are dry for most of the growing season.

Seasonal shallow wetlands ranging from 50 m² to 0.5 ha, flooded during the winter and spring and dry in summer and fall, are called **vernal pools.** Associated with a mediterranean-type climate with wet winters and dry summers, vernal pools are most common in the United States west of the Sierra Nevada Mountains and in the Central Valley of California. Vegetation is dominated by annuals and seasonal invertebrate life. Virtually ignored as wetlands in other parts of the continent, especially the northeast, are similar vernal pools, usually called temporary ponds. In winter these depressions may hold ice-covered water collected from melt-

(a)

(b)

**Figure 15.25** (a) A cypress deep-water swamp in southern United States. (b) A scrub-shrub swamp dominated by alder with a herbaceous understory of skunk cabbage.

**Figure 15.26** A riparian woodland of bottomland hardwoods along a small stream in western United States.

ing snow, but most fill in early spring with snowmelt and rainwater and dry up in summer. Invertebrate life is dominated by fairy shrimp and other small crustaceans that survive the dry summer and winter as eggs or cysts (Kenk 1949). These pools are important breeding sites for frogs, toads, and salamanders.

Wetlands in which considerable amounts of water are retained by an accumulation of partially decayed organic matter are **peatlands** or **mires.** Mires fed by water moving through mineral soil, from which they obtain most of their nutrients, and dominated by sedges are known as **fens** (Figure 15.27a). Mires dependent largely on precipitation for their water supply and nutrients and dominated by *Sphagnum* moss are **bogs** (Figure 15.27b). Mires that develop on upland situations where decomposed, compressed peat forms a barrier to the downward movement of water, resulting in a perched water table above mineral soil, are **blanket mires** (Figure 15.28). Blanket mires are more popularly known as **moors.** Because bogs depend upon precipitation for nutrient inputs, they are highly deficient in mineral salts and low in pH.

Bogs also develop when a lake basin fills with sediments and organic matter carried by inflowing water. These sediments divert water around the lake basin and raise the surface of the mire above the influence of groundwater. These are **raised bogs** (Figure 15.28). Other bogs form when a

(a)

(b)

**Figure 15.27** (a) Morning fog rises over this fen in Michigan. (b) An upland black spruce-tamarack bog in the Adirondack Mountains of New York.

lake basin fills in from above rather than from below (Figure 15.29), producing a floating mat of peat over open water. Such bogs are often termed **quaking bogs.**

## Structure

The structure of a wetland is influenced by the phenomenon that creates it—its hydrology. Hydrology has two components. One is the physical aspects of water and its movement: precipitation, surface and subsurface flow, direction and kinetic energy of water, and the chemistry of the water. The other is the **hydroperiod,** the duration, frequency, depth, and season of flooding. Length of the hydroperiod varies among types of wetlands. Basin wetlands have a longer hydroperiod and usually experience flooding during periods of high rainfall and drawdown during dry periods. Both phenomena appear to be essential to the long-term existence of wetlands. Riverine wetlands have a short period of flooding associated with peak stream flow. The hydroperiod of fringe wetlands, influenced by wind and lake waves, may be short and regular, and does not undergo the type of seasonal fluctuations characteristic of many basin marshes.

Hydroperiod influences plant composition, for it affects seedling germination, survival, and mortality at various stages of the plant's life cycle. The effect of hydroperiod is most pronounced in basin wetlands, especially those of the prairie regions of North America. In basins (called *potholes*

in the prairie region) deep enough to have standing water throughout periods of drought, the dominant plants will be submergents. If the wetland goes dry annually or during a period of drought, tall or midheight emergent species such as cattails will dominate the marsh. If the pothole is shallow and flooded only briefly in the spring, then grasses, sedges, and forbs will make up a wet meadow community.

If the basin is sufficiently deep toward its center and large enough, then zones of vegetation may develop, ranging from submerged plants to deep-water emergents such as cattails and bulrushes, shallow-water emergents, and wet ground species such as spikerush. Zonation reflects the response of plants to hydroperiod. Those areas of wetland subjected to a long hydroperiod will support submerged and deep-water emergents; those with a short hydroperiod and shallow water are occupied by shallow-water emergents and wet ground plants (van der Valk 1981).

Periods of drought and wetness can induce vegetation cycles associated with changes in water levels. Periods of above-normal precipitation can raise the water level and drown the emergence to create a lake marsh dominated by submergents. During a drought the marsh bottom is exposed by receding water, stimulating the germination of seeds of emergents and mudflat annuals. When water levels rise again, the mudflat species drown, and the emergents survive and spread vegetatively (van der Valk and Davis 1978).

Peatlands differ from other freshwater wetlands in that their rate of organic production exceeds the rate of decom-

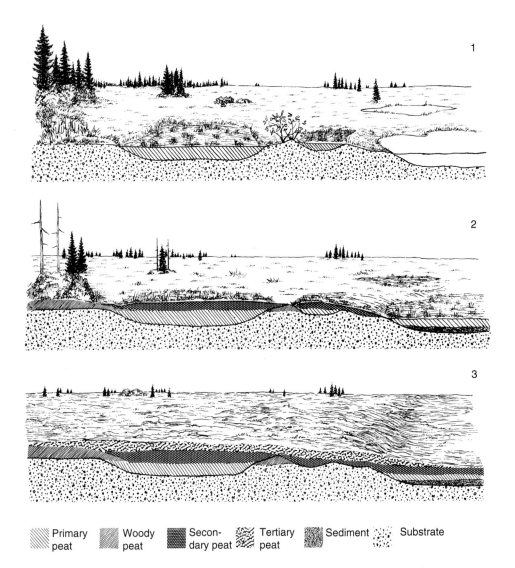

| ⧄ Primary peat | ⧄ Woody peat | ▦ Secon- dary peat | ⬚ Tertiary peat | ▨ Sediment | ⋰ Substrate |

**Figure 15.28** Development of a raised bog and blanket mire. In the first stage primary peat forms in basins and depressions. It reduces the surface retention of the reservoir. After peat fills the basin, it may continue to develop beyond the confines of the basin to form a raised bog. The raised bog consists of secondary peat, which acts as a reservoir, increasing surface retention of water on the area. Peat that develops above the physical limits of the groundwater and blankets even terrestrial situations is tertiary peat. It acts as a reservoir, holding a volume of water by capillary action above the level of the main groundwater drainage. The landscape acts as a perched water table, fed by precipitation falling on it. As the area becomes more wet and acidic, tree growth dies out and the landscape is covered by sphagnum and sedges.

position, and much of the production accumulates as peat. Peatlands are more characteristic of northern regions where inflow and precipitation balance outflow and retention, or where permafrost impedes the drainage of water and the bulk of summer precipitation is retained.

Upland bogs may begin when sphagnum moss invades higher ground surrounding a lake basin, when beaver back up water over streamside forests, or when forests are cut on wet sites. Sphagnum moss has the ability to absorb ions out of nutrient-poor water by exchanging hydrogen ions in its tissues for cations in solution. This exchange, which results in an increase of hydrogen and sulfate ions in the water, increases the acidity of the system, further decreasing the rate of decomposition and increasing the net accumulation of organic matter. Sphagnum moss adds new growth on top of the accumulating remains of past moss generations; and its spongelike ability to hold water increases water retention on the site. As the peat blanket thickens, the water-saturated mat of moss and associated vegetation is raised above and insulated from mineral soil. The peat mat

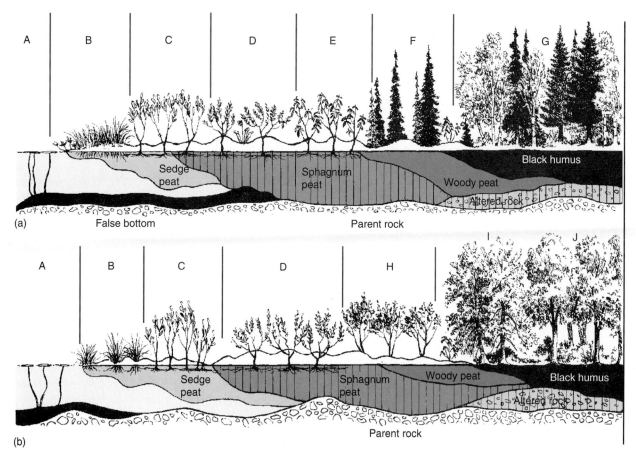

**Figure 15.29** (a) Transect through a quaking bog, showing zones of vegetation, sphagnum mounds, peat deposits, and floating mats. A, pond lily in open water; B, buckbean (*Menyanthes trifoliata*) and sedge; C, sweetgale (*Myrica gale*); D, leatherleaf (*Chamaedaphne calyculata*); E, Labrador tea (*Ledum groenlandicum*); F, black spruce; G, birch-balsam fir-black spruce forest. (b) Alternative vegetational sequence: H, alder; I, aspen, red maple; J, mixed deciduous forest. (Adapted from Dansereau and Segadas-Vianna 1952.)

then becomes its own reservoir of water, creating a perched water table.

Such upland bogs or plaudified landscapes are not uniform but rather are a mosaic of vegetation types ranging from blanket mires and raised bogs, fed by precipitation (**ombrotrophic**), to fens fed by water moving through mineral soil (**minerotrophic**). Although usually associated with and most abundant in boreal regions of the Northern Hemisphere, peatlands also exist in tropical and subtropical regions. They develop in mountainous regions or in lowland or estuarine regions where hydrological situations encourage an accumulation of partly decayed organic matter. Examples are the tropical coal swamps of the Carboniferous period, the Everglades in Florida, the pocosins of the southern United States coastal plains, and small bogs in the northern Appalachian Mountains.

Biologically, wetlands are among the richest and most interesting of ecosystems. They support a diverse community of benthic, limnetic, and littoral invertebrates, especially crustaceans and insects. These invertebrates, along with

small fishes, provide a basic food base for waterfowl, herons, gulls, and other birds, and supply the fat-rich nutrients needed by ducks for egg production and the growth of young. Wetlands support a diversity of amphibians and reptiles, notably frogs, toads, and turtles, and such mammalian inhabitants as beaver, muskrats (*Ondatra zibethicus*), mink (*Mustela vison*), and otter (*Lutra canadensis*).

Herbivores make up a conspicuous component of animal life. Microcrustaceans filter algae from the water column. Snails eat algae growing on the leaves and litter; geese graze on new emergent growth; coots and mallards and other surface-feeding ducks feed on algal mats. The dominant herbivore in the prairie marshes is the muskrat. During population highs, muskrats can eliminate emergent vegetation, creating "eat-outs," and transform an emergent-dominated marsh into an open-water one. Introduced into Eurasia, the muskrat has become the major herbivore in many marshes on that continent. Conversely, the nutria (*Myocaster coypus*), introduced into Louisiana in the 1930s as a furbearer, has spread throughout the southeast into the coastal marshes

of Maryland and New Jersey. Its numbers, size, and herbivorous feeding habits make it a major threat to coastal marshes by destroying marsh vegetation. Muskrats are the major prey for mink, the dominant carnivore on the marshes. Other predators include raccoon, fox, weasel, and skunk, which can seriously reduce the reproductive success of waterfowl on small marshes surrounded by agricultural land.

## Function

Only in recent years have ecologists given serious attention to the function of freshwater wetlands (Good, Whigham, and Simpson 1978; Greeson, Clark, and Clark 1979; Weller 1981; Gore 1983; Odum et al. 1984; Mitsch and Gosselink 1993). Much of this work has concerned glacial prairie wetlands, freshwater tidal wetlands, and southern deep-water swamps. Two wetland types will serve as examples, freshwater marshes and peatlands.

Freshwater Marshes Productivity of freshwater marshes is influenced by hydrological regimes: groundwater, surface runoff, precipitation, drought cycles, and the like. It is also affected by the nature of the watershed in which the wetland lies, soils, nutrient availability, types of vegetation, and the life history of the plant species.

Annual and perennial vegetation differ in their contributions to biomass through the growing season (Figure 15.30). Annual emergents exhibit a linear increase in biomass, reaching a maximum in late summer. Perennials increase their biomass the first part of the growing season, then decline or level off as leaves become senescent (Whigham et al. 1978). However, the average maximum standing crop of biomass usually matches annual aboveground productivity. For example, a sedge wetland had a maximum standing crop biomass greater than 1000 g/m² and a net primary productivity greater than 1000 g/m²/yr. However, if the life history of the plants—including mortality, regeneration, and winter activity—was considered, net productivity was 1600 g/m²/yr (Bernard and Gorham 1978).

Belowground production, much more difficult to estimate, appears for some species to be highest in summer, at the same time the peak aboveground biomass is achieved. Others, such as *Typha* and *Scirpus,* reach peak production in the fall, when nutrients are stored in the roots. Such species may have minimal root biomass in the summer because of nutrient transfer to aboveground biomass.

The cycling of phosphorus in a cattail marsh on Lake Mendota, Wisconsin (Prentki et al. 1978) provides an example (Figure 15.31). In spring the growing new shoots of cattail draw on the phosphorus reserves in the rootstalks for their initial growth before mobilizing P from the soil. By June cattails are accumulating P at a rate higher than they are accumulating biomass. By midsummer *Typha* has accumulated over 4 g/P/m² in the shoots, 78 percent of total P in biomass. At the same time the belowground pool is minimal. In late summer and fall P is remobilized in the rhizomes be-

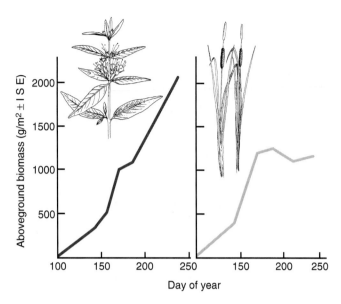

**Figure 15.30** Pattern of aboveground biomass accumulation through the growing season for a freshwater annual, loosestrife (*Lythrum*), and a perennial, cattail (*Typha*). Note the linear increase in biomass in the annual and the sigmoid growth of the perennial. (From Whigham et al. 1978:12.)

fore the shoots die, but at a rate slower than accumulation of belowground biomass. In fall a large amount of P is lost through leaching and death of the shoots. Only about 28 percent of the summer accumulation is returned to the rhizomes, to be rapidly depleted again by spring growth.

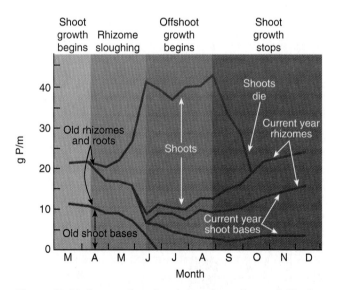

**Figure 15.31** Seasonal stocks of phosphorus in cattail (*Typha latifolia*) plant parts and belowground deficit in a marsh at Lake Mendota, Wisconsin. Note the sharp rise in P in the shoots during the summer months and its just as rapid decline in the fall when the shoots die. Much P is lost through leaching, but note the accumulation of some P in the current-year rhizomes. Belowground deficit is highest during the summer, when most of the P is in the shoots, but the deficit is erased by winter. (After Prentki, Gufason, and Adams 1978:176.)

To balance losses, cattails and other emergent plants must draw on a P supply from the soil. By doing so, the plants act as a nutrient pump, drawing nutrients from the soil, translocating them into the shoots, and then during the growing and post-growing season releasing them to the surface soil by leaching and death of the shoots. In this way marsh plants make nutrients sequestered in the soil available for growth.

Nutrients accumulated in plant biomass become available for use through decomposition. Initial decomposition in wetlands is high, as leaching removes soluble compounds that enter DOM (Figure 15.32). After initial leaching, decomposition proceeds more slowly. Permanently submerged leaves decompose more rapidly than those on the marsh surface, because they are more accessible to detritivores (Smock and Harlowe 1983) and because the stable physical environment is more favorable for microbes.

**Peatlands** In a bog, because vegetation is not in contact with mineral soil and because inflow of groundwater is blocked, nutrient input comes largely from precipitation. Nutrient availability, especially nitrogen, phosphorus, and potassium, is low, and most of the nutrients fixed in plant tissue are removed from circulation in the accumulation of peat. The amount of nutrients received by a bog depends upon the location of peatland. Bogs near the sea, such as English blanket mires, receive considerably more nutrients, especially magnesium and potassium, than inland bogs. Few data exist on nutrient cycles in ombrotrophic bogs, but evidence suggests that some bog plants possess mechanisms to conserve nutrients.

In general nitrogen is available from three sources: (1) precipitation, the major source; (2) nitrogen fixation by cyanobacteria living in close association with bog mosses and by bog myrtle, if the pH of the substrate is at least 3.5, below which nitrogen-fixing root nodule bacteria are inhibited; and (3) carnivory, by which certain plants such as sundews and pitcher plants extract nitrogen from captured and digested insects. Cyanobacteria may fix 0.23 to 0.9 $g/N/m^2/yr$. Rosswall has calculated a nitrogen budget and nitrogen cycle for an ombrotrophic bog dominated by the trailing ground plant *Rubus chamaemorus* (Figure 15.33). Yearly demand is about 1 to 2 $g/m^2$.

The phosphorus cycle in a bog holds losses to a minimum and is more closed than the nitrogen cycle. *Rubus chamaemorus* increases its uptake of phosphorus prior to bud break. After bud break the plant increases phosphorus in stem, leaf, and root. This increase correlates with an increase in peat temperature. After the plant completes shoot growth in summer, its phosphorus level declines in the shoots as it mobilizes phosphorus in developing fruits and in roots and rhizomes. In advanced senescence the plant rapidly loses phosphorus from the shoots and accumulates it in roots and winter buds.

Energy flow in the mire system differs from that of other systems, because decomposition is impaired. In most ecosystems material that enters the detrital food web eventually is recycled, and energy is liberated or stored in living material. In mire systems about 10 percent of primary production accumulates in an undecomposed state, and energy is locked up in peat until environmental conditions change, favoring decomposition, or the material is burned.

Because of low temperatures, acidity, and nutrient immobilization primary production in peatlands is low. In the Stordalen mire in Sweden primary production amounted to 70 $g/m^2/yr$ by bryophytes and 83 $gm/m^2/yr$ by dwarf shrubs, forbs, and monocots collectively (Rosswall 1975). In an English blanket mire average production was 635 $g/m^2/yr$, with sphagnum on wet sites contributing 300 $g/m^2/yr$ (Heal et al. 1975).

Herbivore utilization of bog vegetation is low, in part because of unpalatability. Herbivores may include red grouse, willow grouse, hares, bog lemmings, and sheep. With these herbivores consumption is selective, usually confined to current shoots, opening buds, and fruits. In most bogs the dominant herbivores are insects (Heal et al. 1975, Moore et al. 1975). Predators include rodent-consuming weasels, harriers, short-eared owls, insect-consuming frogs, shrews, pipits (Europe), Nashville warblers (North America), and invertebrates such as spiders and ground beetles.

Decomposer organisms are low in number, and the dominant species vary from year to year. In a Swedish mire total fungal biomass was 58 $g/m^2$ and the bacterial biomass, largely anaerobic, was 22 $g/m^2$ (Rosswall et al. 1975).

**Figure 15.32** Decomposition of leaves of arrow arum in a tidal freshwater marsh, as measured by the percentage of original ash-free dry weight remaining in litter bags under three conditions: irregularly flooded high marsh exposed to alternate wetting, creek bed flooded two times daily, and permanently submerged. Note that the detrital material consistently wet showed the highest rate of decomposition, although the overall pattern of decomposition is similar. (From Odum and Haywood 1978:92.)

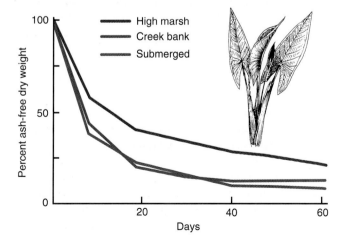

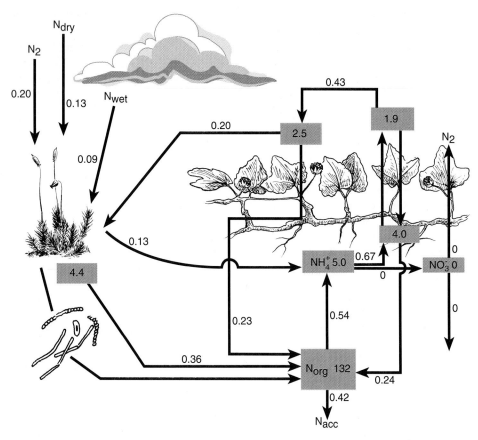

**Figure 15.33** Nitrogen budget for an ombrotrophic mire in Stordalen, Sweden. Quantities of nitrogen are expressed as g/N/m², flows as g/N/m²/yr. The vegetation has been divided into aboveground parts exemplified by *Rubus chamaemorus* and lichens (1.6 g N/m²), belowground parts of vascular plants (3.0 g N/m²), aboveground litter (3.2 g N/m²), and mosses (4.4 g N/m²). $N_{org}$ = organic nitrogen; $N_{acc}$ = nitrogen accumulated. (After Rosswall 1975.)

Invertebrate detritivores include rotifers, tardigrades, mites, and nematodes. In an English mire a single species of enchytraeid, *Cognettia sphagnetorum,* accounted for 70 to 75 percent of the total energy assimilated by the detritivores (Heal et al. 1975).

Rates of decomposition vary among bog vegetation. Litter of *Rubus chamaemorus* lost 2 percent of its weight in one year and 50 percent in three years. Shrub litter decomposes more slowly, while sphagnum decomposes hardly at all.

In an English mire (Moor House National Nature Reserve) the vegetational standing crop has an annual turnover of about 0.3 percent, and only about 1 percent of production is consumed by herbivores. The remainder of the vegetation enters the decomposer food web, where about 5 percent is assimilated by decomposers and 10 percent passes below the water table. About 85 percent of the production is decomposed by microflora; but the rate of decomposition is slow, with 95 percent turnover time of about 3000 years. In the top 20 cm, 95 percent turnover time is about 70 years (Heal et al. 1975).

# SUMMARY

Freshwater ecosystems are lentic or still water, or lotic or flowing water. Lentic ecosystems are lakes and ponds, standing bodies of water that fill a depression in the landscape. Geologically speaking, lakes and ponds are ephemeral features of the landscape. In time they fill in, draw smaller, and finally are replaced by a terrestrial community.

A lake exhibits gradients in light, temperature, and dissolved gases, resulting in seasonal stratification. In summer a top layer of warm, light water floats on top of a middle mass of water, the metalimnion, characterized by a steep decline in temperature. Below the metalimnion is a bottom layer of cold water, the hypolimnion. These gradients influence biological stratification. The area where light penetrates to the bottom of the lake, a zone called the littoral, is occupied by rooted plants. Beyond is the open water or limnetic zone, inhabited by plant and animal plankton and fish. Below the depth of effective light penetration is the

profundal region, where the diversity of life varies with temperature and oxygen supply. The bottom or benthic zone is a place of intense biological activity, for here decomposition of organic matter takes place. Anaerobic bacteria are dominant on the bottom beneath the profundal water, whereas the benthic zone of the littoral is rich in decomposer organisms and detritus feeders. Although lake ecosystems are often considered as autotrophic systems dominated by phytoplankton and the grazing food web, lakes are strongly dependent on the detrital food web. Much of that detrital input comes from the littoral zone.

Lakes may be classified as eutrophic or nutrient-rich, oligotrophic, or dystrophic, acidic and rich in humic material. Most lakes are subject to cultural eutrophication, which is the rapid addition of nutrients, especially nitrogen and phosphorus, from sewage and industrial wastes. Cultural eutrophication has produced significant biological changes, mostly detrimental.

Lotic ecosystems are characterized by inputs of detrital material from terrestrial sources and currents of varying velocities that carry nutrients and other materials downstream. Lotic ecosystems exhibit a continuum of physical and ecological variables from source to mouth. There is a longitudinal gradient in temperature, depth, and width of the channel, velocity of the current, and nature of the bottom. Changes in physical conditions are reflected in biotic structure. Headwater streams in forested regions are shaded and strongly heterotrophic, dependent on inputs of detritus. This detrital material is processed by a number of invertebrates, classified functionally as shredders, collectors, grazers, and gougers. These organisms, algae, and detritus slow the downstream movement of nutrients, known as spiraling. Larger streams, open to sunlight, shift from heterotrophic to autotrophic. Primary production from algae and rooted aquatics becomes an important energy source. Large rivers return to a heterotrophic condition. They are dependent on fine particulate organic matter and dissolved organic matter as sources of nutrients and energy. Downstream systems depend upon the inefficiencies of energy and nutrient processing upstream.

Closely associated with lakes and streams are wetlands, areas where water is at, near, or above the level of the ground and occupied by hydrophytic vegetation. Wetlands dominated by grasses are marshes; those dominated by woody vegetation are swamps. Wetlands characterized by an accumulation of peat are mires. Mires fed by water moving through the mineral soil (minerotrophic) and dominated by sedges are fens; those dominated by sphagnum moss and dependent on precipitation for moisture and nutrients (ombrotrophic) are bogs. Bogs are characterized by blocked drainage, accumulation of peat, and low productivity. A significant portion of nutrients fixed in plants is removed from circulation and stored in accumulated peat. The stored energy remains locked in peat until environmental conditions change to favor decomposition or until the material is burned. In contrast, marsh ecosystems are heavily influenced by nutrients stored in the soil organic material. Marsh vegetation acts as a nutrient pump, drawing nutrients from the substrate, translocating them to the shoots, and depositing them on the surface or losing them through leaching to the water.

## REVIEW QUESTIONS

1. Compare and contrast the roles of phytoplankton and macrophytes in the function of lentic ecosystems.
2. Compare and contrast nutrient cycling in a lake, a stream, a freshwater marsh, and a bog.
3. Characterize the major functional groups of stream invertebrates and describe their role in the flowing water ecosystem.
4. What is spiraling, and how does it affect nutrient cycling and retention in streams?
5. How do downstream lotic systems relate to upstream systems?
6. What is the weakness of defining a wetland by vegetation alone?
7. What is hydroperiod? How does it relate to the structure of wetlands?
8. Consider the concept of a wetland (refer to Tiner 1992, Lyon 1993, Mitsch and Gosselink 1993). Then relate how the following would interpret a wetland from their own perspective and why each has a vested interest in how wetlands are defined: a wildlife biologist and conservation biologist, a developer, a midwest agriculturalist, the Army Corps of Engineers, the Soil Conservation Service.
9. Study a topographic map of your area. Locate and mark drainage patterns and watershed or basin boundaries, starting with first-order streams. Then visit them in the field to study their gradient, size, rate of flow, and condition. Look for pollution, siltation, streambank vegetation, and proximity to human development. What major changes do you observe?

## CROSS-REFERENCES

Structure of water, 64; specific heat of water, 65; water cycle, 67–69; plant responses to flodding, 73–75; light, 100–105; nutrients and nutrient cycles, 116–120; gley soil, 142–143; decomposition, 159–165; energy flow, 189–192; food webs, 192–193, 336, 621–623; oxygen cycle, 196–199; nitrogen cycle, 210–212; phosphorus cycle, 212–213.

# Saltwater Ecosystems

## Outline

## Concepts

1. Temperature, salinity, tides, waves, and currents dictate the structure and function of marine ecosystems.
2. At the base of food webs in relatively unproductive open seas are phytoplankton, nanoflagellates, and cyanobacteria.
3. The most productive areas of the sea are coastal waters and coral reefs.
4. Rocky and sandy intertidal ecosystems are shaped by tidal cycles and the action of waves.
5. Coral reef ecosystems are created by coral organisms that make massive deposits of calcium carbonate in shallow tropical seas.
6. Estuarine ecosystems are strongly influenced by the inflow of fresh water from rivers.
7. The high production of salt marshes and mangrove swamps results from tidal subsidies of nutrients and energy.

Freshwater rivers eventually empty into the oceans, and terrestrial ecosystems end at the edge of the sea. For some distance beyond there is a region of transition at the land and water interface. Rivers enter the saline waters of the ocean, creating a gradient of salinity. Coastal regions exposed to the open sea are dominated by tides. Beyond them lies the open ocean—shallow seas overlying continental shelves and the deep oceans.

# PHYSICAL FEATURES

## Stratification

Just as lakes exhibit vertical stratification, so do the seas. The ocean is divided into two main layers, the **pelagic region** or whole body of water, and the **benthic region** or bottom (Figure 16.1). Within the pelagic region there are three vertical layers. From the surface to about 200 meters is the **photic** zone, in which there are sharp gradients of light, temperature, and salinity. Ranging from 200 to 1000 m, where little light penetrates and where the temperature gradient is more gradual and without much seasonal variation is the **mesopelagic** layer. It contains an oxygen-minimum layer and often a maximum concentration of nitrate and phosphate. Below the mesopelagic is the **bathypelagic,** where darkness reigns, except for bioluminescence, the temperature is low, and pressure is great. Depending upon its global location, it lies between 100 to 700 m and 2000 to 4000 m.

Horizontally, the pelagic region is subdivided into the **neritic province,** water that overlies the continental shelf, and the **oceanic province** (Figure 16.1). Lying over the major plains of the ocean, down to about 6000 m, is the **abyssal pelagic** or **benthipelagic province.** Water of the deep oceanic trenches is called the **hadalpelagic.**

The sea bottom is called the benthic region. It, too, can be divided into three zones, which differ in the nature of their bottom life. The **bathyal zone** covers the continental shelves and down to 4000 m. The **abyssal zone** embraces the broad ocean plains down to 6000 m, and the **hadal zone** is the benthic zone of oceanic trenches between 6000 and 10,000 m. The benthic zone underlying the neritic pelagic zone on the continental shelf is the **sublittoral** or **shelf zone.** The intertidal zone, like the margins of lakes, is called the **littoral.**

**Figure 16.1** Regions of the ocean.

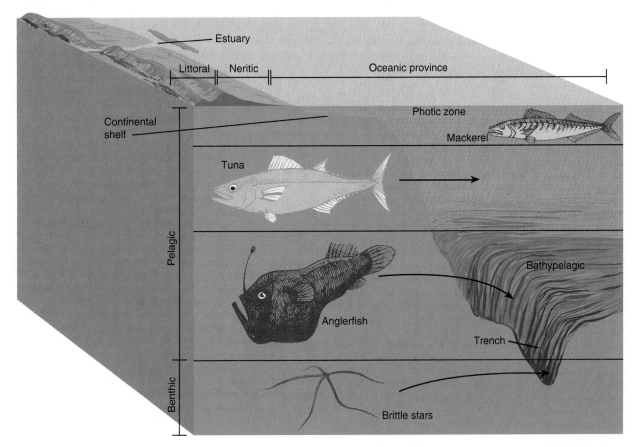

## Temperature

The photic layer exhibits a stratification of temperature (Figure 16.2). In polar regions conditions remain fairly constant throughout the year. In middle latitudes temperatures vary with the season, as the climate changes. In summer the surface waters become warmer and lighter, forming a strong thermocline that decays with coming of winter. Tropical oceans, whose surface waters are constantly heated, have a strong permanent thermocline between 100 and 200 m.

## Salinity

Two elements, sodium and chlorine, make up about 86 percent of sea salts. Along with other such major elements as sulfur, magnesium, potassium, and calcium they comprise 99 percent of sea salts. Seawater, however, differs from a simple sodium chloride solution in that the amounts of cations and anions are not balanced against each other (Table 16.1). The cations exceed the anions by 2.38 milliequivalents/kg. As a result, seawater is weakly alkaline (pH 8.0 to 8.3) and strongly buffered, a condition that is biologically important.

The amount of dissolved salt in seawater is usually expressed as chlorinity or salinity. Because oceans are usually

**Figure 16.2** Temperature stratification in the oceans at high, middle, and low latitudes. (After R. V. Tait 1968.)

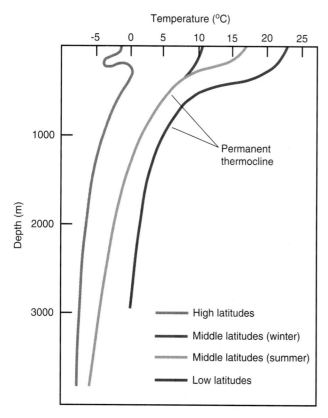

**Table 16.1  Composition of Seawater of 35‰ Salinity**

| Elements | g/kg | Millimole/kg | Milliequivalent/kg |
|---|---|---|---|
| **Cations** | | | |
| Sodium | 10.752 | 467.56 | 467.56 |
| Potassium | 0.395 | 10.10 | 10.10 |
| Magnesium | 1.295 | 53.25 | 106.50 |
| Calcium | 0.416 | 10.38 | 20.76 |
| Strontium | 0.008 | 0.09 | 0.18 |
| Total | | | 605.10 |
| **Anions** | | | |
| Chlorine | 19.345 | 545.59 | 545.59 |
| Bromine | 0.066 | 0.83 | 0.83 |
| Fluorine | 0.0013 | 0.07 | 0.07 |
| Sulphate | 2.701 | 28.12 | 56.23 |
| Bicarbonate | 0.145 | 2.38 | — |
| Boric acid | 0.027 | 0.44 | — |
| Total | | | 602.72 |

*Note:* Surplus of cations over strong anions (alkalinity): 2.38.
*Source:* Kalle 1971.

well mixed, sea salt has a constant composition; that is, the relative proportions of major elements change little. Thus the determination of the most abundant element, chlorine, can be used as an index of the amount of salt present in a given volume of seawater. Chlorine, expressed in parts per thousand (o/oo) is the amount of chlorine in grams in a kilogram of seawater. The relationship of salinity to chlorinity is expressed by

$$S(‰) = 1.80655 \times \text{chlorinity}$$

The salinity of parts of the ocean is variable because physical processes change the amount of water there. Salinity is affected by evaporation, precipitation, movement and mixing of water masses of different salinities, formation of insoluble precipitates that sink to the ocean floor, and diffusion of one water mass to another. Salinities are most variable near the interface of sea and air.

The salinity of the ocean has been ascribed to the accumulation of salts derived from the weathering of Earth's crust. This source can account for much of the sodium, but it cannot account for chlorine or salts such as bromides and iodides and boron. A major source of chlorine is hydrochloric acid (HCl) from submarine volcanoes in the midocean ridges. The hydrochloric acid reacts with bicarbonates to form water and carbon dioxide, freeing the chloride ion. The chloride ion then replaces the bicarbonate ion as the anion, balancing the sodium cation. (For the chemistry of seawater see Macintyre 1970.

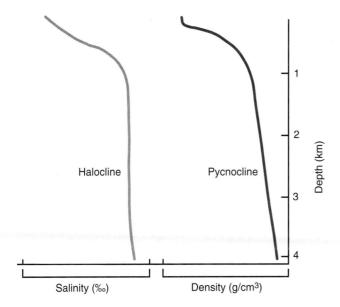

**Figure 16.3** The halocline, the stratification of salinity in the ocean, and the pycnocline, stable stratification of density. (After M. Gross 1972.)

Salinity, like temperature, exhibits a vertical gradient in the sea (Figure 16.3). This gradient or **halocline** is especially pronounced at higher latitudes. There abundant precipitation reduces surface salinity, whereas in tropical regions, evaporation increases salinity. In the middle latitudes, salinity together with temperature produces a marked gradient in density. Because the density of seawater increases with depth but does not reach its greatest density at 4° C, as with fresh water, there is no seasonal overturn. However, the rapid change in temperature, which produces a thermocline, is paralleled with a rapid change in the density of seawater. This zone of rapid change in density is the **pycnocline.**

Salinity imposes restrictions on life that inhabits the oceans (see Kinne 1971). Fish and marine invertebrates in marine, estuarine, and tidal environments have to maintain osmotic pressure, often under conditions of changing salinities. Most marine species are osmoconformers, lacking the ability to move into less saline water. Others, such as striped bass and eels, are osmoregulators and have the ability, at least during some part of their life cycle, to control the concentration of salts or water in their bodies. They are able to move into waters of lower than normal oceanic salinity.

## Pressure

Another aspect of the marine environment is pressure. Pressure in the ocean varies from 1 atmosphere at the surface to 1000 atm at its greatest depth. Pressure changes are many times greater in the sea than in terrestrial environments and have a pronounced effect on the distribution of life. Certain organisms are restricted to surface waters,

whereas others are adapted to pressure at great depths. Some marine organisms, such as sperm whales and certain seals, can dive to great depths and return to the surface without difficulty.

## Waves

Waves are generated on the surface of seas (and lakes) by pressures of wind on their surfaces. The frictional drag of the wind on the surface of smooth water ripples it. As the wind continues to blow, it applies more pressure to one side of the ripple, and wave size grows. As the wind becomes stronger, short, choppy waves of all sizes appear; and as they absorb more energy, they continue to grow in size. When the waves reach a point at which the energy supplied by the wind is equal to the energy lost by the breaking waves, they become the familiar whitecaps.

Up to a certain point, the stronger the wind, the higher the waves and the longer the fetch (the distance the waves can run without obstruction under the drive of the wind in a constant direction). As waves travel out of the fetch (or if the wind dies), sharp-crested waves change into smooth, long-crested waves or swells, characterized by troughs and ridges. Swells can travel great distances because they lose little energy as they travel.

The waves that break on beaches are generated in and transported from distant seas. Each particle of water remains largely in the same place and follows an elliptical orbit with the passage of the wave form. As the wave moves forward with a velocity that corresponds to its length, the energy of a group of waves moves with a velocity only half that of individual waves. The wave in front loses energy to the waves behind and disappears, its place taken place by another. Thus the swells that break on the beach are distant descendants of waves generated far out at sea.

As waves approach land, they advance into increasingly shallow water. The height of the waves rises until the wave front grows too steep and topples over. As the waves break on the shore, they dissipate their energy against it, pounding rocky shores or tearing away at sandy beaches at one point and building up new beaches elsewhere.

## Langmuir Cells

Wind blowing across the surface may stir up more than surface waves. Winds blowing above three meters per second generate circulating vertical cells of water spinning just below the surface. These are called **Langmuir cells** after the physical chemist I. Langmuir (1938), who studied them in Lake George, New York. These cells of water, which may range from a few centimeters to many meters in diameter, are parallel to the water's surface and the direction of the wind (Figure 16.4). Each circulates in a direction opposite to the adjacent one. One circulates clockwise, the adjacent one

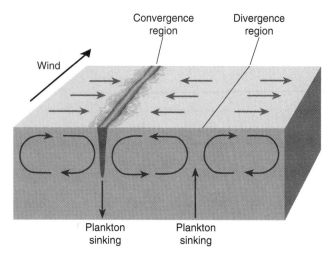

**Figure 16.4** Langmuir cells in a body of water. Note that these cells are marked by accumulated materials on the surface where the adjacent cells upwell. (After Nybakken 1988:63.)

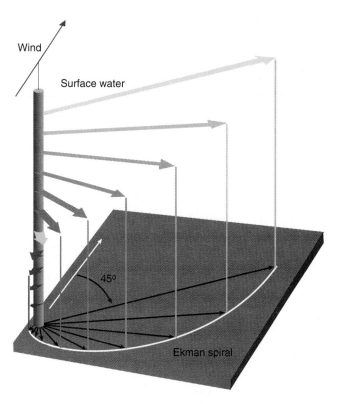

**Figure 16.5** The Ekman spiral. The wind-driven surface current moves at an angle of 45° to the direction of the wind, to the right in the Northern Hemisphere, to the left in the Southern. Successively deeper water layers are deflected even further than those immediately above them and move at slower speeds. Net water movement is at 90° to the wind. (Adapted from Strahler 1971:258.)

counterclockwise, so that between one pair there are up-welling or divergent currents and the between the next pair, downwelling or convergent currents. The position of the downwelling currents is marked on the surface by streaks of accumulated materials, such as leaves and strands of sub-merged aquatic vegetation lifted to the surface. Where up-welling with its divergent currents occurs, no such streaks appear. These internal currents mix heat and nutrients in the surface water during the summer and circulate phytoplank-ton, accounting for its often patchy and changing distribu-tion in lakes and ponds.

## Ekman Spiral and Upwelling

Also influencing subsurface currents in the oceans is the **Ekman spiral** (Figure 16.5). Ocean currents do not flow parallel to the wind; they are deflected by the Coriolis force. As wind moves the surface water, some of the energy passes down through the water column and sets each successive layer into motion. Energy, however, decreases with depth, so each layer moves more slowly than the layer above it. Like the surface water, each layer in motion is deflected relative to the one above. Ultimately the water mass at the base of the spiral moves counter to the flow on the surface, although it is at right angles to the wind. In coastal regions, Ekman spiral can influence upwelling, the bringing of deep water to the surface.

Coupled with horizontal motions of water are other verti-cal water motions. Wind blowing parallel to continental margins causes surface water to be blown offshore. This loss of water is replaced by water moving upward from the deep. Although cold and containing less dissolved oxygen, up-welling water is rich in nutrients that support abundant growth of phytoplankton. For this reason regions of up-

welling, such as the west coasts of South America and North America, are highly productive.

## Tides

The most familiar feature of marine ecosystems, especially along the coasts, is the tide. Tides are periodic and pre-dictable rising and falling sea levels over a given interval. They result from a combination of gravitational action of sun and moon on Earth and centrifugal forces of the Earth-moon system.

Earth and moon form a single system revolving about a common center of gravity. Because Earth is so much larger than the moon, the common center of the mass is within Earth. Gravitational force is a function of the mass of each body and the distance between them. Although the moon is much smaller than the sun, it is also very much closer, so its gravitational force on Earth is two times that of the sun. The centrifugal force within this revolving system more or less balances the gravitational force acting between these two bodies. The interaction of these two forces pulls the water in the ocean basin into bulges. On the side of Earth facing the

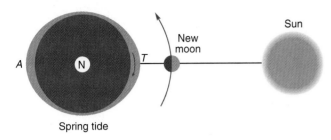

Spring tide

**Figure 16.6** Tides result from the gravitational pull of the Moon on Earth. Centrifugal force applied to a kilogram of mass is 3.38 milligrams. This centrifugal force on a rotating Earth is balanced by gravitational force, except at those (moving) points on Earth's surface that come into direct line with the Moon. Thus the centrifugal force at point N, the center of the rotating Earth, is 3.38 milligrams. Point T on Earth is in direct line with the Moon. At this point the gravitational force of the Moon is 3.49 milligram, a difference of 0.11 mg. Because the Moon's gravitational force is greater than the centrifugal force at T, the force is directed away from the Earth and causes a tidal bulge. At point A on the opposite side of the Earth from T, the Moon's gravitational force is 3.27 mg, 0.11 milligram less than the centrifugal force at N. This difference causes a tidal bulge on the opposite side of Earth. (See text for further details.)

moon, the gravitational force exceeds the centrifugal force, and the water is pulled into a bulge, the high tide. On the opposite side of Earth, away from the moon, the centrifugal force exceeds the gravitational force, and it, too, pulls the water into a bulge, creating another high tide (Figure 16.6). As Earth rotates on its axis, the position of the moon relative to Earth changes during the 24-hour day. High tides follow the changing position of the moon around Earth. As Earth rotates eastward on its axis, the tides advance westward. Thus any given place on Earth will in the course of one daily rotation pass through two of the lunar tidal bulges, or high tides, and two of the lows or low tides, at right angles to the high. Because the moon revolves in a 29.5-day orbit around Earth, the average period between successive high tides is approximately 12 hours, 25 minutes.

The sun also causes two tides on opposite sides of Earth, but because they are less than half as high as lunar tides, solar tides are usually partly masked by the lunar tides. Twice during the month, when the moon is full and the sun and moon are opposite each other, and when it is new and the sun is behind the moon, Earth, sun, and moon are nearly in line. Then the gravitational pulls of the sun and moon are additive. The intensified force causes what are known as **spring tides,** tides of maximum rise and fall. When the moon is at either quarter, the gravitational pulls of sun and moon interfere with each other, creating **neap tides** of minimum difference between high and low tides.

Tides are not entirely regular, nor are they the same all over Earth. They vary from day to day in the same place, following the waxing and waning of the moon. They may act differently in several localities within the same general area.

In the Atlantic semidaily tides are the rule. In the Gulf of Mexico flood and ebb tides follow one another at about 24-hour intervals to produce one daily tide. Mixed tides, combinations of semidaily and daily tides, are common in the Pacific and Indian oceans, with different combinations at different places.

Local inconsistencies of tides are due to many variables. The elliptical orbits of Earth about the sun and of the moon about Earth influence the gravitational pull, as does the declination of the moon—the angle of the moon in relation to the axis of Earth. Latitude, barometric pressure, offshore and onshore winds, depth of water, contour of the shore, and internal waves modify tidal movements. Islands in the middle of a tidal basin may experience minimal tides. Geological structures of some basins funnel tidal water into restricted channels, such as the Bay of Fundy and Cook Inlet of Alaska. They experience extremely high tides.

## THE OPEN SEA

The oceans occupy 70 percent of Earth's surface. All of the seas are interconnected by currents, dominated by waves, influenced by tides, and characterized by saline waters. They are deep, in some places 7 km. The volume of surface area lighted by the sun is small compared to the total volume of water. Dimness and the dilute solution of nutrients limit production.

Compared to terrestrial ecosystems, the pelagic ecosystem lacks well-defined communities, largely because of the lack of the supporting structures and framework of large dominant plant life. Exceptions are the subtidal kelp forests, dominated by brown algae, found throughout the cold temperate regions of the world, and the Sargasso Sea, dominated by floating sargassum weed, also a brown algae. However, differences based on physical characteristics and life forms permit a division of the ocean into different regions.

The Arctic Ocean comprises marine waters that lie north of the land masses in the Northern Hemisphere and is open only to the Atlantic Ocean. The Southern or Antarctic Ocean lies about the continent of Antarctica and is open to three oceans, the Atlantic, Pacific, and Indian. Warm oceanic waters making up the Atlantic and Pacific have some of their own distinctive communities. Deep sea benthic ecosystems are quite distinctive from those of lighted waters. Other important distinctive marine ecosystems include coral reefs, upwelling systems off the coasts of California, Peru, Northwest Africa, Southwest Africa, India and Pakistan. Important and distinctive are the shelf-sea ecosystems. Shallow, productive, and nutrient-rich, they support a diversity of fish and invertebrate marine life.

### Structure

Because the global regions vary in their diversity and nature, ranging from oligotrophic to eutrophic and from cold to tropical, a brief discussion of structure must be general and

emphasize the major structural groups. (For a good introduction to the structure and function of marine ecosystems see Nybakken 1988.)

**Phytoplankton**  As in lakes, the dominant form of plant life or primary producer in the open sea is phytoplankton. Seawater is so dense that oceanic phytoplankton do not need well-developed supporting structures. In coastal waters and areas of upwelling phytoplankton 100 microns or more in diameter may be common, but in general plant life is much smaller and widely dispersed.

Because of its requirement for light, phytoplankton is restricted to the upper surface waters. The depth of its occurrence is determined by the depth of light penetration and so many range from tens to hundreds of meters. Because of seasonal, annual, and geographic variations in light, temperature, nutrients, and grazing by zooplankton, the distribution and composition of phytoplankton change with time and place.

Each ocean or region within an ocean appears to have its own dominant forms. Littoral and neritic waters and regions of upwelling are richer in plankton than midoceans. In regions of downwelling, dinoflagellates concentrate near the surface in areas of low turbulence. They attain their greatest abundance in warmer waters. In summer they may so concentrate in the surface waters that they color it red or brown. Often toxic to other marine life, such concentrations of dinoflagellates are responsible for red tides.

In regions of upwelling the dominant forms of phytoplankton are diatoms. Enclosed in a silica case, diatoms are particularly abundant in arctic waters.

Smaller than diatoms are the Coccolithophoridae, so small they pass through plankton nets (and so are classified as nannoplankton). Their minute bodies are protected by calcareous plates or spicules embedded in a gelatinous sheath (Figure 16.7). Universally distributed in all waters except the polar seas, the Coccolithophoridae possess the ability to swim. Droplets of oil aid in buoyancy and storage of food.

In the equatorial currents and in shallow seas the concentration of phytoplankton is variable. Where both lateral and vertical circulation of water is rapid, the composition reflects in part the ability of the species to grow, reproduce, and survive under local conditions.

**Zooplankton**  Grazing on the phytoplankton is the herbivorous zooplankton, consisting mainly of copepods, planktonic arthropods that are the most numerous animals of the sea, and the shrimplike euphausiids, commonly known as krill (Figure 16.8). Other planktonic forms are the larval stages of such organisms as gastropods, oysters, and cephalopods. Feeding on the herbivorous zooplankton is the carnivorous zooplankton, which includes such organisms as the larval forms of comb jellies (Ctenophora) and arrowworms (Chaetognatha).

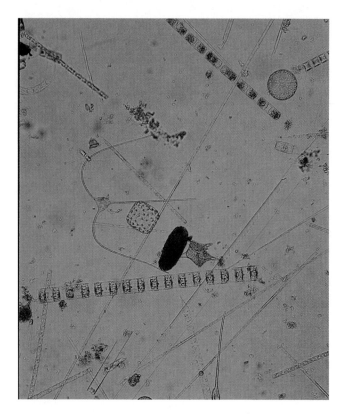

**Figure 16.7** An array of marine phytoplankton including diatoms and dinoflagellates.

Because its food in the ocean is so small and widely dispersed, zooplankton has evolved ways more efficient and less energy-demanding to harvest phytoplankton than filtering water through pores. It has webs, bristles, rakes, combs, cilia, sticky structures, and even bioluminescence.

**Figure 16.8** Small euphausiid shrimps called krill are eaten by baleen whales.

The composition of zooplankton, like that of phytoplankton, varies from place to place, season to season, year to year. In general zooplankton falls into two main groups, the larger forms characteristic of shallow coastal waters and the smaller forms characteristic of the deeper open ocean. Zooplankton forms of the continental shelf contain a large proportion of larvae of fish and benthic organisms. They have a greater diversity of species, reflecting a greater diversity of environmental and chemical conditions. The open ocean, being more homogeneous and nutrient-poor, supports less diverse zooplankton. In polar waters zooplankton species spend the winter in a dormant state in the deep water and rise to the surface during short periods of diatom blooms to reproduce. In temperate regions distribution and abundance depend upon temperature conditions. In tropical regions, where temperature is nearly uniform, zooplankton is more abundant, and reproduction occurs throughout the year.

Also like phytoplankton, zooplankton lives mainly at the mercy of the currents; but many forms possess sufficient swimming power to exercise some control. Most species of zooplankton undertake a daily vertical migration from as deep as 800 m to 100 m or to the surface. As darkness falls, zooplankton rapidly rises to the surface to feed on phytoplankton. At dawn the forms move back down to preferred depths.

By feeding in the darkness of night and hiding in the darkened waters by day, zooplankton avoids heavy predation; and by remaining in cooler water by day, it conserves energy during the resting period. Surface currents move zooplankton away from its daytime location during feeding, but it can return home by countercurrents present in the deeper layers. Response to changing light conditions is useful in another way. As clouds of phytoplankton pass over the water above, zooplankton responds to the shadow of food by moving upward. This motion takes it out of the deep current drift and nearer the surface. At night it can move directly up to the food-rich surface water.

Zooplankton that lack a vertical migration, and even some of those that have it, drift out of the breeding area with surface currents. Survival of a breeding population is assured by a complex cycle of seasonal migration.

Nekton  Feeding on zooplankton and passing energy along to higher trophic levels is the **nekton,** swimming organisms that can move at will in the water column. They range in size from small fish to large predatory sharks and whales, seals, and marine birds such as penguins. Some of the predatory fish, such as herring and tuna, are more or less restricted to the photic zone. Others are found in the deeper mesopelagic and bathypelagic zones or can move between them, as does the sperm whale (*Physeter catadon*). Although the ratio in size of predator to prey falls within certain limitations, some of the largest nekton organisms in the sea, the baleen whales

(Mysticeti), feed on disproportionately small prey, euphasiids (popularly called krill). The sperm whale, 13 to 18.5 m in length, attacks very large prey, the giant squid (*Architeuthis*), whose overall length (tentacles extended) may be 18 m or more.

Living in a world that lacks any sort of refuge against predation or site for ambush, inhabitants of the pelagic zone have evolved various means of defense and of securing prey. Among them are the stinging cells of jellyfish, the remarkably streamlined shapes that allow speed both for escape and for pursuit, unusual coloration, advanced sonar, a highly developed sense of smell, and a social organization involving schools or packs. Some animals, such as the baleen whales, have specialized structures that permit them to strain krill and other plankton from the water. Others, such as sperm whale and certain seals, have the ability to dive to great depths to secure food. Phytoplankton lights up darkened seas, and fish take advantage of that bioluminescence to detect their prey.

The dimly lighted regions of the mesopelagic and the dark regions of the bathypelagic zone depend upon a rain of detritus as an energy source. Such food is limited. The rate of descent of organic matter, except for larger items, is so slow that it is consumed, decayed, or dissolved before it reaches the deepest water or the bottom. Other sources include saprophytic plankton, which exist in the darker regions, particulate organic matter, and such material as wastes from the coastal zone, garbage from ships, and dead whales and fish.

Residents of the deep have special adaptations for securing food. Some, like the zooplankton, swim to the upper surface to feed by night. Others remain in the dimly lighted or dark waters. Darkly pigmented and weak-bodied, many of the deep-sea fish depend upon luminescent lures, mimicry of prey, extensible jaws, and expandable abdomens (which enable them to consume large items of food) as means of obtaining sustenance in this zone. Although most fish are small (usually 15 cm or less in length), the region is inhabited by rarely seen large species such as the giant squid.

In the bathypelagic region bioluminescence reaches its greatest development. Two-thirds of the species produce light. Bioluminescence is not restricted to fish. Squid and euphausiids possess searchlightlike structures complete with lens and iris, and squid and shrimp discharge luminous clouds to escape predators. Fish have rows of luminous organs along their sides and lighted lures that enable them to bait prey and recognize other individuals of the same species.

Benthos  There is a gradual transition of life from the benthos on the rocky and sandy shores to that in the ocean's depths. From the tide line to the abyss, organisms that colonize the bottom are influenced by the nature of the substrate. If the bottom is rocky or hard, the populations consist

largely of organisms that live on the surface of the substrate, the epifauna and the epiflora. Where the bottom is largely covered with sediment, most of the inhabitants, chiefly animals, are infauna and live within the deposits. Particle size of the substrate determines the type of burrowing organisms in an area, because the mode of burrowing is often specialized and adapted to a certain type of substrate.

The substrate varies with the depth of the ocean and the relationship of the benthic region to land areas and continental shelves. Near the coast bottom sediments are derived from the weathering and erosion of land areas along with organic matter from floating marine life. Sediments of deep water are characterized by fine-textured material, which varies with depth and the type of organisms in the overlying waters. Although these sediments are termed organic, they contain little decomposable carbon, consisting largely of skeletal fragments of planktonic organisms. In general, with regional variations, organic deposits down to 4000 m are rich in calcareous matter. Below 4000 m hydrostatic pressure causes some forms of calcium carbonate to dissolve. At 6000 m and lower sediments contain even less organic matter and consist largely of red clays, rich in aluminum oxides and silica.

Within the sediments are layers that relate to oxidation-reduction reactions. The surface or oxidized layer, yellowish in color, is relatively rich in oxygen, ferric oxides, nitrates, and nitrites. It supports the bulk of the benthic animals, such as polychaete worms and bivalves in shallow water, flatworms, copepods, and others in deeper water, and a rich growth of aerobic bacteria throughout. Below the oxidized layer is a grayish transition zone to the black reduced layer, characterized by a lack of oxygen, iron in the ferrous state, nitrogen in the form of ammonia, and hydrogen sulfide. It is inhabited by anaerobic bacteria, chiefly reducers of sulfates and methane.

In the deep benthic regions variations in temperature, salinity, and other conditions are negligible. In this world of darkness there is no photosynthesis, so the bottom community is strictly heterotrophic, depending entirely for its source of energy on what organic matter finally reaches the bottom. Estimates suggest that the quantity of such material amounts to only 0.5 g/m²/yr (H. B. Moore 1958). Bodies of dead whales, seals, and fish may contribute another 2 or 3 g.

Bottom organisms have four feeding strategies: (1) they may filter suspended material from the water, as do stalked coelenterates; (2) they may collect food particles that settle on the surface of the sediment, as do the sea cucumbers; (3) they may be selective or unselective deposit-feeders, such as the polychaetes; or (4) they may be predators, as are the brittle stars and the spiderlike pycnogonids.

Important in the benthic food web are bacteria of the sediments (Figure 16.9). Common where large quantities of organic matter are present, bacteria may reach several tens of grams per square meter in the topmost layer of silt. Bacteria synthesize protein from dissolved nutrients and in turn become a source of protein, fats, and oils for deposit-feeders.

In spite of living in a world of darkness and pressure, the benthic fauna exhibits a surprising diversity. Faunal composition and diversity increases with depth down to the mid or lower bathyal region and decreases toward the abyssal plain (Rex 1981). Diversity along a gradient of depth appears to be related to the productivity of the waters above, competition among bottom fauna for food, and predation.

Hydrothermal Vents In 1977 oceanographers discovered along volcanic ridges in the ocean floor of the Pacific near the Galagapos Islands high-temperature deep sea springs never known to exist before. These springs vent jets of hydrothermal fluids that heat the surrounding water to 8° to 16° C, considerably higher than the 2° ambient water. Since then oceanographers have discovered similar vents on volcanic ridges along other fast-spreading centers of ocean floor, particularly in the eastern Pacific (for the geology and chemistry of such vents see Hayman and McDonald 1983, van Dover 1990, Tunnicliffe 1992).

Vents form when cold seawater flows down through fissures and cracks in the basaltic lava floor deep into the underlying crust, specifically in areas where continental plates are separating from each other. The waters react chemically with the hot ballast, giving up some minerals but becoming enriched with others such as copper, iron, sulfur, and zinc. Heated to a very high temperature, the water reemerges through the floor through mineralized chimneys rising up to 13 m above the floor. Among the types of chimneys are white smokers and black smokers (Figure 16.10). White smoker chimneys rich in zinc sulfides issue a milky fluid under 300° C. Black smokers, narrower chimneys rich in copper sulfides, issue jets of clear water 300° to over 450° C that is soon blackened by precipitation of fine-grained sulfur-mineral particles.

Associated with these vents is a rich diversity of newly discovered deep sea forms of life confined within a few meters of the vent system. Of the 293 species of vent animals described, 97 percent are endemic to the hydrothermal vent habitat. There is one new class, Vestimentifera, the giant tubeworms, which lack a digestive system. There are three new orders and 22 new families (van Dover 1990, Tunnecliffe 1992). These include giant clams, mussels, polychaete worms that encrust the white smokers, and crabs. The primary producers are chemosynthetic bacteria that oxidize the reduced sulfur compounds, such as $H_2S$, to release energy, which they use to form organic matter from carbon dioxide. Primary consumers, the clams, mussels, and worms, filter bacteria from water and graze on bacterial film on rocks. The giant clam *Calyptogena magnifica* and the large vestimentiferan worm *Riftia pachyptila* contain symbiotic chemosynthetic bacteria in their coelomic tissue. These bacteria need a reduced sulfide source, which is carried to them by the blood of these animals. Aix and Childress (1983) report that *Riftia* has in its blood a sulfide-binding protein that concentrates sulfide from the environment and

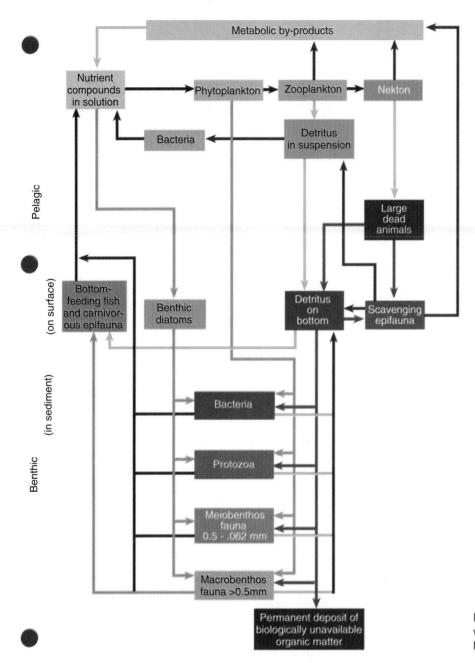

**Figure 16.9** Simplified marine food web. (After Raymont 1963, from M. Gross 1972.)

transports it to the bacteria. Such concentrations would poison normal animals, but the sulfide-bearing protein of the worm and apparently the clam has a high affinity for free sulfides, preventing them from accumulating in the blood and entering the cells (Power and Somber 1983).

## Function

The oceans occupy 70 percent of Earth's surface and their average depth is around 4000 m; yet their primary production is considerably less than that of the land. They are less productive because only a superficial illuminated area up to 100 m deep can support plant life; and that plant life, largely phytoplankton, is patchy because most of the open sea is nutrient-poor. A limited to almost nonexistent nutrient reserve can be recirculated. Phytoplankton, zooplankton, and other organisms and their remains sink below the lighted zone into the dark benthic water. While this sinking supplies nutrients to the deep, it robs the upper layers.

The depletion of nutrients is most pronounced in tropical waters. The permanent thermal stratification, with a layer of warmer, less dense water lying on top of colder, denser deep water, prevents an exchange of nutrients between the surface and the deep. Thus in spite of high light intensity and warm temperatures, tropical seas are the lowest in production, about 18 to 50 g C/m$^2$/yr.

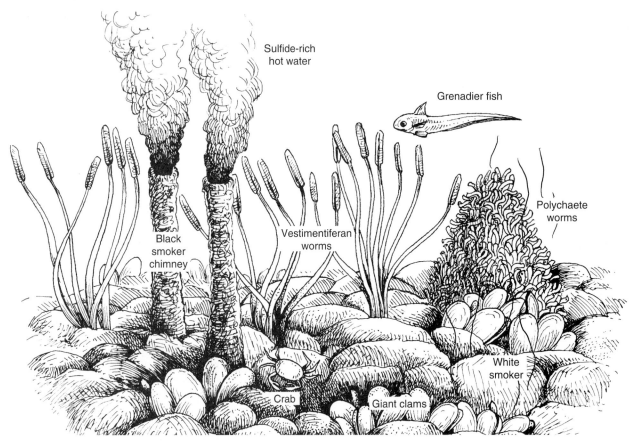

**Figure 16.10** A typical hydrothermal vent mound resting on flows of black basaltic lava. (Adapted from Hayman and McDonald 1985 and other sources.)

The temperate oceans are more productive, largely because a permanent thermocline does not exist. During the spring and to a limited extent during the fall temperate seas, like temperate lakes, experience a nutrient overturn. The recirculation of phosphorus and nitrogen from the deep stimulates a surge of spring phytoplankton growth. As spring wears on the temperature of the water becomes stratified and a thermocline develops, preventing a nutrient exchange. The phytoplankton growth depletes the nutrients and the phytoplankton population suddenly declines. In the fall a similar overturn takes place, but the rise in phytoplankton production is slight because of decreasing light intensity and low winter temperatures. Reduced production in winter holds down annual productivity of temperate seas to a level a little above that of tropical seas, 70 to 110 g C/m²/yr.

Most productive are coastal waters and regions of upwelling, whose annual production may amount to 1000 g C/m²/yr. Major areas of upwelling are largely on the western sides of continents: off the coasts of southern California, Peru, northern and southwestern Africa, and the Antarctic. Upwellings result from the differential heating of polar and equatorial regions that produces the equatorial currents and the winds. As the water is pushed northward or southward toward the equator by winds, it is deflected away from the coasts by the Coriolis force. As the deflected surface water

moves away, it is replaced by an upwelling of colder, deeper water that brings a supply of nutrients into the sunlit portions of the sea. As a result regions of upwellings are highly productive and contain an abundance of life. Because of their high productivity, upwellings support (or did support until overfished) important commercial fisheries such as the tuna fishery off the California coast, the anchoveta fishery off Peru, and the sardine fishery off Portugal.

Other zones of high production are coastal waters and estuaries, where productivity may run to 380 g C/m²/yr. Turbid, nutrient-rich waters are major areas of fish production. Thus between upwellings and coastal waters, the most productive areas of the seas are the fringes of water bordering the continental land masses. A great deal of the measured productivity of the coastal fringes comes from the benthic as well as the surface waters because these seas are shallow. Benthic production, largely unavailable, is not considered in the productivity of the open sea. Recent estimates (Koblentz-Mishke et al. 1970) of the total production for marine plankton is 50 Gt dry matter per year; if benthic production is considered, total production may be 55 Gt of dry matter per year.

Carbohydrate production by phytoplankton, largely diatoms, is the base upon which life of the seas exist. Conversion of primary production to animal tissue is accom-

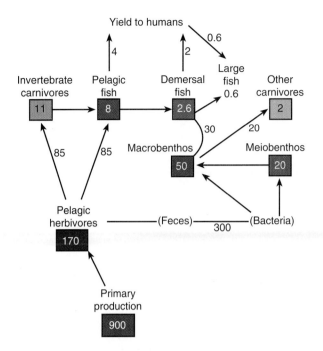

**Figure 16.11** Food web and energy flow from an area of the North Sea. Note the low energy yield to humans, 6.6 kcal/m²/yr from a primary production of 900 kcal/m²/yr. (From Steele 1971.)

cyanobacteria (1–2μ) are responsible for a large part of photosynthesis in the sea. These phytoplankton cells excrete a substantial fraction of their photosynthate in the form of dissolved organic material that is utilized by heterotrophic bacteria.

Populations of such bacteria are dense, around 1 million cells per milliliter of sea water. These heterotrophic bacteria account for about 20 percent of primary production. Bacterial growth efficiency does not exceed 50 percent, so one-half of phytoplankton primary production in the form of dissolved organic material is consumed by bacteria. Bacterial numbers in the sea remain relatively stable, suggesting predation; but filter-feeding zooplankton cannot retain particles of bacterial size. Therefore, the consumption of bacteria by heterotrophic nanoflagellates, experimentally demonstrated, accounts for the disappearance of bacterial production. This uptake by heterotrophic bacteria of dissolved organic matter produced by the plankton, and the subsequent consumption of bacteria by nanoplankton introduces a feeding loop, termed the **microbial loop** (Azam et al. 1983) (Figure 16.12). Nanoplankton, a newly discovered group of phytoplankton unicellular cyanobacteria (less than

**Figure 16.12** A microbial loop that feeds into the classic food web. On the right are representative photoautotrophs, nanoflagellates, and cyanobacteria and on the left are associated heterotrophs, bacteria, and heterotrophic nanoflagellates. (Adapted from Fenchel 1988:24.)

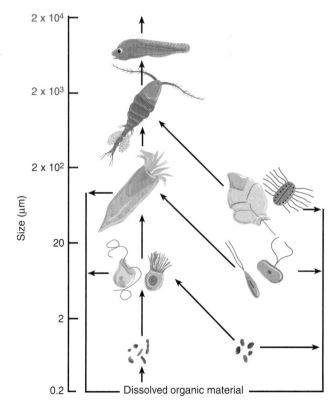

plished by zooplankton, the most important of which are the copepods. To feed on the minute phytoplankton, most of the grazing herbivores must also be small, measuring between 0.5 and 5.0 mm. Most of the grazing herbivores in the oceans are the genera *Calanus, Acartia, Temora,* and *Metridia,* probably the most abundant animals in the world. The single most abundant copepod is *Calanus finmarchicus* and its close relative *Calanus helgolandicus.* In the Antarctic krill, fed upon by the blue whale, are the dominant herbivores.

The copepods then become the link in the classic food chain between the phytoplankton and the second-level consumers, illustrated in the North Sea food web (Figure 16.11). The dominant primary producer is the diatom *Skeletoma,* which is grazed by *Calanus.* The major predator on *Calanus* is the semiplankton sand eel *Ammodytes,* which in turn is food for herring. The herring, however, can shorten the food chain by bypassing the sand eel and feeding directly on *Calanus.* The herring is also involved in a number of side food chains that add stability to its food supply.

Part of the food chain begins with organisms even smaller. Recent investigations show that bacteria and protists, both heterotrophic and photosynthetic, make up one-half of the biomass of the sea, and are responsible for the largest part of energy flow in pelagic systems (Fenchel 1987, 1988). Photosynthetic nanoflagellates (2–20μ) and

5 µm in size), and other abundant types of small marine phytoplankton are preyed upon by phagotrophic protozoa, adding several more trophic levels in the plankton food chain (Sherr and Sherr 1991).

## ROCKY SHORES

All ocean shores have one feature in common. They are alternately exposed and submerged by tides. Roughly, the region of the seashore is bounded by the extreme high water mark and extreme low water mark. Within this intertidal zone conditions change from hour to hour with the ebb and flow of tides. At high tide the seashore is a water world; at ebb tide it belongs to the terrestrial environment, with its extremes of temperature, moisture, and solar radiation. Nevertheless the intertidal inhabitants are essentially marine, able to withstand some degree of exposure to air for varying periods of time.

As the sea recedes at ebb tide, life hidden by the tidal water emerges, layer by layer. The uppermost layers, those near the high water mark are exposed to air, wide temperature fluctuations, intense solar radiation, and desiccation for a considerable period of time, whereas the lowest fringes on the intertidal shore may be exposed only briefly before the tide submerges them again.

**Figure 16.13** The broad zones of life exposed at low tide on the rocky shore of the Bay of Fundy.

### Structure

Along rocky shores varying exposures to environmental conditions result in pronounced zonation (Figure 16.13). Although this zonation may be strikingly different from place to place as a result of local variations in aspect, substrate, wave action, light intensity, shore profile, exposure to prevailing winds, climatic differences, and the like, it possesses everywhere the same general features. All rocky shores have three basic zones, characterized by the dominant organisms occupying them (Figure 16.14).

Where land ends and seashore begins is difficult to fix. The approach to a rocky shore from the landward side is marked by a gradual transition from lichens and other land plants to marine life, dependent in part on tidal waters (Figure 16.15). The first major change from land shows up on the **supralittoral fringe,** where salt water comes only once every two weeks on the spring tides. It is marked by a black zone, a patchy or beltlike incrustation of Verrucaria-type lichens and crustose algae such as *Calothrix.* Capable of existing under conditions so difficult that few other plants could survive, these algae, enclosed in slimy, gelatinous sheaths, and their associated lichens represent a nonmarine community, on which graze basically marine mammals, the periwinkles (Doty 1957). Common to this black zone on American and European shores is the common periwinkle

(*Littorina littorea*) (Figure 16.16) that grazes on wet algae covering the rocks. On European shores lives the small periwinkle (*L. neritiodes*), highly adapted to the driest conditions of the black zone. Often living for days, or even weeks, without being wetted by the spray, this periwinkle is the most highly resistant to desiccation of all shore animals.

Below the black zone lies the **littoral zone,** a region covered and uncovered daily by tides. The littoral tends to be divided into subzones. In the upper reaches, barnacles are most abundant. The oyster, blue mussel, and limpet appear in the middle and lower portions of the littoral, as does the common periwinkle.

Occupying the lower half of the littoral zone (midlittoral) of colder climates and in places overlying the barnacles is an ancient group of plants, the brown algae, more commonly known as rockweeds (*Fucus*) and wrack (*Ascophyllum*) (Figure 16.17). These intertidal brown algae attain their finest growth on protected shores, where they may grow 2 m long; on wave-whipped shores they are considerably shorter.

The lower reaches of the littoral zone may be occupied by blue mussels (*Mytilus edulis*) instead of rockweeds, particularly on shores where hard surfaces have been covered in part by sand and mud. No other shore animal grows in such

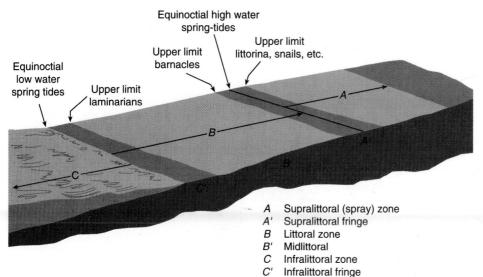

A   Supralittoral (spray) zone
A'  Supralittoral fringe
B   Littoral zone
B'  Midlittoral
C   Infralittoral zone
C'  Infralittoral fringe

**Figure 16.14** Basic or universal zonation on a rocky shore. Use this diagram as a guide when studying the drawing in Figure 16.15. (Adapted from Stephenson 1949.)

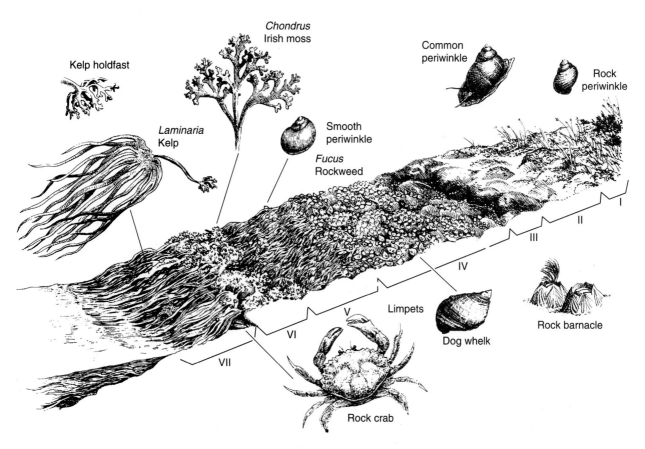

**Figure 16.15** Zonation on a rocky shore along the North Atlantic. Compare with the diagram in Figure 16.14. I, land: lichens, herbs, grasses; II, bare rock; III, zone of black algae and rock periwinkles; IV, barnacle zone: barnacles, dog whelks, common periwinkles, mussels; V, fucoid zone: rockweed (*Fucus*) and smooth periwinkles; VI Irish moss (*Chondrus*) zone; VII, kelp (*Laminaria*) zone. (Based on data from Stephenson and Stephenson 1954; author's photographs and observations.)

**Figure 16.16** Low tide reveals common periwinkles clinging to the rocky substrate.

abundance; the blue-black shells packed closely together may blanket the area (Figure 16.18).

Near the lower reaches of the littoral zone, mussels may grow in association with the low-growing red algae *Gigartina.* Algae and mussels together often form a tight mat over the rocks. Here, well protected in the dense growth from waves, live infant starfish, sea urchins (Echinoidae), brittle stars (Stelleroidae), and bryozoan sea mats or sea lace (Membranipora).

The lowest part of the littoral zone, uncovered only at the spring tides and not even then if wave action is strong, is the **infralittoral fringe.** This zone, exposed for short periods of time, consists of forests of the large brown algae *Laminaria,* one of the kelps, with a rich undergrowth of smaller plants and animals among the holdfasts.

Beyond the infralittoral fringe is the **sublittoral zone,** part of the open sea. This zone is principally neritic and benthic and contains a wide variety of fauna, depending upon the substrate, the presence of protruding rocks, gradients in turbulence, oxygen tensions, light, and temperature.

Although waves, tides, exposure, desiccation, and temperature influence the intertidal zonation, the pattern of

**Figure 16.17** Knotted wrack (*Ascophyllum nodosum*) exposed at low tide on the Bay of Fundy.

**Figure 16.18** A tidal flat covered with blue mussels.

## Function

The rocky shore is both autotrophic and heterotrophic. Many organisms, such as barnacles, depend upon tides to bring them food; others, such as periwinkles, graze on algal growth on the rocks. In fact, the functioning of the rocky seashore ecosystem involves complex interactions between physical and biotic aspects of the ecosystem. Like a salt marsh, which receives a daily energy subsidy from tidal flooding, rocky shores receive an energy subsidy, although indirectly, from the waves. Leigh et al. (1987) estimate that wave-generated energy impinging on a rocky shore is roughly twice that of solar radiation, amounting to 0.045 watts/cm$^2$, compared to 0.017 to 0.025 watts/cm$^2$ of solar radiation. Although this energy is not directly stored as carbohydrates, its input, nevertheless, is reflected in productivity. For example, the productivity of kelp beds of the Pacific Northwest exceeds that of the tropical rain forest. Part of this productivity relates to the much higher leaf surface per square meter—about 20 m$^2$ per m$^2$ of growing surface, compared to about 8 m$^2$ for a tropical forest. The highest standing crops of kelp, mussels, and other organisms are restricted to areas receiving the heaviest wave action.

Intertidal organisms do not use wave energy directly. Rather the force of the waves favorably affects the environmental conditions that promote increased productivity. For one, heavy wave action reduces the activity of such predators of sessile intertidal invertebrates as starfish and sea urchins (see Chapter 24). Waves bring in a steady supply of nutrients and carry away products of metabolism. They keep in constant motion the fronds of various seaweeds, moving them in and out of shadow and sunlight, allowing for more even distribution of incident light and thus more efficient photosynthesis. By dislodging organisms, both plants and invertebrates, from the rocky substrate, waves open up space for colonization by algae and invertebrates and reduce strong interspecific competition. In effect disturbance (see Chapter 30), which influences community structure, is the root of intertidal productivity.

## Tide Pools

The ebbing tide leaves behind pools of water in rock crevices, in rocky basins, and in depressions (Figure 16.19). They are tide pools, "microcosms of the sea," as Yonge (1949) describes them. They represent distinct habitats, which differ considerably from the exposed rock and the open sea, and even differ among themselves.

At low tide all the pools are subject to wide and sudden fluctuations in temperature and salinity, but these changes are most marked in shallow pools. Under the summer sun the temperature may rise above the maximum many organisms can tolerate. As the water evaporates, especially in the smaller and more shallow pools, salinity increases and salt crystals may appear around the

life on rocky shores is heavily influenced, too, by biotic interactions of grazing, predation, and competition, and by larval settlement (see Underwood, Denley and Moran 1983). Where wave action is heavy on New England intertidal rocky shores, periwinkles are rare, which permits more vigorous growth of algae. Lack of grazing favors ephemeral algal species such as *Ulva* and *Enteromorpha*, whereas grazing allows the perennial *Fucus* to become established (see Lubchenco 1980, 1983; Chapter 30). On the New England coast the blue mussel outcompetes barnacles and algae; but predation by starfish (*Asterias* spp.) and the snail *Nucella lapillus* prevents dominance by mussels except on the most wave-beaten areas. A similar situation exists on the Pacific coast. Barnacles of several species tend to outcompete and displace algal species, but in turn the competitively dominant mussel *Mytilus californianus* destroys the barnacles by overgrowing them (Dayton 1971). Where present the predatory starfish *Pisaster ochraceus* prevents the mussel from completely overgrowing barnacles (Paine 1966). The end result of such interactions of the physical and the biotic is a patchy distribution of life across the rocky intertidal shore. (For a good overview see Nybakken 1988:255–283.)

**Figure 16.19** Tidal pools fill depressions on this piece of rocky coast in Maine.

edges. When rain or land drainage brings fresh water to the pool, salinity may decrease. In deep pools such fresh water tends to form a layer on the top, developing a strong salinity stratification in which the bottom layer and its inhabitants are little affected. If algal growth is considerable, oxygen content of the water varies through the day. Oxygen will be high during the daylight hours but will be low at night, a situation that rarely occurs at sea. The rise of carbon dioxide at night means a lowering of pH.

Pools near low tide are influenced least by the rise and fall of the tides; those that lie near and above the high-tide line are exposed the longest and undergo the widest fluctuations. Some may be recharged with seawater only by the splash from breaking waves or occasional high spring tides. Regardless of their position on the shore, most pools suddenly return to sea conditions on the rising tide and experience drastic and instantaneous changes in temperature, salinity, and pH. Life in the tidal pools must be able to withstand wide and rapid fluctuations in the environment. Among tidal pool inhabitants are green seaweeds (**Enteromorpha** spp.), filamentous seaweeds (**Cladophera** spp.), small sea anemones, acorn barnacles, and periwinkles.

# SANDY SHORES AND MUDFLATS

Both the sandy shore and the mudflat at low tide appear barren of life, a sharp contrast to the life-studded rocky shores; but beneath the wet and glistening surface life hides, waiting for the next high tide.

Both in some ways are harsh environments, but the sandy shore is especially so. The very matrix of this seaside environment is a product of the harsh and relentless weathering of rock, both inland and along the shore. Through eons the ultimate products of rock weathering are carried away by rivers and waves to be deposited as sand along the edge of the sea.

The size of the sand particles deposited influences the nature of the sandy beach, water retention during low tide, and the ability of animals to burrow through it. Beaches with steep slopes usually are made up of larger sand grains and are subject to more wave action.

Beaches exposed to high waves are generally flattened, for much of the material is transported away from the beach to deeper water, and fine sand is left behind (Figure 16.20). Sand grains of all sizes, especially the finer particles in which capillary action is greatest, are more or less cushioned by a film of water about them, reducing further wearing away. Retention of water by the sand at low tide is one of the outstanding environmental features of the sandy shore.

## Structure

Life on the sand is almost impossible. Sand provides no surface for attachments of seaweed and their associated fauna. The crabs, worms, and snails so characteristic of rocky crevices find no protection here. Most life, then, is forced to live beneath the sand.

Life on sandy and muddy beaches consists of the **epifauna**, organisms living on the surface, and the **infauna**, organisms living within the substrate. Most infauna either occupy permanent or semipermanent tubes within the sand or mud, like the lugworm (*Arenicola* spp.), or are able to

**Figure 16.20** A long stretch of sandy beach washed by waves on the southern Australian coast. Although the beach appears barren, life is abundant beneath the sand.

burrow rapidly into the substrate, like the coquina bivalve (*Donax variabilis*). Multicellular infauna obtain oxygen either by gaseous exchange with the water through their outer covering or by breathing through gills and elaborate respiratory siphons.

Within the sand and mud live vast numbers of **meiofauna** with a size range between 0.05 and 0.5 mm, including copepods, ostracods, nematodes, and gastrotrichs. These interstitial fauna are generally elongated forms with setae, spines, or tubercles greatly reduced. The great majority do not have pelagic larval stages. These animals feed mostly on algae, bacteria, and detritus. Interstitial life, best developed on the more sheltered beaches, shows seasonal variations, reaching maximum development in summer months.

Sandy beaches also exhibit zonation related to tides (Figure 16.21), but it must be discovered by digging. Sandy and muddy shores can be divided into supralittoral, littoral, and infralittoral zones, based on animal organisms, but a universal pattern similar to that of the rocky shore is lacking. Pale, sand-colored ghost crabs (*Ocypode quadrata*) and

**Figure 16.21** Life on a sandy ocean beach along the mid-Atlantic Coast. Although strong zonation is absent, organisms still change on a gradient from land to sea. I, supratidal zone: ghost crabs and sand fleas; II, flat beach zone: ghost shrimp, bristle worms, clams; III, intratidal zone: clams, lugworms, mole crabs; IV, subtidal zone. The dashed line indicates high tide.

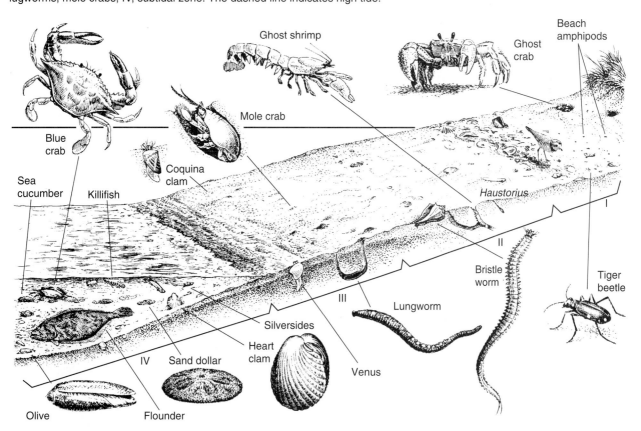

beach fleas (not true fleas) (*Talorchestia* and *Orchestia* spp.) occupy the upper beach, the supralittoral. The intertidal beach, the littoral, is a zone where true marine life appears. Although sandy shores lack the variety found on rocky shores, the populations of individual species of largely burrowing animals often are enormous. An array of animals, among them starfish and the related sand dollar, can be found above the low-tide line and in the infralittoral.

Organisms living within the sand and mud do not experience the same violent fluctuations in temperature as those on the rocky shores. Although the surface temperature of the sand at midday may be 10° C or more higher than the returning seawater, the temperature a few inches below remains almost constant throughout the year. Nor is there a great fluctuation in salinity, even when fresh water runs over the surface of the sand. Below 25 cm, salinity is little affected.

## Function

For life to exist on the sandy shore, some organic matter has to accumulate. Most sandy beaches contain a certain amount of detritus from seaweeds, dead animals, feces, and material blown in. This organic matter accumulates within the sand, especially in sheltered areas. An inverse relationship exists between the turbulence of the water and the amount of organic matter on the beach, with accumulation reaching its maximum on the mudflats. Organic matter clogs the space between the grains of sand and binds them together. As water moves down through the sand, it loses oxygen from both the respiration of bacteria and the oxidation of chemical substances, especially ferrous compounds. The point within the mud or sand at which water loses all its oxygen is a region of stagnation and oxygen deficiency, characterized by the formation of ferrous sulfides. The iron sulfides cause a zone of black whose depth varies with the exposure on the beach. On mudflats such conditions exist almost to the surface.

The energy base for sandy beach and mudflat fauna is organic matter. Much of it becomes available through bacterial decomposition, which goes on at the greatest rate at low tide. The bacteria are concentrated around the organic matter in the sand, where they escape the diluting effects of water. The products of decomposition are dissolved and washed into the sea at each high tide, which, in turn, brings in more organic matter for decomposition. Thus the sandy beach is an important site for biogeochemical cycling, supplying offshore waters with phosphates, nitrogen, and other nutrients.

Energy flow in sandy beaches and mudflats differs from that of terrestrial and aquatic systems because the basic consumers are bacteria. In other systems bacteria act largely as reducers responsible for conversion of dead organic matter into a form that can be used by producer organisms. In sandy beaches and mudflats bacteria not only feed on detrital material and break down organic matter, but they are also a major source of food for higher level consumers.

A number of deposit-feeding organisms ingest organic matter largely as a means of obtaining bacteria. Prominent among them on the mudflats are numerous nematodes and copepods (Harpacticoida), the polychaete clam worm (*Nereis*) and the gastropod mollusk (*Hydrobia*). Deposit-feeders on sandy beaches obtain their food by actively burrowing through the sand and ingesting the substrate to obtain the organic matter it contains. The most common among them is the lugworm, which is responsible for the conspicuous coiled and cone-shaped casts on the beach.

Other sandy beach animals are filter feeders, obtaining their food by sorting particles of organic matter from tidal water. Two of these "surf fishers" who advance and retreat up and down the beach with the ebb and flow of the tides are the mole crab (*Emerita talpoida*) and coquina.

Associated with these essentially herbivorous animals are the predators, always present whether the tide is in or out. Near and below the low tide line live predatory gastropods, which prey on bivalves beneath the sand. In the same area lurk predatory crabs such as the blue crab (*Callinectes sapidus*) and green crab (*Carcinus maenas*) which feed on mole crabs, clams, and other organisms. They move back and forth with the tides. The incoming tides also bring other predators such as killifish (*Fundulus heteroclitus*) and silversides (*Menidia menidia*). As the tide recedes, gulls and shorebirds scurry across the sand and mudflats to hunt for food.

Because of their dependence upon imported organic matter and their essentially heterotrophic nature, sandy beaches and mudflats should be considered not as separate ecosystems, but as part of the whole coastal ecosystem (Figure 16.22). Except in the cleanest of sands, some primary production does take place in the intertidal zone. The major primary producers are the diatoms, confined mainly to fine-grained deposits of sand containing a high proportion of organic matter. Productivity is low. One estimate places productivity of moderately exposed beaches at 5 g C/m$^2$/yr. Production may be increased temporarily by phytoplankton carried in at high tide and left stranded on the surface. Again, these organisms are mostly diatoms. More important as producers are the sulfur bacteria in the black sulfide or reducing layer. These chemosynthetic bacteria use energy released as ferrous oxide is reduced. Some mudflats are covered with the algae *Enteromorpha* and *Ulva,* whose productivity can be substantial.

In effect, then, sandy shores and mudflats are part of a larger coastal ecosystem involving the salt marsh, the estuary, and coastal waters. They act as sinks for energy and nutrients because the energy they utilize comes not from primary production, but from organic matter that originates outside the area. Many of the nutrient cycles are only partially contained within the borders of the shores.

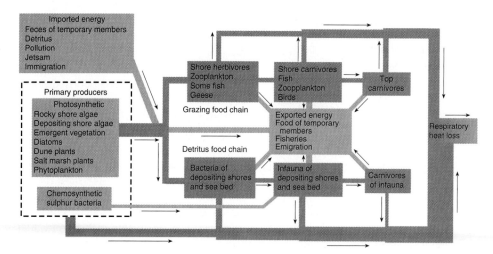

**Figure 16.22** Diagram of the coastal ecosystem, a supraecosystem consisting of the shore, the fringing terrestrial regions, and the sublittoral zones. It connects two food webs: that of the rocky shore with its algae, herbivores, and zooplankton, and the detrital food webs involving the bacteria of the depositing shore and sublittoral muds and the dependent detritivores and carnivores. Coastal ecosystems are extremely productive; because energy imports exceed exports, the system is continuously gaining energy. (From S. K. Eltringham 1971.)

# CORAL REEFS

In subtropical and tropical waters around the world are structures of biological rather than geological origin: coral reefs (Figure 16.23). Coral reefs are built by carbonate-secreting organisms, of which coral (Cnidaria, Anthozoa) is the most conspicuous, but not always the most important. Also contributing heavily are the coralline red algae (Rhodophyta, Corallinaceae), Foraminifera, and mollusks. Built only underwater at shallow depths, no deeper than 50 to 70 m and preferably 25 m, coral reefs need a stable foundation upon which to grow. Such foundations are provided by shallow continental shelves and submerged volcanoes.

Coral reefs are of three types with many gradations among them. (1) **Fringing reefs** project directly seaward from the shore. (2) **Barrier reefs** parallel shorelines and are separated from land by a lagoon. (3) **Atolls** are coral islands that begin as horseshoe-shaped reefs surrounding a lagoon. Such lagoons are about 40 m deep and are usually connected to the open sea by breaks in the reef. Reefs build up to sea level. To become atolls, the reefs have to be exposed by a lowering of the sea level or built up by the action of wind and waves.

## Structure

Coral reefs are complex ecosystems involving close relationships between coral organisms and algae. In the tissues of the gastrodermal layer live **zooxanthellae,** symbiotic endozoic dinoflagellate algae. On the calcareous skeleton live still other kinds of algae, both the encrusting red and green coralline species and filamentous species, including turf algae. Associated with coral growth are mollusks such as giant clams (*Tridacna, Hippopus*), echinoderms, crustaceans, polychaete worms, sponges, and a diverse array of fishes, both herbivorous and predatory.

Zonation and diversity of coral species are influenced by an interaction of depth, light, grazing, competition, and disturbance (Huston 1985). The basic diversity gradient is established by light. Diversity is lowest at the crest, where only species tolerant of intense or frequent disturbance by waves can survive. Diversity increases with depth to a maximum of about 20 m and then decreases as light becomes attenuated, eliminating shade-intolerant species (see Wellington 1982a). Imposed upon this condition are a variety of biotic and abiotic disturbances that vary in intensity and decrease with depth. Growth rates of photosynthetic coral are highest in shallow depths, and a few species, especially the branching corals, can easily dominate the reef by overgrowing and shading the crustose corals and algae. Disturbances by wave action, storms, and grazing reduce the rate of competitive displacement among corals. Heavy grazing of overgrowing algae by sea urchins and fish, such as parrotfish (Scaridae), increases encrusting coralline algae. Light grazing allows rapidly growing filamentous and foliose algal species to eliminate crustose algae (see Wellington 1982b).

## Function

Coral organisms are partially photosynthetic and partially heterotrophic. During the day zooxanthellae carry on photosynthesis and directly transfer organic material to coral tis-

**Figure 16.23** Coral reef viewed from the air.

sue. At night coral polyps feed on zooplankton, securing phosphates and nitrates and other nutrients needed by the anthozoans and their symbiotic algae. Thus nutrients are recycled in place between the anthozoans and the algae (Pomeroy and Kuenzler 1969, Hatcher 1988). In addition, carbon dioxide concentrations in animal tissue enable the coral to extract the calcium carbonate needed to build the coral skeletons. Adding to the productivity of the coral are crustose coralline algae, turf algae, macroalgae, seagrass, sponges, phytoplankton, and a large bacterial population.

Coral reefs are among the most highly productive ecosystems on Earth. Net productivity ranges from 1500 to 5000 g C/m²/yr, compared to 15 to 50 g C/m²/yr for the surrounding ocean. Because of the ability of the coralline community to retain nutrients within the system and to act as a nutrient trap, coral reefs are oases of productivity within a nutrient-poor sea.

This high productivity and the diversity of habitats within the reef support a high diversity of life (Figure 16.24). There are thousands of kinds of invertebrates, some of which, such as sea urchins, feed on coral animals and algae, hundreds of kinds of herbivorous fish that graze on algae, and many predatory species (see Yonge 1963, Sale 1980). Some of these predators, such as the puffers (Tetraodontidae) and filefishes (Monacanthidae), are corallivores, feeding on coral polyps. Others lie in ambush for prey in coralline caverns. In addition, there is a wide array of symbionts, such as cleaning fish and crustaceans, that pick parasites and detritus from larger fish and invertebrates.

**Figure 16.24** Colorful fish life in a coral reef.

## ESTUARIES

Water of all streams and rivers eventually drains into the sea. The place where this fresh water joins and mixes with the salt is called an **estuary.** Estuaries (for example, the Chesapeake Bay and the Humboldt Bay of California) are semienclosed parts of the coastal ocean where river water mixes with and measurably dilutes the seawater (Ketchum 1983).

Estuaries differ in size, shape, and volume of water flow, all influenced by the geology of the region in which they occur. As the river reaches the sea, it drops its sediments in quiet water. They accumulate to form deltas in the upper reaches of the mouth and shorten the estuary. When silt and mud accumulations become high enough to be exposed at low tide, tidal flats develop, which divide and braid the original channel of the estuary. At the same time, ocean currents and tides erode the coastline and deposit material on the sea-

ward side of the estuary, also shortening the mouth. If more material is deposited than is carried away, barrier beaches, islands, and brackish lagoons appear.

## Structure

Current and salinity, both complex and variable, shape life in the estuary. Estuarine currents result from the interaction of a one-direction stream flow, which varies with the season and rainfall, with oscillating ocean tides and with the wind (Ketchum 1951, 1983, Burt and Queen 1951, Smayda 1983). Because of the complex nature of the currents, generalizations about estuaries are difficult to make (see Lauff 1967).

Salinity varies vertically and horizontally, often within one tidal cycle. Vertical salinity may be the same from top to bottom, or it may be completely stratified, with a layer of fresh water on top and a layer of dense saline water on the bottom. Salinity is homogeneous when currents, particularly eddy currents, are strong enough to mix the water from top to bottom. The salinity in some estuaries is homogeneous at low tide but stratified at high tide. As the tide floods, a surface wedge of seawater moves upstream more rapidly than the bottom water, creating a density inversion of salinity stratification. Seawater on the surface tends to sink as lighter fresh water rises, and mixing takes place from the surface to the bottom. This phenomenon is known as tidal overmixing. Strong winds, too, tend to mix salt water with the fresh (Barlow 1956) in some estuaries; but when the winds are still, the river water flows seaward on a shallow surface over an upstream movement of seawater that only gradually mixes with the salt.

Horizontally, the least saline waters are at the river entrance, and the most saline at the mouth of the estuary (Figure 16.25). The configuration of the horizontal zonation is determined mainly by the deflection caused by the incoming and outgoing currents (see Officer 1983). In all estuaries of the Northern Hemisphere, outward-flowing fresh water and inward-flowing seawater are deflected to the east because of Earth's rotation. As a result, salinity is higher on the western side.

Salinity also varies with changes in the quantity of fresh water pouring into the estuary through the year. Salinity is highest during the summer and during periods of drought, when less fresh water flows into the estuary. It is lowest during the winter and spring, when rivers and streams are discharging their peak loads. This change in salinity may happen rather rapidly. For example, early in 1957 a heavy rainfall broke the most severe drought in the history of Texas. The resultant heavy river discharge reduced the salinities in Mesquite Bay on the central Texas coast by over 30 ppt in a two-month period. At the height of the drought salinities ranged from 35.5 to 50.0 ppt, but after the break in the drought they ranged from 2.3 to 2.9 ppt (Hoese 1960). Such rapid changes have a profound impact on the life of the estuary.

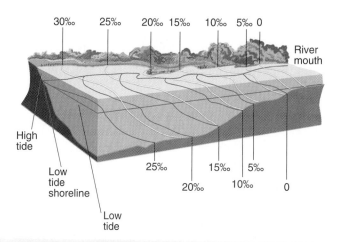

**Figure 16.25** Vertical and horizontal stratification of salinity from the river mouth to the estuary at both low and high tide. At high tide the incoming seawater increases the salinity toward the river mouth; at low tide salinity is reduced. Note also how salinity increases with depth, because lighter fresh water flows over denser salt water.

The salinity of seawater is about 35[‰]; that of fresh water ranges from 0.065 to 0.30 [‰]. Because the concentration of metallic ions carried by rivers varies from drainage to drainage, the salinity and chemistry of estuaries differ. The proportion of dissolved salts in the estuarine waters remains about the same as that of seawater, but the concentration varies in a gradient from fresh water to sea.

Exceptions to these conditions exist in regions where evaporation from the estuary may exceed the inflow of fresh water from river discharge and rainfall (a negative estuary). This situation causes the salinity to increase in the upper end of the estuary, and horizontal stratification is reversed.

Temperatures in estuaries fluctuate considerably diurnally and seasonally. Waters are heated by solar radiation and inflowing and tidal currents. High tide on the mudflats may heat or cool the water, depending on the season. The upper layer of estuarine water may be cooler in winter and warmer in summer than the bottom, a condition that, as in a lake, will result in a spring and autumn overturn.

Mixing waters of different salinities and temperatures acts as a nutrient trap (Officer 1983). Inflowing river waters more often than not impoverish rather than fertilize the estuary, except for phosphorus. Instead, nutrients and oxygen are carried into the estuary by the tides. If vertical mixing takes place, these nutrients are not soon swept back out to sea, but circulate up and down among organisms, water, and bottom sediments (Figure 16.26).

Organisms inhabiting the estuary are faced with two problems—maintenance of position and adjustment to changing salinity (Vernberg 1983). The bulk of estuarine organisms are benthic and are securely attached to the bottom, buried in the mud, or lodged in crevices and crannies about sessile organisms. Motile inhabitants are chiefly crustaceans

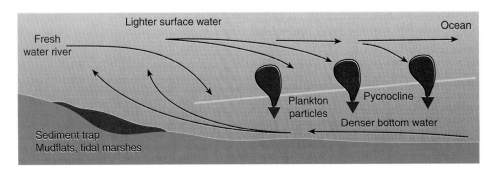

**Figure 16.26** Circulation of fresh and salt water in an estuary creates a nutrient trap. The trap develops because of a salt wedge of intruding seawater on the bottom, a surface flow of lighter fresh water and a counterflow of heavier brackish water. This countercurrent serves to trap nutrients, recirculating them toward the tidal marsh. The same countercurrent also sends phytoplankton up the estuary, repopulating the water. When nutrients, especially in the form of particulate matter, are high in the upper estuary, they are taken up rapidly by tidal marshes and mudflats, converted to soluble forms, and exported back to the open waters of the estuary. (From Correll 1978.)

and fish, largely young of species that spawn offshore in high-salinity water.

Planktonic organisms are wholly at the mercy of the currents. Because the seaward movement of stream flow and ebb tide transports plankton out to sea, the rate of circulation or flushing time determines the nature of the plankton population. If the circulation is too vigorous, the plankton population may be small. Phytoplankton in summer is most dense near the surface and in low-salinity areas. In winter phytoplankton is more uniformly distributed. For any planktonic growth to become endemic in an estuary, reproduction and recruitment must balance the losses from physical processes that disperse the population (Barlow 1955).

Changing salinity dictates the distribution of life in the estuary. Essentially, the organisms of the estuary are marine, able to withstand full seawater. Except for anadromous fishes, no freshwater organisms live there. Some estuarine inhabitants cannot withstand lowered salinities, and these species decline along a salinity gradient. Sessile and slightly motile organisms have an optimum salinity range within which they grow best. When salinities vary on either side of this range, populations decline. Two animals, the clam worm and the scud, illustrate this situation. Two species of clam worm, *Nereis occidentalis* and *Neanthes succinea,* inhabit the estuaries of the southern coastal plains of North America. *Nereis* is more numerous at high salinities and *Neanthes* at low salinities. In European estuaries the scud *Gammarus* is an important member of the bottom fauna. Two species, *G. locusta* and *G. marina,* are typical marine species and cannot penetrate far into the estuary. Instead they are replaced by a typical estuarine species, *G. zaddachi.* This species, however, is broken down into three subspecies, separated by salinity tolerances. *G. zaddachi* lives at the seaward end, *G. z. salinesi* occupies the middle, and *G. z. zaddachi,* which can penetrate up into fresh water for a short time, lives on the landward end (Spooner 1947, Segerstrale 1947).

Stage of development influences the range of motile species within the estuarine waters. This influence is particularly pronounced among estuarine fish. Species such as the striped bass spawn near the interface of fresh and low-salinity water (Figure 16.27). The larvae and young fish move downstream to more saline waters as they mature. Thus for the striped bass the estuary serves both as a nursery and feeding ground for the young. Anadromonous species, such as shad (*Alosa sapidissima*), spawn in fresh water; the young fish spend the first summer in the estuary, then move out to the open sea. Other species, such as the croaker (*Micropogon undulatus*), spawn at the mouth of the estuary, and the larvae are transported upstream to feed in the plankton-rich, low salinity areas. Still others, such as the bluefish (*Pomatomus saltatrix*), move into the estuary to feed. In general, marine species drop out toward fresh water and are not replaced by freshwater forms.

The oyster bed and the oyster reef are the outstanding communities of the estuary. The oyster (*Crassostrea virginica*) is

**Figure 16.27** Relationship of a semianadromous fish, the striped bass, to the estuary. Adults live in the marine environment, but young fish grow up in the estuary. (From Cronin and Mansueti 1971.)

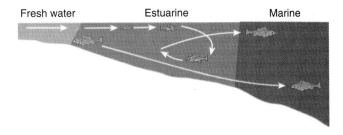

the dominant organism about which estuarine life revolves. Oysters may attach to every hard object in the intertidal zone or form reefs, areas where clusters of living oysters grow cemented to the almost buried shells of past generations. Oyster reefs usually lie at right angles to tidal currents, which bring planktonic food, carry away wastes, and sweep the oysters clean of sediment and debris. Closely associated with oysters are encrusting organisms such as algae, sponges, barnacles, and bryozoans. Beneath and between the oysters live polychaete worms, decapods, pelecypods, and a host of other organisms.

## Function

Estuarine systems function on both plankton-based and detrital-based food webs. The producer component, particularly in the middle and lower estuary, consists of dinoflagellates and diatoms. The latter convert some of the carbon intake to high-caloric fats and lipids, rather than low-energy carbohydrates typical of green plants. This fat provides a high-energy food base for higher trophic levels.

Inflowing water from rivers and coastal marshes carries nutrients, especially nitrogen and phosphorus, into an estuary and can stimulate an increase in phytoplankton production. Phytoplankton production, however, is regulated more by internal nutrient cycling than by external sources. This internal cycling involves excretion of mineralized nutrients by herbivorous zooplankton, release of nutrients remineralized by invertebrates of the bottom sediments, stirring of sediments, and steady-state exchanges between nutrients present in the particulate and dissolved phases (see Smayda 1983).

Nutrient buildup over winter in temperate estuaries stimulates a winter-spring bloom. As the nutrients become depleted and the phytoplankton experience intensive predation by zooplankton, the bloom collapses and falls to the bottom sediments. There it is fed upon by bivalves and other filter-feeding invertebrates (see Wolff 1983). In well-mixed estuaries nutrients remineralized in the benthos are released to the water column, stimulating a summer bloom.

Estuarine zooplankton, dominated by copepods (for ecology and life history, see Miller 1983), undergo their own seasonal fluctuations. Although expansive growth and subsequent declines in zooplankton populations can be associated with phytoplankton blooms, population dynamics are determined by many physical and biotic influences, including flushing rates of the estuary. The rate of increase for many zooplankton populations must balance the rate of loss from river flood and tidal flushing. Because of the unstable physical environment, zooplankton never have evolved species endemic to the estuary.

In shallow estuarine waters, rooted aquatics, such as widgeon grass (*Ruppia maritima*) and eelgrass (*Zostera marina*) assume major importance (for review see Phillips and McRoy 1980, Thayer, Adams and LaCroix 1984, Zieman

1982). They support complex systems with a large number of epiphytic and epizootic organisms. Such communities are important to certain vertebrate grazers, such as brant (*Branta bernicla*), Canada geese (*Branta canadensis*), snow geese (*Chen caerulescens*), the black swan (*Cygnus atratus*) in Australia, and sea turtles, and provide a nursery ground for shrimp and bay scallops.

Data for Pamlico River estuary in North Carolina (Copeland et al. 1974) provide an example of energy flux in an estuary. The estuary is characterized by low salinity, high turbidity, and shallow water. The shallow water supports dense stands of widgeon grass, with associated attached algae and animals such as scuds and grass shrimp (*Palaemonetes pugio*). The benthos is dominated by the clams *Rangia* and *Macoma*. The euphotic zone, the upper 2 m, is dominated by dinoflagellates. Grazing on the phytoplankton are the zooplankters (*Acartia tonsa*) and the harpacticoid copepods that move up to the surface at night. Each day ctenophores crop 30 percent of the zooplankton population. At the same time considerable detrital material enters the estuarine system as dissolved and particulate organic matter. Because phytoplankton depends heavily on dissolved organic matter, the complex food web can be considered primarily detrital. Both detritus and phytoplankton support a number of trophic levels that lead eventually to fish and humans (Figure 16.28).

Rivers flowing into estuaries carry along with the sediment a load of inland pollutants: domestic wastes, drainage from agricultural lands with nutrients and pesticides, and industrial effluents carrying toxic elements. Industrial, shipping, and housing developments along the estuarine coast add their own load of contaminants, alter flow of tidal water, and increase the anoxic condition of the estuarine bottom sediments. These conditions favor only a few bottom organisms such as polychaete worms and reduce the production of fish, clams, and oysters or make them inedible because of the accumulation of toxic elements and pesticides in their tissues. At the same time withdrawal of water from coastal aquifers and streams allows salt water to intrude into fresh water aquifers, destroying the water supply.

## SALT MARSHES

On the alluvial plains about the estuary and the in shelter of offshore bars and islands exists a unique community, the salt marsh (Figure 16.29). Although at first glance a salt marsh appears as waving hectares of grass, it is a complex of distinctive and often clearly demarked plant associations. The nature of the complex is determined by tides and salinity. Tides play a most significant role in plant segregation. Twice each day at high tide the outermost tidal flats up to mean high water are submerged in salt water; then at low tide they

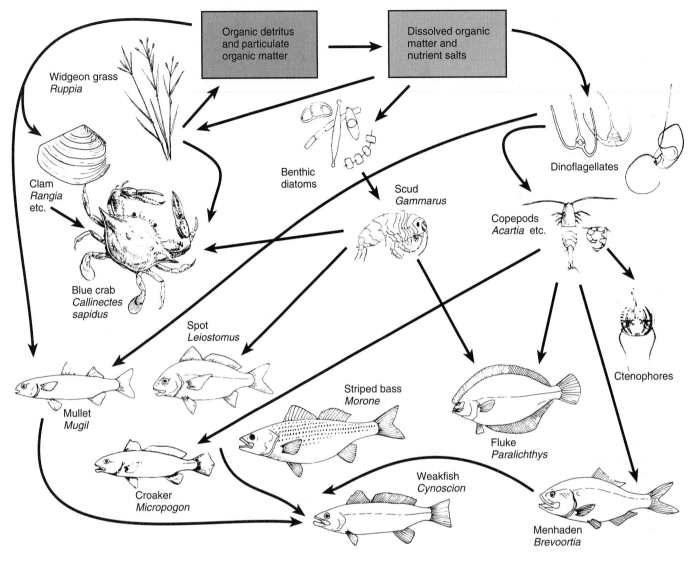

**Figure 16.28** Simplified estuarine food web based on an estuary in the southeastern United States. The energy base is largely heterotrophic, supported by particulate matter and dissolved organic matter. (Adapted from Copeland et al. 1974.)

assume a semiterrestrial existence. These flooded tidal flats make up the low marsh. At mean high water level is a region flooded irregularly at high tides. This area is the high marsh. The low marsh and high marsh form marked zones and patches of vegetation distinctive in color and texture (Figure 16.30).

Salt marshes develop over time by accretion, subsidence, and rising sea levels. Low marshes develop as sediments build up along the coast or in areas flooded by a rising sea. In eastern North America these new flats of accumulated sediments are invaded by the seaward encroachment of the dominant plant of the low marsh, smooth cordgrass (*Spartina alternifolia*). High marshes are formed over areas of low marsh where silt deposition raises the level of the

marsh, or where high marsh vegetation moves over low upland areas as sea levels rise.

## Structure

**Plants** Salt marshes begin in most cases as tidal sands or mudflats colonized first by algae and, if the water is deep enough, by eelgrass. As organic debris and sediments accumulate, eelgrass is replaced by the first salt marsh colonists—sea poa (*Puccinellia*) on the European coast or smooth cordgrass on the eastern coast of North America. Stiff, leafy, up to 3 m tall, and submerged in salt water at every high tide, *Spartina alternifolia* forms a marginal zone between the open mudflat to the front and the high marsh

**Figure 16.29** A salt marsh on the Bay of Fundy, Nova Scotia. Here river and tidal erosion are high because of the great tidal change. The marsh, not well-developed, is dominated by *Puccinellia* rather than *Spartina.* Compare this figure with views of a salt marsh on the Virginia coast, Figures 16.32 and 16.33.

**Figure 16.30** A stylized transect of part of the salt marsh, showing the relationship of plant distribution to microrelief and tidal submergence.

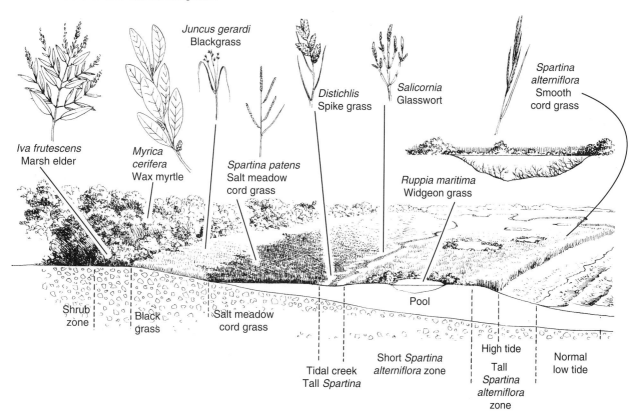

behind. No litter accumulates in the stand. Strong tidal currents sweep the floor of the low marsh clean, leaving only thick, black mud.

*Spartina alternifolia* is well adapted to the intertidal flats to which it has sole possession. It has a tolerance for salt water and it is able to live in a semisubmerged state. It can live in the saline environment by selectively concentrating sodium chloride at a higher level in its cells than in the surrounding seawater, thus maintaining its osmotic integrity. To rid itself of excessive salts, *Spartina alternifolia* has special salt-secreting cells in its leaves. Water excreted with the salt evaporates, leaving behind sparkling crystals of salt on the leaves to be washed off by tidal water and rain. To get air to its roots buried in anaerobic mud, *Spartina* has hollow tubes (aerenchyma) leading from the leaf to the root, through which oxygen diffuses.

The low marsh is monospecific, dominated by tall *Spartina* that quickly fills in any disturbance. It gives way sharply to the high marsh, marked by one of its most conspicuous plants, a short form of *Spartina alternifolia* (Figure 16.31). Yellowish, almost chloritic in appearance, short *Spartina* contrasts sharply with the dark green tall form. Why such a difference should exist has been the subject of study and debate. Although there may be some genetic difference in the two forms (Gallagher et al. 1988), the short form reflects a deficiency of nitrogen caused by salinity stress and sulfide concentrations that inhibit its uptake (Chalmers 1982, Haines and Dunn 1985).

The high marsh with its lower tidal exchange rates, a shorter, more open canopy, higher soil temperatures, and higher evaporation rates, and thus a higher salinity, provides

**Figure 16.32** The cowlick sweep of salt meadow cordgrass is distinctive next to a stand of salt grass. Bayberry in the background marks the shrub zone.

**Figure 16.31** The high marsh holds a greater diversity of species than the low marsh.

the opportunity for other plants to grow among the short *Spartina.* Here are the fleshy glassworts (*Salicornia*) that turn bright red in the fall, sea-lavender (*Limonium carolininum*), and spearscale (*Atriplex patula*).

Above mean high water the short *Spartina* and its associates are replaced by *Spartina patens,* salt meadow cordgrass, and an associate, spike grass, *Distichlis spicata* (Figure 16.32). *Spartina patens* is a fine, small grass that grows so densely and forms such a tight mat that few other plants can grow with it. Dead growth of the previous year lies beneath the current growth, shielding the ground from the sun and keeping it moist. Where the soil is more saline or waterlogged, *Spartina patens* is replaced or shares the site with *Distichlis spicata.*

In back of this zone, where the marsh level is somewhat higher or where some fresh water intrudes on the marsh, *Spartina* and *Distichlis* may be replaced by two species of black grass (*Juncus roemerianus* and *Juncus gerardi*), so called because their dark green color becomes almost black in the fall. Rarely are these rushes covered by ordinary high tides, but they may be submerged by the spring tides of spring and fall.

Beyond the rushes and often replacing them are shrubby growths of marsh elder (*Iva frutescens*) and groundsel (*Baccharis halimifolia*). These shrubs tend to invade the high marsh where a slight rise in elevation exists, but such invasions are often short-lived, as strong tides sweep in and kill the plants. On the upland fringe grow other salt marsh shrubs, particularly bayberry (*Myrica* spp.) and sea hollyhock (*Hibiscus palustris*).

Draining and supplying the high marsh are tidal creeks, bordered with growths of tall *Spartina alternifolia*. In estuarine marshes, the river itself forms the main channel, joined by numerous braiding tributaries. The creeks are deepened by scouring and heightened by a steady accumulation of organic debris. At the same time, the heads of the creeks erode backward and more small branch creeks develop. The distribution and pattern of the creeks influence the movement and drainage of water from the high marsh.

The exposed banks of tidal creeks support a dense population of mud algae, the diatoms and dinoflagellates, photosynthetically active all year. Some of the algae are washed out at ebb tide and become part of the estuarine plankton available to such filter feeders as oysters and mussels.

Across the high marsh are many circular and ellipical depressions and flats called **salt pans** or **pannes** (Figure 16.33). These pans come about in several ways. Many are formed as the marsh develops. Bare spots on the marsh become surrounded by vegetation.

As the level of the marsh rises, the bare spots lose their water outlet and become shallow ponds. Others are derived from creeks across which growth of vegetation or organic debris blocks the channel. Death of small patches of vegetation from one cause or another creates rotten spots in the peat that develop into ponds.

Such pans support distinctive plant life, which varies with the depth of the water and salt concentration. Pools with a firm bottom and sufficient depth to retain water support dense growths of widgeon grass, with long, threadlike leaves and small black, triangular leaves relished by waterfowl. The pools are usually surrounded by forbs such as sea-lavender.

Shallow depressions or pans in which water evaporates are covered with a heavy algal crust and crystallized salt. The edges of the salt flats may be invaded by *Salicornia*, *Distichlis*, and even short *Spartina alternifolia*.

The salt marsh described is typical of the North American Atlantic Coast, but many variations exist locally and latitudinally around the world (Chapman 1976). North America has several distinctive types: Arctic salt marshes with few species, East and Gulf Coast marshes dominated by *Spartina* and *Juncus*, and Pacific Coast marshes whose low marshes are dominated by *Spartina foliosa* and high marshes by *Salicornia virginica*. Tidal marshes of Europe, similar to those of the east coast of North America, support *Salicornia angelica*.

**Consumers** The three dominant animals of the low marsh are ribbed mussel (*Modiolus dimissus*), fiddler crab (*Uca pugilator* and *Uca pugnax*), and marsh periwinkle (*Littorina* spp.). The marsh periwinkle moves up and down the stems

**Figure 16.33** A salt pan or pool in the high marsh.

(a)

(b)

**Figure 16.34** (a) A ribbed mussel half-buried in the mud at low tide. (b) A fiddler crab emerges from its burrow to forage across the tidal mud at low tide.

of *Spartina alterniflora* as the tidal cycle changes. At low tide the periwinkle moves onto the mud to feed on algae and detritus.

Buried halfway in the mud is the ribbed mussel (Figure 16.34a). At low tide the mussel is closed; at high tide the mussel opens to filter particles from the water, accepting some and rejecting others in a mucous ribbon known as pseudofeces.

Running across the marsh at low tide are fiddler crabs (Figure 16.34b). They are omnivorous feeders, consuming plant and animal remains, algae, and small animals. Fiddler crabs live in burrows, marked by mounds of freshly dug, marble-sized pellets. The burrowing activity of crabs is similar to that of earthworms. In overturning the mud the crabs mix the soil and bring nutrients to the surface (Bertness 1985).

Prominent about the base of Spartina stalks and under debris are sandhoppers (*Orchestia*). These detrital-feeding amphipods may be very abundant and are important in the diet of some of the marshland birds. Grazing on tall *Spartina* leaves is the salt marsh grasshopper (*Orchelium*) and sucking the plant juices is the plant hopper (*Prokelsis*). Neither harvests significant portions of the standing crop (Teal 1962).

Two conspicuous vertebrate residents of the intertidal marshes of eastern North America are the diamondback terrapin (*Malaclemys terrapin*) and the clapper rail (*Rallus longirostris*).

In the salt marshes of the southern United States they are joined by the crab-eating marsh rat (*Oryzomys palustris*). The diamondback terrapin feeds on fiddler crabs, small mollusks, and dead fish. The clapper rail finds its diet of fiddler crabs and sandhoppers along the creek banks and in the tall *Spartina* at low tide. Less conspicuous is the seaside sparrow (*Ammospiza maritima*), which eats sandhoppers and other small invertebrates.

On the high marsh animal life changes almost as suddenly as the vegetation. The pulmonate marsh snail (*Melampus*) replaces the marsh periwinkle of the low marsh. Within the matted growth of *Spartina patens* is a maze of runways made by the meadow mouse, which feeds heavily on the grass. Replacing the clapper rail and seaside sparrow are the willet (*Catoptrophorus semipalmatus*) and the seaside sharp-tailed sparrow (*Ammospiza caudacuta*).

Low tide brings a host of predacious animals into the marsh to feed. Herons, egrets, willets, ibis, raccoons, and others spread over the exposed marsh floor and the muddy banks of tidal creeks to feed. At high tide the food web changes. Such fish as the killifish, silversides, and the four-spined stickleback (*Apeltes quadracus*), restricted to creek waters during low tide, spread over the marsh at high tide, as does the blue crab (*Callinectes sapidus*). During the winter migrant waterfowl claim the high marsh, especially snow geese, which graze on *Salicornia*.

## Function

The salt marsh is one of the most productive ecosystems. Aboveground production of salt marshes of eastern North America ranges from 650 to 2800 g dry weight/$m^2$/yr; belowground production is two times greater. Aboveground production of British salt marshes (Ranwell 1961, Jefferies 1972) and California *Salicornia* marshes (Mahall and Park 1976) falls with the same range of values.

The reason for high production is a tidal subsidy. Tidal flushing brings in new nutrients, sweeps out accumulated salts, metabolites, sulfides, and toxic wastes, and replaces anoxic interstitial water with oxygenated water. Added to this is a rather tight internal nutrient cycling. Algal and bacterial populations turn over rapidly, and detrital material is fragmented and decomposed. Up to 47 percent of net primary production is respired by microbes; part is grazed by nematodes and microscopic benthic organisms, both of which are consumed by deposit-feeders. Depending upon the nature of the salt marsh, between 11 and 66 percent of decomposed marsh grass is converted to microbial biomass. The rest is lost to the sea. Belowground production enters an anaerobic food web of fermentation, reducing sulfates and producing methane.

Sulfur, present in tidal water, appears to be significant in energy flow in the salt marsh (Howarth and Teal 1979, Howarth et al. 1983). In the anaerobic environment of the salt marsh, soil bacteria convert the seawater sulfates to sulfites by oxidizing organic compounds. In doing so, bacteria trap a portion of the energy, with the remainder residing in sulfide radicals. These stored sulfides are reoxidized over the year by oxygen diffusing from the roots of salt marsh plants and become available for further growth of sulfur-oxidizing bacteria. An estimated 70 percent of the aboveground net primary production in a New England salt marsh flows through reduced inorganic sulfur compounds, and an equivalent of 20 percent of aboveground net primary production is exported from the marshes as reduced sulfur compounds.

What happens to excess carbon production in a tidal marsh is not well understood (Table 16.2). Each salt marsh apparently differs. What happens to its excess carbon production is influenced by its route of transformation though a food web, and its route and importance of export. Some salt marshes are dependent on tidal exchanges and import more than they export (Woodwell et al. 1979), whereas others export more than they import (Valiela et al. 1978, Valiela and Teal 1979). Some of the excess production goes into sediments; some may be transformed microbially in the water in the marsh and tidal creeks. A portion may be exported to the estuary physically as detritus, as bacteria, or as fish, crabs, and intertidal organisms by way of the food web.

The importance of tidal marshes as a nutrient source and sink for the estuary has been a question of long standing. An insight into this relationship is partially provided by phosphorus and nitrogen cycling in the marsh.

**Table 16.2  Simulated Annual Carbon Budget for the Duplin River Marshes, Sapelo Island, Georgia**

| Source or Process | | Net Balance (g C/$m^2$/yr) |
|---|---|---|
| Production | | |
| *S. alternifolia* | 1575 | |
| Algae | 131 | |
| **Total production** | | 1706 |
| Loss | | |
| Respiration ($CO_2$ + $CH_4$) | | |
| in soil | − 623 | |
| in water | − 222 | |
| in air | − 68 | |
| Tidal Movement | − 586 | |
| **Total loss** | | − 1499 |
| **Net change** | | 207 |
| Sedimentation | 29 | |
| Unexplained | 178 | |

*Note:* All values prorated over the entire marsh system—soil, creeks, and Duplin River.
*Source:* Pomeroy and Wiegert 1981:225.

The major flux of phosphorus in and out of salt marsh and estuarine waters is dissolved phosphate. Over the year input equals output and usually phosphorus is surplus to the requirements in both soil and water. Excess phosphorus is mineralized and utilized by autotrophs and moves back through the system by the way of heterotrophs.

Nitrogen is a different story (Table 16.3). Denitrification often exceeds nitrogen fixation, and the marsh depends upon inputs into the system. The amount of nitrogen cycled is determined by tidal input, physical and chemical exchanges

**Table 16.3  Provisional Nitrogen Budget for the Duplin River Watershed at Sapelo Island, Georgia**

| Flux Components | Nitrogen Flux (g N/$m^2$/yr) |
|---|---|
| Inputs | |
| Rain | 0.3 |
| Sedimentation | 3.3 |
| Nitrogen fixation | 14.8 |
| Tidal exchange | 46.6 |
| Losses | |
| Denitrification | 65.0 |
| Internal Cycles | |
| Primary production | 70.0 |
| Soil remineralization | 70.0 |

*Note:* Tidal exchange is depicted as seeking a steady state, offsetting differences in nitrogen fixation and denitrification.
*Source:* Pomeroy and Wiegert 1981:180.

with air and water, and biological fluxes. In some salt marshes, exemplified by Great Sippewissett Marsh in Massachusetts, groundwater inflow brings in nitrates, some of which percolate through the peat and are exported to the ocean as organic nitrogen, ammonium, and nitrates. Of the total influx, one-third is exported by denitrification and two-thirds by tides (Valiela et al. 1978). The much larger salt marshes of Sapelo Island, Georgia, depend upon inputs of nitrogen from the associated river and tides. Only in summer does the system appear to export any nitrogen. Mostly the marsh is a net sink for nitrogen.

The nitrogen cycle in the salt marsh is not well understood, but a study of nitrogen cycling in *Spartina alternifolia* by Hopkinson and Schubauer (1984) in the Sapelo marshes, Georgia (Figure 16.35) provides some insight into the mechanisms. The aboveground pool of total nitrogen was greater than the belowground pool. The concentration was highest in young stems, but it decreased with age. Maximum accumulation was above ground in midsummer and below ground in midwinter, when 83 percent of nitrogen was in roots and rhizomes. Uptake of nitrogen was 34.8 g/m²/yr. Total transfer of nitrogen from belowground to aboveground tissue was 33 g/m²/yr. Forty-six percent of new nitrogen was taken up from the soil. Of N transferred to aboveground biomass, 14.4 g/m²/yr was lost as detritus upon culm death, 0.7 g was leached from the living culm, and 17.9 g was translocated back to rhizomes. The movement of N from aboveground to belowground biomass during the period of active growth and the dependence of new growth in spring on N stored in the rhizomes suggests strong nutrient conservation and recycling.

**Figure 16.35** Nitrogen cycle (g/m²/yr) in a Georgia salt marsh. Note both the degree of internal cycling and the amount exported to the mud and to the estuary. (From Hopkinson and Schubauer 1984:966.)

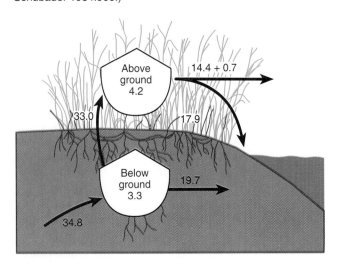

Maintenance of this cycling of N within *Spartina* involves an interaction with the plant's animal associate, the ribbed mussel, whose strong byssal threads bind sediments and prevent their erosion. Jordan and Valiela (1982) found that ribbed mussels pump tidal water in excess of the tidal volume of their New England salt marsh and deposit as much particulate matter in the form of feces and pseudofeces on the surface mud as is exported. This nitrogen is immediately available for use by *Spartina*. Investigating further, Bertness (1984), by manipulating density of ribbed mussels, experimentally demonstrated that this invertebrate stimulated production of *Spartina*. Increased height, biomass, and flowering of *Spartina* and soil nitrogen levels were positively associated with mussel density.

# MANGROVE WETLANDS

Replacing salt marshes on tidal flats in tropical region are **mangrove swamps** or **mangal** (Figure 16.36a). The term *mangrove* describes a group of halophytic species belonging to twelve genera in eight different families. The term also refers to forest plant communities fringing sheltered tropical shores (Lugo and Snedaker 1974). Mangrove forests develop in tropical coastal areas where wave action is absent, sediments accumulate, and the muds are anoxic. Mangrove forests are dominated by *Rhizophora, Avicennia, Brugivera, Sonneratia,* and *Laguncularia.* Growing with them are other halophytic species, mostly shrubs.

Mangrove trees are characterized by shallow, widely spreading roots, or by prop roots coming from the trunk and branches (Figure 16.36b). Many species have root extensions or pneumatophores that take in oxygen for the roots. The leaves although tough, are often succulent and may have salt glands. Red mangroves (*Rhizophora*) have a unique method of reproduction. Seeds germinate on the tree, grow into a seedling with no resting stage between, drop to the water, and float upright until they reach water shallow enough for their roots to penetrate the mud.

## Structure

The formation and physiognomy of mangrove swamps is strongly influenced by the range and duration of tidal flooding and surface drainage (Lugo and Snedaker 1974, Lugo 1990). One of the features of this response is zonation, the changes in vegetation from seaward edge to true terrestrial environment. Although often used as an example, the mangrove swamps of the Americas, particularly Florida, have the least pronounced zonation, largely because of the few species involved. The pioneering red mangrove (*Rhizophora mangle*) occupies the seaward edge and experiences the deepest tidal flooding. Red mangroves are backed by pneu-

(a)

(b)

**Figure 16.36** (a) Mangrove forests replace tidal marshes in tropical regions. (b) Shallow-rooted mangroves often have prop roots descending from the trunk and branches.

matophore-possessing black mangroves (*Avicennia germinanas*), shallowly flooded by high tides. The landward edge is dominated by white mangroves (*Laguncularia racemosa*) along with buttonwood (*Conocarpus erectus*), a nonmangrove species that acts as a transition to terrestrial vegetation.

In the Indo-Pacific region mangrove forests are much better developed, contain up to 30 to 40 species, and have a more pronounced zonation. The seaward fringe is dominated by one or several species of *Avicennia* and perhaps trees of the genus *Sonneratia,* which do not grow well in the shade of other mangrove species. Behind the *Avicennia* is a zone of *Rhizophora,* which grow in areas covered by daily high tide up to the point covered only by the highest spring tides. At and beyond the level of high spring tides is a broad zone of *Bruguiera.* The final and often indefinite mangrove zone is an association of small shrubs, mainly *Ceriops.*

Associated with mangroves are a mix of marine organisms that occupy prop roots and the mud, and those that live in the trees. As in the salt marsh, *Littorina* snails live on the roots and trunks of mangrove trees. Attached to stems and prop roots are barnacles and oysters. Fiddler crabs burrow into the mud during low tide and live on prop roots and high ground during high tide. In the Indo-Pacific mangrove swamps live mudskippers, fish of the genus *Periophthalmus* with modified eyes set high on the head, which live in burrows in the mud, spend time out of the water crawling about the mud, and in many ways acting more like amphibians than fish. Closely associated with these mangrove forests are commercially important species of penaeid prawns and shrimp that exist on mangrove detritus, phytoplankton, and associated microbial populations (Sasekumar and Ching 1987). Herons and other wetland birds nest in the mangrove trees, and alligators, crocodiles, bears, pumas, and wildcats inhabit the forest interior.

## Function

Net productivity of mangrove swamps is variable, ranging in Florida from about 450 to 2700 g C/m$^2$/yr, about the productivity of salt marshes. Productivity is influenced by tidal inflow and flushing, water chemistry, salinity, and soil nutrients, much as in the salt marsh. Highest rates of productivity occur in the mangrove forests that are under the influence of daily tides (Lugo and Snedaker 1974, Careter et al. 1973). For example, gross primary productivity of red mangrove, flooded daily by high tides and receiving some fresh water, is greatest at salinities below that of seawater, whereas gross primary productivity of white and black mangroves is positively correlated with salinity. These two species predominate at higher elevations, zones of increasing salinity (Hicks and Burns 1975). In areas of intermediate salinity, white mangroves have twice the productivity of red mangroves. Zonation of mangrove swamps appears to reflect the optimal productivity niches of the species rather than physical or successional conditions (Lugo 1980).

Nutrient cycling is regulated by hydroperiod, which affects the amount and turnover of litter on the mangrove swamp floor and the ability of plants to translocate and take up nutrients. Data on nutrient cycling in mangrove forests are minimal. Recycling of calcium appears to be low, except on calcium-poor waters. White mangroves (*Laguncularia racemosa*) appear to have a high recycling efficiency for nitrogen. Australian mangroves seem to be more efficient in recycling phosphorus than the ones in the Malaysian Peninsula (Lugo et al. 1992).

Like the salt marsh, mangrove swamps export a considerable portion, up to 50 percent, of their aboveground primary production to the adjacent estuary. This detrital input is important to the commercial and sport fisheries of the Gulf of

Mexico (W. Odum 1970) as well as those of the mangrove inlets and estuaries of Indo-Pacific regions. In mangrove-fringed peninsular Malaysia, the high commercial harvest of prawns is positively correlated with the presence of mangroves. The prawns assimilate an average of 65 percent of their carbon from the mangroves. The mud substrates are rich in food for benthic meiofauna, which feed on phytoplankton, benthic flora, and mangrove detritus. Prawns in turn feed on mangrove detritus, assimilating its leachable organic matter and microbial populations associated with it (Sasekumar and Ching 1987).

## SUMMARY

The marine environment, which occupies 70 percent of Earth's surface, is characterized by high salinity, waves, currents, tides, and vastness. Border ecosystems between terrestrial and marine environments are estuaries, rocky, sandy, and muddy shores, salt marshes, and mangrove wetlands.

Estuaries, where fresh water meets the sea, and their associated tidal marshes and swamps are a unit in which the nature and distribution of life are determined by salinity. As salinity declines from the mouth up through the river, so do estuarine fauna, chiefly marine species. The estuary serves as a nursery for marine organisms, for here the young can develop protected from predation and competing species unable to withstand lower salinity. If vertical mixing between fresh and salt water occurs in the estuary, the nutrients circulate up and down between the organisms and bottom sediments. Tidal marshes are dominated by salt-tolerant plants and flooded by daily tides. They are highly productive because tidal flushing brings in new nutrients and sweeps away wastes. Most of the primary production goes unharvested by herbivores. A large part is consumed by various heterotrophs or through sulfate reduction. Some accumulates as peat, and another portion is exported to the estuary. Mangrove forests replace salt marshes on tidal flats in tropical regions. Mangrove trees have shallow, widely spreading roots with additional roots arising from trunk and branches. Like salt marshes, mangrove forests have pronounced zonation from the seaward fringe to the level of high spring tides. Associated with and depending on the mangroves are wide diversity of species from barnacles, snails, and crabs to reptiles and birds. Closely associated with mangroves are commercially important penaeid prawns and shrimp.

Sandy shore and rocky coast are places where sea meets the land. The drift line marks the furthest advance of tides on the sandy shore. On the rocky shore the tide line is marked by a zone of black algal growth. The most striking feature of the rocky shore, its zonation of life, results from alternate exposure and submergence of the shore by the tides. The black zone marks the supralittoral, the upper part of which is flooded only every two weeks by spring tides. Submerged daily by tides is the littoral, characterized by barnacles, periwinkles, mussels, rockweed, and knotted wrack. Uncovered only at spring tides is the infralittoral, which is dominated by large brown laminarian seaweeds, Irish moss, and starfish. Although the basic pattern of zonation may result from tidal flooding and exposure, distribution and diversity of life across the rocky shore is influenced by disturbance from wave action, competition, herbivory, and predation. Left behind by outgoing tides are tidal pools. These distinct habitats are subject over a 24-hour period to wide fluctuation in temperature and salinity. They are inhabited by varying numbers of organisms, depending upon the amount of emergence and exposure.

By contrast, sandy and muddy shores appear barren of life at low tide, but beneath the sand and mud conditions are more amenable to life than on the rocky shore. Zonation of life is hidden beneath the surface. The energy base for sandy and muddy shores is organic matter made available by bacterial decomposition. They are important sites for biogeochemical cycling, supplying nutrients for offshore waters. The basic consumers are bacteria, which in turn are a major source of food for both deposit-feeding and filter-feeding organisms. Sandy shores and mudflats are a part of the larger coastal ecosystem including the salt marsh, estuary, and coastal waters.

Beyond the estuary and rocky and sandy shores lies the open sea. There the dominant plant life is phytoplankton and the chief consumers are zooplankton. Also important in the marine food web are photosynthetic nanoflagellates and cyanobacteria, and heterotrophic bacteria. Depending upon this base are the nekton organisms dominated by fish.

The open sea can be divided into three main regions. The bathypelagic is the deepest, void of sunlight and inhabited by darkly pigmented, weak-bodied animals characterized by luminescence. Above it lies the dimly lit mesopelagic region, inhabited by species such as certain sharks and squid. Both the mesopelagic and bathypelagic regions depend upon a rain of detritus from the upper region, the epipelagic, for their energy source.

Along the volcanic oceanic ridges, especially in the mid-Pacific, are hydrothermal vents inhabited by unique and newly discovered forms of life, including clams, worms, and crabs. The source of primary production for these hydrothermal vent communities are chemosynthetic bacteria that use sulfides as an energy source.

The impoverished nutrient status of ocean water results in low productivity. Impoverishment comes about because nutrient reserves in the upper layer of water are limited, phytoplankton and other life sink to the deep water, and a thermocline, permanent in deep water, prevents the recirculation of deep water to the upper layer. Most productive are shallow coastal waters and upwellings, where nutrient-rich deep water comes to the surface. The most productive fisheries

are confined to these areas, but pollution and overexploitation are reducing their productivity.

Nutrient-rich oases in the nutrient-poor tropical seas are coral reefs. Coral reefs are complex ecosystems based on anthozoan corals and their symbiotic endozoan dinoflagellate algae and coralline algae. Recycling nutrients within the system and functioning as nutrient sinks, coral reefs are among the most productive ecosystems in the world. Their productivity and varied habitats support a high diversity of colorful invertebrate and vertebrate life.

## REVIEW QUESTIONS

1. How does the tidal cycle relate to the structure, function, and productivity of intertidal ecosystems?
2. Give some possible explanations for the high diversity of the ocean benthos.
3. Why are hydrothermal vents unique ecologically?
4. What is the relationship between photosynthetic nanoflagellates and cyanobacteria and associated heterotrophic bacteria and nanoflagellates in the marine food web?
5. Contrast the energy source of a sandy or muddy shore with that of a rocky shore. What groups of organisms are the basic consumers in each?
6. How does the fact that corals are partially photosynthetic and partially heterotrophic contribute to the high productivity of coral reefs?
7. How is the estuary a nutrient trap?
8. What contributes to the high productivity of salt marshes and mangrove wetlands?

## CROSS-REFERENCES

Coriolis effect, 39; ocean currents, 42–43; characteristics of water, 64–66; halophytes, 123–124; productivity of oceans, 172, 175–176; energy flow, 189–191; nitrogen cycle, 210–212; phosphorus cycle, 212–213; plant-herbivore interactions, 522–524; mutualism, 582–596; disturbance and community structure on rocky shores, 621, 622.

Part *5*

Population Ecology

# Properties of Populations

## Concepts

1. Populations may be unitary or modular.
2. Individuals within a population and populations may be distributed randomly, uniformly, or in clumps.
3. Mortality may be expressed as survivorship and delineated in a life table.
4. The life table is a basic tool in the study of population dynamics.
5. Fecundity may be expressed as a net reproductive rate that combines survivorship with an age-specific schedule of births.
6. Age structure influences the survivorship and fecundity characteristics of a population.

If you have been caught in a mass of surging people on a crowded city street or at an athletic event, if you have watched yellow sulfur butterflies congregating at a tiny pool or a flock of robins foraging on a lawn, you have seen an aggregation of individuals of the same species called a population.

A population is not just an aggregation of individuals, however. More precisely, a **population** is a group of interbreeding, or potentially interbreeding, organisms of the same species occupying a particular space at the same time. Because populations are aggregations, they have properties not associated with individuals. First, they are characterized by density, the number of organisms occupying a definite unit of space. Second, they have an age structure, the ratio of one age class to another. Third, populations grow by adding new members through birth and immigration, or shrink by losing members through death and emigration. The difference between gains and losses is the rate of population growth.

The study of populations is a major component of ecology, essential to the investigation and understanding of other aspects of ecology. One part of population ecology is demography, basic to all population studies. **Demography** deals with the vital statistics of a population: age structure, density, births, deaths, growth, and reproduction. Knowledge of these aspects of a population is basic to the study of behavior and population interactions such as predation, competition, and mutualism. Another part of population ecology, concerned with the breeding structure and genetic composition of a population, is **population genetics.** Whereas demography deals with individual members, population genetics is concerned with individuals only to the extent that, as packages of genetic samples, their survival and reproductive success influence the genetic constitution of the population.

# DEFINING POPULATIONS

## Unitary and Modular Populations

When we try to define the properties of a population, we are faced with two problems. The first is defining the boundaries of a population. Most populations know no boundaries other than those drawn by the ecologist studying them. The second problem is defining the number of individuals present.

If you are studying animals, typically there is no major problem in defining individuals. You can recognize a deer as an individual. It consists of morphological parts: head, four legs, a tail, hair, and other sharply defined characteristics. Other deer, regardless of age, have a similar appearance. Thus individual deer are easy to identify and count.

But what about the trees among which the deer are standing (Figure 17.1)? Each trunk appears to belong to an individual tree, but does it? If you were to do some digging, you

**Figure 17.1** Populations consist of either unitary or modular organisms. The white-tailed deer is a unitary organism with a discrete growth form. The surrounding trees appear as individuals, but some of them are probably modules of the original parent tree, arising from buds on the horizontal roots to form clones. In addition, the tree is made up of iterated growth forms of buds, leaves, stems, and twigs.

might discover that some of the trees are connected by the same root structure. Further, no tree is quite the same as others in shape, number of leaves, size of crown, and amount of branching. Each is made up of populations of smaller units: leaves, buds, twigs, branches, and seasonally, flowers and fruits. The deer obviously is a real individual, a unit. Some trees, such as aspen, probably are not; each stem is a part of a larger unit, a clump of closely associated trees. Each is a **module,** a part of the whole. This problem was not recognized until the late 1960s.

An individual, strictly speaking, is derived from a zygote. The deer is such an individual, and many trees can also be traced back to sexual reproduction, when wind-carried pollen fertilized ovules in the female catkins. But once established, some species of these sexually produced trees, such as aspen (*Populus* spp.), and many species of herbaceous plants, such as strawberries (*Fragaria*) and mayapples (*Podophyllum peltatum*), produce new "individuals" or modules asexually by means of buds on shallow horizontal roots or stems that touch the ground (Figure 17.2). Individual plants or zygotes produced by sexual reproduction are termed **genets.** They are the genetic, evolutionary

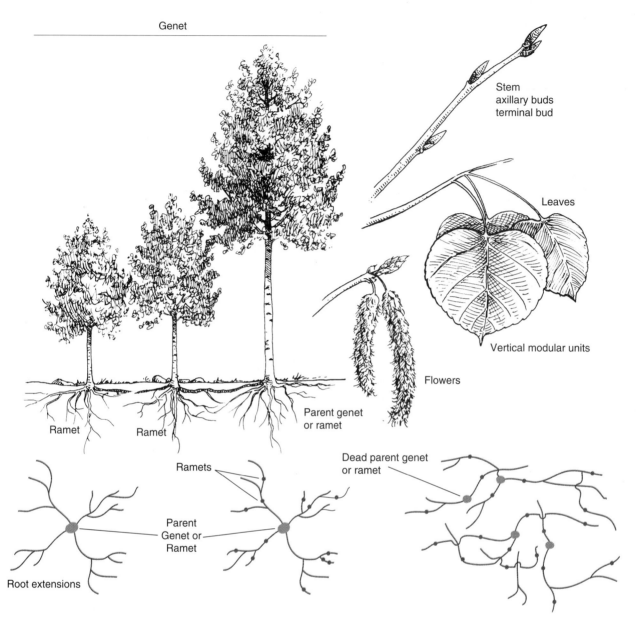

**Figure 17.2** Horizontal and vertical modular growth in plants, as exemplified by aspen. Vertical modular growth involves iterations of units of leaf, axillary bud, and associated stem (internode). Horizontal growth involves iterations of roots and root buds, which give rise to clones of the genet. The clones or ramets are various ages, with the youngest individuals forming the leading edge. The clones and vertical modular units remain in close connection through maintenance of woody tissue. Connections between the original genet and its ramets are lost in time, and the ramets lead independent existence. Because all the ramets from the parent genet are identical genetically, the ramets collectively make up the genet.

units. The asexually produced individuals derived from the parent zygote or genet are called **ramets.** Because they arise over a period of time, ramets are of different ages and different sizes. They can exist independently of the parent genet and can give rise to their own ramets; yet all possess the same genetic constitution as the parent genet. By producing ramets, the parent genet can extend over a considerable area. Usually the connections between the original stem and its

clones may die, resulting in separate ramets that produce their own lateral extensions or ramets.

A population of ramets arising from the same parent genet makes up a **clone.** Because all individuals of the clone have arisen from the same zygote and have the same genotype, they now collectively make up a single organism or genet, no matter how many "individuals" the clone may possess. Because they are a part of the original growth,

iterations from the parent plant, the genet may be very old, although the "individual" trees or plants, the ramets, may be very young. As a result the life spans of genets are indeterminate for many plants.

This feature is not restricted to plants. Some animal groups are also modular, including sponges, corals, and hydroids, all of them sessile organisms.

There is more to modular growth in plants than ramets. Woody plants exhibit vertical growth of modules. These modules give rise to buds, leaves, stems, and other structures one above the other while maintaining strong woody connections to older modules, rather than rotting away. Stems eventually develop into branches, limbs, and vertical extensions of trunks (Figure 17.2). Some modules develop into flowers, producing potential new genets.

Such modular growth in itself is a demographic process (see Harper 1977, White 1979, Harper and Bell 1979). Leaves, buds, stems, and roots are populations in their own right. They compete with neighboring modules of the same type for moisture, light, and nutrients, and respond in their own way to environmental conditions; yet they may supply carbon (energy) to neighboring modules. They experience their own birth rates, death rates, and changes in age structure. The distribution of birth rates and death rates of the modules influences the growth form the plant will take, including the ordering of branches in trees.

## Populations as Genetic Units

Because it is composed of interbreeding organisms, a population is a genetic unit. Each individual carries a certain combination of genes, a sample of the population's total genetic information. The sum of all genetic information carried by all individuals of an interbreeding population is the **gene pool.** Gene flow, the exchange of genetic information between populations, comes about through immigration and emigration.

Populations may also be considered as evolutionary units. Evolution results from changes in gene frequency in a given gene pool over a period of generations. An outcome may be changes in the physical expression of organisms in the population, reflecting its genetic constitution. These changes in the genetic constitution often result from selective pressures brought to bear by the environment on individuals of the population.

## Populations as Demes

Populations of species consist of numerous subpopulations or **demes** of varying densities. These subpopulations are both ecological and genetic, but an ecological population is not necessarily the same as the genetic population. To define this incongruity, Antonovics and Levin (1980) have proposed the neighborhood concept of population density and gene flow.

To visualize the concept, consider a large number of even-aged plants distributed across a landscape. If individual plants are so spaced that each plant does not affect the growth of any other plant, that spacing is the **ecological effective distance.** Draw a circle about each plant. Within the circle the plant is free from any interference from neighboring plants. The area within the circle represents the plant's **ecological neighborhood.** The number of plants in each such neighborhood is one, and all plants within the idealized population have the same neighborhood area (Figure 17.3). The density of such a population as a whole is the **ecological effective density,** defined by the ecological effective distance. The ecological effective density is the highest density the population can attain before plants have some crowding effect on the others. If we increase the density of plants by some magnitude, we will arrive at a point where ecological neighborhoods will be adjacent, but density effects are still absent. If we further increase the density of the plants, ecological neighborhoods will overlap. Although the ecological neighborhood of each plant remains the same, it has to share its neighborhood with others. The plants compete for light, moisture, and nutrients. The interactions are now affected by the density of the plant population. Thus the ecological neighborhood is defined in terms of distances over which density-dependent effects do or do not operate.

The deme also represents a genetic population, but it is defined in terms of the distances genes travel away from the reproductive individuals. This distance may have little relationship to the ecological neighborhood. For example, pollen carried by the wind may travel distances measured in kilometers away from the plant that produced it. The pollen may arrive on the stigma of an individual of the same plant species far removed from the ecological neighborhood of the pollen producer. Thus the ecological neighborhood of such plants is smaller than their genetic neighborhood. Plants that compete for animal pollinators, however, may have similar genetic and ecological neighborhoods. Animals whose foraging distances are about the same as their search distances for mates have ecological and genetic neighborhoods of similar sizes. Thus ecological effective distances may be smaller, larger, or about the same as the genetic neighborhood. This point is important to remember when considering the interrelations of individuals and demes at both ecological and genetic levels.

# DENSITY AND DISPERSION

## Crude Versus Ecological Density

The size of a population in relation to a definite unit of space is its **density.** Every ten years the Census Bureau counts the number of people living in the United States. Often it presents the data as the number of individuals per square mile. Wildlife biologists estimate the number of game animals in

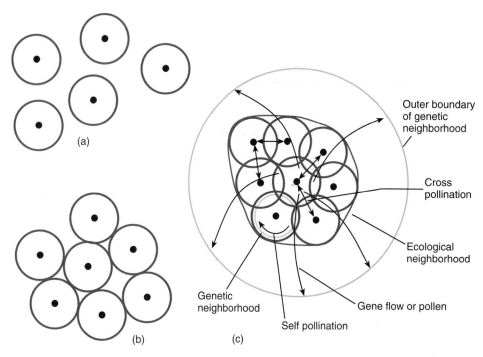

**Figure 17.3** Ecological and genetical neighborhoods. (a) Individuals are so widely spaced that the effects of density are absent. The circle about each plant indicates its ecological neighborhood. (b) The highest density possible before the effects of crowding take place. This density forms a neighborhood continuum. (c) Plants are more crowded and the number of plants per ecological neighborhood increases, indicated by the overlapping circles. The dotted lines indicate the genetic neighborhood of the population, and the arrows show gene flow. If plants are self-pollinated, the genetic neighborhood is smaller than the ecological neighborhood. If gene flow remains in the within the total ecological neighborhood, the genetic and ecologic neighborhoods are the same. If gene flow occurs over much larger distances, the genetic neighborhood is larger than the ecological neighborhood.

a particular area; a forester determines the number and volume of trees in a timber stand. The measure of the number of individuals per unit area is called **crude density.**

Populations do not occupy all the space within a unit, because it is not all a suitable habitat. A biologist might estimate the number of deer per square mile, but the deer might avoid half the area because of human habitation and land use practices, lack of cover, or lack of food. A soil sample may contain 2 million arthropods per square meter, but these arthropods inhabit only the pore spaces in the soil. Goldenrods inhabiting old fields grow in scattered groups or clumps because of soil conditions and competition from other old field plants. No matter how uniform a habitat may appear, it is not uniformly habitable because of microdifferences in light, moisture, temperature, or exposure, to mention a few conditions. Each organism occupies only areas that can adequately meet its requirements, resulting in patchy distribution. Density measured in terms of the amount of area available as living space is **ecological density.**

Attempts have been made to make such ecologically realistic measurements. For example, one study in Wisconsin expressed the density of bobwhite quail as the number of birds per mile of hedgerow rather than per acre (Kabat and Thompson 1963). However, ecological densities are rarely estimated, because it is difficult to determine what portion of a habitat represents living space, especially if organisms require different habitats during the year or during developmental change.

The density of organisms in any one area varies with the seasons, weather conditions, and food supply, to name only a few factors. However, an upper limit to the density is imposed by the size of the organism and its trophic level. Generally the smaller the organism, the greater its abundance per unit area. A 40 ha forest will support more woodland mice than deer and more trees 5.1 to 7.6 cm inches dbh (diameter breast height) than trees 30 to 35 cm inches dbh.

From a practical point of view, density is one of the more important parameters of populations. It both determines in part and is determined by energy flow, resource availability and utilization, physiological stress, dispersal, and productivity of a population. The density of human populations, for example, relates to economic growth and the expansion and management of towns, cities, regions, states, and nations. Increasing or decreasing populations can place strains on economic and social institutions. The distribution of humans in a given region affects land use and pollution problems.

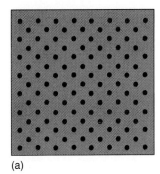
(a)

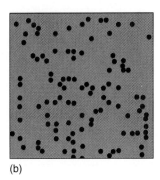
(b)

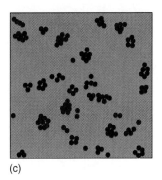
(c)

**Figure 17.4** Patterns of distribution: (a) uniform, (b) random, and (c) clumped.

Wildlife biologists need to know about densities of game populations to regulate hunting and manage habitats. Foresters base timber management and evaluation of site quality in part on the density of trees.

## Patterns of Dispersion

Crude density also tells us nothing about how evenly individuals within the population are distributed over space and time. Determining this dispersion is a major field problem.

Spatial Dispersion Individuals may be distributed randomly, uniformly, or in clumps (Figure 17.4). Distribution is considered random if the position of each individual is independent of the others. Random distribution is rare, for it can occur only where the environment is uniform, resources are equally available throughout the year, and interaction among members of the population produces no patterns of attraction or avoidance. Some invertebrates of the forest floor, particularly spiders (Cole 1946, Kuenzler 1958), the clam *Mulinia lateralis* of the intertidal mudflats of the northeastern coast of North America, and certain forest trees (Pielou 1974) appear to be randomly distributed.

Uniform or regular distribution is the more even spacing of individuals than would occur by chance. Regular patterns of distribution result from intraspecific competition among members of a population. For example, territoriality under homogeneous environmental conditions can produce uniform distribution (Figure 17.5). In plants it may result from severe competition for crown and root space among forest trees (see Gill 1975) and for moisture among desert plants (Beals 1968) and savanna trees (Smith and Goodman 1986). Autotoxicity (the production of exudates toxic to seedlings of the same species), a characteristic common among plants of arid country, is another mechanism for achieving uniform distribution.

The most common type of distribution is clumped, also called clustered, contagious, and aggregated. This pattern of dispersion results from responses by plants and animals to habitat differences, daily and seasonal weather and environmental changes, reproductive patterns, and social behavior.

**Figure 17.5** Golden eagle territories in Scotland, showing the even dispersion of breeding sites. Territoriality in birds and other animals, a function of intraspecific competition, usually results in a fairly relatively uniform distribution over an area of suitable habitat. (From Brown 1976:109.)

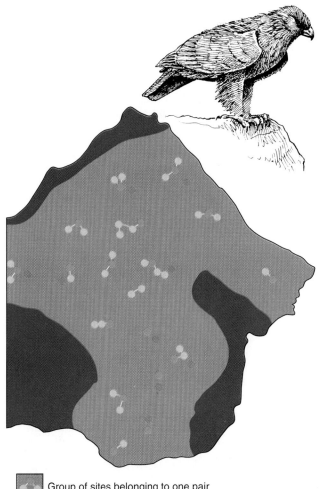

Group of sites belonging to one pair

Single site

Marginal site not regularly occupied

Breeding, year of survey 1967

Low ground unsuited to breeding eagles

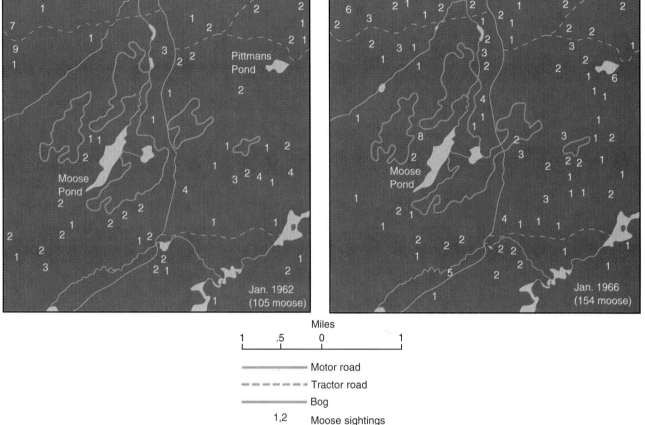

**Figure 17.6** Randomly distributed aggregate distribution of moose in the Lower Noel Paul River, Newfoundland. The numerals indicate the number of moose in each aggregation. The small aggregates reflect the low sociality of moose. Individuals and groups of moose are randomly distributed over a rather homogeneous habitat. (From Bergerud and Manuel 1969:912.)

The distribution of human beings is clumped because of social behavior, economics, and geography.

There are various degrees and types of aggregated distribution. Groups of varying sizes and densities may be randomly or nonrandomly distributed over an area; individuals within the clumps may be distributed randomly or uniformly. Aggregations may be small or large. Population clusters may tend to concentrate around a geographical feature that provides nutrients or shelter (Figure 17.6).

Aggregations among plants are often influenced by the nature of propagation and specific environmental requirements. Poorly dispersed seeds, such as those of oaks and cedar, are clumped near the parent plant or where they are placed by animals. Well dispersed seeds are more widely distributed, but even they tend to concentrate near the parent plant. Vegetative propagation produces clumping. Seed germination and survival of seedlings also affect the degree and type of aggregation.

Some animal aggregations represent individual responses to environmental conditions. Individuals may be drawn together by a common source of food, water, or shelter. Moths attracted to light, earthworms congregated in a moist pasture field, barnacles clustered on a rock, all have little or no social interaction. The individuals do not aid one another and only passively prevent other members of the same species from sharing the condition that brought each of them to the same location.

Aggregations on a higher social level reflect some degree of interaction. Prairie chickens congregate for communal courtship; elk band together in herds with some social organization, usually with a cow as the head (Altmann 1952); birds congregate on feeding grounds away from territorial sites, yet show intolerance for each other near the nest. Aggregations of the highest social structure are found among insect societies, such as ants and termites, and in the naked mole rat. Here individual members are organized into social castes according to the work they perform.

The pattern of local distribution has another feature of note, its **grain** (Figure 17.7) (Pielou 1974). A pattern is *fine-grained* if each individual has an equal likelihood that its neighbors will belong to any other species in proportion to their relative abundance. A pattern is *coarse-grained* when each individual is likely to have as neighbors members of its own species. Coarse-grained distribution is clumped with relatively large spaces between groups; in other words, the distribution is patchy. (Those terms also relate to habitat variability and size of patches.)

A change in density in the local population can affect grain. As density increases, fine-grained species spread randomly (Figure 17.7). Coarse-grained species, preferring certain patches of habitat, become more aggregated as density increases; but as density becomes very high, individuals may be forced to occupy marginal patches in a fine-grained manner. As density increases, dispersion may become more uniform.

On a regional scale, individual populations of a species are not distributed continuously over the landscape but in clusters or patches and exhibit some degree of variability in the nature of the habitats they occupy. Losses and gains of habitat over a period of time influence abundance and distribution within a region. Regional distributions of populations make up the **range** of the species (Figure 17.8). The boundaries of a range or biogeographical distribution are not fixed; they may fluctuate greatly. Habitat changes, competition, predation, and climatic changes can influence the extent of a species range, expanding it one year and contracting it another.

**Temporal Dispersion**  Organisms in populations are distributed not only in space, but also in time. Temporal distribution can be circadian, relating to daily changes in light and dark. The environmental rhythm of daylight and dark

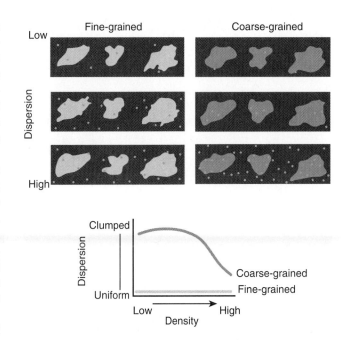

**Figure 17.7** Habitat or patch occupancy for a fine-grained and a coarse-grained population as density increases from low to high levels.

(see Chapter 7) is responsible for the daily movement of some animal populations, such as nectar-feeding insects seeking patches of open flowers, the daily movement of plankton from deeper to upper layers of water, and the withdrawal and emergence of nocturnal and diurnal animals.

Other temporal distributions relate to changes in humidity and temperature, seasons, lunar cycles, and tidal cycles. Seasonal changes are reflected in the sequential blooming of wildflowers in forest and field and in the return and departure of migrant animals. The populations of forests and fields are quite different in spring, summer, fall, and winter. Distributions may also be related to longer periods of time, which encompass annual cycles, successional stages, and evolutionary changes.

**Dispersal Movements**  Most organisms disperse at some stage of their life cycle. They leave their immediate environments either permanently or seasonally for more favorable habitats. Such movements are essential for individual survival, especially of the young, the group most prone to disperse, for there is no room for all in any one immediate environment. Insufficient resources, deteriorating habitats, and alleviation of inbreeding are major impetuses for dispersal. Whatever the reasons, the dispersers improve their potential fitness by moving to a new territory, in spite of the risks of travel. Dispersal leads to colonization of suitable new areas, expansion of a species range, and the spread of genes. Isolated pockets of colonizers, subject to genetic drift, establish locally adapted populations that serve as focal points for further colonization.

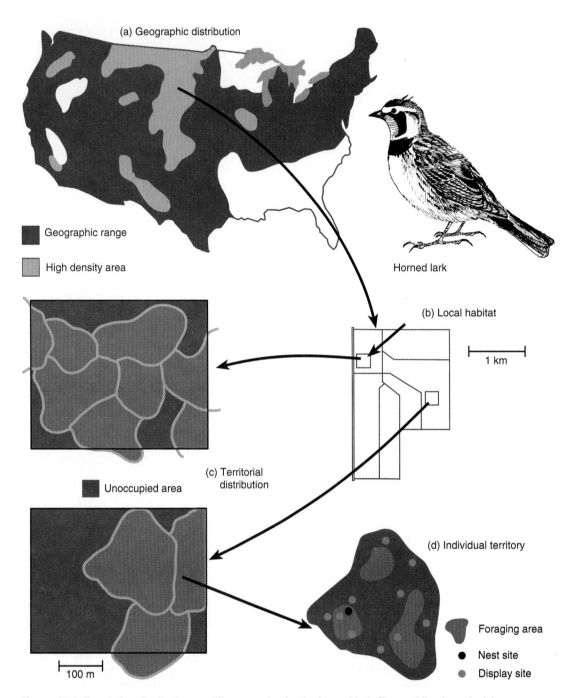

**Figure 17.8** Population distribution on different scales for the horned lark *Eromophila alpestris.* (a) Although the horned lark is distributed over a wide biogeographical breeding range, its regional distribution ranges from areas of low to high densities. (b) On a local scale the distribution of the bird is influenced by the availability of habitat. (c) Within a plot of given habitat, the bird's distribution is influenced by territorial behavior. (d) Within each territory the bird allocates space to different activities. (From Wiens 1973:237.)

Dispersal movements may be one-way out of one habitat and one-way into another with no return trip. The former is termed **emigration** and the latter **immigration.** Both involve the same movement; the distinction is one of viewpoint. Emigrants from one area become the immigrants to another. Dispersal with a return to the place of origin is termed **migration.**

For mobile animals dispersal is active; but many sessile organisms, particularly plants, depend on passive means of dispersal, involving gravity, wind, water, the coats of

mammals, the feathers of birds, or the guts of both. The distance these organisms travel depends upon the quality of their dispersal agent. Seeds of most plants fall near the parent, and their density falls off quickly with distance (Figure 17.9). Heavier seeds, such as those of oaks, have a much shorter dispersal range than the lighter wind-carried seeds of maples, birch, milkweed, and dandelion. Wind also is the means of dispersal for many animals, such as the young of some species of spider, larval gypsy moths, and cysts of brine shrimp. In streams the larval forms of some invertebrates disperse downstream in the current to suitable microhabitats; but larval offspring of sessile marine organisms such as barnacles become active planktonic swimmers in the marine environment and determine where they will settle. Many plants depend upon some active carrier, such as birds and mammals, to disperse their seeds, which in some cases has evolved to mutual advantage. Seeds of some plants are armed with spines and hooks that catch on the fur of mammals, feathers of birds, and clothing of humans. Other plants, such as the cherries and viburnums, depend upon birds and mammals to consume their fruits and carry the seeds to some distant point in their guts. The seeds of such passive dispersers usually are heavy and few in number, but they have higher chances of being deposited in suitable places for germination than wind-dispersed seeds.

Migratory movements fall into three categories, the most familiar of which is the repeated return trips made by individuals. Such round-trip migrations may be daily or annual, short-range or long-range. Zooplankton in oceans move down to lower depths by day and move up to the surface water at night. The movement appears to be a response to light intensity. Bats leave their roosting places in the evening hours to travel to their feeding ground and return by daybreak.

Other migrations are seasonal. Annually earthworms make a vertical migration deeper into the soil to spend the winter and move back to the upper soil in spring and summer. Mule deer in the western mountains move from their summer ranges on north-facing slopes to wintering grounds on south-facing slopes (Figure 17.10). Similarly, elk move down from their high mountain summer ranges to lowland winter ranges. On a larger scale caribou move from summer calving ranges in the taiga to the arctic tundra for the winter, where lichens comprise their major food source (Figure 17.11). Gray whales move down from Arctic waters in summer to their wintering waters off the California coast, and humpbacked whales migrate from northern oceans to the central Pacific off the Hawaiian Islands. The most familiar of all migrations is the annual spring return and fall departure of birds.

A second type of migration involves only one return trip. Such migrations are common to some species of Pacific salmon. Young hatch and grow in headwaters of coastal streams and rivers. The young move downstream and out to

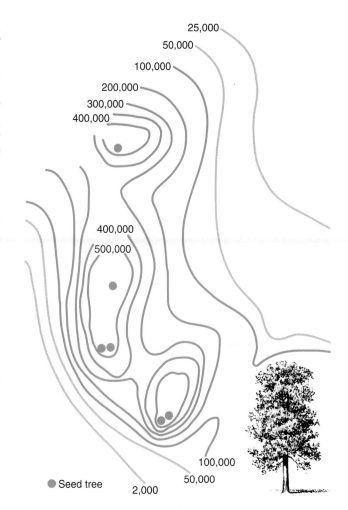

**Figure 17.9** Pattern of annual seedfall of yellow-poplar, *Liriodendron tulipifera*. Lines show equal seeding density. With this wind-dispersed species, seedfall drops off rapidly away from the parent tree. (After Engle 1960.)

the open sea, where they reach sexual maturity. At this stage they return to their home streams to spawn and die.

A third type of migration, exemplified by the monarch butterfly, is unusual because the fall migrants do not return north but their offspring do. About 70 percent of the last generation of monarch butterflies in summer moves south in noticeable flights to their wintering grounds in the highlands of Mexico, a trip that covers about 14,000 km. From the wintering grounds monarchs undertake a northward movement in January and arrive in the deep southern United States in early spring, where they start a new generation. The first spring generation continues the northward trek, following the milkweed north with the spring. One generation succeeds another until the monarchs that finally arrive on the northern breeding ground are several generations removed from the ancestors that migrated south the previous fall. Such migrations are not confined to the monarch butterfly. Similar but less extensive migrations are undertaken

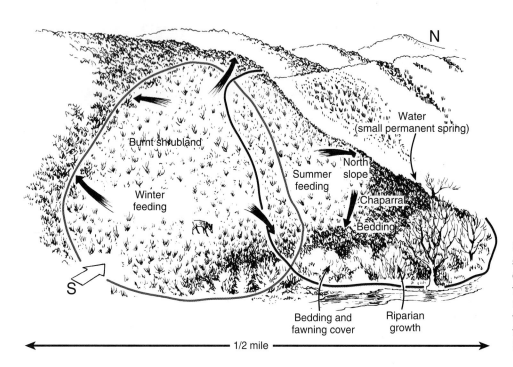

1/2 mile

**Figure 17.10** The seasonal short-range migration of mule deer in California. In summer the deer keep to the cooler, moister north-facing slopes. In winter they move to the south-facing slopes. (From Taber and Dasmann 1958:34.)

**Figure 17.11** Migratory pathways of four vertebrates. (1) Ring-necked ducks (*Aythya collaris*) breeding in the northeast migrate in a corridor along the coast to wintering grounds in South Carolina and Florida. (2) The canvasback duck (*Aythya valisinera*), whose major breeding areas are in the prairie pothole region, has a number of corridors; but most canvasbacks diverge to the Atlantic or Pacific Coast. (3) The gray whale summers in the Arctic and Bering Seas; it winters in the Gulf of California and the waters off Baja California. (4) The barren-ground caribou winters in the taiga and spends the summer on the arctic tundra.

by other insects, including two leafhoppers (*Macrosteles fascifrons* and *Empoasca fabae*), the harlequin bug (*Murgantia histrionica*), and the milkweed bug (*Oncopeltus fasciatus*).

# AGE STRUCTURE

When you attend events, you may notice the differences in age of those attending. For example, rock concerts are dominated by younger people, symphonic concerts by older ones. Parades and fairs have a greater diversity of age classes. Once aware of these differences, you can appreciate how the nature and character of a population are influenced by its age structure. The age structure of plant, animal, and human populations determines, in part, population reproductive rates, death rates, vigor, survival, and other demographic attributes.

The age structure of populations can be characterized by specific age categories, such as years or months. We may use other categories: (1) life history stages, such as prereproductive, reproductive, and postreproductive stages in birds and mammals; (2) eggs, pupae, larvae, and instars in insects; and (3) size classes in plants, such as heights of herbaceous plants or seedlings and diameters of trees.

## Age Structure in Animals

Theoretically, all continuously breeding populations should tend toward a **stable age distribution.** The ratio of each age group in a growing population remains the same if the age-specific birth rate and the age-specific death rate do not change. If the stable age distribution is disrupted by any cause, such as natural catastrophe, disease, starvation, or emigration, the age composition will tend to restore itself upon return to the previous conditions, provided, of course, the rates of birth and death are still the same.

When deaths balance births and the population is closed—that is, it experiences movements neither into nor out of the population—the population has reached a constant size. It assumes a special form of stable age distribution, known as a **stationary age distribution.** This distribution exists when a population is not growing and its age structure remains the same. Thus all stationary age distributions are stable, but only some stable age distributions are stationary.

A population can do three things: increase, decrease, or remain stable. The ratio of young to adults in a relatively stable population of most mammals and birds is approximately 2:1 (Figure 17.12). A normally increasing population should have an increasing number of young, whereas a decreasing population should have a decreasing number of young. Within this framework are a number of variations. A popula-

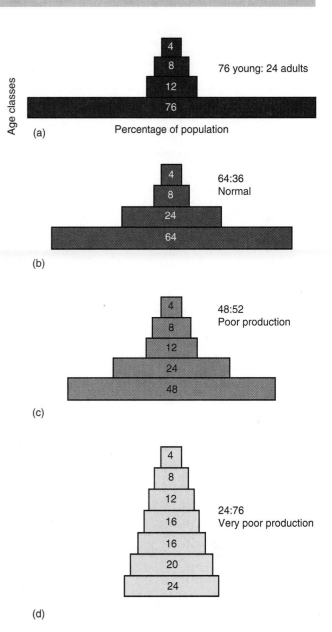

**Figure 17.12** Theoretical age pyramids, especially applicable to mammals and birds. (a, b) Growing populations in general are characterized by a large number of young, giving the pyramid a broad base. (c) A population with a high proportion of individuals in the older age classes. Such a population is aging. (d) If the ratio of young to adults declines, the population is dominated by older individuals and the production of young is low. The pyramid of such a population is a narrow one with a narrow base of young. (After M. Alexander 1958.)

tion, for example, may be decreasing yet show an increasing ratio of young to adults. Whatever the relationships, the number in one age class that enters the next age class influences the age structure of a population from year to year. A deficiency of young can lead to an aging population, but a high proportion of old individuals may not be a cause of a

deficiency of young. By combining information on population density, age ratios, and reproduction, a biologist can correlate changes in population structure with habitat changes and ecological and human influences.

The history of a population can be detected in a series of age distributions. Grant and Grant (1992) provide the age distribution of the cactus ground finch on Isla Daphne Major, Galapagos, for the years 1983 and 1984 (Figure 17.13). Reproductive activity and success are governed by rainfall. In 1982 rainfall was negligible and reproduction failed. No young entered year class 1 in 1983. Heavy rainfall marked 1983, a year of El Niño (see Chapter 4), so reproduction was highly successful. This success was marked by the presence of year class 1 in 1984, but no young were available to enter year class 2. The very small age class 3 in 1983 results in a very small age class 4 in 1984. Note the shrinkage of other age classes as they move up the age ladder.

The loss of age classes can have a profound influence on a population's future. Consider an exploited fish population in which the older reproductive age classes are re-moved. If the population experiences reproductive failure for one or two years (as in the ground finch age pyramids), there will be no young fish to move into the reproductive age class to replace the fish removed, and the population can collapse.

Although useful in looking at a population's history and trends, age pyramids, especially for wild populations subject to exploitation, rarely indicate whether the rate of increase is positive, negative, or zero. When age distributions are not stable, change in age distribution does not imply changes in survivorship or fecundity (see Caughley 1977) and cannot be safely used alone to predict population trends.

## Age Structure in Plants

The use of age structure in demographic studies of plant populations is more recent. Modular structure and asexual reproduction in plants impose difficulties on research. Determining age in plants involves following a cohort of marked individuals over time, destroying a sample of the population, and determining the ages of individuals by growth rings, bud scars or other indicators (Figure 17.14).

Among many plant species, especially woody ones, age may not tell enough. In even-aged stands of trees, the bulk of individuals falls into a very few age classes because they dominate the site, competitively excluding young age classes. Because of competition among individuals, some trees will achieve greater size than others, giving the impression of age differences, when in fact the trees are the same

**Figure 17.13** Age structure of the cactus ground finch (*Geospiza scandens*) on Isla Daphne Major for 1983 and 1984. Note the missing year classes and the shrinkage of a particular age class from one year to the next. (From Grant and Grant 1992:775.)

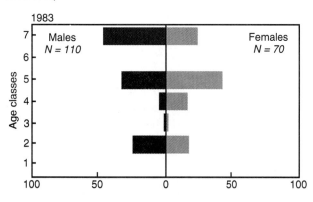

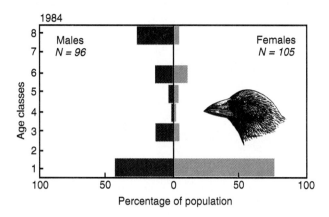

**Figure 17.14** Age classes in a colony of common plantain (*Plantago rugelii*). The age structure is based on marked individuals of known ages. (From Hawthorne and Cavers 1976:516.)

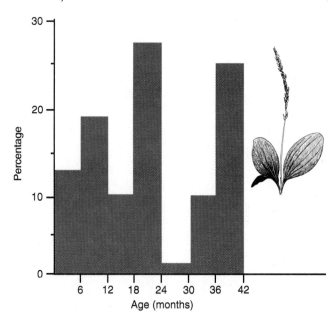

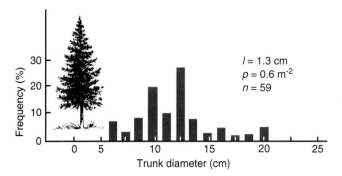

**Figure 17.15** Size classes in an even-aged 54-year-old stand of balsam fir (*Abies balsamea*). Although all trees are the same age, note the difference in the size classes, as measured by diameter at breast height. The smaller trees give the impression of being younger individuals, when instead they are suppressed individuals. (*l* = interval of trunk diameter between successive bars determined by dividing the range of observed diameters into 12 equal intervals; *p* = stand density; *n* = number of trees in sample.(From Mohler et al. 1978.)

**Table 17.1** Yield Table for Douglas-fir on Fully Stocked Hectare

| Age (Years) | Trees | Av. DBH (cm) | Basal Area (m²) |
|---|---|---|---|
| 20 | 1427 | 14.5 | 9.4 |
| 30 | 875 | 29.9 | 14.3 |
| 40 | 600 | 31.0 | 18.1 |
| 50 | 440 | 38.8 | 20.8 |
| 60 | 345 | 46.2 | 23.1 |
| 70 | 283 | 53.0 | 24.9 |
| 80 | 243 | 59.1 | 26.5 |
| 90 | 210 | 65.0 | 27.8 |
| 100 | 188 | 70.1 | 29.0 |
| 110 | 172 | 74.7 | 30.7 |
| 120 | 158 | 79.0 | 30.9 |
| 130 | 148 | 83.0 | 31.7 |
| 140 | 138 | 87.1 | 32.6 |
| 150 | 128 | 90.9 | 33.2 |
| 160 | 120 | 94.5 | 33.8 |

*Source:* Derived from McArdle et al. 1949.

age. Age structure in such stands is replaced by some size criterion, such as diameter classes (Figure 17.15).

A similar approach is necessary for uneven-aged stands, because older individuals are not necessarily the larger dominant ones. What appear to be smaller young trees are in fact often the same age as the dominant individuals. One or two age classes dominate and hold onto the site until they die or are removed. Then new young age classes develop. Size classes and numbers within each class change as the population ages and mortality increase. In such a population age alone is a poor criterion of population dynamics. In Table 17.1, for example, as the basal area (or size) of the individual trees increases, the number of trees that can occupy a given space decreases.

Plant ecologists have also applied analysis of age and size to modular units of plants, particularly leaves, twigs, and stems. Age distributions influenced by growth and senescence of leaves have an important influence on the structure, growth form, and productivity of plant populations. For example, Dickerman and Wetzel (1985) studied the clonal growth and ramet demography of the cattail *Typha latifolia* in south-central Michigan, providing a detailed picture of the shoot emergence, mortality, density, height growth, and effects of intraspecific competition on the population. They followed the fate of marked individual shoots over a two-year period and were able to develop both age and height classifications of the population (Figure 17.16). Shoots emerged in three main pulses each year, resulting in three major cohorts. The first cohort emerged in early spring and was gone by late autumn. The second cohort appeared in midsummer, and about 80 percent of these shoots died by autumn. The remainder resumed growth the following spring. The third cohort emerged in late summer, and about 90 percent of its shoots resumed growth the following spring.

## SEX RATIOS

An almost universal characteristic of eukaryotes is sexual reproduction, although some organisms reproduce asexually at times. Even these species have some provision for sexual reproduction in their life cycle, for only by mixing and recombining genes can a population maintain genetic variability.

Populations of most organisms tend toward a 1:1 sex ratio (the proportion of males to females). The **primary sex ratio** (the ratio at conception) also tends to be 1:1. (This may not be universally true and is, of course, difficult to determine.)

The **secondary sex ratio** (the ratio at birth) among mammals is often weighted toward males, but the population shifts toward females in older age groups. For example, the ratio among fetuses of elk in western Canadian national parks was 113 males to 100 females (Flook 1970). Between ages 1½ and 2½ the ratio of males to females dropped abruptly, until it remained at about 85:100, although in certain areas it dropped as low as 37:100. The greatest decline in the number of males occurred between the ages of 7 and 14. The decline of females was much less rapid. The difference in the rate of decline has an additional effect on the higher ratio of females to males. The loss of males allows more food and space for the females and young, further increasing the female rate of survival.

Among humans, too, males exceed females at birth, but as age increases the ratio swings in favor of females. In 1965 the ratio of males to females based on a stable age distribution in the United States was age 0 to 4, 104:100; age 40 to 44, 100:100; age 60 to 64, 88:100; age 80 to 84, 54:100.

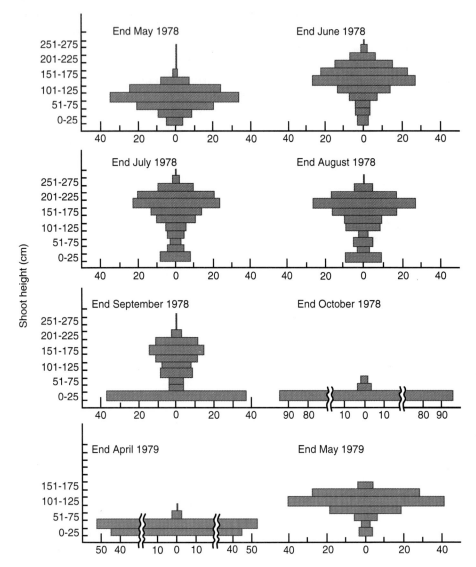

**Figure 17.16** Age classes, as indicated by size and growth, of shoots of cattail *Typha latifolia* at Lawrence Lake, Michigan over two years. Note that as the plants grow taller, smaller shoot size classes make up a proportionately smaller segment of the population. By summer's end the overstory shoots have died and smaller shoots, the young segment of the population, becomes proportionately large. (From Dickerman and Wetzel 1985:542.)

Lower mortality among females is characteristic of both advanced and underdeveloped countries.

Among birds the sex ratios tend to remain weighted toward the males (Bellrose 1961). Fall and winter sex ratios of prairie chickens show a preponderance of males in adult groups; similar ratios are characteristic of other gallinaceous birds and some passerine birds.

It is not easy to explain why sex ratios should shift from an equal ratio at birth to an unequal one later in life. Perhaps a partial answer lies in factors related to the genetic determination of sex and the physiology and behavior of the two sexes.

Physiological and behavioral patterns affect mortality of the sexes differently. For example, during the breeding season the male elk battles other males for dominance of a harem, defends his harem from rivals, and mates with the females. These activities not only consume considerable energy, but leave little time for feeding, and the male often ends the breeding season in poor physical condition. Among birds the female may help the male defend territory, builds

the nest, lays and incubates the eggs, broods the young, and, often with the help of the male, feeds the young. While incubating the eggs and brooding the young, the female is much more vulnerable to predation and other dangers than is the male. Thus adult males in mammals and adult females in birds apparently expend more energy, are subject to greater stress, and are more vulnerable to predation and other dangers. Their higher vulnerability and mortality is consistent with the imbalance in the sex ratio in the older age cohorts.

## MORTALITY AND NATALITY

The age structure of a population reflects two continuous processes in a population: deaths and births. Death is a great leveler of populations. Typically, it concentrates on the very young, reproducing a potential abundance of new individuals entering a perhaps already crowded population, and on

the old, removing senescent individuals to make room for more vigorous young.

Mortality is often expressed as a **crude death rate,** usually the number of deaths per 1000 in a given period. However, it is more precise to express mortality in a population either as the probability of dying or as a death rate. The **death rate** is the number of deaths during a given time interval divided by the average population; it is an instantaneous rate. For example, if the population size at the beginning of the period is 1000 and the number alive at the end of the period is 600, the average size of the population for the period would be 1600/2 or 800. The number of deaths is 400, so the death rate is 400/800 or 0.50. The **probability of dying** is the number that died during a given time interval divided by the number alive at the beginning of the period, 400/1000 or 0.40. The complement of the probability of dying is the **probability of surviving,** the number of survivors divided by the number alive at the beginning of the period. Because the number of survivors is more important to a population than the number dying, mortality is better expressed in terms of **life expectancy,** the average number of years to be lived in the future by members of a given age in the population.

The greatest influence on population increase is the birth of new individuals, called **natality.** Natality may be described as maximum or physiological natality and as realized natality. **Physiological natality** represents the maximum possible number of births under ideal environmental conditions, the biological limit. Because this number is rarely achieved in wild populations, its measure is of little value to a field biologist, but it is a useful yardstick against which to compare realized natality. **Realized natality** is the amount of successful reproduction that actually occurs over a period of time. It reflects the type of breeding season (continuous, discontinuous, or strongly seasonal), the number of litters or broods per year, the length of gestation or incubation, and so on. It is influenced by environmental conditions, nutrition, and density of the population.

Natality, measured as a rate, may be expressed either as crude birthrate or specific birthrate. **Crude birthrate** is expressed in terms of population size, for example 50 births per 1000 population. **Specific birthrate,** a more accurate measure, is expressed relative to a specific criterion, such as age. It is usually expressed as an **age-specific schedule of births,** the number of offspring produced per unit time by females in different age classes.

## The Life Table

A clear and systematic picture of mortality and survival in a population is best provided by a life table. It is a useful device to analyze probabilities of survivorship of individuals in a population, to determine ages most vulnerable to mortality, and to predict population growth. This device was developed by students of human populations and is widely used by life insurance companies. For them it is a critical tool in determining probabilities of survivorship upon which to base life insurance premiums. Population ecologists have adopted the life table for the study of natural populations.

The life table consists of a series of columns, each of which describes an aspect of mortality statistics for members of a population according to age. (For details of construction see Appendix B.) Figures are presented in terms of a standard number of individuals all born at the same time, called a **cohort.** By convention, the initial number of individuals in a cohort is set at 1000. The columns include $x$, the unit of age or age level; $l_x$, the number of individuals in a cohort that survive to that particular age level; $d_x$, the number of a cohort that die in an age interval $x$ to $x + 1$ (from one age level listed to the next); and $q_x$ the probability of dying or age-specific mortality rate. This rate is determined by dividing the number of individuals that died during the age interval by the number alive at the beginning of the age interval. Another column, $s$, or survival rate may be added. It can be calculated from $1 - q$. In order to calculate life expectancy, given in column $e_x$, two additional statistics are needed, $L_x$ and $T_x$. $L_x$, is the average years lived by all individuals in each category in the population. It is obtained by summing the number alive at age interval $x$ and the number at age $x + 1$, and dividing the sum by 2. $T_x$ is the number of time units left for all individuals to live from age $x$ onward. It is calculated by summing all the values of $L_x$ from the bottom of the table upwards to the age interval of interest. Life expectancy, $e_x$, is obtained by dividing $T_x$ for the particular age class $x$ by the survivors for that age, as given in the $l_x$ column.

Data for life tables are easy to obtain for laboratory animals and for humans. Laboratory animals are confined populations of known ages. For humans data on births and deaths are reported to and recorded by state governments and life insurance companies. Data on age, mortality, and survivorship for organisms in the wild are much more difficult to obtain. Mortality ($d_x$) can be estimated by determining the ages at death of a large number of animals born at the same time, if they were marked or banded. We can record the ages at death of animals marked at birth. Ideally we would like to follow a group born at a given season in a particular year. However, to obtain a sufficiently large sample we may need to combine the fates of individuals born in different seasons or different years.

Another approach to gathering data is to determine the age at death of a representative sample of carcasses of the species concerned collected in the same year. Age can be determined by examining wear and replacement of teeth in deer, growth rings in the cementum of teeth of ungulates and carnivores, annual rings in the horns of mountain sheep, and weight of the lens of the eye. This information also goes into the $d_x$ column. Recording ages at death of a sample of a population wiped out by some catastrophe could provide data for the $l_x$ series, because all animals were alive just prior to sampling and experienced death in the same short time period. Similarly, determining the age of animals shot during a

hunting season provides information for the $l_x$ column, because the sample is from a living population.

There are three basic types of life tables: horizontal, vertical, and dynamic-composite. **Horizontal life tables,** also called **cohort** or **dynamic life tables,** are constructed by following a cohort of individuals, a group all born within a single short span of time, from birth to the death of the last member. Construction of such life tables is most easily accomplished when the species is short-lived (one to two years), so that the generations are discrete.

To develop a cohort life table for organisms with overlapping generations, we have to mark or in some manner follow all the individuals born in one year through to the death of the last survivor. Because many animals are long-lived and get thoroughly mixed with individuals of other age classes, following them becomes a real problem. Grant and Grant (1992) constructed such tables for populations of the medium ground finch (*Geospiza fortis*) and the cactus ground finch (*Geospiza scandens*) on Isla Daphne Major in the Galapagos archipelago. They followed four color-banded cohorts born in the years 1975 and 1978 until nearly every individual died. Table 17.2 provides an example of a cohort life table for the cactus ground finch in 1975. The survivorship of males and females is combined.

The second type of life table is the **vertical, time-specific,** or **static life table** (Table 17.3). It is constructed by sampling the population in some manner (such as hunting takes, or core samples of trees) and aging the organisms to obtain a distribution of age classes during a single time period. This life table involves the assumptions that each age class is sampled in proportion to its numbers in the population and ages at death, that the birthrate and death rate are constant, and that the population is neither increasing nor decreasing. It assumes, for example, that survivors of one year class were survivors from the year before and so on, the same as they would have been if they were a single cohort. Such assumptions, of course, are false because when data are so collected, the number of survivors in one year may be greater than the year previous. In this case the data have to be adjusted or smoothed to get rid of this anomaly (see Caughley 1977).

The third type is the **dynamic-composite life table.** It records the same information as the dynamic life table, but it takes as the cohort a composite of a number of animals marked over a period of years rather than at just one birth period. For example, wildlife biologists may mark or tag a number of newly hatched young birds or newly born young mammals each year over a period of several years. After following the fate of each year's group, they pool the data and treat all of the marked animals as one cohort (see Barkalow et al. 1970). Or biologists may record the ages at death of animals found over a series of years and pool those data to construct a life table. Houston (1982) used this method to construct a life table for Yellowstone elk (Table 17.4).

Both the static (time-specific) and dynamic-composite life tables are inaccurate. Mortality and reproduction vary from year to year over which the data are collected. The data reflect standing age distributions; yet the life table is based on a stable age distribution. If age distributions are unstable, populations are changing continuously

**Table 17.2** Life Table of the 1975 Cohort of *Geospiza scandens*

| Age class (x) | Number alive each year ($N_x$) | Proportion surviving at start of age x ($l_x$) | Proportion dying between x and x + 1 ($d_x$) | Mortality rate ($q_x$) | Mean expectation of further life ($e_x$) |
|---|---|---|---|---|---|
| 0 | (82) | 1.000 | 0.488 | 0.488 | 2.63 |
| 1 | 42 | 0.512 | 0.207 | 0.404 | 3.64 |
| 2 | 25 | 0.305 | 0.098 | 0.301 | 4.78 |
| 3 | 17 | 0.207 | 0.036 | 0.174 | 5.79 |
| 4 | 14 | 0.171 | 0.000 | 0.000 | 5.93 |
| 5 | 14 | 0.171 | 0.012 | 0.070 | 4.93 |
| 6 | 13 | 0.159 | 0.000 | 0.000 | 4.27 |
| 7 | 13 | 0.159 | 0.000 | 0.000 | 3.27 |
| 8 | 13 | 0.159 | 0.074 | 0.465 | 2.27 |
| 9 | 7 | 0.085 | 0.037 | 0.435 | 2.79 |
| 10 | 4 | 0.048 | 0.000 | 0.000 | 3.50 |
| 11 | 4 | 0.048 | 0.000 | 0.000 | 2.50 |
| 12 | 4 | 0.048 | 0.024 | 0.500 | 1.50 |
| 13 | 2 | 0.024 | 0.000 | 0.000 | 1.50 |
| 14 | 2 | 0.024 | 0.012 | 0.500 | 0.50 |
| 15 | 1 | 0.012 | 0.012 | 1.000 | 0.50 |
| 16 | 0 | 0 | . . . | . . . | . . . |

*Source:* Grant and Grant 1992.

*Note:* Age classes are designated in years. The starting number in the cohort ($N_0$) is in parentheses because it was estimated.

**Table 17.3** Time-Specific Life Table for Red Deer on Isle of Rhum, 1957

| x | $l_x$ | $d_x$ | 1000 $q_x$ | $e_x$ |
|---|-------|-------|------------|-------|
| Stags | | | | |
| 1 | 1000 | 282 | 282.0 | 5.81 |
| 2 | 718 | 7 | 9.8 | 6.89 |
| 3 | 711 | 7 | 9.8 | 5.95 |
| 4 | 704 | 7 | 9.9 | 5.01 |
| 5 | 697 | 7 | 10.0 | 4.05 |
| 6 | 690 | 7 | 10.1 | 3.09 |
| 7 | 684 | 182 | 266.0 | 2.11 |
| 8 | 502 | 253 | 504.0 | 1.70 |
| 9 | 249 | 157 | 630.6 | 1.91 |
| 10 | 92 | 14 | 152.1 | 3.31 |
| 11 | 78 | 14 | 179.4 | 2.81 |
| 12 | 64 | 14 | 218.7 | 2.31 |
| 13 | 50 | 14 | 279.9 | 1.82 |
| 14 | 36 | 14 | 388.9 | 1.33 |
| 15 | 22 | 14 | 636.3 | 0.86 |
| 16 | 8 | 8 | 1000.0 | 0.50 |
| Hinds | | | | |
| 1 | 1000 | 137 | 137.0 | 5.19 |
| 2 | 863 | 85 | 97.3 | 4.94 |
| 3 | 778 | 84 | 107.8 | 4.42 |
| 4 | 694 | 84 | 120.8 | 3.89 |
| 5 | 610 | 84 | 137.4 | 3.36 |
| 6 | 526 | 84 | 159.3 | 2.82 |
| 7 | 442 | 85 | 189.5 | 2.26 |
| 8 | 357 | 176 | 501.6 | 1.67 |
| 9 | 181 | 122 | 672.7 | 1.82 |
| 10 | 59 | 8 | 141.2 | 3.54 |
| 11 | 51 | 9 | 164.6 | 3.00 |
| 12 | 42 | 8 | 197.5 | 2.55 |
| 13 | 34 | 9 | 246.8 | 2.03 |
| 14 | 25 | 8 | 328.8 | 1.56 |
| 15 | 17 | 8 | 492.4 | 1.06 |
| 16 | 9 | 9 | 1000.0 | 0.50 |

*Source:* V. P. W. Lowe, 1969.

**Table 17.4** Dynamic-Composite Life Table for Northern Yellowstone Female Elk

| x | $l_x$ | $d_x$ | $q_x$ | $e_x$ |
|---|-------|-------|-------|-------|
| 0 | 1000 | 323 | .323 | 11.8 |
| 1 | 677 | 13 | .019 | 16.2 |
| 2 | 664 | 2 | .003 | 15.5 |
| 3 | 662 | 2 | .003 | 14.6 |
| 4 | 660 | 4 | .006 | 13.6 |
| 5 | 656 | 4 | .006 | 12.7 |
| 6 | 652 | 9 | .014 | 11.7 |
| 7 | 643 | 3 | .005 | 11.0 |
| 8 | 640 | 3 | .005 | 10.0 |
| 9 | 637 | 9 | .014 | 9.0 |
| 10 | 628 | 7 | .001 | 8.1 |
| 11 | 621 | 12 | .019 | 7.3 |
| 12 | 609 | 13 | .021 | 6.4 |
| 13 | 596 | 41 | .069 | 5.5 |
| 14 | 555 | 34 | .061 | 4.9 |
| 15 | 521 | 20 | .038 | 4.2 |
| 16 | 501 | 59 | .118 | 3.3 |
| 17 | 442 | 75 | .170 | 2.7 |
| 18 | 367 | 93 | .253 | 2.1 |
| 19 | 274 | 82 | .299 | 1.7 |
| 20 | 192 | 57 | .297 | 1.2 |
| 21+ | 135 | 135 | 1.000 | 0.5 |

*Source:* Adapted from Houston 1982.

developmental stages rather than discrete time intervals. The $l_x$ values are obtained by observing a natural population over the annual season and by estimating the size of the surviving population at each stage of development from eggs, larvae, pupae, and adults. If we keep records of weather conditions, abundance of predators, parasites, and diseases, we can also estimate how many die from various causes.

An example is a life table for a sparse stable population of gypsy moths in Connecticut (Table 17.5). A brief description of the life cycle will give the table more meaning. In April and May gypsy moth larvae emerge from eggs deposited by females the previous summer. In response to light, the larvae climb to the tops of trees and spin down from unfolding leaves by a silken thread. Because the thread is easily broken, larvae may be carried by the wind for long distances, which often results in a mass redistribution of the population.

The larvae go through a number of developmental stages called instars. Male larvae pass through five instars, females through six. Each instar, particularly the last three, is characterized by different feeding behaviors. Larvae reach full size by late June or July, stop feeding, and spin a few threads that will hold the pupae on the tree. The insects spend about two weeks as pupae and then emerge as adult moths. Males emerge first, several days ahead of females. Females, who do not fly, attract strong flying males by means of a powerful sex attractant. After mating, the females lay an egg mass

and the data from which life tables are constructed do not reflect the true nature of the population. In spite of these shortcomings, such life tables may present a reasonable assessment of average conditions in the population, useful to compare life history trends within and between populations.

The life tables described are typical of long-lived species in which generations overlap and in which different age groups are alive at the same time. However, tremendous numbers of organisms have one annual breeding season and generations do not overlap. All individuals belong to the same cohort or age class. Many insects and annual and biennial plants follow this pattern.

Some organisms, notably insects, have several stages to their life cycle. In this case, the life table is best divided into

**Table 17.5** Life Table Typical of Sparse Gypsy Moth Populations in Northeastern Connecticut

| x | $l_x$ | $d_{xf}$ | $d_x$ | $100\, q_x$ |
|---|---|---|---|---|
| Eggs | 550.0 | Parasites | 82.5 | 15 |
| | | Other | 82.5 | 15 |
| | | Total | 165.0 | 30 |
| Instars I–III | 385.0 | Dispersion, etc. | 142.4 | 37 |
| Instars IV–VI | 242.5 | Deer mice | 48.5 | 20 |
| | | Parasites and disease | 12.1 | 5 |
| | | Other | 167.3 | 69 |
| | | Total | 227.9 | 94 |
| Prepupae | 14.6 | Predators, etc. | 2.9 | 20 |
| Pupae | 11.7 | Vertebrate predators | 9.8 | 84 |
| | | Other | 0.5 | 4 |
| | | Total | 10.3 | 88 |
| Adults | 1.4 | Sex (SR = 30:70) | 1.0 | 70 |
| Adult, female | 0.4 | — | — | — |
| Generation | — | — | 549.6 | 99.93 |

*Source:* R. W. Campbell 1969.

containing a variable number of eggs (the sparser the population, the greater the number of eggs).

This life table represents the fate of a cohort from a single egg mass. The age interval or x column indicates the life history stages of unequal duration. The $l_x$ column indicates the number of survivors at each stage. The $d_x$ column gives the breakdown of deaths by causes in each stage. Note that life expectancy is not calculated because there is none. All the adult population will die in late summer.

Applying the life table concept to plants is an active area in the study of plant demography. One example is the life table (Table 17.6) for the 1978–1979 generation of the annual plant *Phlox drummondii* at Nixon, Texas, developed by Leverich and Levin (1979). On a study plot they located, mapped, and censused at intervals a cohort of individuals in the population from their emergence as seedlings in November 1978 to their death in 1979. In this table the $N_x$ column records the actual number of individuals. The $l_x$ is the probability of survivorship.

Plant ecologists have adapted the life table to study the population dynamics of plant modules such as leaves, buds, and stems that exhibit age specific mortality. Life tables are a useful tool in studying the role of plant modular growth in primary production and life history traits of plants.

McGraw (1989) constructed a life table for shoots of *Rhododendron maximum* (Table 17.7). A shoot is an apical meristem and the live leaves produced by that meristem. McGraw was interested in whether shoot life histories were patterned more clearly by age or size and whether interactions between age and size determined shoot fate (death, growth, flowering, and branching). He was able to use a life table approach because shoots are analogs of individuals in

**Table 17.6** Life Table for *Phlox drummondii* at Nixon, Texas

| Age Interval (days) x – x' | No. Surviving to Day x $N_x$ | Survivorship $l_x$ | No. Dying During Interval $d_x$ | Average Mortality Rate per Day $q_x$ | Mean Expectation of Life (days) $e_x$ |
|---|---|---|---|---|---|
| 0–63 | 996 | 1.0000 | 328 | .0052 | 122.87 |
| 63–124 | 668 | .6707 | 373 | .0092 | 104.73 |
| 124–184 | 295 | .2962 | 105 | .0059 | 137.59 |
| 184–215 | 190 | .1908 | 14 | .0024 | 137.05 |
| 215–231 | 176 | .1767 | 2 | .0007 | 115.72 |
| 231–247 | 174 | .1747 | 1 | .0004 | 100.96 |
| 247–264 | 173 | .1737 | 1 | .0003 | 85.49 |
| 264–271 | 172 | .1727 | 2 | .0017 | 68.94 |
| 271–278 | 170 | .1707 | 3 | .0025 | 62.71 |
| 278–285 | 167 | .1677 | 2 | .0017 | 56.78 |
| 285–292 | 165 | .1657 | 6 | .0052 | 50.42 |
| 292–299 | 159 | .1596 | 1 | .0009 | 45.19 |
| 299–306 | 158 | .1586 | 4 | .0036 | 38.46 |
| 306–313 | 154 | .1546 | 3 | .0028 | 32.36 |
| 313–320 | 151 | .1516 | 4 | .0038 | 25.94 |
| 320–327 | 147 | .1476 | 11 | .0107 | 19.55 |
| 327–334 | 136 | .1365 | 31 | .0325 | 13.85 |
| 334–341 | 105 | .1054 | 31 | .0422 | 9.90 |
| 341–348 | 74 | .0743 | 52 | .1004 | 5.58 |
| 348–355 | 22 | .0221 | 22 | .1428 | 3.50 |
| 355–362 | 0 | .0000 | | | |

*Source:* Leverich and Levin 1979.

**Table 17.7** Life Table for Shoots of *Rhododendron maximum*

| Age | $l_x$ | $d_x$ | $q_x$ | $e_x$ |
|-----|-------|-------|-------|-------|
| 0 | 1.000 | 0 | 0 | 5.60 |
| 1 | 1.000 | 0.016 | 0.016 | 4.60 |
| 2 | 0.984 | 0 | 0 | 3.67 |
| 3 | 0.984 | 0.075 | 0.077 | 2.67 |
| 4 | 0.909 | 0.185 | 0.024 | 1.85 |
| 5 | 0.724 | 0.346 | 0.477 | 1.19 |
| 6 | 0.378 | 0.270 | 0.714 | 0.82 |
| 7 | 0.108 | 0.095 | 0.882 | 0.62 |
| 8 | 0.013 | 0.013 | 1.000 | 0.50 |

*Source:* Adapted from McGraw 1989.

traditional populations. They grow, reproduce (branch), and die. They are easily aged by counting annual bud scars along the stem; new leaves and stems are produced annually. McGraw sampled the entire shoot population on a large rhododendron, judged median for the rhododendron population based on previous studies. McGraw divided the shoots into age classes, noted shoot death by counting those that no longer supported any live green leaves, and determined fertility by the number of daughter branches produced over a period of one year.

## Survivorship and Mortality Curves

The life table is a very important tool in the analysis of population dynamics. From it we can derive survivorship curves based on the $l_x$ column and mortality curves based on the $q_x$ column. These curves enable us to determine the ages at which a particular organism most often dies. This information provides some leads in determining the causes of death and ultimately the processes that affect the population dynamics of a given species. They enable biologists to compare survival between the sexes, between cohorts arising in different years, between populations, and between species.

Survivorship Curves The survivorship curve depicts age-specific mortality through survivorship. It is obtained by plotting the number of individuals of a particular age cohort against time. The usual form is to plot the logarithms of the numbers of survivors (usually on semilogarithmic graph paper) against age.

Survivorship curves may be classified into at least three hypothetical types (Figure 17.17) (Deevey 1947). These curves are conceptual models only, against which real-life survivorship curves can be compared. The Type I curve is convex. It is typical of populations whose individuals tend to live out their physiological life span; they exhibit a high degree of survival throughout life and experience heavy mortality in old age. Such a curve is typical of some plants, such as *Phlox drummondii* (Figure 17.18), and many mammals,

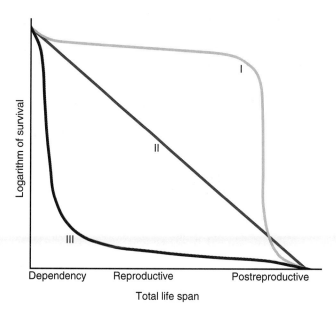

**Figure 17.17** Three basic types of survivorship curves. The vertical axis may be scaled arithmetically or logarithmically. If it is logarithmic, the slope of the lines will show the following rates of change: Type I, curve for populations in which the survival of juveniles is high and mortality is concentrated among old individuals. Type II, curve for populations in which the rate of mortality is fairly constant at all age levels, so that there is more or less a uniform percentage decrease in survivorship over time; Type III, curve for a population in which high mortality is concentrated among juveniles.

**Figure 17.18** Type I survivorship curve for the annual *Phlox drummondii* at Nixon, Texas. (From Leverich and Levin 1979:885.)

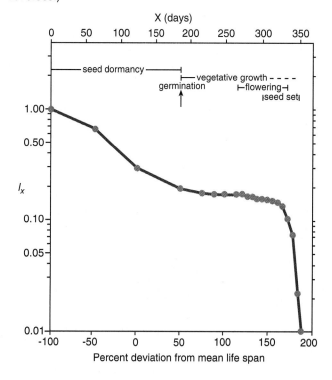

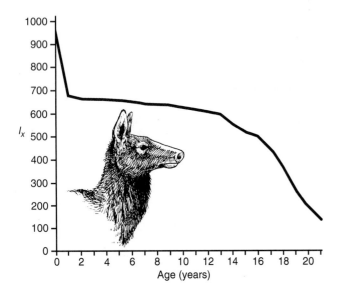

**Figure 17.19** Type I survivorship curve for female elk of Yellowstone National Park.

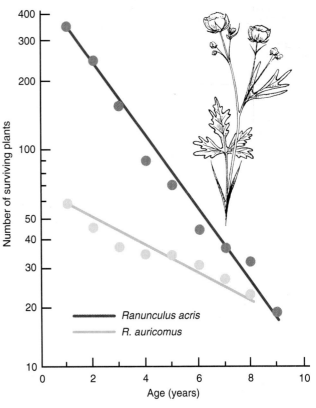

**Figure 17.21** Types II survivorship curves of buttercups *Ranuculus acris* and *Ranuculus auricomus*. (From Sarukhan and Harper 1974.)

such as elk (Figure 17.19). The Type II curve is linear and typical of organisms with constant mortality rates. Such a curve is characteristic of the adult stages of many birds (Figure 17.20), rodents, and some plants (Figure 17.21). Type III is concave and is typical of organisms with extremely high mortality rates in early life, such as many species of invertebrates (Figure 17.22), fish, and some plants.

Among plants, life table attributes and survivorship apply not only to individuals but also to the modular units: the

**Figure 17.22** A Type III survivorship curve exhibited by the oystershell scale insect (*Lepidosaphes ulma*). (Adapted from Price 1975.)

**Figure 17.20** Type II survivorship curves for the 1975 cohorts of the cactus ground finch (*Geospiza scandens*) and the medium ground finch (*G. fortis*). (Adapted from Grant and Grant 1992:771.)

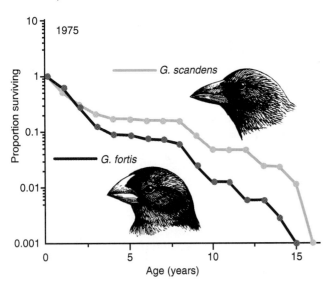

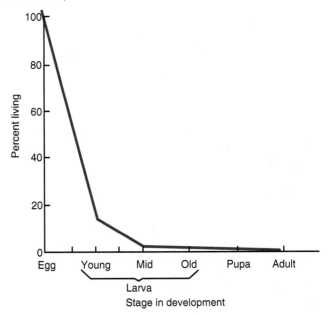

populations of leaves, buds, flowers, fruits, and seeds, as well as clones (Figure 17.23). Mortality and survivorship of these subpopulations influence the individual growth form of the plant, its ability to compete, and its photosynthetic and reproductive performance.

The validity of the survivorship curves depends upon the validity of the life table and the $l_x$ column. Life tables and thus survivorship curves are not typical of some standard population, but depict instead the nature of the population at different places at different times under different environmental conditions, and assume stable age distribution. Their greatest usefulness rests in the comparison of populations of one area, time, sex, or species with populations of another. For example, Figure 17.24 shows the variation in survivorship of the small-mouthed salamander (*Ambystoma texanum*) larvae over several years. It emphasizes the effects of variations in the environment on survivorship. Low survivorship in 1983 and 1984 resulted from severe flooding of the stream habitat. Similarly, survivorship curves can point

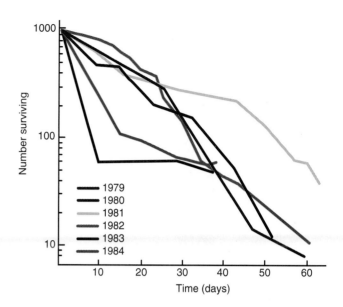

**Figure 17.24** Variation in survivorship of small-mouthed salamander (*Ambostyma texanum*) larvae over a period of five years. Low survivorship in 1983 and 1984 resulted from severe flooding of their stream habitat. (From Pefranka and Sih 1986:731.)

**Figure 17.23** Survivorship curve of plant modules: shoots of *Rhododendron maximum*. The curve is Type I. (Based on data from McGraw 1989.)

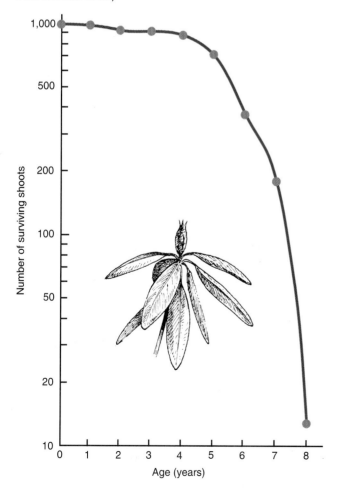

out the effect of environmental changes on plant populations. Klemow and Raynal (1983) followed the fate of a population of the summer annual garden rocket (*Erucastrum gallicum*) for five years. Survival of plants that emerged in the spring varied markedly with rainfall. Cohorts in a year with abundant rainfall exhibited a Type I survivorship; cohorts in a year of intermittent rainfall and drought had a Type II survivorship; and cohorts experiencing a spring drought had a Type III survivorship curve (Figure 17.25).

Mortality Curves If you plot the data in the $q_x$ or mortality rate column of the life table against age, you get a mortality curve. It consists of two parts: (1) the juvenile phase, in which the rate of mortality is high; and (2) the post-juvenile phase, in which the rate first decreases as age increases, and then increases with age after a low point in mortality (Figure 17.26). Most populations have a roughly J-shaped curve.

Because $q_x$ is the ratio of the number dying during an age interval to the number alive at the beginning of the period (or surviving the previous age period), that parameter is independent of the frequency of each of the previous age classes. Therefore the parameter is free of the biases inherent in the $l_x$ column and survivorship curves. Most life tables of wild populations are subject to bias because the first-year age class is not adequately represented. This error distorts each succeeding $l_x$ and $d_x$ value. If the first values are inaccurate, all succeeding ones are inaccurate. But if the first values of $q_x$ are wrong, the error does not affect the other values. For this reason mortality curves, which indicate the

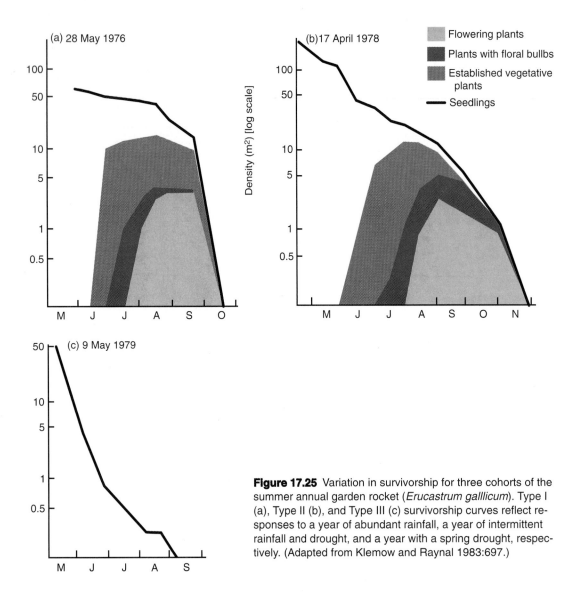

**Figure 17.25** Variation in survivorship for three cohorts of the summer annual garden rocket (*Erucastrum galllicum*). Type I (a), Type II (b), and Type III (c) survivorship curves reflect responses to a year of abundant rainfall, a year of intermittent rainfall and drought, and a year with a spring drought, respectively. (Adapted from Klemow and Raynal 1983:697.)

rate of mortality indirectly by the slope of the line, are more informative than survivorship curves.

## Animal Natality

As pointed out earlier, (page 376) natality is measured as a birthrate. In animals, especially, it is expressed as age-specific schedule of births, the number of offspring produced per unit time by females in different age classes.

Because population increase depends mostly on the number of females in the population, the age-specific birth schedule usually counts only the females giving rise to females. The age-specific schedule is obtained by determining the mean number of females born in each group of females, designated as $m_x$. If we take the $l_x$ or survivorship column from the life table and age-specific $m_x$ values, we can con-

struct a fecundity table and calculate the **net reproductive rate,** $R_o$, the number of females left during a lifetime by a newborn female. When the reproductive success of known males can be determined, a fecundity table can be constructed for that sex also.

An example of a fecundity table for both males and females, constructed by Grant and Grant (1992) for the 1975 cohort of the cactus ground finch, is given in Table 17.8. In this table Grant and Grant assumed that the production of male and female fledglings is the same for both sexes. Therefore $m_x$ is the average number of daughters produced by mothers, or sons produced by fathers at a given age.

Both male and female cactus ground finches produce no young at age 0, their first year of life; therefore the $m_x$ values for that age are 0. Neither did this cohort produce young at ages 10 and 11 because of a protracted drought. To adjust for

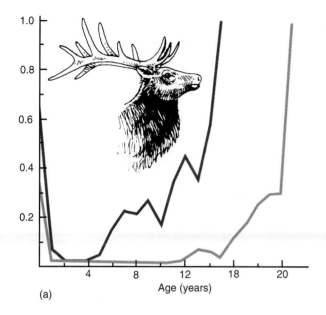

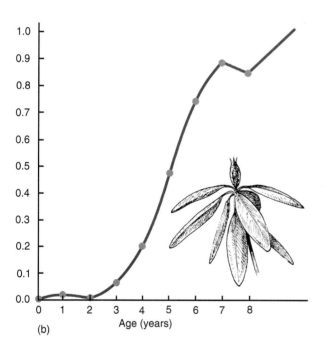

**Figure 17.26** (a) Elk or wapiti (*Cervus canadensis*) of Yellowstone. (From Houston 1982:57). (b) The mortality curve of modular *Rhododendron maximum* shoots also assumes a strong J-shape, with maximum mortality late in life. (Data from McGraw 1989.)

which reproduction occurs, the sum represents the number of females that will be left during a lifetime by a newborn female, and the number of males left by a newborn male, or $R_o$, the net reproductive rate. For the 1975 cohort of male and female cactus ground finches, the net reproductive rate for males was 0.762, and for females, 2.101. The birds replace themselves when $R_o$ equals 1. Thus in these birds, the males of the 1975 cohort were not replacing themselves, while the females were doubling each generation.

## Plant Natality

The distinct demographic differences between plants and most animals create problems in determining natality in plants. Plants accumulate structural units: new buds, leaves, flowers, fruits, new roots below ground, and new clones or ramets, extensions of the parent plants. Each of these modular populations has its own natality rate. At the same time, plants produce new individuals through sexual reproduction. Sexual reproduction in plants involves both the production and germination of seeds. Except for annuals and biennials, it is difficult to estimate seed production by individual plants and by populations. Perennial and woody plants, even within a population, vary in longevity and in seed production, which is not necessarily an annual event. Individual plants vary widely in seed production from year to year and from age class to age class, and the amount of seed produced over a lifetime is largely unknown. Moreover, seeds usually undergo a period of dormancy, which in some species may last for a number of years until conditions are right for germination. Germination is the formal equivalent of birth in plants.

The simplest situation exists with annual and biennial plants, because both produce seeds in one reproductive effort at the end of their individual life spans. By censusing the number of seeds produced by the different individuals, we can arrive at some idea of fecundity. Leverich and Levin (1979) did this for the annual Drummond's phlox, arriving at a fecundity schedule for both ovules and seeds (Table 17.9 seeds only). From this fecundity schedule they were able to determine $R_o$, the net rate of increase. In this case, involving an annual species with no overlapping generations $R_o$ not only gives us the average number of offspring produced per individual over a lifetime, but also provides the factor with which to multiply to obtain the size of the next generation, a point to be considered later.

## Fecundity Curves and Reproductive Value

The $m_x$ column of the fecundity table provides the data needed to plot a fecundity curve, which depicts the number of offspring per individual per age class (Figure 17.27). It suggests that individuals of certain age classes contribute

mortality in each age group, the $m_x$ value for each age class is multiplied by the corresponding $l_x$ or survivorship value. The resulting $l_x m_x$ value gives the mean number of females or males born in each age group adjusted for survivorship. For example, for female age class 3, the $m_x$ value is 1.44, but when adjusted for survivorship, the value drops to 0.40. When these adjusted values are summed over all ages in

**Table 17.8 Fecundity and Reproductive Values for the Cactus Ground Finch**

| Age class (x) | Probability of surviving to x ($l_x$) | Fledglings per season ($m_x$) | Product of survival and reproduction ($l_x m_x$) | Reproductive value ($V_x$) |
|---|---|---|---|---|
| **Males** | | | | |
| 0 | 1.000 | 0.0 | 0.0 | 0.762 |
| 1 | 0.512 | 0.0 | 0.0 | 1.561 |
| 2 | 0.317 | 0.0 | 0.0 | 2.404 |
| 3 | 0.293 | 0.104 | 0.030 | 2.601 |
| 4 | 0.268 | 0.250 | 0.067 | 2.730 |
| 5 | 0.268 | 0.250 | 0.067 | 2.480 |
| 6 | 0.244 | 0.475 | 0.116 | 2.450 |
| 7 | 0.244 | 0.125 | 0.030 | 1.975 |
| 8 | 0.244 | 1.350 | 0.330 | 1.850 |
| 9 | 0.122 | 0.600 | 0.073 | 1.000 |
| 10 | 0.073 | 0.0 | 0.0 | 0.667 |
| 11 | 0.073 | 0.0 | 0.0 | 0.667 |
| 12 | 0.073 | 0.667 | 0.049 | 0.667 |
| 13 | 0.049 | 0.0 | 0.0 | 0.0 |
| 14 | 0.049 | 0.0 | 0.0 | 0.0 |
| 15 | 0.024 | 0.0 | 0.0 | 0.0 |
| 16 | 0 | . . . | . . . | . . . |
| | | $R_o = \Sigma l_x m_x$ | = 0.762 | $\Sigma = 21.814$ |
| **Females** | | | | |
| 0 | 1.000 | 0.0 | 0.0 | 2.101 |
| 1 | 0.512 | 0.364 | 0.186 | 4.294 |
| 2 | 0.279 | 0.187 | 0.052 | 6.873 |
| 3 | 0.279 | 1.438 | 0.401 | 6.686 |
| 4 | 0.209 | 0.833 | 0.174 | 7.006 |
| 5 | 0.209 | 0.500 | 0.104 | 6.173 |
| 6 | 0.209 | 0.833 | 0.174 | 5.673 |
| 7 | 0.209 | 0.250 | 0.052 | 4.839 |
| 8 | 0.209 | 3.333 | 0.696 | 4.588 |
| 9 | 0.139 | 0.125 | 0.017 | 1.888 |
| 10 | 0.070 | 0.0 | 0.0 | 3.500 |
| 11 | 0.070 | 0.0 | 0.0 | 3.500 |
| 12 | 0.070 | 3.500 | 0.245 | 3.500 |
| 13 | 0 | . . . | . . . | . . . |
| 14 | . . . | . . . | . . . | . . . |
| 15 | . . . | . . . | . . . | . . . |
| | | $R_o = \Sigma l_x m_x$ | = 2.101 | $\Sigma = 60.621$ |

*Source:* Grant and Grant 1992: 773.

more to population growth than those of other age classes. This average contribution to the next generation that members of a given age group make between their current age and death is termed reproductive value. Examples of reproductive values are given in the $V_x$ column of the fecundity table for the cactus ground finch in (Table 17.8).

The concept of reproductive value was developed by R. A. Fisher (1930). In a population of constant size, he defined the reproductive value as the age-specific expectation of future offspring:

$$V_x = \sum_{y=x}^{\infty} \frac{l_y}{l_x} m_y$$

where $l_y/l_x$ = probability of living from age $x$ to age $y$, $m_y$ = average reproductive success of an individual at age $y$, and $V_x$ = reproductive value.

For newborn individuals in a population that is neither increasing nor decreasing, the reproductive value is the same as the net reproduction rate, $R = 1$. Populations change in size, however, so the definition of reproductive value becomes the present value of future offspring. This value represents the

**Table 17.9** Fecundity Schedule for *Phlox drummondii* at Nixon, Texas, Based on Seed Production

| $x - x^1$ | $N_x$ | $l_x$ | $m_x^{seed}$ | $l_x m_x$ |
|---|---|---|---|---|
| 0–299 | 996 | 1.0000 | .0000 | .0000 |
| 299–306 | 158 | .1586 | .3394 | .0532 |
| 306–313 | 154 | .1546 | .7963 | .1231 |
| 313–320 | 151 | .1516 | 2.3995 | .3638 |
| 320–327 | 147 | .1476 | 3.1904 | .4589 |
| 327–334 | 136 | .1365 | 2.5411 | .3470 |
| 334–341 | 105 | .1054 | 3.1589 | .3330 |
| 341–348 | 74 | .0743 | 8.6625 | .6436 |
| 348–355 | 22 | .0221 | 4.3072 | .0951 |
| 355–362 | 0 | .0000 | .0000 | .0000 |
| | | | | $\Sigma = 2.4177$ |

$R_o = \Sigma l_x m_x = 2.42$ (per capita)

$R = \dfrac{ln\ R_o}{365} = 0.0024$ (per capita per day)

*Note:* $x - x' =$ age interval; $N_x =$ no. surviving to day $x$; $m_x^{seed} =$ average no. of seeds per individual during interval; $l_x =$ survivorship. $l_x m_x =$ contribution to net reproductive rate during interval.
*Source:* From Leverich and Levin 1979.

number of offspring an individual dying at age $x + 1$ would have to produce at age $x$ to leave behind as many offspring had it not died at age $x + 1$ and survived to produce young thereafter. The general equation for reproductive value is:

$$\frac{v_x}{v_o} = \frac{e^{rx}}{l_x} \sum_{y=x}^{\infty} e^{-ry} l_y m_y$$

(For calculation of reproductive value see Appendix B.)

The reproductive value of a newborn individual is influenced by the state of the population. In an expanding population the reproductive value of young is low for two reasons. First, in an increasing population the probability of death before reproduction may also increase. Second, because the future breeding population of which the young will be a part will be larger, they will contribute less to the overall gene pool than offspring currently being born. Conversely, young born into a declining population will contribute more to the future than present progeny, so are worth more.

Reproductive values are usually calculated for females because reproductive values for males are difficult to obtain, but calculations can be done if we know the number and fate of the offspring sired by the male. Grant and Grant have done this calculation for the cactus ground finch (Table 17.8 and Figure 17.28). Reproductive values for females are higher those of males. Generally reproductive values rise to a peak at an early age and then decline. However the reproductive value of the cactus ground finch varies, depending upon annual fluctuations in rainfall. In the polygamous red deer (*Cervus elaphus*) on the Isle of Rhum, the reproductive value of the hinds (plotted in Figure 17.29) declined gradually from four years of age. This decline reflects the physiological drain of reproduction on the females. The reproduc-

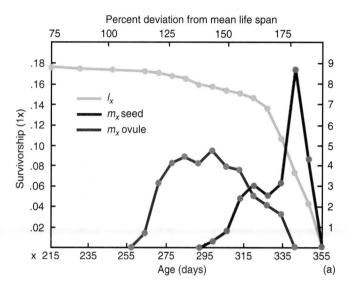

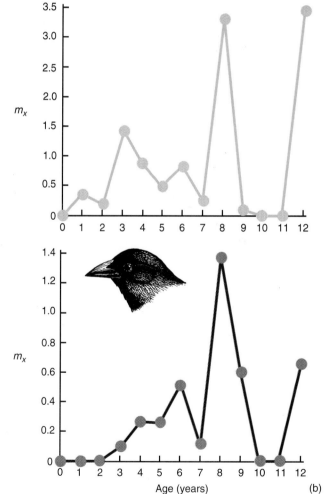

**Figure 17.27** Fecundity curves. (a) *Phlox drummondii*. (After Leverich and Levin 1979:888). (b) Male and female cactus ground finch. (Data from Grant and Grant 1992.) In *Phlox* fecundity is the highest in the old age classes, which we would expect in an annual plant.

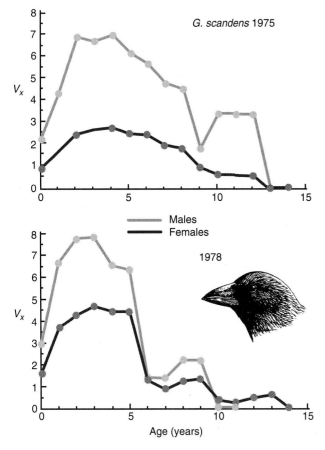

**Figure 17.28** Reproductive value as a function of age in two cohorts of the cactus ground finch. (From Grant and Grant 1992:772.)

**Figure 17.29** Plot of reproductive values for red deer hinds and stags. Reproductive value for both sexes was calculated in terms of the number of female offspring surviving to one year old that parents of different ages can be expected to produce in the future. (From Clutton-Brock 1984:154.)

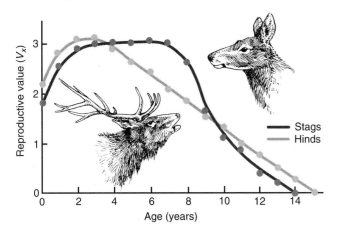

tive value of the stag remained constant from three to seven years, then declined sharply, indicating the loss of vigor after several years of intense reproductive activity. Unlike old hinds, stags apparently do not invest heavily in breeding activities during their declining years. The determination of reproductive values, a measure of fitness, provides a better understanding of life history patterns, and indicates which age classes contribute most to population growth, useful information in the management of wildlife populations and endangered species.

## SUMMARY

Living organisms exist in groups of the same species. These groups, or populations, occupy a particular space at a given time. The size of a population in relation to a defined unit of space is its density.

Populations are distributed in some kind of pattern over an area. Some are uniformly distributed, a very few are randomly distributed, and most exhibit a clumped distribution, which results in aggregations. Some aggregations reflect a degree of sociality, which may lead to cooperative or competitive situations. The pattern of clumped distribution is called coarse-grained when the clumps and areas between them are large and fine-grained when the clumps and areas between them are small.

Movements within and among population involve emigration, the permanent movement of individuals out of a population; immigration, the movement of individuals into a population or vacant habitat; and migration, a daily or seasonal round trip movement between two areas.

Age and sex ratios affect the structure of a population and influence the rates of births and deaths. Reproduction is limited to certain age classes; mortality is most prominent in others. Changes in age class distribution bring about changes in the production of young and mortality. The sex ratio tends to be balanced between males and females at birth, but the ratio changes to favor females in mammals and males in birds as age cohorts grow older.

Population density depends upon the number of individuals added to the group and the number leaving—the difference between the birthrate and the death rate and the balance between immigration and emigration. Birthrate usually has the greatest influence on the addition of new individuals. Mortality is the reducer and is greatest among young and old. Mortality and survivorship are best analyzed by means of a life table, which is an age-specific summary of mortality in a population. From the life table we can derive mortality curves that are useful in comparing demographic trends within and between populations living under different environmental conditions. Reproductive rates of populations and reproductive values of individuals within a population can be best analyzed by means of a fecundity table that relates age-specific reproduction to age structure and survivorship.

# REVIEW QUESTIONS

1. What characteristics are unique to populations?
2. Distinguish between unitary and modular populations.
3. Distinguish among uniform, clumped, and random distribution of individuals within a population.
4. Contrast emigration, immigration, and migration.
5. What is an age pyramid? What can it tell about a population's history? Its future potential?
6. Define death rate, mortality rate, probability of dying, and life expectancy.
7. What is a life table? Explain each of the major columns.
8. What are three basic types of survivorship curves?
9. What is a fecundity table? How is it used to obtain the net reproductive rate? What is reproductive value?
10. Using the following data for a population of squirrels, construct a life table and a fecundity table. First convert the raw population data, $N_x$, to proportionality to get the $l_x$ column.

$x = 0, 1, 2, 3, 4, 5, 6, 7, 8, 9$

$N_x = 410, 164, 85, 44, 21, 12, 8, 4, 2, 1$

$m_x\ 0.0, 1.0, 2.2, 2.4, 3.2, 1.8, 1.7, 2.0, 1.4, 1.4$

11. Determine the net reproductive rate of the ground squirrel population.

# CROSS-REFERENCES

Adaptation, 30; population growth, 392–396; dispersal, 413–417; mating systems, 428–430; reproductive costs, 436–446.

# Population Growth and Regulation

## Concepts

1. The difference between the instantaneous birthrate and the instantaneous death rate is the rate of population increase, $r$.
2. Geometric increase of a population is characterized by a constant schedule of births and deaths.
3. As population density increases, population growth slows and fluctuates about some point of equilibrium with the environment.
4. Population fluctuations are influenced by density-dependent mechanisms.
5. Major density-dependent regulating mechanisms involve intraspecific competition.
6. Density-independent influences are not regulatory, but they can greatly influence both density-dependent mechanisms and population growth.
7. With $k$-factor analysis ecologists recognize the key factors determining population trends.
8. When populations become very small, chance events alone can lead to extinction.

Populations are dynamic. Depending upon the nature of the organism, population numbers may change from hour to hour, day to day, season to season, year to year. In some years certain organisms are abundant; in other years they are scarce. Local populations appear, expand, decline, or become extinct. Eventually the area may be recolonized by individuals moving in from other populations. Such changes come about as the interaction of organisms and their environment affects birthrates, death rates, and the movement of individuals.

# RATE OF INCREASE

Possessing a sufficiently large sample of individuals of all ages in a population to construct a life table and knowing the age-specific fecundity (discussed in Chapter 17), we can determine some characteristics of population growth. By multiplying age-specific survivorship, $(l_x)$ by age-specific fecundity, $(m_x)$, and summing all the $l_x m_x$ values for the entire lifetime, as we did in Chapter 17, we obtain the net reproductive rate $R_O$, defined as

$$\sum_{x=0}^{\infty} l_x m_y$$

If $R_O$ equals 1, the birthrate equals the death rate; individuals are replacing themselves and the population is remaining stable. If the value is greater than 1, the population is increasing; if it is less than 1, it is decreasing.

With the survivorship schedule $(l_x)$ and the fecundity schedule $(m_x)$, and an additional parameter, $p_x$, the proportion of animals surviving in each age class, we can chart the growth of a population by constructing a **population projection table.** The parameter $p_x$ is the complement of $q_x$ and is expressed as $1 - q$.

The construction of a population projection table can be illustrated using a hypothetical population of ten female squirrels, all age 1, introduced into an oak forest at initial year 0. The survivorship schedule, fecundity schedule, and $p_x$ are given in the life table shown in Table 18.1. For ease in calculation the $l_x$ column is expressed as a decimal fraction

**Table 18.1 Life Table for a Hypothetical Squirrel Population**

| $x$ | $l_x$ | $q_x$ | $p_x$ | $e_x$ | $m_x$ | $l_x m_x$ | $x l_x m_x$ |
|---|---|---|---|---|---|---|---|
| 0 | 1.0 | 0.7 | 0.3 | 1.09 | 0 | 0 | |
| 1 | 0.3 | 0.5 | 0.5 | 1.47 | 2.0 | 0.60 | 0.60 |
| 2 | 0.15 | 0.4 | 0.6 | 1.43 | 3.0 | 0.45 | 0.90 |
| 3 | 0.09 | 0.55 | 0.45 | 1.05 | 3.0 | 0.27 | 0.81 |
| 4 | 0.04 | 0.75 | 0.25 | 0.75 | 2.0 | 0.08 | 0.32 |
| 5 | 0.01 | 1.0 | 0.00 | 0.5 | 0.00 | 0.00 | 0.00 |
| $\Sigma$ | | | | | | 1.40 | 2.63 |

of 1 rather than in terms of 1000. These 10 females in the year of introduction (year 0) will give birth to 20 females, which now make up age class 0 in the year 0 (Table 18.2). The total population of female squirrels for year 0 stands at 30 squirrels, 10 of which are age 1 and 20 of which are age 0. Not all of these squirrels will survive into the next year. The survival of these two age groups is obtained by multiplying the number of each by the $p_x$ value. Because the $p_x$ of the females of age 1 is 0.5, we know that 5 individuals ($10 \times 0.5 = 5$) survive to year 1 (age 2). The $p_x$ value of age 0 is 0.3, so only 6 of the 20 in this age class in year 0 survive ($20 \times 0.3 = 6$) to year 1 (age 1). In year 1 we now have 6 one-year-olds and the 5 two-year-olds, with both age classes reproducing. The $m_x$ value of the 6 one-year-olds is 2.0, so they produce 12 offspring. The 5 two-year-olds have an $m_x$ value of 3.0, so they produce 15 offspring. Together the two age classes produce 27 young that now make up age class 0. The total population for year 1 is 38. Survivorship and fecundity are determined in a similar manner for each succeeding year. Survivorship is tabulated year by year diagonally down the table to the right through the years, while new individuals are being adding each year to age class 0.

The steps for determining the number of offspring in year $t$ $(N_t)$ are given by the equation

$$N_{tx} = \sum_{x=1}^{\infty} N_{tx} m_x$$

where $N_{t0}$ is the number in age class 0 at the given year $t$ and $N_{tx}$ is the number of age $x$ at year $t$. For year 0 the calculation is

$$N_0 = (10)(2) = 20$$

and for succeeding years

$$N_1 = (6)(2) + (5)(3) = 27$$
$$N_2 = (8.1)(2) + (3)(3) + (3)(3) = 34.1$$

and so on. Thus the number of offspring added to age 0 each year is obtained by multiplying the number in each age group by the $m_x$ value for that age and summing these values over all ages.

From such a population projection table we can calculate **age distribution,** the proportion of individuals in the various age classes, for any one year by dividing the number in each age group by the total population size for that year. The general equation is

$$C_{tx} = \frac{N_{tx}}{\sum_{y=0}^{\infty} N_{ty}}$$

where $C_{tx}$ is the proportion of age group $x$ at year $t$, $N_{tx}$ is the number in each age group $x$ at year $t$, and $N_{ty}$ is the number in age group $y$ at year $t$.

Comparing the age distribution of the hypothetical squirrel population in year 3 with that of the population in year 7,

**Table 18.2** Population Projection Table, Hypothetical Squirrel Population

| Age | \multicolumn{11}{c}{Year} |
|---|---|---|---|---|---|---|---|---|---|---|---|
| | 0 | 1 | 2 | 3 | 4 | 5 | 6 | 7 | 8 | 9 | 10 |
| 0 | 20 | 27 | 34.1 | 40.71 | 48.21 | 58.37 | 70.31 | 84.8 | 101.86 | 122.88 | 148.06 |
| 1 | 10 | 6 | 8.1 | 10.23 | 12.05 | 14.46 | 17.51 | 21.0 | 25.44 | 30.56 | 36.86 |
| 2 | 0 | 5 | 3.0 | 4.05 | 5.1 | 6.03 | 7.23 | 8.7 | 10.50 | 12.72 | 15.28 |
| 3 | 0 | 0 | 3.0 | 1.8 | 2.43 | 3.06 | 3.62 | 4.4 | 5.22 | 6.30 | 7.63 |
| 4 | 0 | 0 | 0 | 1.35 | 0.81 | 1.09 | 1.38 | 1.6 | 1.94 | 2.35 | 2.83 |
| 5 | 0 | 0 | 0 | 0 | 0.33 | 0.20 | 0.27 | 0.35 | 0.40 | 0.49 | 0.59 |
| Total | 30 | 38 | 48.2 | 58.14 | 68.93 | 83.21 | 100.32 | 120.85 | 145.36 | 175.30 | 211.25 |
| Lambda | $\lambda$ | 1.27 | 1.27 | 1.21 | 1.19 | 1.21 | 1.20 | 1.20 | 1.20 | 1.20 | 1.20 |

we observe that the population attains a stable or unchanging age distribution by year 7 (Table 18.3). From that year on, the proportions of each age group in the population and the rate of growth remain the same year after year, even though the population is steadily increasing.

The construction of the population projection table demonstrates some concepts of population growth. The constant rate of increase of the population from year to year and stable age distribution depend upon survivorship, $(l_x)$ from each age class to the next and upon fecundities of each age class $(m_x.)$ Both factors were used in the development of the population projection table. By dividing the total number of individuals in year $x + 1$ by the total number of individuals in the previous year $x,$ one can arrive at the finite multiplication rate, $\lambda$ (lambda), for each time period.

The method of determining $\lambda$ from changes in the size of a natural population from one year to another is useful in the study and management of natural populations. However, the construction of a population projection table is not always necessary to determine $\lambda$. By obtaining reasonably accurate counts of a population size from one year to another, we can mimic what we did on the population projection table. We can divide the current population size by the population size of the previous year and thus gain some rough idea of the current rate of population increase.

Lambda can be used as a multiplier to predict population size some time in the future:

$$N_t = N_0 \lambda^T$$

For our hypothetical population, we can multiply the population size, 30, at time (year) 0 by $\lambda = 1.20$, the value derived from the population projection table, to obtain a population size of 36 for year 1. If we multiply 36 again by 1.20, or the initial population size 30 by $\lambda^2$ ($1.20^2$), we get a population size of 43 for year 2; and if we multiply the population at $N_0 = 30$, by $\lambda^{10}$ we arrive at a projected population size of 186 for year 10. These population sizes do not correspond to the population sizes early in the population projection table, because in the lower years $\lambda$ is higher and the population has not reached a stable age distribution. Only after the population achieves a stable age distribution does the $\lambda$ value of 1.20 predict future population size. For example, if population size $N$ in year 7 is used as $N_0$, then $142 \times 1.20^3$ predicts a population size of 245 in the year 10. This fact emphasizes that population projections using $\lambda$ assume a stable age distribution.

Generations in the squirrel population are not discrete, as the population projection table makes evident; they overlap. The female squirrels in the initial population at time 0 continue to contribute to population growth, although at a reduced rate, at the same time as their offspring are adding their own reproductive efforts. In Table 18.1, the net repro-

**Table 18.3** Approximation of Stable Age Distribution, Hypothetical Squirrel Population

| Age | \multicolumn{11}{c}{Proportion in Each Age Class for Year} |
|---|---|---|---|---|---|---|---|---|---|---|---|
| | 0 | 1 | 2 | 3 | 4 | 5 | 6 | 7 | 8 | 9 | 10 |
| 0 | .67 | .71 | .71 | .71 | .69 | .70 | .70 | .70 | .70 | .70 | .70 |
| 1 | .33 | .16 | .17 | .17 | .20 | .17 | .17 | .18 | .18 | .18 | .18 |
| 2 | | .13 | .06 | .07 | .06 | .07 | .07 | .07 | .07 | .07 | .07 |
| 3 | | | .06 | .03 | .03 | .04 | .04 | .03 | .03 | .03 | .03 |
| 4 | | | | .02 | .01 | .01 | .01 | .01 | .01 | .01 | .01 |
| 5 | | | | | .01 | .01 | .01 | .01 | .01 | .01 | .01 |

ductive rate, $R_0$, assumes a discrete generation time. If generations are discrete, as they are in annual plants and many insects, then the unit of time, ($t$) and generation time, ($T$) are the same. In populations with overlapping generations the value of $T$ becomes $T_c$, the **mean cohort generation time.** $T_c$ is the mean period of time elapsing between the birth of the parents and the birth of the offspring. Mean cohort generation time is computed from the fecundity schedule by multiplying each $l_x m_x$ by its appropriate age $x$; all values are summed and divided by the sum of the $l_x m_x$'s or $R_0$ to obtain $T_c$:

$$T_c = \sum \frac{x l_x m_x}{R_0}$$

The net reproductive rate $R_0$, provided it is near 1, can be used to estimate the **intrinsic rate of natural increase** ($r$), a measure of the instantaneous rate of change of a population size per individual, by the formula:

$$r \approx \log_e R_0 / T_c$$

For the squirrel population in described in Table 18.1 the cohort generation time is $\Sigma x l_x m_x / R_o = 2.63/1.40 = 1.87$. The intrinsic rate of increase can be approximated by $\log_e 1.40/1.87 = 0.336/1.87 = 0.1797$.

The intrinsic rate of increase (or decrease) can be converted to the finite rate of increase by the formula $\lambda = e^r$, where $e$ is the base of natural logarithms (approximately 2.7183). By determining $e^r$ with an $r$ value of 0.1797 or 0.180 we obtain $\lambda = 1.198$ or 1.20. This value agrees with that obtained from the population projection after stable age distribution was achieved.

The intrinsic rate of increase can be obtained more precisely from the Euler equation:

$$\sum_{\lambda}^{\infty} e^{-rx} l_x m_x = 1$$

which can be solved only by using a trial and error method (iteration, see Appendix B). **Iteration** is accomplished by introducing different values of $r$ and $rx$ above and below the estimated value of $r$ until the solution approximates 1. The iterated value of $r$ for the squirrel population is 0.186, which gives $\lambda$ a value of 1.203.

Because $r$ is derived from $R$, which involves schedules of survivorship ($l_x$) and fecundity ($m_x$), $r$ includes both births and deaths. In a closed population, one in which no individual enters or leaves, the intrinsic rate of increase is the instantaneous birth rate minus the instantaneous death rate, or $r = b - d$. In an open population $r$ also must include immigration ($i$) and emigration ($e$) or $r = (b + i) - (d + e)$.

Deriving any sort of value for $r$ for natural populations can be very difficult. Yet in the management of wild populations, whether game, endangered, or pest species, some estimate of $r$ is needed.

Knowing the value of $r$ for a population allows the calculation of **doubling time** of population growth. The doubling time of a population at some time $t$ results when $N_t/N_O = 2$. Therefore $N_O e^{rt}/N_t = 2$. Thus $e^{rt} = 2$, and $rt = \log_e 2 = 0.693$; then $t = .693/r$. The doubling time of the hypothetical squirrel population is $0.693/0.186 = 3.72$ years.

However useful $r$ may be, the parameter has certain limitations. It refers to that particular period of time during which the data for the determination of $r$ were obtained. It is free from the effects of the population's own density and its own age structure. The rate of increase is influenced by life history events, such as frequency of reproduction and the number of young born in each reproductive period. Local environmental conditions also affect mortality and the production of young. In effect, values of $r$ relate to specific populations under specific conditions.

## POPULATION GROWTH

### Exponential Growth

If a population were suddenly presented with an unlimited environment, as can happen when a small number of bacteria, nonnative plants, or animals are introduced into a suitable but unoccupied habitat, it would tend to expand geometrically. Assuming there were no movement in or out of the population and no mortality, then birthrate alone would account for changes in population numbers. Under this condition population growth would simulate compound interest, a continual increase called **exponential growth.** But growth of populations is tempered by death, so a death rate must be factored in with the birthrate.

Exponential growth is described by the differential equation

$$dN/dt = (b - d)N$$

or

$$dN/dt = rN$$

This differential equation states that the rate of increase $dN/dt$ is directly proportional to population size and that the growth rate $r$ (the difference between instantaneous birthrates and instantaneous death rates) expresses growth or decline per individual or per capita. The time intervals between geometric growth increments are reduced to zero (Figure 18.1), so theoretically change in population numbers occurs instantaneously.

A more useful equation for calculating exponential growth is the integrated form

$$N_t = N_o e^{rt}$$

where $e$ is the base of natural logarithms 2.71828 . . . , $r$ is the rate of increase, and $t$ is the unit of time. This exponential equation is equivalent to the equation for geometric growth:

$$N_t = N_o \lambda^t$$

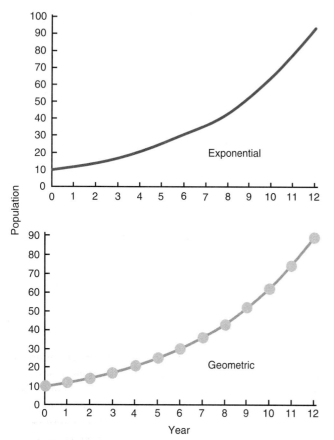

**Figure 18.1** Geometric growth curve plotted for λ = 1.20 and exponential growth curve plotted for $r$ = 0.186. The curves are similar except that the geometric curve has discrete points.

In this equation λ takes the place of $e^r$. Provided that the geometric growth rate and the exponential growth rate are equivalents, the two equations will produce the same growth curves, except that the geometric curve will have discrete points (Figure 18.1). If the populations are increasing, $r$ will be greater than 0, and λ will be greater than 1. If the populations remain the same, $r$ = 0 and λ = 1. If the population has a negative growth rate, $r$ will have negative exponential growth rate, and λ will be between 0 and 1.

Compare the exponential growth of the hypothetical squirrel population with its geometric growth rate:

| | Population Growth | |
|---|---|---|
| Year | Exponential ($r$ = 0.186) | Geometric (λ = 1.20) |
| 0 | 10 | 10 |
| 1 | 12 | 12 |
| 3 | 17 | 17 |
| 5 | 25 | 25 |
| 7 | 37 | 36 |
| 9 | 53 | 52 |
| 12 | 93 | 89 |

The population's growth described is for specific values of $r$, λ, and initial population size. The exponential growth of any population depends upon and varies with the value of $r$ (Figure 18.2).

The rate of growth at first is influenced by heredity or life history features, such as the age at beginning of reproduction, the number of litters produced during the lifetime of each female, the number of young produced, survival of young, and length of the reproductive period. Regardless of the initial age of the colonizers, the number of animals in the prereproductive age class would increase because of births, whereas those in the older age categories for a time would remain the same. As the young mature, more would enter the reproductive stage and more young would be produced. A J-shaped growth curve is characteristic of many organisms introduced into a new and unfilled environment. An example of exponential growth is the rise of a reindeer herd on St. Paul, one of the Pribilof Islands in Alaska (Figure 18.3). Introduced in 1910, reindeer expanded from 4 males and 22 females to a herd of 2000 in only 30 years. Exceeding the ability of the range to support it, the population crashed in an exponential fashion. Such curves are typical of species that grow rapidly, exceed available resources, and crash in a boom-and-bust cycle.

## Logistic Growth

Exponential growth is not biologically realistic: no population can grow indefinitely. The environment is not constant, and resources such as food and space are limited. As population density increases, competition for available resources among its members also increases. Eventually the detrimental effects of increased density—increased mortality from disease, starvation, and predation, decreased fecundity, or both, and emigration—begin to slow population growth until it ceases. This level, at which the population theoretically is in equilibrium with its environment, is called *carrying capacity*. It is expressed in the logistic equation as $K$.

This concept of the decline of the exponential growth rate as the size of the population increases was described by the mathematician Pierre-Francois Verhulst in 1838. He called this equation logistic because of its logarithmic-exponential form. His paper, however, was buried in a seldom-read journal. In 1920 Raymond Pearl and I.J. Reed of the Institute of Biological Research, Johns Hopkins University published a nearly identical version of the equation in a paper on the growth of the population of the United States since 1790. They predicted a regular pattern of decline in exponential growth as the population size increased. They modified the exponential equation $dN/dt = rN$ by adding a variable to describe the effect of density:

$$\frac{dN}{dt} = rN\left(\frac{K - N}{K}\right)$$

where $dN/dt$ is the instantaneous rate of change, $K$ is carrying capacity, and $(K - N)/K$ or $(1 - N/K)$ is the unutilized

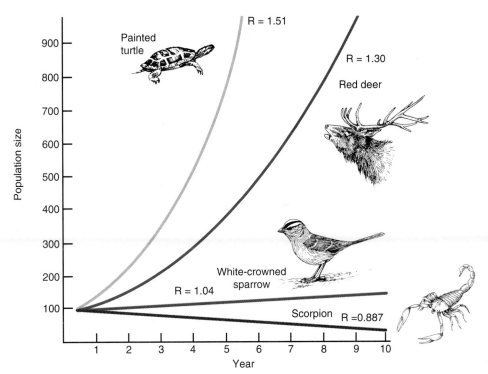

**Figure 18.2** Exponential growth curves for four different values of R as represented by data for the painted turtle (*Chrysemys picta*), red deer (*Cervus elaphus*), white-crowned sparrow (*Zonotrichia leucophrys*), and scorpion (*Paruroctonus mesaensis*). The initial population for each is 100. Note that the scorpion has an R value less than 1.0; its population is declining. (Data from Tinkle, Condon, and Rosen 1981 for the painted turtle; Lowe 1969 for the red deer; Baker, Mewaldt, and Stewart 1981 for the white-crowned sparrow; and Polis and Farley 1980 for the scorpion.)

opportunity for population growth. The equation says that the rate of increase of a population over a unit of time is equal to the potential increase of the population times the unutilized portion of the resources. When $N$ is low, $K − N/K \approx 0$ (thus most of the resources are unutilized). When $N \approx K − N/K \approx 0$ (thus most of the resources are utilized). If $N > K$, then $dN/dt$ is negative and $N$ declines toward $K$.

Let us use our hypothetical squirrel population as an example of logistic growth. The population will have a starting size of 10 with a rate of increase of 0.186. We will assume the squirrels inhabit an imaginary oak woodlot with a carrying capacity of 50.

| Year | Growth |
|------|--------|
| 0    | 10     |
| 1    | 11     |
| 5    | 19     |
| 8    | 26     |
| 10   | 30     |
| 15   | 40     |
| 25   | 48     |
| 30   | 49.2   |
| 40   | 49.8   |
| 50   | 49.98  |
| 70   | 49.99  |

Note how the rate of increase is slow at first, accelerates, and then slows. The point in the logistic growth curve where population growth is maximal is $K/2$ (in this example, 25),

**Figure 18.3** Exponential growth of the St. Paul reindeer herd is illustrated by the left side of the graph. When the herd outstripped its resources, the population crashed. (From V. C. Scheffer 1951.)

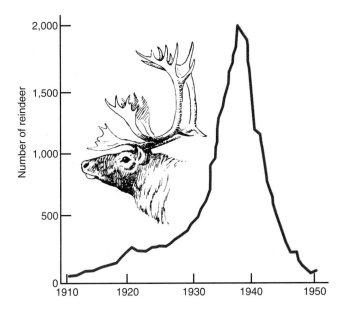

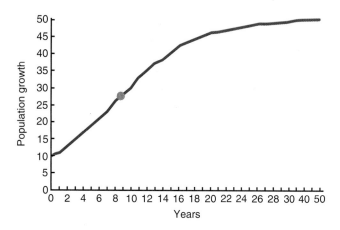

**Figure 18.4** The logistic growth curve for the hypothetical squirrel population with an *r* value of 0.186. The curve is sigmoidal. Its upper limit is termed the carrying capacity, *K*. The inflection point at which density begins to slow population growth is located at *K*/2. In this example the inflection point stands at population size 27.

known as the **inflection point.** From this point on, population growth slows (Figure 18.4). As population density approaches carrying capacity, *N* approaches *K*, and the rate of growth declines. Thus $dN/dt$ is low when $N \approx 0$ and $N \approx K$, and highest when $N = K/2$.

The logistic equation involves several assumptions. Age distribution is stable, initially at least. No immigration or emigration takes place. Increasing density depresses the rate of growth instantaneously without any time lags in reproductive responses. The relationship between population size and the rate of growth is linear. When plotted logarithmetically, the rate of growth declines directly as population increases. We also have to assume a predetermined level for *K*.

We can view the logistic equation in two ways. One is to consider it a law of population growth. That was the original proposition elaborated by Pearl (1927) (see Kingsland 1985). A second, preferable view is to regard the logistic equation as a mathematical description of how populations might grow under most favorable conditions and as a yardstick against which to measure actual population growth. Although some natural populations appear to grow logistically, rarely, if ever, does their growth curve match the predicted logistic curve, because the assumptions are too simplistic.

For example, consider the growth of the human population in Monroe County, West Virginia, whose economic base has been agriculture and small industry. It was settled by Europeans in the early 1700s, and the population was well established in 1800, the year for which U.S. Census data are first available. The population reached 13,200 in 1900 and has fluctuated about that number since then. The population grew most rapidly from 1800 to 1850, so growth

during that period provide the data to estimate *r,* the rate of increase, as 0.074. *K,* based on the population levels since the year 1900, is 13,200. With these values of *r* and *K* the logistic equation predicts that the population would reach the asymptote around 1870, 30 years before it actually did. The predicted growth curve rises much more steeply than the actual growth curve (Figure 18.5). The human growth curve approaches but does not conform to the logistic curve. The reasons for nonconformity are obvious. The age structure was not stable, birth rates and death rates varied from census period to census period; immigration and emigration were common to the population. The most surprising feature of the population is its relative stability after reaching *K*.

## Time Lags

The logistic equation suggests that populations function as systems, regulated by positive and negative feedback. Growth is stimulated by positive feedback (as illustrated by the exponential growth curve), then slowed by the negative feedback of competition and dwindling. As *N* approaches *K*, the population theoretically responds instantaneously as density-dependent reactions set in.

Rarely does such feedback work as smoothly in real life as the equation suggests. Often adjustments lag, and available resources may allow the population to overshoot equilibrium. Unable to sustain itself on the available resources, the population declines to some point below carrying capacity, but not before it has altered resource availability for future generations. Its recovery as determined by reproductive rates is influenced by the density of the previous generation and the recovery of the resources, especially food supply. These factors build a time lag into population recovery.

To make the logistic equation more realistic, we need to factor into the equation a **reaction time lag** (*W*), a lag

**Figure 18.5** Actual and logistically predicted population growth of the white European population in Monroe County, West Virginia. (Data from U. S. Census Bureau.)

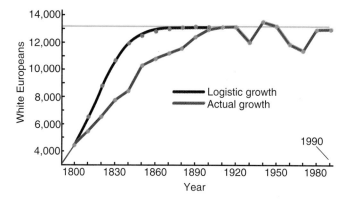

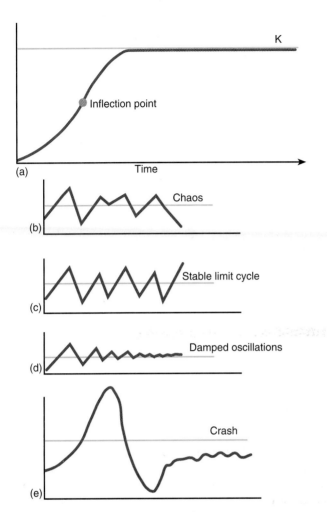

**Figure 18.6** (a) A stylized logistic growth curve and (b–e) types of fluctuations that take place when the population overshoots *K*. (b) Chaotic fluctuation. The population fluctuates wildly with no regulation. Such fluctuations can lead to sudden extinction. (c) Stable-limit cycles. The population fluctuates about some equilibrium level, with fluctuations having a certain period and amplitude. (d) Oscillations decrease over time. After overshooting *K*, the population levels off and maintains itself at *K* through compensating birth and death rates. (e) The population strongly overshoots *K* and then crashes. It may drop very low, recover, and return to some lower equilibrium level, or it may go extinct.

between environmental change and corresponding change in the rate of population growth:

$$\frac{dN}{dt} = rN\left(\frac{K - N_{t-w}}{K}\right)$$

Another factor is a **reproductive time lag** (*g*), a lag between environmental change and change in the length of gestation or its equivalent:

$$\frac{dN}{dt} = rN_{t-g}\left(\frac{K - N_{t-w}}{K}\right)$$

Means of incorporating time lags into the logistic equation are detailed by Krebs (1985) and Berryman(1981).

Time lags result in fluctuations in populations (Figure 18.6). The population may fluctuate widely without any reference to equilibrium size. Such populations may be influenced by some powerful outside, or extrinsic, force such as weather or by some chaotic changes inherent in the population. (For an introduction to chaos in populations, in which simple processes can lead to complex behaviors, see R. May 1976, Schaffer 1985, Schaffer and Kot 1985, Yodzis 1989.) A population may fluctuate about the equilibrium level, *K,* rising and falling between some upper and lower limits. Such fluctuations are called **stable limit cycles.**

# DENSITY-DEPENDENT POPULATION REGULATION

Sooner or later all populations have to confront the carrying capacity of the environment in which they live. Although explicitly defined in the logistic equation, a constant equilibrium level rarely exists. Because the environment is variable, *K* also is variable. Carrying capacity is the level at which available resources can sustain individuals in a population at survival level. The availability of resources, however, varies from season to season and year to year. While one resource may be adequate or abundant, another resource may be lacking. Thus carrying capacity is influenced by the most limiting resource. If that resource increases, then the carrying capacity increases. Further, populations over time are limited by disease, predation, unfavorable weather, and habitat quality. For this reason the population level at any one time may not reflect the potential carrying capacity of the environment. The best estimate we can obtain of *K* is to average the population size over time, which provides a mean population size. Populations tend toward this equilibrium through density-dependent regulation.

Such population regulation increases or decreases mortality and reproduction. Both tend to bring the population back to the equilibrium set by limiting factors. As a population increases, competition among its members and a scarcity of resources result in increased mortality, decreased natality, or both for a species so regulated. If a population drops to some lower level and resources become more abundant, it increases through some combination of decreasing mortality and increasing natality. This relationship is illustrated in Figure 18.7. For population regulation the birthrate, death rate, or both must be density-dependent.

Through most of the sigmoid growth curve, both positive and negative feedback operate, with a change in the relative importance of each. In the early stages of population growth, positive feedback dominates. As population reaches *K*/2,

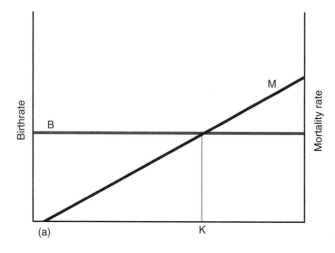

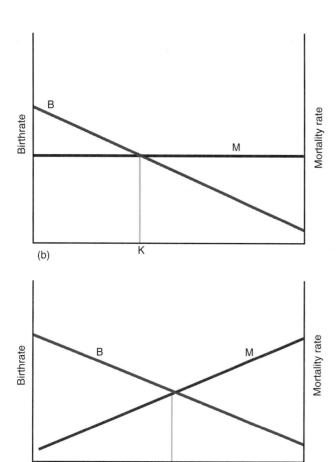

**Figure 18.7** Population regulation requires density-dependent birthrates, mortality rates, or both. In (a) the birthrate is independent of population density but mortality increases with density. In (b) the mortality rate is independent of population density, but the birthrate declines as density increases until the population reaches *K*. In (c) both birthrate and death rate are density-dependent, and the population reaches equilibrium when the birthrate equals the death rate.

negative feedback overcomes positive feedback. How rapidly these two responses function relates to the population's impact on resources and future growth. If individuals remove a resource faster than it is replaced, then the present population impoverishes the environment for the next generation, slowing population growth.

Density-dependent effects influence a population in proportion to its size. At some low density there is no interaction, and mortality and natality are independent of population size. As population density increases, density-dependent effects set in. Mortality increases, but the number dying is less than the number surviving. Density-dependent effects undercompensate for increased density, so the population still increases. Eventually there comes a point where mortality exceeds survival. Now density-dependent mechanisms overcompensate for increased density, and the population declines (see Bellows 1981 for mathematical models).

## Intraspecific Competition

Density-dependent regulation acts through intraspecific competition when a needed resource such as food, nesting sites, and mates is short relative to the number of individuals of the same species seeking it. As long as resources are abundant enough to allow each individual a sufficient amount for survival and reproduction, no competition exists. When resources are insufficient to satisfy the needs of all individuals, the means by which they are allocated has a marked influence on the welfare of the population.

Nicholson (1954) demonstrated how intraspecific competition for a limited resource, such as food, affects population numbers. Although his long-term experiments on sheep blowflies (*Lucilia cuprina*) lacked all the complex interactions we would expect to find in nature, his work does show what might happen and established some basic concepts in the study of competition.

In one of his many experiments Nicholson fed to a culture of blowflies containing both adults and larvae a daily quantity of beef liver for the larvae and an ample supply of dry sugar and water for the adults. The number of adults in the experimental cages varied, with pronounced oscillations. When the population of adults was high, the flies laid such a great number of eggs that the resulting larvae ate all the food before they were large enough to pupate. As a result, no adult offspring came from the eggs laid during that period. The number of adults progressively declined through natural mortality, and fewer eggs were laid. Eventually the intensity of larval competition was so reduced that a fraction of the larvae obtained sufficient food to grow to a size large enough to pupate. These larvae in turn gave rise to egg-laying adults. Because of the developmental time lag between the survival of larvae and an increase in egg-laying adults, the population

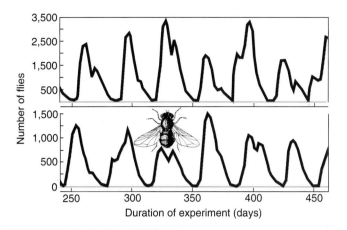

**Figure 18.8** Fluctuations in the number of adult blowflies in two cultures subjected to the same constant conditions, but restricted to different daily quotas of food: (a) 50 gm; (b) 25 gm. Although a greater food supply permitted a greater increase in density, the final outcome in both situations was similar. The adults, experiencing scramble competition, increased rapidly then declined sharply as they reached the limits that the food could support. (From Nicholson 1957.)

continued to decline, further reducing the intensity of larval competition and permitting an increasing number of larvae to survive. Eventually the adult population again rose to a very high level and the whole process started again.

Competition for limited food held this blowfly population in a stage of stability and prevented a continuing increase and decrease. The time lag involved in the addition of egg-laying adults to the declining population resulted in an alternate overshooting and undershooting of the equilibrium position, causing an oscillating population (Figure 18.8).

In this and the other experimental competitive situations the larvae and adults were seeking food, the rate of supply of which was not influenced by the activity of the flies. In effect the resource, the available food, was subdivided into many small parts to which the competitors, the larvae and adult flies, had general access. The individuals "scrambled" for their food. Gross crowding resulted in wastage because each competitor got such a small fraction of the food that it was unable to survive.

Nicholson called this type of competition **scramble.** In its purest form, all competing individuals garner such a small share of the resources that none survive. Among some populations outcomes are less severe; competition is scramble-like rather than pure scramble. Such competition, in which each individual is affected by the amount of shared resource remaining, can be called **exploitative competition.** Competing individuals do not necessarily react to each other, only to the level of resources. Scramble competition tends to produce sharp fluctuations in a population over time. It limits the average density of the population below that which the

resources could support if an adequate amount of resources were obtained by only a part of the population.

That is exactly what takes place with **contest competition.** The deleterious effects of limited resources are confined to a fraction of the population, and members of the population interact directly. For that reason contest competition can also be called **interference competition.** Once a population characterized by contest competition passes the point at which resources become limiting, a fraction of the individuals obtains all the resources it needs. The remaining individuals get less and produce no offspring or die.

Intraspecific competition influences births, deaths, and growth of individuals in a density-dependent manner. Its effects come slowly, involving at first the general welfare rather than survival of individuals. Later, as its impacts become accentuated, intraspecific competition affects individual fitness.

## Growth and Fecundity

When a population reaches a point at which the resources, particularly food, are insufficient to meet the needs of individuals something has to give. In populations characterized by scramble or exploitative competition, individuals respond to lowered level of food by reducing growth. Examples of this inverse relationship between density and rate of body growth are found among poikilothermic vertebrates. Dash and Hota (1980) discovered that frog larvae reared experimentally at high densities grew slower, decreasing their chances of successfully transforming from tadpoles to frogs, and took longer to reach the minimum threshold size for metamorphosis to frogs, 0.75 g. Tadpoles held at lower densities grew more rapidly and larger and transformed at an average size of 0.889 g (Figure 18.9). In

**Figure 18.9** Influence of density on the growth rate of the tadpole *Rana tigrina.* Note how rapidly the growth rate declines as as density increases from 5 to 160 individuals confined in the same space. (After Dash and Hota 1980:1027.)

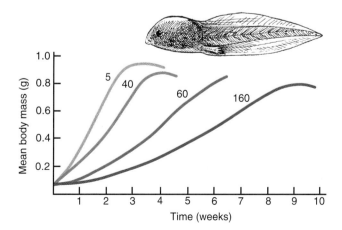

this situation intraspecific competition had minimal influence on population size, and total biomass remained approximately the same, but it was apportioned among many small individuals at high densities.

Relationships among density, growth, and fecundity extend to other vertebrate groups. Harp seals (*Phoca groenlandica*) become sexually mature when they reach 87 percent of their mature body weight of about 120 kg. At low population densities young animals  ̇ ʹain this weight at a much faster rate than when population densities are high (Figure 18.10). Fertility in harp seals, as measured by the number of females giving birth to young, is density dependent, too (Figure 18.11). Some birds exhibit a similar density-dependent relationship to fecundity. Roseberry and Klimstra (1984) found that the higher the breeding population of bobwhite quail relative to carrying capacity on their southern Illinois study area, the lower was the rate of summer population gain (Figure 18.12.)

These examples suggest that vertebrate populations do respond to increasing numbers in a density-dependent fashion through intraspecific competition. The timing of the response depends upon the nature of the population. Fowler (1981) has hypothesized that among large mammals with a long life span and low reproduction, regulating mechanisms do not function until the population approaches carrying capacity. In other words, mortality and natality are more or less in balance, births compensating for deaths, until the

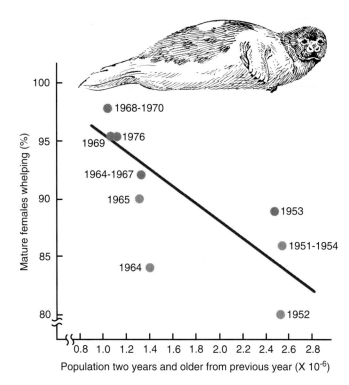

**Figure 18.11** Fertility, too, is density-dependent in harp seals. As the population of seals (measured by including only animals 2 years and older from the previous year) increases, the percentage of females giving birth to young decreases markedly. (From Lett et al. 1981:146.)

**Figure 18.10** The mean age of maturity of harp seals (and other marine and terrestrial mammals) is related not so much to age as to weight. Seals arrive at sexual maturity when they reach 87 percent of average adult body weight. Seals attain this weight at an earlier age when population density is low. (From Lett et al. 1981:144.)

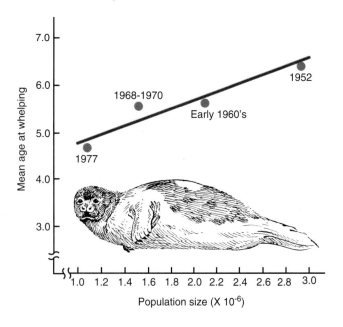

**Figure 18.12** The bobwhite quail population shows a curvilinear density-dependent response to natality. Natality declines at a slower rate in early stages of population growth and continues the decline much more rapidly as density increases. The breeding population of bobwhite quail on the Southern Illinois Carbondale Research area, 1954–1979, was adjusted to account for changes in land use and thus carrying capacity of the area. (From Roseberry and Klimstra 1984:96.)

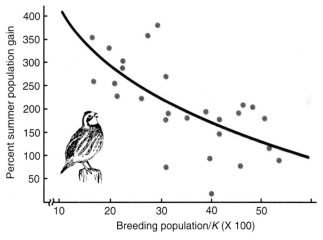

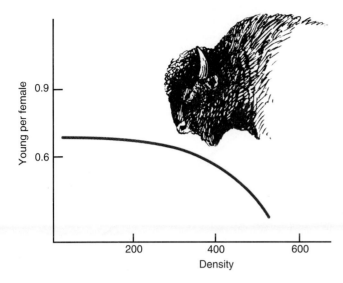

**Figure 18.13** Nonlinear density-dependent change in a large mammal population, the American bison. The birth rate of the bison, expressed as young per female, is independent of density until the population reaches a certain size. Then birth rates overcompensate for increased density. (From Fowler 1981:607.)

population is close to $K$, at which point density-dependent mechanisms set in and tend to overcompensate. The birthrate of bison (Figure 18.13) shows such a response, but birthrates of other large mammals, such as elk (Figure 18.14), appear to be linear.

## Plant Biomass

If you walk into two pine plantations of the same age but at contrasting densities, one thinned artificially and the other

**Figure 18.14** Linear relationship between calf recruitment and population size in Yellowstone elk (From Houston 1982:45).

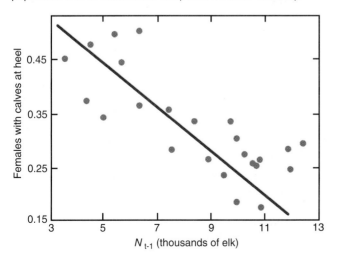

unthinned, you will be aware of differences. The trees of the artificially thinned stand are fewer, larger in diameter, and taller. Those in the unthinned stand are fewer, and smaller in diameter than those in the thinned stand. If you were to cut and weigh trees from equal-sized plots from both stands, you would find that the total biomass was similar, but mean weight of the plants was greatest on the thinned plots with the lower density. These observations would suggest that a close relationship exists between plant density and the growth of individual plants as measured by biomass accumulation.

Suppose a group of seedlings occupies the same habitat but starts out at different densities. The pattern of growth among all will be similar. At first all individuals will grow at about the same rate, accumulating biomass in a density-independent fashion. In time, all populations, regardless of initial density, will reach a point at which the canopy closes. Because only so much canopy area is available, individuals begin to compete for light and space. Growth of individual plants slows. However, some individuals gain more of the available resources than others and grow faster than other plants of equal size. Eventually some individuals lack the ability to compete and die. The result is the gradual thinning of a stand and developing of a hierarchy of size with few large and many small individuals. This progressive decline in density in a population of growing individuals is known as **self-thinning.**

As density declines, both the mean weight of individual plants and the total yield (as measured by biomass) increase. The total biomass per unit area will increase only as the number of individuals declines. How soon plant populations experience competition and depressing effects on the increase in the mean weight of plants depends upon the initial density. If the density of seedlings is high, that population will experience the effects of competition, such as reduction in growth and increased mortality, much sooner than less dense populations. Regardless of their starting densities, populations eventually will converge upon a common density that will decrease through time. They will reach a point at which growth exactly compensates losses from mortality and a reduction in mean plant weight exactly compensates for increased density. The final yield from all starting densities as measured by plant biomass becomes constant. This relationship is known as the **law of constant final yield.**

This relationship between mean plant weight and density was demonstrated by Yoda et al. in 1963 in experimental monocultures of horseweed (*Erigeron canadensis*) with a fixed sowing density and wide range of fertilities. Prompted by the work of Yoda, others have carried out similar experiments, including Lonsdale and Watkinson (1982). They sowed ryegrass (*Lolium perenne*) at densities of 1000, 5000, 10,000, 50,000, and 100,000 individuals at two different light intensities, high and low (17 percent of natural light). They harvested samples of the population at 14, 35, 76,

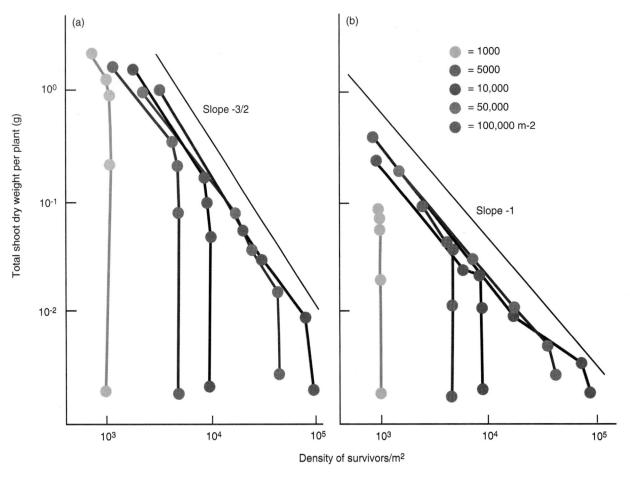

**Figure 18.15** Self-thinning slopes for perennial ryegrass (*Lolium perenne*) sown at five different densities under two light conditions: (a) 0 percent shade; (b) 83 percent shade. The populations were harvested on five successive occasions, indicated by the vertical lines. Note that the denser populations reached and followed the −3/2 slope of self-thinning much sooner than the less dense populations. The last to reach it was the population with the thinnest density, but regardless of starting density all lines converged at a similar final density. (From Lonsdale and Watkinson 1982: 433.)

104, and 146 days. Plant weight at first increased independent of density, but at some point plant weight increased as density decreased because of increasing mortality (Figure 18.15). The populations with highest density experienced increased mortality much sooner than the low-density populations. In time all populations, regardless of initial density, reached the same position where density declined as plant weight increased. All populations moved along the same straight line described by the logarithm of average plant weight plotted against the logarithm of density of survivors, the slope of which is approximately −3/2. Plant ecologists call this relationship the **−3/2 power law** of self-thinning (Yoda et al. 1963).

The interactions observed are described by the equation $w = cd^{-a}$, where $w$ is mean plant weight, $d$ is plant density, and $a$ and $c$ are constants. The value $c$ has the dimensions of $g/m^2$, and $a$ changes with time. In the absence of competi-

tion, the value of $a$ is 0; under intense competition $a$ is 1. At $a = 1$ growth of survivors balances the deaths of other individuals, total weight per unit area remains constant, and biomass accumulation reaches the maximum for a particular species under given environmental conditions. This is what happened with the ryegrass grown under low light conditions, the slope of which line was −1. In this case $w- = cd^{-1}$, in contrast to $w = cd^{-3/2}$. The position of the −3/2 line can be altered by light condition, but usually not by nutrients. If the nutrient supply is increased, the line is reached more quickly and populations move along it faster because of an increased growth rate.

The −3/2 power law describes the reciprocal changes in mean plant biomass $w$ and the number of plants per unit area $d$, during the development of a monospecific even-aged stand with complete crown closure. It is based on the geometry of space occupied by a plant. The law assumes that all

plants of any given species grow to a similar shape, regardless of size and growing conditions, and that the combined action of crown growth and self-thinning maintains a complete crown closure. The average ground area or space (basal area) occupied by a plant will be proportional to the square of the linear dimension of the plant. Its weight will be proportional to the cubic volume of space that the plant occupies. The canopy volume relates to a cube of some linear dimension, particularly height; thus the −3/2 relationship. Mortality in the population occurs when the percentage of canopy cover relative to ground area approaches 100 percent, blocking light to plants below.

Because data for a number of plants from herbaceous to woody seem to obey it, the −3/2 power law has been accepted as an ecological law (D. E. Weller 1987a, 1987b, 1989). However, not all plants necessarily conform to the law, woody plants in particular. Plant shapes are variable, even within a species. Because thinning exponents vary according to plant geometry, they are not always near the idealized value of −3/2. Deviations in woody plants are related to changes in plant height, average mass, average bole diameter at breast height, and average bole basal area. In fact for trees the use of diameter at breast height, which is correlated with crown diameter and tree number (for examples see Lampson 1987) rather than tree mass, weakens its application as an ecological law.

# DENSITY-INDEPENDENT INFLUENCES

During the late 1950s and early 1960s not all ecologists accepted the idea that populations were regulated by density-dependent influences. Many argued that density-independent mechanisms were more important. (For density-dependent viewpoints, see Nicholson 1954, 1956; M. E. Solomon 1957; Lack 1966; and Hairston, Smith and Slobodkin 1960. For density-independent viewpoints see Andrewartha and Birch 1954; Milne 1957. For a historical review see Krebs 1985.) The arguments were largely semantic, stemming from different approaches, different philosophies, and experiences with different taxons, all infused with a dose of advocacy. The density-independent school, for example, was dominated by insect population ecologists interested in proximate causes of population fluctuations and densities. The density-dependent side was dominated by vertebrate ecologists more interested in natural selection and in evolutionary problems. Most ecologists now agree that the numbers of organisms are determined by an interaction between density-dependent and density-independent influences, which may vary among and within populations.

By themselves, density-independent influences do not regulate population growth. Regulation implies a homeo-static feedback that functions with density. However, they can have considerable impact on population size and they can affect birthrates and death rates. Density-independent influences may so affect a population that they completely mask any effects of density-dependent regulation. A cold spring may kill the flowers of oaks, causing a failure of the acorn crop. Because of the failure squirrels may experience widespread starvation the following winter. Although the proximate cause of starvation is the density of squirrels and the meager food supply, weather is the ultimate cause. In general, population fluctuations influenced by annual and seasonal changes in the environment tend to be irregular and correlated with variations in temperature and moisture. Conditions beyond the organisms' limits of tolerance can have a disastrous impact, affecting growth, maturation, reproduction, survival, movements, and dispersal of individuals within a population and even eliminating local populations.

In general the influence of weather is stochastic: it is irregular and unpredictable and it functions largely by influencing the availability of food. Pronounced changes in population growth often can be correlated directly with variations in moisture and temperature. For example, outbreaks of spruce budworm (*Choristoneura fumiferana*) are usually preceded by five or six years of anticyclonic weather characterized by low rainfall and high evaporation and end when wet weather returns. Such density-independent effects can take place on a local scale where topography and microclimatic conditions influence the fortunes of local populations.

Consider the San Francisco Bay checkerspot butterfly (*Euphydryas editha bayensis*). Adults emerge from pupae from mid-March to early May and lay their eggs at the base of the plantain (*Plantago erectus*). Within two weeks the eggs hatch, and for three weeks the larvae feed on the plantain and owl's clover (*Orthocarpus densiflorus*), when they reach the third instar. Now the dry season sets in and the larvae go into diapause until December. Diapause is broken by the winter rains, which also stimulate the germination and growth of the food plants. The larvae go through three more instars before they pupate and emerge as adults. During a two-year period, 1983 and 1984, record rainfalls and cool weather retarded the development of the larvae, delaying the flights of the adults and the deposition of eggs. Meanwhile the food plants *Plantago* and *Orthocarpus* were developing normally. By the time the eggs hatched the host plants were already senescent or dying. Lacking food, the prediapause larvae faced heavy mortality from starvation. Normally the south-facing slope experiences earlier plant growth and faster development of larvae because of the warmer microclimate, while the north-facing slope experiences about a two-week delay in growth of food plants. Because of the delay in their development, the larvae faced senescent plants on the south-facing slope, while the host plants on the north-

facing slope were not available for another week. If they had been available, the plants would have died before the larvae reached the third instar. The following year only 18 adults were captured (Dobkin et al. 1987). In such a manner microclimatic changes can lead to sharp declines or extinctions in local populations.

Deer in the northern part of their range are sensitive to severe winters. In Minnesota Mech et al. (1987) found a significant relationship between snow accumulation over the previous three years and viability of offspring, as indicated by fawn-doe ratios and percent change in the deer population (Figure 18.16). The data show a winter-to-winter carryover. Snow limits food, affecting prenatal nutrition over winter and in utero development in two-year old deer. When the sum of a three-year average of snow accumulation exceeds 10.2 m and the average single year accumulation exceeds 340 cm, one can expect fewer fawns and a decline in deer population compared to winters with less snow.

In desert regions a direct relationship exists between precipitation and rate of increase in certain rodents and birds (Figure 18.17). Merriam's kangaroo rat (*Dipodomys merriami*) occupies lower elevations of the Mojave Desert. The kangaroo rat has the physiological capacity to conserve water and survive long periods of aridity. However, it does require in its environment a level of moisture sufficient to stimulate the growth of herbaceous desert plants in fall and winter. The kangaroo rat becomes reproductively active in January and February when plant growth, stimulated by fall rains, is green and succulent. Herbaceous plants provide a source of water, vitamins, and food for pregnant and lactating females. If rainfall is scanty, annual forbs fail to develop and the production of kangaroo rats is low (Beatley 1969, Bradley and Mauer 1971). This close relationship to seasonal rainfall and relative success of winter annuals is also apparent in other rodents occupying similar desert habitats and in Gambel's quail and scaled quail (Francis 1970).

## POPULATION FLUCTUATIONS AND CYCLES

As you may have observed, the abundance of some species of insects, birds, and mammals seems to remain the about the same from one year to the next. Other species may be noticeably abundant some years and noticeably scarce in others. Why the difference?

Population fluctuations are mostly local phenomena. A species may show stability over the whole of its range, but be highly variable locally. Populations fluctuate because time lags in density-dependent mechanisms, particularly birthrates and death rates, tend to either undercompensate or overcompensate for population size. In addition, populations are affected by changes in carrying capacity and other

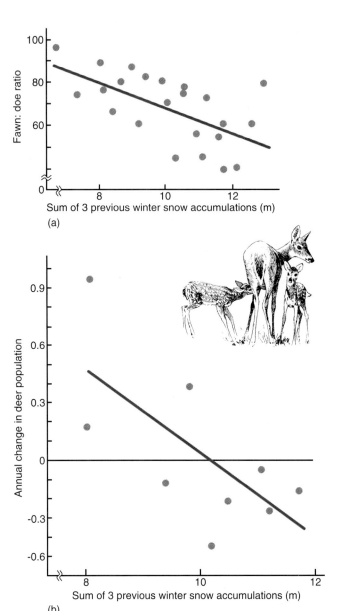

**Figure 18.16** The relationship between the sum of the previous three winter monthly snow accumulations in northeastern Minnesota and the population of white-tailed deer. (a) Fecundity (fawn:doe ratio). (b) Percent annual change in next winter's population. (From Mech et al. 1987:619, 622.)

extrinsic influences, especially weather, predation, and competition (Figure 18.18).

The pattern of the fluctuations, based on census data, is colored by timing, whether the populations are censused in fall, winter, or spring. Population trends in winter may reflect neither the true carrying capacity of the breeding habitat nor the breeding population. The carrying capacity of the environment during the most critical time of the year can influence population size during the breeding season. Fall populations of some species, reflecting the success of

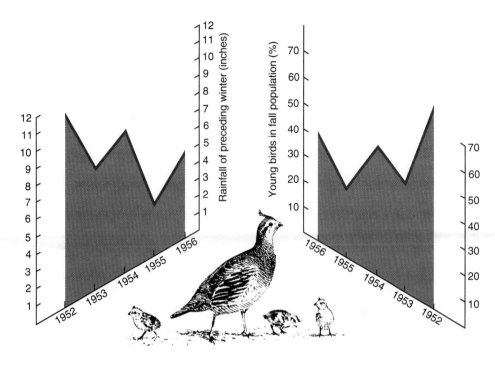

**Figure 18.17** Relationship of winter rainfall to the percentage of young the following year in a Gambel's quail population in southern Arizona. Note how the production of young follows rainfall. (After Sowls 1960:187.)

recruitment, may be larger or more variable from year to year than spring populations (Figure 18.19).

The nature of the fluctuation reflects the population's **resilience.** Resilience is the rate at which a population returns to equilibrium after a disturbance takes it away from equilibrium (Pimm 1991). In other words, resilience is a measure of how fast the population declines from above and how quickly it increases from below equilibrium. The resilience of a population is strongly influenced, if not determined, by its reproductive rate.

Size provides a clue. Small-bodied animals, such as meadow mice, fluctuate more widely than large-bodied ones, such as deer. Small animals have shorter lives, die more quickly, and thus decrease more dramatically from year to year. However, they reproduce fast, and can recover from their losses quickly. Such species have a high resilience. Large-bodied animals possess more stability about an equilibrium level because they live longer and are less subject to environmental vagaries; but they reproduce slowly and may require a very long period of time to return to equilibrium. Consider, for example, the slow population growth of whales. Such species possess low resilience. The measure of resilience, then, is the time required for the pop-

**Figure 18.18** Fluctuation in a wintering population of black-capped chickadees in northwestern Connecticut about a mean long-term density of 160 birds. The short-term decline in the population in 1968-1969 was attributed to an influx of competitive tufted titmice. In spite of fluctuations the population exhibited no sustained increase or decrease. (Data from Loery and Nichols 1985.)

**Figure 18.19** Trends of spring and fall populations in the bobwhite quail in Wisconsin. Note the seasonal fluctuations of population density. Fall densities fluctuate more than spring densities. (After Kabat and Thompson 1963:78.)

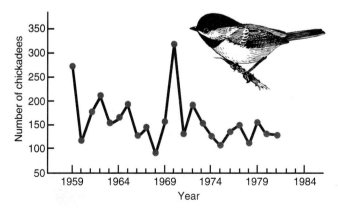

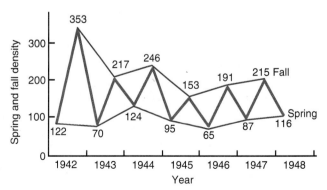

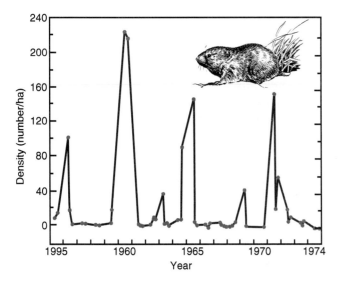

**Figure 18.20** The lemming cycle over a 20-year period in the coastal tundra at Barrow, Alaska. Three to six years elapse between peaks. Whereas the period is regular, the amplitudes vary greatly. (From Batzli et al. 1978:338.)

ulation to return toward equilibrium. The longer the time, the less the resilience.

The return time can be influenced by interactions with other species. A population of a given species does not live alone. If you disturb one species, you affect others as well. If species A depends on species B for food, both have experienced disturbance. Species A cannot return to equilibrium until species B has done so (for a good discussion see Pimm 1991). This point will be emphasized throughout later chapters.

Population fluctuations that are more regular than we would expect by chance are called oscillations or cycles. The fluctuations of Nicholson's blowfly populations in the laboratory, for example, are oscillations (see Figure 18.8) brought about by delayed density responses. In natural populations the two most common intervals between peaks are three to four years, typified by lemmings (Figure 18.20), and nine to ten years, typified by the snowshoe hare and the lynx, which preys on it (Figure 18.21). These cyclic fluctuations are largely confined to simpler ecosystems, such as the boreal forest and tundra. Usually only local or regional populations are affected, although there is some evidence to suggest broader synchronies.

A number of theories have been advanced to explain cycles. One theory, proposed by Cole (1951, 1954) and Palmgren

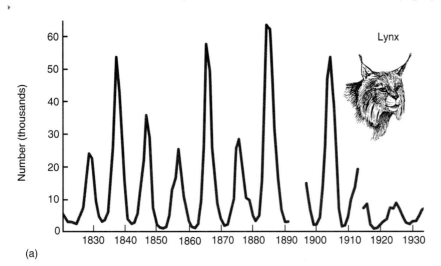

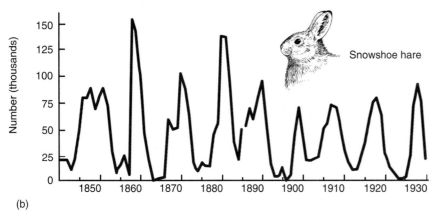

**Figure 18.21** (a) The nine- to ten-year cyclic fluctuation of lynx populations in the northwest territories of Canada. Data are based on lynx fur returns of the Northern Department, Hudson's Bay Company, from 1821 to 1913, and an equivalent area from 1915 to 1934. (From Elton and Nicholson 1942.) (b) The nine- to ten-year snowshoe hare cycle in eastern Canada. The lynx preys on snowshoe hare. (From MacCulich 1937.)

(1949), holds that cycles cannot be distinguished statistically from random fluctuations. Populations reflect a variety of random oscillations or fluctuations in environmental conditions. However, the statistical reliability of cycles has been demonstrated by Bulmer (1974, 1975) for a number of northern animals (red fox, lynx, muskrat, snowshoe hare, horned owl, and ruffed grouse) during the period 1951–1969 and by R. M. May (1976) for lemmings. The main feature of these cycles is the regularity of the period and the irregularity of the amplitude.

Another group of biologists maintains that something in the physical environment, in the ecosystem, or in the population itself causes cycles. Predation has been singled out as a cause. Malfunction of the animal's endocrine system has been cited, as well as changes in the frequencies of genes that make much of the population less resistant to environmental changes, aggressive behavior (C. J. Krebs 1985), dispersal (Stenseth 1983), and food shortages. Food shortages have been implicated in snowshoe hare decline (Keith 1974) (see Chapter 24) and in cycles of lemmings in the arctic tundra (Pitelka 1973, Batzli and others 1980). Lynx populations track snowshoe hare populations, the basis of the classic snowshoe-lynx cycle (see Chapter 25).

L. Keith and his associates (Meslow and Keith 1968; Keith and Windberg 1978; Keith et al. 1984) followed snowshoe hare populations through two periods of increase growth and three of decline in the Rochester district of central Alberta. These studies provide insights into the demographic features of the ten-year snowshoe hare cycle. The decline, which set in prior to the peak winter populations, was characterized by a high winter-to-spring weight loss, decrease in juvenile growth rate, decreased juvenile overwinter survival, reduction of adult survival beginning one year after the population peak and continuing to the low, and decreased reproduction (characterized by reductions in ovulation rates, third and fourth litter pregnancy rates, and length of breeding season).

Although the decline was attributed to food shortages, an intensive study involving radio-collared hares revealed new insights (Keith et al. 1984). Starvation in winter is a short-term phenomenon, confined largely to the first few months of the major decline. It is followed by one to two winters in which malnutrition (determined by sampling bone marrow) and low temperatures (below −30° C) interact strongly with heavy predation on nutritionally stressed hares by coyote, lynx, horned owl, and goshawk. Predation was the immediate cause of mortality, accounting for the death of 80 to 90 percent of the radio-collared hares. The upswing of the cycle, which set in about three years after the peak winter, was characterized by a lower winter-to-spring weight loss, increased juvenile growth rate, increased overwinter survival, and increased reproduction. Increased reproductivity activity and survival were stimulated by the recovery of vegetation during the low of the hare cycle.

## KEY FACTOR ANALYSIS

How do we determine what density-dependent influences are at work in a given population? One method is **key factor analysis.** A key factor is a biological or environmental condition associated with mortality that causes major fluctuations in population size. Key factor analysis is based on a $k$ value derived from the life table. Related to the mortality rate $q$, $k$ (sometimes called killing power) is defined as $\log_{10}l_x - \log_{10}l_{x-1}$. The $k$ value has the advantage over $q$ because it is additive. The summation of $k$ values over age classes provides $K$, the total killing power, which reflects the rate or intensity of mortality. Like $l_x$, it is comparable between populations.

Key factor analysis has been useful in the study of insect populations that have discrete generations and life stages to which mortality can be assigned. It is more difficult to use in populations with overlapping generations. However, with modifications it can be used to detect when regulation may be occurring in the life cycle and to aid in the search for the causes of mortality.

To demonstrate how key factor analysis works, we can use the data from the life table of the gypsy moth. The first figure in the $k$ factor table (Table 18.4) is the maximum potential natality for each generation. It is determined by multiplying the number of females of reproductive age by the maximum number of eggs per female. For our example we will simply consider the maximum fecundity for one female gypsy moth in a sparse population as 800 eggs. The $l_x$ value of each successive stage of the life cycle as it appears in the life table is entered, and the values are converted to logarithms. Each logarithm is subtracted from the previous one to give a $k$ or mortality value for each age class. These values are added to give a total generation mortality $K$. This is done for a number of successive generations. To identify the key factor which influences trends in adult populations, the $k$ values for each successive gen-

**Table 18.4** $k$ **Values for a Gypsy Moth Population Based on Life Table for a Sparse Population in Connecticut**

| $x$ | $l_x$ | Logarithm of $l_x$ | $k$ Value |
|---|---|---|---|
| Maximum natality | 800.0 | 2.903 | — |
| Eggs laid | 550.0 | 2.740 | 0.163 |
| Larvae I–III | 385.0 | 2.585 | 0.155 |
| Larvae IV–VI | 242.0 | 2.384 | 0.201 |
| Prepupae | 14.6 | 1.164 | 01.22 |
| Pupae | 11.7 | 1.068 | 0.096 |
| Adult | 1.4 | 0.146 | <u>0.922</u> |
| | | | $K = 2.757$ |

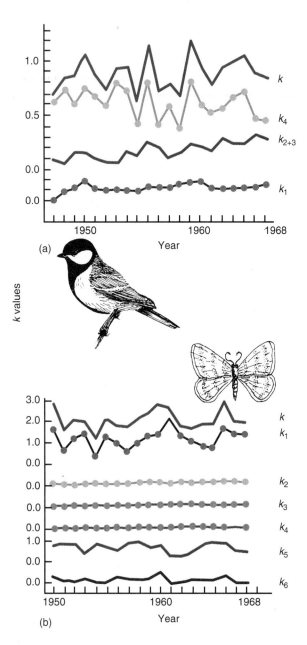

**(a)**

**(b)**

**Figure 18.22** Key factor analysis applied to the great tit (*Parus major*) and the winter moth (*Operophtera brumata*). (a) The key factor in the life cycle of the great tit in Marley Woods, Oxford, is $k_4$, mortality outside of the breeding season. In the rest of the annual life cycle variations in clutch size, ($k_1$) and hatching success ($k_2$) are density-dependent and are sufficient to regulate the population. (b) Key factors in the life cycle of the winter moth are $k_1$, overwintering loss of winter moth eggs and larvae before the first larval census in spring, and $k_5$, a density-dependent loss of the pupae in the soil due to predation in the spring. (From Podoler and Rogers 1975:97–101.)

eration are plotted along with $K$. The plot shows whether the mortality rate for one particular stage or age class consistently displays over the generations a strong correlation with total mortality. Such graphs are shown for the great tit and winter moth in Figure 18.22. If there is some correlation between the $k$ value of a particular stage and total $K$, then the analysis can be carried further to determine the $k$ factor within that stage.

Houston (1982) used that approach in his study of the population dynamics of the northern Yellowstone elk. It was obvious from the life table (Table 17.3) that juvenile mortality was the major cause of population loss. Houston used five sets of data in his key factor analysis: annual census of elk, reduction in fertility ($k_0$) from potential maximum, neonatal mortality ($k_1$), calf hunting mortality ($k_2$), and overwinter mortality of calves ($k_3$). Reduction in fertility was measured as the difference between the number of calves that would be born if all mature females produced calves and observed pregnancy. Neonatal mortality involved calves alive at 6 months; hunting mortality, calves alive at 9 months; and winter mortality, calves alive at 12 months. Total reduction in number of young, $K$, is equal to $k_0 + k_1 + k_2 + k_3$. Values for each were obtained by subtracting log (calves produced) from log (maximum potential calves) and so on. Houston assumed that each reduction acted in sequence with negligible overlap in time. Because of censusing and other problems, the estimates of $k$ were crude, but nevertheless reflected the true situation. Reduction in fertility $k_0$ essentially was a constant and hunting mortality was low. The major source of loss in calves turned out to be neonatal mortality ($k_1$). (For details on the use of key factor analysis see Dempster 1975.)

# EXTINCTION

When deaths exceed births and emigration exceeds immigration, then populations decline. $R$ becomes less than 1; $r$ becomes negative. Unless the population can reverse the trend then it at worst faces extinction or at best increases its probability of becoming extinct.

Vulnerability to extinction varies widely among species. Some species are common—widely distributed across their range and occupying a variety of habitats. Most species are relatively rare. Some occupy a wide natural range, but within it they are restricted to certain habitats. Others have a narrow range within which they are restricted to a very narrow habitat. Such species are endemic, typical of island situations. They are much more vulnerable to extinction than the common ones; but even common species are not immune, as exemplified by the passenger pigeon and the black rhinoceros.

408 PART 5 POPULATION ECOLOGY

There are several causes for the decline of sparse populations. When only a few individuals are present, females of reproductive age may have small chance of meeting a fertile male. Many females remain unfertilized, reducing average fecundity. A small population suffers more from predation and sudden environmental changes, because there are fewer individuals to survive. Losses feed upon losses until the population disappears.

Extinction is a natural process. Through millions of years of Earth's history, species have appeared and disappeared, leaving a record of their existence as fossils and trails in sedimentary rock. Some species could not adapt to geological and climatic changes. Others diverged into new species while the parent stock disappeared. Still others could not withstand the predatory pressures of a relatively new species, *Homo sapiens,* who appeared to have the first great impact in the Pleistocene. Massive extinctions have occurred at several points in Earth's history: the late Ordovician; the late Devonian; the late Permian, which witnessed the extinction of up to 96 percent of species; and the Cretaceous-Tertiary, which saw the end of the dinosaurs.

Mass extinction is happening today at an accelerated pace. Some estimates place it at 100 species a day, many of which are not yet known to science. The greatest number of extinctions has taken place since A.D. 1600. Well over 75 percent of modern-day extinctions have been caused by humans through the alteration and destruction of habitat, introduced predators and parasites, predator and pest control, competition for resources, and hunting of various types.

Despite popular impressions, extinction does not take place simultaneously over the full range of a species. It begins with isolated local extinctions when environmental conditions deteriorate or the population is unable to replace itself. Local extinctions often begin when habitats are destroyed and the dispossessed find remaining habitats filled. Restricted to marginal habitats, the individuals may persist for a while as nonreproducing members of a population or succumb to predation and starvation. As the habitat becomes more and more fragmented, the species is broken down into small isolated or "island" populations out of contact with other populations of its species. As a result the population is subject to inbreeding and genetic drift, reducing the ability of the small population to withstand environmental changes (see Chapter 21).

The maintenance of local populations often depends heavily on the immigration of new individuals of the same species. As the distance between local populations or islands increases and as the size of the local population declines, their continued existence becomes more precarious. As the number falls below some minimum level (see Chapter 21), the local population may become extinct simply through random fluctuations. Although we equate such situations mostly with rarer species, even the more common species experience local extinctions. These often go unnoticed because the loss is masked by the influx of immigrants from surrounding areas. One study of a suburban population of robins showed that because of the losses of nests and young through predation by cats and interference by humans, the robins were not replacing themselves. Robins sang each spring only because new birds moved into the area. Thus suburbia became a population sink rather than a population source.

In fact, most local populations do not thrive for long. They are revived fast enough by new immigrants that replace the losses and keep the population going. As one local population slides down the slope toward extinction, a local population somewhere else is experiencing overcrowding and supplies new recruits for depleted habitats.

Extinctions are of two sorts: deterministic and stochastic. **Deterministic extinction** comes about through some force or change from which there is no escape. The Cretaceous-Tertiary extinctions are an example. So is destruction of habitat on a local or regional scale. Habitat destruction rarely causes the extinction of a species except when a species is a local endemic or it is already on the verge of extinction. Examples are the recently extinct dusky seaside sparrow (*Ammodramus maritimus nigrescens*) of the Merritt Island and St. Johns River marshes of southern Florida. and the endangered spotted owl (*Strix occidentalis*) of the Pacific northwestern United States.

Small localized populations of a species are more often subject to stochastic extinction. **Stochastic extinction** comes about from normal random changes within the population or environment. Such changes normally do not destroy a population but merely thin it out, but a smaller population faces an increased risk of extinction from some decimating event.

Stochastic events may be demographic or environmental. Demographic stochasticity is chance variations in individual births and deaths. Demographic stochasticity results from habitat deterioration and loss through normal successional processes, reducing population size and restricting them to local patches of habitat. In a small population a high death rate or low birthrate can lead to a random or accidental extinction. When a population falls below a minimal viable size, it faces very great risk of going extinct. Environmental stochasticity is a random series of adverse environmental changes which comes about mostly through deterioration in environmental quality. If all members of a local population are affected equally by an adverse environmental change, the population may be reduced to a level at which demographic stochasticity takes over. It can be accompanied by a reduced effective population size and an increase in genetic drift and inbreeding. How long a population can exist at a low level depends upon the size of

individuals, longevity, mode of reproduction, and seed banks in plants.

A classic example of an extinction is that of the heath hen (*Tympanuchus cupido cupido*). Formerly abundant in New England, the heath hen, an eastern form of the prairie chicken, was driven by excessive hunting and habitat destruction to the island of Martha's Vineyard off the Massachusetts Coast and to the pine barrens of New Jersey. By 1880 it was restricted to Martha's Vineyard. At this point the population was subjected to deterministic extinction over most of its range. The small population, confined to a small island, was highly vulnerable to stochastic events. At first the population prospered, growing from a population of 200 birds in 1890 to over 2000 in 1920. Then a major stochastic event, a combination of fire, winter gales, and cold weather, reduced the population to 50. The heath hen never recovered, and the last bird died in 1932.

# SUMMARY

Populations increase when births and immigration exceed deaths and emigration. The difference between the two (when measured as an instantaneous rate) is the population's rate of increase, *r*. In an unlimited environment a population expands geometrically, a phenomenon that may occur when a small population is released in an unfilled habitat. Geometric increase is characterized by a constant schedule of birth and death rates, an increase in numbers equal to the intrinsic rate of increase, and the assumption of a fixed or stable age distribution, which is maintained indefinitely. Because the environment is limited, such growth is not maintained indefinitely. Population growth eventually slows and arrives at a point of equilibrium with the environment's carrying capacity, *K*. However, natural populations rarely achieve such equilibrium levels; instead they fluctuate in numbers. When populations become quite small, chance events, demographic or environmental, alone can lead to extinction.

Fluctuations about the equilibrium point suggest that populations possess some form of regulatory mechanism. Some ecologists have maintained that population fluctuations are most affected by density-independent influences. Others have argued that populations are regulated by density-dependent influences that relate to optimum population size. A more general idea is that local populations fluctuate between upper and lower levels and the fluctuations are brought about by an interaction of population density and influences extrinsic to the population, such as weather.

Density-independent influences affect but do not regulate populations. They can reduce local populations, even to the point of extinction, but their effects do not vary with population density. Regulation implies a homeostatic feedback that functions with density.

Intraspecific competition for resources in short supply is a density-dependent mechanism in the regulation of population numbers. There are two basic types of competition. In scramble competition all individuals have equal access to the resource, and each attempts to get a part of it. In extreme cases scramble competition results in each individual obtaining insufficient amounts to survive or reproduce. In contest competition successful individuals divide the resource and the unsuccessful are denied access to it. Contest competition is characteristic of species whose individuals are able to defend a resource from others.

Intraspecific competition can result in density-dependent reduced growth, delayed maturity, lessened fecundity, and increased mortality, especially of the young. In plants response to density results in a decrease in the number of individuals and an increase in the mean weight of individuals. Regardless of starting densities, monocultural populations of a plant will arrive at a point where growth compensates for losses through mortality, and reduction in mean plant weight compensates exactly for mean plant density, a constant final yield. This relationship is expressed in the $-3/2$ power law, $w = cd^{-3/2}$.

One approach to the study of density-dependent mortality in animal populations is *k*-factor analysis. A key factor is a biological or environmental condition associated with mortality that causes major fluctuations in population size.

# REVIEW QUESTIONS

1. What is the difference between lambda $\lambda$, *R*, and *r?* What is the relationship among them?
2. Distinguish between exponential growth and logistic growth. Give the equation for each.
3. What are the weaknesses and limitations of the logistic equation? Why is the equation useful?
4. What is carrying capacity (*K*)? Why do populations fluctuate about some estimated value of *K?*
5. Why do species go extinct?
6. Distinguish between density-dependent regulation and density-independent influences.
7. Define competition. What is scramble competition? Contest competition?
8. How can density-independent influences affect density-dependent mechanisms?
9. Explain key-factor analysis. Why is it useful?
10. The role of density dependence and density independence in population growth was a major controversy among ecologists in the 1960s. Read Davidson and Andrewartha (1948) and F. E. Smith (1961) and

summarize the ideas. Was the basic argument over density dependence or over the role of physical versus biological factors?

11. Has the human population in some parts of the world reached or exceeded carrying capacity (economic as well as ecological)? Consider Rwanda, Somalia, and India. What are the signs of population stress?

## CROSS-REFERENCES

Plant responses to moisture, 70–75; animal responses to moisture, 75–77; genetic drift, 471–472; viable populations, 475–476; plant-herbivore interactions, 522–524; predator-prey cycles, 543–545; habitat fragmentation, 614–618.

# Intraspecific Competition

## Concepts

1. High population density can cause social stress, resulting in increased mortality and decreased fecundity.
2. Dispersal is the movement of an organism from its natal site to its first breeding site.
3. Motivation to disperse can be intrinsic or extrinsic.
4. Social dominance, expressed as territoriality and social hierarchy, can regulate population.
5. Territoriality has costs as well as benefits.

What form does intraspecific competition take, especially in situations where resources such as food, space, and mates are not shared equally? How are resources allocated? What happens to individuals who do not receive their share? Intraspecific competition can operate in subtle and not-so-subtle ways. The outcome of competition affects the fitness of the individuals and can influence population density.

# DENSITY AND STRESS

How do individuals respond as their population increases, as their living space becomes more crowded, as their food, shared by more and more hungry members of the group, becomes less available? One response is increased social stress, especially among vertebrates.

Stress in vertebrates, evidence suggests, can act on the individual through a physiological feedback involving the endocrine system. This feedback is most closely associated with the functioning of the pituitary and adrenal glands (Christian 1963, 1978; Christian and Davis 1964; D. E. Davis 1978). Stress triggers hyperactivation of the hypothalmus-pituitary-adrenocortical system, which in turn alters the secretion of growth and sex (gonadotrophic) hormones. Profound hormonal changes suppress growth, curtail reproductive functions, and delay sexual activity. Further, these hormonal changes may suppress the immune system and cause breakdowns in white blood cells, increasing an individual's vulnerability to disease (Sinclair 1977). Social stress among pregnant females may increase intrauterine mortality (spontaneous abortion) and cause inadequate lactation and subsequent stunting of nurslings. Thus, stress can result in decreased births and increased mortality.

Such population-regulating effects have been confirmed in confined laboratory populations of several species of mice and to a lesser degree in enclosed wild populations of woodchucks (*Marmota monax*) (Lloyd et al. 1964) and Old World rabbits (*Oryctolagus cuniculus*). K. Myers and his associates (1967, 1971) held such rabbits at several densities in different living spaces within confined areas of natural habitat. Those living in the smallest space, in spite of a decline in numbers, suffered the most debilitating effects. Rates of sexual and aggressive behavior increased, particularly among females. Reproduction declined, fat about the kidneys (an indicator of physical condition) decreased, and the kidneys exhibited inflammation and pitting on the surface. The weight of liver and spleen decreased and adrenal size increased.

Young rabbits born to stressed mothers were stunted in all body proportions and in organs. As adults, they showed behavioral aberrations such as a high rate of aggressive and sexual activity and physiological aberrations such as abnormal adrenal glands, low body weight, and poor survival. Rabbits from low and medium densities had excellent health and survival.

**Pheromones** (chemicals released by animals that serve as communication among individuals of the same species) present in the urine of adult rodents may inhibit reproduction among members of a population. Such a function is suggested in a study involving wild female house mice (*Mus musculus*) living in high-density and low-density populations confined to grassy areas within a highway cloverleaf. Urine from females of a high-density population was absorbed onto filter paper. The paper was placed with juvenile wild female mice held individually in laboratory cages. Similarly, urine from females in low-density and sparse populations was placed with other juvenile test females. Juvenile females exposed to urine from high-density populations experienced delayed puberty, whereas females exposed to urine from low-density populations did not. The results suggest that pheromones present in the urine of adult females in high-density populations may delay puberty in juveniles and help slow population growth (Massey and Vandenberg 1980). Juvenile female house mice from low-density populations exposed to urine of dominant adult males accelerate the onset the puberty (Lombardi and Vandenbergh 1977).

Plants growing under crowded conditions, in which competition is high, respond in various ways to conditions of low nutrients, low moisture, and other environmental stress, such as too little or too much light (see Chapter 7). Individual plants react to increased density with decreased growth, as expressed by reduction in mean plant weight, loss of leaves and branches, and changes in growth form. In spite of this reduction, yield per unit area is constant over a wide range of densities (see Chapter 18). Genets may respond to high density by reducing the number of ramets, as perennial ryegrass (*Lolium perenne*) does, resulting in a lower density of tillers than genets in the population (Kays and Harper 1974). Plants also modify their morphology by reducing the number of nodes per stem, internode length, number of flowers and seeds, leaves per stem, and branches (for examples see White 1984, Sarukhan et al. 1984). Such reductions, especially in leaf area, can increase mortality (Fowler and Antonovics 1981, Antonovics and Primack 1982).

Ruderal plants, those adapted to persistent and severe disturbance, such as ragweed and other annual weeds, respond to low moisture and nutrient stress by producing seeds at the expense of vegetative development. Individual plants are small and poorly developed, yet the number of seeds relative to individual plant biomass is high. Seeds of such plants can survive buried in soil for long periods of time. They can germinate quickly when disturbance exposes the seeds to light and fluctuating daily temperatures.

When subjected to competitive stress, biennials, such as fetterbush (*Pieris hieracioides*) and bluethistle (*Echium vulgare*), may delay reproduction for three to five years, even though they can reproduce in two years under ideal conditions. Under conditions imposed by dense populations, a lower proportion of plants survives to maturity, reproduction is delayed, and fewer seeds are produced by mature plants

(Klemow and Raynal 1985). Plants growing on infertile, drought-prone sites exhibit a similar behavior, responding to unfavorable conditions rather than to high densities.

Such responses to stress influence individual fitness among plants and the maintenance and expansion of populations. However, there is still insufficient evidence that density-dependent stress in plants acts in any sort of regulatory way. (For detailed review of density-dependent regulation in plants see Antonovics and Levin 1980.)

# DISPERSAL

By late summer the reproductive season for animals is over and juveniles have swelled the population. General belief is that the young will leave their natal area, driven out by competitive interactions with the adults. Actually, dispersal is more complex. Why disperse, when the animal might be better off staying put? Who disperses? When do dispersers leave, and where do they go? What motivates dispersers to leave, and what effect does dispersal have on the fitness of the dispersers and on population regulation? Answers to such questions provide insights into one of the most common phenomena in the natural world.

## Who Disperses?

There is no hard and fast rule about who disperses. Howard (1960) defined dispersal as "the permanent movement an individual makes from its birth site to the place where it reproduces or would have reproduced if it had survived and found a mate." This definition suggests that juveniles and subadults are the dispersers.

Young are the major dispersers among birds. Greenwood (1980) suggests their leaving be called **natal dispersal,** and that the dispersal of adults from poor to better reproductive sites be termed **breeding dispersal** (see also Greenwood and Harvey, 1982).

Among rodents subadult males and females make up most of the dispersing individuals (Tamarin 1978, C. J. Krebs et al. 1976, Beacham 1980). At times dispersing females outnumber males, and at other times the opposite is true; or there may be no difference at all (M. C. Johnson and Gaines 1987). Many dispersers, up to 40 percent in some cases (Beacham 1980), are in breeding condition. They tend to be lighter in weight than stay-at-homes, and reach sexual maturity at a lower weight and younger age (C. J. Krebs et al. 1976).

Among many groups of insects, dispersal is undertaken by a polymorphic component of normally flightless insects that acquire wings, notably aphids (Aphidoidae), leafhoppers (Cicadellidae), and by long-winged (versus short-winged) water striders (Gerridae). Crowding, temperature change, quality and abundance of food, and photoperiod all have been implicated in stimulating the development of winged individuals. These winged forms have the option of flying to new habitats or not flying, depending upon environmental cues. Within insect populations in which all individuals are capable of flight, a proportion may disperse and colonize new habitats; others may not. (For review, see Harrison 1980.)

**When and How Far?** The prereproductive period is the usual time of dispersal. Many species of birds disperse in spring, following the return of migrants or breakup of winter flocks among nonmigratory resident species. In some resident species, such as the ruffed grouse, young disperse in late fall and are settled on breeding areas by early winter. Rodents leave their home place mostly when the population is increasing and at the peak phase of population growth. Dispersal decreases during the periods of population decline (C. J. Krebs et al. 1976, Beacham 1980).

Because dispersers are seeking vacant habitat, the distance they travel will depend in part on the density of surrounding populations and the availability of suitable unoccupied areas. Murray (1967) stated a rule of dispersal: move to the first uncontested site you find and no further. Animals generally will either disperse in a straight line from their natal area or make exploratory forays into surrounding areas before leaving the natal site. They should settle in the first empty site.

For five years Martin and Bucher (1993) studied natal dispersal of the colony-nesting monk parakeet (*Myiopsitti monoachus*) in the province of Cordoba, Argentina. Ninety-one percent of the banded young remained with the parents until three months old. Then the young birds left gradually and intermittently in random directions, until early in the next breeding season (Figure 19.1a). Their dispersal distance ranged between 300 and 2000 m, with a mean of 1230 m. Fifty percent of the breeding adults moved an average of 500 m when they changed nesting sites between census periods.

Larsen and Boutin (1994) monitored the individual fates and documented the movements of 205 (94 percent) of 219 young red squirrels (*Tamiasciurus hudsonicus*) at Fort Assiniboine, Alberta, Canada. Prior to settlement, the young squirrels made exploratory forays of up to 900 m from their natal territory (Figure 19.1b), which they did not abandon until they settled on their own territory. Just under one-half of the 73 offspring that acquired territories settled either on or adjacent to their mothers' territories. The farthest distance a young squirrel settled was only 323 m from the natal territory, or about the distance of three territory widths.

In developing a model of dispersion based on straight-line movements, P. M. Waser (1985) measured dispersal distance in terms of the number of home ranges (based on average size of home range for the species) the disperser has to travel from its natal site (Figure 19.2). His model predicts the distribution of dispersal as a function of turnover of home ranges or territories vacated by the death of previous owners, and of the distance the disperser has to travel to find them.

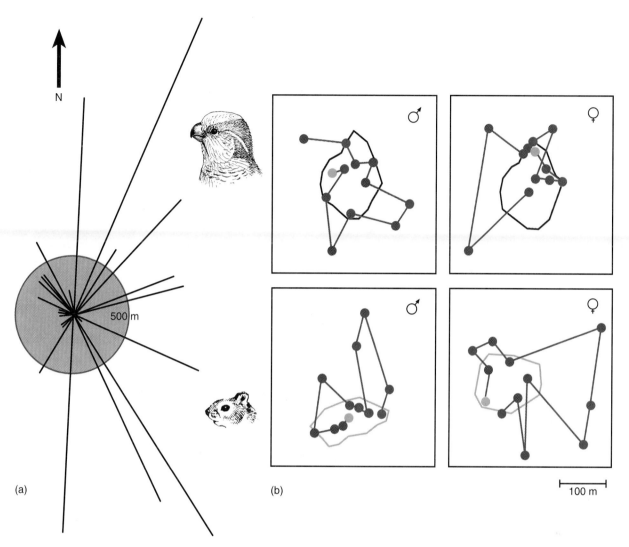

**Figure 19.1** Direction of dispersal. (a) Directions moved by young monk parakeets from their natal nest to their first occupied nest in Cordoba Province, Argentina. (After Martin and Bucher 1993.) (b) Samples of forays made by young radio-collared red squirrels. o = initial observation; [sb] = subsequent observations. (After Larsen and Boutin 1994:218.)

## Why Disperse?

Why should a young animal disperse at all, and leave its natal area for some strange place? Some do not. Whatever the choice, it carries certain costs and benefits (Table 19.1). A few dominant juveniles may be able to establish themselves reproductively when the natal area is vacated through mortality of the adults. Because there is little probability they will reproduce on their natal area, most juveniles or subadults can maximize their fitness only if they leave their birthplace, in spite of the risks. Where intraspecific competition is intense, dispersers can locate habitats where resources are more accessible, breeding sites are available, and competition is less. Dispersers also increase the probability of encountering more new individuals with which to mate, reducing the prob-

ability of inbreeding and increasing fitness of offspring because of heterozygosity (see Chapter 21).

An animal has to have some motivation to disperse. Intraspecific competition of some sort may be a driving force behind some dispersals. Christian (1971) hypothesized that increased population density in rodents increased levels of aggression. Aggressive individuals force subdominant ones to disperse. Under these conditions maximum dispersal would be at peak densities and involve social subordinates, mostly males.

Evidence obtained from fenced populations seems to point out that highest rates of dispersal occur before peak densities are reached. Not all dispersers are necessarily subdominants, or the less aggressive, and a large fraction of dispersing animals are females. Some animals may be born to disperse.

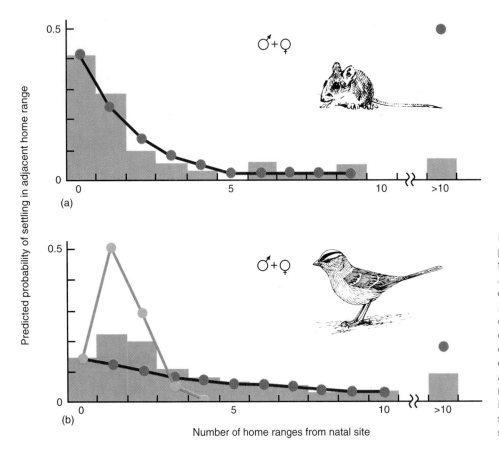

(a)

(b)

Number of home ranges from natal site

Predicted probability of settling in adjacent home range

**Figure 19.2** Observed and expected dispersal distances (a) in the deer mouse (*Peromyscus maniculatus*) and (b) in white-crowned sparrow (*Zonotrichia leucophrys*). The bar graphs indicate the observed dispersal distances. The reddish line indicates the dispersal distance expected if the search were along a radius away from the natal site. The yellow line in (b) is the expected pattern if the search were to the nearest empty site. (From P. W. Waser 1985:1173.)

**Table 19.1  Potential Costs and Benefits of Dispersal Choices (G = Genetic, S = Somatic)**

| Stay at Home: Philopatry | |
| --- | --- |
| *Costs* | *Benefits* |
| Inbreeding depression (G) | Optimal inbreeding: maintain locally adapted genes (G) |
| Reduced fitness because of resource shortage (S) | Reduced physical risks: increased survivorship (S) |
| Reduced indirect fitness: competition with kin (S) | Familiarity with local terrain: security (S) |
| | Familiar social environment (S) |
| | Adaptive local traditions (S) |
| | Maintain kin association (S) |

| Disperse | |
| --- | --- |
| *Costs* | *Benefits* |
| Outbreeding depression: disrupt coadapted genes (G) | Outbreeding enhancement (G) |
| Hybrid young not well adapted (G) | Avoid overcrowding (S) |
| Alleles less suited to the environment (G) | Avoid competing with kin (S) |
| Greater risk in movement: predators, local diseases, unfamiliarity with terrain (S) | Improve fecundity (S) |

*Source:* Adapted from Shields 1987.

W. E. Howard (1960) suggested a genetic basis for dispersal, and Myers and Krebs (1971) and C. J. Krebs et al. (1976) found that dispersers were not a random subsample from their control populations of voles, *Microtus pennsylvanicus* and *M. ochrogaster*. Certain genotypes were more prone to dispersal than others when populations were increasing. However, there is little evidence on the heritability of tendencies to disperse.

Bekoff (1977) suggested that social interaction prior to dispersal rather than aggressiveness is the mechanism behind dispersal. Asocial individuals, either dominant individuals avoided by their sibs or subdominants avoiding their sibs, are the most likely to disperse because they fail to develop social ties. Because aggressive behavior is not an adequate stimulus for dispersal, no relationship exists between population density and dispersal.

Lidicker (1975) has hypothesized two types of dispersal, presaturation and saturation. **Presaturation dispersal** takes place during the increase phase of population growth before population reaches a peak or carrying capacity and before resources are depleted. The dispersers are in good condition, consist of any sex or age group, have a good chance of survival, and have a high probability of settling in a new area. Such dispersals seem to be density-independent. **Saturation dispersal** occurs when carrying capacity has been exceeded. The individuals, mostly juveniles and subdominants, have two options: to stay and either perish or not breed, or to leave the area. If they move out, the odds are the dispersers will perish, although a few may arrive at some suitable area and settle down. The mass movements of squirrels in the 1800s (D. L. Allen 1962) and muskrat dispersal described by Errington (1963) are examples of saturation dispersal. Such dispersal is density-dependent.

Dispersal requires a source and a **sink,** an empty or unfilled habitat, or even marginal or unsuitable habitat in which the animals can survive for a time. The dispersal sink must permanently remove animals from the resident or source population. Many dispersers die during their travels. Some dispersers discover and settle into patches of optimal habitat. Others move into areas where conditions (predation, poor nesting sites, lack of protective cover) preclude successful reproduction (Figure 19.3). In such habitats reproduction does not balance mortality. Even though sink habitats often support very large populations, sometimes populations would disappear without the immigration of surplus individuals from more productive source areas (Pulliam 1988). Because a species may actually be more abundant in a sink habitat than in the source habitat, the sink appears to be optimal. In reality, the habitat may be luring dispersers into areas subject to high predation or other causes of reproductive failure.

Such population sinks can be distinguished from optimal habitats only with some knowledge of the demographics of the species in each. Such knowledge is critical in the conservation of species. Because of population abundance, we may mistakenly select a sink habitat as an optimal source habitat

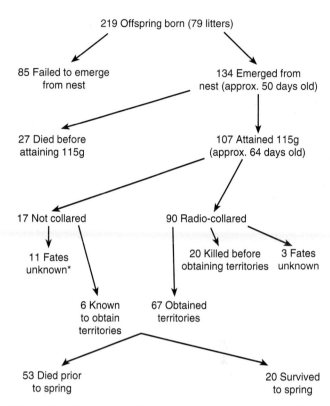

**Figure 19.3** The fate, starting at birth, of 219 red squirrel offspring (79 litters) from the 1988, 1989, and 1990 cohorts at Fort Assiniboine, Alberta, Canada. Of the 90 radio-collared young dispersers, only 20 survived on new territories. The rest either died during dispersal or were lost to various dispersal sinks. (From Larsen and Boutin 1994:217.)

when selecting refuges and unintentionally lead both sink and source populations to local extinction.

## Does Dispersal Regulate Populations?

Does dispersal play any role in population regulation? This question has especially intrigued microtine biologists and ecologists, who have been seeking explanations for population cycles of voles and other rodents ever since they were described by Elton in 1926. Much of what we have learned about dispersal and population interactions is based on comparisons of fenced rodent populations in which dispersal could occur with those in which dispersal was prevented. The magnitude of dispersal was estimated by comparing the numbers of marked animals disappearing from a control grid to the number of those animals reappearing in grids from which all animals are continuously removed (for reviews see Gaines and Johnson 1987, Gaines and McClenaghan 1980). For example, in an early experiment C. Krebs et al. (1969) enclosed three populations of meadow voles (*Microtus pennsylvanicus*) in southern Indiana in such a way that individuals could not immigrate or emigrate, but predators had access. They compared these populations with

a control population whose members were able to disperse. The enclosed populations increased in size; in one grid the population was three times as high as in the control. Overpopulation in the enclosures resulted in overgrazing, habitat deterioration, and starvation. Physiological stress was relatively unimportant because *Microtus* can exist at densities several times higher than those normally experienced by other voles. This experiment led to a long series of studies involving fenced populations to study emigration from growing populations.

One of the first of these studies was done by Krebs and his associates (1976) in British Columbia. They set up control and experimental grids populated with *Microtus townsendii.* They established control populations whose members were allowed to emigrate through exit tubes. They cleared experimental grids of voles, continually kept the areas vacant, and monitored recolonization. Colonization of experimental areas was most rapid when populations in the control areas were increasing. In declining populations on the control grids, in which losses were due to death and not emigration, very little dispersal occurred.

The outcome of these experiments, reviewed by Gaines and McClenaghan (1980) and Gaines and Johnson (1987), show that dispersal in fluctuating populations of voles is density-independent. Most dispersals take place during times of population increase. Although dispersal is positively correlated with population density and with the rate of population increase, there is no association between the proportions of the population leaving the area and the rate of population increase and decrease. The dispersers are not a random subset of the set of the resident population, but are mostly younger males.

Although dispersal may not function as a regulatory mechanism in a traditional sense, it can expand populations, aid in the persistence of local populations, and function as a form of natural selection by sorting out phenotypes and genotypes.

# SOCIAL INTERACTIONS

A flock of mourning doves, two pairs of adults and their young, are feeding on grain you have scattered for them. Each maintains some distance from the others and protects that distance by slight movement toward any individual that comes too close. Among them a dominant male commands the largest space, driving all others away. Several of the young are harassed, chased away from the grain. Noting their frustration, you scatter more seed over a wider area, a space too large to be controlled by the dominants. The subdominant birds feed on the periphery, and peace has settled on the flock for a time. A resource in a limited space caused intraspecific conflict among the birds. An increase in both available food and feeding space reduced competitive interactions.

Aggressive and submissive interactions (called agonistic) are the basis of social organization, which takes two forms, dominance and territoriality. The difference between the two involves not only individual interactions but also the utilization of space. Social organization based on individual distance (distance from another individual which provokes aggressive or avoidance behavior) and dominance relationships among members of a social unit that shares space is **social dominance.** Social organization involving the division and exclusive occupation of space by a social unit or individual with a defended boundary is **territoriality.**

## Social Dominance

Social dominance is based on intraspecific aggressiveness and intolerance and on the dominance of one individual over another. Two opposing forces are at work simultaneously: mutual attraction versus social intolerance, a negative reaction against crowding. Each individual occupies a position in the group or local population based on dominance and submissiveness.

In its simplest form an alpha individual is dominant over all others, a beta individual is dominant over all but the alpha, and so on to the omega, which is totally subordinate. This relationship was first described by Schjelderup-Ebbe (1922) for the domestic chicken. It is a straight-line or linear peck order, so called because pecking follows dominance—that is, birds peck at others of lower rank (Figure 19.4). Even within a peck order complexities may exist, such as triangular or nonlinear hierarchies. In these triplets, the first individual is dominant over the second, the second is dominant over the third, and the third is dominant over the first. In such a situation, an individual of a lower rank can peck an individual of a higher rank.

Among some animals, birds in particular, peck order is replaced by peck dominance, in which social rank is not absolutely fixed. Threats and pecks are dealt by both members during encounters, and the individual that pecks the most is regarded as dominant. The position of the individual in the social hierarchy may be influenced by levels of male hormones, strength, size, weight, maturity, previous fighting experience, previous social rank, injury, fatigue, close associates, and environmental conditions.

In mixed groups, males and females may have separate hierarchies with males dominant over females; or females may be equal to the males or dominant over them. In other species, dominance is unrelated to sex. Once social hierarchies are well established within a group, newcomers and subdominant individuals rise in rank with great difficulty. Strangers attempting to join the group are either rejected or relegated to the bottom of the social order.

Rise in hierarchy often is related to sexual activity and hormones. This is particularly true among those species that remain in flocks throughout the year. Individuals, male or female, that come into breeding condition early, even though subdominant in the winter group, rise in hierarchy through increased aggressiveness.

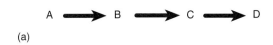

(a)

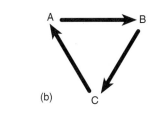

(b)

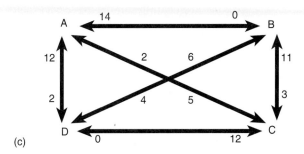

(c)

**Figure 19.4** Examples of peck orders. (a) A straight-line peck order in which one animal is dominant over the animal below it. A is the alpha individual; D is the omega. (b) Triangular peck order in which A is dominant over B, B is dominant over C, yet C is dominant over A. (c) A more complex triangular peck order. The double arrows indicate encounters between individuals. The numbers represent the number of wins of one individual over another. For example, A is clearly dominant over B with 14 wins. B never dominated A. D is the omega individual.

To the dominant individuals in such a social structure go most of the resources. Among many species the dominant male secures the most mates, thereby insuring greater fitness at the expense of subdominant males. Dominant individuals have first choice of food, shelter, and space, and subdominant individuals obtain less. When shortages are severe, low-ranking individuals may be forced to wait until all others have fed, to take the leavings, face starvation, or disperse.

Such social interactions cannot be called regulatory. Rather, the outcome of this type of contest competition simply ensures that the dominant animals in the population will continue to reproduce successfully. The fitness of such individuals is secured at the expense of subdominant individuals.

Social dominance, however, can influence population regulation if it affects reproduction and survival in a density-dependent manner. An example is the wolf. Wolves live in small groups of 6 to 12 or more individuals called packs. The pack is an extended kin group consisting of a mated pair, one or more juveniles from the previous year who do not become sexually mature until the second year, and several nonbreeding related adults.

The pack has two social hierarchies, one headed by an alpha female and one headed by an alpha male, the leader of the pack to whom all other members defer. Below the alpha male is the beta male, closely related, often a full brother, who has to defend his position against pressures from other males below.

Mating within the pack is rigidly controlled. The alpha male (occasionally the beta male) mates with the alpha female. She prevents lower-ranking females from mating with the alpha and other males, while the alpha male inhibits mating attempts by other males. Thus each pack has one reproducing pair and one litter of pups each year. These pups are reared cooperatively by all members of the pack. At low wolf densities, some packs may rear two litters of pups a year (Ballard et al. 1987).

The size of a wolf population in a region is governed by the size of the packs, which hold exclusive areas. Regulation of pack size is achieved by events within the pack that influence the amount of food available to each wolf. The food supply itself does not affect births and deaths, but the social structure that leads to an unequal distribution of food does. The reproducing pair, the alpha female and the alpha male, has priority for food; they, in effect, are independent of the food supply. The subdominant animals, male and female, with little reproductive potential, are affected most seriously. At high densities the alpha female will expel other adult females from the pack. Other individuals may leave voluntarily. Unless these animals have an opportunity to settle successfully in a new territory and form a new pack, they fail to survive.

The social pack, then, becomes important in population regulation. As the number of wolves increases, the size of the pack increases. Individuals are expelled or leave and the birthrate relative to the population declines because most sexually mature females do not reproduce. Overall the percentage of reproducing females declines. When the population of wolves is low, sexually mature females and males leave the pack, settle in unoccupied habitat, and establish their own packs with one reproducing female. More rarely, the pack may produce two litters instead of one litter in a year (Ballard et al. 1987, Van Ballenberghe 1983). But at very low densities, females may have difficulty locating males to establish a pack and so fail to reproduce or even survive. (For details on social regulation of population size in wolves see Mech 1970, Zimen 1978, Fritts and Mech 1981, Ballard et al, 1987.)

## Territoriality

The flock of mourning doves feeding on the ground will retain that social organization through the winter. In early spring the flock will break up and scatter across the countryside, form pairs, and establish another type of social organization, territoriality. A **territory** is a defended area, more or less fixed and exclusive, maintained by an individual or by a social group, such as a wolf pack. Some would give the term an even broader definition and say that a territory exists when individuals or groups are spaced out more than you

would expect from a random occupation of suitable habitat (Davies 1978).

Often it is difficult to draw a sharp distinction between social dominance and territoriality. Depending upon the season, the degree of crowding, and the distribution of resources, territory can grade into social hierarchy and vice versa. For example, winter flocks of chickadees and ring-necked pheasants have a social hierarchy. As the breeding season approaches the hierarchy breaks down into territoriality. Dominant males in the winter flocks may be the first to establish territories in the spring.

**Types of Territory.** Types of territories vary according to the needs of the animals that defend them (Nice 1941). One type is a general purpose territory, established during the breeding season. This type of territory is common among songbirds and some mammals, such as muskrats. Within it all activities from feeding, to mating and rearing the young take place. Late in the breeding season or soon after, territorial defense breaks down.

A second type, common among hawks, is a mating and nesting territory with feeding done elsewhere. Mating territories are exemplified by leks of prairie grouse and the singing grounds of woodcock (*Philohela minor*). Swallows and many colonial birds defend only a nesting territory, the size of which is often determined by the distance the bird can strike from its nest. Other animals, such as hummingbirds and some squirrels, defend only a food resource, a feeding territory. Some birds defend a winter roosting territory when adequate roosting sites are scarce.

**Territorial Defense** Once an animal has established a territory, the owner must defend it against intruders. At first conflicts may be numerous. Birds, frogs, and insects usually defend their claims vocally by singing from some conspicuous spot. Songs and calls advertise the fact that the area is already occupied. They are a long-distance warning that potential trespassers should not waste their energies trying to settle there. Birds may shift their song perches throughout the territory (Figure 19.5) or vary their song patterns, perhaps in an effort to suggest that more than one male is in the area.

Song can be effective in maintaining space between individuals. If a bird is removed from its territory, the space is quickly claimed by another deprived of a territory outright or forced to settle in some suboptimal area. If a territorial male is removed but his song has been recorded and is played in the territory, other males will stay away (Carrick 1963).

If songs and calls fail and another individual does move into the area, the owner may confront the trespasser with visual display. That display may involve raising the crest, fluffing the body feathers, spreading the wings and tail, and waving wings among birds, erecting the ears and baring the fangs among mammals (Figure 19.6). Such displays usually intimidate the invader, encouraging it to leave.

If intimidation displays fail, then the territory owner is forced to attack and chase the intruder, an activity easily observed among many birds in spring and among dragonflies

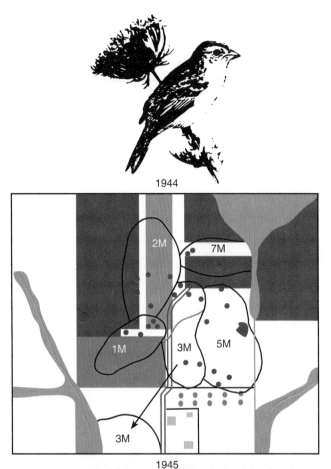

1944

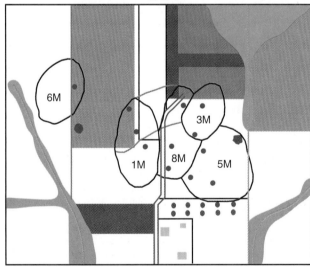

1945

**Figure 19.5** Mating territories of six banded male grasshopper sparrows (*Ammodramus savannarum*). Dots indicate song perches. Note how they are distributed near the territorial boundaries. The shaded areas represent crop fields, the white areas hayfields. Dashed lines indicate boundary shifts in territory prior to second nesting. Males returned to nearly the same territorial area the second year. Such behavior is philopatry. (From R. L. Smith 1963:160.)

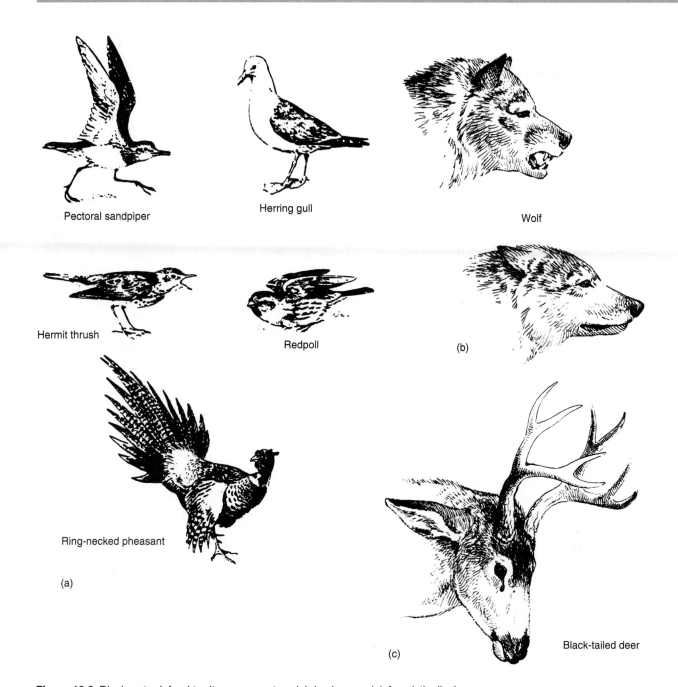

**Figure 19.6** Displays to defend territory or assert social dominance. (a) Agonistic displays among birds. (Pectoral sandpiper, (*Calidris melanotos*) after W. J. Hamilton 1959; herring gull (*Larus argentatus*) based on photographs in N. Tinbergen 1953; hermit thrush (*Catharus guttatus*) after Dilger 1956; redpoll (*Carduelis flammea*) after Dilger 1960; ring-necked pheasant (*Phasianus colchicus*) after Collias and Taber 1951.) (b). Aggressive (upper) and submissive (lower) expressions in the American wolf, typical of canids. (c) Aggressive display in the black-tailed deer (*Odocoileus hemionus sitkensis*). The illustration shows details of the head during the snort that occurs when the buck is circling in a crouch position. Note the widely opened preorbital gland near the eye, curled upper lip, and bulged neck muscles. The snort is a sibilant expulsion of air through the closed nostrils, causing them to vibrate. (After Cowan and Geist 1961.)

about a pond's edge. The gall-forming aphid *Pemphigus betae* (about which more will be said later) defends her territory by engaging in end-to-end kicking-shoving contests with the intruder, which may last more than two days and may result in the death or one or both aphids (Whitham 1987).

Some animals defend a territory by the use of scent markers. Wolf packs and coyotes mark territories with well-placed scent posts, frequently renewed by urine (Peters and Mech 1975, Barrette and Messier 1980). These scent marks warn neighboring members of their species about boundary rights. Just as important, these scent posts tell members of a wolf pack that they are within their own territory and prevent accidental straying into hostile territory.

The use of scent and other chemical releasers, or pheromones, is widespread among animals, especially mammals and insects. They are secreted from endocrine glands, transmitted as a liquid or a gas, and smelled or tasted by others. They are important not only to mark territory and trails but also to convey such information as the identity of an individual, its sex and social rank, the location of food, and the presence of danger or a potential mate (see E. O. Wilson 1971, Whittaker and Feeney 1971).

**Why Defend a Territory?** Why should an insect, bird, or mammal defend a territory? The reasons vary among animals. For some it is the acquisition and protection of a needed resource such as food, or a reduction in the risk of predation. For others it is the attraction of a mate. The basic benefit is always an increased probability of survival and improved reproductive success—in short, increased fitness. By defending a territory, the individual forces others into suboptimal habitat, reducing their fitness and increasing the proportion of its own offspring in the population.

Consider the territorial behavior of the gall-forming aphid *Pemphigus betae* on the leaves of its host plant, narrowleaf cottonwood (*Populus augustifolia*), which grows along streams in western North America. The female sexual aphid deposits a single overwintering egg in the deeply fissured bark of the main branches and trunk of the tree. From the egg a wingless stem mother, who will give rise to new offspring, emerges in spring at the time leaves are unfolding. The stem mothers move en masse from the trunk and branches and establish themselves on immature leaves. They probe the expanding tissues to form a small depression. In a short time leaf tissue will envelop the aphid, forming a hollow gall in which she will parthenogenetically produce up to 300 progeny. The number of aphids invading the tree may be high, up to 850 per 1000 leaves.

The aphids have a preferred site, the base of the largest leaves (Whitham 1987). These leaves have the lowest concentration of phenolics, complex molecules that inhibit feeding activities of herbivorous insects (see Chapter 24), and are the least susceptible to leaf fall. Such leaves are rare, representing only about 1.6 percent of available leaves, so there is a great deal of competition for them. Kicking-and-shoving contests between prospective colonists settle disputes for microterritories 3–5 mm in length. The winners usually are the largest stem mothers, who may or may not share a leaf with another. Stem mothers who have a leaf to themselves have the highest reproductive success. When two aphids share a leaf, the stem mothers that secure the basal portion of the leaf produce on the average 56 percent more progeny than those displaced to the distal portion of the leaf. These differences in reproductive success result from microhabitat variations in the quality of the leaf. Only a few millimeters' difference in leaf position affects aphid reproduction (Figure 19.7).

To determine what would happen to reproductive success, Whitham removed one member of a competing pair. Immediately the remaining aphid crossed the former territorial boundary and enlarged her own. If the remaining member was the distal stem mother, she increased the number of her progeny on an average of 48 percent. If she was the basal stem mother, she improved her reproductive success by only 18 percent.

His study points out several aspects of territoriality. The territorial animal that claims the best territory has the

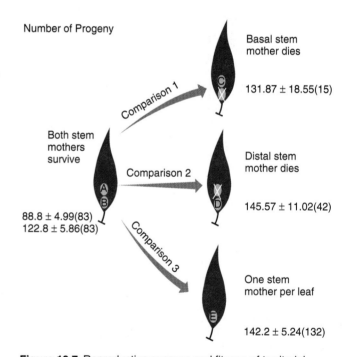

**Figure 19.7** Reproductive success and fitness of territorial stem mothers of the gall-forming aphid *Pemphigus betae.* When either stem mother of a competing pair suffered an early death, the remaining stem mother produced significantly more progeny (A<C and B<D). Stem mothers occupying leaves singly from the beginning of gall formation produced more progeny than either member of a competing pair (A<B<E). When released from competition early in development, the surviving stem mothers on the average produced the same number of progeny as stem mothers solitary from the beginning of gall formation (C = D = E) Mean number of progeny from leaves of same quality and size ± SE (*n*) is indicated for 355 surviving stem mothers. (From Whitham 1986:140.)

highest reproductive success. The reproductive cost of losing is lower fecundity, if the losers reproduce at all. The winners are the most dominant individuals. Among aphids dominant individuals usually are the largest. Solitary stem mothers achieved the highest reproductive success, followed by the basal stem mother of a competing pair and then by the distal stem mother. Smaller stem mothers are forced to share leaves or have to settle on inferior smaller leaves. Thus the cost of losing a competitive interaction is reduced fitness.

Defending a territory can be a costly business, especially if optimal resources are limited. When the stakes of winning or losing in the aphids are high, contests over a position on a leaf may last two days and result in the death of one or both contestants. Even on a less intense scale, territorial defense uses energy, consumes time, and interferes with feeding, courtship, mating, and rearing of young.

Like all economic endeavors, territorial ownership has costs and benefits, and the owner has to balance the two (Figure 19.8) (Davies and Houston 1984). Some territories are economically defendable and some are not. A general prerequisite is a predictable resource somewhat dispersed. Then acquisition of an area insures its owner of resources, reduces foraging costs, and allows time for other activities. If resources are unpredictable and patchy, it may be advantageous for individuals to belong to a group and cooperatively seek needed resources without being restricted to one area. Spotted hyenas (*Crocuta crocuta*), for example, live in clan territories in the Ngorongora Crater of Kenya, where resources are predictable; whereas on the Serengeti Plains, where food is seasonal, they range over wide areas and do not defend a territory (Kruuk 1972).

**Territorial Size**   Closely associated with the cost-benefit ratio of territorial ownership is territorial size. As the size of a territory increases, the cost of territorial defense increases. Many male birds in spring attempt to claim more ground than they can economically defend. They are forced to draw in the boundaries to make the area more manageable. However, there is a minimum size below which they cannot go, because it will be too small to meet their needs. (For models see Stamps and Krishnan, 1990, Adams 1994). The number of territory owners an area can hold is the total area divided by the minimum size of the territory. Minimum and maximum sizes, however, can change from year to year. Size is influenced by resource availability, habitat changes, adult mortality, and settlement patterns. Somewhere along the gradient of too small and too large is an optimal size for a territory, one from which the owner gains maximum benefits for the costs incurred (Figure 19.8).

Some animals, notably certain spiders, have fixed territories (Riechert 1981). The size of the territories remains the same, regardless of spider density or resource availability. Such fixed sizes may be transmitted as part of an inherited behavior, or the fixed-size territories may be transferred from one owner to the next, and between generations.

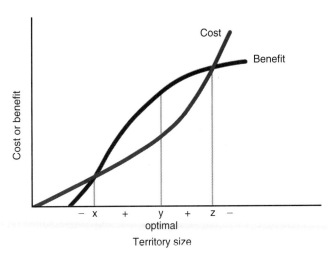

**Figure 19.8** Model of cost–benefit ratio curves as they might apply to territorial defense. Costs of territorial defense increase as the size of territory increases. Defense is profitable only between X and Z. A maximum cost–benefit ratio is at Y in each case.

Other animals are more flexible. Optimal size will vary from year to year and from locality to locality. If a resource such as food is abundant, the territory may be small; and if resources are less abundant, it may be larger. In general, territory size tends to be no larger than required to provide needed resources. For example, the territories of the golden-winged sunbird (*Nectarina reichenowi*) vary greatly in size and in floral composition, but each territory contains just enough of a nectar supply to meet an individual's daily energy requirement (Gill and Wolf 1975). Such flexibility in the size of territories has been likened by J. Huxley (1945) to an elastic disk compressible to a certain size. Territory size decreases as density increases, but when the territory compresses to a certain size, the resident resists further compression and denies access to additional settlers (see Getty 1981, Stamps and Krishnan 1990). Because aggressive behavior varies among individuals, the most aggressive have the advantage and the less aggressive are forced to settle elsewhere.

For some animals—birds, in particular—it is not the size of the territory that counts, but its quality (Figure 19.9). Some males are successful in claiming the best territories, superior nesting sites or food sources. For example, hummingbirds and Hawaiian honeycreepers seem to assess differences in standing crops of floral nectar and defend areas with the richest nectar supplies, but cease to defend territories when food is superabundant (F. L. Carpenter 1987). The ovenbird (*Seiurus aurocapillus*), a ground-nesting forest warbler, selects and defends areas with greatest prey abundance per unit area (Smith and Shugart 1987). Optimal territory for the dickcissel (*Spiza americana*), a finch of the midwestern grasslands of the United States, is measured by litter depth and vegetation density (Zimmerman 1971). Less suc-

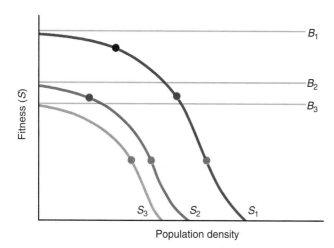

**Figure 19.9** Relationship among habitat suitability, population density, and fitness. The suitability of habitat is ranked from $B_1$, the best, to $B_3$, the poorest. The curves from $S_1$, the highest fitness, to $S_3$, the lowest, show the fitness of individuals in habitats at different population densities. At low population density all individuals will be able to settle in optimum habitat $B_1$. At intermediate density some individuals will be forced to settle in habitat $B_2$. At high density all three habitats will be settled. The model implies that as density increases, habitat suitability for part of the population decreases. (After Fretwell and Lucas 1969:24.)

cessful males occupy suboptimal territories, and many males fail to secure a territory.

To successful males come the females, and the "wealthiest" males always obtain a mate. In fact, among some species of birds, such as the dickcissel (Zimmerman 1971) and some of the marshland blackbirds (Orions 1969), females will accept a male already mated if he holds an outstanding territory. Apparently the odds of a successful nesting are greater there, even though the female has to share the area with another. Lenington (1980) found that early-arriving female redwinged blackbirds (*Agelaius phoeniceus*) chose males based on territory quality. Late-arriving females faced aggression from resident females. They had to settle for males holding territories in poorer habitat.

Floaters As a result of contest competition among males for space, some individuals are denied territory. Thus a portion of the population does not reproduce because they are excluded from the suitable breeding sites by territorial individuals. They make up a surplus population, a **floating reserve** that would be able to reproduce if a territory became available to them.

Such a floating reserve of potentially breeding adults has been described for a number of species. The number of floaters may be great. Studies of a banded white-crowned sparrow population (*Zonotrichia leucophrys nuttalli*) in California indicated a nonbreeding surplus of potentially breeding birds (Petrinovich and Patterson 1982). In fact, 24 percent of the territory holders entered the breeding population two to five years after banding, and 25 percent of the nestlings that acquired territories did so two to five years after their birth. Territory holders that disappeared during the breeding season were quickly replaced.

Although the existence of floating reserve populations is acknowledged, few data exist on the social organization and behavior of surplus birds. Floaters may live singly off the territories as white-crowned sparrows do; or the floaters may form flocks with a dominance hierarchy on areas not occupied by territory holders, as the floating populations of red grouse do on the heather-dominated moors of Scotland (Watson and Moss 1971, 1972).

Floaters in some species live on the breeding territories of other individuals. An example of this strategy is provided by the detailed studies by S. M. Smith (1978) of the rufous-collared sparrow (*Zonotrichia capensis*) in Costa Rica Figure 19.10). By observing banded birds, both territorial and nonterritorial, and by selectively removing certain individuals, Smith was able to determine the role of the floater or "underworld" bird. Territorial sparrows on her study area occupied small territories ranging from 0.05 to 0.40 ha and made up 50 percent of the total population. The other 50 percent, underworld birds consisting of both males and females, lived in well-defined restricted home ranges within other birds' territories. Male home ranges, often disjoined, embraced three or four territories. Female home ranges were usually restricted to a single territory. Because home range boundaries of both sexes coincided with territorial boundaries, each territory held two single-sex dominance hierarchies of floaters, one male and one female. When a territory owner, male or female, disappeared, it was quickly replaced by a local underworld bird of appropriate sex on the territory. These floaters usually entered the territories as young birds hatched some distance away and were tolerated by the owners.

A similar situation exists among some territorial spiders (Reichert 1981) and the aphid *Pemphigus* (Whitham 1987). They, too, have a floating reserve of individuals who quickly claim vacated sites. Among the spiders, the floaters live in cracks and crevices within occupied territories. Periodically, floaters will unsuccessfully attempt to take over an occupied habitat.

Population Regulation A consequence of territoriality can be population regulation. If no limit to territorial size exists and all pairs that settle on an area get a territory, then territoriality results only in spacing out the population. No regulation of the population results. If territories have a lower limit in size, then the number of pairs that can settle on an area is limited. Those that fail to get a territory have to leave. Thus territoriality might regulate population density, but only under certain conditions.

For example, in the arctic ground squirrel (*Spermophilus undulatus*) all females are allowed to nest. However,

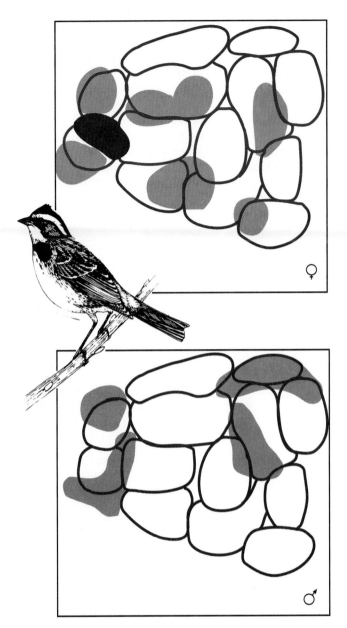

**Figure 19.10** Territorial boundaries of rufous-collared sparrows. Home ranges of "underworld" females and males are superimposed on occupied territories. Eventually some of the females occupied a territory by replacing a missing bird. The male home ranges *a* and *b* are disjoined, and each includes several territories. (Adapted from S. Smith 1978:570.)

territorial polygamous males drive excess males from the colony into submarginal habitat, where they exist as a nonbreeding floating population. The number of breeding males remains constant because losses are continually replaced from the floating population. This floating population decreases drastically over the year, largely from predation and weather. In this species, territoriality stabilizes the number of breeding males but does not regulate the population, because only males appear to be surplus.

Krebs (1971) removed breeding pairs of great tits from their territories in an English oak woodland. The pairs were replaced by new birds, largely first-year individuals, that moved in from territories in hedgerows, considered suboptimal habitat (Figure 19.11). The vacated hedgerow territories, however, were not filled, suggesting that a floating reserve of nonterritorial birds did not exist. In this case territorial behavior limits the density of breeding birds in optimal habitat, but does not regulate the population, because all birds are breeding in some habitat.

To demonstrate that territorial behavior limits a population, we have to show that a portion of a population, including males and females, does not breed because they are excluded from suitable breeding sites by dominant or territorial individuals. These surplus individuals should be able to breed if territories became available to them. Such is the case with the white-crowned sparrows, rufous-collared sparrows, and red grouse. The size of the reproductive population is limited by territoriality, and density-dependent population regulation results.

## Home Range

Territory should not be confused with home range. **Home range** is an area in which an animal normally lives. It is not necessarily associated with any type of aggressive behavior. A home range may or may not be defended in part or in whole (in which case the home range is identical with a territory) and may or may not overlap with those of other individuals of the same species (Figure 19.12). However, dominance may exist among individuals with overlapping home ranges, and subordinate individuals tend to avoid contact with dominant individuals. Both use portions of the same area by avoiding contact both spatially and temporally.

**Figure 19.11** Replacement of removed individuals and settlement of vacated territory. Six pairs of great tits (*Parus major*) were removed between March 19 and March 24, 1969 (left stippled area). Within three days, four new pairs had taken up territories (right stippled area) and some residents had expanded their territories, so that territories again formed a complete mosaic over the woods. (From Krebs 1971:7.)

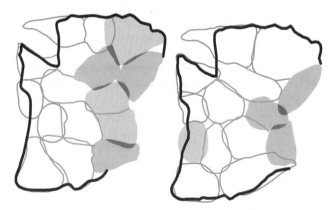

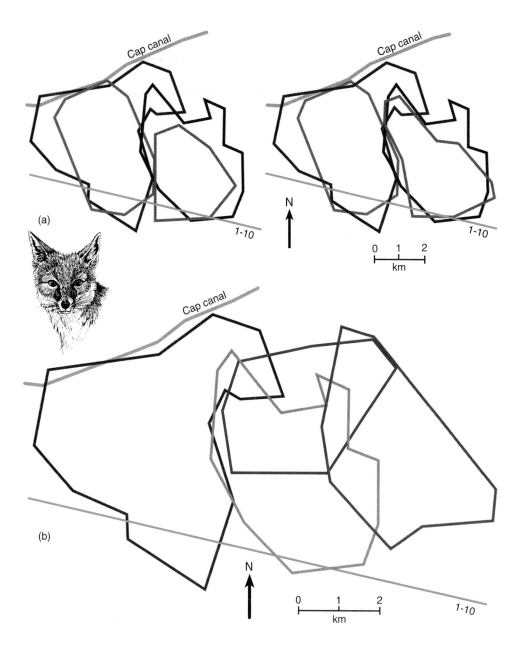

**Figure 19.12** (a) Home ranges of paired female and male kit foxes (*Vulpes velox macrotis*) in western Arizona, determined by observations of marked animals. Each pair of foxes shares much the same home range, which suggests a degree of territoriality. (b) Home ranges of two unpaired kit foxes overlap the home ranges of two paired males. Much of the overlap resulted from the movement of non-paired males during the breeding season to dens used by females of the other pairs. (From Zoellick and Smith 1992:86.)

Generally, home ranges do not have fixed boundaries. Seldom is a home range rigid in its use, size, and establishment. It may be compact, continuous, or broken into two or more discontinuous parts reached by trails and runways. Irregularities in distribution of food and cover produce corresponding irregularities in home range and in frequency of animal visitation. The animal does not necessarily visit every part daily. Its movements may be restricted to trails and most of its activities may be concentrated in a smaller core area used more intensively than other parts.

The size of a home range relates to body size (McNab 1963, Harestad and Bunnell 1979). Large mammals have larger home ranges than smaller ones, and carnivores generally have larger home ranges than herbivores and omnivores

of similar size. Males and adults have larger home ranges than females and subadults. Difference in weight alone is sufficient to account for this within a species without invoking any competitive interactions. The home range of herbivores and omnivores increases at a nearly constant rate as body weight increases; among carnivores the home range increases at a greater rate as body weight increases (Figure 19.13).

Like territoriality, possession of a home range confers certain advantages. The animal becomes familiar with the local area, knowing where to find food, shelter, and cover from enemies with a minimum expenditure of energy. It can define a series of escape routes to cover and travel routes to food sources throughout the year.

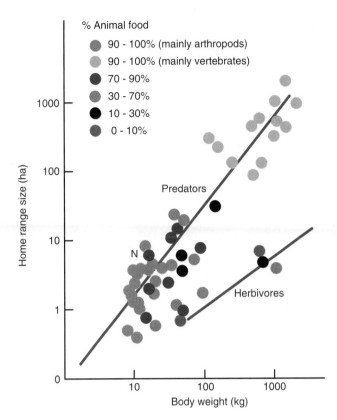

% Animal food
- 90 - 100% (mainly arthropods)
- 90 - 100% (mainly vertebrates)
- 70 - 90%
- 30 - 70%
- 10 - 30%
- 0 - 10%

**Figure 19.13** 13. Relationship between the size of home range and body weight of North American mammals. (Adapted from Harestad and Bunnell 1979:390.)

## SUMMARY

Some mechanisms of intraspecific competition involve physiological and behavioral interactions among members of the population. Increased density may affect population growth through physiological responses of individuals. Crowding may produce stresses that can result in endocrine imbalances, especially in the adreno-pituitary complex, and other responses, which result in abnormal behavior, abnormal growth, degeneration, and infertility.

Social pressure and crowding may also induce emigration or dispersal. However, dispersal appears to be density-independent except at the highest population levels; most dispersal appears to take place when populations are increasing rather than at the peak. Dispersal may function in population regulation by encouraging mostly subadults and some, but not necessarily all, subdominant individuals to leave their natal area and occupy vacant habitats. Although risks of moving may be great, successfully dispersing individuals improve their fitness.

Density influences social behavior of individuals in a population. Basic social behavior involves aggressive and submissive behavior expressed as social dominance and ter-

ritoriality. Both act as spacing mechanisms. Territoriality, the defense of a fixed area, divides space among some individuals and excludes others. Territoriality is expressed only when a resource such as food is contained within an economically defensible space. Owners defend their space by aggressive signals such as song, display, or chemical marking. In contrast, social dominance results in the sharing of space and maintenance of individual distances, but with some individuals dominant. Subordinate individuals avoid contact with dominant ones by using different parts of a shared area or home range and by using the area at different times. Territoriality can function in population regulation only if it creates a surplus population consisting of sexually mature individuals prevented from breeding by territory holders.

## REVIEW QUESTIONS

1. How does stress affect population growth?
2. Define dispersal. What types of animals are most prone to disperse?
3. What are some costs and benefits of dispersal?
4. Under what conditions can dispersal aid in population regulation?
5. Distinguish among social dominance, territoriality, and home range.
6. How can social hierarchy and territoriality function in population regulation?
7. What are the advantages and costs of territoriality?
8. What is a floating reserve, and how does it function?
9. Watch the social interactions of birds or squirrels at a winter feeding station over a period of a week and report on your observations. Observe the territorial interactions among robins or another common species and record behavioral interactions, including song, chase, and other conflicts. Attempt to determine territorial boundaries and nest locations.
10. Some plant ecologists have suggested that even spacing among some desert plants and the crowns of tropical trees suggests territoriality among plants. Discuss this idea. Given the definition of territoriality as observed in animals, can plants be territorial?

## CROSS-REFERENCES

Plant responses to moisture, 72–75; plant responses to light, 102–105; –3/2 power law, 400–402; genetic drift, 471–472; chemical defense in plants, 527–529; chemical defense in animals, 536; habitat fragmentation, 614–618.

# Life History Patterns

## Outline

## Concepts

1. Reproduction may be asexual or sexual. Sexual reproduction involves the production of gametes by males and females, either as separate individuals or as hermaphrodites.
2. Mating systems are broadly defined as monogamous and polygamous.
3. Sexual selection, in contrast to natural selection, involves intrasexual and intersexual components.
4. Parental allocation of reproductive costs balances selection for fecundity with selection for survival of parents and offspring.
5. Equal investment in sons and daughters can result in a biased sex ratio in favor of the less expensive sex.
6. *r*-traits and *K*-traits describe differences in life history patterns.
7. Animals use environmental cues to select habitat.

Reproduction is the major vocation of all living things. By transmitting genetic characteristics from one generation to another, individuals maximize the numbers of their descendents. All their activities, their life history patterns, and traits are directed toward that one end. It colors the way in which individuals select and acquire a mate, their mode of reproduction, the number of young produced, and how often. It is the underlying reason for bright colors and fragrances of flowers, the songs of birds, the flashing of fireflies by night, the antlers of deer.

This concentration on activities and morphologies, which most of us rarely relate to sexual activity and reproduction, is metabolically expensive to individuals. Parents have to make "decisions" about the amount of energy to allocate to growth, defense, and reproduction. How organisms budget these energy expenditures has intrigued behavioral ecologists. How do individuals meet these costs and allocate them to reproduction and survival? How has sexual selection and the drive to contribute to a new generation affected the fitness of individuals? These are questions explored in this chapter.

## PATTERNS OF REPRODUCTION

How organisms form new individuals falls into two categories: asexual and sexual reproduction. Asexual reproduction, which takes many forms, results in the formation of new individuals genetically the same as the parents (see Chapter 17). Paramecia reproduce by dividing in two. Hydras reproduce by budding; strawberry plants spread by runners. Aphids produce eggs by normal cell division or mitosis that develop into female adults without fertilization (parthenogenesis). However, organisms that rely heavily on asexual reproduction turn on occasion to sexual reproduction. Hydras at some time in their life cycle produce egg and sperm. Plants produce flowers. At the end of the summer, aphids resort to sexual reproduction, halving the chromosomes and producing males.

Sexual reproduction is common to eucaryotic organisms. It involves the production of haploid gametes, egg and sperm that combine to form a diploid cell or zygote. This halving and recombination of genes allows the gene pool to become mixed. Mixing supplies the genetic variability necessary to meet changing selective pressures, and to prevent an accumulation of harmful mutations (see Chapter 21). For the individual, however, sexual reproduction is expensive. Each individual can contribute only one-half of its genes to the next generation. Because the success of that contribution depends upon the contribution gained from a member of the opposite sex, it follows that each individual should acquire the best possible mate.

Sexual reproduction can take a variety of forms. The most familiar involves separate male and female individuals. It is common to most animals. Plants with that characteristic are called **dioecious;** examples are holly trees and stinging nettle (*Urtica* spp.). In such plants fertilization takes place with the fusion of a large nonmobile female egg and a small mobile male gamete. Some individual organisms possess both male and female organs. They may be **monoecious** or **hermaphroditic.** In plants individuals can be hermaphroditic either by having bisexual flowers with both male (stamens) and female (ovules) organs, as in lilies and buttercups, or by being monoecious (Figure 20.1). In monoecy, the individual plant possesses separate male and female flowers, as do birch (*Betula*) and hemlock (*Tsuga* spp.) trees. Among animals hermaphroditic individuals possess both testes and ovaries, a condition common to some invertebrates, such as earthworms, and some fish. Some hermaphrodites are simultaneous, as earthworms are; others are sequential. The latter type are one sex when young or small and develop into the opposite sex when mature or large, as snails and jack-in-the-pulpit do.

## MATING SYSTEMS

On a brushy rise of ground at the edge of a woods, a pair of red fox has a deeply-dug burrow within which are the female and her litter of pups. Outside at the burrow entrance are scattered bits of fur and bones, the leftovers of meals carried to the den by the male for his mate and pups. Back in the woods a doe has a young dappled fawn hidden in a patch of ferns on the forest floor. The fawn's father is nowhere near and has no knowledge of this offspring, sired during a short encounter of a few days the previous fall, along with several others.

The monogamous fox and the promiscuous deer represent two means by which organisms obtain a mate. A **mating system** includes such aspects as the number of mates males and females acquire, the manner in which they are acquired, the nature of the pair bond, and the pattern of parental care provided by each sex. The structure of mating systems ranges from monogamy through many variations of polygamy. Mating systems may even vary within a species, involving different strengths of pair bonds (Table 20.1).

The nature and evolution of male-female relationships are influenced by ecological conditions, especially the availability and distribution of resources and the ability of individuals to control access to mates or resources. If the male has no role in the feeding and protection of young, and defends no resources available to them, the female gains no advantage by remaining with the male; and it makes no sense to the male to remain with her. If the habitat is sufficiently

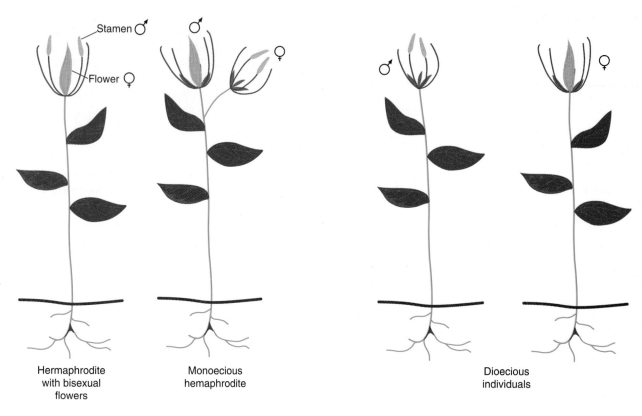

**Figure 20.1** Floral structure in dioecious and monoecious (hermaphroditic) plants.

**Table 20.1  An Ecological Classification of Mating Systems**

| | |
|---|---|
| Monogamy | Neither sex has opportunity of monopolizing additional members of the opposite sex. Fitness often maximized through shared parental care. (Fox, beaver.) |
| Polygyny | Individual males frequently control or gain access to multiple females. |
| Resource defense polygyny | Males control access to females *indirectly,* by monopolizing critical resources. (Hummingbirds, redwinged blackbird.) |
| Female (or harem) defense polygyny | Males control access to females *directly,* usually by virtue of female gregariousness. (Elk, wild horses, seals.) |
| Male dominance polygyny | Mates or critical resources are *not economically monopolizable.* Males aggregate on leks during the breeding season and *females select mates* from these aggregations. (Sage grouse, prairie chicken.) |
| Explosive breeding assemblages | Both sexes converge for a short-lived, highly synchronized mating period. The operational sex ratio is close to unity and sexual selection is minimal. (Frogs, insects.) |
| Rapid multiple clutch polygamy | Both sexes have substantial but relatively *equal* opportunity for increasing fitness through multiple breedings in rapid succession. Males and females each incubate separate clutches of eggs. (Some shorebirds.) |
| Polyandry | Individual females frequently control or gain access to multiple males. |
| Resource defense polyandry | Females control access to males *indirectly,* by monopolizing critical resources. (Spotted sandpiper, jacana.) |
| Female access polyandry | Females do not defend resources essential to males but, through interactions among themselves, may limit access to males. Among phalaropes, both sexes converge repeatedly at ephemeral feeding areas where courtship and mating occur. The mating system most closely resembles an explosive breeding assemblage in which the operational sex ratio may become skewed with an excess of females. (Phalaropes, small arctic shorebirds.) |

*Source:* Emlen and Oring 1977.

uniform so that little difference in territorial quality exists, the number of young raised in the poorer habitat is only slightly less than the number reared in the best. Selection would favor monogamy because female fitness in both would be nearly the same. However, if the habitat is diverse, with some parts more productive than others, competition may be intense and some males will settle on poorer territories. Under such conditions, a female may find it more advantageous to join another female in the territory of the male defending a rich resource than to settle alone with a male on a poorer territory. Selection under those conditions will favor a bigamous situation, even though the male may provide little aid in feeding the young.

## Monogamy

**Monogamy** is the formation of a pair bond between one male and one female. It is most prevalent among birds and rare among mammals, except several carnivores, such as the fox and mustelids, and a few herbivores, such as the beaver.

Monogamy occurs mostly among those species in which cooperation by parents is needed to rear young successfully. Nearly 90 percent of all species of birds are seasonally monogamous, because in most birds the young are helpless at hatching and need food, warmth, and protection. The avian mother is no better suited to provide these needs than the father. Instead of wandering off to other females and seeking other mates, the male can increase his fitness more by continuing his investment in the young. Without him, the young carrying his genes may not survive. Among mammals the situation is different. The females lactate, providing food for the young. Males often can contribute little or nothing to the survival of the young, so it is to their advantage to mate with as many females as possible. Among the exceptions are the canids, among which the male provides for the female and young and defends the territory. Both male and female are able to regurgitate food for the weaning young.

Monogamy, however, has another side. Among some 100 species of monogamous birds, the female cheats on her mate. The male too engages in extra-pair copulations while maintaining his reproductive relationship with his primary mate and young (Mock and Fujioka 1990).

## Polygamy

**Polygamy** means acquiring two or more mates, none of which is mated to other individuals. It can involve one male and several females or one female and several males. A pair bond exists between the individual and each mate. When one member of the pair is freed from parental duty, partly or wholly, the emancipated member of the pair can devote more time and energy to intrasexual competition for more mates and resources. The more unevenly such critical resources as food or quality habitat are distributed, the greater is the opportunity for a successful individual to control the resource and thus available mates.

The number of members of the other sex an individual can monopolize depends upon the degree of synchrony in sexual receptivity. For example, if females in the population are sexually active for only a brief period, as with the white-tailed deer, the number a male can monopolize is limited. However, if females are receptive over a long period of time, as with elk, the number a male can control depends upon the availability of females and the number of mates the male can energetically defend.

Such variability in environmental and behavioral conditions results in various types of polygamy. There are two basic forms: **polygyny,** in which an individual male gains control of or access to two or more females; and **polyandry,** in which an individual female gains control of or access to two or more males. A special form of polygamy is **promiscuity,** in which males and females copulate with one or many of the opposite sex and form no pair bonds. Emlen and Oring (1977) have classified these types of polygamous relationships according to the means by which individuals gain access to the limiting sex (Table 20.1).

In polyandry the female rather than the male is the competitive individual. In resource defense polyandry, best developed in two orders of shorebirds, Gruiformes (cranes and rails) and Charadriiformes (shorebirds), the female competes for and defends resources essential for the male. As in polygyny, this mating system depends upon the distribution and defensibility of resources, especially quality habitat. The production by females of multiple clutches, all of which require the brooding services of a male, leads to competition among females for access to available males. After the female lays a clutch, the male begins incubation. Then he becomes sexually inactive and is effectively removed from the male pool, resulting in a scarcity of available males.

## SEXUAL SELECTION

Choosing a proper mate is essential if a plant or animal is to contribute genetically to the next generation. The tumult of bird song in spring, the frenzy of frogs calling in ponds, and the clashes among stags in the season of rut—all focus on sex. All this activity seems to involve males attempting to attract females; but at the same time, more quietly perhaps, females are also seeking out males. In any population there are just so many males and females. Because males are not as selective with whom they mate, females usually have no trouble finding a partner. Females, however, are selective, and males must prove their fitness; those who cannot go mateless. The result is intense rivalry among males for female attention (Figure 20.2). The outcome of male rivalry, called **intrasexual competition,** may indeed give the victors access to females. However, success depends not so much on who wins as on how the competition affects their acceptance by choosy females. In the end the female determines which male will sire her offspring.

**Figure 20.2** This bull elk is bugling a challenge to other males in a contest for control of a harem.

Supposedly, males (and in some situations females) that win the intrasexual competition are the fittest. By selecting the fittest mate from among competing males, based on some specific characteristic during courtship, females attempt to assure their own fitness. Making such a choice is known as **sexual selection.** Ultimately sexual selection results in the evolution of morphological and behavioral traits that influence both competitive mating and mating systems.

## Models of Sexual Selection

How does a female select a mate with greatest fitness? That question still intrigues behavioral ecologists. Darwin (1871) attempted to answer it first. Because the elaborate and often outlandish plumages of birds, and the horns, antlers, and large size of polygamous males seemed incompatible with natural selection, Darwin developed the concept of sexual selection. He hypothesized that competition among males selected for weapons, body size, and striking plumage patterns. Females then selected males from among the winners, based on their appearance and behavior. Supposedly these traits reflected the highest fitness.

How does sexual selection square with natural selection when the traits selected by females seem to expose males to predation? The geneticist R. A. Fisher (1930) attempted to provide an explanation. He hypothesized that some novel genetic characteristic, such as a plumage pattern, that indirectly indicates maleness becomes the object of a female mating preference. Females who select males with this trait produce fitter progeny. The original selective advantage of the preferred characteristic paired with female preference for the characteristic increases the advantage accrued to the male possessing the trait. Positive feedback results in an accelerated increase or "runaway" both in the male trait and in the female preference for the trait.

Both of these hypotheses assume that mate competition is a male trait and that mate choice is a female trait. Trivers (1972), and earlier, J. Maynard Smith (1956) suggested a more balanced view of sexual selection, based on the reproductive interests of the female (see Small 1992). The basic strategy for both male and female is to ensure their own maximum fitness. What increases male fitness is not necessarily what improves female fitness. Sperm is cheap. With little investment, males should mate with as many females as possible to achieve maximum fitness. By contrast, females invest considerably more in reproduction. It is to their advantage to be selective in choosing a mate, one who will pass on the best genes to the next generation.

This concept led to another explanation of sexual selection, advanced by the Israeli behavioral ecologist A. Zahavi (1975, 1977), the **handicap hypothesis.** It postulates the evolution of three characteristics: a male handicap, a female mating preference for the handicap, and a general viability trait. The handicap is a secondary male characteristic, such as bright plumage, that could reduce the male's survival. The viability trait is one that affects an individual's ability to survive, as by escaping a predator. If a male can carry handicaps and survive, it is proof of a superior genotype. Females showing preference for the handicapped male receive a selective advantage because their offspring carry genes for high viability. The handicap principle has sparked a good deal of controversy and research in sexual selection (Kodric-Brown and Brown 1984, Kirkpatrick 1986, 1987).

## Processes of Selection

J. Maynard Smith (1991) has nicely classified the processes of selection into two major types: intrasexual selection and

intersexual selection. Male-to-male (or in some cases female-to-female, as in some shorebirds) competition for the opportunity to mate with females is **intrasexual selection.** It leads to exaggerated secondary sexual characteristics, such as bright or elaborate plumage in male birds and antlers in deer. **Intersexual selection** is mostly female choice of a mate. Intersexual selection breaks down into parts, the female choice of a conspecific mate and the choice between conspecifics. Choice between conspecifics may be one of two types. In one the female makes a choice based on resources such as territory or access to food, which will improve her fitness. The other choice may involve genes only, as with many polygamous species. By selecting a male with exaggerated traits, the female will also be acquiring genes for highest fitness. The problem facing behavioral ecologists is finding out what females look for in a mate in each of these situations, and how they make their final choice.

**Resource-Based Selection**  For monogamous females the criterion for mate selection appears to be acquisition of a resource, usually a high-quality territory (for example, see J. L. Zimmerman 1971, R. D. Howard 1978b, Searcy 1979). The question is whether the female selects the male and accepts the territory that goes with him or whether she selects the territory and accepts the male that goes with it. There is some evidence from the laboratory and the field that female songbirds base their choice, in part, on the complexity of the male's song. This evidence is based on studies in which hormone-implanted females (the hormone maintains the female in a sexually receptive condition during the experimental period) are exposed to the playback of actual and simulated conspecific bird song of varying complexity of males of their own species. These females responded with stronger solicitation displays to the more complex song types (Searcy et al. 1985, Searcy and Andersson 1986, Baker et al. 1986, Catchpole 1987) (Figure 20.3). In the field males with more complex song repertoires appeared to hold the higher-quality territories, so the more complex song may convey that fact to the female (Figure 20.4). None of these studies, however, determined the fitness of females attracted to these males.

Another criterion of fitness in birds may be plumage. Female large cactus finches (*Geospiza conirostric*) on Isla Genovesa, Galapagos, select a mate on the basis of courtship performance and black plumage of males. Both black plumage and courtship behavior indicate older males with past breeding experience and proven viability (Grant and Grant 1987). By mating with experienced males, females gain a fitness advantage in production of young and their recruitment into the breeding population.

Such behavior relates to the bright plumage hypothesis advanced by Hamilton and Zuk (1982). This hypothesis proposes that only healthy males can develop bright plumage. Females use differences in the brightness of male plumage as a criterion of health. There is some evidence, based on the many studies stimulated by this model, that a relationship

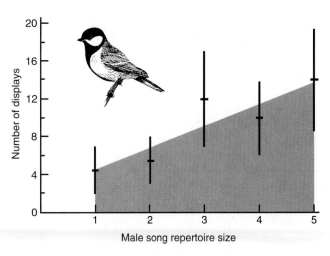

**Figure 20.3** Mean number of copulation-solicitation displays given by 11 female great tits as a function of male song presented in repertoire sizes ranging from one to five song types. (From M. C. Baker et al. 1986:495.)

**Figure 20.4** Male sedge warblers (*Acrocephalus schoenobaenus*) with more complex songs (larger repertoire of syllable types) attracted females for pairing before their rivals with simpler songs. Songs were recorded from a marked population. (K. K. Catchpole 1980:150.)

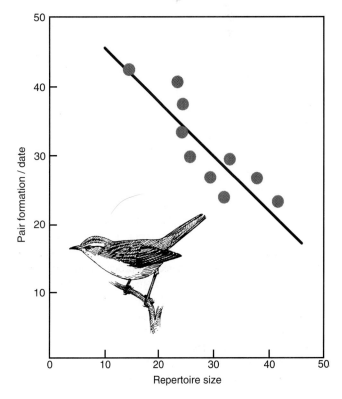

exists between infection of the male bird by coevolved endoparasites and the brightness of his plumage (Borgia and Collis 1989, Moller 1991, Zuk 1991). Males with lowest parasitic infection have brightest plumage. Females pick these males because they are the most disease-resistant. This hypothesis, controversial to say the least, still awaits an accumulation of data, intraspecific tests, and tests of alternative hypotheses (Zuk 1991).

Among polygamous species, the question becomes more complex. In those cases in which females acquire a resource along with the male, the situation is similar to that of monogamous relationships. Among birds, for example, the females show strong preference for males with high-quality territories (Plesczynska and Hansell 1980, Lenington 1980). On such territories with superior nesting cover and an abundance of food, females can attain reproductive success, even if this choice involves sharing the territory and the male with other females.

Among those fish species in which parental care of the eggs is an exclusive male function, females prefer to spawn with males already caring for another female's eggs, although the males give less care to adopted eggs (Sargent 1989). Males of some fish, such as the fantail darter (*Etheostoma flabellare*), capitalize on such female preferences by resorting to egg-mimicry. Territorial males excavate nests beneath flat rocks and care for eggs attached by the female to the ceiling of the nest. Male darters have each of the seven to eight spines of the distal end of the first dorsal fin tipped with fleshy knobs that appear as eggs when the males are resting beneath the rocks. Female darters preferred males with eggs over males without eggs, and males with egg-mimics over males without them (Knapp and Sargent 1989).

**Genes-Only Selection**    That females select males on the basis of resources they offer requires little explanation. What puzzles behavioral ecologists most is the basis of female choice in polygamy situations in which the males offer only genes and no resources. How does the female select the male with greatest fitness? This question has stimulated a body of growing research.

It would seem that the female has limited information with which to select a mate. She might select a winner from among males that best others in combat, as in bighorn sheep, sheep, elk, and seals. She might select mates based on intensity of courtship display or some morphological feature that may reflect a male's genetic superiority and vitality. Whatever the situation, the selection process comes down to salesmanship on the part of the male and sales resistance on the part of the female (G. C. Williams 1975).

Among some polygamous species, such as elk and seals, a dominant male does control a harem of females. Even in this situation the females are not placid and do have some choice. Protestations by female elephant seals over the attention of a dominant male may attract other large males nearby, who may attempt to dislodge the male from the group. Such behavior ensures that females will mate only with the highest-ranking male (Cox and LeBoeuf 1977).

In most cases females seem to control mate choice. Consider the long-tailed widowbird (*Euplectes progne*) of the African savanna. It is one of several species of widowbirds and whydahs (*Steganura* spp.) whose males possess highly elongated tails in their breeding plumage; but none can equal that of the long-tailed widowbird. The males are black with a red shoulder patch and a tail 50 cm long (Figure 20.5) which they display in courtship flight. A male may attract as many as six females (mottled brown, short-tailed birds the size of a sparrow) to their territories. A behavioral ecologist, M. Andersson (1982), investigated female choice in this species in an experiment involving males. From some males he cut the tail feathers to shorten them and glued these feathers onto the tails of other males to greatly lengthen them. On some males he clipped the feathers and glued them back in place, and others he left unclipped as controls. He quickly discovered that the females had a decided preference for the males with the artificially exaggerated tail feathers. These males attracted more mates and had a higher reproductive success (Figure 20.6) than both short-tailed and normal birds. Although the long tails did not seem to impair the males in territorial defense, Andersson did not determine whether the extra-long tails reduced the males' survival over time.

Such advertising can be costly for courting polygamous males, however. In the breeding season male three-spined sticklebacks (*Gasterostens aculeatus*), a small fish of marine and freshwater habitats in North America, Europe, and Pacific Asia, are polymorphic in the coloration of their throats. Some males have red throats and some do not.

**Figure 20.5** The long-tailed widowbird (*Euplectes progne*), unmistakable with its extra long tail and bright red shoulder bordered with white, inhabits the open grasslands of eastern South Africa.

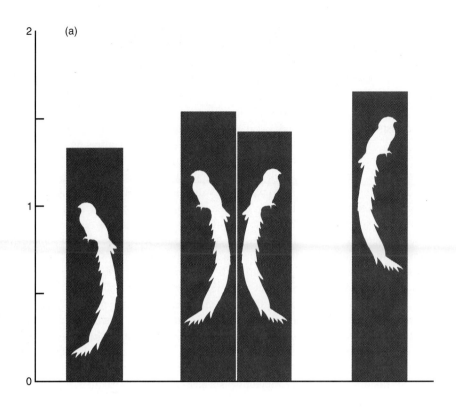

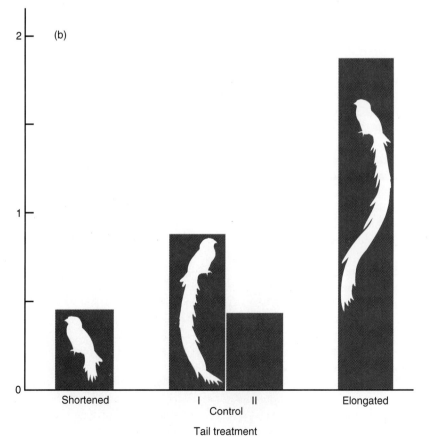

**Figure 20.6** Mating success in male long-tailed widowbirds subjected to different tail treatments. (a) Mean number of active nests for the nine males of the four treatment categories before the experiment. (b) Number of new active nests in each territory after treatment of males. (From Andersson 1982:819.)

Females seem to prefer to mate with red-throated males. Semler (1971) presented females a choice between two genetically non-red males, one of which had been given an artificial red throat painted with nail polish. Females selected artificially-colored males, even though they had no genetic differences with the plain-colored males. The polymorphism is probably maintained because red-throated males are subject to intense predation by rainbow trout. Thus a cost accrues to the red-throated males that acts against their reproductive advantage.

Lek Behavior Extreme examples of genes-only female choice appear in the lek species. These animals aggregate into groups on communal courtship grounds called **leks** (or arenas). Males on the lek defend very small territories that hold no resources and advertise their presence by colorful vocal and visual displays. Females visit the leks of displaying males, select a male, mate, and move on. Although few species engage in this type of mating system, it is widespread in the animal world, from insects and frogs to birds and mammals. The lek mating system is characterized by male defense of small clustered mating territories in an area where females have large overlapping ranges that the males cannot economically defend.

At least three hypotheses have been advanced to explain lek behavior (Davies 1991). One hypothesis, **female choice**, is that females show preference for a courtship arena because it is the safest place to mate or it forces males to cluster. Leks provide an unusual opportunity for females to choose a mate among the displaying males (Bradbury 1981). This advantage holds especially among lekking insects, such as certain tropical Drosophila, in which females are widely dispersed.

A related hypothesis is the **"hotspot" model** (Bradbury and Gibson 1983), which says males cluster in places where encounters with females are potentially high. If females have large overlapping ranges, males can ensure encounters with them by aggregating in overlap zones. For example, sage grouse (*Centrocercus urophasianus*) males display in leks near the females' wintering ranges (Figure 20.7). As females abandon the winter range for nesting areas, the males establish new lek areas near them (Bradbury et al. 1989).

A third hypothesis is the **"hotshot" model** (Beehler and Foster 1988). This model emphasizes an inequality of mating success. It envisions a strong hierarchy among the males with the dominant male, in smaller leks at least, displacing all others and leaving no opportunity for female choice. In other words, the visiting females select an arena and mate with the dominant males. Indeed, among some species this situation may be true, as in the case of the tropical manakins and birds of paradise. Among these birds one dominant male on the lek may perform over 90 percent of successful copulations (Beehler 1983). In spite of the odds of not mating, males congregate on the lek because by displaying together they draw in females from a larger area. By associating with

**Figure 20.7** Male sage grouse on a strutting ground or lek strut inflate the air sacs of the breast and spread their pointed tail feathers in courtship displays before the females.

and congregating about "hotshot" males with the most effective displays, subdominant and satellite males may be able to steal mating opportunities. A majority of matings on the lek, however, are done by a small percentage of the males in the male dominance hierarchy formed in the absence of females.

The hotshot model and the hotspot models may not describe the real situation for many species. Consider the fallow deer (*Dama dama*), a lekking ungulate of Europe, whose reproductive behavior has been followed by Clutton-Brock and his associates (1988, 1989). The fallow deer may defend and mate in resource territories, especially at low population densities. At high densities the males commonly defend small territories about 15 to 20 m in diameter on traditional mating grounds. Females visit and move through several territories before settling in one. The buck collects a harem of females, although each doe moves frequently among territories, rarely staying in one for more than several hours. The investigators found that bucks holding central territories possessed larger harems and had a higher mating success than those with holding edge territories. On the leks studied, the most successful male mated with 15 percent of 200 does and about 50 percent of the bucks on the lek did not mate at all. Thus it would appear that positions or hotspots on the lek could explain females' choices.

To test that hypothesis, the investigators pinned sheets of black polyethylene on the territories of the most successful bucks. Females refused to visit such territories, forcing the males to shift their territories and set up new display sites. The females moved with them, suggesting a strong choice of these particular males by the females. According to the hotshot model, males in territories adjacent to the original display site should shift their territories also, but they remained on the old lek. The results of these studies suggest that female fallow deer do choose their mates on the basis of male appearance and behavior and that territory position may not influence their choice. It is difficult, however, to separate territory position and mate choice, if the best males happen also to occupy the best sites.

# REPRODUCTIVE EFFORT

Natural selection favors individuals that produce the maximum number of reproducing offspring in a lifetime. The size and number of offspring vary greatly among plants and animals. Whether the offspring are large or small, few or many, parents allocate a certain amount of resources and time to fecundity. In doing so, they have to make tradeoffs among growth, maintenance, protection, and reproduction. The nature and amount of allocations to reproduction over a period of time are the organism's **reproductive effort.**

## Parental Care

Caring for young is a major reproductive expenditure: providing food, shelter, and protection from predators, brooding, grooming, and any related activity that increases the fitness of the offspring. The kind and amount of care, which vary widely among species, is influenced by the maturity of young at birth. Some birds incubate eggs for a long period of time, investing considerable energy prior to hatching. Some mammals have much longer gestation periods than others and give birth to young in a more advanced stage of development.

The degree of development at hatching or birth among birds and placental mammals, the eutherians, translates into the two major types of young produced: precocial and altricial. **Precocial** young are able to move about at or shortly after birth, although they may have a long infancy and grow slowly. **Altricial** animals are born helpless, naked or nearly so, and often blind. They grow rapidly and mature early. Between the two extremes is a wide variation in the degree of development (Nice 1962, Derrickson 1992) (Table 20.2). Among noneutherian mammals (monotremes and marsupials), the marsupials give birth to tiny and poorly developed young that possess the minimal anatomical development needed to survive outside of the uterus. Attached to a nipple in the pouch or marsupium, the young marsupial continues its development and leaves the pouch at about the same weight as a newborn placental mammal of the same relative size.

Among birds, the most altricial offspring occur among the order Passeriformes, the perching birds. The most precocial are the mound-building megapods (Megapodiidae), such as the malleefowl (*Leipoa ocellata*) of southern Australia. They lay their eggs in a nest mound and allow heat from the sun and from decaying vegetation within the mound to incubate them. Upon hatching, the well-developed young burrow to the surface and run into the bush.

**Table 20.2  Classification of Maturity at Hatching in Birds**

| Type | Characteristics | Example |
|---|---|---|
| **Feed Selves** | | |
| Precocials | Eyes open, down-covered, leave nest first day or two | |
|   Precocials 1 | Independent of parents | Megapods |
|   Precocials 2 | Follow parents but find own food | Ducks, shorebirds |
|   Precocials 3 | Follow parents and are shown food | Quail, chickens |
| **Fed by Parents** | | |
|   Precocials 4 | Follow parents and are fed by them | Grebes, rails |
| Semiprecocials | Eyes open, down-covered, stay at nest though able to walk | Gulls, terns |
| Semialtricials | Down-covered, unable to leave nest | |
|   Semialtricials 1 | Eyes open | Herons, hawks |
|   Semialtricials 2 | Eyes closed | Owls |
| Altricials | Eyes closed, little or no down, unable to leave nest | Passerines |

*Source:* Adapted from Nice 1962.

Within several hours they are able to fly and to lead fully independent lives.

Most mammals are altricial or semiprecocial. Among the most precocial mammals are the seals. Not only is their delivery extremely rapid, approximately 45 seconds in the gray seal (Bartholomew 1959), but movements and vocalizations appear shortly after birth. Newly born fur seals are able to rise up and call from 15 to 45 seconds after birth and are capable of shaky but effective locomotion a few minutes later. Even while the umbilical cord is still attached, the pups are able to shake off water, nip at each other, and scratch dog-fashion with the hind flippers. The cows are protective and attentive only between parturition and estrus, about a week's time, although they nurse the young much longer.

Many species of fish, such as the cod (*Gadus morhua*) lay millions of floating eggs that drift free in the ocean with no parental care. Other species, such as bass, lay eggs in the hundreds and provide some degree of parental care. The quality of parental care appears to be related to egg size (Sargent et al. 1987). The larger the egg size relative to the size of the fish, the greater the amount of parental care. This extra care apparently relates to the longer time offspring take to develop and absorb the yolk sac and become juveniles. Juveniles that hatch from larger eggs experience lower mortality, faster growth, and earlier sexual maturity.

Among amphibians, parental care is most prevalent among tropical anurans (frogs and toads) and is poorly developed in salamanders. Females of some species of salamanders, such as the dusky salamander (*Desmogathus fucus*) of the eastern United States, remain with the eggs until they hatch. Males of some tropical anurans carry tadpoles in their vocal pouches or on their backs. In other species the eggs develop on the backs of females or in a brood pouch formed in the dorsal skin (Figure 20.8). Male midwife toads (*Alytes obstetricans*) of Europe carry eggs and developing young on their backs. All of these species eventually place them in a suitable environment for further growth.

Internal fertilization and terrestrial reproduction are fully developed among reptiles. Few eggs well supplied with yolk may be carried inside the mother's body until they hatch; or they may be placed in nests buried in the ground and given little subsequent care. Crocodiles, however, are the exception in reptiles. They actively defend the nest and young for a considerable period of time.

Except for the social insects, parental care is not well developed among invertebrates. Some retain eggs within the body until they hatch; others, such as crayfish, carry eggs externally. Invertebrate parental care is most highly developed in social ants, bees, and wasps. Social insects provide all five functions of parental care: food, defense, heat, sanitation, and guidance. (See Tallamy 1984 on insect parental care.)

We may be stretching the imagination a bit to consider maternal care in plants; yet the plant does have an influence on its offspring beyond its genetic contribution (Schall

**Figure 20.8** The South American hylid frog *Gastrotheca cornuta* carries the developing eggs in a brood pouch of skin on her back.

1984). The seed-producing parent has a nurturing role on the developing seeds, which influences the offspring's fitness. This relationship was demonstrated in intensive greenhouse studies of the seed size and germination and seedling development in the Texas bluebonnet (*Lupinus texensis*). When released from the parent plant, the larger seeds, supplied with greater nutrient resources, had a higher germination rate and higher seedling survivorship and seedling biomass than did small seeds and seedlings.

## Parental Investment

Costs of reproduction are high in terms of energy expenditure and survival (Clutton-Brock and Godfray 1991). The physical drains and energy costs of brooding, gestation, birth, feeding the young, lactation, and defense of the females are great. For polygamous males, in particular, the strain of defending females, fighting off male competitors, and sexual activity shortens life expectancy and future reproductive activity.

Parents have only a limited amount of energy. From this amount they have to allocate a certain amount to their own maintenance and survival. To make its maximum contribution to future generations, an organism must balance the profits of immediate reproductive investments of time and energy against the costs to future prospects, including fecundity and its own survival.

Suppose you were given $1000 to spend. You could spend the entire $1000 at once on some item. You would

have made a one-time investment of your money; you would have nothing left for future expenditure. As another alternative, you might spend part of the money on one item and place the balance in a savings account. Some time later you might withdraw some of the money for another expenditure, which would reduce the amount you have to spend in the future. At each withdrawal, you further reduce the amount for future expenditure, until the whole $1000 is gone. How long the money lasts depends on the amount and number of withdrawals for present expenditure.

In a similar manner organisms work under a budget. Because the organism has to allocated its energy budget to different expenditures, it will have to make some tradeoff. The organism can expend all its energy in one major reproductive effort in its lifetime, or it can opt for repeated reproductive bouts, producing fewer young each time and saving some of its energy for future reproduction.

Those organisms that go for one major reproductive effort in a lifetime are termed **semelparous.** They invest all their energy in growth, development, and energy storage, and then expend that energy in one massive suicidal reproductive effort. Such a reproductive strategy is employed by most insects and other invertebrates, some species of fish, notably salmon, and by many plants—annuals, biennials, and some bamboos. Many semelparous plants, such as ragweed, are small, short-lived, and occupy ephemeral or disturbed habitats. For them it would not pay to hold out for future reproduction, for their chances of future reproduction are slim. They gain their maximum fitness by expending all their energies in one bout of reproduction.

Other semelparous organisms, however, are long-lived and delay reproduction. Mayflies (Ephemeroptera) may spend several years as larvae before emerging from the surface of the water for an adult life of several days devoted to reproduction. Periodical cicadas spend 13 to 17 years below ground before they emerge as adults to stage an outstanding exhibition of single-term reproduction. The Hawaiian silverswords (*Argyroxiphium*) live 7 to 30 years before flowering and dying. Some species of bamboo delay flowering for 100 to 120 years, produce one massive crop of seeds, and die (Janzen 1976). The Haleakala silversword (*A. sandwicense*) (Figure 20.9), found on the summit of Haleakala volcano on Maui in Hawaii, lives under environmental conditions that are extreme: intense light, periods of strong heat and cold, and little moisture. The plant accumulates energy and moisture within its leaves, which it expends in a very short period of flowering once in its lifetime. Environmental conditions preclude repeated reproduction. In general, for a species to evolve semelparity, that mode of reproduction has to increase fitness enough to compensate for the loss of repeated reproduction.

Organisms that produce fewer young at one time and repeat reproduction throughout their lifetime are termed

**Figure 20.9** The Haleakala silversword stays in a rosette stage for an unknown number of years before forming a massive flowering stalk up to 2 m tall covered with maroon-colored flowers in large heads. After the seeds form, the plant dies.

**iteroparous.** Iteroparous organisms include most vertebrates, perennial herbaceous plants, shrubs, and trees. For an iteroparous organism the problem is timing reproduction—early in life or at a later age. Whatever the choice, it involves tradeoffs. Early reproduction means less growth, earlier maturity, reduced survivorship, and reduced potential for future reproduction. Later reproduction allows for increased growth, later maturity, and increased survivorship, but decreased fecundity after each bout of reproduction. In effect, if an organism is to make a maximum contribution to future generations, it has to balance the profits of immediate reproductive investments against costs to future prospects, including fecundity and its own survival. It has to make a tradeoff between present progeny and future offspring (G. B. Williams 1966). The situation is defined by partitioning the equation for reproductive value, discussed in Chapter 18,

into two parts, one for present reproduction and one for residual future reproduction:

$$V_x = m_x \sum_{t=x+1}^{\infty} \frac{l_t}{l_x} t$$

In words, reproductive value at age $x$ equals present progeny (current investment) plus expected future progeny (future investment). Energy expended in repeated reproduction weighs against future prospects, as measured by declining fecundity and reduced potential for survival (Figure 20.10). Thus a parent that has invested heavily in early reproduction will have less to invest in later fecundity and survival. Increased early investments will require that the organism increase its level of present investment in reproduction, because of decreased probability of survival. Somewhere in between lies the best strategy for **optimal lifetime reproductive success,** the sum of present and future reproductive success.

Why some plants are annuals or semelparous and others perennial or iteroparous has intrigued ecologists. What are the evolutionary advantages of semelparity and iteroparity, of reproducing and dying in a single season or reproducing over a number of seasons? What would an annual plant gain by becoming a perennial? L. C. Cole (1954) assessed this question mathematically. He calculated that the gain would be equivalent to adding one more offspring to an annual's output. Because it is easier to make one more sperm, egg, or seed than to survive for an additional year, Cole wondered why an organism should be a perennial. The answer, Charnov and Schaffer (1973) proposed, is that survivorship

varies from one life cycle phase to another, and the young have lower survivorship than older individuals. A perennial life cycle or iteroparity is favored in those situations in which the ratio of juvenile survival to adult survival is low. An annual life cycle or semelparity is favored where the ratio of juvenile survival to adult survival is high.

For an iteroparous organism to achieve high fitness, it has to settle for neither maximum present reproduction nor maximum future reproduction, a chancy gamble. An optimal level of parental investment balances the rate of return from investment in the present with the rate of return from investment in the future. Present investment in fitness should continue to grow through reproductive success of offspring while future investments in fitness continue to decline (see Sargent and Gross 1985; Coleman et al. 1986) (Figure 20.11).

Empirical evidence from natural populations comparing growth, survival, and future reproduction of reproductive and nonreproductive individuals is weak and confusing. There have been manipulative experiments with plants to

**Figure 20.11** Effects of present investment on reproductive success. Present reproductive success (P) is assumed to increase with diminishing returns (reproductive success of adult offspring) on present investment, while future reproductive success declines with present investment. Lifetime reproductive success is the sum of present and future reproductive success (P + F). The optimal level of present investment (I) is that which brings about maximum lifetime reproductive success (P + F). Although theoretical, these models and that of Figure 20.10 provide some insights on why birds abandon a nest after some effort in defense or why grizzly bears abandon young as a reproductive tactic (see Tait 1980). (After Coleman et al. 1985:60.)

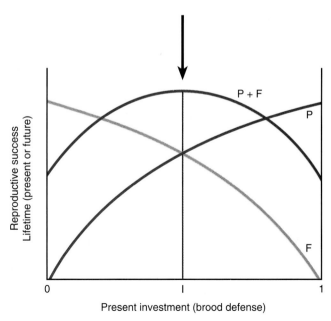

**Figure 20.10** A graphic analysis of C. B. Williams' (1966) tradeoff model between fecundity and adult survival. Parental investment in offspring is measured by brood defense. As investment increases, offspring survival(s) increases and parental survival (S) decreases. The point of intercept represents optimal parent investment for survivorship and future fecundity. Beyond this point parents are sacrificing future investment for present investment. (After Sargent and Gross 1985:44.)

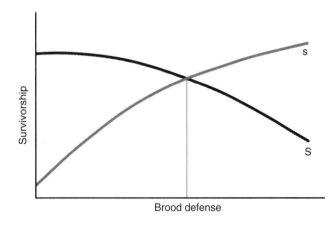

increase and decrease the costs of reproduction in plants by removing leaves (which decreases the leaf area of the plant) or pollinating the flowers (which increases reproductive output). In some of these studies, plants manipulated for one to two years show lower survival, growth, and reproduction the following year; others do not. For four years Primack and Hall (1990) followed the costs of reproduction for randomly selected hand-pollinated pink lady's slipper orchids (*Cypripedium acaule*) compared to insect-pollinated plants. Because the lady's slipper depends upon bees for pollination, plants growing in the wild rarely develop fruit. Costs of annual reproduction in hand-pollinated plants did not show up until the third and fourth years, when the fruiting plants had lower rates of growth and flowering than the control plants. An average-size fruiting

plant experienced a 10 to 13 percent decrease in leaf area and a 5 to 16 percent decrease in the probability of flowering the following year.

Demonstrating the costs of reproduction is no easier in animals. In some mammals it appears that energy expended in repeated reproduction weighs against future prospects as measured by declining fecundity and reduced potential for survival. Red deer females debilitated by intense reproduction in one year perform poorly the next year. Milk hinds, ones who suckled a calf the previous year, have higher reproductive costs and higher mortality than yeld hinds, ones that did not produce a calf the previous year (Figure 20.12) (Clutton-Brock 1984). In the American bison (*Bison bison*) rearing sons is more debilitating to the mother than rearing daughters, because the

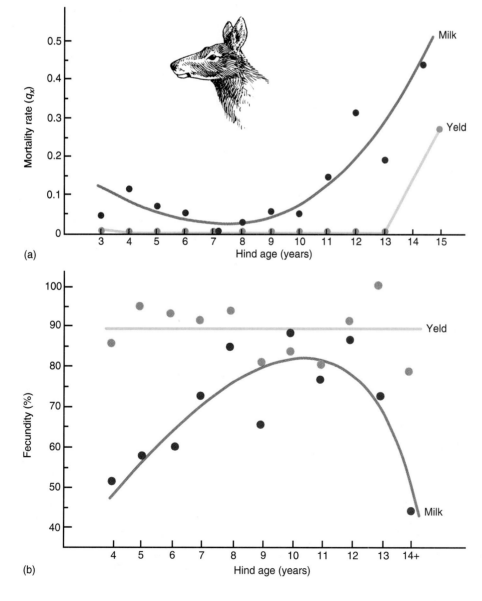

**Figure 20.12** (a) Age-specific mortality of red deer milk hinds (who have reared a calf to weaning age) and yeld hinds (who have not). Milk hinds have a higher reproductive cost, which increases mortality. (b) The effect of that mortality and past investment in young is decreased future fecundity. (From Clutton-Brock 1984:224, 216.)

sons suckle longer—up to 15 months (Wolff 1988). Cows that have produced sons breed later than other cows and and are more likely to be barren than cows that have produced daughters the previous year.

## Parental Energy Budgets

How organisms allocate energy to growth and reproduction is central to their reproductive strategy. If an organism directs more energy to reproduction, it has less energy to allocate to growth and maintenance. As a result the individual may grow more slowly to the next age or fail to reproduce or to survive. For example, Lawler (1976) found that reproductive females of the terrestrial isopod *Armadillidium vulgare* had a lower rate of growth than nonreproductive females. The nonreproductive females devoted as much energy to growth as the reproductive females devoted to both growth and reproduction. Tinkle (1969) demonstrated that among lizards, species with larger fecundity have poorer individual survival. They channeled more energy into egg production than to growth over a short period, resulting in heavy physiological stress. Many North American freshwater fish experience a considerable loss of size or growth from reproduction, up to 25 percent or more in some trout (*Salmo*) (Bell 1980). Energy costs are more difficult to determine in homeotherms, although the costs show up in survival and future reproduction. Birds expend about four times the basal metabolic rate in feeding their young (Robbins 1983), and lactating mammals use up 2.5 to 5 times basal metabolic energy (Gittleman and Thompson 1988).

Plants, too, must allocate a portion of their net annual assimilation to reproductive effort. The amount plants allocate to vegetative reproduction can influence seed production and the growth of the genet (Fitter 1986). Annual species of plantain (*Plantago*) had higher reproductive costs, based on milligrams of seeds produced per square centimeter of leaf area, than perennial species (Primack 1979). Perennials, however, can allocate a high amount of yearly biomass to reproduction. For example, the meadow buttercup (*Ranunculus acris*) allocates up to 60 percent of its biomass to reproduction, and the showy goldenrod (*Solidago speciosa*) of old fields and woods allocates up to 35 percent (Abrahamson and Gadgil 1973). Many plants that grow under a range of environmental conditions are flexible in adjusting energy allocation to reproduction. Plants that grow in more adverse environments allocate more energy to reproduction than those in more moderate habitats.

The shortcomings of such estimates are many. In plants the estimates fail to take into account belowground energy allocations, including the production of bulbs, corms, and roots. Simply comparing aboveground biomass of vegetative parts to biomass of reproductive output may not be an accurate assessment of the energy costs of reproduction, as

Reekie and Bazzaz (1987) found in intensive studies of reproductive costs in quack grass (*Agropyron repens*). They point out that resource allocation is not necessarily a reliable indicator of costs of reproduction, because plants experience different costs. A plant with a low cost of reproduction could afford to allocate a larger proportion of its resources to reproduction.

**Clutch Size**   To produce viable offspring, organisms need a certain minimal amount of energy to be apportioned among a certain number of young. As parents divide available energy among an increasing number of offspring, the fitness of individual offspring declines. Eventually parents reach a point at which increased expenditure or investment per offspring from birth to independence results in declining payoff in the fitness of their offspring (Figure 20.13) (C. C. Smith and Fretwell 1974).

Two basic choices are to apportion parental investment among many small young or to concentrate their investment in a few larger ones. Within that range of choices the parents have to adjust the number of young they can rear with the resources available without significantly reducing their own fitness and survival, as well as the fitness of their offspring.

One method by which parents can optimally apportion investment in offspring is adjustment of brood or clutch size.

**Figure 20.13** The Smith-Fretwell model of tradeoffs between size and number of offspring. The fitness curve shows how the success of an offspring varies with the amount of parental care the offspring receives. The parent obtains maximum return on its investment at the point where the line from the origin is tangent to the offspring fitness curve. (C. C. Smith and Fretwell 1974:501.)

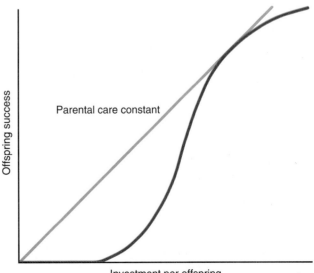

David Lack (1954) originally hypothesized that among birds clutch size evolved through natural selection to correspond to the average largest number of young the parents can feed. There is indeed evidence that the optimal clutch size should be lower than that which would produce the maximum number of young each season. This hypothesis can and has been tested experimentally in birds simply by manipulating clutch size, especially by adding additional eggs, usually one, to the nest.

What happens if birds attempt to rear more than an average-sized brood? If Lack's hypothesis is correct, an experimental increase in brood size at hatching should result in fewer young surviving to reproductive age. About one-third of such studies involving birds support Lack's hypothesis (Lessels 1986); mean clutch size was consistently less than most productive brood size in those studies. In 35 out of 60 studies, however, increase in brood size resulted in an increase in the number of young fledged, without necessarily affecting parental survival (Lessels 1991). In one such study Gustafsson and Sunderland (1988) experimentally modified the clutch size of altricial young of the collared flycatcher (*Muscicapa latirostris*) by increasing or reducing the natural clutch size by one or two eggs. The unmanipulated clutch size had no effect on adult survival, juvenile survival, or future reproduction. The large-sized clutches produced more young but at the cost of lower juvenile survival, subsequent reduced fecundity of the parents, and lower fecundity in the offspring. Reduced clutches produced fewer young. Natural clutch size produced more recruits than the enlarged or reduced clutches. C. M. Lessels (1986) manipulated the brood size of monogamous Canada geese, a species with precocial young that do not need to be fed by the parents. Because gosling survival is independent of brood size, larger broods produced more surviving young. The larger broods, however, had an adverse effect on weight and timing of molt in the nesting females, but not on the males. Birds handling larger clutches bred later the following year, but the clutch size was unaffected. Such studies suggest that physiological and ecological factors probably set clutch size among birds at an optimum level that results in maximum lifetime reproduction, as Lack postulated.

A similar relationship holds among some insect parasitoids, but because no parental care is involved, the constraints involve hosts in which the insects lay their eggs. Two models are proposed (Parker and Courtney 1984): (1) individuals should adjust their clutch size so that large clutches are laid on hosts that can support the growth of more offspring; and (2) as time between bouts of egg-laying increases, usually when hosts are rare, large clutches should be laid in higher quality hosts. The swallowtail butterfly that lays its eggs on the leaves of the pipevine (*Aristolochia reticulata*) follows this pattern (Pilson and Rausher 1988). The females lay larger clutches on younger plants with more edible leaves and terminal buds and consistently discriminate between high- and low-quality plants. Hardy and his associates (1992) manipulated the clutch size of an ectoparasitoid wasp (*Goniozus nephantidis*), which lays a clutch of up to 20 eggs on small caterpillars. They artificially created clutches of different sizes on hosts of approximately the same weight. Increasing clutch size had little effect on juvenile survival, but it did result in smaller adult females. In nature clutch size of such parasitic wasps is strongly influenced by host size. The larger hosts support larger clutches.

When parents cannot predict available food resources at the time of nesting (which they usually cannot), they may have to adjust to the situation by brood reduction. The decision is when to withdraw investment from one or more offspring. Theoretically parents should attempt to rear all of the offspring until they arrive at the point where the potential fitness of the discontinued offspring (if cared for) is less than the augmented fitness of the remaining offspring.

Decisions in brood reduction, of course, are not consciously made by the parents. It comes about largely through asynchronous hatching and siblicide. In asynchronous hatching the young are of several ages. The older siblings may beg more vigorously for food, forcing the harried parents to ignore the calls of the younger, smaller sib, who perishes, or the older or more vigorous young may simply kill their weaker sib. For example, the common grackle (*Quiscalus quiscula*) begins incubation before its clutch of five eggs is complete, a pattern of hatching ensuring survival of some young under adverse conditions (Howe 1976). The eggs laid last are heavier and the young from them grow fast. However, if food is scarce, the parents fail to feed these late offspring because of more vigorous begging by the siblings. The last-hatched young then die of starvation. Thus asynchronous hatching favors the early-hatched young at the expense of those hatched later. A similar situation exists in western gulls (*Larus occidentalis*) (Sydeman and Emslie 1992). Although parents attempt to ensure the survival of all young, their investment in older young is protected by starvation of late-hatched birds if a food shortage arises.

A number of birds, including raptors, herons, egrets, gannets, boobies, and skuas, practice siblicide. The parents normally lay two eggs, possibly to insure against infertility of a single egg. The larger of the two hatchlings kills the smaller sibling or runt, and the parents redirect all resources to the surviving chick (Godfray and Harper 1990). These birds are not alone in siblicidal tendencies. The females of some parasitic wasps lay two or more eggs in a host and the larvae fight each other until only one survives (Godfray 1987).

Lack (1954) also proposed that clutch size is an adaptation to food supply. According to Lack's hypothesis, temperate species have larger clutches because increasing

daylength allows parents a longer time to forage for food to support larger broods. In the tropics, where daylength does not change, food becomes a limiting resource. Cody (1966) modified this concept by employing the principle of allocation of energy. He proposed a second hypothesis, that clutch size results from different allocations of energy to egg production, avoidance of predators, and competition.

A third hypothesis, proposed by Ashmole (Ashmole 1963, Ricklefs 1980), states that clutch size varies in direct proportion to the seasonal variation in resources, largely food supply, used by a population. Population density is regulated primarily by winter mortality, when resources are scarce. Available resources during the breeding season relative to breeding population density influence clutch size and number of young reared. Greater winter mortality makes more food available to the survivors during the breeding season. This resource availability is reflected in larger clutches. Thus geographical variation in mean clutch size is more strongly correlated with winter food availability and production, which directly influences the size of the breeding population, than with summer food production.

All three hypotheses predict that clutch size is higher in high latitudes than low. A number of field examples agree. Birds in temperate regions have larger clutch sizes than birds in the tropics (Figure 20.14), and mammals at higher latitudes have larger litters than those at lower latitudes (Lord 1960). Lizards exhibit a similar pattern. Those living at lower latitudes have smaller clutches, have higher reproductive success, reproduce at an earlier age, and experience higher adult mortality than those living at higher latitudes (Tinkle and Ballinger 1972, Andres and Rand 1974).

Insects, too, such as the milkweed beetle (*Oncopeltus*) support the hypotheses (Landahl and Root 1969). Temperate and tropical milkweed beetles have a similar duration of the egg stage, egg survivorship, developmental rate, and age at sexual maturity. Although the clutch sizes are the same, the temperate species lay a larger number of eggs because they lay more clutches. Total egg production of the tropical species is only 60 percent of that of the temperate species.

Plants follow the general principle of allocation on a latitudinal basis. McNaughton (1975) investigated the allocation of resources to reproduction in a series of greenhouse studies of cattail (*Typha*). He measured reproductive effort by the growth of rhizomes, the cattails' principal means of expansion within a given habitat. Cattails shed their very small seeds to the wind as means of colonizing a new habitat. McNaughton considered three species of cattails on a climatic gradient: the common cattail (*Typha latifolia*), a climatic generalist that grows from the Arctic Circle to the equator; the narrowleaf cattail (*T. augustifolia*), restricted to the northern latitudes of North America; and *T. domingensis*, restricted to the southern latitudes of North America. The cattails exhibited a climatic gradient in the allocation of energy to reproduction. *T. angustifolia* and northern populations of *T. latifolia* grew earlier and faster and produced a greater number of rhizomes than the *T. domingensis* and southern populations of *T. latifolia*, although the southern cattail produced larger rhizomes. Faced with stronger intra- and interspecific competition, the southern cattail invested more energy in vegetative growth. Likewise, tropical members of the sunflower tribe Heliantheae in the aster family (Asteraceae) have significantly smaller clutch sizes (ovules per flowering head) than those of temperate and alpine regions, with an average of 51.2, 82.7, and 89.6 ovules respectively (Levin and Turner 1977).

Although reproductive output among organisms does appear to be greater at higher latitudes, the number of comparable species for which we have data is few, too few to consider the hypotheses as a general law. Many more studies along a latitudinal gradient from the tropics to arctic regions are needed.

**Hermaphrodism and Gender Change** The discussion of reproductive costs relates mostly to species among which males and females are separate individuals. Cost allocation among hermaphrodites, particular sequential hermaphrodites, has interesting differences. Hermaphrodism, especially

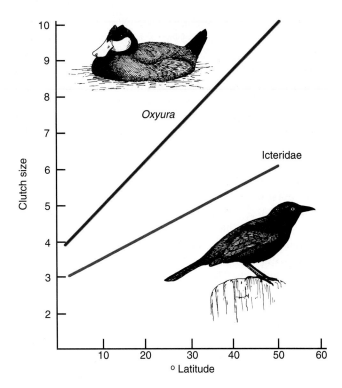

**Figure 20.14** Relationship between clutch size and latitude in birds. Represented as examples are the subfamily Icterinae (blackbird, orioles, and meadowlarks) in North and South America and the worldwide genus *Oxyura* (ruddy and masked ducks) of the subfamily Anatinae. (Adapted from Cody 1966.)

common in plants, probably evolved among those organisms in which the female could take on the male function by relinquishing only a small amount of female function. Similarly, the male could add a female function without seriously compromising sperm production (Charnov 1982). In hermaphrodism the contribution of combined male and female exceeds the contributions that would be made by separate sexes.

Mating in hermaphroditic organisms can involve either outcrossing or self-fertilization. Apparently most hermaphroditic animals are not self-fertilized. Earthworms mate with other individuals, as do hermaphroditic coral reef fish. Some animal hermaphrodites are completely self-sterile, and only a few, like certain land snails, are self-fertilized.

Among animal hermaphrodites, how is reproductive effort allocated between egg and sperm and between male and female behaviors? The male side of behavior in a hermaphrodite attempts to fertilize the eggs of other hermaphrodites, increasing the reproductive success of the male. The female side attempts to secure fertilization of her eggs to enhance her reproductive success.

Once the eggs are laid, should the hermaphrodite male defend the eggs of its partner or those of its female aspect? Fischer (1981) investigated this problem in the hermaphroditic coral reef fish *Hypoplectrus nigracans,* the black hamlet. He found that the reproductive success of the hamlet as a male depends upon its ability to reproduce as a female, because spawning partners "trade eggs," giving up eggs to be fertilized in exchange for the opportunity to fertilize those of another individual. Courtship is largely a female function rather than a male function; it advertises that an individual has eggs. Each fish parcels out its daily clutch in four to five spawns, and it alternates sex roles with each spawn, taking turns giving up parcels to be fertilized.

Many hermaphroditic plants are self-compatible but have evolved different means to prevent self-fertilization. Anthers and pistils may mature at different times, or the pistil may extend well above the stamens. Other plants have evolved more effective means. One is a genetic mechanism that prevents the growth of a pollen tube down the style of the same individual. Some hermaphroditic species are divided into two or three morphologically different types between which pollination takes place.

Although many hermaphroditic organisms possess mechanisms to reduce self-fertilization, the capacity for self-fertilization does carry advantages, especially among plants. A single self-fertilized individual is able to colonize a new habitat and then reproduce itself, establishing a new population. Other hermaphrodites produce self-fertilized flowers under stressed conditions, ensuring a new generation. Jewelweed, for example, under normal and optimal environmental conditions produces cross-pollinated (chastogamous) flowers. Under adverse environmental conditions or after the chastogamous flowers have set seed, jewelweed (as well as violets and other species) produces tiny, self-fertilized (cleistogamous) flowers that never open (Figure 20.15).

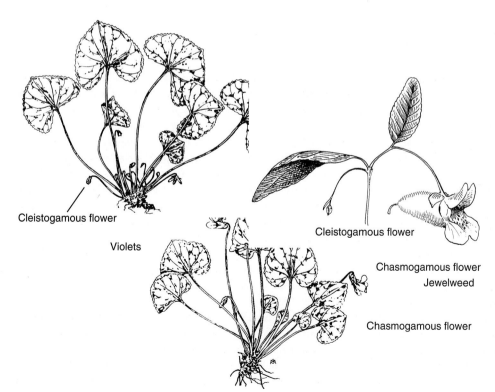

Cleistogamous flower

Violets

Cleistogamous flower

Chasmogamous flower
Jewelweed

Chasmogamous flower

**Figure 20.15** The normal cross-pollinated flowers of jewelweed and violets are termed chasmogamous. They are conspicuous and produce obvious seeds. But hidden beneath the leaves of jewelweed and at ground level in violets are tiny self-fertilized cleistogamous flowers that never open but also produce seed. The seeds of cleistogamous flowers are important to the maintenance of local adapted populations of these plants.

They have vestigial petals, no nectar, and few pollen grains and remain in a green, budlike stage. Thus, if outcrossing fails or never develops, the plants have ensured a next generation by self-fertilization.

The above discussion applies to simultaneous hermaphrodism. More interesting is sequential hermaphrodism, in which individuals budget their reproductive costs by changing gender as they grow during their life cycle. Gender change, in which the individual switches between a costly female reproductive function and a less expensive male, maximizes individual reproductive success.

Gender change among animals, notably fish species, appears to be stimulated by a social change involving sex ratios in the population. Sex reversal among some species of marine fish can be initiated by the removal of one or more individuals of the other sex. Among some coral fish, removal of females from a social group stimulates an equal number of males to change sex and become females (Friche and Friche 1977). In other species, removal of males stimulates a one-to-one replacement of males by sex-reversing females (see Shapiro 1979, 1987). Among the mollusks, the Gastropoda (snails and slugs) and Bivalvia (clams and mussels) have sex-changing species, and almost all of them change from male to female (W. G. Wright 1988). For example, if a young slipper limpet (*Crepidula* spp.) of rocky shorelines settles on its adult habitat without others of its own kind, it grows to a large size and develops into a female. If it settles on an occupied rock, it will develop into a small male. The male, however, changes rapidly into a functional female if larger occupants are removed.

Plants also exhibit a gender change (Freeman et al. 1980). One such plant is jack-in-the-pulpit (*Arisaema triphyllum*) (Policansky 1981, Lovett, Doust and Cavers 1982, Bierzychudek 1982a, 1984), a clonal woodland herb whose genet is a perennial corm and whose ramet is a single annual shoot (Figure 20.16). Jack-in-the-pulpit produces staminate (male) flowers one year, an asexual vegetative shoot the next, and a carpellate (female) or monoecious shoot the next. Over its life span a jack-in-the-pulpit may produce both genders as well as an asexual vegetative shoot, but in no particular sequence. The sequence may be from asexual to male or female, from male to asexual or female, or from female to asexual to male. Usually an asexual stage follows a gender change. Gender change in jack-in-the-pulpit appears to be triggered by an excessive drain on the photosynthate by female flowers. If the plant is to survive, one carpellate flowering could not follow another. To avoid death, the plant reduces its reproductive effort the next year by changing its gender or becoming vegetative. Among jack-in-the-pulpits, the size of the plant is more important than its sexual state in determining its future sex. Large males are more likely to be female the next spring than are small females (Policansky 1981).

**Figure 20.16** The jack-in-the-pulpit can change from asexual to male or female, from female to asexual to male, or from male to asexual to female, depending upon energy allocations and reserves. The plant gets its name from the spadix enclosed in a hoodlike sheath. This fruiting plant is the female stage.

Age and Size For many species clutch size and fecundity is related to age and size of the parent. This relationship is especially strong in plants and poikilothermic animals. Among plants perennials delay flowering until they have attained a sufficiently large leaf area to support seed production. Many biennials living in poor environments also delay flowering beyond the usual two-year life span, until environmental conditions become more favorable. Annuals show no such relationship between leaf area and the percentage of energy devoted to reproductive output, but size differences among annuals do result in differences in the number of seeds produced (Primack 1979). Small plants produce fewer seeds, even though the plants themselves may be contributing the same proportionate share to reproductive effort as larger plants.

Age at maturity and seed production vary among trees. Some, like Virginia pine (*Pinus virginianus*) start to produce seed at five to eight years when growing in open stands. In dense stands, Virginia pine may delay seed production for 50 years. Quaking aspen (*Populus tremuloides*) produces

seeds at the age of 20 years and white oak not until 50 years. Acorn production varies with size. Trees 16 inches in diameter at breast height produce about 700 acorns; those of 24 to 26 dbh produce 2000 or more acorns (Downs and McQuilkin 1944).

Fecundity in fish increases with size, which in turn increases with age. Because early fecundity reduces both growth and later reproductive success, fish obtain a selective advantage by delaying sexual maturation until they grow larger. Gizzard shad (*Dorosoma cepedianum*) reproducing at two years of age produce about 59,000 eggs. Those delaying reproduction until the third year produce about 379,000 eggs. Among the gizzard shad only about 15 percent spawn at two years of age and about 80 percent at three years.

The optimal clutch size of the loggerhead sea turtle (*Caretta caretta*) is constrained by the female's egg-carrying capacity, which is related to body size (Hays and Speakman 1991). Because of the high energy sea turtles expend on land, they may be under selective pressure to maximize clutch size to minimize the energy expended per egg.

An apparent relationship also exists between body size and fecundity among homeotherms. Heavier females are more successful in reproduction and experience increased survival (Sadlier 1969). For example, body weight of female European red squirrels (*Sciurus vulgaris*) in Belgian forests was strongly correlated with lifetime reproductive success (Wauters and Dhondt 1989) (Figure 20.17). Squirrels weighing less than 300 g did not come into estrus. Among red deer, young hinds and old hinds enjoyed the highest survival of their calves and middle-aged hinds the poorest (Clutton-Brock et al. 1982). This difference probably re-

flects the better body condition of young hinds and a greater energy investment in offspring by experienced old hinds. Such an investment by old hinds, however, is not reflected in another measure of fecundity, reproductive value, which usually declines with age.

## GENDER ALLOCATION

Another significant aspect of reproduction is that perennial question: is it a girl or a boy? Because gender in many animals is randomly determined by the X and Y chromosomes (or some variation thereof), we expect on the average a 50:50 ratio. R. A. Fisher (1930) carried the explanation one step further to explain why sex ratios tend to be equal. Natural selection favors those parents who invest equally in their sons and daughters. When an average son and an average daughter are equally expensive to produce, the sex ratio of the offspring at the end of parental care should be 1:1. The costs of rearing each sex, however, may not be equal. One sex may cost parents less to produce and rear than the other because it requires less food or other resources, has a higher mortality during the period of parental care (thus freeing additional resources for the remaining siblings), or imposes less stress on the mother. Eventually the sex ratio will become skewed toward the less expensive sex. If the sex ratio becomes skewed toward females, any factor that leads the bearer to produce more sons than daughters would be selected for. The reason is that sons would mate, on the average, with more than one female and thus more efficiently transmit genes to future generations. If the population sex ratio, however, is biased toward males, sons will mate, on the average, with less than one female, so any gene that favors female offspring will be selected for. When the sex ratio is at equilibrium, then selection for either sex ceases.

Among the higher vertebrates there is virtually no genetic variability of the sex ratio near fertilization; it is close to 1:1. Any alteration of sex ratio follows fertilization. What physiological mechanism determines sex of individual embryos is still a mystery.

Much of the theory of sex ratio adjustment applies to polygamous mammals. It is based on the premise that variations in physical condition or social rank of the mother result in differential investment in sons and daughters. One model (Trivers and Willard 1973) is based on three assumptions: (1) females in the best condition produce the healthiest offspring; (2) differences in offspring condition follow into adulthood; and (3) differences in body condition affect the reproductive success of males more strongly than the reproductive success of females. The reason is that the physically strong males assume positions of dominance in mating. This model predicts that a female in good condition would leave more surviving grandchildren if she produced

**Figure 20.17** Lifetime reproductive success of the European red squirrel according to body weight in their first winter as an adult. Coniferous habitat (yellow), deciduous habitat (brown). (From Wauters and Dhondt 1989:645.)

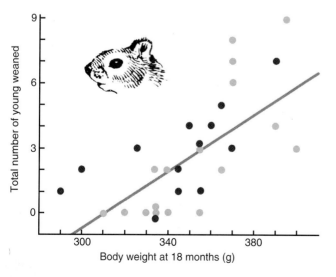

more sons than daughters, whereas a female in poorer condition would maximize her fitness by producing more daughters.

A second model, local resource competition (Clark 1978), is based on the predominance of male dispersal and female philopatry (remaining at home) in mammal populations. This model predicts the opposite of the Trivers-Willard model. Its premise is that mothers of high social rank and associated access to resources should produce mostly daughters, because they would inherit the mother's rank and privileges. Low-ranking females should produce sons that could disperse to areas where they could maximize their reproductive success (Silk 1983).

In polygamous mammals, males invest most of their reproductive effort in securing as many matings as possible and little, if any, in the rearing of offspring; females invest most of their reproductive effort in the production and rearing of offspring. For this reason, males have a greater variability in reproductive success than females (Clutton-Brock, Guinness, and Albon 1982). Some males—the strong, dominant ones—do extremely well; other males do poorly in fathering offspring. Females, however, do about equally well. In many species of mammals the sex ratio at birth is biased toward males; however, between birth and independence, male mortality is higher than that of females. Because of this higher mortality and because local resource competition exists among mother and daughters whereas males disperse, males may be the less expensive sex.

How do these predictions hold up in natural populations? In the red deer (*Cervus elaphus*) social rank and body size affected the breeding success of both sexes. Dominant hinds were heavier, conceived earlier, and produced larger calves and significantly more sons than subordinate hinds. Stags born to females above median rank experienced higher levels of reproductive success than hinds, whereas hinds born to subordinate females accrued greater reproductive success than stags (Figure 20.18) (Clutton-Brock et al. 1986). In the white-tailed deer (*Odocoileus virginianus*) the opposite was true. Sex ratios of offspring were related to the nutritional status and age of the doe, which reflect habitat quality. Fawns, yearling does, and undernourished does produced a preponderance of males, whereas heavier and older does produced significantly more daughters than sons (Ozoga and Verme 1986). The preponderance of females produced by older, well-nourished does fits the local resource competition model. Because daughters remain with their mothers until three years old, when they move to an area adjacent to the mother's home range and form a family group of their own (Ozoga et al. 1982), local competition for resources could develop, whereas males offspring disperse as yearlings. Does occupying poor range would benefit little by producing potentially competing philopatric offspring and would maximize their fitness by producing males.

Also conforming to the local resource competition model are the dominant females of the Cape Mountain zebra (*Equus zebra zebra*) of southern Africa (Lloyd and Rasa 1989). They produce significantly more daughters than sons, a trend not found in the subordinate females. The mother's status was positively correlated with the dominance status of her female offspring, but was not related to the subsequent status of her sons. Dominant mares gained maximum fitness by producing daughters because they had more than twice the chance of breeding than sons.

Among some species of primates, high-ranking females have access to more resources than low-ranking females,

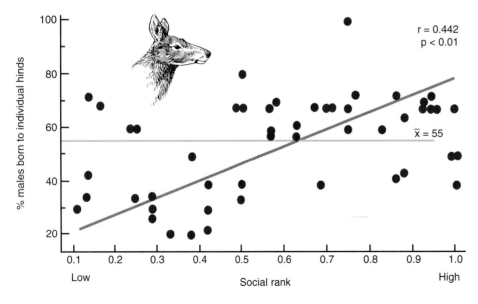

**Figure 20.18** Percentage of male offspring born to individual red deer hinds differing in social rank. High-ranking females tend to have sons. (Clutton-Brock et al. 1986:464.)

and males disperse. Because daughters inherit their mother's rank and its associated access to resources, high-ranking females produce more daughters, whereas low-ranking females, with access to fewer resources, produce more sons.

If access to resources resulting in better-nourished, heavier, and larger females biases sex ratios, then perhaps sex ratios are influenced by environmental conditions. Such influences appear in some of the parasitoid wasps that lay a single egg per host. Host size correlates with the size of the wasp that emerges from it. Sex determination in wasps is through haplodiploidy. Fertilized eggs produce females; unfertilized eggs produce males. Thus females can control the sex of the offspring by deciding to deposit a fertilized or unfertilized egg. Because female size and fecundity are closely related, female wasps lay fertilized or female-producing eggs in large hosts and male eggs in small hosts. The relationship is somewhat more complex, but the point is made. A clearer relationship exists among many of those reptiles that bury their eggs in the ground. The sex of their offspring depends upon temperatures during incubation. Among some species males develop at cool temperatures, among others at high temperatures (Charnov and Bull 1977). How and why this type of gender allocation occurs has no satisfactory explanation yet (Bull and Charnov 1989).

# *r*-SELECTION AND K-SELECTION

There are obvious broad differences among reproductive patterns in organisms. Some species, such as weeds and insects, are small, have high reproductive rates, and live short lives. Others, like trees and deer, are large and have low reproductive rates and long lives. Ecologists call the former *r* species and the latter *K* species, after the two terms on the logistic equation. Populations of the former grow rapidly and do not seem to reach or remain at carrying capacity. The populations of the latter attain and more or less remain around carrying capacity.

The concept of *r* and *K* species originated with MacArthur and Wilson (1967). They used the terms *r*-selection and *K*-selection in their theory of island biogeography to distinguish between types of selection pressures on island colonists. The distinction was elaborated by Pianka (1970). Empty islands would be colonized by a variety of immigrants, the most successful of which would be species with the best mechanisms for dispersal and the best ability for rapid population growth in unfavorable habitats. Natural selection would favor individuals able to achieve rapid population growth in uncrowded resource-rich environments. However, as populations increased, conditions became crowded, and individuals competed for resources, natural selection would favor the most competitive individuals, those

that were able to continue population growth at high sustained densities near carrying capacity. MacArthur and Wilson considered the former as *r*-selected, because environmental conditions keep growth of such populations on the rising part of the logistic curve. Mortality in these species is largely density-independent. They considered the latter as *K*-selected because they are able to maintain their densest populations at equilibrium (asymptote) or carrying capacity (*K*). *K* species were are able to compete effectively for food and other resources in a crowded environment. Mortality in these species results mostly from density-related factors.

The theory of *r*- and *K*-selection predicts that species in these different environments will differ in life history traits such as size, fecundity, age at first reproduction, number of reproductive events during a lifetime, and total life span. Species popularly known as **r-strategists** are typically short-lived. Among *r* species selection favors those genotypes that confer high reproductive rate at low population densities (Figure 20.19), early and single-stage reproduction, rapid development, small body size, large number of offspring (but with low survival), and minimal parental care. They have the ability to make use of temporary habitats. Many inhabit unstable or unpredictable environments where catastrophic mortality is environmentally caused and relatively independent of population density. For them environmental resources are rarely limiting, and they are able to

**Figure 20.19** A model of *r*- and *K*-selection involving the rates of increase of two genes, $X_1$ and $X_2$. Under the environmental conditions at point A, $X_2$ increases faster than $X_1$ and continues to do so up to point C. This population is *r*-selected. Under the environmental conditions at point B, $X_1$ increases faster than $X_2$. This population is *K*-selected. Where the lines cross at point C, *r*-selection switches to *K*-selection. (After MacArthur 1972:228.)

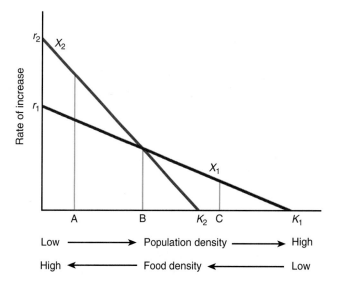

exploit relatively uncompetitive situations. Tough and adaptable, *r*-strategists, such as weedy species, have means of wide dispersal, are good colonizers, and respond rapidly to disturbance.

**K-strategists** are competitive species with stable populations of long-lived individuals. Among them selection favors genotypes that confer a slower growth rate at low populations, but the ability to maintain that growth rate at high densities (Figure 20.19). *K*-strategists have the ability to cope with physical and biotic pressures, possess both delayed and repeated reproduction, and have a larger body size and slower development. They produce few seeds, eggs, or young. Among animals, parents care for the young; among plants seeds possess stored food that gives the seedlings a strong start. *K*-strategists exist in environments in which mortality relates more to density than to unpredictability of conditions. They are specialists, efficient users of a particular environment, but their populations are at or near carrying capacity and are resource-limited. These qualities, combined with their lack of means of wide dispersal, make *K*-strategists poor colonizers.

The original concept of *r*-selection and *K*-selection applied to the natural selection of those traits among individuals in populations. Later the concept was applied to species characteristics. Small species with rapid reproductive rates, such as insects and mice, were categorized as *r* species, and long-lived species like oak trees, deer, and elephants were considered *K* species. *r* and *K* species occupied end points on a continuum from *r*-selection to *K*-selection. Such a view tempts a classification of species either as *r*-selected or *K*-selected, but it is difficult to force species into such a classification. Under certain conditions individuals or populations will exhibit *r*-selected traits or *K*-selected traits. Meadow mice living in environments where dispersal can take place easily exhibit characteristics of *r*-selection, whereas those living under conditions in which there is no dispersal sink assume *K*-selected characteristics (Tamarin 1978). White-tailed deer exhibit *r*-selected traits when the species spreads into new habitat or is greatly reduced by hunting. However, at high densities, especially where females are not reduced, the population remains at *K* and the population exhibits *K*-selected traits with low overall recruitment (McCullough 1979). Thus the concept of *r*-selection and *K*-selection is most useful in its original meaning, comparing organisms of the same type, individuals within a population, or populations within a species.

The concept of *r*-selection and *K*-selection assumes deterministic environments: one that is unpredictable and another that is stable. Environments, however, randomly fluctuate over the life of an organism. Such conditions favor adult survival at the expense of present fecundity. Schaffer (1974), Stearns (1976) and others have called this adjustment "bet-hedging." According to this hypothesis, in a variable environment in which adult mortality is high and juve-

nile mortality is low, selection should favor early maturity, larger reproductive effort, and more young to replace the adults. Under the same conditions if adult mortality is low and juvenile survival is high, then natural selection should favor late maturity, for no need exists for either early maturity or high production of young. Thus a variable environment that affects juvenile survival more than adult survival should lead to lower reproduction each year and increased iteroparity (*K*-traits). Conversely, if the variable environment affects adult survival more than juvenile survival, then it should lead to higher reproduction each year and reduced iteroparity (*r*-traits). Thus one can arrive at the same strategies by considering the response of individuals to selection pressures only in terms of *r*.

In contrast to the use of *r*- and *K*-strategies in plants, J. Grime (1977, 1979) has proposed a three-endpoint system consisting of ruderal or *R*-strategists, competitive or *C*-strategists, and stress-tolerant or *S*-strategists (Figure 20.20). *R*-strategists, typically weedy species, occupy uncertain or disturbed habitats, have a short growth form, reproduce early in life, possess high fecundity, experience one lethal reproduction (semelparity), and have well-dispersed seeds. *C*-strategists and *S*-strategists occupy more stable environments and are long-lived, often drastically reducing the opportunity of seedling establishment, resulting in high juve-

**Figure 20.20** Equilibrium points among competition (*C*), stress (*S*), and disturbance (*R*) strategies. Strong competitors occupy the upper part of the triangle; the stress tolerators occupy the lower left-hand part of the triangle, and the ruderals, opportunistic species adapted to disturbance, occupy the lower right-hand part of the triangle. Species within the triangle are intermediate in their responses, depending upon environmental conditions and life history traits. (Grime 1977:1187.)

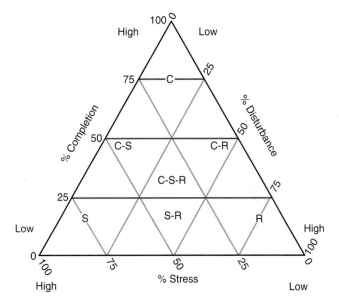

nile mortality. Beyond these two characteristics, the two have evolved quite different life-history strategies. *C*-strategists, such as grasses in an ungrazed grassland, live in competitive but productive and relatively undisturbed environments, attain maximum vegetative growth, reproduce early, and repeatedly utilize an annual expenditure of energy stored prior to seed production. *S*-strategists live in stressed environments, such as highly disturbed sites or forest understory, have delayed maturity, intermittent reproductive activity, and long-term energy storage. Within these three end-points are intermediate types: competitive ruderals (*C-R*), living in environments of low stress and moderate disturbance, such as fertilized moderately grazed grasslands; stress-tolerant ruderals (*S-R*), living in extreme environments such as rock crevices; and *C-S-R* plants, found in habitats where competition is reduced by the combined effects of stress and disturbance, such as old fields. *R* species are equivalent to species on the *r* end of the *r–K* continuum, *C* species occupy the middle, and *S* species fill the *K* end of the continuum.

# HABITAT SELECTION

Reproductive success depends heavily upon choice of habitat. Settling on less than optimal habitat can result in reproductive failure. How are organisms able to assess the quality of the area in which they settle? What do they seek in a living place? Such questions have been intriguing ecologists for many years. Answering them is important in these times of diminishing habitats. To maintain, improve, and restore habitats, we need to know what features are important to the species occupying them (see Verner et al. 1986).

Habitat selection among vertebrates is partly genetic (Wecker 1963, Klopfer 1963) and partly psychological. Lack and Venables (1939) and Miller (1942) suggested that birds recognize their ancestral habitat by conspicuous though not necessarily essential features. The Nashville warbler (*Vermiforma ruficapilla*), a typical inhabitant of open heath edges of northern bogs, selects open stands of aspen and balsam fir and forest openings of blackberry and sweet fern (*Comptonia peregrina*) in the southern part of its range in New York. These habitats are visually suggestive of bog openings (R. L. Smith 1956). MacArthur and MacArthur (1961) and later many others (for example, MacArthur 1972, James 1971, Balda 1975) demonstrated a strong correlation between structural features of vegetation and the species of birds present.

Habitat selection probably involves a hierarchical approach (Hilden 1965, Wiens and Rotenberry 1981, Hutto 1985, Orians and Whittenberger 1991). Birds appear to assess initially the general features of the landscape—the type of terrain; presence of lakes, ponds, streams, and wetlands; gross vegetational features such as open grassland, shrubby areas, types and extent of forests; and homogeneous or patchy vegetational distribution. Once in a broad general area, the birds respond to more specific features of habitats, such as the structural configuration of vegetation, particularly the density of leaves at various elevations above the ground and degree of vegetational patchiness (Figure 20.21). James called this vegetational profile associated with the breeding territory of a particular species the "niche gestalt." Although once regarded as unimportant in habitat selection and utilization, floristics (the individual plant species present) also appear to be important. The structural characteristics of the trees and shrubs may affect the foraging activities of birds. Various species of plants influence the type of herbivorous arthropods, and thus the levels of prey abundance (Holmes and Robinson 1981, Robinson and Holmes 1984, Wiens 1985).

Still other structural features determine a habitat's suitability (see Fish and Wildlife Service 1980, Verner et al. 1986). The lack of song perches may prevent some birds from colonizing an otherwise suitable habitat. Their introduction can stimulate the colonization of that area. For example, when telephone lines were strung across a treeless heath, tree pipits, birds that require an elevated singing perch, moved into the area (Lack and Venables 1939). Woodcock (*Philohela minor*) will not utilize a singing ground, an opening in shrubby fields and forest from which the bird performs its courtship flights, unless the opening allows sufficient room for flight (Sheldon 1967). A small opening surrounded by tall trees is not suitable, but an opening of the same size surrounded by low shrubs is.

An adequate nesting site is another requirement. Animals require sufficient shelter to protect parents and young against enemies and adverse weather. Selection of small island sites, such as muskrat houses, by geese provides protection against predators. Cavity-nesting animals require suitable cavities, dead trees, or other substrate in which they can construct such cavities. In areas where such sites are absent, populations of birds and squirrels can be increased dramatically by providing nest boxes and den boxes.

There appears to be a relationship between food availability and habitat selection (Hutto 1985, T. M. Smith and Shugart 1987, Orians and Whittenberger 1991). This is particularly true in habitat selection in migratory birds as they arrive on their nesting grounds. For them the availability of food is more important than sharply defined structural features (Hutto 1985). Because food resources are not available at the time the birds have to make their decision, they do not know exactly what the habitat has to offer in the future. They need to rely on direct or indirect cues to assess quality of habitat and the use of space within a habitat.

Female yellow-headed blackbirds (*Xanthocephalus xanthocephalus*) of western North American marshes are a polygamous species in which three to four females nest in the territory of a male. Females select the territory in which

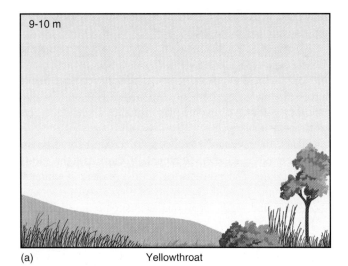

(a)                    Yellowthroat

(b)                    Hooded warbler

(c)                    Ovenbird

**Figure 20.21** The vegetation structure that characterizes the habitat of three neotropical warblers. (a) Yellowthroat (*Geothlypis trichas*), a bird of shrubby margins of woodland and wetlands and brushy fields. (b) Hooded warbler (*Wilsonia citrina*), a bird of small forest openings. (c) Ovenbird (*Seiurus aurocapillus*), an inhabitant of deciduous or mixed conifer deciduous forests with open forest floor. (After F. James 1971.)

they will nest and the male simultaneously, using vegetation density as a cue (Orians and Whittenberger 1991). Because the females themselves are not territorial, they can range across the marsh, including the territories of adjacent males (with whom they sometimes mate). The females seem to settle on marshes that will have the highest emergence rates of damselflies and dragonflies (Odonata), their major food item. Essential cues may be the vegetational patchiness of the marsh and the amount of shoreline along open water without emergent vegetation, places where odonatates may emerge.

Being polygamous and nonterritorial, female yellow-headed blackbirds can use cues on a larger scale than territorial monogamous species that must forage within their territories. Among territorial monogamous species, males select the territory. What cues do they use? That question was explored by Smith and Shugart (1987) in a study of habitat selection by the ovenbird (*Sieirus aurocapillus*), a ground-foraging warbler of the eastern North American deciduous forest. They hypothesized that structural habitat cues are the proximate factor determining territory settlement and size. To test the hypothesis they examined the relationships among habitat structure, prey abundance, and intrapopulation variation in the ovenbird in a deciduous forest in Tennessee. They assessed vegetation structure, plotted territories, and determined prey abundance by sampling forest floor litter invertebrates from each of 115 sample vegetation plots contained within 23 territories of ovenbirds and from 100 plots outside of ovenbird territories during late April and early May (the time when arriving ovenbirds select their habitat within the woods). T. M. Smith and Shugart found a significantly higher prey abundance per unit area within the territories than outside the territories. Seventy-five percent of the variation in prey abundance was correlated with habitat structure, particularly the nature of the canopy. Forest litter invertebrates were most abundant in those areas of the forest with large trees, a closed canopy, and a sparse understory. In contrast, areas of more open canopy and dense understory had fewer invertebrates. The microclimate of the forest floor under a closed canopy, with its lower temperatures, less drastic diurnal fluctuations in temperature and moisture in forest litter, and increased relative humidity, provided superior habitats for forest litter invertebrates, major food of the ovenbird. The ovenbird apparently cued in on areas of closed canopy and sparse understory as microhabitats within the woods offering the greater abundance of food. This hypothesis was supported by a significant correlation between predicted prey abundance and territory size. However, variations in territory size were related to structural features of the habitat rather than to prey abundance or intraspecific competition.

Habitat selection is a common behavioral characteristic of vertebrates; fish, amphibians, reptiles, birds, and mammals furnish most of the examples. Garter snakes (*Thamnophis elegans*) living along the shores of Eagle Lake in

sagebrush-ponderosa pine country in northeastern California selected as their retreat sites rocks of intermediate thickness (20 to 30 cm) over thinner and thicker rocks. Shelter under thin rocks became lethally hot; shelter under thicker rocks would not allow the snakes to warm to their preferred range of body temperature ($T_b$). Under the rocks of preferred thickness, the snakes would never overheat, and they would achieve and maintain their preferred body temperature for a long period (Huey et al. 1989, Huey 1991). Insects, too, cue in habitat features. Whitham (1980) found that the gall-forming aphid *Pemphigus,* which parasitizes the narrowleaf cottonwood (*Populus angustifolia*), selects the largest leaves to colonize and discriminates against small leaves. Beyond that they select the best positions on the leaf (see Chapter 19). Occupancy of this particular habitat, which provides the best food source, produces individuals with the highest fitness.

Even though a given habitat may provide suitable cues, it still may not be selected. The presence of others of the same species may be necessary to attract more individuals. In social or colonial species like herring gulls (Dorst 1958), an animal will choose a site only if others of the same species are already there. On the other hand, the presence of predators and human activity may discourage a species from occupying otherwise suitable habitat. Human activities on northern lakes inhibit the nesting of loons (Reem 1976).

Most species exhibit some plasticity in habitat selection. Otherwise these animals would not settle in what appears to us as less suitable habitat or colonize new habitats. Often individuals are forced to make this choice. Available habitats range from optimal to submarginal; the optimal habitats, like good seats at a concert, fill up fast. The marginal habitats go next, and the latecomers or subdominant individuals are left with the poor habitats where they may have chance of reproducing successfully (Fretwell and Lucas 1969). The ability of some members of a particular species to select habitats that deviate from those of others must exist on both a phenotypic and genetic level. Ovenbirds in the northern part of their range select woodland habitats that are exactly the opposite of those in Tennessee. The Ontario birds prefer more a open canopy and denser understory (Stenger and Falls 1959). The differences in habitat may result from differences in plant species associated with structural gradient and from their influence on composition, microclimate, and chemistry of the forest floor that affects prey abundance. Late successional stages in Ontario forests are dominated by mixed conifers and hardwoods, whereas those in Tennessee are deciduous. The contrast in ovenbird habitats suggests that the species may have evolved patterns of habitat selection to match patterns of productivity over its geographical range (T. M. Smith and Shugart 1987). Another example of strong contrast in choice of habitat types is found in the black-throated green warbler (*Dendroica virens*). It is associated with coniferous forests in the northern part of its range and with drier oak forests in the middle and southern Appalachians (Collins 1983).

Do plants select habitats, and if so how? F. A. Bazzaz (1991) explored some answers. Plants can hardly get up and move about to find a suitable site. The only recourse plants have in habitat selection is to send out seeds in anticipation that they will arrive at some place suitable for seedling germination and survival.

Plants, like animals, fare better in certain habitat types, characterized by such environmental factors as light, moisture, nutrients, and presence of herbivores and symbionts. Habitat choice involves the ability of plants to disperse with the aid of wind, water, or animal agents to preferred patches of habitat, which more often than not involves an element of chance. The problem for plant dispersers is that many plants have the same set of requirements for light, nutrients, and the like, so they may face competitors once the propagule arrives. A plant has a better chance of survival and growth if it is a habitat specialist. Once established, such plants can expand into surrounding areas by modular growth.

Consider the cattail, a habitat specialist inhabiting marshy places. Each year a cattail stand sends out billions of seeds that are carried miles by the wind. Most settle on highly unsuitable places. Just a few seeds drop into very wet roadside ditches where they quickly germinate, take root, and within a few years spread their rhizomes into surrounding wet ground.

The rapid loss and degradation of habitats that is causing the decline of many species of animals has stimulated intensive research since the 1970s into the relationship between organisms and their specific habitat requirements (see Verner et al. 1986.) Such knowledge is imperative if we are to maintain and restore wildlife habitats. Habitat assessment and habitat modeling are two approaches to determining essential features of an organism's habitat. Unfortunately, a model that predicts habitat features for a species in one part of its range may not apply in other parts of its range. We must incorporate habitat assessment information into land use management, development, and zoning plans. Failure to understand the habitat requirements of animals and plants and to use such knowledge in land use planning leads to species extinctions.

## SUMMARY

Life history patterns encompass traits and behaviors that help individuals to achieve fitness. These activities range from sexual reproduction and sexual selection to components of an individual's reproductive effort including parental care, investments, and expenditures.

Sexual reproduction requires the acquisition of mates, which involves both mating systems and sexual selection.

Mating systems include two basic types, monogamy and polygamy. Two general kinds of polygamy are polygyny, in which the male acquires more than one female, and polyandry, in which the female acquires more than one male. The potential for competitive mating and sexual selection is higher in polygamy than in monogamy.

Sexual selection (as opposed to Darwinian natural selection involving differential survival) usually involves intrasexual selection or male competition, which increases selective pressures for the evolution of horns, large size, or exaggerated morphological features such as elaborate plumages and bright colors that in some way influence female choice. Intersexual selection involves female choice among males offering resources or offering only genes. Sexual selection favors traits that enhance mating success even if they handicap the male by making him more vulnerable to predation.

Optimal fitness in sexual reproduction can be achieved only if parents balance energy allocated to present reproduction with that allocated to survival and future fecundity. One alternative is to invest a maximum amount of energy into a single reproductive effort in a lifetime, as exemplified by annual plants and many insects. The other alternative is to allocate less energy to each reproduction and repeat reproductive efforts through a lifetime. Organisms may invest reproductive effort into many small offspring and provide a minimal amount of parental care; or they may invest a similar amount of energy into fewer, larger individuals and extended parental care. A single reproductive effort (semelparity) or production of many young with minimal energy invested in each is characteristic where mortality of adults is high. Because little difference in potential for population increase exists between single and repeated reproduction, repeated reproduction (iteroparity) may be a response to an unpredictable survival of individuals from young to adult. By reproducing several times, an organism is more likely to assure reproductive success.

Reproduction has costs. Individuals with repeated reproduction face decreased future survival and decreased future fecundity with each present investment in reproduction. Heavy early investment in reproduction can reduce individual survival and thus future fecundity. Light investment can improve survival and improve future fecundity, but the individual still has potential lower life expectancy. Somewhere in between lies the best strategy for optimal lifetime reproductive success, the sum of present and future reproductive success.

Resource allocation may influence sex ratios in the offspring because of the differential costs of producing males and females. Natural selection often favors those parents who invest equally in sons and daughters. When both are equally expensive, the sex ratio will be 1:1. The costs of rearing, however, may not be equal, and the sex ratio will become skewed toward the less expensive sex.

Allocation of energy to reproduction is related in part to mortality of parents and offspring, its relation to the population, and predictability of the environment. Organisms living in an unpredictable environment or subject to heavy environmentally induced mortality tend to allocate a greater proportion of energy to reproduction, and to expand rapidly when conditions are favorable. Such organisms are said to be $r$-selected because selection favors high productivity. Organisms that occupy a more predictable environment and are more subject to density-related mortality tend to allocate less energy to reproduction. They are said to be $K$-selected because selection favors efficient use of the environment.

Because quality of habitat is essential to reproductive success, habitat selection is an important part of an organism's life history pattern. How animals cue in on habitat quality is not clearly understood, but features such as structure and diversity of vegetation cover, particularly as they relate to food, appear to be important. Quality may be modified by the availability of nesting sites, song perches, escape cover, and the like. Individuals across a species' range exhibit regional plasticity in selection of habitats.

# REVIEW QUESTIONS

1. What forms might sexual reproduction take? What are the stimuli for and the advantages of sexual reversals?
2. What are the advantages and disadvantages of monogamy and polygamy for males and females? What are the several types of polygamy?
3. What is sexual selection and how does it function? Which sex does the selecting? What are some cues?
4. What is the relationship between intrasexual selection and intersexual selection?
5. Distinguish between resource-based and genes-only sexual selection. Which one imposes the greater selection pressure on males?
6. What is unique about lek behavior? Discuss the three models: female choice, hotspot, and hotshot. Does one model really explain all lek behavior?
7. Distinguish between altricial and precocial young.
8. Define reproductive effort.
9. What are the costs of reproduction? What is the significance in the manner in which these costs are apportioned? How does brood reduction enter the picture?
10. What is the relationship between clutch size and fitness?
11. What is the relationship among the parent's age, size, and fecundity?
12. Report on the lek behavior of other species such as prairie chicken, black grouse, bird of paradise, Uganda kob, and hammer-headed bats. What model do they fit, if any?

13. In a randomly mating population, male and female offspring should be produced in equal numbers, and parents should invest equally in both sexes. Why are sex ratios in many natural populations skewed toward one sex or the other?

14. In the light of this chapter, discuss such human behavior as mate selection, family size, investment in young, age and fecundity. (See Buss 1994.)

# CROSS-REFERENCES

Adaptation, 30–31; energy allocation, 170–172; modular populations, 362–364; reproductive values, 384–387; population genetics, 456–477; natural selection, 461–463; fitness, 461; intraspecific predation, 541–542; parasitism, 558–577.

*Chapter* **21**

# Population Genetics

*Concepts*

1. Genetic information is coded in the DNA molecule.
2. Major sources of genetic variation are mutations and the segregation and recombination of genes.
3. Natural selection can change gene frequency.
4. Natural selection may be stabilizing, directional, or disruptive.
5. Maintenance of altruistic traits may involve group or kin selection.
6. Small populations are vulnerable to inbreeding and genetic drift, which reduce genetic variability.
7. The maintenance of genetic variability depends upon effective population size.
8. A minimum viable population size is necessary to ensure the persistence of a population over time.

As a rapidly expanding human population steadily encroaches on natural areas, large, expansive populations of many plants and animals species are being reduced to small isolated populations. These populations are forced to exist in fragments of their former habitat. Not only are these small populations now vulnerable to chance demographic changes in birthrates and death rates and chance variations in the environment; they are also subject to genetic deterioration. How much genetic variation have these small populations lost or retained from the original population? What is the probability of the loss of genetic diversity? How would such a loss affect the fate of a small population? What must be done to maintain genetic diversity in small populations? How can genetic variation be maintained in small captive populations of endangered species that will provide the source of stock for reintroduction? These and related questions are explored in this chapter.

## GENETIC VARIATION

Wherever you go along the seashore, whether it be on long stretches of beach or harbors and docks, you see and hear the gulls, especially the ubiquitous herring gull. Even a moderately alert observer will detect differences among herring gulls. Most conspicuous are the adults with their bluish-gray back, their white head, neck, underparts, and tail, their black-tipped primary wing feathers, and their yellow bill with a bright red spot near the tip of the lower mandible. Among the adult gulls are younger birds with a different pattern. Some are darkish brown-gray, mottled and barred on the back with white and grayish-buff. Others are lighter in tone with some gray on the back. Still others are similar to adults but with some dusky spottings on the tail and wings. Their bills may have only a suggestion of the red spot.

If you examine the gulls more closely, you will detect other subtle differences. Size, shades and patterns of gray on the back, length of the bill, shape of the red spot, length of the wing, and other characteristics vary among the birds. In fact, so widespread are these smaller differences that if you look carefully at the birds and become acquainted with a colony, you can distinguish one bird from another, just as you can tell one person from another.

### Types of Variation

The most obvious variations among the members of a population are **discontinuous,** that is, variations in a specific character or sets of characters that separate individuals into discrete categories, such as male or female or age classes. Thus differences in patterns of plumage enable ornithologists to classify individuals in the gull population as first-year birds, second-year birds, third-year birds, and mature

adults. Another type of discontinuous variation is morphological, such as male and female or the red and gray phase of the screech owl. Other discontinuous differences are biochemical, such as blood groups in humans, or even behavioral, such as song dialects in birds.

A second type of variation is **continuous,** a variation in a character that can be placed along a range of values. Characters subject to continuous variation can be measured—for example, tail length of a species of mouse, number of scales on the belly of a snake, rows of kernels on an ear of corn, and shapes and sizes of sepals and petals. The measurements of such a character or set of characters for several individuals in a population can be tabulated as a frequency distribution and arranged graphically as a histogram (Figure 21.1).

## Genotypes and Phenotypes

Inherited characteristics of a species and variations in individuals are transmitted from parent to offspring. The sum of the hereditary information carried by the individual is the **genotype.** The genotype directs the development of the indi-

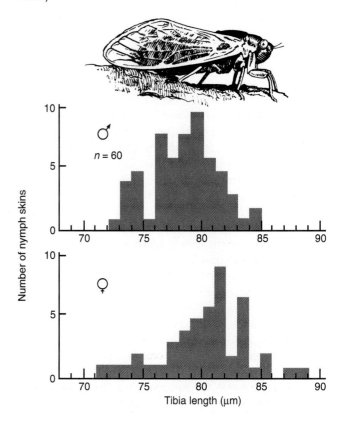

**Figure 21.1** Histogram showing the frequency distribution of the hind tibia lengths of nymphal exuvia (shed skin) of the periodical cicada *Magicicada septendecim.* (After Dybas and Lloyd 1962.)

vidual and underlies the morphological, physiological, and behavioral characteristics of the individual. The external or observable expression of the genotype is the **phenotype.** Some phenotypic expressions result from an interaction of the genotype and the environment. For example, a seedling with a gene for the formation of chlorophyll will develop the normal green color if germinated in the light, but will be white if germinated in the dark. The gene directs the character of the green color, but its expression is affected by environmental conditions, because chlorophyll develops only in the light.

The ability of a genotype to give rise to a range of phenotypic expressions under different environmental conditions is known as **phenotypic plasticity.** Some genotypes have a narrow range of reaction to environmental conditions and therefore give rise to a fairly constant phenotypic expression. However, many plants and animals that can survive under a wide range of environmental conditions possess variable and diverse phenotypic responses. Some of the best examples of such phenotypic plasticity are found among plants. The size of plants, the amount of reproductive tissue, and even the shape of the leaf may vary widely at different levels of nutrition, light, and moisture (Figure 21.2). Lacking the mobility of animals, plants must possess more flexibility in their response to environmental conditions to survive. Phenotypic plasticity represents nongenetic variation. An environmentally induced modification of a character is not inherited. However, the ability to modify such a character under certain environmental conditions are inherited.

## Sources of Variation

The primary genetic control mechanism, found within the nucleus of every cell in the organism, is deoxyribonucleic acid, DNA. DNA, the information template from which all cells in the organism are copied, is a complex molecule in the shape of a double helix, resembling a twisted ladder. The long strands, comparable to the uprights of the ladder, are formed by an alternating sequence of deoxyribose sugar and phosphate groups. The connections between the strands, or the rungs, consist of pairs of the nitrogen bases adenine, guanine, cytosine, and thymine. In the formation of the rungs adenine is always paired with thymine and cytosine is always paired with guanine. The DNA molecule is divided into smaller units, called nucleotides, consisting of three elements: phosphate, deoxyribose, and one of the nitrogen bases bonded to the strand at the deoxyribose. The information of heredity is coded in the sequential pattern in which the base pairs occur. Each species is unique in that its base pairs are arranged in a different order and probably in different proportions from every other species.

In eukaryotic cells DNA is present in larger units called chromosomes, which are found in most living organisms.

**Figure 21.2** Environmental plasticity in the growth of leaves of the yellow water-buttercup (*Ranunculus flabellaris*). The submerged leaves are divided into threadlike segments. The floating leaves are much broader and less divided.

Each species has a characteristic number of chromosomes in every cell, and the chromosomes occur in pairs. When cells reproduce (a process of division called **mitosis**), each resulting cell nucleus receives the full complement of chromosomes, or the **diploid** number (for example, 46 in humans). In organisms that reproduce sexually the germ cells or gametes (sperm and egg) result from a different process of cell division, **meiosis,** in which the pairs of chromosomes are split, so that each resulting cell nucleus receives only one-half the full complement, or the **haploid** number (23 in humans). When egg and sperm unite to form a new individual, the diploid number is restored. The chromosomes recombine in a great array of combinations. This segregation and recombination of chromosomes and the hereditary information they carry are the primary sources of variation.

Each chromosome carries units of heredity called **genes,** the informational units of the DNA molecule. Because chromosomes are paired, genes are also paired in the body cells. The position a gene occupies on a chromosome is known as a *locus.* Genes occupying the same locus on a pair of chromosomes are termed **alleles.** If each member of the pair of alleles affects a given trait in the same manner, the two alleles are called **homozygous.** If each affects a given trait in a different manner, the pair is called **heterozygous.** During meiosis the alleles are separated as the chromosomes separate. At the time of fertilization the alleles, one from the sperm and one from the egg, recombine as the chromosomes recombine.

Major interest lies in sources of new genetic variations. One source is the reassortment and recombination of existing genes, both at the level of the gene and at the level of the

chromosomes. The other source is mutation, or change in the gene or chromosome.

Recombination of Genetic Material The major source of genetic (and thus phenotypic) variation among individuals in a population is the recombination of genetic material during sexual reproduction. Each egg and sperm, being haploid, carries one of the two members of a pair of alleles. When the two gametes unite to form a zygote, the alleles are recombined. The number of possible recombinations is extremely large. Recombination does not result in any change in genetic information, as mutation does, but it does provide different combinations of genes upon which selection can act. Some combinations of genes are more adaptive than others. Selection determines the variations or new types that survive in the population. The poorer combinations are eliminated.

The amount of recombinations in a population is limited by a number of characteristics of the species. One limitation is the number of chromosomes and the number of genes. Another is the frequency of crossing-over, the exchange of corresponding segments of homologous chromosomes during meiosis. Others include gene flow between populations, the length of generation time, and the type of breeding—for example, single versus multiple broods in a season among animals and self-pollination versus cross-pollination in plants.

Mutation A **mutation** is an inheritable change of genetic material in the gene or chromosome. Organisms that possess such changes are called mutants.

Mutations come in two broad types: macromutations and micromutations. **Macromutations** are chromosomal mutations that result from a change in the number of chromosomes or a change in the structure of the chromosome. Of greatest evolutionary interest is a mutation resulting from a change in the number of chromosomes. Such a change can arise in two ways: the complete or partial duplication of the diploid number rather than the transmission of the haploid number, or a deletion of some of the chromosomes.

**Polyploidy** is the duplication of entire sets of chromosomes. It can arise from an irregularity in meiosis or from the failure of the whole cell to divide at the end of the meiotic division of the nucleus. Forms of polyploidy are triploid ($3n$ or three haploid sets), tetraploid ($4n$), and so on.

Polyploidy exists mostly in plants. The condition is rare in animals, because an increase in sex chromosomes would interfere with the mechanism of sex determination and the animal would be sterile. Polyploid plants, on the other hand, can reproduce and spread asexually. Polyploid plants differ from the normal diploid individuals of the same species in appearance and are usually larger, more vigorous, and occasionally more productive.

Another form of macromutation involving sets of chromosomes is duplication or deletion of part of a normal complement of chromosomes. Such deletions or duplications result in abnormal phenotypic conditions. (One such condition is Down's syndrome in humans, in which there are three copies of a particular chromosome.)

Macromutations also arise with a change in the physical structure of a chromosome involving the deletion, duplication, inversion, or translocation of segments of the chromosome (Figure 21.3). **Deletion** is the loss of a part of a chromosome; a definite segment and the genes thereon are missing in the offspring cell. **Duplication** is an addition to the chromosome due to internal doubling. **Inversion** is an alteration of the sequence of genes in the chromosomes. It may occur when a chromosome breaks in two places and the segment between the breaks becomes turned around. Such a break reverses the order of genes in respect to an unbroken chromosome, and interferes with pairing in a heterozygous individual. **Translocation** is the exchange of segments between two nonpaired (nonhomologous) chromosomes in a heterozygous individual. The genes in the translocated segment become linked to those of the recipient chromosome. Most mutations arising from physical changes in chromosomes are lethal, produce abnormal, nonviable gametes, or have a marked effect on the development of an individual.

**Micromutations** are gene mutations. They are alterations in the DNA sequence of one or a few nucleotides. During meiosis the gene at a given locus usually is copied exactly and eventually becomes part of the egg or sperm. On occasion the precision of this duplication process breaks down, and the offspring DNA is not an exact replication of the parent DNA. The alteration may be a change in the order of nucleotide pairs, the substitution of one nucleotide pair for another, the deletion of a pair, or various kinds of transpositions.

The rate of mutation in general is low. Most common mutations involve the change of one allele into another. Consider a pair of alleles, one dominant $A$ and the other recessive $a$. In a population homozygous for gene $A$, for example, $A$ eventually will mutate to $a$ in some of the gametes; and in a population having both genes, mutations may be forward to $a$ or backward to $A$. If $A$ mutates to $a$ faster than $a$ to $A$, the frequency of $A$ decreases. Rarely is one of the alleles lost to the population, for reversibility prevents a long-term or permanent loss. Eventually such mutations arrive at an equilibrium. Even if one allele is lost from the population, it may reappear by mutation.

## Hardy-Weinberg Equilibrium

If a gene occurs in two forms, $A$ and $a$, then any individual carrying it can fall into three possible diploid classes: $AA$, $aa$, and $Aa$. Individuals in which the alleles are the same, $AA$ or $aa$, are called **homozygous;** and those in which the alleles are different, $Aa$, are called **heterozygous.** The haploid gametes produced by homozygous individuals are

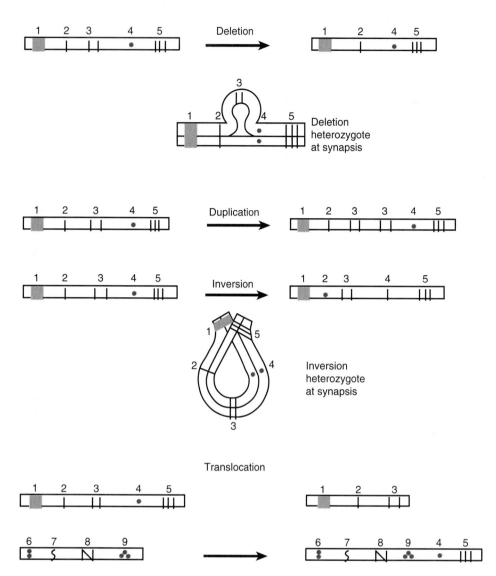

**Figure 21.3** Types of chromosomal aberrations. When these altered chromosomes join with normal homologues during the first meiotic division, they assume characteristic configurations that allows locus-by-locus matching. Synaptic configurations of deletion and inversion are shown. (Diagrams after Wilson and Bossert 1971.)

either all *A* or all *a;* and those by the heterozygous, half *A* and half *a*. They can recombine in sexual reproduction in three possible ways: *AA, aa,* and *Aa*. The proportion of the gametes carrying *A* and *a* is determined by the individual genotypes, the genes received from the parents. Eggs and sperm unite at random, enabling the prediction of the proportion of offspring of different genotypes based on parental genotypes.

Assume that a population homozygous for the dominant *AA* is mixed with an equal population homozygous for the recessive *aa*. Their offspring, the $F_1$ generation, will consist of 0.25 *AA*, 0.25 *aa*, and 0.50 *Aa* (Figure 21.4). These proportions are called **genotypic frequencies.** The **allele frequencies,** of course, are 0.50 of *A* and 0.50 of *a*.

Will this proportion be maintained through successive generations of a bisexual population? We can phrase this question in the form of a null hypothesis that states the pro-

**Figure 21.4** Mixing two homozygous populations.

|  |  | Males | |
|---|---|---|---|
|  |  | *AA* .50 | *aa* .50 |
|  |  | A | a |
| *AA* .50 | A | *AA* .25 | *Aa* .25 |
| Females |  |  |  |
| *aa* .50 | a | *Aa* .25 | *aa* .25 |

portions will be maintained through successive generations of a bisexual population under certain conditions: (1) mating is random; (2) mutations do not occur, or if they do, the rate of mutation from A to a is the same as a to A; (3) the population is closed (gene flow from one population to another does not exist); (4) population size is infinite; and (5) no natural selection occurs.

Let us go back to that hypothetical population of homozygous males and females that produced the $F_1$ generation. The frequency of allele A can be designated as p (whose value is between 0 and 1) and the frequency of a as q (whose value is also between 0 and 1). If mating is random, then the probability that an offspring will receive an A allele from the father is p and an A allele from the mother is also p, so the probability that the offspring will be homozygous for AA is $p^2$. The same argument holds true for the probability that the offspring will be homozygous for aa, or $q^2$. But the probability also exists that the offspring will receive an A allele from its father and an a allele from its mother: $p \times q$. Or it could receive an a allele from the father and an A allele from its mother: $q \times p$. The combined probability is $2pq$. Thus the predicted genotypic frequency in the offspring in the population can be expressed as

$$p^2 + 2pq + q^2 = 1$$

This expression is known as the Hardy-Weinberg law.

Simply stated, the Hardy-Weinberg law says that the allele frequency is $p + q = 1.0$, where p is frequency of allele A, the dominant, and q is frequency of allele a, the recessive. Then the genotypic frequency will be $p^2 + 2pq + q^2 = 1.0$, where $p^2$ = frequency of homozygous individuals AA, $q^2$ is frequency of homozygous individuals aa, and $2pq$ is fre-

quency of heterozygous individuals Aa. In the hypothetical population, the proportion of the genotypes in the $F_1$ generation will be $(0.5)^2 + 2(0.5 \times 0.5) + (0.5)^2$. The same genotypic frequency will be maintained into the $F_2$ generation if the conditions of the Hardy-Weinberg law hold (Figure 21.5).

The same tendency can be demonstrated even if the ratio is not the classical Mendelian ratio. Normally both p and q = 0.5. Imagine a population in which the ratio of A alleles (p) to a alleles (q) is 0.6 to 0.4 (Figure 21.6). The frequency of the genotypes in the $F_1$ generation will be 0.36 AA, 0.48 Aa, and 0.16 aa; and the allele frequency will be $(0.6)^2 + 2(0.6 \times 0.4) + (0.4)^2$. We can conclude that all succeeding generations will carry the same proportions of the three genotypes, provided the assumptions mentioned earlier are met.

The stated assumptions, of course, are never met perfectly in any real population. If not, then what is the value of the Hardy-Weinberg law? First, it serves as a null hypothesis against which to test the departure of frequencies away from the Hardy-Weinberg equilibrium. A departure indicates that one or more of the assumptions is invalid, a starting point for further investigation.

In certain cases the Hardy-Weinberg equation can be used to calculate genotypic frequencies, especially when the character is one for which mating is usually random, such as certain blood types. For example, about 1 in every 10,000 newborn Caucasians suffers the effects of phenylketonuria, a severe form of mental retardation caused by the inability to to metabolize phenylalanine. The gene causing the condition obviously is the recessive q, so the genotypic frequency $q^2 = 1/10,000$; therefore $q = (1/10,000)^{1/2} = 0.01$, and $p = 1.00 - 0.01 = 0.99$. Carriers for the disease would be the het-

**Figure 21.5** Proportions in the $F_2$ generation.

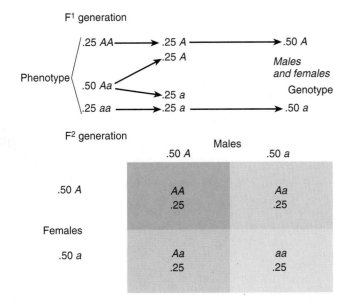

**Figure 21.6** The Hardy-Weinberg law applied to a hypothetical case.

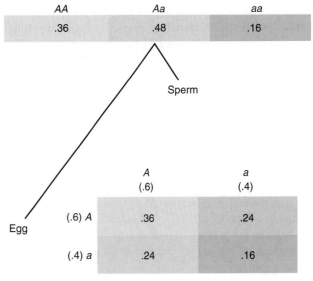

erozygotes $2pq = 2(0.01 \times 0.99) = 0.0198$. Thus about 1 person in 50 ($50 \times 0.0198$) would carry the allele. These heterozygotes are phenotypically indistinguishable from the noncarriers.

# NATURAL SELECTION

## Nonrandom Reproduction

Variation in a population seldom is constant from generation to generation. One reason is gene mutation, the ultimate source of genetic variation. Of more immediate consequence is nonrandomness of reproduction within a population. Not all members contribute their genetic characteristics to the same degree to the next generation. It is this selectivity that is natural selection.

Before a given individual in a population can contribute to the succeeding generation, it must first survive to reproduce. Survival begins from the time of fertilization through the periods of development, growth, and sexual maturation. Fertilized eggs may fail to develop fully and die from physiological or environmental causes. Disease, predation, and accidents eliminate those young not quite as swift or as strong as their siblings. In such survival genetic variation plays a key role, for natural selection influences the frequency of alleles in a population. If a mutation arises that places its carrier at a disadvantage when expressed, selective pressures eliminate the individual; on the other hand, and advantageous mutation is retained.

An example of such selection can be found among the flies. When DDT was first used as an insecticide against houseflies, the chemical was highly effective, destroying the bulk of local populations. Among the flies were a few that did not die, that carried an allele or a combination of genes that made them resistant to the spray. Resistance in one strain of flies was due to a recessive gene. Flies homozygous for this gene tolerated a high concentration of DDT, while homozygous dominants and heterozygotes were killed. These flies survived to multiply. Many of their offspring were as resistant to the sprays as the parents; some were even more resistant. The least resistant were selected against; the most highly resistant were retained in the reproductive population. Later applications of DDT continually selected for a combination of genes most resistant to the insecticide. As a result DDT became ineffective in fly control, and newer, stronger sprays were required. Eventually even these sprays select resistant strains of flies, which will become adapted to the new environmental conditions. However, to acquire this resistance the flies pay a price. In the absence of DDT the resistant flies are inferior competitors to the nonresistant flies, which have a shorter development time (Pimentel et al. 1951). If the spraying is stopped,

evolution will be reversed and the resistance will largely disappear from the fly population.

Similar selective processes operate on all organisms. Once surviving young reach reproductive age, more individuals are eliminated from the parental population. The maintenance of genetic equilibrium is based on random mating, but mating is not random. Many species of animals, particularly among birds, fish, and some insects, have elaborate courtship and mating rituals (Chapter 20). Any courtship pattern that deviates from the commonly accepted pattern is selected against, and the deviating individual and its genes are eliminated from the reproductive population. On the other hand, animals possessing a color pattern or movement that accents the typical pattern are selected for. Any new mutations that improve on courtship, mating signals, and ritual would possess a favored position in subsequent generations. Among polygamous species, in particular, the majority of males go mateless, for the females mate with dominant males that tolerate no interference from younger or less aggressive males. States of psychological and physiological readiness also are involved in mate selection. Unless both male and female are at the same state of sexual readiness, mating will not occur.

Neither is fecundity random. Some families or lines increase in number through time; others fade away. Obviously those who produce more offspring increase the frequency of their genes in a population and affect natural selection. For example, if individuals with allele *A* produce ten offspring to every one produced by those with allele *a,* the proportion of *A* in the population will increase. There is a limit, however, for natural selection does not always favor fecundity. If an increased number of young per female results in reduced maternal care, survival of offspring may be reduced, particularly among those animals whose chances of individual survival are high. Those organisms whose chances of individual survival are low—for example, ground-nesting game birds, oceanic fish, parasites, and marine invertebrates—have become very fecund.

## Fitness and Modes of Selection

If an organism can tolerate a given set of conditions so that it can leave fertile progeny, thus contributing its genetic traits to the population gene pool, it can be said to be adapted to its environment. If an organism survives only as an individual and leaves few or no mature, reproducing progeny, thus contributing little or nothing to the gene pool of the population, it is poorly adapted. Those individuals that contribute the most to the gene pool are said to be the most fit, and those that contribute little or nothing are the least fit. The **fitness** of the individual is measured by its reproducing offspring. Natural selection is not a measure of individual survival, but of differential reproduction, the ability to leave the most offspring capable of further reproduction.

In simplest terms fitness is measured by comparing the number of reproducing offspring produced by one genotype to the number produced by another. Suppose genotype *AA* produces 250 reproducing offspring and genotype *BB* produces 200. The reproductive success of genotype *BB* compared to *AA* is reduced by 50; or expressed in fractional terms, 50/250 = 0.20. Obviously *AA* is the more successful genotype.

In measuring selection or the adaptative value of a genotype, fitness is frequently designated as *W*, the value of which ranges from 1.00 for the most productive genotype to 0 for no reproduction (lethal genes). In our simple example the value of *W* for *AA* would be designated as 1.00; the value of *W* for *BB* would be 1.00 − 0.20, or 0.80.

The selective pressure acting on a genotype is designated as a **selection coefficient,** *s*. It can be stated as the difference between 1.00 and the fitness value. In the example the selection coefficient for *AA* is 0; for *BB* it is 0.20. Thus, *W* (fitness) = 1 − *s;* similarly, *s* = 1 − *W*. For *BB*, *W* = 1 − 0.20 = 0.80; *s* = 1 − 0.80 = 0.20. This example is simplified. The calculations used to determine fitness values and selection coefficients are more complex, but not difficult. Good discussions are given in Haldane (1954), Wallace (1968), and Ricklefs (1979).

Within a population selection may act in three ways. Given an optimum intermediate genotype, **stabilizing selection** favors the average expression of the phenotype at the expense of both extremes. This type of selection takes place in all populations. **Directional selection** favors one extreme phenotype at the expense of all others. The mean phenotype is shifted toward the extreme, provided that heritable variations of an effective kind are present. **Disruptive selection** favors both extremes over a range of phenotypes, although not necessarily to the same extent, at the expense of the average type (Figure 21.7).

Disruptive selection is most apt to occur in a population living in a heterogeneous environment in which there is a strong selection for adaptability to a developing environmental condition. As a result the population may subdivide into two phenotypes. This division would give rise either to polymorphism or to separation into populations with different characteristics. The latter is most likely to take place in areas where selection is intense and where optimum habitat adjoins or is penetrated by less than optimum habitat. Organisms settling in these habitats will adapt to the local environment. If disruptive selection is strong enough, it will lead to perferential mating and eventually to genetic divergence of two or more groups.

An example of the development of a polymorphic species through disruptive selection is the peppered moth (*Biston betularia*) in England. Before the middle of the nineteenth century the moth, as far as is known, was always white with black speckling in the wings and body (Figure 21.8). In 1850 near the manufacturing center of Manchester, a

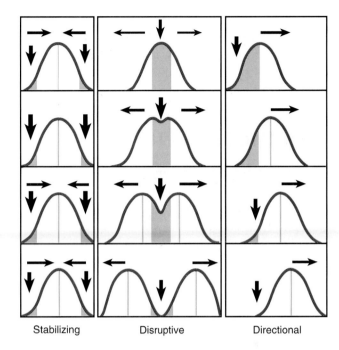

Stabilizing     Disruptive     Directional

**Figure 21.7** Three modes of selection. Stabilizing selection favors organisms with values close to the population mean. Consequently, little or no change is produced in the population. Directional selection accounts for most of the change observed in evolution. Disruptive selection increases frequencies in the extremes. The curves represent the frequency of organisms with a certain range of values. Upward arrows favorable selection and downward arrows adverse selection. Shaded areas represent the phenotypes being eliminated by selection.

melanistic form of the species was caught for the first time. The black form, *carbonaria,* increased steadily through the years until it became extremely common, often reaching a frequency of 95 percent or more in Manchester and other industrial areas. From these places *carbonaria* spread mostly westward into rural areas far from the industrial cities. The black form came about by the spread of dominant and semidominant mutant genes, none of which are recessive. This increased frequency and spread was brought about by natural selection. The normal form of the peppered moth has a color pattern that renders it inconspicuous when it rests on lichen-covered tree trunks; but the grime and soot of industrial areas carried great distances over the English countryside by prevailing westerly winds killed or reduced the lichen on trees and turned the bark of the trees a nearly uniform black. The dark form is conspicuous against the lichen-covered trunk, but inconspicuous against the black.

A British biologist, H. R. D. Kettlewell (1961; see also Kettlewell 1965) experimentally demonstrated the role of natural selection in the spread of the dark form. He reared, marked, and released melanistic and light forms in polluted woods. The melanistic form had a better survival rate than

**Figure 21.8** Normal and melanistic forms of the polymorphic peppered moth *Biston betularia* at rest on a lichen-covered tree. The spread of the melanistic form *carbonaria* in the industrial areas is associated in part with improved concealment of black individuals on soot-darkened, lichen-free tree trunks. Away from industrial areas the normal color is most frequent because black individuals resting on lichen-covered trunks are subject to heavy predation by birds.

the light form and therefore left more offspring. To confirm the role of natural selection, Kettlewell released light and dark forms in unpolluted woods. There the light form survived better. The reason was selective predation. In woods with lichen-covered trees the melanistic form was more easily seen by several species of insect-feeding birds and was therefore subject to heavier predation. In polluted woods the light form bore the brunt of predation. For this reason the normal form has virtually disappeared from polluted country but it is still common in the unpolluted areas in western and northern Great Britain.

Although selective predation is considered to be the major influence, melanistic polymorphism in the peppered moth may also be maintained by another, independent factor (Lees and Creed 1975). The dark-colored individuals apparently have a physiological advantage over the nonmelanistic form in withstanding the effects of air pollution. Predation aside, the dark form will increase in those areas where it has the physiological advantage. In some cases the visual advantage is less than the physiological advantage.

## Group and Kin Selection

According to evolutionary theory, selection operates on the level of the individual and acts on phenotypes. Certain traits, however, such as warning calls, warning coloration, and behaviors that benefit a group are often at the expense of the individual. Such acts and traits are called **altruistic.** Altruism, strictly defined, is the sacrifice of one's own well-being in the service of another. In genetic terms, the altruist contributes to the genetic fitness of another individual, the recipient of the altruistic act, while decreasing its own fitness (Michod 1982).

For many years, evolutionary ecologists have debated this puzzling question. How can such traits be maintained in a population when the altruist and thus the altruistic gene are selected against, whereas the nonaltruistic or selfish gene is at a selective advantage? It is possible to pass on a trait that benefits the species but harms the organism that carries it?

Group Selection   To explain such traits, some population and behavioral ecologists have suggested that **group selection** operates on the differential productivity of local populations (Wilson 1980). The characteristic selected improves the fitness of the group, though it may decrease the fitness of any individual in the group. Any genetic differences among local populations that decrease the likelihood of local extinction or increase the likelihood that one local population will produce emigrants or colonists who will affect the genetic composition of another will favor that group.

The idea of group selection was suggested by Darwin in *The Origin of Species* (1859) and by Sewall Wright (1931a, b, 1935), who called it *interdemic selection*. The concept began to receive serious and controversial attention when a Scottish ecologist, V. C. Wynne-Edwards, published in 1962 *Animal Dispersion in Relation to Social Behavior.* In that book Wynne-Edwards advanced the idea that animals tend to avoid overexploitation of their habitat, especially the food supply, by altruistic restraint in population growth, either by reducing or refraining from reproduction. Restraint was achieved through the mechanism of social behavior in which displays provided information about the local population. Local populations that restrained reproduction were more likely to survive than populations that grew beyond the ability of the resource to support them. Such populations would decline or go extinct, leaving empty habitats. These habitats would then be colonized by dispersers from altruistic populations.

The idea of group selection was vigorously challenged by many evolutionary ecologists (see G. B. Williams 1966, Wines 1966, Lack 1966, Ghiselin 1974, J. Maynard Smith 1976), who argued that natural selection could act only on individuals. It was accepted in modified forms by others (Lewontin 1965, J. Emlen 1973, Gilpin 1975, Alexander and Borgia 1978). D. S. Wilson (1975, 1977, 1979, 1980, 1983) framed the idea of group selection in a somewhat different context. He views group selection as a component of natural selection that operates on the differential productivity of local populations within a global system. Local populations or **demes** are groups of individuals that interact with one another sometime in their life cycle, giving any two of them an equal opportunity of becoming neighbors (Mayr 1963, D. S. Wilson 1977). Within demes are genetically distinct individuals that form trait groups (D. S. Wilson 1979, 1980) for a part of a generation or for many generations. These trait groups interact in a number of ways, such as mating, competition, defense against predation, and the like. We can

consider a deme not only as a population of individuals but also as a population of groups isolated relative to certain traits. Thus the population of a given organism has levels. Total population over a large given area, the global population, consists of numerous local populations, or demes, isolated in varying degrees from one another; and each deme consists of a number of trait groups. Within these trait groups intrademic group selection as well as individual selection takes place.

Individual selection is that component of natural selection that operates on the differential fitness of individuals within local populations. Traits promoted by individual selection are considered selfish, and those promoted by group selection as altruistic. Altruistic or group-benefit traits increase the relative productivity of local populations within a global system. Altruistic traits may be weak, that is, not strongly sacrificial, or strong. Selection for an altruistic trait might take place in species with small freely interbreeding local populations sufficiently isolated to allow some differentiation in gene frequency, and with little gene flow between other local groups.

A model of intrademic group selection illustrates how an altruistic trait, although selected against in an individual, may be selected for in a group (Wilson 1983). Assume that an altruistic allele $A$ benefiting the group at the expense of itself exists in a global population with a frequency of $p = 0.5$. Individuals, however, are distributed in local populations in which $p$ varies from 0.3 to 0.7, and selection operates within each group as outlined in Figure 21.9. Note that in each local population, the $A$ allele declines in frequency because of individual selection, but at the same time populations $N'$ increase relative to $p$. When we weigh the new global frequency of $A$ by group size, the frequency of $p$ actually increases, suggesting that an altruistic allele can evolve, even though it is selected against within each group. As with individual selection, the model suggests that group fitness involves some heritable phenotypic variation that influences group productivity or persistence. Natural selection will favor that characteristic, just as genetic variation influences individual selection. Only within groups can an allele can have low relative individual fitness and still be selected. Thus an altruistic allele in a population selected against in mixed groups can persist in a population by intrademic group selection.

Consider a local population or deme possessing a high frequency of a gene that decreases mortality or increases reproduction that is surrounded by demes possessing the gene at lower frequencies. The deme possessing a high frequency of the adaptive gene will develop some selective advantage over neighboring demes. It will produce a greater surplus population that will emigrate to surrounding demes or into empty habitat patches (created perhaps by local extinctions of populations possessing the adaptive gene in low frequencies). In fact, group selection may be particularly important for traits affecting dispersal (see Slatkin 1987). Group selec-

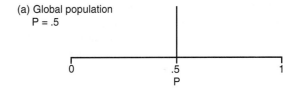

(a) Global population
P = .5

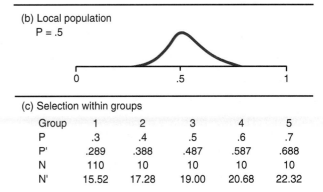

(b) Local population
P = .5

(c) Selection within groups

| Group | 1 | 2 | 3 | 4 | 5 |
|-------|------|-------|-------|-------|-------|
| P | .3 | .4 | .5 | .6 | .7 |
| P' | .289 | .388 | .487 | .587 | .688 |
| N | 110 | 10 | 10 | 10 | 10 |
| N' | 15.52 | 17.28 | 19.00 | 20.68 | 22.32 |

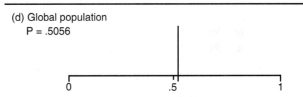

(d) Global population
P = .5056

**Figure 21.9** A numerical example of interdemic selection. The $x$ axis of the graph shows the given frequencies ($p$) of the $A$ allele and the $y$ axis the relative abundance $s$ of groups within a given frequency $p$. (a) Global population represented by a single frequency of $A_1$. (b) Local groups that vary in $p$ values. (c) Table showing selection within five representative groups, using Wright's model with $b = 2$ and $s = 0.05$. (d) Global population after selection. Wright's model for the socially advantageous $A_1$ allele (frequency = $p$) that benefits the group ($b$ term) at the expense of itself ($s$ term) is:

| Genotype | Frequency | Selective value |
|----------|-----------|-----------------|
| AA | $p^2$ | $(1 + bp)(1 - 2s)$ |
| Aa | $2p(1 - p)$ | $(1 + bp)(1 - s)$ |
| aa | $(1 - p)^2$ | $(1 + bp)$ |

tion can then be defined as changes in gene frequencies resulting from differential extinction or productivity of groups. Because the groups themselves are short-lived, extinction of groups is not as relevant as differential productivity of groups or the contribution of groups to the mating pool (Michod 1982).

**Kin Selection** Evolution of an altruistic trait may take place more easily when groups involved are made up of closely related individuals sharing genes by common descent. In such a setting an individual may increase its own fitness and the fitness of close relatives in the long run by decreasing its fitness in the short run. This idea, advanced by W. D. Hamilton in 1964, was called **kin selection** by Maynard Smith (1964).

The cost and benefit of an altruistic trait depends upon the closeness of the relationship. Genes for altruism can be selected only when the benefit to the recipient is greater than the reciprocal of the coefficient of relationship between the altruist and the recipient. This requirement was defined by Hamilton:

$$k > 1/r$$

where $r$ is the proportion of the genes of two individuals identical because of common descent and $k$ is the relationship of recipient's benefit ($b$) to the altruist's cost ($c$) $k = b/c$. The value of $r$ varies from 1 for identical twins to 0 for no relationship. For parent–offspring the value is 0.5, because the offspring receives one-half of its genes from each parent; and conversely, a parent contributes one-half of its genes to an individual offspring. The value for a full sibling would be 0.5; for a half-sib, 0.25; for an uncle, aunt, niece, or nephew, 0.25; and for a first cousin, 0.125. The closer the relationship, the higher the value of $r$. If the value of $r$ is large, that is, the relationship is close, the value of $1/r$ is small. If $k$, the ratio of fitness to the loss of fitness, exceeds $1/r$, then the altruistic gene would be selected for. If the value of $1/r$ is small, then the ratio of benefits $b$ to cost $c$, or $k$, can be large and still favor the altruistic gene. In other words, an altruist can take greater risk to help a close relative than a distant one. Even in that situation the benefit/cost ratio $k$ would have to exceed a value of 2 (1/0.5).

Consider an individual A who, facing 100 percent chance of death, gives a warning call to a full sibling. If this alarm warns only a single sibling, then the benefit (survival of 1) relative to the cost (death of 1), would be 1/1 < 1/.05, and altruism would be selected against. If the warning saves four full sibs, then 4/1 > 1/.5, and the trait would be selected for. If the act incurs only 50 percent chance of death, then the benefit, 1/0.5, would be the same as 1/r; 1/0.5, and there would be no benefit from the act. By the same token, the altruist in the initial situation above would have to save over eight first cousins before the trait would be selected for.

Two aspects of fitness are involved in kin selection. One is individual fitness, which involves the genes in the individual's own offspring (classic selection). The other is the additional fitness acquired by improving the fitness of very close relatives, especially parents and sibs (kin selection). An individual's own direct reproductive success added to that of its close relatives possessing replication of its own genes is **inclusive fitness** (W. D. Hamilton 1964), a concept used to explain many aspects of social behavior.

Kin selection, then, is the evolution of a genetic trait expressed by one individual that affects the genotypic fitness of one or more directly related individuals (Michod 1982). It is favored when an increase in fitness of closely related individuals is great enough to compensate for the loss in fitness of the altruistic individual. If the dispersers and helpers are very close kin, as they usually are, those individuals serve to increase the fitness of the genetic traits they hold in common.

An altruistic act need not mean sacrifice of life. Any act counts that improves reproductive opportunities for the remaining sibs or parents. An example is dispersal from a kin group, as in black bear (see Rogers 1987) and ground squirrel (Holekamp and Sherman 1989), or helping parents raise full or half sibs.

The most widespread examples of altruism are eusociality and cooperative breeding. **Eusociality** is the extreme in social living (Andersson 1984). It is characterized by cooperative caring for the young, division of labor with more or less sterile workers caring for individuals engaged in reproduction, and an overlap of at least two generations of life stages able to contribute to colony labor. Such eusociality is found among bees, wasps, and termites, and one species of mammal, the naked mole rat.

Among the Hymenoptera there are various levels of eusociality. It is most highly developed among bees and termites, where it gives real meaning to the term kin selection. In the bees the workers are nonreproducing or sterile females waiting on a sister, the reproductive queen. In these insects males arise from unfertilized eggs and possess one set of chromosomes from the mother (haploidy). The females have one set from each parent (diploidy). Such sex determination is called haplodiploidy. The females have all the genes from their father ($r = 1$) and one-half their mother's genes ($r = 0.5$), so the female offspring have three-quarters of their genes in common. Thus the females are more closely related to full sisters than to their mother and their own offspring. Nonreproductive females can propagate their own genes most effectively and increase their own fitness best by helping create more sisters.

The most remarkable example of eusociality among vertebrates is the naked mole rat, a member of the family Bathyergidae, the mole rats of Africa. The eusocial naked mole rat, restricted to the hot, arid Horn of Africa, lives in colonies ruled by a queen waited upon by nonreproducing individuals, both male and female. The queen, with two or three large male consorts, breeds year-long and produces litters of up to 12. The queen and the breeding males are concerned with the handling, care, and grooming of the young. Larger nonbreeding males and females defend the colony and remove dirt from the tunnels; smaller nonbreeders transport food, gather nest material, and clear tunnels (Honeycutt 1992, Sherman, Jarvis, and Alexander 1991).

A much less complex form of reproductive altruism is cooperative breeding behavior, expressed in 220 species of birds (Brown 1987) and 120 species of mammals (Riedman 1982). In cooperative breeding, adult individuals regularly help the genetic parents in the rearing of the young. The cooperative breeding groups are typically family groups formed through the retention of grown offspring. Such breeding systems can develop complex extended families involving more than one breeding pair (Emlen 1991, Koenig

**Figure 21.10** The Florida scrub jay inhabits the thickets of sand pine, scrub oaks, and palmettos along the east and west coasts of Florida. Young scrub jays help their parents feed and care for young for several years.

and Mumme 1987). Two of the simpler ones are those of the Florida scrub jay (*Aphelocoma coerulescens*) and the black-backed jackal (*Canis mesomelas*) of Africa.

The basic social unit of the Florida scrub jay (Figure 21.10) is a monogamous breeding pair together with some of their young from the previous one or two years (Woolfenden 1975, Woolfenden and Fitzpatrick 1984). These mature young help in group defense of the territory (McGowan and Woolfenden 1989). During the breeding season they bring food to the nestlings, tend and guard the nest, and continue to provide food and protection to the fledgling out of the nest. Female helpers rarely remain for more than two years because they join the breeding population. Unmated males may be around longer.

The black-backed jackal (Figure 21.11) has a similar social structure (Moehlman 1979). Some of the young of the year remain with their parents through the following breeding season, forming groups of three to five adults. The mature offspring help the parents rear the next year's litter. They bring food to the nursing female, regurgitate food for the young, play with and groom them, and guard the pups when the parents are away from the den.

In these cooperative breeding systems, the helpers do enhance the fitness of the breeders and indirectly their own fitness (Emlen 1991). Pairs with helpers have higher reproductive success than those without them (Figure 21.12). Helpers increase the success of reproductive attempts by the mated pair by helping to feed and care for the young. Helpers also reduce the workload of the parents, enabling them to produce a second litter. Because they defend against predators

**Figure 21.11** The black-backed jackel inhabits savannas and open woodlands of southern Africa. Some young remain with the monogamous parents, do not breed, and assist with the next litter.

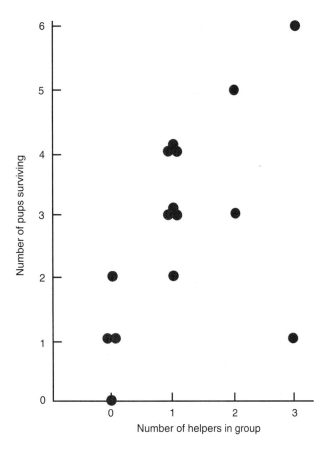

**Figure 21.12** Correlation between the number of parental helpers and reproductive success in the black-backed jackal. (From Moehlman 1979.)

and reduce reproductive stress, helpers improve the breeders' survivorship, increasing their inclusive fitness.

If these helpers are so successful at aiding the parental pair, why don't they go out on their own? Several hypotheses have been advanced to explain why young adults stay at home. One hypothesis is that the habitat is saturated and suitable territories are limited. The mature offspring have to wait for a suitable territory to open, and waiting is best done at home in familiar surroundings. Another hypothesis is that although vacant territories are available, they are at best marginal, so why not wait for a high quality opening? Because of the helpers' assistance, the family size of the pair increases. As the family group expands, its territory expands. Eventually a dominant male helper may claim a part of the enlarged territory as his own, or inherit it and become a breeder, or hold a competitive edge for nearby openings. By helping, the male has come out ahead. He has increased his inclusive fitness and improved the chances of his own direct fitness by increasing his opportunity to become a breeder. A third hypothesis is that dispersal into unknown situations is risky, especially if the only habitat available is marginal. Early dispersers face an increased risk of mortal-

ity, a reduced probability of securing a mate, and the high costs of independent reproduction, especially in a fluctuating environment.

In the long run, then, what does a helper gain by staying at home for a while? First, it increases the probability of its own survival. By belonging to a larger group it has increased vigilance against predators and access to social and physical resources of the natal territory. Second, the helper enhances its own likelihood of becoming a successful breeder by gaining access to a superior territory and thus increased probability of obtaining a mate. Third, the helper can increase its reproductive success because of its experience at parenting, and it will gain a similar set of helpers. Finally, helping increases the survival of related breeders whose production of young will increase the helper's own inclusive fitness through their reproductive efforts. Thus in a way, a helper may not be altruistic at all, but practicing a form of selfish individual selection under the guise of cooperation (Lignon 1981).

## INBREEDING

Although a local population of a species consists of all individuals occupying a given area, it really is made up of a number of semi-isolated subpopulations, often with minimal interchange (Ralls, Harvey, and Lyles 1986). Isolation becomes most pronounced when subpopulations occupy patches or islands of habitats, as do house mice restricted to individual barns or white-footed mice confined to islands of forest growth scattered through agricultural or suburban landscapes. Such small populations may be subject to inbreeding.

**Inbreeding,** simply defined, is breeding between relatives. With inbreeding, mates are more closely related on the average, than they would be if they had been chosen at random from the population. Some reasons for inbreeding are small isolated populations, close proximity of potential mates, ecological preferences, morphological resemblances among individuals, and the like. The principal effect of inbreeding is an increased frequency of homozygous genotypes (Figure 21.13).

The extreme in inbreeding is self-fertilization, which may occur in plants. Such inbreeding provides a measure of comparison for degrees of inbreeding. Consider the population described by the Hardy-Weinberg equilibrium earlier, in which $p(A) = q(a)$ and both are equal to 0.5. Thus $p^2 = .25$, $2pq = .50$, and $q^2 = .25$. In our hypothetical population, all breeding involves selfing from the start. Within the population all homozygotes $AA$ and $aa$ breed true. Offspring from the heterozygotes will be one-half heterozygotes and one-half homozygotes each generation. The homozygotes produced will be added to the pool of homozygotes in the population; the remaining heterozygotes in the next generation

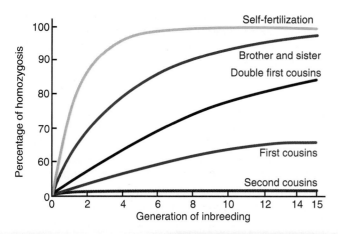

**Figure 21.13** The percentage of homozygous offspring from systematic matings with different degrees of inbreeding. Note how rapidly homozygosity declines as the relationship of offspring becomes further removed from the original parent stock.

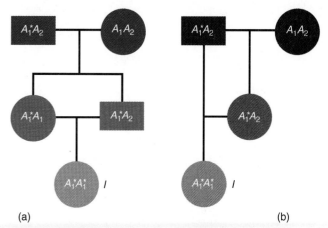

(a)　　　　　　　　　　　　　　　　　　(b)

**Figure 21.14** Development of autozygosity, possession of alleles identical by descent. The pedigrees chart inbreeding due to (a) sib mating and (b) parent-offspring mating. The A* allele is obtained by direct descent. The inbreeding coefficient of *I* is 1/4 in both cases.

again will produce offspring one-half homozygotes and one-half heterozygotes. Eventually the self-fertilizing population will become exclusively homozygous.

Within one generation the gene frequencies will be [1/4 (1) + (1/2)(1/4)], (1/2) (1/2), [1/4 (1) + (1/2)(1/4)], or 3/8 *AA*, 2/8 *Aa*, 3/8*aa* (Table 21.1). Already the degree of heterozgosity has declined from 1/2 to 1/4. The expected frequency of heterozygosity in a self-fertilizing population is $2pq \times (1/2)^n$, where $2pq$ is the initial frequency of heterozygotes and $n$ is the number of consecutive generations. Note that with inbreeding, the frequency of alleles remains the same, as predicted by the Hardy-Weinberg equilibrium, but homozygosity increases at the expense of heterozygosity. In effect, variation originally partitioned within individuals becomes partitioned among individuals.

## The Inbreeding Coefficient

Exclusively self-fertilizing populations, the extremes of inbreeding, are uncommon. Even populations of self-fertilizing plants experience some periods of cross-fertilization or outbreeding. Close inbreeding usually involves mating between brother and sister, parent and offspring, or more frequently cousins. Sometimes a homozygote's alleles are identical because of independent mutations. Such individuals are **allozygous** for the alleles involved. Inbred individuals possess alleles that are identical by descent; they can be traced back to a common ancestor (Figure 21.14). Such homozygotes are **autozygous.** Thus one outcome of inbreeding is an increase in the frequency of autozygous individuals.

The degree of inbreeding depends upon relationship. Brothers and sisters and parents and offspring share one-half of their genes, so the relationship is very close. First cousins share one-quarter of their genetic heritage. As relationships diverge, the frequency of shared genes diverges (Figure 21.15). The amount of inbreeding between relatives is measured by the coefficient of inbreeding, *F. F* is the probability that an individual receives at a given locus two genes that are identical by descent; stated differently, it is the amount of heterozygosity that has been lost. For a self-compatible population the inbreeding coefficient is

$$F = \frac{H_0 - H}{H_0}$$

where $H_0 = 2pq$ and $H$ is the actual frequency of heterozygous genotypes in the population. $F$ is a measure of the fractional reduction of heterozygosity in an inbreeding population relative to reduction in a randomly mating population with the same frequency of alleles.

**Table 21.1** Decrease in Heterozygosity under Systematic Self-fertilization Starting with an Equilibrium Population ($p = q = \frac{1}{2}$)

| | Genotypic Frequencies | | | | |
| Generations | A/A | A/a | a/a | F | q |
|---|---|---|---|---|---|
| 0 | ¼ | ½ | ¼ | 0 | ½ |
| 1 | ⅜ | ¼ | ⅜ | ½ | ½ |
| 2 | ⁷⁄₁₆ | ⅛ | ⁷⁄₁₆ | ¾ | ½ |
| 3 | ¹⁵⁄₃₂ | ¹⁄₁₆ | ¹⁵⁄₃₂ | ⅞ | ½ |
| 4 | ³¹⁄₆₄ | ¹⁄₃₂ | ³¹⁄₆₄ | ¹⁵⁄₁₆ | ½ |
| n | $\dfrac{1-(\frac{1}{2})^n}{2}$ | $(\frac{1}{2})^n$ | $\dfrac{1-(\frac{1}{2})^n}{2}$ | $1-(\frac{1}{2})^n$ | ½ |
| ∞ | ½ | 0 | ½ | 1 | ½ |

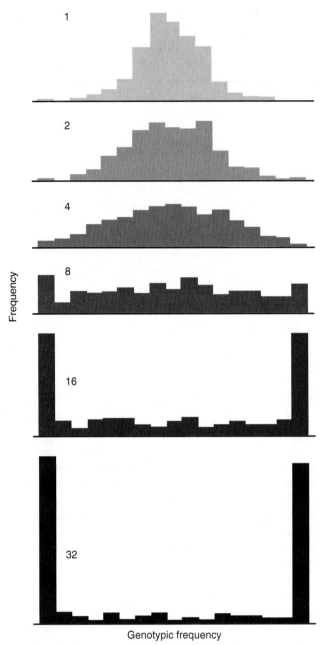

**Figure 21.15** A computer simulation of the dispersal of gene frequencies among 400 hypothetical populations over 32 generations. The genetic model had the following conditions: (1) each population consisting of 8 diploid individuals—four randomly formed pairs—is constant from generation to generation; (2) each individual mates but once and the number of offspring produced by each mating varies—a Poisson distribution with a mean of two; (3) selection and mutation are absent; (4) each population started with 2 *AA*, 4 *Aa*, and 2 *aa*, so the initial gene frequency was 0.5 for each allele at a single autosomal locus. The figure shows the populations classed according to gene frequencies in generations 1, 2, 4, 8, 16, and 32. Because of chance variations—random genetic drift—in such small populations, gene frequencies show an increasing spread toward fixation of one allele or the other in most of the populations. This simulation demonstrates why small populations exhibit fixed genes and a lack of genetic diversity. (From Mettler and Gregg

Consider a heterozygous population possessing two kinds of gametes in which no two alleles are alike. The total number of kinds of gametes produced in the population is $2N_0$. When the gametes unite to form the zygotes of the next generation, a probability of $1/2N_0$ exists that two identical gametes will unite to form a homozygote. Within an inbreeding population the probability an individual in the first filial or F1 generation will have two alleles identical by descent is the same as the probability of having two identical alleles, $1/2N_0$. Thus the inbreeding coefficient of the first generation is $F_1 = 1/2N_0$.

The second inbreeding generation will not consist only of homozygotes produced by union of two gametes with alleles identical by descent from the first generation. The probability also exists that two gametes from different homozygous individuals in the $F_1$ generation but descending from the same ancestors in generation 0 will also be present. Thus the population consists not only of homozygotes produced by new breeding but also homozygotes attributed to previous inbreeding (Table 21.2):

$$F_1 = \frac{1}{2N}$$

$$F_2 = \frac{1}{2N_1} + \left(1 - \frac{1}{2N}\right)F_1$$

$$F_3 = \frac{1}{2N_2} + \left(1 - \frac{1}{2N_2}\right)F_2$$

$$F_n = \frac{1}{2N_{n-1}} + \left(1 - \frac{1}{2N_{n-1}}\right)F_{n-1}$$

The inbreeding coefficient consists of two parts. The first part, $1/2N_{n-1}$ is derived from new inbreeding. The second part, $(1 - 1/2N_{n-1})F_{n-1}$, is attributed to previous inbreeding.

The above coefficients relate to an idealized population. This constraint can be removed if the term $1/2N_{n-1}$ is replaced by $F$, the rate at which heterozygosity is lost or alternatively the rate at which genes become fixed:

$$F = \frac{F_n - F_{n-1}}{1 - F_{n-1}}$$

When $F = 0$ no inbreeding occurs, and when $F = 1$ complete inbreeding occurs. In summary, upon inbreeding the Hardy-Weinberg frequencies change to:

$$[p^2(1 - F) + pF] + 2pq(1 - F) + [q^2(1 - F) + qF]$$

Models of inbreeding have been the domain of plant and animal breeders. They apply these concepts in the development of new varieties of horticultural plants and crops and in the improvement of livestock breeds from broiler chickens to racehorses. Geneticists also use such models in the study of genetically carried defects in humans, such as susceptibility to certain diseases. Now conservation geneticists are using the same models to study ways of conserving genetic

**Table 21.2** Genotypic Frequencies with Inbreeding

| | Frequency in Population | | | |
|---|---|---|---|---|
| Genotype | With Inbreeding Coefficient F | | With F = 0 (Random Mating) | With F = 1 (Complete Inbreeding) |
| | Allozygous genes | Autozygous genes | | |
| AA | $p^2(1-F)$ + | $pF$ | $p^2$ | $p$ |
| Aa | $2pq(1-F)$ | | $2pq$ | $0$ |
| aa | $q^2(1-F)$ + | $qF$ | $q^2$ | $q$ |

diversity in captive breeding populations of endangered species.

The use of inbreeding models requires the analysis of pedigrees. From pedigrees geneticists can determine lineage and thus the degree of inbreeding among related individuals. Conservation geneticists have been developing pedigrees of individuals in the zoo populations of many endangered species. They use this information to plan breeding programs designed to increase genetic diversity in the captive populations that they hope to reintroduce to the wild.

Obtaining pedigrees of individuals in natural populations is almost impossible without following marked breeding individuals and their offspring. However, DNA "fingerprinting" enables conservation geneticists to study the degree of inbreeding within local populations of a particular species and to assess the degree of genetic diversity among separate populations of that species.

## Consequences of Inbreeding

In normally outcrossing populations, close inbreeding is detrimental. It increases autosomal homozygosity with all its attendant problems. Rare recessive deleterious genes become expressed in a homozygous state. This condition can cause premature death, decreased mating success, decreased fertility, decreased fecundity, small body size, loss of vigor, reduced pollen and seed fertility in plants, and various meiotic abnormalities, such as poor chromosomal pairing. These consequences are termed **inbreeding depression.**

The effect of inbreeding depression is evident in the lions of Ngorongoro Crater (O'Brian and Evermann 1988, Packer et al. 1988). The present population is about 100 lions, 30 of which are adults, in six prides. All of them descended from 15 survivors of an outbreak of a fly plague. They have lost 10 percent of their genetic diversity over the past 20 years, and possess only one-half of the genetic diversity of Serengeti lions. In addition, they exhibit a genetically weakened immune system, making them more vulnerable to disease. Because resident males are strong enough to deter the few interlopers that do venture into the crater, no new male

lions have entered Ngorongoro Crater breeding population since 1969, resulting in inbreeding for five generations. There is no indication, however, that these physiological problems as yet have reduced reproductive fitness.

Inbreeding is not always bad for a population. Occasional inbreeding will maintain rare alleles that otherwise might be lost. For example, descendants of the Ngorongoro lions might be resistant to future bloodsucking fly plagues. Close inbreeding under certain artificial situations is used in the breeding of domestic plants and animals to fix certain desirable genes that will breed true, in spite of the expression of deleterious ones. These inbred lines, however, are then outcrossed to produce "hybrid" vigor. Paradoxically, the outcrossing of mildly inbred populations in natural populations (as happens when new individuals are introduced into local populations to augment small populations) can result in **outbreeding depression.** Outbreeding depression causes a loss of fitness through the contamination of a gene pool with new alleles that produce offspring poorly adapted to the local environment.

Close inbreeding in nature is rare, less than 2 percent in natural populations of vertebrates for which there are data (Ralls, Harvey, and Lyles 1986). Populations of three species exhibit low levels of genetic variability: seals (Bonnell and Selander 1974), sea otter (Ralls et al. 1983), and Indian rhinoceros (Dinerstein and McCracken 1990). Natural mechanisms exist to reduce inbreeding. One is spatial separation or differences between the sexes in dispersal of young (Greenwood 1980, Holmes and Sherman 1983). One sex stays behind; the other leaves. Among birds, juvenile females most frequently leave the home area; the males tend to return to the vicinity of their birth. Separation is further enhanced in many species by monogamous mating habits and the frequent loss of a mate during the nesting season or between years. Among mammals young females stay close to the home place, while young males, often driven away by the females, seek new places to live. Adult male lions defend groups of females from competitors, but in a few years they are forced to relinquish the pride to vigorous younger males from the outside (Pusey and Packer 1986).

Sex-biased dispersal among black-tailed prairie dogs (*Cynomys ludovicianus*) reduces breeding between first-degree relatives by 90 percent (Hoogland 1982). Dispersal involves virtually all males. Young males leave the family group or coterie before breeding, and young females remain behind. Adult male prairie dogs are more likely to move to new breeding groups if adult daughters are in the home colony. Although dispersal does reduce inbreeding, sex-biased dispersal may have evolved for other reasons, such as enhanced reproductive success, increased access to food resources and space, and other advantages (Greenwood 1980).

A second mechanism in reducing close inbreeding is kin recognition (Holmes and Sherman 1982, 1983). Because of their close association during early life, siblings recognize one another later on. Females of both the Beldings ground squirrel and Arctic ground squirrel can distinguish between full and half sisters, although this discrimination fades over long absence. Females mate with unrelated males; they leave the group if a related male returns, or fail to come into estrus, especially if the father is in the group (Hoogland 1982).

# GENETIC DRIFT

In sexual reproduction only a few of the gametes produced actually form a new generation. In general, all an individual's genes will be represented somewhere among its gametes, but not in any two of them. Under conditions of stable population size, two gametes are about all that an average individual can leave behind. For a heterozygote *Aa*, there is a 50:50 chance that the two gametes will either be both *A* or both *a*, assuming no natural selection. Thus a 50:50 chance exists that a heterozygote will fail to pass on one of its genes. In a whole population these losses tend to balance each other, so the gene frequencies of the filial generation are a replica, but never an exact one, of the parents' gene frequencies. This is simply the familiar law of averages at work. The larger the population, the more closely the gene frequencies of each generation will resemble those of the previous generation.

However, large populations consist of numerous subpopulations, often more or less isolated from one another, like white-footed mice in a small woodlot. Because of habitat fragmentation, more and more populations of species are being isolated to varying degrees. In these cases, the subpopulations represent random sampling among gametes not representative of the gene pool of the larger population. Each subpopulation has its own distinct sample. Chance fluctuation in allele frequencies in these small populations as a result of random sampling is called **genetic drift.** Over time some genes will continue to segregate, while others will become fixed. After some time, the population will become fixed or homozygous for some alleles, and other alleles will be lost. Over time, allele frequencies spread out progressively as the proportion of fixed genes steadily increases, ultimately resulting in homozygous populations (Figure 21.15).

How rapidly this outcome occurs depends upon the size of the population. Consider the data in Table 21.3, which compares the variances over generations for small and large populations when $p = q = 0.5$, the number of individuals is constant, and the generations are discrete. The variance at any given time depends upon the size of the population, the number of generations elapsed, and initial gene frequencies. Note that a population of 500 shows little variance over 50 generations and minimal variance after 100 generations. But allele frequencies in natural populations behave so erratically that one cannot predict the probability of ultimate fixation.

Genetic drift mimics inbreeding. Inbreeding comes from nonrandom mating, whereas genetic drift comes from random mating. However, both processes increase homozygosity.

The measure of genetic drift is the **fixation index,** the reduction of heterozygosity of a subpopulation due to random genetic drift. It measures the amount of inbreeding due solely to population subdivision. It is similar to the inbreeding coefficient presented earlier:

$$F_{ST} = \frac{H_T - H_S}{H_T}$$

where $H_S$ represents the heterozygosity of a randomly mating subpopulation and $H_T$ represents the heterozygosity in an equivalent randomly-mating total population. Thus the fixation index $F_{ST}$ is the probability that two alleles chosen at random in the same subpopulation are identical by descent. (For a full discussion of the derivation of the fixation index, see D. Hartyl 1988.)

As in inbreeding, the value of $F$ will change as genetic drift continues generation after generation. The average

---

**Table 21.3 Variances over Generations for Populations of Different Sizes with Discrete Generations and a Constant Number of Individuals, all with $p = q = 0.5$**

| Number of Generations | Size of Population | | | | | |
|---|---|---|---|---|---|---|
| | 6 | 10 | 50 | 100 | 500 | 1000 |
| 1 | 0.02 | 0.01 | | | | |
| 2 | 0.04 | 0.02 | 0.01 | | | |
| 3 | 0.06 | 0.04 | 0.01 | 0.01 | | |
| 4 | 0.07 | 0.05 | 0.01 | 0.01 | | |
| 5 | 0.09 | 0.06 | 0.01 | 0.01 | | |
| 10 | 0.15 | 0.10 | 0.02 | 0.01 | | |
| 50 | 0.25 | 0.23 | 0.10 | 0.06 | 0.01 | |
| 100 | 0.25 | 0.25 | 0.16 | 0.10 | 0.05 | 0.01 |

*Source:* Mettler and Gregg 1969:51.

value of the fixation index in subpopulations in generation $t$, or $F_t$, is:

$$F_t = 1 - \left(1 - \frac{1}{2N}\right)$$

The value of $F_t$ will range from 0 for no homozygosity to 1 for complete homozygosity.

## Effective Population Size

The values of $F_t$ are based on an ideal population characterized by a constant population size, equal sex ratios, equal probability of mating among all individuals, and a constant dispersal rate. Most populations are not "ideal." There are age-related differences in reproduction. Particularly in polygamous populations the ratio of breeding males to females is unequal. In such populations the number of males is more important than the number of females in determining the amount of random drift. For these reasons the actual size of a small subpopulation is of little meaning. Of greatest importance is the genetically **effective population size**, $N_e$.

$N_e$ is not the same as the actual number of breeding individuals. The effective population size equals the number of adults contributing gametes to the next generation. If the sexes are equal in number and all have an equal probability of producing offspring, $N_e$ equals the number of breeding adults in the population. If the number of breeding males and females in the population is not equal, then $N_e$ is less than $N$. $N_e$ is defined as the size of an ideal population subject to the same degree of genetic drift as a particular real population. The ideal population is a randomly breeding one with a 1:1 sex ratio and with the number of progeny per family randomly distributed (Poisson).

**Unequal Sex Ratios** In a monogamous population in which one male mates with one female, all offspring are less closely related than in a polygamous population with, say, a ratio of one breeding male to five females. In the latter situation the offspring would be half or full sibs. The chance of an allele becoming lost or fixed (and thus the amount of genetic drift) is much greater in such a population.

The effective population size is given by

$$N_e = \frac{4 N_m N_f}{N_m + N_f}$$

where $N_m$ and $N_f$ are the numbers of breeding males and females respectively. As the disparity in the ratio of males to females widens, the effective population diminishes (Figure 21.16).

Consider a population of white-tailed deer consisting of 100 adult does and 50 adult bucks. The actual size of the population is 150. Because of the unequal sex ratio we might assume that the effective population size is $4(50 \times 100)/150 = 133$. However, the white-tailed deer is a polygamous species with a dominance hierarchy among the

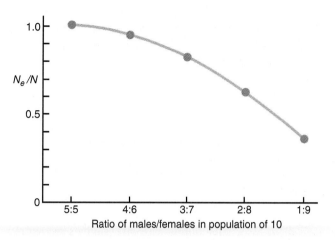

**Figure 21.16** The depression of effective population size $N_e$ due to disparity in the sex ratio (m/f) of reproducing animals. Note that as the ratio widens with fewer males relative to the number of females, as it may in polygamous species, especially where males are hunted, the effective population size drops dramatically. (From Foose and Foose 1984.)

males. Only the dominant males breed, and subdominant, potentially breeding males cannot contribute their genes to the next generation during any particular breeding season. Thus the effective population size is even smaller. Assume that 25 dominant males mate with 100 does, an average of four does per breeding male, not an unrealistic breeding ratio. Under these conditions the actual breeding population consists of 100 does and 25 bucks. The effective breeding population, $N_e$, now is $4(25 \times 100)/150 = 66$. What $N_e$ tells us is that the sampling error or genetic drift in our deer population of 150 animals with a sex ratio of 1:4 is equal to that of a population of 66 with an equal number of males and females. In other words, in our deer population the 25 males are not genetically equivalent to the 100 females. The males contribute disproportionately to the gene pool of the next generation. Each male contributes $1/2 \times 1/25 = 0.02$ of the genes to the next generation, whereas each female contributes $1/2 \times 1/100 = 0.005$ to the next generation.

Effective population size becomes more significant in exploited populations, especially those in which trophy animals, usually dominant males, are selected by hunters and removed from the population. For example, antler size in the white-tailed deer is genetic (Harmel 1983, M. H. Smith et al. 1983), controlled by a single set of dominant-recessive alleles (Templeton et al. 1983). A single dominant allele has a major effect on the phenotypic expression of five to ten points and a single recessive determines the phenotypic expression of two to four points. If these males are selected against by sport hunting (as they are in North America), then subdominant males become the dominants and genes for smaller antlers could become fixed in the local population.

**Isolation by Distance** The concept of effective population size also applies to a population whose individuals are more or less continuously distributed over a large area, larger than the greatest dispersal distance of the species. This is known as the neighborhood area for the species. An estimate of the neighborhood area is the root mean square of dispersal distance of individuals about their natal origin (Dobzhansky and Wright 1947). The effective breeding size of such populations (often called **neighborhood size**) depends upon the number of breeding individuals per unit area and the amount of dispersion between an individual's birthplace and the birthplace of its offspring. The latter is denoted by $\sigma^2$, the one-way variance (half the standard deviation squared) of the distance between birth and the site of first breeding, or the root mean square dispersal distance. The variance is estimated as

$$s^2 = \frac{1}{N} \sum_{i=1}^{N} \left( d - d_i \right)^2$$

where $d_i$ is the location of individual's breeding site, $d$ is the individual's birthplace, and $N$ is the number of individuals in the sample. Since $d$, the birthplace, is at distance 0, the mean-square distance is

$$s^2 = \frac{1}{N} \sum_{i=1}^{N} d_i^2$$

In terms of dispersal distance, $\sigma$, and the number of breeding individuals per unit area, $\delta$, the effective or neighborhood population size is given by $N_e = 4\pi\sigma^2\delta$ (where $\pi$ is 3.14).

The equation can be applied to data on the bannertail kangaroo rat (*Dipodomys spectabilis*) obtained by W. T. Jones (1987), who studied the dispersal of this rodent on a 36-ha study area in southeastern Arizona. The kangaroo rat lives in mounds that persist for and are occupied over many generations. Dispersing juveniles settle in vacant mounds. Jones determined that the adult breeding population consisted of 70 males and 72 females for a density of 2.27/ha. Two hundred and eighteen juveniles (107 males, 111 females) moved a mean distance of 50 m from natal site to breeding site. The estimated mean dispersal distance $d$ for the juvenile population

$$\sum_{t=1}^{n} d_i f(d_i)$$

where $d$ is dispersal distance and $f(d_i)$ is the fraction of individuals dispersing distance $d_i$ (see Moore and Dolbeer 1989) is 140.25 m, and the mean square dispersal distance $\sigma^2$,

$$\sum_{t=1}^{n} d^2 f(d_i)$$

is 10943 m²/2 = 5472 m² or .5472 ha. The estimated effective population size is $N_e = 4\pi(.5476)(2.27) = 15.75$. The neighborhood size is small because of the very limited dispersal of the kangaroo rat.

If dispersion follows a normal curve, then 39 percent of all individuals will have their offspring within a radius of $\sigma$ centered at their own birthplace; 87 percent will have their offspring within a radius of $2\sigma$; and 99 percent will have their offspring within a radius of $3\sigma$ (S. Wright, 1978). Therefore even within large areas of habitat, the overall population is a mosaic of subpopulations restricted by dispersal distance, which may promote a degree of inbreeding. Even northern white-tailed deer subpopulations sharing winter deer yards inhabit largely exclusive summer ranges that rarely overlap with other subpopulations (Nelson and Mech 1981).

**Population Fluctuations** Actual populations are dynamic; they fluctuate through time. In fact, numbers may change dramatically. Under adverse environmental conditions or sudden loss of habitat, the population may decline sharply or "crash" (Figure 21.17). The survivors of the crash, the progenitors of future populations, possess only a sample of the original gene pool. Suppose that the effective size of a particular population for generation 1 is $N_1$, for

**Figure 21.17** A population faced with an environmental catastrophe or overexploitation enters a "bottleneck" in which the surviving population consists of only a sample of the total gene pool. If the population makes a complete recovery, it may lack the genetic diversity found in the original population. If the population remains small, as in captive or isolated situations, it is subject to random genetic drift. (From Frankel and Soulé 1981:32.)

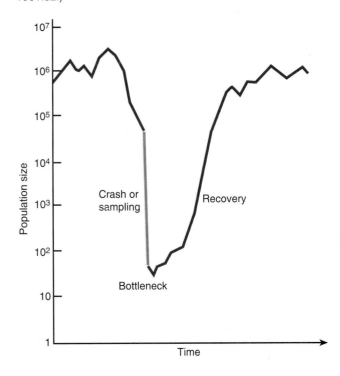

generation 2, $N_2$, and so on. The overall increase in $F_{ST}$ is given as

$$1 - F_{ST} = \left(1 - \frac{1}{2N_e}\right)^t$$

where

$$\frac{1}{N_e} = \left(\frac{1}{t}\right)\left(\frac{1}{N_1} + \frac{1}{N_2} + \ldots \frac{1}{N_t}\right)$$

The effective size of a population fluctuating over time is the harmonic mean or reciprocals of the effective number of each generation. This harmonic mean is strongly influenced by the smaller values (Crow and Kimura 1970). Thus a population crash would tend to reduce the average effective size over time. The sharp reduction in numbers creates a severe population **bottleneck.** During a bottleneck some genes may be lost from the gene pool as a result of chance. This loss, which can severely reduce genetic diversity in the remaining population, will be carried through into future generations.

There are several examples from natural populations. One is the cheetah (*Acinonyx jubatus*), the populations of which are sparse and isolated. S. T. O'Brien and associates (1983) sampled the blood of 55 cheetahs from two geographically isolated populations in South Africa and found them genetically the same (monomorphic) at each of 44 allozyme loci. Analysis of 155 abundant soluble proteins from cheetah fibroblasts (cells found in vertebrate connective tissue that form and maintain collagen) revealed a low frequency of polymorphism. The average heterogeneity was only 0.013. Compare this with an average heterogeneity of 0.036 for mammal populations studied so far. The homozygosity of the cheetah populations suggests that the species experienced a severe bottleneck perhaps 100 generations ago, resulting a severe range contraction. A similar situation exists with the elephant seal (Bonnell and Selander 1974).

A bottleneck also occurs when a small group of emigrants from one subpopulation founds a new subpopulation and when small populations of animals, such as wild turkey or otter, are introduced into new or empty habitats. The emigrant or introduced population carries only a sample of genes from the parent population, so it, too, is subject to random genetic drift. This drift is known as **founder's effect** (Figure 21.18). The gene pool of the future population is derived from the genes in the original founding population.

## Gene Flow, Mating Strategies, and Genetic Drift

Unless they are separated by a wide expanse of inhospitable habitat, some interchange takes place among subpopulations. A few white-footed mice move from one woodlot to another, and home ranges of subpopulations of deer overlap at the peripheries. If immigrants or dispersers become part of the breeding population, they introduce a different ge-

netic sample that tends to reduce or slow random genetic drift and helps maintain genetic diversity.

The degree to which dispersal influences genetic drift depends upon effective population size, amount of dispersal or immigration, and the degree to which individuals in the population are monogamous or polygamous. In a small population of constant size, monogamous species should experience inbreeding depression more slowly than polygamous species, because of their maximum population size. The greater the sex disparity in polygamous species and the larger the harem, the greater the rate of inbreeding. Where immigration or dispersal brings genetic interchange, loss in homozygosity declines (Figure 21.19). Little genetic interchange is needed to prevent significant random genetic drift among subpopulations and to maintain genetic diversity. Change in homozygosity in subpopulations experiencing some immigration is given by

$$F = \frac{1}{4N_m + 1}$$

where $N_m$ is the actual number of migrants per generation.

$F_{ST}$ decreases as the number of immigrants increases. For example, one immigrant per generation would give a value of $F = 1/4(1) + 1 = 1/5 = 0.2$. Five immigrants per generation would greatly reduce $F$: $F = 1/4(5) + 1 = 1/21 = .05$. Conservation geneticists believe that one reproduc-

**Figure 21.18** Genetic diversity in founder populations of various sizes. Founder populations of 10 may hold 90 percent of the genetic diversity found in the parent population. Populations of 50 may contain a nearly complete sample. That genetic diversity can be maintained only if the population expands. These facts are important in the introduction of a species into vacant or new habitat. (From Foose 1983:388.)

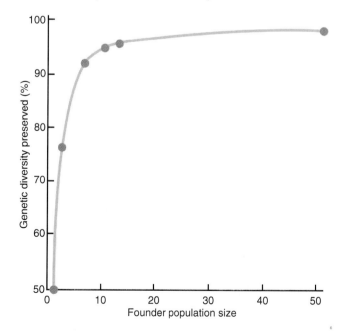

tively successful immigrant per generation is the minimal number needed to slow genetic drift and five immigrants the maximum (Frankel and Soulé 1981:129), although the percentage of immigrants needed may depend more upon whether the population is monogamous or polygamous. Too

**Figure 21.19** Alleviation of inbreeding over time. (a) Monogamous populations with either sex dispersing. (b) Polygamous populations with males dispersing. Immigrant males were always considered successful in polygamous matings. Numbers on the curves represent dispersal rates. Note that when dispersal rates in monogamous populations are low relative to population size, inbreeding continues to increase rapidly because of the slow diffusion of new alleles into the population. Contrast the rapid alleviation of inbreeding depression in polygamous matings with male dispersal and a harem of five. Even in polygamous populations that reach large breeding coefficients, only a few dispersing males are necessary to reduce genetic drift. (From Chesser 1983:74, 75.)

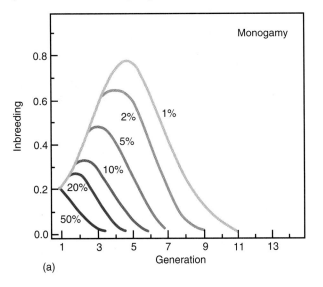

(a)

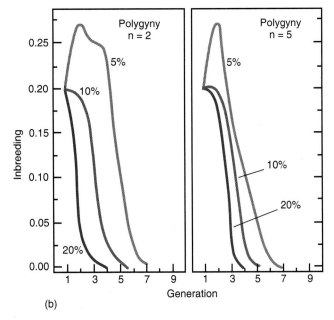

(b)

many immigrants could swamp the genetic character of the subpopulation.

Information from the field on genetic effects of immigration is lacking, because it is difficult to collect data. However, R. K. Chesser (1983) provides a simulation model, which suggests that in small monogamous subpopulations a slow immigration rate does not alleviate inbreeding depression, because of the slow diffusion of unrelated alleles into the gene pool and the equally slow decline of inbreeding among kin. In polygamous populations, however, inbreeding is quickly alleviated when the harem size is large and immigrant males are involved in breeding. Dominant immigrant males with a harem of five (Figure 21.19) can rapidly reduce inbreeding coefficients.

Remember that certain advantages and disadvantages derive from monogamy and polygamy. Although inbreeding may be more severe in small monogamous populations, the effective size is maximized (assuming equal numbers of both sexes). Monogamous populations do retain rare alleles and great qualitative genetic variation. The largely monogamous sea otter, for example, was hunted to virtual extinction in the late 1800s. With protection, populations have increased to over 100,000, retaining about 80 percent of their genetic variability (Ralls et al. 1983). The polygamous seals, however, have low levels of genetic variability (Testa 1986). Polygamous populations may retain greater quantitative genetic variation, but they are subject to the loss of rare alleles and the potential genetic variation carried by nonbreeding males (Testa 1986).

Given sufficient numbers and some immigration, cycles of inbreeding and outbreeding should retain genetic variation. As populations become divided into subpopulations and their habitats become increasingly fragmented, the need for research into the role of mating strategies and gene flow between populations in the maintenance of genetic variation in small populations increases.

## MINIMUM VIABLE POPULATIONS

The effective population statistic tells us something about the nature of the breeding population relative to the total population; but we need to know more. Breeding populations experience mortality and dispersal and these lost individuals must be replaced by younger animals. For this reason, the total population must have an adequate age structure and be above a critical size to maintain itself. Below that size and structure inbreeding and the loss of selectable genetic variation become a problem for continued survival.

The threshold number of individuals that will insure the persistence of a subpopulation in a viable state for a given interval of time, say a hundred years, is the **minimum**

**viable population (MVP)** (Schaffer 1981, Gilpin and Soulé 1986). The minimum viable population has to be large enough to cope with chance variations in individual births and deaths, random series of environmental changes, and random changes in allele frequency or genetic drift.

How large must a viable population be to balance the rate of loss of genetic variability by the rate of gain by new mutations? Based on studies of the mutution-genetic drift equilibrium in fruitflies (*Drosophila*), conservation geneticists estimate that the rate of added genetic variance by mutation that balances the genetic variance reduced by genetic drift balance for a given trait is on the order of $10^{-3}$ (0.001) (Lande and Barrowclough 1987, Stacey and Taper 1992). This rate translates into a minimum effective population size of 500 individuals that is needed to retard the effects of genetic drift. An absolute minimum $N_e$ of 50 is needed to avoid inbreeding depression (Franklin 1980). Any smaller population is subject to serious genetic drift, to loss of the genetic variability necessary to track environmental changes, and to high probability of stochastic extinction. These estimates (see Table 21.3 and Figure 21.20) have given rise to the 50/500 rule (Franklin 1980).

This 50/500 rule has created a number of problems in the management of endangered species and has built-in dangers and inefficiencies. Relying on variations in mutation rates is risky for two reasons: we do not know the reliability of those rates; and many mutations are deleterious. Moreover, we rarely know the effective population size for the population concerned; and genetic formulas estimating the MVP are appropriate only for populations with discrete generations.

Applying the 50/500 rule is a dangerous numbers game for species survival. If we consider a population of 500 individuals as sufficient to maintain all species, we may be condemning certain species to extinction. If we consider a population of 50 as minimal for sustaining subpopulations, we may also be condemning them to the same fate. In assessing the fate of 120 bighorn sheep populations over a 70-year span, Joel Berger (1990) found that 100 percent of the populations with fewer than 50 individuals went extinct within 50 years, but populations with more than 100 individuals went extinct within 70 years. He concluded that 50 individuals is not a viable population size for bighorn sheep. On the other hand, if we had considered a population of 50 as minimal for a species' recovery, we would have given up on the whooping crane, peregrine falcon, black-footed ferret (*Mustela nigripes*), European bison (*Bison bonatus*), Pere David deer (*Elaphurus davidianus*), and other comeback species whose remnant populations were considerably below 50. The point to remember is that MVP is only a general guideline for the genetic management of endangered species (Hedrick and Miller 1992).

Immediate efforts to save endangered species depend more on demography than on genetics. What is important initially is the size of the remnant population. Sizes of MVP vary among species (Pimm 1991). Those species in which

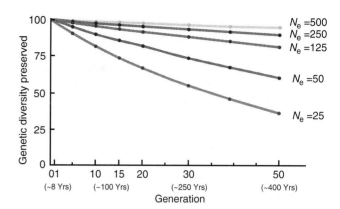

**Figure 21.20** Loss of genetic diversity as measured by heterozygosity due to random genetic drift for various effective population sizes based on a rate of decline in heterozygosity of $1/2N_e \times 100\%$ per generation. Note how rapidly genetic diversity declines when $N_e$ is small. Even with a $N_e$ of 500 some loss of genetic diversity takes place. (From Foose 1983:376.)

reproduction is highly density-dependent and which live in a more or less constant environment may persist for a long time in spite of a decline in genetic diversity. Species with highly variable population sizes, such as lagomorphs, may need large populations to counteract environmental stochastic effects on population growth. Thus MVP should be a guide to the size of a population that should be maintained, but it should never be used as a precise rule in deciding a species' fate.

# SUMMARY

The raw material of evolution and adaptation to local environments is the genetic variability of individuals in local populations. Natural selection acts upon this variability, reducing the influence of less fit and favoring the more fit, with fitness measured by the number of reproducing offspring contributed to the next generation. Natural selection works in three ways: nonrandom mating, nonrandom fecundity, and nonrandom survival. Certain genetic combinations are more fit than others, and these transmit more genes to future generations than less fit combinations. Two sources of genetic variation acted upon by natural selection are mutations and recombination of genes provided by parents in bisexual populations.

Theoretically, variations in biparental populations, as reflected in gene frequencies and genotypic ratios, remain in Hardy-Weinberg equilibrium, $p^2 + 2pq + q2 = 1$, which gives the expected genotypic frequencies of *AA, Aa,* and *aa,* respectively, if the conditions of random mating, equilibrium in mutation, lack of selection, and a relatively large, closed population exist. In nature such conditions do not oc-

cur, and there is a departure from genetic equilibrium. The direction evolution takes depends upon the genetic characteristics of those individuals in the population that survive and leave behind viable progeny. Natural selection may be stabilizing, maintaining the current genetic equilibrium and favoring intermediate phenotypes; directional, favoring phenotypes at one extreme of the range; or disruptive, favoring genotypes at both extremes of the range at the same time.

Although natural selection applies only to individuals, the evolution of altruistic traits seems to require some form of group and/or kin selection. Group selection, which operates on the differential production of local populations, favors characteristics that improve the fitness of the group, but may decrease the fitness of any individual within the group. Kin selection is the evolution of a genetic trait expressed by one individual that increases the genotype fitness of one or more directly related individuals. Individual fitness plus fitness acquired by improving the fitness of close relatives is inclusive fitness.

Because of habitat fragmentation and human exploitation of the landscape, populations of many species of plants and animals are being reduced to isolated or semi-isolated small populations. These small populations carry only a sample of the genetic variability of the total population. These situations can lead to inbreeding and genetic drift.

Inbreeding, mating between relatives, brings out hidden genetic variation, increases homozygous genotypes at the expense of heterozygous ones, and reveals the effects of homozygous rare alleles that often result in reduced fecundity, viability, and even death. The degree of inbreeding is given by the inbreeding coefficient, $F$, which measures the probability that two alleles at a locus in an individual are identical by descent.

Small populations are subject to random genetic drift, chance fluctuations in allele frequency as a result of random sampling among gametes. Drift results in the fixation of alleles and populations lacking in genetic variability. Genetic drift mimics inbreeding by increasing homozygosity, but it is the result of random mating and the alleles involved are not necessarily identical by descent. The effects of random drift are measured in terms of heterozygosity of individuals, subpopulations, and the total population. The value $F_{ST} = (H_t - H_s)/H_t$ measures the degree of genetic structuring among subpopulations.

Influencing the degree of random genetic drift is the effective population size, $N_e$, the size of an ideal population having the same rate of increase in $F_{ST}$ as the population in question. In monogamous populations with an equal number of breeding males and females, the actual population size and the effective population size are the same. In polygamous populations a wide disparity in sex ratios can strongly reduce the effective population size and increase genetic drift. The effective population size is given by $4N_m N_f/(N_m + N_f)$ where $N_m$ and $N_f$ are the number of males and females, respectively. If the population is spread out more or less uni-

formly across the landscape, the effective size becomes $4\pi\sigma^2 d$, where $d$ is the density of breeding individuals per unit area and $\sigma^2$ is the root-mean-square of dispersal distance. If the population fluctuates over time, the effective population is given by the harmonic mean of the various population values.

Exchange of individuals among populations (emigration and immigration) can reduce genetic drift and inhibit genetic divergence among subpopulations. The equilibrium value of $F_{ST}$ for immigration or dispersal is given by $1/(4N_m + 1)$ where $m$ is the number of migrants per generation. Only a few per generation are needed to reduce random genetic drift and keep $F_{ST}$ below 0.10.

If a subpopulation is to persist over time, it has a threshold number below which the population must not fall to maintain genetic diversity. This is called the minimum viable population, the size at which the rate of loss of genetic variance is balanced by the rate of mutation. Although a figure of 500 is generally given, this may be much too low. Such a value fails to consider age structure and sex ratios of the population and chance variations in the environment. Population genetics is becoming crucial in the preservation and management of wild species in the face of human development.

## REVIEW QUESTIONS

1. Distinguish between genotype and phenotype. Upon which does natural selection work?
2. Define allele, locus, homozygous, heterozygous, diploid, haploid, gene pool.
3. What are the major sources of variation in the gene pool?
4. What is the Hardy-Weinberg law? What is its significance?
5. Contrast stabilizing, directional, and disruptive selection.
6. What is meant by fitness, direct fitness, indirect fitness, and inclusive fitness?
7. What is altruism? Relate this trait to the theories of group selection and kin selection.
8. What is inbreeding? Why does inbreeding increase homozygosity? Contrast allozygous with autozygous genes.
9. Define the coefficient of inbreeding. What does it measure?
10. What are the end values of $F$, the inbreeding coefficient, and what do the values tell us?
11. What is inbreeding depression? Outbreeding depression? How do they affect small populations?
12. To what degree and under what conditions does inbreeding occur in natural populations?
13. What is genetic drift? What is the difference between inbreeding and genetic drift?

14. What is effective population size? How does it relate to inbreeding and genetic drift? To the sex ratios of reproducing animals in a population?

15. What is the relationship between dispersal distance and effective population size in a continuously distributed population?

16. What is a population bottleneck? What is its significance genetically?

17. What is the founder's effect?

18. What is meant by minimum viable population?

19. Show that a breeding population with a ratio of 2 males to 10 females is genetically less variable than a population with a ratio of of 6 males to 10 females.

20. Argue why it is unwise to base the minimum size of a population necessary for species survival on genetics alone.

21. Suggest ways in which gene flow between isolated populations of a species could be maintained.

## CROSS-REFERENCES

Adapation, 30; hypothesis testing, 18–21; age structure, 372–374; population growth, 392–396; extinction, 407–409; demographic stochasticity, 408; environmental stochasticity, 408; habitat fragmentation, 614–618.

# Interspecific Competition

*Concepts*

1. Interspecific competition results when two or more different species seek the same resource that is in short supply.
2. In exploitative competition, both species consume the resource; in interference competition, one competitor denies another access to a resource.
3. Interspecific competition may result in stable or unstable coexistence or in exclusion of a competitor.
4. Competition is reduced by the differential use of resources by potential competitors or resource partitioning and by environmental variability.
5. The niche is the functional role of an organism in a community. It may or may not be constrained by interspecific competition.

Up to this point the emphasis has been on single-species populations. Now we must broaden our scope. Individuals within a single-species population interact not only with others of their own kind but with individuals of other species populations as well. The effects of these interactions on population growth can be positive, negative, or neutral (Table 22.1).

Neutral interactions (designated 0 0) have no effect on the growth of interacting populations. Positive interactions (+ +) benefit both populations. Such relationships are termed **mutualism** (Chapter 27). In some situations a one-sided relationship develops in which one species benefits and the other is neither benefited nor harmed (+ 0). This relationship is known as **commensalism.** Examples of commensalism are epiphytes, plants that grow on the branches of trees. They depend on trees for support only; their roots draw nutrients from the humid air. In another one-sided relationship between two species, one population is negatively affected while the other remains unaffected (– 0). This relationship is called **amensalism.** A nebulous relationship, amensalism probably involves a chemical interaction such as the production of an antibiotic or allelochemical agent by one of the organisms. Other relations positively affect one population and are detrimental to the other (+ –). Such relationships are predation and parasitism. **Predation** (Chapter 23) is the killing and consumption of prey. **Parasitism** (Chapter 26) is an interaction in which one usually small organism (the parasite) lives in or on another (the host) from which it obtains food. Finally, relationships can have adverse effects on both populations (– –). Such relationships develop when a needed resource is in limited supply relative to the number seeking it. This interaction between species is **interspecific competition.**

Individuals within a population experience intraspecific competition with members of their own kind (Chapter 19). At the same time they also experience to a greater or lesser degree interspecific competition with individuals of other species. To demonstrate that interspecific competition is important, we have to show that one species uses a resource to the extent that it limits the population size of another.

Interspecific competition may be exploitative or interference, two terms somewhat akin to the scramble and contest types of intraspecific competition. In **exploitative competition** the species use the same resource. Use by one reduces availability for another. The outcome is determined by how effectively each of the competitors uses the resource. It often results in reduced growth of all competitors. **Interference competition** is a direct interaction between competitors in which one interferes with access to the resource by another. In animals interference usually involves aggressive behavior.

Among individual plants and sessile animals, both fixed in space, interspecific competition is influenced by the degree of proximity. Each individual affects the environment

**Table 22.1  Population Interactions, Two-Species System**

|  | Response | |
| --- | --- | --- |
| Type of Interaction | A | B |
| Neutral | 0 | 0 |
| Mutualism | + | + |
| Commensalism | + | 0 |
| Amensalism | – | 0 |
| Parasitism | + | – |
| Predation | + | – |
| Competition | – | – |

*Note:* 0 = no direct effect; + = positive effect on growth of population; – = negative effect on growth of population.

of its neighbor by consuming resources in limited supply, by modifying environmental conditions (for example, shading and protecting plants from wind and predators), and by producing toxins. These changes can alter the rate of growth, biomass accumulation, and the growth form of individual plants.

# CLASSIC COMPETITION THEORY

## The Lotka-Volterra Model

The mathematicians Lotka and Volterra, who independently formulated the logistic growth equation (Chapter 18) also independently modified the logistic equation to describe the relationship between two species using the same resource. They added to the logistic equation for the population of each species a constant to account for the interference of one species on the population growth of another. This constant, in effect, converts the number of members of one species population into the units of the other:

$$\textbf{Species 1:}\quad \frac{dN_1}{dt} = r_1 N_1 \left( \frac{K_1 - N_1 - \alpha N_2}{K_1} \right)$$

$$\textbf{Species 2:}\quad \frac{dN_2}{dt} = r_2 N_2 \left( \frac{K_2 - N_2 - \beta N_1}{K_2} \right)$$

where $r_1$ and $r_2$ are the rates of increase for species 1 and 2, respectively; $K_1$ and $K_2$ are equilibrium population size for each species in the absence of the other; $\alpha$ is a constant characteristic of species 2, a measure of the inhibitory effect of one $N_2$ individual on the population growth of species 1; and $\beta$ is a constant characteristic of species 1, a measure of the inhibitory effect of one $N_1$ individual on the population growth of species 2.

Some major assumptions lie behind the Lotka-Volterra model: (1) the environment is homogeneous and stable,

without any fluctuations; (2) migration is unimportant; (3) the effect of competition is instantaneous; (4) coexistence requires a stable equilibrium point; and (5) competition is the only important biological interaction (see Schoener 1982; Chesson and Case 1986; Roughgarden 1986).

In the absence of any interspecific competition—either $\alpha$ or $N_2 = 0$ in equation 1 and $\beta$ or $N_1 = 0$ in equation 2—the population of each species grows logistically to equilibrium at carrying capacity. Inherent in the logistic equation is the inhibitory effect of each individual on its own species' population growth. This effect is represented by $1/K_1$ for species 1 and $1/K_2$ for species 2. In competing populations, the inhibitory effect of each $N_2$ individual on $N_1$ is $\alpha/K_1$. Similarly, the inhibiting effect of each $N_1$ individual on the population growth of species 2 is $\beta/K_2$. The outcome of competition depends upon the relative values of $K_1$, $K_2$, $\alpha$, and $\beta$. If $N_2 = K_1/\alpha$, $N_1$ can never increase; and if $N_1 = K_2/\beta$, $N_2$ can never increase.

What the equations describe can be better understood with some graphic models (Figure 22.1). In each case the ordinate will represent the population size of species 1 and the abscissa the population size of species 2. The two extreme cases, no competition and complete competition, are the ends of the diagonal line, $K_1$ and $K_2$, for carrying capacity of species 1 and 2 respectively, and $K_1/\alpha$ and $K_2/\beta$ for the competitive effects on species 1 and species 2, respectively. The diagonal line represents equilibrium conditions; the space below it represents the area within which the population can increase. For species 1 in Figure 22.1a, equilibrium conditions are represented by the line $K_1/\alpha$, $K_1$ obtained by plotting $N_1$ against $N_2$. The line represents a set of joint values of $N_1$ and $N_2$ along which the number of individuals in species 1 is neither increasing nor decreasing. This set of joint values is represented by $dN_1/dt = 0$. Populations of species 1 inside the line, indicated by the shaded area, will increase in size until they reach the diagonal line, which represents all points of equilibrium. Actual values can fall anywhere along this line. All populations outside the line (to the right) will decrease to the equilibrium points. A similar set of joint values exists in which $N_2$, the number of individuals in species 2, reaches equilibrium level (Figure 22.1b). Outside the diagonal equilibrium line, species 2 decreases; inside the line it increases. Values of $N_1$ and $N_2$ are considered jointly because of the effect of competition. An increase in $N_1$ diminishes the growth rate of species 2, and an increase in $N_2$ decreases the growth rate of species 1. In the presence of species 2, the higher the value of $N_2$, the lower is the value of $N_1$ at which species 1 stops growing.

What happens when species 1 and species 2 occupy the same space simultaneously as competitors? According to the Lotka-Volterra model, there are four possible outcomes: (1) species 1 wins and species 2 becomes extinct; (2) species 2 wins and species 1 becomes extinct; (3) either species can win, depending upon the ecological variables operative at

any one time; or (4) neither species wins, and they eventually coexist, dividing the resources between them in some manner.

To illustrate the first two outcomes, if species 1 and species 2 occupy the same space as competitors and species 2 is the stronger competitor, species 2 slows down the population growth of species 1 and eventually wins, leading to the extinction of species 1 ($\alpha > K_1/K_2$ and $\beta < K_2/K_1$). In Figure 22.1c the plot of species 1 moves upward because the area on or above the line $K_1,K_1/\alpha$ (the carrying capacity of species 1) and the area below $K_2,K_2/\beta$ is below the carrying capacity of species 2. Species 2 will increase until it arrives at $K_2$; at that point only species 2 remains. Under a different set of conditions species 1 will win, as illustrated in Figure 22.1d. In each case no equilibrium exists because one species is able to increase in an area where the other species must decrease. Equilibrium can come about only when the diagonal lines cross one another.

In the third outcome (Figure 22.1e) the diagonal equilibrium lines cross each other. The equilibrium point is represented at their crossing, but it is unstable. The vectors are directed away from the equilibrium point, indicating that the true equilibrium points are $K_1$ and $K_2$. In this situation equilibrium between competing species is unstable and either of the two species can win. Above the line $K_2,K_2/\beta$, species 2 is unable to increase; and above $K_1,K_1/\alpha$, species 1 is unable to increase. If the mix of the species is such that the point $N_1,N_2$ falls within the triangle $K_2,E/K_1/\alpha$, species 1 is above its carrying capacity and species 2 is not. Species 2 will continue to increase and species 1 will decrease until it is gone. The reverse situation occurs in triangle $K_1,E,K_2/\beta$. What happens in parts of the diagram outside the triangles depends upon whether the starting value of $N_1$ is larger or smaller than that of $N_2$.

Finally, the two species might coexist with their populations in equilibrium (Figure 22.1f). As species 1 increases, species 2 may decrease and vice versa. Each species inhibits its own growth through intraspecific competition more than it inhibits the growth of the other species through interspecific competition. Neither species reaches a high enough density to bring about any serious competition between them, and the population growth of each is not strongly controlled by the same limiting conditions. As long as each species is limited by a different resource and both are only weakly competitive, then the two species will continue to coexist. Thus, in Figure 22.1f species 2 has the advantage in the area $K_1,E,K_2/\beta$ and the plot moves up and to the left to the equilibrium point $E$. The two competing species will reach a stable equilibrium point and persist indefinitely when

$$\alpha < \frac{K_1}{K_2} \text{ and } \beta < \frac{K_2}{K_1}$$

The Lotka-Volterra equation and graphic models based on them apply well to animal populations, but not to plants.

The models are based on an animal's potential for increase in numbers. In many plant populations the potential is for increase in biomass. Accordingly, competitive relations among plants may be examined in terms of the influence of one species on the growth of another. Models developed from experimental studies of mixtures of two species sown at a variety of proportions detect changes in yield of dry matter, number of tillers, production of seed, and so on after a lapse of time. Proportions of the two species at the end of the period are plotted against proportions at the beginning. Such studies arrive at five basic types of interactions (Figure 22.2), suggestive of those predicted by the Lotka-Volterra equations (Harper 1977):

1. Neutral interaction. Proportion of two species remains unaltered after a period of growth together. The balance of the two species is subject only to random variation. Such interactions are rarely, if ever, seen in nature.
2. Directional interaction in favor of species 1. Species 1 has the competitive advantage in the mixture at all proportions. If the advantage is carried from planting to harvest and back to planting again, generation after generation of species 1 ultimately would bring about the extinction of species 2.
3. Directional interaction in favor of species 2. Species 1 is eliminated by species 2.
4. Stabilizing interaction. In this frequency-dependent situation the minority component is always at an advantage.
5. Disruptive interaction. In this frequency-dependent situation the majority component is at an advantage. This nonequilibrium mixture is unstabilized.

## Competitive Exclusion

The Lotka-Volterra equations predict, as graphs (c) and (d) in Figure 22.1 illustrate, that if one species in a competitive situation grows rapidly enough to prevent the population increase of another, it can reduce that population to extinction or exclude it from the area. This model led to **Gause's principle,** named for the Russian ecologist who demonstrated the concept experimentally. It states that two species with identical ecological requirements cannot occupy the same environment.

The idea was far from original with Gause (and he laid no claim to it). For example, the ornithologist Joseph Grinnell wrote in 1904: "Two species of approximately the same food habits are not likely to remain long evenly balanced in numbers in the same region. One will crowd the other out. The one longest exposed to local conditions, and hence best fitted, though ever so slightly, will survive to the exclusion of any less-favored would-be invader."

The concept eventually gained the name **competitive exclusion principle.** Hardin (1960) wrote: "Complete competitors cannot coexist. Two competing species with identical ecological requirements cannot occupy the same area." However, this concept is not really a principle (Cole 1960). It is little more than an ecological definition of a species. A corollary of the statement is that if two species coexist, they must possess ecological differences. Obviously, two separate species cannot have identical requirements; being different species, they must have somewhat different ecologies. However, two or more species can compete for some essential resource without being complete competitors.

Competition, for simplicity, is usually considered on a one-dimensional gradient such as food size on water availability; but in natural situations competition is spread over a number of resources. A high competitive interaction for one resource may be counterbalanced by low competitive interactions for other resources. In such situations, minimal competitive inhibitions on several gradients among a number of species can for some individual species be equivalent to strong competitive interaction for one resource from a single competing species. This relationship has been termed **diffuse competition** by MacArthur (1972). Theoretically, diffuse competition can exclude a species or greatly reduce its numbers through competitive interactions with a specific combination of other species, rather than just one strong competitor.

Pielou (1974) provides a set of conditions in addition to the utilization of resources in short supply that should be met for competitive exclusion to take place: (1) competitors must remain genetically unchanged for a sufficiently long period of time for one species to exclude the other; (2) immigrants

**Figure 22.1** (opposite)  The Lotka-Volterra model of competition between two species. In (a) and (b) populations of species 1 and 2 in the absence of competition will increase in size and come to equilibrium at some point along the diagonal line or isocline of zero growth ($r = 0$). In the shaded area below the line $r$ is positive and the population increases (as indicated by arrows), whereas above the line the population decreases. In (c) species 1 and 2 are placed in competition. Because the isocline, the zero growth curve, of species 2 falls outside the isocline of species 1, species 2 always wins, leading to the extinction of species 1. In (d) the situation is reversed, and species 1 wins, leading to the exclusion of species 2. In (e) and (f) the isoclines cross. Each species, depending upon the circumstances, is able to inhibit the growth of the other. In (e) each species inhibits the growth of the other species more than it inhibits its own growth. Which species wins often depends upon the initial proportion of the two species. In (f) neither species can exclude the other. Each by intraspecific competition inhibits the growth of its own population more than it inhibits the growth of the other population.

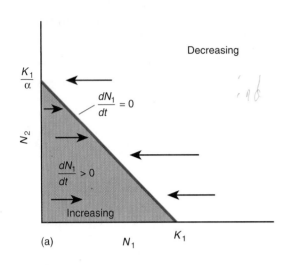

(a)

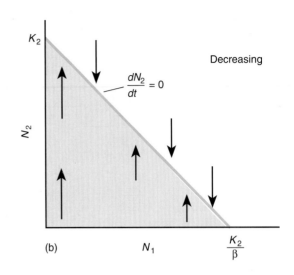

(b)

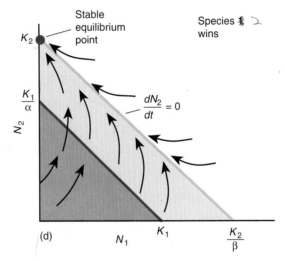

(c)

(d)

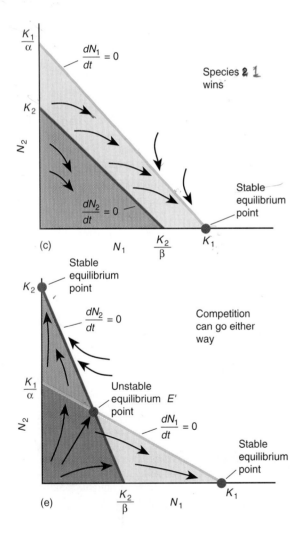

(e)

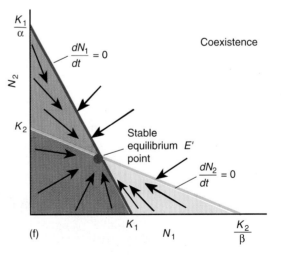

(f)

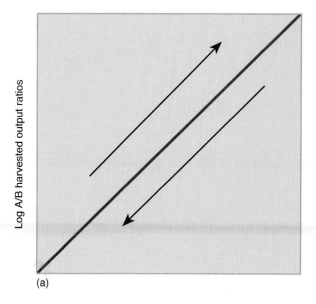

(a)

*Log A/B harvested output ratios*

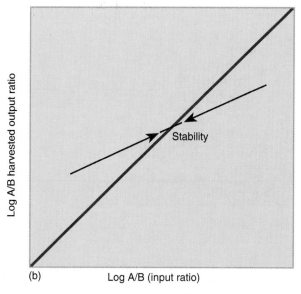

(b)

Stability

*Log A/B harvested output ratio*

Log A/B (input ratio)

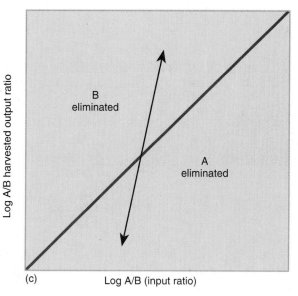

(c)

B
eliminated

A
eliminated

*Log A/B harvested output ratio*

Log A/B (input ratio)

from areas with different conditions cannot move into the population of the losing species; (3) environmental conditions must remain constant; and (4) competition must continue long enough for equilibrium to be reached. In the absence of any of these requirements, species usually coexist.

# STUDIES OF COMPETITION

## Laboratory Studies

The best place to observe interspecific competition is in laboratory cultures of small invertebrates and microorganisms and in the greenhouse, isolated from environmental fluctuations and outside interference. Gause (1934) set out to test the Lotka-Volterra equations experimentally. He used two species of *Paramecium, P. aurelia* and *P. caudatum. P. aurelia* has a higher rate of increase than *P. caudatum.* When both were introduced into one tube containing a fixed amount of bacterial food, *P. caudatum* died out. The population of *P. aurelia* interfered with the population growth of *P. caudatum* because of its higher rate of increase (Figure 22.3). (Compare Figure 22.1c, d.) In another experiment Gause used *P. caudatum* and *P. bursaria.* Both species reached stability, because *P. bursaria* confined its feeding to bacteria on the bottom of the tube, whereas *P. caudatum* fed on bacteria suspended in solution. Although the two used the same food supply, they occupied different parts of the culture. In effect, each utilized food unavailable to the other. (Compare Figure 22.1f.) Park (1948) and Crombie (1947) carried out competition experiments involving several species of flour beetle and obtained results similar to those of Gause.

Tilman and associates (Tilman et al. 1981) grew laboratory populations of two species of diatoms, *Asterionella for-*

**Figure 22.2** Input/output diagrams showing four possible outcomes of competition among plants. The ratio of two species present after a period of time is plotted against the ratio of the species sown or planted. (a) The 45° slope represents a stable mixture in which the proportion of the species remains unaltered after a period of growth together. To the left of the 45° slope species 1 (B) gains advantage over species 2 and species 2 goes extinct. To the right the situation is reversed and species 1 moves to extinction. (b) Stabilizing interactions. In this frequency-dependent situation the minority species is at an advantage. If a high proportion of species 1 is sown, species 2 gains in the mix. The mix tends toward a stable equilibrium. (c). Disruptive interaction. In this frequency-dependent situation the major species in the mix is at an advantage. If a high proportion of species 1 is sown, species 2 will go extinct. There is no equilibrium. (Based on Etherington 1976 and Harper 1977.)

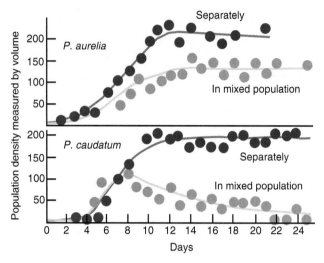

**Figure 22.3** Competition experiments with two related ciliated protozoans, *Paramecium aurelia* and *P. caudatum,* grown separately and grown in a mixed culture. In a mixed culture *P. aurelia* outcompetes *P. caudatum,* and the result is competitive exclusion. (From Gause 1934.)

*mosa* and *Synedra ulna,* which require silica for the formation of cell walls. They monitored not only population growth and decline but also the level of silica. When grown alone in a liquid medium to which resource (silica) was continually added, both species kept silica at a low level (Figure 22.4). When grown together, *S. ulna* took silica to a level below which *A. formosa* could not survive and reproduce (Figure 22.4c). In this experiment *S. ulna* competitively excluded *A. formosa* by reducing resource availability.

Such laboratory studies of two interacting species under controlled and manipulated conditions demonstrate the reality of interspecific competition. They provide the experimental data needed to test and modify mathematical models of competition and refine the concepts of competition. However, do the findings of laboratory experiments apply to the more complex competitive interactions in the real world of nature?

## Field Studies

A major tenet of ecology holds that competitive interactions among taxa, genera, and trophic levels determine the nature and structure of natural communities (see Chapter 28). In a classic paper, Hairston, Smith, and Slobkin (1960) (HSS) hypothesized that herbivores do not compete because they are predator-limited, whereas producers and carnivores experience strong competition. In 1976 Menge and Sutherland (MS) proposed that plants compete least for resources, predators compete the most, and herbivores are intermediate. These propositions opened up more questions. What is the intensity and overall effect of competition? Do the ef-

fects of competition differ among taxa and between trophic levels?

Such questions cannot be answered in the laboratory, where life for the experimental organisms is akin to being in a jail or compound. In the real world organisms encounter many competitors and predators. Questions on competition have to be explored by quantified, manipulative experiments in the field.

Under natural conditions, we can observe fairly easily what appear to be competitive interactions. One species of bird replaces another at a winter feeding station; bluebirds, tree swallows, and starlings compete for a limited number of nest boxes or tree cavities. In these cases we have observations but not experimental evidence that competition has a pronounced effect on population growth and on the survival of individuals and populations.

Attempting to demonstrate interspecific competition under truly natural conditions is difficult. The experimental populations must live naturally, yet be held under some type of control amenable to experimental manipulations. Most field experiments involve enclosed or caged populations of assumed competitors. These enclosed populations, usually held above or below natural densities, may be forced to compete more actively than they would if they were free. Further, the organisms may enjoy more protection from predators that, under uncaged conditions, would alleviate the effects of competition.

Field experiments have their own sets of problems. Experimenters have difficulty establishing adequate controls and sufficient replicates and manipulating the populations. They are hampered by poor experimental design and by confounding interspecific competition with environmental effects. However, the task is not impossible. The hypotheses to be tested must be well-defined, the experimental procedures must represent the natural situation, and the experiments must be replicated.

Well over 100 field studies involving several hundred experiments have attempted to determine the effects of interspecific competition on species assumed to be competitors and the relative importance of competition at different trophic levels. Most of these studies have involved organisms that have similar resource requirements, including fish, amphibians, marine mollusks, aquatic insects and snails, and marine and terrestrial plants. In one survey Schoener 1983b found that terrestrial and freshwater species were supportive of the HSS theory; marine competition was only weakly supportive. Competition was demonstrated in 90 percent of the studies and 76 percent of the species involved. Connell (1983) in his survey found competition demonstrated in most studies, in 50 percent of the species, and in two-fifths of the experiments. Many studies, however, involved intraspecific competition. Goldberg and Barton (1992) found that among plants interspecific competition was stronger than intraspecific competition, and that competition did affect distribution patterns, relative abundance,

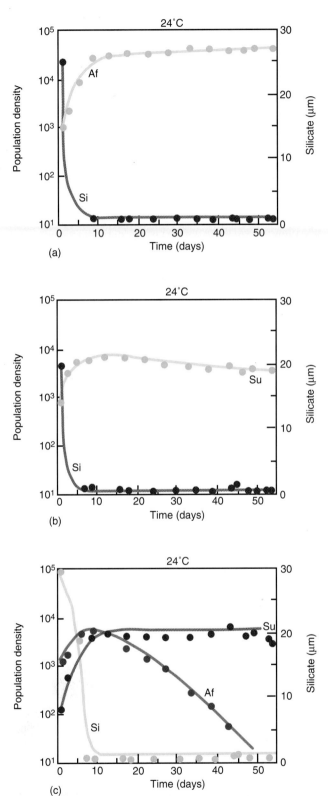

(a)

(b)

(c)

**Figure 22.4** Competition between two species of diatoms, *Asterionella formosa* and *Synedra ulna,* both of which require silica for the formation of cell walls. (a) When grown alone in a culture flask *A. formosa* reaches a stable population level at which it keeps silica at a constant low level. (b) *S. ulna* does the same, except that it draws the silica down to an even lower level. (c) When the two species are grown together in a culture flask, *S. ulna* drives *A. formosa* to extinction, because *S. ulna* reduces the silica level to a point below which the other species cannot exist. (From Tilman et al. 1981:1025, 1027.)

and diversity. Most studies they reviewed, however, were short-term, lasting from several months to three years.

Surveys of individual studies considered independently tell us little about wider effects. Gurevitch and associates (1992) took a different approach to their analysis of studies covering 93 species in a variety of habitats. They employed a statistical test then new to ecology but used in medicine and social sciences—meta-analysis (Hunter et al. 1982). Meta-analysis is a statistical synthesis of results from a set of primary studies. It employs differences between means of two groups, control and experimental, and does not reanalyze original data. Meta-analysis tests whether the outcomes of different studies are consistent with one another, whether together they demonstrate an effect that is large, moderate, small, or not much different from zero.

This approach yielded better answers about the organization of nature, some of which contradict predictions of ecological theory. Interspecific competition does have a large overall effect, but this effect varies widely among organisms. Large differences exist among trophic levels, but the magnitude of competition within them is contrary to predictions. The effects of competition among carnivores and primary producers are small-to-moderate, not large as predicted by HSS or MS. Effects among small herbivores vary. Frogs and toads and arthropods of flowing waters experience large effects; marine mollusks and echinoderms experience small effects. Effects of interspecific competition among terrestrial arthropods are not significantly different from zero. Among herbivores interspecific competition is less than intraspecific competition. Among most organisms the effects of intraspecific and interspecific competition do not differ, nor are there differences in effects between terrestrial and aquatic organisms. As we might expect, caged organisms experience greater competition than organisms living free. Among plants competition does not differ between highly productive and poorly productive ecosystems.

**Coexistence** Most interspecific competitive relations are probably expressed as stable or unstable equilibrium, if indeed competition occurs at all. Classical competition theory

assumes that the environment is stable and competition is continuous, but in reality interspecific competition is probably discontinuous (see Wiens 1977, Wiens et al. 1986, Chesson 1986), because environments are variable and populations are patchily distributed in space and time. Organisms using identical but limited resources coexist because of different responses to a fluctuating environment and differing life history traits.

In variable environments resource levels vary between superabundance and scarcity. Periods of scarcity, which occur irregularly, create ecological crises that can result in intense interspecific competition and act as a major selective force. On the small Isla Daphne in the Galapagos Islands, Grant and his associates (Grant 1986) followed the populations of two species of Darwin's finches—*Geospiza fortis,* the medium ground finch, and *G. scandens,* the cactus ground finch–over ten years. In that time there was one long dry period, from May 1976 to January 1978. During that period seed production declined precipitously (Figure 22.5). Many seed-bearing plants died and were not replaced until 1983, the next wet year, keeping seed availability low for several years. This period of food scarcity was accompanied by a population crash of *G. fortis* and a less drastic decline in the population of *G. scandens.* Foraging diets and behavioral changes reduced overlaps in food. The diet of *G. scan-*

*dens,* which specializes in the seeds of cactus, became narrower; and that of the generalist *G. fortis,* which feeds on a variety of seeds including those of cactus, became broader. During the drought the diets of the two birds diverged and the overlap diminished, possibly because of the decline in jointly exploited foods.

Exclusion Because competitive relationships among individuals and species interact with the physical and biological environments, including physiological tolerances and predation, ecologists have a difficult time determining degrees of competitive exclusion. Connell (1961a, 1961b) demonstrated some degree of competitive exclusion among two species of barnacles on the Scottish coast. Heller and Gates (1971) found some evidence of competitive exclusion among four species of chipmunks on an altitudinal gradient in the eastern slopes of the Sierra Nevada mountains of California.

In recent years the black duck (*Anas rubripes*) has been declining in northeastern North America and is being replaced by the mallard (*Anas platyrhynchos*) (Figure 22.6). To examine the hypothesis that mallards first invaded and then replaced black ducks on fertile wetlands in southern Ontario, Merendino, Ankney, and Dennis (1993) studied 131 wetlands. They surveyed occupancy of the wetlands for

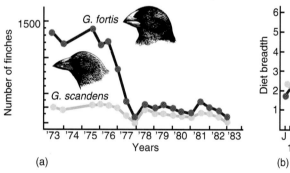

(a)

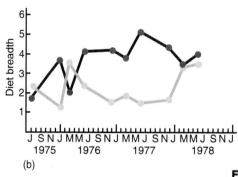

(b)

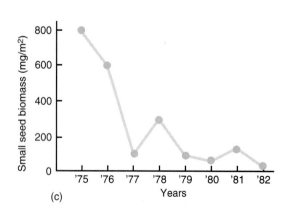

(c)

**Figure 22.5** Change in (a) the population sizes and (b) breadth of diets of two of Darwin's finches, *Geospiza fortis* and *G. scandens,* in response to a drought on Isla Daphne. (c) The precipitous decline in the biomass of small seeds at the beginning of the dry season from 1976 to 1982. In response to this change in food availability, *G. scandens* decreased and *G. fortis* increased its diet breadth. After rainfall broke the drought in 1983, the changes were reversed. (From Grant 1986:180, 186.)

(a)

(b)

(c)

**Figure 22.6** The mallard (a, b), the Northern Hemisphere's most abundant duck, appears to be a strong competitor of the black duck (c). The mallard is increasing and expanding its range into the northeastern North American coastal habitat of the black duck.

breeding populations of the two species over the previous 20 years; and they measured the characteristics of the wetlands, including size, shape of shoreline, percentage of emergent vegetation, distance from disturbance, and water chemistry.

The researchers divided the wetlands into seven categories: (1) wetlands invaded by mallards in the 1970s but never recorded as holding black ducks; (2) wetlands invaded by mallards in the 1980s but not used by black ducks; (3) wetlands where mallards replaced black ducks in the 1970s; (4) wetlands where mallards replaced black ducks in the 1980s; (5) wetlands inhabited by both species; (6) wetlands used only by black ducks; and (7) unused wetlands. Wetlands where the mallards first appeared were more fertile than the wetlands occupied later; and wetlands where mallards first replaced black ducks were more fertile than those occupied later. Black ducks currently are restricted to wetlands with very low fertility. The biologists concluded

that by competitively excluding black ducks from productive wetlands (accompanied by some hybridization), mallards have contributed to the decline of the black duck in southern Ontario.

Competitive exclusion is most conspicuous in places invaded by exotic weedy plant species that displace native plants. The colorful purple loosestrife (*Lythrum salicaria*), a wetland perennial from Europe, has invaded prime wetlands throughout the temperate regions of the United States and Canada (Malecki et al. 1993) (Figure 22.7). It replaces native wetland species and eliminates natural food and cover essential to wetland wildlife. Its life history characteristics, including the production of 2.5 million long-lived, easily dispersed seeds per plant and growth rates of established seedlings exceeding 1 cm/da, make it a formidable competitor to wetland plants. Similarly the Australian pine (*Casuarina*), introduced in south Florida, is invading and displacing vegetation of the Everglades.

**Figure 22.7** The colorful growth of purple loosestrife in this marsh belies its aggressive exclusion of native wetland plants.

**Allelopathy** A particular form of interference competition among plants is *allelopathy,* the production and release of chemical substances by one species that inhibit the growth of other species. These substances range from acids and bases to simple organic compounds that reduce competition for nutrients, light, and space. Produced in profusion in natural communities as secondary substances, most compounds remain innocuous, but a few may influence community structure. For example, broomsedge (*Andropogon virginicus*) produces chemicals that inhibit the invasion of old fields by shrubs and thus maintains its dominance (Rice 1972). Bracken fern (*Pteridium aquilinum*), the most widely distributed vascular plant in the world (Page 1982), produces plant poisons that accumulate in the upper surface of the soil. These phytotoxins kill the germinating seeds of many plants, especially conifers, and reduce growth of seedlings. These allelopathic effects along with the heavy, smothering overwinter accumulation of dead fronds allow bracken ferns to dominate competitively large areas of ground (Ferguson and Boyd 1988). Likewise, the black walnut (*Juglans nigra*) of the eastern North American deciduous forest is antagonistic to many plants (Brooks 1951).

In desert shrub communities a number of shrubs (*Larrea, Franseria,* and others) release toxic phenolic compounds to the soil through rainwater. Under laboratory conditions, at least, these substances inhibit germination and growth of seeds of annual herbs (McPherson and Muller 1969). Other desert shrubs (*Artemisia* and *Salvia*) that commonly invade desert grasslands release aromatic terpenes such as camphor to the air. These terpenes are adsorbed from the atmosphere onto soil particles. In certain clay soils these terpenes accumulate during the dry season in quantities sufficient to inhibit the germination and growth of herb seedlings. As a result, invading patches of shrubs are surrounded by belts devoid of herbs and by wider belts in which the growth of grassland plants is reduced (Muller et al. 1968). Allelopathy may not be the only reason for belts devoid of vegetation. Studies of plant-animal interactions suggest the impact of predation by hares and consumption of seeds by rodents, birds, and ants (Bartholomew 1970).

**Diffuse Competition** If interspecific competition between two species is difficult to establish, it is more difficult to demonstrate diffuse competition, the sum of weak competitive interactions among ecologically related organisms. Davidson (1985) carried out a five-year experiment with colonies of three species of harvester ants, the large *Pogonomyrmex rugosa,* the intermediate-sized *P. desertorum,* and the small *Pheidole xerophila.* The intermediate *P. desertorum* increased on plots from which its interference competitor, the large species *P. rugosa,* was removed; and the small species, *P. xerophila,* its exploitative competitor, declined. Davidson attributed the decline in the small species to the absence of the large harvester ant. The large harvester ants apparently aided the small species indirectly by suppressing populations of the intermediate-sized ants.

# RESOURCE PARTITIONING

Observations of a number of species sharing the same habitat suggest that they co-exist by utilizing different resources. Animals eat different sizes and kinds of food, or feed at different times or in different areas. Plants occupy a different position on a soil moisture gradient, require different proportions of nutrients, or have different tolerances for light and shade. Each species exploits a portion of the resources, which becomes unavailable or is unusable to others. Such resource partitioning or differential resource utilization is often regarded as an outcome of interspecific competition.

## Theoretical Considerations

Consider an animal species A, which in the absence of any competitor utilizes a range of different-sized food items (Figure 22.8). We can picture that utilization as a bell-

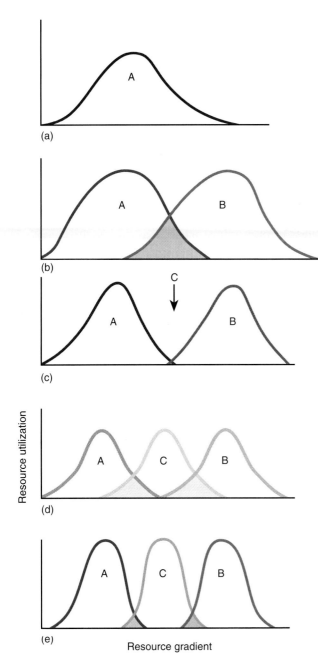

Resource utilization

Resource gradient

**Figure 22.8** Theoretical resource gradient utilized by three competing species, A, B, and C. (a) A as the only species occupying the resource gradient. (b) Species B invades the resource gradient and partially competes with A. (c) In response to selective pressures both A and B narrow their range of resource use to optimum, and C invades that portion used at less than an optimal level. (d) C competes with both A and B on parts of the resource gradient. (e). In response to selection pressure A, B, and C narrow their range of resource use to optimum.

shaped curve on a graph, with food as the ordinate and fitness as the abscissa. Most individuals feed about the optimum. Individuals at either tail feed on larger or smaller food items, respectively. As population size increases, the range of food taken may increase, as intraspecific competition forces some individuals to seek food at two extremes. Such intraspecific competition fosters increased genetic variability in the population.

Now allow a second species, B, to enter the area. When its resource use curve is superimposed on the curve of species A, B shows considerable overlap. Selective pressure from interspecific competition forces both species A and species B to narrow their range of resource use. Natural selection will favor those individuals living in areas of minimal or no overlap. Ultimately the two species will narrow their ranges of resource use. They will diverge, moving to the left and the right on the graph. Direct interspecific competition will be reduced, and the two species will coexist. Thus, while intraspecific competition favors expansion of the resource base, interspecific competition narrows the range. The populations involved have to arrive at some balance between the two.

Now allow a third species, C, to invade this resource gradient at a point between the utilization curves of A and B. Species C can successfully invade if A and B are rare, if they are below carrying capacity, and if resources are abundant. Under these conditions competition will force each of the three to become more specialized in their resource utilization, to utilize optimal resources, and to space themselves more narrowly on the resource gradient (for theory, see MacArthur and Levins 1967).

## Field Examples

Intensive field studies have turned up numerous examples of presumed resource partitioning. Lack (1971) noted that nine species of the genus *Parus* living in the broadleaf woods and coniferous forests of Europe and Great Britain feed in different parts of the tree canopy and consume different food throughout the year. Three of them, the blue tit, the great tit, and the marsh tit, inhabit the broadleaf woods. The blue tit, the most agile of the three, works high up in the trees gleaning insects, mostly 2 mm in size or smaller, from leaves, buds, and galls. The great tit, which is large and heavy, feeds mostly on the ground and seeks prey in the canopy only when taking caterpillars to feed its young. Its food consists of large insects 6 mm and over, supplemented with seeds and acorns. The marsh tit feeds largely on insects around 3 to 4 mm, which it gleans in the shrub layer and in twigs and limbs below 6 m above the ground. It, too, feeds extensively on seeds and fruits. In the northern coniferous forests live the coal tit, the crested tit, and the willow tit. The more agile coal tit forages high up in the trees among the needles. There it seeks and feeds on aphids

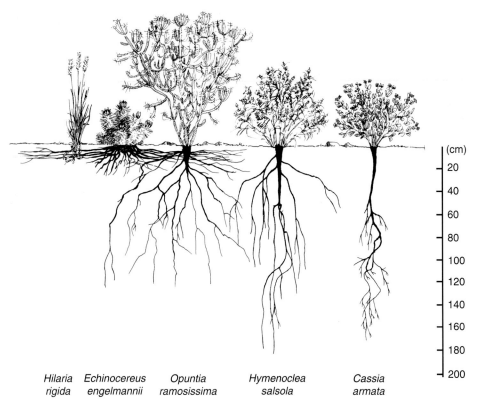

| Hilaria | Echinocereus | Opuntia | Hymenoclea | Cassia |
| rigida | engelmannii | ramosissima | salsola | armata |

**Figure 22.9** Partitioning of the soil resource by a group of Mojave Desert plants. Root system morphology is species-specific. Species such as *Hilaria rigida* and *Echinocereus engelmannii* are shallow surface rooters, able to take up moisture quickly during occasional rains. *Opuntia* and *Hymenoclea* employ more spreading roots at various intermediate depths. Plants such as *Cassia* have deep taproots. Plants with the same root morphology are not near neighbors. (After Cody 1986:386–387.)

and spruce seeds. The willow tit consumes a high proportion of vegetable matter and feeds in the few available broadleaf trees. When in the conifers, it spends most of its time in the lower parts and on the branches rather than on the twigs. The crested tit is confined mostly to the upper and lower parts of the trees and the ground, but the bird does not feed in the herb layer. Thus by feeding in different areas and on different size insects, as well as different types of vegetable matter, these species divide the resources among them.

MacArthur (1958) observed a similar partitioning among five species of warblers inhabiting the spruce forests of the northeastern United States. Each fed in a different part of the canopy, and each was specialized behaviorally to forage in a somewhat different manner.

Intraspecific rather than interspecific competition appears to be most important among phytophagous (leaf-eating) insects. Although the assemblages or guilds of such insects are similar to those of other groups of species, there is no evidence that they result from interspecific competition (Lawton and Strong 1981). Interspecific competition for food is weak (except, perhaps, in outbreaks of gypsy moths, which consume most or all of the foliage in oak forests). Instead, environmental pressures such as harsh climatic conditions, phenology of host plants, seasonal changes in chemical composition of plant tissues (see Chapter 24), and

patchy distribution of food plants overshadow the relationships among leaf-eating insect species.

A similar partitioning of resources exists among plants. Plants experience strong abiotic and biotic selective pressures on gross morphology, both above ground and below ground. These pressures result in differing methods of exploiting light, water, and nutrients. Once plants are committed to a life form, they are committed to a particular mode of resource utilization. For this reason intraspecific competition can be more influential than interspecific competition.

This point is well illustrated in Cody's (1986) studies of life forms of desert plants. He found that belowground root morphologies were much more important than aboveground structures to coexistence. Conspecifics with similar root structures tended to be widely spaced, and their nearest neighbors were species with different and complementary root systems (Figure 22.9). Species such as *Echinocereus,* the hedgehog cactus, and *Yucca,* the yuccas, have spreading roots within 15 cm of the surface. Others have deep taproots that extend 2 m or more below the surface to reach water and nutrients. Still others, such as the *Opuntia* cactus, have deep spreading roots rather than taproots. The deep-rooted species would conflict with conspecific near neighbors but not with shallow-rooted species. These shrub species coexist because they have different root systems that allow them to exploit water and nutrients in separated areas.

# DIFFERENTIAL RESOURCE UTILIZATION

The usual approach to the study of competition along a resource gradient emphasizes population responses. However, as populations use resources they are also depleting the resource base, which may or may not be renewed. What becomes important in interspecific competition, then, is the rate of consumption versus the rate of renewal. Among plants, in particular, competition is not for just one resource, such as food among animals, but for several resources simultaneously, such as light, soil nutrients, and water.

Tilman (1980, 1982, 1986) has presented a model of plant competition that considers both species growth and resource levels. It is based on the theory that each plant species is a superior competitor along a resource gradient of light and nutrients or light and moisture. Changes in these resources should result in changes in competitive interaction between plant species.

The theory is best illustrated graphically, first for a single species then for two species. Consider a single species using two essential resources, soil nitrogen and light. The resource availability can be plotted on two axes, with light on the $x$ axis and soil nitrogen on the $y$ axis (Figure 22.10). The population response of the species at zero growth on the two gradients is plotted as a solid line with a right-angle corner where the two gradients intersect. This line represents the resource-dependent zero net growth isocline. For all points along the isocline population growth is zero. Above and to the right of the zero net growth isocline populations increase. Below and to the left of the isocline populations decline or fail to survive.

For any one point along the zero net growth isocline, there is only one point, the supply point, where resource supplies are constant. If a species uses a resource at a rate faster than it can be renewed, then the population will decline along the zero isocline (toward the bottom left of the diagram) until it establishes a new equilibrium point. Conversely, if renewal of the resource exceeds consumption, then the population will grow along the isocline until it establishes a new equilibrium point (Figure 22.11).

Now consider the interaction of two species, A and B, each of which has different requirements for light and soil nitrogen (Figure 22.12). The zero net growth isocline of A lies closer to the axes than does that of B. Thus A can survive and grow at lower supply levels of both light and nitrogen than B. If the resource supply point (S) falls outside the isoclines of both A and B, neither species can exist. If the supply point falls in the region above and to the right of A's zero net growth isocline, but outside of B's isocline, only A can exist. If the supply point falls within the region above the isoclines of both species, then A reduces the resource

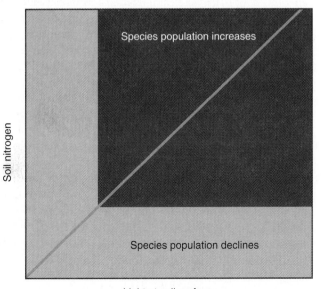

**Figure 22.10** Zero net growth isocline for a plant species potentially limited by two resources, light at soil surface and available soil nitrogen. The zero net isocline is rectangular because both resources are essential. The species population increases within the shaded area; it declines in the area outside. (After Tilman 1986:362.)

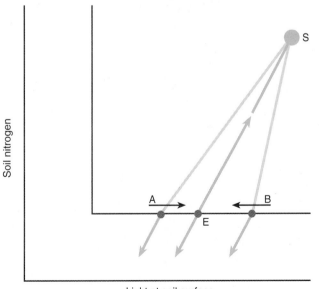

**Figure 22.11** What happens if consumption vectors veer away from the equilibrium point? At A the consumption rate is less than the renewal rate, so the population will grow toward the equilibrium point E. If the consumption rate is greater than the renewal rate, as at B, then the population will decline to the equilibrium point.

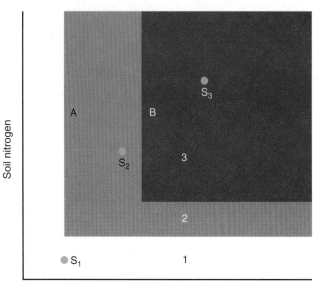

**Figure 22.12** Competitive exclusion. The resource-dependent zero net growth isocline of species B lies farther away from the resource axes than that of species A. In area 1 neither species can exist because $S_1$ the supply point of light and nitrogen, lies outside the isoclines of both species. In area 2 only A can exist because the supply point $S_2$ falls within the isocline of species A but outside that of species B. In area 3 A also wins because the supply point $S_3$ falls within its isocline, which is closer to the resource axes than that of species B. (After Tilman 1986.)

species of the resource that limits its own growth. In an intermediate habitat, point 5, each species is limited by a different resource, B by nitrogen and A by light. Each species uses more of the resource that limits its growth and will be limited by the resource for which it is the poorer competitor. The equilibrium point of stable coexistence in habitat 5 will be that at which combined resource consumption by the two species equals resource supply. This concept relates to plant succession, discussed in Chapter 30.

This model of differential resource utilization appears to be more realistic than the classic competition model, because it takes into account not only population growth but also depletion and renewal of resources that affect growth. Like other ecological models, though, it will require extensive field testing under natural conditions.

The seedlings occupy niches different from the mature plants. This fact is an important aspect of niches. The fundamental and realized niches of an organism can change with its growth and development. Insects with a complex life history may occupy one niche as larvae and an entirely different niche as adults. As a larva, a butterfly feeds on green foliage; as an adult it feeds on nectar. For other organisms, such as many fish, the niche space changes with size as the organism matures (see Figure 11.15). As the fish grows larger, it may switch from feeding on plankton to eating small fish. With this change in diet comes a change in its interspecific relationships.

concentrations along its own isocline to a point where B cannot exist, and B is competitively excluded.

In the usual situation, each species has a greater requirement for one resource than for another. In this case the isoclines will intersect (Figure 22.13). In our example A is the superior competitor for nitrogen and B is the superior competitor for light. The point at which their isoclines cross is the equilibrium point at which both species can coexist. Again, if the supply point falls outside either of the two isoclines, as at point 1, neither species can survive. If the supply point falls at 2 between A and B, then A wins and B loses. If it falls at 3 between B and A, then B wins. If the supply point falls somewhere within the common region enclosed by both, then the outcome depends upon the position of the supply point. If it falls at 4, then B, with lower light requirements, should displace A from habitats with low light availability. B does so because it reduces the light at the soil surface to a level below that required for species A to survive. At point 6, which has a low supply of nitrogen, both species will be nitrogen-limited. A, the superior competitor for nitrogen, should displace species B. In both situations, the reason for displacement is the greater use by each

**Figure 22.13** Competition for soil nitrogen versus light in two hypothetical species with a different need for each resource. (After Tilman 1986:363.)

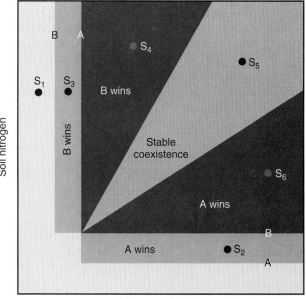

# THE NICHE

Closely associated with interspecific competition is the concept of the niche. **Niche** is one of those nebulous terms in ecology, its meaning colored by various interpretations that equate it with habitat, functional roles, food habits, and morphological traits. In everyday terms, a niche is a recess in a wall where you place something, usually an ornamental object; or it is a position in life suitable for a person. In ecology it means an organism's place and function in the environment—or does it?

One of the first to propose the idea of the niche was the ornithologist Joseph Grinnell (1917, 1924, 1928). In his study of the California thrasher (*Toxostoma redivivum*) and other birds, he suggested that the niche be regarded as a subdivision of the environment occupied by a species, "the ultimate distributional unit within which each species is held by its structural and functional limitations." Essentially Grinnell was describing the habitat of the species.

Charles Elton (1927), in his classic *Animal Ecology*, considered the niche as the fundamental role of the organism in the community—what it does, its relation to its food and enemies. This idea stresses the occupational status of the species in the community.

Other definitions are variations on the same theme. Odum (1971) considers the habitat as the animal's address and the niche as its occupation. Whittaker, Levin, and Root (1973) consider the niche only as a functional position, whereas the niche and habitat combined comprise the *ecotope*, "the ultimate evolutionary context of the species." Pianka (1978) regards the niche as embracing all the ways in which a given individual, population, or species conforms to its environment.

The definition that links the niche to competition was proposed by G. E. Hutchinson (1957). It is based on the competitive exclusion principle. According to this concept, an organism's niche consists of many physical and environmental variables, each of which can be considered a point in a multidimensional space. Hutchinson called that space the **hypervolume.**

We can visualize a multidimensional niche to a certain extent by creating a three-dimensional one. Consider three niche-related variables for a hypothetical organism: food size, foraging height, and humidity (Figure 22.14). Suppose the animal can handle only a certain range in food size. Food size, then, is one dimension of the niche. Add the foraging height, the area to which it is limited seeking food. If we graph that on the second axis and enclose the space, we have a rectangle, representing a two-dimensional niche. Suppose, too, that the animal can survive and reproduce only within a certain range of humidity. Humidity can be plotted on a third axis. Enclosing that space, we come up with a volume, a three-dimensional niche. Of course, many more variables, both biotic and abiotic, influence a species' or an individ-

ual's fitness. A number of these dimensions, *n*—difficult to visualize and impossible to graph—make up the *n*-dimensional hypervolume that would be the species' niche. An individual or a species free from the interference of another could occupy the full hypervolume or range of variables to which it is adapted. That is the idealized **fundamental niche** of the species.

The fundamental niche of the species assumes the absence of competitors, but rarely is this the case. Competitive relationships may force the species to constrict a portion of the fundamental niche it could potentially occupy. In those

**Figure 22.14** Models of niche dimension for a hypothetical animal. (a) A one-dimensional niche involving food size. The animal can live on food of intermediate size but not on food that is large or small. (b) A second dimension has been added, foraging height. Enclosing that space, we obtain a two-dimensional niche. (c) Suppose the organism can survive and reproduce only within a certain range of humidity, graphed as a third axis. By enclosing all those points, we arrive at a three-dimensional niche space or volume.

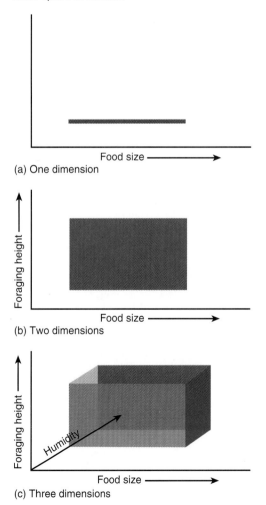

(a) One dimension

(b) Two dimensions

(c) Three dimensions

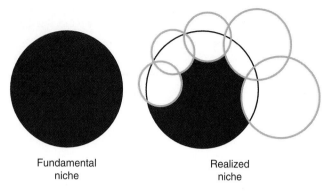

Fundamental
niche

Realized
niche

**Figure 22.15** Fundamental and realized niches. The fundamental niche of a species is the full range of environmental conditions, biological and physical, under which it can exist. Under pressure of superior competitors whose niches overlap, the species may be displaced from part of its fundamental niche and forced to retreat to that portion of the fundamental niche hypervolume to which it most highly adapted. The portion it occupies is its realized niche.

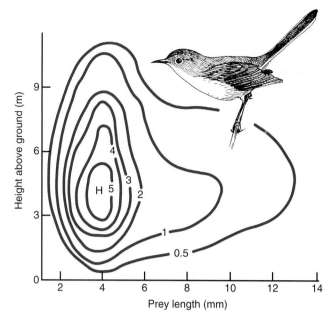

**Figure 22.16** The feeding niche of the blue-gray gnatcatcher (*Polioptila caerulea*), based on of two variables, size of prey and feeding height. The contour lines map the feeding frequencies for adult gnatcatchers during the incubation period during July and August in California oak woodlands. The maximum response level is at H. Contour lines spreading out from this optimum represent decreasing response levels. The outer contour line probably represents the outer boundary of the realized niche for these two variables. (For discussion of such an analysis, see Maguire 1973, Hutchinson 1978.) (Diagram from Whittaker et al. 1974, based on data from Root 1967.)

parts its fitness might be reduced to zero. The conditions under which an organism actually exists are its **realized niche** (Figure 22.15). The niche may be further restricted by the absence of certain features of the niche at any given point in time and space. Like the fundamental niche, the realized niche is an abstraction. In their studies, ecologists usually confine themselves to one or two niche dimensions, such as a feeding niche, a space niche, or a tolerance niche.

Consider two examples. Root (1967) studied the exploitation of the feeding niche by the blue-gray gnatcatcher in California oak woodlands. He characterized the niche of the bird by the size of its food and by the height above the ground at which it captured food (Figure 22.16). Simplified for the sake of example, the bird's fundamental niche could be described by a maximum range of size of prey between 1 and 14 mm in length and by a foraging area of ground level to 10 m. The gnatcatcher's niche center, indicated by frequency of capture and stomach content analysis, consists of insects 3 to 5 mm long taken 2.4 to 8.5 m above the ground. The further the height and food dimensions diverge from this center, the more the gnatcatcher's niche may overlap those of other species.

Putwain and Harper (1970) studied the population dynamics of two species of dock, *Rumex acetosa* and *R. acetosella,* each growing in hill grasslands in North Wales. *R. acetosa* grew in a grassland community dominated by velvet grass (*Holcus lanatus*) and red and sheep fescues (*Festuca rubra* and *F. ovina*); *R. acetosella* grew in a community dominated by sheep fescue and bedstraw (*Galium saxatile*). To determine interference and niches of the two dock species, Putwain and Harper treated the flora with specific herbicides to remove selectively in different plots (1)

grasses, (2) forbs except *Rumex* species, and (3) the *Rumex* species. All species except *R. acetosella* spread rapidly after the grasses were removed, but *R. acetosella* increased only after both grasses and nongrasses were removed.

The niches of these two plants are diagramed in Figure 22.17. The fundamental niche of *R. acetosella* (R) overlaps the fundamental niches of both grasses (G) and other forbs (D). Only when these competitors are eliminated does this dock realize its fundamental niche. *R. acetosa,* however, overlaps only with the grasses and only their removal is necessary to permit expansion of this dock throughout its fundamental niche.

## Niche Overlap

The example of *Rumex* brings up the question of niche overlap. What happens when two or more species use a portion of the same resource, such as food, simultaneously? The theoretical model of the niche assumes that competition is intense, that only one species can occupy a niche space, and

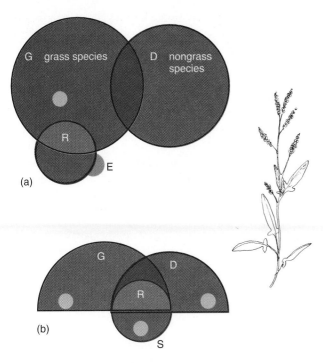

(a)

(b)

**Figure 22.17** Niche relationships of (a) *Rumex acetosa* and (b) *R. acetosella* in mixed grassland swards. In each diagram the fundamental niches of grass species (G) and nongrass species (D) overlap. The fundamental niche of *Rumex* species (R) is shown as a continuous line, and the realized niche is shaded. E is that part of the fundamental niche of *R. acetosa* which is expressed in the presence of nongrass species and does not overlap the fundamental niches of G and D. The fundamental niches of seedlings (S), shown by the small colored circles, are contained within the fundamental niches of grasses, nongrasses, and mature *Rumex*. (From Putwain and Harper 1970.)

that competitive exclusion takes place in areas of overlap. The amount of niche overlap is assumed to be proportional to the degree of competition for that resource. In a condition of minimal or no competition, niches may be adjacent to one another with no overlap, or they may be disjunct (Figure 22.18). At the other extreme, in a condition of intense competition, the fundamental niche of one species may be completely within or correspond exactly to another, as in the case of the seedling *Rumex*. In such instances there can be two outcomes. If the niche of species 1 contains the niche of species 2 and species 1 is competitively superior, species 2 will be eliminated entirely. If species 2 is competitively superior, it will eliminate species 1 from the part of the niche space species 2 occupies. The two species then coexist within the same fundamental niche.

When fundamental niches overlap, some niche space is shared and some is exclusive, enabling the two species to coexist (Figure 22.18). Considerable niche overlap does not necessarily mean high competitive interaction. In fact, the reverse may be true. Competition involves a resource in short supply. Extensive niche overlap may indicate that little competition exists and that resources are abundant. Pianka (1972, 1975) has suggested that the maximum tolerable overlap in niches should be lower in intensely competitive situations than in environments with low demand/supply ratios.

In fact, both high niche overlap and the absence of overlap may reflect other environmental and behavioral influences and not interspecific competition at all. To attribute niche overlap or the lack of it to interspecific competition may be to ignore real reasons. Consider a study of habitat partitioning by three species of grassland sparrows: the grasshopper sparrow (*Ammodramus savannarum*), Henslow's sparrow (*A. henslowii*), and the savannah sparrow

**Figure 22.18** Niche relationships visualized as graphs on a resource gradient and as Venn diagrams. Species A and B have overlapping niches of equal breadth but are competitive at opposite ends for the resource gradient. B and C have overlapping niches of unequal breadth. Species C shares a greater proportion of its niche with B than B does with C. (However, B shares its niche also with A at the other end.) C and D occupy adjacent niches with little possibility of competition. D and E occupy disjunct niches and no competition exists. Species F has a niche contained within the niche of E. If F is superior to E competitively, it persists and E shares that part of its niche with F. Compare Figure 22.17. (Adapted from Pianka 1978.)

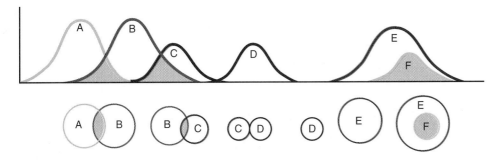

(*Passerculus sandwichensis*), on broad expanses of reclaimed surface mines in north-central Pennsylvania by Piehler and Whitmore (Piehler 1987). They mapped 65 sparrow territories (20 grasshopper, 23 savannah, and 22 Henslow's sparrows) and measured 14 structural variables of the vegetation, including basal and overhead cover, height, density, and litter depth. Territorial boundaries of all three species overlapped to a degree. Except for the semicolonial Henslow's sparrow, territorial boundaries within species were contiguous. Territorial boundaries among the three species showed varying degrees of overlap, the greatest being a 31 percent overlap between the grasshopper and savannah sparrows. The partitioning of the grassland habitat

and thus habitat niche configurations could be explained by a gradient of vegetational structure (Figure 22.19) alone, based on increasing habitat richness (vegetation density, vegetation height, bare ground cover, and litter depth). Henslow's sparrows inhabited areas of tall, dense, thick grass. On the other end of the gradient, with some bare ground, thinner and shorter forb and grass cover, and less litter depth, was the grasshopper sparrow. Intermediate between the two was the savannah sparrow. In addition to the wide differences in the vegetational cover they inhabited, the two congeneric species, grasshopper sparrow and Henslow's sparrow, were very similar in morphological characters of bill width and bill length, whereas the savannah

**Figure 22.19** Niche overlap and habitat partitioning among three species of grassland sparrows on a Pennsylvania study site. (a) A two-dimensional plot of the first (habitat) and second (territory size) discriminant axes. Note the wide separation between the two congeneric species, the grasshopper and Henslow's sparrows. (From Piehler 1987.) (b) The three sparrows arranged along a habitat resource gradient. There is a 1 percent overlap in habitat between Henslow's and grasshopper sparrow, a 14 percent overlap between Henslow's and savannah sparrows, and a 35 percent overlap between savannah and grasshopper sparrows.

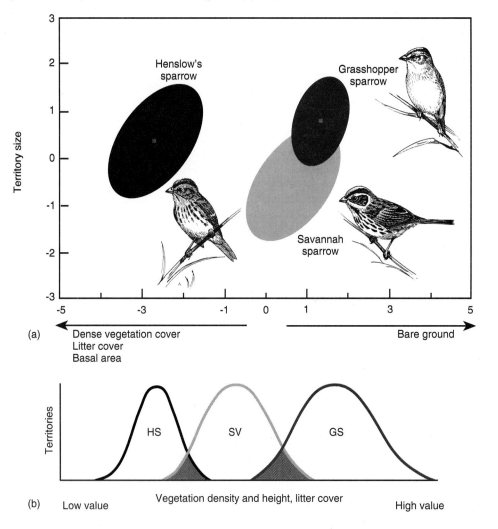

sparrow had significantly smaller ($p < .05$) bill width and length.

For simplicity, niche overlap is usually considered one-dimensional or two-dimensional. In reality, a niche involves many types of resources: food, a place to feed, cover, space, and so on. Rarely do two or more species possess exactly the same requirements. Species overlap on one gradient, but not on another; thus total competitive interactions may be less than the competition or niche overlap suggested by one gradient alone (Figure 22.20).

## Niche Width

If we plotted the range of resources—for example, food size—used by an animal or the range of soil moisture conditions occupied by plants, the length of the axis intercepted by the curve would represent **niche width** (Figure 22.21). Theoretically, niche width (also called niche breadth and niche size) is the extent of the hypervolume occupied by the realized niche. A more practical definition is the sum total of the different resources exploited by an organism (Pianka 1975). Measurements of a niche usually involve the measure of some ecological variable such as food size or habitat space.

Niche widths are usually described as narrow or broad. The wider the niche, the more generalized the species is considered to be. The narrower the niche, the more specialized is the species. Most species have broad niches and sacrifice efficiency in the use of a narrow range of resources for the ability to use a wide range of resources. As competitors they are superior to specialists if resources are somewhat un-

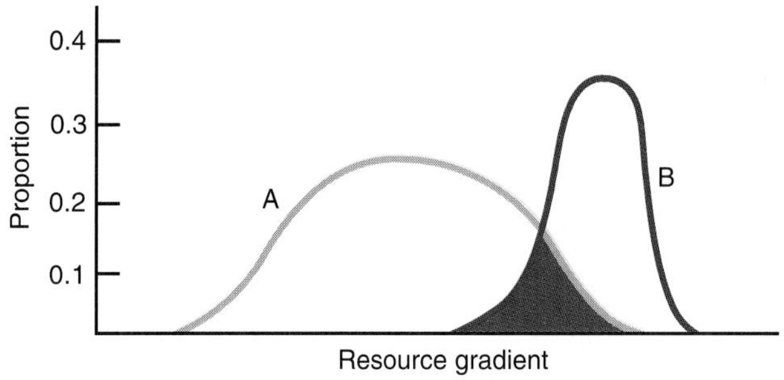

**Figure 22.21** Hypothetical distribution of a species with a broad niche (A) and a species with a narrow niche (B) on a resource gradient. The niches overlap (shaded area). Species A overlaps species B more than species B overlaps species A.

dependable. Specialists, equipped to exploit a specific set of resources, occupy narrow niches. As competitors they are superior to generalists if resources are dependable and renewable. A dependable resource is closely partitioned among specialists with low interspecific overlap (Roughgarden 1974). If resource availability is variable, generalist species are subject to invasion and close packing with other species during periods of resource abundance.

## Niche Change

Niche width provides some indication of resource utilization by a species. If a community made up of a number of species with broad niches is invaded by competitors, intense competition may force the original occupants to restrict or compress their utilization of space and to confine their feeding and other activities to those patches of habitat providing optimal resources. Competition that results in the contraction of habitat rather than a change in the type of food or resources utilized is called **niche compression** (MacArthur and Wilson 1967).

Conversely, if interspecific competition is reduced, a species may expand its niche, utilizing space previously unavailable to it. Niche expansion in response to reduced interspecific competition is called **ecological release.** Ecological release may occur when a species invades an island that is free of competitors, moves into habitats it never occupied on the mainland, and increases in abundance (Cox and Ricklefs 1977). Such expansion may also follow when a competing species is removed from a community.

Associated with compression and release is another response, niche shift. **Niche shift** is the adoption of changed behavioral and feeding patterns by two or more competing populations to reduce interspecific competition. The shift may be a short-term ecological response or a long-term evolutionary response involving some change in a basic behavioral or morphological trait. With the exception of evolution-

**Figure 22.20** Niche relationships based on two gradients. Species may exhibit considerable overlap on one gradient and little or none on another. When niche dimensions are added, niche overlap may be reduced considerably. (Adapted from Pianka 1978.)

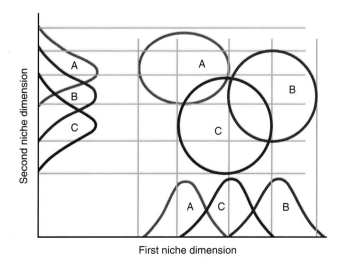

First niche dimension

ary change, niche shift does not involve the establishment of a new niche, as the term might imply. Rather niche shift refers (or should refer) to a shifting of the realized niche within the range of the fundamental niche.

Werner and Hall (1976, 1977, 1979) demonstrated niche shift in three cogeneric species of sunfish (Centrarchidae): the bluegill (*Lepomis maerochirus*), the pumpkinseed (*L. gibbosus*), and the green sunfish (*L. cynellus*), occupying experimental ponds 30 m in diameter supporting natural stands of emergent and submerged vegetation and their associated prey populations. Although this experiment lacked replication and the fish were stocked at higher than natural densities, it does provide an example of processes involved in niche shift. When stocked in the ponds alone, each species inhabited the vegetation zones supporting the largest prey. When the three species were stocked together in equal densities, all initially occupied the vegetated zones. As food resources declined, the bluegill and the pumpkinseed left for more open water, leaving the green sunfish, a more efficient forager, in the vegetation. Bluegills and pumpkinseeds concentrated their foraging efforts on bottom invertebrates, mainly Chironominae. The pumpkinseed, with its short, widely spaced gill rakers that do not become fouled when the fish sorts through the bottom sediments, exploited that food supply. The bluegill, with long, fine gill rakers that retain small prey, foraged in the open water.

The fish exhibited shifts in habitat utilization as well. When bluegill and green sunfish were confined together in equal densities in a pond, the bluegill shifted to open water; but in the absence of green sunfish, bluegills invaded the dense vegetation. Bluegills and pumpkinseeds have a dietary overlap of 50 to 55 percent. When the two are stocked together in a pond, the bluegill again moves to open water. In the end, however, the bluegill may be the superior competitor. Young sunfish of all three species feed on zooplankton in open water. This fact places the young of green sunfish and pumpkinseeds in direct competition for small-sized food with both young and adult bluegill. A more efficient open-water forager, the bluegill probably affects the recruitment of the other two species. Perhaps its generalized diet and its ability to move into different habitats as the situation requires accounts for the bluegill's position as the most common sunfish in ponds.

## ■ SUMMARY

Relations between species may be positive (+) or beneficial; negative (−) or detrimental; or neutral (0). There are six possible interactions: (0 0), neutral; (+ +), in which both populations mutually benefit (mutualism); (− −), in which both populations are affected adversely (competition); (0 +), in which one population benefits and the other is unaffected (commensalism); (0 −), in which one population is harmed and the other is unaffected (amensalism); and (+ −), in which one population benefits and the other is harmed (predation, parasitism).

Interspecific competition—the seeking of a resource in short supply by individuals of two or more species, reducing the fitness of both—may be one of two kinds, interference and exploitative. Exploitative competition depletes resources to a level of little value to either population. Interference involves aggressive interactions (passive in plants, active in animals). A particular form of interference competition is allelopathy, the secretion of chemical substances that inhibit the growth of other organisms.

As described by the Lotka-Volterra equations, four outcomes of interspecific competition are possible. Species 1 may win over species 2, or species 2 may win over species 1. Both of these outcomes represent competitive exclusion. A third possibility is unstable equilibrium, in which the potential winner is the one most abundant at the outset or most able to respond to a changing environmental condition. A final possible outcome is stable equilibrium, in which two species coexist, but at lower population levels than if each existed in the absence of the other.

The competitive exclusion principle—two species with exactly the same ecological requirements cannot coexist—has conceptual difficulties. It has, however, stimulated critical examinations of other competitive relationships, especially how species coexist and how resources are partitioned. One way of looking at plant competition is differential resource utilization. It is based on the premise that species compete simultaneously for several resources, such as light and nutrients, but differ in their requirements for two limiting essential resources. Coexistence occurs when the combined resource consumption of the two species equals the resource supply.

Closely associated with interspecific competition is the concept of the niche. A niche is the functional role of an organism in the community. It might be constrained by interspecific competition. In the absence of competition, an organism occupies its fundamental niche. In the presence of interspecific competition, the fundamental niche is reduced to a realized niche, the conditions under which an organism actually exists. When two different organisms use a portion of the same resource, such as food, these niches are said to overlap. Overlap may or may not indicate competitive interaction.

The range of resources used by an organism suggests its niche width. Species with broad niches are generalists, whereas those with narrow niches are specialists. Niche compression results when competition forces an organism to restrict its type of food or constrict its habitat. In the absence of competition, the organism may expand its niche and experience ecological release. Organisms may also undergo niche shift by changing their behavioral or feeding patterns to reduce interspecific competition.

# REVIEW QUESTIONS

1. Distinguish among the following population interactions: neutral interactions, mutualism, commensalism, amensalism, parasitism, competition, predation.
2. According to the Lotka-Volterra theory of competition, what are the four possible outcomes of competition?
3. What conditions are necessary for competitive exclusion? Is allelopathy a mechanism for competitive exclusion among plants?
4. Why is it more difficult to demonstrate interspecific competition in the field than in the laboratory? Can you relate laboratory studies to field situations?
5. What is resource partitioning? How might it relate to interspecific competition?
6. Define the niche. What is a functional niche? A realized niche? Relate niche to interspecific competition and resource partitioning.

7. Read an experimental study of competition in the field critically. Analyze the experiment, considering experimental design, application of laboratory studies to the field situation, demonstration of interspecific competition, and applicability of the results to the natural community. Good choices are Kross 1980, Brown 1982, Wilbur and Alford 1985, Gurevitch 1985, and McGraw and Chapin 1989.

# CROSS-REFERENCES

Trophic levels, 191–192; food chain, 185–187, 192–193, 336, 621–623; population growth, 392–396; intraspecific competition, 412–424; social dominance, 417–418; territoriality, 418–425; chemical defense by plants, 527–529; parasitism, 558–573; competition and community structure, 620–621; predation and community structure, 621–623; plant succession, 656–669.

## Predation

### Concepts

1. Predator-prey models predict oscillations in predator and prey populations.
2. Functional response relates the amount of prey taken by predators to prey density.
3. Increase in number and fecundity of predators in relation to changes in prey density is called numerical response.
4. Optimal foraging is a strategy to maximize food intake and minimize time and energy spent in acquiring it.
5. Risk-sensitive foraging relates to the forager's response to the mean and variance of its food distribution.

No type of population interaction is more misunderstood or stirs more emotional reaction than predation. Sympathy goes to the prey. When we look at it objectively, though, predation becomes a step in the transfer of energy in the ecosystem.

Predation is commonly associated with the strong attacking the weak, the lion pouncing on the deer, the hawk upon the sparrow. However, considered more broadly, predation also includes **parasitoidism,** a case of the weak attacking the strong. In this situation, one organism, the parasitoid, attacks the host (the prey) by laying its eggs in or on the body of the host. After the eggs hatch, the larvae feed on the tissues of the host until it dies. The effect is the same as that of predation. Another special form of predation is **cannibalism,** in which the predator and the prey are the same species. The concept of predation has been extended still further to include **herbivory,** in which grazing animals of all types feed on plants. Herbivores kill their prey when consuming seeds or the whole plant, or they function as parasites when they consume only part of the plant but do not destroy it. Thus predation in its broadest sense can be defined as one organism feeding on another living organism, or **biophagy.**

Ecologically, predation is more than just a transfer of energy and nutrients. It represents a direct and often complex interaction of two or more species, of the eaters and the eaten. The numbers of some predators may depend upon the abundance of prey, and predation may be involved in the regulation of prey populations.

## MODELS OF PREDATION

In 1925 A. J. Lotka, a mathematician and physical scientist, proposed the first model of predator-prey interactions in *Elements of Physical Biology.* In 1926 the Italian mathematician A. Volterra independently came up with a similar model. Neither of the two extended the logistic equation to the two-species system. Instead, Lotka adapted the chemical principle of mass action (Berryman 1992). Mass action assumes that individual predators and individual prey encounter each other randomly in the same way that molecules interact in a chemical solution. The responses of predator and prey populations are assumed to be proportional to the products of their population densities.

The Lotka-Volterra model involves paired equations, one for the prey population and one for the predator population. The prey growth equation has two components, the maximum rate of increase per individual and the predatory removal of prey from the population:

$$dN/dt = aN - bNP$$

For the predator population:

$$dP/dt = cNP - dP$$

where $N$ and $P$ are the densities of the prey and the predator, respectively, and $a$ and $d$ are the per capita rates of change in

absence of each other, and $b$ and $c$ are the rates of change for prey and predator resulting from the interaction of the two populations.

The Lotka-Volterra model is based on a number of underlying assumptions: (1) in the absence of predation, the prey experiences exponential growth; (2) the predator population declines exponentially in the absence of prey; (3) predators move at random among randomly distributed prey; (4) the proportion of encounters that result in the capture and consumption of prey are constant at all predator and prey densities; (5) the number of prey taken increases in direct proportion to the number of predators, a linear response; (6) all responses are instantaneous with no time lag for handling and ingesting prey; and (7) energy input to predators is immediately converted to the birth of more predators. The model makes no allowance for age structure, for interaction of prey with their own food supply, or for density-dependent mortality of the predator.

The Lotka-Volterra model is depicted graphically in Figure 23.1. The ordinate $P$ is the number of predators; the abscissa $N$ is the number of prey. The horizontal straight line is the zero growth curve of the prey, and the vertical line is

**Figure 23.1** Lotka-Volterra model of predator-prey interactions. (a) The abundance of each population is plotted as a function of time. (b) The joint abundance of species. The zero growth curve or isocline of the predator is vertical and the isocline of the prey is horizontal. They intersect at right angles. A minus sign shows population decline and a plus sign population increase. Predators increase to the right of the vertical line; prey increase below the horizontal line.

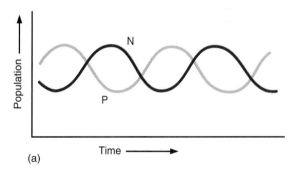

(a)

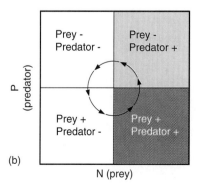

(b)

the zero growth curve of the predator. In the area to the right of the vertical line, predators increase; to the left, they decrease. In the area below the horizontal line, the prey increase, and above it they decrease. The circle of arrows represents the joint population of predator and prey and the way it changes. If a point or arrow falls in the region to the left of the vertical line, the prey population is not large enough to support the predators and the predator population declines. If an arrow falls to the upper left, both populations are declining. The predator population decreases enough to permit the prey population to increase, moving the arrow to the lower left. The increase in the prey population now permits predators to increase, and the arrow moves to the lower right. As the predator population increases, depressing the prey population, the arrow moves to the upper right. This interaction between predator and prey results in reciprocal oscillations of predator and prey with some time delay in the predator's response. These regular oscillations or cycles will continue ad infinitum.

A decade later an ecologist, A. J. Nicholson, and a mathematician, W. Bailey, recognized the deficiencies in the Lotka-Volterra model. They developed a model for a host-parasitoid relationship (Nicholson and Bailey 1935). Parasitoids differ from predators in that their attacks do not remove the host from the prey population. In fact, one host may be attacked many times. The number of parasites reared on the host depends upon the number of attacks, not the number of hosts encountered.

Nicholson and Bailey based their model on a set of assumptions somewhat different from the Lotka-Volterra model. Predators search randomly for static prey uniformly dispersed over a homogeneous landscape. The predators have a constant area of discovery and an insatiable appetite. The predators "sample" a certain proportion of the prey population. The time element is discrete, not continuous. The generations of both predator and prey have the same time span and are of the same length. Predator mortality is density-independent. The conversion of energy input by the predator into the birth of more predators is not immediate but delayed by one generation.

Like the Lotka-Volterra model Nicholson and Bailey's model consists of two equations, with the exception that they are difference equations:

$$H_{t+1} = \lambda H_t \exp(-aP_t)$$
$$P_{t+1} = H_t[1 - \exp(-aP_t)]$$

where $H$ is the host, $P$ the predator or parasitoid, $t$ is time, $\lambda$ is the finite rate of increase for the host population, and $a$ is the rate of parasitism per parasitoid.

These features of the Nicholson-Bailey model allow an estimate of the prey in the next generation. If the number of hosts that the parasitoid removes within its sampling area is equal to the fraction of the prey that represents recruitment, then the base parental stock remains. If the parental stock remaining is sufficient to replace the individual prey taken and

if it is sufficient to maintain the density of the parasitoids, then the two populations remain stable indefinitely. However, if the parasitoid removes part of the parental stock along with recruitment, the prey and ultimately the parasitoid populations decline. If much of the recruitment is left untouched, prey increase and the predators may not be up to the task of removing increased recruitment. In either of these two cases, the two interacting populations will undergo oscillations with increased amplitude, according to the Lotka-Volterra equations.

The Nicholson-Bailey model does not respond in exactly the same manner. If parental stock of the prey is sufficient to replace the prey taken as well as to maintain the predator population, then the two populations will remain stable as in the Lotka-Volterra model. If, however, predators remove too many individuals or if the growth of the prey population outstrips the ability of the predators to reduce the prey population, the results are unstable populations with ever-increasing amplitudes and the possible extinction of one population or the other (Figure 23.2).

The Lotka-Volterra and Nicholson-Bailey models emphasize the influence of predators on prey populations. Genetic changes, stress, emigration, aggression, availability of cover and hiding places, difficulty of finding prey as they become more scarce, and other attributes also influence predator-prey relationships. To add more realism to predation models, Rosenzweig and MacArthur (1963) developed a series of graphic models that consider a wider range of predator-prey interactions (Figure 23.3). Modifying the zero growth isocline of the prey to account for a low rate of growth at low and high densities, they plot the growth curve of the prey as convex rather than horizontal. The isocline of the predator levels off at high prey density, reflecting nonlinearity, compared to the linear response of the Lotka-Volterra

**Figure 23.2** Population oscillations of host and parasitoid predicted by the Nicholson-Bailey equations. (After Nicholson 1933.)

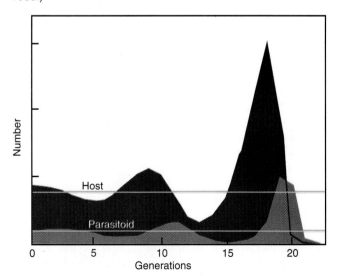

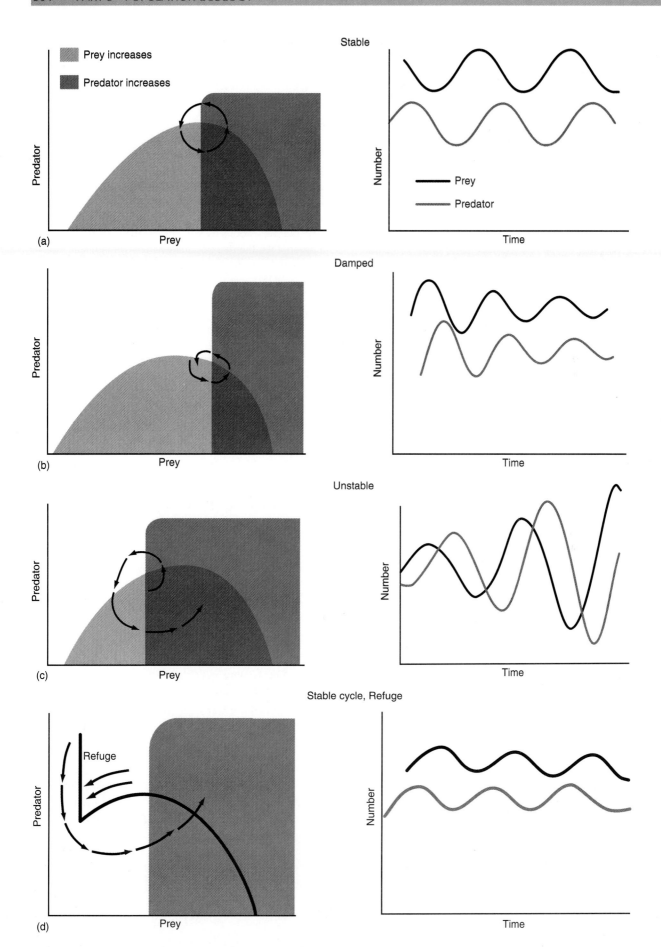

**Figure 23.3** (opposite)  Rosenzweig-MacArthur model of predator-prey interactions. The prey curve is convex rather than straight, involving density-dependent growth, unlike the predator curve. (a) Stable cycle. The prey population can increase if the joint abundances of predator and prey fall inside the area below the isocline of the prey. Prey population will decrease outside this area. The growth curves intersect at right angles as in the Lotka-Volterra model (Figure 23.1). (b) If the predator isocline is moved to the right or to the left it will no longer intersect the prey isocline at right angles. When the predator's isocline is moved to the right, it intersects the descending part of the prey isocline, which dampens oscillations. (c) If the predator isocline intersects the ascending part of the prey isocline, it increases oscillations, producing an unstable system. Vectors spiral outward and the amplitude of the population increases steadily until a limit cycle is reached. Such situations can lead to the extinction of either the predator or the prey. (d) If prey have access to a refuge, a portion of the population escapes predation and supplies immigrants to the exploited area. This immigration limits oscillations and leads to a stable cycle. (See MacArthur and Connell 1966:154–156.)

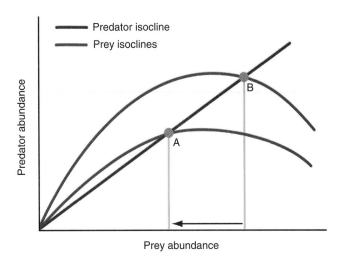

**Figure 23.4**  Predator-prey interaction when a ratio-dependent functional response curve is incorporated into the logistic model. The predator zero isocline is slanted, rather than vertical. The prey zero isocline is convex. Prey increases below its zero isocline. The vertical lines indicate increased prey vulnerability to predation or a lower rate of increase. The points of intersection represent points of equilibrium relative to the predator populations. Prey equilibrium changes as its vulnerability to predation changes. (After Arditi and Berryman 1991.)

equations. The Rosenzweig-MacArthur model of a stable cycle of interaction is similar to the Lotka-Volterra model. Predator and prey populations increase if the joint abundance of predator and prey falls inside the convex curve to the right of the vertical. If the joint abundance falls outside the curve, populations of both decline. Because the prey curve intersects the predator curve at right angles, the curve produces neutral, stable cycles in populations of predator and prey. By moving the predator curve to the right or to the left so that the curves no longer intersect at right angles, we can produce damped and unstable cycles.

These two models represent situations in which the prey has no refuge from the predator. Growth rates of both predator and prey populations are a function of the frequency with which the predator comes into contact with the prey. In nature a fraction of the prey escapes predation by hiding in refuges. Predators then decline from emigration, starvation, or failure to reproduce. As the prey population increases in the refuge area, the surplus repopulates the surrounding area.

In these models the equilibrium density of the prey population does not increase, even if the prey population is increasing. All the gain in production goes to the predator (which you can demonstrate by moving the vertical predator isocline to the right and left). Thus the Rosenzweig-MacArthur models consider that prey equilibrium is independent of prey density and that prey stability depends upon characteristics of the predator. Ignored is the ratio of prey numbers to predator numbers. To take this ratio into account, Arditi and associates (Arditi et al. 1991, Arditi and Berryman 1991) proposed a ratio-dependent model in which the predation rate depends upon the ratio of predators to prey (Figure 23.4). As prey abundance increases, predator abundance increases; and conversely, as prey abundance decreases, predator abundance decreases. Thus both predator and prey equilibria rise and fall with prey productivity.

## PREDATOR-PREY SYSTEMS

How well does the Lotka-Volterra model predict the behavior of predator-prey systems? The question stimulated experimental investigations into systems involving a single predator species and a single prey species.

The Russian biologist G. F. Gause (1934) reared together under constant environmental conditions a predatory ciliate, *Didinium nasutum,* and its prey, another ciliate, *Paramecium caudatum* (Figure 23.5). In one experiment Gause introduced the predator into a clear nutritive medium in test tubes supporting a population of *Paramecium.* Unable to escape the predatory ciliates, the *Paramecium* succumbed to predation. With their food supply gone, the *Didinium* starved. Gause discovered that regardless of the density of the two populations, the predator always exterminated the prey. Only by periodically introducing prey (immigrants) to the medium was Gause able to maintain the predator population. In this manner he was able to maintain populations together and produce the regular fluctuations in both predicted by the Lotka-Volterra equations.

In another experiment, Gause introduced sediment into the bottom of the tubes. Here some of the *Paramecium* could find refuge from their predator. When the predators ate all the available prey, they died from the lack of food. Meanwhile the *Paramecium* that took refuge in the sediment

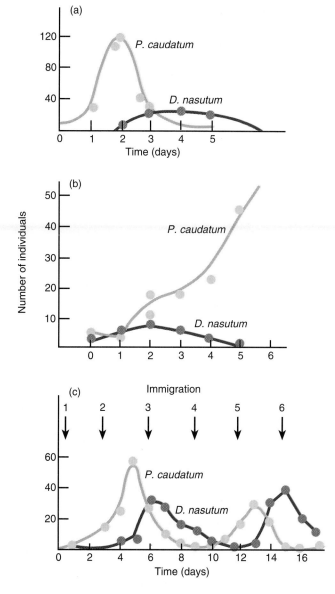

**Figure 23.5** Outcome of Gause's experiments of predator-prey interactions between the protozoans *Paramecium caudatum* and *Didinium nasutum* in three microcosms: (a) oat medium without sediment; (b) oat medium with sediment; (c) with immigration in oat medium without sediment. (After Gause 1934.)

multiplied, and eventually they took over the medium. Gause could maintain both populations only when a portion of the prey population could escape predation by retreating to a refuge and when he periodically introduced a small number of predators to the medium. Gause's laboratory experiments did not support the predictions of the Lotka-Volterra predation model. Rather, his results suggested that the predator-prey relationship was one of overexploitation and annihilation of the prey followed by the collapse of the predator population. Only immigration from other prey populations could maintain the predator-prey system.

The Gause experiments took place in a simple environment. This fact prompted C. Huffaker, a University of

California entomologist studying the biological control of crop pests, to investigate whether he could establish an adequately large and complex laboratory environment in which a predator-prey system would not be self-exterminating (Huffaker 1958). He chose as the prey the six-spotted mite, *Eotetranychus sexmaculatus,* which feeds on oranges and as predator another mite, *Typhlodromus occidentalis.* Both reproduce parthenogenetically. Huffaker established experimental populations on trays holding 40 cells in four rows of ten each. He placed oranges in some of the positions and rubber balls of the same size in the remaining cells. By changing the number and placement of oranges on the tray, and by covering the oranges with paper and sealing wax to vary the amount of exposed surface, he was able to control both the total food available and the pattern of food dispersion. Thus he manipulated the size and distribution of the prey population. In most of his experiments he established a prey population with 20 females and introduced a predatory population with 2 females 11 days later.

In one experiment Huffaker created a simple environment by concentrating the food. He placed four oranges in adjacent positions. The concentrated prey populations grew up to 650 individuals and survived 25 to 30 days before being eliminated by the predaceous mites. In another experiment Huffaker alternated 20 small areas of food with 20 foodless positions. The prey population grew to 2000 individuals (Figure 23.6). Because the scattered oranges provided a refuge that took the predators a longer time to discover, the population survived for 36 days.

In still another experiment Huffaker established a complex environment where food was widely dispersed. He used three 40-cell trays for a 120-cell universe in which he placed 120 oranges. Each orange had 1/20 of its surface exposed as a feeding area for the mites. Thus the available feeding area was equivalent to six oranges. Huffaker added small wooden pegs within the trays to speed the dispersal of the six-spotted mites. The wooden pegs provided dispersal launching points for six spotted mites that float to new areas on strands of silk. He established a maze of Vaseline barriers across the trays to slow the dispersal of the predatory mites, which could travel only by foot. Huffaker then stocked the trays with 120 six-spotted mites, placing one mite on each of the 120 oranges, and added 27 female predatory mites five days later. He placed one predator on each of 27 oranges across the major sections of the trays.

In this experiment and the one using the 20 small areas of food, the prey and predator found plenty of available food for population growth. Density of the predators increased as the prey population increased. In the experiment in which the food was concentrated and dispersion of the prey population was minimal, predators readily found the prey, quickly responded to changes in prey density, and destroyed the prey rapidly. In fact, that situation was self-annihilative. In the experiment in which the primary food supply and the prey were dispersed, predator and prey went through three oscillations before the system died out (Figure 23.7).

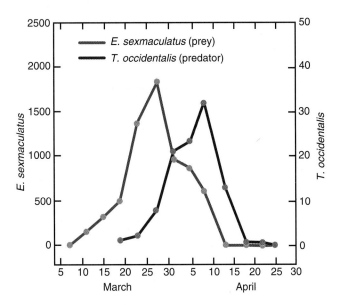

**Figure 23.6** Densities per orange area of the prey, *Eotetranychus sexmaculatus,* and the predator, *Typhlodromus occidentalis,* with 20 small areas of food for the prey alternating with 20 foodless positions. (After Huffaker 1958:365.)

Several important conclusions resulted from the study. First, predators cannot survive when the prey population is low for a prolonged period relative to the longevity of the predators. Second, a self-sustaining predator–prey relationship cannot be maintained without immigration of prey. Third, the complexity of prey dispersal and predator-searching relationships, combined with a period of time for the prey to recover from the effects of predation and to repopulate the area, had more influence on the period of oscillation than did the intensity of predation. Thus Huffaker's experiments emphasized that the Lotka-Volterra model says nothing about the need for heterogeneity of the environment to maintain a predator–prey system. Only by introducing heterogeneity was Huffaker able to obtain empirical results that even approached the Lotka-Volterra model.

The degree of dispersion and the area employed were too restricted in Huffaker's experiment to perpetuate the system. Pimentel, Nagel, and Madden (1963) attempted to provide an environment with a space-time structure that would allow the existence of a parasite-host system. They chose as subjects a parasitic wasp (*Niasonia vitripennis*) and a host fly (*Musca domestica*) and provided for the environment a special population cage, a group of interconnected cells. A predator-prey system living in 16 cells died out, but a 30-cell system persisted for over a year. Increasing the system from 16 to 30 cells decreased the average density of parasites and hosts per cell and increased the chances for survival of the system. The lower density was due to the breakup and sparseness of both parasite and host populations. The greater number of individual colonies that remained following a severe decline of the host assured survival of the system, because these colonies provided a source of immigrants to repopulate the environment. Moreover, amplitude of the fluctuations of the host did not increase with time, as proposed by the Nicholson-Bailey model. Apparently the fluctuations were limited by intraspecific competition.

These laboratory experiments support studies made in the field. Sometime around 1839 prickly pear cactus (*Opuntia*) was introduced from America into Australia as an ornamental. As is often the case with introduced plants and animals, the cactus escaped from cultivation and rapidly spread to cover 60 million acres in Queensland and New South Wales. To combat the cacti, a South American cactus-feeding moth (*Cactoblastis cactorum*) was liberated. The moth multiplied, spread, and destroyed the cacti until plants existed only in small, sparse, widely distributed colonies.

The decline of the prickly pear also meant decline of the moth. Most of the caterpillars coming from moths that had bred on prickly pear in the previous generation died of starvation. In areas where only a few moths survived, not many plants were parasitized. As prickly pear increased, so did the moth, until the cactus colony was again destroyed. In areas where no moths survived the colony spread once more, but sooner or later it was found by moths from other areas and

**Figure 23.7** Three oscillations in predator and prey populations when the predatory mite *Typhlodromus occidentalis* preyed upon the orange-feeding six-spotted mite, *Eotetranychus sexmaculatus,* in a complicated environment. (After Huffaker 1958:370.)

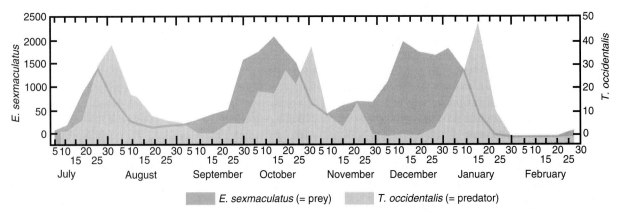

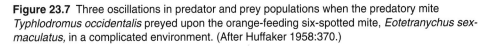

destroyed. However, seed scattered into new areas established new colonies that maintained the existence of the species and thereby maintained the predator-prey system.

The rate of establishment of prickly pear colonies is determined by the time available for the colonies to grow before they are found by moths. As a result an unsteady equilibrium exists between cactus and moth. Any increase in the distribution and abundance of the cactus leads to an increase in the number of moths and subsequent decline in the cactus. The maintenance of this predator-prey, or more accurately herbivore-plant, system depends upon environmental discontinuity. The relative inaccessibility of host or prey in time and space limits the number of parasites and predators.

In further investigations of the moth-cactus relationship, J. Monro (1967) found that the moth may conserve food for succeeding generations of moths by limiting its own numbers. At high densities the moth clumps its egg sticks rather than laying them randomly on prickly pear. The clustering overloads certain plants of prickly pear with eggs, resulting in the destruction of the plants. In dense stands of prickly pear clustering initially has little influence on larval survival, for as an overloaded plant collapses, it falls on its neighbor and larvae can move to a new source of food. Later, as dense stands become broken up into isolated plants, the sedentary larvae are unable to cross the wide gaps and die of starvation. As mean density increases, the proportion of eggs wasted by clumping increases.

However, because the eggs are clustered rather than widely distributed, more plants escape infestation altogether or are subject to a lighter infestation than would be expected if eggs were laid at random. Monro found that this mechanism, which is employed most in the center of the range of the moths, acts to conserve the food supply for succeeding generations and to maintain a constant level of both the food resource and the moth population.

These examples to some extent illustrate predatory-prey interactions, both at the plant-herbivore and at the herbivore-carnivore level. Although for simplicity they are considered separately, predator-prey interactions at one trophic level influence predator-prey interactions at the next trophic level. Interactions at two or more trophic levels are often involved in predator-prey stability.

# FUNCTIONAL RESPONSE

The Lotka-Volterra equations of predation hint at two distinct responses of predators to changes in prey density. As prey density increases, each predator may take more prey or take them sooner, a **functional response;** or predators may become more numerous through increased reproduction or immigration, a **numerical response.**

The idea of a functional response was introduced by Solomon (1949) and explored in detail by Holling (1959, 1961, 1966). Holling recognized three types of functional response (Figure 23.8): Type I, in which the number of prey eaten per predator increases linearly to a maximum as prey density increases (the Lotka-Volterra assumption); Type II, in which the number of prey eaten increases at a decreasing rate toward a maximum value; and Type III, in which the number of prey taken is low at first and then increases in a sigmoid fashion, approaching an asymptote.

## Type I Response

Type I response is a specialized one, the sort assumed in the simple predation models and best demonstrated in the laboratory. Predators of any given abundance capture food at a rate proportional to their encounter with prey items up to the point of satiation. There is density-independent mortality of the prey up to that point. In northern Finland, where *Microtus* vole populations are cyclic, the European kestrel (*Falco tinnunculus*), short-eared owl (*Asio flammeus*), and long-eared owl (*Asio otus*) exhibit a linear Type I functional response (Figure 23.9 ) (Korpimaki and Norrdahl 1991), except that the horizontal saturation level is never reached. The three raptors take voles in proportion to their availability. As the density of the voles increases, the kill rate increases but does not level off. This failure to level off suggest that the populations of the voles do not become so abundant that the predators become satiated.

## Type II Response

The Type II response is generally but not exclusively associated with invertebrate predators. It is described by the disk equation, named for an element in the experiment from which it was derived. In Holling's experiment, the predator was represented by a blindfolded person and the prey by sandpaper disks 4 cm in diameter thumbtacked in different densities to a 1 m square table. The "predator" tapped the table with a finger until a prey was found and then removed the disk. The "predator" continued to search and encounter (tapping, discovery, and removal) for one minute. Holling found that the number of disks the predator could pick up increased at a progressively decreasing rate as the density of disks increased. Predator efficiency rose rapidly as the density of the disks increased, up to the point where the predator spent so much time picking up and laying aside disks that the predator could handle only a maximum number at a time.

Holling repeated this experiment using insect predators such as the praying mantis and insect prey such as flies. He found the same relationship between prey density and handling time as he did with the disk-handling experiment. Handling time increased as the density of the prey increased. Although search time was short, the predator could handle only a limited number of prey in a given time. These experiments demonstrated several important components of predation: density of prey, attack rate of the predator, and han-

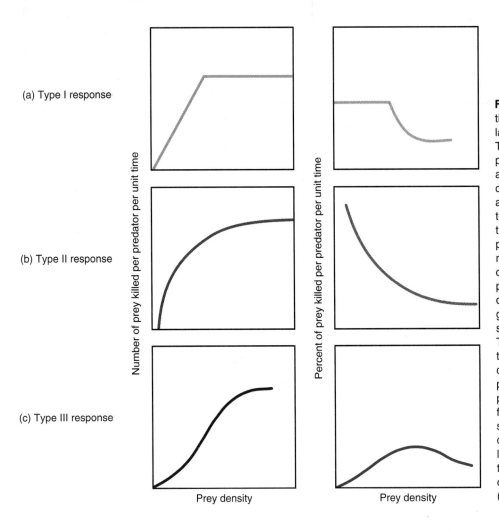

Number of prey killed per predator per unit time

Percent of prey killed per predator per unit time

(a) Type I response

(b) Type II response

(c) Type III response

Prey density

Prey density

**Figure 23.8** Three types of functional response curves, which relate predation to prey density. (a) Type I. The number of prey taken per predator increases linearly to a maximum as prey density increases. Graphed as a percentage, predation declines relative to the growth of the prey population. (b) Type II. The number of prey taken rises at a decreasing rate to a maximum level. When considered as a percentage of prey taken, the rate of predation declines as the prey population grows. Type II predation cannot stabilize a prey population. (c) Type III. The number of prey taken is low at first, then increases in a sigmoid fashion approaching an asymptote. When plotted as a percentage, the functional response still retains some of the sigmoid features, but declines slowly as the prey population increases. Type III functional response has the potential of stabilizing prey populations. (After Holling 1959.)

dling time, including time spent pursuing, subduing, eating, and digesting prey.

Type II functional response is described by the disk equation

$$\frac{N_a}{P} = \frac{aNT}{1 + aT_hN}$$

where $N_a$ is the number of prey or hosts killed or attacked, $P$ is the number of predators or parasitoids, $N$ is the number of prey, and $N_a/P$ is number of prey eaten per predator; $a$ is a constant representing the attack rate of the predator or the rate of successful search, $T$ is total time predator and prey are exposed, $T_h$ is handling time, and $T_s$ is time spent by predator in search of prey, $T$ is determined by the equation

$$T = T_s + T_hN_a$$

The Type II functional response curve derived by Holling is identical to the Michaelis-Menten equation of enzyme kinetics:

$$b(N) = m(N/(w + N))$$

**Figure 23.9** Type I linear functional response of a pair of European kestrels to *Microtus* vole densities during the breeding season. The curve did not reach a horizontal level, presumably because the vole population was not high enough for the predator to reach saturation level. (After Korpimaki and Norrdahl 1991.)

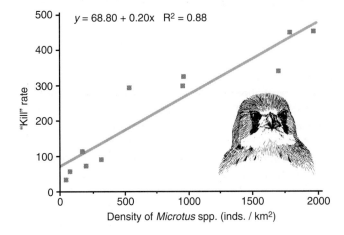

$y = 68.80 + 0.20x$   $R^2 = 0.88$

"Kill" rate

Density of *Microtus* spp. (inds. / km²)

where $m$ is the maximum predator attack rate, $w$ is prey density at which the attack rate is half saturated, and $N$ is prey density (Real 1977). This equation describing enzymes acting on a substrate is analogous to predators feeding on a prey. Although the Holling disk equation describes the Type II functional response, it is difficult to apply in Type III functional responses. The Michaelis-Menten equation is not. We will point out the use of this equation later.

Because handling time is the dominant component, rise in the number of prey taken per unit time decelerates to a plateau (Figure 23.10) while the number of prey is still increasing. For this reason Type II functional response cannot act as a stabilizing force on a prey population unless the prey occurs in patches. Thus Type II response is destabilizing (see Murdock and Oaten 1975).

## Type III Response

Type III functional response is more complex than Type II. It has been associated with vertebrates that can learn to concentrate on a prey when it becomes more abundant, but studies by Hassell et al. (1977) show that Type III is found among invertebrate predators as well. Because some vertebrate predators, especially feeding specialists, may also show Type II response, it is much wiser not to attempt to assign types either to invertebrate or vertebrate predators.

In Type III response the number of prey taken per predator increases with increasing density of prey and then levels off to a plateau where the ratio of prey taken to prey available declines (Figure 23.10). Because the amount of prey taken is density-dependent, Type III functional response is potentially stabilizing.

## Threshold of Security

Type II responses occur in situations of varying densities of one prey species. Type III responses invariably involve two or more prey species; the predator has a choice of prey. In the presence of several prey species, the predator may distribute its attacks among the prey in response to the relative density of the prey species. Predators may take most or all of the individuals of a prey species that are in excess of a certain minimum number, determined, perhaps, by the availability of prey cover and the prey's social behavior. Errington (1946), drawing on his studies of predation in muskrat and bobwhite quail populations, called this minimal prey population level, at which the predator no longer finds it profitable to hunt the prey species, its **threshold of security.** Type III responses have been called compensatory because as prey numbers increase above the threshold, the "doomed surplus" becomes vulnerable to predation through intraspecific competition. Below the threshold of security the prey species compensates for its losses through increased litter size and greater survival of young.

Although intuitively the concept of the threshold of security appears to be sound, it has never been rigorously tested in the field. Based on intrinsic density-dependent population regulation through intraspecific competition, the threshold of security concept minimizes the role of predators as an important extrinsic force in population regulation. This concept has been uncritically accepted as dogma by some wildlife biologists. Upon it they have based their beliefs (1) that in good habitats wildlife populations cannot be overhunted in normal seasons and (2) that hunting take is compensatory and not additive to other population losses. What hunters remove is supposedly the "doomed surplus." Recently there is increasing evidence that this concept does not hold true and that even in good habitats some wildlife species can be reduced below the presumed threshold of security.

## Aggregative Response

Although the functional response equations assume predators searching at random in a uniform prey population, the usual situation in nature is an uneven distribution of prey. This setup, in turn, may attract a number of predators to an area. One or two members of the predator species discover and begin to feed on the prey item; other members of the species observe the feeding response and follow suit (see Curio 1976). In this **aggregative response,** predators tend to congregate in patches of high prey density.

Intermediate levels of aggregation may increase the efficiency of predation, because a number of predators foraging together can locate areas of prey abundance more quickly than a few. At high levels of aggregation in areas of high prey density, interference among predators may be so great that the efficiency of predation declines. Encountering an individual of its own species, a predator may temporarily cease hunting or leave the area. Interference among predators reduces the proportion of total prey or hosts the predator or parasite encounters, because the predator's search time is reduced (Hassell et al. 1976).

Aggregative responses of predators to areas of high prey density may have a pronounced influence on stability of predator-prey interactions. Hassell and May (1974) presented an idealized general aggregative response curve for predator-to-prey distribution (Figure 23.11). The response curve exhibits a lower plateau of low prey density and an upper plateau of high prey density where predators do not distinguish between prey areas. The model predicts that predators discriminate markedly in areas of intermediate prey density and tend to congregate in areas of higher density. An example of this type of distribution is the response curve for the redshank (*Tringa totanus*), a shorebird that tends to concentrate in areas of its preferred food, the amphipod crustacean *Corophium volutator* (Figure 23.12). On the other hand, predators tend to avoid areas of low prey density, making those individuals much less vulnerable. Low density areas provide the prey species with partial refuges against predation.

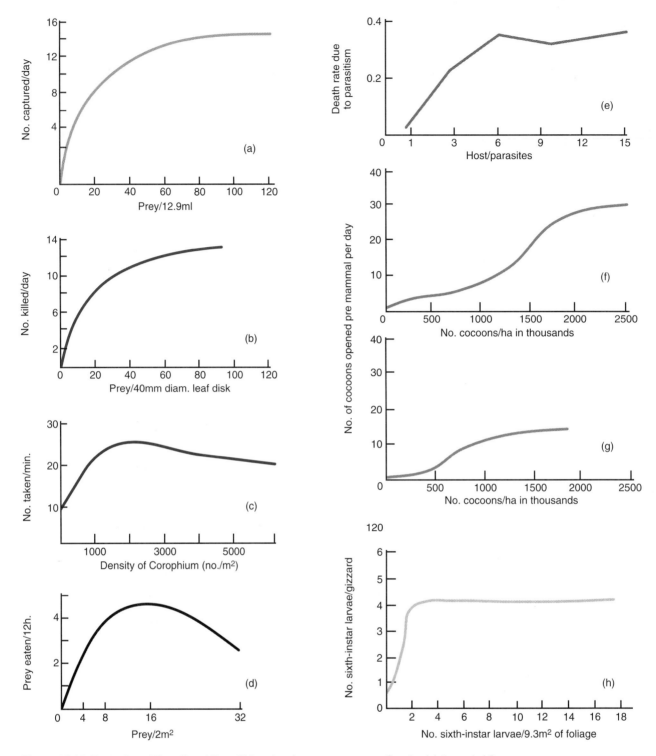

**Figure 23.10** Examples of Type II and Type III functional response curves. Graphs (a) through (d) are Type II curves. (a) First instar of the spider *Linyphia triangulaus* feeding on *Drosophila*. (b) Second instar of the coccinellid *Harmonia axyridis* feeding on aphids. (c) Adult female mites of *Phytoseiulus persimilis* feeding on nymphs of *Tetranychus urticae*. (d) Numbers of amphipod crustaceans *Corophium* taken per minute by a shorebird, the redshank. The decline of prey eaten in at high densities in (c) and (d) probably results from interference among predators. Female parasitoids discover many hosts are already parasitized, or predators are discouraged from feeding in areas where a large number of individuals are already congregated and leave the area.

Type III functional response curves are graphed in (e) through (h). (e) *Encarsia formosa* parasitizing *Trialeurodes vaporariorum*. (f) The shrew *Sorex* feeding on sawfly larvae. (g) *Peromyscus* (deer mice) preying on sawfly larvae. (h) Bay-breasted warbler (*Dendroica castanea*) feeding on spruce budworm (*Choristoneura fumiferana*) larvae. (a, b, c, e, f from Hassell et al. 1976 and Beddington et. al. 1976; d from Goss-Custard 1977b; g, h from Holling 1964.)

## Switching

Involved in the Type III response curve is the role of the facultative predator and alternate prey. Although the predator may have a strong preference for a certain prey, it can turn to an alternate, more abundant prey species that provides more profitable hunting. If rodents, for example, are more abundant than rabbits and quail, foxes and hawks will concentrate on the rodents. This idea was advanced early by Aldo Leopold in *Game Management* (1933), in which he described alternate prey species as buffer species because they stood between the predator on one hand and game species on the other. If the population of buffer prey is low, the predators turn to the game species; the foxes and hawks will concentrate on the rabbits and quail. This turning by a predator to an alternate, more abundant prey such as mice was later termed **switching** by Murdoch (1969). In switching the individual predator concentrates a disproportionate amount of attacks on the more abundant species and pays little attention to the rarer species (Figure 23.13). As the relative abundance of the two prey species changes, the predator changes its diet.

Switching, according to Murdoch and Oaten (1975), is caused by three behaviors of the predator: (1) changing its preference toward the more abundant prey as it eats it more frequently by choice; (2) ignoring rare prey; or (3) concentrating search in more rewarding areas. Any one of these three behaviors results in Type III response curves.

## Search Image

The reason for the sigmoidal shape of Type III response is the subject of much study and debate (see Royama 1970, Croze 1970, Murdock and Oaten 1975, Curio 1976). One explanation was advanced by L. Tinbergen (1960), based on

**Figure 23.11** Model of general aggregative response. At the lower plateau of prey density predators do not distinguish among low (unprofitable) prey areas; at intermediate densities (shaded area) predators discriminate markedly; at the upper plateau predators do not discriminate among high density (profitable) areas. (Hassel and May 1974.)

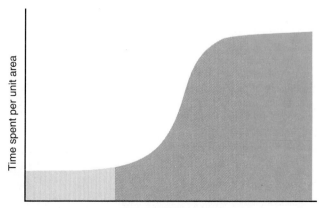

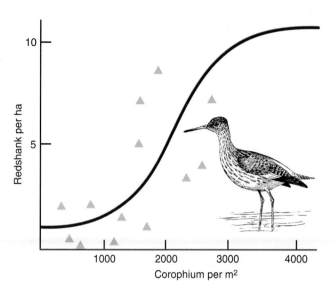

**Figure 23.12** Aggregative response in the redshank. The curve plots the density of the predator (the redshank) against the average density of arthropod prey (*Corophium*). Compare this curve with the profitability curve in Figure 19.5. (After Hassel and May 1974.)

his studies of the relation between woodland birds and insect abundance. According to Tinbergen's hypothesis, when a new prey species appears in a given area, its risk of becoming prey is low. The birds have not yet acquired a **search image** for the species. A search image is a perceptual change in the ability of a predator to detect a familiar cryptic prey. Once the predator has secured a palatable item of prey, the predator finds it progressively easier to find others of the same kind. The more adept the predator becomes at securing a particular prey item, the longer and more intensely it concentrates on the item. In time the numbers of the prey species become so reduced or its population so dispersed that encounters between it and the predator lessen. The search image for that species begins to wane, and the predator begins to react to another species. The combination of increasing density of prey and establishment of a search image results in a sudden increase of the perceived prey species in the predator's diet, giving a sigmoid functional response curve.

Studies have shown that predator can acquire a search image from remarkably few experiences (Croze 1970, Dawkins 1971, Curio 1976). In losing an image the predator may simply not respond to the perceived stimulus or may in fact no longer perceive it, that is, no longer distinguish the properties of the prey from the background. The search image is maintained by rewards in the form of the acquisition of food. When rewards are no longer there, the bird turns to another image. In effect, the predator responds to changes in rewards. The extinction of an image tends to occur more slowly than its acquisition, and among some predators the search image may be retained for some time, even in the absence of rewards. Croze (1970) found that carrion crows

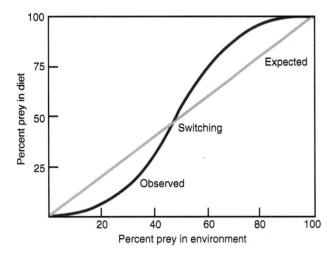

**Figure 23.13** A model of switching. The straight line represents a situation in which the theoretical amount of prey is consumed at a rate proportional to its availability in the environment. It represents a constant preference with no switch. The curved line represents the proportion of prey species actually taken. At low densities the proportion taken is less than expected. Switching occurs at the point where the curved line crosses the straight.

(*Corvis corone*) retained their search image for eight days without reward. After that time the search image declined rapidly, but was still retained.

## NUMERICAL RESPONSE

In addition to functional responses, predators may exhibit numerical responses. One type is the immigration or emigration of predators in response to changes in prey density. Such movements, however, can be considered a true numerical response only when predators move into an area from some distance. For example, in their long-term study of the European kestrel, short-eared owl, and long-eared owl inhabiting a 47 km² farmland in western Finland, Korpimaki and Norrdahl (1991) found that the nesting pairs of these raptors fluctuated in close accordance with the spring density of *Microtus* voles (Figure 23.14). The raptors were able to track the vole populations without a time lag because the birds were mobile and the vole cycles were asynchronous in the region. An increase in the vole population stimulated a rapid immigration of raptors to the area. There was just as rapid an emigration when the vole population declined.

A second type of numerical response is an increase or decrease in the rate of predator natality and mortality in response to changing prey density. This type is influenced by the kind of predator involved: parasitoid or true predator. For true predators that require several prey to complete their development, numerical response depends upon the rate at which predators are able to locate and consume suitable prey (Lawton et al. 1975).

Survival rates of arthropod instars and the young of nonarthropod predators are directly dependent on the density, size, and availability of prey. A scarcity of prey means a lack of food. A lack of food results in poor survival of young or instars. A poor survival rate has a direct bearing on numerical response of a predator population.

Some parasitoids feed only on one host. Among them adult fecundity is limited by the number of hosts the female can find. With them the relationship between prey density and fecundity is linear (Hassel and May 1973).

In all other situations nutrition, controlled by the amount of prey eaten during the adult stage, affects fecundity. Predators can use the energy remaining after meeting maintenance demands for reproduction. If predators face limited food and therefore have low energy, they experience low fecundity and thus a low numerical response. With increasing prey density, the predator's fecundity increases and with it the numerical response. For example, the population of the European kestrel in western Finland increased from 2 during a period of low density of voles to 46 at a high density. The mean number of young produced ranged from 4.6 at high vole densities to 0 at low densities. Fecundity of the kestrel, however, made only a minimal contribution to its numerical response.

Numerical response, positive or negative, is not immediate in most cases, especially in situations where it depends upon increased fecundity. There is necessarily a time lag between adequate nutritional intake, development and birth of young, and their maturation to reproducing individuals.

The "fugitive" Cape May and bay-breasted warblers of the northeastern North American forests provide examples of delayed numerical response. Outbreaks of spruce budworm apparently dictate the abundances of those birds (MacArthur 1958, Morris et al. 1958, Mook 1963). During outbreaks the two species increase more rapidly than other warblers because of extra-large clutches. During years between outbreaks they decline and even become extinct locally.

In general, then, numerical response takes three basic forms (Figure 23.15): (1) direct or positive response, in which the number of predators per unit area increases as the prey density increases; (2) no response, in which the predator population remains proportionately the same; (3) inverse or negative response, in which the predator population declines in relation to prey population (Hassell 1966).

## FORAGING THEORY

The Type III response curve may reflect the acquisition of a search image in some predators, but the same response curve could result from the manner in which predators must allocate energy to secure prey (Royama 1970). It is energetically unprofitable for predators to spend time where prey density is low. Predators must discover the most productive

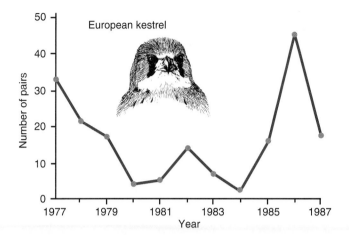

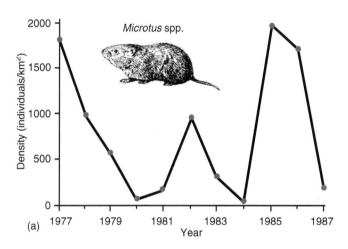

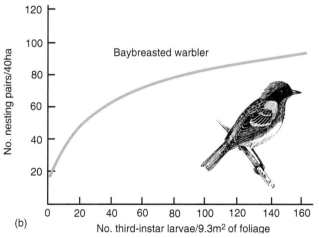

**Figure 23.14** (a) Numerical response of nesting pairs of European kestrels to changes in density of *Microtus* voles at Alajaki, western Finland, during 1977–1978. (Adapted from Korpimaki and Norrdahl 1991.) (b) Numerical response of nesting bay-breasted warblers to density of spruce budworm. (After Mook 1963.)

way to allocate their hunting time among different prey species of different abundances in different patches. Profitability is measured not by prey density, but by the amount of prey (preferably measured in terms of biomass) that a predator can harvest in a given time. This profitability of hunting is sufficient to produce Type III curves.

The profitability of hunting by a predator relates to the manner in which its prey is distributed. If prey were distributed in a fine-grained manner (see Chapter 17), the predator could pick and choose with a search image. Prey, however, is distributed in a coarse-grained manner in patches across the landscape. These patches vary in size and in the quality and quantity of resource, so the predator must be able to locate profitable patches. This fact gave rise to the concept of the optimal use of patchy environments advanced by MacArthur and Pianka (1966), which later evolved into optimal foraging theory. This theory forms a basis against which actual foraging strategies can be compared.

A foraging animal wants to obtain the most energy from food intake relative to the energy expended in securing and eating the food. The difference is net energy gain. Its **optimal foraging strategy** provides a maximum net rate of energy gain, endowing the animal with the greatest fitness. It involves two separate but related components. One is optimal diet; the other is foraging efficiency. Theoretical ecolo-

**Figure 23.15** Basic forms of numerical response. No response means that the number of predators remains the same in the face of increasing prey density. A direct response implies that predators increase in response to an increasing prey density. Inverse response means that the number of predators per unit area declines as prey density increases. (From Hassell 1966.)

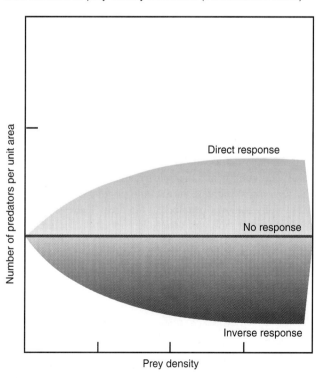

gists have come up with certain rules (hypotheses) for optimal foraging.

## Optimal Diet

Suppose you scatter black oilseed (a medium-sized sunflower seed) and white millet (a small seed) for winter birds. According to studies of preferred food of winter birds, black oilseeds are preferred by cardinals, house finches, and nuthatches, who extract the meat. The preferred food of mourning doves supposedly is millet, a close relative of its natural food. You soon discover that the doves choose oilseed over the millet. They will not crack the seeds as finches do—their thin pointed bills are not adequate to the task—but will eat them whole. Only after the day's allotment of oilseed is consumed will the mourning doves turn to their supposedly preferred millet.

Obviously, the mourning doves found it energetically more profitable to take large oilseeds, rich in carbohydrates, over small millet seeds. Because of oilseed's larger size, the doves could acquire more energy with less handling time, which meant remaining for a much shorter period in the food patch. When feeding on millet, the doves had to handle many more seeds providing much smaller packets of energy per unit effort. In effect, the doves made an optimal economic decision. They chose larger, more profitable seeds over smaller ones.

According to the "decision rules," a consumer should: (1) prefer the most profitable prey (items that yield the greatest net energy gain); (2) feed more selectively when profitable prey or food items are abundant; (3) include less profitable items in the diet when the most profitable foods are scarce; and (4) ignore unprofitable items, however common, when profitable prey are abundant. The mourning doves made all the "right" decisions, but they were operating under ideal conditions unwittingly provided: an abundance of food with a choice of only two items. Natural conditions in which they had to locate food patches that provided much smaller and more diverse of prey items might have produced a different outcome.

Although the theory of optimal diet makes practical sense, it is difficult to test under field conditions. Not only would we have to know exactly what items the animals were consuming; we would also have to know the relative availability of all potential food items in the habitat, as well as the profitability index of each. Even then we would not necessarily know enough. Consumers might make their decisions on criteria other than relative availability (for a short discussion of this point see Taylor 1984:93–94). Some studies have been done under controlled conditions involving birds (great tit, J. R. Krebs et al. 1978), fish (brown trout *Salmo trutta*, Ringler 1979) and invertebrates (shore crabs *Carcinus maenas*, Elner and Hughes 1978).

Werner and Hall (1974) presented groups of ten bluegill sunfish with three sizes of *Daphnia* in a large aquarium. They allowed the fish to forage for a period of time, then killed them and examined their stomachs to determine the number and size of *Daphnia* taken. When the density of the prey presented was low, the fish consumed the three sizes according to the frequency encountered. They showed no preference for any size category (Figure 23.16). When the prey population was dense, the fish consumed the largest prey items. When presented with an intermediate number, the fish took the two largest size classes. The results of these feeding trials support the optimal foraging theory.

Feeding trials involving the shore crab and its prey, the blue mussel (*Mytilus edulis*), gave much the same results (Elner and Hughes 1978). The crab's diet extended to smaller mussels as the preferred size became scarce. But at no time did the crabs exclude small mussels completely from their diet, even when larger mussels were more than abundant enough to fill them. The crabs did not pass up good food when encountered, even through it did come in smaller packages. That is probably the way animals in the wild respond.

Two field studies provide some insight into the way animals forage under natural conditions. Goss-Custard (1977a, 1977b) studied food selection by redshanks (*Tringa totanus*) on mudflats containing different sizes of polychaete worms. He found that as the number of large worms increased, redshanks became more selective. They tended to ignore small worms, regardless of how common they were, as long as the density of large worms remained high. These field observations were supported by laboratory studies.

Davies (1977) studied the feeding behavior of the pied wagtail (*Montacilla alba*) and yellow wagtail (*M. flava*) in a pasture field near Oxford, England. The birds fed on various dung flies and beetles attracted to droppings. They had access to prey of several sizes: large, medium, and small flies and beetles. The wagtails showed a decided preference for medium-sized prey (Figure 23.17). The size of the prey corresponded to the optimum-sized prey the birds could handle profitably (Figure 23.18). The birds ignored small sizes. Although easy to handle, small prey did not return sufficient energy, and large sizes required too much time and effort to handle.

## Foraging Efficiency

Most animals live in a heterogeneous or patchy environment. In feeding, they have to concentrate on the most productive food patches. This fact has given rise to another set of decision rules in optimal foraging theory. The consumer should: (1) concentrate foraging activity in the most productive patches; (2) stay with those patches until their profitability falls to a level equal to the average for the foraging area as a whole (Figure 23.19); (3) leave the patch once it has been reduced to a level of average productivity; and (4) ignore patches of low productivity.

These rules are covered by the **marginal value theorem** (Charnov 1976, Parker and Stewart 1976), which gives the length of time a forager should profitably stay in a resource

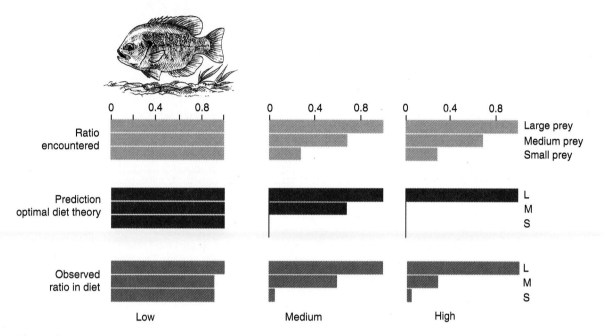

**Figure 23.16** Optimal choice of diet in the bluegill sunfish preying on different sizes of *Daphnia*. The histograms show the ratio of encounter rates with each size class at three different densities, the prediction of optimal ratios in the diet, and observed ratios in the diet. Note the bluegill's preference for large prey. (After Werner and Hall 1974:1048.)

patch before it seeks another. The length of stay relates to the richness of the food patch, the time required to get there, and the time required to extract the resource. When a forager arrives on a patch, it initially has a high rate of extraction and energy gain (Figure 23.19); but as time progresses, the abundance of the resource and the rate of extraction decline, until on the average it is no longer profitable for the forager to remain. Too long a stay depletes the resource. Conversely, if the forager leaves a patch too soon, it does not utilize the resource efficiently. Ideally the forager should leave for another patch at the point (indicated by the intersect of the straight line tangent to the curve in Figure 23.19b) where energy gains start to diminish. The model predicts that foragers should remain in a rich food patch longer than in a poor one, and that as travel time between patches increases, it should remain in the patch longer to balance energy loss in travel. Overall the forager should leave all patches, regardless of their profitability, when they have been reduced to the same marginal value that is average for the environment as a whole.

Whether animals go by these rules has also been the object of experimentation both in the laboratory and in the field. Hubbard and Cook (1978) studied the foraging behavior of a parasitoid, the ichneumon wasp *Memeritis canescens,* in a laboratory arena containing patches of the host, larvae of *Ephestia cantella.* Hosts in densities of 64, 32, 16, 8, and 4 were placed in petri dishes filled to the brim with hardened plaster of Paris, the hosts' substrate. Space between the larvae was filled with wheat bran. One result of

**Figure 23.17** Pied wagtails show a definite preference for medium-sized prey, which are taken in amounts disproportionate to sizes of prey available in the environment. (Davies 1977:48.)

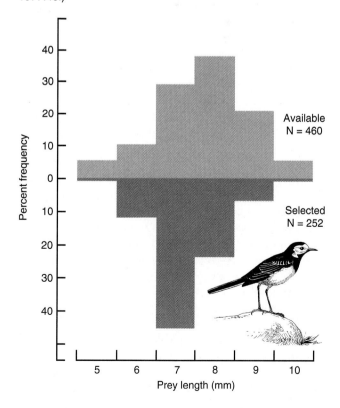

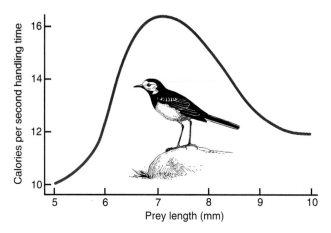

**Figure 23.18** Prey size chosen by pied wagtails is the optimal size for maximum energy per handling time. Small sizes provide too few calories. Large sizes require too much handling time. (Davies 1977:48.)

the experiment was that all patches were depleted of larvae to a common level of host abundance. The richest patches suffered the greatest depletion. As exploitation proceeded, the amount of time spent by the wasp in patches of highest density declined, and the proportion of time spent in the next-richest patches increased.

Zack and Falls (1976a, 1976b, 1976c) studied foraging strategy under more natural conditions. They exposed captive ovenbirds (*Seiurus aurocapillus*) individually to a patchy food supply (mealworms) presented in natural outdoor pens in typical habitat. The ovenbird is a ground-nesting, ground-foraging neotropical warbler of eastern deciduous and northern mixed deciduous-conifer forests. While foraging, it walks slowly and scans continuously for invertebrates in the leaf litter and low vegetation.

In one experiment Zack and Falls presented four fixed patch locations in which they interchanged prey densities. They found that the ovenbirds increased their search path exponentially with prey density. The birds rapidly shifted their search efforts as prey densities were interchanged. Because less search path was required per prey found in patches of dense prey, the birds concentrated their efforts in areas of high profitability and took a higher percentage of prey available on the sites. Ovenbirds did not always visit every patch location during the observation periods, but they always visited the high-density prey patches. This finding suggests that the birds' discovery of one or more profitable feeding areas discourages the sampling of other patches to assess their profitability. The ovenbird's tendency to quit searching after encountering one or more profitable patches and its ability to learn the location of and return to patches of high prey density may limit the number of patches the bird will exploit. The ovenbird may use other patches only if it discovers them by chance.

In another set of experiments Zack and Falls (1976b) exposed ovenbirds to various sets of patches of food. When

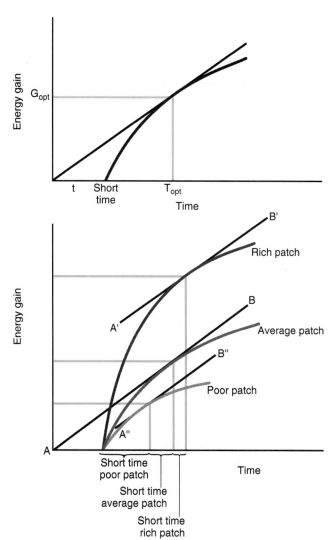

**Figure 23.19** How long should a predator remain in a habitat patch seeking food? This graph provides a theoretical answer. Time spent in travel and time spent in a habitat patch are plotted against energy gain, which declines as food is depleted. The curve represents the cumulative amount of food harvested relative to time in the patch. The straight line represents average food intake per unit time for the habitat as a whole. Where the line touches the curve, the predator has reached average cumulative net food gain for the habitat as a whole. Beyond this point net food gain declines below average. Thus the point represents the optimal time ($T_{opt}$) for the predator to seek a more profitable food patch. (After Krebs 1978:42.)

they exposed birds to a single patch of mealworms, the ovenbirds quickly concentrated their foraging there. The birds took equal amounts of food at different prey densities and did not vary the length of search path per visit. However, after Zack and Falls increased the prey density in the patches, ovenbirds decreased the number of visits to the patches. The birds also reduced the total search path per prey located and exploratory searches outside the high prey-density patches. That finding suggests that with low depletion

and high renewal rates, a single profitable feeding site might be enough.

In a third experiment Zack and Falls presented two prey patches successively. Prey in patch 1, presented on day 1, was not renewed on day 2. Instead a new patch with prey was presented. After an initial visit to patch 1, the birds quickly abandoned the first patch location, and after some exploratory behavior, rapidly concentrated their search efforts in the new location.

Zack and Falls (1976c) also investigated when ovenbirds would give up the search in one area and move on to the next. They discovered that the birds did not take some optimal number based on previous experience. Instead the birds learned rapidly to find patches of prey; they chose feeding sites nonrandomly and avoided areas of no food and patches visited previously. Ovenbirds improve their foraging efficiency by searching nonrandomly within the patches, avoiding areas already exploited. By doing so, ovenbirds were less likely to deplete their prey. If the birds followed a systematic search pattern, they gave up and left when a patch was completely covered. The time at which they gave up the search was unrelated to prey density.

Finally, Zack and Falls (1979) observed the foraging behavior of free-ranging male ovenbirds in a deciduous forest in Ontario with no control over the birds or their prey (Figure 23.20). They sampled prey density in the foraging areas of each bird and found the litter invertebrates to be patchily distributed. This study yielded three main results: (1) the areas used for foraging consistently had higher prey densities than areas not used for foraging; (2) the birds concentrated their search paths in areas of high prey density and returned to those areas; (3) search paths typically were directional; and (4) the birds foraged systematically and avoided inefficient random searching. However, the investigators could not determine whether the birds were foraging optimally because of the complexity of the situation and the lack of detailed information on prey types. Zack and Falls (1981) concluded that the results of their series of controlled laboratory experiments were of slight value

in evaluating wild birds' performance, because the laboratory situation was much simpler than the complex world of the forest.

These and other experimental studies, however, support the hypothesis of optimal foraging up to a point. It is not surprising that they should. Much optimal foraging theory concerns actions you would expect any mobile animal to take: forage in areas where food is abundant, leave when searching is no longer rewarding, select the larger and most palatable items of food, leave the poor items until last, and travel no farther than necessary to feed. Where the theory breaks down is in the expectation that animals will choose patches in the order of profits or take only optimal food items first and ignore the rest. Such choices may be characteristic of animals foraging in a stable laboratory environment. It is not necessarily the way animals behave in the wild.

This observation is recognized in the concept of **satisficing,** in which the decision-maker is satisfied after meeting some minimal requirement (Simon 1956, Stephens and Krebs 1986). An analogy is your seeking a good restaurant in a new town. If you "foraged" optimally, you would wait to dine until you had checked out all the restaurants in town and found what you considered to be the best one. Rarely, however, would you "forage" in that manner. Instead, you would probably look at several restaurants, but not all, and settle on one that is acceptable. Animals appear to do the same.

Being opportunists, animals will take some less than optimal food items upon discovery, and they may quit before food items are reduced to some minimal level. Nor will they pass up certain profitable patches because they do not meet some theoretical expectation. Animals quickly learn where food is and where food is not, and they do not waste much time on a patch after it is depleted. Foragers, however, will stay with a patch as long as the rate of replenishment exceeds the rate of depletion.

Some animals are highly restricted in their choices. Sedentary animals, such as corals, barnacles, and blackfly larva, filter feeders all, have to take what food flows past

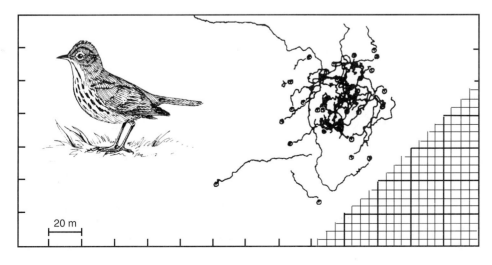

**Figure 23.20** Search paths of a free-ranging foraging male ovenbird (*Seiurus aurocapillus*) in a deciduous forest in Ontario. (After Zack and Falls 1979.)

20 m

them. Others have severely restricted foraging patterns that limit their choices of food.

## Risk-Sensitive Foraging

The marginal value theorem model assumes that the animal knows the quality of food patches and expects a constant reward on each visit. In reality the quality of the patches varies randomly over space and time. In this situation the animal has to decide whether to go back to a patch that gives it a constant rate of return or visit a new patch where the return is unknown. The choice is important if the animal is to avoid a shortfall in energy needs. Animal behaviorists call such decision-making **risk-sensitive foraging.**

The word "risk" here is borrowed from economics. Financial investors may have some knowledge about the probability of an event (such as a rise in interest rates or increased demand for a product). On the other hand, they may lack such knowledge and face a degree of uncertainty. This uncertainty or risk is what foragers face in their variable environment. The foragers must first learn the probability distribution of rewards derived from certain behaviors. Then they have to seek a strategy to exploit those distributions. That is where the risk comes in.

How animals make such decisions has been the subject of a number of behavioral experiments in the laboratory (Real and Caraco 1986). For example, Caraco and associates (1980) determined the daily energy requirements of a captive flock of yellow-eyed juncos (*Junco haeonotuis*). They provided food (millet) at two feeding stations separated by a partition in their aviary cage. The experimenters could manipulate the energy budgets of the birds by depriving the birds of food prior to any trials. In a given experiment one feeding station always offered a constant reward (risk-averse or low risk). The other feeding station offered an unpredictable reward—no seeds half the time; some seeds the other half of the time (risk-prone or constant risk). Thus the birds faced choices between a constant number of seeds and a random number of seeds; but always the mean of the variable reward equaled the mean of the constant reward.

When deprived of food for one hour in experimental tests and still in a positive energy balance, the juncos avoided risk by preferring the predictable site. When deprived of food for four hours, the birds switched their preference for the variable reward. They changed from being risk-averse to risk-prone. Under energy stress the variable site offered the possibility of providing 50 percent more food, whereas the constant site would not provide sufficient food to meet energy needs. Of course, there was the 50 percent risk of finding no food. Nevertheless, in the face of high energy demand, risk-prone behavior maximized daily survival.

Animals living in natural conditions face such choices each day. They may start out risk-prone and as time goes on become risk-averse. This behavior has given rise to the **expected energy budget rule:** be risk-prone if the daily energy budget is negative; be risk-averse if it is positive (Stephens 1981).

Animal behaviorists may demonstrate risk-sensitive foraging in the laboratory, but do foragers in the wild practice it? Wild foragers do have a number of choices denied captive experimental animals. A few behaviorists have undertaken experimental studies of risk sensitivity under natural conditions (for example, Wunderle and Cotto-Navarro 1988, Barkan 1990, Cartar 1991). Cartar (1991) investigated the risk-sensitive foraging of three species of colonial bumblebees in coastal southwest British Columbia—*Bombus melanopygus, B. mixus,* and *B. sithensis.* These bees feed on the nectar and pollen of their two most common food plants, seablush (*Plectritis congesta*) and dwarf huckleberry (*Vaccinium caespitosum*). Both plants offer foraging bumblebees equivalent expected rates of return, but the huckleberry is more variable. The bees store the nectar they collect in open-topped honey pots. The nectars they accumulate during the day they use during the night.

Cartar manipulated the energy requirements of the bees by depleting the honey pots or enhancing them with 50 percent sucrose solution. He then censused the foragers from depleted and enhanced colonies visiting the two flower species. He hypothesized that if the bees were risk-sensitive, they should increase their relative use of the more variable huckleberry when the colonies were depleted of energy than when they were enhanced. The bees did so. Because the huckleberry had the higher probability of greater nectar returns, the bees accepted the gamble. They made their foraging decisions based on the energy requirements of the colony relative to expected intake of energy. Their behavior suggested that the bees were sensitive to the mean and variance of the energy rewards offered by the two plants.

Another, wholly unrelated type of risk-sensitive foraging relates to **predation risk.** Habitat cover and foraging areas both vary in their foraging profitability and predation risk. In deciding where it will feed, the forager must balance its energy gains against the risk of being eaten. If predators are about, then it may be to the forager's advantage not to visit a most profitable but predator-prone area and to remain in a less profitable but more secure part of the habitat. Ecologists have done many studies on how the presence of predators affects foraging, mostly in aquatic invertebrates and fish (for a review see Lima and Dill 1990).

In the coniferous forests of central Finland flocks of willow tits (*Parus montanus*) and crested tits (*P. cristatus*) forage in spruce, pine, and birch trees in winter. Their major threat to survival is the pigmy owl (*Glaucidium passerinum*). The owl is a diurnal ambush or sit-and-wait hunter that pounces downward on its prey. Its major food is voles, and when vole populations are high, usually every three to five years, the predatory threat to these small birds declines. When vole populations are low, however, the small birds become the owl's major prey. During these lows the willow and crested tits forsake the outer branches and more open parts of the trees. They restrict their foraging to the dense

inner parts of spruce and to the tops of the more open pine and the leafless birches (Suhonen 1993).

Foraging theory is an area of active interest among theoretical behavioral ecologists. They are attempting to develop models incorporating animal foraging decisions, resource quantity and quality, and environmental constraints to explain and to predict foraging behavior. Because of an animal's own individuality in its decision-making, it is doubtful that foraging behavior can be reduced to sets of predictive mathematical equations. Foraging models, however, can provide valuable insights into ways in which animals utilize their environment. (For discussions see J. Krebs 1987, Kamil, Krebs, and Pullium 1986, Real and Caraco 1986.)

# SUMMARY

Predation is the consumption of one living organism by another, a relationship in which one organism benefits at the other's expense. In its broadest sense predation includes herbivory and parasitism.

Interactions between predator and prey have been described by the mathematical models of Lotka and Volterra, modified by others. Essentially all of these models predict oscillations of predator and prey populations. The oscillations may be stable, damped, or unstable. Relationships between predator and prey populations result in two distinct responses. As density of prey increases, predators may take more of the prey, a functional response, or predators may become more numerous, a numerical response.

There are three types of functional responses. In Type I the number of prey taken per predator increases linearly to a maximum as prey density increases. In Type II the number of prey taken rises at a decreasing rate toward a maximum. In Type III the number of prey taken is low at first, then increases rapidly to an asymptote, resulting in a sigmoidal pattern.

Both invertebrate and vertebrate predation may exhibit Type II and Type III response curves. Type II occurs in situations of varying densities of one species prey. Type III involves two or more species of prey. Inherent in Type III responses are a search image, in which the predator develops a facility for finding a particular prey item, and switching, in which the predator turns to an alternate, more abundant prey species for more profitable hunting. It takes that prey in a disproportionate amount relative to other prey species. Functional response relates predator attack rates to prey density. Numerical response refers to the increase of predators resulting from an increased food supply. Numerical response may involve an aggregative response, the influx of predators to a food-rich area, or more importantly, a change in the rate of growth of the predator population through changes in developmental time, survival rates, and fecundity. Such changes produce a delayed numerical response, for a time lag necessarily exists between birth of young and maturation of reproducing individuals.

Because prey occurs in patches, the predator finds it most efficient to spend time not necessarily where prey is most abundant, but where hunting is most profitable in terms of time allocated relative to net energy gained. Study of such behavior has given rise to the concept of optimal foraging, a strategy that obtains for the predator a maximum rate of net energy gain. There is a break-even point above which foraging in a particular patch is profitable and below which it is not. Optimal foraging involves an optimal diet, one that includes the most efficient size of prey for both handling and net energy return. Optimal foraging efficiency involves the concentration of activity in the most profitable patches of prey and the abandonment of those patches when they are reduced to the average profitability of the area as a whole. Based on the mean and variance of foraging rewards, decisions about patch choice are risk-sensitive. Depending upon their energy status, foragers choose between a constant source of reward or a variable one with the probability of much greater reward. When confronted with an energy shortfall, foragers seem to gamble on the latter.

# REVIEW QUESTIONS

1. What do the Lotka-Volterra and Nicholson-Bailey models of predation predict? What are some weaknesses of these models?
2. How does the Rosenweig-MacArthur model of predation differ from the Lotka-Volterra and Nicholson-Bailey models?
3. How does the ratio-dependent predation theory modify predation models?
4. What is functional response in predation? Distinguish between Type I, Type II, and Type III responses.
5. What is switching? How might search image relate to switching?
6. What is numerical response?
7. What is optimal foraging? How might risk-sensitive foraging relate to optimal foraging?
8. The models of predation and functional response are theoretical. How can we apply these models to the management of species populations, their food resources including vegetation, and pest management?

# CROSS-REFERENCES

Population growth, 392–396; logistic equation, 393–396; carrying capacity, 393; intraspecific competition, 412–417; exploitative and interference competition, 398, 480.

# Plant-Herbivore Systems

## Concepts

1. Herbivores are predators on plants.
2. Herbivory can reduce plant fitness, or it can stimulate biomass production.
3. Plants affect herbivore fitness through the quantity and quality of plant tissue and through defensive mechnisms.
4. Many herbivores have evolved ways of breaching plant defenses.
5. Plant production has a major effect on herbivore production.

Predator-prey systems involve three trophic levels: producers, herbivores, and carnivores. Interactions between plants and their herbivorous predators make up plant-herbivore systems. These systems have a direct influence on the next level systems, herbivores and their predators, the carnivores.

# PREDATION ON PLANTS

Predation on plants by herbivores includes both defoliation and consumption of fruits and seeds. The results of the two forms of predation are different.

Defoliation is the destruction of plant tissue (leaf, bark, stem, sap, and roots). Some plant predators, such as aphids, do not eat tissue directly, but, acting as parasites, tap plant juices without killing the plant. Other herbivores consume tissue directly, destroying part or all of the plant. If grazers eat seedlings, they kill the plant. If they remove only part of the plant, its survival depends upon the amount and continuation of grazing. Continued grazing may eventually kill the plant, but if grazing ceases, the plant may regenerate. Although grazed plants may persist and regenerate, defoliation still has an adverse effect. Grazing decreases plant biomass. Removal of leaves changes the plant's competitive position in the stand. Loss of foliage and subsequent death of some roots (root pruning) reduce the vigor of the plant, its competitive ability, and its fitness (Harper 1977).

The impact of seed predation is difficult to assess. If density-dependent processes let few individuals survive, seeds removed by predators represent that portion of the population that has no future. In such instances seed predation has no real impact. If predators remove seeds from an expanding population or from areas being colonized, predation reduces the rate of increase. On the other hand, if consumption of seeds is a mechanism for seed dispersal, as when seeds are contained in a palatable fruit and then carried in the gut of a fruit-eating herbivore, predation can be to the plant's advantage.

## Effects on Plant Fitness

Removal of plant tissue—leaves, flowers, bark, stems, roots, sap—affects a plant's fitness and its ability to survive, even though it may not be killed outright (see Dirzo 1984). Loss of foliage and loss of roots decrease plant biomass, reduce the vigor of the plant, place it at a competitive disadvantage to surrounding vegetation, and lower its reproductive effort or fitness (Figure 24.1). These effects are especially strong in the juvenile stage, when the plant is most vulnerable.

Although the plant may be able to compensate for the loss of leaves by increasing photosynthetic assimilation in the remaining leaves, it may be adversely affected by the loss of nutrients, depending on the age of the tissues removed. Young leaves are dependent structures, because

**Figure 24.1** Intense predation on oaks by gypsy moths. Such defoliation can kill weaker trees in the forest and reduce the growth of others. Increased light and nutrient input from droppings of caterpillars can increase understory growth.

they import and consume nutrients drawn from reserves in roots and other plant tissues. As the leaf matures, it becomes a net exporter, reaching its peak before senescence sets in. Grazing herbivores such as sawfly and gypsy moth larvae, deer, and rabbits concentrate on more palatable, more nutritious leaves. They tend to reject older leaves because they are less palatable, being high in lignin and other secondary compounds (tannin, for example). If grazers concentrate on young leaves, they remove considerable quantities of nutrients.

Plants respond to defoliation with a flush of new growth that drains nutrients from reserves that otherwise would have gone into growth and reproduction. Defoliation also draws on the plants' chemical defenses, a costly response. Often the withdrawal of nutrients and phenols from roots exposes them to attack by root fungi while the plant marshals its defenses in the canopy (Parker 1981). If defoliation of trees is complete, as often happens during an outbreak of gypsy moths or fall cankerworms (*Alsophila pometaria*), replacement growth differs from the primary canopy removed. The leaves are smaller, and the total canopy area may be reduced by as much as 30 to 60 percent (Heichel and Turner 1976). Defoliation in a subsequent year may cause an even further reduction in leaf size and number. Some trees may end up with only 29 to 40 percent of the original leaf area to produce food in a shortened growing season.

Severe defoliation and subsequent regrowth alter the tree physiologically. Removal of leaves changes hormonal growth regulators that control bud dormancy. The plant uses up reserve food to maintain living tissues until new leaves

are formed. Buds for the next year's growth are late in form-ing. The refoliated tree is out of phase with the season, and the drain on nutrient reserves adversely affects the tree over-winter because twigs and tissues are immature at the onset of cold weather. Such weakened trees are more vulnerable to attacks by insects and disease the next year. Also, because the plant used nutrient reserves for regrowth and mainte-nance, it has no resources available for reproduction.

Defoliation of coniferous trees results in their death. Conifers do not have the physiological traits that allow them to recover by forming new needles. That is why outbreaks of spruce budworm in spruces and adelgids in balsam fir devas-tate large areas dominated by these species.

Some plant predators, such as aphids, tap plant juices on new growth and young leaves rather than consume tissue di-rectly. Sap suckers can decrease growth rates and biomass of woody plants by 25 percent.

Damage to the cambium and growing tips (apical meris-tem) is more destructive in some plants (Figure 24.2). Deer, mice, rabbits, and bark-burrowing insects feed on those parts, often killing the plant or more likely changing its growth form.

Moderate grazing, even in a forest canopy, can have a stimulating effect, increasing biomass production, but at some cost to vigor and at the expense of nutrients stored in the roots. The degree of stimulation depends upon the nature of the plant, nutrient supply, and moisture. Although many biennial and perennial herbaceous species are adversely af-fected by herbivory, some species respond by producing more branch tissue than shoots.

For grasses, moderate grazing typically increases bio-mass, but severe grazing causes a decline in biomass pro-duction. Adverse effects are greatest when new growth is de-veloping. Defoliation then, as in deciduous trees, results in a loss of biomass, decreased growth, and delayed maturity (Andrzejewska and Gyllenberg 1980). Grasses, however, are well adapted to grazing, and may benefit from it. Because the meristem is close to the ground, older rather than more expensive young tissue is consumed first. Grazing stimu-lates production by removing older tissue functioning at a lower rate of photosynthesis. It reduces the rate of leaf ag-ing, prolonging active photosynthetic production, and it in-creases light intensity on underlying young leaves, among other things. Some grasses can maintain their fitness only under the pressure of grazing, even though defoliation re-duces sexual reproduction (McNaughton 1979, Owen 1980, Owen and Wiegert 1981).

Some ecologists, however, challenge the idea that her-bivory benefits grazed plants (Belsky 1986). Although abun-dant evidence exists that herbivory affects all aspects of plant growth, little evidence exists to show beneficial effects on fitness and population dynamics (Crawley 1989). Herbivory, especially grazing by large herbivores and out-breaks of foliage-consuming insects such as gypsy moth and spruce budworm, can eliminate or sharply reduce popula-tions of certain trees and shrubs and stimulate the growth and spread of previously nonselected species. These cases, however, are exceptional. Can herbivory on a less dramatic scale improve the survival and reproductive success of plants?

There are few experimental studies following the effects of herbivory on seed production and seedling survival. Paige and Whitman (1987) experimentally demonstrated the re-sponse of scarlet gilia (*Ipomopsis aggregata*), a tall showy biennial growing in extensive patches in Rocky Mountain meadows, to grazing by elk and mule deer. Fifty-six percent of the experimental population lost 95 percent of its above-ground biomass. The gilia overcompensated for this loss by producing multiple-flowering stems and up to three times as many flowers, fruits, and seeds as the nongrazed controls. The grazed and ungrazed populations did not differ in the number of seeds produced per fruit, seed weight, and germi-nating success and survival. The grazed plants, however, av-eraged a 2.4-fold increase in fitness over uneaten plants. These results suggest that for this plant there was a repro-ductive advantage to having parts removed by hebivores. A similar concept applies to the pinching back of horticultural plants such as chrysanthemums and certain annuals to stim-ulate flower production.

**Figure 24.2** Heavy browsing on low woody growth and herba-ceous plants by white-tailed deer prevents any plant from es-caping predation and achieving significant growth.

However, compensatory response of plants to grazing may be temporary. Doak (1991) investigated the impact of herbivory on a long-lived herbaceous species, dwarf fireweed (*Epilobium latifolium*), in Alaska. Often growing in extensive pure stands on highly disturbed sites, dwarf fireweed is an early successional species found in the arctic and alpine regions of the Northern Hemisphere. Although it spreads by small horizontal roots, dwarf fireweed is characterized by persistent pseudorhizomes that produce clumps of annual shoots in the same location for years. Clumps of shoots arise as suppressed branches from buds at or below ground level on the shoots of the previous year. Dwarf fireweed is attacked by a species-specific herbivore, a small leaf-rolling lepidopteran caterpillar (*Mompha albapalpella*). The damaged individual shoots respond by increasing branch growth on the stem. This branching increases biomass growth 80 to 500 percent over undamaged shoot growth, but branches rarely flower. Whereas damage to individual shoots appears to stimulate growth, a single year's damage to a clump of shoots reduces net growth by more than 75 percent over a three-year period. A year after the attack, the number of shoots is reduced by over 50 percent while mean shoot size is unaffected. Thus compensatory response to grazing observed in a single season may be misleading. We must consider effects over a longer time span.

## Effects on Herbivore Fitness

Herbivory is a two-way street. Plants, which support herbivores, have a pronounced effect on the fitness of herbivores. Herbivore biomass, consumption, and productions are correlated with primary production (Figure 24.3) (McNaughton et al. 1989). Thus plants may have more effect on herbivore fitness and population dynamics than herbivores have on plant populations (Crawley 1991). Although seemingly passive in the process, plants have a pronounced effect on the fitness of herbivores.

For herbivores it is not the quantity of food that is critical—usually there is enough plant biomass available—but the quality. Because of the complex digestive process needed to break down plant cellulose and convert plant tissue into animal flesh, high-quality forage rich in nitrogen is necessary. Without that, herbivores can starve to death on a full stomach. Low-quality foods are tough, woody, fibrous, and indigestible. High-quality foods are young, soft, and green, or are storage organs such as roots, tubers, and seeds. Most food is low-quality, and herbivores forced to live on such resources experience high mortality or reproductive failure (Sinclair 1977). Added to the problem of quality is the task of overcoming the various defenses of plants that make food unavailable, hardly digestible, unpalatable, or even toxic.

Secondary plant substances (discussed later) can affect the reproductive performance of some mammals. Isoflavonoids in plants—usually concentrated in legumes,

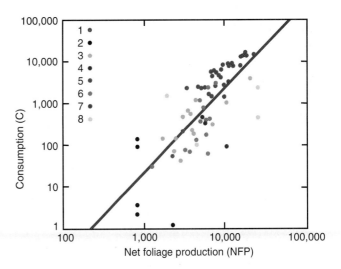

**Figure 24.3** Relationship between net foliage production (NFP) and consumption by herbivores (C) for various ecosystems: (1) desert; (2) tundra; (3) temperate grassland; (4) temperate successional old field; (5) unmanaged tropical grassland; (6) temperate forest; (7) tropical forest; (8) salt marsh. Units are kJm²/yr by type of ecosystem. (Adapted from McNaughton et al. 1989:143.)

particularly alfalfa and ladino clover—mimic estrogenic hormones, especially progesterone. When consumed, these isoflavonoids induce a hormonal imbalance in grazing herbivores that results in infertility, difficult labor, and reduced lactation.

Secondary compounds also serve as reproductive cues for some voles (Berger et al. 1981). A particular compound, 6-methoxybenzoxaxolinone (6-MBOA), found in grass, rapidly stimulates reproductive effort in montane voles (*Microtus montanus*). When voles feed on grass, they stimulate the injured plant tissue to release an enzyme that converts a precursor compound abundant in young growing tissue to 6-MBOA. The ingested chemical serves as a cue to the voles that the vegetative growing season has begun. Such a chemical cue allows the voles to produce offspring when food resources will be available to them. These chemical signals are important to the voles because they live in an environment where food resources are unpredictable and depend upon the timing of snowmelt and other environmental conditions. Yearly differences in the appearance of new vegetative growth and of 6-MBOA may influence population fluctuations (Negus, Berger, and Forslund 1977).

## PLANT DEFENSES

Because plants are fixed in place, they are at a distinct disadvantage in any predator-prey relationship. To counteract predatory attack, plants have arrived at some modes of de-

fense. These defenses range from chemical methods, widespread among plants, to mimicry and structural features.

## Mimicry

Mimicry is usually considered an evolutionary response of animals (see Chapter 25), but animals in search of food may have stimulated mimicry in the plant kingdom. L. E. Gilbert (1975) found evidence of plant mimicry in his study of the passionflower butterfly (*Heliconius*) (Figure 24.4) and its food plant, the passionflower (*Passiflora*). *Passiflora,* a vine of the New World tropics, comprising around 350 species, has a wide range of intraspecific and interspecific leaf shapes. The number of *Passiflora* species found in any one area is about 2 to 5 percent of the 350 species. Some 45 species of highly host-specific *Heliconius* butterflies use *Passiflora* species as an egg-laying site and as a source of larval food. Each species of *Heliconius* uses a limited group of plants. Visually sophisticated butterflies learn the position of the vines and return to them on repeated visits. Within a habitat, the leaf shapes of each passionflower species vary. Under visual selection by butterflies passionflowers apparently evolved leaf forms that make them more difficult to locate. Because the larval food niche is broader than that of the ovipositing females, there has been selective pressure for divergence among *Passiflora* species (Figure 24.5). Probably because of these selection pressures, *Passiflora* leaf shapes converge with those of associated tropical plants that *Heliconius* finds inedible. So close are the convergences that plant taxonomists have named some *Passiflora* species after the genera they resemble.

In addition, two *Passiflora* species, *P. cyanea* and *P. auriculata,* have evolved glandular outgrowths on the stipules (structures near the leaf bases) that mimic the size, shape, and golden color of *Heliconius* eggs at the point of hatching. Because *Heliconius* females reject shoots that carry eggs and young of other females, *Passiflora* achieves a measure of protection by egg mimicry (Williams and Gilbert 1981).

## Structural Defenses

Some of the least costly defenses available to plants are structures that make penetration by predators difficult, if not impossible. They include tough leaves, spines, and epidermal hairs on leaves, which may trap, impale, or fence out insects and discourage browsing by vertebrate herbivores. These structures may have evolved early in the history of the plants when they might have been subject to even greater predatory pressure. Because they represent little investment, plants still retain them. Many seeds have thick, hard seed coats that provide protection from seed-eating animals. The problem with such seed defense is that the seeds need to be scarified—the hard seed coat softened—so that the seedling itself can escape. If the seed is not scarified, the seedling is sealed in, never to germinate. Many plants, however, have turned seed predation into a mechanism for seed dispersal.

The role of structural defenses in plants is mostly presumed. Little experimental evidence exists to demonstrate the effectiveness of such apparent defensive structures against grazing herbivores. Cooper and Owen-Smith (1986) investigated experimentally the effects of plant spinescence on the feeding habits of three large browsing mammalian herbivores. These were the kudu (*Tragelapus strepsiceros*), a large African antelope attaining female weights of 180 kg.; the impala (*Aepyceros melampus*), a medium-sized African antelope attaining a female body weight of 50 kg; and the Boer goat, a domestic ungulate weighing about 35 kg. The experimenters hand-reared the antelopes from calves to allow close observation of feeding habits at close range. In the rearing pens the antelopes and goats were introduced to plant species from the study area, the Nylsvley Nature Reserve in the northern Transvaal bushveldt of South Africa. When the antelope were 6 months of age, the experimenters released them into a 213 ha enclosure where they ranged freely and secured food from natural vegetation. Cooper and Owen-Smith observed the feeding animals from distances of 1 to 5 m. They determined the biting rates by counting the number of sequential plucking actions the animals made while feeding on a particular plant. They converted the bites to dry biomass by collecting samples of leaves and shoots of a size similar to those eaten and drying them to a constant weight. They calculated eating rate as the product of bite size (dry mass) and biting rate.

Cooper and Owen-Smith carried out two additional experiments to examine further the influence of spinescence. They selected ten plants each of five species of trees at a height accessible to impalas outside the enclosure. On each tree two branches were matched for size, shape, density of

**Figure 24.4** *Heliconius vicina,* one of the *Heliconius* butterflies that use *Passiflora* species as egg-laying sites and a source of larval food.

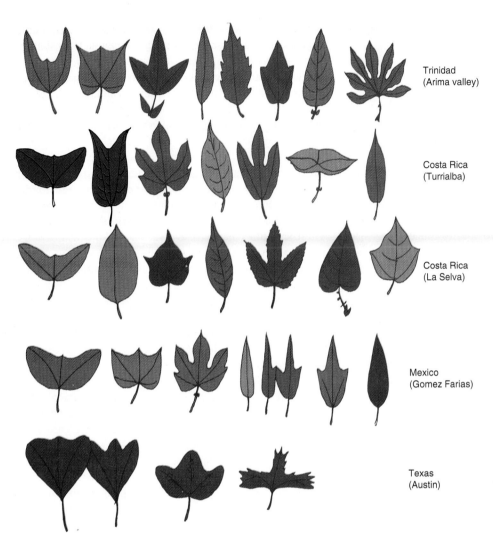

Trinidad
(Arima valley)

Costa Rica
(Turrialba)

Costa Rica
(La Selva)

Mexico
(Gomez Farias)

Texas
(Austin)

**Figure 24.5** Variation in leaf shape among groups of sympatric species of *Passiflora*. The leaf shapes tend to converge with those of other common species of a number of genera, inedible to *Heliconius* butterflies. (After Gilbert 1975.)

leaf cover, and ease of access to impala. They labeled the paired branches and removed the thorns from one of the pairs. Two months later, they visually estimated the relative loss of foliage from browsing. The other experiment involved feeding goats clipped and unclipped branches.

The woody plants exhibited three basic types of spinescence: (1) paired prickles or thorns situated in or close to the leaf axils; (2) short, sharp-tipped branchlets or spines, sometimes carrying small leaves; and (3) prickles of various kinds on leaves. Thorns were either straight and long, up to 70 mm, or short and sharply curved (hooked). Results clearly showed that thorns and spines affected the feeding behavior of the three ungulates. These structures restricted bite sizes to mostly single leaves or leaf clusters, and hooked thorns retarded biting rates. Acceptability of leaves of those plant species offering small leaf size along with prickles was lower, at least for kudu, than those of other palatable plant species. The inhibitory effect of prickles was less for impala and goats than for kudu. Kudu bit off the shoot ends despite the prickles (Figure 24.6). For certain straight-thorned species kudu compensated partially for their slow eating rates by spending more time gathering the leaves. Most spinescent species were similar to unarmed palatable species in their acceptability to the ungulates, even though the armed species had higher crude protein in their foliage. Probably these spinescent species, especially of *Acacia,* would be preferred over unarmed species but for the thorns.

The main effect of these structural defense features is to restrict bite size, increasing handling time. Thorns, spines, and prickles restrict foliage losses to large herbivores. In addition, the animals may incur scar tissue in the esophagus and scratches in the buccal and esophageal mucosa.

## Predator Satiation

A more subtle defense among plants and animals is the physiological mechanism of timing reproduction so that a maximum number of offspring is produced within one short period of time. The great abundance of prey satiates

**Figure 24.6** Kudu browsing on *Acacia*.

decline rapidly when the food supply is depleted. This strategy reduces the number of predators available to exploit the next seed crop.

The production of a periodic seed crop depends upon synchronization of seed production among individuals. Weather events, such as late frosts or protracted dry spells, and internal physiological constraints usually bring about this synchronization (Janzen 1971). As individuals of a tree species in a community become synchronized, strong selection pressures build up against nonsynchronizing individuals, because they experience heavy seed predation between peak years. Such individuals either drop out of the community over evolutionary time or become synchronized.

A fourth approach is synchronization across species. Predator satiation may be further assured if during any fruiting season the timing of the seed crop of one species is the same as another, and both share seed predators. Seed predators may be attracted away from one species to another, reducing predatory pressure on both species.

## Chemical Defenses

Chemical defense is another first line of defense by plants against herbivores (see Levin 1976). The basis of chemical defense is an accumulation of secondary products (metabolic products not directly related to plant metabolism) ranging from alkaloids to terpenes, phenolics, steroidal, cyanogenic, and mustard oil glycosides and tannins, and resins. Phenolics, a by-product of amino acid metabolism, are ubiquitous in seed plants. Alkaloids, also amino acid derivatives, occur in several thousand species, and cyanogenic glycosides in a few hundred species. The secondary products may be stored within the cells and released only when cells are broken, or they may be stored and secreted by epidermal glands to function as a contact poison or a volatile inhibitor.

Production and storage of such metabolites are expensive to the plant. They seem to require a tradeoff between defense and reproductive effort, but there is little evidence that plants evolved these defensive metabolites for such a purpose. During the evolutionary history of plants, these secondary compounds may have been metabolic by-products or have served some past (or even present) physiological function (see Futuyma 1983). In time these compounds became useful deterrents to herbivore predation, although they are not able to defend against a full suite of enemies.

Chemical resistance to attack falls into two general types (Levin 1976). One involves accumulations and changes in metabolites of the host plant that act as toxins at the wound site. This response is commonly used to resist attacks from bacteria, fungi, and nematodes. The other type of resistance is based on the presence of inhibitors prior to attack. This approach is commonly employed against animals feeding on plants as well as against fungi.

the predators and allows a percentage of the reproduction to escape.

Predator satiation is a major strategy against predation in plants and is most prevalent in those species lacking strong chemical defenses. It involves four approaches (Janzen 1971). The first approach is to distribute seeds so that all of a seed crop is not equally available to seed predators. Seeds of most trees fall near the parents, and the number of seeds declines rapidly as the distance from the tree increases. Seed predators concentrate about the parent plant. Many of the scattered seeds are missed by the searching predators, in part because of search image and unprofitability. These survivors must produce most of the recruitment. A second approach is to shorten the time of seed availability. If all seed matures and is available at one time, seed predators will not be able to use the entire crop before germination. A third approach is to produce a seed crop periodically rather than annually, like oaks. The longer the time between seed crops, the less opportunity seed-dependent predators have to maintain a large population between crops. Seed predators often experience local increases in density during good seed years, but

Inhibitors may function as warning odors, repellents, attractants, or in some cases direct poisons. Volatile components advertise substances that insects and other herbivores would find repellent if they touched the plant. Bitter tastes imparted by tannins and cardiac glycosides can deter further consumption of both seeds (Janzen 1971) and foliage. Metabolites such as phenolic terpenes and saponins may be toxic and cause illness and occasionally death. Some, such as tannins, reduce digestiblility of the plant materials consumed and depress growth rates (Robbins et al. 1987, Bernays et al. 1989). Such repellents not only inhibit feeding on the plant possessing them, but also add a measure of protection to associated plants. For example, grazing by cattle on bent grass (*Agrostis*) and fescue (*Festuca*) is reduced considerably in the presence of buttercup (*Ranunculus bulbosus*), which contains a powerful irritant of skin and mucous membranes (Phillips and Pfeiffer 1958). The presence of such plants can cause the herbivore to fail to locate the palatable plants or to reject them along with the repellent plant (Atsatt and O'Dowd 1976).

Some mutualistic relationships between plant and fungus also lead to chemical defense. A mutualistic relationship has evolved between tall fescue (*Festuca arundinacea*) and an endophytic fungus, *Acremonium ceonophialum*. The mycelium of this fungus grows as the leaves of the grass grow and extends through the intercellular space between plant cells (Figure 24.7). It derives its nourishment from the intercellular fluids at little cost to the plant. In return the fungus provides resistance to drought and strong toxic defenses against grazing (Ball et al. 1993). The toxic effects of this fungus on large herbivores, especially cattle and horses, include intolerance to heat, poor weight gain, and reproductive failure in mares. Similar complex plant-fungus-animal relationships, yet to be discovered, may exist among other grasses and herbaceous plants.

Plants containing secondary metabolites may also function as attractant-decoys that cause the herbivore to feed on alternate prey. Many attractant plants are not what they advertise. They function as decoys and cause mortality or reduced fecundity because of the presence of toxins or deficiency in certain nutritive materials (Atsatt and O'Dowd, 1976). Coexisting toxic and nontoxic plants with similar attractant chemistry present a problem for host-specific herbivores. Some insects, for example, may be stimulated to lay eggs on a "wrong" but closely related plant, which results in larval deaths.

The mode of defense varies with the nature of the plants. Feeney (1975) divided plants into two groups, apparent and unapparent; and Rhodes and Cates (1976) divided them into available and predictable and unavailable and unpredictable (Rhodes and Cates 1976). The two sets of terms are synonymous—the former from the viewpoint of the herbivore, the latter from the viewpoint of the plant.

**Apparent plants** are large, easy to locate, available to herbivores, usually long-lived, and woody. They possess the most expensive type of defense—quantitative or dosage-dependent. Such a defense, not easily mobilized, is most effective against herbivore specialists. The secondary compounds involved are mostly tannins and resins concentrated near the surface tissues of leaves, in bark, and in seeds. They form in-

**Figure 24.7** Endophytic fungi in a blade of fescue.

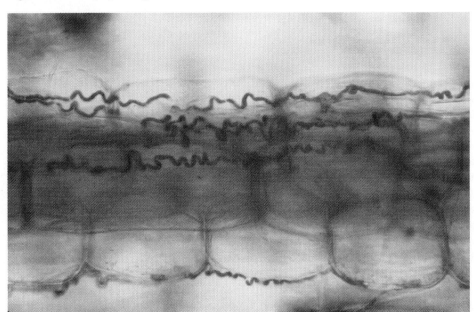

digestible complexes with leaf proteins, reduce the rate of assimilation of dietary nitrogen, reduce the ability of microorganisms to break down lead proteins in herbivore digestive systems, and lower palatability. The problem with such a defense is slow response. One year after defoliation by gypsy moths, oaks increased tannin and phenolic content of their leaves and increased their toughness (Schultz and Baldwin 1982).

**Unapparent plants** are short-lived, mostly annuals and perennials, and scattered in space and time. The plants employ a qualitative, highly toxic defense involving secondary substances such as cardiac glycosides and alkaloids that interfere with metabolism or disrupt development of non-adapted insects. These substances can be synthesized quickly at little cost, are effective at low concentrations, are readily transported to the site of attack, and work quickly. They can be shuttled about the plants from growing tips to leaves, stems, roots, and seeds; and they can be transferred from seed to seedling. These substances protect mostly against generalist herbivores.

However, the distinction between quantitative and qualitative defense or between apparency and unapparency is not absolute. Although testing causal relationships of apparency is difficult (Fox 1981, Courtney 1985), the concept is useful in understanding some aspects of herbivore-plant relationships.

# HERBIVORE COUNTERMEASURES

Although plants possess powerful chemical defenses that work well against generalist herbivores, they can be breached, especially by specialists. The main mechanism is detoxification of secondary compounds. The major detoxifying system is mixed function oxidase, MFO. Possessed by all animals, MFO metabolizes foreign, potentially toxic substances. In vertebrates the MFO activity is located in the liver; in insects it is in the gut, fat bodies, and Malpighian tubules. By oxidation, reduction, and hydrolysis MFO converts fat-concentrating (lipophilic) foreign chemicals into water-soluble molecules that can be eliminated by the excretory system.

The MFO system is a general detoxifying agent, nonspecific in character and induced into activity by a wide array of toxic compounds. It probably evolved in animals to degrade toxic by-products of animal metabolism and harmful compounds ingested. Thus animals, especially the insects, are preadapted to handle many toxic chemically unrelated compounds. Because of the ubiquitous occurrence of MFO, adaptations to new specific toxic compounds require little genetic change, as witnessed by the rapidity with which insect pests become adapted to new insecticides. Thus insects discovering an abundant new source of food can adapt

quickly to novel toxic compounds and become feeding specialists on certain families of plants. For example, some butterfly species of the family Pieridae, notably the cabbage butterfly (*Pieris rapae*), the large white, and small white, and cabbage aphids (*Brevicoryne brassicae*) feed on members the of Cruciferae family. Its allyl glucosinolate is toxic to all noncruciferous feeders. Larvae of the monarch butterfly feed on the highly toxic milkweed (*Asclepias syriaca*), and sequester its cardiogenic glycosides in their bodies as a chemical deterrent to predation. For such specialists, the volatile chemicals of the host plants may act as an attractant rather than a deterrent. Thus the females of specialists are programmed to seek out and lay eggs on plants on which the larvae are able to overcome chemical defense.

Other insects get around chemical defenses by stopping the flow of toxic sap to the leaf on which they will feed (Carroll and Hoffman 1980). The plants they feed on possess toxic latex and resins under pressure in secretory canals associated with leaf veins. In milkweeds and sumacs the secretory canals follow major leaf veins. On these plants beetles, katydids, and certain caterpillars cut leaf veins to sever the secretory canals and block the flow of latex to the intended feeding sites (Figure 24.8) (Dussourd and Eisner 1987, Dussourd and Denno 1991). Among plants in the aster and cucumber families the canals are arranged in networks instead of only along the leaf veins. On these plants, caterpillars cut trenches across the leaf, to sever all strands of the network. Among toxic plants in the morning glory family, the secretory canals are confined to the major leaf veins. Herbivores feeding on these plants dispense with cutting and feed between the major veins.

# FUNCTIONAL RESPONSE OF HERBIVORES

The concept of functional response developed around the consumption of animal prey by predators and parasitoids. What about herbivorous predation on plants? Does this relationship involve a functional response? Until recently ecologists gave little attention to the relationship between the abundance and spacing of plants and the eating rates of herbivores. They have discovered fundamental differences between the functional responses of carnivores and herbivores.

Consider the fox as the carnivore, the rabbit as the mammalian herbivore, and various grasses and herbs as the rabbit's food. The fox searches for, attacks, kills, eats, and digests the rabbit. Each process is discrete and on a one-to-one basis. The fox will not hunt for another rabbit until it has processed the one it has killed. The plant-eating rabbit, however, uses different processes. The rabbit's prey is a patch of plants, not an individual. This patch may be a clump of

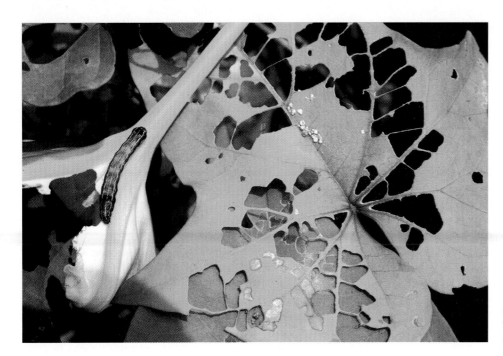

**Figure 24.8** Caterpillars avoid toxins by severing leaf veins and cutting trenches.

plants or a part of an individual plant. The rabbit does not completely eat and rarely kills the plant, so it is available for additional attacks. In this way the rabbit acts more like a parasite than a predator. In seeking food in a given area, the rabbit commits itself to the plants available. Within its area of search, the rabbit may encounter three conditions of food availability that influence its search time. It may find some plants, such as plantain, scattered and hidden beneath grass. The rabbit cannot find more of these plants without moving to a new location in its search area. It may find other plants, such as clumps of red clover, scattered but easily located. The rabbit can move quickly from clump to clump. Third, it may come across a concentrated growth of clover. Then it can crop the plants with little need to move. Thus the rabbit's choice of food and the distribution and visibility of those plants influence the cropping rate.

Having located a patch of food, the rabbit eats by cropping or taking bites of the plant. With each bite it takes a certain amount of plant tissue. How big a bite it takes depends upon the size of its mouth and the nature of the plants (see Shipley et al. 1994). The rabbit can take a larger bite of dense growth of clover than it can of the single leaves of plantain. There is a point at which to eat a given amount of plant tissue, the rabbit will have to take two bites instead of one. As bite size declines, the intake rate declines. Bite size, then, regulates the intake of food. (Spalinger and Hobbs 1992).

Food intake, in turn, depends upon how fast a grazing herbivore, in our case the rabbit, can chew and swallow food. As the rabbit feeds, one process affects the other. The rate at which the rabbit crops food is influenced by the spacing and density of plants. Because the rabbit cannot crop and chew at the same time, in dense growth chewing competes with cropping and vice versa, which reduces the intake rate. However, as it moves between clumps, the rabbit chews what it has eaten, which can increase its intake rate. Because the rabbit may chew while searching for new plants, searching and handling time overlap.

What separates the functional responses of mammalian herbivores from those of other predators are the spacing and apparency of plants and variations in search and handling times. In some studies the Michaelis-Menton function describes and predicts the intake of food. In other cases a poor relationship exists. To account for the factors controlling food intake, ecologists may have to modify the classic functional response equation. (See Spalinger and Hobbs 1992.)

Fryxell and Doucet (1993) investigated the functional responses of beaver (*Castor canadensis*) to various densities of trembling aspen (*Populus tremuloides*) saplings, a preferred food, in three kinds of stand: (1) pure stands of saplings of the same size; (2) pure stands with different sizes of saplings; and (3) mixed stands of aspen, alder (*Alnus rugosa*), and red maple (*Acer rubrum*) of various sizes. Beaver had the highest consumption or cutting rates in stands of aspen of the same size, especially at high densities, producing a classic Type II response curve (Figure 24.9). They had intermediate consumption rates in stands of aspen of different sizes, partly because of handling time. The beaver had the lowest consumption rate in mixed stands of multiple-sized

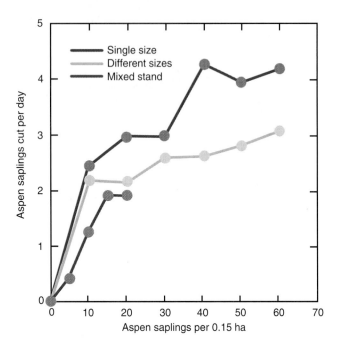

**Figure 24.9** Mean daily rates of aspen sapling cutting in relation to density in three experimental treatments. (After Fryxell and Doucet 1993.)

saplings, partly because of increased search time for aspen growing among nonpreferred foods.

## MODELS OF INTERACTION

The interrelations of plants and herbivores have been examined theoretically by May (1973), Caughley (1976a, 1976b), and Noy-Meir (1973), all of whom present mathematical analyses.

The growth of vegetation as a function of plant biomass can be described by an expression comparable to the logistic growth equation:

$$\frac{dV}{dt} = aV\left(1 - \frac{V}{K}\right)$$

where $V$ is biomass of vegetation, $K$ is maximum sustained biomass (carrying capacity), and $a$ is rate of increase. The rate of increase slows as competition for sunlight, moisture, and nutrients increases and interference increases (Figure 24.10).

When ungrazed vegetation is subject to grazing by a herbivore population, the vegetation's rate of growth is slowed by an amount proportional to the intensity of grazing or predation (number of herbivores consuming plants multiplied by the rate at which vegetation is consumed). When vegetation is at maximum sustained biomass ($K$), herbivores can eat all they want, although the quantity is limited by the her-

bivore's intake capacity. If the vegetation increases while the herbivore population remains the same, grazers increase consumption up to a point of satiation. If the vegetation declines, the amount herbivores consume also declines, because the intake is limited by the forage available. These conditions represent a Type II functional response curve (Figure 24.10a).

If herbivores increase, a numerical response, they reach a level where they overgraze the vegetation, as frequently happens with deer, snowshoe hare, and lemmings. If the vegetation has no ungrazable reserve biomass or if it is grazed to a point where the reserve is too sparse to maintain production, the plant population may go extinct (Figure 24.10c, d). If the vegetation has an ungrazed reserve, the reserve may be used by the plants to attain a low biomass steady state (Figure 29.10b). Depending upon the population density of the herbivore, the vegetation may stabilize at a high biomass ($V$) or reach an unstable equilibrium point at which the vegetation may be able to restore itself if grazing pressure lets up, or it may slip to extinction. In some situations, especially where a Type III functional response is involved, the vegetation may exhibit two stable steady points, one at a high plant biomass and another at a low plant biomass (Figure 24.10e). Between the two is an unstable equilibrium point.

Interactions between various vegetation growth curves and herbivore densities can result in a number of plant-herbivore relations (for examples and discussion, see Noy-Meir 1975). As herbivores increase, vegetation declines (Figure 24.11). Then herbivores decline, the vegetation recovers, the herbivore population increases once again, and the two populations approach equilibrium, the vegetation with grazing pressure and the herbivore with its food supply (Caughley 1976b).

A notable plant-herbivore interaction is the ten-year cycle of the snowshoe hare and woody browse in the Yukon (Smith et al. 1988) (Figure 24.12). As snowshoe hare populations increased toward peak densities, they depleted their winter food supply, dominated by the preferred woody browse species, bog birch (*Betula glandulosa*) and willow (*Salix glauca*). The hares had to subsist on large twigs and less preferred browse, such as white spruce (*Picea glauca*). The situation became critical when essential browse fell below that needed to support the population overwinter, approximately 300 g per individual per day of stems 3 to 4 mm in diameter (Keith 1974, Pease, Vowles and Keith 1979, Wolf 1980). This shortage of food brought on malnutrition (J. M. M. Smith et al. 1988). Malnutrition at peak densities in a snowshoe hare population results in a high winter mortality of juvenile hares, a high overwinter loss of weight, low bone marrow fat, low levels of liver glycogen, and a late onset of reproduction the following summer. Malnourished hares became highly vulnerable to predation, and the population crashed.

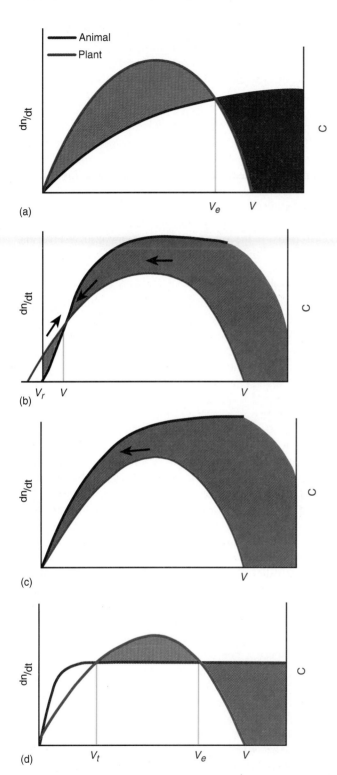

**Figure 24.10** Logistic plant growth as a function of plant biomass, over which is imposed consumption per animal as a function of plant biomass. In dark areas plant growth exceeds consumption; in light areas consumption exceeds growth. (a) Consumption curve is below the growth curve at all biomass levels. Intersection of the two curves indicates the point of stable equilibrium between plant growth and herbivore consumption. Deviation in either direction will cause net changes in $V$, tending to restore stable equilibrium. It is an undergrazed state as plant growth, animal consumption, and secondary production are below maximum. (b) Overgrazing to low biomass steady state. Vegetation has some ungrazable reserve biomass that prevents complete extinction. (c) Overgrazing to extinction. The consumption curve exceeds the plant growth curve at all levels of $V$. If no inaccessible plant reserves exist, the plant population becomes extinct. (d) Steady state and unstable turning point to extinction. This situation occurs if the consumption curve is steeper than in (a). The two curves intersect at two points, one at a steady state at high biomass and the other at low biomass. Any deviation can lead to extinction if $V$ becomes lower than $V_t$. (e) Two steady states, one at high biomass and the other at low biomass. This situation occurs where a plant is ungrazable. The two curves intersect three times, producing a stable steady state at high plant biomass ($V_e$) and at a low plant biomass ($V_l$) and an unstable equilibrium or turning point between them ($V_t$). (After Noy-Meir 1976.)

In other regions the decline in hares is also related to chemical defenses of alder, birches, and some willows, which strongly influence the selection of winter forage among many subarctic browsing vertebrates (Bryant and Kuropat 1980). Hares avoid these more nutritious plants, especially juvenile growth and buds, because they contain more resin and phenolic glycosides (Reichardt et al. 1984,

Palo 1984, Sinclair and Smith, 1984, Bryant et al. 1985). For example, hares avoid the juvenile internodes of Alaskan green alder (*Alnus crispus*). They contain three times the concentration of two deterrent secondary metabolites, pinosylvin and pinosylvin methyl ether, found in mature twigs which the hares do consume (Bryant et al. 1983, Clausen et al. 1986). Mountain (arctic) hares (*Lepus*

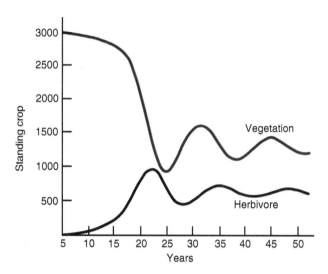

**Figure 24.11** Trend of vegetation density and animal numbers after an herbivore eruption. Note that as the herbivore population increases, vegetation biomass decreases; and as herbivores decline, vegetation increases. Eventually vegetation growth and herbivore consumption reach a steady state. (After Caughley 1976.)

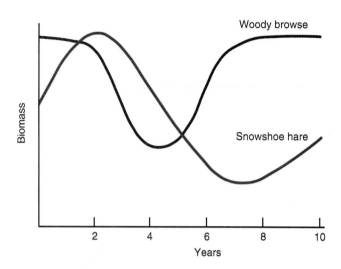

**Figure 24.12** A vegetation-herbivore cycle involving woody vegetation, particularly aspen, and the snowshoe hare. Note the time lag between the cycle of vegetation recovery and the growth and decline of the snowshoe hare population. (Adapted from Keith 1974.)

*timidus*) selectively feed on mature over juvenile twigs of a number of species of willow and avoid low-growing species, which have high levels of secondary metabolites (Takvanainen 1985). The decline in secondary metabolites in mature woody growth suggests that juvenile resistance can be an adaptation against mammal browsing at the ground level. This adaptation could further reduce available winter food and help trigger cycling in hares.

# SUMMARY

Predators interacting with their prey make up a predator-prey system. Predator-prey interactions at one trophic level influence predator-prey interactions at the next trophic level. Involved are plant-herbivore systems and herbivore-carnivore systems.

Plant predation includes defoliation by grazers and consumption of seeds and fruits. Interactions between changes in plant biomass and herbivores result in oscillations of both plant and herbivore populations or in an equilibrium situation, the vegetation with grazing pressure and herbivores with the vegetation.

Herbivory affects plant fitness by reducing the amount of photosynthate and the ability to produce more. Plants respond by denying herbivores palatable or digestible food or by producing secondary compounds that interfere with growth and reproduction.

In response to the selection pressures of herbivory, plants have evolved measures of defense. These include

mimicry among plants to hide from specialized herbivores or to attract seed dispersers, structural defense involving hairs, thorns, and spines, and predator satiation. Reproduction is timed so that fruits and seeds are so abundant that seed predators can take only a fraction of them, leaving a number to escape and germinate. Widespread is chemical defense. It involves distasteful or toxic substances that repel, warn, or inhibit would-be attackers. These substances are secondary metabolic products such as alkaloids, phenolics, and cytogenic glucosides. Chemical defense is most successful against generalist herbivores. Certain specialists breach the chemical defense and detoxify secretions or sequester the toxins in their own tissues as defense against predators.

# REVIEW QUESTIONS

1. Differentiate between the forms of predation on plants.
2. What effect does herbivory have on plant fitness? How do plants respond to defoliation?
3. Why do grasses withstand herbivory much better than woody plants?
4. In what ways do plants affect herbivore fitness?
5. How do the following mechanisms allow plants to defend themselves against herbivores: mimicry, structural defense, predator satiation?
6. What two types of chemical resistance are employed by plants, and how do some herbivores breach them?

7. What is the difference between apparent and unapparent plants?

8. How do vegetation and herbivores interact over time? Discuss this in relation to the woody browse-snowshoe hare cycle. How does chemical defense by plants enter the picture?

9. Schultz (1988) makes the statement, "So little is known (especially by ecologists) of the regulation of plant development and biosynthesis of secondary compounds, that is it reasonable to ask whether 'induced' responses represent anything other than the inevitable and incidental consequence of tissue loss."

Relate this statement to evolution or coevolution of defensive responses of plants to herbivory.

# CROSS-REFERENCES

Herbivores, 183; seed dispersal, 310, 590; logistic growth, 393–396; Type II functional response, 508–509; search image, 512–515; optimal foraging, 515–519; defensive mutualism, 528, 587; functional response in herbivores, 529–531; herbivore-carnivore cycles, 543–545.

# Herbivore-Carnivore Systems

## Outline

## Concepts

1. Prey and predators have evolved various tactics of defense and attack that vary according to their costs and benefits in time and energy.
2. The ability of predators to regulate prey populations depends upon the population dynamics of the prey.
3. Predator-prey cycles seem to be driven by interactions of food shortages, malnutrition, and predation.
4. Cannibalism, the killing and eating of conspecifics, can have pronounced demographic effects within a population.
5. Intraguild predation, the killing and eating of species using the same resources, results in an energy gain and a reduction of competition for the predator.
6. Successful exploitation of a natural population depends upon sustained yield, in which the yield per unit time balances production per unit time.
7. Sustained yield is rarely achieved because economic pressures override ecological considerations.

Herbivory supports carnivory. Unlike herbivores, carnivores are not faced with a lack of quality in their food. It is quantity that is frequently lacking. That dictates a somewhat different relationship between the eater and the eaten. Numbers of prey become important. Fitness of the predator depends upon its ability to capture prey; and fitness of the prey depends upon its ability to elude predation and, if a herbivore, at the same time to overcome plant defenses. That combination puts a squeeze on herbivores.

In an evolutionary context, predator and prey play a sort of game. Prey evolve often elaborate means of defense. To capture the prey, the predator must come up with a way to breach the defense. The relationship between the two involves a flux of adaptive genetic change in each. All of these changes are variations within the context of several simple tactics with different time and energy costs and benefits (Malcolm 1990) (Figure 25.1). For the prey, the choice is hide, run, or fight. For the predator, the choice is sit and wait, stalk and trap, or active search and pursuit.

Over evolutionary time, predators do not seem to track closely the defensive changes in their prey. They seem to leave an adaptive gap between themselves and their prey (Bakker 1983). Predators, as suggested by fossil records, did not evolve rapidly enough to track the escape adaptations of their prey. Thus predators possess a suboptimal and hardly adequate efficiency in predation.

# PREY DEFENSE

Prey have evolved an array of defensive tactics ranging from chemical defenses to simply overwhelming the predator. They affect the predator's breadth of diet and its ability to secure prey.

## Chemical Defenses

Chemical defenses are widespread among many groups of animals; and as with plants, they may have been borrowed from some other use. Venom, for example, protects snakes from enemies, but it is also the means by which the snakes capture prey.

There is an array of chemical defenses. Some species of fish release pheromones from the skin into the water that act as alarm substances and induce fright in other members of the same or related species (Pfeiffer 1962). The fish produce the pheromone in specialized cells in the skin that do not open to the surface, so the pheromone is released only when the skin is broken. Fish in the vicinity receive the stimulus through the olfactory organs. Such alarm substances are most common among fish that are social, nonpredaceous, and lack defensive structures.

Arthropods, amphibians, and snakes employ secretions to repel predators. Many arthropods produce, often in copi-

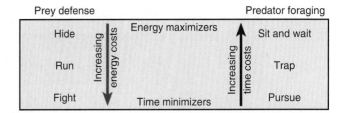

**Figure 25.1** Basic categories of prey defense and predator offense and the costs and benefits of time and energy. (After Malcolm 1990:58.)

ous amounts, strongly odorous, easily detected substances (Eisner and Meinwald 1966, Eisner 1970). They produce the secretions in glands with large saclike reservoirs that are essentially infoldings of the body wall and discharge it through small openings. The secretions may ooze on the animal's body surface, as in millipedes; be aired by the evagination of the gland, as in beetles; or be sprayed for distances of up to a meter, as in grasshoppers, earwigs, and stink bugs. These secretions effectively repel birds, mammals, and insects alike by their effect on the predator's face and mouth. Some mammals, such as shrews and skunks and other mustelids, also possess secretions that discourage attacks by would-be predators.

Active components in the defensive secretions of many arthropods occur as toxic secondary substances, such as saponins, glossypol, and cyanogenic glycosides, used as chemical defenses by plants. Although these toxins inhibit herbivores from feeding on the plants (see Chapter 24), some arthropods can incorporate toxic substances ingested from the plants into their own tissues. In turn, the toxin protects the herbivore from its enemies. The monarch caterpillar, for example, feeds on milkweeds (page 521) that contain a cardiac glycoside, a substance that causes illness in birds that eat the monarch (Figure 25.2) (Brower 1984, 1988; Brower and Fink 1985).

## Warning Coloration and Mimicry

Animals that possess pronounced toxicity and other chemical defenses often possess warning coloration, bold colors with patterns that serve as warning to would-be predators. The black-and-white stripes of the skunk, the bright orange of the monarch butterfly, and the yellow-and-black coloration of many bees and wasps serve notice of danger to their predators (Figure 25.3). All their predators, however, must have some unpleasant experience with the prey before they learn to associate the color pattern with unpalatability or pain.

The association of conspicuous coloration with unpalatability or other averse qualities has for a long time been attributed to the evolution of warning signals advertising that fact. However, unpalatability can evolve independently of warning coloration, as evidenced by shrews and other un-

**Figure 25.2** Highly unpalatable and relatively invulnerable to vertebrate predation, the migratory monarch butterflies congregate in large masses in wintering areas.

palatable animals. Were animals conspicuously colored for some other reason highly vulnerable to predation? Could the cost of production of toxins, which might have also evolved for different reasons, be outweighed by the protective advantage it confers together with conspicuous coloration?

Similarly, animals living in the same habitats as inedible species sometimes evolve a similar mimetic or false warning coloration. That phenomenon was described some 100 years ago by the English naturalist H. W. Bates in his observations of tropical butterflies. The type of mimicry he described, now called Batesian, is the resemblance of an edible species, the mimic, to an inedible one, the model. Once the predator has learned to avoid the model, it avoids the mimic also.

**Figure 25.3** The warning coloration of the poisonous coral snake (*Micurus fulvius*) is mimicked by nonvenomous milk snakes (*Lampropeltis*).

Batesian mimicry is disadvantageous to the model because the predator will encounter a number of tasty mimics and thus take longer to avoid the model. The model, as a result, will suffer greater losses in the learning process. The greater the proportion of mimics to models, the longer the learning time of the predator. Usually the number of mimics is fewer than the number of models.

Mimicry of various sorts is common among animals (Pasteur 1982). A familiar example among North American butterflies, the palatable viceroy butterfly (*Basilarchia archippus*), mimics the monarch (*Danaus plexippus*), most of which are distasteful to birds (Brower 1958) (Figure 25.4). Both model and mimic have an orange ground color with white and black markings. They are remarkably alike; yet the viceroy's nonmimetic relatives are largely blue-black in color. Mimicry is widespread among butterflies, tropical ones in particular, but mimicry is not confined to mimicking models within a taxon. Some butterflies and butterfly larvae possess eyespot patterns that suggest the eyes of snakes or the eyes of large avian predators that attack small insectivorous passerine birds. Juvenile lizards and snakes mimic highly unpalatable large millipedes (Vitts 1992), insect larvae mimic snakes, and snakes even mimic snakes (see Pough 1988).

A less common type of mimicry is Mullerian mimicry, described in 1879 by Fritz Muller. In this type of mimicry one unpalatable or venomous species mimics another. Such mimicry is advantageous to both. The pooling of numbers between the model and the mimic reduces the losses of each, because the predator associates distastefulness with the pattern without having to handle both species. Mullerian mimicry differs from Batesian in that feedback from handling either species is negative, reinforcing the learning process on

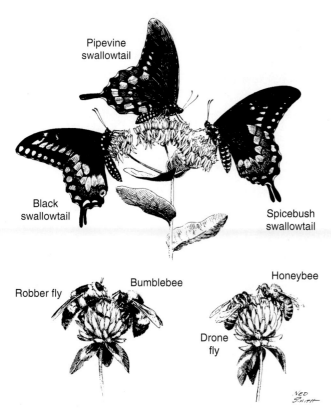

**Figure 25.4** Mimicry in insects. The model, the distasteful pipevine swallowtail, has as its mimics the black swallowtail and the spicebush swallowtail. The black female tiger swallowtail is a third mimic. All these butterflies are found in the same habitat. The robber fly, a mimic of the bumblebee, illustrates aggressive mimicry. The drone fly is a mimic of the bumblebee.

**Figure 25.5** The walking stick (Phasmatidae), which feeds on the leaves of deciduous trees, strongly resembles a twig.

the part of the predator. Batesian mimics and models belong to different phylogenetic lines, whereas Mullerian mimics include members of the same genus and family. This type of mimicry, may have become fixed in an evolutionary line because of the ability of species within a taxon to use and store poisons from plants.

## Cryptic Coloration

Another defense makes locating prey more difficult. Certain color patterns and behaviors evolved by prey enable them to hide from predators (Figure 25.5). Such cryptic colorations involve patterns, shapes, postures, movements, and behaviors that tend to make the prey less visible.

Some animals are protectively colored, blending into the background of their normal environment. Such protective coloration is common among fish, reptiles, and many ground-nesting birds. Countershading or obliterative coloration, in which the lower part of the body is light and the upper part is dark, reduces the contrast between the unshaded and shaded areas of the animal in bright sunlight.

Object resemblance is common among insects. For example, walkingsticks (Phasmatidae) resemble a twig, and katydids resemble leaves. Some animals possess eyespot markings, which intimidate potential predators, attract their attention away from the animal, or delude them into attacking a less vulnerable part of the body.

Associated with cryptic coloration is flashing coloration. Certain butterflies, grasshoppers, birds, and ungulates, such as the white-tailed deer, display extremely visible color patches when disturbed and put to flight. The flashing coloration may distract and disorient predators; or as in the case of the white-tailed deer, it may serve as a signal to promote group cohesion when confronted by a predator (W. Smith 1991). When the animal comes to rest, the bright or white colors vanish, and the animal disappears into its surroundings (see Harvey and Greenwood 1978 for review).

## Armor and Weapons

Some of the most effective means of defense involve protective armor (Figure 25.6). Clams, armadillos, turtles, and nu-

**Figure 25.6** The armor-bearing armadillo (*Dasypus novem-cinctus*) has strong bony plates overlain with horns that develop from the skin.

merous beetles all withdraw into armor coats or shells when danger approaches. The associated problem is the animal's inability to assess the external environment. Is the predator large or small, still present or departed? How much foraging time should an animal sacrifice before daring to open up its defenses? Porcupines, hedgehogs, and echidnas have quills (modified hairs), which effectively discourage predators.

## Behavioral Defenses

Some animals' defenses are behavioral. One is the alarm call, given at the moment of potential danger when a predator is sighted. High-pitched alarm calls are not species-specific. They are recognized by many different animals close by. But an unanswered question is to whom the calls are directed—the predator or the conspecific prey. If directed toward potential prey, the alarm call could be either altruistic or selfish (Chapter 21). If the alarm exposed the caller's position to the predator, the caller could draw the predator's attention away from conspecifics, including kin, or it could attract more conspecifics for cooperative defense and lower its risk of being taken. Alarm calls do function to warn close relatives, as in the case of Belding ground squirrels (Sherman 1977). Highly sedentary, closely related females live in close proximity to each other. Adult and one-year-old females living with relatives respond quickly to danger and give most of the alarm calls, which warn offspring and other relatives. Beyond that there are few conclusive studies on the evolution and function of alarm calls.

Alarm calls often bring in numbers of potential prey that respond by mobbing or harassing the predator. An example is the harassment of an owl perched in a tree by many small birds attracted to the scene by general alarm calls. Mobbing may involve harassment at a safe distance or direct attack.

The outcome for the prey is a reduction in the risk of predation to themselves and their offspring. But like alarm calls, the adaptive and evolutionary significances of mobbing are still obscure.

A distraction display diverts the attention of predators away from eggs or young. Distraction displays are most common among birds. Birds with precocious young, such as the killdeer, usually exhibit the most vehement distraction displays at the time the eggs hatch, and altricial birds, such as the vesper sparrow, at the time the young fledge. Because the beneficiaries of distraction display are the immediate offspring, the behavior probably evolved through kin selection.

Living in groups may be the simplest defense for some prey species. Groups, especially in mobbing situations, would probably deter a predator that would not be so inhibited when facing only one or two prey individuals. Sudden explosive group flight can confuse a predator, unable to decide which individual to follow. A predator is less likely to find prey when individuals are grouped than if an equal number of individual prey were distributed as solitary prey. By keeping to a group, an individual reduces its chances of being taken. Collectively these two antipredator advantages of group living (called the **attack-abatement effect** by Turner and Pitcher 1986) reduce the risk to group prey.

Among some animals the size of the group may show a direct relationship to the density of specialized predators. For example, in winter, when the availability of forage was less abundant and more patchy, the group size of muskox in the northwestern Yukon was about 1.7 times as large as in the summer (Heard 1992). During both seasons, however, variation in group size was directly related to wolf densities (Figure 25.7). The group size of muskox appeared to be a tradeoff between decreased predation risk for individuals in larger groups at high wolf densities in winter and a decreased benefit of group foraging in summer.

## Predator Satiation

A more subtle defense is the timing of reproduction so that most of the offspring are produced in a very short period of time. A good deal of this restricted seasonality of birth can be attributed to the selective advantage of producing young when food will be sufficiently abundant to support them. The other advantage is the synchronization of births to reduce predation on the newborn. This strategy is employed by such ungulates as the caribou and wildebeest (Bergerud 1971, Schaller 1972, Rutberg 1987). Such reproductive synchrony can reduce predation in three ways: (1) the collective defense of the young by breeding adults; (2) interference with the predator's ability to pick out a prey individual; and (3) production of such an abundance of prey that predators can take only a fraction of them. The remaining young quickly grow beyond a size easily handled by predators (Schaller 1972, Bergerud 1971). Such reproductive synchrony functions best against specialized predators with a

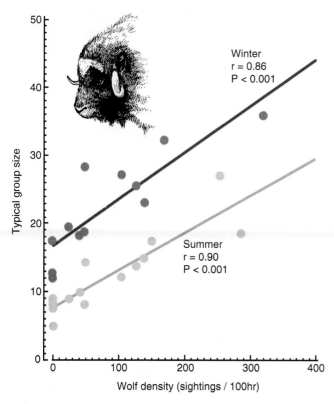

**Figure 25.7** Increase in muskox group size in winter ($y = 0.063x + 16.76$) and in summer ($y = 0.0546x + 7.76$) in the Yukon. (After Heard 1992.)

Type II functional response. For prey faced with generalized predators, asynchronous reproduction may be the best strategy (Ims 1990).

The 13-year and 17-year appearances of the periodical cicadas (*Magicicada* spp.) function in much the same manner. By appearing suddenly in enormous numbers, they quickly satiate predators and do not need to evolve costly defensive mechanisms (Figure 25.8). Although huge numbers of adults succumb to predators, the losses hardly dent the total

population (Williams et al. 1993). In other years, predators must seek alternative prey. Thus the cicadas' major defense is to prevent predators from ever evolving any dependence upon them.

# PREDATOR OFFENSE

As prey evolved ways of avoiding predators, predators by necessity had to evolve better ways of hunting and capturing prey (Bakker 1983).

## Hunting Tactics

Predators have three general methods of hunting: sit-and-wait or ambush, stalking and trapping, and search and pursuit (Figure 25.1). Ambush hunting involves lying in wait for prey to appear. This method is typical among certain insects and some frogs, lizards, and alligators. Ambush hunting has a low frequency of success, but it requires a minimal expenditure of energy. Stalking, typical of herons and some cats, is a deliberate form of hunting with quick attack. The predator's search time may be great, but pursuit is minimal. Therefore it can afford to take smaller prey. Pursuit hunting, typical of many hawks, involves minimal search time. Because pursuit time is great, these predators must secure relatively large prey to compensate for the energy expended. Searchers spend more time and energy to encounter prey. Pursuers, theoretically, spend more time to capture and handle prey once they notice it.

Predators have energy requirements that can be met only by profitable foraging. Predators cannot afford to pursue prey too small to meet their energy requirements unless that prey is abundant and can be captured quickly. Predators also have a limit on upper size. The prey may be too large to consume or too difficult or dangerous to handle. In fact, some prey species become invulnerable to predation through body growth.

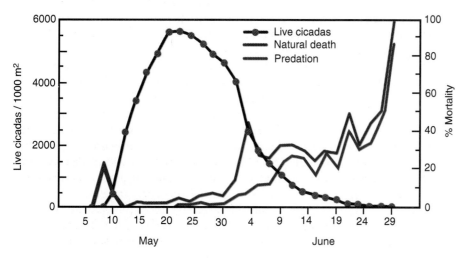

**Figure 25.8** Estimated daily population density of periodical cicadas on a study site in Arkansas (left) and estimated daily bird predation rates based on cicada wing counts (right). Maximum cicada density occurred around May 24, and maximum predation occurred around June 10. Predatory pressures built up as birds apparently acquired a search image for the cicadas. At the height of predation, most of the cicadas had already emerged and escaped bird predation. (From Williams et al. 1993:1148.)

Some predators have evolved methods of killing prey much larger than themselves. Certain snakes, arthropods, and shrews use venom to kill large prey. Predators may hunt in groups to lessen the risks they face in attacking prey and to increase foraging efficiency. Wolves, African hunting dogs, jackals, African lions, and Harris' hawks (*Parabueto unicinctus*) (Bednarz 1988) are examples of predators that take large prey by hunting cooperatively. True cooperative hunting involves at least two members of a stable social unit pursuing, capturing, and sharing prey in a coordinated manner. Individual participation in group hunting, however, varies considerably (Packer and Ruttan 1988, Scheel and Packer 1991). Some get into the thick of the action; others refrain or hold back. More individuals tend to refrain from the final pursuit if the prey is small; most join in if the prey is large and dangerous.

Conversely, predators much larger than their prey have evolved ways of filtering organisms from their environments, particularly in aquatic communities. Examples are net-spinning caddisflies that feed on drift and baleen whales that feed on krill. Intermediate-sized predators are usually hunters, whereas very small-sized predators are usually parasitoids.

## Cryptic Coloration and Mimicry

If prey can use cryptic coloration, concealing coloration, and mimicry to their advantage, predators can do the same. Cryptic and concealing coloration enable them to blend into the background or break up their outlines. Predators can deceive prey by resembling the host or prey, a deception called **aggressive mimicry** (Pasteur 1986). An example is the model bumblebee and the mimic robber fly (*Mallophora bomboides*) (see Figure 25.2). Not only does the robber fly benefit from reduced predation, but it also exploits the model for food. The robber fly preys on Hymenoptera by preference, and its resemblance to its prey allows it to escape notice until the bee finds it too late to flee or defend itself (Brower and Brower 1962). The females of certain species of fireflies imitate the flashing of other species, attracting to them males of those species, which they promptly kill and eat (Lloyd 1980). Among birds, the zone-tailed hawk (*Buteo albonotatus*) soars with groups of vultures that scavenge for food as it searches for prey on the ground below (Willis 1963). By mimicking nonpredatory vultures, the zone-tailed hawk can deceive its live prey to false security.

## Adaptations for Hunting

Predators have acquired various adaptations that improve their hunting ability in addition to such weapons as fangs and claws. Bats, for example, produce ultrasonic sounds through the nose and mouth that enable them to detect prey by echolocation (for good summary see Vaughan 1986). Night-hunting owls can locate prey by hearing rather than by sight. Their feathered facial disks reflect the sounds of prey and direct them to the ears (Konishi 1973). The owl's large ear openings are positioned asymmetrically, enabling the bird to detect differences in the elevation of the prey. The owl's ability to fly noiselessly allows it to come upon the prey without alerting the victim. Day-hunting northern harriers (*Circus cyaneus*) flying over densely grown grass fields have similar facial disks and placement of ears that enable the hawks to locate by sounds voles hidden in the grass (Rice 1982).

## CANNIBALISM

A special form of predation is cannibalism, more euphemistically called intraspecific predation. Cannibalism, more widespread and important in the animal kingdom than many ecologists admit, is killing and eating an individual of the same species. Cannibalism is common to a wide range of animals, aquatic and terrestrial, from protozoans and rotifers through centipedes, mites, and insects to frogs, birds, and mammals, including humans. About 50 percent of terrestrial cannibals, mostly insects, are normally herbivorous species, the ones most apt to encounter a shortage of protein. In freshwater habitats, most cannibalistic species are predaceous, as they are in all marine ecosystems (Fox 1975a, 1975b, 1975c).

Cannibalism has been found mainly in stressed populations, particularly those facing starvation. Some animals do not become cannibalistic until other food runs out; others do so when the availability of alternative foods declines and individuals in the population become nutritionally disadvantaged (Alm 1952). Cannibalism is probably initiated when hunger triggers search behavior, lowers the threshold of attack, increases the time spent foraging, and expands the foraging area. It is consummated when the individual encounters vulnerable prey of the same species (see Polis 1981). Other conditions that may promote cannibalism are: (1) crowded conditions or dense populations, even when food is adequate; (2) stress, relegating some members to vulnerable low social rank; and (3) the presence of other vulnerable individuals, such as nestlings, eggs, or runts, even though food resources are adequate.

Whatever the cause, local conditions and the nature of local populations influence the intensity of cannibalism. In general, cannibalism fluctuates greatly over both long and short periods of time. Among some predaceous fish, such as walleye (*Stizostedium vitreum*) (Fortney 1974), and insects, such as freshwater backswimmers (*Notonecta hoffmanni*) (Fox 1975b, 1975c), cannibalism is most prevalent in summer. This season coincides with a decrease in normal prey and a reduction of spatial refuges for the young.

Not all individuals in a population become cannibals (see Polis 1981). Intraspecific predation is usually confined to older and larger individuals. Those receiving the brunt of

cannibalism are the small and the young, but not always. In some situations groups of smaller individuals will attack and devour larger individuals.

Demographic consequences of cannibalism depend upon the age structure of the population and the feeding rates of various age classes. Even at very low rates, cannibalism can produce demographic effects. Three percent cannibalism in the diet of walleyes could account for 88 percent of mortality among young (Chevalier 1973). Cannibalism can account for 23 to 46 percent of the mortality among eggs and chicks of herring gulls (Parsons 1971), 8 percent of young Belding ground squirrels (Sherman 1981), and 25 percent of lion cubs (Bertram 1975). If a large proportion of either an entire population or a vulnerable age class is eaten, cannibalism can cause violent fluctuations in recruitment.

Cannibalism can become a mechanism of population control that rapidly decreases the number of intraspecific competitors as food becomes scarce. It is unlikely to bring about extinctions of local populations because of its short-term nature. It decreases as resources become more available to survivors and as vulnerable individuals become scarcer. By reducing intraspecific competition at times of resource shortages, cannibalism may actually reduce the probability of local extinction of a population. In the long term, however, cannibalism can be self-defeating because it runs counter to the second law of thermodynamics and trophic level dynamics. The exceptions are among those animals whose young feed at lower trophic levels than the adults. Then cannibalism would involve the harvesting of young grazers.

Cannibalism can provide a selective advantage to survivors. Survivors gain a meal and eliminate a potential competitor for food as well as a potential conspecific predator. With the population reduced, the survivor has more food, enhancing its chances of longer survival, rapid growth, and increased fecundity. Cannibalism may also be rewarding from a nutritional standpoint, leading to increased growth rates and reproduction of cannibalistic individuals over noncannibalistic ones. This may come about in part because the individuals consumed contain the proper proportion of nutrients necessary for growth, maintenance, and reproduction.

Cannibals can also increase their own fitness by reducing the fitness of competitors. By killing and eating other individuals of the same sex they reduce competition for mates. They can eat the offspring of a competitor, as adults of the Belding ground squirrels do. Among some animals, insects and spiders in particular, the females will kill and eat the male after mating, reducing the probability that other females will encounter a mate (Jackson 1992).

Cannibalism can be a selective disadvantage if individual survivors become too aggressive and destroy their own progeny or genotype completely, reduce their genotype faster than the genotypes of conspecific competitors, or reduce the chances of successful reproduction by eliminating suitable mates.

Selection can balance advantages against disadvantages. In some situations the disadvantages of cannibalism are less severe than starvation and reproductive failure caused by inadequate nutrition. For example, parents cannibalizing some of their own offspring can increase the probability of survival and fitness of either parents or surviving offspring or both (Polis 1981, Rohwer 1978) and use rather than waste energy already invested in them. If starvation reduces a population, the survivors may be nutritionally stressed; but if cannibalistic individuals remove conspecifics early, they reduce population density early, and per capita food supply remains high. Survivors have improved their fitness because, being well-fed as juveniles, they grow faster, survive better, and produce more young.

Cannibalism may be less costly to individuals, but is disadvantageous from an evolutionary viewpoint. For this reason it is highly improbable that strong selection exists for the trait. With a few exceptions cannibalistic individuals do not distinguish between conspecifics and other prey, but rather are opportunistic predators. Rarely is cannibalism a distinct behavioral trait.

## INTRAGUILD PREDATION

Typical food web construction shows predators occupying specific links. Species A is herbivorous; species B feeds on species A; species C and D both feed on species B. What if D also feeds on species C? That situation, fairly common to many communities, is **intraguild predation (IGP)**. Intraguild predation is the killing and eating of a species that uses similar resources and thus is a potential competitor (Polis and Holt 1990) (Figure 25.9). What makes intraguild predation unique is that it combines elements of predation and competition. By feeding on a competitor the predator acquires an energy gain. Simultaneously, the predator has reduced potential competition.

The simplest form of intraguild predation is a three-species system in which one of the competitors is also a

**Figure 25.9** Food webs. (a) Web without intraguild predation (IGP). (b) A three-species food web with IGP. (c) A three-species food web with an age-structured species whose juveniles compete with the consumer and whose adults eat the consumer. (After Polis and Holt 1992:152.)

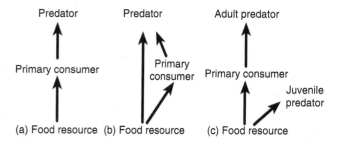

predator and the other is its prey (Polis et al. 1989, Polis and Holt 1992). It occurs among species that eat the same foods. The potential prey species or age class has smaller body size than the predator, and it falls into its normal prey size. The young of many species are particularly vulnerable to species with whom the adults compete. An IG prey species may compete heavily for resources shared with the young of an IG predator; or the IG predator may prey heavily on the young of the IG competitor. Where the IG predator feeds on the young of competing IG prey, it can greatly reduce or even eliminate the local abundance of the IG prey. If such predation excludes or decreases the population of a more efficient IG competitor, the predator increases the supply of the resource for itself.

Intraguild predation can be severe enough to reduce local populations and to regulate or affect community structure. In the Caribbean islands lizards are the dominant vertebrate insectivores. They feed extensively on web spiders as well as on the prey taken by the spiders. Lizards are both predators on and competitors with the spiders. Populations of web spiders are about ten times as dense on islands without lizards as on islands with lizards (Schoener and Toft 1983).

To discover the process underlying this effect, Schoener and Spiller (1987) experimentally removed lizards from randomly selected plots on a very large island in the Bahamas. They found that spider densities in the removal plots were 2.5 times higher than on the control plots. The effects of removing the lizards showed down through the food web. Abundance of arthropod prey increased, as did the consumption of prey. This study, coupled with one by Palaca and Roughgarden (1984), demonstrated that lizards can significantly reduce spider populations on tropical areas. Contrary to most studies, in which predation increases species diversity, predator removal resulted in an increase in the number of spider species.

Intraguild predation has important implications in resource management, especially fisheries management. Fisheries management has a long history of introducing exotic species into lakes and streams either for sport fishery or for forage production. An example is the introduction of opossum shrimp (*Mysis relicta*), a voracious predator of zooplankton, especially cladocerans, as a forage species into the Flathead River-Lake ecosystem in Montana (Spencer et al. 1991). It was stocked to stimulate the production of kokanee salmon, a landlocked species of sockeye salmon. Kokanee salmon (*Oncorhynchus nerka*) were also introduced to the system in 1916, when they replaced native populations of cutthroat trout (*Salmo clarki*). Shortly after the introduction of the shrimp populations, various species of cladocerans declined sharply or became extinct. Instead of increasing with an increased food supply, the salmon population declined, in part because the shrimp retreated to the benthic region during the day, escaping predation by the diurnal feeding salmon. Rather than feeding on shrimp, the salmon preferred cladocerans, which were also the major prey of young salmon. The shrimp, however, were superior competitors with the young salmon for the cladocerans. The result was a collapse of the kokanee salmon population in the lake. The demise of the salmon population reduced the autumnal congregation of bald eagles on the lake, as well as other wildlife attracted to the spawning run.

The moral of these studies is two-fold. One, look beyond the current management concept that predator-prey relations, especially in fisheries, are on a one-to-one basis. Two, never introduce a new or exotic species into an ecosystem in which you do not understand trophic structure. To do so can destroy its evolved predator-prey relationship, eliminate native species, and reduce biodiversity.

## PREDATOR-PREY CYCLES

Cycles in population density over time are a pervasive idea in population ecology, one that has risen to the level of ecological dogma. The concept of predator-prey cycles is an outgrowth of the neutral Lotka-Volterra predation equations, later modified as stable limit cycles, in which the prey cycle is driven by predation. The classic example of a predator-prey cycle is that of the snowshoe hare and lynx. By using data obtained from the fur returns of the Hudson Bay Company, MacLuich (1937) first plotted the snowshoe hare-lynx cycle. Elton and Nicholson (1942) further analyzed the cycle, and that analysis has become enshrined in the ecological literature. As first described, the snowshoe hare and lynx cycled in tandem with a time lag between the two. The lynx population peaked as the snowshoe population declined (Figure 25.10), and the snowshoe hare population recovered before that of the lynx. Although this cycling appeared to be an excellent example of a long-term predator-prey interaction, a more critical examination of the data suggested otherwise. The classic cycle was based on hare data from the Hudson Bay region of eastern Canada, whereas the lynx data were from western Canada (Finerty 1980). The two cycles were not exactly coupled.

As we saw in Chapter 24, the nine- to ten-year snowshoe hare cycle appears to be related in part to a vegetation-hare interaction. As the number of snowshoe hares increases, they experience an increasing shortage of food over winter that leads to malnutrition. Malnutrition and low winter temperatures weaken the hares, making them extremely vulnerable to predation (Keith et al. 1984, Sinclair et al. 1988). Intense predation by lynx and other predators causes a rapid decline in the number of hares and for several years holds the population at a level much lower than the habitat could actually support. Facing a shortage of food, the predators (mostly lynx) fail to reproduce or to rear their young. With a decline in predatory pressure and a growing abundance of winter food, the hare population begins to rise sharply, starting another cycle (Figure 25.11).

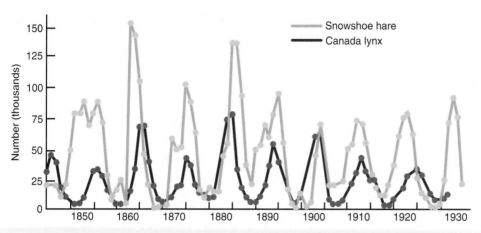

**Figure 25.10** The classic nine- to ten-year snowshoe hare-lynx cycle in northern North America. Cycling is evident in the fur returns from the snowshoe hare and the lynx. Although the cycles are real, the interaction between predator and prey is an artifact. The lynx fur returns are from western Canada and the snowshoe hare returns are from the Hudson Bay region. The two cycles are not coupled. For this reason in some years the lynx decline precedes the hare decline. If the predator were truly attacking the prey, the lynx decline would always follow the hare decline. (From MacLuilch 1937.)

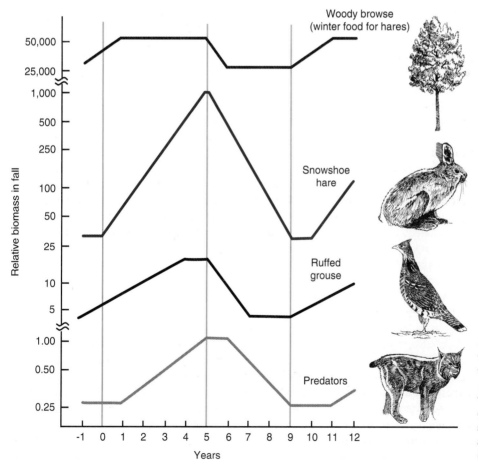

**Figure 25.11** Model of the vegetation-herbivore-predator cycle involving the snowshoe hare and lynx. Note the time lag between the cycle of vegetation recovery, growth and decline of the snowshoe hare population, and the rise and fall of the lynx population. (Adapted from Keith 1974.)

Seeking causes of cycles, especially the ten-year cycle of hares and lynx, has long intrigued ecologists and wildlife biologists. The causes are especially puzzling because the ten-year cycle is synchronized across the boreal regions of North America. Suggested reasons have ranged from random variations in population oscillations (Cole 1951) to sunspots. In 1930 Delury (1930) suggested that cycles are caused by sunspot variations in solar radiation and weather. Now the idea that sunspots affect the snowshoe hare cycle has emerged again, not as a cause, but as a synchronizer.

Sinclair et al. (1993) reexamined the relationship of sunspot activity to snowshoe hare cycles. They first looked for clues in the growth rings of white spruce (*Picea glauca*) in the Yukon. At the highs of their cycle, snowshoe hares, deprived of birch and willow browse, turn to eating the nutrient-poor apical stems of white spruce. Loss of the stems affects the growth of spruce. These periods of stress are marked in cross sections of the trunk by dark marks in the growth rings of the trees. The experimenters cleared a five-kilometer strip of forest three meters wide in 1987 at Kluane, southwest Yukon. They determined the number of trees available for browsing as those less than 50 years old. They found that among these trees the presence of tree ring marks correlated with the number of stems browsed. The frequency of the marks correlated with the density of hares in the same region over one well-documented hare cycle. With this information the biologists were able to determine the frequency of dark marks in tree rings within the first 50 years of life of trees germinating between 1751 and 1983. They cross-correlated this data with the hare fur records of the Hudson Bay Company. They found that the tree ring marks of low growth and highs of hare cycles were locked in phase. Next the biologists cross-correlated both tree ring marks and snowshoe hare cycle with sunspot numbers. The resulting correlogram showed a ten-year periodicity. Phase analysis of the data revealed that tree marks and sunspot numbers have periods of nearly constant phase difference during the years 1751–1787, 1838–1870, and 1948 to the present. These periods coincided with the time spans of sunspot maxima. The biologists then examined the data from a 102.5 m ice core from Mt. Logan in Kluane National Park, Yukon, to determine the net annual snow accumulation over the years. They found a nearly constant phase relationship between net annual snow accumulation (an indicator of climatic conditions), tree mark ratios, hare fur records before 1895, and sunspot number during high amplitudes in the hare cycle. These phase relationships suggest links among the solar cycle, climate, tree marks, and hare populations.

In the intervening time between the years of nearly constant phase differences, the hare cycles drift out of phase. They are synchronized again by a period of sunspot maxima. Sunspot cycles do not cause snowshoe hare cycles. The correlations among climate, food, and hare populations with sunspots suggest that solar activity acting through amplified climate cycles synchronizes hare cycles over the boreal region.

The ten-year cycle of the snowshoe hare is characteristic of the boreal region. South of the boreal region, some populations of snowshoe hares exhibit cyclic fluctuations and some do not. Dolbeer and Clark (1975), Tanner (1975), and Wolff (1980) have studied why. In coniferous forests and associated regions south of the boreal forests, snowshoe hares exhibit cyclic fluctuations only in a uniform environment of spruce and fir (which is also characteristic of the boreal forest). In regions where the environment is very patchy, where many kinds of vegetation patterns exist, cycles do not occur. There are several possible reasons. Fragmented habitats support a greater diversity of prey species, which adds stability to populations of facultative predators. These predators are able to maintain sustained predation on hares (Keith 1983). Hares occupying high-quality habitats are protected from predation, whereas hares living in areas of poor cover are subject to predation. These patches of high-quality habitat that provide excellent food and cover act as refuges from which surplus animals repopulate poorer habitat patches. Predators in turn eliminate these hares. Such dispersal and predation tend to hold the population of hares in better habitats at a level at which they do not overutilize their food supply, thus damping cyclic behavior.

The other major cycle is the two- to five-year cycle of microtine voles. Like the snowshoe hare, some vole populations are cyclic; others are not. For example, voles cycle in northern Fennoscandia, but not in the south (Figure 25.12). (Hanski 1987). Paralleling the rise and fall of the voles are similar trends occurring in populations of shrews (*Sorex* ssp.). Predation by least weasels (*Mustela nivalis*) appears to be the common factor synchronizing the crash of voles and shrews (Henttonen 1985). Following the crash of voles and shrews, the weasels also decline. During this time of food scarcity, the weasels turn to cavity-nesting passerine birds, ignored as prey when rodent densities are high (Järvinen 1985). Failure of the vole and shrew populations to cycle in southern Fennoscandia is probably related to the diversity of general predators exhibiting a Type III functional response. In the north the few specialized predators exhibit a numerical response.

## REGULATION

That predators have an adverse impact on the abundance of prey species is an ingrained idea hard to dislodge. Humans quickly blame predators for the decline in any species in which they have a vested interest, such as game species. However, long-term studies on the effect of predation on vertebrate populations by Paul Errington (1943, 1945, 1946, 1963) suggest that predators feed on the "doomed surplus" and have little impact on the productivity of prey populations. The unstated premise in Errington's studies was that

the predator would have to reduce *r*, the rate of increase of the prey population, to regulate it. To accomplish that, predators must remove a portion of the reproductive age classes (see R.J. Taylor 1984:125–139), which is not part of the doomed surplus.

**Figure 25.12** Two examples of fluctuations in co-occurring species of voles in northern and southern regions of Eurasia. (a) Synchronous cycles in *Clethrionomys glareolus* (solid line) and *Microtus oeconomus* in the north (Fennoscandia). (b) Nonsynchronous and noncyclic fluctuations in *C. glareolus* and *M. arvalis* in the south (Tula region near Moscow). (After Hanski 1987:55.)

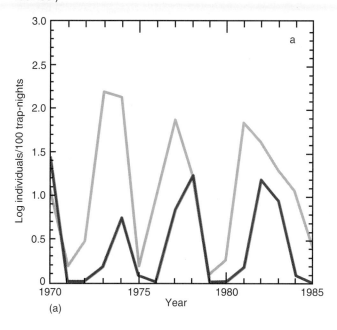

(a)

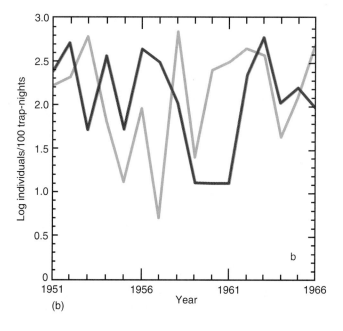

(b)

There is some evidence that natural predation accomplishes such regulation, especially where specialized predators are involved. Larger predators, such as African lions and wolves, tend to take the most vulnerable individuals—the young, the old, and those in poor condition—and few of the reproductive age classes (Figure 25.13). Such predation is compensatory; that is, the losses are made up by future reproduction. Declines for various reasons in prey populations force large predators to take increased numbers of young as well as reproductive individuals, in spite of the greater energy costs incurred. Such predation can drive prey to low levels (Newsome et al. 1989). The lack of prey then reduces the predator population through malnutrition, reproductive failure, and social interactions. Freed from intensive predation and experiencing an abundance of food, prey becomes abundant again. Such loosely regulatory feedback in predator populations and the changing impact of predation on the prey result in wide fluctuations in both predator and prey populations.

Theory holds that predators interact with prey to produce either stable equilibria or cycles, but field evidence is scarce. Erlinge et al. (1984) studied the interactions of nine vertebrate predator species and their major prey, largely rodents and rabbits, in central Sweden. They found that the feeding habits of generalist predators in combination with their territorial behavior did prevent significant annual fluctuation in rodent numbers. Their field studies, supplemented by simulation models, suggest that generalist predators can maintain stable populations by shifting their diets in response to changing prey densities. Such regulation can occur only if alternative prey is abundant and predator populations are intrinsically regulated.

The role of predators and alternate prey in the regulation of prey populations is difficult to assess under natural conditions. However, Lindstrom and associates (1994) were provided a fortuitous opportunity to study the effects of red fox predation in Sweden on cyclic voles and their alternate prey, mountain hare (*Lepus timidus*) and forest grouse (black grouse *Tetrao tetrix,* capercaillie *T. urogallus,* and hazel grouse *Bonasa bonasia*). Fox prey heavily on voles, but during the low of the three- to four-year vole cycle, fox turn to their alternate prey and induce cyclic fluctuations in hare and grouse. If fox predation limited the populations of hare and grouse, then a decrease in the fox population should result in an increase in the prey and reduce or eliminate cycles in the alternate prey. In the late 1970s and 1980s an epizootic of sarcoptic mange (*Sarcoptes scabiei*), swept through fox populations in Sweden. The mange mite, discovered for the first time in Sweden in 1975, causes hair loss, skin deterioration, and the ultimate death of the host. The epizootic decimated fox populations and provided the opportunity for the Swedish biologists to monitor the populations of the fox and its prey over a 20-year period, from 1972 to 1992 (Figure 25.14). Even though the voles were the

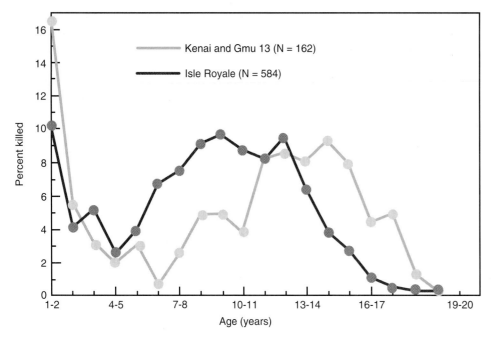

**Figure 25.13** The ages of adult moose killed by wolves on the Kenai, Alaska and on Isle Royale, Michigan. The yearlings and older individuals are most vulnerable, especially on the Kenai. A smaller moose population on Isle Royale apparently increased predatory pressure on younger moose, but heaviest predation still fell on the old individuals. (Data from Peterson et al. 1984 and Ballard et al. 1987.)

major prey of the fox, they were not affected by the loss of the fox and continued to exhibit cyclic behavior. Populations of hare and grouse, however, increased by 40 to 100 percent. The results suggest that predation by the red fox does limit the number of hares and grouse during the down side of the vole cycle. As the voles declined in their cycle, the fox turned to alternate prey and in effect transferred the cycle to the hare and grouse. Thus predatory pressure on alternate prey varied with the cyclic fluctuations of the primary prey species, the voles.

By contrast, in predator-ungulate systems the predators are loosely regulated by social interactions but lack any alternative prey. One prey species has to bear the brunt of a specialized predator. This situation can result in wide fluctuations in prey populations. Extensive studies of moose-wolf relationships point out that the role of predation depends on the ratio of prey to predator (Gasaway et al. 1983, Peterson and Page 1983, Ballard et al. 1987). At a ratio of more than 30 moose per wolf, predation can be significant. The moose population is likely to remain stable or increase if it is below carrying capacity, unless some other source of mortality such as a severe winter intervenes. At a ratio of 20 to 30 moose per wolf, predation can control a moose population. Whether the population remains stable or declines depends upon the combined effects of hunting, food supply, and winter snows. At a ratio of 20 moose per wolf, predation can cause a decline in the moose population.

Assessing the role of predation in the regulation of prey populations is also difficult because human intrusion into predator-prey systems has so altered the relationships between predator and prey that few resemble the system under natural conditions. For example, human settlement in the prairie pothole region of North Dakota severely reduced waterfowl breeding habitat and greatly increased the red fox population. A study of the decline of nesting waterfowl showed that predation by red fox was more intense on nesting females than on males. It distorted the sex ratio strongly toward males, 128 to 100, compared to the presumed pristine ratio of 110 males to 100 females. A reduction in the fox population resulted in a more even sex ratio on an experimental area, and a higher fox population shifted the ratio more strongly toward males (D. H. Johnson and Sargeant 1978). Fox predation accompanied by noncompensatory hunting mortality further distorts the sex ratio.

## EXPLOITATION BY HUMANS

A form of highly selective and intensive predation, often not related to the density of either predators or prey, is exploitation by humans. Overexploitation of wild populations, especially when coupled with loss of habitat, has resulted in a serious decline and local or global extermination of some species. The overharvesting of buffalo, great auk, African ungulates, whales, and many pelagic fish are examples of short-sightedness on the part of humans. On the other hand, such wildlife populations as the white-tailed deer are underharvested in many places, particularly since their natural predators have been eliminated. In contrast to these examples of destructive exploitation, the objective of the wise exploitation of any natural population is the maintenance of equilibrium between recruitment and harvest.

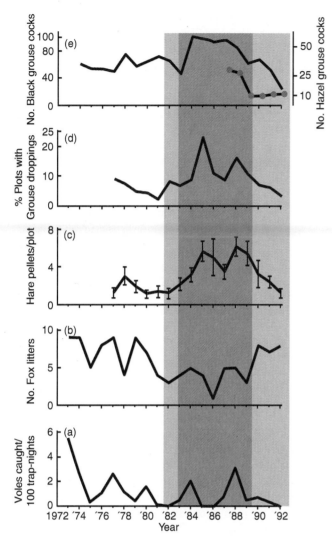

**Figure 25.14** Population fluctuations in the Grimso Research Area in south-central Sweden: (a) voles; (b) red fox litters; (c) mountain hare; (d) grouse; (e) displaying black grouse cocks and hazel grouse cocks. Shaded area indicates years of mange. Darker shading indicates a period of low fox densities as interpreted from the variation in the number of fox litters. (After Lindstrom et al. 1994:1046.)

## Sustained Yield

Although some of the terms used in defining exploitation of populations are similar to those used in productivity, the meanings are somewhat different. When fishery and wildlife biologists speak of **yield,** they refer to the individuals or biomass removed when the population is harvested. **Biomass yield** is the product of the number harvested times the average weight. (Yield may indicate weight without numbers or vice versa.) The **standing crop** is the biomass present in a population at the time it is measured. **Productivity** is the difference between the biomass left in the population after harvesting at time $t$ and the biomass present in the population just before harvesting at some subsequent time $t + 1$.

The objective of regulated exploitation of a population is **sustained yield:** the yield per unit time is equal to productivity per unit time. In its simplest form sustained yield is described by an equation first proposed by E. S. Russell for fishery exploitation. Although the equation was specifically developed for fisheries, it is applicable to any exploitable population. With some minor modifications this equation is

$$B_{t+1} = B_t + (A_{br} + G_{bi} - (C_{bf} + M_b)$$

where $B_{t+1}$ is total biomass of exploitable stock just before harvesting, at time $t + 1$; $B_t$ is total biomass of exploitable stock just after the last harvest, at time $t$; $A_{br}$ is biomass gained by the younger recruits just grown to exploitable stock; $G_{bi}$ is biomass added by the growth of individuals in both $B_t$ and $A_{br}$; $C_{bf}$ is any biomass exploitatively removed during the harvest period; and $M_b$ is biomass lost from exploitable stock by natural causes during the time $t$ to $t + 1$. The equation stresses the fact that productivity also includes individuals that were born and individuals that died during the time interval from the end of one harvest period to the beginning of the next.

The equation is highly simplified. In an unexploited fish population $A_{br}$, $C_{bf}$, and $M_b$ are interdependent. For example, in a stable environment largely undisturbed by humans fish populations appear to be dominated by large species. In turn each species population appears to be dominated by large old fish. When humans start to exploit such a population, significant changes take place. To compensate for exploitation directed first toward the largest members of the population (organisms that under natural conditions are normally secure from predation), the population exhibits an increased growth rate, a reduced age of sexual maturity, increased number of eggs per unit of body weight, and reduced mortality of small members of the population (Regier and Loftus 1972). Populations of other vertebrates react in a similar way.

Exploitation may also influence behavior of the species. Fishing or hunting techniques that are employed constantly because of their initial success gradually become less effective with time because of conditioning or learning by members of the population. As harvest of the species begins to decline, the exploiters are forced to improve or change their methods of fishing and hunting. Also involved may be an interspecific competition. As both the numbers and larger members of a population decline, the niche may be occupied by unexploited, highly competitive, and closely related sympatric or introduced species. Thus, as the Pacific sardine populations declined, their place was taken by anchovies.

If exploitation is carried far enough, then the age classes in the population are too young to carry on reproduction and the population collapses. The principle behind sustained yield is to avoid that collapse of the population.

Sustained yield is clearly dependent on the rate of increase. Sustained yield does not imply holding a population at ecological carrying capacity ($K$), for at that level the rate

of increase equals zero. A population stable in the absence of harvesting can be harvested under sustained yield only after the plant-herbivore or the herbivore-carnivore system has been manipulated to raise $r$. This change can be made in two ways: (1) improve food and cover to increase the carrying capacity by increasing available resources, fecundity, and survival; or (2) lower the density by removing a certain number and then stabilize the population at some lower density (Figure 25.15). Within limitations the lower the density of a population is below the carrying capacity, the higher is the rate of increase. Thus a higher rate of harvest is needed to hold the reproductive population stable at some desired lower density.

The idea is to have the rate of harvest $H$ equal to the rate of increase $r$. The rate of harvest should hold the rate of growth at zero. $H$ would have to equal the rate at which the population would increase if the harvest were stopped (Caughley 1976).

Consider a deer population increasing at a rate of 20 percent a year. It has a finite rate of increase, $e^r = 1.20$ and $r = 0.182$. To hold the population stable, the herd must be harvested at the instantaneous rate of $H = 0.182$. If the population is harvested only during a certain season of the year, as is usual with deer, then the sustained yield is calculated from

an isolated rate of harvest $h$, defined as $h = 1 - e^{-H}$, in this case 0.167. This isolated rate of harvest would have taken into account natural mortality to the population occurring between the period of birth and the time of harvest. Assuming a deer population after the fawning season of 1000 animals and allowing a natural finite rate of mortality of 0.25 per year, we could remove 151 animals in the fifth month. This action would allow the number at the next fawning season to climb back to 1000. The addition of 375 young would compensate for hunting and natural mortality (for details on carrying out calculations, see Caughley 1976).

Although often considered as such, sustained yield is not a particular value for a given population. There may be a number of sustained yield values corresponding to different population levels and different management techniques. The level of sustained yield at which the population declines if exceeded is known as **maximum sustained yield.** Maximum sustained yield is not always the most efficient harvest level, because of other considerations such as species interactions, esthetics, land use problems, and the like. Harvesting may be aimed at the **optimum sustained yield,** the level of sustained yield determined by consideration of these other factors as well as maximum sustained yield.

The higher the $r$ of a species, the higher will be the rate of harvest that produces the maximum amount of biomass production. Species characterized by scramble competition ($r$-strategists) have a high wastage of production. To manage a population influenced by density-independent variables, such as climate or temperature, the objective is to reduce wastage by increasing the rate of exploitation. The role of harvesting is to take all individuals that otherwise would be lost to natural mortality. This type of exploitation is described by the expression

$$\text{maximum yield } B_t - \min (R_t)$$

where $B_t$ is biomass at time $t$ and $\min (R_t)$ is the minimum number of reproducing individuals left at time $t$ in order to ensure replacement of maximum yield at time $t + 1$ (Watt 1968).

Such a population is often difficult to manage because the stock can be severely depleted unless there is repeated reproduction. An example is the Pacific sardine (G. I. Murphy 1966, 1967), a species in which there is little relationship between breeding stock and the subsequent number of progeny produced. Exploitation of the Pacific sardine population in the 1940s and 1950s shifted the age structure of the population to younger age classes. Prior to exploitation 77 percent of the reproduction was distributed among the first five years. In the fished population, 77 percent of the reproduction occurred in the first two years of life. The population approached that of single-stage reproduction subject to pronounced oscillations (Figure 25.16). Two consecutive years of reproductive failures resulted in a collapse of the population from which it never has recovered.

**Figure 25.15** Effect of harvesting on a plant-herbivore system. Superimposed on the interaction diagram in Figure 24.11 is the path of vegetation and herbivore standing crops after harvesting of herbivores over the intervals and at the rates per year in the rectangles above. As a result of management, fluctuations are reduced and the system moves to equilibrium. Such equilibrium results if harvest is initiated at a rate of about one-half of the population's intrinsic rate of increase when the animal population is well below the peak. This rate of harvest must be maintained until plant density levels off and begins to increase. (After Caughley 1976a.)

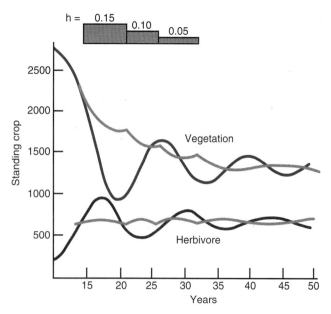

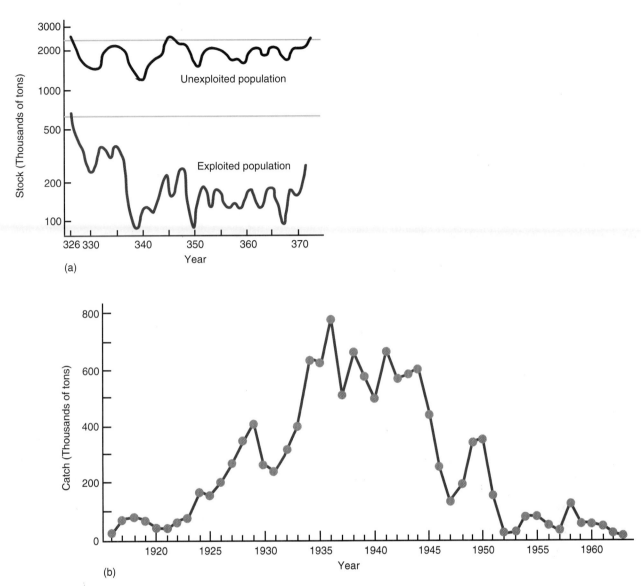

**Figure 25.16** (a) Simulation of an exploited and an unexploited population of sardines, both subject to random environmental variation in reproductive success. The dotted line indicates the asymptotic population size. Note how exploitation adds to instability and how dangerously low the population can get. (From Murphy 1967.) (b) Compare this simulation with the annual catch of the Pacific sardine along the Pacific coast of North America. The population collapsed from overfishing, environmental changes in the Pacific Ocean, and the increase in a competing fish, the anchovy, following a population decline of the sardine. (After G.I. Murphy 1966.)

In populations characterized by density-dependent regulation ($K$-strategists) the maximum rate of harvest depends on age structure, frequency of harvest, number left behind after harvest, fluctuations in environment, and fluctuations in fecundity. It also depends on density of the population to be harvested and the rate of harvest needed to stabilize the density at that level.

This type of harvest is described by the expression

$$P_b = \max B_{t+1} - B_t(X)$$

where $P_b$ is biomass productivity from $t$ to $t+1$, $B_t$ biomass at time $t$, $B_{t+1}$ is biomass at time $t+1$, and $X$ is the several variables that influence biomass production over time $t$ to $t+1$.

This relation is illustrated by Figure 25.17. For any position of the stock to the left of the 45° line there is a rate of exploitation that will maintain the stock at that position. Maximum sustained yield does not necessarily require a large standing crop. Let $a$ be any position on the curve and $c$ a perpendicular line that cuts the 45° line at $b$. At equilibrium the portion $bc$ of the recruitment must be used for the maintenance of the stock, for $bc = ac$; $ab$ can be harvested. There is, however, a limit to exploitation, a limit that is influenced by the inflection point of the curve. For

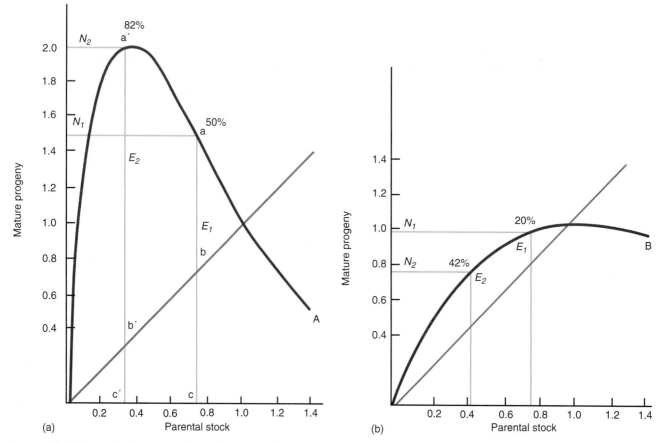

**Figure 25.17** Reproduction curves illustrating rates of exploitation. Reproduction curves diagram the relationship between recruitment or net reproduction, a function of number of mature progeny, and the density of parental stock. The 45° line represents the replacement level of the stock, along points where density dependence is absent. The dome-shaped curve is the plot of actual recruitment in relation to size of parental stock. The apex of the curve, lying above and to the left of the diagonal line, represents maximum replacement reproduction. The curve must cut the diagonal line at least once and usually only once. Where the curve and diagonal line intersect is the point at which parents are producing just enough progeny to replace current losses from reproductive units.

(a) The reproduction curve typical of *r*-strategists suggests that low parental stock can be very productive. On curve A a perpendicular line *ac* cuts the 45° line at *b*. Segment *ab* ($E_1$) is harvest; *bc* is stock left for recruitment. Line *ac* represents the point on the curve at which 50 percent of the mature population is harvested each period. The line *a'c'* is the 82 percent point, the maximum surplus reproduction and the maximum rate of exploitation possible for this population. (b) Curve B is typical of *K*-strategists, among which a low density of parental stock is not very productive. The two points represent levels at which 20 and 42 percent, the maximum, can be harvested.

Note the difference between curves A and B. In curve A, $E_2 < E_1$ and $N_2 < N_1$. Under these conditions the greater the standing crop, the greater is sustained yield. In curve B, where $E_2 < E_1$ yet $N_1 < N_2$, a high standing crop does not result in greater sustained yield. A knowledge of parent-

curve A, the maximum rate of harvest is about 82 percent, for curve B 42 percent. In these curves the size of the reproductive stock that will give maximum sustained yield will not be greater than half of the replacement of the reproductive population. The greater the area of the reproduction curve above the 45° line, the greater the optimum rate of reproduction.

In summary, Caughley (1976) gives six points applicable to harvesting of populations:

1. A population stable in numbers must be reduced below a steady density to obtain a croppable surplus.
2. There is an appropriate sustained yield for each density to which a population is reduced.
3. For each level of sustained yield there are two levels of density from which this sustained yield can be harvested (Figure 25.18).
4. Maximum sustained yield can be harvested at only one density (Figure 25.18).

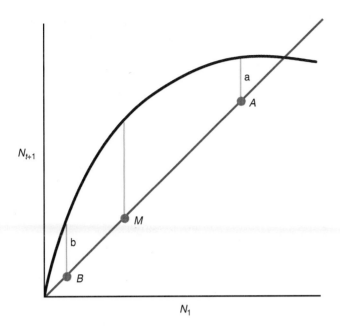

**Figure 25.18** A reproduction diagram of an idealized population of *K*-strategists harvested for sustained yield under three regimes. In the first the population is harvested down from a steady state to a size $N_t = A$. The dashed line *a* represents the number that must be harvested or replaced each year to hold the population size stable at $N_t = A$. In the second region the population is reduced to $N_t = B$, and a sustained yield represented by line *b* could be harvested each year to hold the population stable at $N_t = B$. The yield in this case is a large proportion of a small population. Maximum sustained yield, *M*, is obtained at the population size where the diagonal line and the curve have maximum separation, $N_t = M$. (From Caughley 1976a.)

5. If a constant number is harvested from a population each year, the population will decline from steady density and stabilize at the upper population size for which that number is sustained yield. If this number exceeds the maximum sustained yield, the population declines to extinction.

6. If a constant percentage of the population is harvested each year (the percentage applying to the standing crop of that year), the population will decline and stabilize at a level at equilibrium with the rate of harvesting. This level may be above or below that generating maximum sustained yield.

One group of animals, notably fish and whales, is exploited commercially; another group is hunted for sport. Most game animals are characterized by contest competition and are largely, but not always, density-regulated. It has been assumed by wildlife managers that regulated hunting and compensatory predation have similar effects on game animals. Just as a predator turns to another source of prey when the first source demands too great an expenditure of energy to hunt, so sport hunters abandon the field when

hunting success (animals taken per unit time) drops below a certain level. Hunting mortality supposedly replaces natural mortality. If the surplus were not harvested, the animals would succumb to disease, exposure, and the like. Thus for each individual removed by hunting natural mortality is reduced by one. It is assumed that population stability and sustained yield can be maintained if the population is harvested at the rate represented by the percentage of the young of the year or in some cases percentage of year-old animals in the population.

The weakness of this assumption is that if hunting mortality and natural mortality replace each other and if the rate of harvest applied to the population equals the rate of mortality in the absence of hunting, then the only cause of death in the post-young population is hunting. Similarly, if the nonhunted population is below carrying capacity, no adult dies until *K* is reached (Caughley 1976). Obviously other mortality does take place among some hunted populations, and hunting and natural mortality are not compensatory. Part of the hunting mortality may be an addition to rather than a replacement of nonhunting mortality, as seems to be the case in some waterfowl populations (Geis et al. 1971).

## Problems with Management

Although sustained yield management (SYM), particularly multiple sustained yield (MSY), looks good on paper, it depends too heavily on the logistic equation. The problem is SYM views management as a numbers game. The number or biomass removed theoretically is replaced by an equal number of new recruits to the harvestable component. Under SYM the "surplus," based on density compensation, supposedly is reduced; future production is optimized by optimizing harvest. Thus SYM assumes, as the models indicate, that a relationship exists between stock and recruitment. No proof of this theory exists. The usual approach to SYM fails to consider adequately: size and age classes, differential rates of growth among them, sex ratio, survival, reproduction, and environmental uncertainties, all data difficult to obtain. To add to the problem is the common property nature of the resource—since it belongs to no one, it belongs to everyone to use as each sees fit. To attempt MSY without such information is to balance the population on the edge of catastrophe. Overestimate the potential harvest in any one year and the population is headed for a sharp decline.

Several approaches to management of exploitable populations are in current use. One is the **fixed quota,** in which a certain percentage is removed each harvest period based on MSY estimates. Harvesting matches recruitment. Often used in fisheries, such an approach is risky because a fixed quota can drive a population to commercial if not actual extinction. Because populations fluctuate from harvest to harvest, the MSY will vary from year to year. If such fluctuations are not taken into account, there will be times of overharvest. Overharvest combined with environmental

changes have been responsible for the demise of some fisheries, such as the Pacific sardine.

A second approach is **harvest effort,** often used in establishing seasons for sport hunting and fishing. The number of animals killed is manipulated by controlling hunting effort: the number of hunters in the field, the number of days of hunting (season length), and the size of the bag limit. Hunting effort is decreased by reducing the allowable kill, shortening the season, or closing the season entirely. The reverse approach is used to increase the kill. In general such a rule-of-thumb approach has been more successful in managing exploitable populations than the fixed quota approach.

The permit system is a special variant of the harvest effort approach, used to control more tightly the number of animals killed. It is based on the maximum number of animals to be removed from any one defined area and the amount of hunting effort needed to achieve that goal. For example, it is well-established that approximately 15 percent of deer hunters are successful in any one hunting season. To achieve a desired level of kill the number of deer to be removed from any one area is used as the expansion ratio to set the number of permits issued. For example, if 100 animals is the desired harvest level, then approximately 700 permits should be issued.

A third approach is the **dynamic pool model.** It is essentially the one described on page 548. It assumes a constant natural mortality rate that is independent of density and is the same for all age classes. Growth rates are age-specific but unrelated to density. Animals removed replace density-related natural mortality and are not added to it. Most fishery biologists, for example, believe that density-related mortality is concentrated in the early life-stages of the fish. The flaws in such an approach are many. Fishing mortality can be additive to natural mortality; growth rates are affected by population density, as is recruitment to the population. In practice the dynamic pool model translates fishing mortality into fishing effort, based on type of equipment such as size selective gill-nets that sort out age (size) classes, efficiency of equipment, and season of exploitation. Few dynamic pool models have been developed, let alone put into practice. A general weakness of the model is the inability to estimate natural mortality accurately.

All of the three models are biological, based on logistic growth, and have a major flaw. They fail to incorporate the most important component of population exploitation, economics (Figure 25.19) (see C. A. Hall et al. 1986, Walters 1986). Once exploitation of a natural resource becomes a commercial enterprise, the pressure is on to increase that exploitation to maintain the economic infrastructure built upon it. Once in place, any attempt to reduce the rate of exploitation meets strong opposition, supported by arguments that reduction will result in unemployment and industrial bankruptcy. This, of course, is a short-term argument, because in the long run the resource will be depleted and the economic collapse will occur anyway. That fact is written across the country in abandoned fishery processing plants, rusting fishing fleets, and deserted logging towns. With conservative exploitation on lower economic and biological scales, the resource could still be exploited. Because of the failure of harvest regulations, nationally and internationally, and the commonality of the resource, exploitation efforts increase, even in the face of declining stocks. Instead of reducing harvest effort, we invest in technological improvements in finding and harvesting the remaining resource to a point where it collapses.

An example is the whaling industry (Figure 25.20). The early whaling industry centered around the extraction of oil from blubber to use in lamps. Early whalers with hand lances and harpoons sought only those whales that could be overtaken and killed from small boats. As whales became scarce along the shore, whalers took to the high seas. In the

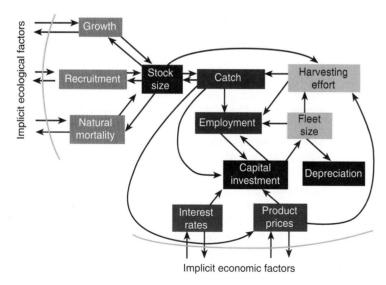

**Figure 25.19** The relationship between sustained yield models and economic factors. Sustained yield models ignore economics which has the greatest impact on resource exploitation, however carefully planned. (From Walters 1984:37.)

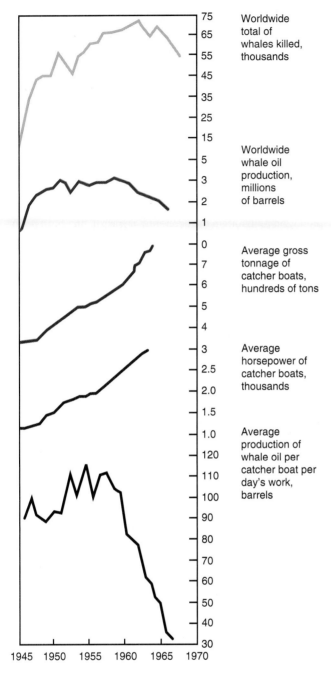

**Worldwide total of whales killed, thousands**

**Worldwide whale oil production, millions of barrels**

**Average gross tonnage of catcher boats, hundreds of tons**

**Average horsepower of catcher boats, thousands**

**Average production of whale oil per catcher boat per day's work, barrels**

1945  1950  1955  1960  1965  1970

**Figure 25.20** Relationship between economics, intensity of harvesting, and the declining stock of whales. These graphs pick up the data on the whaling industry following its resurgence after World War II. From 1945 on more and more whales were killed to produce less and less oil. Boats became larger and more powerful, but their efficiency dropped greatly. Today by international agreement whaling has virtually ceased, although several nations, notably Norway and Japan, are attempting to resume whaling. (From R. Payne 1968.)

sixteenth century whalers hunted off the coasts of Newfoundland and Iceland until stocks failed. The populations of Spitzenbergen and Davis Strait were next to go. Then the colorful New England whaling industry exploited the stock of right whales (*Eubalaena glacialis*). (Right whales were so named because the carcasses floated on the surface; thus they were the right whales to kill.) It peaked in the first half of the nineteenth century, but as stocks of slow-moving, easily exploited species were depleted and as petroleum replaced whale oil as fuel, the New England whaling industry died.

Then two developments put international whaling into business. One was the invention of the explosive harpoon in 1865; the other was the development of more powerful, faster boats that could overtake the swifter whales. The revitalized industry began to concentrate on the plankton-feeding blue whale (*Balaenoptera musculus*) and its relatives, the finback (*B. physalus*) and sei (*B. borealis*) whales. Again stocks were overexploited. The blue whale fishery in Norway ended in 1904, followed by a decline of the species off Iceland, the Faeroes, the Shetlands, the Hebrides, and Ireland. When these areas failed, whalers sailing free-ranging factory ships turned to the Antarctic and the Falklands to concentrate on the small fin whales. The catches rose and the stocks again collapsed. Antarctic whaling was over. Japan and the former USSR developed a whaling industry and hunted the sei, sperm, and other whales wherever they were found. They made investments in large factory ships equipped with helicopters and accompanied by catcher boats that captured whales but with diminishing returns, as measured by less oil and meat per effort and per ship.

Overexploited populations exhibit certain easily discernible symptoms (K. E. F. Watt 1968). Up to a certain rate of exploitation, the stock is able to replace itself. Beyond this critical point, certain changes point to impending disaster. Exploiters experience decreased catch per unit effort as well as a decreasing catch of one species relative to the catch of related species. There is a decreasing proportion of females pregnant, due both to sparse populations and to a high proportion of young nonreproducing animals. The species fails to increase its numbers rapidly after harvest. A change in productivity relative to age and age-specific survival shows that the ability of the population to replace harvested individuals has been impaired. An outstanding example is the blue whale (Figure 25.21). After 1860 the blue whale became the most important commercial species. Catches peaked in 1931 at 50,000 animals and declined to 1000 to 2000 in 1963. For the past 40 years the average age of the blue whale caught in the Antarctic has been six years, mostly immature females or females carrying their first calf. The species is near extinction, but there are indications that with international sanctions, the species is slowly recovering.

Another problem is that traditional population management, especially fisheries, considers stocks of individual

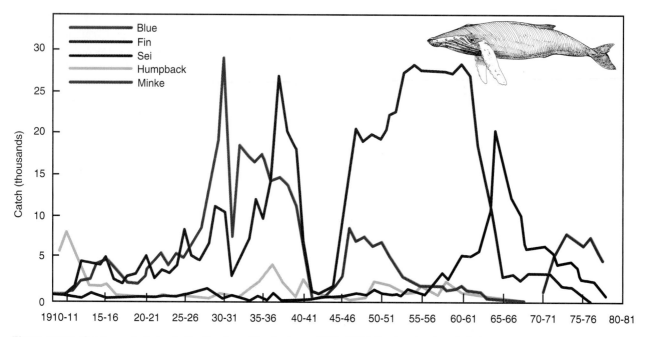

**Figure 25.21** Catches of whales in the Southern Hemisphere, 1910–1977, the focal point of whaling after stocks in the Northern Hemisphere had been depleted. Note the virtual cessation of whaling during World War II years, 1941–1945. The precipitous decline in blue whales (*Balaenoptera musculus*) began before 1940. After World War II the fin whale (*Balaenoptera physalus*) bore the heaviest exploitation, but for a while blue whale take increased. The increase in harvesting intensity on the blue whale points out a truism in resource exploitation. A high harvesting effort on a declining stock can continue if some alternative resource is abundant enough to support that effort. The stock of fin whale supported the incidental harvesting of blue whales. Then the story was repeated with the fin whale. As that stock declined, whalers turned to the smaller sei and minke whales in an effort to maintain their investment in boats and equipment. Finally the whaling industry collapsed, but not until whales were close to extinction. (After Allen 1980.)

species as single biological units rather than as components of a larger ecological system. Each stock is managed to bring in a maximum economic return, overlooking the need to leave behind a certain portion to continue its ecological role as predator or prey. This attitude encourages a tremendous discard problem, euphemistically called "bypass." Employing large drift nets that encompass square kilometers of ocean, fishermen haul in not only commercial species they seek, but a range of other marine life as well, from sea turtles to dolphins, and scores of other species of fish. Fishermen bypass this unwanted catch by dumping it back into the sea. The ecological effects of bypass can be enormous. Because much of the bypass consists of juvenile and undersized fish of commercial species, the practice can seriously affect the future of those fisheries (Bricklemyer et al. 1989). The removal or reduction of other species can interfere with predator-prey interactions in the sea, the dynamics of interspecific competition, and intraguild predation. Such disturbances can alter food webs of ocean ecosystems.

An economic attitude also prevails in the management of some game species. In too many instances biologists have emphasized the increase of recreational opportunities for hunters over the welfare of the species they hunt. For example, rather than restrict human exploitation on moose populations, the approach is to kill wolves to reduce natural predatory loss (see Gasaway et al. 1983).

This reluctance to reduce seasons or to restrict bag limits relates in part to the economics of hunting. Most wildlife programs, directly or indirectly, depend upon hunting license revenue. Loss of such revenue reduces income to wildlife agencies. This reduction has a two-edged effect. It may reduce pressure on some hunted species, but it also reduces money to support wildlife habitat restoration and acquisition, which benefits all species, including endangered ones, and other programs essential to wildlife welfare. A different means of financing these programs is needed.

Management of exploitable populations depends too much on crisis management. No steps are taken to rescue a species until its population has fallen so low that the species becomes endangered. Then it is protected and expensive recovery programs are initiated with the hope that populations will recover. The simpler solution is to manage the population more judiciously in the first place.

# SUMMARY

Herbivore-carnivore systems involve interactions between the second and third trophic levels. These interactions influence the fitness of both predator and prey.

In response to selective pressures of predation, prey species have evolved measures of defense. Chemical defense involves a distasteful or toxic secretion that repels, warns, or inhibits a would-be attacker. Cryptic coloration and behavioral patterns enable the prey to escape detection or inhibit predators. Warning coloration signals to a predator that the intended prey is distasteful or disagreeable in some manner. Usually the predator has to experience at least one encounter with such a prey to learn the significance of a color pattern. Some palatable species mimic unpalatable species, thus acquiring some protection from predators. Another form of defense is predator satiation. Reproduction in a prey population is timed to produce so many young that predators can take only a fraction of them. Those left escape by growing to a size too large for the predator to handle.

Although warning and cryptic coloration and mimicry are usually associated with prey species, such mechanisms are also employed by predators to increase hunting efficiency. Predators also use special hunting tactics and adaptations.

Predation may stabilize prey populations, but more often the relationship results in unstable fluctuations. The ten-year snowshoe hare-lynx cycle involve interactions between vegetation and hare and between hare and lynx. Malnourished hares fall quickly to predators. Recovery of the hares follows the recovery of vegetation.

An unusual form of predation is cannibalism, in which predator and prey are the same species. Often associated with stressed populations, cannibalism can result in pronounced demographic effects within a population, including the loss of younger age classes and lowered reproduction. It can become an important form of population control.

Some predators kill and eat species that use similar resources. This type is called intraguild predation. Although it combines elements of predation and competition, intraguild predation differs from competition because the predator gains energy and nutrients. Intraguild predation differs from classical predation because it reduces potential competition.

Commercial and sport hunting represents predation by humans on natural populations. Some attempt is made at sustained yield, in which the yield per unit time equals production per unit time. Such an approach to management of exploited populations necessarily differs in its effect on $K$-selected and $r$-selected species. Based on the logistic equation, sustained yield models fail to take into consideration all aspects of population dynamics, including natural mortality and environmental uncertainty. Also ignored is the economics of population exploitation. Economic pressures on a common resource encourage overexploitation, resulting in the commercial and even biological extinction of species.

# REVIEW QUESTIONS

1. Discuss the effectiveness of the various kinds of defenses employed by prey species against predators. Do the same for the countermeasures used by predators. Use specific examples.
2. What is mimicry? What is the model? The mimic? Contrast Batesian with Mullerian mimicry.
3. How can predator satiation reduce the impact of predation? What behavioral and physiological characteristics of a species can lead to this type of defense?
4. Can predators regulate prey species? If so, under what conditions?
5. What is the relationship between predator and prey cycles? Why are populations of similar or same species not cyclic in southern regions?
6. What seems to trigger cannibalism in a species? What are some of the selective advantages? Disadvantages?
7. What is intraguild predation? How does this concept relate to the overexploitation of species by humans and to the introduction of an exotic species into the ecosystem?
8. What is the relationship between economics and the management of exploited populations?
9. What are some signs of overexploitation of a population? Choose a particular species endangered by overexploitation, such as haddock or the rhino, and relate those indicators to them.
10. Choose one of the three scenerios for the evolution of cryptic coloration and mimicry and present arguments for its plausibility. (Refer to Guilford 1988, Turner 1988).
11. What ethical, economic, social, and political controversies arise when a major prey species, such as deer or moose, happens also to be a major game species, preyed upon by both wolves and humans?

# CROSS-REFERENCES

Animal nutrition, 124; logistic growth, 393–396; population regulation, 396–402; population fluctuations, 403–406; altruistic behavior, 463; plant mimicry, 525; chemical defense in plants, 527–529; plant-herbivore interactions, 522; herbivore-vegetation cycles, 531–533; parasitism, 558–573.

# Parasitism

## Concepts

1. Parasitism is an association between two organisms in which one, the parasite, benefits at the expense of the other, the host.
2. Parasites may be facultative or obligate.
3. Parasites can be grouped into two types, microparasites and macroparasites.
4. Transmission from host to host may be direct or indirect.
5. Host response to parasites ranges from biochemical to behavioral changes.
6. The population dynamics of hosts and parasites are interactive.
7. Brood parasitism is dependent upon the social structure of both host and parasite.

**Parasitism** is a condition in which two organisms live together but one derives its nourishment at the expense of the other. Parasites, strictly speaking, draw nourishment from the live bodies of another species. Typically, parasites do not kill their hosts as predators do, although the host may die from secondary infection, suffer stunted growth, emaciation, or sterility, or in a weakened condition fall to predators.

Many species of insects have a parasitic larval stage that draws nourishment from the tissues of their hosts. By the time the parasitic larvae have reached metamorphosis, they have completely consumed the soft tissues of the host. These parasites, known as **parasitoids,** essentially act as predators and were considered Chapter 23.

For decades, ecologists paid scant attention to the role of parasites in population dynamics and community structure. A major reason was that in most natural populations the effects of parasitism are difficult to detect. Bodies of victims that die are quickly eaten by detrivores and scavengers, eliminating traces of disease.

However, over 60 years ago Aldo Leopold, in his classic *Game Management* (1933), remarked that the role of disease in wildlife populations was radically underestimated. He hypothesized that parasites control both predators and prey, that they may delimit the range of some species, limit population density, and become involved in population fluctuations. "Disease," he wrote, "does not yield to observational methods of study." Understanding begins only when field observations are combined with experimental study and laboratory techniques. In recent years ecologists have been undertaking the kind of studies that Leopold suggested. As a result we know more about disease in natural populations.

These studies are revealing that parasites do have a pervasive influence on plant and animal populations, just as Leopold surmised. Parasites can increase the death rates and decrease the birthrates of their hosts, and affect their nutritional status, energy demands, and growth rates. Parasites can alter host behavior, affect the outcome of intraspecific and interspecific competition, and increase the susceptibility of the host populations to predation. Parasites may influence mate choice, alter the sex ratios of host populations, and induce sterility (Minchella and Scott 1991). In effect, parasitism may have a greater influence on community structure and population dynamics than predation or interspecific competition.

# CHARACTERISTICS OF PARASITES

Parasites include viruses, many bacteria, fungi, and an array of invertebrate taxonomic groups, including the arthropods. The presence of a heavy load of parasites is considered an infection, and the outcome of an infection is a disease. A **disease** is any condition of a plant or animal that deviates from normal well-being. Not all parasites, however, are agents of disease.

Parasites exhibit a tremendous diversity in the ways they exploit their hosts (for overview see Croll, 1966). They may parasitize plants or animals or both. They may be **ectoparasites** that live on the outside of the host or **endoparasites** that live within the body of the host. Some are full-time parasites, others only part-time. Part-time parasites may be parasitic as adults and free-living as larvae or the reverse.

Parasites have developed numerous ways to gain entrance to their hosts, even to the point of using several hosts as dispersal agents. They have evolved various means and degrees of mobility, ranging from free-swimming ciliated forms to ones totally dependent upon other organisms for transport. They have developed diverse ways of securing themselves to the host to maintain position and means of surviving the biochemical hazards of living inside a host.

Parasites may be restricted to one or a limited number of species or genera of host. A number of parasites of birds, especially certain tapeworms, live only in one particular order or genera (see Baer 1951). Some parasites live their entire life cycle on one host, whereas others require several hosts.

The usual approach to typing parasites has been taxonomic: tapeworms, roundworms, and the like. May and Anderson (1979) have suggested that parasites be distinguished on the basis of size, as microparasites and macroparasites. **Microparasites** include the viruses, bacteria, and protozoans. They are characterized by very small size and a short generation time. They develop and multiply rapidly within the host and tend to induce immunity to reinfection in hosts that survive initial infection. The duration of the infection is short relative to the expected life span of the host. Transmission of these parasites from host to host is direct, although they may involve some other species as a vector.

**Macroparasites** are relatively large in size. In animals they include parasitic worms, the platyhelminths, acanthocephalans, roundworms, flukes, lice, fleas, ticks, mites, and fungi (Figure 26.1). Parasites of plants include fungi, rusts,

**Figure 26.1** Lungworm in the air passages of a white-tailed deer.

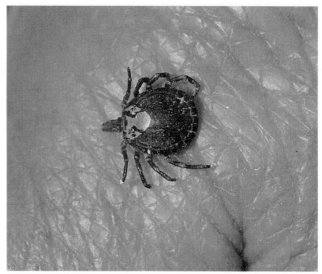

**Figure 26.3** A tick (*Dermacentor*) on a human. Ticks are carriers of such diseases as Rocky Mountain spotted fever and Lyme disease.

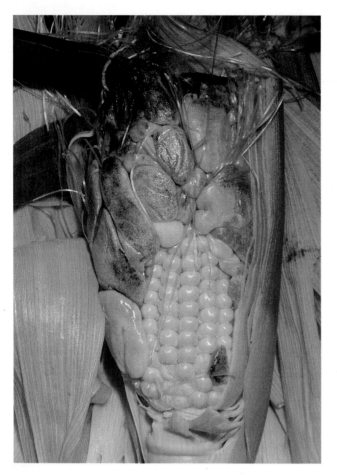

**Figure 26.2** Smut, a fungal infection, on ears of corn.

smuts, and plants such as dodder, broomrape, and mistletoe (Figure 26.2). Macroparasites have a comparatively long generation time, and direct multiplication in the host is rare. The immune response macroparasites stimulate is of short duration and depends upon the number of parasites in the host. Macroparasites persist in hosts by continual reinfection. They may spread by direct transmission from host to host or by indirect transmission, involving intermediate hosts and vectors.

## HOSTS AS HABITAT

Hosts are homes for parasites, and parasites have exploited every conceivable habitat on and within them. In fact, they represent the extreme in specialization for resource exploitation (Price 1980). Among animals some parasites live on the skin hidden within the protective covers of feathers and hair (Figure 26.3). Some burrow beneath the skin. Others live in the bloodstream, in the heart, brain, digestive tract, liver, spleen, mucosal lining of the stomach, spinal cord, or brain, in the nasal tract and lungs, in the gonads, bladder, muscle tissue, pancreas, eyes, or gills of fish, to mention some sites

among many. Parasites of insects live on the legs, on the upper and lower body surfaces, and even on mouthparts.

Parasites within the host will colonize different sites within the same organ system. For example, different species of the coccidian protozoans of the genus *Eimeria* inhabit different regions in the guts of mammals: one in the duodenum, another in the lower duodenum and upper small intestine, a third in the lower small intestine, and a fourth in the caecum and rectum.

Plant parasites, too, divide up the habitat. Some live on the surface of roots and stems; others penetrate the roots and bark to live in woody tissue beneath. Some live at the base of the stem (root collar) where the plant emerges from the soil. Others live within the leaves, on young leaves, on mature leaves, on flowers, pollen, or fruits.

A major problem for parasites, especially parasites of animals, is gaining access to and escape from the host. Parasites of the alimentary tracts of vertebrates enter the host orally and escape by the way of the rectum, a path used by other parasites. Parasites of the lungs enter orally or penetrate the skin and travel to the lungs by the pulmonary system. They escape mainly by being coughed up and swallowed into the alimentary tract. Liver parasites, exploiting one of the nutritionally richest habitats in the animal body, arrive there by way of the circulatory system, bile duct, and hepatic portal systems and escape by the same route. Parasites of the urogenital systems enter orally and travel through the gut to the site of residency and exit through the urinary system. Blood parasites enter and escape by way of the skin, but always with the aid of some obliging vector such as mosquitoes and ticks. Parasites then end up in muscle tissues, where they usually exit in capsules, reaching a blind end. For them the only way out is for their host to be killed and eaten by a predator.

# LIFE CYCLES

Parasites face unique problems moving from host to host. For the parasite, the host is both habitat and food. Many parasites have evolved complex life cycles geared to escape and relocation. For most parasites the hosts function as islands from which they must escape; then they must survive in a generally hostile external environment while locating other host islands, some of which are occupied and some of which are not. For animal parasites the host islands are movable.

Parasites cannot move from host to host at will. They can escape only during a transmission stage, no matter whether the means of dispersal is direct or indirect. Successful transmission depends upon contact between the host and the infective stage. The infective stage is essential.

All parasites reach a stage in their life cycle within the host when they can develop no further. **The definitive host** is the one in which the parasite becomes an adult and reaches maturity. All others are **intermediate hosts,** which harbor some developmental phase. Parasites may require one, two, or even three intermediate hosts. Each infective stage can develop only when it is independent of the definite host, and it can continue its development only if it can find another intermediate or its definitive host. To this end many parasites exploit the hosts' feeding habits or behavioral patterns. Thus the population dynamics of a parasite population is closely tied to the population dynamics of the host.

## Direct Transmission

Direct transmission is the transfer of the parasite from one host to another by direct contact or through a carrier or vector. The life cycle of the parasite does not involve intermediate stages in a secondary host. Most microparasites are transmitted directly.

One such disease of wild mammals and humans is rabies, transmitted through the saliva by the bite of a rabid animal. The rabies virus follows the nerves from the infection point to the spinal column and brain before symptoms appear. The symptoms are both behavioral and physical. Infected animals are excitable and restless, exhibit convulsions, wander aimlessly, and show no fear of humans. Physically they are emaciated, exhausted, and partially paralyzed. Among wild animals, rabies occurs most commonly in coyotes, foxes, skunks, raccoons, and bats (Figure 26.4). Foxes, dogs, and raccoons are the primary transmitters of the disease.

Microparasites that infect plants are also transmitted directly when the plant comes into contact with the virus or resting spores in the soil. In the presence of suitable hosts the spores germinate and penetrate the roots. Or spores may be wind-carried and come to rest on and infect leaves of the plant. Others are carried by insect vectors. One example is elm phloem necrosis, caused by a mycoplasmal organism. It is transmitted by root grafts (a situation in which roots of one tree become grafted onto the roots of a neighbor) or by the widely distributed white-banded elm leafhopper

● Arctic Fox
● Red Fox
● Insectivorous Bats
● Raccoon
● Striped Skunk
● Grey Fox
● Vampire Bat

**Figure 26.4** The distribution of wildlife rabies in North America. The many vector species include the fox, raccoon, skunk, and bat. In Europe the only wildlife vector is the fox. In South and Central American the major vector is the vampire bat. (From Steck 1982:58.)

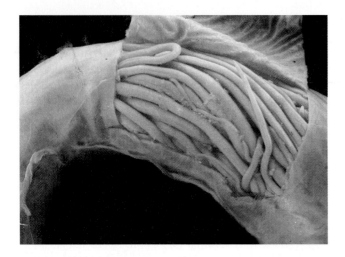

**Figure 26.5** Roundworms (*Ascaris*) in the intestine of a pig.

(*Scaphoideus luteolus*). This insect ingests the mycoplasma while feeding on the sap of a diseased tree and transmits it while feeding on a new host tree.

Many important macroparasites of animals and plant move directly from infected to uninfected host by direct contact. Parasitic nematodes (*Ascaris*) live in the digestive tracts of mammals (Figure 26.5). There female roundworms lay thousands of eggs, which are then expelled with the feces. If they are swallowed by the host of the correct species, the eggs hatch in the intestines, bore their way into the blood vessels, and come to rest in the lungs. From here they ascend to the mouth, usually by causing the host to cough, and are swallowed again to reach the stomach, where they mature and enter the intestines.

Nematodes are important macroparasites of plants. The female of one species, the eelworm, *Heterodera schachti,* lives a parasitic existence in the roots of such plants as tomatoes and beets. Fertilized by the free-living male, the female sends thousands of larvae into the soil, ready to take up a parasitic existence in other plants.

Most important external and debilitating parasites of birds and mammals are spread by direct contact. They include lice, mange mites, ticks, fleas, and botfly larvae. Many of these parasites lay their eggs directly on the host, but fleas lay their eggs and their larvae hatch in the nests and bedding of the host (even in the shag rugs of homes with dogs and cats), and eventually leap onto nearby hosts.

Some fungal parasites of plants spread through root grafts, and others are carried by insect vectors. For example, an important fungal infection of white pine (*Pinus strobus*), *Fomes annosus,* spreads rapidly through pure stands of the tree by root grafts. The devastating Dutch elm disease caused by the introduced fungus *Ceratocystis ulmi* is spread from tree to tree by spore-carrying elm beetles, *Scolytus multistriatus* and *Hylurgopinus rufipes.*

Plant macroparasites even include a number of flowering plants themselves. One group are **holoparasites,** plants that lack chlorophyll and draw their water, nutrients, and carbon from the roots of host plants. Notable among them are members of the broomrape family, Orobanchaceae. Two are the familiar squawroot (*Conopholis americana*), which parasitizes the roots of oaks, and beechdrops (*Epifagus virginiana*), which parasitizes mostly beech trees (Figure 26.6).

Another group are the **hemiparasites.** They are photosynthetic, but they draw water and nutrients from their host plant. The most familiar hemiparasites are mistletoes, whose

**Figure 26.6** Squawroot, a member of the broomrape family, is a hemiparasite on the roots of oak.

sticky seeds attach to limbs (see Chapter 27), send out roots that embrace the limb, and send roots into the sapwood. Mistletoe can reduce the growth of its host.

## Indirect Transmission

Many parasites, both plant and animal, utilize indirect transmission. During different stages of the life cycle they require different hosts. The brainworm (*Parelaphostrongylus tenuis*) (in spite of its name the parasite is a lungworm) of the white-tailed deer has as its intermediate host during its larval stage a snail or slug that lives in the grass (Figure 26.7) (R. C. Anderson 1963, 1965). The deer ingests the infected snail while grazing. In the deer's stomach, the larvae leave the snail, puncture the deer's stomach wall, enter the abdominal membranes, and follow the lumbar and other nerves to the vetebral canal carrying the spinal cord and on to the spaces surrounding the brain. Here, about 40 days after infection, the subadults mature and migrate forward to the cranium,

where they deposit eggs in the venous circulatory system. Eggs pass through the bloodstream to the lungs, where they develop into first-stage larvae. The larvae move into the air sacs of the lungs, are coughed up by the deer, swallowed, and passed out through the feces. The larvae invade the foot of snails as the snails move across the feces. In the snails the larvae grow, molt twice, and give rise to the infective stage that the deer acquire as they ingest the snails clinging to the vegetation. In white-tailed deer, infestations of this parasite produce minor or no symptoms. In moose and elk, infestations cause serious and often fatal neurological symptoms.

Other parasites need two intermediate hosts. An example is the black grub or black spot infection of minnows and sunfish (Figure 26.8), caused by a small white larvae of a fluke (*Trematoda*), *Uvulifer ambloplitis*. Once infected, the fish host lays down a black pigment around the thick-walled cysts, causing black spots on the body wall. The adult stage is attached by suckers to the mucosa of the intestine of the primary host, the kingfisher. The eggs pass through the in-

**Figure 26.7** Life cycle of a macroparasite, the brainworm *Parelaphostrongylus tenuis,* with indirect transmission in white-tailed deer, moose, and elk. (After Strelive in Davidson 1981:141.)

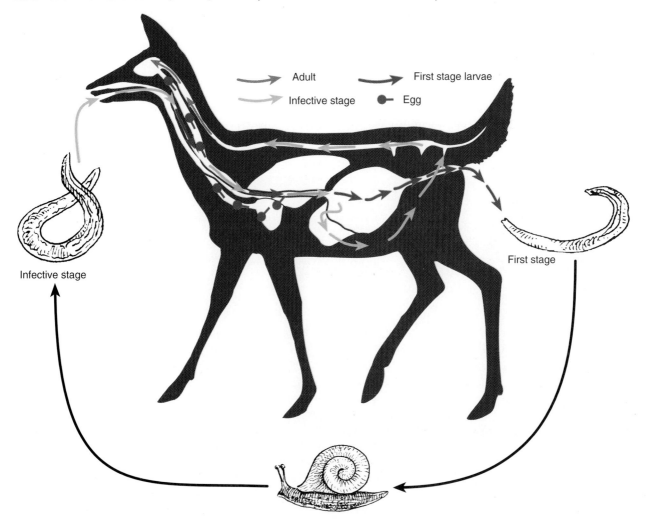

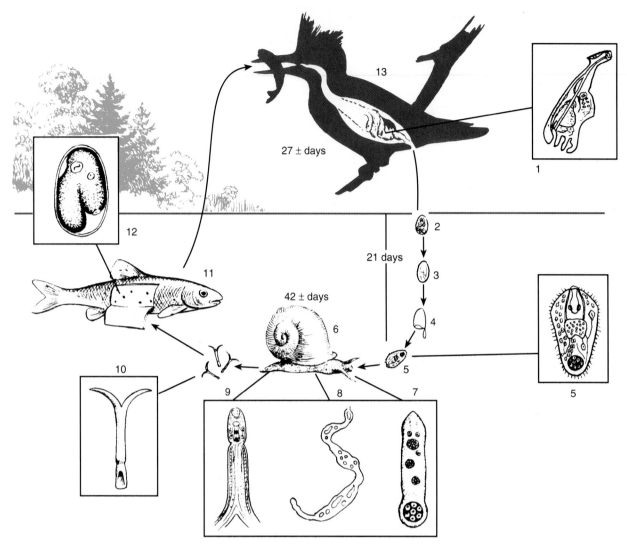

**Figure 26.8** The life cycle of the black grub, *Uvulifer ambloplitis*. (1) Adult trematode in the intestine of the kingfisher; (2) immature egg; (3) mature egg; (4) empty shell; (5) miracidium, the ciliated larva of the fluke that parasitizes the snail; (6) the first intermediate host, the snail; (7) and (8) sporocysts; (9) cercaria, larval form of the fluke produced asexually in the snail; (10) free-swimming cercaria, which penetrates beneath the scales or into the fins of fish and encysts, producing black spots in about 22 days; (11) minnow with skin cut away to show black grubs; (12) sketch of parasite within inner cyst; (13) kingfisher. (Adapted from Hunter and Hunter 1934, New York State Conservation Department.)

testine of the kingfisher into the water, where the miracidia hatch. The miracidia, moving through the water by cilia, seek an intermediate host, a snail, to which they gain entrance by secreting a tissue-dissolving substance. Within the snail, the miracidia transform into saclike sporocysts, which eventually give rise to tailed larvae called cercaria. These free-swimming larvae emerge from the snail and seek the next intermediate host, a fish, which they must find within two days or die. The cercaria penetrate the fish and encyst. The fish, in turn, is eaten by the kingfisher, the cercaria become free and mature in the bird's intestinal tract. The new adults start laying eggs to repeat the cycle.

Indirect transmission among plant macroparasites is uncommon except among the rusts, which employ the wind to carry the infective stages from primary and intermediate hosts. An example is white pine blister rust (*Cronartium ribicola*) (Figure 26.9). The rust produces spores in the diseased bark of pine in the spring. These spores, which cannot reinfect pines, are carried by the wind. Some may land on species of *Ribes,* or gooseberry shrubs. On the infected leaves of *Ribes* the rust produces two kinds of spores, early and late summer forms. The early summer form infects other *Ribes* only, intensifying infection on these plants during the summer. Late summer spores infect only white pine

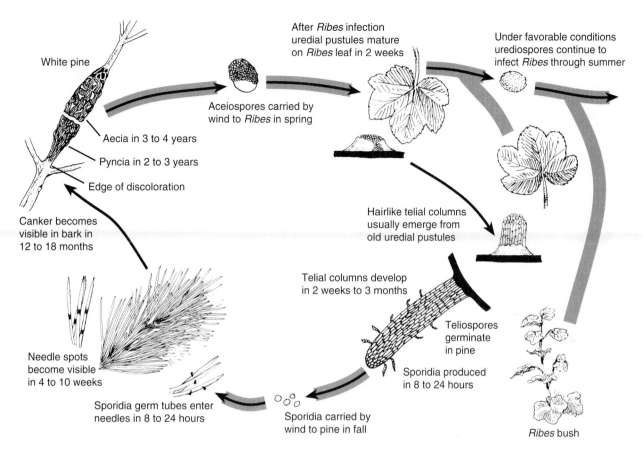

**Figure 26.9** Life cycle of the white pine blister rust. The rust infection on the shrub gooseberry, *Ribes,* appears as yellowish to orange pustules of spores. An annual infection, it has little effect on *Ribes.* On the white pine the infection is perennial and appears as large orangish blisterlike cankers on the bark.

needles. Carried by the wind, the spores germinate on the needles and grow until they reach the bark. There the rust produces spindle-shaped diseased areas or cankers. Two to five years after infection, and annually thereafter, spores produced in the bark are carried by the wind to nearby *Ribes.* Like other parasites that require intermediate hosts, conditions for transmission are critical. These include a widespread infection of *Ribes,* an abundant production of pine-infecting spores, favorable temperature, moisture, and wind conditions, and hosts within 280 meters of the infected *Ribes,* the maximum carrying distance of spores on the wind. The disease has been controlled by eliminating *Ribes* from within and near white pine stands.

## Dynamics of Transmission

Transmission from host to host is the key to parasitic existence. It can take place only with the dispersal of an infective stage independent of the definitive host. Parasites requiring more than one host reach only a certain stage in their life cycle in each. They can complete the life cycle only if they can infect another host. The sexual form of the malarial parasite cannot survive in the mosquito. It must transfer to a vertebrate host to develop into an adult stage. The brainworm of deer has to locate a snail as an intermediate host to continue its development to an infective stage that can be transmitted back to the deer. Even more complex is the trematode infecting the kingfisher and other fish-eating birds. It has to develop three infective stages. Each stage has to be uniquely adapted to its host. It has to be able to access the host, overcome the immune response of the host, and escape from the host. For many animal parasites this process means exploiting the feeding habits of the definitive host and adapting to the habits of the intermediate hosts. The brainworm of the deer exploits the snail's habit of crawling up grass stems where it risks being eaten along with the grass by a grazing deer. Unless the snail is swallowed by the deer, the parasite will perish.

Transmission of parasites is further complicated by the patchy or clumped distribution of the hosts. Only a few members of the host population harbor major parasite loads and act as reservoirs of the infection. Uninfected hosts are widely scattered or intermixed among populations of other species, so the probability of the parasite or its carrier com-

ing into contact with the susceptible individuals is low. In other words transmission is highly dependent on both the density of hosts and the distance between parasite and potential host. Transmission of parasites is most successful when the population of potential hosts is dense, particularly if the parasites depend upon direct contacts among hosts. The rapid spread of viral and bacterial diseases in dense populations is called **epizootics** in animal populations and **epidemics** in humans. Such epizootics involve virulent forms that sweep through susceptible populations as an advancing front, exemplified by the spread of rabies through Europe (Figure 26.10).

Some parasites depend on predator-prey relationships. The tapeworm *Echinococcus* (Cestoda) lives in the intestines of carnivores. It spends its larval stage as hydatid cysts in the muscles and tissues of any organ in its immedi-

ate hosts, herbivores, and under certain conditions humans closely associated with them. When a carnivore feeds on the flesh of infected herbivore prey, the hydatid cysts gain entrance to the digestive tract of the carnivore and develop into adult tapeworms. One species, *E. granulosus,* the dog tapeworm, is found in dogs and wild canids, especially the wolf. A second species, *E. multilocularis,* has a restricted range in the northern regions of Asia, Europe, and North America. Its definitive host is the Arctic fox and its intermediate host several species of voles. Both species of tapeworm appear to change the behavior of its intermediate host to ensure transmission back to its definitive host. Wolves preferably select moose infected with hydatid larvae, which concentrate in their lungs; their presence may be indicated by some odor on the breath of infected moose, by excretions, or by some aberrant behavior. Voles heavily infected with hydatid larvae

**Figure 26.10** The spread of rabies in Europe. Originally diagnosed in Poland in 1939, it has spread westward over 1400 km of Europe at a rather constant rate of about 20–60 km per year. This epizootic developed independently of dogs and has spread across the continent with foxes as the principal vector. (From Steck 1982:62.)

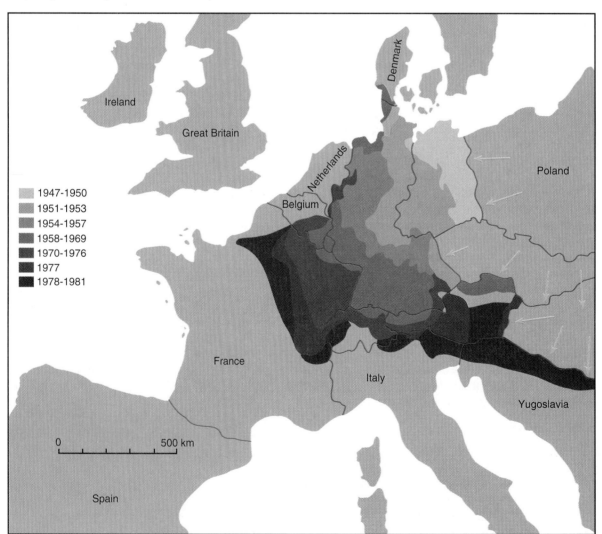

of *E. multilocalaris,* which rapidly grow in and destroy the liver, appear large and fat, but they have difficulty moving and fall easy prey to the fox.

# HOST RESPONSES

Just as prey respond to predators, so do hosts react to invasions of parasites. Some responses are defensive, attempts to prevent parasitic invasion. Other responses follow parasitic infection.

## Biochemical Responses

A major form of defense among animals is biochemical and cellular response. One response is inflammation, provoked by the death or destruction of host cells. The infection stimulates the secretion of histamines and increased blood flow to the site, bringing in phagocytes, lymphocytes, and leucocytes to attack the infection. Scabs form on the skin by fibrosis, as in the case of heavy mange mite infestations on red fox and other canids (Figure 26.11). Internal reactions can produce calcareous cysts in muscle or skin that imprison the parasite. Examples are the cysts of the infective stage of the trematode in its intermediate fish host (see Figure 26.8) and the cysts of the roundworm *Trichinella spiralis* (Nematoda), which causes trichinosis in humans, in the muscles of pigs and bears.

The ultimate biochemical response involves the immune system. Depending upon the parasite, this response may produce short-term or life-long immunity to the infection. When a foreign protein or antigen enters the bloodstream, it is taken up by lymphocytes, which produce a molecular template corresponding to the antigen. This template is known as an antibody. The antibodies in the bloodstream neutralize invading antigens.

The immune response, however, can be breached. Antibodies specific to an infection normally are composed of proteins. If the animal suffers from poor nutrition and its protein deficiency is severe, its normal production of antibodies is inhibited. This depletion of energy reserves breaks down the immune system and allows the virus or other parasites to become pathogenic. The ultimate breakdown in the immune system occurs in humans infected with the human immunodeficiency virus (HIV), the causal agent of AIDS. The virus attacks the immune system itself, exposing the host to a range of deadly infections.

## Abnormal Growths

The host may also react to parasitic infections with abnormal growth. Plants respond to the invasion of their tissues by bacteria and fungi by cyst formation in the roots; scab formation in fruits and roots, which cuts off contact of the fun-

**Figure 26.11** Encrustations of mange, caused by a mite on fox.

gus with healthy tissue; root knot, a reaction to nematodes in tomatoes and other plants; black knot, a limb canker in black cherry (*Prunus serotina*) in response to the fungus *Apiosporina morbosa;* and root nodules in legumes. They react to attacks on leaf, stem, fruit, and seed by gall wasps, bees, and flies by forming abnormal growth structures unique to the particular gall insect (Figure 26.12). Production of galls does not benefit the plant. In fact, a heavy infestation of gall insects, particularly leaf galls, can weaken and kill the plant. However, gall formation can expose the larvae of some galls to predation. The conspicuous, swollen knobs of the goldenrod stem gall, for example, attract the downy woodpecker (*Picoides pubescens*), who excavates and eats the larva within the gall.

Malarial parasites in vertebrates can stimulate the spleen into producing many red blood cells and antibodies, resulting in its enlargement. Certain helminth parasites can produce cancerous tumors in their mammalian hosts. Larvae of warble flies (*Oestridae*) develop in boil-like swellings (warbles) just under the skin (Figure 26.13) These growths subside after the transformed larvae escape through a hole cut in the skin. Invertebrates, especially mollusks, respond to irritants or parasitic larvae by pearl formation. Pearl formation is usually associated with some inanimate foreign body, but often it is the response of an invertebrate host to a parasitic infection.

## Sterility

Parasitic infection of plants and certain invertebrates can destroy or alter a host's reproductive system. Consider the ex-

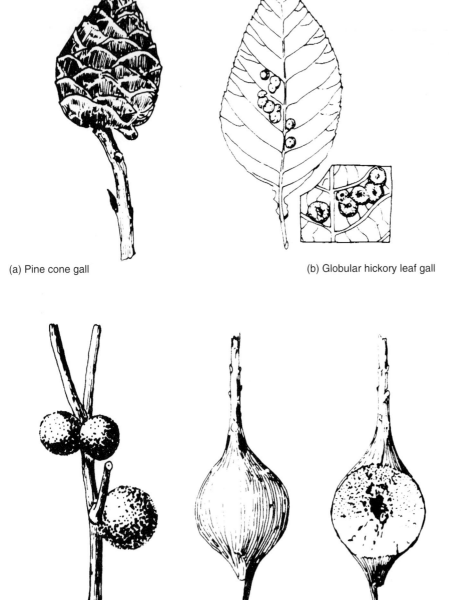

(a) Pine cone gall

(b) Globular hickory leaf gall

(c) Oak bullet gall

(d) Goldenrod ball gall

**Figure 26.12** Galls are a morphogenic response by plants to an alien substance in their tissues, in this case the presence of a parasitic egg. The response involves a genetic transformation of the host's cells. (a) The pine cone gall, a bud gall caused by the gall midge, *Rhabdophaga strobiloides*. (b) The hickory leaf gall, induced by a gall aphid, *Phylloxera canyaeglobuli*. (c) Oak fig root gall, a fleshy, fig-shaped white gall caused by a gall wasp, *Belanocnema treatae*. (d) The familiar goldenrod ball gall, a stem gall induced by a gallfly, *Eurosta solidagin*.

ample of four fungus-eating *Drosophila* flies of eastern North America: *D putrida, D. falleni, D. recens, and D. testaca.* These flies are parasitized, often heavily, by the gut nematode *Howardula aoronymphium* (Allantonematidae). The female nematode, living in a mushroom, infects *Drosophila* larvae on the mushroom by penetrating the cuticle of the larvae. The female grows in the larvae, produces eggs, and releases larval nematodes in the hemocoel of adult flies. The larval nematodes exit through the anus or ovipositor when the fly visits a mushroom. Although the nematode does not

interfere with male mating success, it does cause sterility in female flies. Of the infected females in the populations of several species studied by J. Jaenike (1992), one hundred percent of *D. putrida*, 90 percent of *D. testaca*, and about 50 percent of female *D. falleni* are completely sterile. This parasite-induced sterility can have a pronounced effect on the host populations. Nematode parasites can reduce the potential growth rate of a population of *D. testaca* by 24 percent, of *D. putrida* by 11 percent, and of *D. falleni* by 5 percent. Because all three of these flies feed on the same species of

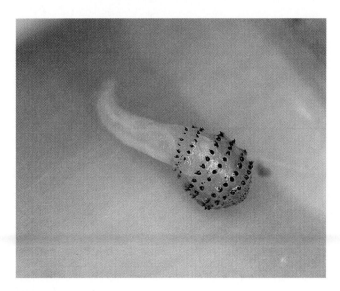

**Figure 26.13** Larvae of the botfly (*Cuterebra emasculator*) develop beneath the skin of an infected mammal, producing swellings or warbles. The larvae emerge through the skin.

mushrooms (Jaenike and James 1991), parasitism can influence the outcome of interspecific interactions among them.

Infection of plants, notably grasses and sedges, by systemic endophytic fungi that grow within the leaves and stems and by nonsystemic smuts and rusts that grow externally can prevent the plants from reproducing sexually by "castration"—inhibiting floral development or aborting developing flowers (Clay 1991). Nonsystemic fungal parasites, the smuts and rusts, which comprise most of the parasitic fungi of plants, colonize new hosts by contagious spread of spores, as in white pine blister rust. Most castrating fungi are systemic endophytes that spread by vegetative growth into tillers and tubers. Although inhibiting sexual reproduction by "castration," these endophytic fungi stimulate vegetative reproduction or cloning. By doing so, such fungi increase the availability of new hosts and prevent the evolution of resistant progeny through sexual reproduction. By interfering with sexual reproduction, such endophytic parasites run the risk of reducing the genetic diversity of their hosts, especially if those hosts are annual plants, making them vulnerable to adverse environmental changes.

## Behavioral Changes

Heavily parasitized animals often behave abnormally. Rabbits infected with the bacterial disease tularemia (*Pasteurella tularensis*), transmitted by the rabbit tick (*Haemaphysalis leporis-paulstris*), are sluggish and difficult to move from cover. Rabid foxes and raccoons may be overly aggressive or overly tame and unafraid of human contacts.

Horsehair or gordiid worms (Nematomorpha) live in and lay their eggs in ponds and livestock troughs. The tiny motile armed larvae that hatch from the egg invade host insects, including wasps, bees, and beetles that visit the water. Approaching maturity, the larvae must return to water. The parasitic larvae seemingly force their hosts to return to water, into which the hosts often dive suicidally. Then the gordiid larvae burst out of the insect and swim free in the water to live a brief life as a reproducing adult (Croll 1968).

Birds and mammals respond defensively to ectoparasites by grooming. Among birds the major form of grooming is preening, which involves manipulation of plumage with the bill and scratching with the foot. Both activities remove adults and nymphs of lice from the plumage (Clayton 1991).

## Mate Selection

Parasites may affect mate selection in birds. The hypothesis has been advanced that parasitic infections or resistance to such infections can influence which mate a female chooses (Hamilton and Zuk 1982, Zuk 1991). Supposedly males possessing the brightest plumage or other ornaments are the most free of of parasites. By choosing males with the brightest colors, females are selecting the healthiest males. This hypothesis is highly controversial (Real 1988). Evidence used to support it comes from correlation studies relating mating success with parasitic loads (Zuk 1991, Hudson and Dobson 1991). The hypothesis has yet to be tested rigorously with experimental studies involving mate choice and controlled parasitic infection of males.

# POPULATION DYNAMICS

By now it should be apparent that interactions of parasites and hosts involve three or more populations. One is the host, the growth rate of which is influenced by intensity of parasitic infection. The second is the adult parasite in the definitive host, the growth rate of which is influenced by intraspecific competition within the host and the dispersion of parasites among the host individuals. The third is the free-living infective transmission stage, which experiences many of the same population limitations as the adult parasites. The growth of the adult population of the parasite in the definitive host is distinct from the production of the infective stage (or stages), which has its own birthrate and growth rate.

## Models of Parasitism

Anderson and May (1978, 1979) and May and Anderson (1978, 1979) provide theoretical models for the growth of the three different populations. The assumptions are that parasites increase the rate of mortality of the host and that the rate of mortality is a function of the number of parasites per host. The basic models can be modified to consider direct and indirect transmission of parasites, parasites that

reproduce within the definite host, time lags, and the like. The basic models are:

$$\text{Host: } dN/dt = (a - b)N - \alpha P$$

$$\text{Parasite: } dP/dt = BwN - (\mu + b + \alpha)P - \alpha(k - 1)P^2/kN$$

$$\text{Free-living infective stage: } dw/dt = \lambda P - cw - \beta wN$$

where $N$ is the number of hosts, $t$ is time, $a$ is the instantaneous host birthrate, and $b$ the instantaneous host death rate; $\alpha$ is the instantaneous parasite-induced host death rate, $P$ the number of adult parasites, $\beta$ the rate of parasite transmission (hosts acquire individual parasites at a rate proportional to the number of contacts between host and parasitic infective stages $\beta wN$), $w$ the number of free-living infective stages, and $k$ the parasites distributed as a negative binomial (degree of overdispersion); $\mu$ is the natural mortality rate of adult parasites, $\lambda$ is rate of production of infective stages by adult parasites, and $c$ is the death rate of infective stages.

The host model states that the growth of the host population $dN/dt$ is determined by its own rate of increase, $(a–b)N$, reduced by parasite-caused mortality, $\alpha P$. The parasite model shows the relationship among the population size and transmissivity of the infective stage, reduction in adult parasite numbers caused by parasite-induced host mortality, natural host death rate, natural mortality of adult parasites, and parasite-induced host death rate as influenced by dispersion or intensity of infection of hosts by adult parasites. The infective stage model relates to production of infective or intermediate stages that is decreased in part by loss of population due to invasion of the definitive host and transformation into adults.

## Parasitism as a Regulatory Mechanism

The impact of parasites on host populations relates in part to the nature of transmission and the density and dispersion of the host population. Microparasites, dependent for the most part on direct transmission, require a high host density to persist. For them ideal hosts live in groups or herds. To persist the parasites need a long-lived infective stage that does not ensure long-term immunity in the host population. Immunity reduces populations of a parasite, if indeed it does not eliminate them. An example of a parasite in wild populations that does not confer long-term immunity is rabies; one that does confer immunity to animals that survive the disease is distemper.

Indirect transmission, typical of macroparasites, is more complex. To persist the parasite requires a highly effective transmission stage, which often involves a close association with food webs. Parasites with indirect transmission exist at low population levels, but because of efficient transmission they do well and persist for long time in low density populations of hosts. The longevity of each parasitic stage varies in different hosts. Longevity is higher in the definitive host than in the intermediate hosts.

To regulate a host population the parasite has to increase the host's death rate or decrease its reproductive capability. Parasites can do so only when they can maintain a reservoir of infection in sparse populations of the host. When host populations become dense and direct contact is frequent among individuals carrying the disease and susceptible ones, the parasite spreads rapidly through the population with high mortality. Such effects are most evident when parasites are introduced into a population with no evolved defenses. In such cases the disease may be density-independent. It can reduce populations, exterminate them locally, or restrict the distribution of the host. For example, the chestnut blight fungus, *Endothia parasitica,* spread rapidly through the American chestnut (*Castanea dentata*), eliminating it as a commercial timber species and changing the composition of the eastern deciduous forest. Rinderpest, a viral disease of cattle, introduced into Somalia by cattle importations from either Arabia or India between 1884 and 1889, spread swiftly through populations of native cattle, wildebeest, and African buffalo, decimating those species. After several periodic epizootics, the disease was removed by vaccination of cattle. Rinderpest has had a pronounced effect on the ecology of the East African savanna ecosystem that extended far beyond the ungulate species involved (see Sinclair 1977, 1979).

These are extreme cases and hardly regulatory. What about directly transmitted endemic diseases maintained in the population by a small reservoir of infected carriers? Outbreaks of these diseases appear to occur when the density of the host population is high, and they tend to sharply reduce host populations. Examples are viral distemper in raccoons, which can be significant in controlling their populations (Gorham 1966, Haberman et al. 1958), and a macroparasite, the mange mite, *Sarcoptes scabiei,* in foxes (H. J. Smith 1978, Tullar 1979).

The distribution of macroparasites, especially those with indirect transmission, is highly clumped or overdispersed (Figure 26.14) Some individuals in the host population carry a higher burden of parasites than others. These individuals are the ones that are most likely to succumb to parasite-induced mortality, suffer reduced reproductive rates, or both (Figure 26.15). Such parasitically induced deaths often are not caused directly by parasites but by secondary infections. Herds of bighorn sheep (*Ovis canadensis*) may be infected with up to seven different species of lungworms (Nematoda), involving 100 percent of the animals. Highest infections occur in the spring when the lambs are born. Heavy lungworm infestations in the young bring about a secondary infection, pneumonia, that kills the lambs. Such infections tend to stabilize or sharply reduce mountain sheep populations by reducing reproductive success. Wildlife biologists successfully reduce the parasite load and increase lamb survival with chemotherapy by treating pregnant ewes with an anthelmintic drug disguised in highly palatable fermenting apple mash. Indiscriminate use of anthelmintic drugs has

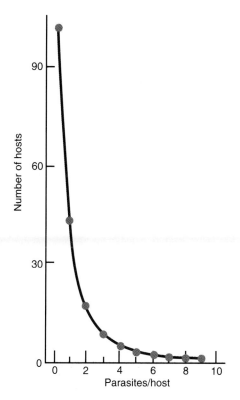

**Figure 26.14** Overdispersion of the tick *Ixodes trianguliceps* (Birula) on a population of the European field mouse, *Apodemus sylvaticus*. Most of the individuals in the host population carry no ticks. A few individuals carry most of the parasite load. (From Randolph 1975:454.)

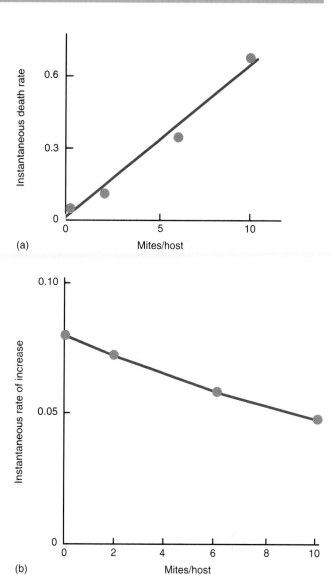

**Figure 26.15** The influence of parasitism by the mite *Hydryphantes tenuabilis* on the water measurer *Hydrometra myrae*, a common hemipteran on ponds. (a) The effects of parasite density in a host on the rate of parasite-induced mortality of the host. (b) The effect of parasitic infection on fecundity. (From Lanciana 1975:691.)

been criticized. Although the drugs reduce parasite-induced mortality, they can increase herbivore growth rates and feeding rates, intensifying the impact on the grazing system (Grenfell 1992).

High populations of hosts may result in a more uniform distribution of parasites (underdispersions), which may have some significance in population regulation. Carroll and Nichols (1986) monitored a high population of meadow voles (*Microtus pennsylvanicus*) and its important ectoparasite, the larvae and nymphs of the American dog tick (*Dermacentor variabilis*). During high vole density nearly all voles examined were parasitized. There was no significant difference ($P > 0.05$) in tick burdens between sexes on an annual basis, but juvenile and subadult voles harbored significantly fewer larval and nymphal ticks than adults. When the vole population declined over a two-year period— 1983 and 1984—adult ticks also declined sharply. Adult ticks in 1983 would have fed as nymphs on a declining population in the spring of the year when the vole population was its lowest. The greater scarcity of adult ticks in 1984 was the result of low populations of both adult ticks and voles in 1983. This study suggests that the dynamics of host and parasite populations are closely related.

Impacts of parasitism on host populations may be the outcome of the combined effects of infestations of a number of different species of both ectoparasites and endoparasites, a case of "diffuse" parasitism. The eastern cottontail rabbit (*Sylvilagus floridanus*) has been declining over parts of its range, in spite of no apparent changes in its habitat. Jacobson, Kirkpatrick, and McGinnes (1978) investigated two local rabbit populations in Virginia occupying habitats similar in structure and vegetational composition, but differing in temperature, humidity, and soil conditions. One population was abundant and the other, inhabiting a locality with

higher summer temperatures and humidity, conditions favorable for many parasites, was chronically low for a number of years. The low population carried a high parasitic load. These parasites included a high infestation of rabbit ticks (carriers of the fatal bacterial disease tularemia), botflies, nematodes, a very high load of intestinal flukes, stomach worms causing hemorrhagic gastritis, anemia, and weight loss, and staphylococcus and subcutaneous infections. Highest parasitic infections occurred in spring, resulting in low rabbit reproduction. The other population carried a much lower infestation of a somewhat different array of parasites. Field evidence seems to point to the role of parasites in holding a local population of rabbits at a level well below the carrying capacity of the habitat.

Evidence is growing that parasites, once overlooked as a major force in community population dynamics, function like top predators (Grenfell 1992). This function is most evident in the dynamics of plant-herbivore systems involving directly transmitted parasites. In these systems the transmission of parasites is coupled with host food consumption. The levels of infection and pathogenicity increase as the density of the host population increases and the availability of food and level of nutrition decrease. There is evidence that overutilization of vegetation and accompanying increases in parasitism alter the stability of plant-herbivore interactions, causing cycles.

Cyclic outbreaks of forest lepidopteran insects, especially the tent caterpillars (*Malacosoma* spp.), may be induced by nuclear polyhedrosis viruses (*Baculovirus* spp.) (Myers 1993). The virus is so named because its DNA is surrounded by a polyhedral-shaped crystal during much of its life cycle. Caterpillars pick up this virus, which replicates in the nucleus of host cells, as they feed on virus-contaminated leaves. It takes four days to three weeks for the infected caterpillar to succumb to the virus. When it does, the caterpillar ruptures, releasing the polyhedra onto leaves and bark. Although the virus is killed by ultraviolet light, it can persist in protected areas, such as bark, for a number of years and spread to new localities by rain, wind, and insects. Gradual buildups of the virus accompanied by increasing densities of the caterpillars generate population cycles (Myers 1988, 1990, 1993).

## Parasite Population Dynamics

There is a similarity between predator-prey relationships (Chapter 25) and parasite-host relations. The parasite acts out in its own way the role of a functional predator, but with a major difference. The host is at once food and habitat, so the parasite is much more intimately related to its "prey" than the predator. Its fate is closely related to the density of the host population and the distribution of the individual parasites among the host. A host can support only so many adult parasites before the parasites experience intraspecific

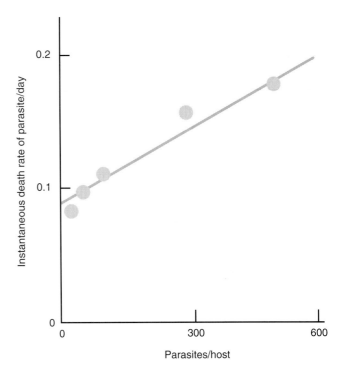

**Figure 26.16** The relationship between parasite density and instantaneous parasite death rate within a host. The hosts are chickens and the parasite is the gut nematode *Ostertagia ostertagi*. The rate of mortality is linear with density. (From Ackert et al. 1931.)

competition or parasite-induced mortality that takes the parasites along with their host.

The situation becomes magnified when the parasites possess two or more infective stages. Each stage behaves as an individual population subject to its own population stresses and the immune responses and longevity of the host. Premature death of a host results in the death of the parasites. Thus the parasite is faced with both the natural and parasite-induced mortality of the host.

Within the host the parasite faces its own intraspecific interactions, especially density-dependent interference competition. Growth of a parasitic population may be constrained by space within the host (Figure 26.16) as well as its own fecundity (Figure 26.17). Ackert and associates (1931) found a relationship between the amount of larvae and eggs taken in by chicks and growth rates of the parasite. An intake of a smaller amount of eggs gave a higher percentage of hatching and a greater growth rate of worms, whereas a high infestation resulted in a low percentage of hatching and slow growth rates.

Immune responses to high infestation can increase parasite mortality and reduce hyperinfestation. The nematode *Haemoinchus contortus* occurs as an adult in the stomach of sheep and is transmitted in the larval stage found on grass. Entering the stomach, the infecting larvae burrow into the

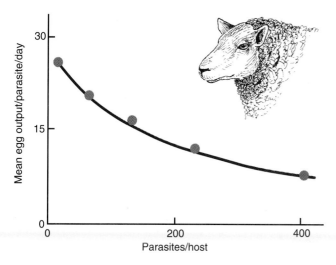

**Figure 26.17** High density of parasites in a host, probably resulting in strong intraspecific competition, sharply reduces the production of transmission stages. The hosts are sheep infected with the liver fluke *Fasciola hepatica*. (From Boray 1969.)

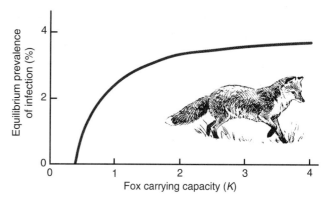

**Figure 26.18** Theoretical relationship between the density of a host population, the fox, and the prevalence of infection of a microparasite, the rabies virus. The equilibrium prevalence of the infection does not rise until the host density rises above the threshold level $N_T$. After this point the level of infection increases rapidly and then tends to slow down and level off at high values of $N$. At high levels of infection and high levels of host mortality, the relationship can result in population regulation of the host by the parasite. Rabies outbreaks in high populations of foxes does reduce the fox population to much lower levels and then periodically reduces them again as the population rises above the threshold level. The same type of curve is applicable to infections of macroparasites. (From A. D. M. Smith 1982:137.)

mucosal lining of the host's gut. They remain there for a short time before returning to the lumen of the stomach as adults. Before their stay ends, the larvae induce a rise in the histamine concentration in the sheep's blood and evoke an antibody response. In the stomach the adult nematodes take up feeding on blood. Continuous reinfection of the sheep brings in more larvae, which attach themselves to the mucosa. Their presence induces further production of histamine and antibodies. They circulate in the blood and are taken up by the blood-feeding adults, killing them. With the expulsion and death of the adult nematodes, the newly invading larvae find a place to live. Such a reaction by the host prevents an overpopulation of parasites within it.

As in the study of the population dynamics of nonparasites, the most important parameter is $R_0$, the basic reproductive rate. The reproductive rate of parasites can be described by an model similar to that derived from the life table in Chapter 18. Among microparasites, $R_0$ is the average number of newly infected hosts that arise from each infected host during its lifetime when introduced into a large susceptible population. The population cannot grow until it reaches some transmission threshold $N_T$ at which $R_0 = 1$. If $R_0$ is less than 1, the parasite population will die out; if $R_0$ is greater than 1, the parasite population increases (Figure 26.18) (Anderson and May 1979). A large value of $R_0$ means a high infection rate. Among macroparasites $R_0$ is the average number of offspring produced by a mature parasite throughout its reproductive lifespan that successfully complete their life cycle and attain reproductive maturity.

$R_0$ is given as $\beta N/(a + b + v)$, where $N$ is host population size, $\beta$ is transmission rate of the parasite, $a$ is the instantaneous host death rate when mortality is due to the influence of the parasite, $b$ is the instantaneous host natural death rate, and $v$ is the instantaneous birthrate of the parasitic transmission stage.

The transmission threshold $N_T$ ($R_0 = 1$) must be crossed if the disease or parasite is to spread. This threshold is given as $N_T = (a + b + v)/\beta$. Thus $R_0 = N/N_T$.

This model is a starting point. The model becomes more complex as variables are introduced, depending upon the parasites and hosts involved. Some of these variables are the expected life span of mature adult parasites, expected life span of the infective stage of the parasite, proportion of mature parasites within the host, and the proportion of the transmissive stage that become infective. The model is basic to studying and understanding the spread of disease. Determining the role of parasites and disease in the population dynamics of host populations is one of the most challenging areas of ecology (Price et al 1986).

## EVOLUTIONARY RESPONSES

A premise in ecology is that parasitic species that do not harm their hosts have the best chance for long-term survival. A parasite that confers high virulence causes a high mortality of hosts. The parasite gains no advantage if it kills its

host and thus itself. Parasites with low virulence ensure a long duration of infection. Thus natural selection would favor adaptive responses between parasite and host. As an outcome host and parasite evolve a mutual tolerance or coadaptation.

Disease represents a lack of adaptation between parasite and host. Although the death of a host may appear to be detrimental to the parasite, there is evidence that natural selection may actually favor parasites with genes that confer virulence, especially among certain microparasites transmitted by vectors (Ewald 1983). Evolutionary pressure may favor increased severity in situations where competition exists between genetically different parasites of the same species. Highly productive, and therefore highly virulent, parasites would reach susceptible hosts much faster than those of lower productivity. Rapid transmission would occur, especially if arthropod vectors were involved. Once these virulent forms entered new hosts, the immune responses of the host might reduce the entry of less virulent forms. Thus highly virulent forms would outcompete the less virulent ones, but reduce the fitness of the host. As the host built up immune responses, it would reduce the fitness of the parasite. Then parasites with genes for lower virulence in the surviving resistant hosts would have the selective advantage. Reduced virulence would bring about increased survival of both parasites and host and ensure long-term propagation.

The parasite-host interaction between the European rabbit and the viral infection myxomatosis is an example of evolutionary forces at work, the adaptation and counteradaptation of parasites and hosts (see J. Holmes 1983, R. H. May and Anderson 1983). The myxomatosis virus is transmitted from host to host by an arthropod vector, mosquitoes.

To control the introduced rabbit, the Australian government introduced the rabbit's viral parasite into the population. Initially, the highly virulent strain had the competitive advantage and spread rapidly. The first epidemic of myxomatosis was fatal to 97 to 99 percent of the rabbits. The second resulted in a mortality of 85 to 95 percent; the third, 40 to 60 percent (Fenner and Ratcliffe 1965). The effect on the rabbit population was less severe with each succeeding epizootic, suggesting that the two populations were becoming more coadapted.

In this adjustment, attenuated genetic strains of virus, intermediate in virulence, tended to replace highly virulent strains. Too high a virulence killed off the host; too low a virulence allowed the rabbits to recover before the virus could be transmitted to another host. Also involved was a passive immunity to myxomatosis conferred upon the young born to immune does. Finally a genetic strain arose in the rabbit population providing an intrinsic resistance to the disease.

The transmission of myxomatosis depends upon *Aedes* and *Anopheles* mosquitoes, which feed only on living animals. Rabbits infected with the more virulent strain lived for a shorter period than those infected with a less virulent strain. Because the latter live for a longer period, the mosquitoes have access to that virus for a longer time. That gives the less virulent strain a competitive advantage over the more virulent.

In those regions where the less virulent strains have the competitive advantage, the rabbits are more abundant because fewer die. That means more total virus is present in those regions than in comparable areas where the more virulent strains exist. Thus, the virus with the greatest rate of increase and density within the rabbit population is not the one with the selective advantage. Instead the virus whose demands are balanced against supply has the greatest survival value.

# SOCIAL PARASITISM

Another form of parasitic relationship is **social parasitism,** in which one organism is parasitically dependent upon the social organization of another. Two forms are brood parasitism and kleptoparasitism. **Brood parasitism** is foisting incubation of eggs or care of young onto surrogate parents. Brood parasitism may be temporary or permanent, facultative or obligatory (E. O. Wilson 1975). Facultative brood parasitism may occur within or among species. **Kleptoparasitism** is the forcible theft of prey by the parasite from the host.

## Brood Parasitism

Temporary facultative brood parasitism within a species is well developed among ants and wasps. For example, a newly mated queen of the wasp genus *Polistes* or *Vespa* will attack established colonies of her own species and displace the resident egg-carrying queen (E. O. Wilson 1975). Females of a number of species of birds, notably some of the Galliformes, ostriches, swallows, starlings, bluebirds, American robin (*Turdus migratorius*), moorhens (*Gallinula chloropus*), American coots (*Fulica americana*), and waterfowl, will lay some of their eggs in the nests of others of the same species (Petrie and Møller 1991).

An example of temporary facultative brood parasitism among species occurs within the ant genus *Lasius*. A newly mated queen of the species *L. reginae* will enter the nest of host species *L. alienus* and kill its queen. The *L. alienus* workers care for the *L. reginae* queen and her brood. In time the *L. alienus* workers, deprived of their own queen and thus of replacements, die out, and the colony then consists of *L. reginae* workers.

Some species of birds exhibit this type of parasitism. Twenty-one species of ducks are known to lay eggs in nests other than their own (Weller 1959). The redhead duck

(*Aythya americana*) is an example. Approximately 5 to 10 percent of female redhead ducks are nonparasitic and nest early. Over 90 percent lay eggs parasitically at one time or another. More than half of these ducks are semiparasites and also build nests of their own. The remainder build no nests, and thus become temporary obligatory parasites.

Although common in ants, the most outstanding examples of temporary, obligatory brood parasitism occur in birds. Brood parasitism has been carried to the ultimate by the cowbirds and old world cuckoos, both of which have lost the acts of nest building, incubating the eggs, and caring for the young. They are obligatory parasites who pass off these duties to the host species by laying eggs in their nests. The brown-headed cowbird (*Molothrus ater*) of North America removes one egg from the nest of the intended host and the next day lays one of her own as replacement (Figure 26.19). Some host birds counter by ejecting the eggs from the nest. Others hatch the egg and rear the young cowbird, usually to the detriment of their own offspring. The host's young may be pushed from the nest or die from lack of food because of the more aggressive nature and larger size of the young cowbird.

A final type of brood parasitism is permanent and obligatory between species. The parasitic form spends its entire life cycle in the nest of the host (E. Wilson 1975). This type of social parasitism is common among ants and wasps. In most cases the species are workers and queens have lost the ability to build nests and care for the young. The queen gains entrance to the nest of the host and either dominates the host queen or kills her outright and takes over the colony.

For obligatory brood parasites, such as the cowbird and European cuckoo, the advantages of social parasitism are obvious. Foisting parental care of young on other species is

the only way they can perpetuate their own. The hosts of such parasites are adversely affected, especially if they are experiencing their first contact with the parasite. Such a situation has developed with Kirtland's warbler (*Dendroica kirtlandi*), a relict species that inhabits extensive jack pine stands in a compact central homeland of about 250 km$^2$ in northern lower Michigan (Mayfield 1960). Before European settlers arrived, Kirtland's warbler apparently was isolated by a belt of unbroken forest 320 k wide from the parasitic brown-headed cowbird of the central plains, a bird closely associated with grazing animals. When settlers cleared the forest and brought grazing animals with them, cowbirds spread eastward and northward into jack pine country. Never associated with the cowbird, Kirtland's warbler did not evolve defenses against brood parasitism, such as ejecting an egg, building a new nest over the old, or rearing young successfully along with the cowbird. This parasitism has resulted in an alarming reduction in warblers fledged, a trend that is being reversed by a strong cowbird control program in this warbler's breeding range. However, with increasing fragmentation of the eastern deciduous forest, warbler species of the forest interior are experiencing similar parasitism by the cowbird.

Although obligatory brood parasitism has attracted the most attention, nonobligatory brood parasitism, especially intraspecific parasitism, is widespread among birds (Rohwer and Freeman 1989). From an evolutionary viewpoint, it would seem that nonobligatory brood parasitism, in which individuals "decide" to be parasitic or not, would be maladaptive. The hosts are forced into misdirected parental care. Parasites capable of rearing their own broods risk lower reproductive success by entrusting the rearing of their offspring to others. Further, an evolutionary conflict of inter-

**Figure 26.19** A newly hatched cowbird crowds the nestling of its barn swallow host. Note the two swallow eggs still unhatched.

est arises in such populations: the parasite can become a host and the host a parasite. Brood parasitism would seem to hold little advantage for either the host or the parasite.

What advantages do accrue for the nonobligatory brood parasite? There is no single or obvious answer. Parasitic behavior may be strongest in those populations of a species occupying areas of uncertain environmental conditions, such as lack of nesting sites or food. Under those conditions the potential brood parasite has four options: (1) become a nonbreeder and join the floating population; (2) become a typical nester; (3) become a parasitic egg layer; or (4) adopt a combined strategy of both parasitic egg laying and nesting. In the last option the individual runs the risk of becoming a host itself (Sorensen 1991).

Brood parasitism could reduce the individual fitness of a host because part of the host's energy goes to rearing young of others. Under certain circumstances, however, intraspeciifc brood parasitism may not affect the host's overall fitness. Being indeterminate layers (they have no upper limit to clutch size), female goldeneye ducks (*Bucephala clangula*) who have their nests parasitized by females of their own species accept the eggs as their own. They adjust their own final clutch accordingly (Andersson and Eriksson 1982). The reproductive success of the host is reduced by the proportion of parasitized eggs. However, because female waterfowl have a strong tendency to return to their own natal place (philopatry), the host and the parasitic bird are probably genetically related, so the effects on inclusive fitness may not be great.

Being parasitized may improve the host's reproductive success, because it spreads the risk of partial predation on the brood among unrelated young, especially in precocial species (Eadie and Lumsden 1985), whereas the parasite is adversely affected. Thus the improved survival of the host's own young may compensate for a reduced clutch and the cost of hatching and caring for the parasitic young.

A major reason for brood parasitism among nonobligates simply is to make the best of a bad situation (Yom-Tov 1980, Petrie and Møller 1991, Sorensen 1993). If the prospects for nesting are poor, the individual may find it advantageous to reduce its own reproductive effort and become parasitic. Parasitic egg laying becomes a low-cost alternative, even though reproductive failure may be high. For the parasitic individual it is better to deposit eggs in the nest of a conspecific with the potential of having some offspring reared to independence than to lay no eggs at all.

Lack of suitable nesting sites can be a major stimulus for an individual to choose parasitism. Among waterfowl drought and drainage of wetlands may eliminate nesting cover. Among secondary cavity-nesting birds (those that do not excavate their own nesting cavities, including waterfowl such as goldeneyes, hooded mergansers, and wood ducks, and passerine birds such as bluebirds), the lack of suitable nesting cavities in trees (and nest boxes) forces unsuccessful females to risk incubation of their eggs by others. An experimental study with bluebirds has shown that when available nest boxes exceeded the number needed by the bluebird population, bluebirds were nonparasitic. When the boxes were reduced by one-third, bluebirds deprived of nesting sites resorted to nest parasitism (Gowaty and Bridges 1991).

Brood parasitism may be induced by reproductive failure. Females who lose their first nest before they begin incubation may drop eggs into nests of others rather than on the ground. Because parasitic egg laying may not be in synchrony with the incubation of the host's eggs, the parasitic eggs may fail to hatch (Sorensen 1993).

Some species adopt the strategy of being both a parasite and a nester. By undertaking both, the birds spread the risk of reproductive failure and ensure that some of their offspring will survive. Such parasitic activity may take place concurrently with their own nesting effort, called a mixed strategy (Brown and Brown 1991), or prior to their own nesting activity, a dual strategy (Sorensen 1990).

A classic example of mixed strategy is found in the European barn swallow (*Hirundo rustica*) (Møller 1987) and the cliff swallow (*H. pyrrhonota*) (Brown and Brown 1991) (Figure 26.20). Uncertain environmental conditions that stimulate mixed strategy among the colonial nesting cliff swallows include heavy ectoparasitic infections by nest mites, loss of poorly attached nests, and extreme weather conditions. Female cliff swallows monitor unattended nests and lay one egg in it during the owner's absence. In response to such parasitism the host lays a smaller clutch than normal, so the total clutch size does not change. In cases of heavy mite infestation of the nest, cliff swallows will carry an egg to a less infested nest of another.

By becoming parasitic, a pair of cliff swallows may also become hosts. When individuals are away from their own nest seeking an unguarded one to parasitize, they leave their own nest unguarded and vulnerable to parasitism. By distributing eggs among several nests, the parasites/hosts are minimizing the risks of an uncertain environment and ensuring the survival of at least one offspring. By employing such a mixed strategy, the swallows apparently improve their reproductive success. Pairs that engage in brood parasitism as well as rearing broods of their own have a higher reproductive success than nonparasitic pairs (Petrie and Møller 1991).

The redhead duck is a notorious parasitic egg layer. About 50 percent of its offspring hatch from parasitically laid eggs. In achieving that distinction, the redhead engages in a dual strategy. Under favorable environmental conditions, redhead females engage in sequential egg laying, depositing two to three eggs in nests either of the same species or of canvasback ducks (*Aythya valisineria*) prior to their own nesting. Under adverse environmental conditions, such as drought, the redheads either lay parasitically or nest; they do not do both (Sorensen 1992). Although annual reproductive success of parasitic ducks may be low, their lifetime reproductive success may be equal to or greater than that of nonparasitic ducks, in part because of increased adult survivorship.

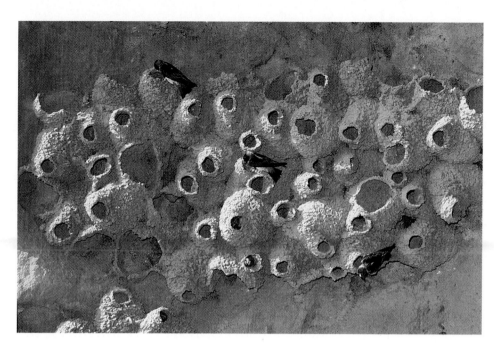

**Figure 26.20** The colonial nesting habit of cliff swallows provides an ideal opportunity for intraspecific nest parasitism.

Many hosts have evolved some antiparasitic strategies. Among these is ejecting of strange eggs, especially if they are recognizable by different markings, guarding the nest, or even deserting it. The last is especially true with the wood duck (*Aix sponsa*), in which nest parasitism markedly increases the size of the clutch (Semel and Sherman 1986). In such situations the cost of rebuilding a nest and relaying eggs is less than that of rearing parasitic young. Because of the apparent ease with which hosts of some species become parasitized, the question arises why potential hosts have not evolved stronger defensive tactics. It could be that the cost of defending the nest is greater than that of rearing parasitic young.

## Kleptoparasitism

In kleptoparasitism the parasites obtain a substantial portion of their food by stealing it from the host. A bald eagle, for example, forces an osprey to drop the fish it caught, which the eagle retrieves and eats. Although kleptoparasitism is widespread among various taxa, it is most prevalent among two groups of birds, the Falciformes (eagles, falcons, and hawks) and the Charadriiformes (skuas, gulls, and waders). As with other forms of parasitism, a close relationship has evolved between the kleptoparasite and its host. Kleptoparasites depend upon other species to locate and obtain prey that can then be stolen. The food item is costly to obtain, but it becomes conspicuous and profitable once acquired by the host. Although kleptoparasitism involves interaction between two individuals, the kleptoparasites often invade the social structure of the hosts. It is most common in situations where potential hosts are aggregated into breeding colonies or feeding groups (Barnard 1984), the availability of food is both temporally and spatially predictable, and the hosts offer little defense.

Behavior of kleptoparasites is illustrated by the black-headed gull (*Larus ridibundus*), which parasitizes feeding flocks of golden plovers (*Pluvialis apricaria*) and lapwings (*Vanellus vanellus*) in the English countryside in winter (Barnard and Thompson 1985). Black-headed gulls exploit the social structure of plovers and lapwings to gain their food. The gulls use the plovers as a quick means of locating profitable flocks of lapwings to parasitize. Typically the gulls join preforaging flocks of plovers in early morning before they fly off. The golden plovers increase their feeding efficiency by joining lapwings feeding on worms in grassy fields. Lapwings are adept at locating rich patches of food. Once settled among the lapwings and plovers, the gulls more or less evenly disperse themselves throughout the flock. The gulls, maintaining some distance between themselves and their intended hosts, closely monitor the lapwings and to a lesser extent the plovers for their foraging success. Their host choice is not random. The gulls select only unwary birds that have extracted a large worm from the soil. The gull then chases or harasses the bird until it drops the worm, which the gull catches or picks up from the ground. In general gulls are more successful against lapwings than against plovers, because the latter are more vigilant.

The kleptoparasite, nevertheless, faces problems. Only a portion of the items procured by the host are worth stealing. The profitability of an item relates to the energy costs of time spent in watching the hosts, of the chase, of chances of success, and of antitheft responses of the host. Nevertheless,

black-headed gulls steal enough worms to meet their daily energy requirements.

Kleptoparasitism, however, has an adverse impact on the hosts. Black-headed gulls impact the time budgets and feeding efficiency of the lapwings and to a lesser extent the plovers. As a result lapwings, unable to meet their energy budgets by day, must also forage in late evening and early night.

Hosts do have some defensive tactics. They can engage in evasive flight. They can shift to smaller, less visible prey, and when foraging maintain a greater distance between themselves and the kleptoparasite. They can attempt to recover the prey. Or they can tolerate the situation and compensate by feeding at a higher rate, as the lapwings do.

# SUMMARY

Parasitism is a situation in which two organisms live together, but one derives its nourishment at the expense of the other. A parasitic infection can result in disease, a state or condition of a plant or animal that deviates from normal well-being.

Parasites may be divided into microparasites and macroparasites. Microparasites include the viruses, bacteria, and protozoa. They are small in size, have a short generation time, multiply rapidly in the host, tend to produce immunity, and spread by direct transmission. They are usually associated with dense populations of the host. Macroparasites, larger in size, include parasitic worms, lice, ticks, fleas, rusts, smuts, fungi, and other forms. They may be ectoparasites, living outside the body of the host, or endoparasites, living inside the body of the host. Macroparasites have a comparatively long generation time, rarely multiply directly in the host, are persistent with continual reinfection, and spread by direct and indirect transmission.

Hosts are the habitat of parasites. The problem faced by parasites is to gain entrance into and escape from the host. The life cycles of parasites revolve about these two problems. The adult stages live in the definitive hosts, from which they escape by means of direct contact with other hosts or by means of vectors. Vectors are organisms than carry or transmit the parasite from one organism to another. Many vectors are hosts for some developmental or infective stage of the parasite. These vectors become intermediate hosts of the parasite. Transmission from definitive to intermediate and back to definitive hosts is considered indirect. Indirect transmission often involves the food chain.

Transmission of parasites, direct or indirect, is complicated by patchy or clumped distribution of the hosts. As a result parasites may become overdispersed: the greatest load of parasites is carried by a few individuals in the population, and most remain free of infection. Hosts respond to parasitic infection by biochemical responses, including inflammatory processes and immune reactions. Parasitic infections result in abnormal growth, sterility, and behavioral changes. There is evidence that parasitic infections also influence mate selection in birds.

Interactions of parasites and host involve populations of up to three stages of parasites, each subject to its own rate of growth and fecundity. Because of the close relationship between parasite and host, the population dynamics of one is influenced by the other. A heavy parasitic load can increase mortality and decrease fecundity of the host population. Under certain conditions, parasitism can function as a population regulatory mechanism. Conversely, a high population of parasites within an individual can experience both intraspecific competition and immune responses from the host. These interact to increase mortality of parasites and reduce their fecundity. Because the death of a host does not benefit a parasite that depends on its host for both food and shelter, natural selection favors less virulent forms of the parasite that can live in the host without killing it. Both hosts and parasites develop a mutual tolerance with a low-grade widespread infection.

Mathematical models of parasite-host relationships are important in studying the spread and effects of parasites and associated diseases. These models suggest those describing predator-prey relationships, except that they are more complex. Depending upon the parasite involved, these models must incorporate such parameters as parasite density in the host, density of the host, proportion of parasites in the transmissive stage, rate of production of eggs and several life stages, and influence of the parasite on host survival.

Another type of parasitism is social parasitism, which includes brood parasitism and kleptoparasitism. In brood parasitism one organism depends upon the social structure of another to rear its young. This arrangement may be temporary or permanent, facultative or obligatory, and interspecific or intraspecific. Brood parasitism occurs most commonly among ants, wasps, and birds. Kleptoparasitism is the piracy of food by one individual from another. Although occurring among many taxa, kleptoparasitism is most prevalent among hawks and gulls, skuas, and wading birds. It is most common where potential hosts are aggregated in breeding colonies and feeding groups. Kleptoparasitism is energetically profitable for the parasite, but adversely affects the foraging behavior and energy budgets of the host.

# REVIEW QUESTIONS

1. What is parasitism, and how does it relate to disease?
2. Characterize microparasites and macroparasites. Cite the groups involved and the types of immune response each stimulates.
3. Distinguish between a definitive host and an intermediate host.

4. What is the relationship between population density and predation in the transmission of parasites?

5. From an evolutionary viewpoint, why should parasites and their hosts develop a mutual tolerance?

6. How would an approach to the control of a parasite with indirect transmission differ from one direct transmission?

7. What is kleptoparasitism? What is its effect on the host?

8. Robertson, Watson, and Cook (1992) commented that with the common eider (*Somateria mollissima*), intraspecific brood parasitism "is a well defined behavior and may be a viable means of obtaining reproductive output." Petrie and Møller (1991) suggest that intrabrood parasitism depresses the average fitness of individuals in a population. Discuss the apparent contradiction.

# CROSS-REFERENCES

Population growth, 392–396; intraspecific competition, 412–414; sexual selection, 430–436; natural selection, 451–463; interspecific competition, 480–513; predator-prey relationships, 536–547; coevolution, 582; defensive mutualism, 587; parasites and community structure, 613–624.

# Mutualism

## Outline

## Concepts

1. Coevolution is reciprocal selection pressure on interacting populations.
2. Mutualism is a positive relationship between individuals or populations of different species.
3. Some mutualisms are a form of symbiosis, in which an individual of one species lives permanently on or within another.
4. Mutualism may be obligate symbiotic, obligate nonsymbiotic, facultative, direct, indirect, or defensive.
5. Mutualism may have evolved from predator-prey, parasite-host, or commensal relationships.
6. Because of positive relationships between populations, the population consequences of mutualism are difficult to define and to model.

In the previous chapters we have explored the relationships between plants and herbivores, predators and their prey, and parasites and their hosts. These relationships share one characteristic: selective pressures on one species affect selective pressures on the other. Pressures of herbivory on plants, of predators on prey, and of parasites on hosts result in counteradaptations. Herbivory on plants is met by structural and chemical defenses. Predation is countered by increased speed, warning coloration, and chemical defense on the part of the prey. Parasites acquire ways of gaining access to and exiting from their hosts and means of counteracting immune responses, whereas hosts acquire improved immune responses. In this chapter we meet another relationship, *mutualism*, in which reciprocal selection pressures benefit two species.

# COEVOLUTION

Reciprocal selection pressure on two interacting populations is called **coevolution.** Certain traits of each species evolve in response to the traits of the other (Ehrlich and Raven 1965, Janzen 1980). Any evolutionary change in one member may change the selective forces acting on the other member. The species play a game of adaptation and counteradaptation.

Such a restricted definition seems to imply that interacting species grow up together over evolutionary time and experience complementary gene-to-gene coevolution. For example, a gene that confers a defensive capability upon a host organism is countered by a distinct corresponding gene that confers the ability upon a parasite to evade the defense. Consider the interaction of winter wheat (*Triticum aestivum*) and its destructive pest, the Hessian fly (*Mayetiola destructor*) (Gallun 1977). The fly overwinters in a pupal case underneath the leaves of winter wheat. In spring the adult emerges and lays eggs on the wheat. The larvae emerge and feed between the leaf sheaths and the stem, killing or weakening the plant. Winter wheat has seven dominant genes and one recessive gene that confer resistance to herbivory by the fly. The fly, in turn, has four homozygous sets of recessive genes that confer resistance to the wheat's defenses. The fly retains this homozygosity because of a unique meiotic condition in which the male transfers only maternal chromosomes to his offspring. This trait ensures retention of the homozygous condition and allows the Hessian fly both to retain and to evolve resistance to wheat defenses. Thus alleles in the wheat that confer resistance are selected for by the attacks of pest, and alleles in the fly that counter wheat defenses permit the fly to reproduce on selected plants. Such gene-to-gene evolution may also take place among highly specialized parasites and their hosts.

However, whether mutualistic species have interacted over evolutionary time is one of the highly debated areas in ecology (Howe and Westley 1988, Futuyma and Slatkin 1983). Some scientists argue that no proof exists that coevolution between species occurs on a one-to-one basis. The

chances are that many supposedly coevolved pairs did not grow up together and that today they are not inhabiting the environments in which they evolved (Howe 1985, Herrara 1985). Instead the organisms probably evolved in different types of habitats through time, each with somewhat different selection pressures. When these plants or animals invaded new habitats, they adjusted to the organisms at hand. When the traits they had already acquired fit the situation, then the two interacted in a manner that suggested long-term coevolution. If the relationship meshed, then further evolutionary changes were minor. What appears to be a highly coevolved system may not have involved an evolutionary change in either partner. The relationship would then continue to be selected for by current interaction.

Another approach to coevolution is to consider it a less restrictive, more general response of one group of species to another. A particular trait may evolve in several species in one taxon in response to the selective pressures of a trait or a suite of traits in several species in another taxon. Plants might have evolved chemical and physical defenses against a diverse array of herbivorous insects. In turn, many insects might have evolved the ability to detoxify a wide range of plant chemicals. Similarly, animals might have evolved a generalized immune system in response to a wide range of parasites. Plants adapt to nonspecific pollinators that visit their flowers. Such interactions are termed **diffuse coevolution** because the adaptive responses are spread over many interacting species. Such interactions contrast with a pairwise response between one species and another, such as a specialized parasite and its host. Among ecologists the most intense interest revolves about the coevolution of mutualistic relationships among species from the polar regions to the tropics.

# TYPES OF MUTUALISM

**Mutualism** is a positive, reciprocal relationship at the individual or population level between two different species (Boucher et al. 1982). Out of this relationship, most obvious at the individual level, both species enhance their survival, growth, and fitness (J. Holmes 1983). Evidence suggests that many mutualistic relationships are more reciprocal exploitations than cooperative efforts between individuals, which the definition of mutualism seems to imply (Barrett 1983).

Mutualism may be symbiotic or nonsymbiotic, obligate or facultative. In **symbiosis** two organisms live together in close physical association, from which one or both derive benefit, and at least one member of the pair cannot lead an independent life. Their relationship is obligate. Intestinal parasites of mammals and birds are examples. In **nonsymbiotic mutualism** both members of the pair benefit each other, but they do not live together. Their relationship is often facultative or opportunistic. They can survive without each other, but benefits accrue from the relationship. (In older literature this relationship was called protocooperation.)

## Obligate Symbiotic Mutualism

Some forms of relationship are so permanent and obligatory that the distinction between the two interacting populations becomes blurred. A good example is the fungi-algae symbiosis in the lichens. The basic structure of the lichen is a mass of fungal hyphae. Within this formation is a thin zone of algae that usually forms colonies of 2 to 32 cells (Figure 27.1). Some 27 different genera of algae have been associated with lichens, about 90 percent of them green algae and 10 percent cyanobacteria. The most common genus involved is *Trebouxia*. It is the only one not found in a free-living state (Holmes 1983). The fungi involved belong to diverse taxonomic groups, most of them being close relatives of free-living ascomycetes that are either parasites of plants and animals or decomposers. Many lichens produce and disperse spores, which form mats of mycelia that may live an independent, saprobic existence for a short time, until they capture algal cells. Depending upon the species involved, the algae change the morphology of the fungi affected. So specific are these morphological changes to a particular algal-fungal relationship that it can be classified as distinct species of lichen.

Algae and fungi supposedly live together for each other's benefit. The algae gain the protection of the fungal thallus; the lichens derive nutrition from the photosynthetic algae. Except for the genus *Trebouxia*, the relationship may not be exactly mutualistic. The lichen may be parasitizing the algae for nutritional gain, since cultured free-living forms of the lichen grow more slowly than those associated with algae. However, algal cells may leak metabolites to the surrounding soil rather than passing them on to the fungus. There is no evidence that the fungus provides anything for the algae other than protection from damaging solar radiation and desiccation (Ahmadjian 1970).

A more convincing example of mutualism is the intimate association of a plant's roots with fungal hyphae, called *mycorrhizae*. The plant supplies energy to the fungi and the fungal hyphae take up mineral nutrients from the soil and transport them into the host's roots.

**Figure 27.1** Section through a typical lichen body, showing the algae within the fungal mass. Chapter opening photo shows some examples of lichens.

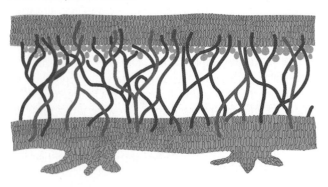

One form common to many trees of temperate and tropical forests is **ectomycorrhizae** (EMC). It consists of a well-developed fungal sheath or mantle around the root (Figure 27.2). This mantle is connected to the inside of the root by a network of hyphae called Hartig's net (Figure 27.3). The outside sheath develops into another network of hyphae that produces shortened and thickened roots that suggest coral. This outside network functions as extended root hairs (Harley and Smith 1983).

Another form is **endomycorrhizae,** notably the vesicular arbuscular mycorrhizae (VAM) (Figure 27.4). They are associated with a wide range of agricultural and native plants. The fungi involved belong to the genus *Endogone;* they have not been cultured independently of the host. The roots are infected by hyphae in the soil or from germ tubes that develop from spores. The hyphae penetrate the cells of the host to form a finely bunched network, called an arbuscle, and oil-rich vesicles. The hyphae act as extended roots for the plant, drawing in phosphorus at distances beyond those reached by the roots and root hairs. The arbuscles have a definite life span. Their disappearance may be due to their digestion by the host plant. Unlike EMC, VAM do not change the shape or structure of the root.

Mycorrhizae, especially important in nutrient-poor soils, aid in the decomposition of litter and translocation of nutrients, especially nitrogen and phosphorus, from the soil into root tissue (Mosse, Stribley, and LeTacon 1981, Newman 1988). Mycorrhizae increase the capacity of roots to absorb nutrients, provide selective ion accumulation and absorption, mobilize nutrients in infertile soil, and make available certain nutrients bound up in silicate minerals (Voigt 1971). In addition, mycorrhizae reduce susceptibility of their hosts to invasion of pathogens by utilizing root carbohydrates and other chemicals attractive to pathogens (Marx 1971, Mosse, Stribley, and LeTacon 1981). They provide a physical barrier to pathogens and stimulate the roots to elaborate chemical inhibitory substances. In return, the roots of the host provide support and a constant supply of carbohydrates. The association between the plant and the fungus can be tenuous. Any alteration in the availability of light or nutrients for the host creates a deficiency of carbohydrates and thiamine for the fungi. Interruption of photosynthesis causes a cessation of fruiting by mycorrhizae. So important are mycorrhizae to the growth of plants that foresters, restoration ecologists, and nursery operators inoculate nursery-grown shrubs and trees, especially pines to be planted in reforestation and restoration of highly disturbed areas. Lack of appropriate mycorrhizae in soil may inhibit the colonization of disturbed areas and early succession sites by forest trees.

Similar mutualistic relationships permitting organisms to exploit nutrient-poor environments are found in the sea. The oceans offer innumerable examples of all sorts of mutualistic interactions, especially among the coral reefs. Coral reefs, self-formed calcareous substrates occupied by anthozoans, are found largely in warm nutrient-poor tropical waters. Living within the cells of the endoderm layer of the oral

(a)

(b)

(c)

**Figure 27.2** Types of mycorrhizae. (a) Coral-like growth of endomycorrhizae on the root of a woody plant. (b) Ectomycorrhizae in the root cells of an orchid. (c) Dye-maker's puffball (*Pisolithus tinctorius*) is the mushroom or spore-producing body of a common and highly important endomycorrhizae of pines.

cavity of the coralline anthozoans are photosynthetic dinoflagellate algae (zooxanthellae) (Figure 27.5). The heterotrophic anthozoans utilize the photosynthetic products of the algae. In turn the coral anthozoans remove, retain, and recycle essential nutrients from the water used by the zooxanthellae (Muscatine and Porter 1977). Although they are carnivorous suspension feeders capturing zooplankton from the surrounding water, anthozoans derive only about 10 percent of their daily energy requirement from zooplankton. They obtain 86 percent of their energy and caloric requirements from algal fixation of C and N and are able to survive and flourish in their nutrient-poor environment by recycling nutrients with their symbiont algae. In addition, algal photosynthesis improves the ability of the anthozoans to lay down calcified coral structures, enabling them to build reefs fast

enough both to counteract destruction and to increase their benthic cover.

## Obligate Nonsymbiotic Mutualism

More common is nonsymbiotic obligate mutualism, in which the mutualists live physically separate lives yet cannot survive without each other. Pollination and seed dispersal systems offer many examples. Other obligatory relationships involve shelter, protection against predators, and reproduction.

Some of the most interesting cases exist between ants and fungi and ants and plants. Fungus-growing attine ants depend upon a slow-growing fungus that they "farm." Neither the ants nor the fungi can survive without each other (M. H. Martin 1970). A classic ant-plant relationship involves the

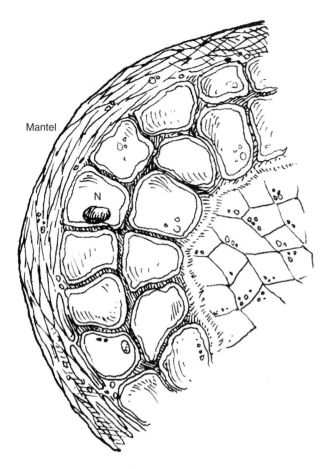

Mantel

N

**Figure 27.3** Ectomycorrhizae form a mantle of fungi about the tips of rootlets. Hyphae invade the tissues of rootlets between the cells. This network is called Hartig's net.

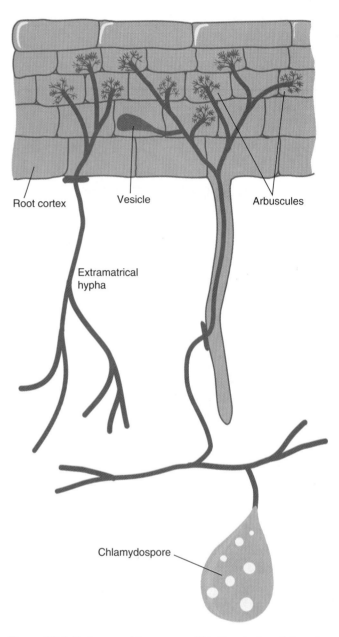

Root cortex Vesicle Arbuscules

Extramatrical hypha

Chlamydospore

**Figure 27.4** Endomycorrhizae grow within the plant rootlet, and fungal hyphae enter the cells.

ant-acacia mutualism (Janzen 1966, Hocking 1975). The Central American ants live in the swollen thorns of acacia (*Acacia* spp.), from which they derive shelter. They feed on protein-rich nodules growing at the tips of pinnate leaves, called Beltian bodies (named after the tropical naturalist Thomas Belt, who discovered their role). These nodules and sugar-secreting nectaries growing on the vegetative parts (extrafloral nectaries) provide an almost complete diet for all stages of the ants' development. In turn the ants protect the plants from herbivorous animals such as cattle. At the least disturbance the ants swarm out of their shelters, emitting repulsive odors and attacking the intruder until it is driven away. Experimental data demonstrate the effectiveness of extrafloral nectaries and the ants they attract in *Acacia* and other plants possessing them in confering protection against herbivores. Among other plants with extrafloral nectares the protective relationship is not so clear. (See Bentley 1977 for a review.)

Some mutualistic relationships involve a third member. Some ectomycorrhizae are epigeous—that is, they produce their sporocarps above ground and forcibly discharge their spores to the air. Hypogeous mycorrhizae produce their sporocarps below ground in structures popularly known as

truffles. In the coniferous forests of western North America, these mycorrhizae depend upon chipmunks and voles to disperse their spores (Maser, Trappe, and Nussbaum 1978). Attracted to the fruiting bodies by species-specific odors, rodents eat the sporocarps, which make up a significant part of their diet. When they defecate, the rodents spread the viable spores necessary for the survival and health of conifers throughout the forest (Figure 27.6)

## Facultative Mutualism

Most mutualisms are nonobligatory and facultative (opportunistic), at least on one side. Such mutualisms are diffuse,

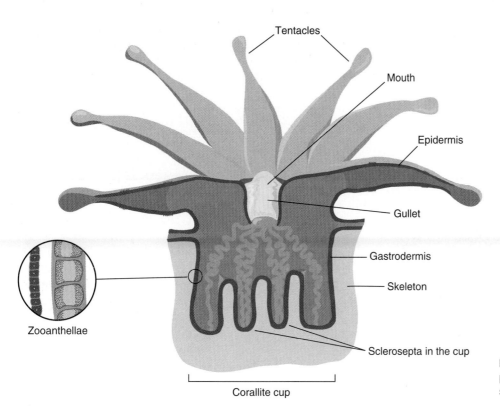

**Figure 27.5** Anatomy of a coral polyp, showing location of the symbiotic zooxanthellae.

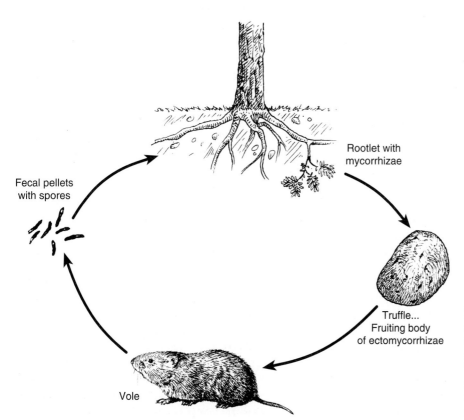

**Figure 27.6** Three-way relationship among hypogeous ectomycorrhizae, trees, and small mammals. The ectomycorrhizae need the tree for energy. The tree needs the ectomycorrhizae for uptake of nutrients from the soil. Small mammals, needed to disperse the spore, feed heavily on the truffles.

involving interactions among guilds of species. They are widely involved in seed dispersal and pollination. The benefits are spread over many plants, pollinators, and seed dispersers. Many of the mutualistic relations discussed later are of this type.

## Defensive Mutualism

A major problem faced by many livestock producers is the toxic effect of certain grasses, particularly perennial ryegrass and tall fescue, on grazing cattle. These grasses are infected by certain fungal endophytes that live inside plant tissue (see Figure 27.4). The fungi (Clavicipitaceae ascomycetes) produce physiologically active alkaloids in the tissue of the host grasses. These alkaloids, which impart a bitter taste to the grass, are toxic to grazing mammals, particularly domestic animals, and a number of insect herbivores. In mammals the alkaloids constrict small blood vessels to the brain, causing convulsions, tremors, stupor, gangrene of the extremities, and death. At the same time these fungi seem to stimulate plant growth and seed production. This symbiotic relationship suggests a defensive mutualism between plant and fungi, in which the fungi defend the host plant against grazing (Clay 1988). There are costs to the plant. The fungal infection causes sterility in the host plant by inhibiting flowering or aborting seeds. Some plants have a few counteradaptations that restore fertility; but in most plants the loss of sexual reproduction is balanced by the greater vegetative growth of the infected plants and enhanced growth in the absence of herbivory.

## Indirect Mutualism

Mutualistic relationships described so far are direct: one species immediately benefits the other. Other mutualistic-type relationships are indirect. For example, suppose that prey species A, eaten by predator X, strongly inhibits prey species B, eaten by predator Y. If the effect of predator X on prey A permits the expansion of prey B, X in effect provides more food for Y. The activities of X then indirectly benefit Y.

A similar mutualistic effect may occur in competitive situations. Suppose species A competes strongly with species B, but mildly with C; and that species C competes strongly with B, but mildly with A. The combined effects of A and C are to their mutual advantage against B (see Chapter 24). A number of such indirect mutualisms could influence community organization. (For more discussion of indirect mutualism see Vandermeer 1980, D. S. Wilson 1980, Waser and Real 1979, Lane 1985.)

There are few empirical examples demonstrating indirect mutualism. How indirect mutualism might function appeared in experimental aquatic food webs studied by H. Wilbur and J. Fauth (1990). This elaborate experiment involved the construction of 16 different food webs with four replicates in an array of 64 experimental ponds. The two predators were the larvae of the green darner dragonfly (*Anax junius*), voracious predators of anuran larvae, and a newt (*Notophthalmus viridescens*). The two prey species were tadpoles of the pickerel frog (*Rana palustris*) and the American toad (*Bufo americanus*), two competitors. The model for each one-predator, two-prey system predicted low and nearly equal survival of the two prey (4 percent for *Bufo* and 7 percent for *Rana*). At the end of the experiment, however, the observed survival in these systems for the toad was 50 percent and 33 percent and for the frog 75 percent and 35 percent. In the four-species system, 35 percent of the toads and 59 percent of the frogs survived. Even though the two prey species were competitors, the minor effects of competition were swamped by the beneficial effects of sharing the risks of predation. Under the pressure of predation the two anurans became mutualists and experienced higher survival than if each was the only prey. The results of this experiment point to the subtle interrelations among species. If one member of an indirect relationship should go extinct, the other species would decline.

## POLLINATION

The "goal" of pollination is specific and direct. The plant must transfer its pollen from the anthers of one plant to the stigma of a conspecific. Some plants simply disperse their pollen to the wind. This method of pollination works well and costs little when the plants grow in large homogeneous stands, as grasses and pine trees do. Wind dispersal is unreliable when conspecifics are scattered individually or in patches across a field or forest. These plants depend upon animals for pollen transfer, mostly insects with some assistance from nectar-feeding birds and bats.

A number of plant-animal pollination systems involve obligates. Examples are the mutualistic relationship of several species of *Yucca* with associated species of yucca moths, *Tegeticula,* and of roughly 600 species of *Ficus* with their own specialized pollinators, minute fig wasps (Agaonidae) (Wiebes 1979). Yuccas depend exclusively upon yucca moths for pollination, so the relationship is a nonsymbiotic obligate mutualism (Figure 27.7). The larvae of yucca moths live symbiotically in the yucca seed heads as obligate predators of their seeds. The adult moths gather pollen; yet they neither take nectar from the flowers nor use the pollen to feed their larvae. At the same time the females lay eggs in the ovaries of newly opened flowers and then deposit the pollen on the stigma. After the seeds have fully developed, each developing larva eats several seeds before dropping to the ground to form a cocoon.

In return for pollination, yuccas must pay out a certain amount of seeds as larval food, without which the yucca moth could not survive. Overall, yucca moth larvae decrease viable seed production by as much as 19 percent, and yuccas expend up to 30 percent of the benefits gained from pollination to support yucca moth larvae (Addicott 1986).

(a)

**Figure 27.7** (a) Yuccas are conspicuous plants of the deserts of southwestern North America. (b) Yucca moth, on which the yucca solely depends for pollination.

(b)

Some other pollination systems involve less complex nonsymbiotic obligate mutualism. Well-known examples are orchids in lowland neotropical forests. Scattered widely through the forests, these orchids depend entirely on male euglossine or golden bees (tribe Euglossini in the honeybee family), specialists in the pollination of orchids (Figure 27.8). Extremely fast fliers, these bees cover long distances between orchids in the tropical forests. However, the male bees obtain no food from the visit. The orchids could not afford to meet the bees' energy demands. Instead the male bees collect fragrances from secretory cells on the lips of the flowers. The male bees use these fragrances to develop their own pheromones to attract females. Female bees ignore the orchids entirely.

Most orchids offer no reward at all. To attract the euglossines, such orchids mimic other flowers that do; or they have flowers with the scent, shape, or color patterns that mimic female bees. When males attempt to copulate with the flower mimic, they pick up pollen.

The orchids contain their pollen in a single mass, the pollinium, with an attachment device. When the bee brushes against it, the pollinium becomes attached to the bee at a specific location on its body. It remains there, for days or weeks if necessary, until the bee visits another orchid of the same species, whose stigma retrieves it. Bees

**Figure 27.8** Tropical orchids depend on male euglossine bees for pollination.

may visit more than one species of orchid and pick up polliniums from each of them. To prevent wrong deliveries, each species of orchid has its pollinium so located within the flower that it adheres at its own specific location on the bee's body. It becomes detached only when the bee enters the correct orchid. This fact suggests that coevolution has occurred between bees and orchids and that the bees have been a strong selection force in the evolution of orchids (see Heinrich 1979, van der Pijl and Dodson 1966, Feinsinger 1983, Faegri and van der Pijl 1979).

Like the fragrance-collecting bees, nectar-feeding animals visit flowering plants to exploit them, not to pollinate them. Most nectivores are generalists. They find little advantage in specializing, except as temporary facultative specialists. Because of the short seasonal flowering of each species, often shorter than the availability of fruits, nectivores depend on a progression of flowering plants through the season. Nectivores cannot afford to commit themselves to one flower, but they do concentrate on one species while its flowers are available.

Rather, plants are the ones which have to specialize, to entice animals by color and odor, dust them with pollen, and then reward them with a rich source of food: sugar-rich nectar, protein-rich pollen, and fat-rich oil. Providing such rewards is expensive for plants. Nectar and oils are of no value to the plant except as attractants for potential pollinators. They represent an expenditure of energy that the plant otherwise might use in growth.

Many species of plants, such as blackberries, elderberries, cherries, and goldenrods, are generalists themselves. They flower profusely and provide a glut of nectar that attracts a diversity of pollen-carrying insects, from bees and flies to beetles (Figure 27.9). Other plants are more selective, screening their visitors to ensure efficiency in pollen transfer. These plants may have long corollas allowing access only to insects and hummingbirds with long tongues and bills and keeping out small insects that eat nectar but do not outcross the plants. Some, such as the closed gentian, have petals that only large bees can pry open.

In addition to nectar, some plants provide oil as a floral reward (Vogel 1969, Buchmann 1987). Many genera in a number of families, including Iridaceae, Orchidaceae, Scrophulariaceae, Concurbitaceae, Solanaceae, and Primulaceae, mostly in neotropical savannas and forests, have specialized oil-secreting organs, called **elaiophores.** One type is epithelial elaiophores. They consist of small areas of secretory epidermal cells beneath a protective cuticle on the petals in which secreted lipids accumulate. A second type is trichome elaiophores, made up of hundreds to thousands of glandular trichomes that secrete lipids in a thin film of oil exposed to the air. In some plants, however, the lipids are protected within deep floral spurs. The flowers are visited by highly specialized bees in four families that use the energy-rich floral oils in place of or along with pollen as provisions for developing larvae. These bees possess modified cuticular and setal

**Figure 27.9** Bees and bumblebees visit the flowers of many species. They are not pollination specialists.

structures designed for mopping up, storing, and transporting oil to the nest (see Buchmann 1987).

The relationship between the common Central American plants *Heliconia* and hummingbirds illustrates the many factors that may be involved in such mutualisms. Growing in the openings of tropical forests or along the forest edge, *Heliconia* propagates vegetatively by rhizomes and usually forms large clumps. When two years of age or older, each individual *Heliconia* plant in the clump blooms. The bloom consists of several showy bracts, each of which encloses several flowers. The bracts open one after another over a period of days or weeks. Each flower within a bract lasts only a day. The flowers are tubular and vary in length and curvature depending upon the species. Some have long curvaceous corollas, 32 mm or less (Stiles 1975). Some species bloom either in the wet season or the dry season, whereas others bloom throughout the year but have a wet or dry seasonal peak in flowering. All are pollinated by insect or birds and offer a supply of sugar-rich nectar as an inducement to their pollinators.

Stiles found that in his Costa Rican study area, nine species of hummingbirds visit nine species of *Heliconia*. Just as the flowers of *Heliconia* have straight or curved corollas, Stiles found that the hummingbirds, too, could be divided into two groups, hermit hummingbirds with long, curved bills and nonhermits with shorter, straight bills. Stiles observed that the five *Heliconia* species with the long, curved corollas are visited to a significantly greater extent by hermits than by nonhermits, whereas three of the species with short corollas are visited commonly by straight-billed nonhermits.

In return for nectar, hummingbirds pollinate the respective flowers. *Heliconia* depend upon hummingbirds for pollen transfer. Hermit hummingbirds, probing long, curved corollas, carry pollen at the base of the bill or on the head. The nonhermits, or straight-billed hummingbirds, carry pollen on the chin or mandibles. If the short corolla flowers are somewhat curved, but the path to the nectar is short and straight, allowing easy access, the hummingbird has pollen deposited on the bill.

Because of the number and types of plants and birds involved, some isolating mechanisms are essential. *Heliconia* select against hybridization by sequential and nonoverlapping peaks in flowering, by spatial isolation, and by promoting behavioral isolation in hummingbirds. The behavioral differences include responses of the hummingbird to visual cues of flowers and to caloric content of nectar. The mutualism thus depends not only on the morphological fit between bird bill and flower corolla, but also on flower phenology, energy content of nectar, and energy demands and behavioral responses of hummingbirds.

Although the relationship of plants and their pollinators has been investigated for years (Real 1983) and the literature is large, many questions remain. Do plants and pollinators coevolve? To get at that question, researchers need to incorporate systematics, morphology, and behavior into their studies. For example, we know little about the evolution of flower choice and flower visiting by animals. Can coevolution take place if plants benefit from constant visits by animals and animals benefit from exclusive access only over a short period of time? Does an exclusive, obligate coevolving relationship increase the fitness of the plant and animal? Do plants experience simultaneous selection from a suite of pollinators? What are the effects of different pollinators on seed set?

The answers are basic to conserving many species and maintaining the integrity of ecosystems, especially tropical ones. If essential relationships between plants and their pollinators are severed by deforestation, loss of habitat, or extinction of one member, then we may experience the loss of associated species and witness a sharp decline in biodiversity. Once any of the participants are gone, there is no way to restore the relationships.

# SEED DISPERSAL

Plants with seeds too heavy to be dispersed by wind depend upon animals to carry the seeds some distance from the parent plant and deposit them in sites favorable for seedling establishment. Some seed-dispersing animals are seed predators, consuming the seeds for their own nutrition. Plants depending on such animals must produce a tremendous number of seeds over their reproductive lifetimes and sacrifice most of them to ensure that a few will survive, come to rest on a suitable site, and germinate.

An example is the relationship between the Clark nutcracker (*Nucifraga columbiana*) and the whitebark pine (*Pinus albicaulis*) of western North America (Figure 27.10). Whitebark pine and a few other pines such as piñon pine, (*Pinus edulis*) produce large wingless seeds that can be dispersed away from the parent trees only by animals. The seeds of whitebark pine are eaten and hoarded by several species of rodents, including chipmunks, and by Steller's jays (*Cyanocitta stelleri*) and Clark nutcrackers. Only the nutcracker possesses the behavior appropriate to disperse the seed systematically and successfully away from the tree (Hutchins and Lanner 1982, Tomback 1982). Typical of jays, the bird carries seed in cheek pouches and caches the seeds deep enough in the soil of forests and open fields to reduce predation by rodents. The number of seeds cached per individual per year is enormous, about 98,800. The nutcrackers fail to retrieve enough of these seeds to allow a large number of them to establish seedlings. Although the cost is high, whitebark pine is virtually dependent on the bird for seed dispersal, and the ranges of the two species roughly coincide.

Some plants use a seed predator not only to disperse the seeds but also to protect them from other predators. In the deserts of the southwestern United States (O'Dowd and Hay 1980), in the sclerophyllous shrublands of Australia (Berg 1975), and in the deciduous forests of eastern North America (Beattie and Culver 1981, Handel 1978) a number of herbaceous plants, including many violets (*Viola* spp.), depend upon ants to disperse their seeds. Such plants, called **myrmecochores,** have an ant-attracting food body on the seed coat, called an *elaiosome*. Appearing as shiny tissue, the elaiosome contains lipids and sterols essential to certain physiological functions in insects.

The ants carry seeds to their nests, where they sever the elaiosome and eat it or feed it to their larvae. The ants discard the intact seed within abandoned galleries of the nest. Ant nests, richer in nitrogen and phosphorus than surrounding soil (Culver and Beattie 1978), provide a suitable substrate for seedling emergence and establishment. Further, by taking seeds away from the parent plant, the ants significantly reduce rodent predation on them (O'Dowd and Hay 1980, Heithaus 1981). Heithaus (1981) found that rodents, particularly the white-footed mouse (*Peromyscus leuco-*

*pus*), removed 84 percent of the seeds of bloodroot (*Sanguinaria canadensis*) when ants were excluded from the parent plants, but only 13 to 43 percent when ants were allowed access.

Plants have an alternate approach to seed dispersal: to enclose the seed in a nutritious fruit attractive to fruit-eating animals, the frugivores. Frugivores are not seed predators but consume only the endocarp surrounding the seed. With some exceptions, they do not impair the vitality of the seed. Most frugivores do not depend exclusively on fruits, because fruits tend to be deficient in certain nutrients such as protein, and because they are only seasonally available.

To use frugivorous animals as agents of dispersal, plants must attract them but discourage the consumption of unripe fruit. Plants protect unripe fruit by cryptic coloration, such as green fruit among green leaves, and by unpalatable texture, repellent substances, and hard outer coats. When seeds mature plants attract fruit-eating animals by presenting attractive odors (Howe 1980), altering the texture of fruits and seeds, improving succulence, acquiring a high content of sugar and oils, and "flagging" their fruits with colors—red, black, blue, yellow, white—to catch attention (Stiles 1982) (Figure 27.11).

Plants have two alternative approaches to seed dispersal by frugivores. One is to become opportunistic and evolve fruits that can be exploited by a large number of dispersal agents. Such plants opt for quantity dispersal, the scattering of a large number of seeds with the chance that a diversity of consumers will drop some seeds in a favorable site. Such a strategy is typical of but not exclusive to plants of the temperate regions. There most fruit-eating birds and mammals are opportunistic consumers, rarely specializing in any one kind of fruit and not depending exclusively on fruit for their basic sustenance. The fruits are usually succulent and rich in sugars and organic acids and contain small seeds that pass through the digestive tract unharmed (Stiles 1980). Large numbers of small seeds are so dispersed, but few are deposited on suitable sites. The length of time such seeds remain within the digestive tracts of some small birds is no more than 30 minutes, so the distance dispersed depends on how far the birds go right after eating (Stiles 1980). Such dispersal is a lottery in the truest sense.

In temperate regions fruits ripen in early and late summer, when the young of the year are no longer dependent on highly proteinaceous food and both adults and young can turn their attention to a growing abundance of fruits. Fruits of late summer and fall ripen when migrant birds come through. They congregate in flowering dogwoods, spicebush, and wild grape to feed on high quality fruits with nutrient-rich flesh (Figure 27.12). Such fruits do not last long, and their seeds are scattered widely. Fruits of lower quality, with less fats and sugars, hang on well into winter (E. Stiles 1980). They are available to birds over a longer period of time and are consumed when more palatable, short-lived

**Figure 27.10** The Clark nutcracker, a jay of the high mountains of western North America, stores and eats the seeds of the white-barked pine in which it is perched.

succulent fruits are gone. In such a manner those plants avoid competition for dispersers in early fall.

The second approach is to depend upon a small number of birds and mammals that are exclusively consumers of fruit. Such plants are mostly tropical forest species, 50 to 75 percent of which produce fleshy fruits whose seeds are dispersed by animals. Rarely are frugivores obligates of the fruits on which they feed. Exceptions include the oilbirds and a large number of tropical fruit-eating bats. Among these the flying foxes of the Old World eat the fruits in place and drop the seeds beneath the tree (Figure 27.13). The smaller spear-nosed fruit bats of the New World pluck the fruits and fly to a safe perch some distance away to avoid predators waiting for them in the fruiting trees.

Most trees attract frugivores that consume many different fruits. Dispersers of the seeds of one plant are also dispersers for others, for several reasons. Fruits vary widely in their nutritional value; by eating a variety of them, frugivores tend to balance their diets. Plants have few means available to restrict consumption of their fruits to a certain few frugivores. However, because plants do have fruits of various sizes, shapes, colors, aromas, nutrient contents, and palatability, some are consumed chiefly by mammals and others by birds (for review see Howe 1986).

Like other areas of mutualism, the study of the relationships among plants and their seed dispersers is still embryonic. It lacks strong empirical approaches in spite of the

(a)

(b)

**Figure 27.11** Color attracts fruit-eating birds and mammals to plants that depend on them for seed dispersal. (a) White berries of poison ivy (*Rhus radicans*). (b) Red fruits or hips of multiflora rose (*Rosa muliflora*), an exotic that has reached a pest status. (c) Blue fruits of wild grape (*Vitus* sp.). (d) Black fruit of Hercules club (*Aralia spinosa*).

(c)

(d)

**Figure 27.12** The frugivorous cedar waxwing (*Bombycilla cedrorum*) feeds on the red berries of mountain ash (*Sorbus*).

relatively large literature available (see Howe 1986). Much needs to be learned. What makes the task more difficult is the fact that seed dispersal is less highly linked to animals than pollination. One reason is that, aside from ants, most seed dispersers are long-lived vertebrates that cannot depend exclusively on one species of a fruit-producing plant. Another is the random fluctuations in fruit production and animal populations over time. Most fruit-producing plants evolved to exploit interchangeable sets of animals for seed dispersion. Nevertheless, many questions need to be investigated. Are all dispersers of the seeds of a particular plant or set of plants equally effective? What happens if the more effective dispersers disappear? Where do most of the seeds go? How is dispersion influenced by the digestive morphology and physiology of the animals involved? Even though they are less specialized than pollinators, the loss of animal dispersers could result in the loss of associated plant species.

## ORIGINS OF MUTUALISM

How did mutualism arise? Population ecologists speculate that mutualism may have evolved from predator-prey, parasite-host, or commensal relationships. Initially one member of the relationship increased the stability of a resource level for the second. In time energy benefits accrued to the second member, and perhaps its activities began to improve the

**Figure 27.13** Fruit-eating bats like the epulated fruit bat (*Epomorphus wahlbergi*) of Africa and the little tent-building bat (*Uroderma bilobatum*) of the neotropical rain forest are important seed dispersers of tropical trees.

fitness of the first. For example, suppose a host tolerant of a parasitic infection exploits the relationship for the host's own benefit. In time the two exploit each other, as in plant-mycorrhizal mutualism. Selection then favors mutual interaction to the point that the two become totally dependent on the other, as in obligate symbiotic mutualism. At the extreme the two function as one individual, as the lichens do.

Among nonsymbiotic mutualists, obligate or facultative, the relationship may have begun with exploitation. Birds or insects came to the plants to feed on pollen or nectar and frugivores to feed on fruits. In the process they accidentally carried pollen to similar plants or dispersed seeds away from the parent plant. Such plants experienced improved fitness and ultimately adapted to exploit the visitors as a means of dispersing pollen or seeds.

Such mutualistic relationships evolved indirectly. Pollinators came to the plant to collect pollen or tap nectar supplies for food, not to aid the plant in completing its life cycle. Fruit-eating animals visited plants for fruits, not to disperse seed. Neither did the mutualisms arise as one-to-one relationships. Plants were visited by an array of hungry insects, birds, and mammals. Except in rare situations, these animals did not, and still do not, specialize on one plant: so groups of plants were visited by groups of various taxa of animals, which differed locally. The result was diffuse coevolution. Guilds of related or unrelated taxa evolved the capacity to use a range of plant resources, resulting in novel ecological associations and diffuse rather than paired coevolution.

A number of constraints favor diffuse coevolution. One is multiple relationships. A flower may be served best by one or two species of pollinators, but it is visited by many. An insect, such as a bee, may visit only one or two species of flowers, but that same flower will be visited by other insects as well. A plant may have its seeds dispersed most efficiently by one type of frugivore, but the frugivore will eat a variety of different fruits seasonally. Therefore it is unlikely that close evolution occurs between just one plant and one pollinator or seed disperser. A second constraint is asymmetrical evolution. For instance, woody plants are much more ancient and have evolved more slowly than any of their pollinators or seed dispersers (Herrera 1985). As one animal group disappeared from the scene, its place was taken by new groups that exploited the plants in a similar manner. Plants, however, retained any traits that encouraged pollination or fruit consumption. Further restricting tight coevolution are variations in the distribution and abundance of plants and animals, seasonal differences in flowering and fruiting, and seasonal changes in animal abundance and distribution.

# POPULATION EFFECTS

Mutualism is most easily appreciated at the individual level. It is fairly easy to comprehend the interaction between ectomycorrhizal fungi and their oak or pine host, to observe and quantify the dispersal of pine seeds by squirrels and jays, and to measure the cost of dispersal to plants in terms of seeds consumed. Mutualism may improve the fitness of the pine and the consumption of pine seeds may improve the fitness of the seed predators; but what about the consequences at the level of the population?

The population consequences of mutualism are considerably more difficult to define than those of predation and parasitism, because the relationship is more nebulous and harder to model. Mutualism exists at the population level only if the population growth rate of species A increases with the increasing density of species B, and vice versa.

The question is most relevant to obligate and facultative nonsymbiotic mutualists. For obligate symbiotic mutualists the relationship is straightforward. Remove species A and the population of species B no longer exists. If ectomycorrhizal spores fail to infect the rootlets of young pine, they will not develop. If the young pine invading a nutrient-poor old field fails to acquire its mycorrhizal symbiont, it will not grow well.

For nonsymbionts, obligate or facultative, the effect on populations may be limited to the extent that one species benefits another and to that part of each other's life history cycle involved in the mutualistic relationship. Consider the yucca and yucca moth mutualism in which the yucca depends upon the moth for pollination. Aker (1982) studied the relationship of *Yucca whipplei* and its moth *Tegeticula maculata*. Throughout the flowering season adult yucca moths were distributed evenly among the available flowers. The number of pollinators on the flower heads was directly proportional to the number of flowers available. In turn the number of fruits set on the plant was directly proportional to the number of flowers produced. If too few moths were available, many flowers would go unpollinated. Most flowers were not fertilized and the fruit production of some plants was limited by pollination. In general, the relative abundance of moths to flowers was low enough that most plants were pollinator-limited.

Individual yuccas regulate mature fruit production by aborting excess fruit. Fruit abortion affects the survival of the seed-consuming moth larvae and their emergence from the capsules at the end of the summer. Thus the size of the adult moth population the following year is determined by the number of plants that flowered and matured fruit. The dropping of immature fruit (along with the larvae) matches the moth larval abundance to the number of plants reproducing in the current year.

These observations suggest that the yucca and the yucca moth have reciprocal influences on each other's population; but whether the yucca moth limits or increases population recruitment and growth of yucca is questionable. The yucca moth may limit seed production (Addicott 1985), but recruitment of yucca seedlings may be influenced more by such extrinsic conditions as insufficient rainfall for seed germination, by seed predation, and by animal browsing on seedlings. The number of moths visiting

the yuccas may be influenced by weather conditions and by the spatial distribution of the plants. The vigor of the yucca as evidenced by the size of the basal rosettes influences the total number of flowers produced (Aker 1982). In other words, the population growth and density of yuccas and moths may be influenced by situations unrelated to their strong mutualistic relationship.

A more definitive example of the population consequences of mutualism is provided by a demographic analysis of an ant-seed mutualism by Hanzawa, Beattie, and Culver (1988). It involves a guild of ants and golden corydalis (*Corydalis aurea*), an annual or biennial widely distributed in open or disturbed sites in the northeastern and western continental United States, Canada, and Alaska. The three researchers compared the survivorship of both seeds and plants, fecundity, reproduction, and growth rates of two seed cohorts of the plant. They relocated one cohort to ant nests undisturbed by ant foragers and hand-planted a control cohort of equal numbers near each nest. The ant-handled cohort had significantly higher survivorship than the control (Figure 27.14). The ant-handled cohort produced 90 percent more seeds than the control cohort; its net reproductive rate $R_0$ was 8.0 and that of the control 4.2. The finite rate of increase of the ant-handled cohort was 2.83 per year compared to 2.05 per year for the control. The ant-handled cohort experienced greater reproductive success, not because of any great difference in the fecundity of the plants but because of a significantly higher survival to reproductive age. Its higher survival was due largely to dispersal and to the protected microsites of the ant nests, and not to the removal of seeds from the vicinity of parent plants or the distance moved.

Theoretical ecologists have been attempting to model the population dynamics of mutualism. Mutualism, like competition, involves interactions between two populations, but the interactions are positive rather than negative. The general approach to modeling mutualism has been a modification of the terms of the Lotka-Volterra equations for competition (Chapter 22) in which the negative alphas of competition become positive. An increase in one population directly influences and is directly influenced by an increase in the other. Because each species benefits the other, the population growth equation of each must include a term for the rate of increase of the other. The maximum values of $K$ do not enter into the relationship, so $K$ is changed to $X$ (Pianka 1994):

$$\frac{dN_1}{dt} = r_1 N_1 \left( \frac{X_1 - N_1 + \alpha_{21}N_2}{X_1} \right)$$

$$\frac{dN_2}{dt} = r_2 N_2 \left( \frac{X_2 - N_2 + \alpha_{12}N_1}{X_2} \right)$$

where $X_1$ and $X_2$ are the equilibrium density of species 1 and species 2 in the absence of the other species, ($\alpha_{12}$ is the beneficial effect of one individual of $N_1$ on $N_2$, and ($\alpha_{21}$ is the beneficial effect of one individual of $N_2$ on $N_1$.

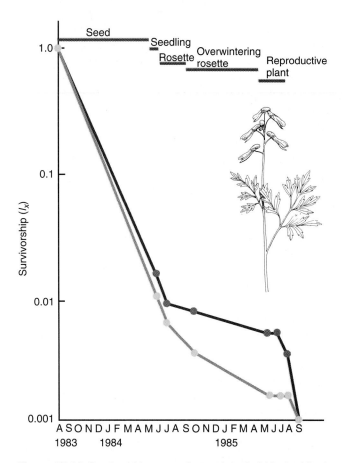

**Figure 27.14** Survivorship curves for ant-handled (dashed line) and control (solid line) cohorts of golden corydalis (*Corydalis aurea*).The mutualism between ants and the plant results in a higher survivorship of overwintering rosettes and reproductive plants. (After Hanzawa, Beattie, and Culver 1988:7.)

The pair of equations predicts that the population equilibrium density of each species is increased by the density of the other species. Such equations suggest that mutualism could produce a runaway positive feedback that would be unstabilizing for both populations unless a strong negative feedback of density-dependent population regulation checked the positive feedback of mutualism.

Mutualistic relationships obviously are much more complex than the two basic equations describe. They fail to include any variable to stop unbounded growth. This weakness has led to the development of a number of more complex mathematical formulations, especially ones relative to stabilizing the population dynamics of the mutualistic species (see Vandermeer and Boucher 1979, Vandermeer 1980, Travis and Post 1979, Heithaus, Culver, and Beattie 1980, Wolin and Lawlor 1984, Post, Travis, and DeAngelis 1985).

To cap growth and to stabilize the system, a third species that is a competitor or a predator of one of the mutualists must be brought into the system, as often happens in real-life

mutualisms involving seed dispersal. In a plant-pollinator system a certain population density of both interacting species is necessary before any equilibrium is possible and both populations reach maximum stable densities. If plant density is too low and pollinators have difficulty finding plants, the pollinators may decline below replacement level. Further, the environment may impose its own limits on population growth. The nature of many mutualistic systems—diffuse and involving arrays of species in very different taxa—makes it difficult to develop a realistic two-species model analogous to the predator-prey and interspecific competition models. Nevertheless, an understanding of the role of mutualism in population dynamics and community structure must rest on a stronger empirical foundation than we now have. The future of many species and communities depends on that understanding.

# SUMMARY

When predators and parasites seek prey and hosts, prey and host seek escape or defense. When one individual derives a benefit from another, it may confer a benefit too. In these situations each individual of the interacting pair is imposing selection pressure on the other. These reciprocal selection pressures result in coevolution. Traits of each interacting species evolve in response to traits of the other. A positive reciprocal relationship between two species that may have evolved from predator-prey, host-parasite, and commensal relationships is mutualism.

Many coevolved relationships are symbiotic. Symbiosis is a permanent association in which one organism lives within or on the other. The symbiosis is parasitism if the guest exploits its host, commensalism if it has no effect on its host, and mutualism if both guest and host benefit.

Mutualism may be symbiotic or nonsymbiotic and obligate or facultative. Obligate symbiotic mutualists are physically dependent on each other, one usually living within the tissues of the other, such as lichens. Obligate nonsymbiotic mutualists depend upon each other, but they lead independent lives, such as certain plants and their pollinators. Nonobligate facultative mutualists include interactions among guilds of species involved in seed dispersal and pollination. In exchange for dispersal of pollen and seeds, plants reward animals with food—fruit, nectar, and oil. To reduce wastage of pollen, some plants possess morphological structures that permit only certain animals to reach the nectar.

Population effects of mutualism are difficult to model because so many mutualistic systems are diffuse, involving arrays of species in different taxa rather than one-to-one relationships. Because an increase in the population of the one species results in an increase in the population of the other, models must incorporate some variables to halt unbounded growth.

# REVIEW QUESTIONS

1. What is coevolution? How does the concept relate to mutualism, predator-prey relationships, and parasite-host relationships?
2. What is mutualism? Symbiosis? How are they different?
3. Distinguish between facultative and obligate mutualism, symbiotic and nonsymbiotic mutualism, and direct and indirect mutualism.
4. What are mycorrhizal mutualisms? What is the importance of these mutualisms? Why are they obligate?
5. In what ways are pollination and seed dispersal forms of facultative mutualism?
6. How would selective pressures differ between a plant visited by several species of bumblebees and a plant visited by a single species of bee?
7. Why do fruit-eating animals not specialize on one species of fruit? What is an advantage of seed dispersal by way of the animal gut?
8. Why is it difficult for coevolution to take place on a one-to-one species basis? Are there any examples of one-to-one coevolution? If so, present arguments for its existence.

# CROSS-REFERENCES

Adaptation, 30–31; nutrient cycling, 116–120; rumen metabolism, 183; nitrogen fixation, 206–207; function of temperate forests, 277–280; function of tropical forests, 285–287; coral reefs, 344; natural selection, 461–463; interspecific competition models, 480–484; predator-prey interactions, 536–547; parasite-host relationships, 560–568.

*Part 6*

## The Community

# Community Structure

## Concepts

1. The community is an assemblage of interacting species occupying a particular position in the landscape
2. Communities possess a vertical structure and a form horizontal pattern across the landscape.
3. Emergent properties of the community include species richness, species diversity, and species abundance.
4. Edge, which develops where two or more vegetation types meet, increases local species diversity.
5. Island theory states that species equilibrium, influenced by island size, is attained when immigration rates balance extinction rates.
6. Fragmentation of the environment results in habitat islands of various sizes.
7. Community organization is influenced by competition, predation, parasitism, and mutualism.
8. Functional structure of the community can be defined by feeding relationships or food webs.

A saying goes, "You can't see the forest for the trees." When we walk through a forest, an oak, maple, or pine catches the eye; we notice one squirrel, deer, or bird. As we walk on, we realize that collections of such individuals make up populations of species in the area. It is harder to see that these populations in the forest are all in some way influencing each other. In doing so, they form another recognizable ecological unit, the community. Forest, grassland, hedgerow, lawn, stream, marsh—these and other units in the landscape contain unique groupings of organisms known as biotic communities.

# THE COMMUNITY DEFINED

A **community** is a naturally occurring assemblage of plants and animals living in the same environment and interacting in many ways, including mutualism, predation, and competition. Interactions may be direct or indirect, but in some way each population affects the well-being of the others. For example, the assemblage of organisms occupying a fallen log on the forest floor interacts closely, whereas its interaction with organisms living in the upper canopy of the forest is minimal. The organisms within the rumen of deer make up still another community. However, life in the fallen log and in the rumen of the deer ultimately depends upon the functional roles of organisms in a broader context.

Like many ecological terms, community has several meanings. Some ecologists use it to describe groups of similar organisms occupying the same habitat, such as a community of desert lizards or grassland birds. Others reserve the term for specialized groups within a specific habitat, such as insect-feeding birds in the forest canopy or hummingbirds feeding on meadow flowers. Then ecologists are referring to guilds. **Guilds** are groups of species that forage or feed similarly in a given habitat. In these cases, the term community does refer to an assemblage of interacting species, but in a restricted sense.

A community may be autotrophic in the sense that it includes photosynthetic plants and gains its energy from the sun. Other communities, such as those in springs and caves, are heterotrophic. They depend upon the input of fixed energy, such as organic material, from the outside. Autotrophic communities usually contain a number of heterotrophic microcommunities, such as fallen logs.

Just as populations have attributes beyond those of the individuals that comprise them, so a community is more than the sum of its parts. The community has attributes or emergent properties. The nature of the community is influenced by the interactions of species and their adaptations to the physical environment. These adaptations and interactions are reflected in such attributes of the community as structure, dominance, diversity, and niches.

# PHYSICAL STRUCTURE

## Life Forms

Differences in terrestrial communities are defined by the form and structure of vegetation. Vegetation may be classified according to growth form. Plants may be tall or short, evergreen or deciduous, herbaceous or woody. We might speak of trees, shrubs, and herbs, and then further subdivide the categories into needle-leafed evergreens, broad-leafed evergreens, evergreen sclerophylls, broad-leafed deciduous trees, thorn trees and shrubs, dwarf shrubs, ferns, grasses, forbs, mosses, and lichens.

A more useful system is the one designed in 1903 by the Danish botanist Christen Raunkiaer. Instead of considering growth form, he classified plants by life forms. He defined these life forms in terms of the height of the plants' **perennating tissue** above ground. Perennating tissue is the embryonic or meristemic tissue of buds, bulbs, tubers, roots, and seeds. It survives from one growing season to the next by remaining inactive over winter or the dormant season.

Raunkiaer recognized six principal life form classes (Table 28.1 and Figure 28.1) into which all species in a region can be grouped: therophytes, cryptophytes, hemicryptophytes, chamaephytes, phanerophytes, and epiphytes. The ratio among these life form classes expressed as a percentage provides a life form spectrum for the area that reflects the plants' adaptations to the environment, particularly climate (Figure 28.2). A community with a high percentage of perennating tissue well above the ground (phanerophytes) would be characteristic of warm climates. A community consisting mostly of chamaephytes and hemicryptophytes would be characteristic of cold climates. A community dominated by therophytes would be characteristic of deserts and frequently disturbed sites.

**Table 28.1  Raunkiaer's Life Forms**

| Name | Description |
|---|---|
| Therophytes | Annuals survive unfavorable periods as seeds. Complete life cycle from seed to seed in one season. |
| Geophytes (Cryptophytes) | Buds buried in the ground on a bulb or rhizome. |
| Hemicryptophytes | Perennial shoots or buds close to the surface of the ground; often covered with litter. |
| Chamaephytes | Perennial shoots or buds on the surface of the ground to about 25 cm above the surface. |
| Phanerophytes | Perennial buds carried well up in the air, over 25 cm. Trees, shrubs, and vines. |
| Epiphytes | Plants growing on other plants; roots up in the air. |

**Figure 28.1** Raunkiaer's life forms: (a) phanerophytes; (b) chamaephytes; (c) hemicryptophytes; (d) geophytes (cryptophytes); (e) therophytes; (f) epiphytes. The parts of the plant that die back are unshaded; the persistent parts with buds (or seeds in the case of therophytes) are dark.

## Vertical Stratification

A distinctive feature of the community is vertical stratification of plant and animal life (Figure 28.3). Stratification of the community is determined largely by the form of plants—their size, branching, and leaves—which in turn influences and is influenced by the gradient of light. The vertical structure of the plant community provides the physical structure in which many forms of animal life are adapted to live.

A well-developed forest community has a highly stratified structure with a large variety of components. It consists of several layers of vegetation, each of which provides a habitat for animal life. From top to bottom these layers are the canopy, the understory tree layer, the shrub layer, the herbaceous or ground layer, and the forest floor or litter layer. The tropical rain forest has one additional stratum, emergents, trees that rise above the general canopy of the forest. The canopy is the major site of primary production and has a pronounced influence on the structure of the rest of the forest. If the canopy is fairly open, considerable sunlight reaches the lower layers, and the shrub and understory tree strata are well developed. If the canopy is closed, understory trees, shrubs, and even the herbaceous layer are poorly developed. The nature and development of the several layers are influenced by soil moisture, density of the

**Figure 28.2** Life form spectra of a tropical rain forest (adapted from P. W. Richards 1952), a Minnesota hardwood forest (data from Buell and Wilbur 1948), and a New Jersey pine barren (data from W.I. Stern and Buell 1951). Note the absence of hemicryptophytes, geophytes, and therophytes from the tropical rain forest and the prominence of epiphytes. The pine barrens are dominated by phanerophytes.

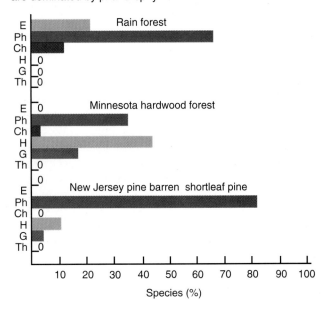

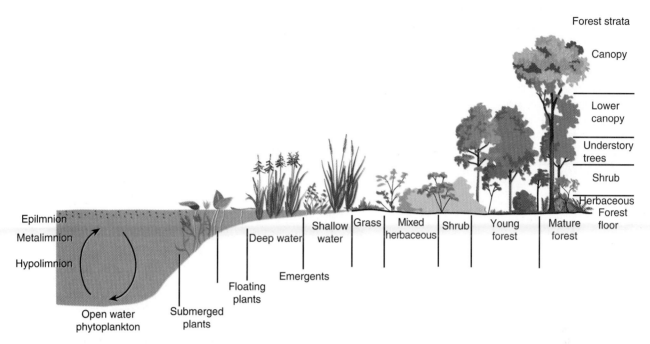

**Figure 28.3** Vertical structure of communities from aquatic to terrestrial. All are structurally similar in that the zone of decomposition and regeneration is the bottom stratum and the zone of energy fixation is the upper strata. In the sequence from the aquatic to terrestrial, stratification and complexity of the community become greater. Stratification in aquatic communities depends on physical factors, the gradients of oxygen, temperature, and light. Stratification in terrestrial communities is largely biological. Dominant vegetation affects the physical structure of the community and microclimate conditions of temperature, moisture, and light. Because the forest often has five or six strata, it can support a greater diversity of life than a grassland with three strata. Floating and emergent aquatic plant communities can support a greater diversity of life than open water.

overstory, slope position in hilly and mountainous country, aspect, and disturbance. All of these conditions can vary from place to place throughout the forest. The final layer, the forest floor, depends on all these factors and in turn determines in part how rapidly nutrients are recycled within the community.

Other communities have a similar, if not as highly stratified, structure. Grasslands have a herbaceous layer that changes through the seasons, a ground or mulch layer, and a root layer. The root layer is more highly developed in grasslands than in any other community, and the mulch layer has a pronounced influence on plant development and animal life, especially insects and small mammals.

The strata of aquatic communities are affected heavily by vertical gradients of light, temperature, and oxygen. Well-stratified lakes in summer contain the epilimnion, a layer of freely circulating surface water; the metalimnion, characterized by a thermocline; and the hypolimnion, a deep layer of dense water about 4° C, often low in oxygen; and a layer of bottom mud (see Figure 28.3). In addition, two structural layers based on light penetration are recognizable—an upper zone roughly corresponding to the epilimnion, dominated by autotrophic phytoplankton, and a lower layer in

which decomposition is most active. The lower layer roughly corresponded to the hypolimnion and bottom mud.

Communities, whether terrestrial or aquatic, have similar biological structures. They possess an autotrophic layer, which fixes the energy of the sun and manufactures food from inorganic substances. It consists of the area where the light is most available: the canopy of the forest, the herbaceous layer of grassland, and the upper layer of water of lakes and seas. Communities also possess a heterotrophic layer that utilizes the food stored by autotrophs, transfers energy, and circulates nutrients by predation in the broadest sense and by decomposition.

Each vertical layer in the community is inhabited by characteristic organisms. Although considerable interchange takes place among several strata, many highly mobile animals confine themselves to only a few layers, particularly during the breeding season. Occupants of a vertical stratum may change during the day or season. Such changes reflect daily and seasonal variations in humidity, temperature, light, oxygen content of water, and other conditions or the different requirements of organisms for the completion of their life cycles. For example, D. L. Pearson (1971) found that birds occupying the upper strata of a tropical dry forest in

Peru moved to the lower strata during the middle of the day for several reasons: to secure food (insects move to lower levels), to escape heat stress, to escape a high degree of solar radiation, and to conserve moisture.

In general, the greater the vertical stratification of a community, the more diverse its animal life. The variety of life in a terrestrial community is heavily influenced by the number and development of layers of vegetation. If a certain stratum is absent, the animal life it normally shelters and supports is also missing. Therefore grassland, with few strata, is poorer in species than a highly stratified forest ecosystem (see Karr and Roth 1971). Likewise in aquatic communities, the greater the variation along vertical gradients of light, temperature, and oxygen, the greater is the diversity of life.

## Horizontal Structure

Walking across a typical old field, we move through patches of open grass, clumps of goldenrods, tangles of blackberry, and small thickets of sumac and other tall shrubs. Continue into an adjacent woodland and we cross through an open understory, and patches of shade-tolerant undergrowth of laurel and viburnum. Then we come upon open gaps in the canopy where dense thickets of new growth have claimed the sunlit openings. We continue our walk to the top of an open hill and view the pattern of vegetation across a larger landscape. We note the quilt-like patches of forests, fields, croplands, roads, human settlements. These large patches, spatially separated from one another, produce a horizontal pattern that increases the physical and ecological complexity of the environment.

The patchiness of vegetation across the landscape exists on different scales: we discovered within-patch heterogeneity in the old field, between-patch heterogeneity when we passed from field to forest, and finally the patterned land-scape. At all levels, size, shape, and dispersion of patches affect their colonization by individuals, the persistence of these individuals on the patch, the number of breeding units, and the number of species in an area. Horizontal patterns influence dispersal and foraging by animals. Within the old field the size and number of patches of goldenrods are important to the goldenrod beetles. On a larger scale the presence of patches of low woody vegetation is essential to the rabbit. For the deer heterogeneity on an even larger scale, involving large patches of woods, old fields, and grassland, is essential. Each species and individual requires its own scale of patchiness within the matrix of the landscape.

For plants opportunities for recruitment, growth, reproduction, and survival change with spatial variations in soils, nutrients, moisture, light, and competition from other plants. The type of plant reproduction and the timing and patterns of seed production over time influence vegetation patchiness. Plants with wind-dispersed and animal-dispersed seeds have a wider distribution across the landscape and exhibit fewer patches than plants with poor dispersal mechanisms. Vegetative or clonal reproduction produces distinctive clumps of certain plants in an otherwise homogeneous environment. Allelopathic effects and shading lead to the suppression of some plant species and to the development and growth of others.

The patterns we see at all scales in the landscape are the product of an array of environmental and biological influences (Figure 28.4). Soil structure, soil fertility, moisture conditions, and aspect influence the microdistribution of plants. Patterns of light and shade shape the development of understory vegetation. Runoff and small variations in topography and microclimate produce well-defined patterns of plant growth. Grazing animals have subtle but important effects on the spatial patterning of vegetation, as do wind-

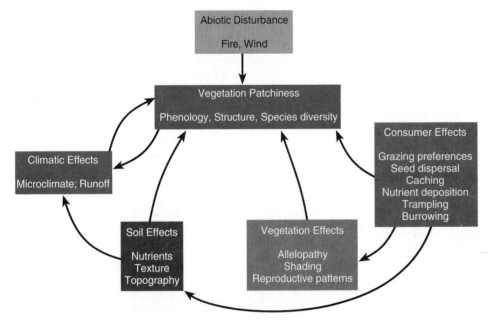

**Figure 28.4** General relationships among some of the major influences that govern vegetational patchiness in terrestrial environments. (From Wiens 1976.)

throw, fire, and human disturbances (see Chapter 29). A patchy environment in turn influences the distributional pattern of animal life across the landscape (Wiens 1976).

# BIOLOGICAL STRUCTURE

The physical structure of the community sets the stage for even more powerful influences: the abundance and the diversity of species, and the interactions among them. Interactions among individuals in some communities result in dominance by one, several, or a group of prominent species. These organisms are called **dominants.**

## Species Dominance

It is not easy to determine what constitutes a dominant species or, in fact, to determine which are the dominant species. Dominants in a community may be the most numerous, possess the highest biomass, preempt the most space, make the largest contribution to energy flow or mineral cycling, or by some other means control or influence the rest of the community.

You can assign dominance to numerically superior organisms, but numerical abundance alone is not sufficient. A species of plant, for example, may be widely represented yet exert little influence on the community as a whole. In the forest small or understory trees may be numerically superior; yet the nature of the community is controlled by the fewer large trees that overshadow them. In such a situation dominance is measured not by number but by biomass or basal area. Local differences in the environment, such as moisture availability, nutrient levels, and topographic positions, can create patches of different dominants within the community, further complicating the dominance relationships within it.

The dominant organism may be scarce yet by its activity control the nature of the community. The predatory starfish *Pisaster,* for example, preys on a number of species similar in habit and thereby reduces competitive interactions among them, so these different prey species coexist (Paine 1966). If the starfish is removed, a number of prey species disappear and one of them becomes dominant. In effect, the predator controls the structure of the community and so must be regarded as the dominant or **keystone species.** Keystone species are those whose presence is critical to the integrity of the community.

To determine dominance ecologists have used several approaches. They can measure **relative abundance** of species, comparing the numerical abundance of one species to the total abundance of all species. They can measure **relative dominance,** a ratio of basal area occupied by one species to total basal area. It is best used when species involved are approximately the same size. They can use **relative frequency** as a measure. It is best used when the species involved are very different in size. Often all three measurements are combined to arrive at an **importance value** for each species. Most species do not arrive at a high level of importance in the community, but those that do serve as **index species.** Once species have been assigned importance values, the stands can be grouped by their leading dominants according to those values. Such techniques are useful in the study and placement of communities on some environmental gradient, such as moisture.

## Species Diversity

Species Richness and Evenness Among the array of species that comprise a community, few are abundant. Indeed, individuals of most species make up only a small proportion of the total population in the community. Consider the structure of the tree component of a mature woodland consisting of 24 species over 10 cm dbh (diameter breast height), as presented in Table 28.2. Individuals of two tree species, yellow-poplar and white oak, make up nearly 44 percent of the stand. The next most abundant trees—black oak, sugar maple, red maple, and American beech—each make up from under 7 percent to just over 5 percent of the stand. Nine species range from 1.2 to 4.7 percent, and the nine remaining species as a group represent less than 5 percent of the stand. A second woodland, described in Table 28.3, has a different composition. That community consists of ten species of which two, yellow-poplar and sassafras, make up nearly 84 percent of the stand.

These two forest communities illustrate a typical pattern in temperate forests—a few common species associated with less abundant ones. They also illustrate two other characteristics of the distribution of species within a community—**species richness,** the number of species, and **evenness,** the relative abundance of individuals among the species. The more equitable the distribution, the greater is evenness. **Species diversity,** which considers both the species richness and evenness, increases as the numbers of individuals in the total population are more equitably distributed among the species. The stand described in Table 28.2 is richer in species (24 versus 10) and has a greater evenness than the stand described in Table 28.3.

In order to quantify species diversity for comparison, a number of indexes have been proposed (Pielou 1975). One of the most widely used is the Shannon Index (also called Shannon-Weiner), which considers both richness and evenness. The Shannon Index measures diversity by the formula

$$H = -\sum_{i=1}^{s} (p_i)(\log_2 p_i)$$

where $H$ is the diversity index, $s$ is the number of species, $\log_2$ is the natural log, and $p_i$ is the proportion of individuals of the total sample belonging to the $i$th species.

**Table 28.2  Structure of Vegetation of a Mature Deciduous Forest in West Virginia**

| Species | Number | Percentage of Stand | Species | Number | Percentage of Stand |
|---|---|---|---|---|---|
| Yellow-poplar (*Liriodendron tulipifera*) | 76 | 29.7 | Bitternut hickory (*Carya cordiformis*) | 5 | 2.0 |
| White oak (*Quercus alba*) | 36 | 14.1 | Pignut hickory (*Carya glabra*) | 3 | 1.2 |
| Black oak (*Quercus velutina*) | 17 | 6.6 | Flowering dogwood (*Cornus florida*) | 3 | 1.2 |
| Sugar maple (*Acer saccharum*) | 14 | 5.4 | White ash (*Fraxinus americana*) | 2 | .8 |
| Red maple (*Acer rubrum*) | 14 | 5.4 | Hornbeam (*Carpinus caroliniana*) | 2 | .8 |
| American beech (*Fagus grandifolia*) | 13 | 5.1 | Cucumber magnolia (*Magnolia grandiflora*) | 2 | .8 |
| Sassafras (*Sassafras albidum*) | 12 | 4.7 | American elm (*Ulmus americana*) | 1 | .39 |
| Red oak (*Quercus rubra*) | 12 | 4.7 | Black walnut (*Juglans nigra*) | 1 | .39 |
| Mockernut hickory (*Carya tomentosa*) | 11 | 4.3 | Black maple (*Acer nigrum*) | 1 | .39 |
| Black cherry (*Prunus serotina*) | 11 | 4.3 | Black locust (*Robinia pseudoacacia*) | 1 | .39 |
| Slippery elm (*Ulmus rubra*) | 10 | 3.9 | Sourwood (*Oxydendrum arboreum*) | 1 | .39 |
| Shagbark hickory (*Carya ovata*) | 7 | 2.7 | Tree of heaven (*Ailanthus altissima*) | 1 | .39 |
| | | | | 256 | 100.00 |

This index, based on information theory, is a measure of uncertainty. The higher the value of *H*, the greater is the uncertainty, or the probability that the next individual chosen at random from a collection of species containing *N* individuals (in our example the species of tree) will not belong to the same species as the previous one. The lower the value of *H*, the greater the probability that the next individual encountered will be the same species as the previous one. In the forest described in Table 24.3, whose diversity index is 1.87, the probability is high that in sampling the trees, the next tree picked at random will be a yellow-poplar or a sassafras. In the forest described in Table 28.2, the diversity index, 3.59, is higher and the chance that the next tree encountered at random will be a yellow-poplar is considerably less.

We can gain a better appreciation of evenness or equability by comparing the proportion of individuals in the community to the maximum possibility of evenness if *N* individuals were evenly distributed among *s* species, that is, if each $n_i = N/s$. For the Shannon Index evenness is given by

$$J = H/H_{max} = -\Sigma p_i \ln p_i / \ln s$$

The stand described in Table 28.2 has an evenness index of 0.78 (H maximum 3.17) where *J* ranges from 0 to 1.0. The

**Table 28.3  Structure of Vegetation of a Second Deciduous Forest in West Virginia**

| Species | Number | Percentage of stand |
|---|---|---|
| Yellow-poplar (*Liriodendron tulipifera*) | 122 | 44.5 |
| Sassafras (*Sassafras albidum*) | 107 | 39.0 |
| Black cherry (*Prunus serotina*) | 12 | 4.4 |
| Cucumber magnolia (*Magnolia grandiflora*) | 11 | 4.0 |
| Red maple (*Acer rubrum*) | 10 | 3.6 |
| Red oak (*Quercus rubra*) | 8 | 2.9 |
| Butternut (*Juglans cinerea*) | 1 | .4 |
| Shagbark hickory (*Carya ovata*) | 1 | .4 |
| American beech (*Fagus grandifolia*) | 1 | .4 |
| Sugar maple (*Acer saccharum*) | 1 | .4 |
| | 274 | 100.0 |

stand in Table 28.3 has an evenness index of 0.56 (H maximum 2.26).

The more abundant species are not necessarily the most influential members of the community. In communities embracing organisms with a wide range of sizes, an index may lead us to underestimate the importance of fewer but larger individuals and overestimate that of the more common species. One of the distinctive failures of the indexes is the inability to distinguish between the abundant and the rarer species, which contribute little to the index. Nevertheless, diversity indices do provide one measure of community differences. (For a discussion of indexes, see Hurlbert 1971, Peet 1974.) Diversity indexes may be used to compare species diversity within a community (alpha diversity, $\alpha$), between communities or habitats (beta diversity, $\beta$), and among communities over a geographical area (gamma diversity, $\gamma$) (Whittaker 1972).

**Local and Regional Diversity** The richness of species in a local area reflects the uniqueness of the local habitat, which is exposed to a variety of conditions. The presence or absence of a species is influenced by the availability of and competition for nutrients, moisture, and food. It is further complicated by natural and human-induced disturbances, the availability of seeds relative to safe sites for successful colonization, and the physical condition of the environment. All of these factors change over space and time.

The death of one or several trees in a forest, tropical or temperate, can open up a microsite and bring in shrubby growth and associated wildlife. The influence of that disturbance is temporary and will decline as vegetation grows. Certain species may persist in a locality only because immigrants move in from more productive areas. Some plant species may occur locally only because persistent disturbances maintain potential habitats.

Local diversity reflects and is influenced by regional diversity, which is a product of climatic history, historical accidents, and the geographical position of dispersal barriers (Ricklefs 1987). The plowing of the Great Plains in the late 1800s and early 1900s, a historical event, reduced diversity of grassland plants and animals. The massive clearing of the eastern deciduous forest in North America, which reduced a dispersal barrier, and the subsequent conversion of much of the region to farmland allowed midwestern grassland species to move eastward, increasing species diversity in the deciduous forest region.

**Global Diversity** The variety of life existing worldwide emphasizes diversity on a global scale. The diversity of life becomes more abundant at lower latitudes of Earth and at lower elevations in mountainous country. Diversity of life is lowest in high latitudes, the polar regions, and high altitudes in the mountains. Species of nesting birds (Fischer 1960), mammals (Simpson 1964), fish (Lowe-McConnell 1969), lizards (Pianka 1967), and trees (Monk 1967) decrease latitudinally from the tropics to the polar regions. The richness of marine life increases from the continental shelf, where food is abundant but the environment is changeable, to deep water where food is less abundant but the environment is more stable. Mountainous regions generally support more species than flatlands because of topographic diversity, and peninsulas have fewer species than adjoining continental areas. From east to west in North America the number of species of breeding land birds (MacArthur and Wilson 1967) and mammals (Simpson 1964) increases (Figure 28.5a). This increased diversity on an east-west gradient relates to an increased diversity of the environment both horizontally and altitudinally. Currie (1991) demonstrates a correlation of regional species richness of vertebrates across North America with annual potential evapotranspiration (PET) and the species richness of trees to actual evapotranspiration (AET). Because of lower PET conditions, amphibians are more abundant and diverse in eastern North America (Figure 28.5b) than in the western part of the continent. Reptiles, on the other hand, are more diverse in the hot arid regions of western North America (Kiester 1971) (Figure 28.5c), where PET is low. Diversity of trees in North America shows a strong correlation with AET (Currie and Paquin 1987) (Figure 28.5d).

**Species Diversity Hypotheses** Why should such latitudinal and altitudinal gradients in diversity exist? Why should the tropics hold a greater abundance of species than temperate regions? Why should one locality hold more species than another? These questions have intrigued ecologists for years. Although they have generated a number of hypotheses, answers to these questions are elusive.

Two hypotheses involve time, so they might be considered historical hypotheses. One is the **evolutionary time hypothesis** (Fischer 1960, Simpson 1964). It proposes that diversity relates to the age of the community. Old communities (in an evolutionary sense) hold a greater diversity than young communities. Tropical communities are older and evolve and diversify faster than temperate and arctic communities.

Considering a shorter time scale, the **ecological time hypothesis** is based on the time needed for a species to disperse into unoccupied areas of suitable habitat. Because not enough time has passed since the glacial period for many species to move into temperate zones, these areas are unsaturated by the species they now support. Many cannot move until barriers to dispersal are broken; others are moving out of the tropics into temperate regions. Examples are the spread of the cattle egret (*Bubulcus ibis*) from Africa by way of South America and the northward spread of the armadillo (*Dasypus novemcinctus*).

The **spatial heterogeneity hypothesis** (Simpson 1964) holds that the more complex and heterogeneous the physical environment, the more complex will its flora and fauna be. The greater the variation in topographic relief, the more

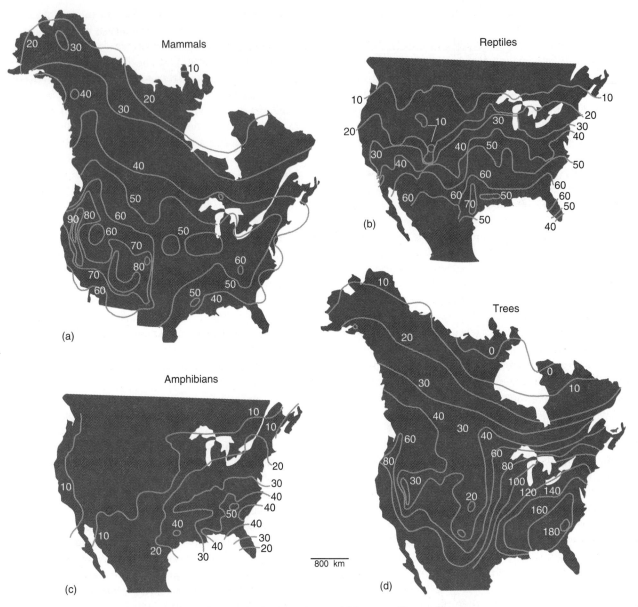

**Figure 28.5** Latitudinal variation in the distribution of mammals, amphibians, reptiles and trees. Contour lines connect points with about the same number of species. (a) Diversity of mammals reflects both latitudinal and altitudinal variations across the continent. (After Currie 1991.) The most pronounced latitudinal variations occur among amphibians and reptiles. (b) Poikilothermic and ectothermic reptiles have their greatest diversity in hot desert regions and lower latitudes of North America. (c) Being not only poikilothermic but also highly sensitive to moisture conditions, amphibians reach their greatest diversity in the central Appalachians and decrease northward, southward, and westward. Species numbers are lowest in the dry and in the cold regions of the continents. (From Kiester 1971.) (d) Distribution of trees reflects actual evapotranspiration. (From Currie and Paquin 1987.)

types of habitats the community contains ($\beta$ diversity); the more complex the vertical structure of the vegetation ($\alpha$ diversity); the more kinds of species it will hold. The fact that communities with marked vertical structure hold more species of birds supports this theory (MacArthur and MacArthur 1961, MacArthur 1972, Pearson 1971).

Several hypotheses relate to climate. The **climatic stability hypothesis** (Fischer 1960, Connell and Orias 1964) states that because a stable climate, one that does not change much with the seasons, provides a more favorable environment, the species richness will be high. Organisms living in a constant benign climate, typical of tropical regions, would

not require the broad tolerance limits needed by species living in a more variable climate. Constant environment would favor specialization in feeding niches and microhabitats. Such specialization would increase species diversity, typical of the tropics. In the higher latitudes and altitudes, where seasonal climatic conditions range from arctic cold to tropical heat, selection would favor those organisms able to tolerate climatic variability. Species tolerant of such climatic regimes are less restricted in their habitat use and are able to occupy a wider latitudinal range (Stevens 1989).

The **climatic predictability hypothesis** relates species diversity in temperate and polar regions to a variable but predictable climate. In such climates organisms have evolved some dependence on regularly occurring changes and specialize on conditions that recur every year. Desert annuals respond to brief seasonal periods of winter and summer rainfall. Migratory birds respond to seasonal climatic changes by arriving at the most favorable periods for nesting and leaving before adverse climate changes return. These responses result in temporal diversity in the region.

The great regional variability in species diversity in temperate and polar regions may also be related to energy balance. The **energy hypothesis,** originally advanced by J. Brown (1981), predicts that in regions of roughly equal area, energy flux per unit area should be the major determinant of species diversity (Currie 1991). For trees primary production represents realized capture of solar energy, the best indicator of which is actual evapotranspiration. Among vertebrates, the regulation of body temperature is closely linked to atmospheric energy. It is best correlated with potential evapotranspiration, which is closely related to latitude and variability in solar radiation.

Sanders (1968) modified the climatic stability hypothesis into the **stability-time hypothesis.** This hypothesis assumes that two contrasting types of communities exist: the physically controlled and the biologically controlled. In physically controlled communities organisms are subjected to physiological stress by fluctuating physical conditions. Organisms in time evolve adaptive mechanisms to meet these conditions. However, at least some of the time, organisms face severe physiological stress, which increases the probability of low reproductive success and survival. These conditions result in low diversity.

In biologically controlled communities, physical conditions are relatively uniform over long periods of time and are not critical in controlling species. Evolution proceeds along lines of interspecific competition, one species adapting to the presence of the other and sharing resources with it. The environment is more predictable, the physiological tolerances are low, and diversity is high. However, no community is wholly physically controlled or biologically controlled.

Another interesting hypothesis also related to climatic stability is the **productivity hypothesis** advanced by Connell and Orias (1964). In essence, this hypothesis proposes that the level of diversity of a community is deter-

mined by the amount of energy flowing through the food web. The rate of energy flow is influenced by the limitation of the ecosystem and by the degree of stability of the environment. The productivity hypothesis in effect states that the more nutrients available and the greater the productivity, the greater the diversity.

Although this statement may be true in a general sense, there are too many exceptions. Nutrient-rich and highly productive salt marshes are species-poor compared to a nutrient-poor old field; and an old field is more species-rich than highly fertile grassland. In some aquatic systems increased nutrient enrichment from sewage and other nutrient sources decreases diversity. Marine benthic regions of low productivity have a higher abundance of species than areas of high productivity (Sanders 1968). In tropical bird communities, the diversity of prey rather than the total available energy is best correlated with diversity (Karr 1975).

A **competition hypothesis,** originally proposed by Dobshansky (1951) and C. B. Williams (1964), contains some elements of the climatic stability hypothesis. In tropical regions, where the climate is benign and stable, populations of species reach near maximal size. As a result both intraspecific and interspecific competition are high. Under competitive pressures, selection favors specialization in foods and microhabitats. Species occupy narrow niches, which makes for high diversity. In temperate and polar regions, where the climate is less benign and more variable, populations rarely reach their maximal size. Because competition for resources is relatively low and tolerance limits are broad, the species are not strongly specialized. As a result, niches are broad and species diversity is relatively low. In such regions selection is controlled largely by physical variables.

A **predation hypothesis** also has been proposed to account for species diversity, particularly on a local and regional basis (Paine 1966). This hypothesis proposes a higher species diversity in those communities in which predators reduce prey species to a numerical level where interspecific competition among them is greatly reduced. The reduction in competition allows the coexistence of a number of prey species.

Predation, herbivory, fluctuations in the physical environment, and all sorts of density-independent mortality keep natural communities in a state of disturbance or nonequilibrium. Huston (1979) proposed a **dynamic equilibrium hypothesis** based on the differences in the rates at which populations of competing species reach competitive equilibrium. The major determinant of diversity in nonequilibrium situations is the population growth rate of competitors. Most communities fail to achieve equilibrium because of a fluctuating environment and periodic reductions in populations. In the absence of disturbance, an increase in the population growth of major competitors results in low diversity. Disturbances so frequent that many populations fail to increase also cause low diversity. It is at the intermediate level of disturbance that greatest diversity occurs. At this level a dynamic balance becomes established between competitive

displacement and frequency of population reductions due to disturbance.

However intriguing these hypotheses may be, they are difficult to test experimentally. They present ecologists with some formidable challenges. Among them all lie parts of an explanation for species diversity: habitat structure, diversity of microhabitats, physical environment, climate and protection from its adverse effects, competition, predation, availability of food and nutrients, disturbances, time, historical accidents, and geographical barriers.

## Species Abundance

As the discussion of species evenness suggested, few species in a community are common and most species are rare. Species diversity indexes tell us little about species abundance. To determine species abundance, we need to plot the proportion of species present into abundance classes. Theoretically, abundance classes represent the manner in which species divide the niche space or the environmental space. Three statistical models are used to describe species abundance: the random niche boundary model, the niche preemption model, and the log-normal hypothesis. All involve plotting the relative abundance of each species in the community in order of its rank from the most to the least abundant (Figure 28.6).

The **random niche model** or broken stick model (MacArthur 1960) views abundance as a random partitioning of resources distributed along a continuum. The analogy is that of a stick upon which are randomly marked a number of points that represent niche boundaries. Then the stick is broken at each point into segments. The length of each segment represents the abundance of species. If the segments, representing the importance value of the species (the percentage of total density, biomass, and so on of all species in a community that a single species represents), are plotted in a sequence from the longest to the shortest on a log scale, then a curve like A in Figure 28.6 will result. The model assumes that species in the community use the critical resource with no overlap between species. This model is realistic only on rare occasions (Hairston 1969). It produces the highest evenness of the three models. The curve is approached only by some small samples of taxonomically related animals with stable populations and long life cycles occupying a small homogeneous community, such as nesting birds in a forest.

The **niche preemption hypothesis** or **geometric distribution hypothesis** supposes that the most successful or dominant species preempts the most space. The next most successful claims the next largest share of space, and the least successful occupies what little space is left. If the relative importance of each species is plotted in species sequences on a log scale, the result is a straight line (curve B in Figure 28.6), and the distribution of the species forms a geometric series. This model produces the highest dominance and lowest evenness of the three models. Such a distribution is achieved only by a few plant

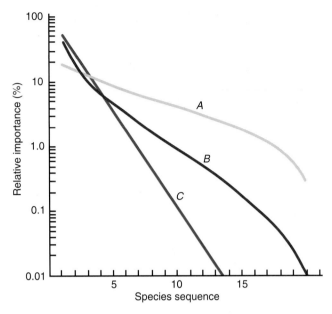

**Figure 28.6** Graphic representation of species abundance hypotheses. In these graphs, the importance values (the percentage of all species that a particular species represents), expressed as total density, total biomass, total frequency, total productivity, or some other measurement, are plotted against a ranked sequence of species. According to the random niche hypothesis (curve A), the boundaries are located at random positions along the line. By the niche preemption hypothesis or geometric distribution hypothesis (curve B), the relative importance of each species is plotted in sequence on a log scale, resulting in a straight line. The log-normal hypothesis (curve C) supposes that species distribution is determined by a large number of variables that affect the competitive abilities of the species involved. (After Whittaker 1965.)

communities containing few species and occupying severe environments such as a desert. In most plant and animal communities, species overlap in the use of space and resources.

The **log-normal hypothesis** (Preston 1962) supposes that niche space occupied by a species is determined by a number of conditions, such as food, space, microclimate, and other variables that affect the success of one species in competition with another. The relative importance of each species is determined by the way the variables affect it (see May 1981). A plot of the importance values results in a bell-shaped or normal curve (Figure 28.7). As in the broken stick model, the line can be divided into segments. In this case each segment represents a range in importance values grouped into frequency distribution classes called octaves (for methodology see Ludwig and Reynolds 1988, C. J. Krebs 1989, Poole 1974). The octaves represent a doubling of individuals—1–2, 2–4, 4–8, and so on. This doubling translates into taking logarithms of abundance to the base 2. The octave with the greatest number of individuals is equal to 0. The rest of the octaves are numbered in plus and minus directions away from the mode. If this distribution is plotted

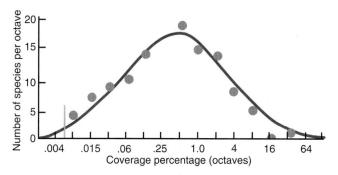

**Figure 28.7** A bell-shaped curve of importance values resulting from the log-normal distribution of plant species. The importance value in this example is determined by the percent ground surface covered by the species. It is represented on the horizontal scale (logarithmic) by octaves in which the species are grouped by doubling. (After Whittaker 1965.)

logarithmically, the curve produced will fall somewhere between the random niche and geometric distribution curves (curve C in Figure 28.6).

The log-normal distribution most closely approximates the distribution of importance values obtained from communities rich in species. It is most useful in summarizing observed abundance relationships within and among communities. All of these distributions describe species abundances, but they are of little value in determining the underlying causes for the observed abundance relationships.

## EDGE COMMUNITIES

Edges are one of the most conspicuous features of the landscape. They mark the boundaries between a housing development and forest from which it was carved, between a roadside and a woods, and between a pond and a field. The boundaries where these blocks of vegetation meet are called **edges.** Some edges between vegetation types are abrupt or sharp. Others are diffuse. The vegetation of one patch blends with the other to form a transition zone called an **ecotone** (Figure 28.8)

Some edges result from abrupt changes in soil type, topography, geomorphic features (such as rock outcrops), and microclimate. Under such conditions adjoining vegetation types are determined by long-term natural features. Such edges, usually stable and permanent, are considered **inherent edges** (Thomas et al. 1979). Other edges result from such natural disturbances as fire, storms, and floods, or

**Figure 28.8** Types of ecotones. (a) Abrupt, narrow edge with no development of an ecotone. (b) Narrow ecotone developed by advancement of community Y into community X to produce ecotone YX. (c) Community X advances into community Y to produce ecotone XY. (d) Ideal ecotone development in which plants from both communities invade each other to create a wide ecotone X2Y2. This type of ecotone will support the most edge species.

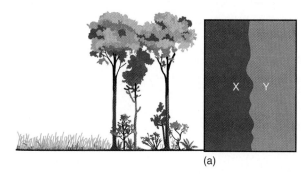

(a)

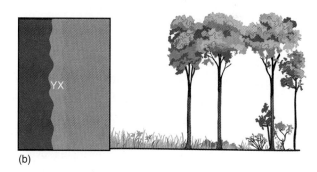

(b)

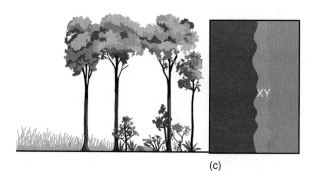

(c)

(d)

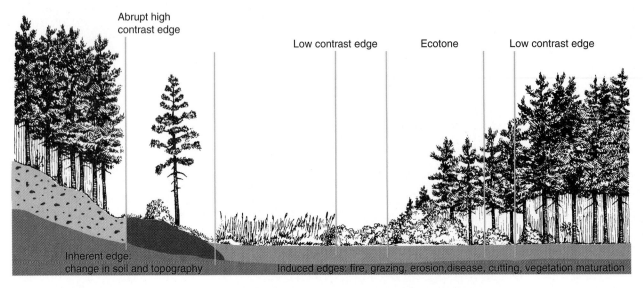

**Figure 28.9** Types of edges: inherent and induced, high contrast and low contrast. Inherent edges are mostly abrupt. Edges of high contrast exist widely different adjacent communities, such as shrub and mature forest. Edges of low contrast involve two closely related successional stages, such as shrubs and sapling growth. (After Thomas et al. 1979.)

from such human-induced disturbances as livestock grazing, timber harvesting, land clearing, and agriculture. The adjoining vegetative types are successional or developmental and will change or disappear with time. Such edges, maintained only by periodic disturbances, are called **induced edges** (Figure 28.9). Induced edges may be abrupt, too, or transitional, resulting in an ecotone.

Ecotones arise from the blending of two or more vegetational types. Plants competitively superior and adapted to environmental conditions existing in the edge advance as far into either community as their ability to maintain themselves will allow (Figure 28.8). Thus in the ecotone species common to each type mingle with species common to the edge. Plants of the edge, such as blackberries, sumac, and hawthorn, are opportunistic and shade-intolerant. They grow well in a xeric environment, with its high rate of evapotranspiration, reduced soil moisture, and fluctuating temperatures.

Animals characteristic of the edge are those that require two or more vegetational communities. For example, one edge species, the ruffed grouse (*Bonasa umbellatus*), requires forest openings with an abundance of herbaceous plants and low shrubs for summer food, dense sapling stands to shelter drumming or display logs, small timber for nesting cover, and mature forests for winter food and cover. Because the ruffed grouse spends its life in an area of 5 to 10 hectares, this amount of land must provide all of its seasonal requirements. Some species, such as the indigo bunting, are restricted exclusively to edge situations (Figure 28.10). Because of species responses, the variety and density of life are often greatest in and about edges and ecotones. This phenomenon has been called the **edge effect** (Leopold 1933).

Edge effect is influenced by the amount of edge available—its length, width, and degree of contrast between adjoining vegetational communities (Patton 1975). The greater the contrast between adjoining plant communities the greater species richness should be (Figure 28.8). An edge between a forest and grassland in general supports more species than an edge between a young and a mature forest (McElveen 1977, Harris and McElveen 1981, Harris 1989). The larger the adjoining communities, the more opportunity exists for plants and animals of each of those communities to occupy the area along with the species of the edge. If patches of vegetation are too small to support species characteristic of larger patches of that habitat, then the area becomes a homogeneous community dominated by edge species.

The edge effect comes about because environmental conditions differ from those of adjacent vegetational communities, especially forests (Ranney 1977). Environmentally, such edges reflect steep gradients of wind flow, moisture, temperature, and solar radiation between extremes of open land and forest interior. Wind velocity is greater at the forest's edge than within the forest, creating xeric conditions in and around the edge. Evaporation increases, so plants place greater demands on soil moisture. Variations in solar radiation, both direct and reflected, probably have the strongest physical influence on forest edge, especially north-facing and south-facing edges (Figure 28.11). A south-facing edge may receive three to ten times more hours of sunshine a month during midsummer than a north-facing edge, making it much warmer and drier. Also affecting the physical environment of the edge is the depth to which sunlight penetrates the vertical edge of the forest. This penetration depends upon solar angle, edge aspect, density and height of

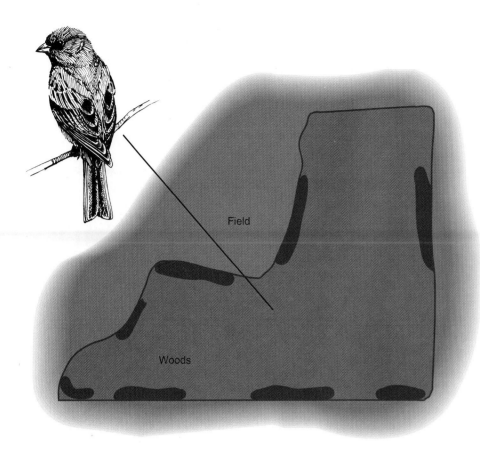

**Figure 28.10** Map of territorries of a true edge species, the indigo bunting (*Passerina cyanea*) that inhabits woodland edges, larger gaps in the forest creating edge conditions, hedgerows, and roadside thickets. The male requires tall, open song perches and the female a dense thicket in which to build a nest. (After Whitcomb et al. 1981:143.)

vegetation, latitude, season, and time of day. In general, the edge effect extends about 50 m into the forest.

Because of increased exposure to wind and direct sunlight, vegetation of the edge is subject to different environmental stresses than forest interior species. Trees on the forest edge experience greater heat stress and dissipate proportionately more heat by evaporation than by radiation; interior trees dissipate more heat by radiation. Higher light intensities cause sun scald in some tree species, such as yellow-poplar, and increase crown expansion and epicormic branching (development of new limbs on the bole of a tree) in others, such as oaks. Some mesic, shade-tolerant trees succumb. High light intensity and xeric conditions favor those plant species, such as blackberries and sumac, that are tolerant of high light and dry soil, that can compete successfully for moisture and nutrients, and that can reproduce vegetatively. Density and frequency of saplings and sprout stumps are greater in border areas. As the edge canopy thickens, shade-tolerant plants appear. Eventually the edge becomes a mixture of shade-tolerant and shade-intolerant species (Wales 1972, Ambuel and Temple 1983, J. G. Blake and Karr 1984, Freemark and Merriam 1986).

The concept of the edge and its relationship to a heterogeneous environment was formalized by Aldo Leopold (1933) in his classic text *Game Management,* as the law of the edge. He stated that the potential abundance of wildlife

(game) species with small home ranges that require two or more vegetation types is roughly proportional to the sum of the edges. Thus density of an edge species varies as a constant proportion of edge. The amount of edge depends upon the intermingling of different vegetation types that provide habitat and the dispersion of those types across the landscape. Thus heterogeneity of the landscape determines vegetational richness and the availability of habitat patches for edge species.

Leopold applied this principle only to resident, edge-obligate species with restricted home ranges, not to mobile species. However, wildlife biologists and ecologists recognized that animal and plant life, in general, were more abundant about the edges than in the interior of forests and grassland. Soon they were extending the idea beyond Leopold's original concept. The way to increase wildlife diversity was to increase edge habitat. This idea became a tenet of ecology and wildlife and forest management. Application of this concept involved creating openings in the forest and breaking up large open fields with plantings of shrubs and trees, all to increase the edge effect. Carried to an extreme, this management for diversity results in breaking up a large homogeneous stand of vegetation into smaller and smaller patches or fragments, eliminating wildlife requiring larger areas of nonedge habitat. At some point edge becomes redundant (Guthery and Bingham 1992), in excess of the

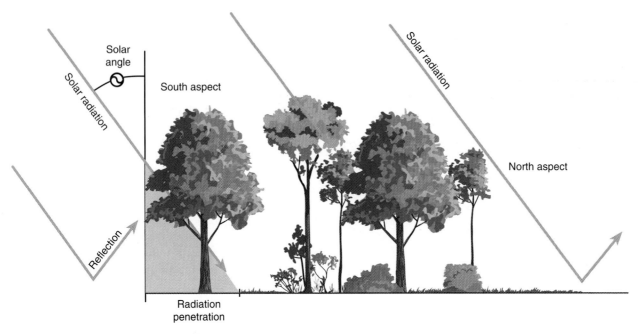

**Figure 28.11** Influence of solar radiation on the edge between a field and a forest. Solar radiation does not affect all edge aspects equally. North-facing edges receive almost no direct sunlight and limited reflection from nearby fields, whereas south-facing edges receive both. The depth to which solar radiation penetrates forest edge on level ground depends upon the height of the canopy and solar angle. Within the zone of radiation penetration light intensities and summer daytime temperatures are higher than in forest interiors sheltered by tree canopies. (After Ranney 1977.)

amount required for the expected population response relative to the density-imposed limits. The heterogeneity once provided by edge becomes itself a homogeneous community dominated by edge and early and mid-successional species.

Edge effect may increase abundance and diversity of species, but it also creates severe ecological problems only recently addressed (Temple 1986, Reese and Ratti 1988, Yahner 1988). Edge vegetation attracts generalist species, such as grackles, that compete with habitat specialists, such as wood thrushes, for food in forest litter. More importantly edges, especially abrupt ones, attract both mammalian and avian predators. Raccoon, opossum, and fox use edges as travel lanes, and avian predators, especially crows and jays, find easy access to birds nesting within the habitat block. Edges also provide the brown-headed cowbird easy access to new and inexperienced hosts, especially neotropical warblers. Edges can become a population sink for some species attracted to it. They experience reproductive failure because of nest predation and parasitism.

## ISLAND COMMUNITIES

The various patches, large and small, that form the vegetation patterns across the landscape suggest islands of different sizes. Some are near each other; others are remote and isolated. A grassland, for example, is surrounded by forest, or a patch of woods sits in a sea of cropland or housing developments. Other patches are connected by a corridor of similar vegetation. The sizes of these patches and their distances from each other have a pronounced influence on the nature of the patches. The role of size and distance relates to island biogeography.

## Island Biogeography Theory

Large areas hold more species than small areas. This fact was not overlooked by early naturalist explorers and biogeographers in the late 1700s. Much later, the zoogeographer P. Darlington (1957) suggested a rule of thumb: a tenfold increase in area leads to a doubling of the number of species. In 1962 F. W. Preston formalized the relationship between the area of an island and the number of species present. When the two values are plotted as logarithms, the number of species changes linearly with island size (Figure 28.12):

$$S = cA^z \text{ or } \log S = \log c + z \log A$$

where $S$ is the number of species, $A$ is the area of the island, $c$ is a constant measuring the number of species per unit area, and $z$ is a constant measuring the slope of the line relating $S$ and $A$. The slope, $z$, is a measure of how fast species change with a unit change in area. The steeper the slope of

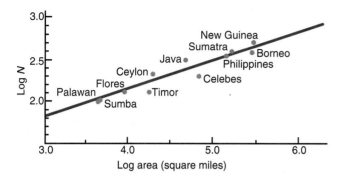

**Figure 28.12** Number of bird species on various islands of the East Indies in relation to area. The abscissa gives areas of the islands. The ordinate is the number of bird species breeding on each island. (From Preston 1962:195.)

the line, the faster the number of species decreases as island size decreases.

MacArthur and Wilson (1963) extended this concept into a theory of island biogeography. Briefly, it states that the number of species of a given taxon established on an island represents a dynamic equilibrium between the immigration of new colonizing species and the extinction of previously established ones (Figure 28.13).

Consider an island empty of life some distance from a mainland. The mainland is a source of potential colonists for the islands. Some mainland species are better dispersers and colonists than others. They will rapidly occupy the island at first, followed eventually by species with poorer dispersal abilities. As the number of colonizing species increases, the number of new immigrant species arriving on the island decreases, because fewer new potential immigrants are available from the source pool.

At the same time some island inhabitants go extinct. As more species arrive, the extinction rate increases. At equilibrium between immigration and extinction the number of species remains stable, although the composition of species may change. The rate at which one species is lost and a replacement gained is the **turnover rate.**

All islands are not created equal. Some are smaller than others; some are more distant from the mainland pool of colonists. These conditions strongly influence immigration, extinction, turnover rates, and equilibrium size of species on islands. According to the theory, the equilibrium level is influenced by the size of the island and the distance of the island from a pool of potential immigrants (Figure 28.14).

Extinction rates are higher on smaller islands than on larger ones. Large islands have lower extinction rates than small islands because they hold larger populations. Because immigration rates decrease with increasing distance from a source area, distant islands attain species equilibrium with fewer species than near islands, all else being equal. Considering both immigration and extinction rates, one can hypothesize that near islands reach equilibrium with more species than distant islands. Small islands reach equilibrium with fewer species than large islands, and large, far islands have an equilibrium density equal to that of near, small islands.

Turnover rates are greater for near, small islands and near, large islands than for distant small and large islands because they are closer to a source of replacement immi-

**Figure 28.14** Graphical representation of the island biogeography theory, involving both distance and area. Immigration rates decrease with increasing distance from a source area. Thus distant islands attain species equilibrium with fewer species than near islands, all else being equal ($S_3 > S_2$ for large islands; $S_2 > S_1$ for small islands). Extinction rates increase as the size of the island becomes smaller.

**Figure 28.13** Equilibrium model of species on a single island. The point at which the curve for rate of immigration intersects the curve for rate of extinction determines the equilibrium number of species in a given taxon on the island. $S$ represents the equilibrium number of species.

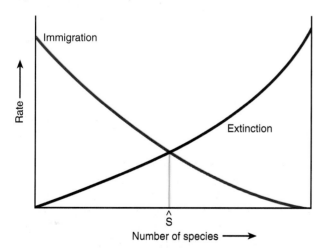

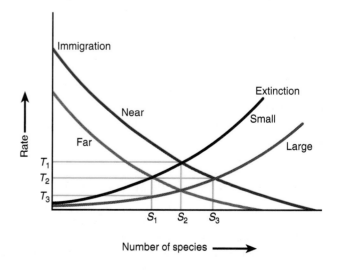

grants. However, turnover rates for near, small islands are greater than for near, large islands because extinction rates are greater on small islands.

Verifying this model is difficult. Several intensive studies on the defaunation of insects and recolonization of small islets of mangroves off the Florida Keys (Simberloff and Wilson 1969, 1970; Wilson and Simberloff 1969) and small islets of salt marsh at Oyster Bay, Florida (Ray 1981, Strong and Ray 1982) were undertaken. The mangrove experiments tended to verify the MacArthur-Wilson dynamic equilibrium model. The islands richest and poorest in species prior to defaunation were also the richest and poorest after defaunation. Islands with the greatest number of species were those closest to the main stands of mangroves. However, species found to immigrate frequently were not the same as those present before the experiment, and species turnover was high.

In the salt marsh experiments, recolonization and extinction rates corresponded to the island biogeography curves in general, with immigration rates negatively correlated and extinction rates positively correlated with species number. However, great variability in both rates, especially extinctions, made correspondence to the MacArthur-Wilson curves statistically insignificant. This result suggests that both immigration rates and extinction rates are influenced by life history and trophic characteristics of the species, predation, host plant chemistry, and other factors.

Equilibrium size and turnover rates on already occupied islands can be assessed only by long-term annual censuses on each island. An example is the seasonal census of confirmed nesting birds over a 26-year period in the 16 ha Eastern Wood of Bookham Common, an isolated woodland near Surrey, England. Over the 26-year period, 44 species of birds appeared in the woods. Of these, six apparently did not nest, leaving 38 breeding species. Of these breeders, four species had territories extending beyond the woods, and nine species never had more than two pairs nesting in the woods itself. Eleven more species nested in fewer than 5 years during the 26 years. They had to be considered casual species, counted as immigrants and extinctions in calculating annual turnover. Only 14 species were regular breeders. During the 26 years Eastern Wood experienced a considerable turnover of species with an average of three immigrations and three extinctions annually. Species equilibrium as determined by immigration and extinction graphs (Figure 28.15) is 32 species, somewhat higher than the 27 species that on the average Eastern Woods supported over the years (Williamson 1981).

The equilibrium theory of island biogeography deals with numbers of species only. The model makes no assumptions about species composition; it treats all species in a taxon as equals. It assumes that the probability of extinctions and immigrations is the same for all species. It makes no assumptions about the life history or habitat requirement of the species involved. Species richness may remain constant, but species composition changes. However, the degree of isolation is relative, depending upon the taxon involved.

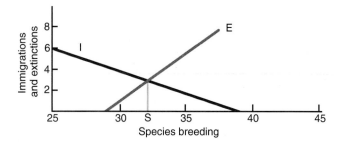

**Figure 28.15** Immigration and extinction curves for Eastern Wood, Bookham Common, Surrey, England, a 16 ha oak woodland. The immigration curve is a regression line (the points have been omitted) which cuts the abscissa at 39 species. The maximum number of species that bred at one time or another in Eastern Wood was actually 44. The extinction line is at 45°, indicating one extinction for every species present over 29.

What is a short distance for a bird may be an insurmountable distance for a mammal or a lizard. Moreover, the assumption of the model that extinctions relate to an island's area and distance does not consider the fact that immigrations and extinctions may not be independent. Extinction of a dwindling population may be slowed or even halted by an influx of immigrants, the **rescue effect** (Brown and Kodrich-Brown 1977). Extinctions are influenced by life history traits and are not due only to isolation. Thus the equilibrium theory is neutral to species interactions and habitat heterogeneity. In a way it is analogous to the Hardy-Weinberg equilibrium in population genetics.

Another approach to island biography is the habitat diversity theory. There is considerable evidence that habitat heterogeneity overrides island size in some situations (Rigby and Lawton 1981, Blake and Karr 1984, Freemark and Merriam 1986). Larger islands may have lower extinction rates and support more species than smaller islands because they hold larger populations and have more diverse habitats.

Even though the theory of island biogeography has its deficiencies, conservation ecologists have recognized its applicability to fragmented habitats. A major debate has developed over the value of a *single large or several small preserves* (SLOSS) to guarantee the survival of most species. One group (Simberloff and Abele 1976, 1982, 1984, Abele and Connor, 1979, Higgs 1981, Järvinen 1982, 1984) argues that two or more smaller reserves will hold more species than one large reserve of the same area. Data on species richness seem to support that argument. Others (Diamond and May 1976, Wilson and Willis 1971, E. D. Willis 1984, Whitcomb et al. 1976) argue that a large island is preferable because that area will not only support more rare species but also large populations of other species, making them less vulnerable to random extinction. This argument has abated somewhat as conservation biologists have turned their attention to the application of island biogeography theory to the effects of fragmentation on terrestrial habitats (Harris 1984).

# Habitat Fragmentation

Not all islands are oceanic. Mountain tops (J. H. Brown 1971, 1978), bogs, ponds, dunes, areas fragmented by human land use, host plants and their insects (Janzen 1968, Strong et al. 1984), and animals and their parasites—all are essentially island habitats. As Simberloff (1974:162) put it: "Any patch of habitat isolated from similar habitat by different, relatively inhospitable terrain traversed only with difficulty by organisms of the habitat patch may be considered an island." Investigations of these island situations are leading to new insights and modification of island theory, as it applies to terrestrial situations.

Considerable differences exist between habitat islands and oceanic islands, on which the theory was developed (Wilcox 1980). Oceanic islands are isolates. They are ecological units surrounded by an aquatic barrier to dispersal. They are inhabited by organisms of various taxa that arrived there by chance dispersal over a long period of time or represent remnant populations that existed on the area long before isolation. By contrast, habitat islands are samples of populations of a much larger area. These samples contain fewer species, fewer individuals within a species, and more species represented by only a few individuals.

Fragmentation in large tracts of forest and grassland may begin intrusively with clearing within the tract, such as clearcuts, wildlife food plots, and housing development (Harris and Silva-Lopez 1992). Although considerable forest or grassland remains, the structural integrity of the forest declines. Carried far enough, such intrusion results in a fragmented forest, grassland, or shrubland (Figure 28.16). A fragmented forest is reduced but retains ecological characteristics and organisms of the forest. As a large area of habitat becomes even more fragmented, all that is left are disjointed fragments of varying size (Figure 28.17). These fragments are separated by other types of habitats, particularly urban and suburban developments and agricultural lands. When such fragments become isolated from others in a matrix of a nonforest or grassland landscape, they become habitat islands or insular habitat. The flows and interactions of forest are dominated by the nonforested landscape. The same holds true for native grassland and shrublands.

Not only do areas surrounding the habitat fragment present barriers to dispersal, but they are also terrestrial habitats in their own right, with their own sets of species, including domestic animals such as cats, dogs, sheep, and cattle, and species of wildlife highly adaptable to human habitations. Some species, like cats, weasels, raccoons, and crows, invade the edges and move into the interior, increasing predatory pressure on interior species (Gates and Gysel 1978, Chusko and Gates 1982, Wilcove 1985). The brown-headed cowbird encounters new naive hosts for nest parasitism (Brittingham and Temple 1983). Exposure to wind and solar radiation causes mortality among mesic and shade-tolerant plant species on the edge, further reducing the integrity of the fragment (Wales 1972, Ranney 1977, Lovejoy et al. 1986).

Insular habitats may be further reduced by fragmentation pressures encroaching from one, several, or all sides. Increasing contraction has severe impacts on the fragment by reducing size and increasing isolation and edge effects. Causes of such fragmentation are highways, roads, power-line rights-of-way, and ski slopes that bisect a habitat and create a barrier that effectively divides populations of many organisms into fractions on either side. The impact of such division varies with the species. For some large or highly mobile animals, a highway is not uncrossable but dangerous, as evidenced by road kills. For most small mammals, reptiles, amphibians, and invertebrates of low mobility, roads impose a barrier individuals cannot cross (Mader 1984). Even in large tracts of seemingly unfragmented forests, dirt roads and hiking trails can have a divisive effect on forest floor inhabitants. The endangered Cheat Mountain salamander (*Plethodon nettingi*), endemic to West Virginia, exists in isolated populations across its restricted range in forested areas above 960 m. These populations are further subdivided in their high elevation forest habitat by hiking trails that individuals will not cross, restricting gene flow within the population (T. Pauley 1992).

Up to a point in fragmentation, no species are lost. As the process continues, the remaining area is reduced to a critical size below which it will not support many of the original species, and a number of them disappear (Whitcomb et al. 1981, Robbins et al. 1989) (Figure 28.18). The first to go are the susceptible species, those on the higher trophic levels, habitat and food specialists, and ones that need large areas of habitat to maintain viable populations. The latter are **interior species** whose habitat begins some distance within the forest. The others are **area-sensitive species** that require large territories or foraging areas but may or may not be interior species (Figure 28.19). Many interior species are area-sensitive to the degree that they will disappear when the habitat, even though much larger than their activity space, declines to some critical minimum. As these species disappear, edge and area-insensitive species, ones at home in small to large units of habitat, increase in abundance. Although the number of species in a fragmented habitat may increase initially, the number of species it contains eventually will decline to a new steady state, called **faunal relaxation.** If the habitat island continues to shrink or becomes more isolated, faunal relaxation will lead to **faunal collapse.**

What sizes of habitat fragments are needed to maintain regional populations and satisfy habitat requirements of the species concerned? What size of island maintains greatest species diversity? At what size of habitat patch do interior and area-sensitive species disappear? Such questions have stimulated a number of studies (see Wilcove et al. 1986,

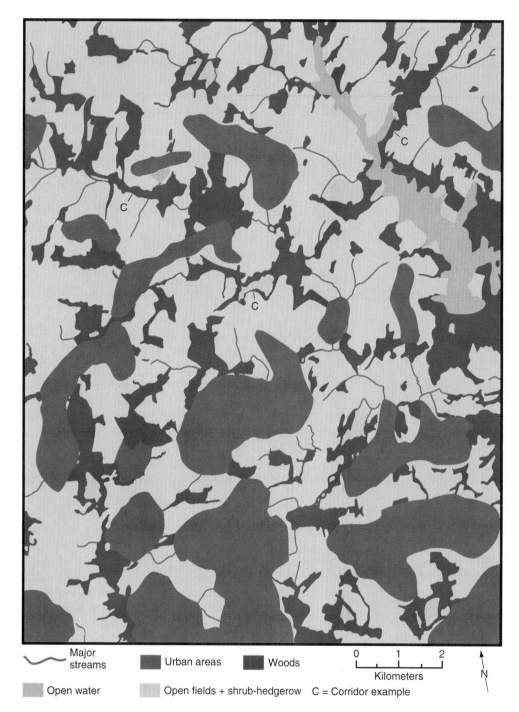

Major streams

Open water

Urban areas

Open fields + shrub-hedgerow

Woods

C = Corridor example

0    1    2
Kilometers

N

**Figure 28.16**  At the time of European setlement, this section of eastern Maryland was completely forested. Over the course of several hundred years, much of the land was cleared for agriculture, fragmenting the remaining forest land into isolated patches. In the past quarter century urban and suburban developments have encroached upon the area, further fragmenting and isolating the forest, and doing the same to agricultural lands. Note the corridors that still remain, connecting various forest islands. (Adapted from Whitcomb et al. 1981:127.)

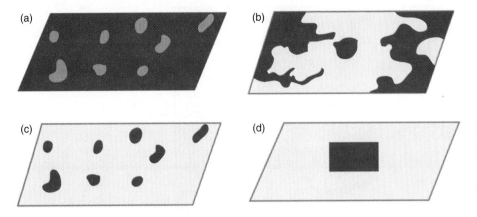

**Figure 28.17** (a–c) How a forest becomes fragmented. (d) A larger forest tract becomes isolated when the surrounding land has been cleared, a situation common to parks and reserves.

Lovejoy et al. 1986, McLellan et al. 1986, Higgs 1981, for example).

The minimum size of habitat needed to maintain interior species differs with plants and animals. For plants, patch size is not as important in species persistence and extinction as environmental conditions (M. Weaver and Kellman 1981). For forest interior plants, the minimum area depends upon the patch size at which moisture and light conditions become adequate to support shade-tolerant species. That area depends in part on whether the edge of the stand allows light and wind to penetrate (Levenson 1981, Raney et al.

1981), on canopy closure, and on the ratio of edge to interior. If the stand is too small or too open, the interior environment becomes so xeric that mesic species, both herbaceous and woody, cannot reproduce. When mature residual mesic species such as sugar maple and beech die, they are replaced by xeric species such as oak.

Levenson (1981) found that the species richness of plants was the greatest in edge situations where xeric species coexist with some interior species. The total number of woody species in Wisconsin woodlots increased with woodlot size up to approximately 2.3 ha. At that point vegetation achieved a maximum balance between edge and residual interior species. Beyond that size species richness declined and leveled off at 9.3 ha (23 acres), as mesic conditions and shade-tolerant species persisted. A similar situation occurs in islands of prairie vegetation (Simberloff and Gotelii 1984), Yorkshire limestone pavements, chalk quarry reserves, and lowland heaths in Great Britain (Higgs and Usher 1980). There appears to be a negative correlation between edge species of plants and the size of forest islands and a positive correlation between interior species and increased area.

Among birds a similar pattern exists. In general, larger forests have more species than smaller forests. The latter are occupied by edge and ubiquitous species at home in any size forest tract (N. W. Moore and Hooper 1975, Galli et al. 1976, Ambuel and Temple 1983, J. G. Blake and Karr 1984, Freemark and Merriam 1986). A study in New Jersey suggested that maximum bird diversity was achieved with woodlands 24 ha in size. However, those woodlands held no true forest interior species such as the worm-eating warbler (*Helmitheros vermivorus*) and the ovenbird (*Seiurus aurocapillus*), which are highly sensitive to forest fragmentation and require extensive areas of woods (Whitcomb et al. 1981, Lynch and Whitcomb 1977) (Figure 28.19). Both Blake and Karr (1984) and Freemark and Merriam (1986), investigated bird species composition of large and small forest habitat islands (from 3 to 7620 ha) in agricultural regions in Illinois and Ontario. They found that although two or more smaller forest habitats supported more species, long-distance mi-

**Figure 28.18** A model of the number of species remaining in each species pool as fragmentation proceeds. Large, solid circles show the pool of species with larger area requirements and low freedom of movement; open circles show the species pool with less stringent area requirements. The small solid dots depict the proportion of the first species pool that would be present when the habitat is minimally fragmented. (From McLellen et al. 1986.)

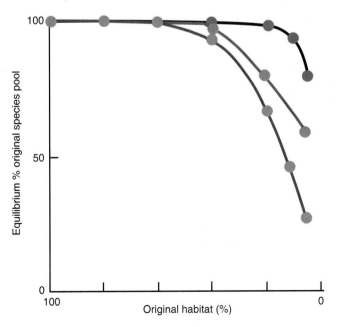

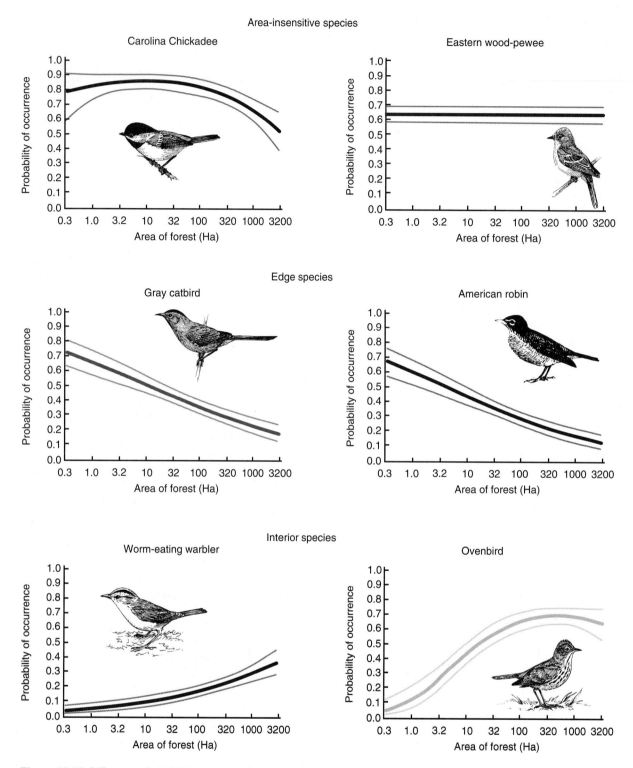

**Figure 28.19** Differences in habitat responses between edge, area-insensitive, and area-sensitive or interior bird species. The graphs indicate the probability of detecting these species from a random point in forests of various sizes. The dotted lines indicate 95 percent confidence intervals for the predicted probabilities. (Adapted from Robbins, Dawson, and Dowell 1989:18.)

grants and interior species, typical of larger tracts, were missing or poorly represented. Large forest tracts were important for increasing the number of forest interior neotropical migrants and certain resident species, such as the hairy woodpecker. Large forest tracts with a high degree of heterogeneity held the most species, supporting birds of both edge and interior. The presence of forest interior species in smaller woodlands depends upon the nearness of those habitat islands to a pool of replacement individuals.

Area alone is not the full story. What is important is the ratio of edge (or perimeter) to area. The length of perimeter is directly proportional to the square root of the area. At some small size islands are all edge, as illustrated in Figure 28.20a. If we allow the depth of the edge to remain constant while increasing area, the ratio of edge to interior decreases as the size of habitat island size increases. When the island size becomes large enough to maintain mesic and interior conditions, an interior begins to develop. However, size is not the sole determinant of edge-interior conditions. The shape of the island is also critical. In Figure 28.20b the rectangular wooded island contains 39 ha; yet because of its shape it is entirely edge. Its width does not exceed the depth of its edge, or to put it differently, it has a high ratio of edge to area. Out of its 16 species, none are interior. By contrast, the square woodland of 47 ha has a core interior area of 20 ha, and 6 of its 16 species are interior species sensitive to fragmentation. This relationship is further illustrated in Figure 28.20c. Any island to the left of point A, the size at which interior species can exist, is all edge. As island size increases, the interior area increases and the ratio of edge to interior decreases. This relationship of area to edge holds for circular or square islands, but not for irregular or rectangular islands.

The perennial argument over size of habitat islands and the maintenance of species diversity is not as simple as it appears. Given that small islands hold a higher species diversity than large islands, an overemphasis on managing for species diversity, so prevalent today, can lead to decline and extinction of interior and area-sensitive species that need larger or more homogeneous habitats. A high diversity of species is not as important as the kinds of species present.

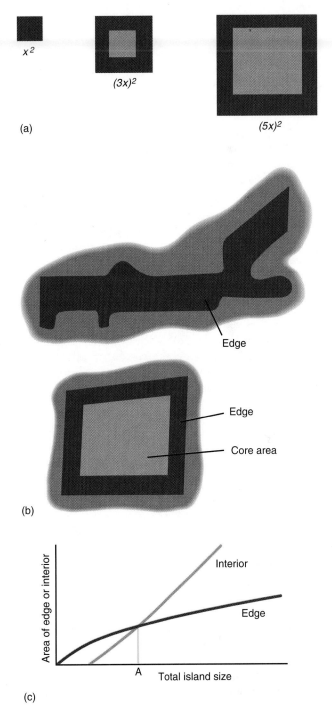

**Figure 28.20** Relationship of island or fragment size to edge and interior conditions. (a) Functionally all small islands of habitat are edge. Allowing the depth of edges to remain constant, the ratio of edge to interior decreases as island size increases. When the island size is large enough to maintain mesic and interior conditions, an interior begins to develop. (b) Size alone is not a determinant of interior-edge conditions. Shape of the island is also critical. A long or rectangular habitat patch is all edge. (c) A graph shows the relationship between edge and interior as island size increases. Below point A, the woodland is all edge. As its size increases, interior area increases and the ratio of edge to interior decreases. This relationship of size to edge holds for circular or square islands. Long narrow woodland islands as in B, whose width does not exceed the depth of the edge, would be edge communities, even though their area might be the same as that of square or circular ones (and c after Levenson 1981:32; b after Temple 1986:304.)

## Corridors

Important to the integrity of fragmented habitats are corridors linking one with the other. Corridors are strips of vegetation similar to the fragments but different than the landscape matrix in which they are set. (Various corridors appear in Figure 28.16.) Narrow line corridors include lines of trees planted as windbreaks, hedgerows, roads and roadside strips, and drainage ditches (Forman and Godron 1981, 1986). Such corridors usually originate from human disturbance or activity: remnants of forest are left between fields and developments, volunteer shrubs grow along boundaries between fields, and the like. Such corridors are usually inhabited by edge species, although in Europe the long history of hedgerows in the landscape has allowed the evolution of typical hedgerow animal and plant communities. Strip corridors are wider bands of vegetation with an edge effect on each side; they may hold interior species. Such corridors may be wide strips of woodland or powerline rights-of-way. Belts of vegetation along streams and rivers form stream corridors; many of these corridors form important riparian habitat for animal life.

Corridors have important functional roles in the landscape. They provide dispersal routes for species between habitat islands, especially when corridors interconnect to form networks, and they offer stopover habitat for migrating birds. They facilitate gene flow between habitat patches, and the reestablishment of species in habitats that have experienced local extinctions. Corridors provide protective cover for prey and lookout cover for predators hunting adjacent vegetation patches. Finally, corridors are habitats in their own right for edge species and, if sufficiently wide, for interior species. However, corridors can also facilitate the spread of disease between habitat blocks, expose prey species to predation, and increase the vulnerability of large mammals to poaching.

## Assembly Rules

Islands of fragmented habitats hold a remnant of the larger community of which they were once a part. Communities that have developed on some oceanic islands and large areas of disturbance started at one time from scratch. The proportion of species on those areas differ from those of the mainland and from each other. What factors are involved in assemblage of species and how those species fit together to form community structure is one of the more intriguing problems in ecology.

Ecologists have used several approaches to investigate this question. One, already alluded to, is to remove a keystone species and study how the community restructures itself—what species become the new dominants, which species increase, and which species decrease. Another approach is to attempt to reconstruct the sequence in which species were added when the community was formed and to figure out what colonization sequences may or may not be possible according to some assembly rules (Diamond 1975, Drake 1990). **Assembly rules** are the mechanisms of how species fit together in a community. It is much like fitting together a jigsaw puzzle, with one significant difference: a jigsaw puzzle has a single solution, fixed by the shapes of the pieces. Assembly rules for a community involve alternative pieces that fit but change the configuration of the puzzle (Figure 28.21).

Consider species A that initially colonizes a site and multiplies to dominate it. Then two other species, B and C, invade the area. Both of these species can compete with and share the site with A, but B and C are strong competitors. Each can beat the other, with the advantage going to the earlier arriving species. Depending on their arrival, the community now will consist of AB or AC. Now a fourth species, D, arrives on the site. It occupies a somewhat different niche, but still competes with C to the point that if the community it invades is AC, then D cannot succeed. If D invades AB, it can find a place in the community, which becomes ABD.

This example is simplistic. More species are involved in a natural community. The species assemblage of the community is influenced by trophic relationships, competitive interactions, old species lost, and new species added. Some assemblages of species are vulnerable to invasion by other species; other assemblages resist invasion. Although the

**Figure 28.21** Pieces of a jigsaw puzzle that can be fitted together in two combinations represent an analogy of community assembly. If A colonizes first, then either B or C, competing species, can colonize. If B is the successful colonizer, then D can also fit in. If C is the successful colonizer, D will not be able to colonize, because the community will resist invasion; piece D will not fit. (After Drake 1990.)

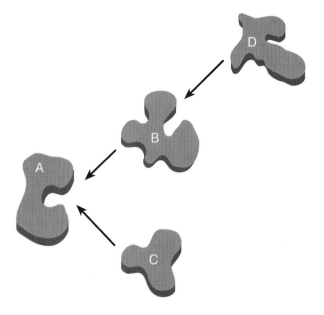

same species may be involved in similar communities, each may have colonized the area at different times and in different sequences, which would change species relationships. Each community, then, develops within a historical context that will influence species interaction. Thus different assembly routes produce differences in community organization.

This concept of species assemblages is not without its strong critics, who offer an alternative approach, the null model (Simberloff 1978, 1983). The null model approach is to draw species at random from a species pool to create an assembled community. This randomly generated community then forms a "null" hypothesis against which the distribution of species in a natural community can be compared. The idea is that community assemblages are based on chance alone. The null hypothesis ignores the historical context within which the assemblage of species operates.

Of what value is this highly theoretical concept? Assembly rules become quite important in the restoration of ecosystems. If you are attempting to reconstruct a prairie, for example, you will want to give some thought to the species of plants to be introduced and what animals might colonize the area naturally. Should you introduce prairie dogs or buffalo? What impact might the introduction of wolves into a wolfless wilderness area have on the current community? How will certain species of fish fit into a restored stream ecosystem? Too often assembly rules are given little consideration in such restoration work.

# POPULATION INTERACTIONS

## Influence of Competition

Since Darwin (1859) ecologists have considered interspecific competition, especially competitive exclusion, as the cornerstone of community structure. Lack (1954, 1971), Hutchinson (1959), and MacArthur (1960, 1972), among others, have stressed the role of competition in shaping community organization, including species distribution, resource allocation, niche segregation, and the like. Salisbury (1929) early pointed out the role of competition in shaping plant communities.

Numerous studies have been undertaken to demonstrate the role of interspecific competition in determining community structure, with emphasis on the animal component (see Connell 1983, Schoener 1983). Many of these studies suggest, if they do not prove with certainty, that competition exists between certain species pairs and within guilds of species. However, do these studies demonstrate that interspecific competition exerts a strong influence on community organization? The answer depends on the definition of community, which as we saw is colored by many interpretations. Is the community in question limited to members of a particular guild of organisms, or to a part of a food web? Do the interventions affect only a small proportion of the species

present, or do they reverberate through all species on the site? Determining the effects of interspecific competition on shaping the whole community is difficult at best.

Nevertheless, interspecific competition does impact to some extent the organization of some communities. Anyone who has a garden experiences the effects of competition between garden plants and weeds. A great deal of competition, both exploitative and interference, for light, moisture, nutrients, and space exists among plants from forests to deserts. The outcome of competitive interaction among plants, which will vary from community to community and is often influenced by herbivory, has a pronounced influence on the physical structure of the terrestrial community in which they belong and in turn on distribution and abundance of animal life.

A great deal of theoretical literature attempts to demonstrate or determine the role of competition in structuring communities on a scale that goes beyond local communities (for example, see Cody and Diamond 1975, Diamond and Case 1986). However, we are concerned here with the effects of competition on local community structure. At this level competition is for food and space. Such exploitative competition is influenced by the nature of the niche each species occupies and is usually most evident at the level of guilds. For example, in autumn many animals, especially jays, wild turkey, deer mice, squirrels, black bear, and white-tailed deer feed on acorns. If the acorn crop is poor, these animals compete strongly for it. Because squirrels depend heavily on acorns, they may feel the effects of exploitative competition more keenly than the more generalist consumers (Allen 1943).

When one species is placed at a competitive disadvantage by some environmental or biological disturbance, competitors can use available resources to the detriment of the other. Mortality of oak trees defoliated by gypsy moths is greater on nutrient-rich mesic sites than on xeric sites (Statler and Serro 1983, Hicks and Fosbroke 1987). Oaks and other trees growing on mesic sites grow large crowns relative to their root systems (low root-to-shoot ratio). On xeric sites oaks, the dominant species, have small crowns and relatively large root systems (high root-to-shoot ratio). One hypothesis for oak mortality on mesic sites is that yellow-poplar and sugar maple, usually ignored by oak-feeding gypsy moth larvae, maintain their canopy and use the moisture and nutrient resources that normally would be used by the now defoliated oaks. On mesic sites the small root systems of the oaks are unable to compete strongly for moisture and to supply sufficient moisture and nutrients to refoliate a large crown. Unable to refoliate successfully, many oaks succumb. On more xeric sites, moisture-demanding competing trees are absent and the large root systems can often supply sufficient resources for summer refoliation.

Competition for space can occur through the mechanisms of dominance and founder control (Yodzis 1988). In the former, some organisms become dominant and control the re-

source. Among plants interspecific competition interacts with resource abundance, local environmental heterogeneity, and disturbances to control community structure. Fast-growing trees, for example, intercept light and shade out plants beneath their canopy. Mussels completely cover rocky substrates along the shore. In contrast, founder control involves no clear dominants. Space belongs to those species that colonize the area first and hold onto it. In this case, disturbance may be important in determining colonization by different species (see Chapter 29).

The clearest examples of competition influencing community structure are the effects of introductions or invasions of exotic species. Exotic species of birds introduced into the Hawaiian islands have not only excluded native species with similar ecological requirements but have also excluded or prevented the successful establishment of other exotics with similar morphological characteristics (Moulton and Pimm 1986). When two introduced species of plankton-feeding fish, the alewife and rainbow smelt, proliferated in Lake Michigan, seven native species of fish with similar food habits declined drastically (Crowder et al. 1981). Japanese honeysuckle, a garden escapee, and multiflora rose, widely planted in the past for soil conservation purposes, have invaded old fields and forest edges, crowding out native plants and affecting the structure and composition of animal life. Since arriving in the ballast tanks of freighters from Europe in 1985, zebra mussels have spread from the Great Lakes to the Mississippi River and its tributaries. Growing on the tops of native freshwater mussels and killing them, zebra mussels threaten the extinction of native species.

The importance of competition in community structure will vary from community to community. It is more pronounced among some groups, especially sessile ones such as plants and mussels, or members of the same guilds. Other groups, such as phytophagous insects, experience no interspecific competition (Strong et al. 1984). Competition among plants may have the greatest influence, because its outcome can affect the structural aspects of the community. At best it is difficult to determine the role of competition, because its apparent outcomes may have alternative explanations, such as weather, climate, and especially predation (Andrewartha and Birch 1984).

## Influence of Predation

Although the influence of competition on community structure is somewhat obscure, the influence of predation is more demonstrable. Because predation affects one or more trophic levels, its influence can more readily be noticed throughout a community, particularly on competitive relationships.

In 1955 Robert MacArthur suggested that communities with more complex food webs are more stable than those with simple ones. The reasoning was that in food webs with more trophic links, predators have access to alternative prey. Should the population of one prey species become rare, the predators would switch their diet to more abundant species, allowing the less abundant prey to recover. Such interactions within the food web would reduce predator and prey fluctuations.

Hairston, Smith, and Slobodkin (1960) proposed that within trophic levels of a community, carnivores compete strongly among themselves and severely exploit their resources. Herbivores are regulated by predators and have little impact on vegetation. Plant populations, not greatly reduced by herbivory, compete for light, water, and nutrients. The food chain is so structured that energy flows from the base to the top, but the abundance of populations in the lower trophic levels above plants is determined by predation.

If predation has a major influence on trophic structure, then the removal of a major predator should reveal a pattern in the food web. R. T. Paine (1966, 1969) experimentally removed the top carnivore, the starfish *Pisaster ochraceus,* from a rocky intertidal community on the Pacific coast of Washington. This community, in addition to the starfish, consisted of four species of algae, a sponge, filter-feeding barnacles and mussels, browsing limpets and chitons, and a predatory whelk. Both barnacles and mussels, when given the opportunity, aggressively and tenaciously take over space and exclude other sessile organisms. The starfish fed on sessile barnacles and mussels as well as on limpets and chitons. For two years Paine excluded the starfish from a swath 8 m long and 2 m deep on a rocky intertidal area. In the undisturbed community, the control, the community remained intact. On the swath where the starfish were removed, barnacles settled successfully but were soon crowded out by the mussels. All but one species of algae disappeared for lack of space, browsers moved away for lack of food, and the number of species dropped from 15 to 8. Apparently the predaceous starfish, by feeding on barnacles and mussels, made space available for the competitively subdominant species and helped to maintain species diversity and complex intertidal trophic structure.

*Pisaster* functioned as a keystone species in the food web; it affected the pattern and trophic structure of the community. Without it certain species disappeared in the face of strong competitors and the stability of the food web was impaired. Thus certain species have a manipulative influence on the rest of the community and the effects of their removal cascade down through the community.

Keystone species are not necessarily top predators. Herbivory affects the structure of plant populations and influences competition and productivity. A classic example is the influence of rabbits on the diversity of species in English pastures (Tansley and Adamson 1925, Zeevalking and Fresco 1977). A sharp reduction in the rabbit population in southern England from myxomatosis resulted in an aggressive growth of meadow grass in fields inhabited by the spectacular large blue butterfly, *Maculinea arion.* The heavy growth of grass resulted in the extinction of open-ground ant colonies, the nests of which were utilized by large blue

caterpillars. As a result the large blue is nearly extinct. The loss of one keystone grazing herbivore, the rabbit, resulted in the local extinctions of two other species as well.

Grazing herbivores, from prairie dogs (Whicker and Detling 1988) and pocket gophers (Huntley and Inouye 1988) to large ungulates (McNaughton et al. 1988), influence the structure of grassland communities from the North American plains to East Africa. Grazing ungulates, in particular, reduce canopy height, stimulate tiller growth, and favor low-growing grasses and herbaceous plants at the expense of taller grasses. Intensive grazing favors plants that are prostrate, have small leaves, or grow in rosettes close to the ground and selects against taller forms. When grazing pressure is relaxed, tall-growing grasses assume dominance at the expense of lower-growing forms, reducing plant diversity (McNaughton 1984, 1985).

Woody vegetation is not exempt from the effects of herbivores, notably browsing ungulates. Browsing by white-tailed deer in many forests of eastern North America has eliminated understory herbaceous and woody plants, including seedlings and sprouts of trees. It has converted the understory to ferns and created browse lines on the trees (Marquis 1981, Marquis and Grisez 1978). Heavy browsing by moose on Isle Royale has favored spruce over deciduous trees. The change in quantity and nature of litterfall has resulted in a decrease in microbial biomass and nitrogen availability (McInnes et al. 1988).

The rocky intertidal regions of the New England coast are inhabited by a number of marine algae, including the ephemeral species *Enteromorpha, Ulva,* and *Porphyra* and the perennial fucoid algae *Fucus vesticulosus* and *F. distichus,* and by a grazing herbivore, the snail *Littorina littorea.* Lubenchenco (1983, 1986) excluded snails from rocks sheltered against wave action. She discovered that the ephemeral species of algae colonized and become established on the rocks. They monopolized the space and inhibited colonization by the perennial fucoid algae. Where snails were present, grazing pressure reduced competition for space among the three ephemerals. The snails prevented any one of them from becoming dominant, although each has the potential of excluding the others. In the absence of grazers in the herbivore exclusion cages, the perennial *Fucus* was able to colonize the rocks and assume dominance. Although *Littorina* snails eat small *Fucus,* they avoid larger individuals if ephemerals are present. *Fucus* has chemical and structural defenses against snails, but the ephemerals have no known defenses. They persist where grazers are absent, exist at very low levels, or are inactive (during winter months). Because snails are rare at sites exposed to heavy wave action, exposed rocky intertidal sites become havens for ephemerals. These experiments suggest that by controlling competitive interactions, herbivory increases species richness and influences the community structure of rocky intertidal shores.

Such studies, and there are many of them, suggest some basic principles about the influence of grazing on plants. If the herbivore selectively grazes the major dominant, then plant diversity increases. Overgrazing in winter and spring followed by undergrazing in summer produces maximum floral richness. (This is usually the case in natural situations because deer, elk, and others can be pressed for food in winter, but find plenty in summer.) The most species-rich communities are developed by continuous grazing with a maintained population of generalist herbivores. If the dominant plant is highly palatable, overgrazing will reduce it and allow other, less palatable species to occupy the area; but if the dominant species are unpalatable, then grazing only serves to increase their dominance. If the herbivore is regulated by food supply and not by predators, then it can reduce a plant species to a minor component and allow invasion by other species. As the intensity of grazing increases, more and more unpalatable species are grazed, until eventually only unpalatable species remain or move into the area. This condition is exemplified by the late-season overgrazed pasture fields with their conspicuous growths of ironweed and thistle. Thus the complex balance of species in a plant community is sensitive to control by feeding activities of grazing animals.

By selectively eating seeds of certain plants over others, seed-eating predators influence plant community composition. Using exclosures and plots sown with known quantities of seeds of annual plants, Borchert and Jain (1978) found that meadow mice and house mice consumed 75 percent of wild oats (*Avena fatua*) seed, 44 percent of wild barley (*Hordeum leporinum*) seed, and 37 percent of ripgut brome (*Bromus diandrus*) seed. Showing a strong preference for *Avena* and *Hordeum,* the mice reduced the numbers of these plants by 62 percent and 30 percent respectively, stimulating competitive release of *Bromus.* Mice also influence the populations of plants by cutting mature plants and eating seedlings (Batzli and Pitelka 1970).

Such studies and computer simulations of food webs allow us to arrive at some general hypotheses about the effects of species removal on community structure and stability (Figure 28.22). The removal of a plant species from a simple food chain can cause the loss of the whole chain. For example, the loss of bamboo results in the extinction of its specialist herbivore, the panda. The removal of a top predator from a simple food chain, however, may have little or no effect. The removal of a single plant species from a more complex food web may have little effect, but if a top predator is removed, the effects may cascade down through the food web to the basal or primary trophic level.

This cascading affects not only production and consumption at the lower trophic levels, but also the availability of nutrients by altering available biomass (Carpenter, Kitchell, and Hodgson 1985). Consider a top predatory fish, such as bass. An increase in its population reduces the population of

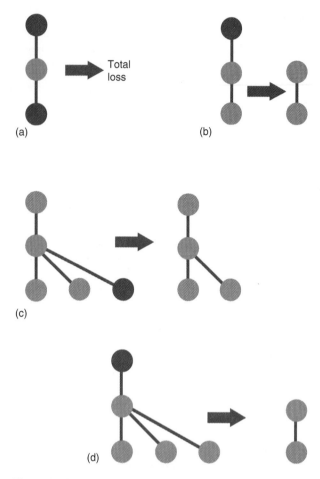

**Figure 28.22** Theoretical effects of species removals on the compexity of a food web. (a) Removal of a plant from a simple food chain may destroy the entire chain. (b) Removal of a predator from the same food chain may have little effect. (c) A plant removed from a complex web may have little effect. (d) The loss of a predator from the same food web may allow herbivores to increase and lead to loss of several plant species. (After Pimm 1987:106.)

*Cichlasoma maculicauda,* whose juveniles formed the prey of some of the species eliminated, increased. With plankton-feeding fish gone, zooplankton that had been their preferred prey increased, and with their prey species largely eliminated, herons, terns, kingfishers, and tarpon decreased (Zaret and Paine 1973).

Sih and associates (1985) summarized an array of studies on the effects of the interaction of predation and competition on prey communities and deduced certain trends. Herbivores have a stronger effect on plant communities than carnivores have on animal communities. Overall effects of predators on a prey community depend upon how the predators impact competitive dominants among the prey species. The effects may be direct or indirect. One exploitative competitor affects another by reducing the level of a shared resource, a third species. Predatory reduction of one or both competitors of the third species allows it to increase, an indirect effect. Predation may also influence lifestyles of the prey, including morphology, physiology, and behavior, such as habitat use, foraging mode, and the like. This in turn determines prey encounter rates with both competitors and predators. In all, the interactions among predation and competition and community structure involve complexities at which experimental studies to date can only hint.

## Effects of Parasites and Diseases

Parasites are an integral part of natural communities. Their overall effect on community structure becomes most apparent when an outbreak of disease decimates or reduces an affected population. With the clearing of forests, the brown-headed cowbird, a nest parasite, has expanded its range through North America. It has invaded forest edge and even the forest interior where it parasitizes birds, such as hooded warblers, that were never exposed to nest parasitism in their evolutionary history. Lacking antiparasitic behavior, these birds are experiencing parasite-induced declines in their populations (Goldwasser et al. 1980, Walkinshaw 1983). The introduction of rinderpest, a viral disease, to East African ungulates (see Chapter 26) decimated herds of African buffalo and wildebeest. The introduction of avian malaria carried by introduced mosquitoes eliminated most native Hawaiian birds below 1000 meters in altitude; above that altitude is the mosquito-free zone. Introduced bird species are resistant to malaria (Warner 1968). Outbreaks of mange periodically reduce red fox populations in New York and other areas.

How parasitism can influence interactions within a community beyond a simple host-parasite relationship is illustrated in the moorland ecosystem of Scotland. Heather (*Calluna vulgaris*) and bracken (*Pteridium aquilinum*) (see Chapter 13) dominate the moorlands of Scotland, the major habitat of the red grouse (*Lagopus scoticus*) and grazing land for domestic sheep. The red grouse, which feeds on

zooplankton-feeding fishes. A decrease of these predators allows an increase in herbivorous zooplankton, which in turn leads to a decrease in phytoplankton and primary production, but an increase in available nutrients. If the top predator is reduced, vertebrate planktivores increase, invertebrate planktivores and large herbivorous zooplankton decrease, and small herbivore zooplankton and primary production increase.

What happens when an introduced top predator invades a complex food web? When the peacock bass, *Cichla ocellaris,* native to the Amazon River, was introduced as a game fish into a tributary of Panama's Lake Gatun around 1967, this voracious predator eliminated 8 of the 11 principal native fish species and reduced 3 others by 75 to 90 percent. With these fish gone, another species, the algae-feeding

young shoots of heather, is the host for several parasites, internal and external. The parasitic nematode *Trichostongyle tenuis* reduces the breeding success of grouse and may account for the short-term population cycles of the bird (Potts et al. 1984, P. J. Hudson 1986). Red grouse, the chicks in particular, are also parasitized by the tick *Ixodes ricinus,* which carries a virus, louping ill. The virus normally is associated with sheep, but ticks transmit the disease to highly susceptible grouse chicks (Duncan et al. 1978). The problem is intensified by management schemes. Sheep grazing on the moorlands provide the reservoir of the virus. Bracken, once managed and harvested for livestock bedding, is spreading into and replacing heather. Mature heather resists invasion; but burning of moorlands to provide young nutritious shoots for the grouse reduces mature heather and its ability to resist invasion of bracken with its fast-growing rhizome system. Bracken produces a mat of damp litter that provides an ideal habitat for ticks. Spread of bracken has increased exposure of the grouse to the tick and consequently to the louping ill virus that is associated with the long-term decline in grouse densities (Dobson and Hudson 1986). Because the virus is so pathogenic in grouse, the birds cannot maintain the virus without the presence of sheep, the vaccination of which should theoretically eliminate the disease.

Virulent tree diseases have markedly changed the composition of North American forests. The chestnut blight, introduced into North America from Europe, nearly exterminated the American chestnut and removed it as a major component of the forests of eastern North America. With its demise oaks and birch increased. Dutch elm disease has nearly removed both the American elm from North America and the English elm from Great Britian.

## Effects of Mutualism

Mutualism is too often overlooked as a mechanism in community structure. Although it is difficult to demonstrate, it may be more significant than either competition or predation. The importance of mutualistic relationships among conifers, mycorrhizae, and voles in the forests of the Pacific northwest is one example (see Chapter 27). The conifers depend upon mycorrhizal fungi for nutrient acquisition from the soil. The mycorrhizae depend upon the conifers for energy, and upon mycorrhizal-feeding voles for dispersal of spores. Temple (1977) suggests that the tree *Calvaria major* on the island of Mauritius has failed to produce seedlings in spite of adequate seed production because its seeds had to pass through the gut of the now-extinct dodo, *Raphus cucullatus.*

Relationships between pollinators and certain flowers are so close that loss of one could result in the extinction of the other (Pimm and Pimm 1982). There is some evidence that a relationship exists between mutualism and interspecific competition, just as there seems to be a relationship between interspecific competition and predation. This relationship

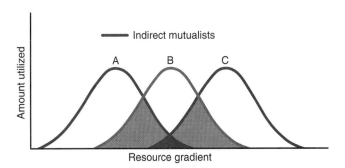

**Figure 28.23** How indirect mutualism might work. In this hypothetical situation species A and species C compete moderately, and both compete strongly with B. When all three species occur together, A and C together exert strong competition on B, reducing the intensity of competition by B. Thus in the presence of B, A and C have an indirect beneficial effect on each other. This model could also be an example of diffuse competition on B. (From Pianka 1981.)

could come about through indirect mutualism, in which the affected species never come into contact but affect each other's fitness or population growth rate (Boucher et al. 1982). Suppose that species A and species C are weak competitors, but both compete strongly with species B for resources partially shared. The joint pressure of A and C in effect reduces the intensity of competition from B for the resources they share (Figure 28.23). The interaction among the three competitors reduces niche overlap, an effect similar to that induced by a keystone predator. A and C benefit each other indirectly by reducing the fitness of a strong competitor of the other species. In another situation species A and C may be mutualistic with B, in which case A and C benefit each other indirectly.

Boucher et al. (1982) provide some possible examples of indirect mutualism. One is the relationships among two species of herbivorous *Daphnia* and their predators, a midge larva (*Chaoborus*) and a larval salamander (*Ambystoma*) in subalpine ponds in Colorado (Dodson 1970). *Ambystoma* eats large species of *Daphnia*, the midge larva small species of *Daphnia*. Where salamander larvae were present, the number of large *Daphnia* was low and the number of small *Daphnia* high. But in ponds in which the salamander larvae were absent, small *Daphnia* were absent and midges could not survive. The two species of *Daphnia* apparently compete for the same resources, while the midge and the salamander use different resources. The midge apparently depends upon the presence of salamander larvae for its survival in the pond. This possible indirect mutualism can be demonstrated only under controlled experiments involving population manipulations of the four species.

The idea of indirect mutualism is highly speculative, requiring strong experimental demonstration in natural com-

munities. It does suggest that mutualism, as well as competition and predation, can be an integrating force in the structuring of natural communities. Mutualism, long overshadowed by interspecific competition, is now receiving the attention it deserves (see Howe and Westley 1988).

# COMMUNITY PATTERNS

The nature of the community has been the object of study and dispute for years. Is a community such as an oak-hickory forest a real entity that is definable, describable, and constant from one stand to another? Or is it an abstraction, a collection of populations of different species that exist together because they have similar environmental requirements? To the last question one group of ecologists says yes, and another group says no.

## Organismic Versus Individualistic

Involved in the concept of the community are two opposing philosophies. One is the *organismic concept* advanced by F. E. Clements (1916). This concept had a powerful influence on the development of ecological thinking in the United States (McIntosh 1976) and persisted as a dominant concept until recent years. The organismic concept regards the community as a superorganism, the highest stage in the organization of the living world—rising from cell to tissue, organ, organ system, organism, population, community. Just as tissues have certain characteristics and functions above and beyond those of the cells that comprise them, so the community has characteristics and functions above and beyond the various populations it embodies. The distribution and abundance of one species in the community are determined by the species' interaction with others in the same community. Species making up the floristic community are organized into discrete closed groups. Groups of stands similar to one another form associations. Stands of one association are clearly distinct from stands of other associations. The community acts as a unit in seasonal activity, in trophic functions, and in succession. Clements (1916) considered the mature form of the community or the climax formation to be an "organic entity":

> As an organism the formation arises, grows, matures, and dies. Its response to the habitat is shown in the processes or functions and in structures which are the record as well as the results of these functions. Furthermore, each climax formation is able to reproduce itself, repeating with essential fidelity the stages of its development. The life history of a formation is a complex but definite process, comparable in its chief features with the life history of an individual plant. . . . Its most striking feature lies in the movement of populations, the waves of invasion which rise and fall through the habitat from initiation to climax. . . . All of these viewpoints are summed up in that

which regards succession as the growth or development of the climax.

The organismic approach was challenged by H. L. Gleason, who advanced an individualistic approach to the community. He argued:

> the association represents merely the coincidence of certain plant individuals and is not an organic entity of itself. While the similarity of vegetation in two detached areas may be striking, it is only an expression of similar environmental conditions and similar surrounding plant populations. If they are for convenience described under the same name, the treatment is in no wise comparable to the inclusion of several plant individuals in one species (1917).

> The sole conclusion we can draw from all the forgoing considerations is that the vegetation of an area is merely the resultant of two factors, the fluctuating and fortuitous immigration of plants and an equally fluctuating and variable environment (1926).

The individualistic approach was ignored by ecologists until the 1940s, when the concept was advanced quantitatively by John Curtis and his students (1951, 1959) as the vegetation continuum and by Robert Whittaker (1962, 1965) as gradient analysis (see McIntosh, 1975). The reintroduction of Gleason's individualistic concept coincided with the development of quantitative methods for analyzing vegetation. The individualistic concept emphasizes the species rather than the community as the essential unit in the analysis of relationships and distribution. Species respond independently to the physiological and biotic environment according to their own genetic characteristics. They are not bound together into groups of associates that must appear together. Instead, when species populations are plotted along an environmental gradient, long or short, the resulting graph suggests a normal or bell-shaped curve (Figure 28.24). The curves of many species overlap in a heterogeneous fashion. Thus the vegetation and its associated animal life exhibit a gradient or continuum from one extreme (for example, dry conditions) to another (wet conditions). In this view, the community is a collection of populations of species requiring the same environmental conditions. It is a continuous variable, not an integrated unit.

The individualist proponents grant that community discontinuities do exist, but that these reflect environmental discontinuities, such as breaks in soil type or sharp changes in moisture and salinity. If stands are as distinct as the organismic group considers them to be, then the boundaries between associations should be distinct, but they rarely are. Except where humans or some environmental catastrophe such as fire have interfered, it is rare for the vegetation to be a mosaic of discontinuous units. Instead, boundaries between units are more or less diffuse. If separate stands of an association are similar, all such associations should have similar distributions, and plants that comprise such associations should have distributions that coincide locally and over

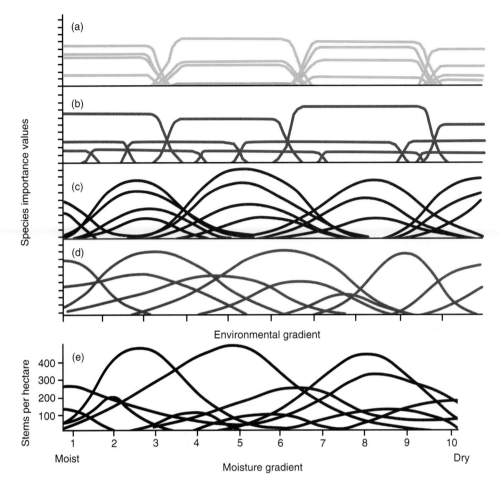

Species importance values

Stems per hectare

400
300
200
100

1   2   3   4   5   6   7   8   9   10
Moist                                          Dry

Environmental gradient

Moisture gradient

**Figure 28.24** (a–d) Four models of species distribution along environmental gradients. Each curve in each part of the figure represents one species population. (a) Competing species exclude one another along sharp boundaries. The species have evolved adaptations for living with one another, resulting in a distinct zone with its own assemblage of species. (b) Competing species exclude one another at some point along the gradient but do not form distinct groups with parallel distribution. (c). Competition does not result in sharp boundaries between species populations. Species so distributed on this gradient have evolved adaptions to one another, resulting in the appearance of groups of species with similar distribution. Although each group characterizes different communities, they intergrade along the environmental gradient. (d) Competition does not produce sharp boundaries between species populations. Because the distribution of one species on an environmental gradient is independent of the others, they do not form well-defined groups of species with similar distribution. (From Whittaker 1975.) (e) Actual distribution of species along a topographic moisture gradient from moist environments of ravines to dry south-west facing slopes in the Santa Catalina mountains of Arizona. (From Whittaker 1967.)

continental limits. If the association is a natural unit, then the species that comprise it should be bound together by obligate relationships. But most species relationships are not obligatory.

The organismic school argues that stands studies by gradient analysis are disturbed stands or ones that are not in equilibrium. If undisturbed, all stands develop to an endpoint in a few hundred years; and stands heading for an endpoint will naturally show a continuum of species. Besides, the technique of gradient analysis forces data into a continuum. These gradients are usually based on one variable such as moisture, but it is impossible to restrict a continuum to one variable alone because many interacting variables influence plant distribution. The continuum approach assumes that all species are equal when in fact some species are dominant.

Although the individualistic concept is widely accepted, the organismic concept is still evident in much ecological thinking, especially in regard to succession. The ecosystem is still considered as a functional, biological unit that modifies its organization to store, conserve, and manage information.

## Gradients and Continua

During the 1960s and 1970s our concept of the community underwent refinement. The current view is that the composition of any one community is determined by the species that happen to be distributed in the area and can survive and grow under prevailing conditions. Seeds of many plants may be carried by wind and animals, but only those adapted to grow and compete successfully in the habitat where they are deposited will take root and thrive. The element of chance is also involved. One adapted species may colonize an area and prevent others equally well adapted from entering. Wind direction and velocity, size of the seed crop, disease, insect and rodent damage, all influence the establishment of vegetation. Thus the exact species that settle an area and the number of individual species that succeed are situations that seldom, if ever, are repeated. Nevertheless, there is a certain pattern, with more or less similar groups recurring from place to place.

Plant communities are often regarded as distinct, closed natural units or associations in the landscape, especially for the practical purposes of description and study. More often than not, however, community boundaries are hard to define. Some communities, such as ponds, tidal beaches, grassy balds, islands of spruce and fir within or adjacent to a deciduous forest, old fields, and burns, have sharply defined boundaries, although even that may be an illusion. Here the vegetational patterns are discontinuous, resulting in edge. Most often, however, one community type blends into another. The species comprising the community do not necessarily associate only with one another, but are found with other species where their distribution overlaps. Some organisms will succeed only under certain environmental conditions and tend to be confined to certain habitats. Others tolerate a wider range of environmental conditions and are found over a wider area. Species shift in dominance and abundance, because of change in altitude, moisture, temperature, and other physical conditions. One species may be dominant in one group, and an associated species in another. The sequence of communities showing a gradual change in composition is a called a **continuum** (Curtis 1959, Whittaker 1956, 1960, 1967). Each community occupying a distinct position or point on the gradient is similar to, but slightly different from its neighbor. Even when the dominant plants change completely, the understory plants may be similar (Austin 1985).

Because the continuum seems to provide a physiognomic view of vegetation, we too often tend to equate the position of a species or a community on the continuum or environmental gradient with its location on a transect. In reality there is no necessary spatial relation between sites with similar values on the gradient, because the continuum is an abstract dimension of ecological space. It is much like the position of an animal niche on a resource gradient (Austin and Smith 1989, Smith and Huston 1989).

Figure 28.24 illustrates four models and an example of actual species distribution along an environmental gradient in which each curve represents a single species. In A the abundance of one species is associated with the abundance of other species. Each zone has its own assemblage of co-adapted species, which gives way at a sharp boundary to another group of species. This discontinuous distribution represents the supraorganismal community concept of Clements. In B the abundance of one species on an environmental gradient is independent of the others. The association of several species along the gradient changes with the differential response of the individual species to environmental changes. The continuum in C shows species always found in association with each other, but the communities intergrade continuously. In D the distribution of a species in each stratum is restricted at some point on the resource gradient, but species in each stratum are independent of those in the other strata. This distribution suggests the relationship of the lower strata in a forest relative to the upper stratum of canopy species. The continuum in E is a plot of the actual distribution of species along an environmental gradient in the Santa Catalina mountains of Arizona.

Ecologists have a tendency to confuse the two types of gradient involved: a physiological gradient, reflecting the physiological optimum of a species, and an environmental gradient, reflecting the environmental optimum. The physiological gradient of a species is evident only when it is grown in monoculture (Ellenberg 1988). Under those conditions many species share a physiological optimum (Figure 28.25). In natural multispecies associations, however, growing conditions are quite different. The position where species achieve their maximum biomass can be quite different from the physiological optimum (Figure 28.26). There are more variations in conditions under which plants are found than in conditions under which they grow best. Competition displaces species toward environmental conditions they are able to tolerate, but which species that outcompete them under optimal conditions cannot tolerate. Thus the ecological optima may be closer to the species' physiological limits or tolerance.

## CLASSIFICATION SYSTEMS

To give some order to the study of communities, we need a system of classification, even though we may not be able to sort the communities of a region into discrete categories. There are a number of approaches to community

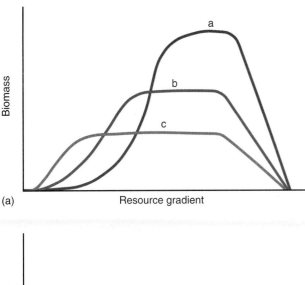

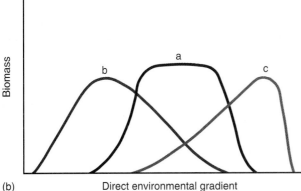

**Figure 28.25** Differences in fundamental responses of species on two types of gradient: (a) a resource gradient and (b) an environmental gradient. (After Austin and Smith 1989:39.)

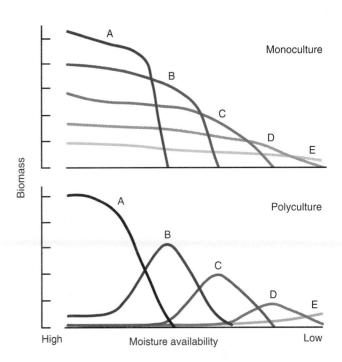

**Figure 28.26** Simulated patterns of plant abundance of five hypothetical functionally different species along a moisture gradient when grown in a monoculture and in a mixed species or polyculture. All five have the same growth rates but differ in maximum height, maximum age, drought tolerance, and shade tolerance. (After Smith and Huston 1989:58.)

| Species | Maximum Height | Maximum Age | Tolerance Index | |
|---------|---------|---------|---------|---------|
| | | | *Drought* | *Shade* |
| A | 36 | 500 | 0.2 | 1 |
| B | 31 | 296 | 0.3 | 2 |
| C | 15 | 175 | 0.4 | 3 |
| D | 14 | 103 | 0.5 | 4 |
| E | 05 | 61 | 0.6 | 5 |

classification, each arbitrary and each suited to a particular need or viewpoint (Mueller-Dombois and Ellenberg 1974). The most commonly used classification systems are based on physiognomy, species composition, dominance, and habitat.

Physiognomy (the general appearance, vertical structure, and growth form of vegetation) is a highly useful method of naming and delineating communities, particularly in surveying large areas and in subdividing major types into their component communities. Because animal distribution appears to correlate with vegetational communities, classification by physiognomy will relate to both plant and animal life. Communities so classified are often named after the dominant form of life, usually plants, such as coniferous or deciduous forest, sagebrush, or shortgrass prairie. A few are named after animals, such as the barnacle-blue mussel (*Balanus-Mytilus*) community of the rocky tidal zone. One community, of course, may grade into another. In these cases the classification may be based on arbitrary, although specific, criteria. Where habitat boundaries are well defined,

communities may be classified by physical features such as tidal flats, sand dunes, cliffs, ponds, and streams.

Finer subdivisons may be based on species composition, a system which works much better with plants alone than with animals. Such a classification requires a detailed study of the individual community (see Mueller-Dombois and Ellenberg 1974), covering frequency, dominance, constancy, presence, and fidelity.

In this system a group of stands in which similar combinations of species occur can be classified as the same community type. The type is named after the dominant organisms, or the ones with the highest frequency. Examples are the *Quercus-Carya* association, or oak-hickory forest, the *Stipa-Bouteloua* association, or mixed prairie, and the animal-dominated *Balanus-Mytilus,* or barnacle-blue mussel community.

European ecologists have developed a floristic classification with emphasis on dominance, constancy, and diagnostic species (Braun-Blanquet 1965). They group communities into classes, orders, alliances, and associations (see Poore 1962, Whittaker 1962, Mueller-Dombois and Ellenberg 1974). Such a classification involves **fidelity,** the faithfulness of a species to a community type. Species with low fidelity occur in a number of different communities and those with high fidelity in only a few. Seldom, if ever, are the latter found away from certain closely associated plants and animals. The greater **constancy,** the ratio of species always associated with the community type to the total number of species, the more homogeneous is the community and the more sharply it can be delineated. Often, however, this close association merely reflects a group of species unable to grow successfully under a wide range of ecological conditions or with other species. Species can be grouped as *exclusive,* those completely or nearly confined to one type of community; *characteristic,* those most closely identified with a certain community; and *ubiquitous,* those with no particular affinity for any community. The species grouped as characteristic, high in constancy and dominance, are the ones that define the community type.

A different approach to vegetation involves ordinational techniques. **Ordination** is the arrangement of stands by along a linear axis according to their similarity. An early version was the **continuum index** (Curtis and McIntosh 1951, Curtis 1959), a synthetic index ordering species relative to changing vegetational composition. It is based on a linear, one-dimensional ordination of stands from early to late stages of succession (Figure 28.27).

The continuum index involves vegetative changes relative to a successional sequence. However, an objective in vegetation analysis and classification is the correlation of vegetation with the environment. One approach is **gradient analysis** (Whittaker 1956, 1960, 1967). It plots the pattern of plant responses to changes along a particular environmental gradient such as soil moisture, temperature, and elevation. Direct gradient analysis uses a graph for displaying patterns of species distribution. The patterns are based on data obtained by sampling the vegetation along some environmental gradient. The resulting bell-shaped curves of plant species on the gradient (Figure 28.28) are suggestive of the utilization curves or niche dimensions on the resource gradients of animal ecologists (Austin 1985).

An advance over the continuum index and direct gradient analysis was the development of **community ordination,** a form of indirect gradient analysis. In community ordination vegetational samples or species are arranged relative to one or more ecological gradients or to one or more abstract coordinate axes that represent such gradients. The relative position of the sample units on the axis and to each other provides maximum information about their ecological similarities (Austin et al. 1984, Ludwig and Reynolds 1988).

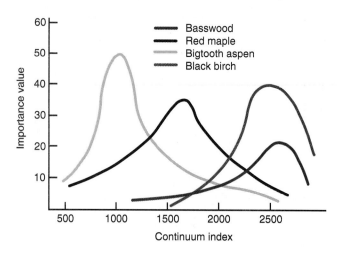

**Figure 28.27** Distribution of some forest trees on a continuum index. The smaller the index value, the earlier the species appears in forest succession. (Adapted from Curtis and McIntosh 1951.)

Ordination is an exploratory data-analysis technique designed to seek patterns or trends in the data.

The earliest of these ordination methods in plant ecology was the polar ordination of Bray and Curtis (1957), a technique specifically developed to analyze plant community data. It involves the use of community similarities or dissimilarities to determine endpoints on the axis, followed by a geometric positioning of the remaining stands between them (see Ludwig and Reynolds 1988, Beals 1984). Although subjective, polar ordination remains a useful method.

Since then more mathematically sophisticated and complex multivariate methods have been developed to analyze the community. One is **principal components analysis.** Although developed for vegetational studies, the method has become widely used in other areas of ecology. Other methods include cluster analysis, discriminate analysis, correspondence analysis, and nonlinear ordinations. The value of these methods is their ability to summarize a large, complex set of variables. They enable ecologists to develop new hypotheses about environmental effects on species distribution and community patterns.

## SUMMARY

A biotic community is a naturally occurring assemblage of plants and animals living in the same environment, mutually sustaining and interacting directly or indirectly with one another.

Communities exhibit some form of layering or vertical stratification. In terrestrial communities stratification is determined by the life forms of plants and influences the nature and distribution of animal life. In addition to vertical

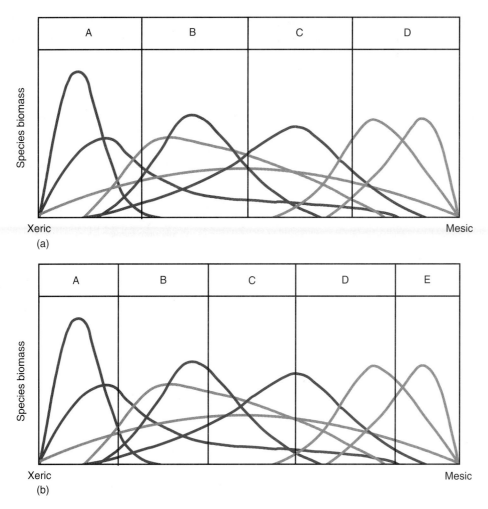

**Figure 28.28** (a) Vegetation distributed along an environmental moisture gradient from xeric to mesic. Each species responds in its own way to the moisture environment; yet sufficient overlap in response allows a number of them to associate with each other on the environmental gradient. The nature of the community, its dominants and associated species, depends upon the point at which community boundaries are placed. In the hypothetical gradient, each demarked community is characterized by its own dominants, although some species are shared with other communities. The communities at either end of the gradient are distinct, although they share one ubiquitous species. (b) Artificially shifting the community boundaries along the gradient (as often happens in sampling communities on a gradient) would result in changes in community composition.

stratification, terrestrial and shallow freshwater communities may also exhibit horizontal structure produced by clumped distribution of vegetation, resulting in a patchy environment. The communities that are highly stratified vertically and are patchy hold the richest variety of animal life, for they contain a greater assortment of microhabitats and available niches.

Dominant species may control the character of communities, especially in the temperate zone. The dominant species may be the most numerous, possess the highest biomass, preempt the most space, or make the largest contribution to energy flow. It is not necessarily the most important species in the community. Other species may be critical in cycling of nutrients or other processes.

Communities differ in their diversity of species. Species diversity implies both a richness in the number of species and the evenness of the distribution of individuals among the species. When evenness is high, species diversity is also high. Several hypotheses have been proposed to explain why species diversity is higher in some regions than in others, especially on a latitudinal basis. These hypotheses are difficult to test. Ecologists have developed mathematical indexes such as the Shannon index that considers both species richness and evenness. A species diversity index is useful

only on a comparative basis, either within a single community over time or among communities. Species diversity within a community is called alpha diversity and between communities beta diversity, and on a geographical area, gamma diversity.

The place where two different communities meet is an edge. The area where two communities blend is an ecotone. An edge may be inherent, produced by a sharp environmental change such as a topographical feature; or it may be induced, created by some form of disturbance that is limited in extent and changes through time. Ecotones typically have a high species richness, because they support not only selected species of adjoining communities but also a group of opportunistic species adapted to edges.

A positive relationship exists between species diversity and area. Generally, large areas support more species than small areas. This species-area relationship is the basis for the theory of island biogeography. It proposes that the number of species an island holds represents a balance between immigration and extinction. Immigration rates in an island are influenced by the distance of an island from a mainland or source of potential immigrants. Islands distant from a mainland receive fewer immigrants than islands closer to the mainland. Extinction rates are influenced by the area of the island. Because small islands hold smaller numbers of individuals of a species and have less variation in habitat, they have higher extinction rates than large islands.

The theory of island biogeography is relevant to the fragmentation of natural habitats such as forests that become isolated by surrounding agricultural and urban lands. Although smaller fragments hold fewer species than larger fragments, size is not the only criterion for the value of such fragments for wildlife. Important is the ratio of interior to edge. Many species are area-sensitive; that is, they require large unbroken blocks of habitat. Unless the habitat parcel is large enough to hold interior species, the area is inhabited only by edge and area-insensitive species.

The structure or organization of a community may also be influenced by interactions among its member species. Interspecific competition may exclude species, reduce the abundance of species, or limit the number of individuals within a species. Predators may reduce the population sizes of competing prey species, permitting their coexistence. When key predators are eliminated, a certain few prey species may assume dominance. Predation in its broadest sense involves feeding relations or food webs that structure the relations among organisms in the community. Parasites and disease may reduce or eliminate certain species from the community, opening it for invasion by others. Finally, mutualistic relationships, both direct and indirect, may integrate community structure in ways that are now only beginning to be appreciated.

Historically, plant ecologists debated whether plant communities are tightly integrated units (superorganisms) or chance assemblages of species with similar physiological tolerances. The composition of a community is determined in part by the species that happen to disperse to and colonize an area and that can survive and grow under prevailing conditions. The exact species that happen to settle on an area and the number that survive are rarely repeated. There is, however, a certain recurring pattern of more or less similar groups.

Rarely can communities be sharply delimited, because they tend to form a continuum along a gradient of environmental resources. Nevertheless, there is a need to classify communities by some criteria for field work and other purposes. The simplest classification scheme is based on physiognomy and dominant species. More detailed schemes involve detailed studies of floristics. Direct gradient analysis is used when a strong environmental gradient is evident. Indirect gradient analysis, which involves such multivariate statistical approaches as principal components analysis and discriminant analysis, is used when such a gradient is not evident.

## REVIEW QUESTIONS

1. What is the basis of vertical stratification in aquatic and terrestrial communities?
2. Why is stratification an important structural component of the community?
3. How does horizontal structure influence the community?
4. Distinguish among species dominance, species richness, and species diversity. How are the concepts related?
5. Distinguish between an edge and an ecotone.
6. Explain the edge effect. How is it related to horizontal structure of communities?
7. What is island biogeography theory?
8. What is habitat fragmentation? How do some aspects of island biogeography relate to it?
9. How do habitat patches in the landscape differ structurally and functionally from oceanic islands?
10. Show the relationship among habitat fragment size, interior species, and species diversity.
11. What is the major weakness of managing for species richness in a given area?
12. Explain how overgrazing can affect community structure. What if the herbivores are removed?
13. Contrast the community concept with the continuum concept.
14. Fishery biologists commonly stock new reservoirs with a top predator, for example bass, and an assortment of prey or forage fish. What are the problems in attempting to construct such a food web?

How should such restoration relate to assembly rules?

15. Secure older and recent air photos of a local area, especially one that has seen development. Compare the landscape patterns. What has been the extent of habitat fragmentation and loss? How might these changes relate to species diversity?

16. E. A. Read (1994) wrote: "Community classification has become more than an academic subject—it plays a key role in the formation of mitigation and restoration plans that are intended to reduce environmental impacts." Discuss.

## CROSS-REFERENCES

Adaptation, 30–31; energy flow and food webs, 189–193; grassland structure, 231–233; forest structure, 268, 276–277, 283–285; lakes and ponds, 294–297; effective population size, 472–474; viable populations, 475; vegetation-herbivore interactions, 522–524; mycorrhizae, 528, 584; predator-prey interactions, 536–547; population effects of parasitism, 569–572; mutualism, 582–596; moisture gradients, 627; succession, 656–667.

# Disturbance

## *Concepts*

1. Disturbance is any physical force that damages natural systems and results in mortality of organisms.
2. Disturbances vary in intensity, frequency, and spatial scale.
3. Small-scale disturbances promote community diversity.
4. Large-scale disturbances can alter the character and functioning of ecosystems.
5. Community stability depends on resistance and resilience.
6. Communities rarely achieve equilibrium because of environmental disturbances.

We think of famous natural landscapes as frozen in time, to remain unchanged like the image in a photograph. To preserve an area as we see it, we protect it against fire, insect attack, and other events we consider harmful. In spite of our efforts, the landscape changes anyhow. We fail to comprehend that nature is not constant, that disturbance is the means by which landscape diversity is maintained.

Consider Figure 29.1a, a photograph of a western Montana landscape in the late 1800s. This scene is one of vegetational diversity with patches of grassland, aspen, Douglas-fir, and pine, showing the influence of periodic fires, set by lightning and native Americans. Protected from fire and grazed for 90 years, the vista has changed. The open country is grown to Douglas-fir, and the diverse patches of vegetation are gone. A patchy landscape protected from disturbance by fire has changed to a homogeneous one.

In Switzerland the scenic Alpine landscape associated with the story of Heidi changed when cattle and sheep no longer grazed the high mountain meadows. Protected from grazing, spruce forests began to claim the meadows, changing the landscape familiar to tourists. Only by reintroducing grazing on a subsidy basis could the traditional Alpine landscape be maintained (Wiegandt 1976). The point of both ex-

amples is that diversity in the landscape comes about only through some form of disturbance.

**Disturbance** is any physical force, such as fire, wind, flood, extremely cold temperature, and epidemic, that damages natural systems and results in the mortality of organisms or loss of biomass (Huston 1994, Pickett and White 1985). These disturbances in themselves do not create diversity; rather they provide the opportunity for colonization of the disturbed site by new species.

# CHARACTERISTICS OF DISTURBANCE

Disturbances have both spatial and temporal characteristics. They include intensity, frequency, and area or scale.

## Intensity

Intensity of a disturbance is measured by the proportion of the total biomass or of the population of a particular species that is killed or removed. Intensity is influenced by at least

(a)

(b)

**Figure 29.1** Two photos of the same area show the effects of fire suppression and grazing disturbance on landscape diversity. (a) A photo taken in 1888 shows a north-northwest view in the vicinity of Fort Maginnis (U. S. Army troops in the foreground), located in the Judith Mountains about 20 miles northeast of Lewistown, Montana. Ground cover on the ridge and lower slopes in the distance apparently is dominated by perennial grasses. Conifers on the far slopes are confined largely to localized areas, suggesting that wildfires burned the slopes several decades earlier. (b) The same site in 1980, 92 years later. The foreground shows the effects of current livestock grazing. Shrubs, including currant, chokeberry, rose, shrubby cinquefoil, and common juniper, are more conspicuous among the rocks, as are shrubs and conifers in the left and central midground. The stream course in the distance at the center left of the photo, which was formerly treeless, now supports large cottonwoods, Douglas-fir, and ponderosa pine. Absence of fire has allowed a profuse growth of Douglas-fir and ponderosa pine on the far slopes.

three factors. One is the magnitude of the physical force, such as the strength of the wind. A second is the morphological and physiological characteristics of the organisms that influence their response to the disturbance. Tall, mature trees are less resistant to wind than small, young trees. A third is the nature of the substrate, especially as for sessile organisms on a rocky shore. The amount of aboveground biomass accumulation and the height of vegetation influences the intensity of wind damage and fire. Intensity of fire is also related to the amount of energy released during burning.

## Frequency

It is difficult to separate the intensity of a disturbance from frequency of disturbance. **Frequency** is the mean number of disturbances that occur within a particular time interval (Figure 29.2). The **return interval** or turnover time is the inverse of frequency, or the mean time between disturbances on the same piece of ground. If the intervals between disturbances are long, that is, the frequency is low, biomass accumulates during the interval between disturbances and the intensity of disturbance can be high.

In temperate and tropical forests, in which natural disturbances are on a small scale, the frequency of disturbance within a given stand is high. The rate of disturbance for all stands, however, is low, between 0.5 to 2.0 percent per year. The return interval is between 50 and 200 years (Brokaw 1985, Runkle 1985). Runkle (1985) estimates that 4 to 14 percent of the total land area of the Great Smoky Mountains National Park is subject to disturbance over a decade. This slow rate of disturbance and replacement maintains diversity in a mature forest over many years and allows the coexistence of species with different life history characteristics. Such disturbances give the time-locked observer the impression that the forest is unchanging.

In such communities the return interval of disturbance is roughly proportional to the life span of the dominant organisms. In forest communities with trees that potentially live for several hundred years, return intervals of disturbance are shorter than the life span of the trees (see Figure 29.2). The frequency of disturbance then is low, but the intensity of the disturbance is high.

Consider a boreal forest of black spruce (*Picea mariana*). Longevity of black spruce is about 200 years, but few stands ever survive beyond the age of 70 years. In fact, old unburned stands are difficult to find (Yarie 1981). In the boreal forest black spruce, which grows on cool, wet ground, experiences fire frequencies of about 40 to 45 years. Dense stands of spruce are prone to lightning fires, which burn off all vegetation, but revegetation is rapid. Moss and light-seeded herbs provide the seed bed and cover for black spruce seedlings, which grow above the shrubs and establish closed dense stands in 40 to 60 years. At about this age productivity of the stand declines, and the microenvironment becomes cooler and wetter. Lacking a means of self-perpetuation, old spruce stands that do not experience a fire degenerate into a treeless moss and lichen bog.

How frequently a fire burns over a given area—its return rate—is influenced by the occurrence of droughts, accumulation and flammability of the fuel, the resulting intensity of the burn, and human interference. In grasslands of presettlement North America, fires occurred about every three years. This time was needed for sufficient mulch, dead stems, and leaves to accumulate. In forest ecosystems the frequency of fires varies greatly, depending upon the type of forest (Heinselman 1981a). Frequent light surface fires may have a return interval of 1 to 25 years, whereas crown fires may have a return interval of 25, 100, or even 300 years. Usually light and severe fires occur in combination. A red pine forest may experience a light to moderate surface fire every 5 to 30 years and a crown fire every 100 to 300 years.

Various forest ecosystems appear to burn and develop under certain fire frequencies. Frequent low-intensity surface fires every 5 to 20 years were typical in presettlement forest of ponderosa pine (*Pinus ponderosa*) in western North America. Such fires prevented the buildup of a heavy fuel load, thinned the stand, eliminated the incoming shade-tolerant conifers that could carry fire up into the crown, and encouraged an open, grassy understory. Red pine and white pine forests of the Great Lakes region experienced infrequent

**Figure 29.2** (pages 636–637) Frequency of disturbance in a community as illustrated by the fire history of the Tolan watershed in the Bitterroot National Forest between 1734 and 1900. Twelve of the 23 fires are illustrated in this series. No fires burned after 1900 because of fire suppression. The watershed supports ponderosa pine, Douglas-fir, western larch, lodgepole pine, and white-barked pine.

The general pattern was one of frequent fires leaving substantial remnants of older trees. Most fires burned lightly on the drier slopes at lower elevations, perpetuating ponderosa pine as the dominant species in open stands. The fires killed much of the ponderosa pine regeneration and most of the invading Douglas-fir. Fires burned with greater intensity on north-facing slopes, where dense young growth of Douglas-fir provided more opportunity for crown fires. However, even here, much of the old growth survived. In the lower subalpine forest, dominated by lodgepole pine, fires often were of low or medium intensity, spreading mostly on the forest floor. (From Arno 1976:23–29.)

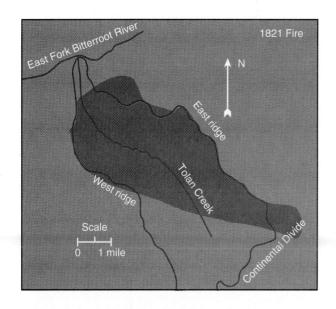

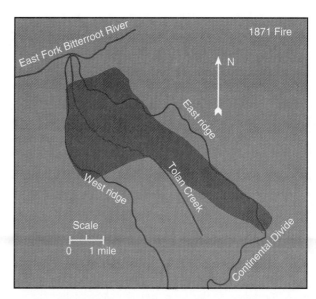

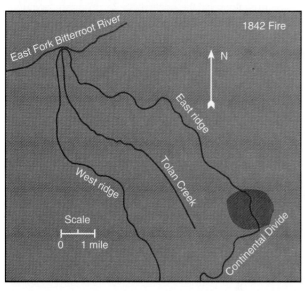

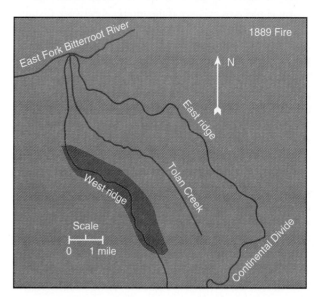

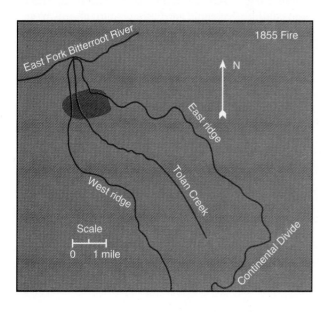

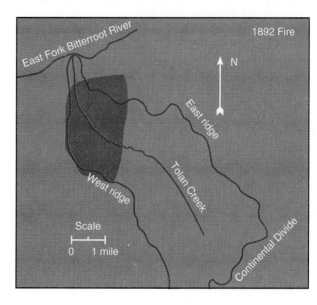

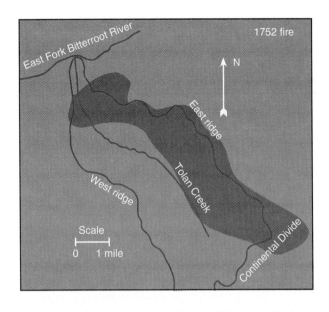

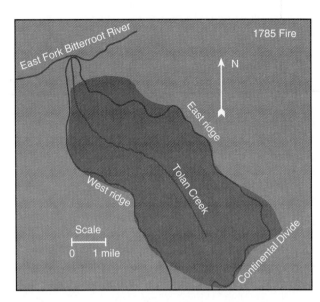

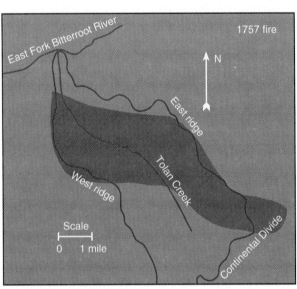

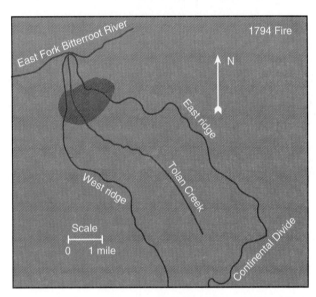

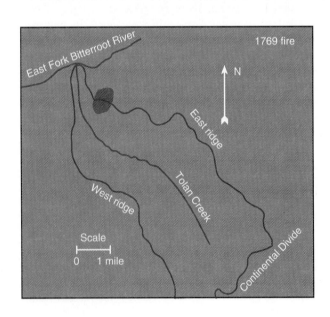

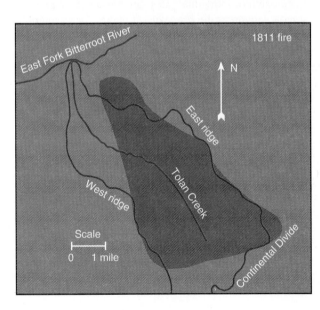

surface fires of low intensity with a return interval of about 28 to 35 years. They were punctuated at longer intervals of 150 to 300 years by severe surface and crown fires that destroyed the stand.

Systems dependent on infrequent large-scale disturbances for regeneration respond negatively to disturbances that come at too frequent intervals. Under such conditions, the system becomes degraded or converted into some other community. For example, Zedler and associates (1983) report on the effects of short intervals between fires on chaparral vegetation in California. In 1979 a fire burned over a large area of chaparral. Although such burned-over areas are quickly recolonized by seedlings and sprouts, they typically are reseeded artificially to ryegrass to control erosion (Taskey et al. 1989). Fine-leafed ryegrass itself is fire-prone. In the following year a portion of the same area was burned again, which resulted in a drastic change in vegetation. The two most abundant shrubs, *Ceanothus olignathus* and *Adenostoma fasciculatum,* were reduced in density by 97 percent. Because the seed reserves were depleted by the previous fire, the area changed to a relatively permanent, degraded shrubland and annual grassland. The increased frequency of fires brought about by the planting of grass and by arson is having a similar effect on large areas of California chaparral.

Suppression of disturbance can lead a disturbance-controlled system into a more fragile, less resilient one, susceptible to destruction. In a lodgepole pine or mixed coniferous forest of western North America, a natural fire regime involves recurring ground fires at intervals of 7 to 25 or more years. These fires have a significant role in maintaining conditions for tree regeneration and nutrient cycling. They also create and maintain openings in the forest canopy, in effect producing natural fire breaks. Fire suppression allows the accumulation of fuel, crown closure, development of an understory that can carry fire to the crown, and senescence and death of individual trees. Such protection against fires greatly lengthens the fire cycle. It makes for an intense fire that can cause extensive tree damage and mortality over a very large area (Holling 1980, Kilgore 1973). The long years of fire suppression in Yellowstone National Park until 1975 (Romme and Knight 1982), accompanied by an unexpected period of drought, set the stage for extensive fires in Yellowstone National Park in 1988 (Figure 29.3).

A similar effect occurs when natural insect predators of trees are suppressed to protect stands of susceptible timber. Spruce budworm infestations, like short-interval ground fires, create a patchy environment. However, protection of spruce and fir forest, from short-term timber losses sets the stage for a widespread outbreak of the insects, covering hundreds or thousands of hectares.

Kessell and Fischer (1981) developed a model that predicts the effects of high and low frequency disturbances by fire in Douglas-fir stands (Figure 29.4). High frequency fires reduce the system to one dominated by opportunistic species. Low frequency fires result in the establishment of Douglas-fir forest. Intermediate frequency fires, characterized by ground fires and low intensity surface fires, result in more open stands with greater species diversity and a mix of tolerant and intolerant species.

A hurricane is another powerful recurring event of high intensity but low frequency. The frequency of hurricanes with the intensity of Hurricane Hugo in 1989, with winds over 166 km/hr, is once every 50 to 60 years in the Caribbean. The return interval of all hurricanes there is about 21 years (Scatena and Larsen 1991).

## Scale

The impact of a disturbance on community structure is also a matter of scale. Disturbances range from very small, frequent ones such as the death of a single tree in a forest to large-scale, infrequent disturbances that embrace extensive areas swept by fire, buried under volcanic ash, torn by landslides, or denuded by human land-clearing schemes.

What constitutes a small disturbance depends upon the scale of the landscape in which it occurs. Small-scale disturbances that cause the death of an individual tree or group of trees, opening the canopy, have more impact on a small woodland than on a large forest.

The abrasive action of waves tears away mussels and algae from tidal rocks. In grasslands digging by badgers and groundhogs exposes small patches of mineral soil that are colonized by herbaceous plants. A dominant tree falls in a forest. The outcome of such disturbances is the creation of a **gap,** a term originally applied by A. S. Watt (1947) to openings that become localized sites of vegetative regeneration and growth.

Consider a forest. A single tree struck by lightning or killed by a fungal infection remains standing as a snag, creating a simple gap. A dominant tree uprooted by a windstorm creates a larger, more complex gap (Figure 29.5). When it falls, it opens a space in the canopy. Its upturned roots tear up the soil and form a pit and a mound of exposed mineral soil open for colonization. Its falling trunk slices through its neighbors, snapping the limbs of nearby canopy and understory trees. The fallen crown crushes the understory beneath it. Once created, some gaps in the forest continue to grow as peripheral trees succumb to winds and insect damage.

Within the gaps the microclimate differs from the rest of the forest. Light and soil temperature increase, while soil moisture and relative humidity decrease. How woody plants respond to these changes is suggested by studies on the effects of different methods of timber harvesting on forest regeneration. Removing single trees or small groups of trees (selection cuts) simulates natural gap formation. In small gaps, the response is typically vegetational reorganization. The crowns of trees about the edge of the gap expand to fill the opening. Their closure of the canopy inhibits any initial

**Figure 29.3** In 1985 massive crown fires swept through Yellowstone National Park. The fire was intense in part because of the buildup of fuel brought about by fire suppression in the past. Compare the lower photo with Figure 29.8 (p. 643) showing ecosystem recovery.

growth response understory plants may have made (Trimble and Tyron 1966). Larger gaps in group selection cuts provide environmental conditions and opportunities for tolerant species in temperate hardwood forests (Trimble 1973) and primary or mature phase species in tropical forests (Brokaw

1984). In these cases replacement vegetation is likely to be similar to that of the canopy. Few incoming trees in small gaps in the forest reach the canopy without the aid of another disturbance (Runkle and Yetter 1987, Tryon and Trimble 1969). These disturbances need not occur within the

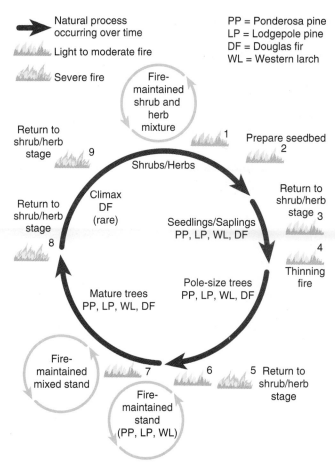

Natural process occurring over time

Light to moderate fire

Severe fire

PP = Ponderosa pine
LP = Lodgepole pine
DF = Douglas fir
WL = Western larch

Fire-maintained shrub and herb mixture

Prepare seedbed 2

1

Return to shrub/herb stage 9

Shrubs/Herbs

Return to shrub/herb stage 3

Climax DF (rare)

Seedlings/Saplings PP, LP, WL, DF

Return to shrub/herb stage 8

4

Thinning fire

Mature trees PP, LP, WL, DF

Pole-size trees PP, LP, WL, DF

Fire-maintained mixed stand

7

6

5 Return to shrub/herb stage

Fire-maintained stand (PP, LP, WL)

**Figure 29.4** A model of forest succession influenced by fire in a moist Douglas-fir habitat type in the western United States. (1) Following a fire disturbance, secondary succession begins with a mixture of shrubs and herbs. (2) Fire creates a seed bed conducive to seedling germination. However, frequent fires will kill off conifer regeneration and maintain the shrub and herb community. (3) If a fire should recur during the seedling-sapling stage the site will return to the shrub-herb community. (4) In somewhat older stands fire will eliminate accumulated fuel and thin the conifer stand. (5) If a severe fire occurs in a pole-stage stand, the site will go back to the shrub-herb stage. (6) However, surface fires of lower intensity maintain open stands, often dominated by ponderosa and lodgepole pines and larch. (7) Where Douglas-fir is more aggressive, it will dominate the mixture of other trees. Periodic light surface fires maintain this mixed stand. Long periods of fire suppression permit the establishment of a dense understory of Douglas-fir, which acts as a fuel ladder to the overstory. (8) A severe stand-destroying crown fire in a closed mature stand will recycle the site to a shrub-herb stage. (After Kessell and Fisher 1981:4.)

**Figure 29.5** Trees uprooted by ice and windstorms create new microreliefs on the forest floor to be colonized by herbaceous woodland plants.

gap but only among neighboring trees. In such gaps tolerant understory trees such as flowering dogwood may respond more quickly and successfully than replacement canopy saplings (Ehrenfeld 1980). They fill in the lower canopy and for a time, at least, inhibit the growth of young canopy-species trees.

Large gaps encourage the growth of intolerant species. Such species require openings of at least 0.1 ha or one in which the minimum width is two times the height of the surrounding trees. Even then the most responsive growth is in the center of the gap (Tyron and Trimble 1969). In eastern deciduous forests of North America, intolerant species such as yellow-poplar and black cherry respond to larger gaps. Their ability to do so accounts for their conspicuous and continued presence in mature forests dominated by tolerant species. Gaps up to 0.4 ha provide the greatest diversity in the forest. Overall, openings of various ages in the forest result in patches of different stages of successional or compositional maturity. Chronic small-scale disturbances are important in the maintenance of species richness and structural diversity within a mature forest ecosystem (see Shugart 1984, Platts and Strong 1989).

A vigorous response of herbaceous plants and ferns can adversely affect the growth of seedlings of forest trees. The density of tree seedlings can be much greater in places on the forest floor with little herbaceous cover than in patches of herbaceous growth (Maguire and Forman 1983). The inhibitory effect of dense patches of herbs and ferns on forest tree regeneration is well known among foresters (Horsley 1977, Ferguson and Boyd 1988).

Large-scale disturbances induced by fire, logging, land-clearing and other such events to be discussed result in responses that go beyond vegetational reorganization and involve colonization by opportunistic species. Some of these species may already be on the area as seeds of woody and herbaceous plants, root stocks, stump sprouts, and surviving seedling and saplings. Other colonizing species are wind and animal dispersed. Long-term recovery involves a series of successional stages in which short-lived opportunistic species are replaced by long-lived species characteristic of the original community. (See Chapter 30.)

## SOURCES OF DISTURBANCE

Many of the most powerful sources of disturbance are natural, such as lightning-set fires, hurricanes, and floods. Some of the most lasting disturbances are human-induced, such as land-clearing for crops and development.

### Fire

Fire, a recurring event, is a major natural disturbance that has had an important role in vegetation development world-wide. Globally, large regions are characterized by vegetation that evolved under fire: the grasslands of North America, the mediterranean-type shrublands of the southwestern United States and the Mediterranean region, the African grasslands and savannas (Batchelder 1967), the southern pinelands of the United States, and even-aged stands of coniferous forests of western North America. Up to 95 percent of the virgin forests of Wisconsin were burned during the five centuries before the land was settled by white Europeans (Curtis 1959). These fires not only enabled such species as yellow birch, hemlock, pines, and oaks to persist, but also were normal and necessary to perpetuate those forests (Maisurow 1941, Curtis 1959). In Alaska fires have converted white spruce stands into treeless herbaceous and shrub communities of fireweed and grass or dwarf birch and willow (Lutz 1956). Their growth is so thick that forest trees cannot become reestablished. Thus fire is a powerful selective and regulatory force on the evolution and maintenance of many types of ecosystems.

Certain regions of Earth possess conditions conducive to the start and spread of fires. One condition is a fire climate. It involves extended dry periods during which fuel accumulated during wetter periods can burn and the prevalence of dry lightning storms, ones unaccompanied by precipitation. Such conditions prevail in certain parts of North America, Africa, Mediterranean regions, and Australia. In the western United States 70 percent of forest fires are caused by dry lightning during the summer. Because of the seasonal nature of lightning, fires so caused in western North America are more numerous during the growing season, from April through August. At that time fires are the least severe but have the greatest impact as a selective force.

Fires fall into three types: surface, ground and crown. Type and behavior depend upon the kind and amount of fuel, moisture, wind, and other meteorological conditions, season of the year, and the nature of the vegetation. **Surface fire,** the most common type, feeds on the litter layer. In grasslands it consumes dead grass and mulch, converting organic matter to ash. Usually surface fire does not harm the basal portions of roots, stalks, tubers, and underground buds, but it does kill most of the invading woody vegetation. In the forest, surface fires consume leaves, needles, woody debris, and humus. They kill herbaceous plants and seedlings and scorch the bases and occasionally the crowns of trees. Damage to trees depends upon the intensity of the fire and the susceptibility of trees to heat. Surface fires may kill thin-barked trees like maples by scorching the cambium layer. Thick-barked trees, like oaks and pines, are better protected, but they can be scarred, allowing fungal infection (Figure 29.6).

If the fuel load is high and the wind strong, surface fires may leap into the forest canopy to cause a **crown fire,** one that sweeps through the canopy of the forest. Crown fires are most prevalent in coniferous forests because of the flammability of the foliage. If the canopy is unbroken, the fire may sweep across it, and tops and branches may fall to the

**Figure 29.6** This (species) has experienced several surface fires in its lifetime, as evidenced by the fire scars at its base. The tree was relatively fire-resistant, but the scars, although walled off by subsequent growth, did allow fungal infection to gain entrance into the heartwood.

ground to further feed the fire. A crown fire kills most above-ground vegetation through which it burns. It skips, and hops, leaving patches unburned.

**Ground fire** that consumes organic matter down to the mineral substrate or bare rock is the most destructive (Figure 29.7). It is most prevalent in areas of deep, dried-out peat and of extremely dry, light organic matter such as an accumulation of conifer needles. Such a fire is flameless, extremely hot, and persistent, until all available fuel is consumed. In spruce and pine forests, with their heavy accumulation of fine litter, a ground fire can burn down to expose rocks and mineral soil, eliminating any opportunity for that vegetation type to return.

Fire induces certain synergistic effects not common to other types of large-scale disturbances. It raises the temperature of the soil, but the degree depends upon soil moisture present. Soil temperature does not rise above 100° C until all the moisture is evaporated, and even under hot fires in semi-arid country temperatures rarely exceed 200°C at depths of 2.5 cm. However, heat breaks down soil aggregates, increasing bulk density of soil and decreasing its permeability. This change reduces infiltration of water into the soil, increases surface runoff, and promotes erosion and soil slippage on steep slopes.

As an agent of disturbance, fire sets the process of stand regeneration into motion by stimulating sprouting from roots and germination of seeds (Figure 29.8). Fire prepares the seedbed for some species of trees by exposing mineral soil, eliminating competition for soil moisture and nutrients from fire-sensitive and shade-tolerant species. Periodic surface fires thin some coniferous stands, such as ponderosa and longleaf pine. Importantly, fire acts as a sanitizer, terminating outbreaks of insects and such parasites as mistletoe by destroying senescent stands and deadwood and providing conditions for regeneration of vigorous young trees. In grassland fire consumes the standing dead material and accumulation of litter. It exposes the soil to the heat of the sun and speeds the recycling of nutrients, both of which stimulate the growth of new grasses.

**Figure 29.7** Fires of great intensity can have a profound influence on ecosystems. After the spruce forest located in the Allegheny Plateau area in West Virginia known as Dolly Sods was cut in the 1860s, intense ground fires burned, fed by piles of logging debris. Fire consumed the peatlike ground layer to bedrock and mineral soil. The forest never recovered and the plateau is a now a boulder-strewn landscape with intermittent patches of blueberry, dwarfed birches, mountain ash, and bracken fern.

**Figure 29.8** After the Yellowstone fire in 1988, the burned-over areas responded with a diversity of new vegetation.

Some vegetation types require the rejuvenating effects of periodic fires to persist. Such vegetation possesses characteristics that enhance fire spread and increase flammability (Mutch 1970). Some of these characteristics are chemical, physical, or physiological (Philpot 1977). They include secondary metabolites like flammable resins, waxes, terpenes, and other volatile products. Other characteristics are morphological, like finely branched needle-like leaves that carry fire from plant to plant, and accumulation of dead material as the plants age.

Plants of fire-prone ecosystems, such as chaparral, lodgepole pines, and eucalyptus, also possess adaptive traits that permit them to respond to fire in one of three general ways. First, mature adults develop defenses. One is bark thick enough to insulate the cambium from the heat of surface fires. Because such protection is not 100 percent effective and the heat of fire is not uniform about the tree, one side of the tree may burn. This injury creates a permanent fire scar. A second defense is an accumulation of a mat of needles that supports frequent low-intensity surface fires. These frequent surface fires prevent the buildup of a heavy fuel load that once ignited could destroy the forest. A third defense is rapid growth that results in the self-pruning of the lower branches, raising the canopy to reduce the danger of surface fires leaping into the crown.

A second adaptation is to use the death of mature plants as a means of regenerating and perpetuating the stand. In forests such destruction results in even-aged stands in which all the trees arise and become senescent simultaneously, promoting severe but infrequent fires. Such a response works only if frequency between fires is long enough to allow the plants to mature and produce seed. The seed may be stored in the soil or on the plant, awaiting the fire to release seeds or stimulate germination.

Jack pine and lodgepole pine are two coniferous species that retain unripened cones for many years on trees. Seeds remain viable within the cones until a crown fire destroys the stand. Then the heat opens the cones and releases the seeds (**serotiny**) to a newly prepared seedbed well fertilized with ash.

Other species rely on stand destruction and fire-stimulated germination of seeds stored in the soil. These seeds are hard. The seed coat is impervious to water and other softening agents. Only the high temperature of fire-heated soil cracks the seed coat or releases the seed from soil-stored chemical inhibitors deposited by living overhead vegetation. Some of the most abundant shrubs in the chaparral of California, ceanothus (*Ceanothus*) and manzanita (*Arctostaphylos*), are obligate seeders. These plants regenerate by seeds only rather than by root or stem sprouting. When mature and senescent, these shrubs possess an abundance of dead stems that feed intense fires and few potential resprouters.

A third adaptation to fire is resprouting. Although fire kills the tops and foliage, new growth appears as bud sprouts and root sprouts. Certain trees, particularly a number of *Eucalyptus* species in Australia, possess buds protected beneath the thick bark of larger branches. The buds survive crown fires and break out to develop new foliage (Figure 29.9). Other plants sprout from buds on roots, rhizomes, root collars, and specialized structures called lignotubers. These are basal burls on which latent axillary buds are released from inhibition when fire removes the region of growing cells at their tips (apical meristems). Ferns have subterranean buds on rhizomes that respond to loss of aboveground foliage. Shrubs such as blackberries and blueberries and trees such as aspen (*Populus*) sprout vigorously from roots. Trees such as oaks and hickories sprout from buds that develop at the root collar just below the ground. Certain species of mediterranean-type shrubs, including the North American chamise (*Adenostoma fasciculatum*) as well as species in at least seven genera in Australia, have lignotubers. Such sprouting is an effective means of survival and increase.

**Figure 29.9** Eucalyptus forests in Australia are subject to periodic fires. The bark may be scorched and charred and the trees defoliated by the heat, but the trees are not killed. The fire-scarred trees respond by resprouting not only from roots but from trunk and branches, as in this recently burned area. New foliage regenerates from dormant shoots, either along the trunk, on main branches, only in the upper branches, or in some combination, depending upon the species.

## Wind

Wind can inflict a major disturbance on vegetation. It shapes the canopies of trees exposed to prevailing winds, affects their growth of wood, and uproots them from the ground (windthrow). Mature trees, whose trunks lack the suppleness of youth, are especially vulnerable to windthrow. So, too, are trees growing on shallow and poorly drained soils in which the roots, spreading along the ground, are not well anchored. Trees weakened by fungal disease, insect damage, and lightning strikes and tropical forest trees carrying a heavy load of epiphytes in their crowns are also candidates for windthrow. The impact of wind is accentuated when strong winds accompany heavy snowfall that weighs down trees or heavy rains that soften the soil about the roots of trees. The position trees occupy in the forest also affects their vulnerability to wind damage. Trees bordering ragged forest gaps and those growing along forest edges, roads, and power lines are more likely to blow down than trees in the forest interior.

Hurricanes, especially those with wind speeds in excess of 166 km/hr and associated rainfall in excess of 200 mm, have a devastating impact on ecosystems. The force of a hurricane extends about 40 km about the center of the storm (Scatena and Larsen 1991). Hurricanes cause landslides in hilly and mountainous country and massive defoliation and blowdowns of timber. They greatly increase the total nutrient input to the forest floor, and they alter nutrient cycling, especially nitrogen and phosphorus (Lodge et al. 1991). In addition, hurricanes are a major force in maintaining diversity in montane rain forests of the Caribbean (Doyle 1981, Weaver 1989). Hurricane Hugo, which swept through the southern United States and Puerto Rico in 1989, and Hurricane Andrew in 1992 devastated much of the vegetation in their paths and adversely affected certain animal species (Figure 29.10). For example, Hurricane Hugo destroyed much of the remaining old growth stands of longleaf pine, habitat of the endangered red-cockaded woodpecker (*Picoides borealis*). This bird depends completely on old-growth pines affected with a fungus-caused soft rot of the heartwood for nesting sites.

## Moving Water

Moving water is a powerful agent of disturbance. Storm floods scour stream bottoms, cut away banks, change the courses of streams and rivers, move and deposit sediments, and bury or carry away aquatic organisms. Strong waves on

**Figure 29.10** Hurricane Hugo destroyed many pine forests on the coastal plain in South Carolina.

rocky intertidal and subtidal shores overturn boulders and dislodge sessile organisms. This action clears patches of hard substrate for recolonization and maintains local diversity (Sousa 1979, 1985; Connell and Keough 1985). High storm tides break down barrier dunes, allowing seawater to invade behind the dunes, and change the geomorphology of barrier islands.

## Drought

Prolonged drought can have a pronounced effect on vegetation composition and structure. On the grasslands of western Kansas during the drought years of the 1930s, blue grama with its physiological ability to resist dry conditions became two times as dense as the drought-sensitive buffalo grass. When the rains came, buffalo grass quickly responded; in two years it reversed its position and became five times as dense as blue grama. After ten years without drought conditions, the two species were codominants (Coupland 1958). In temperate forests prolonged drought can result in heavy mortality of shallow-rooted tree species, such as hemlock and yellow birch, as well as understory trees and shrubs (Hough and Forbes 1943, Bjorkbom and Larson 1977). Drought dries up wetlands, causing crisis conditions among waterfowl and other wetland birds, muskrats, and amphibians, greatly reducing their populations.

## Animals

A walk into a forest inhabited by a high population of deer or across an overgrazed grassland provides visual evidence of the impact that herbivorous animals can have on communities. Overgrazing in rangelands of the southwestern United States, for example, has reduced the organic mat and thus the incidence of fire. By dispersing seeds of mesquite and other shrubs through their droppings, cattle have encouraged the invasion of these woody plants onto overgrazed rangelands (Phillips 1965, Box et al. 1967). In many parts of eastern North America, large populations of white-tailed deer have eliminated certain trees, such as white cedar (*Thuja occidentalis*) and American yew (*Taxus canadensis*), from the forest. They have destroyed forest reproduction, and developed a browse line—the upper limits on a tree at which deer can reach foliage (Figure 29.11). In cutover areas of hardwood forests on the Alleghany Plateau in Pennsylvania, deer greatly reduced pin cherry and blackberry. They selectively reduced sugar maple and favored the expansion of ferns and grass which inhibits the regeneration of oaks and other trees (Marquis 1974, 1981; Marquis and Grisez 1978).

The African elephant has long been considered a major influence on the development of savanna vegetation. When their numbers are in balance with the vegetation and their movements are not restricted, elephants have an important role in creating and maintaining the woodlands. When ele-

**Figure 29.11** An overpopulation of deer results in the elimination of forest understory plants and the creation of a browse line in the forest trees. The browse line is the highest point that hungry deer can reach for food.

phants exceed the capacity of their habitat to support them, their feeding habits combined with fires devastate flora, fauna, and soils. Elephant destruction of trees (Figure 29.12) acts as a catalyst to fires, which are the primary cause of converting woodland to grassland (Wing and Buss 1970).

However, interactions between key herbivores and community structure are not quite that simple. Vegetation, climate, fire, and large herbivores all interact to reach long-term equilibrium. Consider the *Acacia* woodlands of the central Savuti channel of the Botswanas' Chobe National Park in South Africa. There the *Acacia* woodlands are declining as mature, even-aged trees die from old age and elephant damage. The *Acacia* woodlands became established in the late 1800s, when the viral disease rinderpest killed off many of the large ungulate species, elephants were decimated by ivory hunters, and the Savuti channel had dried up from drought. The channel refilled in the 1950s only to dry again in the 1980s. A return of the large woods will require another combination of events that reduces browsing pressure long enough to allow the seedlings to reach a safe size, perhaps 10 to 15 years (Walker 1989). Such a combination of events might have occurred again in Africa during recent prolonged drought of the 1990s with its great losses in ungulates.

**Figure 29.12** Consuming great quantities of woody vegetation and uprooting trees, elephants of the African savanna have an important influence on ecosystem succession and stability. The life cycles of certain trees and the maintenance of the savanna ecosystem depend in part upon the disturbance regime of elephants. Too many elephants destroy the ecosystem; too few elephants allow bush encroachment.

Beaver modify many forested areas in North America and Europe. By damming streams they alter the structure and dynamics of flowing water ecosystems (Naiman et al. 1986). Pools behind dams become catchments for sediments and sites for organic decomposition. By flooding lowland areas, beaver convert forested stands into wetlands. By feeding on aspen, willow, and birch, beaver maintain stands of these trees, which otherwise would be replaced by later successional species.

Birds may appear to be unlikely to cause major vegetation changes. In the lowlands along the west coast of Hudson Bay, however, large numbers of the lesser snow goose (*Chen caerulescens caerulescens*) have impacted the brackish and freshwater marshes. Snow geese grub for roots and rhizomes of graminoid plants in early spring and graze intensively on leaves of grasses and sedges in summer. With a dramatic increase in their number, geese have stripped large areas of their vegetation, resulting in the erosion of peat and the exposure of underlying glacial gravels. There is little likelihood that the vegetation that reestablishes will closely resemble the original (Kerbes et al. 1990).

Outbreaks of insects such as gypsy moth and spruce budworm defoliate large areas of forest, killing or reducing growth of affected trees. The degree of mortality may range from 10 to 50 percent in hardwood forests infested by gypsy moth to 100 percent in spruce and fir stands. Outbreaks of bark beetles have much the same effect in pine forests. The impact of spruce budworm, bark beetles, and other major

forest insects is most intense in large expanses of homogeneous forested landscapes where natural fires have been suppressed for long periods of times, allowing late-stage stagnated stands, highly susceptible to spruce budworm outbreaks, to develop (Wolf and Cates 1987). In their own way these insects act to regenerate senescent or stagnated forests.

## Timber Harvesting

One of the major large-scale disturbances to the world's forests is timber harvesting. Disturbance by logging depends upon the method: selection cutting, (uneven-aged management) or some form of clear-cutting (even-aged management). In **selection cutting** mature single trees or groups of trees scattered through the stand are removed. The cuts per hectare are relatively light, and new trees to replace those cut are grown as understory to the older remaining trees. Selection cutting produces only gaps in the forest canopy and favors reproduction of shade-tolerant over shade-intolerant trees. Forest composition essentially remains unchanged (Trimble 1973, J. Johnson 1984, Lorimer 1989).

**Even-aged management** is removal of the forest and reversion to an early stage of succession (Figure 29.13a). Unless followed by fire or badly disturbed by erosion and logging activities (Figure 29.13b), the cutover area fills in rapidly with herbs, shrubs, sprout growth, and seedlings of trees present as advanced regeneration in the understory. The area passes quickly through the shrub stage to an even-aged young forest. Because many of the most valuable timber trees are shade-intolerant to mid-tolerant species, they can be regenerated only by removal of mature trees, exposing the ground to sunlight.

There are three approaches to even-aged management (D. Smith 1986). One is **clear cutting** 11 to 44 ha blocks of timber within large forest tracts. A second method is **strip cutting,** the removal of all merchantable timber and remaining trees in strips 15 to 30 m wide. Every third strip is removed, followed by removal of the remaining strips in two cuttings two to four years apart. A third method is **shelterwood cutting,** which leaves 10 to 70 percent of the stand after initial cutting. When new growth is well under way, the remaining trees are removed. The first two methods of even-aged management favor regeneration of intolerant tree species. Shelterwood cutting retains some of the characteristics of the original forest yet permits intolerant species to regenerate.

Foresters often modify the regenerating forest to meet their requirements, introducing another form of disturbance during the development of the stand. Early in the life of a new forest, foresters may remove tree species not desired for timber or individuals of poor form. This improves, by economic but not ecological standards, the composition of the stand and the quality of the trees. The maximum growth of crop trees can be encouraged by thinning. Increased space between the trees stimulates crown expansion and increases growth.

Response of vegetation to the sudden removal of a forest canopy in timber harvesting is often rapid, as seeds and

(a)

(b)

**Figure 29.13** (a) Block clearcutting in a western coniferous forest. Such cutting fragments the forest. (b) Unless carefully managed, clearcutting can cause severe disturbance to a forest ecosystem.

seedlings of intolerant woody plants take advantage of changed environmental conditions (Bormann and Likens 1979). One example in the northern hardwood forests is pin cherry (*Prunus pensylvanica*). Its seeds are carried to a forest by birds and small mammals or are deposited on the forest floor by an earlier stand of cherry. Pin cherry seeds can remain dormant for up to 50 years. When the forest canopy is removed and moisture, temperature, and light conditions become favorable, pin cherry seeds germinate. Young trees quickly dominate the site, crowding out the associated blackberry that also colonizes the area mostly from resident seeds (Marks 1974). If the seedling growth is dense, a pin cherry canopy can close in four years. Canopy closure eliminates other species except highly shade-tolerant seedlings of sugar maple or beech (Figure 29.14) If seedling growth is moderately dense, species with wind-disseminated seeds, such as yellow birch and paper birch, will also occupy the site. Within 30 to 40 years, pin cherry dies out, allowing birch, sugar maple, and beech to dominate the gap. However, during its period of tenure, pin cherry contributes numerous seeds to the forest floor, ready to reclaim the site when another disturbance provides the opportunity.

Sites may be colonized by species seeding from outside the area. Yellow-poplar and black birch are two such species in the eastern deciduous forest. The nature and success of their colonization depend upon a number of conditions, including distance from the site, size of the seed crop, timing of seed arrival, and exposed mineral soil for a seed bed.

Although the response of woody vegetation to large-scale removal of timber is rather well understood, little attention has been given to the response of the shade-tolerant understory herbaceous plants to disturbance. Ash and Barkham (1976) studied the response of the herbaceous understory layer of an English coppice forest after cutting. (A coppice forest is one in which the stump sprouts or root suckers are maintained as a main source of regeneration, with cutting rotations between 20 and 40 years.) Cutting coppice involves complete canopy removal, resulting in increased surface temperature on the forest floor and full exposure to light. Typically a number of open habitat or opportunistic species germinate and become established (Figure 29.15) but are soon excluded by the developing canopy cover. In spite of the disturbance, characteristic woodland species persist throughout the cycle. Adapted to a high light regime in spring before the leaves are out, these plants have the ability to tolerate high light intensity. At the same time they are able to coexist with annuals and open-habitat perennials, which cast a shade on the ground like a tree cover. As opportunistic species disappear, woodland herbs again assume dominance, often developing into monospecific stands. There is evidence that logging practices in the broadleaf deciduous forests of the eastern United States, involving clearcutting, frequent cutting, soil compaction, and microclimatic changes on the forest floor, adversely affect survival and recolonization of forest understory species (Duffy and Meier 1992, Bratton 1994, Matlack 1994).

## Cultivation

Human activity has a more profound impact on ecosystems than natural disturbances. We have the ability to change the natural environment radically. One of the more permanent and extreme changes in vegetation communities comes about when we remove natural systems and replace them with cultivated cropland (Figure 29.16).

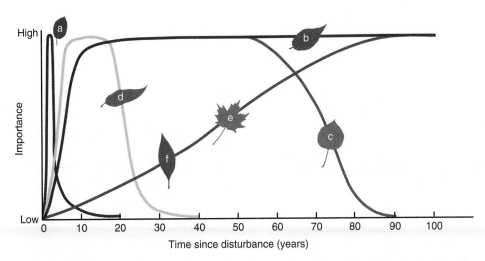

**Figure 29.14** The importance of different species along a gradient of time following logging of a typical northern hardwoods forest. Immediately after disturbance blackberries (a) dominate the site, but they quickly give way to yellow birch (b), quaking aspen (c), and pin cherry (d). Intolerant pin cherry assumes dominance early but within 30 years fades from the forest. Yellow birch, an intermediate species, assumes early dominance which it retains into mature or climax stand. Quaking aspen, an intolerant species, begins to drop out after 50 years. Meanwhile sugar maple (e) and beech (f), highly tolerant species, slowly gain dominance through time. In about 100 years the mature forest is dominated by beech, sugar maple, and birch. (After Marks 1974:75.)

Humans began to convert natural ecosystems to cultivation long ago. In fact, prehistoric human populations converted land to pasture and cultivation as long as 5000 years ago. Their activities changed the pattern of the landscape, extended or reduced the ranges of woody and herbaceous plants, allowed the invasion and spread of opportunistic weedy species, and changed the dominance structure in woodlands (Delcourt 1987). Although the greatest changes took place in Europe, the same effects occurred in North America and South America. Prehistoric native Americans in the Tennessee River Valley, for example, intensively cultivated the bottomlands (B. Smith 1978).

Cultivated plant communities are simple and highly artificial. They consist mainly of introduced genetically altered species adapted to grow on disturbed sites. Because of the simple and homogeneous ecosystem, tillage brings with it new pests destructive to both cultivated and natural vegetation. Tillage disturbs the structure of the soil and exposes it to water and wind erosion. In temperate forest regions, abandoned and degraded agricultural lands eventually return to some form of forest. In tropical regions the degraded cultivated lands may never do so. Intermediate-size disturbances, typified by slash-and-burn or tropical swidden agriculture, in which intensive cultivation rarely lasts more than three years, return to forest, provided the plots are close to a seed source and the soil has not been heavily eroded. Where slash-and-burn is replaced by more intensive cultivation, abandoned lands rarely return to any semblance of the original forest (see Kowal 1966, C. F. Jordan 1986). After high intensity disturbances involving large areas of land, as exemplified by clearing and conversion to grassland in the Amazon, the areas are so degraded that the abandoned sites may be permanently altered to heathlike communities.

## Surface Mining

Surface mining accounts for a high percentage of coal production and for extraction of most other minerals, such as iron, copper, gold, gravel, and limestone. The impact and magnitude of damage vary with the region and the degree and success of reclamation efforts. The effects are most pronounced in mountainous regions, as in the Appalachians. There surface mining follows the contour of the mountain slopes or involves mountaintop leveling and valley fill (Figure 29.17). Whatever the method, surface mining does violence to the land. Deep, unweathered rock strata are broken and brought to the surface, where the material is subject to rapid weathering, releasing toxic nutrients required in small amounts. Carried away in high concentration by water coming off mined sites, these elements reduce water quality downstream for both aquatic life and humans. Unless expensive precautions are taken, sediment deposition in stream can be extremely high.

Surface mining alters the groundwater regime. Water tables once deep in the underlying rock strata are exposed and flow freely to the newly created surface. Large quantities of water that would have been taken up by trees and lost to the atmosphere are added to the amount of runoff. During heavy storms this runoff intensifies the height and damage of flash floods.

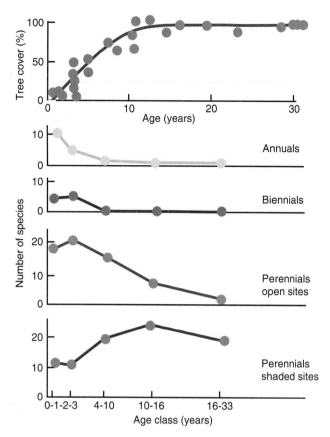

**Figure 29.15** Response of understory herbaceous vegetation to disturbance in a coppice stand in England. The graphs show changes in the percentage of cover produced by growth of the canopy and the number of herb and shrub species in the field or ground layer at different times after coppicing. Not all species were present during the 30-year period of growth. Total numbers for each type are indicated. Note the rapid decline of annuals, biennials, and open-site perennials as the canopy closes. Open-site perennials dominate the field layer shortly after coppicing. Perennials of shaded sites show a sigmoidal growth response as the canopy closes. (From Ash and Barkham 1976:706.)

# EFFECTS ON NUTRIENT CYCLING

Major disturbances such as fire, timber harvesting, and insect outbreaks cause a sudden release of the nutrient capital and provide the opportunity for reestablishing a cycle of change, initiating succession, and increasing productivity. The key is the retention of nutrients. Without uptake by rejuvenating vegetation, by the activity of soil microorganisms, and by the colloidal properties of soil, nutrients could be lost to the system.

Normal populations of foliage-consuming insects account for up to 40 percent of the input of nitrogen and phosphorus to the litter, strongly influencing nutrient cycling in the forest (Kitchell et al. 1979). During an outbreak of

**Figure 29.16** Most of the tallgrass prairie and oak woodland has been converted to fields of corn and soybeans.

foliage-consuming insects such as gypsy moth, fall cankerworm (*Alsophila pometaria*), and elm spanworm (*Ennoomos subsignarius*), great amounts of nutrients contained in the leaves are returned to the litter in the form of frass, the fecal material of caterpillars.

Swank and associates (1981) investigated the effects of an epidemic outbreak of fall cankerworm (a defoliator of such hardwood trees as maple, hickory, oak, and basswood), on nitrate export from three mixed hardwood forests in North Carolina. The larvae consumed approximately 33 percent of the foliage biomass and produced considerable quantities of frass. This deposition resulted in an accelerated leaching and stream export of nitrate nitrogen ($NO_3$–N), although there were no changes in the concentrations of other nutrients. The overall effect of such defoliation was a large increase in leaf litterfall, total litter, soil metabolism, the standing crop of total microbes and nitrifying bacteria, and pools of available nutrients. The trees, in response to defoliation, increased uptake of nutrients made accessible by frass and more rapid litter turnover and shifted production from wood to leaf to compensate for the loss of leaves to insects.

Fire is the great regenerator of nutrients (Ahlgren and Ahlgren 1960), provided they stay in place and the ash is not eroded away. Nutrient recycling by fire is influenced by a number of factors. One is the rate and temperature at which the litter burns. Others are the nutrient content of the fuel, how rapidly nutrients are leached from the ash, nutrient release from the ash, nutrient storage capacity of the soil, surface runoff, soil erosion, and the response of rejuvenating vegetation (Stark 1976). Burning of heather in the Scottish highlands reduced annual primary production and tended to

**Figure 29.17** Contour multiple-seam mining in the steep-sided Appalachians mountains, forever changes the mountains, destroys the forests, eliminated many forms of wildlife, and disrupts the natural water regimes.

make calcium and magnesium less soluble. Young growth of heather, experiencing a postburn flush, contained increased concentrations of K, Mg, P, and N for the first several years; then the concentration declined rapidly back to the prefire level (Miller and Watson 1978b).

Fire reduces soil organic matter to ash. Temperatures of 200° to 300° C destroy 85 percent of organic matter; release $CO_2$, nitrogen, and some ash to the atmosphere; and deposit minerals in the form of ash on the soil. Temperatures over 200° C volatilize nitrogen and release potassium; temperatures under 200° C cause little loss (Knight 1966). Much of the nitrogen lost, however, is in a form unavailable to plants. Some of it is replaced by nitrogen-fixing legumes, the growth of which is stimulated by fire and by increased activity of soil microorganisms, including free-living nitrogen-fixing bacteria. Left behind are cation nutrients such as calcium, magnesium, and potassium in both mobile and plant-available forms. Rainfall following a burn moves these cations through the soil, where they displace hydrogen ions, increasing soil pH.

Timber harvesting, cultivation, and surface mining differ from other types of disturbance because we remove biomass and the nutrients it contains from the ecosystem. In timbering, the quantity of nutrients removed depends upon the type of harvest. Stem-only harvesting leaves tops and foliage behind. Whole tree harvesting leaves only some foliage, with all woody material being reduced to chips. Most exploitative of all is complete tree harvesting, the removal of the total tree, including roots. Block clearcutting involving stem-only removal on Hubbard Brook Forest removed about 28 percent of the calcium contained in aboveground bio-

mass, 28 percent of the potassium, 24 percent of nitrogen, 34 percent of sulfur, and 19 percent of phosphorus (Hornbeck et al. 1987).

Nutrient cycling within the disturbed forest is altered almost immediately after timber harvesting. Loss of the trees severs the uptake of nutrients and adds large amounts of detritus to the forest floor. The loss of canopy cover allows increased precipitation and sunlight to reach the forest floor, increasing decomposition and accelerating nitrification. Increased concentrations of nutrients in the dissolved organic and inorganic fractions are easily leached and exported by surface water and groundwater to streams. This flux of nutrients represents a loss to the disturbed site and a gain to the stream ecosystem. Such an input to a stream may have positive effect if the stream is nutrient poor or negative effect if the stream is already nutrient-rich (Bormann and Likens 1979, Hornbeck et al. 1987, Waide et al. 1988, Vitousek and Melillo 1979, Martin et al. 1986). Input of nutrients into receiving streams is greater in the second year following cutting than during the first, and by the fourth year stream ion concentration usually returns to precutting levels.

The second year increase and subsequent return to original status may be affected by the means of nutrient conservation available to disturbed ecosystems. One is immobilization of nutrients by decomposer organisms. Bormann and Likens (1979) suggest that in the first year after harvest, when the C:N and element:P ratios of decaying organic matter are high, dissolved nutrients in the soil are taken up and immobilized by increased numbers of microorganisms. As the ratios decrease, the demands by microorganisms decline, and export of nutrients to streams increases. During this

time, however, rapidly growing, early succession vegetation begins to recycle nutrients efficiently, retaining most of the nutrients on site. Boring and associates (1981) found that incoming woody and herbaceous plants in a clear-cut Appalachian hardwood forest during the first year held 29–44 percent of the amount of N, P, K, Mg, and Ca found in the net primary production of the control. Uhl and Jordan (1984) found similar responses in a cutover tropical forest. By the end of the fifth year, live plant regeneration held 15 percent of precut levels of N, 23 percent of P, 39 percent of K, 48 percent of Mg, and 45 percent of Ca. Thus microbial activity, soil processes, and rapid recovery of vegetation contribute to the resilience of forest ecosystems following logging.

## ANIMAL RESPONSE TO DISTURBANCE

How does animal life respond to disturbances of various magnitudes? Over a short term the effect may be negative; over the long term it may be positive, depending upon the species. With disturbance it is clearly a case of "One man's meat is another man's poison." Small-scale disturbances, particularly treefall gaps, result in a positive response. The new vegetative growth filling in the gap provides low ground cover attractive to such gap species as hooded warblers (*Wilsonia citrina*) and Kentucky warblers (*Oporornis formosus*). The canopy gaps provide open areas needed by flycatchers (Tryannidae) for hawking insects; yet the openings have no effect on canopy-dwelling species. Clearcut areas in a forest create the early successional habitats needed by opportunistic or ephemeral species such as prairie warblers (*Dendroica discolor*), chestnut-sided warblers (*Dendroica pensylvanica*), and woodcock. Such species require recurring disturbances to maintain their habitat. Large clearcut areas, on the other hand, eliminate habitat of canopy-dwelling vertebrates and invertebrates that will not return for several decades. Generally, ground-dwelling small mammals are little affected by timber harvesting. Cutting of old-growth forest can permanently eliminate old-growth dependent species such as spotted owls (*Strix occidentalis*) and red tree voles (*Phenacomys longicaudus*) of the Pacific Northwest.

A view of its immediate aftermath may give the impression that fire is a major agent of destruction to wildlife. On a short-term basis fire destroys or partially destroys habitat. It may cause some injury and death, either directly by fire or indirectly by predators who take advantage of prey suddenly driven from or deprived of cover. On the African savanna, kites and other birds follow grass fires, hunting insects driven to flight by the advancing fires. Many flying insects, such as grasshoppers and moths, fly in front of the flames and are often engulfed by wind-driven gas clouds, but a surprising number go through the fire unscathed. Unless nesting, birds are rarely directly affected by fire other than

short-term loss of habitat. Some species of birds may decrease following fires, but ground-foraging birds seem to increase (Wright and Bailey 1982).

Many mammals, especially those that live in burrows, survive fires (Bradley et al. 1992). Large mammals are adept at keeping ahead of flames and working their way back through gaps and unburned patches to burned-over areas behind the flames (Main 1981). The major problem faced by these mammals is the short-term lack of food and cover. High populations of post-fire grazing herbivores feeding on newly regenerating plants may overgraze or eliminate palatable species.

For fire-intolerant species, such as tree-dwelling squirrels, severe fires destroy habitat and eliminate for some time those species dependent on it. At the same time fire improves habitat for fire-impervious species, especially those that favor open and shrubby land. Many animals favor both pre- and post-fire conditions and in fact are dependent on such fluctuations in habitat. Fire produces a mosaic of shrubs, timber, and open land (see Figure 29.1). Such patchy environment is essential for such species as snowshoe hare, black bear, white-tailed deer, and ruffed grouse.

A few species are fire-dependent. They require the periodic disturbance of fire to maintain their habitat. One is the endangered Kirtland's warbler (*Dendroica kirtlandii*) (Figure 29.18). Restricted to the jack pine forests of the lower peninsula of Michigan, the warbler requires large blocks (40+ ha) of even-aged stands of pine 1.5 to 4.5 m tall with branches close to the ground. Smaller or larger trees are unacceptable. Intervals of fire are needed to maintain blocks of young jack pine habitat.

Another group of fire-dependent birds is the *Silva* warblers that occupy the Mediterranean shrublands of Sardinia (Walter 1977). Of the five species, two are fire-dependent.

**Figure 29.18** The endangered Kirtland's warbler, is a fire-dependent species inhabiting the jack pine forests of Michigan.

One species, the Dartford warbler (*S. sarda*), can occupy a habitat patch that has been burned within six years. Another species, Marmora's warbler (*S. undata*), occupies only 18- to 20-year old tall, shrubby growth. Two of the remaining species are fire-adapted, and one is fire-tolerant.

# COMMUNITY STABILITY

Community stability has been one of the tenets of ecology. Ecologists have regarded communities as highly organized assemblages of coevolved species. Each species is competitively superior in its own niche in the habitat. Competitive interactions among these species maintain a state of equilibrium and high diversity without continual change in species composition. After a disturbance the species eventually reoccupy their former positions; the community arrives again at equilibrium or some degree of stability.

## Equilibrium Communities

Community equilibrium or stability can be viewed in two ways: resistance and resilience. **Resistance** is the ability of a system to withstand or resist variation. It is measured by the degree to which the system is changed from an equilibrium state following a disturbance. Communities most resistant to change characteristically have a large biotic structure, as trees do, and nutrients and energy stored in standing biomass. A forest community is relatively resistant. It can withstand such environmental disturbances as sharp temperature changes, drought, and insect outbreaks because the system is able to draw on stored reserves of nutrients and energy. For example, a late spring frost may kill the new leaves of forest trees, but they are able to draw on energy reserves in the roots to replace leafy growth. If the forest is highly disturbed by fire or logging, its return to the original condition is slow. The system exhibits low *resilience,* which is the inverse of resistance.

Resilience is the speed with which a disturbed system returns to equilibrium or the same general state after being changed. A rapid return is evidence of high resilience, and a slow return indicates low resilience. For example, in the spruce-fir forests of northern North America under certain environmental conditions the spruce budworm population increases rapidly and escapes control of predators and parasites. It feeds heavily on balsam fir, killing many trees and leaving only the less susceptible spruce and birch. After the spruce budworm population collapses because of the exhaustion of the food supply, young balsam fir grows back in thick stands with spruce and birch. Between budworm outbreaks balsam fir outcompetes spruce and birch, but during outbreaks spruce and birch are favored over balsam fir. Thus some time after the outbreak of budworm, the system re-

turns to balsam fir. The system is resilient even though some of the interacting populations have low resistance.

Aquatic ecosystems, which lack any long-term storage of energy and nutrients in biomass, exhibit little resistance but are resilient. Frequent flooding dislodges stream invertebrates, creating moaics of empty patches. These patches are recolonized by different subsets of stream invertebrates available for recruitment. Although the recovered patches may hold a different set of organisms, the stream has returned to original conditions. An influx of pollutants such as sewage effluents disturbs the system, adding more nutrients and organic matter to the ecosystem than it can handle; but because the system is limited in its capacity to retain and recycle nutrients, it returns to its original condition soon after the inflow of nutrient-rich pollutants is reduced or removed. In both cases the systems have high resilience but low resistance.

For example, Lake Washington, near Seattle, Washington was used as a basin for sewage disposal. It received a large input of nutrients, especially phosphorus. The input killed diatom and algal populations and shifted the community from green algae to cyanobacteria (blue-green algae), especially the filamentous forms. It clouded the water and changed the structure of the lake ecosystem. Once the sewage input was diverted from the lake, phosphorus levels in the lake declined, filamentous "algae" declined, and the lake returned to its clear condition.

A very strong disturbance may carry an ecosystem into a different level of stability. The system may be so greatly disturbed that it is unable to return to its original state, and a different ecosystem with a different domain of stability takes its place. Perry and associates (1989) provide an excellent example of such a response to disturbance in 10 to 15-ha clearcuts in high elevation white fir (*Abies concolor*) forests of the Siskiyou Mountains of southwestern Oregon and northern California. Following clearcutting, the areas were sprayed with herbicides to eliminate incoming hardwood growth preparatory to replanting the sites with conifers. Hardwood growth, however, is a natural successional recovery process after a disturbance, such as fire, in these forests. The rapid response of hardwood growth shades the soil, ameliorates surface temperatures and moisture regimes, and maintains the integrity and stability of the organisms of the soil.

The elimination of the natural incoming hardwood growth greatly weakened the close link between plants and soils and altered the belowground system. Rapid growth of certain soil bacteria, mostly actinomycetes, inhibited the growth of other microbes and plants and reduced the belowground mutualists. Loss of the belowground microbial community in turn caused the loss of soil structure and a reduction in soil pore space and water holding capacity. The original white fir forest was replaced by brome grass, bracken fern, and manzanita, species that have a competitive

advantage when mycorrhizal fungi and rhizosphere organisms decline.

According to equilibrium theory, stability is the tendency of an ecosystem to reach and maintain an equilibrium condition of either a steady state or a stable oscillation (Holling 1973, 1984). If the system is highly stable, it exhibits high resistance to departure from that condition. If the system is directed away from that condition, it is resilient. It returns rapidly to its initial condition with the least fluctuation. This stability may be local or global. *Local stability* is the tendency of a system to return to its original state from a small disturbance. Forest gaps filling in with tree species are examples of local stability. *Global stability* is the tendency of a community to return to its original condition from all possible disturbances. It implies that without disturbance no species will ever be lost. Chaparral or eucalyptus forest returning quickly to its original condition and species composition after a fire represents global stability. Such systems exhibit low resistance to change.

## Nonequilibrium Communities

Although equilibrium theory has long dominated ecological thought, most ecologists now agree that ecological communities seldom attain equilibrium. Disruptions are common, preventing species assemblages from reaching any highly ordered state. Contrary to the equilibrium theory, communities exist at some level of nonequilibrium, held in that state by environmental disturbances (Connell 1978).

Communities are subject to all sorts of disturbances—storms, wind, waves, ice, heavy snow, drought, fire, insect defoliation, grazing and browsing by herbivores. All of these disturbances affect populations of species within the community differently. Although competition for resources is potentially intense, disturbance prevents competitive interactions from being played out to the end. If they were, then the community eventually would be dominated by a few of the most competitive species and diversity would be low. Any disturbance that would occur before competitive exclusion could take place would result in higher diversity (Huston 1979). In nonequilibrium communities highest diversity is maintained at an intermediate level of disturbance, a concept known as the **intermediate disturbance hypothesis** (Connell 1978).

This hypothesis is summarized in Figure 29.19. At a high frequency of disturbance, there is no time for certain species to mature. The community is dominated by species with a rapid growth rate, a short life span, and a strong ability to colonize disturbed areas. At a low frequency of disturbance, short-lived species (*r* species) are outcompeted by long-lived species that come to dominate the community (*K*-species). Diversity is low: However, within the two extremes, as the time between disturbances lengthens, species of varying rates of growth, longevity, and competitive abili-

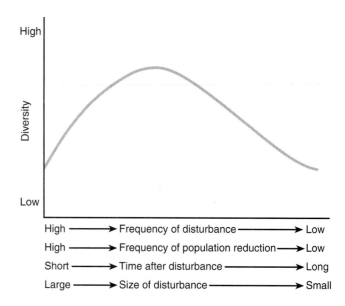

**Figure 29.19** A model for the intermediate disturbance hypothesis.

ties can persist. Further influencing the process is the scale of disturbance. Small-scale disturbances favor higher diversity than large-scale ones. Overall, small-scale environmental disturbances at intermediate frequencies best maintain diversity in communities.

## SUMMARY

Disturbance is any physical force, such as fire and wind, that damages natural systems and kills organisms. It influences community structure and contributes to biodiversity. Disturbances vary in intensity, frequency, and spatial scale. Intensity is measured by the proportion of the biomass or populations killed or removed. Frequency is the mean number of disturbances that take place within a given time interval. Related to frequency is the return interval, the mean time between disturbances on the same piece of ground.

Important to the nature of disturbances is spatial scale. Small-scale disturbances are typical of rocky intertidal shores and temperate and tropical forests. Wave action and moving water create gaps among sessile organisms on rocky substrates. Treefall and removal of individual trees by logging create small gaps in forests. Responses to gap formation are canopy closure, invasion by opportunistic species, and growth of tolerant species, depending upon biotic conditions. Large-scale disturbances induced by major events such as fire, hurricanes, logging, insect outbreaks, and land clearing for cultivation allow colonization by opportunistic species. Such disturbances can modify the system by favoring certain species and eliminating others, or they can ensure regeneration of the system itself.

Of great ecological importance are the frequency and return interval of disturbances. Small-scale disturbances have a high frequency within the system, but the rate of disturbance is low and the return interval is between 50 and 200 years. Natural large-scale disturbances have low frequency and a return interval of 25 to several hundred years. Frequency of large-scale disturbances relates to the life span of the longest living species. Too frequent disturbances can eliminate certain species, destroying plants before they have had time to mature and seed. Too long a time between disturbances can eliminate midtolerant species, reduce system diversity, and set the stage for highly destructive disturbances.

Fire is the major natural large-scale disturbance to terrestrial ecosystems. It has both beneficial and adverse effects. It results in loss of soil nutrients but also makes nutrients available. It sets into motion regeneration of fire-adapted systems by stimulating root sprouting and germination of seeds. It can favor fire-resistant species and eliminate fire-sensitive ones.

Disturbances release nutrients locked up in biomass. Although considerable amounts of nutrients may be lost to the system following major disturbances, much of the nutrient pool is immobilized by increased populations of soil microorganisms and incorporated into plant biomass by the resurging growth of vegetation.

Response of the animal component of the systems depends upon the species. Short-term impacts include loss of food and cover. Long-term effects may be the loss of habitat for some species and the gain of habitat for others. Some species depend upon disturbance for the maintenance of their habitat, especially those associated with the more ephemeral stages of early succession. Other species depend upon periodic fires to provide the mosaic of vegetation types required in their life cycle. A few fire-dependent species would go extinct without periodic fires to maintain their habitat.

Disturbance affects stability of communities. Stability is the tendency of a system to return to and maintain an equilibrium condition after a disturbance. If the system is highly stable, it resists departure from or returns rapidly to equilibrium. Stable systems are highly resistant to disturbance but not necessarily resilient. Resilience is the time required for a system to return to equilibrium, absorbing changes.

The concept of ecosystem stability implies that the species composition of the community is in a state of equilibrium. However, most communities are in a state of nonequilibrium brought about by environmental disturbances. The degree of nonequilibrium is influenced by the frequency and scale of disturbances. Both very frequent and highly infrequent disturbances result in low diversity. Communities kept in a moderate state of disequilibrium by small-scale intermediate disturbances support the greatest diversity.

# REVIEW QUESTIONS

1. Define disturbance.
2. How do intensity and frequency of a disturbance influence its effects?
3. How do small-scale disturbances relate to the diversity and long-term stability of a community?
4. What constitutes large-scale disturbance? What are some major natural forces behind such disturbances?
5. How does fire differ in its ecological effects from other disturbances?
6. What major microclimatic changes from surrounding forest would you expect to discover in a small gap in the forest? Consider light, air temperature, soil temperature, and humidity. Discuss how these changes affect opportunistic species. (Refer to Chapters 4, 6, 7, and 8.)
7. Controlled burning of the understory of longleaf pine at regular intervals is a common management practice. What is the purpose of such burns? What regulates the frequency? What effect do they have on understory vegetation? What would be the outcome if such disturbances were eliminated?
8. How do resistance and resilience relate to ecosystem stability?
9. What is the intermediate disturbance hypothesis? Discuss its relationship to frequency and scale of disturbances.
10. Refer to the pattern of forest fires in Figure 29.2. Assume that the 12 fires diagrammed rather than the actual 23 represent the total occurrence of fire from 1750 to 1900. Select several points at random in the watershed and determine the frequency and return interval of fire for those points.
11. Read John Muir's essay, "Fire in the Sequoias" (John Muir, *Our National Parks,* Houghton Mifflin, 1904, pp. 307–314). Contrast the effects of natural fires in the sequoias to complete fire exclusion imposed by the National Park Service.

# CROSS-REFERENCES

Plant response to drought, 72–73; nutrient cycling, 116–120; r and K selection, 448; herbivory and community structure, 622; predation and community structure, 621–623; parasitism and community structure, 623–624; secondary succession, 659–660; cyclic succession, 666–667.

# Succession

## Concepts

1. Succession is a change in species composition and community structure and function over time and space.
2. Succession may be primary, starting on sites unoccupied by plants, or secondary, starting on sites where plants have grown.
3. Succession results from different combinations of life history traits and physiological properties of plants interacting under changing environmental conditions.
4. The climax, a terminal point in succession, is maintained by small disturbances through time and space.
5. Cyclic replacement of vegetation contributes to community persistence.
6. Succession involves changes in biomass accumulation, nutrient cycling, and community structure.
7. Animal life changes as plant communities change. Some animals depend upon certain successional stages.
8. Patterns of vegetation observed today are products of evolution and plant succession over geological time.

Abandoned cropland is a common sight in agricultural regions in once forested areas in eastern North America. No longer tended, the land quickly grows up in grasses, goldenrod, and weedy herbaceous plants. In a few years these same weedy fields are invaded by shrubby growth—blackberries, sumac, and hawthorn. These shrubs are followed by fire cherry, pine, and aspen. Many years later this abandoned cropland supports a forest of maple, oak, cherry, or pine. Thus over a period of years one community replaces another until what appears to be a stable forest occupies the area (Figure 30.1).

## SUCCESSION DEFINED

The changes in the return of the forest to these crop fields are not haphazard, but orderly, and barring disturbance by humans or natural events, the reappearance of a forest is predictable. This change in species composition and community structure and function over time is **ecological succession.** More precisely, succession may be defined as a continuous unidirectional, sequential change in the species composition of natural communities. Succession is usually set in motion by some sort of disturbance.

The sequence of communities from grass to shrub to forest that terminates in a relatively stable community historically has been called a **sere** (from the word *series*), and each of the changes is a seral stage. Although each **seral stage** is a point in a continuum of vegetation through time, it is recognizable as a distinct community. Each has its characteristic structure and species composition, especially at the point of optimal development. A seral stage may last only one or two years or several decades. Some stages may be missed completely, or they may appear only in abbreviated or altered form. For example, when an abandoned field grows up immediately in forest trees (Figure 30.1e), the shrub stage appears to have been bypassed; but structurally its place is taken by the incoming young trees.

Eventually succession slows and the plant community achieves some degree of equilibrium or steady state with the environment. This mature, relatively self-maintaining seral stage traditionally has been called the **climax** community, and the vegetation supporting it the climax vegetation. The term, however, has fallen into disrepute with some ecologists.

Succession that begins on areas unoccupied or unchanged by organisms is called **primary succession** (Figure 30.2). Succession that proceeds on areas where other organisms were present is called **secondary succession** (Figure 30.1). Secondary succession arises on sites where the vegetation cover has been disturbed by humans, animals, or natural forces such as fires, wind storms, and floods. Its development may be controlled or influenced by the activities of humans or domestic and wild animals.

Barren areas, whether they are natural primary sites, such as rock outcrops, sand dunes, and alluvial deposits, or disturbed areas, such as abandoned cultivated fields or roadbanks, are a natural biological vacuum eventually filled by living organisms. Organisms that colonize such sites comprise the **pioneer species.**

Succession is self-driven or **autogenic** when changes in the environment are brought about by the organisms themselves. In other cases changes in the environment are caused by some external factor unaffected by the organisms. Then succession is called **allogenic.** Such environmentally induced succession, often long-term, may involve climate change, or gradual changes in soil over tens or hundreds of years. However, such succession still has a short-term autogenic component.

## A DESCRIPTIVE APPROACH

Most successional studies describe the process rather than explaining the forces that drive it. Descriptive approaches have their place, because they set the pattern of succession, illustrate the process, and provide useful points from which to discuss what happens during succession.

### Terrestrial Primary Succession

Primary succession begins on sites that never have supported life. Such places are rock outcrops and cliffs (Burbanck and Platt 1964, Shure and Ragsdale 1977), sand dunes (Cowles 1899, Olson 1958), and newly exposed glacial till (Crocker and Major 1955, Lawrence 1958). Succession begins with the colonization of a site by early successional species that both alter and improve conditions, allowing new species to invade.

For example, consider primary succession on a very inhospitable site, a sand dune. A product of pulverized rock, sand is deposited by wind and water. Where deposits are extensive, as along the shores of lakes and oceans and on inland sand barrens, sand particles may be piled in long windward slopes to form dunes (Figure 30.3). Such dunes move

**Figure 30.1** (opposite) Successional changes in an old field in western Pennsylvania over 50 years. (a) The field as it appeared in 1942, when it was moderately grazed. (b) The same area in 1963. (c) A close view of the rail fence in the left background of (a). (d) The same area 20 years later. The rail fence has rotted, and white pine and aspen grow in the area. (e) The field in 1972, when aspen has claimed much of the ground. (f) In 1992, the field is covered with a young forest dominated by quaking aspen and red maple.

(a)

(b)

(c)

(d)

(e)

(f)

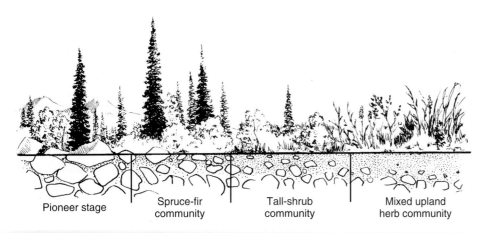

Pioneer stage    Spruce-fir community    Tall-shrub community    Mixed upland herb community

**Figure 30.2** Primary succession in the subalpine zone of the Wasatch Mountains, Utah. Here the early stages include trees; the climax is a mixed herb community. Note the changes in soil depth from a rocky surface with fine soil only in crevices to a well-defined deep soil essentially free from rocks. Such succession may be considered retrogressive, because it leads to community with less vertical structure dominated by vegetation traditionally associated with earlier stages of succession. (Based on data from Ellison 1954.)

before the wind and often cover forests and buildings in their path until stabilized by plants. Colonization of dunes and the progressive development of vegetation was originally described by H. C. Cowles (1899) in a pioneering classic study of plant succession on the dunes about Lake Michigan.

Grasses, especially beach grass (*Ammophila breviligulata*), are the most successful pioneering and binding plants. When they and such associated plants as beach pea (*Lathyrus japonicus*) stabilize the dunes at least partly, mat-forming shrubs invade the area. From this point the vegeta-

tion may pass to pine and then oak, or to oak directly without an intervening pine stage. The low fertility of the dunes favors plants with low nutrient requirements. Because these plants are inefficient in cycling nutrients, especially calcium, soil fertility remains low. Because of this infertility and low moisture reserves in the sand, oak is rarely replaced by more mesophytic and moisture-demanding trees (Olson 1958). Only on the more favorable leeward slopes and in depressions, where microclimate is more moderate and where moisture can accumulate, does succession proceed to more

**Figure 30.3** Primary succession on a coastal sand dune colonized by beach grass.

mesophytic trees such as sugar maple, basswood, and red oak. Because these trees recycle nutrients more efficiently, effectively shade the soil, and add to litter accumulation, they aid in the rapid improvement of nutrients and moisture conditions. On such sites a mesophytic forest may become established without going through the oak and pine stages. This example emphasizes one aspect of primary succession: the colonizing species ameliorate the environment, paving the way for invasion of other species.

Newly deposited alluvial soil on a floodplain represents another barren primary site. Walker and associates (1986) studied primary succession on an alluvial floodplain in Alaska. Seeds of all colonizers—willow, alder, balsam poplar (*Populus balsamifera*), and white spruce—arrived on the site more or less simultaneously and gave rise to seedlings. Their appearance, however, was determined by life history traits. Willow with its light, wind-dispersed seed was most abundant, and the three other species were less widely distributed. Willow and alder grew rapidly, but the willow, naturally short-lived, was heavily browsed by snowshoe hare. Nitrogen-fixing alder then became dominant, eventually to be replaced by balsam poplar and long-lived white spruce. Thus primary succession may be influenced by such events as seed production and by the life history traits of the colonists.

Primary succession does not always involve colonization of the site by species with the highest environmental tolerances and their improvement of the site for the growth of later arriving species. Wood and del Moral (1987) followed early primary succession in subalpine habitats on Mount St. Helens for the first six years following its eruption. They found that species of high environmental tolerance, such as *Aster ledophyllus,* dispersed only short distances, and most seedlings were within 3 m of the conspecific adults. These species were unable to colonize much of the barren substrate. Species that dispersed the farthest were incapable of growing on barren ash. They could survive only in patches of original substrate that remained free of volcanic ash. Thus primary succession may be impeded by the poor dispersal of potential colonists or by the inability of some pioneering species to grow under prevailing conditions.

## Terrestrial Secondary Succession

Secondary succession is most commonly encountered on abandoned farmland and noncultivated ruderal sites (waste places) such as fills, spoil banks, railroad grades, and roadsides. These artificially disturbed areas are frequently subject to erosion and settling.

Species most likely to colonize such places are the so-called weeds, species out of place from a human perspective. Although hard to define, weeds have two characteristics in common. First, they invade areas modified by human action; in fact, a few are confined to such artificially modified habitats. Some, such as ragweed, are native (to North America). They evolved in highly disturbed habitats and thus are already adapted to invade human disturbed sites. Many, such as kudzu, are exotics, not native to the region. As competitors on disturbed sites they are superior to native species.

Second, whether annual, biennial, or perennial, all plants that successfully colonize disturbed areas possess tolerance for soil disturbance. Their seeds remain viable for a long time. They may remain in the soil for years until conditions are right for germination. Some weeds require an open seedbed and exposed mineral soil for germination. Their rapid and successful colonization is aided by an efficient means of dispersal. Some have light seeds that are carried by the wind; others spread by underground rhizomes. These vigorous pioneer plants (*r*-selected species) grow rapidly under favorable conditions; in less favorable habitats they set seed even when small. In spite of their vigor, these plants cannot maintain dominance for long—two or three years at the most—if all disturbance ceases. Short life cycles, advantageous at first, are not adaptable to conditions imposed by incoming plants with longer life cycles (*K*-selected species), ones that begin growth early in the spring and persist throughout the summer (see Peterson and Bazzaz 1978).

One of the classic examples of secondary succession is Keever's (1950) study of old-field succession in the Piedmont of North Carolina. The year a crop field is abandoned, the ground is claimed by annual crabgrass (*Digitaria sanguinalis*), whose seeds, lying dormant in the soil, respond to light and moisture and germinate. But the crabgrass' claim to the ground is short-lived. In late summer the seeds of horseweed (*Lactuca canadensis*), a winter annual, ripen. Carried by the wind, they settle on the old field, germinate, and by early winter have produced rosettes. The following spring horseweed, off to a head start over crabgrass, quickly claims the field. During the summer the field is invaded by other plants—white aster (*Aster ericoides*) and ragweed (*Ambrosia artemissifolia*). Competition from aster and inhibiting effects of decaying horseweed roots on horseweed itself allows aster to achieve dominance.

By the third summer broomsedge (*Andropogon virginicus*), a perennial bunchgrass, invades the field. Abundant organic matter and ability to exploit soil moisture efficiently permit broomsedge to dominate the field. About this time pine seedlings, finding room to grow in open places among the clumps of broomsedge, invade the field. Within five to ten years the pines are tall enough to shade the broomsedge. A layer of poorly decomposed pine needles (duff) that prevents most pine seeds from reaching mineral soil, dense shade, and competition for moisture among successfully germinating seedlings and shallow-rooted parent trees inhibit pines from regenerating on the site. Hardwoods, such as oaks and ash, grow up through the pines, and as the pines

die (if they are not cut) take over the field (if it is not burned). Development of the hardwood forest continues as shade-tolerant trees and shrubs—dogwood, redbud, sourwood, hydrangea, and others—fill the understory. The sere has arrived at the mature or tolerant stage, in which only the dominant species of the crown can replace themselves in their own shade.

As plants change, so do the animals. Early stages are characterized by such arthropods as crickets, grasshoppers, and spiders and by such seed-eating birds as mourning doves. Broomsedge brings in meadowlarks and meadow mice, and low pine growth brings in rabbits. Mature pines shelter pine warblers and sparrows. As pines decline and hardwoods claim the area, downy woodpeckers, flycatchers, and hooded and Kentucky warblers appear.

## Aquatic Succession

Succession in aquatic environments is primary. Succession starts with open water and a bottom barren of life. The first organisms to colonize the pond or lake are plankton, which may become so dense it clouds the water. If the plankton growth is rich enough, the pond may support other forms of life—caddisflies, bluegills, green sunfish, and large-mouthed bass.

At the same time the pond acts as a settling basin for inputs of sediment from the surrounding watershed. These sediments form an oozy layer that provides a substrate for rooted aquatics such as the branching green algae, *Chara*, and pondweeds. These plants bind the loose matrix and add materially to the accumulation of organic matter. Rapid addition of organic matter and sediments reduces water depth and increases the colonization of the basin by emergent and submerged vegetation. That, in turn, enriches the water with nutrients and organic matter. This enrichment further stimulates pelagic production and sedimentation, and expands the surface area available for colonization by macrophytes (see Carpenter 1981). Thus aquatic succession goes from an oligotrophic to a eutrophic state. Eventually the substrate, supporting emergent vegetation such as sedges and cattails, develops into a marsh. As drainage improves and the land builds higher, emergents disappear. Meadow grasses invade to form a marsh meadow in forested regions and wet prairie in grass country. Depending upon the region, the area may pass into grassland, swamp hardwoods or conifers, or peat bogs; or succession may remain stalled at the grassland stage, resisting invasion by hardwood species.

Is the sequence of aquatic succession self-driven or allogenic? Paleoecological investigations of aquatic succession on the Indiana sand dunes by Jackson and associates (1988) suggest that the classical sequence of supposed autogenic succession observed today is the result of the differential effects of disturbances rather than gradual, self-induced successional changes. The dune ponds remained more or less in floating and submerged vegetation until a time of human disturbance, when changes were most rapid. Increased sedimentation and changes in water chemistry probably hastened the extinction of certain macrophytes and permitted the invasion of the ponds by others. Thus observed differences in open water succession over a short time reflect differences in disturbances that distort vegetation patterns rather than autogenic successional changes.

## Intertidal Succession

The pattern of changes in intertidal vegetation is evidence of succession. The subtidal kelp forests off the California coast, for example, exhibit three major vertical layers: a *Mycrocystis pyrifera* surface canopy, a dense subsurface canopy of *Pterygophora californica*, and an understory of coralline algae. When Reed and Foster (1984) experimentally removed the canopy and subcanopy layers, annual brown algae (*Desmarestia ligulata*) and a moderate recruitment of the canopy species responded to increased light. This response suggests that, unless disturbed, these perennial kelp species inhibit their own recruitment as well as the invasion of other species.

Disturbance alters the situation. G. G. Harris and associates (1984) observed that a severe storm denuded a southern California marine reef, exposing large areas of virgin rock, and decimated the herbivorous sea urchins that graze on the kelp. The ecologists followed the sequence of recolonization. Immediately after the storm, diatoms colonized the surface and were soon followed by filamentous and leafy seaweeds. Dense stands of filamentous brown algae preceded the appearance of small sporophytes of kelp. Where growing in the open, small kelp experienced heavy grazing by two herbivorous fishes, whereas kelp growing within stands of brown algae escaped serious predation. The brown algal refuge permitted the kelp to become established and grow large enough to withstand predation.

Predation can have a major influence on successional development in subtidal regions. In the Pacific Northwest sea urchins prefer early successional kelp *Nereocystis leutkeana* over late successional kelp *Agarum cribrosum*. When urchins are present, late successional *Agarum* occurs. If the urchins are naturally rare or removed, the early successional *Nereocystis* inhibits the establishment of *Agarum* (Vadas 1977). In the rocky intertidal zone of New England, littorine snails prefer the early successional green alga *Ulva lactuca*, which can delay the recruitment of less preferred *Fucus vesiculosus*. When the snails are abundant, *Fucus* becomes established much more quickly (Lubchenco 1978). In southern California sea urchins may overgraze the middle and late successional perennial red and brown algae and permit the early successional ephemeral species to persist. The marine successional example emphasizes another important point about succession, whether marine, freshwater aquatic, or terrestrial, that succession can be strongly influenced by the grazing action of herbivores.

# MODELS OF SUCCESSION

Although Henry Thoreau (1860) had described plant succession following logging in New England and Anton Kerner (1863) wrote about plant succession in the Danube Basin in central Europe, it was H. C. Cowles (1899) in his study of vegetational development on the sand dunes of Lake Michigan who first explicitly described the process of plant succession. Later F. E. Clements (1916) developed a descriptive theory of succession and advanced it as a general ecological concept. His theory of succession had a powerful influence (or inhibition, some critics say) on ecological thought.

## Pioneering Concepts

Clements viewed succession as a process involving several phases, to each of which he gave his own terminology. Succession began with the development of a bare site, called **nudation** (disturbance). Nudation was followed by **migration,** the arrival of propagules. Migration was followed by the establishment and initial growth of vegetation. Clements called this phase **ecesis** (growth) As vegetation became well established, grew, and spread, various species began to compete for space, light, and nutrients. This phase Clements called **competition.** It was followed by **reaction,** the self-induced (autogenic) effects of plants on the habitat. The outcome of this reaction was the replacement of one plant community by another, resulting in the persistence of one species complex or community, or **stabilization.** Clements' theory of succession dominated ecology for years, almost becoming dogma, because it provided an orderly, logical explanation for the development of plant communities. In spite of vigorous criticism, the basic processes he outlined are still valid and useful.

The problem critics have with Clements' view of succession lies outside the outline of the process. To Clements, each stage of succession represented a step in the development of a superorganism, the climax. It was analogous to the embryological development of an organism. The climax was an assemblage of vegetation that belonged to the highest type of vegetational community possible under the prevailing climate. The climax, according to Clements, is able to reproduce itself, "repeating with essential fidelity the stages of its development." Each seral stage so modifies the environment that plants of that stage eventually can no longer exist there. Instead they prepare the site for the replacement plants of the next stage. The process continues until the vegetation arrives at the self-reproducing climax. That marks the end of succession.

Henry Gleason (1917) challenged Clements' view of succession. Gleason regarded a plant association or climax not as a superorganism, but as a community consisting of individual species that respond independently to environmental conditions. Succession resulted from these individual responses of different species. Plants involved in succession were those that arrived first on the site and were able to establish themselves under prevailing environmental conditions. Competitive and other interactions among species, the plants' modification of the microenvironment, and the ability of the plants to exploit nutrients, moisture, and other environmental inputs on the site determine the final outcome.

Later Frank Egler (1954) recast these views of succession in terms of relay floristics and initial floristic composition. **Relay floristics** most nearly corresponds to Clements' idea of succession—groups of associated species marching together and disappearing through time as one group replaces another. **Initial floristics,** involved only in secondary succession, proposes that the propagules of most species, both pioneer and late stage, are initially on the site. Which species becomes dominant depends upon life history characteristics and competitive interactions. Short-lived species are eventually replaced by long-lived, but not necessarily climax, species. These views find further expression in two divergent approaches to succession: a holistic or ecosystem approach and a reductionist or population approach.

## Holistic Concept

The holistic approach is a direct descendent of Clements' organismal theory of succession. It views succession as driven by changes in attributes between youthful and mature systems. Succession involves changes in the emergent properties that arise out of the organization of the ecosystem. Succession eventually leads to the formation of an emergent entity with unique characteristics involving nutrient flow, biomass accumulation, and species diversity (E. P. Odum 1969, 1983). Succession begins with the developmental stages of short-lived, shade-intolerant plant species and terminates with a mature stage dominated by long-lived, shade-tolerant species. Young stages so modify the environment that the existing community is replaced by a more mature one, better able to exploit the changed environment.

As succession proceeds, ecosystem attributes change. (See section on Attributes.) For example, succession begins with an unbalanced community metabolism. Net community production is greater than respiration, and biomass accumulates over time. As the ecosystem matures, respiration begins to equal production. Ultimately, in the terminal stage, production balances respiration. This approach views succession as an orderly transition from one seral stage to another driven by self-induced modification of the environment.

## Reductionist Concept

Reductionists reject the holistic concept. They argue that successional changes can be explained by population dynamics, especially competition, regeneration, and mortality (Peet and Christensen 1980, Noble and Slatyer 1980) and by

physiology and life history strategies (Bazzaz 1979). Noble and Slatyer (1980) proposed a set of vital attributes applicable to secondary plant succession. Such attributes include recovery after disturbance and the ability of plants to establish themselves and reproduce on the site. Recovery involves vegetative spread, germination of seeds from a seed bank, and arrival of seeds dispersed to the site by wind and animals. Establishment involves the species' degree of tolerance to light and competition and the time a species needs to reach its critical life stage and replacement. What species become established is largely subject to chance. Species composition over time is determined by development, longevity, and response to competition.

This approach to succession emphasizes the role of population dynamics and life history patterns. Succession in its earliest stages begins with an open site that is eventually colonized by early successional species, variously called pioneer, opportunistic, or fugitive (r species). On primary sites, propagules of colonizing species arrive by means of wind, water, or animals. On secondary sites, residual propagules—seeds lying dormant in the soil, roots, and rhizomes—as well as dispersed seeds are important in colonization. The colonizers establish themselves and grow.

Successful early colonists have characteristics that enable them to establish quickly on open sites. They are generally small and low-growing, have short life cycles, reproduce annually by seeds, or send out new shoots from buds near the ground. They produce large numbers of easily dispersed small seeds. They can remain dormant for a long time in the soil, waiting for favorable conditions to germinate. These plants respond quickly to disturbance, especially exposure of mineral soil. They grow rapidly and put most of their production into photosynthetic tissue. They attain dominance quickly by suppressing for a while the growth of any later-stage plants that might exist as seedlings beneath them. They are tolerant of fluctuating environmental conditions, particularly a wide range of daily temperatures on the soil's surface, alternate wetting and drying, and intense light.

Early colonists soon lose their temporary dominance—in part because they have to renew their photosynthetic biomass each year—to the taller, more vigorous growth of later-stage plants, which carry over biomass from year to year. These plants gradually assume dominance and place a much greater demand on resources such as nutrients, light, and moisture. As availability of resources decreases and demand increases, competition within and among populations of plants becomes more intense. Mortality, reflected in self-thinning, increases. As the population declines, resources are released for use by plants of still later stages (Peet 1981). Species characteristic of later stages of succession grow more slowly and are longer-lived. They are able to dominate the site over a much longer period of time because much of their production goes into storage and maintenance. The species mostly are specialists, adapted to a narrow range of

environmental conditions in which the plants either hoard resources or use them more effectively.

Thus population ecologists regard the successional process as one involving differences in colonizing ability, growth, and longevity of species adapted to grow along a gradient of changing environmental conditions. As environmental conditions such as light and nutrients change, species composition also changes. The replacement of one or several species or groups of species by others results in part from interspecific competition, which permits one group of plants to suppress slower-growing species. As plants of earlier stages are supplanted by species of later stages, the structure of the community, as dictated by growth forms and longevity of plants, also changes. Eventually succession arrives at a point where long-lived species create a relatively stable community.

The major argument between the two approaches, both of which incorporate many of the basic features outlined by Clements, is over the process of succession. The holistic approach holds that succession depends upon emergent properties of plant communities and ecosystems. The reductionists hold that succession is the consequence of the properties of individual plants.

## Connell-Slatyer Models

Holistic and reductionist approaches both appear in models of succession advanced by Connell and Slatyer (1977). They proposed three different models of succession: facilitation, inhibition, and tolerance. The **facilitation model,** basically a Clementsian approach, is autogenic. Changes are brought about from within by the organisms themselves. Early stage species modify the environment and prepare the site for later stage species, facilitating their success. The **inhibition model,** is purely competitive. No species is competitively superior to another. The site belongs to those species that become established first and are able to hold their position against all invaders. They make the site less suitable for both early and late successional species. As long as they live, they maintain their position. The ultimate winners, however, are the long-lived plants, even though early successional species may suppress later stage species for a long time. Such sucession is not orderly and is less predictable. The **tolerance model,** intermediate between the two, involves the interaction of life history traits, especially competition. It suggests that later successional species are neither inhibited nor aided by species of earlier stages. Later stage species can invade a site, become established, and grow to maturity in the presence of those preceding them. They can do so because these later species have a greater tolerance for a lower level of resources than the earlier ones. Such interactions lead to communities composed of those species most efficient in exploiting resources either by interference competition or by using sources unavailable to other species. These three models emphasize some points of Clements' original theory.

# Mechanistic Models

The models presented above explain the shifting patterns of species dominance through succession in terms of features of the ecosystem or community, or of differences in the rates of population recruitment and growth, and competitive ability. These responses reflect the aggregated properties of individual plants interacting with the environment in time and space.

Recent experimental and computer simulation studies have taken a more mechanistic approach to the temporal and spatial dynamics of communities and ecosystems. A mechanistic approach considers both the direct process by which an interaction, such as competition, occurs and how the physiology, morphology, and life history traits of individual species influence those processes (Tilman 1987, 1988; Huston and Smith 1987).

**Resource Ratio Model** One such approach is the **resource ratio model** advanced by Tilman (1985, 1988). This hypothesis has two components: interspecific competition for resources, and limiting resources, particularly soil nutrients and light. Succession comes about as the relative availability of those resources changes through time. The resource gradient ranges from habitats with soils poor in nutrients but with a high availability of light at the soil surface (early sucessional sites) to habitats with nutrient-rich soils and low availability of light (late successional sites). Community composition changes along that gradient as the availability of two or more limiting resources, particularly nitrogen and light, changes. Species reach an equilibrium with the supply rates of the limiting resources. In doing so, they lower the available resources to a point at which other species cannot invade.

In early primary succession, the colonizing species are those adapted to a low soil nutrient and high light regime. As biogeochemical processes make more soil nutrients available, plant growth increases, reducing the availability of light at the soil surface. The changing ratio of soil nutrients to light leads to the replacement of one plant species by another. Over time these changes favor plants adapted to high nutrient and low light availability at the soil surface. Differences in competitive abilities are the result of tradeoffs in characteristics required to exploit aboveground or belowground resources needed to allocate carbon to leaves and roots. Because of the slowness of primary succession, succession along the gradient arrives at various plateaus of equilibrium dominated by species competitively superior at each ratio.

Secondary succession flows in much the same pattern, but at a more rapid rate. Species composition and rapidity of change depend upon at which point on the gradient the species colonize the area. For example, old field succession often involves a low level of soil nutrients and high light on the soil surface that supports early successional herbaceous plants. However, where a seed source is available and a site has both high soil nutrients and high light, the area may be colonized by such pioneering shade-intolerant tree species as pines and yellow-poplar. They exclude early successional herbaceous and woody species by shading.

**Individual-Based Plant Model** This reductionist approach proposes that succession is a population process involving competition. Competition results from inversely correlated traits such as maximum growth rate and tolerance for a low level of light. Those species that do best in high light conditions outcompete shade-tolerant species but lose their competitive advantage under lower light.

Although competition is viewed as a population process, it actually takes place among individuals of the same and other species. The ability of an individual to compete is constrained by individual traits based on a suite of life history and physiological attributes (Table 30.1). There are two components to plant interactions. One is the response of the individual to the prevailing environment, such as light and moisture. The other is the influence of the individual on light and moisture in the environment. Thus the outcome of plant interactions is a result of plants' modifying the environment. These interactions influence the direction of succession.

Huston and Smith (1987) developed an individual-based mechanistic model for plant succession. Their model is

**Table 30.1 Physiological and Life History Characteristics of Early and Late Successional Plants**

| Characteristic | Early Succession | Late Succession |
|---|---|---|
| Photosynthesis | | |
|   Light saturation intensity | high | low |
|   Light compensation point | high | low |
|   Efficiency at low light | low | high |
|   Photosynthetic rate | high | low |
| Respiration rate | high | low |
| Water-use efficiency | | |
|   Transpiration rate | high | low |
|   Mesophyll resistance | low | high |
| Seeds | | |
|   Number | many | few |
|   Size | small | large |
|   Dispersal distance | large | small |
|   Dispersal mechanism | wind, birds, bats | gravity, mammals |
|   Viability | long | short |
|   Induced dormancy | common | uncommon? |
| Resource acquisition rate | high | low? |
| Recovery from nutrient stress | fast | slow |
| Root-to-shoot ratio | low | high |
| Mature size | small | large |
| Structural strength | low | high |
| Growth rate | rapid | slow |
| Maximum life span | short | long |

*Sources:* Huston and Smith 1987.

derived from two widely used individual-based models (Botkin et al. 1972, Shugart and West 1977). The individual-based model simulates annual population dynamics of trees on a defined area by considering birth, growth, and death of individual plants. Each species is defined by given species-specific life history traits.

Huston and Smith gave their species five specific traits: maximum size (height and dbh), maximum age, growth rate, tolerance for shade, and rate of seed production. They incorporated light as the limiting resource. The model is based on competition among individual plants to explain species replacement on a spatial and temporal gradient. It assumes that both relevant environmental resources (in this case light) and the intensity of competition change through time and among communities. The hypothesis is based on three premises: (1) As plants grow, they alter the environment in such a way that the relative availability of resources changes, altering the rules for competitive success. (2) The physiological traits of plants prevent any one species from achieving maximum competitive ability under all circumstances. (3) The interaction between 1 and 2 produces an inverse correlation between certain groups of traits, so species that are good competitors under one suite of environmental conditions are poor competitors under another.

By means of computer simulation, Huston and Smith demonstrated the various patterns of vegetational change that result from different combination of traits. The outcome of a simulation involving five species with assigned life history traits is summarized in Figure 30.4. Note that regardless of the initial combination, the large, slow-growing, long-lived species dominates the late successional stage and that in each case the pattern of vegetation changes through time. When other resources are incorporated into the model, the vegetational pattern becomes more diverse over ecological time (T. M. Smith and Huston 1987).

Both of the above models examine successional dynamics by focusing on the life history and physiological properties of individual organisms interacting under changing environmental conditions with a community. For this reason the models are able to generate the many described successional patterns at the population and community levels. Mechanistic models bring the investigation of succession full circle from the observations of Drury and Nisbet (1973). They pointed out, "The basic cause of the phenomenon of succession is the known correlation between stress tolerance, rapid growth, small size, short life, and wide dispersal of seed."

# THE CLIMAX

According to classical ecological theory, succession stops when the sere has arrived at an equilibrium or steady state with the physical and biotic environment. At this point the community is stable and self-replicating. Barring major disturbances, it will persist indefinitely. This end point of succession is termed the **climax.**

The climax, theoretically, takes on certain characteristics. The vegetation is tolerant of the environmental conditions it has imposed upon itself. It is characterized by an equilibrium between gross primary production and total respiration, between energy used from sunlight and energy released by decomposition, between the uptake of nutrients from the soil and the return of nutrients by litterfall to the soil. It has a wide diversity of species, a well-developed spatial structure, and complex food chains.

Individuals in the climax stage are supposedly replaced by others of the same kind. Thus the species composition maintains an equilibrium. If offspring of the same species are favored over others, then a dead mature individual may be replaced by a plant of its own kind. If the offspring are concentrated about a mature parent, it may be replaced by its own progeny. This situation is most likely if the replacement is essentially the same unit—a root or stump sprout from the dead individual. However, if conditions beneath the mature tree are less favorable for its own species than for other species, it will be replaced by one of another associated species. If conditions are neither more nor less favorable for the offspring than for other species, replacement individuals will be influenced by the relative abundance of propagules arriving on the site, by suppressed individuals already present, and by competitive interactions among them.

There are three theoretical approaches to the climax. One is the **monoclimax** or **climatic climax theory,** advanced by Clements (1916). This theory recognizes only one climax, whose characteristics are determined solely by climate. Successional processes and modifications of the environment overcome the effects of differences in topography, parent material of the soil, and other factors. Given sufficient time, all seral communities in a region will converge to and stabilize at a single climax. The whole landscape will be clothed with a uniform plant community. Communities other than the climax are related to it by successional development and are recognized as subclimax, postclimax, disclimax, and so on.

Tansley (1935) advanced the **polyclimax theory.** It proposes that the climax vegetation of a region consists of a mosaic of vegetation climaxes controlled by soil moisture, soil nutrients, topography, slope exposure, fire, and animal activity (see Daubenmire 1968a, b; Whittaker 1974). The spatial pattern of habitats influences the spatial pattern of climax communities.

Whittaker (1953) proposed a third theory, the **climax pattern theory.** The total environment of the ecosystem determines the composition, species structure, and balance of a climax community. Involved in the environment are the species responses to moisture, temperature, and nutrients, their biotic relationships, availability of flora and fauna to

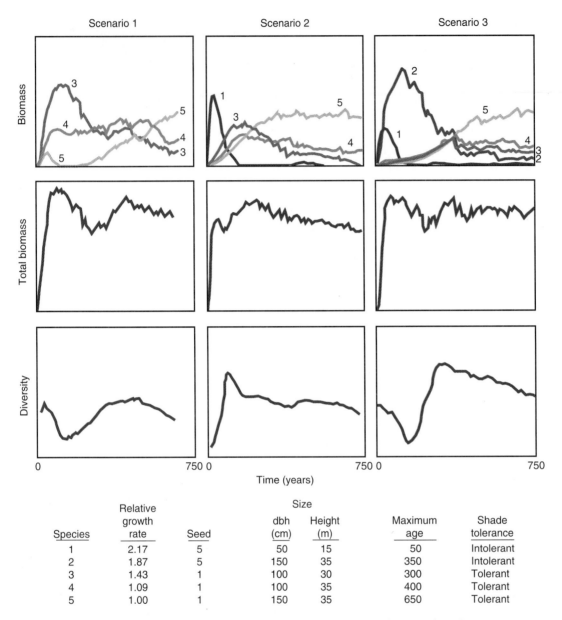

**Figure 30.4** Simulation results for an individual-based model of succession. In scenario 1, the three species (3, 4, and 5) have late successional characteristics but differ sufficiently in relative competitive abilities to produce a typical successional replacement. In scenario 2, an early succes-sional is added to the three species in scenario 1. The early successional species is a stronger competitor early on. In scenario 3, an intolerant "superspecies" (species 2) with a high growth rate and large size is added to the four species in scenario 2. The superspecies is competitively supe-rior for a long period of time, inhibiting the growth of more tolerant species 3, 4, and 5. It retains a position in the community long after it has lost dominance to the tolerant species. An example of such a species is yellow-poplar. Note that the curves of abundance are skewed, not bell-shaped. (From Huston and Smith 1987:184.)

colonize the area, chance dispersal of seeds and animals, soils, climate, and disturbances such as fire and wind. The mosaic of climax vegetation will change as the environment changes. The climax community represents a pattern of pop-ulations that corresponds to and changes with the pattern of environmental gradients to form ecoclines. The central and most widespread community is the prevailing or climatic climax.

The climax pattern theory recognizes a variety of cli-maxes governed by responses of species populations to bi-otic and abiotic conditions along an environmental gradient. Climate does control vegetation over a broad region (such as

the boreal forest and grasslands). However, because environmental conditions, including climate, are variable, rarely do communities achieve equilibrium with their environment. Climax vegetation, then, is an abstract concept seldom realized.

## Cyclic Succession

Vegetation is dynamic and mortality is a feature of the climax. Trees grow old and die and more often than not are replaced by individuals of a different species. Such changes are constantly occurring in patches across the community. Although succession may slow down, it never ceases. This concept was presented by A. S. Watt (1947) in a classic paper, "Pattern and Process in the Plant Community."

Successional stages within a community that appear to be directional are often phases in a cycle of vegetation replacement. Death of vegetation or periodic disturbance starts regeneration again at some particular stage. Such changes usually occur on a small scale within a stable community, and are repeated over the whole of it. Each successive community or phase is related to the others by orderly changes in an upgrade and downgrade series (Figure 30.5). Such cyclic replacements contribute to community persistence.

Cyclic succession is a common and important phenomenon in communities such as old fields (Evans and Cain 1952), desert scrub (Yeaton 1978), coastal tundra in Alaska

(Webber et al. 1980), and dwarf heather in Scotland (Watt 1955). Cyclic replacement, depending heavily upon allogenic succession driven by rainfall as well as on autogenic succession, maintains the long-term stability of pothole marshes in north-central North America (Figure 30.6). During periods of drought—about every 5 to 20 years—shallow marshes dry. Organic debris accumulated on the bottom decays rapidly, releasing nutrients for recycling and stimulating the germination of seeds. The upgrade of the cycle begins with seed germination on exposed mud. That is followed by (1) a newly flooded stage with sparse, often well-dispersed vegetation, dominated by annuals and immature perennials; (2) a flooded, dense marsh dominated by perennials; and (3) a deep open marsh rimmed with emergents, in part caused by the feeding activities of muskrats. The cycle begins anew when the marsh dries.

A similar cyclic succession takes place in the climax forest community. Shade-tolerant sugar maple and beech and the early successional yellow birch (*Betula allegheniensis*) are three dominants in the northern hardwood forests of New Hampshire. Rarely do seedlings and saplings of each of these species grow beneath overstory parents. Beech seedlings and saplings are most commonly associated with a sugar maple overstory, and sugar maple with a yellow birch overstory. Yellow birch seedlings are rather widely distributed throughout the forest. Forcier (1975) has hypothesized that a cyclic succession of disturbance—yellow birch–sugar

**Figure 30.5** Cyclic replacement in an old field community in Michigan. The bare areas at the bottom of the downgrade are invaded by moss to start the upgrade series. Mosses are invaded by Canada bluegrass (*Poa compressa*) and dock (*Rumex acetosella*). The accumulated dead leaves of these plants are covered by lichens that crowd out the grass. Rain, frost, and wind destroy the lichens to start the cycle with bare ground again.

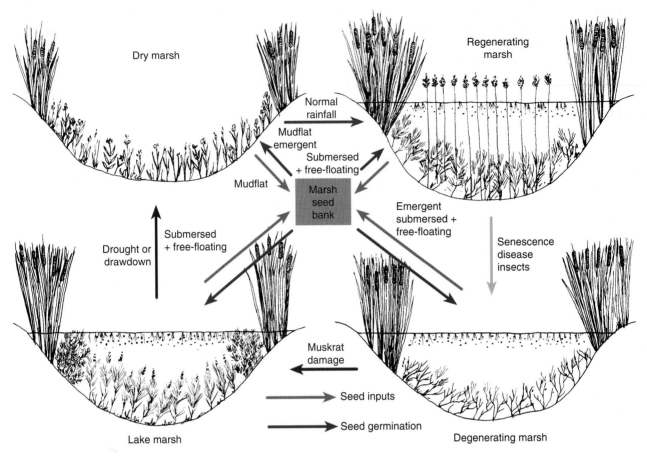

**Figure 30.6** Cyclic replacement of vegetation in a prairie glacial marsh. The cycle is initiated by periods of drought followed by periods of normal rainfall, but the key to replacement is the seed bank in the marsh mud. (After van der Valk and Davis 1978.)

maple–beech—will occur within the forest. When a beech or sugar maple dies, its place in the canopy will not be taken immediately by another beech, nor a sugar maple by another. Instead, yellow birch will fill the opening in the canopy. When the short-lived yellow birch dies, sugar maple will fill the gap. Eventually it will be replaced by beech. Such small-scale succession within a forest has also been called **reciprocal replacement.**

## Shifting Mosaic

Studies of old or mature ecosystems needed to test the climax theory are rare, especially in the eastern deciduous forest, where extremely few old untouched stands exist. One is the Dick Cove Natural Area, a mesic forest on the western slope of the Cumberland Plateau in Franklin County, Tennessee. It is a 100-acre tract surrounded by second growth forest. The forest dominants are northern red oak, white oak, chestnut oak, hickories, sugar maple, and yellow-poplar, associated with some 17 other species. McGee (1984) studied the mortality and succession in this old-growth forest. He found that during a nine-year period from 1972 through 1981, 26 percent of the hickory and 18 percent of white oak and red oak over 43 cm (17 inches) dbh died from a combination of senescence, drought, insect damage, and blowdown. These dead trees ranged in age from 90 to 375 years. Dead red oak averaged 135 years and dead white oak, chestnut oak, and hickory averaged 210 years old. The demise of the oaks after 350 years appears to be accelerating. It is bringing about a change in structure and species composition of the stand from the original apparent climax of oak and hickory to one of the slow-growing, shade-tolerant sugar maple and the fast-growing, shade intolerant yellow-poplar with a strong component of hickory. Red oak lacks replacement trees and yellow-poplar is filling in the gaps. This study emphasizes that even old-growth, apparent climax forests are in a stage of flux observable only over periods of several hundred years. What appears permanent to the observer is slowly transitory.

This point is even more applicable to 450- to 1000-year-old stands of Douglas-fir of the Pacific Northwest. Technically, these old Douglas-fir forests are not climax, because

Douglas-fir is a pioneer species that happens to be very long-lived (Franklin and Hemstrom 1981). The climax species that might replace Douglas-fir in another 500 years are western hemlock (*Tsuga heterophylla*) and Pacific silver fir (*Abies amabilis*). Only by human standards of permanency are these old stands "climax." These old Douglas-fir stands are uneven-aged when, theoretically, they should be even-aged, with all trees in the stand coming in at the same time after a major disturbance. Age classes range from 145 years on up within the stand, suggesting occasional disturbances and replacement, or perhaps failure of canopy closures over a period of time because of variable seed crops.

Although the variation in age classes and the pattern of species composition within both the Tennessee and Pacific Northwest stands change, their average physiognomy remains the same. This observation suggests a **shifting-mosaic steady state,** a concept advanced by Bormann and Likens (1979). Rather than achieving equilibrium, forest vegetation achieves a state in which the standing crop of living and total biomass (living + dead) fluctuates about a mean. Such an ecosystem consists of patches in various stages of seral development, from ones exhibiting high net production and biomass accumulation to ones of senescence and downed timber in which respiration exceeds production. The total production of these patches remains more or less the same throughout time. In the ancient Douglas-fir forest gross production is high, but so is respiration. Although net production is low, it never achieves equilibrium with respiration. Biomass continues to accumulate on individual trees, however slowly. Ratio of live biomass to dead biomass peaks at 300 to 400 years. Total biomass is around its highest at about 750 years, and total dead biomass is greatest at 800 to 1000 years. The forest is a complex system that includes many species of saprophytes and epiphytic lichens, as well as plants and animals. Nutrient retention within the system involves complex detrital pathways. Nitrogen is fixed by lichens in tree crowns and by microflora in decomposing logs. Structural diversity is high, with a large size range of individual living trees and numerous large standing dead trees, large downed trees, and decomposing logs.

# FLUCTUATIONS

Fluctuations are nonsuccessional or short-term reversible changes (Rabotnov 1974). Fluctuations differ from succession in that the floristic composition over time is stable. No new species invade the site, and changes in dominants may be reversible. These changes in floristic composition result from such environmental stresses as soil moisture fluctuations, wind, grazing, and the like.

Fluctuations in forest communities may involve an alternation of species in canopy gap replacements (reciprocal replacement). In such forests there appears to be a tendency for each species to be replaced by its competitor (J. F. Fox 1977). If one species becomes moderately abundant in the canopy, alternate species may be abundant beneath it. Thus over time dominance in the canopy may eventually shift in favor of the temporarily disadvantaged species. For example, J. F. Fox (1977) found that in old northern hardwood stands, sugar maple tends to replace beech in small openings and beech replaces sugar maple (see Forcier 1975, discussed earlier). In general, the tendency is for the dominant tree to be replaced more than half the time by its competitor. Such alternation probably results for two reasons. (1) The dominant tree usurps the site, concentrating the bulk of biomass at one particular place in a single tree. Its conspecifics are thinned out more severely than its competitors. (2) Because of its influence on nutrient regeneration and light and moisture regimes, the canopy tree creates a somewhat species-specific microhabitat for seeds and seedlings beneath it. With an alternate species favored in the understory, the forest maintains a species equilibrium.

Fluctuations may also involve replacement of one age class by another within a species. Such fluctuations are important in maintaining certain forest ecosystems, particularly coniferous forests (Korchagin and Karpov 1974). Sprugle (1976) describes a wave regeneration pattern in balsam fir (*Abies balsamae*) forests in the northeastern United States. Trees die off continually at the edge of a "wave" and are replaced by vigorous stands of young balsam fir (Figure 30.7). Openings in the forest, exposing trees to the wind on the lee-

**Figure 30.7** Cross section through a regeneration wave in a balsam fir forest. The wave is initiated at the location of standing dead trees with mature trees beyond and vigorous reproduction below. In the area where dead trees have fallen, a crop of young fir seedlings is developing. Beyond them is a dense stand of fir saplings, followed by a mature fir forest and then by a second wave of dying trees. (After Sprugle 1976.)

ward side of the opening, initiate the cycle. Desiccation of the canopy foliage by winter winds, the loss of branches and needles in winter from rime frost forming on them, and decreased primary production due to cooling of needles in summer cause the death of trees. Their death exposes the trees behind them to the same lethal conditions and they die. This process continues so that a wave of dying trees through the forest is followed by a wave of vigorous reproduction.

These regeneration waves follow each other at intervals of about 60 years. The process is so regular that all stages of degeneration and regeneration are found in the forest at all times, provided the stand is not cut. The phasic cycle results in a steady state because the degenerative changes in one part of the forest are balanced by regenerative stages in another. The wave regeneration process ensures the stability of the forest and prevents its advancement into a hardwoods stage.

# CHANGES IN ECOSYSTEM ATTRIBUTES

As succession proceeds and species composition and community structure change, functions and other emergent properties also change. In 1942, R. Lindeman postulated that productivity should increase and the efficiency of transfer and use of energy and nutrients improve as succession proceeds. This postulate was modified and expanded by Margelef (1963, 1968) and E. P. Odum (1969) into the concept of maturity: that structural complexity and organization of an ecosystem increase and mature with time (Table 30.2).

Succession may begin with a bare area colonized by small plants and terminate with large plants whose growth form results in increased vertical stratification and increased influence on environmental variables within the community. These changes and others, according to Margelef and Odum, are found during the progression of young or early successional ecosystems to mature or late successional ecosystems.

Early successional stages are characterized by few species, low biomass, and dependence on an abiotic source of nutrients. Net community production is greater than respiration, resulting in increased biomass over time. Energy is channeled through few pathways to many individuals of a few species, and production per unit biomass is high. Food chains are short, linear, and largely grazing. Accompanying these changes is an increase in species diversity.

The mature stage in succession is characterized by high biomass, a nutrient source largely organic in nature, and gross production that about equals respiration. Food chains are complex and largely detrital, as opposed to grazing. Inorganic nutrients accumulate in the soil and vegetation, and considerable quantities are locked or hoarded in plant tissue. Fundamental niches shift from broad and general to narrow and specialized. Accompanying these changes is an increase in species diversity.

How well does succession conform to these generalizations? Studies of ecosystem functions provide some insights. Consider species diversity. Diversity in succession can be considered from two points of view, diversity across an environmental gradient (spatial) and diversity through time (temporal) (Auclair and Goff 1971). Diversity may or

**Table 30.2** Expected Trends in Succession from Development to Mature Systems

| Attribute | Trend | Accept or Reject |
|---|---|---|
| Biomass | Increases | A |
| GPP/ER | Approaches 1 | R (plants) |
| GPP/B | Decreases | A |
| B/ER | Increases | A |
| Net community production | Decreases | R (plants) |
|  |  | A (animals) |
| Total organic matter | Increases | A |
| Inorganic nutrient input | External to internal | ? |
| Species richness | Increases | A |
| Species equitability | Increases | R |
| Stratification | Increases | R |
| Size of organisms | Increases | R |
| Niche specialization | Broad to narrow | ? |
| Role of detritus | Increases in importance | ? |
| Growth form | $r$ to $K$ | R |
| Nutrient conservation | Increases | R |

*Note:* GPP is gross primary production; ER is ecosystem respiration; B is standing crop biomass.
*Accept or Reject:* If trends of attributes are considered hypotheses, they can be tentatively accepted or rejected based on current data. A is Accept; R is Reject.
*Source of attributes:* Odum 1981:446.

may not increase with advancing successional stages (Figure 30.8). Some early stages of succession may have a greater diversity of vegetation than later stages (Bazzaz 1975), because plants with strong allelopathic interference may dominate later stages, reducing species diversity. Sassafras (*Sassafras albidum*), for example, maintains itself in relatively pure stands by releasing into the soil at different times of the year phytotoxins that inhibit germination of seeds and growth of other plants (Gant and Clebsch 1975). Also affecting diversity is the initial state of the site. Auclair

and Goff (1971) concluded that pioneering forest communities on xeric sites increase in diversity through time (Figure 30.9), whereas successional forest communities on mesic sites decline in diversity from late successional to equilibrium forest. On intermediate sites diversity approaches an asymptote late in succession. Even within this framework, soil type and fertility, microtopography, disturbance, and grazing by herbivorous animals will affect diversity on a given site. Diversity of bird species and trees increases with succession but reaches a maximum before the mature stage.

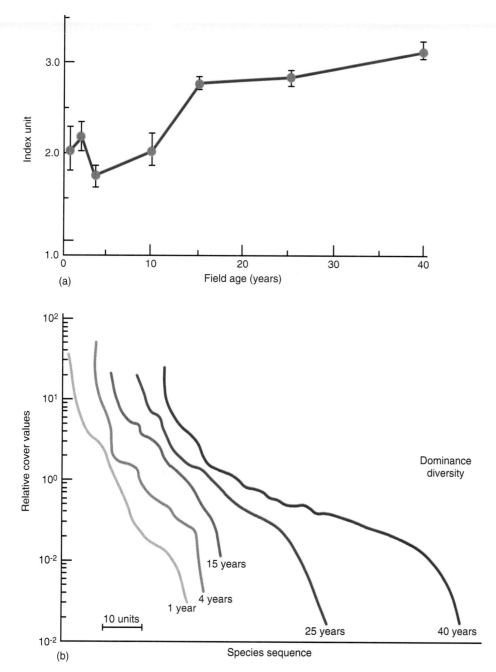

(a)

(b)

**Figure 30.8** Relationship between plant species diversity and succession. (a) Plant species diversity generally increases with succession and may reach a maximum in the forest stage when shade-tolerant and shade-intolerant guilds are present together in the community. Relatively low species diversity in a successional community (as in the example graphed here) may result from development of strong dominance by a species with allelopathic interference or an otherwise strong competitor. (b) Dominance-diversity curves of a successional community are geometric at first, suggesting the niche-preemption theory. Dominance curves become less steep with time as more species are added and gradually a log-normal distribution develops, with an increase in species with intermediate relative importance values. (From Bazzaz 1975.)

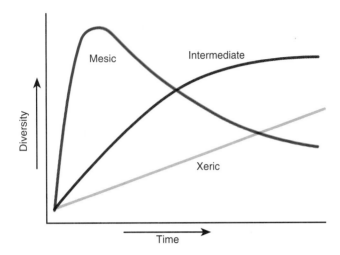

**Figure 30.9** Species diversity in relation to time for xeric, intermediate, and mesic moisture conditions. Note that diversity increases with time on xeric and intermediate sites but decreases on mesic sites, which reach their maximum diversity in earlier stages of succession. Diversity in later stages is maximal on intermediate sites. (From Auclair and Goff 1971.)

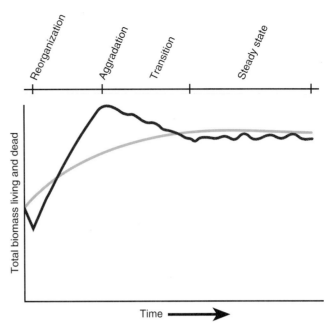

**Figure 30.10** Model of biomass accumulation during successional development of a forest after clearcutting. The yellow line is an asymptotic model of net biomass accumulation (living and dead) until a steady state is achieved, as predicted by the ecosystem theory of succession. The brown line is observed biomass accumulation as predicted by a shifting-mosaic steady-state model. (From Bormann and Likens 1979:166.)

Other observed attributes also fail to match generalizations. This deviation is suggested by results obtained from 15 years of data collected from a developing hardwoods forest at Hubbard Brook, New Hampshire. Bormann and Likens (1979) divided the seral stages of forest recovery following clearcutting into a **reorganization** or young seral stage; an **aggradation** or developmental seral stage in which the growing forest is rapidly accumulating biomass; a **transition** phase, when growth is beginning to slow and a number of trees die from competition and self-thinning; and finally a mature or **steady state** phase. Data for ecosystem function in the reorganization and early aggradation phase are actual; those for later stages are simulations based on actual data using forest growth models.

The ratio of production to respiration (P/R) did not gradually increase through time from less than 1 to 1. Instead, the P-R ratio for the reorganization stage and the transition stage was less than 1 and for the steady state, almost exactly 1. Only in the aggradation phase was the P-R ratio greater than 1. These changes in the P-R ratio reflected the loss of biomass through decomposition of dead organic matter following clearcutting in the reorganization phase and death of trees in the transition phase.

Gross production increased through all stages, and net production was highest in the aggradation phase, which led to a rapid accumulation of biomass. Thus biomass accumulation was highest in the aggradation phase. It declined during the transition stage and leveled off during the steady-stage phase (Figure 30.10). Nutrient cycling was slow and inefficient during the reorganization phase, when quantities of nutrients were being exported from the system. The rea-

son was rapid decomposition of detrital material and the inability of incoming vegetation to use all the resources. Nutrient cycling was most rapid and efficient during the aggradation phase, with little loss from the system. Habitat diversity was highest during the aggradation phase, with its mix of shade-intolerant, shade-tolerant, and intermediate species.

# TIME AND DIRECTION IN SUCCESSION

Time is an integral component of succession. Time, however, is relative, measured in terms of human experience. Theoretically, climax vegetation is regarded as stable with a degree of permanency; but what is permanent? Is it vegetation that remains the same over several human lifetimes, as the old growth forests? By that standard, old field vegetation could be a climax community to ants, birds, and meadow mice, because the vegetation would extend over several lifetimes of those organisms.

Nevertheless, successional communities have their life spans, governed in part by the longevity of the plants that comprise the seral stages. The annual-weed stage in an old

field may last no longer than one or two years. Pioneer lichens and moss on a granite outcrop may remain for hundreds of years. Seral grass stages may last 10 to 15 years or even fewer before being overtaken by woody growth. Woody growth in a shrub stage—whether true shrubs or incoming tree growth below a height of 6 m—may last an additional 10 to 15 years until pioneering trees take over or the canopy closes. If the trees are pioneer or shade-intolerant species such as aspen, the stage may last 25 to 40 years before shade-tolerant species become dominant and hold the site 250 to 500 years.

The time line implies that shade-tolerant species replace shade-intolerant ones and that trees replace shrubs. Not always. A dense growth of shrubs such as meadowsweet, mountain laurel, or Saint-John's-wort may dominate a site for 60, 70, 80 years without any indication of change. Shade-intolerant species such as yellow-poplar and red pine can hang onto a site and maintain their positions for many years. Douglas-fir, a pioneer species in western North America, may dominate a site for more than 1000 years before giving way to western hemlock, which may need another 500 years to achieve dominance.

How long a seral stage lasts could be simply an academic question, except that time has implications in forestry and wildlife management. Certain types of wildlife habitat are seral and therefore ephemeral. Their maintenance may be critical to the welfare of certain species, which requires human interference with the successional process. Certain commercially valuable timber trees are pioneering species that require periodic disturbance to ensure regeneration.

Classical successional theory holds that through time succession is directional and therefore predictable. For succession to operate it has to head somewhere, even if in a circle (cyclic succession). The concept of succession, however, implies unidirection—a one-way trip. In that case, the answer to the question of whether succession is directional would have to be a qualified no.

We could predict with a high degree of probability that an old field in eastern North America will, barring further disturbance, return to forest. We would have much more difficulty predicting the kind of forest, even if we had knowledge of previous vegetation. Each successional community, climax and otherwise, is individualistic, a one-time product of abiotic and biotic forces operating during its development. The exact interaction of these forces will not be repeated again. Any new successional community will be molded by current abiotic and biotic inputs. The exact original composition of species will not be duplicated. Thus, we might predict successfully the physiognomy over a broad region but miss the successional communities on a local level, in spite of recent attempts to do so (see Horn 1975, 1981). There are too many side roads succession can take, influenced by environmental conditions, stochastic dispersal of species onto a site, and biotic interactions that do not lead to the supposed climax (Figures 30.11, 30.12).

# SUCCESSION AND ANIMAL LIFE

As seral stages change, animal life also changes (Figure 30.13). Because animal life is influenced more by structural characteristics of vegetation than species composition, each stage has its own distinctive group of animals. For this reason successional stages of animal life might not correspond to the successional stages identified by plant ecologists. Animals behaviorally classify a young stand of yellow-poplar or balsam fir under 6 m tall as a shrub stage community; a plant ecologist would consider yellow-poplar an intolerant tree stage and fir a tolerant tree stage.

Early terrestrial successional stages in eastern North America support animals of grasslands and old fields, such as meadowlarks, meadow voles, and grasshoppers. Invasion of scattered woody plants adds a new structural element. Meadowlarks decline and field sparrows and song sparrows appear. When tall woody vegetation, whether shrubs or young trees, eventually claims the area, shrubland animals move in. Field sparrows decline and the thickets are claimed by towhees, catbirds, brown thrashers, and goldfinches. Meadow mice give way to white-footed mice. When woody growth exceeds 6 m in height and the canopy closes, shrub-inhabiting species decline, to be replaced by birds and insects of the forest canopy. As the community matures and more structural elements are added, new species appear, such as tree squirrels, woodpeckers, and birds of the forest understory like hooded warblers and wood thrushes.

A number of animal species are highly dependent on specific successional stages. Grasshopper sparrows need large areas of open grassland. The future of this bird depends upon maintenance of grassland. The golden-winged warbler (*Vermivora chrysoptera*) depends upon on early successional shrubland that ultimately converts to forest. As the shrubland disappears, so does the warbler (Confer and Knapp 1981). The American woodcock requires three early successional types: abandoned fields and forest openings, alder, aspen, and young hardwood thckets; and young, open second-growth hardwoods (Sheldon 1967). Decline and loss

**Figure 30.11** (opposite) Soil conditions, particularly moisture, influence species composition and pattern of succession. This diagram illustrates plant succession on moderately drained to well-drained soils in a high elevation valley, Canaan Valley, West Virginia. Succession proceeds along several directions, resulting in different "climax" communities. The original climax in the valley was spruce, to which it probably will never return. (After Fortney 1975.)

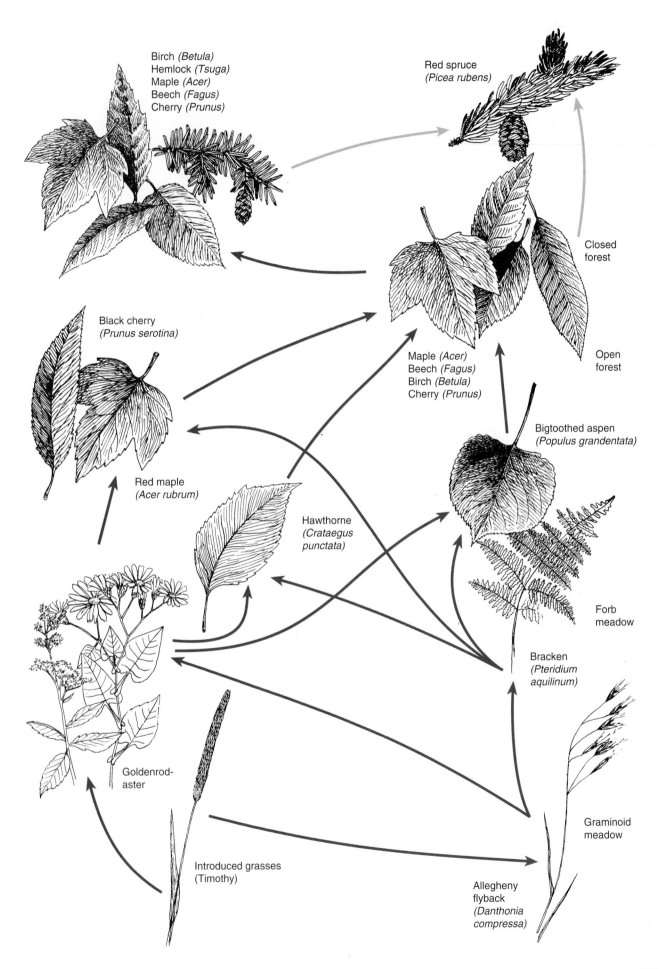

Birch *(Betula)*
Hemlock *(Tsuga)*
Maple *(Acer)*
Beech *(Fagus)*
Cherry *(Prunus)*

Red spruce
*(Picea rubens)*

Closed
forest

Open
forest

Maple *(Acer)*
Beech *(Fagus)*
Birch *(Betula)*
Cherry *(Prunus)*

Black cherry
*(Prunus serotina)*

Red maple
*(Acer rubrum)*

Bigtoothed aspen
*(Populus grandentata)*

Hawthorne
*(Crataegus
punctata)*

Bracken
*(Pteridium
aquilinum)*

Forb
meadow

Goldenrod-
aster

Graminoid
meadow

Introduced grasses
(Timothy)

Allegheny
flyback
*(Danthonia
compressa)*

673

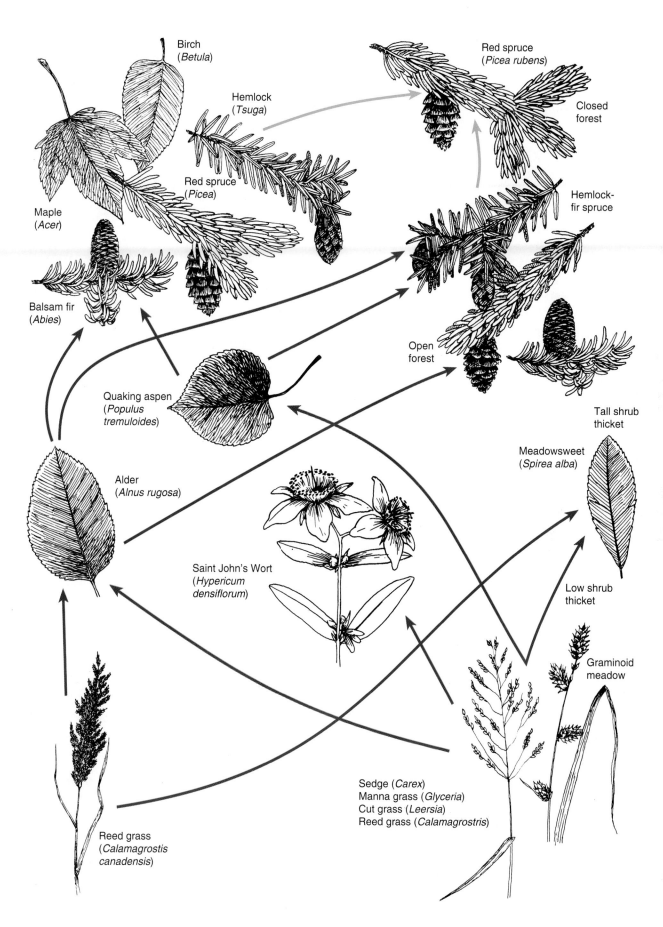

Birch (*Betula*)

Hemlock (*Tsuga*)

Red spruce (*Picea rubens*)

Closed forest

Red spruce (*Picea*)

Maple (*Acer*)

Hemlock-fir spruce

Balsam fir (*Abies*)

Quaking aspen (*Populus tremuloides*)

Open forest

Tall shrub thicket

Meadowsweet (*Spirea alba*)

Alder (*Alnus rugosa*)

Saint John's Wort (*Hypericum densiflorum*)

Low shrub thicket

Graminoid meadow

Sedge (*Carex*)
Manna grass (*Glyceria*)
Cut grass (*Leersia*)
Reed grass (*Calamagrostris*)

Reed grass (*Calamagrostis canadensis*)

**Figure 30.12** (opposite)  Plant succession on wet mineral soil in Canaan Valley, West Virginia, advances along routes different from that of succession on well-drained soils. The terminal community includes a mixed forest of hemlock, balsam fir, red spruce, and maple. Succession to forest can be blocked by terminal shrub communities of either meadowsweet or Saint-John's-wort. (After Fortney 1975.)

of these successional stages is reducing the woodcock population. Other species, such as the spotted owl, inhabit only mature or old growth forests. When they are cut, animal life they support goes, and they are replaced by species associated with early successional stages. Other animals, particularly mammals such as the short-tailed shrew and deer mice, are ubiquitous, inhabiting a range of successional stages.

Diversity of animal life across a range of seral stages varies with the nature of each individual community. In a very general way, shrubland and edge communities and mature stands have a greater diversity of animal life than young forest stands. The key to diversity of wildlife in a given area is the maintenance of a heterogeneous landscape with habitat patches of various successional stages. Each stage must be sufficiently large to support its characteristic species.

In addition to being influenced by successional stages, animal life, particularly grazing and browsing herbivores, influence succession. The exploding population of white-tailed deer in the eastern United States has severely impacted many forested areas. By selectively feeding on seedlings and saplings of oaks, yellow poplar, and maple, and avoiding other species such as black cherry, deer are influencing the future composition of the forest. In other places deer eliminate woody reproduction and covert the site to ferns, grass, and goldenrod (Figure 30.14).

**Figure 30.13** Wildlife succession in large conifer plantations in central New York State. Note how some species appear and disappear as vegetation density and height change. Other species are common to all stages. (After R. L. Smith 1960.)

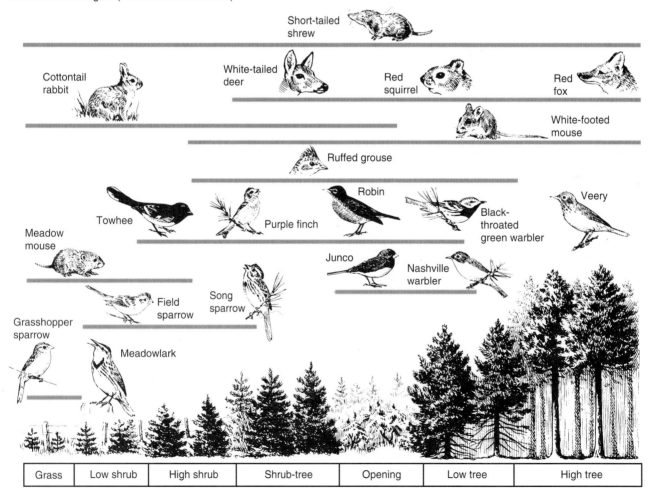

| Grass | Low shrub | High shrub | Shrub-tree | Opening | Low tree | High tree |

**Figure 30.14** An infrared aerial photograph of tree regeneration in a forest clearcut in northwestern Pennsylvania. The area on the right was fenced from deer. It shows a heavy regrowth of forest (red color), whereas the area browsed by deer shows little woody vegetation and much exposed ground (green color).

# DEGRADATIVE SUCCESSION

Within each major community and dependent upon it for an energy source are a number of microcommunities. Dead trees, animal carcasses and droppings, plant galls, tree holes—all furnish a substrate on which groups of plants and animals live, succeed each other, and eventually disappear, becoming in the final stage a part of the nutrient base of the major community itself. In these instances succession is characterized by early dominance of heterotrophic organisms, maximum energy available at the start, and a steady decline or degradation in energy as the succession progresses.

An acorn supports a tiny parade of life from the time it drops from the tree until it becomes a part of the humus (Winston 1956). Succession often begins while the acorn still hangs on the tree. The acorn may be invaded by insects, which carry to the interior pathogenic fungi fatal to the embryo. Most often the insect that invades the acorn is the acorn weevil (*Curculio rectus*). The adult female burrows through the pericarp into the embryo and deposits its eggs. Upon hatching, the larvae tunnel through to the embryo and consume about half of it. If fungi (*Penicillium* and *Fusarium*) invade the acorn simultaneously with the weevil or alone, they utilize the material. The embryo then turns brown and leathery and the weevil larvae become stunted and fail to develop. These organisms represent the pioneer stage.

When the embryo is destroyed, partially or completely, by the pioneering organisms, other animals and fungi enter the acorn. Weevil larvae cut through the outer shell and leave the acorn. Through this exit hole fungi-feeders and scavengers enter. Most important is the moth *Valentinia*

*glandenella,* which lays its eggs on or in the exit hole, mostly during the fall. Upon hatching, the larvae enter the acorn, spin a tough web over the opening, and proceed to feed on the remainder of the embryo and the feces of the previous occupant. At the same time several species of fungi enter and grow inside the acorn, only to be utilized by another occupant, the cheese mites (*Tryophagus* and *Rhyzoglyphus*). By the time the remaining embryo tissues are reduced to feces, the acorn is invaded by cellulose-consuming fungi. The fruiting bodies of these fungi, as well as the surface of the acorn, are eaten by other mites and collembolans and, if moist, by cheese mites too. At this time predaceous mites enter the acorn, particularly *Gamasellus,* which is extremely flattened and capable of following smaller mites and collembola into crevices within the acorn. Outside on the acorn, cellulose and lignin-consuming fungi soften the outer shell and bind the acorn to twigs and leaves on the forest floor.

As the acorn shell becomes more fragile, holes other than the weevil's exits appear. One of the earliest appears at the base of the acorn where the hilum (the scar marking the attachment point of the seed) falls out. Through this hole, larger animals such as centipedes, millipedes, ants, and collembolans enter, although they contribute nothing to the decay of the acorn. The amount of soil in the cavity increases and the greatly softened shell eventually collapses into a mound and gradually becomes incorporated into the humus.

The pattern of succession exhibited in the degradation of the acorn is typical of that found in any type of fresh organic matter, whether it is a log, an animal carcass, or dung. The organisms that first colonize the site are ones that can feed on fresh organic matter. Their feeding activities bring about

physical and chemical changes in the substrate. After they have exploited the energy and nutrients accessible to them, they disappear. Replacing them is a group that is able to extract nutrients and energy left in a less accessible form. One group of organisms follows another. Each group changes the substrate to a point that it can no longer survive there. Eventually the organic matter is degraded. Thus heterotrophic succession follows a pattern similar to that of autotrophic succession, in which changes in the substrate are brought about by the organisms themselves. It differs from from autotrophic succession, however, in that energy is degraded and not accumulated in organic biomass.

# PALEOSUCCESSION

Succession as experienced by humans takes place over a very short period in terms of Earth's history. Changes also take place on a grander time scale. Since its inception some 4.6 billion years ago, Earth has changed profoundly. Land masses drifted northward and broke into continents. Mountains emerged and eroded; seas rose and fell; ice sheets advanced to cover large expanses of the Northern and Southern hemispheres and retreated. All these changes affected climate and other environmental conditions from one region of Earth to another. Many species of plants and animals evolved, disappeared, and were replaced by others. There were major shifts in vegetation patterns. What life was like in the past and under what conditions it existed, we infer from present-day conditions. Conversely, as the geologist Lyell remarked, the past is the key to the present. The key to distributions of animals and plants today may be found in the past.

Records of plants and animals composing past communities lie buried as fossils: bones, insect exoskeletons, plant parts, and pollen grains. These fossils enable us to determine plant and animal associations of the past and in a broad way the climatic changes that brought about the gradual destruction of one type of community and the emergence of another. Such interpretation is based on the assumption that organisms of the past possessed ecological requirements similar to those of related species living today. For example, if modern palms and broadleaf evergreens are tropical plants, we assume that their ancient prototypes also lived in a tropical climate. The study of the relationships of ancient flora and fauna to their environment is **paleoecology.**

## The Pleistocene Epoch

Of particular interest to the paleoecologist are the climatic and vegetation changes that followed the advance and retreat of glaciers during the Pleistocene. Changes in postglacial vegetation and climate are recorded in the bottoms of lakes

and bogs. As glaciers retreated, they scooped out holes and dammed up rivers and streams, which filled with water to form lakes. Organic debris accumulated on the bottom to form peat, marl, and mud. Pollen, spores, and small invertebrates that blew in from adjacent vegetation settled on the water and sank. Microscopic examination of samples of organic deposits obtained at regular intervals reveals the fossil remains of these organisms. Various genera of fossil pollen can be identified by comparison with pollen growing today. Radiocarbon dating reveals the age of the sediments in which the pollen occurs. The relative abundance of pollen of several genera indicates the predominant vegetation at the specified depth of deposition (Figure 30.15).

Pollen investigation can indicate only trends in vegetation and climate through the past. At present it is impossible to determine the exact structure and composition of the prevailing vegetation at any one time period. Tree species that produced more pollen than others will appear more abundant than they really were. Some pollen might have been carried some distance by the wind or perhaps buried deeper by soil invertebrates. Insect-pollinated plants might not be represented at all. Many pollen grains can be identified only to genera and not to species. For example, the pollen of different types of oak cannot be distinguished to give a clue to the particular type of oak forest existing at a particular time. However, modern paleoecological techniques that take these problems into account and improved identification procedures are enabling paleoecologists to provide a rather accurate picture of postglacial vegetation.

The Pleistocene, which began some two million years ago, marked the end of the Tertiary period of the Holocene and the beginning of the Quaternary by ushering in the Ice Age. However, recent studies of deep sea sediments, the geophysics of ocean bottoms, and the Antarctic ice suggest that ice caps have been part of Earth's geological and ecological history at least since the Miocene and even earlier. Thus the Pleistocene is simply a stage in a continuum of ice and vegetation through the Cenozoic. The present distribution of plants and animals can be appreciated only in the context of longer successional development in the past.

## Pre-Pleistocene Development

At the beginning of the Cenozoic era, some 70 million years ago, most of present-day continental North America and Europe was land. By the beginning of the Miocene epoch (see Table 30.3) forests closely related to the present-day deciduous forest existed with little variation across the northern continents. Known as the Arcto-Tertiary forest, it was a mixture of broadleafed and coniferous species, roughly divided into boreal and temporal elements. The boreal elements consisted of pines, spruces, cypress, birches and willows. Because tropical and subtropical climates existed far north of the present positions, neotropical and Paleotropical-Tertiary forest, ancestors of today's tropical forests, covered

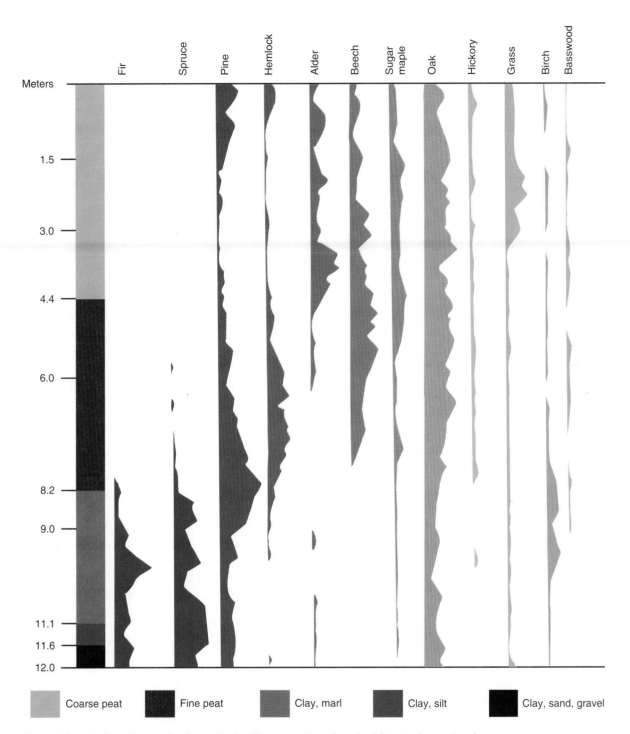

Coarse peat    Fine peat    Clay, marl    Clay, silt    Clay, sand, gravel

**Figure 30.15** Pollen diagram for Crystal Lake, Hartstown Bog, Crawford County, Pennsylvania, at the edge of the maximum advance of the Wisconsin glaciation. The graphs indicate the various genera based on counts of 20 pollen grains for each spectrum level. Grass pollen counts are expressed as percentages of total free pollen. Note the five major forest successions from bottom to top. (1) Initial spruce-fir forest together with some pine and oak invaded as the glacier retreated. Alder, which today precedes spruce, followed it then. (2) As the climate warmed, spruce and fir gave way to jack pine forest with some oak and birch. (3) Later forest surrounding the pond was dominated by oak with hickory, beech, and hemlock. (4) Oaks next dominated the forest, with some hickory, sugar maple, and beech. Grass became an important component in the forest openings. (5) Forests were dominated by oaks and pine (white?) with sugar maple and beech. Hemlock again became important. Compare this diagram with Figures 30.21 and 30.22. (Adapted from P. C. Walker and Hartman 1960:463.)

**Table 30.3  Geological Time Scale**

| Era | Period | Epoch | Age (Millions of Years) | Dominant Life — Plants | Dominant Life — Animals |
|---|---|---|---|---|---|
| CENOZOIC: the age of mammals | Quaternary | Recent | 0.01 | Agricultural plants | Domesticated animals |
| | | Pleistocene | 2 | | Ice Age—first true humans; mixture and then thinning out of mammalian fauna |
| | Tertiary | Pliocene | 10 | Herbaceous plants rise; forests spread | Culmination of mammals; radiation of apes |
| | | Miocene | 25 | First extensive grasslands | |
| | | Oligocene | 35 | | Modernization of mammals; mammals become dominant |
| | | Eocene | 55 | | Mammals become conspicuous |
| | | Paleocene | 70 | | Expansion of mammals; extinction of dinosaurs |
| MESOZOIC: the age of reptiles | Cretaceous | | 135 | Angiosperms rise; gymnosperms decline | Dinosaurs reach peak; first snakes appear |
| | Jurassic | | 180 | Cycads prevalent | First birds and mammals appear |
| | Triassic | | 230 | Gymnosperms rise; seed ferns die out | First dinosaurs; reptiles prominent |
| PALEOZOIC | Permian | | 280 | Conifers become forest trees; cycads important | Great expansion of primitive reptiles |
| | Carboniferous | | | | |
| | Pennsylvanian | | 310 | Lepidodendron, sigillaria, and calamites dominant; the swamp forest | Age of cockroaches; first reptiles |
| | Mississippian | | 345 | Lycopods and seed ferns abundant | Peak of crinoids and bryozoans |
| | Devonian | | 405 | First spread of forests | First amphibians; insects and spiders |
| | Silurian | | 425 | First known land plants | First land animals (scorpions) |
| | Ordovician | | 500 | Algae, fungi, bacteria | Earliest known fishes; peak of trilobites |
| | Cambrian | | 600 | Algae, fungi, bacteria; lichens on land | Trilobites and brachiopods; marine invertebrates |
| PRECAMBRIAN | Late | | | Algae, fungi, bacteria | First known fossils |
| | Early | | 4500 | Bacteria | No fossils found |

most of central North America. Probably in the Eocene, a mixed woodland, the Madro-Tertiary flora, developed on the Mexican plateau.

From the Miocene on, the climate began to cool. The western mountain system in North America rose; climatic zones and their biota were pushed southward, and tropical forests were driven into Central America. The American portion of the Arcto-Tertiary forest was separated from that of Europe and Asia by continental drift, and certain species, such as *Metasequoia, Ailanthus,* and *Ginkgo,* became extinct in North America. As the mountains rose, the broad rain shadow on their lee side wiped out the Arcto-Tertiary forest and stimulated the development of grassland in the central part of North America. A relict Arcto-Tertiary forest, poor in species but including the sequoias and redwoods, was left in the Pacific Northwest. Elements of the Madro-Tertiary forest moved northward to occupy dry lands vacated by the Arcto-Tertiary forest and eastward to form sclerophyllous-pine woodlands ancestral to the Southern oak-pine woodlands of today. In the late Pliocene a continuing climatic cooling accompanied by mountain building brought on continental glaciation.

## Glacial Periods

The Pleistocene was an epoch of great climatic fluctuations throughout the world. At least four times during the Pleistocene ice sheets advanced and retreated in North America (Figure 30.16) and at least three times in Europe (Figure 30.17). Four times in North America and three times in Europe the biota retreated and advanced, each advance having a somewhat different mix of species.

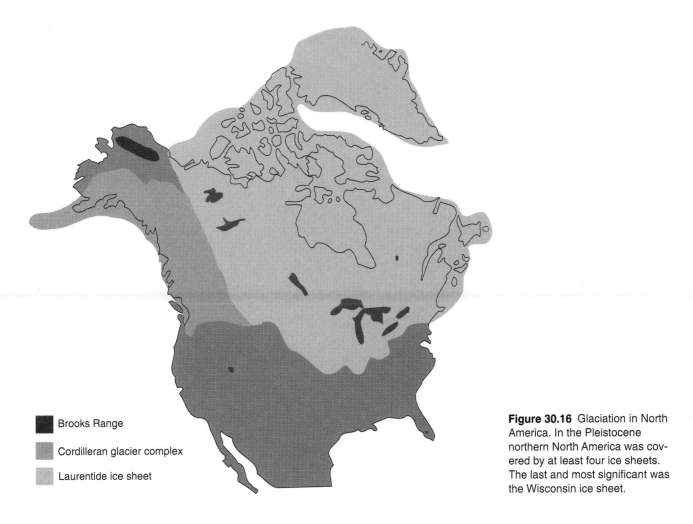

Brooks Range

Cordilleran glacier complex

Laurentide ice sheet

**Figure 30.16** Glaciation in North America. In the Pleistocene northern North America was covered by at least four ice sheets. The last and most significant was the Wisconsin ice sheet.

**Figure 30.17** Glaciation in Europe. In the Pleistocene, northern Eurasia was covered by ice sheets similar to those covering North America. The most important was the last, called the Weichselian. Note the disjunct glacier in the region of the Alps. (After Flint 1971:545.)

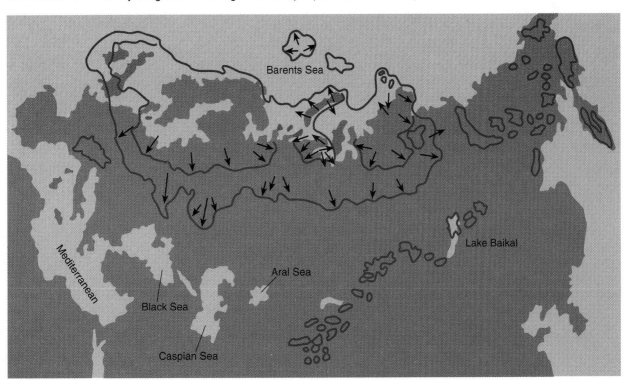

Mediterranean

Barents Sea

Lake Baikal

Aral Sea

Black Sea

Caspian Sea

Each glacial period was followed by an interglacial period (Table 30.4 and Figure 30.18). The climate in each stage oscillated between cold and temperate. During the cold stage tundralike vegetation and boreal species of fir and spruce dominated the landscape. As glaciers retreated and the climate ameliorated, easily dispersed, light-demanding forest trees such as pine and birch advanced northward. As the soil improved and the climate continued to warm, these trees were replaced by the slower dispersing, shade-tolerant trees such as oak and ash. The development of the next glacial period brought in spruce, fir, and tundra vegetation. This vegetation changed the soil from mull to acid mor. As both climate and soil began to deteriorate, heaths dominated the vegetation and forest trees disappeared.

Major differences in vegetation bordering the glacier existed between Europe and North America. In Britain and

**Table 30.4  Glacial and Interglacial Stages**

| Britain | Northern Europe | North America | Climate |
|---|---|---|---|
| Flandrian (postglacial) | Weichselian | | temperate |
| Devensian (last glaciation) | Weichselian | Wisconsin | cold glacial |
| Ipswichian (last interglacial) | | Sangamon | temperate |
| Wolstonian | Saalian | Illinoian | cold glacial |
| Hoxnaian | Holstein | Yarmouth | temperate |
| Anglian | Elsterian | Kansas | cold glacial |
| Comerian | | Aftonian | temperate |
| Beestonian | | Nebraskan | cold |
| Pastonian | | | temperate |
| Baventian | | | cold |

**Figure 30.18** The glacial-interglacial cycle in northwestern Europe. (After van der Hammer, Wijmstra, and Zagwign 1974.)

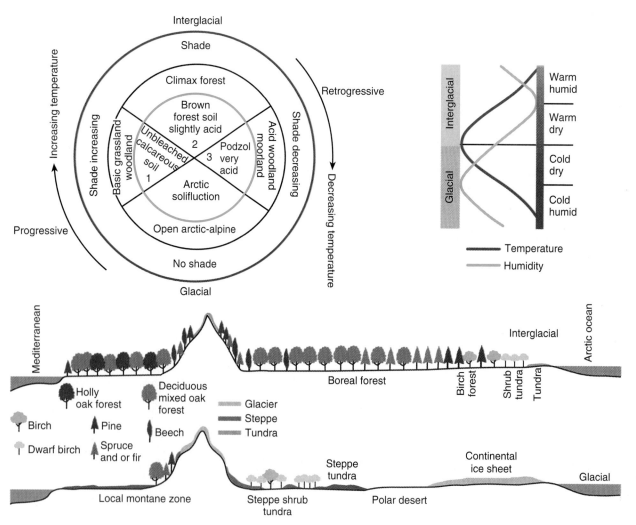

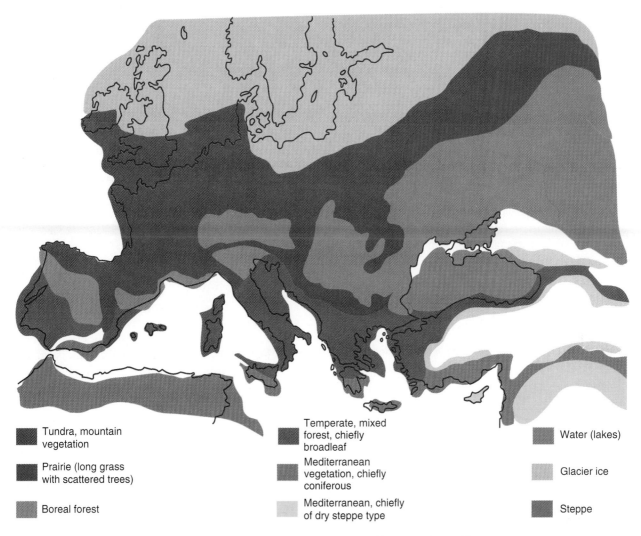

| Tundra, mountain vegetation | Temperate, mixed forest, chiefly broadleaf | Water (lakes) |
| Prairie (long grass with scattered trees) | Mediterranean vegetation, chiefly coniferous | Glacier ice |
| Boreal forest | Mediterranean, chiefly of dry steppe type | Steppe |

**Figure 30.19** Assumed distribution of vegetation in Europe at the Weichsel/Wurm maximum. The Black Sea and the Caspian Sea are interconnected lakes. Note the predominance of tundra vegetation, the patches of boreal forest, and the highly restricted distribution of temperate deciduous forest. (After Flint 1971.)

Europe a wide belt of tundra edged the glacier. The Alps, with their large ice cap, diverted the westerly flow of warm air southward. As a result the Arcto-Tertiary flora was decimated by the early cold stages of the Pleistocene. Temperate genera were forced southward, but the retreat was blocked by mountains, deserts, and seas (Figure 30.19). Only the hardy boreal genera could survive. In North America no such barrier existed. The glacier extended farther south into a warmer zone across which the flow of air was unimpeded (Figure 30.20). As a result spruce forests grew virtually to the edge of the ice and Tertiary precursors of present-day deciduous forest survived to spread north.

The last great ice sheet, the Laurentian, reached its maximum advance about 20,000 BP to 18,000 BP during the Wisconsin glaciation stage in North America. Canada was under ice. A narrow belt of tundra about 60 to 100 km wide bordered the edge of the ice sheet. It probably extended southward into the high Appalachians, where a few relict examples exist today, and into the high peaks of the Adirondacks (H. E. Wright 1970, Delcourt and Delcourt 1981, 1985). Boreal forest, dominated by spruce and jack pine, covered most of the eastern and central United States as far as western Kansas. Its southern limit was about 1200 km south of the modern southern border of boreal forest in Canada. West of Kansas lay uninterrupted sand dunes, a treeless landscape shaped by the intense winds generated by the nearby ice sheet. South of the boreal forest was a transition belt of conifers and mixed hardwoods that separated the boreal forest from the oak-pine forests to the south. Mesic, temperate hardwood species—beech, sugar maple, bass-

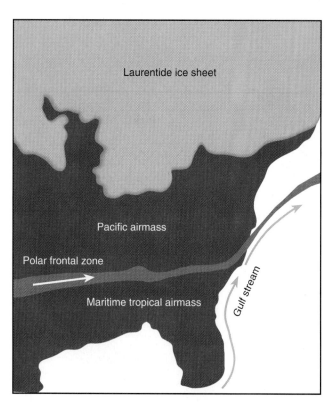

**Figure 30.20** At the full glacial interval at 18,000 BP in North America, the boreal forest was separated from the temperate deciduous forest further south by a narrow belt of cool temperate coniferous and deciduous tree species (Figure 30.21b). This narrow belt of vegetation probably represents the mean annual position of a strong zonal circulation of air, the Polar Frontal Zone, which extended eastward across the western North Atlantic Ocean. The narrow stable climatic boundary separated the Pacific Air Mass immediately to the north of the Polar Frontal Zone from the Maritime Air Mass to the south. (Adapted from Delcourt and Delcourt 1984:277.)

wood, walnut, buckeye, yellow-poplar, chestnut, hickory, and oak—found a refuge in the loess-capped uplands of the Mississippi Valley, in dissected valley slopes along major southern river systems, ravines, and irregular karst terrain, and perhaps along the southern Atlantic and Gulf coastal regions exposed by seas that were 300 m lower than today (Figure 30.21) (Delcourt and Delcourt 1985).

The climatic changes that accelerated the retreat of the Laurentian ice sheet in the late Wisconsin also brought a sudden end to the boreal forest in unglaciated North America. In the western part of the glacial region spruce was replaced by prairie grass. Farther east, closer to the edge of the present prairie region, spruce gave way to pine, birch, and alder (Amundson and Wright 1979). In the southern Appalachians oak and pine replaced spruce, except for relict stands at high elevations. Pines moved northward rapidly from their Appalachian refuge in the Carolinas and domi-

nated much of the region about the newly formed Great Lakes (Figure 30.22) (M. B. Davis 1981, 1983). Hemlock and other species appeared in the southern Appalachians, from which they invaded the deglaciated areas. Other species, such as chestnut, moved much more slowly, taking 3000 years to reach New England from the central Appalachians (M. B. Davis 1981). In the Western Cordillera (parallel chains of mountains) during the Wisconsin glaciation the tree line and tundra vegetation moved downslope 800 to 1000 m lower than today. Many of the modern desert basins were shrub steppes dominated in part by sagebrush. (For an excellent description of the continent after the Ice Age in North America, see Pielou 1992.)

The end of the Pleistocene witnessed a major transition in vegetation and associated animal life. As the climate slowly warmed, southern species advanced slowly northward, overtaking northern species growing along the glacial edge. How far one southern species advanced, and how far south more northern species remained depended upon the physiological adaptations of the species and local climatic conditions.

Now we are in another transition period induced by humans. Increased $CO_2$ and other greenhouse gases in the atmosphere and massive human global environmental changes threaten a rapid rise in Earth's mean temperature. A rise in temperature will endanger the survival of many northern species of plants and animals. Spruce forests would advance into the tundra, and southern plants and exotics overtake spruce forests, changing vegetational patterns and species associations.

However, there are obstacles to this scenario. The northward advance of forest trees in postglacial times was on the order of a fraction of a kilometer a year, as the climate slowly warmed. However, the predicted future global warming would be much more rapid. Temperatures in postglacial times rose about one degree Celsius per 500 years; now the rise may be one degree Celsius per decade. To keep up with this rapidity of warming, tree species would need to advance up to five to ten kilometers a year. Rapidity of range extension would depend upon the abundance, dispersal ability, and physiological abilities of the species involved. Unlike their counterparts in postglacial times, species today face great obstacles to dispersal. Communites are so fragmented that large pools of dispersers no longer exist. Many dispersal agents are gone. Dispersers encounter such formidable barriers as megalopolises of the east and west coast and about the Great Lakes, other urban and suburban developments, dams, highways, and lack of suitable habitats for colonization.

As in the postglacial period, some species will disappear, others will become more common. Species composition of communities as we know them today will change, affecting animal life. What the magnitude of potential future changes will be is purely speculative. One fact, however is obvious. We humans now in control of the planet have the ability to influence the direction the future will take.

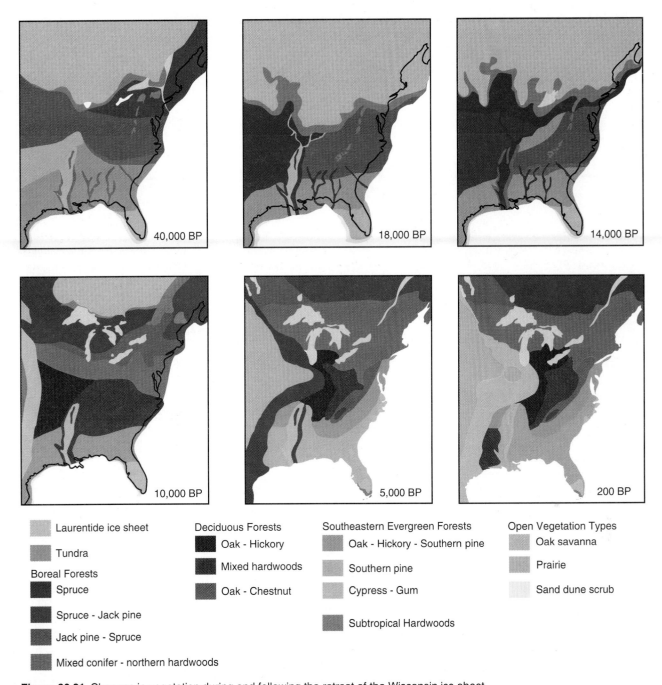

**Figure 30.21** Changes in vegetation during and following the retreat of the Wisconsin ice sheet, reconstructed from pollen analysis at sites throughout eastern North America. (Adapted from Delcourt and Delcourt 1981.)

## SUMMARY

Natural communities change with disturbance and the passage of time. Old fields of today return to forests tomorrow; weedy fields in prairie country revert to grasslands. This gradual sequential change in the relative abundances of dominant species in a community is succession. It is characterized by (1) the replacement of opportunistic, early successional stage species by late stage species; (2) a progressive change in community structure; (3) an increase in biomass and organic matter accumulation; and (4) an approach toward a balance between community production and community respiration. Succession that begins on sites devoid of or unchanged by organisms is termed primary; succession that proceeds from a state in which other organisms were already present is called secondary.

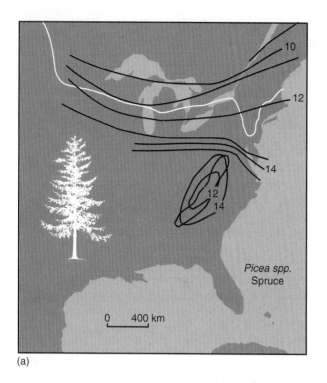

(a)

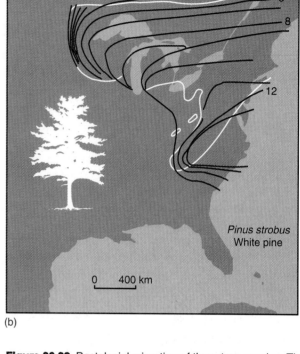

(b)

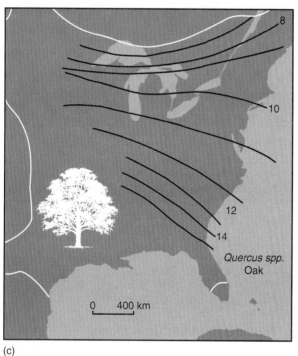

(c)

**Figure 30.22** Postglacial migration of three tree species. The isopleths represent the leading edge of the expanding population. (a) An aggressive pioneer, spruce moved quickly into the tundra and took over the landscape newly exposed by the melting ice sheet. Its speed of northward movement varied. It moved into the Great Lakes region shortly after the ice receded, but 2000 years intervened until it arrived in New England. Spruce preceded alder, the opposite of successional sequences on glacial material today. (b) White pine found a refuge during the height of the ice sheet along the East Coast and in the foothills of the Appalachians, where it was mixed with stands of hardwoods. The species was absent from the full glacial sites in the Mississippi Valley and Florida. Paleoecological evidence of white pine appeared first in Virginia, then expanded rapidly north westward to Minnesota 7000 years ago. Its westward expansion was blocked by dry climate. White pine extended northward, reaching sites north of its present range. There it occurred briefly in large populations before competitive reduction by hardwoods. (c) Oaks, widespread in the southern United States during full glaciation, spread rapidly northward between 10,000 and 9000 BP and reached the full limits of their range by 7000 BP. The northern limit of oak coincides with the northern limit of white pine. (Adapted from M. B. Davis 1981:138, 144, 145.)

The processes and mechanisms involved in succession have been the subject of considerable study, controversy, and review. Several models have proposed to interpret the successional process. Succession typically begins with organisms capable of growing successionally on an open site because of their early arrival as seeds, spores, or residual propagules. According to a facilitation model, succession comes about because of changes induced by the organisms themselves. As they exploit the environment, their life activities make the habitat unfavorable for their own continued survival and create a favorable habitat for other species. An inhibition model has the colonists preempting space and continuing to exclude or inhibit the growth of others until the former colonists die or are damaged. Their death releases resources and allows new, longer-lived species to enter. In the tolerance model, species of all stages colonize

together. "Earlier" species become evident first due to their rapid growth. Later other species outcompete them. All three models apply to most successions.

The resource ratio model and the individual-based plant model are more mechanistic. According to these models, the pattern of species replacement involves competition and the species-specific life history traits and physiological properties of plants interacting under changing environmental conditions, particularly light, nutrients, and moisture. Such environmental changes are brought about in part by individual plants.

Eventually, communities arrive at some form of steady state with the environment. This stage, usually called the climax, is more or less self-sustaining, largely through small-scale disturbances as individuals die and are replaced. The degree of change in a climax depends upon whether the replacements are of the same or different species. If replacements are the same species, stability is assured. Because of different microenvironmental conditions, progeny of the same species are usually less favored than those of other species. As a result the climax is usually a mosaic of regenerating patches.

Some changes are not unidirectional. They include cyclic replacement and fluctuations. In cyclic replacement the destruction of vegetation by some periodic biotic or environmental disturbance starts regeneration again at some particular stage. Each phase is related to the other by orderly changes in the rise and decline of successive communities. Such cyclic succession aids in community persistence. Nonsuccessional, reversible changes or fluctuations in communities result from environmental stresses such as changes in soil moisture, wind, and grazing.

As succession proceeds, emergent properties also change. Community structure becomes more stratified, biomass and organic matter accumulate, and nutrient cycling becomes more internal in the system. Food webs become more complex, species diversity increases up to a point, and net community production and respiration slowly approach equilibrium.

Perhaps the most outstanding characteristic of natural communities is their dynamic nature. They are constantly changing through time—rapidly in early stages of development, more slowly in later stages. Even those communities that are seemingly the most stable, including tropical ones, slowly change through time.

Although we experience succession on a short-term temporal scale, succession and changes in Earth's vegetation and life have taken place on a grander scale over geological time. The distribution of life we observe today has been influenced heavily by long-term dynamic changes in Earth: the breakup of large land masses into continents that drifted apart, the buildup of mountains, the rise and fall of sea levels, the advance and retreat of ice sheets. With these physical changes came climatic changes—warming and cooling, aridity and heavy precipitation. Patterns of vegetation changed as climates changed.

Some of the most pronounced changes, as evidenced by pollen profiles from bogs and pond and lake bottoms, occurred during the Pleistocene, the Ice Age. At that time several advances and retreats of ice sheets eliminated vegetation over much of the northern part of the Northern Hemisphere and pushed plants and animals southward. With each northward retreat of the ice sheet vegetation moved north, producing different forest communities during each glacial interval. The nature and types of vegetational communities today reflect the evolutionary interval. The nature and types of vegetational communities today reflect the evolutionary impact of changing conditions during the Pleistocene.

## REVIEW QUESTIONS

1. Define succession. Distinguish between primary and secondary succession.
2. Contrast autogenic succession with allogenic succession. What is the difference between them and heterotrophic succession?
3. Distinguish between the holistic concept of succession and the reductionist concept.
4. Compare the major models of succession: facilitation, tolerance, inhibition, and mechanistic.
5. Consider early and late successional species terms in terms of $r$ and $K$.
6. What is climax vegetation?
7. How does the concept of shifting-mosiac steady state relate to climax theory?
8. What is cyclic replacement?
9. What are expected trends in ecosystem attributes through succession?
10. Comment on time and direction in succession. How long do various stages last?
11. What major events in the Pleistocene shaped vegetational patterns today?
12. Study the succssional patterns illustrated in Figures 30.11 and 30.12. Then determine some of the life history traits and competitive abilities of the species shown. Based on the information obtained, discuss successional sequence in terms of the mechanistic models.

## CROSS-REFERENCES

Shade tolerance, 102–104; nutrient cycling, 116–120; nutrient budgets, 118; decomposition, 159–165; competition in plants, 400, 484, 490; $r$ and $K$ selection, 448; differential resource utilization, 492–493; vertical structure of communities, 599–601; patch dynamics, 638–641; disturbance by animals, 645–646.

# Sampling Plant and Animal Populations

One of the major field problems in ecology is the determination of population distribution, size, and change in abundance. The problem involves sampling to estimate the true population value of interest, known as the **parameter,** which can be expressed as a number. It may be the total number of plants or animals, population density, average survival rate, proportion of males in a population, seed production per individual plant, and the like. From the sample, taking into account variability within the population, we can make some general inferences about the population as a whole. To be valid the samples must be random; that is, all combinations of sampling units must have an equal probability of being selected.

The object of sampling is to estimate some parameter or a function of some parameter. The mathematical expression that indicates how to calculate an estimation of a parameter from the sample data is the **estimator.** The estimator is usually marked with a "hat" over the parameter to indicate that it is an estimator, not the true parameter. The numerical value obtained by inserting the sample data into the estimator is the **estimate.**

The investigator hopes that the value of the parameter as estimated is **accurate**—that is, close to the true value and free of bias. **Bias** is a systematic distortion due to some flaw in the measurement or in the method of collecting the sample. If the investigator were to repeat the sampling experiment a number of times, each providing a different estimate of the parameter, the average of the estimates should equal the parameter being estimated, but usually it does not. In some statistical procedures bias cannot be avoided, but it is always important to recognize the source of the bias and to take it into account. A biased estimate can never be accurate, although it may be *precise. Precision* is the repeatability of a result, the clustering of sample values about their own mean (Figure A.1). It is measured by the sampling variance and its square root, the **standard error of estimation.**

What the experimenter seeks is a robust estimator. A **robust** estimator has a small bias relative to its standard error, regardless of its ability to detect the true population. An estimator is robust if it is affected very little by the failure of one of the assumptions. A good estimator is (1) robust to critical assumptions; (2) the most precise possible—it exhibits minimal variance; (3) distributed normally; and (4) unbiased (White et al. 1982).

Estimators may be parametric or nonparametric. Parametric estimators involve certain assumptions concerning the specific distribution of the population. Nonparametric estimators involve no specific assumptions about

**Figure A.1** Statistical concepts involved in estimating populations. The solid target represents the parameter being estimated, in this instance population size. The dotted target represents the sampling estimates. The distance of the samples or shots from the true target or bull's-eye is the "deviation." The mean value of estimates gives the expectation of the estimator. The distances of the shots from the mean point of impact are the deviations. The distance between the mean point of impact or expectation of the estimator and the true value, the bull's-eye (which we may never know) is the bias. A tight shot or clumping of estimates around a mean value indicates a high precision, but the estimates are not necessarily accurate. A shot group close to the parameter being estimated increases accuracy. The variance is the mean squared deviation about the mean point of impact. (From Giles 1969, *Wildlife Management Techniques.*)

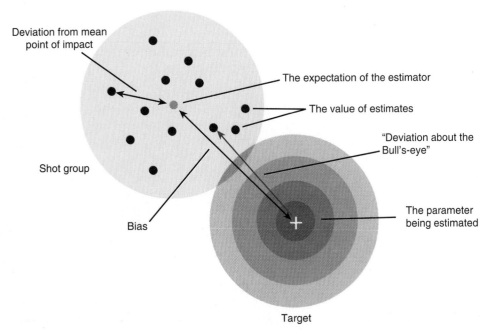

Deviation from mean point of impact

The expectation of the estimator

The value of estimates

"Deviation about the Bull's-eye"

Shot group

The parameter being estimated

Bias

Target

random variables under study. The experimenter hopes to approximate the true parameter. Most field work involving natural populations of plants and animals is nonparametric.

# SAMPLING PLANT POPULATIONS

## Terrestrial Vegetation

Methods of analyzing the vegetation occupying a given site are numerous, and the literature discussing them, the underlying philosophies, and the statistical treatments extensive. Which method to choose for a specific study is a major decision. The basic references at the end of this appendix should help you decide. This appendix describes selected methods with some comments on their advantages and disadvantages.

Quadrats or Sample Plots    Strictly speaking, the **quadrat** is a square sample unit or plot. It may be a single sample unit or it may be divided into subplots. Quadrats vary in size, shape, number, and arrangement, depending upon the nature of the vegetation and the objectives of the study.

The size of the quadrat must be adapted to the characteristics of the community. The richer the flora, the larger or more numerous the quadrats must be. To sample forest trees, the 100 m² plot is a popular size, but it may be too large if trees are numerous or if many species are involved. Smaller 10 m² plots can be used to study shrubs and understory trees. For grass and herbaceous plants, 1 m² is the usual size.

Quadrats may be square, rectangular, or circular. Circular plots are the easiest to lay out, requiring only a center stake and string of desired length. Rectangular plots appear to furnish a more accurate sampling of vegetation composition.

The number of sample units to be employed always presents a problem. The number will vary with the characteristics of the community, objectives of the investigation, degree of precision, and so on. The final number more often than not is arbitrary. By using statistical methods the reliability of the sample and the number of samples needed for any desired degree of accuracy can be determined, once a normal distribution around a mean has been established.

A second approach to this problem is the use of the **species-area curve** (Figure A.2), obtained by plotting the number of species found in plots of different sizes (vertical axis) against the sample size area (horizontal axis). The curve rises sharply at first because the number of new species found is large. As the sample plot size or number is increased, the quantity of new species added declines to a point of diminishing returns, where there is little to be gained by continuing the sampling. This curve can be plotted on an arithmetic or a logarithmic base (see Mueller-Dumbois and Ellenberg 1974). The method can be employed to determine the largest size of a single plot (minimal area) needed to survey the community adequately. In this

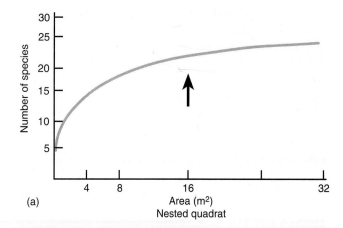

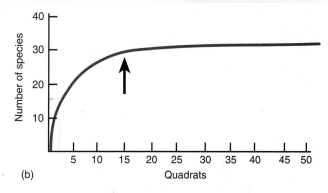

**Figure A.2**  Species-area curves: (a) for minimal area of quadrat; (b) for minimal number of quadrats. Arrows indicate minimums.

case the sampling should be done by using a geometric system of nested plots (Figure A.3). The curve also can be used to determine the minimum number of small multiple plots needed for a satisfactory sample. In addition, the species-area curve can be used to compare one community with another (Figure A.4).

Quadrats fall into four types, according to the type of data recorded.

1. List quadrat. Organisms found are listed by name. A series of list quadrats gives a floristic analysis of the community and allows an assignment of a frequency index, but nothing else.
2. Count quadrat. Numbers as well as names of species encountered are recorded. Quadrats in browse studies fall into this category, which is widely used in forest survey work. In forest studies additional information, such as height, volume, and basal area, is also taken.
3. Cover quadrat. Actual or relative coverage is recorded, usually as a percentage of the area of the ground surface covered or shaded by vegetation.
4. Chart quadrat. A quadrat is mapped to scale to show the location of individual plants. This tedious job provides an overall view especially useful in long-range studies of herbaceous vegetation, mosses, and lichens.

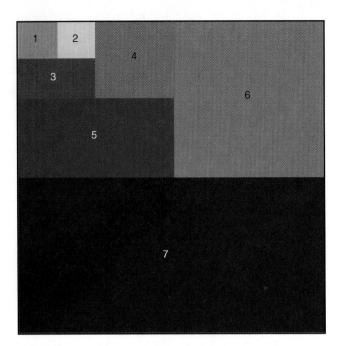

**Figure A.3** An example of nested quadrats.

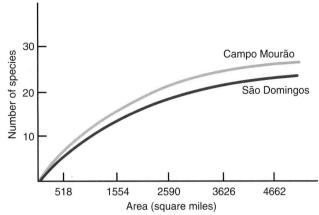

**Figure A.4** Species-area curves compare two stands of *Araucaria* forest on different sites near Campo Mourão, Parana. (From Cain and Castro 1959.)

For statistically reliable estimates, the location of plots must be randomized. This task is rather easy. Numbered grid lines are drawn over an area photo or map of the study area. The numbers of the vertical and horizontal grid lines are written on small squares of paper. To draw random numbers for the two lines, use a table of random numbers, available in statistical tables and statistics texts.

The quadrat method has advantages and disadvantages. It is a popular method, easily employed. If the individual organisms are randomly distributed, then the accuracy of the sample and the estimate of the density depend upon the size of the sample. However, individuals seldom are randomly dispersed, so the accuracy of quadrat sampling may be low, unless a great number of plots is involved. The quadrat method is tedious and time-consuming.

**The Belt Transect** A variation of the quadrat method is the belt transect. A **transect** is a cross section of an area used as a sample for recording, mapping, or studying vegetation. Because of its continuity through an area, the transect can be used to relate changes in vegetation along the transect to changes in the environment. As a sample unit the measurements within a transect can be pooled, so that each transect is treated as a single observation. The belts also can be divided into intervals and each interval treated as a plot.

There are three steps in this method:

1. Determine the total area of the site to be sampled; then divide by 5 or 10 to obtain the total sample area.
2. Lay out a series of belt transects of a predetermined width and length, sufficient to embrace the area to be sampled. Then divide the belts into equal-sized segments. These units are sometimes called quadrats or plots, but they differ from true quadrats in that each represents an observational unit rather than a sampling unit.
3. Measure the vegetation in each unit for some attribute, such as abundance, sociability, frequency, or stem count.

A variation of the segmented-belt transect consists of taking observations only on alternate segments. The precision seems to be affected very little (Oosting 1956). For example, ten segments alternately spaced on a 6-m belt are nearly twice as efficient statistically as ten on a 3-m belt.

The belt transect is well adapted to estimate abundance, frequency, and distribution. For estimating the frequency index, it has the disadvantage that frequency by classes is related to the size of the plot. To compare one area with another, the segment size used in sampling must be the same for both areas.

There are several ways to analyze data from quadrats and belt transects. If we have simply recorded the presence of species, it limits analysis to frequency and relative frequency.

If we have recorded the number of individuals of the various species found in the quadrat, these data give a density figure. Because of the variations in growth forms among the various species, numbers mean little. Counts are most useful in certain situations, such as counting the number of stems of shrubby plants available for deer browse or counting the amount of forest reproduction. The samples are broken down into classes, such as 1 m high, 2 to 12 m, or 1 to 3 cm, 4 to 9 cm, and so on.

A third method is that of Braun-Blanquet (1951). It involves a total estimate based on abundance and cover. If the number of individuals in a plant community is estimated but

**Table A.1  Total Estimate Scale (abundance plus coverage)**

| | |
|---|---|
| + | Individuals of a species sparsely present in the stand; coverage very small |
| 1 | Individuals plentiful, but coverage small |
| 2 | Individuals very numerous if small; if large, covering at least 5% of area |
| 3 | Individuals few or many, collectively covering 6–25% of the area |
| 4 | Individuals few or many, collectively covering 26–50% of the area |
| 5 | Plants cover 51–75% of the area |
| 6 | Plants cover 76–100% of the area |

**Table A.2  Sociability Classes of Braun-Blanquet**

| | |
|---|---|
| Class 1 | Shoots growing singly |
| Class 2 | Scattered groups or tufts of plants |
| Class 3 | Small, scattered patches or cushions |
| Class 4 | Large patches or broken mats |
| Class 5 | Very large mats of stands or nearly pure populations that almost blanket the area |

not counted, the data are referred to as **abundance.** Abundance implies a number of individuals, but number does not necessarily reflect dominance or cover. **Cover** is the result of both numbers and massiveness. Although abundance and coverage are separate and distinct, they can be combined in a community description as the total estimate. For many field studies this method works well, but it is subjective and the data are difficult to handle statistically (see Mueller-Dombois and Ellenberg 1974). However, this method does provide a useful general picture of the plant community and a mechanism for classifying vegetation. The scales are given in Table A.1. Along with total estimates an estimate of sociability of each species should be given (as in Table A.2)—whether the plant grows singly, in clumps, mats, and so on.

The total estimate and the sociability estimate can be expressed together to give a paired value for each species. For example, if a plant species has the value 4.3, the first figure is the total estimate, the second the sociability. Once a number of stands have been surveyed, the community characteristics can be combined in an association table. The plant species usually are listed on the basis of fidelity or presence, the characteristic species of the community often heading the list. A partial example in Table A.3 shows how such ta-

bles are constructed. For details see Mueller-Dombois and Ellenberg (1974).

Data so collected describe individual stands. By the use of another attribute, presence, we can compare stands of a community type or of related types. **Presence** is the degree of regularity with which a species recurs in different examples of a community type. It is commonly expressed as a percentage that can be assigned to one of a limited number of presence classes, given in Table A.4. Presence is determined by dividing the total number of stands in which the species is found by the total number of stands investigated. Species that have a high percentage of presence or that fall within presence class 5 often are regarded as characteristic of that community.

Line Intercept   The line intercept is one-dimensional. Most useful for sampling shrub stands and woody understory of the forest, the line intercept or line transect method consists of taking observations on a line or lines laid out randomly or systematically over the study area. The procedure is as follows:

1. Stretch a metric steel tape, steel chain, or a tape between two stakes 50 to 100 m apart.
2. Subdivide the line into predetermined intervals, such as 1 m.
3. Move along the line, and for each interval record the plant species found and the distance they cover along that portion of the line intercept. Consider only those plants touched by the line or lying under or over it.

**Table A.3  Partial Stand Composition, Cumberland Plateau, West Virginia**

| Herbaceous Species | Plot Number | | | | | | | | | | Frequency (%) |
|---|---|---|---|---|---|---|---|---|---|---|---|
| | 1 | 2 | 3 | 4 | 5 | 6 | 7 | 8 | 9 | 10 | |
| Polystichum acrostichoides | 2·2 | +·1 | 1·2 | 1·2 | 2·2 | 2·2 | 2·2 | 2·2 | 2·1 | 1·1 | 100 |
| Cimicifuga racemosa | 3·2 | 2·2 | | | 2·2 | 3·2 | 2·2 | 1·2 | 2·2 | 2·2 | 80 |
| Geranium maculatum | | +·1 | +·1 | +·1 | 1·2 | 2·2 | 2·2 | +·2 | 1·1 | +·1 | 90 |
| Disporum lanuginosum | 3·2 | 3·3 | | 3·3 | 1·1 | 2·2 | 1·1 | +·1 | +·2 | 2·2 | 90 |
| Galium circaezans | +·2 | +·2 | | | 1·2 | +·1 | 2·2 | +·1 | | | 60 |
| Thalictrum dioicum | | +·2 | | | 1·2 | 1·1 | 2·1 | | | +·2 | 50 |
| Sanicula canadensis | +·1 | 2·2 | 1·1 | | +·1 | | | | | | 40 |

**Table A.4** Presence Classes

| Presence Class | Stands in Which Species Occur (%) |
|---|---|
| 1 | 1–20 |
| 2 | 21–40 |
| 3 | 41–60 |
| 4 | 61–80 |
| 5 | 81–100 |

Treat each stratum of vegetation separately, if necessary. (For grasses, rosettes, and dicot herbs, measure the distance along the line at ground level. For shrubs and tall dicot herbs, measure the shadow or distance covered by a downward projection of the foliage above.)

4. Repeat. Usually 20 to 30 such lines are sufficient.

The data can be summarized as follows:

1. Number of intervals in which each species occurs along the line.
2. Frequency of occurrence for each species in relation to total intervals sampled.
3. Total linear distance covered by each species along the transect.
4. Total length of line covered by vegetation and total "open" length.
5. Total number of individuals, if they can be so recorded. Because of branching and size variations it is difficult to count individual plants on a line transect.

This method is rapid, objective, and relatively accurate. The area may be determined directly from recorded observations. The lines can be randomly placed and replicated to obtain the desired precision. The method is well adapted for measuring changes in vegetation if the ends of the lines are well marked. Generally it is more accurate in mixed plant communities than quadrat sampling and is especially suited for measuring low vegetation.

On the debit side, the method is not well adapted for estimating frequency or abundance, because the probability of an individual's being sampled is proportional to its size. Nor is it suitable where vegetation types are intermingled and the boundaries indistinct.

From line-intercept data the following measurements may be calculated. (Calculation of relative density may not be possible if individual plants can not be identified.)

$$\frac{\text{relative}}{\text{density}} = \frac{\text{total individuals species A}}{\text{total individuals all species}} \times 100$$

$$\frac{\text{dominance}}{\text{(cover)}} = \frac{\begin{array}{c}\text{total intercept length,}\\ \text{species A}\end{array}}{\text{total transect length}} \times 100$$

$$\frac{\text{relative}}{\text{dominance}} = \frac{\begin{array}{c}\text{total intercept length,}\\ \text{species A}\end{array}}{\begin{array}{c}\text{total intercept length,}\\ \text{all species}\end{array}} \times 100$$

$$\text{frequency} = \frac{\begin{array}{c}\text{intervals in which}\\ \text{species occurs}\end{array}}{\begin{array}{c}\text{total number}\\ \text{of transect intervals}\end{array}} \times 100$$

$$\frac{\text{relative}}{\text{frequency}} = \frac{\begin{array}{c}\text{frequency value,}\\ \text{species A}\end{array}}{\begin{array}{c}\text{total frequency value,}\\ \text{all species}\end{array}} \times 100$$

**Point-Frequency Intercept** The point-frequency intercept is useful for grassland and herbaceous vegetation. The method involves the use of a point-frequency frame to sample basal and canopy cover of grassland vegetation.

The point-frequency frame, 0.5 m high and 1 m long, is made of wood or aluminum (Figure A.5). Guide holes spaced 5 cm apart are bored perpendicularly through the horizontally fixed laths. Ten sharp-pointed wire pins or steel rods of the same length as the legs are slid through the holes.

The linear frame is mounted vertically over the herbaceous vegetation to be measured. The pins are lowered vertically one after the other and the first cover and basal interceptions are recorded by species. This procedure gives a measure of both crown or shoot and basal cover. After reading all points in the frame, move the frame to a new location.

The frame may be set at random points throughout the grassland. The most efficient way is to locate a number of random points in the field. From them establish some present number of transect lines radiating from the center point along randomly selected compass bearings (Figure A.6). The number of frames required depends upon the homogeneity of the vegetation. In relatively homogeneous vegetation, 200 points (20 frames) may be sufficient, but usually many more are required.

Although slow, the point-frequency method provides a highly accurate measure of foliar cover of grassland vegetation. However, it will often miss scattered clumps of herbaceous species in the sampling area. The method is most useful in studies of changes in the condition of grassland vegetation over time.

Analysis involves two sampling units—the ten-point frame or "plot" and the ten points within the frame. Both the number of frames and the number of points are utilized.

$$\frac{\text{dominance}}{\text{(coverage)}} = \frac{\begin{array}{c}\text{number of points}\\ \text{with species A}\end{array}}{\begin{array}{c}\text{total number}\\ \text{of points}\end{array}} \times 100$$

$$\frac{\text{relative}}{\text{dominance}} = \frac{\begin{array}{c}\text{number of points}\\ \text{with species A}\end{array}}{\begin{array}{c}\text{total points any species}\\ \text{(excluding empty points)}\end{array}} \times 100$$

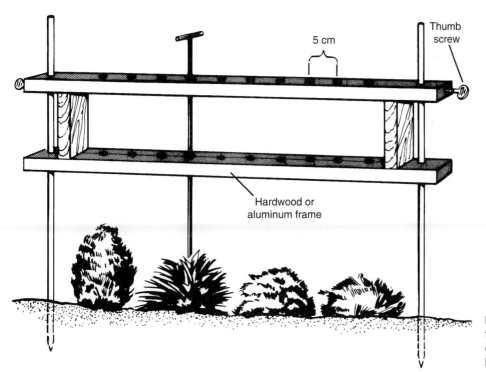

**Figure A.5** A point-frequency-frame sampler. The inside length of the frame is 1 m. The distance between the pins is 5 cm.

$$\text{frequency} = \cfrac{\begin{array}{c}\text{number of frames}\\ \text{with species A}\end{array}}{\begin{array}{c}\text{total number}\\ \text{of frames}\end{array}} \times 100$$

$$\text{relative frequency} = \cfrac{\begin{array}{c}\text{number of frames}\\ \text{with species A}\end{array}}{\begin{array}{c}\text{number of frames}\\ \text{with any species}\end{array}} \times 100$$

**Point-Quarter Method** Several variations of the variable plot or "plotless" method have been developed for ecological work. These methods arose from the variable radius method of forest sampling developed in Germany by Bitterlich. He used it to determine timber volume without establishing plot boundaries. The method was introduced into the United States by Grosenbaugh (1952, 1958).

One of the most useful of the plotless methods is the point-quarter method (see Cottam and Curtis 1956, Greig-Smith 1983). It is most useful in sampling communities in which individual plants are widely spaced or in which the dominant plants are large shrubs or trees.

The procedure is as follows:

1. Locate a series of random points within the stand to be sampled, or pick random points along a line transect passing through the stand.

**Figure A.6** A sampling scheme for use with the point-frequency-frame sampler. From a central stake or point, use random numbers to determine directional and distance coordinates. A minimum of four lines and 16 frames per line for a total of 64 frames and 640 points should be sampled from each point.

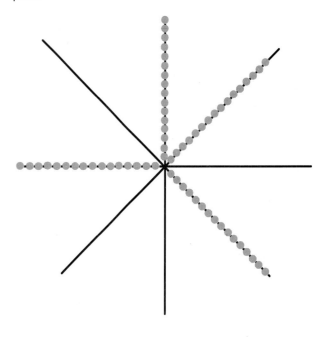

2. At each station mark a point in the ground.
3. Divide the working area into four quarters or quadrants by visualizing a grid line, predetermined by compass bearing, and a line crossing it at right angles, both passing through the point (Figure A.7).
4. Select the tree (or plant) in each quarter that is closest to the point. Record its distance from the point, diameter at breast height, and species. The tally sheet will thus contain data for four trees at each point, one from each quarter.
5. Tally at least 50 such points.

The computations entail several steps. First, add all distances in the samples and divide the total distance by the number of distance to obtain a mean distance of point to plant.

$$\text{mean distance} = \frac{\text{total distance}}{\text{number of distances}}$$

**Figure A.7** The point-quarter method of sampling forest stands.

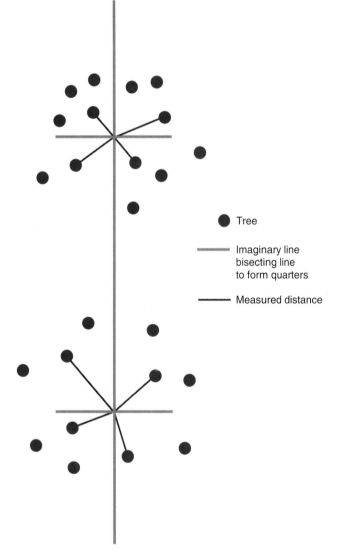

● Tree

━━━ Imaginary line bisecting line to form quarters

━━━ Measured distance

Square the mean distance to obtain the mean area covered on the ground per plant. To obtain the total density of trees, divide the mean area per plant into the unit area on which density is expressed. If the area is in feet, then divide the mean distance squared into 43,560 ft$^2$ to obtain the total density of trees per acre.

Next, determine basal area for each tree (see Table A.5) from diameter measurements. (Basal area is the area of a plane passed through the stem of a tree at right angles to its longitudinal axis at breast height. Because the cross section approximates a circle, its area can be computed from the standard formula for the area of a circle.)

Now these calculations may be made:

$$\frac{\text{relative}}{\text{density}} = \frac{\text{individuals of species A}}{\text{total individuals of all species}} \times 100$$

$$\text{density} = \frac{\text{relative density of species A}}{100} \times \frac{\text{total density}}{\text{of all species}}$$

$$\frac{\text{relative}}{\text{dominance}} = \frac{\text{total basal area of species A}}{\text{total basal area of all species}} \times 100$$

$$\text{dominance} = \frac{\text{density}}{\text{of species}} \times \frac{\text{average dominance}}{\text{value for species}} \times 100$$

$$\text{frequency} = \frac{\text{number of points at which species A occurs}}{\text{total number of points sampled}} \times 100$$

$$\frac{\text{relative}}{\text{frequency}} = \frac{\text{frequency value for species A}}{\text{total frequency value for all species}} \times 100$$

Absolute values for the number of trees per unit area of any species and the basal area per unit area of any species are determined by multiplying the relative figures for density by the total trees per ha to determine density and by the total basal area per ha to determine absolute dominance.

The point-quarter method is simple, rapid, and effective. The underlying assumption is that individuals of all species together are randomly dispersed. Although this assumption may not be true, it does not seem to produce significant error, except where a deviation from overall randomness is obvious. Relative density and relative dominance are valid even if dispersion is not random. Error would appear in the calculation of absolute density and absolute dominance.

**Importance Value** In regions where the plant communities are highly heterogeneous, the classification of communities on the basis of dominants or codominants becomes

## Table A.5  Area of Circles

| Diameter (in./cm) | Circumference (in./cm) | Area (ft²) | (m²) | Diameter (in./cm) | Circumference (in./cm) | Area (ft²) | (m²) |
|---|---|---|---|---|---|---|---|
| 1 | 3.14 | 0.005 | — | 51 | 160.22 | 14.186 | 0.240 |
| 2 | 6.28 | 0.002 | — | 52 | 163.36 | 14.748 | 0.212 |
| 3 | 9.42 | 0.049 | 0.001 | 53 | 166.50 | 15.321 | 0.221 |
| 4 | 12.57 | 0.087 | 0.001 | 54 | 169.65 | 15.904 | 0.229 |
| 5 | 15.71 | 0.136 | 0.002 | 55 | 172.79 | 16.499 | 0.238 |
| 6 | 18.85 | 0.196 | 0.003 | 56 | 175.93 | 17.104 | 0.246 |
| 7 | 21.99 | 0.267 | 0.004 | 57 | 179.07 | 17.721 | 0.255 |
| 8 | 25.13 | 0.349 | 0.005 | 58 | 182.21 | 18.348 | 0.264 |
| 9 | 28.27 | 0.442 | 0.006 | 59 | 185.35 | 18.986 | 0.273 |
| 10 | 31.42 | 0.545 | 0.008 | 60 | 188.50 | 19.635 | 0.283 |
| 11 | 34.56 | 0.660 | 0.010 | 61 | 191.64 | 20.295 | 0.292 |
| 12 | 37.70 | 0.785 | 0.011 | 62 | 194.78 | 20.966 | 0.302 |
| 13 | 40.84 | 0.922 | 0.013 | 63 | 197.92 | 21.648 | 0.312 |
| 14 | 43.98 | 1.069 | 0.015 | 64 | 201.06 | 22.340 | 0.322 |
| 15 | 47.12 | 1.227 | 0.018 | 65 | 204.20 | 23.044 | 0.332 |
| 16 | 50.26 | 1.396 | 0.020 | 66 | 207.34 | 23.758 | 0.342 |
| 17 | 53.41 | 1.576 | 0.023 | 67 | 210.49 | 24.484 | 0.352 |
| 18 | 56.55 | 1.767 | 0.025 | 68 | 213.63 | 25.220 | 0.363 |
| 19 | 59.69 | 1.969 | 0.028 | 69 | 216.77 | 25.967 | 0.374 |
| 20 | 62.83 | 2.182 | 0.031 | 70 | 219.91 | 26.725 | 0.385 |
| 21 | 65.97 | 2.405 | 0.035 | 71 | 223.05 | 27.494 | 0.396 |
| 22 | 69.12 | 2.640 | 0.038 | 72 | 226.19 | 28.274 | 0.407 |
| 23 | 72.26 | 2.885 | 0.042 | 73 | 229.34 | 29.065 | 0.418 |
| 24 | 75.40 | 3.142 | 0.045 | 74 | 232.48 | 29.867 | 0.430 |
| 25 | 78.54 | 3.409 | 0.049 | 75 | 235.62 | 30.680 | 0.442 |
| 26 | 81.68 | 3.687 | 0.053 | 76 | 238.76 | 31.503 | 0.454 |
| 27 | 84.82 | 3.976 | 0.057 | 77 | 241.90 | 32.338 | 0.466 |
| 28 | 87.96 | 4.276 | 0.062 | 78 | 245.04 | 33.183 | 0.478 |
| 29 | 91.11 | 4.587 | 0.066 | 79 | 248.18 | 34.039 | 0.490 |
| 30 | 94.25 | 4.909 | 0.071 | 80 | 251.33 | 34.907 | 0.503 |
| 31 | 97.39 | 5.241 | 0.075 | 81 | 254.47 | 35.785 | 0.515 |
| 32 | 100.53 | 5.585 | 0.080 | 82 | 257.61 | 36.674 | 0.528 |
| 33 | 103.67 | 5.940 | 0.086 | 83 | 260.75 | 37.574 | 0.541 |
| 34 | 106.81 | 6.305 | 0.091 | 84 | 263.89 | 38.484 | 0.554 |
| 35 | 109.96 | 6.681 | 0.096 | 85 | 267.04 | 39.406 | 0.567 |
| 36 | 113.10 | 7.069 | 0.102 | 86 | 270.18 | 40.339 | 0.581 |
| 37 | 116.24 | 7.467 | 0.108 | 87 | 273.32 | 41.282 | 0.594 |
| 38 | 119.38 | 7.876 | 0.113 | 88 | 276.46 | 42.237 | 0.608 |
| 39 | 122.52 | 8.296 | 0.119 | 89 | 279.60 | 43.202 | 0.622 |
| 40 | 125.66 | 8.727 | 0.126 | 90 | 282.74 | 44.179 | 0.636 |
| 41 | 128.81 | 9.168 | 0.132 | 91 | 285.88 | 45.166 | 0.650 |
| 42 | 131.95 | 9.621 | 0.138 | 92 | 289.03 | 46.164 | 0.665 |
| 43 | 135.09 | 10.085 | 0.145 | 93 | 292.17 | 47.173 | 0.679 |
| 44 | 138.23 | 10.559 | 0.152 | 94 | 295.31 | 48.193 | 0.694 |
| 45 | 141.37 | 11.045 | 0.159 | 95 | 298.45 | 49.224 | 0.709 |
| 46 | 144.51 | 11.541 | 0.166 | 96 | 301.59 | 50.266 | 0.724 |
| 47 | 147.65 | 12.048 | 0.173 | 97 | 304.73 | 51.318 | 0.739 |
| 48 | 150.80 | 12.566 | 0.181 | 98 | 307.88 | 52.382 | 0.754 |
| 49 | 153.94 | 13.095 | 0.189 | 99 | 311.02 | 53.456 | 0.770 |
| 50 | 157.08 | 13.635 | 0.196 | 100 | 314.16 | 54.542 | 0.785 |

impractical. Therefore, Curtis and McIntosh (1951) came up with the index of "importance value" to develop a logical arrangement of the stands. This index is based on the fact that most species do not normally reach a high level of importance in the community, but those that do serve as guiding species.

Importance value (IV) is the sum of relative density, relative frequency, and relative dominance for each species. It may be expressed as a range from 0 to 3.00 or 300 percent, or it may be divided by 3 to give importance percentage, which will range from 0 to 1.0 or 100 percent. In situations involving only two values, such as relative frequency and relative density, importance value is expressed in terms of 0 to 2 or 200 percent.

Importance value provides an overall estimate of the influence or importance of a species in a community. Once importance values have been obtained for species within a stand, stands can be grouped by their leading dominants according to importance values, and the groups can be placed in a logical order based on the relationships of several predominant species. In Table A.6, for example, are four species that were the leading dominants in 80 of 95 forest stands in southern Wisconsin. Note that the dominants are arranged in order of decreasing importance value, from stands dominated by black oak to those dominated by sugar maple. Such an arrangement also shows increasing values for sugar maple. Trees intermediate in dominance can be handled in the same way.

## Aquatic Vegetation

In aquatic communities algae are the dominant vegetation. Two kinds of growth are involved: the plankton suspended on the water and the periphyton growing attached to some substrate.

**Phytoplankton** The phytoplankton can be obtained by drawing water samples from several depths. Cell counts of algae present in each sample, either normal or concentrated, can be made with a Sedgewich-Rafter counting chamber and a Whipple ocular. If necessary, the samples can be concentrated by centrifugation in a Forest plankton centrifuge. The centrifuged samples are then diluted to a suitable volume (100 to 200 ml) in a volumetric flask.

As cells are counted, a separate tally is kept for each species to permit an analysis of community structure at each station. The number of cells for single-celled forms and the number of colonies for colonial forms are recorded. The number of colonies is multiplied by an appropriate factor for each species to convert the colonies into cells. These factors are predetermined by averaging the cell counts from a large number of typical colonies from the area in question.

Another method of handling the phytoplankton is by filtration (see McNabb 1960, Clark and Sigler 1963). The organisms in the sample are first fixed by the addition of 4 parts 40 percent aqueous solution of formaldehyde to each 100 parts of the sample. The analysis is as follows:

1. Thoroughly agitate the sample. Withdraw a fraction with a pipette large enough to hold a sample that will provide an optimum quantity of suspended matter on the filter.
2. Place the sample in the tube of a filter apparatus designed to accommodate a 3 cm-diameter membrane filter. Draw the water through the filter with a vacuum pump.
3. Remove the filter and place it on a glass slide. Put two or three drops of immersion oil over the residue, and store the slide in the dark to dry (about 24 hours). The oil replaces the water in the pores of the filter and makes it transparent.
4. Place a cover slip over the transparent filter.
5. Determine the most abundant species by scanning and then choose a quadrat size that will contain individuals of this species approximately 80 percent of the time.
6. Move the mechanical stage so that approximately 30 random quadrats are viewed. Note the presence or absence of individual species. There is no need to count.
7. When 30 quadrats have been surveyed, calculate the percentage frequency.

$$\frac{\text{frequency}}{(\%)} = \frac{\text{total number of occurrences of a species}}{\text{total number of quadrats examined}} \times 100$$

**Periphyton** The periphyton has not received quite the same attention from ecologists as the phytoplankton, partic-

**Table A.6** The Average Importance Value Index of Trees in Stands with Four Species as the Leading Dominants

| Species | Leading Dominant in Stand | | | | Ecological Sequence Number |
|---|---|---|---|---|---|
| | Quercus velutina | Quercus alba | Quercus rubra | Acer saccharum | |
| Black oak (*Quercus velutina*) | 165.1 | 39.6 | 13.6 | 0 | 2 |
| Shagbark hickory (*Carya ovata*) | 0.3 | 8.8 | 5.2 | 5.9 | 3.5 |
| White oak (*Quercus alba*) | 69.9 | 126.8 | 52.7 | 13.7 | 4 |
| Black walnut (*Juglans nigra*) | 1.5 | 1.2 | 2.2 | 1.9 | 5 |
| Red oak (*Quercus rubra*) | 3.6 | 39.2 | 152.3 | 37.2 | 6 |
| American basswood (*Tilia americana*) | 0.3 | 5.9 | 19.0 | 33.0 | 8 |
| Sugar maple (*Acer saccharum*) | 0 | 0.8 | 11.7 | 127.0 | 10 |

*Source:* Adapted from Curtis and McIntosh 1951.

ularly in a quantitative way. Methods for studying the peri-phyton are given in detail by Sladeckova (1962).

Epiphyton, the periphyton growing on living plants and animals, can be observed in place on the organism if the sub-strate is thin or transparent enough to allow the transmission of light. If the leaves are thin and transparent, the task is rel-atively easy, but the growth on one side must be scraped away. If the leaf is opaque, the chlorophyll can be extracted by dipping the leaf in chloral hydrate. Small leaves can be examined over the whole area. Large leaves can be sampled in strips marked by grids on a slide or by an ocular microm-eter. With large aquatic plants a square will have to be cut from the leaf or stem. If the leaf is too thick to handle under the microscope, scrape off the periphyton and mount in a counting cell for examination. The results can be related to the total surface area.

Algae growing on such aquatic animals as turtles or mol-lusks and on stones must be removed for study. Scraping and transfer is difficult, but there are several techniques available.

One method employs a simple hollow, square instrument with a sharpened edge, which is pressed closely on or driven into the substrate. It separates out a small area of given size around which the periphyton is washed away. The instru-ment is then raised and the periphyton remaining in the sam-ple square is scraped into a collecting bottle.

If the stones can be picked up from the bottom, then the periphyton can be removed with an apparatus consisting of a polyethylene bottle with the bottom cut out and a brush with nylon bristles. A section of the stone is delimited by the neck of the bottle held tightly on the surface. The periphyton is scraped loose by the brush and then washed into a collecting bottle with a fine-jet pipette.

The periphyton can be counted in a Sedgewich-Rafter cell recording a predetermined number, usually 100 to 1000, as they appear in the field of view. The results can be ex-pressed as a percentage; or the algae can be checked for fre-quency in the field of view, using the Braun-Blanquet scale of total estimate (see Table A.1).

Some of the difficulties can be avoided by growing the periphyton on an artificial substrate, usually glass or trans-parent plastic slides, attached in the water in a variety of ways. In lentic situations they can be placed on sand or stones in the water. In lotic situations they can be placed in saw-cuts on boards, set in holes in bricks, clipped to a rope, attached to a wooden frame, or tucked into rubber corks.

The usual procedure is to use either a Wildco periphyton sampler, which holds glass slides, or a Hester-Dendy sam-pler (Hester and Dendy 1962), which can be fabricated eas-ily. A Hester-Dendy sampler consists of eight $7\frac{1}{2} \times 7\frac{1}{2}$ cm or $3'' \times 3''$ squares of tempered hardboard (Masonite) with a hole drilled in the center. The boards are held apart by seven spacers, such as $5/8''$ or 17 mm faucet washers. The unit is assembled using a $2\frac{1}{2}''$ or 60 mm stove bolt and tightened with a wing nut. The sampler is attached to a length of stiff

wire with a loop on the end to fit around the head of the bolt. The stake carrying the unit is pushed into the stream bottom.

For algae and protozoans the plates should be exposed for one to two weeks; for hydras, sponges, and the like, about one month. At the time of data collection, place each sampler in a plastic bag underwater, then close the bag with the sampler and water inside. Allow the wire stake to pro-trude from the top of the bag. In the lab, observe the peri-phyton directly under the microscope if glass slides were used. For hardboard samplers, remove the macroscopic or-ganisms in small bowls. Scrape off algal growth and exam-ine it in a Sedgewich-Rafter cell. For further procedures and data sheets see Brewer and McCann (1982:153).

**Stream-Bottom Organisms**   Samples of stream-bottom organisms can be taken with a modified Surber bottom-fauna sampler (Figure A.8). The sampler consists of a box made of a brass frame with stainless steel side pieces and a current baffle. To this box are attached on a removable brass frame two cone-shaped nets. A smaller cone, made of coarse net (19 meshes per linear inch, 7 per linear cm), is fitted within a larger cone, made of fine net (74 meshes per linear inch, 29 per linear cm). Flanges on the insert prevent the coarse net from being forced into the fine net.

This modified sampler picks up many small organisms that might otherwise be lost. In fact, collection of virtually all macroorganisms is assured. In addition we obtain two subsamples with respect to size, and the small organisms are associated with fine detritus only.

The sampler encloses a specified area of stream (500 cm$^2$), which is the sample unit. Organisms, detritus, and trash are scrubbed free from the substrate, and the current washes them into the net. The contents can be transferred to a container and taken back to the lab for examination and sorting.

# Dendrochronology

Dendrochronology is the science of dating past events by the study of the aging of trees. It is a valuable tool for the ecolo-gist. It has been used in a number of studies—to age trees for management information, and to establish dates of past forest fires, insect outbreaks, ice storms, and periods of sup-pression and release in the life history of forest trees. It has been involved in hydrological and archaeological studies and even in legal cases involving boundary disputes, in which specimens are taken from fence posts and witness trees. Examples of growth-ring analysis in ecological inves-tigations are Spenser's study (1964) of population fluctua-tions over the centuries in Mesa Verde National Park and a study by Sinclair and associates (1993) of the relationship among solar cycles, climate, and snowshoe hare cycles.

Dendrochronology is based on the variation of growth rings. Growth rings, despite popular belief, are not regular, nor are they all necessarily laid down annually. Because of

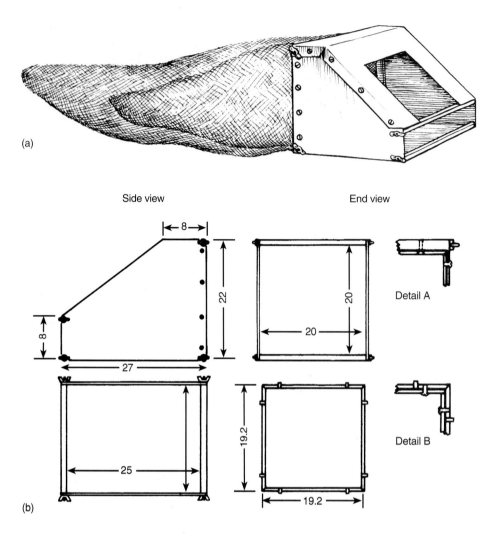

(a)

Side view                End view

Detail A

Detail B

(b)

**Figure A.8** (a) A modified Surber bottom sampler. (b) Construction details for the Surber bottom sampler. (Redrawn from Withers and Benson 1962.)

the failure of the cambium to form a sheath of xylem the entire length of the bole, rings may be omitted, especially near the base. Possible causes are the lack of food manufacture in the crown, drought, fire, extreme cold, insects, and so on. At the other extreme are multiple rings produced by multiple waves of cambial activity during the growing season. They are caused by temporary interruptions in normal growth, such as a late spring frost, or by regrowth after normal seasonal growth has ended. Thus the growth rings reflect the interaction of woody plants and their environment as well as the passage of time.

The fundamental principle of dendrochronology is cross-dating, the correlation of distinctive patterns of growth between trees for a given sequence of years. Because no two plants have exactly the same growing conditions and life history (although the broad features are common to all trees involved), the similarities are relative rather than quantitative. The relative widths of corresponding rings to adjoining rings are the same. By lining up these similarities, the investigator can establish the relative identity of any rings in sequence and aberrant rings in the individual specimen. A great number of specimens must be cross-dated before each ring with a sequence can be dated.

**Collection of Material** A recently logged-over area can provide an abundance of material, but new sections must be cut from the stump. Stump sections should be cut at a 30° angle. The cut may be clean enough for examination or may require smoothing with a carpenter's plane or machine sanding. To study shrubs, cut sample stems close to the ground and use the entire cross section.

The usual method of obtaining samples from forest trees is increment boring. The increment borer, available from forestry supply houses, is an instrument designed to bore a core from a tree. It consists of a T handle, a hollow bit, and an extractor. Increment borers are fairly easy to use, but without care they can be damaged or broken. Here are a few hints:

1. For growth and age studies, remove the core as near to the base of the tree as the instrument handle will allow.
2. Coat the screw with heavy-grade oil.
3. To start the borer, use a strong pushing and twisting motion until the borer is engaged in the wood.

4. Line the borer on the radius, keep the borer straight, and attempt to reach the center of the trunk.
5. When the core is drilled, insert the extractor and press firmly to cut the core from the trunk.
6. Remove the borer with reverse rotation.
7. Paint the wound with tree paint.
8. Store the cores in large diameter soda straws or polyethylene tubing. Be sure to label each sample fully, including the directional side of the tree from which it was removed.

To obtain a freshly cut edge for examination, the core is held firmly in a core holder. The groove in a plastic ruler is fine if the ruler is clamped to a table and the end is stopped. With a razor blade a transverse cut is made the length of the top of the core. It can then be brushed with water or kerosene to make the rings stand out better. When the core is ready, clip it to the stage of a microscope for examination. One-hundred-power magnification usually is strong enough. Looking at the whole core under a dissecting scope is often sufficient.

The width of each ring is measured with a graduated mechanical stage, a stage micrometer, or a dial micrometer. The total distance included in the layers observed can be measured and then compared with the accumulated individual measurements. Any error should be distributed over the individual measurements. For serious research a dendrochronometer, a special instrument with a microscope and precise measuring devices, should be used.

Cross-Dating   The methods and problems of cross-dating are complex. The basic procedure is as follows:

1. On graph paper, write down a series of numbers horizontally from left to right to represent growth layers. You can begin with one or with the years, the first number being the season preceding. This setup gives a series of numbers starting with the present and leading backward through the tree's life. A number of such blanks should be made up.
2. Set up a scale on the graph in millimeters so that the largest bars represent the *narrowest* widths.
3. Make a small bar graph for each year of the tree's life.
4. Make such a coded summary of all wood samples available.
5. Compare these graphs visually, two at a time, sliding them along each other. Keep looking for corresponding groups of years with the same pattern of ring sequences. By such a technique, multiple rings can be checked, or extremely narrow growth rings previously missed can be picked up.

A simpler, but less precise method is to draw a line under the year that has a ring slightly less than the rings adjacent to it; to draw two lines if the decrease is more pronounced; to draw three lines if very narrow; and to draw two lines above the year for very wide rings.

Statistical Analysis   The data can be reduced to average values and then compared to weather data covering the principal growing season for the species. Comparisons can be made between rainfall and the current year's growth, rainfall and the previous year's growth, monthly evapotranspiration deficits and growth, frost-free periods, and so on.

Data will have to be analyzed by simple or multiple regression, depending upon the variables, using partial correlations and standard errors for tests of significance. For example analysis and interpretations, see Fritts (1962) and Cox (1985:83).

## Palynology

Palynology is the study of past plant communities by the analysis of pollen profiles. These studies are especially enlightening if they are coupled with carbon-14 dating.

Peat Collection   Peat cores are bored at one to several stations in a bog. They are taken with a peat borer, available commercially. At each station two separate borings should be drilled, several meters apart. By taking successive samples from alternate borings (for example, first 0.3 m sample from core number one; second 0.3 m sample from core number two) contamination of one sample with another can be prevented. Two 150 cm samples are collected at each boring, one from the lower part of the cylinder, the other from the upper. The samples are placed in glass vials. If the vials are completely filled and tightly sealed, no preservative should be needed (Walker and Hartman 1960).

Treatment of Samples   Back in the laboratory, the samples can be treated as follows:

1. Thoroughly mix each 150 cm sample and remove a pea-size lump for deflocculation.
2. Boil the peat for a few minutes in a dilute solution of NaOH, gently breaking it apart with a wooden cocktail stirrer. Use only one stirrer for each sample to avoid contamination.
3. Add several drops of gentian-violet stain to the boiling mixture.
4. Stir vigorously and strain through fine wire mesh. Then stir again and draw up a 0.5-ml sample into a pipette.
5. Add a very small amount of warm glycerine jelly to the sample and mix.
6. Mount several drops on a slide and add a cover slip.

Examination   The samples, now transferred to slides, should be examined under a microscope equipped with a mechanical stage. Now follow three steps:

1. Tally 100 or 200 pollen grains as they are encountered by systematically moving the slide.
2. Identify each kind of pollen grain, if only by code, and tally the kinds separately. Identification should be made from a reference pollen collection made up beforehand.
3. Record the results from each slide directly as a percentage for each kind of pollen.

Plotting the Pollen Profile The pollen profile can be constructed by plotting a graph for each species or kind of pollen. The vertical scale is set up for depth in meters; the horizontal scale is percentage, based on counts of 100 or 200 pollen grains for each spectrum level.

# SAMPLING ANIMAL POPULATIONS

The study of animals involves considerably more problems than the study of plants. Animals are harder to see and most are not stationary. When it comes to sampling, the animals have something to say about getting caught, and they are more liable to mortality than plants. The following methods of estimating animal numbers, determining age structure, mortality, home range, and so on, enable the field biologist to make some measurements, however rough, of animal populations in the ecosystem.

## Trapping and Collecting

The sampling of an animal population involves collecting animals, either alive, for marking and release, or dead. Detailed information on collecting and trapping is available in other publications (see references at the end of this appendix).

Flying Insects Diurnal insects are collected with aerial nets and heavy-duty sweep nets designed to withstand hard wear when put through grass and woody vegetation. Nocturnal insects may be collected by using traps containing ultraviolet light or a mercury-vapor light or an old sheet fitted on a slant against some support with a strong light above it. Insects can then be picked off the sheet. If the insects are to be killed they are placed in a killing jar containing a layer, either on the bottom or in a deep lid, of plaster of paris and potassium cyanide. Thin layers of tissue or light cloth in the jar prevent damage to moths and butterflies. Another technique is the use of a Malaise trap. This interceptive device uses a series of baffles to herd insects into a closed chamber that may or may not contain a killing fluid.

Aquatic Organisms Aquatic organisms may be collected with dip nets, bottom nets for scraping along the bottom of ponds, wire-basket scraper nets, or plankton towing nets. For collecting from the shore, aquatic throw nets are useful. A bottom dredge lowered from a boat can collect bottom organisms in deep water. Fish, tadpoles, and large crustaceans can be collected with seines. A set of assorted widths will be necessary.

Soil Organisms The most difficult components of soil fauna to study are the soil arthropods. They are the most numerous, the most difficult to identify, and possibly the most difficult to sample accurately. Nematodes, white or pot worms, and protozoans require highly specialized extraction techniques (Murphy 1962).

Soil arthropods can be extracted by means of a Tullgren funnel, an improved version of the Berlese funnel, the construction of which is simple (Figure A.9). Essentially it consists of a heat source, such as a light bulb; a smooth funnel, preferably glass, fitted into a collecting vial; and a screen made of hardware cloth or a sieve meshing inside the funnel. The sample is placed on the screen; the heat and then desiccation drive the arthropods downward, until they fall through the funnel into the collecting bottle.

The procedure is simple:

1. Place the sample of litter or soil on the hardware cloth so fitted in the funnel that air space is present between the wire and the wall of the funnel.
2. To begin extraction, open the lid of the funnel 90° and turn on the 100-watt bulb.
3. After about 16 hours, depending on sample size and moisture content, change to a 15-watt bulb and shut the lid. There will be two periods of arthropod exodus, the first wave due to heat, the second due to desiccation. The collecting bottle beneath the funnel may contain alcohol, formalin, or water. Water may be preferable, because it increases the humidity gradient toward which the animals move.
4. Sort and identify the animals under the microscope.

These funnels are adequate for introductory soil biology. For serious studies in soil zoology, a better extractor is required. An extractor for woodland litter has been described by Kempson, Lloyd, and Ghelardi (1963). The funnels are replaced by wide-mouthed bowls filled with an aqueous solution of picric acid. The acid not only preserves the specimens but also produces by evaporation a high humidity in the air just under the sample. The humid air is cooled by conduction from a cold-water bath in which the bowls are immersed.

Another method of extraction is flotation. Procedures, although simple, are too lengthy to include here. Refer to Jackson and Raw (1966) and Andrews (1972).

Larger soil animals, such as spiders and beetles, can be taken in traps made from funnels and cans set in the soil to ground level. Boards placed on the ground may attract millipedes, centipedes, and slugs. Meat bait in small wire traps will attract scavenger insects.

Sampling earthworm populations presents some difficulties, for no really successful method has been devised to extract the animals from lower layers of the soil. One of the better methods is a combination of formalin and a shovel:

1. Apply a dilute solution of formalin (25 ml of 40 percent formalin to 3.75 liters of water) to a quadrat 0.2 m². Within a few minutes worms will come to the surface.
2. After earthworm movement to the surface stops, pour on a second application.
3. When worms cease to come to the surface the second time, dig out the quadrat as deep as necessary.
4. Hand-sort the soil for maximum recovery. Earthworm cocoons can be extracted by the flotation method.

**Small Animals in Vegetation**  Sweep nets with stout frames to withstand sweeps close to the ground and in woody growth are useful for collecting many types of insects and even some arboreal amphibians and reptiles. Drag nets, consisting of light tubular frames to which are attached canvas bags, are useful on flat ground. Overhead vegetation can be sampled by beating the limbs with sticks to dislodge the animals, which should fall into canvas collecting trays beneath. Other techniques involve traps on which the insects settle on a sticky surface and pit traps in the ground, most useful for spiders and beetles.

**Birds and Mammals**  Birds can be trapped for banding in specially constructed traps, cannon nets for larger game birds, and mist nets. Both federal and state permits are required for such work. Once a permit is granted, the operator of the banding station will be furnished with plans for suitable traps. For mammals, live traps of wood or wire and snap traps are used. Both are available commercially, but live traps are easily constructed. Traps can be baited with natural foods, dripping water, and so on. For small mammals, a mixture of peanut butter and oatmeal works well. Also useful are grain, apple, meat, and appropriate scents.

## Marking Animals

Marking individuals in an animal population is necessary if you wish to distinguish certain members of a population at some future date, to recognize individuals from their neighbors, to study movements, or to estimate populations by the mark-recapture method.

(a)

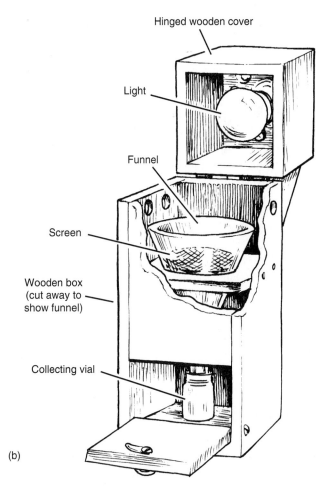

(b)

**Figure A.9** Although the Berlese and other types of funnels can be purchased, they can be constructed easily in the workshop: (a) a simple funnel for introductory work; (b) a more elaborate and efficient design.

Arthropods and snails are best marked with a quick-drying cellulose paint. It is easily applied with any pointed object. Marking butterflies is a two-person operation. One has to hold the wings together with a pair of forceps, while the other marks the side of the wing exposed at rest. For aquatic mollusks and insects better results are obtained through the use of ship-fouling paint, because acetate paints do not hold up well in water.

Fish are usually marked by tagging in several ways. Strap tags of Monel Metal may be attached to the jaw, the preopercle, or the operculum. Stream or pennant tags attached to various parts of the body, usually at the base of the dorsal fin, are used in some studies. Another method is to insert a plastic tag into the body cavity. The tag is inserted through a narrow incision made in the side of the abdominal wall. Once the incision heals, which it does quickly, the tag is carried by the fish for life and is recovered only when the fish is cleaned. Clipping the fins is still another way to mark fish, but it does not permit the individual recognition of a very large number of fish.

Frogs, toads, salamanders, and most lizards can be marked by some system of toe clipping, which involves the removal of the distal part of one or more toes. One method worked out by Martof (1953) is as follows. The toes on the left hind foot are numbered 1 to 5, the toes on the right foot 10 to 50. The left forefoot toes are numbered 100 to 400, and the toes on the right forefoot 800, 1600, 2400, 3200. Thus one can mark up to 6399 individuals by clipping no more than two toes.

Snakes and lizards can be marked by removing scales or patches or scales in certain combinations.

Birds are usually marked by serially numbered aluminum bands and by cellulose and aluminum colored bands. The colored bands are necessary for individual recognition in the field. In some specialized studies, the plumage is dyed a conspicuous or contrasting color.

Small mammals may be marked by toe clipping in combinations similar to those given for amphibians or by notching the ear.

A number of other methods have been devised for marking mammals. Fur clipping and tattooing may be employed. Bear, deer, elk, moose, rabbits, and hares can be marked with strap tags or plastic discs attached to the ear. Aluminum bands similar to those used on birds can be attached to the forearm of bats. Dyes can be used to mark both large and small mammals. Small amphibians and reptiles can be marked by branding (Clarke 1971).

Some species lend themselves to noninvasive means of identification, such as color patterns on the flukes of humpbacked whales and facial patterns and expressions in gorillas. With patience and observation you can identify individuals of other mammals, such as gray squirrels and chipmunks.

Radioactive tracers are particularly useful for studying animals that are secretive in habits, live in dense cover,

spend part or all of their lives underground, or have radically different phases in their life cycle, as the moths and butterflies do. Animals are fed small traces of gamma-emitting radioactive material. The material is metabolically incorporated into the tissue and the tracer becomes a part of the animal. It is passed along to egg or offspring. Radioactive larvae remain so as they transform to adults. The same is true for birds. This technique is useful for studying dispersal, for identifying specific broods or litters, and for obtaining data on population dynamics and natural selection.

Another method involves the application of a radioactive tracer in or on an animal in such a way that the animal is not seriously injured and behaves in a normal way. Usually a radioactive wire is fastened to the animal or inserted under the skin of the abdomen with a hypodermic needle. The movements of the tagged individual are then followed with a Geiger counter.

Although these techniques have their merits, they also have disadvantages. The greatest is the potential radioactive hazard to the investigator, to other humans, and to the ecosystem. Another disadvantage is the impossibility of separating one animal from another. Most work with radioactive tracers requires a federal license. Specific techniques can be reviewed in Tester (1963), Godfrey (1954), Pendleton (1965), Graham and Ambrose (1967), and T. J. Peterle in Schemnitz (1980).

## Aging Animals

Information on the age structure of wild populations is not easily obtained. During the past several decades, a number of aging techniques have been developed, mostly for game and fish.

Fish aging began when Hoffbauer (1898) published his studies on the scale markings of known-age carp. Since then the technique has been refined. It is based on the fact that a fish scale starts as a tiny plate and grows as the fish grows. A number of microscopic ridges, the circuli, are laid down about the center of the scale each year (Figure A.10). When the fish is growing well in summer, the ridges are far apart. During winter, when growth slows down, the ridges are close together. This annual check on growth enables the biologist to determine the age of a fish by counting the number of areas of closed rings, the annuli.

Salmon and some species of trout spend one or two years in streams before migrating out to sea or into lakes. Because stream growth is slower than lake or sea growth, the scales show when the fish migrate. When salmonid fish spawn, reabsorption of scales occurs, eroding the margins of the scales and interrupting the pattern of circular ridges. This erosion leaves a mark that can be detected in later years.

Because the growth of a scale continues throughout the life of a fish, it also provides information on the growth rate. This rate is obtained by measuring the total radius of the scale, the radius to each year's growth ring, and the total

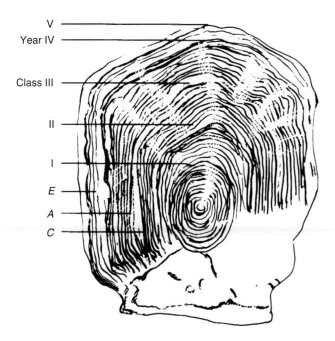

**Figure A.10** The age of a fish can be determined by the growth rings on the scales. C is the circuli; A, the annuli; E, the erosion of the scale from spawning.

body length of the fish. Then by simple proportion, the yearly growth rate can be determined.

Other techniques in aging fish include the length-frequency distribution, vertebral development, and rings or growth layers in the otolith or ear stone (see Jerald 1983).

Because of the large number of year classes (animals born in a population during a particular year) that can be identified, we can determine dominant year classes, learn the age when fish reach sexual maturity, estimate production mortality, and estimate the effects of fish harvest.

Aging techniques for mammals and birds have developed more slowly and are not as refined as those for fish, but a number of methods are in common use. The age of the animal may be indicated by specific characteristics of body parts (Figure A.11). Among birds, plumage development is frequently used. Until molted, the tail feathers of juvenile waterfowl are notched at the tip, in contrast to the normally contoured feather of the winter plumage. The shape of the primary wing feather separates adults from young among many gallinaceous game birds. The presence or depth of the bursa of Fabricus, a blind pouch lying dorsal to the cecum and opening into the cloaca, indicates juvenile birds.

Among mammals the examination of reproductive organs is useful because the majority do not breed until the second year. This method can be used only during the breeding season. The presence of epiphyseal cartilage in rabbits, squirrels, and bats (Figure A.11) identifies juveniles up to 6 or 7 months. Black bars on the pelage of the underside of the

tail of juvenile gray squirrels separate the young from the adults. Primeness of pelt on the inside of skins is a good means of aging muskrat during the trapping season. Dark pigmentation on the flesh side of the pelt indicates areas of growing hair. This pigmentation in adults appears in irregular, scattered dark areas, whereas in immature animals it is more or less symmetrical and linear. Skull measurements are useful in beavers and muskrats. Annual growth rings on the roots of canine teeth indicate age for the first few years of life in the fur seal and other pinnipeds and in canids. Growth rings also show up in the horns of mountain sheep. The wear and replacement of teeth in deer and elk permit the determination of different age classes in these mammals (see Schemnitz 1980).

Because the lens of the eye of most mammals (and possibly birds) grows continuously throughout life and because there is only slight variation among individuals in lens size and growth, the measurement of the lens is a feasible method for aging a number of mammals. It has been done successfully for the cottontail rabbit, raccoon, black bear, and fur seal.

The technique involves comparing the weight of the dry lens with the chart of lens weights of known-age individuals (Figure A.12). The investigator may have to develop the chart by rearing young animals in captivity and sacrificing them week by week for their eyes. A table has been prepared for rabbits by Lord (1959, 1963) and for cotton rats by Birney, Jenness, and Baird (1975).

The technique is as follows:

1. Remove eyes as soon as possible after the animal is killed and place in a solution of 10 percent formalin. The formalin will harden the lens so that it can be removed from the vitreous humor.
2. Fix for a minimum of 1 week, but the longer the better.
3. After fixing, remove the lens from the eye and roll it on a paper towel for a few minutes to remove excess moisture.
4. Place the lens in an oven to dry at 80°C.
5. Lenses are considered dry when repeated weighing after intervals of drying results in no additional loss of weight. This step will usually require 24 to 36 hours.
6. Weigh immediately after removal from oven, because the dried lenses are hygroscopic and take on water. Electronic digital scales that read weights rapidly are preferred over other types of balances.

The lens growth curve permits a close approximation of the age of the mammal. For cottontail rabbits, the method permits the determination of the month of birth of young rabbits and the year of birth of young rabbits over one year of age.

There are statistical and procedural problems in the use of aging techniques, including the lens-weight technique. See Dapson (1980) for guidelines in the use of statistics in age estimation.

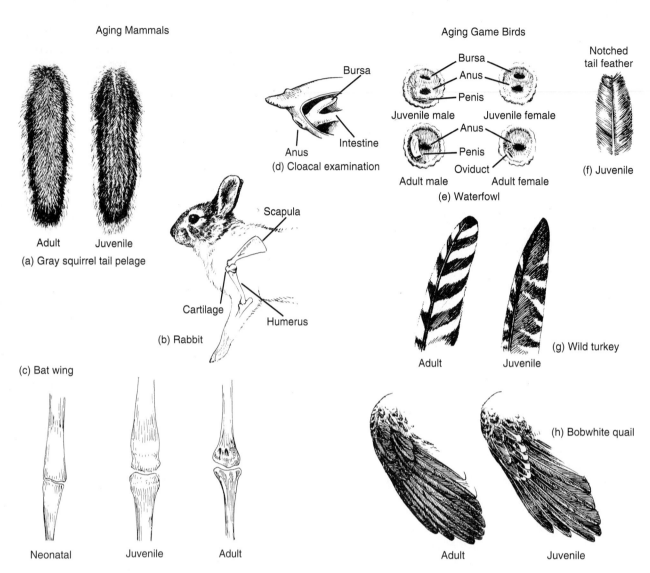

**Figure A.11** Age determination in some mammals and birds. (a) Regular barring on the underside of the tail distinguishes the juvenile gray squirrel from the adult. (b) Epiphyseal cartilage on the humerus of juvenile cottontail rabbits separates that age class from the adult. (c) The epiphyseal cartilage of the fourth metacarpal-phalangeal joint separates juvenile from adult living bats. The degree of ossification in the joint can be determined by extending the wing and holding its ventral surface firmly against the glass stage of a dissecting microscope. Also useful in age determination of young bats are forearm lengths between 14.4 mm and 35.5 mm regressed against age (up to 11–12 days). (See Kunz and Anthony 1982.) (d) The bursa of Fabricus (enlarged). Its presence or greater depth indicates a juvenile bird. The depths vary with the species. This method is useful in both waterfowl and some gallinaceous birds. (e) Sexing and aging waterfowl by examining the cloaca. Note the presence of the bursal opening on juvenile waterfowl and its absence on adult waterfowl. (f) The notched tail feather of juvenile waterfowl. (g) The number X (ten) primary in juvenile gallinaceous birds is sharply pointed; in adults it is rounded. The juvenile wild turkey in addition has its outer primary indistinctly barred. (h) The juvenile bobwhite quail, in addition to having a pointed number X primary, possesses buff-tipped primary coverts. (For a more complete discussion on aging see Schemnitz 1980, DeBlase and Martin 1981.)

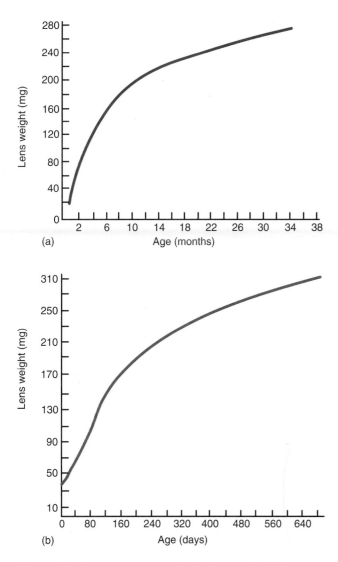

**Figure A.12** Growth rate curves for the lenses of rabbits (a) Cottontail rabbit. (From Lord 1961.) (b) Black-tailed jack rabbit. (After Tiemeier and Plenert 1964.)

## Determining Sex

The sex of mammals in most instances can be determined by examining external genitalia, and the sex of birds by plumage differences (Figure A.13). For example, the male ruffed grouse can be distinguished from the female by the length of the central tail feather; the female prairie chicken can be distinguished from the male by strong mottling or barring of all the tail feathers; the male wild turkey has black-tipped body feathers, whereas the female has brown-tipped feathers. (For detailed information, see Schemnitz 1980.)

## Determining Home Range and Territory

A number of methods are available for obtaining an approximation of the size of home range. Five methods are offered here.

**Home Range Map** On a map, outline and measure the area that includes all the observations made on the movements of individuals. If the observations are obtained by trapping, then assume that the animal could have gone halfway toward an adjacent trap, especially if the traps were set in a regular grid. For techniques and analysis see Anderson (1982), Dixon and Chapman (1980), Samuel and Garton (1985), Samuel et al. (1985).

**Center of Activity** Arrange the recaptures on a grid and determine the values on X and Y axes. An average of these locations will give the center of activity (see Hayne 1949). This method has the advantages that the information is easy to summarize and the calculations are not complicated. However, a map of the area must be made and many recaptures of the same individual are required before the extent of the home range can be obtained. As the number of recaptures increases, the area of the known range increases. At least 15 recaptures or more are necessary. This method is unsuited for mammals that follow paths or tunnel underground.

**Frequency of Capture** Record the distances between captures in live traps set randomly or in grids. Record the number of captures as a frequency distribution according to the distance between them. Distances between captures are then tallied and proportions calculated for each distance by sex or age categories.

The distances can be measured in two ways. They can be taken from the place where the animal was first marked or observed or from each successive location.

This method has the advantage that the traps can be set out haphazardly, avoiding the labor of setting them out in grids. The recaptures of all individuals can be used. Information can be obtained during a short period of time because the data from animals captured only two or three times can be used. The disadvantages of this method are that short movements are favored and that no definite boundaries of home ranges can be given. Home range is described as a frequency of distances observed.

**Territory** Because territorial boundaries are rigidly maintained by birds during the breeding season, territorial boundaries can be mapped by observing the movements of the birds during the day, by plotting singing perches, by observing locations of territorial disputes, and on occasion by chasing the bird (Wakeley 1987c). When the bird arrives at the boundary of its territory, it generally will double back. This technique is most useful in studies of a single species. Do not confuse it with spot mapping (see Verner 1985).

**Radio Tracking.** The development of transistors and other miniature electronic devices has made possible the construction of small transmitters that can be attached to animals, usually by a specially designed collar or harness. Mercury cells are the source of power. The transmitter is a transistor-crystal-controlled oscillator with the tank coil for the oscillator acting as the magnetic dipole transmitting

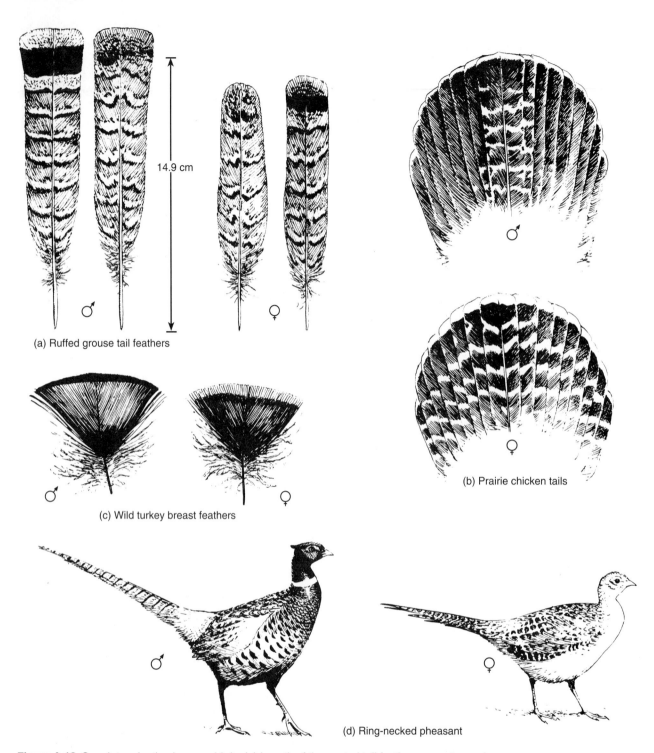

14.9 cm

(a) Ruffed grouse tail feathers

(c) Wild turkey breast feathers

(b) Prairie chicken tails

(d) Ring-necked pheasant

**Figure A.13** Sex determination in game birds. (a) Length of the central tail feather separates male and female ruffed grouse. The male tail feather is 14.9 cm long or more; the female tail feather is shorter. (b) The female prairie chicken has heavily barred tail feathers; barring is absent in the outer tail feathers of the male. (c) The breast feathers of the male wild turkey are black-tipped; those of the female are brown-tipped. (d) Sexual dimorphism distinguishes males from females in many species of birds.

antenna. The antenna is constructed of copper or aluminum and has a figure-8 directional pattern. The receiver is a portable battery-powered unit, whose basic components are the receiver, a radio range filter, and two transistorized radio-frequency converters. Positions of stationary animals can be obtained by a single portable direction-finding receiver. Running animals are best located by triangulation, using at least two direction-finding receivers. Radio tracking equipment is available commercially.

Although highly useful in obtaining data that could not be acquired otherwise, radio tracking has its disadvantages for most investigators. The most serious disadvantage is the need for electronic expertise. Unless you are an electronics whiz or have the assistance of an electronics technician, you should not attempt radio tracking. Other disadvantages are cost and the need for an FCC license.

# Estimating Numbers

Basic to the study of animal populations is the estimation of their numbers, no small task in wild populations. During the past several decades, much work has gone into the development of techniques and statistical methods to arrive at some estimates of animal populations. Basically the methods of estimating the numbers of animals can be put into three categories: true census, a count of all individuals on a given area; sampling estimate, derived from counts on sample plots; and indices, the use of different types of counts, such as roadside counts, animal signs, and call counts, to determine trends of populations from year to year or from area to area.

True Census   A **true census** is a direct count of all individuals in a given area. It is difficult to do for most wild populations, but there are situations where a total count can be made.

Many territorial species are easily seen and heard and can be located in their specific area. Such a census is regularly used for birds. The spot mapping method is probably the best approach (Wakeley 1987b). A sample plot of at least 10 ha is marked out in a grid with numbered stakes or tree tags placed at intervals of 50 m. Five or more daily counts are made throughout the breeding season. Each time a bird is observed, it is marked on a map of the plot. At the end of the census period all the spots at which a species is observed are placed on one map. The spots should fall into groups, with each group indicating the presence of a breeding pair. The groups for each species can then be counted in order to arrive at the total population for the given area. Results are usually expressed as animals per hectare.

Direct counts can be made in areas of concentration. Deer in open country, herds of elk and caribou, waterfowl on wintering grounds, rookeries, roosts, and breeding colonies of birds and mammals permit direct counting either from the air or from aerial photographs. Coveys of bobwhite quail can be located and counted with the aid of a well-trained bird dog.

Sampling Estimates   A **sampling estimate** of population size involves two basic assumptions: (1) mortality and recruitment during the period the data are being taken are negligible or can be accounted for; and (2) all members have an equal probability of being counted—they are not trap-shy or trap-addicted, they are distributed randomly through the population if marked and released, and they do not group by age, sex, or some other characteristic.

Sampling also involves one major general consideration. The method employed must be adapted to the particular species, time, place, and purpose.

Relatively immobile forms, such as barnacles, mollusks, and cicada emergence holes, can be estimated by the quadrat method, similar to that employed for plants. The data can be analyzed for presence, frequency, and so on, or the results can be converted to a density per hectare. The size and shape of the quadrat will depend upon the density of the population, the diversity of the habitat, and the nature of the organism. A few preliminary surveys are made before settling on a quadrat size.

Foliage arthropods may be sampled by a number of strokes with a standard sweep net over a 10 m² area. The number of strokes needed to secure the sample must be predetermined. It will vary with the type of vegetation.

Estimates of zooplankton, obtained by pulling a plankton net through a given distance of water at several depths, can be made by filtering a known volume of sample through a funnel using a filter pump. The filter paper should be marked off in equal squares. With the aid of a hand lens or a binocular microscope, the organisms in each square can be counted. The numbers then can be related back to the total volume of water sampled.

If the organisms are too small to be counted in this manner, a Rafter plankton-counting cell can be used. It consists of a microscope slide base plate ruled into ten 1 cm squares. The slides are made from strips of microscope glass slides cemented to the base with Canada balsam. This cell should hold 1 cc of liquid. After a small volume of water is introduced, the cell is covered with a long cover glass and placed under the microscope. The organisms are counted square by square and the number of each form recorded per square until at least 100 observations have been made. The occurrence of individual species can be recorded as a percentage frequency. (Note: Plankton-counting cells and eyepiece micrometers can be purchased commercially, but they are expensive.)

Capture-Recapture Sampling   **Capture-recapture methods** are based on trapping, marking, and releasing a known number of marked animals into the population, and then later recapturing individuals from the population after an appropriate interval of time (approximately one week for rabbits and mice). An estimate of the total population is then computed from the ratio of marked to unmarked individuals in the sample, which supposedly reflects the ratio of marked to unmarked animals in the population.

Some points about the capture-recapture method need to be emphasized. The size of the population is estimated by the ratio of marked to unmarked animals in the population, not the number of recaptures. The population estimate is for the time the marked animals are released in the population, not the time of recapture. The estimate is biased because one is examining a sample of the population rather than the whole population, and the precision is dependent on the number of marked animals recaptured.

The capture-recapture method involves a number of assumptions:

1. All individuals in the population have an equal chance of being captured.
2. The ratio of marked to unmarked individuals remains the same from the time of capture to the time of recapture.
3. Marked individuals redistribute themselves homogeneously throughout the population with respect to unmarked ones.
4. Marked individuals do not lose their marks.
5. The population is closed. No emigration or immigration takes place within the sampling period.

Because populations are dynamic, the investigator needs to know some features of the population and adjust accordingly. They include reproductive history, mortality patterns, the effects of marking on behavior, the seasonal patterns of movements, and the bias of age and sex on the individual probability of capture. In effect, what the investigator faces is unequal probabilities of capture.

The *single mark-single recapture method* is known as the *Lincoln index* or *Petersen index* of relative population size. The basic model is

$$N : M :: n : R$$

or

$$N = nM/R$$

where $M$ = number marked in the precensus period, $R$ = number of marked animals trapped in the census period, $n$ = total animals trapped in the census period, and $N$ = the population estimate. The probability of capture on any given occasion is $R/n$.

This basic model assumes that every animal has the same capture probability $p$ on every capture occasion. It involves only two parameters, $N$ and $P$, and only two data, the number of captures at each time $t$.

Capture probability can be influenced by effects of time, behavioral response to capture, and variations in probability of capture among individuals (trap-happy, trap-shy). To account for these factors, White and associates (1982) provide

variations on the basic model and a Fortran computer program for them.

As an example, suppose that in a precensus period, biologists tag 39 rabbits. During the census period they capture 15 tagged rabbits and 19 unmarked ones, a total of 34. The following ratio results:

$$N : 39 :: 34 : 15$$

or

$$nM/R = (34)(39)/15 = 88$$

The estimated population is 88 rabbits.

The confidence limits at the 95 percent level may be calculated from

$$\text{S.E.} = N \sqrt{\frac{(N-M)(N-n)}{Mn(N-1)}}$$

(S.E. is Standard Error.)

To determine the limits within which the population lies, add and subtract two standard errors from the estimate. A large standard error and rather wide confidence limits are the result of a small number of recaptures and a small sample size. For this example:

$$\text{S.E.} = 88 \sqrt{(88-39)(88-34)/(34)(39)(87)}$$
$$= 88 \,(0.1513) = 13.31$$

Upper limit: $88 + 26 = 114$

Lower limit: $88 - 26 = 62$

The chances are 95 out of 100 that the population of rabbits lies between 62 and 114. Such results are typical in population studies of animals.

*Repeated mark and recapture* is a variation of the single mark-single recapture method. Investigators capture, mark, and recapture animals over $t$ occasions. The method expands the estimation beyond the $t = 2$ occasions of the Lincoln-Petersen index and provides a series of population estimates, which can be repeated until the investigator is satisfied with the results. The most familiar of these methods is the Schnabel estimate, developed by Zoe Emily Schnabel in 1938.

An example is given in Table A.7. All animals captured are tagged or marked and released daily. A record is kept of the total animals caught each day, the number of recaptures, and the number of animals newly tagged. The assumptions and the calculations are similar to those of the Lincoln-Petersen method. However, in the Schnabel method, $M$ (the number marked in the precensus period) becomes progressively larger. Population estimates can be calculated daily or at intervals during the sampling period, for example weekly. True confidence limits cannot be determined, and the calculation of the standard error becomes complex. (For details, see Ricker 1958.)

*Multiple capture-multiple recapture* is another variation, in which all animals caught on any particular day are marked and released, including the recaptures. Thus an ani-

**Table A.7** Schnabel Method of Estimating Populations

| Period (Date) (P) | Number Trapped (A) | Number Marked | Marked Animals in Area (B) | (A) × (B) | (A) × (B) Sum | Recaptures | Sum of Recaptures (C) | $\frac{(A) \times (B)}{(C)}$ Estimated Population |
|---|---|---|---|---|---|---|---|---|
| 1 | 4 | 4 | — | 00 | 0 | — | — | — |
| 2 | 4 | 4 | 4 | 16 | 16 | 0 | 0 | — |
| 3 | 2 | 2 | 8 | 16 | 32 | 0 | — | — |
| 4 | 6 | 6 | 10 | 60 | 92 | 0 | — | — |
| 5 | 10 | 7 | 16 | 160 | 252 | 3 | 3 | — |
| 6 | 4 | 4 | 23 | 92 | 344 | 0 | 3 | — |
| 7 | 8 | 6 | 27 | 216 | 560 | 2 | 5 | — |
| 8 | 4 | 2 | 33 | 132 | 692 | 2 | 7 | — |
| 9 | 5 | 4 | 35 | 175 | 867 | 1 | 8 | — |
| 10 | 7 | 6 | 39 | 273 | 1140 | 1 | 9 | — |
| 11 | 7 | 6 | 45 | 315 | 1455 | 1 | 10 | 145 |
| 12 | 9 | 7 | 51 | 459 | 1914 | 2 | 12 | 159 |
| 13 | 6 | 3 | 58 | 348 | 2262 | 3 | 15 | 150 |
| 14 | 10 | 6 | 61 | 610 | 2872 | 4 | 19 | 151 |
| 15 | 8 | 5 | 67 | 536 | 3408 | 3 | 22 | 154 |
| 16 | 6 | 1 | 72 | 432 | 3840 | 5 | 27 | 142 |
| 17 | 4 | 2 | 73 | 292 | 4132 | 2 | 29 | 142 |
| 18 | 12 | 7 | 75 | 900 | 5032 | 5 | 34 | 148 |
| 19 | 8 | 4 | 82 | 656 | 5688 | 4 | 38 | 149 |

mal caught on day 1 and again on day 2 will bear the marks of both days. By such a method, one can both keep an account of total marks recaptured each day of trapping and relate recaptures to the initial day of marking (see Jackson 1939; Fisher and Ford 1947; Blower, Cook, and Bishop 1981; Krebs 1989). The basic equation is

$$\hat{P}_i = \frac{n_i \hat{M}_i}{m_i}$$

where $\hat{P}_i$ is the estimate of population on day $i$; $\hat{M}_i$ is the number of marked animals or marks in a sample (number of recaptures or number marked previously); $n_i$ is animals captured, marked, and released on day $i$; and $m_i$ is the number of marked animals or marks in sample on day $i$. The formula is essentially that of the Lincoln index.

For bookkeeping the trellis diagram is most convenient (Figure A.14). There are two types of trellises. In the Type I trellis, employed here, all marks are counted (Fisher-Ford method). The number of recaptured marks is entered into the cells of the trellis. Because some animals will carry more than one mark, a larger number of marks may be entered than marked animals. In the Type II trellis marked individuals are entered according to their marks. The Type II trellis is employed in the Jolly-Seber method (Jolly 1965, Seber 1973, Southwood 1978, Blower, Cook, and Bishop 1981).

In the Type I trellis diagram the marginal column on the left running downward from left to right contains the total animals captured for each day. The marginal column on the right contains the total release each day. In the body of the table are the figures for the recaptures as they relate to the

day they were originally marked. For example, on trapping day 5, 220 captured, 30 had been marked on day 4, 13 on day 3, 8 on day 2, and 2 on day 1. The data can provide two different population estimates. A column starting at any date and running downward to the left gives the necessary information for determining the population on the day of recapture. Columns starting at any date and running downward to

**Figure A.14** Trellis diagram.

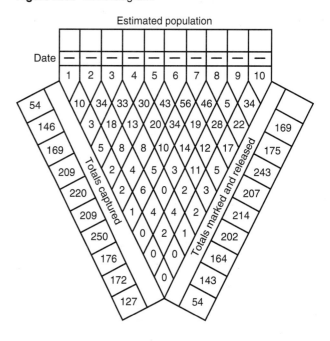

the right give the necessary information to determine the population size on the day of release. To use this method, trapping does not have to be done on consecutive days, but only at equally spaced times.

Although new raw data can be used to estimate population, it is more useful if the recapture values are corrected to the number of marked recaptures per 100 marked on day $i$ and per 100 in the recapture sample. This correction must be done for each recapture value in the body of the table. The corrected values are then substituted for the raw values. The formula is

$$y = \frac{X_i}{1} \times \frac{100}{M} \times \frac{100}{n_i}$$

where $X_i$ = number recaptured on day $n_i$, sample $i$; $M$ = total number of marked animals initially released (prior to time of sample); $n_i$ = total number of marked and unmarked animals in sample $i$; and $y$ = corrected recaptures in sample $i$. For example, on day 5, there were four sets of recaptures: 30, 13, 18, and 2, out of 220 captures. To correct these values:

$$\begin{aligned} \text{recaptures} \\ \text{from day 4:} y \end{aligned} = \frac{30}{1} \times \frac{100}{202} \times \frac{100}{220}$$

$$= 6.7$$

$$\begin{aligned} \text{recaptures} \\ \text{from day 3:} y \end{aligned} = \frac{13}{1} \times \frac{100}{164} \times \frac{100}{202}$$

$$= 3.6, \text{ and so on}$$

The next step is to calculate a weighted ratio, $r$, to show the rate of decrease of recapture values. Determine $r$ for each column running to the right, which gives you $r+$ values, size of the population on the day of release. Do the same for each column running to the left, which gives you $r-$ values for the size of the population on the day of recapture. For three or four values for $y$, the formula is

$$r = \frac{y_2 + y_3 + \ldots \, y_n}{y_1 + y_2 + \ldots \, y_{n-1}}$$

For more than four values for $y$, the formula is

$$r = \frac{y_3 + y_4 + \ldots \, y_n}{y_1 + y_2 + \ldots \, y_{n-2}}$$

With this weighted ratio for each method you can calculate the theoretical number of recaptures that would have been obtained on the day of release:

$$y_0 = \frac{y_1 + y_2 + \ldots \, y_{n-1}}{r}$$
$$- (y_1 + y_2 + \ldots + y_{n-2})$$

With the theoretical values for recaptures at the time of release, you can estimate population size for each day (Table A.8) by the following:

$$N \text{ (on day } i) = \frac{100}{1} \times \frac{100}{y_0}$$

**Table A.8   Fisher-Ford Population Estimates**

| Day | Captured | Released | Recaptured | Survival | Population Size |
|-----|----------|----------|------------|----------|-----------------|
| 2   | 146      | 143      | 10         | 10       | 456.3           |
| 3   | 169      | 164      | 37         | 40       | 460.6           |
| 4   | 209      | 202      | 56         | 84       | 572.1           |
| 5   | 220      | 214      | 53         | 88       | 853.5           |
| 6   | 209      | 207      | 77         | 133      | 659.2           |
| 7   | 250      | 243      | 112        | 210      | 581.1           |
| 8   | 176      | 175      | 86         | 162      | 596.2           |
| 9   | 172      | 169      | 64         | 189      | 725.3           |
| 10  | 127      |          | 84         | 183      | 384.0           |

For convenience and for comparison of population estimates by the two methods, positive and negative, you may tabulate your calculations in a table headed as follows.

| Day | $r+$ | $y_0$ | $N$ | $r-$ | $y_0$ | $N$ |
|-----|------|-------|-----|------|-------|-----|

This method works well where large numbers of animals are involved, such as insect populations.

**Removal Sampling**   The **removal method** of estimating populations, is used in studies of small mammals. The method involves the removal of individuals from the population either permanently or for the duration of the study.

The assumptions underlying this method are: (1) the population is essentially stationary; (2) the probability of capture during the trapping period is the same for each animal exposed to capture; (3) the probability of capture remains constant from trapping to trapping. The last assumption requires that the animals are not trap-shy or trap-prone; and that bait acceptance, weather conditions, and differences in sex and age will not affect the probability of capture. Meeting these assumptions is difficult because dominant mammals, especially males, usually are more trap-prone, and weather can influence mammal movements.

The procedure requires at least two periods of sampling. Sampling requires setting the same number of traps for several days or nights. For small mammals these traps may be snap traps or live traps. If live traps are used, all mammals caught in each sampling period must be marked and not counted if trapped again. The sampling may involve successive removals of animals from the population, the Hayne (1949) method; or it may involve only two periods of sampling, the Zippin (1958) method.

A field procedure is as follows:

1. Set the traps in a grid system, 3 traps to a station; or in two parallel lines 50 meters apart, 20 stations to a line, 3 traps at each station. Space the stations 25 or 50 meters apart, depending upon the vegetation.
2. Prebait for best success.

3. Trap for several successive periods, or for two periods on each of several dates. Depending upon the species, trapping periods may be 24 hours or nighttime only.

A problem with trapping over a number of successive periods is the possibility of attracting new animals into the sampling area as the population is reduced, particularly if the sampling design is a grid. Traps on the outer stations usually capture more animals than those in the center of the grid, because animals in the border zone react to the sudden removal of animals in the center. These immigrants, picked up by the border traps, contribute significantly to the catch on the outer grid lines and influence population estimates.

In the Hayne method the daily catch is plotted against the number of animals previously caught (Figure A.15). A line can be drawn through the data points to cut the horizontal axis. The point at which the horizontal axis is cut represents the population estimate. A more accurate method is to calculate a simple regression line from the catch data. The slope of the line represents the average proportion of the population removed during each sampling period.

In the Zippin method, $N$ is the population size, $n_1$ is the number of animals caught and removed in the first sampling period, and $n_2$ the number caught and removed in the second sampling period. The proportion of the original population captured in the first sampling period is $n_1/N$; $N - n_1$ animals remain; and the proportion of animals caught in the second sampling period is $n_2/(N - n_1)$. An estimate of $N$ is given by

$$N = n_1^2/n_1 - n_2.$$

The two methods give approximately the same answer. In the example given, the Hayne method involving a three-day removal trapline gave an estimated exposed population of 18. The Zippin method, using the catch for the first two days, gives the same answer:

$$N = 36/6 - 4 = 18$$

The removal method is useful when you desire a relative measure or index figure for small mammal populations to compare one habitat with another. Data will be more useful if details on vegetation and litter are recorded for each station. Often some association can be established between vegetation and trapping success.

**Line Transect**   The **line transect** is widely used in terrestrial vertebrate ecology. The method involves walking a line established in an area and noting individual animals observed along that line. The results provide an index rather than an absolute measure of density.

The line transect is most useful in situations in which the animals are difficult to see. There are a number of variations of the line transect, which was developed by R. King to census ruffed grouse and became known as the King census.

The method involves the establishment of a straight line or series of straight line segments in the area to be sampled, marking the line or lines well, walking the line, carefully

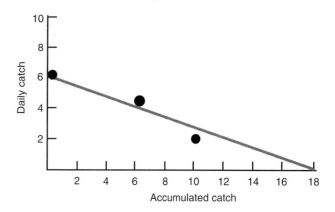

**Figure A.15**  Daily catch on a three-day removal trapline plotted against accumulated catch to estimate the exposed population.

noting any individuals seen or flushed along the line, and measuring the perpendicular distance from the line to the point of sighting or flushing; the sighting distance, the distance of the flushed animal from the observer; or the sighting angle. Preferably all three measurements should be taken (Figure A.16).

The estimator of density from line transect samples is expressed in general as

$$D = nf(0)/2L$$

where $n$ is the number of animals sighted, $L$ is the length of the transect in meters, and $f(0)$ is distances or angles.

The following formula suggested by Haynes (1949) is a modification of the original King formula:

$$D = 10^4 \Sigma(1/d_i)/2L$$

where $d_i$ is distance from the observer to the $i$th animals sighted, measured in meters to the point where the animal was at the time it was flushed; and $10^4$ is a factor for converting m$^2$ to ha.

This line transect method involves four assumptions:

1. Animals on the line are always seen.
2. Animals do not move before being sighted, and none is counted twice.
3. There are no measurement errors and no rounding errors.
4. Sightings are independent events; that is, flushing one animal does not cause another to flush.

As a sampling scheme the line transect is easy to use, but two problems encountered are defining a straight line and holding to it and obtaining accurate measures of distances and angles. Anderson and associates (1979) offer these suggestions:

1. Keep the center line of the transect straight and well marked.
2. Take care that all animals on the center line of the transect are seen with a probability of 1.

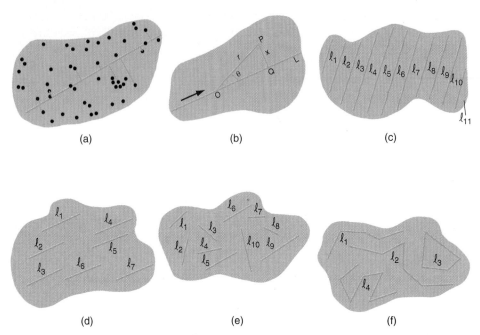

**Figure A.16** (a) Diagram of a line transect sample. The area, with known boundaries, has points distributed within it. Points detected show the perpendicular distance to the line. Points without lines are not detected. Note that points on the line are always detected, whereas those at a greater distance have a low probability of being detected. (b) Possible measurements for a detected object. The observer is at position O when the object is detected at position P, and Q is that point on the line perpendicular to the object. The sighting distance is $r$, the sighting angle is $[\theta]$, and the perpendicular distance from the object to the center line is $x$. The direction of the observer's travel is from O to Q. (c) Transect lines chosen in a systematic, nonrandom fashion. The initial transverse line is at a random distance from the boundary of the area to be surveyed. (d) Individual transect lines dispersed in a random fashion across the study area. All lines are oriented in the same direction. (e) Transect lines need not parallel each other, but can be dispersed across the study area, as long as one line does not meet or cross other lines. (f) In rough terrain transect lines may follow natural features, avoiding boulders, rock outcrops, and other areas difficult to traverse. (Adapted from Burnham, Anderson, and Laake 1980.)

3. Take accurate measurements of distances and angles.
4. Take all three basic measurements: perpendicular distance, sighting (flushing) distance, and sighting (flushing) angle.
5. Record measurements separately for each segment of the total transect length.
6. Record at least 40 animals.

For a successful survey, the observer should take other steps to ensure adequate sampling. Conduct the survey at a time of day appropriate for the species. For songbirds this time would be early morning, for grouse throughout the day, for woodcock in the evening. Conduct surveys under similar weather conditions. Finally, use only well-trained observers, and as few of them as possible to reduce the bias of individual variation in data collecting.

The line transect method may be used with both immobile and mobile populations. Assumptions are most easily met with such immobile populations as dead deer (winter kill), land snails, bird nests, and the like. Most of the problems are encountered with mobile animals—easily flushed animals such as grouse, hares, and rabbits or slow-moving animals such as turtles.

A modification of the transect method is the **strip transect** censusing of songbirds developed by Emlen (1971, 1977 Wakeley 1987a). This method involves tallying birds detected by sight or song on either side of the center transect line. The assumption is that the observer will detect all active birds near the center of the line and for some distance beyond, called the detection distance. Detection distance drops off from the center of the line. The width of the transect line is determined by the full detection distance. For songbirds this distance is about 125 m to either side of the line, assuming the observer's hearing is good. For details consult Emlen (1972, 1977) Cox (1985) and Wakeley (1987a).

Indices **Indices** are estimates of animal populations derived from counts of animals signs, calls, roadside counts, and so on. In this type of estimating, all data are relative and must be compared with data from other areas or other time periods. The results do not give estimates of absolute populations, but they do indicate trends of populations from year to year and from habitat to habitat. Often this type of information is all that is needed.

*Call counts* are used chiefly to obtain population trends of certain game birds, such as the mourning dove, bobwhite quail, woodcock, and pheasant. A predetermined route is established along country roads; it should be no longer than can be covered in one hour's time. Stations are located at quarter- or half-kilometer intervals, depending upon the terrain and the species. The route is run around sunrise for gallinaceous birds and doves and around sunset for woodcock. The exact time to start must be determined for each area by the investigator. The observer stops at each station, listens for a standard period of time (a minute or two), records the calls heard, and goes on to the next station. Routes should be run several times and an average taken. The number of calls divided by the number of stops gives a call-index figure.

The *roadside count* is similar to a call count, with the exception that the number of animals observed along the route is recorded and the results divided by the number of miles. Other variations include counting of animal tracks, browse, signs, active dens and lodges, and so on.

*Counting pellets or fecal groups* is widely used to estimate big-game populations. This method involves the counting of pellet groups in sample plots or transects located in the study area. It may be used to estimate intensity of use of the range by one or more kinds of animals, to determine trends in animal populations, or in rarer cases to estimate the total population. The last is possible only when an entire herd is known to occupy a given area for a definite period. Intensity of use is usually expressed as the number of pellets or pellet groups per unit area.

The accuracy of estimating populations by this method depends upon some knowledge of the rate of defecation. Herein lies the weakness of the technique, because pellet groups vary with the diet, season, age, sex, rate of decomposition, and type of vegetation (plants can cover the pellet groups). Usually rates vary with the region, so preliminary observations have to be made to arrive at some useful figure. For deer, a pellet-group figure of 15 (per deer) is satisfactory on good range, and 13 for poor range. Rabbits vary too widely in their pellet groups for the technique to have much value with them.

The field procedure is as follows:

1. On randomly located transect lines, establish a number of rectangular plots of 100 m² divided in half longitudinally for ease in counting.

2. Count pellet groups at the most favorable time, when plant growth, leaf fall, and so on are least likely to interfere.
3. Mark the plots permanently and clear or paint the pellet groups at the beginning of the study where age determination of pellets is difficult.
4. Then let

$$t = \frac{1}{na^1} y$$

   where $y$ is the sum of pellet groups counted over the plots, $a'$ is the area of one plot, $n$ is the number of plots, and $t$ is pellet groups per unit area.
5. Determine the value of $t$. To translate $t$ to total deer days of use:
6. Assume a defecation rate of 13 pellet groups/deer/day for poor range, 15/day for adequate range.
7. Determine the period, the number of days over which the pellet groups were deposited (for example, since the last count).
8. Divide $t$ by the defecation rate to obtain days of utilization by deer per hectare.
9. Divide the number of days of utilization by the number of days in the period to obtain the number of deer per hectare (assuming a constant population).
10. Multiply the result by 100 ha to obtain the number of deer per square kilometer.

A modification of the pellet-group method for small mammals is the *dropping-board method* (see Emlen et al. 1957).

1. Set out 10 cm squares of weatherproof plywood (in natural color) in lines or grids. Use at least 100 boards. They will cover .64 ha if spaced 10 m apart, 1.9 ha with 15 m spacing, and 7.5 ha with 30 m spacing. Be sure the squares are level and placed firmly on the ground.
2. Number each station.
3. Bait the boards or not, depending upon local conditions.
4. Make a series of at least three visits. The time of day the visits are made and their frequency will depend upon local conditions, such as coprophagous insect activity. Daily visits may be necessary.
5. At each station record the presence of droppings by species, and brush the board clean for the next visit. The droppings of small mammals are distinctive and with some experience can be identified (see Murie 1954).
6. Express results as incidence of droppings for each species.

Figures obtained from the record of usage are indices of the population useful in comparative studies of interspecific, interseasonal, and interregional abundance. The dropping-board technique can be used in studies of population trends and fluctuations, local distribution, species association, activity, rhythms, effects of weather and environmental conditions on activity, and movements if the animal is tagged with a radioactive tracer or with dyed bait.

## ▍REFERENCES

### Sampling Terrestrial Vegetation

ANDERSON, D. R., et al. 1976. *Guidelines for Line Transect Sampling of Biological Populations.* Utah Cooperative Wildlife Research Unit, Logan, UT.

AUBREY, F. T. 1977. Locating random points in the field. *J. Range Management,* 30: 157–158.

BECKER, D. A., AND J. J. CROCKETT. 1973. Evaluation of sampling techniques on tall-grass prairie. *J. Range Management,* 26:61–65.

BORMANN, F. H. 1953. The statistical efficiency of sample plot size and shape in forest ecology. *Ecology,* 34: 474–487.

BRAY, R., AND J. T. CURTIS. 1957. An ordination of upland forest communities of southern Wisconsin. *Ecol. Monographs,* 27: 325–349.

COTTAM, G., AND J. T. CURTIS. 1956. The use of distance measures in phytosociological sampling. *Ecology,* 37: 451–460.

CURTIS., J. T., AND R. P. MCINTOSH. 1951. The upland forest continuum in the prairie-forest border region of Wisconsin. *Ecology,* 32: 476–496.

GROSENBAUGH, L. R. 1952. Plotless timber estimates—new, fast, easy. *J. Forestry,* 50: 32–37.

GROSENBAUGH, L. R. 1958. *Point-Sampling and Line-Sampling: Probability Theory, Geometric Implications, Synthesis.* Southern Forest Expt. Sta. Occ. Paper 160. U.S. Forest Service Experiment Station, New Orleans, LA.

HAWLEY, T. A. 1978. A comparison of line interception and quadrat estimation methods of determining shrub coverage. *J. Range Management,* 31: 60–62.

HAYS, R. L., C. SUMMERS, AND W. SEITZ. 1981. Estimating wildlife habitat variables. U.S.D.I. Fish and Wildlife Service. FWS/OBS-81/47.

HYDER, D. N., AND F. A. SNEVA. 1960. Bitterlich's plotless method for sampling basal ground cover of bunch grasses. *J. Range Management,* 13: 6–9.

LANG, G. E., D. H. KNIGHT, AND D. A. ANDERSON. 1971. Sampling the density of tree species in a species-rich tropical forest. *Forest Science,* 17: 395–400.

LONG, G. A., P. S. POISSONET, J. A. POISSONET, P. M. DAGET, AND M. P. GORDON. 1972. Improved needle point frame for exact line transects. *J. Range Management,* 25: 228.

POISSONET, P. S., J. A. POISSONET, M. P. GORDON, AND G. A. LONG. 1973. A comparison of sampling methods in dense herbaceous pasture. *J. Range Management,* 26: 65–67.

RICE, E. L. 1967. A statistical method for determining quadrat size and adequacy of sampling. *Ecology,* 48: 1047–1049.

U. S. FOREST SERVICE. 1958. *Techniques and Methods of Measuring Understory Vegetation: A Symposium.* Southern Forest Experiment Station, Washington, DC.

U. S. FOREST SERVICE EXPERIMENT STATION, New Orleans, LA. 1962. *Range Research Methods: A Symposium.* Misc. Publ. 940. U.S. Department of Agriculture, Washington, DC.

### Vegetation Analysis

BRAUN-BLANQUET, J. 1951. *Pflanzensoziologie: Grundzuge der Vegetationskunde,* 2nd ed. Springer, Vienna.

BRAUN-BLANQUET, J. 1932. *Plant Sociology.* McGraw-Hill, New York. (An English translation of the 1st edition of above.)

BREWER, R., AND M. T. MCCANN. 1982. *Laboratory and Field Manual of Ecology.* Saunders, Philadelphia.

BROWER, J. E., AND J. H. ZAR. 1984. *Field and Laboratory Methods for General Ecology,* 2nd ed. Brown, Dubuque, IA.

CAIN, S. A., AND G. M. DE. O. CASTRO. 1959. *Manual of Vegetation Analysis.* Harper & Row, New York.

CHAPMAN, S. B. (ED.). 1976. *Methods in Plant Ecology.* Halsted Press, Wiley, New York.

COX, G. 1985. *Laboratory Manual of General Ecology,* 5th ed. Brown, Dubuque, IA.

GOUNOT, M. 1969. *Méthodes d'étude quantitative de la végétation.* Masson, Paris.

GREIG-SMITH, P. 1983. *Quantitative Plant Ecology,* 3rd ed. Blackwell Scientific Publications, Oxford.

KERSHAW, K. A. 1973. *Quantitative and Dynamic Ecology,* 2nd ed. Edward Arnold, London; Elsevier, New York.

KREBS C. J. 1989. *Ecological Methods.* Harper & Row, New York.

MUELLER-DOMBOIS, D., AND H. ELLENBERG. 1974. *Aims and Methods of Vegetation Ecology.* Wiley, New York.

OOSTING, H. J. 1956. *The Study of Plant Communities.* Freeman, San Francisco.

WHITTAKER, R. (ED.). 1973. *Ordination and Classification of Communities.* Vol. 5, *Handbook of Vegetation Science.* W. Junk, The Hague.

### Sampling Aquatic Vegetation

CLARK, W. J., AND W. F. SIGLER. 1963. Method of concentrating phytoplankton samples using membrane filters. *Limnol. Oceanog.,* 8: 127–129.

COWELL, R. R., AND R. Y. MORITA. 1975. *Marine and Estuarine Microbiology Laboratory Manual.* University Park Press, Baltimore.

HARTMAN, R. T. 1958. Studies of plankton centrifuge efficiency. *Ecology,* 39: 374–376.

HESTER, F. E. AND J. S. DANDY. 1962. A multiple-plate sampler for aquatic macroinvertebrates. *Trans. Am. Fish. Soc.* 91: 420–421.

LIND, O. T. 1974. *Handbook of Common Methods in Limnology.* Mosby, St. Louis.

LUND, J. W. G., AND J. F. TALLING. 1957. Botanical limnological methods with special reference to algae. *Botan. Rev.,* 23: 489–583.

MCNABB, C. D. 1960. Enumeration of fresh water phytoplankton concentrated on the membrane filter. *Limnol. Oceanog.,* 5: 57–61.

RODINA, A. G. 1971. *Methods in Aquatic Microbiology.* University Park Press, Baltimore.

SCHWOERBEL, J. 1970. *Methods of Hydrobiology (Freshwater Biology).* Pergamon Press, Elmsford, NY.

SLADECKOVA, ALENA. 1962. Limnological investigation methods for the periphyton (aufwuchs) community. *Botan. Rev.,* 28: 286–350.

WELCH, P. S. 1948. *Limnological Methods.* McGraw-Hill–Blakiston, New York.

WOOD, E. J. F. 1962. A method for phytoplankton study. *Limnol. Oceanog.,* 7: 32–35.

WOOD, R. D. 1975. *Hydrobotanical Methods.* University Park Press, Baltimore.

# Dendrochronology

COX, G. W. 1985. *A Laboratory Manual for General Ecology.* Brown, Dubuque, IA.

CREBER, G. T. 1977. Tree rings: A natural data storage system. *Biological Reviews,* 52: 349–383.

FRITTS, H. C. 1960. Multiple regression analysis of radial growth in individual trees. *Forest Sci.,* 6: 344–349.

FRITTS, H. C. 1962. An approach to dendrochronology—screening by multiple regression techniques. *J. Geophys. Res.,* 67: 1413–1420.

FRITTS, H. C. 1971. Dendroclimatology and dendroecology. *Quarternary Res.,* 1: 419–449.

FRITTS, H. C. 1974. Relationship of ring width in arid zone conifers to variations in monthly temperature and precipitation. *Ecol. Monogr.,* 44: 411–440.

FRITTS, H. C. 1976. *Tree Rings and Climate.* Academic Press, New York.

HUGHES, M. K., P. M. KELLY, J. R. PILCHER, AND V. C. LAMARCHE, JR. 1982. *Climate from Tree Rings.* Cambridge University Press, London.

KIRTPATRICK, M. 1981. Spatial and age dependent patterns of growth in New England black birch. *Amer. J. Botany,* 68: 535–543.

PILCHER, J. R., AND B. GRAY. 1982. The relationship between oak tree growth in Britain and climate in Britain. *J. Ecology,* 70: 297–304.

ROUGHTON, R. D. 1962. A review of literature on dendrochronology and age determination of woody plants. *Tech. Bull. 15,* Colorado Department of Fish and Game, Denver.

SCHULMAN, E. 1956. *Dendroclimatic Changes in Semiarid America.* University of Arizona Press, Tucson, AR.

SINCLAIR, A. R. E., J. M. GOSLINE, G. HOLDSWORTH, C. T. KREBS, S. BOUTIN, J. N. M. SMITH, R. BOONSTRA, AND M. DALE. 1993. Can the solar cycle and climate synchronize the snowshoe hare cycle in Canada? Evidence of tree rings and ice cores. *Am. Nat.,* 141: 173–198.

SPENSER, D. A. 1964. Porcupine fluctuations in past centuries revealed by dendrochronology. *J. Appl. Ecol.,* 1:127–149.

STOCKTON, C. W., AND H. C. FRITTS. 1971. Conditional probability of occurrence of variations in climate based on width of annual tree rings in Arizona. *Tree-Ring Bull.,* 31: 3–24.

TAYLOR, R. F. 1936. An inexpensive increment borer holder. *J. Forestry,* 34: 814–815.

# Palynology

ERDTMANN, G. 1954. *An Introduction to Pollen Analysis.* Ronald Press, New York.

FAEGRI, K., AND J. IVERSEN. 1989. *Textbook of Pollen Analysis,* 4th ed. John Wiley, New York.

FELIX, C. F. 1961. An introduction to palynology. In H. N. Andrews (ed.), *Studies in Paleobotany.* Wiley, New York.

MOORE, P. D., J. A. WEBB, AND M. E. COLLINSON. 1991. *Pollen Analysis,* 2nd ed. Blackwell, Oxford.

WALKER, P. C., AND R. T. HARTMAN. 1960. The forest sequence of the Hartstown bog area in western Pennsylvania. *Ecology,* 41: 461–474.

WODEHOUSE, R. P. 1935. *Pollen Grains.* McGraw-Hill, New York.

# Sampling Animal Populations: Comprehensive References

ANDREWARTHA, H. G. 1970. *An Introduction to the Study of Animal Populations,* 2nd ed. University of Chicago Press, Chicago.

ANDREWS, W. A. 1972a. *A Guide to the Study of Freshwater Ecology.* Prentice-Hall, Englewood Cliffs, NJ.

ANDREWS, W. A. 1972b. *A Guide to the Study of Soil Ecology.* Prentice-Hall, Englewood Cliffs, NJ.

BEGON, M. 1979. *Investigating Animal Abundance.* University Park Press, Baltimore, MD.

BLOWER, J. G., L. M. COOK, AND J. A. BISHOP. 1981. *Estimating the Size of Animal Populations.* George Allen & Unwin, London.

BROWER, J. E., AND J. H. ZAR. 1990. *Field and Laboratory Methods for General Ecology,* 3rd ed. Brown, Dubuque, IA.

CAUGHLEY, G. 1977. *Analysis of Vertebrate Populations.* Wiley, New York.

CORMACK, R. M., G. P. PATIL, AND D. S. ROBSON (EDS.). 1979. *Sampling Animal Populations.* International Cooperative Publishing House, Fairland, MD.

COX, G. W. 1985. *Laboratory Manual of General Ecology,* 5th ed. Brown, Dubuque, IA.

DEBLASE, A. F., AND R. E. MARTIN. 1981. *A Manual of Mammalogy.* Brown, Dubuque, IA.

GILES, R. H., JR. (ED.). 1969. *Wildlife Management Techniques.* The Wildlife Society, Washington, DC.

HOLME, N. A., AND A. D. MCINTYRE. 1971. *Methods for the Study of Marine Benthos.* IBP Handbook No. 19. Blackwell, Oxford, England.

JACKSON, R. M., AND F. RAW. 1966. *Life in the Soil.* St. Martin's Press, New York.

KREBS, C. J. 1989. *Ecological Methods.* Harper & Row, New York.

NIELSEN, L. A., AND D. L. JOHNSON (EDS.). 1983. *Fisheries Techniques.* American Fisheries Society, Bethesda, MD.

PARKINSON, D., T. R. G. GRAY, AND S. T. WILLIAMS. 1971. *Methods for Studying the Ecology of Soil Microorganisms.* IBP Handbook No. 19. Blackwell, Oxford, England.

PHILLIPSON, J. (ED.). 1971. *Methods of Study in Quantitative Soil Ecology.* IBP Handbook No. 18. Blackwell, Oxford, England.

POOLE, R. W. 1974. *An Introduction to Quantitative Ecology.* McGraw-Hill, New York.

RALPH, C. J., AND J. M. SCOTT (EDS.). 1981. Estimating numbers of terrestrial birds. *Studies in Avian Biology,* 6: 1–630.

RICKER, W. E. 1958. Handbook of computations for biological statistics of fish populations. *Fishery Res. Board Can. Bull.,* 119: 1–300.

RICKER, W. E. 1971. *Methods for Assessment of Fish Production in Fresh Waters.* IBP Handbook No. 3. Blackwell, Oxford, England.

SCHEMNITZ, S. D. (ED.) 1980. *Wildlife Management Techniques Manual.* The Wildlife Society, Washington, DC.

SCHWOERBEL, J. 1970. *Methods of Hydrobiology (Freshwater Biology).* Pergamon Press, Elmsford, NY.

SEBER, G. A. F. 1982. *The Estimation of Animal Abundance and Related Parameters.* Macmillan, New York.

SOUTHWOOD, T. R. E. 1978. *Ecological Methods.* Chapman and Hall, London.

VERNER, J., M. L. MORRISON, AND C. J. RALPH (EDS.). 1986. *Wildlife 2000: Modeling Habitat Relationships of Terrestrial Vertebrates.* University of Wisconsin Press, Madison, WI.

WELCH, P. S. 1948. *Limnological Methods.* McGraw-Hill, Blakiston, New York.

## Trapping and Collecting

ANDERSON, R. M. 1948. Methods of collecting and preserving vertebrate animals. *National Museum of Canada Bull.,* 69.

BORROR, D. J., AND R. E. WHITE. 1970. *A Field Guide to the Insects* (pp. 4–28). Houghton Mifflin, Boston.

COOPER, R. J., AND R. C. WHITMORE. 1990. Arthropod sampling methods in ornithology. Pp. 29–37 in *Studies in Avian Biology,* Avian Foraging Theory: Modeling and Application. 13.

DEBLASE, A. F., AND R. E. MARTIN. 1981. *A Manual of Mammalogy.* Brown, Dubuque, IA.

KEMPSON, D., M. LLOYD, AND R. GHELARDI. 1963. A new extractor for woodland litter. *Pedobiologia,* 3: 1–21.

KNUDSEN, J. 1966. *Biological Techniques.* Harper & Row, New York.

MURPHY, P. (ED.). 1962. *Progress in Soil Zoology.* Buttersworth, London.

NEEDHAM, J. G. (ED.). 1937. *Culture Methods for Invertebrate Animals* (reprint). Dover, New York.

OMAN, P. W., AND A. D. CUSHMAN. 1948. *Collection and Preservation of Insects.* Mscl. Publ. 60. U.S. Department of Agriculture, Washington, DC.

SOUTHWOOD, T. R. E. 1980. *Ecological Methods with Particular Reference to Insect Populations.* Chapman and Hall, London.

WAGSTAFFE, R. J., AND J. H. FIDLER. 1955. *The Preservation of Natural History Specimens.* Vol. 1, *The Invertebrates.* Philosophical Library, New York.

WILLIAMS, G. E., III. 1974. New technique to facilitate handpicking macrobenthos. *Trans. Amer. Microscop. Soc.,* 93: 220–226.

## Marking and Aging Animals

BIRNEY, E. C., R. JENNESS, AND D. D. BAIRD. 1975. Eye lens protein as a criterion of age in cotton rats. *J. Wildlife Management,* 39: 718–728.

CLARK, D. R., JR. 1968. Branding as a marking technique for amphibians and reptiles. *Copeia,* 1971: 148–151.

COCHRAN, W. W., AND R. D. LORD, JR. 1963. A radio-tracking system for wild animals. *J. Wildlife Management,* 27: 9–24.

DAPSON, R. W. 1980. Guidelines for statistical usage in age estimation techniques. *J. Wildlife Management,* 44: 541–548.

DAY, G. I., S. D. SCHEMNITZ, AND R. D. TABER. 1980. Capturing and marking wild animals. Pp. 61–88 in S. D. Schemnitz (ed.), *Wildlife Management Techniques Manual.* The Wildlife Society, Washington, DC.

FERNER, J. W. 1979. A review of marking techniques for amphibians and reptiles. *Herptological Circular* No. 9. Society for the Study of Amphibians and Reptiles.

GODFREY, G. K. 1954. Tracing field voles (*Microtus agrestis*) with a Geiger-Muller counter. *Ecology,* 35: 5–10.

GRAHAM, W. J., AND H. W. AMBROSE III. 1967. A technique for continuously locating small mammals in field enclosures. *J. Mammal.,* 48: 639–642.

JERALD, A., JR. 1983. Age determination (fish). Pp. 301–321 in L. A. Nielsen and D. L. Johnson (eds.), *Fisheries Techniques.* American Fisheries Society, Bethesda, MD.

KAYE, S. V. 1960. Gold-198 wires used to study movements of small mammals. *Science,* 13: 824.

KUNZ, T. H., AND E. L. P. ANTHONY. 1982. Age estimation and postnatal growth in the bat *Myotis lucifugus. J. Mamm.,* 63: 23–32.

LARSON, J. S., AND R. D. TAYLOR. 1980. Criteria of sex and age. Pp. 143–202 in S. D. Schemnitz (ed.), *Wildlife Management Techniques Manual.* The Wildlife Society, Washington, DC.

LORD, R. D., JR. 1959. The lens as an indicator of age in cottontail rabbits. *J. Wildlife Management,* 23: 358–360.

LORD, R. D., JR. 1963. The cottontail rabbit in Illinois. *Illinois Dept. Conserv. Tech. Bull.* No. 3.

MADSON, R. M. 1967. *Age Determination of Wildlife: A Bibliography.* Biblio. No. 2. U.S.D.I. Department Library, Washington, DC.

MARION, W. R., AND J. D. SHAMUS. 1977. An annotated bibliography of bird-marking techniques. *Bird-Banding,* 48: 42–61.

MARTOF, B. J. 1953. Territoriality in the green frog *Rana clamitans. Ecology,* 34: 165–174.

NEVILLE, A. C. 1963. Daily growth layers for determining the age of grasshopper populations. *Oikos,* 14: 1–8.

PENDLETON, R. C. 1956. Uses of marking animals in population studies: Labeling animals with radioisotopes. *Ecology,* 37: 686–690.

TABER, R. D. 1956. Marking of mammals: Standard methods and new developments. *Ecology,* 37: 681–685.

TESTER, J. R. 1963. Techniques for studying movements of vertebrates in the field. In *Radioecology.* Van Nostrand Reinhold, New York.

TIEMEIER, O. W., AND M. L. PLENERT. 1964. A comparison of three methods for determining the age of black-tailed jack rabbits. *J. Mammal.,* 45: 409–416.

## Estimating Animal Populations

ANDERSON, D. J. 1982. The home range: A new nonparametric estimation technique. *Ecology,* 63: 103–112.

ANDERSON, D. R., J. L. LAAKE, B. R. CRAIN, AND K. P. BURNHAM. 1979. Guidelines for line transect sampling of biological populations. *J. Wildlife Management,* 43: 70–78.

BURNHAM, K. P., D. R. ANDERSON, AND J. L. LAAKE. 1980. Estimation of density from line transect sampling of biological populations. *Wildlife Monograph* No. 44.

CONNER, R. N., AND J. G. DICKSON. 1980. Strip transect sampling and analysis for avian habitat studies. *Wildlife Society Bull.,* 8: 4–10.

DAVIS, D. E. (ED.). 1982. *Handbook of Census Methods for Terrestrial Vertebrates.* CRC Press, Boca Raton, FL.

DESANTE, D. F. 1986. A field test of variable circular plot censusing methods in a Sierran subalpine forest habitat. *Condor,* 88: 129–142.

DIXON, K. R., AND J. A. CHAPMAN. 1980. Harmonic mean measure of animal activity areas. *Ecology,* 61: 1040–1044.

EBERHARDT, L. L. 1978. Transect methods for population studies. *J. Wildlife Management,* 42: 1–31.

EMLEN, J. T. 1971. Population counts of birds derived from transect counts. *Auk,* 88: 323–341.

EMLEN, J. T. 1977. Estimating breeding season bird densities from transect counts. *Auk,* 94: 455–468.

EMLEN, J. T., ET AL. 1957. Dropping boards for population studies of small mammals. *J. Wildlife Management,* 21: 300–414.

FRANZREB, K. E. 1976. Comparison of various transect and spot map methods for censusing avian populations in a mixed coniferous forest. *Condor,* 78: 260–262.

GRODZINSKI, W., Z. PUCEK, AND L. RYSZKOWSKI. 1966. Estimation of rodent numbers by means of prebaiting and intensive removal. *Acta Theriologica,* 11: 297–314.

HAYNE, D. W. 1949. An examination of the strip census method for estimating animal populations. *J. Wildlife Management,* 13: 145–157.

HEYER, W. R., M. A. DONNELLY, R. W. McDARMID, L. C. HAYEH, AND M. S. FOSTER. 1994. *Measuring and Monitoring Biological Diversity: Standard Methods for Amphibians.* Smithsonian Institute Press, Washington, DC.

JARVINEN, O., AND R. A. VAISANEN. 1975. Estimating relative densities of breeding birds by the line transect method. *Oikos,* 26: 316–322.

MIKOL, S. A. 1980. Field guidelines for using transects to sample nongamebird populations. U.S.D.I. Fish and Wildlife Service. FWS/OBS-80/58.

MURIE, O. J. 1954. *A Field Guide to Animal Tracks.* Houghton Mifflin, Boston.

RALPH, C. J., AND J. M. SCOTT (EDS.). 1981. *Estimating Numbers of Terrestrial Birds. Studies in Avian Biology* 6. Cooper Ornithological Society, Lawrence, KS.

REYNOLDS, R. T., J. M. SCOTT, AND R. A. NUSSBAUM. 1980. A variable circular plot method for estimating bird numbers. *Condor,* 82: 309–313.

ROBINETTE, W. L., R. B. FERGUSON, AND J. S. GASHWEILER. 1958. Problems involved in the use of deer pellet group counts. *Trans. North Amer. Wildlife Conf.,* 23: 411–425.

SAMUEL, M. D., AND E. O. GARTON. 1985. Home range: A weighted normal estimate and tests of underlying assumptions. *J. Wildlife Management,* 49: 513–519.

SAMUEL, M. D., D. J. PIERCE, AND E. O. GARTEN. 1985. Identifying areas of concentrated use within the home range. *J. Anim. Ecol.,* 54: 711–719.

SCHNABEL, Z. E. 1938. The estimation of the total fish population of a lake. *American Mathematics Monthly,* 45: 34–52.

SWIFT, D. M., ET AL. 1976. A technique for estimating small mammal population densities using a grid and assessment lines. *Acta Thierologica,* 21: 471–480.

VERNER, J. 1985. Assessment of counting techniques. Pp. 247–302 in R. J. Johnson (ed.), *Current Ornithology.* Plenum, New York.

WAKELEY, J. S. 1987a. Avian line-transect methods. Section 6.3.2, U.S. Army Corps of Engineers Wildlife Resources Management Manual Tech. Rept. EL-87-5. U.S. Army Engineer Waterways Experiment Station, Vicksburg, MS.

WAKELEY, J. S. 1987b. Avian plot methods. Section 6.3.3, U.S. Army Corps of Engineers Wildlife Resources Management Manual Tech. Rept. EL-87-6. U.S. Army Engineer Waterways Experiment Station, Vicksburg, MS.

WAKELEY, J. S. 1987c. Avian territory mapping. Section 6.3.4, U.S. Army Corps of Engineers Wildlife Resources Management Manual Tech. Rept. EL-87-7. U.S. Army Engineer Waterways Experiment Station, Vicksburg, MS.

WHITE, G. C., D. R. ANDERSON, K. B. BURNHAM, AND D. L. OTIS. 1982. Capture-recapture and removal methods for sampling closed populations. Los Alamos National Laboratory, Los Alamos, NM.

## Analyzing Animal Populations

ADAMS, L. 1951. Confidence limits from the Petersen or Lincoln index in animal population studies. *J. Wildlife Management,* 15: 13–19.

BROWNIE, C., D. R. ANDERSON, K. P. BURNHAM, AND D. S. ROBSON. 1978. *Statistical Inference from Band Recovery Data: A Handbook.* U.S.D.I. Fish and Wildlife Service Resource Pub. No. 131.

BURNHAM, K. P., D. R. ANDERSON, AND J. L. LAAKE. 1980. Estimation of density from line transect sampling of biological populations. *Wildlife Monograph* No. 44.

FISHER, R. A., AND E. B. FORD. 1947. The spread of a gene in natural conditions in a colony of the moth *Panaxia dominula. Heredity,* 1: 143–174.

HAYNE, D. W. 1949. Two methods for estimating populations of mammals from trapping records. *J. Mammal.,* 30: 399–411.

JACKSON, C. H. N. 1939. The analysis of an animal population. *J. Animal Ecol.,* 8: 238–246.

JOLLY, G. M. 1965. Explicit estimates from capture-recapture data with both death and immigration—Stochastic model. *Biometrika,* 52: 225–247.

MARTOF, B. J. 1953. Territoriality in the green frog *Rana clamitans. Ecology,* 34: 165–174.

OTIS, D. L., K. P. BURNHAM, G. C. WHITE, AND D. R. ANDERSON. 1978. Statistical inference from capture data on closed animal populations. *Wildlife Monograph* No. 62.

VAN ETTEN, R. C., AND C. L. BENNET, JR. 1965. Some sources of error in using pellet group counts for censusing deer. *J. Wildlife Management,* 29: 723–729.

ZIPPIN, C. 1958. The removal method of population estimation. *J. Wildlife Management,* 22: 325–339.

## Habitat Analysis

BROWER, J. E., AND J. H. ZAR. 1984. *Field and Laboratory Methods for General Ecology,* 2nd ed. Brown, Dubuque, IA.

COOPERRIDER, A. Y., R. J. BOYD, AND H. R. STUART (EDS.). 1986. *Inventory and Monitoring of Wildlife Habitat.* U.S.D.I. Bur. Land Manage. Service Center, Denver, CO.

FLOOD, B. S., M. E. SANGSTER, R. D. SPARROWS, AND T. S. BASKETT. 1977. *A Handbook for Habitat Evaluation Procedures.* U.S.D.I. Fish and Wildlife Service Resource Publ. No. 132.

MYERS, W. L., AND R. L. SHELTON. 1980. *Survey Methods for Ecosystem Management.* Wiley, New York.

SCHEMNITZ, S. D. (ED.). 1980. *Wildlife Management Techniques Manual,* 4th ed. The Wildlife Society, Washington, DC.

STATES, J. B., P. T. HAUG, T. G. SCHOMAKER, L. W. REED, AND E. B. REED. 1978. *A Systems Approach to Ecological Baseline Studies.* U.S.D.I. Fish and Wildlife Service, Fort Collins, CO.

THOMAS, J. W. (TECH. ED.). 1979. *Wildlife Habitats in Managed Forests: The Blue Mountains of Washington and Oregon.* U.S.D.A. Handbook 553. U.S.D.A. Forest Service, Washington, DC.

U.S. FISH AND WILDLIFE SERVICE. 1980. *Habitat Evaluation Procedures.* U.S.D.I. Fish and Wildlife Service, Washington, DC.

USHER, M. B. (ED.). 1986. *Wildlife Conservation Evaluation.* Chapman and Hall, London.

VERNER, J., M. L. MORRISON, AND C. J. RALPH (EDS.). 1986. *Wildlife 2000: Modeling Habitat Relationships of Terrestrial Vertebrates.* University of Wisconsin Press, Madison, WI.

*Appendix* **B**

# Measuring Community and Population Structure

# MEASURING COMMUNITY STRUCTURE

## Population Dispersion

One of the problems associated with community structure is how the spatial pattern of organisms in the community relates to the interaction of organisms with the environment. Data collected from quadrats, point quadrats, and so on may be used to determine intrapopulation dispersion, as long as we remember that the analysis can be influenced by the size of the sampling unit.

As pointed out in Chapter 17, population dispersion may be uniform, random, or clumped. Where the density of individuals is low for the available surface area or volume, the Poisson method is useful to determine types of dispersion. The Poisson distribution furnishes values expected on the basis of a random dispersion pattern and approximates an extremely asymmetrical distribution. Because the mean of the Poisson is equal to its variance, the Poisson distribution is completely specified by the mean. Thus the theoretical Poisson distribution corresponding to the observed distribution can be constructed from the sample mean μ alone.

To calculate the Poisson we must know the number of sample units, the number of organisms in each sample unit, and the probability that the organism is located in the sample unit or area. The Poisson series is expressed as

$$P_x = e^\mu \left( 1, x, \frac{\mu^2}{2!}, \ldots, \frac{\mu^4}{i!} \right)$$

The steps for setting up the Poisson distribution are as follows:

1. Determine the sample mean obtained by the equation

$$\mu = \frac{\Sigma f(X)}{N}$$

   where $f$ = observed frequencies, $X$ = frequency class, and $N$ = total frequency, individuals, observations. This estimate of the mean density for each sample unit is substituted into the general expression for Poisson probability, $e^{-\mu}$, where $e$ is the base of natural logarithms.
2. Determine from a table of exponential functions (or a scientific calculator) the value of $e^{-\mu}$.
3. Calculate the Poisson probabilities (see example below).
4. Multiply each probability distribution by the total frequency $N$ to convert it to absolute frequency, so that the probabilities are comparable with the observed distribution.

As an example we can use data from Garbutt (1961) on the distribution of fleas on mice. The null hypothesis is that fleas are randomly distributed through a population of mice.

The information for *Microtis* is lumped into five classes as follows:

| Fleas per mouse | ($X$) | 0 | 1 | 2 | 3 | 4+ |
|---|---|---|---|---|---|---|
| Mice | ($f$) | 44 | 8 | 9 | 3 | 4 |

Calculating for the mean,

$$\bar{x} = \frac{\Sigma f(X)}{N}$$

$$= \frac{(0 \times 44) + (1 \times 8) + (2 \times 9) + (3 \times 3) + (4 \times 4)}{68}$$

$$= 0.75$$

Thus the mean, $\bar{x} = 0.75$ and $e^{-\bar{x}} = e^{-0.75} = 0.472$. To determine the Poisson probability:

$$X(0) = e^{-0.075} = 0.472$$

$$X(1) = \bar{x}e^{-0.75} = 0.75 \times 0.472 = 0.354$$

$$X(2) = \bar{x}^2/2!e^{-0.75}$$
$$= (0.75)^2/2 \times 0.472 = 0.133$$

$$X(3) = \bar{x}^3/3!e$$
$$= (0.75)^3/6 \times 0.472 = 0.0332$$

$$X(4) = \bar{x}^4/4!e^{-0.75}$$
$$= (0.75)^4/24 \times 0.472 = 0.0062$$

To obtain the theoretical frequency or distribution, multiply the total number of observations by the Poisson probability for each class (Table B.1). For example, the theoretical distribution for $X$ (0) is $68 \times 0.47 = 31.96$ or 32.

Once the theoretical frequencies have been obtained, the next step is to employ a chi-square goodness-of-fit test to determine how well the data match the Poisson distribution. We set up a chi-square table in which the observed distribution can be compared with the expected distribution (Table B.2).

There are five classes after lumping the data and two constants—$\bar{x}$, the mean, and $N$, the number of cells. In this example the degrees of freedom are $5 - 2 = 3$. The high value of the chi-square (see statistics books for tabled values of

**Table B.1  Poisson Probability and Theoretical Frequency**

| Number of Fleas per Mouse (X) | Observed Frequency | Poisson Probability | Theoretical Frequency |
|---|---|---|---|
| 0 | 44 | 0.47 | 32.0 |
| 1 | 8 | 0.35 | 23.8 |
| 2 | 9 | 0.13 | 8.9 |
| 3 | 3 | 0.04 | 2.7 |
| 4+ | 4 | 0.01 | 0.6 |
| Total | 68 | | 68.0 |

**Table B.2** Comparison of Observed Distribution with Expected Distribution

| Number of Fleas per Mouse (X) | Observed Distribution (O or f) | Expected Distribution (E) | O – E | (O – E)² | $\frac{(O - E)^2}{E}$ |
|---|---|---|---|---|---|
| 0 | 44 | 32.0 | 12.0 | 144 | 4.500 |
| 1 | 8 | 23.8 | −15.8 | 249.64 | 10.480 |
| 2 | 9 | 8.9 | 0.1 | .01 | .001 |
| 3 | 3 | 2.7 | 0.3 | .09 | .033 |
| 4 | 4 | 0.6 | 3.4 | 11.56 | 19.266 |
| Total | 68 | 68.0 | | | $\chi^2 = 34.280$ |

$\chi^2 = 34.28$

variance: $s^2 = \dfrac{f(X - \bar{x})^2}{N - 1} = \dfrac{\Sigma f(X^2) - N(\bar{x}^2)}{N - 1} = \dfrac{135 - 68(0.56225)}{67} = 1.444$

$\dfrac{s^2}{\bar{x}} = \dfrac{1.444}{0.75} = 1.925 \qquad P > .005$

chi-square) with a probability lying well below 0.005 leads to the rejection of the null hypothesis. The fleas are not randomly dispersed among the mice.

A further test for randomness is the ratio of the variance, $s^2$, to the mean, $\mu$. In the Poisson distribution the population mean, $\mu$, is equal to the population variance, $s^2$. A randomly distributed population would have a ratio of its variance to its mean equal to 1.0. A ratio much less than 1.0 would indicate a uniform distribution, and a ratio much greater than 1.0 would indicate clumped distribution. In the example of fleas on mice, the ratio of 1.925 (see Table B.2) suggests contagious or clumped distribution. Significance of departure from randomness may be assessed by the equation

$$t = \frac{|s^2/\bar{x} - 1.0|}{\sqrt{2/n - 1}}$$

Compare the $t$ value to the critical values for $t$ in Student's $t$ table for $n - 1$ degrees of freedom.

For the example of fleas on mice, the $t$ value works out to be

$$t = \frac{1.444/.75}{\sqrt{2/68 - 1}} - 1.0 = .921/.173 = 5.323$$

This value is well above the critical value of 2.660 for 60 DF at the .01 level. It indicates that the frequency distribution is significantly different from random.

Another method is to use the chi-square statistic:

chi-square = sum of squares (SS)/$\bar{x}$

The sum of squares may be computed by

SS = $(n - 1)$ $(s^2)$

The statistical significance may be obtained by using chi-square tables. The degrees of freedom are $n - 1$. In our example   SS = (67)   (1.444) = 96.748   and   chi-square =

96.748/.75 = 128.99. For 60 DF the chi-square value at the 95 percent level is 79.082 and at the 99 percent level, 88.379. The chi-square value is well above critical values, again indicating significant departure from random.

The relationship between observed and expected Poisson distributions can be compared visually by plotting both values on a graph (Figure B.1). The vertical axis can be either the number or proportion of samples containing X. The horizontal axis is X. In general, in a clumped distribution the observed number or proportion of empty cells and aggregations would be higher than predicted random probabilities, so there would be a greater chance of finding many

**Figure B.1** A graph comparing observed distribution with Poisson frequencies for number of fleas per mouse as described in Table B.2.

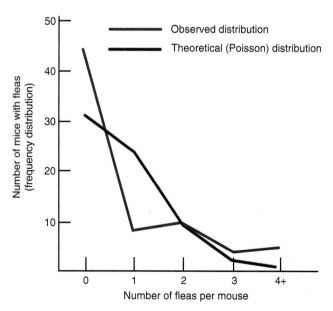

individuals in a few plots. In uniform dispersion, there is a higher than random probability of finding only a few individuals in most plots. In this case the observed proportions or numbers would be lower on both the left- and right-hand side of the mean, but higher near the mean.

In example of the fleas on mice, the number of empty cells is much higher than random probability, but the observed distribution at the higher probabilities $P(3)$ and $P(4)$ do not depart much from random. However, the high number of empty cells, mice without fleas, leads to the clumped distribution of fleas on a small portion of the population.

## Interspecific Association

Some species in a community may occur together more frequently than by chance. This grouping may result from symbiotic relationships, from food chain coactions, or from similarities in adaptation and response to environmental conditions. Some measurement of this association provides an objective method for recognizing natural groupings of species. Negative associations may indicate interactions detrimental to one or both species, such as interspecific competition, or adaptations to different sets of environmental conditions.

The data are obtained by sampling quadrats, point-centered quadrats, trapping stations, and other means. Presence or absence data for pairs are arranged in a $2 \times 2$ or a $2 \times n$ contingency table:

*Species A*

|  |  | + | − |  |
|---|---|---|---|---|
| *Species B* | + | $a$ | $b$ | $a + b$ |
|  | − | $c$ | $d$ | $c + d$ |
|  | + | $a + c$ | $b + d$ | $a + b + c + d = n$ |

where

$a$ = samples containing both species A and B

$b$ = samples containing only species B

$c$ = samples containing only species A

$d$ = samples containing neither species

From these data a coefficient of association, $C$, can be calculated. It will vary from +1.0 for a maximum positive association to −1.0 for a maximum negative association. A value of 0 suggests that the frequency of association is that expected by chance.

If $bc > ad$ and $d \geq a$, then
$$C = ad - bc/(a + b)(a + c) \quad (1)$$

If $bc > ad$ and $a > d$, then
$$C = ad - bc/(b + d)(c + d) \quad (2)$$

If $ad \geq bc$ and $c > b$, then
$$C = ad - bc/(a + b)(b + d) \quad (3)$$

If $ad > bc$ and $c \geq b$, then
$$C = ad - bc/(a + c)(c + d) \quad (4)$$

To learn whether the coefficient of association is significant, we can apply a chi-square test to determine the significant level of the deviations between the observed values of the contingency table and the expected values based on chance association. This test requires a modification of the contingency table as in Table B.3. The number of samples expected in each cell can be determined by

$$a = (a + b)(a + c)/T$$
$$b = (a + b)(b + d)/T$$
$$c = (c + d)(a + c)/T$$
$$d = (c + d)(b + d)/T$$

The chi-square value can be calculated on the basis of observed minus expected differences with the formula

$$\text{chi-square} = \sum \frac{(\text{observed} - \text{expected})^2}{\text{expected}}$$

As an example, consider two species of deer mice, *Peromyscus leucopus*, the white-footed mouse, and *P. maniculatus nubiterrae*, the cloudland deer mouse, inhabiting Appalachian hardwood forests. The data are extracted from a three-year study (Violet 1973) on the relations between the two species. For our example, we use the presence of the two species at live-trap stations located on a mesic north-facing slope. The question is: Are the two species mutually exclusive?

Among the 192 stations, *P. maniculatus* occurred exclusively at 133, *P. leucopus*, 14; both species, 26; and none, 19. Because $bc > ad$ and $a > d$, we will use formula 2 to determine the coefficient of association (Table B.4). The two species of mice show a negative association.

$$c = ad - bc/(b + d)(c + d)$$
$$c = 494 - 1862/(33)(152)$$
$$= -1368/5016$$
$$= -0.273$$

---

**Table B.3  Contingency Table for Determining the Degree of Association**

|  |  | Species A | | | |
|---|---|---|---|---|---|
|  |  | + | | − | |
|  |  | Observed | Expected | Observed | Expected |
| Species B | + |  |  |  |  |
|  | − |  |  |  |  |
|  |  |  |  |  |  |

---

**Table B.4**  Coefficient of Association

|  |  | *Peromyscus maniculatus nubiterrae* | | |
|---|---|---|---|---|
|  |  | + | − |  |
| *Peromyscus leucopus* | + | $a$ 26 | $b$ 14 | $a + b$ 40 |
|  | − | $c$ 133 | $d$ 19 | $c + d$ 152 |
|  | Σ | $a + c$ 159 | $b + d$ 33 | 192 |

To determine whether the coefficient of association is significant, set up a $2 \times 2$ contingency table, Table B.5. Because a $2 \times 2$ contingency table is the smallest possible and has only one degree of freedom, the absolute value of the difference between observed and expected should be reduced by 0.5, an adjustment known as the ***Yates correction for continuity.***

$$\text{chi-square} = (6.5)^2/33 + (6.5)^2/7$$
$$+ (6.5)^2/126 + (6.5)^2/26$$
$$= 1.280 + 6.036$$
$$+ 0.335 + 1.625$$
$$= 9.276$$

The calculated chi-square value of 9.55 exceeds the table value of 3.84, 6.64, and 7.88 for 1 df at the 95, 99, and 99.5 percent levels, respectively. We would reject the hypothesis that the two species are distributed independently and conclude that they are negatively associated.

For a $2 \times 2$ contingency table chi-square can be calculated by an alternative method that already has the Yates correction:

$$\text{chi-square} =$$
$$\frac{(|ad - bc| - 0.5T)^2(T)}{(a + b)(a + c)(b + d)(c + d)}$$

The chi-square value calculated in this manner is 9.74.

**Table B.5**  Chi-Square Contingency Table

|  |  | *Peromyscus maniculatus nubiterrae* | | | |
|---|---|---|---|---|---|
|  |  | + | | − | |
|  |  | O | E | O | E |
| *Peromyscus leucopus* | + | 26 | 33 | 14 | 7 |
|  | − | 133 | 126 | 19 | 26 |

$$\text{chi-square} = \frac{(|1368| - 96)^2(192)}{(40)(159)(33)(152)}$$
$$= \frac{310,652,928}{31,901,760}$$
$$= 9.74$$

In situations having more than two expressions of each variable, the contingency table will have more than two rows and/or two columns. The expected frequencies for a particular cell are given by the expression

$$\frac{(\text{row total})(\text{column total})}{(\text{grand total})}$$

and the chi-square is calculated by the standard formula.

In all chi-square contingency analyses, the degrees of freedom (DF) = (no. of rows − 1)(no. of columns − 1). A $2 \times 2$ contingency table has one degree of freedom. A contingency table with three rows and two columns has two degrees of freedom.

## Community Similarity

When addressing questions of community structure, ecologists often need to compare the species composition of plants or animals of communities over space or time. The similarity of communities can be measured from data as simple as presence or absence of species or as detailed as density, dominance, frequency, and importance value.

**Similarity Coefficients**  Similarity coefficients, often called coefficients of community, are the simplest approaches to comparing community structure. They are based solely on presence (indicated with a 1) and absence (indicated with a 0). Among a number of them appearing in the literature are the Jaccard and Sorensen indexes. The values range from 0 when no species are found in both communities to 1, when all species are found in both communities (complete similarity).

The *Jaccard index* is based on the presence-absence relationship between the number of species in each community and the total number of species:

$$SC_j = \frac{c}{A + B - c}$$

where $c$ is the number of common species, $A$ is the total number of species in stand A, and $B$ is the total number of species in stand B. There are variations to this formula (see Mueller-Dombois and Ellenberg 1974). The coefficient expresses the ratio of common species to all species found in the two vegetational groups. The Jaccard coefficient for the stands described in Table B.6 is

$$SC_j = \frac{6}{10 + 7 - 6} = .5454 \text{ or } 54.54\%$$

**Table B.6** Determination of Percent Similarity

| | Stand 1 | | Stand 2 | |
|---|---|---|---|---|
| Species | Number | % Presence | Number | % Presence |
| Yellow-poplar (*Liriodendron tulipifera*) | 83 | 50.6 | 0 | 0* |
| Red oak (*Quercus rubra*) | 25 | 15.3* | 55 | 34.4 |
| Red maple (*Acer rubrum*) | 19 | 11.6 | 13 | 8.1* |
| Black cherry (*Prunus serotina*) | 25 | 15.3* | 27 | 16.9 |
| Black birch (*Betula lenta*) | 2 | 1.2* | 11 | 6.8 |
| Sugar maple (*Acer saccharum*) | 1 | 0.6* | 2 | 1.3 |
| Sassafras (*Sassafras albidum*) | 0 | 0* | 6 | 3.7 |
| Black locust (*Robinia pseudoacacia*) | 1 | 0.6 | 0 | 0* |
| Black gum (*Nyssa sylvatica*) | 3 | 1.9 | 0 | 0* |
| White ash (*Fraxinus americana*) | 1 | 0.6 | 0 | 0* |
| Chestnut oak (*Quercus prinus*) | 4 | 2.4* | 46 | 28.8 |
| Sum | 164 | | 160 | |

$PS = \Sigma$ (lowest percentage for each species)
  $= 0 + 15.3 + 8.1 + 15.3 + 1.2 + 0.6 + 0 + 0 + 0 + 0 + 2.4$
  $= 42.9$

*Lowest percentage for each species.

The *Sorensen index* differs from the Jaccard by measuring the ratio of the common to the average number of species in the two samples:

$$SC_s = \frac{c}{\frac{1}{2}A + B}$$

where *A* is the total number of species in community A, *B* is the total number of species in community B, and *c* is the number of species common to both communities. In the Sorensen index, theoretically each species has an equal chance of being present in the two communities. The expression $0.5(A + B)$ represents the sum of the theoretically possible coinciding occurrences, and *c* is the expression of the actually coinciding occurrences. The Sorensen formula gives greater weight to species common to both areas and less to species unique to either area and results in a greater similarity.

The Sorensen coefficient for the two stands in Table B.6 is

$$SC_s = \frac{6}{\frac{1}{2}(10 + 7)} = 0.7058 \text{ or } 70.58\%$$

The Jaccard and Sorensen indexes do not take into account the relative abundances of species, limiting their use-

fulness to situations where data on presence and absence of species are sufficient, as in some water pollution studies.

**Percent Similarity**  Another measure of community similarity is the index of percent or proportional similarity, which considers the number of species in each community, the species common to both communities, and the abundance of species. The abundance of a species in a community is tabulated as a percentage of the total species presence in that community (Table B.6):

$$\text{percent presence} = \frac{\text{number of individuals of a species}}{\text{total number of individuals in a community}}$$

The lowest percentage for each species in the communities being compared is identified and used to calculate percent similarity using the equation

$$PS = \Sigma(\text{lowest percentage for each species})$$

An example is considered in Table B.6.

**Coefficient of Community**  A widely used index of similarity between two stands or communities is the coefficient of community. This index ranges in value from 0 to in-

dicate communities with no species in common to 100 to indicate two communities with identical species composition. The proportion of each species can be expressed as density, biomass, frequency, or importance value. The index is calculated using the equation

$$C = \frac{2W}{a+b}(100)$$

where $a$ is the sum of scores for one stand, $b$ the sum of scores for the second stand, and $W$ the sum of lower scores for each species.

An example of the calculations using importance values is given in Table B.8. The two stands involved are Stands 1 and 2, described in Table B.7, which gives stand composition

and importance values for canopy trees in eight selected stands. This calculation is the first step in community ordination.

## Community Ordination

Ordination is the technique of arranging units (for example, forest stands) in a uni- or multidimensional order in such a manner that the position of each unit along the axis or axes conveys the maximum information about its composition or relationship with the other units.

Ordination is based on the assumption that community composition varies gradually over a continuum of environmental conditions. For this reason communities cannot be

**Table B.7** Stand Composition and Importance Values for Canopy Layer

| Species | Importance Values of Stand | | | | | | | |
| --- | --- | --- | --- | --- | --- | --- | --- | --- |
| | 1 | 2 | 3 | 4 | 5 | 6 | 7 | 8 |
| Black oak (Quercus velutina) | 0 | 0 | 0 | 0 | 0 | 0 | 0 | 42.90* |
| Red oak (Quercus rubra) | 10.25 | 0 | 0 | 0 | 23.05 | 0 | 29.98 | 7.81 |
| Chestnut oak (Quercus prinus) | 0 | 0 | 0 | 0 | 4.68 | 0 | 114.46* | 0 |
| Sugar maple (Acer saccharum) | 39.74* | 16.94 | 30.40 | 27.56 | 4.46 | 0 | 25.56 | 21.02 |
| Black maple (Acer nigrum) | 10.16 | 39.37* | 28.95 | 28.95 | 0 | 0 | 0 | 0 |
| Slippery elm (Ulmus rubra) | 20.17 | 78.87* | 0 | 0 | 13.60 | 0 | 0 | 0 |
| American elm (Ulmus americana) | 0 | 30.77 | 0 | 0 | 0 | 0 | 0 | 11.59 |
| Shagbark hickory (Carya ovata) | 26.14* | 0 | 0 | 0 | 0 | 0 | 0 | 36.67* |
| Ironwood (Ostyra virginiana) | 7.02 | 0 | 0 | 0 | 4.46 | 0 | 0 | 40.49* |
| American hornbeam (Carpinus caroliniana) | 0 | 0 | 0 | 0 | 0 | 14.01 | 0 | 0 |
| Beech (Fagus grandifolia) | 29.39* | 0 | 46.86* | 0 | 0 | 161.78* | 0 | 0 |
| Black walnut (Juglans nigra) | 39.85* | 0 | 17.83 | 0 | 0 | 0 | 0 | 0 |
| Yellow-poplar (Liriodendron tulipifera) | 10.44 | 21.49 | 51.86* | 114.10* | 0 | 0 | 0 | 0 |
| White ash (Fraxinus americana) | 0 | 0 | 11.11 | 0 | 106.44* | 0 | 0 | 0 |
| Redbud (Cercis canadensis) | 6.83 | 12.98 | 0 | 0 | 0 | 0 | 0 | 32.86* |
| Flowering dogwood (Cornus florida) | 0 | 0 | 0 | 0 | 0 | 24.21 | 0 | 6.66 |
| Black locust (Robinia pseudoacacia) | 0 | 12.26 | 0 | 0 | 0 | 0 | 0 | 0 |
| Black cherry (Prunus serotina) | 0 | 0 | 0 | 17.75 | 43.31 | 0 | 0 | 0 |
| Sycamore (Platanus occidentalis) | 0 | 0 | 0 | 4.64 | 0 | 0 | 0 | 0 |

*Dominant species.

**Table B.8 Determination of Coefficient of Community**

| Species | Importance Value | |
|---|---|---|
| | Stand 1 | Stand 2 |
| Red oak (*Quercus rubra*) | 10.25 | 0* |
| Sugar maple (*Acer saccharum*) | 39.74 | 16.94* |
| Black maple (*Acer nigrum*) | 10.16* | 39.37 |
| Slippery elm (*Ulmus rubra*) | 20.17* | 78.87 |
| American elm (*Ulmus americana*) | 0* | 30.77 |
| Shagbark hickory (*Carya ovata*) | 26.14 | 0* |
| Ironwood (*Ostyra virginiana*) | 7.02 | 0* |
| Beech (*Fagus grandiflora*) | 29.39 | 0* |
| Black walnut (*Juglans nigra*) | 39.85 | 0* |
| Yellow-poplar (*Liriodendron tulipifera*) | 10.44* | 21.49 |
| Redbud (*Cercis canadensis*) | 6.83* | 12.98 |
| Black locust (*Robinia pseudoacacia*) | 0* | 12.26 |
| Sum | 199.99 | 212.68 |

$$C = \frac{2W}{a+b}(100)$$

$a = 199.99$

$b = 212.68$

$W = 64.54$ (sum of lower scores)

$$C = \frac{2(64.54)}{199.99 + 212.68}(100) = 31.28$$

*Lower score for each species.

classified in discrete units; rather they form a continuum changing in composition and structure over environmental gradients (temperature, elevation, soil, and so on).

Community ordination may be accomplished by two different approaches in deriving the axes. The axes can be based on (1) change in environmental conditions or (2) change in community composition. When the axes represent change in environmental conditions, the position of communities along the axes reflects change in community composition influenced by environmental conditions (gradient analysis). When the axes are based on community composition, the configuration of communities in a geometric space reveals relationship based on similarity in composition.

As an example we will use the Bray-Curtis method, based on the second approach to community ordination. The first step is to determine the degree of similarity among communities or stands using the coefficient of community. The stands involved are described in Table B.7, which gives stand composition and importance values for trees in the canopy layer.

When comparisons are being made among a number of communities, the results are usually presented as a matrix of values representing all pairwise comparisons between communities or stands (see Table B.9).

The next step in the ordination process is to convert the similarity coefficients (Table B.9) to values that express dissimilarity, because the distance between communities in ordination space represents the degree of difference rather than similarity. (Two stands with low values of dissimilarity will appear close together in the ordination arrangement. If the similarity coefficient were used, the greater the similarity, the further apart the two communities would be positioned.) The coefficient of dissimilarity is obtained by subtracting the coefficient of similarity from the highest value of similarity possible. Theoretically this value is 100, but because most replicate samples for a single community show a coefficient of about 85, a more realistic estimate is obtained by using 85 rather than 100 (see Cox 1985). The coefficient of dissimilarity for the stands described in Table B.7 is given in Table B.10.

The position of communities along the ordination axes is determined by calculating values for each stand along the $x$ and $y$ axes (and $z$ axis in the case of three-dimensional ordination). To position the stands along the $x$ axis, terminal points must be determined first. The dissimilarity values between each stand and every other stand are summed (Table B.11). The stand with the highest total dissimilarity values is placed at the 0 point on the $x$ axis and designated as stand A

**Table B.9 Similarity Matrix for Stands in Table B.7**

| Stand | 2 | 3 | 4 | 5 | 6 | 7 | 8 |
|---|---|---|---|---|---|---|---|
| 1 | 31.28 | 50.75 | 24.50 | 16.39 | 14.70 | 19.35 | 31.08 |
| 2 | | 35.20 | 35.20 | 9.13 | 0.00 | 9.26 | 20.97 |
| 3 | | | 57.03 | 8.04 | 24.21 | 14.35 | 10.86 |
| 4 | | | | 11.27 | 0.00 | 13.24 | 10.69 |
| 5 | | | | | 0.00 | 17.40 | 6.13 |
| 6 | | | | | | 0.00 | 3.33 |
| 7 | | | | | | | 15.58 |

**Table B.10**  **Dissimilarity Matrix for Stands in Table B.7**

| Stand | 2 | 3 | 4 | 5 | 6 | 7 | 8 |
|---|---|---|---|---|---|---|---|
| 1 | 55.72 | 34.25 | 60.50 | 68.61 | 70.30 | 65.75 | 53.92 |
| 2 | | 49.80 | 49.80 | 75.85 | 85.00 | 75.74 | 64.03 |
| 3 | | | 27.97 | 76.96 | 60.79 | 70.65 | 74.14 |
| 4 | | | | 73.73 | 85.00 | 71.76 | 74.31 |
| 5 | | | | | 85.00 | 67.60 | 78.87 |
| 6 | | | | | | 85.00 | 81.67 |
| 7 | | | | | | | 69.42 |

(Table B.11). In our example it happens to be stand 6. The stand with the greatest dissimilarity with A is chosen as the end point along the $x$ axis and designated as stand B (Table B.12) Note that in Table B.12 four stands share the highest dissimilarity with 6, with values of 85.00. They are 2, 4, 5, and 7. Because the four stands are equal, one is chosen arbitrarily as stand B. In the example 5 is made the end point. The remaining stands are placed along the $x$ axis a given distance, $D_x$, from stand A using the equation (based on the Pythagorean theorem)

$$D_x = L^2 + (DA^2) - (DB^2)/2L$$

For stand 1:

$$D_x = (85)^2 + (70.30)^2 - (68.61)^2/170$$
$$D_x = 43.88$$

where $L$ is the dissimilarity value between A and B, $DA$ the dissimilarity value between A and the stand in question, and $DB$ the dissimilarity value between B and the stand in question.

The calculation of the $y$ coordinate is designed to account for the greatest amount of remaining between-stand variation (Table B.13). First, the stand with the poorest fit along the $x$ axis is determined by calculating a poorness-of-fit value, $e$, for each stand, using the equation

$$e = \sqrt{DA^2 - x^2}$$

For stand 1:

$$e = \sqrt{(70.3)^2 - (43.88)^2} = 54.92$$

The stand having the largest value of $e$ is designated as stand A' and given the value of 0 along the $y$ axis (Table B.14). The stand showing the greatest dissimilarity with A' and located within (.1) $L$ of A' along the $x$ axis is chosen as the end point along the $y$ axis and designated as stand B'. The value for stand B' along the $y$ axis is its dissimilarity value with A'. In the example, stand 4 (stand 6 is excluded because it is the zero point on the $x$ axis) has the greatest dissimilarity value with stand 8 (74.31) and it is within (0.1) $L$ of stand 8 along the $x$ axis, so it becomes the end point along the $y$ axis. The remaining stands are positioned at a given distance, $D_y$, from A' (as with the $x$ axis), using the equation

$$D_y = \frac{L^2 + (DA'^2) - (DB'^2)}{2L}$$

where $L$ is the dissimilarity value between A' and B', $DA'$ is the dissimilarity value between A' and the stand in question, and $DB'$ is the dissimilarity value between B' and the stand in question.

In our example stand A' is stand 8 and B' is stand 4 (Table B.15).

With the points now determined, they can be plotted on an ordination graph, illustrated in Figure B.2. (For an example of construction of a third axis for a three-dimensional ordination, see Mueller-Dombois and Ellenberg 1974).

**Table B.11**  **Stand Dissimilarity Values**

| Stand | Sum of Dissimilarity Values (from Table B.10) |
|---|---|
| 1 | 409.05 |
| 2 | 455.95 |
| 3 | 394.56 |
| 4 | 443.07 |
| 5 | 526.62 |
| 6 | 552.76 stand A |
| 7 | 518.17 |
| 8 | 496.36 |

**Table B.12**  **Stand Dissimilarities with A**

| Stand | Dissimilarity with A (Stand 6) |
|---|---|
| 1 | 70.30 |
| 2 | 85.00 |
| 3 | 60.79 |
| 4 | 85.00 |
| 5 | 85.00 stand B (arbitrary) |
| 7 | 85.00 |
| 8 | 81.67 |

**Table B.13** Stand Ordination Values, *x* Axis

| Stand | Value along x Axis |
|-------|--------------------|
| 1 | 43.88 |
| 2 | 51.16 |
| 3 | 29.40 |
| 4 | 53.25 |
| 5 | 85.00 |
| 6 | 0.00 |
| 7 | 58.12 |
| 8 | 45.14 |

**Table B.15** Stand Ordination Values, *y* Axis

| Stand | Value along y Axis |
|-------|--------------------|
| 1 | 38.07 |
| 2 | 52.04 |
| 3 | 70.23 |
| 4 | 74.31 |
| 5 | 47.11 |
| 6 | 39.24 |
| 7 | 40.56 |
| 8 | 0.00 |

## Species Diversity

Species diversity includes both the number of species and the number of individuals in a community. We also have to consider how the individuals are apportioned among the species. For example, a community consisting of five species and 100 individuals with the individuals equally divided among all five species would be more equitable than a community in which 80 individuals were of one species and the remaining 20 were allotted to the other four species.

Two approaches to species diversity are widely used today: Simpson's index (Simpson 1949) and the Shannon formula (Shannon and Wiener 1963). Both are sensitive to changes in the number of species and to changes in the distribution of individuals among the species. However, the index value of both is influenced by sample size. If the index is to be used to compare diversity among communities, the sample sizes must be equal. If complete census data are used, the areas sampled must be of equal size.

Simpson's index of diversity considers the number of species, the total number of individuals, and the proportion of the total found in each species. It is based on the number of samples of random pairs of individuals that must be drawn from a community to provide at least a 50 percent chance of obtaining a pair with both individuals of the same species. The index is calculated by the formula

$$\lambda = \frac{\Sigma n_i(n_i - 1)}{N(N - 1)}$$

**Table B.14** Stand Poorness-of-Fit Values

| Stand | e | |
|-------|-------|---------|
| 1 | 54.92 | |
| 2 | 67.88 | |
| 3 | 53.41 | |
| 4 | 66.44 | stand B′ |
| 5 | 0.00 | |
| 6 | NA | |
| 7 | 62.02 | |
| 8 | 68.08 | stand A′ |

where $N$ is the total number of individuals of all species and $n_i$ the number of individuals of a species. This formula actually measures dominance. A group of species with low dominance will have high diversity. To obtain an index of diversity the formula must be changed:

$$D = 1 - \frac{\Sigma n_i(n_i - 1)}{N(N - 1)}$$

However, Simpson's index is usually inverted to obtain a measure of diversity:

$$D = \frac{N(N - 1)}{\Sigma n_i(n_i - 1)}$$

A community containing only one species would have a value of 1.0. Values would increase to infinity, at which every individual belongs to a different species.

The Shannon formula comes from information theory. In ecological use the function describes the degree of uncertainty of predicting the species of a given individual picked at random from the community. As the number of species increases and as the individuals are more equally distributed among the species present, the more the uncertainty increases. This function has been criticized (see Hurlbert 1971). The Shannon formula in a general form is

$$H' = -\Sigma p_i \log p_i$$

where $p_i$ is the decimal fraction of total individuals belonging to the *i*th species.

In words the formula states that the probability that any one individual belongs to species $i$ is $p_i$. This value in turn is equal to the ratio $n_i/n$, where $n_i$ is the number of individuals in the *i*th species and $n$ is the total number of individuals of all species in the sample. Diversity is greatest if each individual belongs to a different species; the least if all individuals belong to one species. A working formula is

$$H' = -\sum_{i=1}^{s} \left(\frac{n_i}{N}\right) \log_2 \left(\frac{n_i}{N}\right)$$

where $s$ is the total number of species collected and $\log_2$ is 3.322 $\log_{10}$. For calculation of the index the equation used is

$$H' = 3.322 \, [\log_{10} n - (1/n \, \Sigma n_i \log_{10} n_i)]$$

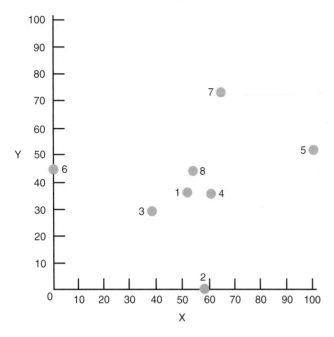

**Figure B.2** An ordination graph for the eight forest stands described in Tables B.7 to B.15.

To compute:

1. Obtain the appropriate $\log_{10}n$.
2. For each species calculate $\log_{10}n_i$
3. Calculate $n_i\log_{10}n_i$, which means multiplying the number of individuals in each species by $\log_{10}n_i$.
4. Sum the $n_i\log_{10}n_i$ and divide the value by $n$, the total number of individuals of all the species.
5. Subtract this value from $\log_{10}n$ and multiply by 3.322 to convert the index value to $\log_2$.

Any log base may be used to calculate diversity as long as it is used consistently. Most commonly used logarithmic bases are 10, $e$, and 2. Because the Shannon formula comes from communication engineering, base 2 is commonly employed and is used here.

If species abundance data are sampled in a nonrandom fashion or if data have been collected for a whole community (such as a total census of organisms), a measure of diversity better suited than Shannon is Brillouin's index:

$$H = \log_{10} \frac{N!}{n_i!}$$

where $N$ is the total number of individuals in all species and $n_i$ the number of individuals in the $i$th species. For calculation the equation used is

$$H = c\,(\log_{10}N! - \Sigma\,\log_{10}n_i!)/N$$

where $c$ is a constant for conversion of logarithms from the base 10 to the base chosen for measure. If the base is 2, $c$ is 3.3219; if the base is $e$, $c$ is 2.3026.

# MEASURING POPULATION STRUCTURE

## Life Table

The life table has been explained in Chapter 17. Procedures for constructing a life table are considered here.

To construct a life table we must be able to determine the age of the organisms in question and distribute the population members into age classes or age intervals. Age intervals can vary according to the longevity of the organism. For small rodents or lagomorphs the age intervals may be one month, for deer one year, for humans five years. For insects age categories may be instars or life history stages. We need information on survival, mortality, or rate of mortality by age classes for a given population. Data on survivorship in each age class provide the information needed for the survivorship column, $l_x$. Data on age-specific mortality provide information for the mortality column, $d_x$. We can use the sum of mortality for each age class over time as the size of the initial population. Thus given information for any one column, we can calculate the others.

To demonstrate the construction of a life table, we will use data for a population of Belding's ground squirrel (*Spermophilus beldingi*) (Sherman and Morton 1984). Data for the life table, Table B.16, were obtained by live-trapping and tagging the entire population of ground squirrels over a period of 11 years. Because the data over the years were pooled, the life table is a dynamic-composite one. The life table considers females only.

These are the steps:

1. Construct a table with the following columns:

    $x =$ age interval or age class

    $n_x =$ number of survivors at start of age interval $x$ (raw field data)

    $l_x =$ proportion of organisms surviving to start age interval $x$

    $d_x =$ number or proportion dying during age interval $x$ to $x + 1$

    $q_x =$ rate of mortality during the age interval $x$ to $x + 1$

    $L_x =$ number of individuals alive on the average during the age interval $x$ to $x + 1$

    $T_x =$ total years to be lived by individuals of age $x$ in the population

    $e_x =$ mean expectation of life for individuals alive at start of age interval $x$

2. Tally raw data for survivorship in the $n_x$ column; adjust this survivorship on the basis of 1000 or as a probability of 1.0. The use of 1000 animals makes the table easier to understand, but later the numbers will have to be converted to probabilities to construct fecundity and other tables. If the frequency of each age class in the raw data is not equal to or greater than $x + 1$, the

**Table B.16** Life Table for a Population of Belding's Ground Squirrel

| $x$ | $n_x$ | $l_x$ | $d_x$ | $q_x$ | $L_x$ | $T_x$ | $e_x$ |
|-----|-------|-------|-------|-------|-------|-------|-------|
| 0–1 | 337 | 1000 | 614 | 0.614 | 693 | 1301 | 1.30 |
| 1–2 | 130 | 386 | 190 | 0.492 | 291 | 608 | 1.58 |
| 2–3 | 66 | 196 | 90 | 0.459 | 151 | 317 | 1.62 |
| 3–4 | 36 | 106 | 53 | 0.500 | 79.5 | 166 | 1.57 |
| 4–5 | 18 | 53 | 24 | 0.452 | 41 | 86.5 | 1.63 |
| 5–6 | 10 | 29 | 15 | 0.517 | 21.5 | 45.5 | 1.57 |
| 6–7 | 5 | 14 | 5 | 0.357 | 11.5 | 24 | 1.71 |
| 7–8 | 3 | 9 | 4 | 0.444 | 7 | 12.5 | 1.38 |
| 8–9 | 2 | 5 | 2 | 0.400 | 4 | 5.5 | 1.10 |
| 9–10 | 1 | 3 | 3 | 1.000 | 1.5 | 1.5 | 0.50 |

age distribution must be smoothed. Caughley (1977:96) provides a method.

3. Determine mortality by subtracting $l_x + 1$ from $l_x$. The difference gives $d_x$.
4. Calculate $q_x$ by dividing $d_x$ by $l_x$.
5. Calculate additional information:

$$L_x = \frac{l_x + l_{x+1}}{2}$$

Do this work for each age interval.

6. Sum the $L_x$ column cumulatively from the bottom up to obtain $T_x$.
7. Calculate the life expectancy for each age class by

$$e_x = T_x/l_x$$

Although the construction of a life table is straightforward, given data on survival of various age classes, the life table may be inaccurate or invalid. Especially questionable are life tables based on capture-recapture of marked or banded individuals (Anderson, Wywialowski, and Burnham 1981). Life tables based on such data involve two assumptions. (1) Annual survival rate is age-specific; it varies only by age and not by year. (2) Recovery rates are constant over all ages and all years. Rarely are these assumptions met. In fact, annual survival rates, especially among birds, do vary by year, often influenced more by weather than by age. (For other assumptions in constructing life tables from various sources of data see Caughley 1977 and Krebs 1989).

## Fecundity Table

If we know the productivity of each age class of females, $m_x$, determined by litter counts, brood counts, placental scars, young fledged, and so on, we can construct a fecundity table (Table B.17). The fecundity table includes the age categories, $x;$ age-specific survivorship from the female life table, $l_x;$ age-specific productivity, $m_x;$ and the mean number of female young produced by each female of age $x$, $l_x m_x$, which is $m_x$ weighted by survivorship. The sum of the $l_x m_x$

column gives the net reproductive rate, $R_0$. Added to the fecundity table is another column, $xl_x m_x$, which records the values obtained by multiplying the $l_x m_x$ by the appropriate age. The sum of this column is used to compute the rate of increase.

In Table B.17, age categories have been converted to mean age to permit the correction calculation of $xl_x m_x$ starting with year class 0.

## Rate of Increase

Given a life table and a fecundity table, we can determine the rate of increase, $r_m$, for a population in a particular environment.

An approximation of $r$ can be obtained by the equation

$$r = \frac{\Sigma l_x m_x \log_e \Sigma l_x m_x}{\Sigma x l_x m_x}$$

For the ground squirrel population an approximation is

$$r = 1.358(0.3060)/3.302 = 0.125$$

**Table B.17** Fecundity Table, Belding's Ground Squirrel

| $x$ | $l_x$ | $m_x$ | $l_x m_x$ | $xl_x m_x$ |
|-----|-------|-------|-----------|------------|
| 0–1 | 1.000 | 0.000 | 0.000 | 0.000 |
| 1–2 | 0.386 | 1.040 | 0.401 | 0.401 |
| 2–3 | 0.196 | 2.171 | 0.426 | 0.852 |
| 3–4 | 0.106 | 2.450 | 0.260 | 0.780 |
| 4–5 | 0.053 | 3.113 | 0.165 | 0.660 |
| 5–6 | 0.029 | 1.875 | 0.054 | 0.250 |
| 6–7 | 0.014 | 1.575 | 0.022 | 0.132 |
| 7–8 | 0.009 | 2.000 | 0.018 | 0.126 |
| 8–9 | 0.005 | 1.440 | 0.007 | 0.056 |
| 9–10 | 0.003 | 1.632 | 0.005 | 0.045 |

$$\Sigma l_x m_x = R = 1.358$$
$$\Sigma x l_x m_x = 3.302$$

Another method requires more calculations but gives the same result. The first step is to determine mean generation time

$$T_c = \frac{\Sigma x l_x m_x}{R_0}$$

For the ground squirrels

$$T_c = 3.302/1.358 = 2.431$$

To find an approximate value of $r_m$ the following formula applies:

$$r_m = \log_e R_0 / T_c = \log_e 1.358/2.431$$
$$= 0.1306/2.431$$
$$= 0.125$$

A more accurate assessment of the rate of increase, also based on life table and fecundity tables at any given density with a stable age distribution, can be obtained by the formula derived by the eighteenth-century French mathematician, Leonard Euler:

$$\Sigma l_x m_x e^{-rx} = 1$$

Because there is no real way to solve for $r$, it can be obtained only by substituting values for $r$ in the equation by trial and error until the right side balances with the left. The ground squirrel population will serve as an example, although we may erroneously assume it has a stable age distribution (Table B.20).

The first step is to obtain some approximate estimate of $r$ to be used in the Euler equation. The estimate of $r = 0.125$, obtained from $\log_e R_0 / T_c$ can be used as the first estimate of $r$. The procedure follows:

1. Multiply the estimated $r$ value by age to obtain $rx$. Thus for the first estimated value 0.125 is multiplied by 0, 1, 2, 3, 4, 5, 6, 7, 8, and 9 respectively to give the following values for $rx$: 0.000, .125, .250, .375, .050, .625, .756, .875, 1.000, and 1.125.
2. In a table of functions look up the tabled value for each of the above values of $e^{-rx}$ in the $e^{-x}$ column, or determine the value on a scientific calculator. Record the values as in Table B.18.
3. Multiply the values of $e^{-rx}$ by the appropriate $l_x m_x$ and record. Sum the column. For the $r$ value 0.125, the sum is 1.0158, not close enough to 1.000 to accept.
4. Proceed by substituting a value higher than 0.125 to attempt to bracket $r = 1.000$ between the first estimate and another. A choice of 0.135 results in a sum of 0.9927, too far on the other side of 1.000. A value of 0.132 results in a sum of 0.9995, a very close approximation.

Such iterations can become time-consuming. Iteration for the Euler equation can be programmed on a computer or a programmable calculator.

## Reproductive Value

The life table and fecundity table provide the data needed to calculate the reproductive values of females age $x$ in the population. The reproductive value is the relative number of female offspring that remain to be born to each female age $x$. To state it differently, it is the number of offspring that will be produced by a female from age $x$ until the end of her life. It can be estimated relative to the reproductive value of the female at birth, $v_0$, which is 1 by the formula given by the geneticist R. A. Fisher:

$$v_x/v_x = r^{rx}/l_x \sum_{y=x}^{\infty} e^{-ry} l_y m_y$$

**Table B.18** Determination of $r_m$, Belding's Ground Squirrel

| $x$ | $l_x m_x$ | $r=0.125$ | | | $r=0.135$ | | | $r=0.132$ | | |
|---|---|---|---|---|---|---|---|---|---|---|
| | | $r_x$ | $e^{-rx}$ | $e^{-rx}l_x m_x$ | $r_x$ | $e^{-rx}$ | $e^{-rx}l_x m_x$ | $r_x$ | $e^{-rx}$ | $e^{-rx+l}l_x m_x$ |
| 0 | 0.000 | 0.000 | 0.0000 | 0.000 | 0.000 | 0.0000 | 0.0000 | 0.000 | 0.0000 | 0.0000 |
| 1 | .401 | 0.125 | 0.8825 | 0.3539 | 0.135 | 0.8737 | 0.3504 | 0.132 | 0.8763 | 0.3514 |
| 2 | .426 | .250 | .7788 | .3317 | .270 | .7634 | .3252 | .264 | .7679 | .3271 |
| 3 | .260 | .375 | .6872 | .1787 | .405 | .6669 | .1734 | .369 | .6730 | .1749 |
| 4 | .165 | .050 | .6065 | .1000 | .540 | .5827 | .0956 | .528 | .5898 | .0973 |
| 5 | .054 | .625 | .5352 | .0289 | .675 | .5091 | .0275 | .660 | .5168 | .0279 |
| 6 | .022 | .756 | .4695 | .0103 | .810 | .4448 | .0098 | .792 | .4529 | .0099 |
| 7 | .018 | .875 | .4723 | .0085 | .945 | .3887 | .0069 | .924 | .3969 | .0071 |
| 8 | .007 | 1.000 | .3678 | .0026 | 1.080 | .3396 | .0024 | 1.056 | .3478 | .0024 |
| 9 | .005 | 1.125 | .3246 | .0016 | 1.215 | .2967 | .0015 | 1.188 | .3048 | .0015 |
| | $\Sigma e^{-rx}l_x m_x =$ | | | 1.0158 | | | 0.9927 | | | 0.9995 |

**Table B.19** Determination of Reproductive Values, Female Belding's Ground Squirrel

| $x$ | $l_x$ | $e^{-rx}$ | $e^{-rx}l_x$ | $e^{-rx}l_xm_x$ | $e^{-rx}l_ym_y$ | $v_x$ |
|---|---|---|---|---|---|---|
| 0 | 1.000 | 0.000 | 0.000 | 0.000 | 0.9995 | 1.00 |
| 1 | .386 | .8763 | .3383 | .3514 | .9995 | 2.95 |
| 2 | .196 | .7679 | .1506 | .3270 | .6481 | 4.30 |
| 3 | .106 | .6730 | .0713 | .1749 | .3210 | 4.50 |
| 4 | .053 | .5898 | .0312 | .0973 | .1461 | 4.68 |
| 5 | .029 | .5168 | .0149 | .0279 | .0488 | 3.28 |
| 6 | .014 | .4529 | .0063 | .0099 | .0209 | 3.12 |
| 7 | .009 | .3969 | .0036 | .0071 | .0110 | 3.05 |
| 8 | .005 | .3478 | .0017 | .0024 | .0039 | 2.29 |
| 9 | .003 | 3048 | .0009 | .0015 | .0015 | 1.67 |

**Table B.20** Stable Age Distribution, Belding's Ground Squirrel

| $x$ | $l_x$ | $\lambda^{-x}$ | $\lambda^{-x}l_xm_x$ | $C_x$ |
|---|---|---|---|---|
| 0 | 1.000 | 1.000 | 1.000 | 0.615 |
| 1 | .386 | .8872 | .3425 | .210 |
| 2 | .196 | .7695 | .1508 | .094 |
| 3 | .106 | .6749 | .0715 | .044 |
| 4 | .053 | .5921 | .0314 | .019 |
| 5 | .029 | .5194 | .0151 | .009 |
| 6 | .014 | .4556 | .0064 | .004 |
| 7 | .009 | .3996 | .0036 | .003 |
| 8 | .005 | .3505 | .0018 | .001 |
| 9 | .003 | .3075 | .0009 | .001 |

$$\Sigma\lambda^{-x}l_x = 1.6240$$

$$\lambda = e^r = e^{0.132} = 1.14$$

Because the reproductive value of a female at birth is 1, $v_0$ can be removed from the formula. The formula can also be written as

$$v_x = \frac{e^{rx}}{l_x} \sum_{y=x} e^{-ry}l_ym_y$$

where $y$ is all ages a female passes through from age $x$ on up. Essentially the formula states that the number of female offspring produced at any one moment of time by females aged $x$ and over is divided by the number of females of age $x$ at any one moment.

To calculate reproductive values for females of each age class (Table B.19):

1. Add the $e^{-rx}l_xm_x$ column from the bottom up to obtain a value for $e^{-rx}l_ym_y$ for each age class.
2. Divide this value by $e^{-rx}l_x$ for each age class. The reproductive values for the ground squirrel population are graphed in Figure B.3. Note that the reproductive values rise and fall with age.

## Age Distribution

A population growing geometrically with constant age-specific mortality and fecundity rates assumes and maintains a stable age distribution (see Chapter 17). From any set of life tables and fecundity tables for a population the stable age distribution can be determined by the equation

$$C_x = \frac{\lambda^{-x}l_x}{\sum\limits_{i=0}^{\infty} \lambda^{-i}l_i}$$

where $C_x$ is the proportion of organisms in age category $x$ to $x + l$ in a population increasing geometrically, $\lambda = e^{rm}$ finite rate of increase, $l_x$ is the survivorship function from the life table, and $x$, $i$ subscripts indicate age.

The ground squirrel population again may serve as the example. In this case $\lambda = e^r = e^{0.132} = 1.14$.

Calculating the proportion of squirrels in each age category gives, for example,

$$C_0 = \frac{1.000}{1.624} = .615$$

$$C_5 = \frac{0.0151}{1.624} = 0.009$$

Compare these calculated age distribution values with the actual age structure of the population given in Table B.20.

**Figure B.3** Reproductive values for the female Belding's ground squirrel population described in Tables B.16–B.19.

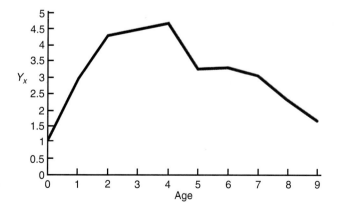

## REFERENCES

ANDREWARTHA, H. G. 1970. *Introduction to the Study of Animal Populations.* University of Chicago Press, Chicago.

BEGON, M., AND M. MORTIMER. 1986. *Population Ecology,* 2nd ed. Sinauer Associates, Sunderland, MA.

BRILLOUIN, L. 1962. *Science and Information Theory,* 2nd ed. Academic Press, New York.

BROWER, J. E., AND J. ZAR. 1984. *Field and Laboratory Methods for General Ecology,* 2nd ed. Brown, Dubuque, IA.

CAUGHLEY, G. 1977. *Analysis of Vertebrate Populations.* Wiley, New York.

COX, G. W. 1985. *Laboratory Manual of General Ecology,* 5th ed. Brown, Dubuque, IA.

GARBUTT, P. D. 1961. The distribution of some small mammals and their associated fleas from central Labrador. *Ecology,* 42: 518–525.

GREIG-SMITH, P. 1984. *Quantitative Plant Ecology,* 3rd ed. Blackwell Scientific Publishers, Oxford. Butterworth, London.

HURLBERT, S. H. 1971. The nonconcept of species diversity: A critique and alternative parameters. *Ecology,* 52: 577–586.

KREBS, C. J. 1985. *Ecology: The Experimental Analysis of Distribution and Abundance,* 3rd ed. Harper & Row, New York.

KREBS, C. J. 1989. *Ecological Methodology.* Harper & Row, New York.

LEVINS, R. 1968. *Evolution in Changing Environments.* Princeton University Press, Princeton, NJ.

LLOYD, M., AND R. J. GHELARDI. 1964. A table for calculating the "equitability" component of species diversity. *J. Animal Ecology,* 33: 217–225.

LUDWIG, J. A., AND J. F. REYNOLDS. 1988. *Statistical Ecology.* Wiley, New York.

MACARTHUR, R. H., AND J. W. MACARTHUR. 1961. On bird species diversity. *Ecology,* 42: 594–598.

MORISITA, M. 1959. Measuring of interspecific association and similarity between communities. *Memoirs of the Faculty of Science, Kyushu University,* Series E (Biology), 3: 66–80.

MUELLER-DOMBOIS, D., AND H. ELLENBERG. 1974. *Aims and Methods of Vegetation Ecology.* Wiley, New York.

PEET, R. K. 1974. The measurement of diversity. *Ann. Rev. Ecology and Systematics,* 5: 285–307.

PIELOU, E. C. 1975. *Ecological Diversity.* Wiley, New York.

POOLE, R. W. 1974. *An Introduction to Quantitative Ecology.* McGraw-Hill, New York.

SHANNON, C. E., AND W. WIENER. 1963. *The Mathematical Theory of Communication.* University of Illinois Press, Urbana.

SHERMAN, P. W., AND M. L. MORTON. 1986. Demography of Belding's ground squirrel. *Ecology,* 65: 1617–1628.

SIMPSON, E. H. 1949. Measurement of diversity. *Nature,* 163: 688.

TRAMER, E. J. 1969. Bird species diversity: Components of Shannon's formula. *Ecology,* 50: 927–929.

VIOLET, R. L. 1973. Microdistribution of small mammals in an Appalachian forest. Unpublished MS thesis, West Virginia University.

ZAR, J. H. 1984. *Biostatistical Analysis,* 2nd ed. Prentice-Hall, Englewood Cliffs, NJ.

# Journals of Interest to Ecologists

*As ecology becomes more specialized, many new journals appear. The following list is by no means complete, but it should call attention to some journals that may be overlooked. The date given is the first year of publication.*

*Acta Botanica Neerlandica,* 1952, Royal Botanical Society of the Netherlands. Wide diversity of papers, mostly in English. Important.

*Agricultural Ecosystems and Environment,* 1981, Elsevier, Amsterdam. Interaction of methods of agricultural production, agricultural ecosystems, and the environment.

*Agroecosystems,* 1974, International Association for Ecology and Elsevier Science Publishers. Ecological interactions within and between agricultural and managed forest ecosystems.

*Agroforestry Systems,* 1982, Martinus Nijhoff/Dr. W. Junk Publishers, The Hague. Application of ecology to land-use systems in agriculture and forestry throughout the world.

*American Birds,* 1946, formerly *Audubon Field Notes.* Devoted to the reporting of distribution and abundance of birds. Important information on continental trends in bird populations.

*American Journal of Botany,* 1914, Botanical Society of America. Technical, of interest chiefly to the professional.

*American Midland Naturalist,* 1909, University of Notre Dame, Notre Dame, IN. Specializes in papers in field natural history.

*American Naturalist,* 1867, American Society of Naturalists. Concerned largely with morphology, evolution, and physiology, but increase in papers on ecology.

*American Scientist,* 1913, Society of Sigma Xi. Contains in every volume several papers of importance to ecologists.

*American Zoologist,* 1916, Quarterly Publication of American Society of Zoologists. Most papers published originate in symposia of the society and its annual refresher course.

*Animal Behavior,* 1952, Baillière, London. A wide range of papers on animal behavior.

*Annales Zoologici Fennici,* 1963, Societas Biologica Fennica Vanamo, Helsinki. An important journal; many papers on ecology.

*Annals of Botany,* 1887, Annals of Botany Co., Clarendon, Oxford. Heavy stress on physiological ecology.

*Annals of the Entomological Society of America,* 1908, Entomological Society of America. The major entomological journal in the United States. Strong on taxonomy and morphology.

*Aquatic Botany,* 1978, Elsevier Science Publishers. Covers submerged, floating, and emergent vegetation of freshwater and marine environments.

*Aquatic Insects,* 1979, International Journal of Freshwater Entomology. Swets Publishing Service, Lisse, Holland.

*Arctic,* 1948, Arctic Institute of North America, McGill-Queens University Press, Montreal. Papers heavily oriented toward natural and human ecology of the arctic region.

*Arctic and Alpine Research,* 1969, Institute of Arctic and Alpine Research, University of Colorado. A major source for information on alpine regions.

*Atmosphere and Ocean,* 1963, Canadian Metero-Ocean Society. Covers meteorological interactions of atmosphere and oceans.

*The Auk,* 1884, American Ornithologists Union. Leading American ornithological journal.

*Australian Journal of Ecology,* 1976, Australian Ecological Society and Blackwell Scientific Publishers. A major journal of general ecology.

*Australian Journal of Marine and Fresh Water Research,* 1952.

*Australian Journal of Zoology,* 1958, Commonwealth, Scientific and Industrial Research Organization. General zoological papers, many of ecological interest.

*Behavior,* 1947, Brill, Leiden, Netherlands. The journal of animal behavior; strongly ethological in viewpoint.

*Behavior Genetics,* 1970, Plenum, New York. Papers on inheritance of behavior in animals and humans.

*Behavioral Ecology and Sociobiology,* 1976, Springer-Verlag, New York. Journal emphasizing ecological and field approach to behavior and sociobiology.

*Biogeochemistry,* 1984, Kluwer Academic Publishers. Papers on all aspects of biogeochemical cycling in the biosphere.

*Biological Bulletin,* 1927, Marine Biological Laboratory, Woods Hole, MA. General papers with emphasis on marine biology.

*Biological Conservation,* 1970, Applied Science Publishers, Barking, Essex, England. Important journal on scientific protection of plants and animals and management of natural resources.

*Bioscience,* 1951, American Institute of Biological Sciences. General biological journal with many papers devoted to ecology and environment.

*Biotropica,* 1969, Association of Tropical Biologists. Major journal on tropical biology.

*Bird-Banding,* 1930, Bird Banding Organizations. Small quarterly noted for its excellent coverage of current literature. (Now *Journal of Field Ornithology.*)

*Botanical Gazette,* 1862, University of Chicago Press, Chicago. Covers all departments of botanical science, including plant ecology.

*Botanical Review,* 1935, New York Botanical Garden. Excellent review papers, often on plant ecology.

*British Birds,* 1897, Witherby, London. An important ornithological journal issued monthly. Contains some first-rate papers on the ecology of birds.

*Brittonia,* 1949, American Society of Plant Taxonomists. Papers deal largely with plant taxonomy.

*The Bryologist,* 1899, American Bryological and Lichenological Society. Only journal devoted to mosses and lichens. Many papers of ecological interest.

*Bulletin of Environmental Contamination and Toxicology,* 1966, Springer-Verlag, New York. Short, current papers dealing largely with pesticides and heavy metal pollution.

*Bulletin of Marine Science,* 1967, Rosensteil School of Marine and Atmospheric Science, University of Miami. Broad aspects of marine science.

*California Fish and Game,* 1914, State of California, Department of Fish and Game. Papers largely devoted to fish and game management. Contains many important papers on western wildlife.

*Cambridge Philosophical Society,* 1923, Biological Reviews, Cambridge University Press. General review papers, many of importance to ecology and behavior.

*Canadian Entomologist,* 1868, Entomological Society of Ontario. A major entomological journal devoted to insects of Canada.

*Canadian Field Naturalist,* 1887, Ottawa Field-Naturalist's Club. Papers on all phases of natural history. Of strong interest to the field biologist.

*Canadian Journal of Botany,* 1921, National Research Council of Canada, Ottawa. General botanical papers, a few of ecological interest.

*Canadian Journal of Fisheries and Aquatic Sciences,* 1942. National Research Council of Canada, Ottawa. A major fisheries journal.

*Canadian Journal of Forest Research,* 1971, National Research Council of Canada. Contains excellent papers on forest ecology.

*Canadian Journal of Soil Science,* 1900, Agricultural Institute of Canada and Canadian Society of Soil Science. Covers various fields of soil science.

*Canadian Journal of Zoology,* 1923, National Research Council of Canada, Ottawa. General zoological papers, many of ecological interest.

*Catena,* Catena Verlag. Interdisciplinary journal of soil science: hydrology, geomorphology, geoecology, landscale evolution.

*Chesapeake Science,* 1960, State of Maryland, Department of Research and Education, Chesapeake Biological Laboratory, Solomons. A regional publication, but papers of wide interest.

*Climate Change,* 1989, Kluwer Academic Publishers. Descriptions, causes, and implications of climate change.

*Condor,* 1899, Cooper Ornithological Society. An ornithological journal with excellent scientific papers.

*Conservation Biology,* 1987, Society for Conservation Biology and Blackwell Scientific Publishers. Emphasis on ecology, population biology, and genetics of fragmented and small populations.

*Copeia,* 1913, American Society of Ichthyologists and Herpetologists. Papers on fishes, amphibians, and reptiles.

*East African Wildlife Journal,* 1963, East African Wildlife Society. Covers general ecology and wildlife and park management in East Africa. (Now *Journal of East African Ecology.*)

*Ecological Applications,* 1990, Ecological Society of America. Journal dealing with applied ecology.

*Ecological Economics,* 1989, International Society for Ecological Economics, Elsevier Science Publishers. Journal dealing with economic aspects of ecological application.

*Ecological Entomology,* 1976, Royal Entomological Society of London, Blackwell, Oxford. Insect ecology.

*Ecological Modeling,* 1975, Elsevier, Amsterdam, The Netherlands. Deals with the use of mathematical models and systems analysis for the description of ecosystems and for the control of environmental pollution and resource development.

*Ecological Monographs,* 1930, Ecological Society of America. Contains papers too long for *Ecology.*

*Ecology,* 1920, Ecological Society of America. Indispensable to ecologists.

*Ecology Law Quarterly,* 1971, School of Law, University of California, Berkeley. Papers on law and the environment.

*Economic Botany,* 1947, Society of Economic Botany. Many interesting papers on human and plant relationships. More ecology than economics.

*Environment,* 1958, Scientists Institute for Public Information, St. Louis. Significant, well-documented articles. Formerly *Scientist and Citizen.*

*Environment and Behavior,* 1969, Sage Publications, New York. Interesting papers on relationship between human environment and behavior.

*Environment and Planning,* 1969, Pion Ltd., London. Papers on environment and ecological approach to urban and suburban planning.

*Environment, Science and Technology,* 1967, American Chemical Society. Environmental articles written from the technological and pro-chemical point of view.

*Environmental Affairs,* 1971, Boston College Environmental Law Center, Brighton, MA. Journal devoted to environmental law. Similar to *Ecology Law Quarterly.*

*Environmental Biology of Fishes,* 1977, Dr. W. Junk, The Hague, Netherlands. Papers on environmental physiology, ecology, life history, and behavior of marine and freshwater fishes.

*Environmental Conservation,* 1974, Elsevier, Sequoia S.A., Lausanne, Switzerland. Articles range from case histories of resource use to environmental policy, education, and management.

*Environmental Entomology,* 1972, Entomological Society of America. Insect interactions with biological, chemical, and physical components of the environment.

*Environmental Law Reporter,* 1971, Environmental Law Institute, Washington, DC. Analysis of laws concerning environmental protection and natural resource use, design of new institutional arrangements to carry out environmental policy.

*Environmental Letters,* 1970, Dekker, New York. Short papers quickly published on current environmental pollution problems.

*Environmental Management,* 1976, Springer-Verlag, New York. New journal focusing on real environmental problems and their solution. Articles deal with practical application of environmental management.

*Environmental Pollution,* 1970, Applied Science Publishers, Barking, Essex, England. Important journal with papers on pollution written from an ecological point of view.

*Environmental Reviews,* 1995, National Research Council of Canada, Ottawa. A new journal emphasizing review papers in environmental sciences.

*Environmental Toxicology and Chemistry,* 1982. Society of Environmental Toxicology and Chemistry and Pergamon Press New York. All aspects of toxic substances in the environment.

*Estuaries,* 1987, Estuarine Research Foundation. A journal covering all aspects of estuarine research.

*Estuarine, Coastal, and Shelf Science,* 1988, Estuarine and Brackish Water Sciences Association and Academic. Covers saline water phenomena in all aspects from continental shelf to tidal zone.

*Evolution,* 1947, Society for the Study of Evolution. Papers on evolution and natural selection.

*Evolutionary Ecology,* 1989, Chapman and Hall, London. Theoretical and empirical papers on the interface of evolution and ecology.

*Fisheries Management and Ecology,* 1994, Blackwell Scientific, New York. Papers concerned with interactions between ecology and management of fish resource.

*Fishery Bulletin,* 1881, U.S. Department of Commerce. Wide range of well-edited papers on ecology and life history of commercial marine species, marine environment, and population dynamics.

*Forest Ecology and Management,* 1976, Elsevier, Amsterdam.

Papers concerned with forest science and conservation and application of ecological knowledge to the management of planted and natural forests. Covers all forest ecosystems.

*Forest Science,* 1955, Society of American Foresters. Technical, but occasionally contains papers of interest to ecologists.

*Forestry,* 1927, Society of Foresters of Great Britain, Oxford University Press, London. Frequently contains good papers on forest ecology.

*Functional Ecology,* 1989, British Ecological Society and Blackwell Scientific Publishers. Emphasis on functional aspects of population ecology, such as foraging, life history patterns.

*The Geographical Journal,* 1893, Royal Geographical Society. Many articles on human ecology.

*Geographical Review,* 1916, American Geographical Society of New York. Outstanding journal with articles of strong ecological interest.

*Geography,* 1916, The Geographical Association, Sheffield, England. Another source of papers relating to ecology and humans.

*Global Biogeochemical Cycles,* 1986, American Geophysical Union, Washington, DC. Encompasses topics related to the earth's atmosphere, oceans, land masses, and biota; interdisciplinary.

*Global Change Biology,* 1995, Blackwell Scientific New York. Deals with interface between all aspects of current environmental change and biological systems.

*Herpetologica,* 1945, Herpetologists League. Papers on all aspects of herpetology. Has improved with age.

*Holarctic Ecology,* 1978, Munksgaard, Copenhagen. Journal covering all aspects of the ecology of the Holarctic region and emphasizing descriptive and analytical studies. Complements *Oikos.*

*Human Ecology,* 1972, Plenum, New York. Journal devoted to the ecology of human beings.

*Hydrobiologica,* 1948, Dr. W. Junk, The Hague. Important journal on freshwater biology.

*Hydrobiological Journal,* 1965, American Fisheries Society and Scripta Publishing Co., New York. Translation of Russian journal. Valuable source of freshwater biology.

*Ibis,* 1859, British Ornithological Society. The leading British ornithological journal.

*Journal of Animal Ecology,* 1933, British Ecological Society, Blackwell, Oxford. Devoted exclusively to animal ecology. Contains many valuable papers.

*Journal of Applied Ecology,* 1966, British Ecological Society, Blackwell, Oxford. Devoted to the applied aspects of Ecology. Many useful papers on wildlife and human-dominated ecosystems.

*Journal of Arid Environments,* 1978, Academic Press, New York. Research work and reviews concerned with problems of arid environments.

*Journal of Biogeography,* 1974, Blackwell Scientific Publishing. Papers on biogeography of plants and animals.

*Journal of Chemical Ecology,* 1975, Springer-Verlag, New York. Major reference for papers on chemical ecology of plants and animals: pheromones, chemical defense, attraction, orientation, etc.

*Journal of Coastal Research,* 1984, The Coastal Education and Research Foundation. Covers social, economic, as well as biological aspects of the coastal environment.

*Journal of East African Ecology,* Ecological Society of East Africa. Continuation of *East African Wildlife Journal.*

*Journal of Ecology,* 1912, British Ecological Society, Blackwell, Oxford. Papers on general ecology with major emphasis on plants. An important reference journal.

*Journal of Economic Entomology,* 1908, Entomological Society of America. Economics of entomology with emphasis on insect damage and control.

*Journal of Entomological Science,* 1966, Georgia Entomological Society. Contains many papers on insect ecology.

*Journal of Environmental Economics and Management,* 1981, Association of Environmental and Resource Economics, Academic Press, New York. Papers on the economic side of ecology and the environment.

*Journal of Environmental Quality,* 1972, American Society of Agronomy, Crop Science Society of America, and Soil Science Society of America, Madison, WI. Papers on environmental quality in natural and agricultural ecosystems. A significant journal.

*Journal of Environmental Systems,* 1971, Baywood. Papers on analysis, design, and management of environmental problems. Applied systems analysis.

*Journal of Experimental Marine Biology and Ecology,* 1963, North Holland. All aspects of marine ecology.

*Journal of Field Ornithology,* 1980. A continuation of *Bird-Banding* with emphasis on field research.

*Journal of Fish Biology,* 1969, Fisheries Society of the British Isles, Academic Press, New York. Papers devoted to fishery biology and fishery management.

*Journal of the Fisheries Research Board of Canada,* 1943, Fisheries Research Board of Canada. Valuable source of papers on fish, fish management, and ecology.

*Journal of Forestry,* 1902, Society of American Foresters. Papers on general forestry. Some of interest to ecologists.

*Journal of Herpetology,* 1967, Society for the Study of Amphibians and Reptiles. General papers on herpetology.

*Journal of Mammalogy,* 1919, American Society of Mammalogists. Only journal in English devoted to mammals.

*Journal of the Marine Biological Association of the United Kingdom,* 1920, Marine Biological Association of the United Kingdom. Contains papers on marine biology.

*Journal of Marine Research,* 1937, Sears Foundation for Marine Research, Bingham Oceanographic Laboratory, Yale University Press, New Haven, CT. A major source of papers on oceanography, heavy on the physical side.

*Journal of the North American Benthological Society,* 1982, North American Benthological Society, Lawrence, KS. Vols. 1 to 4 published as *Freshwater Invertebrate Biology.* An important journal devoted to ecology of bottom organisms, especially in streams; application of ecological theory to study of benthic communities.

*Journal of Parasitology,* 1914, American Society of Parasitology. All aspects of animal parasitism.

*Journal of Range Management,* 1947, American Society of Range Management. Advances in the science and art of grazing land management, understanding of practical and scientific range and pasture problems.

*Journal of Soil Science,* 1949, British Society of Soil Science. Science of soil in its widest aspects.

*Journal of Soil Science of America,* 1936, Soil Science Society of America. All aspects of soil science.

*Journal of Soil and Water Conservation,* 1946, Soil and Water Conservation Society. Covers the basic and applied aspects of soil and water conservation.

*Journal of Tropical Ecology,* 1985, International Association of Ecology. Considers all aspects of tropical ecology.

*Journal of Water Pollution Control Federation,* 1902, Water Pollution Control Federation. Important source of material on water pollution from causes and effects to solutions.

*Journal of Wildlife Management,* 1937, The Wildlife Society. Devoted to wildlife research and management. Contains excellent material for the field biologist.

*Limnology and Oceanography,* 1956, American Society of Limnology and Oceanography. The major journal in the field. All papers of strong interest to ecologists.

*Marine Biology,* 1967, Springer-Verlag, New York. International journal devoted to life in the oceans and coastal waters.

*Marine Ecology,* 1969, Publication della Stazione Zoologice di Napoli, Paul Parier. Ecology of the marine environment.

*Marine Environmental Research,* 1980, Elsevier Applied Science, Essex, England. Papers on chemical, physical, and biological interactions in oceans and coastal waters; strong on marine pollution.

*Microbial Ecology,* 1975, Springer-Verlag, New York. Important journal covering advances in microbiology of natural ecosystems. Contains papers in those branches of ecology in which microorganisms are involved.

*Natural Resources Journal,* 1961, The University of New Mexico School of Law. The original environmental law journal, in print long before a general interest in law and the environment developed.

*Nature,* 1869, Macmillan, London. Contains a number of important papers of interest to ecologists and field biologists.

*New York Fish and Game Journal,* 1953, State of New York Conservation Department. Important regional publication with a number of major papers in the fish and wildlife field.

*North American Journal of Fisheries Management,* 1981, American Fisheries Society. Covers all aspects of fisheries management, including fish population dynamics.

*Northwest Science,* 1927, Northwest Science Association. Includes many papers on ecology, flora, and fauna of the Pacific Northwest.

*Ocean and Shoreline Management,* 1978, Elsevier Scientific Publishers. The applied side of coastal ecology.

*Oecologica,* 1963, Springer-Verlag, New York, in cooperation with International Association for Ecology. International journal containing general ecological papers.

*Oikos, Acta Oecologica Scandinavica,* 1950, Munksgaard, Copenhagen. A major ecological journal emphasizing experimental and theoretical studies. Papers are in English.

*Ornis Scandavia,* 1970, Scandavian Ornithological Union. Experimental studies in ornithology.

*Pacific Science,* 1947, University of Hawaii Press. Physical and biological sciences of the Pacific region.

*Paleobiology,* 1975, Paleontological Society, University of California. Major journal of paleontology.

*Paleogeography, Paleoclimatology, Paleoecology,* 1979, Elsevier Science Publishers. Covers the broad sweep of paleontology.

*Pedobiology,* 1958, Veb Gustav Fisher Verlag Jena. Journal devoted to biology of soil organisms.

*Pesticide Monitoring Journal,* 1967, Federal Workers Group on Pesticide Management, Environmental Protection Agency, Chamblee, GA. Important source of pesticide information.

*Physiological Zoology,* 1927, University of Chicago Press, Chicago. Contains many papers on physiological zoology.

*Plant and Soil,* 1948, Royal Netherlands Society of Agricultural Science. Covers nutrition, plant-microbe associations, soil microbiology.

*Polar Biology,* 1981, Springer-Verlag, New York. Papers on Arctic and Antarctic biology.

*Polar Record,* 1931, Cambridge University Press. Covers all aspects of polar research.

*Proceedings of the Zoological Society of London,* 1822, Zoological Society of London. Contains a number of major papers on ecology and behavior of animals.

*Quarterly Review of Biology,* 1926, American Institute of Biological Sciences. A review occasionally publishing papers of interest to field biologists. Excellent book reviews.

*Quaternary Research,* 1974, University of Washington and Academic Press. Research papers on the ice age.

*Radiation Botany,* 1961, Pergamon Press, Elmsford, NY. International journal with papers on effects of ionizing radiation on plants.

*Regulated Rivers: Research and Management,* 1980, Wiley, NY. interdisciplinary research concerned with river management: physical, biological, ecological.

*Restoration Ecology,* 1993, Blackwell Scientific Publications. Papers on restoration and management of disturbed sites.

*Restoration and Management Notes,* 1981, University of Wisconsin Press. Information and papers on restoration and management of ecological communities on highly disturbed sites.

*Reviews in Aquatic Science,* 1989, CRC Press. Valuable for its synthesis and review papers of aquatic biology.

*Rivers,* 1980, Rivers, Fort Collins, CO. Studies in science, environmental policy, and law of instream flow.

*Science,* 1833, American Association for the Advancement of Science. Covers the whole field of science, but some papers of interest to ecologists.

*The Southwestern Naturalist,* 1956, Southwestern Association of Naturalists. Covers plants, animals, and ecology of southwestern United States, Mexico, and Central America.

*Soviet Journal of Ecology,* 1970 (translated issues from 1972), Consultants Bureau, New York. The major ecological journal of the Soviet Union. Wide range of papers.

*Systematic Zoology,* 1952, Society of Systematic Zoologists. Papers mainly on taxonomy, but many of strong ecological interest.

*Taxon,* 1951, International Association for Plant Taxonomy, Utrecht, The Netherlands. The international journal on plant taxonomy. Papers mostly of a technical nature.

*Theoretical Population Biology,* 1972, Academic Press, New York. Theoretical aspects of population biology, especially ecology, genetics, demography, and epidemiology.

*Transactions of American Fisheries Society,* 1959, American Fisheries Society. Contains a number of papers on fish and freshwater ecology.

*Trends in Ecology and Evolution,* 1986, Elsevier Scientific Publishing. Good review articles of specific topics and commentaries.

*Urban Ecology,* 1975, Elsevier, New York. International journal dealing with ecological processes and interactions within urban areas and between human settlements and surrounding natural systems that support them.

*Vegetatio,* 1950, Institute of Ecological Botany, Uppsala, Sweden. A major journal in plant ecology; synthesis of methods and theories in vegetation science.

*Viltrevy, Swedish Wildlife,* 1964, Swedish Sportsman's Association, Stockholm. Long papers on wildlife, covering both ecology and behavior.

*Wildlife Society Bulletin,* 1973, The Wildlife Society. Articles oriented to field techniques, landscape management, and education.

*Wilson Bulletin,* 1888, Wilson Ornithological Society. Ornithological journal especially oriented to the field.

# Glossary

**A horizon** surface stratum of mineral soil characterized by maximum accumulation of organic matter, maximum biological activity, and loss of such materials as iron, aluminum oxides, and clays

**abiotic** pertaining to the nonliving components of the environment, including soil, water, air, light, nutrients, and the like

**abyssal** relating to the bottom waters of oceans, usually below 1000 m

**acclimation** alteration of an individual's physiological rate or capacity to perform a function through long-term exposure to certain conditions

**acclimatization** changes in physiological state or tolerance that appear in a species after long exposure to different natural environments

**acid deposition** wet and dry atmospheric fallout with an extremely low pH, brought about by a combination of water vapor in the atmosphere with hydrogen sulfide and nitrous oxide vapors released by burning fossil fuels; the sulfuric and nitric acid in rain, fog, and snow, gases, and particulate matter

**active transport** movement of ions and molecules across a cell membrane against a concentration gradient, involving an expenditure of energy, in a direction opposite to simple diffusion

**adaptation** a genetically determined characteristic (behavioral, morphological, physiological) that improves an organism's ability to survive and reproduce under prevailing environmental conditions

**adaptive radiation** evolution from a common ancestor of divergent forms adapted to distinct ways of life

**adiabatic cooling** a decrease in air temperature when a rising parcel of warm air cools by expanding (which uses energy) rather than by losing heat to the surrounding air; the rate of cooling is approximately 1° C/100 m for dry air and 0.6° C/100 m for moist air

**adiabatic lapse rate** rate at which a parcel of air loses temperature with elevation if no heat is gained from or lost to an external source

**adiabatic process** a process in which heat is neither lost to nor gained from the outside

**aerenchyma** plant tissue with large air-filled intercellular spaces, usually found in roots and stems of aquatic and marsh plants

**aerobic** living or occurring only in the presence of free uncombined molecular oxygen, either as a gas in the atmosphere or dissolved in water

**aestivation** dormancy in animals through a drought or dry season

**aggrading** gradually increasing in structure or biomass

**aggregative response** behavior of consumers who spend most time in food patches with the greatest density of prey

**aggressive mimicry** resemblance of a predator or parasite to a harmless species to deceive potential prey

**albedo** proportion of solar radiation reflected by Earth's surface, by the tops of clouds, and by the atmosphere without heating the receiving surface

**alfisols** soils characterized by an accumulation of iron and aluminum in the lower or B horizon

**allele** one of two or more alternative forms of a gene that occupy the same relative position or locus on homologous chromosomes

**allelopathy** effect of metabolic products of plants (excluding microorganisms) on the growth and development of other nearby plants

**allogenic succession** ecological change or development of species structure and community composition brought about by some external force such as fire or storms

**allopatric** having different areas of geographic distribution; possessing nonoverlapping ranges

**allozygous** having two alleles at a locus that are not replicates of a single ancestral allele and therefore are not identical by descent; compare *autozygous*

**alluvial soil** soil developing from recent alluvium (material deposited by running water), exhibiting no horizon development, and typical of floodplains

**alpha diversity** the variety of organisms occupying a given place or habitat

**altricial** among birds and mammals, being hatched or born too weak to support their own weight and usually blind

**altruism** a form of behavior in which an individual increases the welfare of another at the expense of its own welfare

**ambient** referring to surrounding, external, or unconfined conditions

**amensalism** relationship between two species in which one is inhibited or harmed by the presence of the other

**ammonification** breakdown of proteins and amino acids, especially by fungi and bacteria, with ammonia as the excretory byproduct

**anabolism** metabolic reactions requiring energy usually provided by ATP in which molecules are linked to form more complex molecules

**anadromous fish** fish that inhabit seas or lakes but ascend streams to spawn (for example, salmon)

**anaerobic** adapted to environmental conditions devoid of oxygen

**andosols** soils derived from volcanic ejecta, not highly weathered, with a dark upper layer

**antibiotic** a substance produced by a living organism that is toxic to organisms of other species

**anticyclone** an area of high atmospheric pressure characterized by subsiding air and horizontal divergence of air near the surface in its central region

**apparent plants** large, easy to locate plants possessing quantitative defenses not easily mobilized at the point of attack (for example, tannins)

**aridosols** desert soils characterized by little organic matter and high base content

**asexual reproduction** any form of reproduction, such as budding, that does not involve the fusion of gametes

**assimilation** transformation or incorporation of a substance by organisms; absorption and conversion of energy and nutrient uptake into constituents of an organism

**association** a natural unit of vegetation characterized by a relatively uniform species composition and often dominated by a particular species

**atoll** a ring-shaped coral reef that encloses or almost encloses a lagoon and is surrounded by open sea

**aufwuchs** community of plants and animals attached to or moving about on submerged surfaces; also called *periphyton,* but that term more specifically applies to organisms attached to submerged plant stems and leaves

**autogenic** self-generated

**autotrophy** ability of an organism to produce organic material from inorganic chemicals and some source of energy

**autozygous** having two alleles at a locus that have been derived by

replication of a single allele in some ancestral population; having alleles identical by descent

**available water capacity** supply of water available to plants in well-drained soil

**B horizon** soil stratum beneath the A horizon characterized by an accumulation of silica, clay, and iron and aluminum oxides and possessing blocky or prismatic structure

**basal metabolic rate** the minimal amount of energy expenditure needed by an animal to maintain vital processes

**Batesian mimicry** resemblance of a palatable or harmless species, the mimic, to an unpalatable or dangerous species, the model

**bathyal** pertaining to anything, but especially organisms, in the deep sea below the photic or lighted zone and above 4000 m

**benthos** animals and plants living on the bottom of a lake or sea from the high water mark to the deepest depths

**beta diversity** variety of organisms occupying a number of different habitats over a region; regional diversity, as opposed to local or alpha diversity

**biennial** plant that requires two years to complete a life cycle, with vegetative growth the first year and reproductive growth (flowers and seeds) the second

**biochemical oxygen demand (BOD)** a measure of the oxygen needed in a specified volume of water to decompose organic materials; the greater the amount of organic matter in water, the higher the BOD

**biogeochemical cycle** movement of elements or compounds through living organisms and the nonliving environment

**biological clock** the internal mechanism of an organism that controls circadian rhythms without external time cues

**biological magnification** process by which pesticides and other substances become more concentrated in each link of the food chain

**biological species** a group of potentially interbreeding populations reproductively isolated from all other populations

**bioluminescence** production of light by living organisms

**biomass** weight of living material, usually expressed as dry weight per unit area

**biome** major regional ecological community of plants and animals; usually corresponds to plant ecologists' and European ecologists' classification of plant formations and life zones

**biophage** organism that feeds on living material

**biosphere** thin layer about Earth in which all living organisms exist

**biotic community** any assemblage of populations living in a prescribed area or physical habitat

**blanket mire** a large area of upland dominated by sphagnum moss and dependent upon precipitation for a water supply; a moor

**bog** a wetland ecosystem characterized by an accumulation of peat, acid conditions, and dominance of sphagnum moss

**bottleneck** an evolutionary term for any stressful situation that greatly reduces a population

**browse** the part of current leaf and twig growth of shrubs, woody vines, and trees available for animal consumption

**bryophyte** member of the division in the plant kingdom of nonflowering plants comprising mosses (Musci), liverworts (Hepaticae), and hornworts (Anthocerotae)

**buffer** a chemical solution that resists or dampens change in pH upon addition of acids or bases

**C horizon** soil stratum beneath the solum (A and B horizons) little affected by biological activity and soil-forming processes

**$C_3$ plant** any plant that produces as its first step in photosynthesis the three-carbon compound phosphoglyceric acid

**$C_4$ plant** any plant that produces as its first step in photosynthesis a four-carbon compound malic or aspartic acid

**calcicole** plant susceptible to aluminum toxicity, acidity, and other factors influenced by the absence of calcium

**calcification** process of soil formation characterized by accumulation of calcium in lower horizons

**calcifuge** plant with low calcium requirement that can live in soils with a pH of 4.0 or less

**caliche** an alkaline, often rocklike salt deposit on the surface of soil in arid regions, formed at the level where leached Ca salts from the upper soil horizons are precipitated

**calorie** amount of heat needed to raise 1 gram of water 1° C, usually from 15° C to 16° C

**CAM plants (crassulacean acid metabolism)** plants (cactus and other succulents) that separate the processes of carbon dioxide uptake and fixation when growing under arid conditions; they take up gaseous carbon dioxide at night, when stomata are open and water loss is minimal, during the day, when the stomata are closed, $CO_2$ is released and chemically used

**Cambrian** earliest period in geological time in the Paleozoic era, from about 570 to 505 million years ago; algae and many marine invertebrates appeared

**cannibalism** killing and consuming one's own kind; intraspecific predation

**canonical distribution** particular configuration of the log-normal distribution of species abundance

**capillary water** that portion of water in the soil held by capillary forces between soil particles

**Carboniferous** second period of the Upper Paleozoic, from about 360 to 286 million years ago; named after extensive coal deposits formed at that time; warm humid period characterized by forests and swamps, appearance of early amphibians, reptiles, and giant ferns

**carnivore** an organism that feeds on animal tissue; taxonomically, a member of the order Carnivora (Mammalia).

**carr** vegetation dominated by alder and willows and occupying eutrophic peat

**carrying capacity (*K*)** number of individual organisms the resources of a given area can support, usually through the most unfavorable period of the year

**catabolism** metabolic breakdown of complex molecules to simple ones

**catadromous fish** fish that feed and grow in fresh water, but return to the sea to spawn

**catastrophic extinction** a major episode of extinction involving many taxa, occurring suddenly in the fossil record

**catena** a group of related soils

**cation** part of a dissociated molecule carrying a positive electrical charge

**cation exchange capacity** ability of a soil particle to absorb positively charged ions

**Cenozoic** major division of geological time extending from the Mesozoic era some 65 million years ago to the present

**chamaephyte** perennial shoots or buds from the surface of the ground to about 25 cm above the surface

**chaos** apparently random, recurrent, and unpredictable behavior in a simple deterministic system

**chaparral** vegetation consisting of broadleafed evergreen shrubs, found in regions of mediterranean-type climate of hot, dry summers and mild, wet winters

**character displacement** divergence of characteristics in two otherwise similar species occupying overlapping ranges, brought about by the selective effects of competition

**character divergence** evolution of behavioral or morphological differences among species occupying the same area, brought about by the selective pressure of competition

**chasmogamy** production of flowers that open to expose reproductive organs, allowing cross-pollination

**chilling tolerance** ability of a plant to carry on photosynthesis within a range of +5° to +10° C

**chronosequence** a sequence of related soils that differ in the degree of profile development because of age

**circadian rhythm** endogenous rhythm of physiological or behavioral activity, of approximately 24 hours duration

**cleistogamy** self-pollination within a flower that does not open

**climax** stable end community of succession that is capable of self-perpetuation under prevailing environmental conditions

**climograph** a diagram describing a locality based on the annual cycle of temperature and precipitation

**cline** gradual change in population characteristics over a geographical area, usually associated with changes in environmental conditions

**clone** a population of genetically identical individuals resulting from asexual reproduction

**coarse-grained** clumped into large patches affecting the activities of organisms

**coevolution** joint evolution of two or more noninterbreeding species that have a close ecological relationship; through reciprocal selective pressures, the evolution of one species in the relationship is partially dependent on the evolution of the other

**coexistence** two or more species living together in the same habitat, usually with some form of competitive interaction

**cohort** a group of individuals of the same age within a population

**cold resistance** ability of a plant to resist low temperature stress without injury

**colluvium** mixed deposits of soil material and rock fragments accumulated near the base of steep slopes through soil creep, landslides, and local surface runoff

**commensalism** relationship between species that is beneficial to one, but neutral or of no benefit to the other

**community** a group of interacting plants and animals inhabiting a given area

**compensation intensity** light intensity at which photosynthesis and respiration balance each other so that net production is 0; in aquatic systems, usually the depth of light penetration at which oxygen utilized in respiration equals oxygen produced by photosynthesis

**competition** any interaction that is mutually detrimental to both participants, within or between species that share limited resources

**competitive exclusion** hypothesis that when two or more species coexist using the same resource, one must displace or exclude the other

**conduction** direct transfer of heat from one substance to another

**connectance** ratio of potential links or interactions in a food web to those that actually exist

**consumer** any organism that lives on other organisms, dead or alive

**contest competition** competition in which a limited resource is shared only by dominant individuals, resulting in a relatively constant number of survivors, regardless of initial density

**continuum** a gradient of environmental characteristics or changes in community composition

**continuum index** measure of the position of a community on a gradient defined by species composition

**convection** transfer of heat by the circulation of fluids, liquid or gas

**convergent evolution** development of similar characteristics in different species living in different areas but under similar environmental conditions

**coprophagy** feeding on feces

**Coriolis effect** physical consequences of the law of conservation of angular momentum; as a result of the earth's rotation a moving object veers to the right in the Northern Hemisphere and to the left in the Southern Hemisphere relative to the earth's surface

**countercurrent circulation** an anatomical and physiological arrangement by which heat exchange takes place between outgoing warm arterial blood and cool venous blood returning to the body core; important means of maintaining temperature homeostasis in many vertebrates

**Cretaceous** final period of the Mesozoic, from about 144 to 65 million years ago, when much of the present land area was covered by shallow seas; marked by rise of the angiosperms and continued dominance of dinosaurs until mass extinction at the end of the period, by evolution of modern birds and fish, and by appearance of primitive mammals

**critical daylength** the period of daylight, specific for any given species, that triggers a long-day or a short-day response in organisms

**critical thermal maximum** temperature at which an animal's capacity to move is so reduced that it cannot escape from thermal conditions that will lead to death

**crude density** number of individuals per unit area

**cryptic coloration** coloration of organisms that makes them resemble or blend into their habitat or background

**cryptophyte** a bud buried in the ground on a bulb or rhizome

**cyclic replacement** type of succession in which the sequence of seral stages is repeated by imposition of some disturbance so that the sere never arrives at a climax or stable sere

**dark reaction** a group of light-independent reactions following the light reactions in photosynthesis; they reduce carbon dioxide to produce glucose and other carbohydrates

**day-neutral plant** a plant that does not require any particular photoperiod to flower

**death rate** number of individuals in a population dying in a given time interval divided by the number alive at the midpoint of the time interval

**deciduous** (of leaves) shed during a certain season (winter in temperate regions; dry season in the tropics); (of trees) having deciduous parts

**decomposer** an organism that obtains energy from the breakdown

of dead organic matter to more simple substances; most precisely refers to bacteria and fungi

**decomposition** breakdown of complex organic substances into simpler ones

**deductive method** in testing hypotheses, going from the specific to the general

**definitive host** the host in which a parasite becomes an adult and reaches maturity

**deme** a local population or interbreeding group within a larger population

**demographic stochasticity** random variations in population birth and death rates due entirely to chance differences experienced by individuals

**demography** the statistical study of the size and structure of populations and changes within them

**denitrification** reduction of nitrates and nitrites to nitrogen by microorganisms

**density dependence** regulation of size of a population by mechanisms whose effectiveness increases as population size increases

**density independence** condition of a population in which regulation of growth is not tied to population density

**dependent variable** variable *y,* which yields the second of two numbers in an ordered pair (*x, y*); the set of all values taken on by the dependent variable is called the range of the function; compare *independent variable*

**deterministic model** mathematical model in which all relationships are fixed and a given input produces one exact prediction as an output

**detritivore** an organism that feeds on dead organic matter; usually, detritus-feeding organisms other than bacteria and fungi

**detritus** fresh to partly decomposed plant and animal matter

**Devonian** period between the Ordovician and Carboniferous, 408 to 360 million years ago; the age of fish, evolution of early amphibians, and appearance of vascular plants, insects, and plants

**dew point** temperature at which condensation of water in the atmosphere begins

**diameter breast height (dbh)** diameter of a tree measured at 1.4 m (4ft. 6 in.) from ground level

**diapause** a period of dormancy, usually seasonal, in the life cycle of an insect, in which growth and development cease and metabolism is greatly decreased

**diffuse coevolution** coevolution involving the interaction of many organisms in contrast to pairwise interaction

**diffuse competition** competition in which a species experiences interference from numerous other species that deplete the same resources

**dimorphism** existing in two structural forms—two color forms, two sexes, and the like

**dioecious** bearing male and female reproductive organs on separate plants

**diploid** having chromosomes in homologous pairs or twice the haploid number of chromosomes

**directional selection** selection favoring individuals at one extreme of the phenotype in a population

**disease** any negative deviation from a normal state of health

**dispersal** leaving an area of birth or activity for another area

**dispersion** distribution of organisms within a population over an area

**disruptive selection** selection in which two extreme phenotypes in the population leave more offspring than the intermediate phenotype, which has lower fitness

**diversity** abundance of different species in a given location; species richness

**diversity index** the mathematical expression of species richness of a given community or area

**dominance** (ecological) control within a community over environmental conditions influencing associated species by one or several species, plant or animal, enforced by number, density, or growth form; (social) behavioral, hierarchical order in a population that gives high-ranking individuals priority of access to essential requirements; (genetic) ability of an allele to mask the expression of an alternative form of the same gene in a heterozygous condition

**dominant** the population possessing ecological dominance in a given community, thereby governing type and abundance of other species in the community

**dormancy** state of cessation of growth and suspended biological activity, during which life is maintained

**drought avoidance** ability of a plant to escape dry periods by becoming dormant or surviving as a seed

**drought resistance** sum of drought tolerance and drought avoidance

**drought tolerance** ability of plants to maintain physiological activity in spite of the lack of water or to survive the drying of tissues

**dynamic pool model** an optimum yield model, using growth, recruitment, mortality, and fishing intensity to predict yield

**dystrophic** of water, having a high content of humic organic matter, often with high littoral productivity and low plankton productivity

**ecocline** a geographical gradient of communities or ecosystems produced by responses of vegetation to environmental gradients of rainfall, temperature, nutrient concentrations, and other factors

**ecological density** density measured in terms of the number of individuals per area of available living space

**ecological efficiency** percentage of biomass produced by one trophic level that is incorporated into biomass of the next highest trophic level

**ecological release** expansion of habitat or increase in food availability resulting from release of a species from interspecific competition

**ecosystem** the biotic community and its abiotic environment functioning as a system

**ecotone** transition zone between two structurally different communities; see also *edge*

**ecotype** a subspecies or race adapted to a particular set of environmental conditions

**ectothermy** determination of body temperature primarily by external thermal conditions

**edaphic** relating to soil

**edge** place where two or more vegetation types meet

**edge effect** response of organisms, animals in particular, to environmental conditions created by an edge

**effective population size** the size of an ideal population that would undergo the same amount of random genetic drift as the

actual population; sometimes used to measure the amount of in-breeding in a finite, randomly mating population

**egestion**  elimination of undigested food material

**elaiosome**  shiny, oil-containing, ant-attracting tissue on the seed coat of many plants

**emigration**  movement of part of a population permanently out of an area

**endemic**  restricted to a given region

**endogenous**  arising within an organism

**endothermy**  regulation of body temperature by internal heat production, allowing maintenance of appreciable difference between body temperature and external temperature

**energy**  capacity to do work

**entisols**  embryonic mineral soils whose profile is just beginning to develop, common on recent floodplains and wind deposits, and lacking distinct horizons

**entertainment**  synchronization of an organism's activity cycle with environmental cycles

**entropy**  transformation of matter and energy to a more random, more disorganized state

**environment**  total surroundings of an organism, including other plants and animals and those of its own kind

**environmental stochasticity**  unpredictable environmental changes, such as adverse weather or food failure, that affect some aspect of population growth

**epidemic**  rapid spread of a bacterial or viral disease in a human population; compare *epizootic*

**epifauna**  benthic organisms that live on or move across the surface of a substrate

**epilimnion**  warm, oxygen-rich upper layer of water in a lake or other body of water, usually seasonal

**epiphyte**  an organism that lives wholly on the surface of plants, deriving support but not nutrients from the plants

**epizootic**  rapid spread of a bacterial or viral disease in a dense population of animals

**equilibrium species**  a species whose population exists in equilibrium with resources and at a stable density

**equilibrium turnover rate**  change in species composition per unit time when immigration equals extinction

**equitability**  evenness of distribution of species abundance patterns; maximum equitability is the same number of individuals among all species in the community.

**estivation**  dormancy in animals during a period of drought or a dry season

**estuary**  a partially enclosed embayment where fresh water and sea water meet and mix

**eukaryote**  organism whose cells have membranous organelles, notably the nucleus

**euphotic zone**  surface layer of water to the depth of light penetration where photosynthetic production equals respiration

**eutrophic**  of water, high in nutrient content and productivity

**eutrophication**  nutrient enrichment of a body of water; called cultural eutrophication when accelerated by introduction of massive amounts of nutrients by human activity

**evapotranspiration**  sum of the loss of moisture by evaporation from land and water surfaces and by transpiration from plants

**evenness**  degree of equitability in the distribution of individuals among a group of species; see *equitability*

**evolution**  change in gene frequency through time resulting from

natural selection and producing cumulative changes in characteristics of a population

**exothermic**  releasing heat to the environment in a chemical reaction

**exploitative competition**  competition by a group or groups of organisms that reduce a resource to a point that it adversely affects other organisms

**exponential growth**  instantaneous rate of population growth expressed as proportional increase per unit of time

**extinction coefficient**  point at which the intensity of light reaching a certain depth is insufficient for photosynthesis; ratio of intensity of light at a given depth to the intensity of light at a given depth to the intensity at the surface

$F_1$ **generation**  the first generation of offspring from a cross between individuals homozygous for contrasting alleles; the $F_1$ is necessarily heterozygous

$F_2$ **generation**  offspring produced by selfing or by allowing the $F_1$ generation to breed among themselves

**facilitation model**  a model of succession in which the previous community prepares the way for or "facilitates" a succeeding community

**facultative**  ability to adjust optionally to different environmental conditions

**fecundity**  potential ability of an organism to produce eggs or young; rate of production of young by a female

**fell-field**  area within a tundra characterized by stony debris and sparse vegetation

**fen**  a wetland, only slightly acidic, dominated by sedges in which peat accumulates

**fermentation**  breakdown of carbohydrates and other organic matter under anaerobic conditions

**field capacity**  the amount of water held by soil against the force of gravity

**fine-grained**  occurring in patches so small that organisms do not distinguish among them

**fitness**  genetic contribution by an individual's descendants to future generations

**fixation**  process in soil by which certain chemical elements essential for plant growth are converted from a soluble or exchangeable form to a less soluble or nonexchangeable form

**floating reserve**  individuals in a population of a territorial species that do not hold territories and remain unmated, but are available to refill territories vacated by death of an owner

**flux**  flow of energy from a source to a sink or receiver

**foilage height diversity**  measure of the degree of layering or vertical stratification of foliage in a forest

**food chain**  movement of energy and nutrients from one feeding group of organisms to another in a series that begins with plants and ends with carnivores, detrital feeders, and decomposers

**food web**  the interlocking pattern formed by a series of interconnecting food chains

**foraging strategy**  the manner in which animals seek food and allocate their time and effort in obtaining it

**forb**  herbaceous plant other than grass, sedge, or rush

**formation**  classification of vegetation based on dominant life forms

**founder effect**  effect of starting a population with a small number of colonists, which contain only a small and often biased sam-

ple of genetic variation of the parent population, a markedly different new population may arise

**fragipan** a dense, brittle, compact subsoil horizon associated with acid soil conditions

**fragmentation** reduction of a large habitat area into small, scattered remnants; reduction of leaves and other organic matter into smaller particles

**free-running cycle** length of a circadian rhythm in absence of an external time cue

**frost pocket** a depression in the landscape into which cold air drains, lowering the temperature relative to the surrounding area, and which often supports a characteristic group of cold-tolerant plants

**frugivore** an organism that feeds on fruit

**functional response** change in rate of exploitation of a prey species by a predator in relation to changing prey density

**fundamental niche** the total range of environmental conditions under which a species can survive

**gamma diversity** diversity difference between similar habitats in widely separated geographic regions

**gap** opening made in a forest canopy by some small disturbance such as windfall or death of an individual or group of trees that influences the development of vegetation beneath

**gap phase replacement** successional development in small disturbed areas within a stable plant community; filling in of a space left by a disturbance, but not necessarily by the species eliminated by the disturbance

**garrigue** shrub woodland characteristic of limestone areas with low rainfall and thin, poor, dry soils; widespread in the Mediterranean countries of southern Europe

**gene** unit material of inheritance; more specifically, a small unit of DNA molecule coded for a specific protein to produce one of the many attributes of a species

**gene flow** exchange of genetic material between populations

**gene frequency** actually allele frequency; relative abundance of different alleles carried by an individual or a population

**gene pool** the sum of all the genes of all individuals in a population.

**genet** a genetic individual that arises from a single fertilized egg

**genetic drift** random fluctuation in allele frequency over over time due to chance alone without any influence by natural selection; important in small populations

**genetic feedback** evolutionary response of a population to adaptations of predators, parasites, or competitors

**genotype** genetic constitution of an organism

**geometric rate of increase** factor by which the size of a population increases over a period of time

**gley soil** soil developed under conditions of poor drainage, resulting in reduction of iron and other elements and in gray colors and mottles

**global stability** ability of a community to withstand large disturbances and return to its original state

**gouger** member of a group of stream invertebrates that live and feed on woody debris

**granivore** an organism that feeds on seeds

**greenhouse effect** selective energy absorption by carbon dioxide in the atmosphere, which allows short wavelength energy to pass through but absorbs longer wavelengths and reflects heat back to Earth

**gross production** energy fixed per unit area by photosynthetic activity of plants before respiration; total energy flow at the secondary level is not gross production but assimilation, because consumers use material already produced with respiratory losses

**group selection** elimination of one group of individuals by another possessing superior genetic traits; not a widely accepted hypothesis

**growth form** morphological category of plants, such as tree, shrub, and vine

**guild** a group of populations that utilize a gradient of resources in a similar way

**gyre** circular motion of water in major ocean basins

**habitat** place where a plant or animal lives

**hadal** the ocean below 6000 m

**halocline** changes in salinity with depth in the oceans

**haploid** having a single set of unpaired chromosomes in each cell nucleus

**Hardy-Weinberg law** the proposition that genotypic ratios resulting from random mating remain unchanged from one generation to another, provided natural selection, genetic drift, and mutation are absent

**heliothermism** acquisition of heat energy by ectotherms through basking in the sun

**hemicryptophyte** plant with perennial shoots or buds close to the surface of the ground, often covered with litter

**herbivore** an organism that feeds on plant tissue

**hermaphrodite** an organism possessing the reproductive organs of both sexes

**heterogeneity** state of being mixed in composition; can refer to genetic or environmental conditions

**heterotherm** an organism that during part of its life history becomes either endothermic or ectothermic, hibernating mammals become ectothermic, and foraging insects such as bees become endothermic; they are characterized by rapid, drastic, repeated changes in body temperature

**heterotrophic** requiring a supply of organic matter or food from the environment

**heterozygous** containing two different alleles of a gene, one from each parent, at the corresponding loci of a pair of chromosomes

**hibernation** winter dormancy in animals, characterized by a great decrease in metabolism

**hierarchy** a sequence of sets made up of smaller subsets

**histosol** soil characterized by high organic matter content

**home range** area over which an animal ranges throughout the year

**homeohydric** able to maintain a stable internal water balance independent of the environment

**homeostasis** maintenance of nearly constant conditions in function of an organism or in interaction among individuals in a population

**homeotherm** an animal with a fairly constant body temperature; also spelled *homoiotherm* and *homotherm.*

**homeothermy** regulation of body temperature by physiological means

**homologous chromosomes** corresponding chromosomes from male and female parents that pair during meiosis

**homozygous** containing two identical alleles of a gene at the corresponding loci of a pair of chromosomes

**horizon** major zone or layer of soil; each horizon has its own structure and characteristics

**host** an organism that provides food or other benefit to another organism of a different species; usually refers to an organism exploited by a parasite

**humus** organic material derived from partial decay of plant and animal matter

**hybrid** plant or animal resulting from a cross between genetically different parents

**hydrosphere** all water on or near Earth's surface

**hyperthermia** rise in body temperature to reduce thermal differences between an animal and a hot environment, thus reducing the rate of heat flow into the body

**hypha** filament of a fungus thallus or vegetative body

**hypolimnion** cold, oxygen-poor zone of a lake that lies below the thermocline

**hypothesis** proposed explanation for a phenomenon; it should be testable for acceptance or rejection by experimentation

**immigration** arrival of new individuals into a habitat or population

**immobilization** conversion of an element form inorganic to organic form in microbial or plant tissue, rendering the nutrient relatively unavailable to other organisms

**importance value** sum of relative density, relative dominance, and relative frequency of a species in a community

**inbreeding** mating among close relatives

**inbreeding depression** detrimental effects of inbreeding

**inceptisols** mineral soils that have one or more horizons in which mineral materials have been weathered or removed, and that are only beginning to develop a distinctive soil profile

**incipient lethal temperature** temperature at which a stated fraction of a population of poikilothermic animals (usually 50 percent) will die when brought rapidly to it from a different temperature

**inclusive fitness** sum of the total fitness of an individual and the fitness of its relatives, weighted according to the degree of relationship

**independent variable** variable *x*, which yields the first of two numbers of an ordered pair (*x, y*); the set of all values taken on by the independent variable is called the domain of the function; compare *dependent variable*

**individualistic concept** the view, first proposed by H. A. Gleason, that vegetation is a continuous variable in a continuously changing environment; thus no two vegetational communities are identical, and association of species results only from similarities in requirements

**inductive method** in testing hypotheses, going from the specific to the general

**infauna** organisms living within a substrate

**infiltration** downward movement of water into the soil

**infralittoral** region below the littoral region of the sea

**inhibition model** model of succession proposing that the dominant vegetation occupying a site prevents colonization of that site by other plants of the next successional community

**instar** form of an insect or other arthropod between successive molts

**interdemic selection** group selection of populations within a species

**interference competition** competition in which access to a resource is limited by the presence of a competitor

**interglacial period** period between glaciations, during which species recolonize previously occupied areas

**intermediate host** host that harbors the developmental phases of a parasite; the infective stage or stages can develop only when the parasite is independent of its definitive host; compare *definitive host*

**intersexual competition** competition among members of the same sex for a mate, most common among males and characterized by fighting and display

**interspecific** between individuals of different species

**intraguild predation** predation among species occupying the same trophic level and using a similar food resource

**intraspecific** between individuals of the same species

**intrinsic rate of increase** the per capita rate of growth of a population that has reached a stable age distribution and is free of competition and other growth restraints

**introgression** incorporation of genes of one species into the gene pool of another

**inversion** (genetic) reversal of part of a chromosome so that genes within that part lie in reverse order; (meteorological) increase rather than decrease in air temperature with height, caused by radiational cooling of Earth (radiational inversion) or by compression and consequent heating of subsiding air masses from high pressure areas (subsidence inversion)

**island biogeography** study of distribution of organisms and community structure on islands

**isolating mechanism** any structural, behavioral, or physiological mechanism that blocks or inhibits gene exchange between two populations

**isotherm** line drawn on a map connecting points with the same temperature at a certain period of time

**iteroparous** multiple-brooded over a lifetime

**Jurassic** middle period of the Mesozoic 208 to 144 million years ago; dinosaurs abundant; bony fishes evolving rapidly; gymnosperms dominant; warm humid climate

**K-selection** selection under carrying capacity conditions and a high level of competition

**k value** loss of individuals from a given life history stage as expressed by the difference between common logarithms of the numbers at the beginning and end of the stage

**key factor analysis** statistical treatment of population data designed to identify those factors most responsible for change in population size

**keystone species** a species whose activities have a significant role in determining community structure

**kin selection** differential reproduction among groups of closely related individuals

**kinetic energy** energy associated with motion

**kleptoparasitism** the forceable robbing of food acquired by an individual of one species by an individual of another

**Krummholz** stunted trees characteristic of the transition zone between alpine tundra and subalpine coniferous forest

**landscape ecology** study of structure, function, and change in a heterogeneous landscape composed of interacting ecosystems

**Langmuir cell** area of water a few meters wide and hundreds of meters long created by the wind, with a local circulation pattern of upwelling water that diverges from the center, converges, and sinks at the center

**latent heat of fusion** amount of heat given up when a unit mass of a substance converts from a liquid to a solid state, or the amount of heat absorbed when a substance converts from a solid to a liquid state

**laterization** soil-forming process in hot, humid climates, characterized by intense oxidation resulting in loss of bases and in a deeply weathered soil composed of silica, sesquioxides of iron and aluminum, clays, and residual quartz

**leaching** dissolving and removal of nutrients by water out of soil, litter, and organic matter

**leaf area index** ratio of area of canopy foliage to ground area

**lek** comunal courtship area used by males to attract and mate with females

**lentic** pertaining to standing water, such as lakes and ponds

**life table** tabulation of mortality and survivorship schedule of a population; static, time-specific, or vertical life tables are based on a cross section of a population at a given time; dynamic, cohort, or horizontal life tables are based on a cohort of organisms followed throughout life

**life zone** major area of plant and animal life, equivalent to a biome; transcontinental region or belt characterized by particular plants and animals and distinguished by temperature differences; applies best to mountainous regions where temperature changes accompany changes in altitude

**light reaction** light-dependent sequence of photosynthetic reactions

**lignotuber** specialized structure on the roots of certain fire-adapted trees, particularly *Eucalyptus,* from which new growth sprouts following a fire

**limit cycle** stable oscillation in the population levels of a species, usually involving predator-prey interaction

**limiting resource** resource or environmental condition that limits the abundance and distribution of an organism

**limnetic** pertaining to or living in the open water of a pond or lake

**limnetic zone** shallow water zone of a lake or sea in which light penetrates to the bottom

**lithosol** soil showing little or no evidence of soil development and consisting mainly of partly weathered rock fragments or nearly barren rock

**lithosphere** rocky material of Earth's outer crust

**littoral** shallow water of a lake in which light penetrates to the bottom, permitting submerged, floating, and emergent vegetative growth; also the shore zone of tidal water between high-water and low-water marks.

**local stability** ability of a system to return to its initial conditions following a small disturbance

**locus** site on a chromosome occupied by a specific gene

**loess** soil developed from wind-deposited material

**logistic curve** S-shaped curve of population growth, which slows at first, steepens, and then flattens at asymptote, determined by carrying capacity

**logistic equation** mathematical expression for the population growth curve in which rate of increase decreases linearly as population size increases

**log-normal distribution** frequency distribution in which the horizontal or $x$ axis is expressed in a logarithmic scale; in frequency distribution of species, the $x$ axis represents the number of individuals and the $y$ axis the number of species

**long-day organism** plant or animal that requires long days with more than a certain minimum of daylight to flower or come into reproductive condition

**lotic** pertaining to flowing water

**macromutation** mutation at the level of the chromosome

**macronutrients** essential nutrients needed in large amounts by plants and animals

**macroparasite** any of the parasitic worms, lice, fungi, and the like that have comparatively long generation time, spread by direct or indirect transmission, and may involve intermediate hosts or vectors

**mallee** sclerophyllous shrub community in Australia, most of whose species are *Eucalyptus*

**mangal** a mangrove swamp

**maquis** sclerophyllous shrub vegetation in the Mediterranean region

**marginal value theorem** a decision rule model based on economic theory, which holds that a predator should leave a patch of prey when the maximum overall average rate of extraction is the same as that for all other patches in the environment

**marl** unconsolidated deposit formed in freshwater lakes that consists chiefly of calcium carbonate mixed with clay and other impurities

**marsh** wetland dominated by grassy vegetation such as cattails and sedges

**mating system** patterns of mating between individuals in a population

**mattoral** sclerophyllous shrub vegetation in regions of Chile with mediterranean-type climate

**maximum sustainable yield** the maximum rate at which individuals can be harvested from a population without reducing its size; recruitment balances harvesting

**mediterranean-type climate** semiarid climate characterized by a hot, dry summer and a wet, mild winter

**meiofauna** benthic organisms within the size range of 1 to 0.1 mm; interstitial fauna

**meiosis** two successive divisions by gametic cells with only one duplication of chromosomes, so that the number of chromosomes in daughter cells is one-half the diploid number

**meristem** region in a plant containing actively or potentially dividing cells

**mesic** moderately moist

**mesophyll** specialized tissue located between the epidermal layers of a leaf; *palisade mesophyll* consists of cylindrical cells at right angles to upper epidermis and containing many chloroplasts; *spongy mesophyll* consists of interconnecting, irregularly shaped cells next to the lower epidermis with large intercellular spaces

**Mesozoic** middle era in the geological time scale, from 225 to 65 million years ago; the Age of Reptiles, whose three main periods are Triassic, Jurassic, and Cretaceous

**metabolism** the chemical reactions that take place in cells responsible for breaking down molecules to provide energy (catabolism)

and building up more complex molecules from simpler molecules (anabolism)

**metalimnion** transition zone in a lake between hypolimnion and epilimnion; region of rapid temperature decline

**metamorphosis** abrupt transition between life stages

**metapopulation** a set of local populations held together by dispersal

**micelle** soil particle of clay and humus carrying positive electrical charge at the surface

**microbial loop** uptake by bacteria of dissolved organic matter produced by plankton and subsequent consumption of bacteria by nanoplankton, resulting in a feeding loop that adds several levels to the plankton food chain

**microbivore** an organism that feed on microbes, especially in the soil and litter

**microclimate** climate on a very local scale that differs from the general climate of the area and influences the presence and distribution of organisms

**microflora** bacteria and certain fungi inhabiting the soil

**microhabitat** that part of the general habitat utilized by an organism

**micromutation** a mutation at the level of the gene; point mutation

**micronutrients** essential nutrients needed in very small quantities by plants and animals

**microparasite** any of the viruses, bacteria, and protozoans, characterized by small size, short generation time, and rapid multiplication

**migration** intentional, directional, usually seasonal movement of animals between two regions or habitats; involves departure and return of the same individual in a round trip

**mimicry** resemblance of one organism to another or to an object in the environment, evolved to deceive predators

**mineralization** microbial breakdown of humus and other organic matter in soil to inorganic substances

**minimum viable population** size of a population that, with a given probability, will ensure the existence of the population for a stated period of time

**mire** a wetland characterized by an accumulation of peat

**mitosis** cell division involving chromosome duplication, resulting in two daughter cells with a full complement of chromosomes, genetically the same as parent cells

**model** in theoretical and systems ecology, an abstraction or simplification of a natural phenomenon developed to predict a new phenomenon or to provide insights into existing ones; in mimetic association, the organism mimicked by a different organism

**moder** a forest humus layer in which plant fragments and mineral particles form loose netlike structures held together by a chain of small arthropod droppings

**modular organism** an organism that grows by iteration of parts, such as branches or shoots of a plant; some parts may separate and become physically and physiologically independent

**mollisol** soil formed by calcification, characterized by accumulation of calcium carbonate in lower horizons and high organic content in upper horizons

**monoecious** bearing reproductive organs of both sexes on the same plant, either as different flowers or in the same flower (hermaphrodism)

**monogamy** mating of an animal and maintenance of a pair bond with only one member of the opposite sex at a time

**montane** related to mountains

**moor** a blanket bog or peatland

**mor** a forest humus layer of unincorporated organic matter, usually matted or compacted or both, distinct from mineral soil, and low in bases and acid in reaction

**morphology** study of the form of organisms

**mull** humus that contains appreciable amounts of mineral bases and forms a humus-rich layer of forested soil, consisting of mixed organic and mineral matter, blending into the upper mineral layer without abrupt changes in soil characteristics

**Mullerian mimicry** resemblance of two or more conspicuously marked distasteful species, which increases predator avoidance

**mutation** transmissible change in the structure of a gene or chromosome

**mutualism** a relationship between two species in which both benefit

**mycelium** the mass of hyphae that make up the vegetative portion of a fungus

**mycorrhizae** the association of fungus with roots of higher plants, which improves the plants' uptake of nutrients from the soil

**myrmecochory** seed dispersal by ants

**natality** production of new individuals in a population

**natural selection** differential reproduction and survival of individuals that results in elimination of maladaptive traits from a population

**negative feedback** homeostatic control in which an increase in some substance or activity ultimately inhibits or reverses the direction of the processes leading to the increase

**neighborhood size** number of individuals in a population found in an area defined by the average dispersal distance of a single individual

**nekton** aquatic animals that are able to move at will through the water

**neritic** pertaining to the marine environment overlying a continental shelf

**net production** accumulation of total biomass over a given period of time after respiration is deducted from gross production in plants and from assimilated energy in consumer organisms

**net reproductive rate** number of females produced per female per generation

**neutrophilic** preferring a habitat that is neither acid nor alkaline

**niche** functional role of a species in the community, including activities and relationships

**niche breadth** range of a single niche dimension occupied by a population

**niche overlap** the sharing of niche space by two or more species

**niche packing** the tendency of coexisting species to fill all available space along a resource gradient

**niche preemption** procurement by a species of a portion of available resources, leaving less for the next

**nitrification** breakdown of nitrogen-containing organic compounds into nitrates and nitrites

**nitrogen fixation** conversion of atmospheric nitrogen to forms usable by organisms

**nonequilibrium theory** in community ecology, the transient movement of a system away from an equilibrium point relative to time and variation in ecological condition

**null hypothesis** a statement of no difference between sets of values formulated for statistical testing

**numerical response** change in size of a population of predators in response to change in density of its prey

**nutrient** substance required by organisms for normal growth and activity

**nutrient cycle** pathway of an element or nutrient through the ecosystem, from uptake by organisms to release by decomposition

**obligate** having no alternative response to conditions or way of life

**old growth forest** forest that has not been cut or disturbed by humans for hundreds of years

**oligotrophic** of water, low in nutrients and in productivity

**ombrotrophic** in bogs or mires, having water highly deficient in mineral salts and of low pH

**omnivore** an animal that feeds on both plant and animal matter

**opportunistic species** organisms able to exploit temporary habitats or conditions

**optimal foraging** tendency of animals to harvest food efficiently, selecting food sizes or food patches that result in maximum food intake for energy expended

**optimum yield** amount of material that can be removed from a population that will result in production of maximum biomass on a sustained basis

**ordination** process by which communities are positioned graphically on a gradient of one to several axes so that the distance between them reflect differences in composition

**Ordovician** second oldest period of the Paleozoic era, 505 to 438 million years ago, in which marine invertebrates were abundant

**organismic concept of the community** idea that species, especially plant species, are integrated into an internally interdependent unit; upon maturity and death of the community, another identical plant community will replace it

**oscillation** regular fluctuation in a fixed cycle above and below some set point

**osmosis** movement of water molecules across a differentially permeable membrane in response to a concentration or pressure gradient

**osmotic potential** the pressure needed to stop osmosis; the more concentrated a solution, the higher is its osmotic pressure

**outbreeding** production of offspring by the fusion of distantly related gametes

**overdispersion** situation in which the distribution of organisms is not random, but clumped so that areas are both empty and heavily overpopulated; contagious distribution

**overturn** vertical mixing of layers in a body of water brought about by seasonal changes in temperature

**oxisol** soil developed under humid semitropical and tropical conditions, formed by laterization, and characterized by silicates and hydrous oxides, clays, residual quartz, deficiency in bases, and low plant nutrients

**paleoecology** study of ecology of past communities by means of the fossil record

**Paleozoic** geological era of 620 to 245 million years ago, starting with the Cambrian period, during which land plants, insects, amphibians, and the first reptiles evolved

**parapatric** having ranges coming into contact but not overlapping by much more than the dispersal range of an individual in its lifetime

**parasitism** a relationship between two species in which one benefits while the other is harmed (although not usually killed directly)

**parasitoid** an insect larva that kills its host by consuming completely the host's soft tissues before pupation or metamorphosis into an adult

**peat** unconsolidated material consisting of undecomposed and only slightly decomposed organic matter under conditions of excessive moisture

**ped** soil particles held together in clusters of various sizes

**pedalfer** podzolic soil possessing a layer of iron accumulation (hardpan) that impedes free circulation of air and water

**pedon** a three-dimensional sampling unit of soil down to the parent material and wide enough to study all the horizons and their intergradations; the smallest unit in the study of soil

**pelagic** referring to the open sea

**percolation** the movement of water downward and radially through the subsurface soil layer, often continuing down to groundwater

**periphyton** in freshwater ecosystems, organisms that are attached to submerged plant stems and leaves; see *aufwuchs*

**permafrost** permanently frozen soil

**permanent wilting point** point at which water potential and conductivity in the soil assume such low values that the plant is unable to extract sufficient water to survive and wilts permanently

**Permian** most recent period of the Paleozoic, 286 to 245 million years ago, marked by dominance of reptiles, appearance of modern insects, and gymnosperm plants

**phagioclimax** vegetation type maintained as climax over a long period of time by continued human activity

**phanerophyte** a tree, shrub, or vine that bears perennating buds on aerial shoots

**phenology** study of the seasonal changes in plant in animal life and the relationship of these changes to weather and climate

**phenotype** physical expression of a characteristic of an organism as determined by genetic constitution and environment

**phenotypic plasticity** ability to change form under different environmental conditions

**pheromone** chemical substance released by an animal that influences behavior of others of the same species

**photic zone** the lighted water column of a lake or ocean inhabited by plankton

**photoperiodism** response of plants and animals to changes in relative duration of light and dark

**photorespiration** respiration that occurs in light in $C_3$ plants, is not coupled to oxidative phosphorylation, and does not generate ATP; a wasteful process decreasing photosynthetic efficiency

**photosynthate** energy-rich organic molecules produced during photosynthesis

**photosynthetically active radiation (PAR)** those wavelengths in the radiation spectrum used by plants in photosynthesis

**phreatophyte** type of plant that habitually obtains its water supply from groundwater

**physiognomy** outward appearance of the landscape

**physiological longevity** maximum lifespan of an individual in a population under given environmental conditions

**phytoplankton** small, floating plant life in aquatic ecosystems; planktonic plants

**pioneer species** plants that are initial invaders of disturbed sites or early seral stages of succession

**plankton** small, floating or weakly swimming plants and animals in freshwater and marine ecosystems

**Pleistocene** geological epoch from about 2 million to 10,000 years ago, characterized by recurring glaciers; the Ice Age

**pneumatophore** an erect respiratory root that protrudes above waterlogged soils; typical of bald cypress and mangroves

**podzolization** a soil-forming process resulting from acid leaching of the A horizon and accumulation of iron, aluminum, silica, and clays in the lower horizon

**poikilohydric** matching internal water state to that of the environment

**poikilothermy** variation of body temperature with external conditions

**polyandry** mating of one female with several males

**polyclimax theory** idea that the endpoint of succession is controlled by one of many local environmental conditions, such as soil and fire, as well as climate

**polygamy** acquisition by an individual of two or more mates, none of which is mated to other individuals

**polygyny** mating of one male with several females

**polymorphism** occurrence of more than one distinct form of individuals in a population

**polyploid** having three or more times the haploid number of chromosomes

**population** a group of individuals of the same species living in a given area at a given time

**population cycle** oscillations between periods of high and low density in the number of individuals in a population

**population density** the number of individuals in a population per unit area

**population dynamics** the study of the factors that influence variations in the numbers and densities of populations in time and space

**population regulation** mechanisms or factors within a population that cause it to decrease when density is high and increase when density is low

**positive feedback** control in a system that reinforces a process in the same direction

**potential energy** energy available to do work

**potential evapotranspiration** amount of water that would be transpired under constantly optimal conditions of soil moisture and plant cover

**Precambrian** earliest and longest of the geological time periods, between 4600 and 570 million years ago, preceding the Paleozoic era, and marked by appearance of cyanobacteria and fungi

**precocial** young birds hatched with down, eyes open, and able to move about; also young mammals born with eyes open and able to follow their mother after birth (for example fawn deer, calves)

**predation** using one living organism as a food source for another

**preferred temperature** range of temperatures within which poikilotherms function most efficiently

**primary production** production by green plants

**primary productivity** rate at which plants produce biomass

**primary succession** vegetational development starting from a new site never before colonized by life

**producer** a green plant or certain chemosynthetic bacteria that convert light or chemical energy into organismal tissue

**production** amount of energy formed by an individual, population, or community per unit time

**productivity** rate of energy fixation or storage per unit time; not to be confused with production

**profundal** deep zone in aquatic ecosystems below the limnetic zone

**proximate factor** any characteristic of the environment that an organism uses as a cue to behavioral or physiological responses; mechanism behind the operation of an adaptation; compare *ultimate factor*

**punctualism** hypothesis that evolution occurs in a series of sudden bursts with long periods of stability

**pycnocline** the layer of ocean water that exhibits a rapid change in density

**pyramid of biomass** a diagrammatic representation of biomass at different trophic levels in an ecosystem

**pyramid of energy** a diagrammatic representation of the flow of energy through different trophic levels

**pyramid of numbers** a diagrammatic representation of the number of individual organisms present at each trophic level in an ecosystem, the least useful of the pyramids

**Quaternary** the geological period from 2 million years ago to the present, including the Pleistocene and recent epochs

*r*-**selection** selection under low population densities, favoring high reproductive rates under conditions of low competition

**rain shadow** dry area on the lee side of mountains

**raised bog** a bog in which the accumulation of peat has raised the surface above both the surrounding landscape and the water table, so it develops its own perched water table

**ramet** any individual belonging to a clone

**random distribution** distribution lacking pattern or order; distribution of individuals independent of all other individuals

**realized niche** the portion of fundamental niche space occupied by a population in the face of competition from populations of other species; environmental conditions under which a population survives and reproduces in nature

**recombination** exchange of genetic material resulting from independent assortment of chromosomes and their genes during gamete production, followed by a random mix of different sets of genes at fertilization

**recruitment** addition of reproduction of new individuals to a population

**regolith** mantle of unconsolidated material below the soil, from which soil develops

**regular distribution** arrangement of individuals in a pattern that ensures they are more widely separated from each other than would be expected by chance; under dispersion

**relict** surviving species of a once widely dispersed group

**reproductive allocation** proportion of available resource input that an organism expends on reproduction over a given period of time

**reproductive cost** decrease in survivorship or rate of growth experienced by an individual as a result of increasing its current allocation to reproduction, and reflected in its decreased potential for future reproduction

**reproductive effort** proportion of its resources an organism expends on reproduction

**reproductive isolation**  separation of one population from another by the inability to produce viable offspring when the two populations are mated

**reproductive value**  potential reproductive output of an individual at a particular age ($x$) relative to that of a newborn individual at the same time

**residual reproductive value**  reproductive value of an individual reduced by its expected present reproduction

**resilience**  ability of a system to absorb changes and return to its original condition

**resistance**  ability of a system to resist changes from a disturbance

**resource**  environmental component used by a living organism

**resource allocation**  action of apportioning the supply of a resource to a specific use

**respiration**  metabolic assimilation of oxygen accompanied by production of carbon dioxide and water, release of energy, and breaking down of organic compounds

**restoration ecology**  study of the application of ecological theory to the ecological restoration of highly disturbed sites

**rete**  a large network or discrete vascular bundle of intermingling small blood vessels carrying arterial and venous blood that acts as a heat exchanger in mammals and certain fish and sharks

**rheotrophic**  of wetlands (especially bogs and fens), obtaining much nutrient input from groundwater

**rhizobia**  bacteria capable of living mutualistically with higher plants

**rhizome**  a horizontally growing underground stem that gives rise to vegetative structures through branching

**rhizoplane**  root surface

**rhizosphere**  soil region immediately surrounding roots

**richness**  the number of species present in an area, a component of species diversity

**riparian**  along banks of rivers and streams

**root-to-shoot ratio**  ratio comparing the weight of roots to the weight of shoots of a plant

**ruminant**  ungulate with a three- or four-chamber stomach, in the large first chamber of which, the rumen, bacterial fermentation of plant matter takes place

**saprophage**  an organism that feeds on dead plant and animal matter, mainly bacteria and fungi and some invertebrates, such as insect larvae

**savanna**  tropical grassland, usually with scattered trees or shrubs

**scavenger**  an animal that feeds on other dead animals or on animal products, such as dung

**sclerophyll**  a woody plant with hard, leathery, evergreen leaves that prevent moisture loss

**scramble competition**  intraspecific competition in which limited resources are shared to the point that no individual survives

**scraper**  aquatic insect that feeds by scraping algae from a substrate

**search image**  mental image formed in predators enabling them to find prey quickly and to concentrate on a common type of prey

**secondary production**  production by consumer organisms

**secondary substances**  organic compounds produced by plants that are utilized in chemical defense

**secondary succession**  plant succession taking place on sites that have already supported life

**seiche**  oscillation of a structure of water about a point or node

**selective pressure**  any force acting on individuals in a population that determines which individuals contribute more descendants to subsequent generations than others, giving direction to the evolutionary process

**self-thinning**  progressive decline in density of plants associated with the increasing size of individuals

**semelparity**  having only a single reproductive effort in a lifetime over one relatively short period of time

**semiarid**  having a fairly dry climate, with precipitation between 25 and 60 cm a year and with an evapotranspiration rate high enough that potential loss of water to the environment exceeds input

**senescence**  process of aging

**seral**  following one another in succession

**sere**  the series of successional stages on a given site that leads to a terminal community

**serotiny**  retention of seeds in cones until heated, usually by fire

**serpentine soils**  soils that derive from ultrabasic rocks that are high in iron, magnesium, nickel, chromium, and cobalt and low in calcium, potassium, sodium, and aluminum, and that support distinctive communities

**sessile**  not free to move about; permanently attached to a substrate

**sex ratio**  the relative number of males to females in a population

**sex reversal**  a change in functioning so that a member of one sex behaves as the other

**sexual selection**  selection by one sex for an individual of the other sex, based on some specific characteristic or characteristics, usually through courtship behavior

**shade-tolerant**  able to grow and reproduce under low light conditions

**short-day organisms**  plants and animals that come into reproductive condition under short days with less than a certain maximum length

**shredder**  a stream invertebrate that feeds on coarse particulate organic matter

**sibling species**  species with similar appearance but unable to interbreed

**sigmoid curve**  S-shaped curve of logistic growth

**Silurian**  period between the Ordovician and Devonian periods, 438 to 408 million years ago, characterized by early land plants, invertebrates, and primitive jawless fishes

**site**  combination of biotic, climatic, and soil conditions that determine an area's capacity to produce vegetation

**snag**  a dead or partially dead tree at least 10.2 cm dbh and 1.8 m tall, important for cavity-nesting birds and mammals

**social dominance**  physical dominance of one individual over another, usually maintained by some manifestation of aggressive behavior

**social parasite**  an animal that uses other individuals or species to rear its young, for example, cowbirds

**soil association**  a group of defined and named soil taxonomic units occurring together in an individual and characteristic pattern over a geographic region

**soil horizon**  a developmental layer in the soil with characteristic thickness, color, texture, structure, acidity, nutrient concentration, and the like

**soil profile**  distinctive layering of horizons in the soil

**soil series**  basic unit of soil classification, consisting of soils essentially alike in all major profile characteristics except texture

of the A horizon; soil series are usually named for the locality where the typical soil was first recorded

**soil structure** arrangement of soil particles and aggregates

**soil texture** relative proportions of the three particle sizes—sand, silt, and clay—in the soil

**soil type** lowest unit in the natural system of soil classification, consisting of soils alike in all characteristics, including texture of the A horizon

**solar constant** rate at which solar energy is received on a surface just outside of Earth's atmosphere; current value is 0.140 watt/cm$^2$

**speciation** separation of a population into two or more reproductively isolated populations

**species diversity** measurement that relates density of organisms of each type present in a habitat to the number of species in the habitat

**species packing** increases in species diversity within a narrow range of resource variation

**species richness** number of species in a given area

**specific heat** amount of energy that must be added or removed to raise or lower temperature of a substance by a specific amount

**spiraling** mechanism of retention of nutrients in flowing water ecosystems, involving the interdependent processes of nutrient recycling and downstream transport

**spodosol** soil characterized by a horizon in which organic matter and amorphous oxides of aluminum and iron have precipitated; includes podzol soils

**stability** ability of a system to resist change or to recover rapidly after a disturbance; absence of fluctuations in a population

**stabilizing selection** selection favoring the middle in the distribution of phenotypes

**stable age distribution** constant proportion of individuals of various age classes in a population through population changes

**stable equilibrium** ability of a system to return to a particular point if displaced by an outside force

**stable limit cycle** a regular fluctuation in abundance of predator and prey populations, when stabilizing and destabilizing interactions balance

**stand** unit of vegetation that is essentially homogeneous in all layers and differs from adjacent types qualitatively and quantitatively

**standard deviation** statistical measure defining the dispersion of values about the mean in a normal distribution

**standing crop** amount of biomass per unit area at a given time

**static life table** see *life table*

**stationary age distribution** a special form of stable age distribution, in which the population has reached a constant size at which the birthrate equals the death rate and age distribution remains fixed

**stemflow** portion of precipitation that is intercepted by trees and flows down their boles

**stochastic** arising from random factors

**stochastic model** mathematical model based on probabilities, whose predictions are fixed but variable; compare *deterministic model*

**stratification** division of an aquatic or terrestrial community into distinguishable layers on the basis of temperature, moisture, light, vegetative structure, and other such factors, creating zones for different plant and animal types

**sublittoral** lower division of the sea, from about 40 m to 60 m to below 200 m

**subsidence inversion** atmospheric inversion produced by sinking air movement

**subspecies** geographical unit of a species population distinguishable by certain morphological, behavioral, or physiological characteristics

**succession** replacement of one community by another, often progressing to a stable terminal community called the climax

**sun plant** a plant able to grow and reproduce only under high light conditions

**sunspot** relatively dark, sharply defined region on the sun, found mostly in groups of two or more; sunspots are cyclic, with a periodicity of approximately 11 years

**supercooling** in ectotherms, lowering the body temperature below freezing without freezing body tissue, in the presence of certain solutes, particularly glycerol

**survivorship** the probability of a representative newborn individual in a cohort surviving to various ages

**survivorship curve** a graphical description of the survival of a cohort of individuals in a population from birth to the maximum age reached by any one member of the cohort

**sustained yield** yield per unit time from an exploited population equal to production per unit time

**swamp** a wooded wetland in which water is near or above ground level

**switching** a predator changing its diet from a less abundant to a more abundant prey species; see *threshold of security*

**symbiosis** situation in which two dissimilar organisms live together in close association

**sympatric** living in the same area; usually refers to overlapping populations

**synecology** study of groups of organisms in relation to their environment; community ecology

**system** set or collection of interdependent parts or subsystems enclosed within a defined boundary; the outside environment provides inputs and receives outputs

**taiga** the northern circumpolar boreal forest

**temperate rain forest** forests in regions characterized by mild climate and heavy rainfall that produce lush vegetative growth; one example is the coniferous forest of the Pacific Northwest of North America

**territory** area defended by an animal, varying among species according to social behavior, social organization, and resource requirements

**Tertiary** first period of the Cenozoic era, from 65 to 2 million years ago, composed of the Paleocene, Eocene, Oligocene, Miocene, and Pliocene epochs and characterized by the emergence of mammals

**thermal conductance** rate at which heat flows through a substance

**thermal neutral zone** among homeotherms, the range of temperatures at which metabolic rate does not vary with temperature

**thermal tolerance** range of temperatures in which an aquatic poikilotherm is most at home

**thermocline** layer in a thermally stratified body of water in which temperature changes rapidly relative to the remainder of the body

**thermogenesis** increase in production of metabolic heat to counteract the loss of heat to a colder environment

**therophyte** annual or ephemeral plant that survives unfavorable conditions in the form of a seed

**thinning law, 3/2** role that self-thinning plant populations, sown at sufficiently high densities, approach and follow a thinning line with a slope of roughly −3/2; thus in a growing population, plant weight increases faster than density decreases to a point where the slope changes to −1

**threshold of security** point in local population density at which the predator turns its attention to other prey because of harvesting efficiency; the segment of prey population below the threshold is relatively secure from predation

**throughfall** part of precipitation that drips off vegetation and falls to the ground

**tiller** in grasses, lateral shoots arising at ground level

**time lag** delay in a response to change

**time-specific life table** see *life table*

**tolerance model** model of succession that proposes that succession leads to a community composed of those species most efficient in exploiting resources; colonists neither increase nor decrease the rate of recruitment or growth of later colonists

**topography** physical structure of the landscape

**toposequence** a pattern of local soils whose development was controlled by topography of the landscape

**torpidity** temporary condition of an animal involving a great reduction in respiration and loss of power of motion and feeling, usually occurring in response to some unfavorable environmental condition such as heat or cold to reduce energy expenditure

**trace element** element occurring and needed in small quantities; see *micronutrients*

**transient polymorphism** the occurrence of two or more forms of genes in a population during a time when one form is being replaced by another

**translocation** transport of materials within a plant; absorption of minerals from soil into roots and their movement throughout the plant

**transpiration** loss of water vapor by land plants

**Triassic** oldest period of the Mesozoic, 245 to 208 million years ago, marked by an increase in primitive amphibians and reptiles

**trophic** related to feeding

**trophic level** functional classification of organisms in an ecosystem according to feeding relationships from first-level autotrophs through succeeding levels of herbivores and carnivores

**trophic structure** organization of a community based on the number of feeding or energy transfer levels

**trophogenic zone** upper layer of the water column in ponds, lakes, and oceans, in which light is sufficient for photosynthesis

**tropholytic zone** area in lakes and oceans below the compensation point

**tundra** area in arctic, alpine, and high mountain regions characterized by bare ground, absence of trees, and growth of mosses, lichens, sedges, forbs, and low shrubs

**turgor** the state in a plant cell in which the protoplast is exerting pressure on the cell wall because of intake of water by osmosis

**turnover rate** rate of replacement of a substance or a species when losses to a system are replaced

**ultimate factor** survival value of an adaptation; evolutionary reason for an adaptation

**ultimate incipient lethal temperature** upper or lower limit of temperature at which an acclimatized organism will succumb

**unapparent plants** small, short-lived plants that use a qualitative, highly toxic defense involving secondary substances such as cardiac glucosides; see *apparent plants*

**unitary organism** any organism, including arthropods and vertebrates, whose growth to adult form follows a determinate pathway, unlike modular organisms whose growth involves indeterminate repetition of units of structure

**unstable equilibrium** a level of a population, populations, or resources from which slight displacements lead to larger displacements

**upwelling** an area in the ocean where currents force water from deep into the euphotic zone

**vacuole** fluid-filled cavity within the cytoplasm

**vagile** free to move about

**validation** an explicit and objective test of the basic hypothesis

**vapor pressure** the amount of pressure water vapor exerts independent of dry air

**variance** the square of the standard deviation

**vector** organism that transmits a pathogen from one organism to another

**vegetative reproduction** asexual reproduction in plants by means of specialized multicellular organs such as bulbs, corms, rhizomes, stems, and the like

**verification** process of testing whether or not a model is a reasonable representation of a real-life system

**vernal pool** temporary pond of water filled in the spring, an important habitat of many amphibians and aquatic invertebrates

**vertisol** mineral soil that contains more than 30 percent of swelling clays that expand when wet and contract when dry, associated with seasonally wet and dry environments

**vesicular arbuscular mycorrhizae (VAM)** mycorrhizae in which the fungus enters and grows within the host's cells and extends widely into the surrounding soil; a type of ectomycorrhizae

**viscosity** property of a fluid that resists the force within the fluid that causes it to flow

**Wallace's line** biogeographic line between the islands of Borneo and the Celebes that marks the eastward boundary of many landlocked Eurasian organisms and the boundary of the Oriental region

**water potential** measure of energy in an aqueous solution needed to move water molecules across a semipermeable membrane; water tends to move from areas of higher potential to areas of lower potential

**water use efficiency** ratio of net primary production to transpiration of water by a plant

**watershed** entire region drained by a waterway that drains into a lake or reservoir; total area above a given point on a stream that contributes water to the flow at that point; the topographic dividing line from which surface streams flow in two different directions

**weed** a plant possessing a high rate of dispersal, occurring opportunistically on land or water disturbed by human activity, and competing for resources with cultivated plants; a plant growing in the wrong place

**wetfall** component of acid deposition that reaches Earth by some form of precipitation; wet deposition

**wilting point** moisture content of soil, established by oven drying, at which plants wilt and fail to recover their turgidity when placed in a dark, humid atmosphere

**xeric** dry, especially relating to dry soil

**xerophyte** plant adapted to life in a dry or physiologically dry (saline) habitat

**Zeitgeber** the time-setter, usually light, that entrains a circadian rhythm to environmental rhythms

**zero net growth isocline** an isocline along which the population growth rate is zero

**zonation** characteristic distribution of vegetation along an environmental gradient, forming latitudinal, altitudinal, or horizontal belts within an ecosystem

**zoogeography** study of the distribution of animals

**zooplankton** floating or weakly swimming animals in freshwater and marine ecosystems; planktonic animals

# Bibliography

ABELE, L. G., AND E. F. CONNER. 1979. Application of island biogeography theory to refuge design: Making the right decisions for the wrong reasons. In R. M. Linn (ed.), *Proceedings of First Conference on Scientific Research in National Parks,* 89–94, Vol. 1, U. S. Department of Interior, Washington, DC.

ABER, J. D., K. N. NADELHOFFER, P. STEUDLER, AND J. M. MELILLO. 1989. Nitrogen saturation in northern forest ecosystems. *Bioscience* 39:378–386.

ABER, J. S., D. B. BOTKIN, AND J. M. MELILLO. 1978. Predicting the effects of different harvesting regimes on forest floor dynamics in northern hardwoods. *Can. J. For. Res.* 8:305–315.

ABRAHAMSON, W. G., AND M. D. GADGIL. 1973. Growth form and reproductive effort in goldenrod (*Solidago,* Compositae). *Am. Nat.* 107:651–661.

ACKERT, J. E., G. L. GRAHAM, L. O. NOLF, AND D. A. PORTYER. 1931. Quantitative studies on the administration of variable numbers of nematode eggs (*Ascaridia lineata*) to chickens. *Trans. Am. Microscopical Soc.* 50:206–214.

ADAMS, E. S. 1994. Settlement tactics in seasonally territorial animals: Resolving conflicting predictions. *Am. Nat.* 143:939–943.

ADDICOTT, J. F. 1979. A multispecies aphid-plant association: A comparison of local and metapopulations. *Can. J. Zool.* 56:2554–2564.

———. 1985. On the population consequences of mutualism. In T. Case and J. Diamond (eds.), *Community Ecology.* Harper and Row, New York, pp. 425–436.

———. 1986. Variation in the costs and benefits of mutualism: The interaction between yuccas and yucca moths. *Oecologica* 70:486–494.

ADKISSON, P. L. 1966. Internal clocks and insect diapause. *Science* 154:234–241.

AHLGREN, I. F., AND C. H. AHLGREN. 1960. Ecological effects of forest fires. *Bot. Rev.* 26:483–533.

AHMADJIAN, V., AND J. B. JACOBS. 1982. Algal-fungal relationships in lichens: Recognition, synthesis, and development. In L. J. Goff (ed.), *Algal Symbiosis: A Continuum of Interaction Strategies.* Cambridge University Press, Cambridge, England.

AHMED, A. K. 1976. PCBs in the environment. *Environment* 18(2):6–11.

AKER, C. L. 1982. Spatial and temporal dispersion patterns of pollinators and their relationship to the flowering strategy of *Yucca whipplei* (Agavaceae). *Oecologica* 54:243–252.

ALCOCK, J. 1973. Cues used in searching for food by red-winged blackbirds (*Agelaius phoeniceus*). *Behaviour* 17:130–233.

ALEXANDER, M. 1965. Nitrification. In W. V. Barthelomew and F. F. Clark (eds.), *Soil Nitrogen. Am. Soc. Agron. Monogr.* No. 10, pp. 307–343.

ALEXANDER, M. M. 1958. The place of aging in wildlife management. *Am. Sci.* 46:123–131.

ALEXANDER, R. D., AND G. BORGIA. 1978. Group selection, altruism, and levels of organization of life. *Ann. Rev. Ecol. Syst.* 9:449–474.

ALEXANDER, R. D., AND D. W. TINKLE (EDS.). 1981. *Natural Selection and Social Behavior: Recent Research and New Theories.* Chiron, New York.

ALLEE, W. C., A. E. EMERSON, O. PARK, T. PARK, AND K. P. SCHMIDT. 1949. *Principles of Animal Ecology.* Saunders, Philadelphia.

ALLEN, D. L. 1942. *Michigan Fox Squirrel Management.* Michigan Department Conservation, Game Division Publ. 101 Lansing, MI.

———. 1962. *Our Wildlife Legacy,* 2nd ed. Funk and Wagnalls, New York.

ALLEN, L. N., AND E. R. LEMON. 1976. Carbon dioxide exchange and turbulence in a Costa Rican tropical rain forest. In J. C. Monteith (ed.), *Vegetation and the Atmosphere,* Vol. 2 *Case Studies.* Academic Press, London, pp. 265–308.

ALLEN, T. F. H., AND T. W. HOEKSTRA. 1984. The abuse of the concept "disturbance": A scaling problem. Unpublished mss.

ALLEN, T. F. H., AND T. B. STARR. 1982. *Hierarchy: Perspectives for Ecological Complexity.* University of Chicago Press, Chicago.

ALM, G. 1952. Year class fluctuations and span of life of perch. *Rept. Inst. Freshwater Res. Drottningholm* 33:17–38.

ALTMANN, M. 1960. The role of juvenile elk and moose in the social dynamics of their species. *Zoologica* 45:35–40.

AMBUEL, B., AND S. A. TEMPLE. 1983. Area dependent changes in the bird communities and vegetation of southern Wisconsin forests. *Ecology* 64:1057–1068.

AMUNDSON, D. C., AND H. E. WRIGHT, JR. 1979. Forest changes in Minnesota at the end of the Pleistocene. *Ecol. Monogr.* 49:109–127.

ANDERSON, C. C., AND M. TRESHOW. 1980. A review of environmental genetic factors that affect height in *Spartina alternifolia* Loisel (salt marsh cordgrass). *Estuaries* 3:168–176.

ANDERSON, J. G., D. W. TOOKEY, AND W. H. BRUNE. 1991. Free radicals within the Antarctic vortex: The role of CFCs in Antarctic ozone loss. *Science* 251:39–46.

ANDERSON, J. M. 1978. Inter- and intrahabitat relationships between woodland *Cryptostigmata* species diversity and diversity of soil and litter microhabitats. *Oecologica* 32:341–348.

ANDERSON, J. M., AND A. MACFADYIN (EDS.). 1976. *The Role of Terrestrial and Aquatic Organisms in the Decomposition Process.* Blackwell, Oxford.

ANDERSON, M., AND M. O. ERICKSON. 1982. Nest parasitism in goldeneyes, *Bucephala clangula:* Some evolutionary aspects. *Am. Nat.* 120:1–16.

ANDERSON, N. H., AND K. W. CUMMINS. 1979. Influences of diet on the life histories of aquatic insects. *J. Fish. Res. Bd. Can.* 36:335–342.

ANDERSON, R. C. 1963. The incidence, development, and experimental transmission of *Pneumostrongylus tenuis* Dougherty (Metastrongyloidae: Protostrongyliidae) of the meninges of the white-tailed deer (*Odocoileus virginianus borealis*) in Ontario. *Can. J. Zool.* 41:775–792.

———. 1965. Cerebospinal nematodiasis (*Pneumostrongylus tenuis*) in North American cervids. *Trans. N. Am. Nat. Res. Conf.* 13:156–167.

ANDERSON, R. C., AND A. K. PRESTWOOD. 1981. Lungworms. In W. R. Davidson (ed.), *Diseases and Parasites of White-tailed Deer,* Mscl. Publ. no. 7, Tall Timbers Research Station, Tallahassee, FL, pp. 266–317.

ANDERSON, R. M. 1981. Population ecology of infectious diseases. In R. M. May (ed.), *Theoretical Ecology: Principles and Applications,* 2nd ed. Sinauer Associates, Sunderland, MA, pp. 318–355.

ANDERSON, R. M., AND R. M. MAY. 1978. Regulation and stability of host-parasite population interactions. I. Regulatory processes. *J. Animal Ecol.* 47:219–247.

———. 1979. Population biology of infectious diseases. Part I. *Nature* 280:361–367.

ANDERSSON, M. 1982. Female choice selects for extreme tail length in a widowbird. *Nature* 299:818–820.

―――. 1984. The evolution of eusociality. *Ann. Rev. Ecol. Syst.* 15:169–185.

ANDREWARTHA, H. G. 1961. *Introduction to the Study of Animal Populations.* Methuen, London.

ANDREWARTHA, H. G., AND L. C. BIRCH. 1954. *The Distribution and Abundance of Animals.* University of Chicago Press, Chicago.

―――. 1984. *The Ecological Web.* University of Chicago Press, Chicago.

ANDREWS, R., AND A. S. RAND. 1974. Reproductive effort in anoline lizards. *Ecology* 55:1317–1327.

ANDREWS, R. D., D. C. COLEMAN, J. E. ELLIS, AND J. S. SINGH. 1975. Energy flow relationships in a short grass prairie ecosystem. *Proc. 1st Inter. Cong. Ecol.* 22–28. W. Junk Publishers, The Hague.

ANDRZEJEWSKA, L., AND G. GYLLENBERG. 1980. Small herbivore subsystem. In A. I. Bretmeyer and G. M. Van Dyne (eds.), *Grasslands, Systems Analysis and Man.* International Biological Programme no. 19, Cambridge University Press, Cambridge, pp. 201–268.

ANTOINE, L. H., JR. 1964. Drainage and the best use of urban land. *Public Works* 95:88–90.

ANTONOVICS, J. A., A. N. BRADSHAW, AND R. G. TURNER. 1971. Heavy metal tolerance in plants. *Adv. Ecol. Res.* 71:1–85.

ANTONOVICS, J. A., AND D. A. LEVIN. 1980. The ecological and genetical consequences of density-dependent regulation in plants. *Ann. Rev. Ecol. Syst.* 11:411–452.

ANTONOVICS, J. A., AND R. B. PRIMACH. 1982. Experimental ecology and genetics in *Plantago.* VI. The demography of seedling transplants of *P. lanceolata. J. Ecol.* 70:55–75.

ARDITI, R., AND A. A. BERRYMAN. 1991. The biological control paradox. *TREE* 6:32.

ARDITI, R., AND L. R. GINSBERG. 1989. Coupling in predator-prey dynamics: Ratio-dependence. *J. Theor. Biol.* 139:311–320.

ARMOUR, C. L. 1993. *Evaluating temperature regimes for protection of smallmouth bass.* U.S. Fish and Wildlife Service Resource Publication 191, 26 pp.

ARNO, S. F. 1976. *The historical role of fire in the Bitterroot National Forest.* USDA For. Serv. Res. Paper INT 187. Ogden, Utah: U.S. Forest Service.

ARP, A. J., AND J. J. CHILDRESS. 1983. Sulfide binding by the blood of the hydrothermal vent tube worm *Riftia pachyptila. Science* 219:295–297.

ASCHMANN, H. 1973. Distribution and peculiarity of mediterranean ecosystems. In F. di Castri and H. A. Mooney (eds.), *Mediterranean Type Ecosystems: Origin and Structure.* Springer-Verlag, New York, pp. 11–19.

ASCHOFF, J. 1958. Tierische Periodik unter dem Einfluss von Zeitgebern. *Z. F. Tierpsychol.* 15:1–30.

―――. 1966. Circadian activity pattern with two peaks. *Ecology.* 47:657–662.

ASH, J. E., AND J. P. BARKHAM. 1976. Changes and variability in the field layer of a coppiced woodland in Norfolk, England. *J. Ecol.* 64:697–712.

ASHMOLE, N. P. 1963. The regulation of numbers of tropical oceanic birds. *Ibis* 103b:458–473.

ASKENMO, C. 1977. Effects of addition and removal of nestlings and nestling weight, nestling survival, and female weight loss in the pied flycatcher *Ficedula hypoleuca* (Pallas). *Ornis. Scand.* 8:1–8.

―――. 1979. Reproductive effort and return rate of male pied flycatchers. *Am. Nat.* 114:748–752.

ATSATT, P. R., AND D. J. O'DOWD. 1976. Plant defense guilds. *Science* 193:24–29.

AUCLAIR, A. N., AND F. G. GOFF. 1971. Diversity relations of upland forests in the western Great Lakes area. *Am. Nat.* 105:499–528.

AUERBACH, S. I., D. J. NELSON, AND E. G. STRUXNESS. 1974. *Environmental Sciences Division annual progress report period ending Sept. 30, 1973.* Environ. Sci. Div. Pub. No. 57, Oak Ridge National Laboratory.

AUMANN AND EMLEN. 1965. Relationship of population density to sodium availability and sodium selection by microtine rodents. *Nature* 208:198–199.

AUSMUS, B. S., N. T. EDWARDS, AND M. WITKAMP. 1975. Microbial immobilization of carbon, nitrogen, phosphorus, and potassium: Implications for forest ecosystem processes. *Proc. Brit. Ecol. Soc. Symp. on Decomposition,* Blackwell, Oxford.

AUSTIN, M. P. 1985. Continuum concept, ordination methods, and niche theory. *Ann. Rev. Ecol. Syst.* 16:39–62.

AUSTIN, M. P., R. B. CUNNINGHAM, AND R. B. GOOD. 1983. Altitudinal distribution in relation to other environmental factors of several Eucalypt species in southern New South Wales. *Aust. J. Eco.* 8:169–180.

AUSTIN, M. P., R. B. CUNNINGHAM, AND P. M. FLEMING. 1984. New approaches to direct gradient analysis using environmental scalars and statistical curve-fitting procedures. *Vegetatio* 55:11–27.

AUSTIN, M. P., AND T. M. SMITH. 1989. A new model of the continuum concept. *Vegetatio.* 83:35–47.

AZAM, F., T. FENCHEL, J. D. FIELD, L. A. MEYER-REIL, AND F. THINGSTAD. 1983. The ecological role of water-column microbes in the sea. *Mar. Ecol. Prog. Ser.* 10:257–263.

BACON, P. J. (ED.). 1985. *Population Dynamics of Rabies in Wildlife.* Academic Press, London.

BAECKLEN, W. J., AND N. J. GOTELLI. 1984. Island biogeographic theory and conservation practice: Species-area or specious-area relationships. *Biol. Cons.* 29:63–80.

BAER, J. G. 1951. *Ecology of Animal Parasites.* University of Illinois Press, Urbana.

BAES, C. F., JR., H. E. GOELLER, J. S. OLSON, AND R. M. ROTTY. 1977. Carbon dioxide and the climate: The uncontrolled experiment. *Am. Sci.* 65:310–320.

BAILEY, E. P. 1993. *Introduction of foxes to Alaskan islands: History, effects of avifauna, and eradication.* U.S. Fish and Wildlife Service Resource Publication 193, 53 pp.

BAILEY, P. C. E. 1986. The feeding behavior of a sit-and-wait predator *Ranatra dispar* (Heteroptera: Nepidae): Optimal foraging and feeding dynamics. *Oecologica* 68:291–293.

BAILEY, R. W. 1978. *Description of the Ecoregions of the United States.* U.S.D.A. Forest Service Intermountain Region, Ogden, Utah.

BAIN, M. B., J. T. FINN, AND H. E. BOOKE. 1988. Streamflow regulation and fish community structure. *Ecology* 69:382–392.

BAKER, H. G., K. S. BAWA, G. W. FRANKIE, AND P. A. OPLER. 1983. Reproductive biology of plants in tropical forests. In F. B. Golley (ed.), *Tropical Forest Ecosystems: Structure and Function.* Elsevier, Amsterdam, pp. 183–215.

BAKER, M. C., T. K. BJERKI, H. LAMPE, AND Y. ESPMARK. 1986. Sexual response of female great tits to variation in size of males' song repertoires. *Am. Nat.* 128:491–498.

BAKER, M. C., L. M. MEWALDT, AND R. M. STEWART. 1981. Demography of white-crowned sparrows (*Zonotrichia leucophrys nuttalli*). *Ecology* 62:636–644.

BAKKEN, G. S. 1976. A heat transfer analysis of animals: Unifying concepts and the application of metabolism chamber data to field ecology. *J. Theor. Biol.* 60:337–384.

BAKKER, R. T. 1983. The deer flees, the wolf pursues: Incongruencies in predator-prey evolution. In D. Futuyma and M. Slakin (eds.), *Coevolution.* Sinauer Associates, Sunderland, MA, pp. 350–382.

BALDA, R. P. 1975. Vegetation structure and breeding bird diversity. In D. R. Smith (ed.), *Symposium on Management of Forest and Range Habitats for Nongame Birds,* U. S. D. A. For. Serv. GTR WB-1.

BALL, D. M., J. F. PEDERSEN, AND G. D. LACEFIELD. 1993. The tall fescue endophyte. *Am. Sci.* 81:370–379.

BALLARD, W. B., J. S. WHITMAN, AND C. L. GARDNER. 1987. Ecology of an exploited wolf population in south-central Alaska. *Wildl. Monogr.* 98. 54 pp.

BARBOUR, M. M. 1970. Is any angiosperm an obligate halophyte? *Am. Midl. Nat.* 84:106–119.

BARKALOW, F. S., JR., R. B. HAMILTON, AND R. F. SOOTS, JR. 1970. The vital statistics of an unexploited gray squirrel population. *J. Wild. Manage.* 34:489–500.

BARKAN, C. P. L. 1990. A field test of risk-sensitive foraging in black-capped chickadees *(Parus atricapillus). Ecology* 71:391–400.

BARLOW, J. P. 1955. Physical and biological processes determining the distribution of zooplankton in a tidal estuary. *Biol. Bull.* 109:211–225.

———. 1956. Effect of wind on salinity distribution in an estuary. *J. Marine Res.* 15:192–203.

BARNARD, C. 1984. The evolution of food-scrounging strategies within and between species. In C. J. Barnard (ed), *Producers and Scroungers: Strategies of Exploitation and Parasitism.* Croom and Helmt, London, pp. 95–126.

BARNARD, C. J., AND D. B. A. THOMPSON. 1985. *Gulls and Plovers: The Ecology and Behaviour of Mixed-Species Feeding Groups.* Columbia University Press, New York.

BARNHART, R. A., M. J. BOYD, AND J. E. PEQUEGNAT. 1992. *The ecology of Humboldt Bay, California: An estuarine profile.* U.S. Fish and Wildlife Service Biological Report 1. 121 pp.

BARNOLA, J. M., D. RAYNAUD, Y. S. KOROTKEVICH, AND C. LORIUS. 1991. In T. A. Boden, R. J. Sepanski, and F. W. Stoss (eds.), *Trend 91: A Compendium of Data on Global Change.* Oak Ridge National Laboratory, Oak Ridge, TN pp. 4–5.

BARRETT, J. A. 1983. Plant-fungus symbioses. In D. Futuyma and M. Slakin (eds.), *Coevolution.* Sinauer Associates, Sunderland, MA, pp. 137–160.

BARTHOLOMEW, G. A. 1959. Mother-young relations and the maturation of pup behaviour in the Alaskan fur seal. *Anim. Behav.* 7:163–171.

———. 1970. Bare zone between California shrub and grassland communities: The role of animals. *Science* 170:1210–1212.

———. 1981. A matter of size: An examination of endothermy in insects and terrestrial vertebrates. In B. Heinrich (ed.), *Insect Thermoregulation.* Wiley Interscience, New York, pp. 45–78.

BASSAM, J. A. 1965. Photosynthesis. In J. Bonner and J. E. Varna (eds.), *Plant Biochemistry.* Academic, New York, pp. 875–902.

———. 1977. Increasing crop production through controlled photosynthesis. *Science* 197:630–638.

BASSAM, J. A., AND B. B. BUCKANAN. 1982. Carbon dioxide fixation pathways in plants and bacteria. In Govindjee (ed.), *Photosynthesis.* Vol. II, *Development, Carbon Metabolism and Plant Productivity.* Academic, New York.

BATCHELDER, R. R. 1967. Spatial and temporal patterns of fire in the tropical world. *Proc. 6th Tall Timber Fire Ecol. Conf.,* pp. 171–208.

BATZLI, G. O., AND F. A. PITELKA. 1970. Influence of meadow mouse populations on California grassland. *Ecology* 51:1027–1039.

BATZLI, G. O. B. G. WHITE, S. F. MACLEAN, JR, F. A. PITELKA, AND B. D. COLLIER. 1978. The herbivore-based food chain. In J. Brown, F. L. Bunnell, F. S. MacLean, and L. L. Tieszen, (eds). *An Arctic Ecosystem: The Coastal Tundra of Northern Alaska.* US/IBP Synthesis No. 12. Springer Verlag, New York.

BAUMGARTNER, A. 1968. Ecological significance of the vertical energy distribution in plant stands. In F. E. Eckardt (ed.), *Functioning of Terrestrial Ecosystems at the Primary Production Level.* Proc. Copenhagen Symposium Natural Resources Research. UNESCO, Paris, pp. 367–374.

BAWA, K. S. 1980. Evolution of dioecy in flowering plants. *Ann. Rev. Ecol. Syst.* 11:15–39.

BAZZAZ, F. A. 1975. Plant species diversity in old field successional ecosystems in southern Illinois. *Ecology* 56:485–488.

———. 1979. The physiological ecology of plant succession. *Ann. Rev. Ecol. Syst.* 10:351–371.

———. 1991. Habitat selection in plants. *Am. Nat.* 137:S116–S130.

BEACHAM, T. D. 1980. Dispersal during population fluctuations of the vole *Microtus townsendii. J. Anim. Ecol.* 49:867–877.

BEALS, E. W. 1968. Spatial pattern of shrubs on a desert plain in Ethiopia. *Ecology* 49:744–746.

———. 1985. Bray-Curtis ordination: An effective strategy for analysis of multivariate ecological data. *Adv. Ecol. Res.* 14:1–55.

BEATLEY, J. C. 1969. Dependence of desert rodents on winter annuals and precipitation. *Ecology* 50:721–724.

BEATTIE, A. J., AND D. C. CULVER. 1981. The guild of myrmecohores in the herbaceous flora of West Virginia forests. *Ecology* 62:107–115.

BEATTY, S. W. 1984. Influence of microtopography and canopy species on spatial patterns of forest understory plants. *Ecology* 65:1406–1419.

BECK, S. D. 1980. *Insect Photoperiodism,* 2nd ed. Academic Press, New York.

BEDDINGTON, J. R., M. P. HASSELL, AND J. H. LAWTON. 1976. The components of arthropod predation. II. The predator rate of increase. *J. Anim. Ecol.* 45:165–185.

BEDNARZ, J. C. 1988. Cooperative hunting in Harris' hawk (*Parabuteo unicinctus*). *Science* 239:1525–1527.

BEEHLER, B. M. 1983. Lek behavior of the lesser bird of paradise. *Auk* 100:992–995.

BEEHLER, B. M., AND M. S. FOSTER. 1988. Hotshots, hotspots, and female preference in the organization of lek mating systems. *Am. Nat.* 131:203–209.

BEKOFF, M. 1977. Mammalian dispersal and the ontogeny of individual behavioral phenotypes. *Am. Nat.* 111:715–732.

BELL, A. D. 1974. Rhizome organization in relation to vegetative spread in *Medeola virginiana. J. Arnold Arb.* 55:458–468.

BELL, G. 1980. The costs of reproduction and their consequences. *Am. Nat.* 116:45–76.

BELLOWS, T. S., JR. 1981. The descriptive properties for some models for density-dependence. *J. Anim. Ecol.* 50:139–156.

BELLROSE, F. C., T. G. SCOTT, A. S. HAWKINS, AND J. B. LOW. 1961. Sex ratios and age ratios in North American ducks. *Illinois Nat. His. Surv. Bull.* 27:391–474.

BELOVSKY, G. E. 1981a. Food plant selection by a generalist herbivore: The moose. *Ecology* 62:1020–1030.

———. 1981b. A possible population response of moose to sodium availability. *J. Mamm.* 63:631–633.

BELOVSKY, G. E., AND P. F. JORDAN. 1981. Sodium dynamics and adaptations of a moose population. *J. Mamm.* 63:613–621.

BELSKY, A. J. 1986. Does herbivory benefit plants? A review of the evidence. *Am. Nat.* 127:870–892.

BENTLEY, B. L. 1977. Extrafloral nectaries and protection by pugnacious bodyguards. *Ann. Rev. Ecol. Syst.* 8:407–427.

BERG, R. Y. 1975. Myrmecochorous plants in Australia and their dispersal by ants. *Australian J. Bot.* 23:475–508.

BERGER, P. J., N. C. NEGUS, E. H. SANDERS, AND P. D. GARDNER. 1981. Chemical triggering of reproduction in *Microtus montanus. Science* 214:69–70.

BERGERUD, A. T. 1971. The population dynamics of Newfoundland caribou. *Wildl. Monogr.* No. 25.

BERGERUD, A. T., AND F. MANUEL. 1969. Aerial census of moose in central Newfoundland. *J. Wildl. Manage.* 33:910–916.

BERGERUD, A. T., W. WYETH, AND B. SNIDER. 1983. The role of wolf predation in limiting a moose population. *J. Wildl. Manage.* 47:977–988.

BERNARD, J. M., AND E. GORHAM. 1978. Life history aspects of primary production in sedge wetlands. In R. E. Good, D. F. Whigham, and R. L. Simpson (eds.), *Freshwater Wetlands.* Academic Press, New York, pp. 39–51.

BERNAYS, E. A., G. COOPER, DRIVER, AND M. BILGENER. 1989. Herbivores and plant tannins. *Adv. Ecol. Res.* 19:263–302.

BERNHARD-REVERSAT, F. 1975. Nutrients in throughfall and their quantitative importance in rain forest mineral cycles. In F. Golley and E. Medina (eds.), *Tropical Ecological Systems: Trends in Terrestrial and Aquatic Research.* Springer-Verlag, New York, pp. 153–159.

———. 1982. Biogeochemical cycle of nitrogen in a semi-arid savanna. *Oikos* 38:321–332.

BERRY, J. A., AND O. BJORKMAN. 1980. Photosynthetic response and adaptation to temperature in higher plants. *Ann. Rev. Ecol. Syst.* 31:491–453.

BERRY, J. A., AND W. J. S. DOWNTON. 1982. Environmental regulation of photosynthesis. In Govindjee (ed.), *Photosynthesis.* Vol. II, *Development, Carbon Metabolism, and Plant Productivity.* Academic Press, New York, pp. 263–343.

BERRY, J. A., AND J. K. RAISON. 1981. Responses of macrophytes to temperature. In O. Lange, P. S. Nobel, C. B. Osmund, and H. Zeigler (eds.), *Physiological Plant Ecology* I. Vol. 12A, *Encyclopedia of Plant Physiology,* New Series. Springer-Verlag, New York, pp. 237–338.

BERRYMAN, A. A. 1981. *Population Systems: A General Introduction.* Plenum, New York.

———. 1992. The origins and evolution of predator-prey theory. *Ecology* 73:1539–1535.

BERTHOLD, P. 1974. Circannual rhythms in birds with different migratory habits. In E. T. Pengelley (ed.), *Circannual Clocks: Annual Biological Rhythms.* Academic Press, New York, pp. 55–94.

BERTNESS, M. D. 1984. Ribbed mussels and *Spartina alterniflora* production on a New England marsh. *Ecology* 65:1794–1807.

BERTRAM, B. C. R. 1975. Social factors influencing reproduction in wild lions. *J. Zool. Lond.* 177:463–482.

BETZER, P. R., R. H. BYRNE, J. G. ACKER, C. S. LEWIS, R. R. JOLLY, AND R. A. FEELY. 1984. The oceanic carbonate system: A reassessment of biogenic controls. *Science* 226:1074–1077.

BIEL, E. R. 1961. Microclimate, bioclimatology, and notes on comparative dynamic climatology. *Am. Sci.* 49:326–357.

BIERZYCHUDEK, P. 1982a. The demography of Jack-in-the-pulpit, a forest perennial that changes sex. *Ecol. Monogr.,* 52:335–351.

———. 1982b. Life history and demography of shade tolerant temperate herbs: A review. *New Phytol.* 190:757–776.

———. 1984. Assessing optimal life histories in a fluctuating environment: The evolution of sex-changing by jack-in-the-pulpit. *Amer. Nat.* 123:829–840.

BIGALKE, R. C. 1979. Aspects of vertebrate life in fynbos, South Africa. In R. I. Specht (ed.), *Heathlands and Related Shrublands (Ecosystems of the World,* Vol. 9B). Elsevier, Amsterdam, pp. 81–96.

BILBY, R. E. 1981. Role of organic debris dams in regulating the export of dissolved and particulate matter from a forested watershed. *Ecology* 62:1234–243.

BILBY, R. E., AND G. E. LIKENS. 1980. Importance of organic debris dams in the structure and function of stream ecosystems. *Ecology* 1:1107–1113.

BILLINGS, W.D. 1985. The historical development of physiological plant ecology. In B. F. Chabot and H. A. Mooney (eds.), *Physiological Ecology of North American Plant Communities.* New York: Chapman and Hall, pp. 1–15.

BILLINGS, W. D., P. J. GODFREY, B. F. CHABOT, AND D. P. BOURGUE. 1971. Metabolic acclimation to temperature in arctic and alpine ecotypes of *Oxyria digyna. Arctic and Alpine Research* 3:277–289.

BINET, P. 1981. Short-term dynamics of minerals in arid ecosystems. In D. W. Goodall and R. A. Perry (eds.), *Arid Land Ecosystems: Structure, Functioning, and Management,* II. Cambridge University Press, Cambridge, pp. 325–356.

BINKLEY, D., AND R. L. GRAHAM. 1981. Biomass, production, and nutrient cycling of mosses in an old-growth Douglas-fir forest. *Ecology* 62:1387–1389.

BIRCH, L. C., AND D. P. CLARK. 1953. Forest soil as an ecological community with special reference to the fauna. *Quart. Rev. Biol.* 28:13–36.

BIRDSALL, C. W., C. E. GRUE, AND A. ANDERSON. 1986. Lead concentrations in bullfrog *Rana Catesbeiana* and green frog. *R. clamatans* tadpoles inhabiting highway drainages. *Environmental Pollution* (Series A) 43:233–248.

BITMAN, J. 1970. Hormonal and enzymatic activity of DDT. *Agr. Sci. Rev.* 7(4):6–12.

BJORKMAN, O., AND J. BERRY. 1973. High efficiency photosynthesis. *Sci. Amer.* 229(4):80–93.

BLACKBURN, T. H. 1983. The microbial nitrogen cycle. In W. M. Krumbein (ed.), *Microbial Geochemistry.* New York: Blackwell Scientific Publishers, pp. 63–81.

BLACKMAN, F. F. 1905. Optima and limiting factors. *Ann. Bot.* 19:281–298.

BLAIR-WEST, J. R., J. A. COGHLAN, D. A. DENTON, J. F. NELSON,

ET AL. 1968. Physiological, morphological, and behavioral adaptations to a sodium-deficient environment by wild native Australian and introduced species of animals. *Nature* 217:922–928.

BLAKE, D. R., AND F. S. ROWLAND. 1988. Continuing worldwide increase in trophospheric methane, 1978–1987. *Science* 239:1129–1131.

BLAKE, J. G., AND J. R. KARR. 1984. Species composition of bird communities and the conservation benefit of large versus small forests. *Biol. Cons.* 30:193–187.

BLANDIN, W. W. 1992. *Population characteristics and simulation modeling of black ducks.* U.S. Fish and Wildlife Service *Fish and Wildlife Research* 11. 200 pp.

BLISS, L. C. 1956. A comparison of plant development in microenvironments of arctic and alpine tundras. *Ecol. Monogr.* 26:303–337.

———. 1963. Alpine plant communities of the Presidential Range, New Hampshire. *Ecology* 44:678–697.

———. 1975. Devon Island, Canada. In T. Rosswall and O. W. Heal (eds.), *Structure and Function of Tundra Ecosystems.* Swedish National Science Research Council, Stockholm, pp. 17–60.

———. 1981. North American and Scandanavian tundras and polar deserts. In L. C. Bliss, O. W. Heal, and J. J. Moore (eds.), *Tundra Ecosystems: A Comparative Analysis.* Cambridge University Press, Cambridge, pp. 8–24.

———. 1988. Arctic tundra and polar desert biome. In M. G. Barbour and W. D. Billings (eds.), *North American Terrestrial Vegetation.* Cambridge University Press, New York, pp. 1–32.

BLISS, L. C., G. M. COURTIN, D. L. PATTIE, R. R. WIEWE, D. W. A. WHITFIELD, AND P. WIDDEN. 1973. Arctic tundra ecosystems. *Ann. Rev. Ecol. Syst.* 4:359–399.

BLISS, L. C., O. W. HEAL, AND J. J. MOORE (EDS.). 1981. *Tundra Ecosystems: A Comparative Analysis.* International Biological Programme No. 25, Cambridge University Press, Cambridge.

BLONDEL, J. 1981. Structure and dynamics of bird communities in Mediterranean-type habitats. In F. di Castri, D. W. Goodall, and R. Specht (eds.), *Mediterranean-type Shrubland* (*Ecosystems of the World,* Vol. II). Elsevier, Amsterdam, pp. 361–386.

BLUM, J. L. 1960. Algal populations in flowing waters. In *Ecology of Algae.* Spec. Pub. No. 2, Pymatuning Lab. of Field Biology, pp. 11–21.

BOAG, P. T., AND P. GRANT. 1986. Intense natural selection in a population of Darwin's finches. *Science* 241:83.

BOND, R. R. 1957. Ecological distribution of breeding birds in the upland forests of southern Wisconsin. *Ecol. Monogr.* 27:351–384.

BONNELL, M. C., AND R. K. SELANDER. 1974. Elephant seals: Genetic variation and near extinction. *Science* 184:908–909.

BONT, R. G. DE, J. J. VANGELDER, AND J. H. J. OLDERS. 1986. Thermal ecology of the smooth snake *Coronella austriaca Laurenti* during spring. *Oecologica* 69:72–78.

BORAY, J. C. 1969. Experimental Fascioliasis in Australia. *Advances in Parasitology* 7:95–210.

BORCHERT, M. I., AND S. K. JAIN. 1978. The effect of rodent seed predation on four species of California annual grasses. *Oecologica* 33:101–113.

BORGIA, G., AND K. COLLIS. 1989. Female choice for parasite-free male satin bowerbirds and the evolution of bright male plumage. *Behav. Ecol. Sociobiol.* 25:445–454.

BORING, L. R., C. D. MONK, AND W. T. SWANK. 1981. Early regeneration of a clearcut southern Appalachian forest. *Ecology* 62:1244–1253.

BORMANN, F. H., AND G. E. LIKENS. 1979. *Pattern and Process in a Forested Ecosystem.* Springer-Verlag, New York.

BORMANN, F. H., G. H. LIKENS, AND J. H. MELILLO. 1977. Nitrogen budget for an aggrading northern hardwood forest ecosystem. *Science* 196:981–983.

BORNEBUSCH, C. H. 1930. *The Fauna of Forest Soils.* Nielsen and Lydiche, Copenhagen.

BOTKIN, D. B. 1977. Forest, lakes, and the anthropogenic production of carbon dioxide. *Bioscience,* 27:325–331.

BOTKIN, D.B., J.F. JANAK, AND J.R. WALLIS. 1970. Rationale, limitations and assumptions of a northeastern forest growth simulator. *IBM J. Res. Dev.* 16:101–116.

———. 1972a. Some ecological consequences of a computer model of forest growth. *J. Ecol.* 60:948–972.

BOTT, T. L., AND T. D. BROCK. 1968. Bacterial growth rates above 90° C in Yellowstone hot springs. *Science* 164:1411–1412.

BOUCHER, C., AND E. J. MOLL. 1981. South African Mediterranean shrublands. In F. di Castri, D. W. Goodall, and R. L. Specht (eds.), *Mediterranean-type Ecosystems* (*Ecosystems of the World,* Vol. 11). Elsevier, Amsterdam, pp. 233–248.

BOUCHER, D. H., S. JAMES, AND H. D. KELLY. 1982. The ecology of mutualism. *Ann. Rev. Ecol. Syst.* 13:315–347.

BOURGERON, P. S. 1983. Spatial aspects of vegetation structure. In F. B. Golley (ed.), *Tropical Rain Forest Ecosystems* (*Ecosystems of the World,* Vol. 10), Elsevier, Amsterdam, pp. 29–47.

BOURLIERE, F. (ED.). 1983. *Tropical Savannas* (*Ecosystems of the World,* Vol. 13). Elsevier, Amsterdam.

BOURLIERE, F., AND M. HADLEY. 1983. Present-day savannas: An overview. In F. Bourliere (ed.), *Tropical Savannas.* Elsevier, Amsterdam, pp. 1–18.

BOWDEN, W. B. 1986. Nitrification, nitrate reduction, and nitrogen immobilization in a tidal freshwater marsh sediment. *Ecology* 67:88–99.

BOWDEN, W. B., AND F. H. BORMANN. 1986. Transport and loss of nitrous oxide in soil water after clear-cutting. *Science* 233:867–869.

BOX, T. W., J. POWELL, AND D. L. DRAWE. 1967. Influence of fire on south Texas chapparal. *Ecology* 48:955–961.

BRADBURY, J. 1981. The evolution of leks. In R. D. Alexander and D. W. Tinkle (eds.), *Natural Selection and Social Behavior: Research and New Theory.* Chiron Press, New York, pp. 138–169.

BRADBURY, J., R. M. GIBSON, C. E. MCCARTHY, AND S. I. VEHRENCAM. 1989. Dispersion of displaying male sage grouse: The role of female dispersion. *Behav. Ecol. Sociobiol.* 24:15–24.

BRADBURY, J. W., AND R. GIBSON. 1983. Leks and male choice. In D. Bateson (ed.), *Mate Choice.* Cambridge University press, Cambridge, pp. 109–138.

BRADLEY, A. F., N. V. NOSTE, AND W. C. FISCHER. 1992. *Fire ecology of forest and woodland in Utah.* Gen Tech. Rept. INT-287 USDA Forest Service. Intermountain Research Station, Ogden, UT.

BRADLEY, W. G., AND R. A. MAUER. 1971. Reproduction and food habits of Merriam's kangaroo rat, *Dipodomys merriami. J. Mamm.* 52:497–507.

BRAMWELL, A. 1989. *Ecology in the 20th Century.* Yale University Press, New Haven.

BRATTON, S. P. 1994. Logging and fragmentation of broadleaf deciduous forests: Are we asking the right questions? *Cons. Biol.* 8:295–297.

BRAY, J. R., AND J. T. CURTIS. 1957. An ordination of the upland forest communities of southern Wisconsin. *Ecol. Monogr.* 27:325–349.

BRAY, J. R., AND E. GORHAM. 1964. Litter production in the forests of the world. *Adv. Ecol. Rec.* 2:101–157.

BREYMEYER, A. I. 1980. Trophic structure and relationships. In A. I. Breymeyer and G. M. Van Dyne (eds.), *Grasslands, Systems Analysis, and Man.* Cambridge University Press, Cambridge.

BREYMEYER, A. I., AND G. M. VAN DYNE (EDS.). 1980. *Grasslands, Systems Analysis, and Man.* International Biological Programme No. 19. Cambridge University Press, Cambridge.

BRIAND, F. 1983a. Biogeographic patterns in food web organization. In D. L. DeAngelis, W. M. Post, and G. Sugihara (eds.), *Current Trends in Food Web Theory.* ORNL 5983, Oak Ridge National Laboratory, Oak Ridge, TN, pp. 41–44.

———. 1983b. Environmental control of food web structure. *Ecology* 64:253–263.

BRIAND, F., AND J. E. COHEN. 1984. Community food webs have scale-invariant structure. *Nature* 307:264–266.

BRICKLEMYER, E. C., JR., S. IUDICELLO, AND H. J. HARTMANN. 1989. Discarded catch in U.S. commercial marine fisheries. *Audubon Wildlife Report* 1989/1990: 259–295. San Diego, CA: Academic Press.

BRITTINGHAM, M. C., AND S. A. TEMPLE. 1983. Have cowbirds caused forest songbirds to decline? *Bioscience* 33:31–35.

BROCK, T. R. 1967. Life at high temperatures. *Science* 158:1012–1019.

———. 1970. High temperature systems. *Ann. Rev. Ecol. Syst.* 1:191–220.

———. 1979. *Biology of Microorganisms.* Prentice-Hall, Englewood Cliffs, NJ.

BRODSKY, L. M., AND P. J. WEATHERHEAD. 1984. Behavioral and ecological factors contributing to American black duck–mallard hybridization. *J. Wildl. Manage.* 48:846–852.

BROECKER, W. S. 1970. Man's oxygen reserves. *Science* 168:1537–1538.

BROECKER, W. S., T. TAHAKASHI, H. J. SIMPSON, AND T. H. PING. 1979. Fate of fossil fuel carbon dioxide and the global carbon budget. *Science* 206:409–418.

BROKAW, N. V. L. 1985. Gap-phase regeneration in a tropical forest. *Ecology* 66:682–687.

BROOKS, M. B. 1951. Effect of black walnut trees and their products on vegetation. *W. Va. Univ. Agr. Exp. Stat. Bull.* 347:1–31.

———. 1955. An isolated population of the Virginia varying hare. *J. Wildl. Manage.* 19:54–61.

BROWER, J. E., AND J. H. ZAR. 1984. *Field and Laboratory Methods for General Ecology,* 2nd ed. Brown, Dubuque, IA.

BROWER, J. V. Z. 1958. Experimental studies of mimicry in some North American butterflies: 1. The monarch, *Danaus plexippus,* and viceroy, *Limenitis archippus;* 2. *Battus philenor* and *Papilio troilus, P. polyxenes* and *P. glaucus;* 3. *Danaus glippus berenice* and *Limenitis archippus floridensis. Evolution* 12:32–47, 123–136, 273–285.

BROWER, J. V. Z., AND L. P. BROWER. 1962. Experimental studies of mimicry: 6. The reaction of toads (*Bufo terrestris*) to honeybees

(*Apis mellifera*) and their drone mimics (*Eristalis vinetorum*). *Am. Nat.* 97:297–307.

BROWER, L. P. 1984. Chemical defense in butterflies. *Symp. Roy. Entomol. Soc. London* 11:109–134.

———. 1988. Avian predation on the monarch butterfly and its implication for mimicry theory. *Am. Nat.* 131:54–56.

BROWER, L. P., AND L. S. FINK. 1983. A natural toxic defense system in butterflies versus birds. *Ann. N.Y. Acad. Sci.* 443:171–186.

BROWN, C. R., AND M. B. BROWN. 1989. Behavioral dynamics of interspecific brood parasitism in colonial cliff swallows. *Anim. Behav.* 31:777–796.

BROWN, E. R. 1961. The black-tailed deer of western Washington. *Washington State Game Dept. Biol. Bull.* No. 13, Olympia.

BROWN, J. H. 1971. Mammals on mountaintops: Nonequilibrium insular biogeography. *Am. Nat.* 105:467–478.

———. 1978. The theory of insular biogeography on the distribution of boreal birds and mammals. *Great Basin Nat. Mem.* 2:209–227.

———. 1981. Two decades of homage to Santa Rosalia: Toward a general theory of diversity. *Am. Zool.* 21:877–888.

BROWN, J. H., AND A. KODRICH-BROWN. 1977. Turnover rates in insular biogeography: Effect of immigration on extinction. *Ecology* 58:445–449.

BROWN, J. H., AND A. C. GIBSON. 1983. *Biogeography.* C. V. Mosby Co., St. Louis, MO.

BROWN, J. L. 1964. The evolution of diversity in avian territorial systems. *Wilson Bull.* 76:160–169.

———. 1969. Territorial behavior and population regulation in birds: A review and reevaluation. *Wilson Bull.* 81:292–329.

———. 1987. *Helping and Communal Breeding in Birds.* Princeton University Press, Princeton, NJ.

BROWN, J. R., AND E. J. HESKE. 1990. Control of a desert-grassland transition by keystone rodent guild. *Science* 250:1705–1707.

BROWN, K. M. 1982. Resource overlap and competition in pond snails: An experimental analysis. *Ecology* 63:412–422.

BROWN, L. 1976. *British Birds of Prey.* Collins, London.

BROWN, R. J. E. 1970. *Permafrost in Canada.* University of Toronto Press, Toronto.

BROWN, C. R. AND M. B. BROWN. 1989. Behavioral dynamics of interspecific brood parasitism in colonial cliff swallows. *Anim. Behav.* 37:777–796.

BROWN, R. J. E., AND G. H. JOHNSON. 1964. Permafrost and related engineering problems. *Endeavour* 23:66–73.

BROWN, S., AND A. E. LUGO. 1984. Biomass of tropical forests: A new estimate based on forest volumes. *Science* 223:1290–1293.

BRUNIG, E. F. 1983. Vegetation structure and growth. In F. Golley (ed.), *Tropical Rain Forest Ecosystems: Structure and Function.* Elsevier, Amsterdam, pp. 49–75.

BRYANT, J. A. 1987. Feltleaf willow–snowshoe hare interactions: Plant carbon/nutrient balance and flood plain succession. *Ecology* 68:1319–1327.

BRYANT, J. A., AND P. J. KUROPAT. 1980. Selection of winter forage by subarctic browsing vertebrates: The role of plant chemistry. *Ann. Rev. Ecol. Syst.* 11:261–285.

BRYANT, J. P., G. D. WIELAND, T. CLAUSEN, AND P. J. KUROPAT. 1985. Interactions of snowshoe hares and feltleaf willow (*Salix alaxensis*) in Alaska. *Ecology* 66:1564–1573.

BRYANT, J. P., G. D. WIELAND, P. B. REICHERT, V. E. LEWIS, AND M. C. MCCARTHY. 1983. Pinosylvin methyl ether deters snowshoe hare feeding on green alder. *Science* 222:1023–1025.

BRYLINSKY, M. 1980. Estimating the productivity of lakes and reservoirs. In E. D. Le Cren and R. H. Lowe-McConnell (eds.), *The Functioning of Freshwater Ecosystems.* International Biological Programme No. 22. Cambridge University Press, Cambridge England, pp. 411–418.

BRYSON, R. A., D. A. BAERRIS, AND W. M. WENDLAND. 1970. The character of late glacial and post glacial climatic changes. In W. Dort and J. K. Jones (eds.), *Pleistocene and Recent Environments of the Great Central Plains.* Spec. Publ. No. 3, Department of Biology, University of Kansas, Lawrence, pp. 53–74.

BRYSON, R. A., AND R. A. RAGOTZKIE. 1960. On internal waves in lakes. *Limnol. Oceanogr.* 5:397–408.

BUCHMANN, S. L. 1987. The ecology of oil flowers and their bees. *Ann. Rev. Ecol. Syst.* 18:343–369.

BUCKNER, C. H., AND W. J. TURNOCK. 1965. Avian predation on the larch sawfly, *Pristiphora erichsonii* (Hymenoptera: Tenthredinidae). *Ecology* 46:223–236.

BUDYKO, M. I. 1963. *The Heat Budget of the Earth.* Hydrological Publishing House, Leningrad.

BUELL, M. F., AND R. E. WILBUR. 1948. Life form spectra of the hardwood forests of the Itaska Park region, Minnesota. *Ecology* 29:352–359.

BULL, J., AND E. CHARNOV. 1989. Energetic reptilian sex ratios. Evolution. 43:1561–1566.

BULL, J. J., AND R. C. VOGT. 1979. Temperature-dependent sex determination in turtles. *Science* 206:1186–1188.

BULL, K. R., W. J. EVERY, P. FREESTONE, J. R. HALL, AND D. OSBORN. 1983. Alkyl lead pollution and bird mortalities in the Mersey Estuary, UK, 1979–1981. *Environmental Pollution* 31:239–259.

BULMER, M. G. 1974. A statistical analysis of the 10-year cycle in Canada. *J. Anim. Ecol.* 43:701–718.

———. 1975. Phase relations of the 10-year cycle. *J. Anim. Ecol.* 44:609–621.

BUNNELL, F. L., S. F. MCCLEAN, JR., AND J. BROWN. 1975. Barrow, Alaska, U.S.A. In T. Rosswall and O. W. Heal (eds.), *Structure and Function of Tundra Ecosystems.* Swedish Natural Science Institute, Stockholm, pp. 73–124.

BUNNELL, F. L., O. K. MILLER, P. W. FLANAGAN, AND R. E. BENOIT. 1980. The microflora: Composition, biomass, and environmental relations. In J. Brown, P. C. Miller, L. L. Tieszen, and F. L. Bunnell (eds.), *An Arctic Ecosystem: The Coastal Tundra at Barrow, Alaska.* Dowden, Hutchinson, and Ross, Stroudsburg, PA, pp. 255–290.

BUNNING, E. 1964. *The Physiological Clock,* 2nd ed. Academic, New York.

BURBANCK, M. P., AND R. B. PLATT. 1964. Granite outcrop communities of the Piedmont Plateau in Georgia. *Ecology* 45:292–306.

BURGES, A. 1963. The microbiology of a podzol profile. In J. Doeken and J. van der Drift (eds.), *Soil Organisms.* Amsterdam: North Holland Publishing Co., pp. 151–157.

BURGESS, R. L., AND R. V. O'NEILL. 1976. *Eastern deciduous forest biome progress report September 1, 1974–August 31, 1975. EDFB/IBP 76/5.* Env. Sci. Div. Pub. No. 871, Oak Ridge National Laboratory, Oak Ridge, TN.

BURT, W. V., AND J. QUEEN. 1957. Tidal overmixing in estuaries. *Science* 126:973–974.

BUSKIRK, R. E., AND W. H. BUSKIRK. 1976. Changes in arthropod abundance in a highland Costa Rica forest. *Am. Midl. Nat.* 95:288–298.

BUSS, D. M. 1994. The strategies of human mating. *Amer. Sci.* 82:238–249.

BUTCHER, G. S., W. A. NIERING, W. J. BARRY, AND R. W. GOODWIN. 1981. Equilibrium biogeography and the size of nature preserves: An avian case study. *Oecologica* 49:29–37.

BUTLER, P. A. 1964. Commercial fisheries investigations. In *Pesticide-Wildlife Studies,* U.S. Serv. Circ. 226, pp. 11–25.

BUTTEMER, W. A. 1985. Energy relations of winter roost-site utilization by American goldfinches (*Carduelis tristis*). *Oecologica* 18:126–132.

CALDWELL, M. 1985. Cold desert. In B. F. Chabot and H. A. Mooney (eds.), *Physiological Ecology of North American Plant Communities.* Chapman and Hall, New York, pp. 198–212.

CALDWELL, M. M. 1971. Solar ultraviolet radiation as an ecological factor for alpine plants. *Ecol. Monogr.* 38:243–268.

CALDWELL, M. M., R. ROBBERRECHT, AND W. D. BILLINGS. 1980. A steep latitudinal gradient of solar ultraviolet-B radiation in the arctic-alpine life zone. *Ecology* 61:600–611.

CALDWELL, M. M., A. H. TERAMURA, AND M. TEVINI. 1989. The changing solar ultraviolet climate and the ecological consequence for higher plants. *TREE* 4:363–367.

CANHAM, C. D., AND O. L. LOUCKS. 1984. Catastrophic windthrow in presettlement forests of Wisconsin. *Ecology* 65:803–809.

CARACO, T. S., S. MARTINDALE, AND T. W. WHITHAM. 1980. An empirical demonstration of risk-sensitive foraging preferences. *Anim. Behav.* 28:820–830.

CARBYN, L. N. 1974. Wolf predation and behavioral interactions with elk and other ungulates in a high prey diversity. *Can. Wildl. Serv. Rept.,* 233 pp.

———. 1982. Coyote population fluctuations and spatial distribution in relation to wolf territories in Riding Mountain National Park. *Can. Field Nat.* 96:176–183.

———. 1983. Wolf predation on elk in Riding Mountain National Park, Manitoba. *J. Wildl. Manage.* 47:963–976.

CARBYN, L. N., AND M. C. S. KINGSLEY. 1979. Summer food habits of wolves with emphasis on moose in Riding Mountain National Park. *Proc. North Am. Moose Conf. Workshop* 15:349–361.

CAREY, F. C. 1982. A brain heater in the swordfish. *Science* 216:1327–1329.

CARLSON, J. D., JR., W. R. CLARK, AND E. E. KLAAS. 1993. *A model of the productivity of the northern pintail.* U.S. Fish and Wildlife Service Biological Report 7. 20 pp.

CARPENTER, F. L. 1987a. Food abundance and territoriality: To defend or not to defend? *Am. Zool.* 27:387–399.

———. 1987b. The study of territoriality: Complexities and future directions. *Am. Zool.* 27:401–409.

CARPENTER, S. R. 1980. Enrichment of Lake Wingra, Winconsin, by submerged macrophyte decay. *Ecology* 61:1145–1155.

CARPENTER, S. R., AND J. F. KITCHELL. 1984. Plankton community structure and limnetic primary production. *Am. Nat.* 124:159–172.

CARPENTER, S. R., J. F. KITCHELL, AND J. HODGSON. 1985. Cascading trophic interactions and lake productivity. *Bioscience* 35:634–639.

CARRICK, R. 1963. Ecological significance of territory size in the Australian magpie *Gymnorhina tibiten. Proc. Int. Orn. Cong.* 13:740–753.

CARROLL, G. C. 1979. Forest canopies: Complex and independent subsystems. In R. H. Waring (ed.), *Forests: Fresh Perspectives from Ecosystem Analysis.* Proc. Ann. Biol. Coll., Oregon State University Press, Corvallis, pp. 87–107.

CARROLL, J. F., AND J. D. N. NICHOLS. 1986. Parasitization of meadow voles *Microtus pennsylvanicus* (Ord) by American dog ticks *Dermacenter variabilis* (Say) and adult tick movement during high host density. *J. Entomol. Sci.* 21:102–113.

CARROLL, R. C. 1992. Ecological management of sensitive natural areas. In P. L. Feidler and S. K. Jain (eds.), *Conservation Biology: The Theory and Practice of Nature Conservation, Preservation, and Management.* Chapman and Hall, New York, pp. 347–372.

CARSON, R. 1962. *Silent Spring.* Houghton Mifflin, Boston.

CARTER, M. R., L. A. BURNS, T. R. CAVINDER, K. R. DUGGER, K. R. FORE, D. B. HICKS, H. L. REVELLS, AND T. W. SCHMIDT. 1973. *Ecosystem Analysis of the Big Cypress Swamp and Estuaries.* U.S. Environmental Protection Agency 904/9-74-002, Region IV, Atlanta, GA.

CARTER, R. V. 1991. A test of risk sensitive foraging in wild bumblebees. *Ecology* 72:887–895.

CASE, T. J., AND M. L. CODY. 1984. Testing theories of island biogeography. *Am. Sci.* 75:42–411.

CASTRI, F. DI. 1981. Mediterranean-type shrublands of the World. In F. Di Castri, D. W. Goodall, and R. L. Specht (eds.), *Mediterranean-type Shrublands* (*Ecosystems of the World,* Vol. 11). Elsevier, Amsterdam.

CASTRI, F. DI, AND H. A. MOONEY (EDS). 1973. *Mediterranean Type Ecosystems: Origin and Structure.* Springer-Verlag, New York.

CATCHPOLE, C. K. 1987. Bird song, sexual selection, and female choice. *TREE* 2:94–97.

CAUGHLEY, G. 1976a. Wildlife management and the dynamics of ungulate populations. *Appl. Biol* 1:183–246.

———. 1976b. Plant and herbivore systems. In R. B. May (ed.), *Theoretical Ecology: Principles and Applications.* Saunders, Philadelphia, pp. 94–113.

———. 1977. *Analysis of Vertebrate Populations.* Wiley, New York.

CERNUSA, A. 1976. Energy exchange within individual layers of a meadow. *Oecologia* 23:141–149.

CHAFFEE, R. R. J., AND J. C. ROBERTS. 1971. Temperature acclimation in birds and animals. *Ann. Rev. Physiol.* 33:155–202.

CHALMERS, A. G. 1982. Soil dynamics and the productivity of *Spartina alternifolia.* In V. S. Kennedy (ed.), *Estuarine Comparisons.* Academic, New York, pp. 231–243.

CHANGNON, S. A. 1968. La Porte weather anomaly: Fact or fiction? *Bull. Am. Meteorol. Soc.* 49:4–11.

CHAPIN, F. S., P. C. MILLER, W. D. BILLINGS, AND R. I. COYNE. 1980. Carbon and nutrient budgets and their control in coastal tundra. In J. Brown, P. C. Miller, L. L. Tieszen, and F. L. Bunnell (eds.), *An Arctic Ecosystem: The Coastal Tundra at Barrow, Alaska.* Dowden, Hutchinson, and Ross, Stroudsburg, PA, pp. 458–482.

CHAPIN, F. S., III, E. D. SCHULZE, AND H. A. MOONEY. 1990. The ecology and economics of storage in plants. *Ann. Rev. Ecol. Syst.* 21:846–852.

CHAPIN, F. S., L. L. TIESZEN, M. C. LEWIS, P. C. MILLER, AND B. H. MCCOWEN. 1980. Control of tundra plant allocation patterns and growth. In J. Brown, P. C. Miller, L. L. Tieszen, and F. L. Bunnell (eds.), *An Arctic Ecosystem: The Coastal Tundra at Barrow, Alaska.* Dowden, Hutchinson, and Ross, Stroudsburg, PA, pp. 140–185.

CHAPMAN, V. J. 1976. *Coastal Vegetation,* 2nd ed. Pergamon Press, Oxford.

CHARNOV, E. L. 1976. Optimal foraging: The marginal value theorem. *Theor. Pop. Biol.* 9:129–136.

———. 1982. *The Theory of Sex Allocation.* Princeton University Press, Princeton.

CHARNOV, E. L., AND W. M. SCHAFFER. 1973. Life history consequences of natural selection: Cole's results revisited. *Am Nat.* 107:791–793.

CHARNOV, E. L., AND J. J. BULL. 1977. When is sex environmentally determined? *Nature* 266:828–830.

CHASKO, G. G., AND J. E. GATES. 1982. Avian habitat suitability along a transmission line corridor in an oak-hickory forest region. *Wildl. Monogr.* 82. The Wildlife Society.

CHAZDON, R. L. 1988. Sunflecks and their importance to forest understory plants. *Adv. Ecol. Res.* 18:1–63.

CHAZDON, R. L., AND R. W. PEARCY. 1986. Photosynthetic responses to light variation in rainforest species. II. Carbon gain and photosynthetic efficiencies during light flecks. *Oecologica* 69:524–531.

———. 1991. The importance of sunflecks for forest understory plants. *Bioscience* 41:760–765.

CHEATUM, E. L., AND C. W. SEVERINGHAUS. 1950. Variations in fertility of white-tailed deer related to range conditions. *Trans. North Am. Wildl. Conf.* 15:170–189.

CHESSER, R. K. 1983. Isolation by distance: Relationship to the management of genetic resources. In C. Schonewald-Cox, S. Chambers, B. MacBryde, and W. Thomas (eds.), *Genetics and Conservation: A Reference for Managing Wild Animal and Plant Populations.* Benjamin/Cummings, Menlo Park, CA, pp. 51–65.

CHESSON, P. L. 1986. Environmental variation and coexistence of species. In J. Diamond and T. Case (eds.), *Community Ecology.* Harper and Row, New York, pp. 240–256.

CHESSON, P. L., AND T. CASE. 1986. Overview: Nonequilibrium community theories: Chance, variability, history, and coexistence. In J. Diamond and T. Case (eds.), *Community Ecology.* Harper and Row, New York, pp. 229–239.

CHRISTIAN, J. J. 1971. Fighting, maturity, and population density in *Microtus pennsylvanicus. J. Mammal.* 52:556–567.

———. 1978. Neurobehavioral endocrine regulation of small mammal populations. In D. P. Snyder (ed.), *Populations of Small Mammals under Natural Conditions,* Pymatuning Symposia in Ecology, Vol. 5. University of Pittsburg Press, Pittsburgh, PA, pp. 143–158.

CHRISTY, H. R. 1952. Vertical temperature gradients in a beech forest in central Ohio. *Ohio J. Sci.* 52:199–209.

CICERONE, R. J. 1987. Changes in stratosphere ozone. *Science* 237:35–42.

CLARK, A. B. 1978. Sex ratio and local resource competition in a prosimian primate. *Science* 201:163–165.

CLARK, G. A. 1939. Utilization of solar energy by aquatic organisms. *Problems in Lake Biology* (Pub. Am. Assoc. Adv. Science) 10:27–38.

CLARKSON, D. T., AND J. B. HANSON. 1980. The mineral nutrition of higher plants. *Ann. Rev. Plant Physiology* 31:239–298.

CLAY, K. 1988. Fungal endophytes of grasses: A defensive mutualism between plants and fungi. *Ecology* 69:10–16.

———. 1990. Fungal endophytes of grasses. *Ann. Rev. Ecol. Syst.* 21:275–297.

CLAYTON, D. H. 1991. Coevolution of avian grooming and ectoparasite avoidance. In J. E. Loye and M. Zuk (eds.), *Bird-Parasite Interactions: Ecology, Evolution, and Behavior.* Oxford University Press, Oxford, pp. 258–289.

CLEMENTS, F. E. 1916. *Plant Succession: An Analysis of the Development of Vegetation.* Carnegie Inst. Wash. Publ. 242.

CLEMENTS, F. E., AND V. E. SHELFORD. 1939. *Bio-ecology.* McGraw-Hill, New York.

CLOUDSLEY-THOMPSON, J. L. 1956. Studies in diurnal rhythms: VII. Humidity responses and nocturnal activity in woodlice (Isopoda). *J. Exp. Biol.* 33:576–582.

———. 1960. Adaptive functions of circadian rhythms. *Cold Spring Harbor Symp. Quant. Biol.* 255:345–355.

CLUTTON-BROCK, T. H. 1984. Reproductive effort and terminal investment in iteroparous animals. *Am. Nat.* 123:212–229.

CLUTTON-BROCK, T. H., S. D. ALBON, AND F. E. GUINNESS. 1986. Great expectations: Dominance, breeding success, and offspring sex ratios in red deer. *Anim. Behav.* 34:460–471.

CLUTTON-BROCK, T. H., D. GREEN, M. HIRAIWA-HASEGAWA AND S. D. ALBON. 1988. Passing the buck: Resource defense, lekking, and mate choice in the fallow deer. *Behav. Ecol. Sociobiol.* 23:281–296.

CLUTTON-BROCK, T. H., F. E. GUINNESS, AND S. D. ALBON. 1982. *Red Deer: Behavior and Ecology of the Two Sexes.* University of Chicago Press, Chicago.

CLUTTON-BROCK, T. H., M. HIRAIWA-HASEGAWA, AND A. ROBINSON. 1989. Mate choice in fallow deer leks. *Nature* 340:463–465.

CLUTTON-BROCK, T. H., O. F. PRICE, S. D. ALBON, A. ROBERTSON, AND P. A. JEWELL. 1991. Population regulation in Soay sheep. *J. Anim. Ecol.* 60:593–608.

COHEN, J. E., AND C. M. NEWMAN. 1985. A stochastic theory of community food webs. I. Models and aggregated data. *Proc. Roy. Soc. Lond.* 224:421–428.

COHMAR MEMBERS. 1988. Climate changes of the last 18,000 years: Observations and model simulations. *Science* 241:1043–1052.

COKER, R. E. 1947. *This Great and Wide Sea.* University of North Carolina Press, Chapel Hill.

COLE, D. W., S. D. GESSEL, AND S. F. DICE. 1968. Distribution and cycling of nitrogen, phosphorus, potassium, and calcium in second growth Douglas-fir ecosystem. In *Symposium on Primary Productivity and Mineral Cycling in Natural Ecosystems.* Orono, ME, University of Maine Press, pp. 197–232.

COLE, L. C. 1951. Population cycles and random oscillations. *J. Wildl. Manage.* 15:233–252.

———. 1954. The population consequences of life history phenomena. *Quart. Rev. Biol.* 29:103–137.

COLLINS, B. S., AND S. T. A. PICKETT. 1987. Influence of canopy openings on the environment and herb layer in a northern hardwoods forest. *Vegetatio* 70:3–10.

CONFER, J. L, AND K. KNAPP. 1981. Golden-winged warblers and blue-winged warblers: The relative success of a habitat specialist and a habitat generalist. *Auk* 98:108–114.

CONNELL, J. H. 1961. The effects of competition, predation by *Thais lapillus,* and other factors on the distribution of the barnacle *Balanus balanoides. Ecol. Monogr.* 31:61–104.

CONNELL, J. H., AND E. ORIAS. 1964. The ecological regulation of species diversity. *Am. Nat.* 98:399–414.

CONNELL, J. H., AND R. O. SLATYER. 1977. Mechanisms of succession in natural communities and their role in community stability and organization. *Am. Nat.* 111:1119–1144.

CONNELL, J. M. 1978. Diversity in tropical rain forests and coral reefs. *Science* 199:1302–1310.

COOKE, G. D. 1967. The pattern of autotrophic succession in laboratory microcosms. *Bioscience* 17:717–721.

COOPER, J. P. (ED.) 1975. *Photosynthesis and Productivity in Different Environments.* Cambridge University Press, Cambridge.

CORBETT, E. S., AND R. P. CROUSE. 1968. Rainfall interception by annual grass and chaparral. USDA Forest Serv. Res. Paper PSW 48.

COUGHENOUR, M. B. 1985. A mechanistic simulation analysis of water use, leaf angles, and grazing in East African graminoids. *Ecol. Modeling* 6:203–230.

COUGHENOUR, M. B., S. T. MCNAUGHTON, AND L. L. WALLACE. 1984. Modeling primary production of perennial graminoids: Uniting physiological processes and morphometric traits. *Ecol. Modeling* 23:101–134.

COUPLAND, R. T. 1958. The effects of fluctuations in weather upon the grassland of the Great Plains. *Botan. Rev.* 24:273–317.

COURTNEY, S. P. 1985. Apparency in coevolving relationships. *Oikos* 44:91–98.

COUTANT, C. 1970. Biological aspects of thermal pollution: 1. Entrainment and discharge canal effects. *CRC Critical Reviews in Environ. Control,* November 1970, pp. 341–381.

COWAN, I., AND V. GEIST. 1961. Aggressive behavior in deer of the genus *Odocoileus. J. Mammal.* 42:522–526.

COWAN, R. L. 1962. Physiology of nutrition as related to deer. *Proc. 1st Natl. White-tailed Deer Disease Symp.,* pp. 1–8.

COWLES, H. C. 1899. The ecological relations of the vegetation on the sand dunes of Lake Michigan. *Botan. Gaz.* 27:95–117, 167–202, 281–308, 361–391.

COWLES, R. B., AND C. M. BOGERT. 1944. A preliminary study of the thermal requirements of desert reptiles. *Bull. Am. Mus. Natur. Hist.* 83:265–269.

COX, C. R., AND B. J. LEBOEUF. 1977. Female incitation of male competition: A mechanism of mate selection. *Am. Nat.* 111:317–355.

CRAWLEY, M. J. 1989. Insect herbivore and plant population dynamics. *Ann. Rev. Entomol.* 34:531–564.

CRITCHFIELD, W. B. 1971. *Profiles of California Vegetation.* USDA Forest Serv. Res. Paper PSW-76, Berkeley, CA Pacific Southwest Forest & Range Exp. Sta.

CROCKER, R. L., AND J. MAJOR. 1955. Soil development in relation to vegetation and surface age at Glacier Bay, Alaska. *J. Ecol.* 43:427–488.

CROLL, N. A. 1966. *Ecology of Parasites.* Harvard University Press, Cambridge, MA.

CROWLEY, J. 1967. Biogeography. *Can. Geog.* 11:312–326.

CUMMINS, K. W., AND M. J. KLUG. 1979. Feeding ecology of stream invertebrates. *Ann. Rev. Ecol. Syst.* 10:147–172.

CURRIE, D. J. 1991. Energy and large-scale biogeographical patterns of animal and plant species richness. *Am. Nat.* 137:27–49.

CURRIE, D. J., AND V. PAQUIN. 1987. Large-scale biogeographical patterns of species richness in trees. *Nature* (London) 329:326–327.

CURTIS, J. T., AND R. P. MCINTOSH. 1951. An upland forest continuum in the prarie-forest border region of Wisconsin. *Ecology* 32:476–496.

CUSTER, C. M. 1993. Life history traits and habitat needs of the redhead. U.S. Fish and Wildlife Leaflet 13.1.11. 7 pp.

DAHLGREN, R. B., AND C. E. KORSCHGEN. 1992. *Human disturbances of waterfowl: An annotated bibliography.* U.S. Fish and Wildlife Service Resource Publication 188. 62 pp.

D'ANGELO, D. J., J. R. WEBSTER, AND E. F. BENFIELD. 1991.

Mechanisms of stream phosphorus retention: An experimental study. *J. N. Am. Benthol. Soc.* 10:225–237.

DAPSON, R. W., P. R. RAMSEY, M. H. SMITH, AND D. F. URBSTAN. 1979. Demographic differences in contiguous populations of white-tailed deer. *J. Wildl. Manage.* 43:889–898.

DARLINGTON, P. J., JR. 1957. *Zoogeography: The Geographical Distribution of Animals.* Wiley, New York.

DARNELL, R. M. 1961. Trophic spectrum of an estuarine community based on studies of Lake Pontchartrain, Louisiana. *Ecology* 42:553–568.

DARWIN, C. 1859. *The Origin of Species.* Murray, London.

———. 1871. *The Descent of Man, and Selection in Relation to Sex.* Murray, London.

DASH, M. C., AND A. K. HOTA. 1980. Density effects on the survival, growth rate, and metamorphosis of *Rana tigrina* tadpoles. *Ecology* 61:1025–1028.

DAUBENMIRE, R. F. 1968a. Soil moisture in relation to vegetation distribution in the mountains of northern Idaho. *Ecology* 49:431–438.

———. 1968b. Ecology of the fire in grasslands. *Adv. Ecol. Res.* 5:208–266.

DAVIDSON, C. I., J. R. HARRINGTON, M. J. STEVENSON, M. C. MONAGHAN, J. PUDYKEIWICZ, AND W. R. SCHNELL. 1987. Radioactive cesium from the Chernobyl accident in the Greenland ice sheet. *Science* 237:633–634.

DAVIDSON, D. W. 1985. An experimental study of diffuse competition in a desert ant community. *Am. Nat.* 125:500–506.

DAVIDSON, W. R. 1981. *Diseases and parasites of white-tailed deer. Southeastern Cooperative Wildlife Disease Study.* Mscl. Pub. No 7. Tall Timbers Research Station, Tallahassee, FL.

DAVIES, N. B. 1978. Ecological questions about territorial behaviour. In J. R. Krebs and N. B. Davies (eds.), *Behavioural Ecology: An Evolutionary Approach.* Blackwell, Oxford, pp. 317–350.

———. 1991. Mating systems. In J. R. Krebs and N. B. Davies (eds.), *Behavioural Ecology: An Evolutionary Approach.* 3rd ed. Blackwell, London, pp. 263–294.

DAVIES, N. B., AND A. I. HOUSTON. 1984. Territory economics. In J. R. Krebs and N. B. Davies (eds.), *Behavioural Ecology: An Evolutionary Approach,* 2nd ed. Blackwell, Oxford, pp. 148–169.

DAVIS, C. D., AND A. G. VAN DER VALK. 1978. Litter decomposition in glacial prairie marshes. In R. E. Good, D. F. Whigham, and R. L. Simpson (eds.), *Freshwater Wetlands.* Academic Press, New York, pp. 99–144.

DAVIS, D. E. 1978. Physiological and behavioral responses to the social environment. In D. P. Snyder (ed.), *Populations of Small Mammals under Natural Conditions,* Vol. 5. Special Publication Ser., Pymatuning Laboratory of Ecology, University of Pittsburgh Press, Pittsburgh, pp. 84–91.

DAVIS, M. B. 1981. Quaternary history and the stability of forest communities. In D. C. West, H. H. Shugart, and D. B. Botkins (eds.), *Forest Succession: Concepts and Application.* Springer-Verlag, New York, pp. 132–153.

———. 1983. Holocene vegetational history of the eastern United States. In H. E. Wright, Jr. (ed.), *Late Quaternary Environments of the United States.* Vol. II. *The Holocene.* University of Minnesota Press, Minneapolis, pp. 166–188.

DAY, F. P., JR., AND D. T. MCGINTY. 1975. Mineral cycling strategies of two deciduous and two evergreen tree species on a southern Appalachian watershed. In F. C. Howell, J. B. Gentry, and M. H. Smith (eds.), *Mineral Cycling in Southeastern Ecosystems.* National Technical Information Service, U.S. Department of Commerce, Washington, DC, pp. 736–743.

DAY, T. A., T. C. VOGELMANN, AND E. H. DELUCIA. 1992. Are some plant life forms more effective than others in screening out ultraviolet radiation? *Oecologica* 92:513–516.

DAYTON, P. 1971. Competition, disturbance, and community organization: The provision and subsequent utilization of space in a rocky intertidal community. *Ecol. Monogr.* 41:351–389.

———. 1975. Experimental evaluation of ecological dominance in a rocky intertidal algal community. *Ecol. Monogr.* 45:137–159.

DEAN, A. M. 1983. A simple model of mutualism. *Am. Nat.* 121:409–417.

DEAN, R., L., E. ELLIS, R. W. WHITE, AND R. E. BEMERET. 1975. Nutrient removal by cattle from a short-grass prairie. *J. Appl. Ecol.* 12:25–29.

DECOURSEY, P. J. 1960. Phase control of activity in a rodent. *Cold Spring Harbor Symp. Quant. Biol.* 25:49–54.

DEEVEY, E. S. 1947. Life tables for natural population of animals. *Quart. Rev. Biol.* 22:283–314.

DELAUNE, R. D., AND W. H. PATRICK. 1980. Nitrogen and phosphorus cycling in a Gulf coast salt marsh. In U.S. Kennedy (ed.), *Estuarine Perspectives.* Academic Press, New York, pp. 143–151.

DELCOURT, H. R. 1987. The impact of prehistoric agriculture and land occupation on natural vegetation. *TREE* 2:39–44.

DELCOURT, H. R., AND P. A. DELCOURT. 1985. Quaternary palynology and vegetational history of the southeastern United States. In W. M. Bryant, Jr., and R. G. Holloway (eds.), *Pollen Records of Late Quaternary North American Sediments.* American Association of Stratigraphic Palynologists Foundation, Washington, DC, pp. 1–37.

DELCOURT, P. A., AND H. R. DELCOURT. 1981. Vegetation maps for eastern North America, 40,000 yr BP to present. In R. Romans (ed.), *Geobotany.* Plenum Press, New York, pp. 123–166.

———. 1984. Late Quaternary paleoclimates and biotic responses in eastern North America and the western North Atlantic Ocean. *Paleogeography, Paleoclimatology, Paleoecology* 48:263–284.

DELURY, R. E. 1930. Sunspots and living things. *Trans. 17th American Game Conf.* 1930:211–212.

DEMPSTER, J. P. 1975. *Animal Population Ecology.* London: Academic Press.

DERKSEN, D. V., AND D. H. WARD. 1993. *Life history and habitat needs of the black brant.* U.S. Fish and Wildlife Service Fish and Wildlife Leaflet 13.1.15.6 pp.

DERRICKSON, E. M. 1992. Comparative reproduction strategies of altricial and precocial eutherian mammals. *Functional Ecology* 6:57–65.

DESTEVEN, D. 1982. Seed production and seed mortality in a temperate witch-hazel forest shrub (*Hamamelis virginiana*). *J. Ecol.* 70:437–443.

DHYSTERHUIS, E. J., AND E. M. SCHMUTZ. 1947. Natural mulches or "litter of grasslands," with kinds and amounts on a southern prairie. *Ecology* 28:163–179.

DIAMOND, J. 1975. The island dilemma: Lessons of modern biogeographic studies for the design of natural preserves. *Biol. Cons.* 7:129–146.

———. 1986. Overview: Laboratory experiments, field experiments, and natural experiments. In J. Diamond and T. Case

(eds.), *Community Ecology.* Harper and Row, New York, pp. 3–22.

DIAMOND, J., AND R. M. MAY. 1976. Island biogeography and conservation: Strategy and limitations. In R. M. May (ed.), *Theoretical Ecology: Principles and Application.* Saunders, Philadelphia, pp. 163–186.

DIAMOND, J. D., AND T. J. CASE (EDS.). 1986. *Community Ecology.* Harper and Row, New York.

DICKERMAN, J. A., AND R. G. WETZEL. 1985. Clonal growth in *Typha latifolia:* Population dynamics and demography of the ramets. *J. Ecol.* 73:535–552.

DILGER, W. C. 1956. Hostile behavior and isolating mechanisms in the avian genera *Catharus* and *Hylocichla. Auk* 73:313–353.

———. 1960. Agonistic and social behavior of captive redpolls. *Wilson Bull.* 72:115–132.

DINERSTEIN, E., AND G. F. MCCRACKEN. 1990. Endangered greater one-horned rhinoceros carry high levels of genetic variation. *Cons. Biol.* 4:417–423.

DIX, R. L. 1960. The effects of burning on the mulch structure and species composition of grassland in western North Dakota. *Ecology* 41:49–56.

DOAK, D. F. 1991. The consequences of herbivory for dwarf fireweed: Different time scales, different morphological scales. *Ecology* 72:1408–1419.

DOBKIN, D. S. 1985. Heterogeneity of tropical floral microclimates and the response of hummingbird flower mites. *Ecology* 66:536–543.

DOBKIN, D. S., I. OLIVIERI, AND P. R. EHRLICH. 1987. Rainfall and the interaction of microclimate with larval resources in the population dynamics of checkerspot butterflies (*Euphydryas editha*) inhabiting serpentine grassland. *Oecologica* 71:161–166.

DOBSON, A. P., AND P. J. HUDSON. 1986. Parasites, disease, and structure of ecological communities. *TREE* 1:11–15.

DOBSON, F. S., AND J. O. MURIE. 1987. Interpretation of intraspecific life history patterns: Evidence from Columbian ground squirrels. *Am. Nat.* 129:382–397.

DOBSON, F. S., AND W. T. STONE. 1985. Multiple causes of dispersal. *Am. Nat.* 126:855–858.

DOBZHANSKY, T. 1951. *Genetics and the Origin of Species,* 3rd ed. Columbia University Press, New York.

DOBZHANSKY, T., AND S. WRIGHT. 1941. Genetics of natural populations. V. Relations between mutation rates and accumulation of lethals in populations of *Drosophila pseudoobscura. Genetics* 26:23–51.

DODD, R. H. 1973. Insect nutrition: Current developments and metabolic implications. *Ann. Rev. Entomol.* 18:381–420.

DODSON, S. I. 1970. Complementary feeding niches sustained by size-selective predation. *Limno. Oceanogr.* 15:131–137.

DOLBEER, R. A., AND W. R. CLARK. 1975. Population ecology of snowshoe hares in the central Rocky Mountains. *J. Wildl. Manage.* 39:535–549.

DORST, R. 1958. Uber die Ansiedlung von jung ins Binnerien verfrachteten Silbernowesn (*Larus argentatus*). *Vogelwarte* 17:169–173.

DOTY, M. S. 1956. Rocky intertidal surfaces. In J. W. Hedgpeth (ed.), *Treatise in Marine Ecology and Paleoecology: 1. Ecology.* Memoir 67, Geological Soc. Am., pp. 535–585.

DOWNS, A. A., AND W. E. MCQUILKIN. 1944. Seed production of southern Appalachian oaks. *J. For.* 42:913–920.

DOYLE, T. W. 1981. The role of disturbance on the gap dynamics of a montane rain forest: An application of a tropical forest succession model. In D. C. West, H. H. Shugart, and D. B. Botkin (eds.), *Forest Succession: Concepts and Application.* Springer-Verlag, New York, pp. 56–73.

DRAKE, J. A. 1983. Invasability in Lotka-Volterra interaction webs. In S. C. DeAngelis, W. M. Post, and G. Sugihara (eds.), *Current Trends in Food Web Theory.* ORNL 5983, Oak Ridge National Laboratory, Oak Ridge, TN, pp. 83–90.

———. 1990. Communities as assembled structures: Do rules govern pattern? *TREE* 5:159–164.

DREW, M. C. 1979. Root development and activities. In D. W. Goodall and R. A. Perry (eds.), *Arid Land Ecosystems: Structure, Functioning, and Management,* Vol. 1. Cambridge University Press, Cambridge, pp. 573–608.

DREW, M. C., M. B. JACKSON, AND S. GIFFORD. 1979. Ethylene-promoted adventitious roots and development of cortical air spaces (aerenchyma) in roots may be adaptive responses to flooding in *Zea mays. C. Planta* 147:83–88.

DRIFT, J. VAN DER. 1951. Analysis of the animal community in a beech forest floor. *Tijdschrift voor Entomologie* 94:1–168.

———. 1971. Production and decomposition of organic matter in an oak wood in the Netherlands. In P. Duvigneaud (ed.), *Productivity of Forest Ecosystems* (Proc. Brussels Symposium 1969). UNESCO, Paris, pp. 631–634.

DRIZO, R. 1984. Herbivory: A phytocentric overview. In R. Drizo and J. Sarukhan (eds.), *Perspectives in Plant Population Ecology.* Sinauer Associates, Sunderland, MA, pp. 141–165.

DRURY, W. H., JR., AND I. C. T. NISBET. 1973. Succession. *J. Arnold Arboretum* 54:331–368.

DUFFEY, E. 1974. *Grassland Ecology and Wildlife Management.* Chapman and Hall, London.

DUFFY, D. C., AND A. J. MEIER. 1992. Do Appalachian herbaceous understories ever recover from clearcutting? *Cons. Biol.* 6:196–201.

DUGDALE, R. C., AND J. J. GEORING. 1967. Uptake of new and regenerated forms of nitrogen in primary productivity. *Limnol. Oceanogr.* 12:196–206.

DUGGER, K. M., AND L. H. FREDRICKSON. 1992. *Life history and habitat needs of the wood duck.* U.S. Fish and Wildlife Service Fish and Wildlife Leaflet 13.1.6.8 pp.

DUNCAN, J. S., H. W. REED, R. MOSS, J. P. P. PHILLIPS, AND A. WATSON. 1978. Ticks, louping ill, and red grouse on moors in Speyside, Scotland. *J. Wildl. Manage.* 42:500–505.

DUSSOURD, D. E., AND R. F. DENNO. 1991. Deactivation of plant defense: Correspondence between insect behavior and secretory canal architecture. *Ecology* 72:1383–1396.

DUSSOURD, D. E., AND T. EISNER. 1987. Vein-cutting behaviour: Insect counterploy to the latex defense of plants. *Science* 237:898–901.

DUVIGNEAUD, P., AND S. DENAEYER-DESMET. 1967. Biomass productivity and mineral cycling in deciduous forests in Belgium. In *Symposium on Primary Productivity and Mineral Cycling in Natural Ecosystems.* University of Maine Press, Orono, pp. 167–186.

———. 1970. Biological cycling of minerals in a temperate deciduous forest. In D. Reichle (ed.), *Analysis of Temperate Forest Ecosystems.* Springer-Verlag, New York, pp. 109–115.

DWYER, P. D., J. KIKKAWA, AND G. J. INGRAM. 1979. Habitat relations of vertebrates in subtropical heathlands of coastal south-

eastern Queensland. In R. L. Specht (ed.), *Heathlands and Related Shrublands (Ecosystems of the World,* Vol. 9A). Elsevier, Amsterdam, pp. 281–300.

DYBA, H. S., AND M. LLOYD. 1962. Isolation by habitat in two synchronized species of periodical cicadas (Homoptera, Cicadidae, Magicicada). Ecology 43:444–459.

DYER, M. I. 1981. Mammalian epidermal growth factor promotes plant growth. *Proc. Nat. Acad. Sci.* 77:4836–4837.

DYER, M. I., C. L. TURNER, T. R. SEASTEDT. 1993. Herbivory and its consequences. *Ecol. Appl.* 3:10–16.

EADIE, J., AND H. G. LUMSDEN. 1985. Is nest parasitism always deleterious to goldeneyes? *Am. Nat.* 126:859–866.

EATON, J. S., G. E. LIKENS, AND F. H. BORMANN. 1973. Throughfall and stemflow chemistry in a northern hardwood forest. *J. Ecology* 61:495–508.

EDGERTON, P. J., AND B. R. MCCONNELL. 1976. *Diurnal temperature regimes of logged and unlogged mixed conifer stands on elk summer range.* U.S.D.A. Forest Service Res. Note PWN-277. Pac. Northwest For. and Range Exp. Stn., Portland, OR.

EDMONDS, J. 1992. Why understanding the natural sinks and sources of $CO_2$ is important: A policy analysis perspective. *Water, Air, and Soil Pollution* 64:11–21.

EDWARDS, C. A., AND G. W. HEATH. 1963. The role of soil animals in the breakdown of leaf materials. In J. Doeksen and J. van der Drift (eds.), *Soil Organisms.* North Holland, Amsterdam, pp. 76–84.

EGLER, F. E. 1953. Vegetation management for rights-of-way and roadsides. *Smithsonian Inst. Ann. Rept.* 1953, pp. 299–322.

———. 1954. Vegetation science concepts. 1. Initial floristic composition—a factor in old field vegetation development. *Vegetatio* 4:412–417.

EHLERINGER, J. R. 1980. Leaf morphology and reflectance in relation to water and temperature stress. In N. C. Turner and P. J. Kramer (eds.), *Adaptations of Plants to Water and Temperature Stress.* Wiley Interscience, New York, pp. 295–308.

———. 1983. Ecology and leaf physiology in North American desert plants. In E. Rodriguez, P. Heley, and I. Mehta (eds.), *Biology and Chemistry of Plant Trichomes.* Plenum Press, New York, pp. 113–132.

———. 1985. Annuals and perennials of warm deserts. In B. F. Chabot and H. A. Mooney (eds.), *Physiological Ecology of North American Plant Communities.* Chapman and Hall, New York, pp. 166–190.

EHLERINGER, J. R., AND I. FORSETH. 1980. Solar tracking by plants. *Science* 210:1094–1098.

EHLERINGER, J. R., AND R. K. MONSON. 1993. Evolutionary and ecological aspects of photosynthetic pathway variation. *Ann. Rev. Ecol. Syst.* 24:411–440.

EHLERINGER, J. R., R. F. SAGE, L. B. FLANAGAN, AND R. W. PEARCY. 1991. Climate change and the evolution of $C_4$ photosynthesis. *TREE* 6:95–99.

EHRENFELD, J. G. 1980. Understory response to canopy gaps of varying size in a mature oak forest. *Bull. Torrey Bot. Club* 107:29–41.

EHRLICH, P. R., AND P. H. RAVEN. 1965. Butterflies and plants: A study in coevolution. *Evolution* 18:586–608.

EHRLICH, P. R., AND A. H. EHRLICH. 1981. *Extinction: The Causes and Consequences of the Disappearance of Species.* Random House, New York.

EHRLICH, P. R., AND J. ROUGHGARDEN. 1987. *The Science of Ecology.* Macmillan, New York.

EISNER, T. 1970. Chemical defense against predation in arthropods. In E. Sondheimer and J. B. Simeone (eds.), *Chemical Ecology.* Academic Press, New York, pp. 157–217.

EISNER, T., AND J. MEINWALD. 1966. Defensive secretions of arthropods. *Science* 153:1341–1350.

ELLIS, D. H., J. C. BEDMARZ, D. G. SMITH, AND S. P. FLEMING. 1993. Social foraging in raptorial birds. *Bioscience* 43:14–20.

ELLISON, L. 1954. Subalpine vegetations of the Wasatch Plateau, Utah. *Ecol. Monogr.* 24:89–184.

ELNER, R. W., AND R. N. HUGHES. 1978. Energy maximization in the diet of the shore crab *Carcinus maenas. J. Anim. Ecol.* 47:103–116.

ELTON, C. S. 1927. *Animal Ecology.* Sidgwich & Jackson, London.

ELTON, C., AND M. NICHOLSON. 1942. The ten-year cycle in numbers of lynx in Canada. *J. Anim. Ecol.* 11:215–244.

EMANUEL, W. R., H. H. SHUGART, AND M. L. STEVENSON. 1985. Climate change and the broad scale distribution of terrestrial ecosystem complexes. *Climate Change* 7:29–43.

EMLEN, J. M. 1973. *Ecology: An Evolutionary Approach.* Addison Wesley, Reading, MA.

EMLEN, S. T. 1991. Evolution of cooperative breeding in birds and mammals. In J. R. Krebs and N. B. Davies (eds.), *Behavioural Ecology,* 3rd ed. Blackwell, Oxford, pp. 301–337.

EMLEN, S. T., AND L. W. ORING. 1977. Ecology, sexual selection, and the evolution of mating systems. *Science* 197:215–223.

ENGLE, L. G. 1960. *Yellow-poplar seedfall pattern.* Central States Forest. Expt. Stat. Note 143.

ENRIGHT, J. T. 1975. Orientation in time: Endogenous clocks. In O. Kinne (ed.), *Marine Ecology:* Vol. 2, *Physiological Mechanisms,* Part 2. pp. 917–944. Wiley, New York.

ERIKISSON, E. 1963. The yearly circulation of sulfur in nature. *J. Geophys. Res.* 68:4001–4008.

ERKERT, H. G. 1982. Ecological aspects of bat activity rhythms. In T. H. Kunz (ed.), *Ecology of Bats.* Plenum Press, New York, pp. 201–320.

ERKERT, H. G., AND S. KRACHT. 1978. Evidence for ecological adaptation of circadian systems: Circadian activity rhythms of neo-tropical bats and their re-entrainment after phase shifts of the Zeitgeber-LD. *Oecologica* 32:71–78.

ERLINGE, S., G. GORANSSON, G. HOGSTEDT, G. JANSSON, O. LIBERG, J. LORMAN, I. N. NILSSON, T. VON SCHANTZ, AND M. SYLVEN. 1984. Can vertebrate predators regulate their prey? *Am. Nat.* 12:125–133.

ERRINGTON, P. L. 1943. Analysis of mink predation upon muskrats in the north-central U.S. *Iowa Agr. Exp. Sta. Res. Bull.* 320:797–924.

———. 1945. Some contributions of a fifteen-year local study of the northern bobwhite to a knowledge of population phenomena. *Ecol. Monogr.* 15:1–34.

———. 1946. Predation and vertebrate populations. *Quart. Rev. Biol.* 21:144–177, 221–245.

———. 1963. *Muskrat Populations.* Iowa State University Press, Ames, IA.

ETHERINGTON, J. R. 1976. *Environmental and Plant Ecology.* Wiley, New York.

EVANS, F. C., AND S. A. CAIN. 1952. Preliminary studies on the vegetation of an old-field community in southeastern Michigan. *Contrib. Lab. Vert. Biol. U. Michigan* 51:1–17.

EVENARI, M. 1985. The desert environment. In M. Evenari, I. Noy-Meir, and D. W. Goodall (eds.), *Hot Deserts and Arid*

*Shrublands A (Ecosystems of the World,* Vol. 12A). Elsevier, Amsterdam, pp. 1–22.

———. 1986. *Hot Deserts and Arid Shrublands (Ecosystems of the World,* Vol. 12B). Elsevier, Amsterdam.

EWALD, P. W. 1983. Host-parasite relations, vector and the evolution of disease severity. *Ann Rev. Ecol. Syst.* 14:465–485.

EYRE, S. R. 1963. *Vegetation and Soils: A World Picture.* Aldine, Chicago.

FAEGRI, K., AND L. VAN DER PIJL. 1979. *The Principles of Pollination Ecology,* 3rd ed. Pergamon, Oxford.

FARNER, D. S. 1959. Photoperiodic control of annual gonadal cycles in birds. In R. B. Witherow (ed.), *Photoperiodism and Related Phenomena.* American Association for the Advancement of Science, Washington, DC, pp. 717–758.

———. 1964a. The photoperiodic control of reproductive cycles in birds. *Am. Sci.* 52:137–156.

———. 1964b. Time measurement in vertebrate photoperiodism. *Am. Nat.* 98:375–386.

FARNER, D. S., AND R. A. LEWIS. 1971. Photoperiodism and reproductive cycles in birds. In A. C. Giese (ed.), *Photophysiology,* Vol. 6. Academic Press, London, pp. 325–364.

FARNSWORTH, E. G., AND F. B. GOLLEY. 1974. *Fragile Ecosystems: Evaluation of Research and Applications in the Neotropics.* Springer-Verlag, New York.

FEHRIG, L., AND G. MERRIAM. 1985. Habitat patch connectivity and population survival. Ecology 66:1762–1768.

FEINSINGER, P. 1983. Coevolution and pollination. In D. Futuyma and M. Slatkin (eds.), *Coevolution.* Sinauer Associates, Sunderland, MA, pp. 282–310.

FELTON, P. M., AND H. W. LULL. 1963. Suburban hydrology can improve watershed conditions. *Public Works* 94:93–94.

FENCHEL, T. 1987. *Ecology—Potentials and Limitations. (Excellence in Ecology* 1.) Ecology Institute, Oldendorf/Luhe, Federal Republic of Germany.

———. 1988. Marine plankton food chains. *Ann. Rev. Ecol. Syst.* 19:19–38.

FENCHEL, T., AND T. H. BLACKBURN. 1979. *Bacteria and Mineral Cycling.* Academic Press, London.

FENCHEL, T., AND P. HARRISON. 1976. The significance of bacterial grazing and mineral cycling for the decomposition of particulate detritus. In J. M. Anderson and A. MacFayden (eds.), *The Role of Terrestrial and Aquatic Organisms in the Decomposition Process.* Blackwell, New York, pp. 285–321.

FENNER, F., AND F. N. RATCLIFFE. 1965. *Myxamatosis.* Cambridge University Press, Cambridge, England.

FERGUSEN, D. E., AND R. J. BOYD. 1988. Bracken fern inhibition of conifer regeneration in northern Idaho. USDA Forest Service Intermountain Research Station Res. Pap. INT-388, pp. 1–11.

FERRO, D. N. 1987. Insect pest outbreaks in agroecosystems. In P. Barbosa and J. C. Schultz (eds.), *Insect Outbreaks.* Academic Press, San Diego, pp. 195–216.

FINERTY, J. P. 1980. *The Population Ecology of Cycles in Small Mammals.* Yale University Press, New Haven, CN.

FISCHER, A. G. 1960. Latitudinal variation in organic diversity. *Evolution* 14:64–81.

FISCHER, E. A. 1981. Sexual allocation in a simultaneously hermaphroditic coral. *Am. Nat.* 117:64–82.

FISCHER, S. G., AND G. E. LIKENS. 1973. Energy flow in Bear Brook, New Hampshire: An integrative approach to stream ecosystem metabolism. *Ecol. Monogr.* 43:421–439.

FISH AND WILDLIFE SERVICE. 1980. *Habitat as a basis for environmental assessment.* Ecological Services Manual 101, USDI Fish and Wildlife Service, Division of Ecological Services, Washington, DC.

FISHER, R. A. 1930. *The genetical theory of natural selection.* Clarenden, Oxford.

FITTER, A. H. 1986. Acquisition and utilization of resources. In M. J. Crawley (ed.), *Plant Ecology.* Blackwell, Oxford, pp. 375–406.

FLANAGAN, P. W., AND F. L. BUNNELL. 1980. Microflora activities and decomposition. In J. Brown, P. C. Miller, L. L. Tieszen, and F. L. Bunnell (eds.), *An Arctic Ecosystem: The Coastal Tundra at Barrow, Alaska.* Dowden, Hutchinson, and Ross, Stroudsburg, PA, pp. 291–334.

FLINT, R. F. 1970. *Glacial and Quaternary Geology.* Wiley, New York.

FLOOK, D. R. 1970. Causes and implications of an observed sex differential in the survival of wapiti. *Can. Wildl. Serv. Rept. Ser.* No. 11.

FOLTZ, D. W., AND J. L. HOOGLAND. 1983. Genetic evidence of outbreeding in the black-tailed prairie dog (*Cynomys ludovicianus*). *Evolution* 37:273–281.

FOOSE, T. J. 1983. The relevance of captive populations to the conservation of biotic diversity. In C. M. Schonewald-Cox, S. M. Chambers, B. Macbryde, and W. L. Thomas (eds.), *Genetics and Conservation.* Benjamin/Cummings, Menlo Park, CA, pp. 374–401.

FOOSE, T. J., AND E. FOOSE. 1984. Demographic and genetic status and management. In B. Beck and C. Wemmer (eds.), *Pere David's Deer: The Biology and Conservation of an Extinct Species.* Noyes, Park Ridge, NJ.

FORBES, S. A. 1887. The lake as a microcosm. *Bull. Peoria Sci. Assoc* 87; reprinted 1925, *Ill. Natural History Surv. Bull.* 15:537–550.

FORCIER, L. K. 1975. Reproductive strategies and co-occurrence of climax tree species. *Science* 189:808–810.

FOREL, F. A. 1901. *Handbuch der Seenkunde, Allgemeine Limnologie.* J. Engelhorn, Stuttgart.

FORMAN, J. C., B. G. GARDINER, AND J. D. SHANKLIN. 1975. Large losses of total ozone in Antarctica reveal seasonal ClOx/NO$_x$ interaction. *Nature* 315:207–210.

FORMAN, R. T. T., AND M. GODRON. 1981. Patches and structural components for a landscape ecology. *Bioscience* 31:733–740.

———. 1986. *Landscape Ecology.* John Wiley, New York.

FORSETH, I. N., AND J. R. EHLERINGER. 1983. Ecophysiology of two solar tracking desert annuals. IV. Effects of leaf orientation on calculated water gain and water use efficiency. *Oecologica* 58:10–18.

FORTNEY, J. L. 1974. Interactions between yellow perch abundance, walleye predation, and survival of alternate prey in Oneida Lake, New York. *Trans. Am. Fish Soc.* 103:15–24.

FORTNEY, R. H. 1975. The vegetation of Canaan Valley: A taxonomic and ecological study. PhD thesis, West Virginia University, Morgantown.

FOWLER, C. W. 1981. Density dependence as related to life history strategy. *Ecology* 62:602–610.

FOWLER, N. L., AND J. ANTONOVICS. 1981. Small scale variability in the demography of transplants of two herbaceous species. *Ecology* 62:1450–1457.

FOWLER, T. D., AND C. W. SMITH. 1981. *Dynamics of Large Animal Populations.* Wiley, New York.

Fox, B. J. 1981. Niche parameters and species richness. *Ecology* 62:1415–1425.

Fox, F. M., and M. M. Caldwell. 1978. Competitive interaction in plant populations exposed to supplementary ultraviolet-B radiation. *Oecologica* 36:173–190.

Fox, J. F. 1977. Alternation and coexistence of tree species. *Am. Nat.* 111:69–89.

Fox, L. R. 1975a. Some demographic consequences of food shortage for the predator *Notoneta hoffmanni*. *Ecology* 56:868–880.

———. 1975b. Factors influencing cannibalism, a mechanism of population limitation in the predator *Notoneta hoffmanni*. *Ecology* 56:933–941.

———. 1975c. Cannibalism in natural populations. *Ann. Rev. Ecol. Syst.* 6:87–106.

———. 1981. Defense and dynamics in plant-herbivore systems. *Am. Zool.* 21:853–864.

Francis, W. J. 1970. The influence of weather on population fluctuations in California quail. *J. Wildl. Manage.* 34:249–266.

Frankel, O. H., and M. E. Soule. 1981. *Conservation and Evolution*. Cambridge University Press, Cambridge.

Franklin, I. R. 1980. Evolutionary change in small populations. In M. E. Soule and B. A. Wilcox (eds.), *Conservation Biology: An evolutionary-ecological approach*. Sinauer, Sunderland, MA, pp. 135–149.

Franklin, J. F., and C. T. Dyrness. 1973. *Natural Vegetation of Oregon and Washington*. U. S. D. A. Forest Service Gen. Tech. Rept. PNW 8. USDA Forest Service, Corvallis, Oregon.

Franklin, J. F., and M. A. Hemstrom. 1981. Aspects of succession in the coniferous forests of the Pacific Northwest. In D. C. West, H. H. Shugart, and D. B. Botkin (eds.), *Forest Succession: Concepts and Application*. New York: Springer Verlag, pp. 219–229.

Franklin, J. F., and R. H. Waring. 1979. Distinctive features of the northwestern coniferous forest: Development, structure, and function. In R. H. Waring (ed.), *Forests: Fresh Perspectives from Ecosystem Analysis*. Proc. 40th Ann. Biol. Coll., Oregon State University Press, Corvallis.

Franz, H. 1958. *Bodenzoologie als Grundlage der Bodenpflege*. Akademie-Verlag, Berlin.

Freeman, C. L., K. T. Harper, and E. L. Charnov. 1980. Sex change in plants: Old and new observations and new hypotheses. *Oecologica* 47:222–232.

Freemark, K. E., and H. G. Merriam. 1986. Importance of area and habitat heterogeneity to bird assemblages in temperate forest fragments. *Biol. Cons.* 36:115–141.

French, A. R. 1988. The patterns of mammalian hibernation. *Am. Sci.* 76:569–575.

French, C. E., L. C. McEwen, N. C. Magruder, R. H. Ingram, and R. W. Swift. 1955. *Nutritional requirements of white-tailed deer for growth and antler development*. Penn. State Univ. Agr. Exp. Sta. Bull. No. 600.

French, N. R., R. K. Steinhorst, and D. M. Swift. 1979. Grassland biome trophic pyramids. In N. R. French (ed.), *Perspectives in Grassland Ecology*. Springer Verlag, New York, pp. 59–87.

Fretwell, S. D., and H. L. Lucas. 1969. On territorial behavior and other factors influencing habitat distribution in birds. *Acta Biotheoretica* 19:16–36.

Friche, H., and S. Friche. 1977. Monogamy and sex change by aggressive dominance in coral reef fish. *Nature* 266:830–832.

Fritts, S. H. and L. D. Mech. 1981. Dynamics, movements, and feeding ecology of a newly-protected wolf population in northwestern Minnesota. *Wildl. Monogr.* 80.

Fryxell, J. M., and C. M. Doucet. 1993. Diet choice and functional response of beavers. *Ecology* 74:1297–1306.

Futuyma, D. J. 1983. Evolutionary interactions among herbivorous insects and plants. In D. J. Futuyma and M. Slaktin (eds.), *Coevolution*. Sinauer Associates, Sunderland, MA, pp. 207–231.

Gaines, M. M. and M. L. Johnson. 1987. Phenotypic and genotypic mechanisms for dispersal in *Microtus* populations and the role of dispersal in population regulation. In B. B. Chepko-Sade and Z. T. Halpen (eds.), *Mammalian Dispersal Patterns*. University of Chicago Press, Chicago, pp. 162–179.

Galli, A. E., C. F. Leck, and R. T. T. Forman. 1976. Avian distribution patterns in forest islands of different sizes in central New Jersey. *Auk* 93:356–364.

Gallun, R. L. 1977. The genetic basis of Hessian fly epidemics. *Ann. N.Y. Acad. Sci* 287:223–229.

Gates, D. M. 1965. Radiant energy: Its receipt and disposal. *Meteorol. Monogr.* 6:1–26.

———. 1966. Special distribution of solar radiation at the earth's surface. *Science* 151:523–528.

———. 1968. Energy exchange between organisms and environment. In W. P. Lowry (ed.), *Biometeorology*, Proc. 28th Ann. Biol. Colloq. Oregon State University Press, Corvallis.

———. 1972. *Man and His Environment: Climate*. Harper and Row, New York.

———. 1993. *Climate Change and Its Biological Consequences*. Sinauer Associates, Sunderland, MA.

Geiger, R. 1965. *Climate near the Ground*. Harvard University Press, Cambridge, MA.

Geis, A. D., R. I. Smith, and J. P. Rogers. 1971. *Black duck distribution, harvest characteristics, and survival*. U.S. Fish and Wildl. Serv. Spec. Sci. Rept., Wildl. No. 139.

Gerrodette, T. 1987. A power analysis for detecting trends. *Ecology* 68:1364–1372.

Getty, T. 1981. Competitive collusion: The preemption of competition during sequential establishment of territories. *Am. Nat.* 118:426–431.

Ghiselin, M. T. 1974. *The Economy of Nature and the Evolution of Sex*. University of California Press, Berkeley.

Gilbert, L. E. 1975. Ecological consequences of a coevolved mutualism between butterflies and plants. In L. E. Gilbert and P. H. Raven (eds.), *Coevolution of Animals and Plants*. University of Texas Press, Austin, pp. 210–240.

Gill, D. E. 1975. Spatial patterning of pines and oaks in the New Jersey pine barrens. *J. Ecol.* 63:291–298.

Gillon, Y. 1983. The invertebrates of the grass layer. In F. Bourliere (ed.), *Tropical Savannas (Ecosystems of the World*, Vol. 13). Elsevier, Amsterdam, pp. 289–311.

Gilpin, M. E. 1975. *Group Selection in Predator-Prey Communities*. Princeton University Press, Princeton, NJ.

Gimingham, C. H., S. B. Chapman and N. R. Webb. 1981. European healthlands. 365–419. In R. L. Specht (ed.), *Heathlands and Related Shrublands (Ecosystems of the World*, Vol. 9A). Elsevier, Amsterdam.

Gisborne, H. T. 1941. *How the wind blows in the forests of northern Idaho*. Northern Rocky Mt. Forest Range Expt. Sta.

Gist, C. S., and D. A. Crossley, Jr. 1975. A model of mineral-

element cycling for an invertebrate food web in a southeastern hardwood forest litter community. In F. Howell, J. B. Gentry, and M. H. Smith (eds.), *Mineral Cycling in Southeast Ecosystems*. National Technical Information Service, U.S. Dept. Commerce, pp. 84–106.

GITTLEMAN, J. L., AND S. D. THOMPSON. 1988. Energy allocation in mammalian reproduction. *Am. Zool.* 28:863–877.

GLEASON, H. A. 1917. The structure and development of the plant association. *Bull. Torrey Bot. Club* 53:7–26.

———. 1926. The individualistic concept of the plant association. *Bull. Torrey Bot. Club* 53:1–20.

———. 1927. Further views on the succession concept. *Ecology* 8:299–326.

GODFRAY, H. C. J., AND A. B. HARPER. 1990. The evolution of brood reduction by siblicide in birds. *J. Theor. Biol.* 145:163–175.

GOLDBERG, D. E., AND A. M. BARTON. 1992. Patterns and consequences of interspecific competition in natural communities: A review of field experiments with plants. *Am. Nat.* 139:771–801.

GOLLEY, F. B. 1960. Energy dynamics of a food chain of an old field community. *Ecol. Monog.* 30:187–206.

GOLLEY, F. B., AND H. LEITH. 1972. Basis of organic production in the tropics. In P. M. Golley and F. B. Golley (eds.), *Tropical Ecology with an Emphasis on Organic Production*. University of Georgia Press, Athens, GA, pp. 1–26.

GORDON, M. S. 1972. *Animal Physiology*, 2nd ed. Macmillan, New York.

GOSS-CUSTARD, J. D. 1977a. The energetics of prey selection by redshank *Tringa totanus (L)* and a preferred prey *Corophium volutator* (Pallas). *J. Anim. Ecol.* 46:21–35.

———. 1977b. The energetics of prey selection by redshank *Tringa totanus* (L) in relation to prey density. *J. Anim. Ecol.* 46:1–19.

GOSZ, J. R., E. LIKENS, AND F. H. BORMANN. 1976. Organic matter and nutrient dynamics of the forest and forest floor in the Hubbard Brook Forest. *Oecologica* 22:305–320.

GOULD, S. J. 1982. The meaning of punctuated equilibrium and its role in validating a hierarchial approach to macroevolution. In R. Milkman (ed.), *Perspectives in Evolution*. Sinauer, Sunderland, MA, pp. 83–104.

GOULD, S. J., AND E. S. VRBA. 1982. Exaption: A missing term in the science of form. *Paleobiology* 8:4–15.

GOWATY, P. A., AND W. C. BRIDGES. 1991. Nest box availability affects extra-pair fertilization and conspecific nest parasitism in eastern bluebirds *Sialia sialis. Anim. Behav.* 41:661–675.

GRANHALL, R., AND V. LID-TORSVIK. 1975. Nutrition fixation by bacteria and free-living blue-green algae in tundra ecosystem. In F. E. Wielgolaski (ed.), *Fennoscandian Tundra Ecosystems. Part I, Plants and Microorganisms*, Springer Verlag New York, pp. 305–315.

GRANT, P. R., AND B. R. GRANT. 1992. Demography and genetically effective sizes of two populations of Darwin's finches. *Ecology* 73:766–784.

GRANT, R., AND P. R. GRANT. 1987. Mate choice in Darwin's finches. *Biol. J. Linn. Soc.* 32:247–270.

GRENFELL, B. T. 1988. Gastrointestinal nematode parasites and the stability and productivity of intensive ruminal grazing systems. *Phil. Trans. Royal Soc. London, Br. Biol. Sci.* 321:541–563.

———. 1992. Parasitism and the dynamics of ungulate grazing systems. *Am. Nat.* 139:907–929.

GRIME, J. P. 1966. Shade avoidance and shade tolerance in flowering plants. In F. Bainbridge (ed.), *Light as an Ecological Factor.* Blackwell, Oxford, pp. 187–207.

———. 1977. Evidence for the existence of three primary strategies in plants and its relevance to ecological and evolutionary theory. *Am. Nat.* 111:1169–1194.

GRINNELL, J. 1917. The niche relationships of the California thrasher. *Auk* 34:427–433.

———. 1924. Geography and evolution. *Ecology* 5:225–229.

———. 1928. Presence and absence of animals. *U. Calif. Chronicle* 30:429–450.

GROSS, M. 1972. *Oceanography.* Prentice-Hall, Englewood Cliffs, N.J.

GUILFORD, T. 1988. The evolution of conspicuous coloration. *Am. Nat.* 131:S7–S21.

GUREVITCH, J. 1986. Competition and the local distribution of the grass *Stipa neomexicana. Ecology* 67:46–57.

GUREVITCH, J., L. L. MORROW, A. WALLACE, AND J. J. WALSH. 1992. Meta-analysis of competition in field experiments. *Am. Nat.* 140:539–572.

GUSTAFSSON, L., AND W. J. SUTERLAND. 1988. The costs of reproduction in the collared flycatcher *Ficedula albicollis. Nature* 335:813–815.

GUTHERY, F. S., AND R. L. BINGHAM. 1992. On Leopold's principle of edge. *Wildl. Soc. Bull.* 20:340–344.

HABERMANN, R. T., C. M. HERMAN, AND F. P. WILLIAMS. 1958. Distemper in raccoons and foxes suspected of having rabies. *J. Am. Vet. Med. Assoc.* 132:31–35.

HACKETT, C. 1965. Ecological aspects of the nutrition of *Deschampsia flexuosa (L):* The effects of Al, Ca, Fe, K, Mn, P, and pH on the growth of seedlings and established plants. *J. Ecol.* 53:315–333.

HADLEY, E. B., AND L. C. BLISS. 1964. Energy relationships of alpine plants on Mount Washington, New Hampshire. *Ecol. Monogr.* 34:331–357.

HADLEY, E. B., AND B. J. KIECKHEFER. 1963. Productivity of two prairie grasses in relation to fire frequency. *Ecology* 44:389–395.

HADLEY, J. L., AND W. K. SMITH. 1986. Wind effects on needles of timberline conifers: Seasonal influence on mortality. *Ecology* 67:12–19.

HAECKEL, E. 1869. Über Entwichelunge Gang 4. *Aufgabe de Zoologie Jemaische z.* 5:353–370.

HAINES, B. L., AND E. L. DUNN. 1985. Coastal marshes. In B. F. Chabot and H. A. Mooney (eds.), *Physiological Ecology of North American Plant Communities*. Chapman and Hall, New York, pp. 323–347.

HAIRSTON, N. G. 1969. On the relative abundance of species. *Ecology* 50:1091–1094.

———. 1954. The measurement of natural selection. *Proc. 9th Int. Congr. Genetics*, pp. 480–487.

HALDANE, J. B. S. 1932. *The Causes of Evolution.* Longman, Green, London.

———. 1954. The measurement of natural selection. *Proc. 9th Int. Congr. Genetics*, pp. 421–425.

HALL, C. A. S., C. J. CLEVELAND, AND R. KAUFMAN. 1986. *Energy and Resource Quality: The Ecology of the Economic Process.* Wiley, New York.

HALL, R. J., G. E. LIKENS, S. B. FIANCE, AND G. R. HENDREY. 1980. Experimental acidification of a stream in the Hubbard Brook Experimental Forest, New Hampshire. *Ecology* 61:976–989.

HALLE, F. R., A. A. OLDMEMAN, AND P. B. TOMLINSON. 1978. *Tropical Trees and Forests: An Architectural Analysis.* Springer-Verlag, New York.

HALLIDAY, T. R. 1978. Sexual selection and mate choice. In J. R. Krebs and N. D. Davies (eds.), *Behavioural Ecology: An Evolutionary Approach,* 1st ed. Blackwell, Oxford, England, pp. 180–213.

HALVERSON, H. G., AND J. L. SMITH. 1979. *Solar radiation as a forest management tool: A primer and application of principles.* USDA Forest Service Gen Tech. Rept. PSW-33.

HAMILTON, W. D. 1964. The genetical evolution of social behavior I, II. *J. Theoret. Biol.* 82:1–16, 17–52.

———. 1972. Altruism and related phenomena, mainly in insects. *Ann. Rev. Ecol. Syst.* 3:193–283.

HAMILTON, W. D., AND M. ZUK. 1982. Heritable true fitness and bright birds: A role for parasites? *Science* 218:384–387.

HAMILTON, W. J., III. 1959. Aggressive behavior in migrant pectoral sandpipers. *Condor* 61:161–179.

HANDEL, S. N. 1978. The competitive relationship of three woodland sedges and its bearing on the evolution of ant-dispersal of *Carex penduculata. Evolution* 32:151–163.

HANES, T. L. 1981. California chapparal. In F. di Castri and H. A. Mooney (eds.), *Mediterranean-type Shrublands: Origin and Structure.* Springer-Verlag, New York, pp. 139–174.

HANSKI, I. 1987. Populations of small mammal cycles—unless they don't. *TREE* 2:55–56.

HANSON, H. C. 1953. Vegetation types in northwestern Alaska and comparisons with communities in other arctic regions. *Ecology* 34:111–140.

HANSON, W. C. 1971. Seasonal patterns in native residents of three contrasting Alaskan villages. *Health Physics* 20:585–591.

HANSON, W. C., D. G. WATSON, AND R. W. PERKINS. 1967. Concentration and retention of fallout radionuclide in Alaska arctic ecosystems. In B. Aberg and F. P. Hungate (eds.), *Radioecological Concentration Processes.* Pergamon Press, London, pp. 233–245.

HANZAWA, F. M., A. J. BEATTIE, AND D. C. CULVAR. 1988. Directed dispersal: Demographic analysis of an ant-seed mutualism. *Am. Nat.* 131:1–13.

HARDIN, G. 1960. The competitive exclusion principle. *Science* 131:1292–1297.

HARDY, I. C. W., N. T. GRIFFITHS, AND H. C. J. GODFRAY. 1992. Clutch size in a parasitoid wasp: A manipulation experiment. *J. Anim. Ecol.* 61:121–129.

HARESTAD, A. S., AND E. L. BUNNELL. 1979. Home range and body weight—a reevaluation. *Ecology* 60:389–402.

HARLEY, J. L., AND S. E. SMITH. 1969. *The Biology of Mycorrhiza.* Academic Press, London.

HARMEL, D. E. 1983. Effects of genetics on antler quality and body size in white-tailed deer. In R. D. Brown (ed.), *Antler Development in Cervidae.* Caeser Kleberg Wildlife Research Institute, Kingsville, TX.

HARMON, M. E., J. F. FRANKLIN, F. S. SWANSON, ET AL. 1986. Ecology of coarse woody debris in temperate ecosystems. *Adv. Ecol. Res.* 15:133–302.

HARPER, J., AND A. D. BELL. 1979. The population dynamics of growth form in organisms with modular construction. In R. M. Anderson (ed.), *Population Dynamics, 20th Symposium, British Ecological Society.* Blackwell, Oxford, England, pp. 29–52.

HARPER, J. L. 1977. *Population Biology of Plants.* Academic Press, New York.

HARPER, J. S. 1982. After description. In E. I. Newman (ed.), *The Plant Community as a Working Mechanism.* Blackwell, Oxford, pp. 11–25.

HARRIS, G. G., A. W. EBLING, D. R. LAUR, AND R. J. ROWLEY. 1984. Community recovery after storm damage: A case of facilitation in primary succession. *Science* 224:1136–1138.

HARRIS, L. D. 1984. *The Fragmented Forest.* University of Chicago Press, Chicago.

———. 1988. Edge effects and the conservation of biological diversity. *Cons. Biol.* 2:212–215.

———. 1989. The faunal significance of fragmentation of southeastern bottomland forests. In D. D. Hook and L. Ross (eds.), *Proc. Symp. The Forested Wetlands of Southern United States.* USDA Forest Service Gen. Tech. Rept. SE-50. Southeastern Forest Experiment Station, Asheville, NC, pp. 126–134.

HARRIS, L. D., AND J. McELVEEN. 1981. *Effect of forest edge on North Florida breeding birds.* IMPAC Rept. 6. University of Florida School of Forest Resources and Conservation, Gainesville FL.

HARRIS, L. D., AND G. SILVA-LOPEZ. 1992. Forest fragmentation and the conservation of biological diversity. In P. L. Feidler and S. K. Jain (eds.), *Conservation Biology: The Theory and Practice of Nature Conservation, Preservation, and Management.* Chapman and Hall, New York, pp. 197–237.

HARRIS, W. F., P. SOLLINS, N. T. EDWARDS, B. E. DINGER, AND H. H. SHUGART. 1975. Analysis of carbon flow and productivity in a temperate deciduous forest ecosystem. In *Productivity of World Ecosystems.* National Academy of Science, Washington, DC. pp. 116–122.

HARRISON, J. L. 1962. Distribution of feeding habits among animals in a tropical rain forest. *J. Anim. Ecol.* 31:53–63.

HARRISON, R. G. 1980. Dispersal polymorphisms in insects. *Ann. Rev. Ecol. Syst.* 11:95–118.

HART, D. D. 1985. Causes and consequences of territoriality in a grazing stream insect. *Ecology* 66:404–414.

HARTENSTEIN, R. 1986. Earthworm biotechnology and biogeochemistry. *Adv. Ecol. Res.* 15:179–407.

HARTL, D. L. 1988. *A Primer of Population Genetics.* Sinauer Associates, Sunderland, MA.

HARTLEY, J. L., AND S. E. SMITH. 1983. *Mycorrhizal Symbiosis.* Academic Press, London.

HARTSHORN, G. S. 1978. Tree falls and tropical forest dynamics. In P. B. Tomlinson and M. H. Zimmerman, (eds.), *Tropical Trees as Living Systems.* Cambridge University Press, Cambridge, England, pp. 617–638.

HARVEY, P. H., AND P. J. GREENWOOD. 1978. Antipredator defense strategies: Some evolutionary problems. In J. R. Krebs and N. B. Davies (eds.), *Behavioural Ecology: An Evolutionary Approach,* 1st ed. Blackwell, Oxford, England, pp. 129–151.

HARVEY, P. H., AND R. M. ZAMMUTO. 1985. Patterns of mortality and age at first reproduction in natural populations of mammals. *Nature* 315:319–320.

HASLER, A. D. 1969. Cultural eutrophication is reversible. *Bioscience* 19:425–431.

HASSELL, M. P. 1966. Evaluation of parasite or predator response. *J. Anim. Ecol.* 35:65–75.

HASSELL, M. P., J. L. LAWTON, AND J. R. BEDDINGTON. 1976. The components of arthropod predation. *J. Anim. Ecol.* 45:135–164.

————. 1977. Sigmoid functional responses by invertebrate predators and parasitoids. *J. Anim. Ecol.* 46:249–262.

HASSELL, M. P., AND R. M. MAY. 1973. Stability in insect host-parasite models. *J. Anim. Ecol.* 42:693–726.

————. 1974. Aggregation in predators and insect parasites and its effect on stability. *J. Anim. Ecol.* 43:567–597.

HAWKINS, C. P., M. L. MURPHY, AND N. H. ANDERSON. 1982. Effects of canopy, substrate composition, and gradient on the structure of macroinvertebrate communities in the Cascade Range streams of Oregon. *Ecology* 63:1840–1856.

HAWTHORNE, W. R., AND P. B. CAVERS. 1978. Population dynamics of the perennial herbs *Plantago major* and *P. rugelii* Dcne. *J. Ecol.* 64:511–527.

HAYMON, R. M., AND K. C. MCDONALD. 1985. The geology of deep sea hot springs. *Am. Sci.* 73:441–449.

HEAL, O. W., H. E. JONES, AND J. B. WHITTAKER. 1975. Moore House, U.K. In T. Rosswall and O. W. Heal (eds.), *Structure and Function of Tundra Ecosystems.* Swedish Natural Science Institute, Stockholm. pp. 295–320.

HEARD, D. C. 1992. The effect of wolf predation and snow cover on muskox group size. *Am. Nat.* 139:190–204.

HEATH, J. E. 1965. Temperature regulation and diurnal activity in horned lizards. *Univ. Cal. Pub. Zool.* 64:97–136.

HEDRICK, P. W., AND P. S. MILLER. 1992. Conservation genetics: Techniques and fundamentals. *Ecol. Appl.* 2:30–46.

HEICHEL, G. H., AND N. C. TURNER. 1976. Phenology and leaf growth of defoliated hardwood trees. In J. F. Anderson and H. K. Karpus (eds.), *Perspectives in Forest Entomology.* Academic Press, New York, pp. 31–40.

HEINRICH, B. 1976. Heat exchange in relation to blood flow between thorax and abdomen in bumblebees. *J. Exp. Biol.* 64:567–585.

————. 1979. *Bumblebee Economics.* Harvard University Press, Cambridge.

HEINSELMAN, M. L., AND T. M. CASEY. 1981a. Fire and succession in the conifer forests of northern North America. In D. E. West, H. H. Shugart, and D. B. Botkin (eds.), *Forest Succession: Concepts and Applications.* Springer-Verlag, New York, pp. 374–405.

————. 1981b. Fire intensity and frequency as factors in the distribution of northern ecosystems. In *Fire Regimes and Ecosystem Properties,* U. S. For. Serv. Rept. WO-26, U. S. Department of Agriculture, Washington, DC, pp. 7–57.

HEITHAUS, E. R. 1981. Seed predation by rodents on three ant-dispersed plants. *Ecology* 63:136–145.

HEITHAUS, E. R., D. C. CULVER, AND A. J. BEATTIE. 1980. Models of some ant-plant mutualisms. *Am. Nat.* 116:347–361.

HENDERSON, G. S., AND W. H. HARRIS. 1975. An ecosystem approach to the characterization of the nitrogen cycle in a deciduous forest watershed. In B. Bernier and C. W. Wingel (eds.), *Forest Soils and Land Management.* La Presse de l'Université Laval, Quebec.

HENDERSON, L. J. 1913. *The Fitness of the Environment.* Macmillan, New York.

HENDRIKSSON, E., AND B. SIMU. 1971. Nitrogen fixation by lichens. *Oikos* 22:119–121.

HENNTTONEN, H. 1985. Predation causing extended low densities in microtine cycles: Further evidence from shrew dynamics. *Oikos* 45: 156–157.

HERRERA, C. M. 1982. Defense of ripe fruit from pests: Its significance in relation to plant disperser interactions. *Am. Nat.* 120:218–241.

————. 1985. Determinants of plant-animal coevolution: The case of mutualistic dispersal of seeds by vertebrates. *Oikos* 44:132–141.

HETT, J., AND O. L. LOUCKS. 1976. Age structure models of balsam fir and eastern hemlock. *J. Ecol.* 64:1029–1044.

HIBBS, D. E. 1982. Gap dynamics in a hemlock-hardwood forest. *Can. J. For. Res.* 12:522–527.

————. 1983. Forty years of forest succession in central New England. *Ecology* 64:1394–1401.

HICKS, R. R., JR., AND D. E. FOSBROKE. 1987. Stand vulnerability: Can gypsy moth damage be predicted? In S. Fosbroke and R. Hicks (eds.), *Proc. Cooperative Workshop. Coping with the Gypsy Moth in the New Frontier.* West Virginia Office of Publications, Morgantown, WV, pp. 73–80.

HIGGS, A. T. 1981. Island biogeography theory and nature reserve design. *J. Biogeogr.* 8:117–124.

HIGGS, A. T., AND M. B. USHER. 1980. Should nature reserves be large or small? *Nature* (London) 285:568–569.

HILBERT, D. W., AND W. C. OECHEL. 1987. Response of tussock tundra to elevated carbon dioxide regimes: Analysis of ecosystem $CO_2$ flux through nonlinear modeling. *Oecologica* 72:466–472.

HILDEN, O. 1965. Habitat selection in birds: A review. *Ann. Zool. Fennica,* 2:53–75.

HILL, E. P., III. 1972. Litter size in Alabama cottontails as influenced by soil fertility. *J. Wildl. Manage.* 36:1199–1209.

HILL, R. W. 1976. *Comparative Physiology of Animals: An Environmental Approach.* Harper & Row, New York.

HILL, R. W., AND G. A. WYSE. 1989. *Animal Physiology,* 2nd ed. Harper & Row, New York.

HILLBRICHT-ILKOWSKA, A. 1974. Secondary productivity in freshwaters, its value and efficiencies in plankton food chain. In *Proc. 1st Inter. Congr. Ecol.,* pp. 164–167.

HILLS, T. L. 1965. Savannas: A review of a major research problem in tropical geography. *Can. Geog.* 9:216–228.

HOBBIE, J., J. COLE, J. DUNGAN, R. A. HOUGHTON, AND B. PETERSON. 1984. Role of biota in global $CO_2$ balance: The controversy. *Bioscience* 34:492–498.

HOCKING, B. 1975. Ant-plant mutualism: Evolution and energy. In L. E. Gilbert and P. H. Raven (eds.), *Coevolution of Animals and Plants.* University of Texas Press, Austin, pp. 78–90.

HOESE, H. D. 1960. Biotic changes in a bay associated with the end of a drought. *Limnol. Oceanogr.* 5:326–336.

HOFMANN, D. J., AND T. DESHLER. 1991. Evidence from balloon measurements for chemical depletion of stratospheric ozone in the Arctic winter of 1989–1990. *Nature* 349:300–305.

HOFMANN, R. R. 1989. Evolutionary steps of ecological adaptation and diversification of ruminants: A comparative view of their digestive system. *Oecologica* 78:443–457.

HOLDRIDGE, L. R. 1947. Determination of wild plant formations from simple climatic data. *Science* 105:367–368.

————. 1967. Determination of world plant formation from simple climatic data. *Science* 130:572.

HOLDRIDGE, L. R., W. C. GRENKE, W. H. HATHEWAY, T. LIANG, AND J. A. TOSI, JR. 1971. *Forest Environments in Tropical Life Zones: A Pilot Study.* Pergamon Press, New York.

HOLDRIDGE, L. R., AND HOLLING, C. S. 1980. Forest insects, forest fires, and resilience. In H. Mooney, J. M. Boninicksen, N. L. Christensen, J. E. Latan, and W. A. Reiners (eds.), *Fire Regimes and Ecosystem Properties.* USDA For. Serv. Gen. Tech Rept. 20–26.

HOLLING, C. S. 1959. The components of predation as revealed by

a study of small mammal predation of the European pine sawfly. *Can. Entomol.* 91:293–320.

———. 1965. The functional response of predators to prey density and its role in mimicry and population regulation. *Mem. Entomol. Soc. Can.* No. 45.

———. 1966. The functional response of invertebrate predators to prey density. *Mem. Entomol. Soc. Can.* No. 48.

HOLLINGER, D. Y. 1986. Herbivory and the cycling of nitrogen and phosphorus in isolated California oak trees. *Oecologica* 70:291–297.

HOLMES, J. 1983. Evolutionary relationships between parasitic helminths and their hosts. In D. J. Futuyma and M. Slatkin (eds.), *Coevolution.* Sinauer Associates, Sunderland, MA.

HOLMES, R. T., AND S. K. ROBINSON. 1981. Tree species preferences of foraging insectivorous birds in a northern hardwoods forest. *Oecologica* 48:31–35.

HOLMES, W. G. 1984. Sibling recognition in thirteen-lined ground squirrels: Effects of genetic relatedness, rearing association, and olfaction. *Behav. Ecol. Sociobiol.* 14:225–233.

HOLMES, W. T., AND P. W. SHERMAN. 1982. The ontogeny of kin recognition in two species of ground squirrel. *Am. Zool.* 22:491–517.

———. 1983. Kin recognition in animals. *Am. Sci.* 7:46–55.

HOOGLAND, J. C. 1982. Prairie dogs avoid extreme inbreeding. *Science* 215:1639–1641.

———. 1985. Infanticide in prairie dogs: Lactating females kill offspring of close kin. *Science* 230:1037–1040.

HOOK, D. D. 1984. Adaptation to flooding with fresh water. In T. Kozlowski (ed.), *Flooding and Plant Growth.* Academic Press, Orlando, FL, pp. 265–294.

HOPKINSON, C. S., AND J. P. SCHUBAUER. 1984. Static and dynamic aspects of nitrogen cycling in the salt marsh graminoid *Spartina alternifolia. Ecology* 65:961–969.

HORN, H. S. 1975. Markovian properties of forest succession. In M. L. Cody and J. Diamond (eds.), *Ecology and Evolution of Communities.* Harvard University Press, Cambridge, pp. 196–211.

———. 1981. Some causes of variety of patterns of secondary succession. In D. E. West, H. H. Shugart, and D. B. Botkin (eds.), *Forest Succession: Concepts and Applications.* Springer-Verlag, New York, pp. 24–35.

HORN, H. S., H. H. SHUGART, AND D. URBAN. 1989. Simulators as models of forced dynamics. In J. Roughgarden, R. May, and S. Levin (eds.), *Perspectives in Ecological Theory.* Princeton University Press, Princeton, NJ, pp. 256–267.

HORNBECK, J. W., G. E. LIKENS, AND J. S. EATON. 1976. Seasonal patterns in acidity of precipitation and implications for forest-stream ecosystems. In L. S. Dochinger and T. A. Seligar (eds.), *Proc. 1st International Symp. on Acid Precipitation and the Forest Ecosystem.* USDA For. Serv. Gen. Tech. Rept. NE-23.

HORNBECK, J. W., C. W. MARTIN, R. S. PIERCE, F. H. BORMANN, G. E. LIKENS, AND J. S. EATON. 1987. *The northern hardwood forest ecosystem: Ten years of recovery from clearcutting.* U.S.D.A. For. Serv. NERP-576.

HORSLEY, S. B. 1977. Allelopathic inhibition of black cherry by ferns, grass, goldenrod, and aster. *Can. J. For. Res.* 7:205–216.

HOUGH, A. F., AND R. D. FORBES. 1943. The ecology and silvics of forests in the high plateaus of Pennsylvania. *Ecol. Monogr.* 13:299–320.

HOUGHTON, R. A., J. E. HOBBIE, J. M. MELILLO, B. MORE, B. J.

PETERSON, G. R. SHAVER, AND G. M. WOODWELL. 1983. Changes in the carbon cycle in terrestrial biota and soils between 1860 and 1980: A net release of $CO_2$ to the atmosphere. *Ecol. Monogr.* 53:235–262.

HOUSTON, D. B. 1982. *The Northern Yellowstone Elk: Ecology and Management.* Macmillan, New York.

HOWARD, D. V. 1974. Urban robins: A population study. In J. Noyes and D. R. Progulske (eds.), *Wildlife in an Urbanizing Environment.* Massachussetts Cooperative Extension Service, Amherst, MA, pp. 67–75.

HOWARD, R. D. 1978a. The evolution of mating strategies in bullfrogs, *Rana catesbeiana. Evolution* 32:850–871.

———. 1978b. The influence of male defended oviposition sites on early embryo mortality in bullfrogs. *Ecology* 59:789–798.

———. 1979. Estimating reproductive success in natural populations. *Am. Nat.* 114:221–231.

HOWARD, W. E. 1960. Innate and environmental dispersal of individual vertebrates. *Amer. Mid. Natur.* 63:152–161.

HOWARD-WILLIAMS, C. 1970. The ecology of *Becium homblei* in Central Africa with special reference to metaliferous soils. *J. Ecol.* 58:745–763.

HOWARTH, R. W., AND A. GIBLIN. 1983. Sulfate reduction in the salt marshes at Sapelo Island, Georgia. *Limno. Oceanogr.* 28:70–82.

HOWARTH, R. W., AND J. M. TEAL. 1979. Sulfate reduction in a New England salt marsh. *Limno. Oceanogr.* 24: 999–1013.

HOWE, H. F. 1980. Monkey dispersal and waste of a neotropical fruit. *Ecology* 61:944–959.

———. 1985. Gomphothere fruits: A critique. *Am. Nat.* 125:853–865.

———. 1986. Seed dispersal by fruit-eating birds and mammals. In D. R. Murra (ed.), *Seed Dispersal.* Academic Press, Sydney, pp. 123–190.

HOWE, H. F., AND J. SMALLWOOD. 1982. Ecology of seed dispersal. *Ann. Rev. Ecol. Syst.* 13:201–238.

HOWE, H. F., AND L. C. WESTLEY. 1988. *Ecological Relationships of Plants and Animals.* Oxford University Press, New York.

HOWES, B. L., J. W. H. DACEY, AND J. M. TEAL. 1985. Annual carbon mineralization and belowground production of *Spartina alterniflora* in New England. *Ecology* 66:595–605.

HUBBARD, S. F., AND R. M. COOK. 1978. Optimal foraging by parasitoid wasps. *J. Anim. Ecol.* 47:593–604.

HUDSON, H. J., AND J. WEBSTER. 1958. Succession of fungi on decaying stems of *Agropyron repens. Brit. Mycol. Soc. Trans.* 41:165–177.

HUDSON, P. J. 1986. The effect of a parasitic nematode on the breeding production of red grouse. *J. Anim. Ecol.* 55:85–92.

HUDSON, P. J., AND A. P. DOBSON. 1991. The direct and indirect effects of the caecal nematode *Trichostrongylus tenuis* on red grouse. In J. E. Loye and M. Zuk (eds.), *Bird-Parasite Interactions: Ecology, Evolution, and Behavior.* Oxford: Oxford University Press, pp. 49–68.

HUDSON, P. J., A. P. DOBSON, AND D. NEWBORN. 1985. Cyclic and noncyclic populations of red grouse: A role for parasitism? In D. Rollinson and R. M. Anderson (eds.), *Ecology of Host–Parasite Interactions.* Academic Press, New York, pp. 78–89.

HUDSON, J. W. 1973. Evolution of thermal regulation. In G. W. Whittow (ed.), *Comparative Physiology of Thermoregulation,* Vol. 3. Academic Press, New York, pp. 97–165.

HUEY, R. B. 1991. Physiological consequences of habitat selection. *Am. Nat.* 137:S91–S115.

HUEY, R. B., C. R. PETERSON, S. J. ARNOLD, AND W. P. PORTER. 1989.

Hot rocks and not-so-hot rocks: Retreat site selection by garter snakes and its thermal consequences. *Ecology* 70:931–944.

HUFFAKER, C. B. 1958. Experimental studies on predation: Dispersion factors and predator-prey oscillations. *Hilgardia* 27:343–383.

HUMPHREY, R. R. 1958. The desert grassland: A history of vegetational changes and an analysis of causes. *Bot. Rev.* 24:193–252.

HUNGATE, R. E. 1975. The rumen microbial ecosystem. *Ann. Rev. Ecol. Syst.* 6:39–66.

HUNN, J. B., L. CLEVELAND, AND E. E. LITTLE. 1987. Influence of pH and aluminum on developing brook trout in low calcium water. *Env. Pollution* 43:63–73.

HUNTER, G. W., III, AND W. S. HUNTER. 1934. *Further studies on bird and fish parasites.* Supp. 24th Ann. Rept. N. Y. State Dept. Cons., No. 9, Rept. Biol. Surv. Mohawk-Hudson Watershed, pp. 267–283.

HUNTER, J. E., F. L. SCHMIDT, AND G. B. JACKSON. 1982. *Meta-analysis: Cumulating Research Findings across Studies.* Sage, Beverly Hills, CA.

HUNTLEY, B. 1991. How plants respond to climate change: Migration rates, individualism, and the consequences for plant communities. *Am. Bot.* 67 (Supp. 1):15–22.

HUNTLEY, B. J., AND B. H. WALKER (EDS.). 1982. *Ecology of Tropical Savannas.* Springer-Verlag, New York.

HUNTLEY, M., AND R. INOUYE. 1988. Pocket gophers in ecosystems: Patterns and mechanism. *Bioscience* 38:786–793.

HURLBERT, S. H. 1971. The nonconcept of species diversity: A critique and alternative parameters. *Ecology* 52:577–586.

———. 1984. Pseudoreplication and the design of ecological field experiments. *Ecol. Monogr.* 54:187–211.

HUSTON, M. 1979. A general hypothesis of species diversity. *Am. Nat.* 113:81–101.

———. 1985. Patterns of species diversity on coral reefs. *Ann. Rev. Ecol. Syst.* 16:149–177.

———. 1994. *Biological Diversity: The Coexistence of Species on Changing Landscape.* New York, Cambridge University Press.

HUSTON, M., AND T. SMITH. 1987. Plant succession: Life history and competition. *Am. Nat.* 130:168–198.

HUTCHINS, H. E., AND R. M. LANNER. 1982. The central role of Clark's nutcracker in the dispersal and establishment of whitebark pine. *Oecologica* 55:192–201.

HUTCHINSON, G. E. 1957. *A Treatise on Limnology,* Vol. 1. *Geography, Physics, Chemistry.* Wiley, New York.

———. 1959. Homage to Santa Rosalia, or why are there so many kinds of animals? *Am. Nat.* 93:145–159.

———. 1969. Eutrophication, past and present. In (ed.), *Eutrophication: Causes, Consequences, Correctives.* National Academy of Science, Washington DC, pp. 12–26.

———. 1978. *An Introduction to Population Ecology.* Yale University Press, New Haven, CT.

HUTTO, R. L. 1985. Habitat selection by nonbreeding migratory land birds. In M. Cody (ed.), *Habitat Selection in Birds.* Academic Press, Orlando, FL, pp. 455–476.

HUXLEY, J. S. 1934. A natural experiment on the territorial instinct. *British Birds* 27:270–277.

HYDER, D. N. 1969. The impact of domestic animals on the structure and function of grassland ecosystems. In R. L. Dix and R. G. Biedleman (eds.), *The Grassland Ecosystem: A Preliminary Synthesis.* Col. State U. Range Science Dept., Sci. Ser. No. 2, Fort Collins, pp. 243–260.

HYNES, H. 1970. *Biology of Running Water.* University of Toronto Press, Ontario.

IKUSIMA, I. 1965. Ecological studies on the productivity of aquatic plant communities: Measurement of photosynthetic activity. *Bot. Mag. Tokyo* 78:202–211.

IMS, R. A. 1990. On the adaptive value of reproductive synchrony as a predator swamping strategy. *Am. Nat.* 136:485–498.

INGER, R. F. (ED.). 1972. *Man in the Living Environment.* University of Wisconsin Press, Madison.

INGHAM, R. E., J. A. TROFYMOW, E. H. INGHAM, AND D. C. COLEMAN. 1985. Interaction of bacteria, fungi, and their nematode grazers: Effects on nutrient cycling and plant growth. *Ecol. Monogr.* 55:119–140.

INOUYE, R. S., G. S. BYERS, AND J. H. BROWN. 1980. Effects of predation and competition on survivorship, fecundity, and community structure of desert animals. *Ecology* 61:1344–1351.

IRVING, L. 1972. *Arctic Life of Birds and Mammals.* Springer-Verlag, New York.

JACKSON, J. K., AND S. G. FISCHER. 1986. Secondary production, emergence, and export of aquatic insects of a Sonoran desert stream. *Ecology* 67:629–638.

JACKSON, M. B. 1982. Ethylene as a growth promoting hormone under flooded conditions. In P. F. Waring (ed.), *Plant Growth Substances.* Academic Press, London, pp. 291–301.

———. 1985. Ethylene and responses of plants to soil waterlogging and submergence. *Ann. Rev. Plant Physiol.* 36:145–174.

JACKSON, M. B., AND M. C. DREW. 1984. Effects of flooding on growth and metabolism of herbaceous plants. In T. T. Kozlowski (ed.), *Flooding and Plant Growth.* Academic Press, Orlando, FL, pp. 47–128.

JACKSON, R. R. 1992. Eight-legged tricksters. *Bioscience* 42:590–598.

JACKSON, S. T., R. P. FUTUYMA, AND D. A. WILCOX. 1988. A paleoecological test of a classical hydrosere in the Lake Michigan dunes. *Ecology* 69:928–936.

JACOBSON, H. A., R. L. KIRKPATRICK, AND B. S. McGINNES. 1978. Disease and physiologic characteristics of two cottontail populations in Virginia. *Wildl. Monogr.* 60. 53 pp.

JAENIKE, J. 1992. Mycophagous *Drosphila* and their nematode parasites. *Am. Nat.* 139:893–906.

JAENIKE, J., AND A. C. JAMES. 1991. Aggregation and the coexistence of mycophagous *Drosphila.* *J. Anim. Ecol.* 60:913–928.

JAMES, S. W. 1991. Soil, nitrogen, phosphorus, and organic matter processing by earthworms in tallgrass prairie. *Ecology* 72:2101–2109.

JAMES, F. 1971. Ordination of habitat relationships among birds. *Wilson Bull.* 83:215–236.

JANZEN, D. H. 1966. Coevolution of mutualism between ants and acacias in Central America. *Evolution* 20:249–275.

———. 1967. Synchronization of sexual reproduction of trees within the dry season in Central America. *Evolution* 21:620–637.

———. 1968. Host plants as islands in evolutionary and contemporary time. *Am. Nat.* 102:592–595.

———. 1971. Seed predation by animals. *Ann. Rev. Ecol. Syst.* 2:465–492.

———. 1976. Why bamboos wait so long to flower. *Ann. Rev. Ecol. Syst.* 7:347–391.

———. 1979. How to be a fig. *Ann. Rev. Ecol. Syst.* 10:13–51.

———. 1980. When is it coevolution? *Evolution* 34:611–612.

————. 1986. Chihuahuan desert nopaleras: Defaunated big mammal vegetation. *Ann. Rev. Ecol. Syst.* 17:595–636.

JARVINEN, A. 1985. Predation causing extended low densities in microtine cycles: Implications from predation on hole-nesting passerines. *Oikos* 45:157–158.

JARVINEN, O. 1982. Conservation of endangered plant populations: Single large or several small reserves. *Oikos* 38:301–307.

————. 1984. Dismemberment of facts: A reply to Willis on subdivision of reserves. *Oikos* 42:402–403.

JARVIS, R. L., AND M. F. PASSMORE. 1992. *Ecology of band-tailed pigeons in Oregon.* U.S. Fish and Wildlife Service Biological Report 6. 38 pp.

JEFFERIES, R. L. 1972. Aspects of salt marsh ecology with particular reference to inorganic plant nutrients. In R. S. K. Barnes and J. Green (eds.), *The Estuarine Environment.* Applied Science, London, pp. 61–85.

JEFFERS, J. N. R. 1988. *Practitioner's Handbook on the Modelling of Dynamic Change in Ecosystems* (Scope 34). Wiley, New York.

JENNY, H. 1941. *Factors of Soil Formation: A System of Quantitative Pedology.* McGraw-Hill, New York.

————. 1980. *The Soil Resource.* Springer-Verlag, New York.

JOHANNES, R. E. 1968. Nutrition regeneration in lakes and oceans. In M. R. Droop and E. J. Ferguson (eds.), *Advances in the Microbiology of the Sea,* Vol. 1. Academic Press, New York, pp. 203–213.

JOHN-ALDEN, H. A., P. J. MORIN, AND S. LAWLER. 1988. Thermal physiology, phenology, and distribution of tree frogs. *Am. Nat.* 132:506–520.

JOHNSGARD, P. A. 1965. *Handbook of Waterfowl Behavior.* Cornell University Press, Ithaca, NY.

JOHNSON, A. H. 1983. Red spruce decline in the northeastern US: Hypotheses regarding the role of acid rain. *J. Air Pollution Control Assoc.* 33:1049–1054.

JOHNSON, P. L., AND W. O. BILLINGS. 1962. The alpine vegetation of the Beartooth Plateau in relation to cryopedogenic processes and patterns. *Ecol. Monogr.* 32:105–135.

JOHNSON, A. H., AND T. G. SICCAMA. 1983. Acid deposition and forest decline. *Environ. Sci. Technol.* 17:249A–304A.

JOHNSON, D. H., AND A. B. SARGENT. 1977. *Impact of red fox predation on the sex ratio of prairie mallards.* U.S. Fish and Wildlife Service Wildl. Res. Rept. 6.

JOHNSON, D. W., D. W. COLE, S. BLEDSOE, K. CROMACK, R. L. EDMONDS, S. P. BESSEL, C. C. GRIER, B. N. RICHARDS, AND K. A. VOGT. 1982. Nutrient cycling in forests of the Pacific Northwest. In R. L. Edmonds (ed.), *Analysis of Coniferous Forest Ecosystems in the Western United States.* Hutchinson, Ross, Stroudsburg, PA, pp. 186–232.

JOHNSON, D. W., D. O. RICHTER, H. MIEGROET, AND D. W. COLE. 1983. Contribution of acid deposition and natural processes to cation leaching from forest soils: A review. *J. Air Pollution Control Assoc.* 33:1036–1041.

JOHNSON, J. A. 1984. Small woodlot management by single-tree selection: 21-year results. *Northern J. Appl. For.* 1:69–71.

JOHNSON, M. C., AND M. S. GAINES. 1987. The selective basis for dispersal of the prairie vole *Microtus ochrogaster. Ecology* 68:684–694.

JOHNSON, P. L., AND W. D. BILLINGS. 1962. The alpine vegetation of the Beartooth Plateau in relation to cryopedogenic processes and patterns. *Ecol. Monogr.* 32:105–135.

JOHNSON, R. W., AND J. C. TOTHILL. 1985. Definition and broad geographic outline of savanna lands. In J. C. Tothill and J. J. Mott (eds.), *Ecology and Management of the World's Savannas.* Commonwealth Agricultural Bureau, Australian Academy of Science, Canberra, pp. 1–13.

JOHNSTON, J. W. 1936. The macrofauna of soils as affected by certain coniferous and hardwood types in the Harvard forest. PhD Dissertation, Harvard University Library, Cambridge, MA.

JOHNSTON, R. F. 1956a. Predation by short-eared owls in a *Salicornia* salt marsh. *Wilson Bull.* 68:91–102.

————. 1956b. Population structure in salt marsh song sparrows. 2. Density, age structure and maintenance. *Condor* 58:254–272.

JONES, C. G. 1983. Microorganisms as mediators of resource exploitation. In P. W. Price, C. N. Slobodchikoff, and W. S. Gand (eds.), *A New Ecology: Novel Approaches to Interactive Systems.* Wiley, New York, pp. 53–99.

JONES, H. G. 1983. *Plants and Microclimate.* Cambridge University Press, Cambridge England.

————. 1992. *Plants and Microclimate,* 2nd ed. Cambridge University Press, Cambridge, England.

JONES, R. L., AND H. C. HANSON. 1985 *Biogeochemistry, Mineral Licks, and Geophagy of North American Ungulates.* Iowa State University Press, Ames.

JONES, R. L., AND H. P. WEEKS. 1985. Ca, Mg, and P in the annual diet of deer in south-central Indiana. *J. Wildl. Manage.* 49:129–133.

JORDAN, C. F. 1981. Do ecosystems exist? *Am. Nat.* 106:237–253.

————. 1982a. Amazon rain forest. *Am. Sci.* 70:394–401.

————. 1982b. The nutrient balance of an Amazonian rain forest. *Ecology* 63:647–654.

————. 1985. *Nutrient Cycling in Tropical Forest Ecosystems: Principles and Their Application in Management and Conservation.* Wiley, New York.

————. 1986. Local effects of tropical deforestation. In M. Soule (ed.), *Conservation Biology: The Science of Scarcity and Diversity.* Sinauer, Sunderland, MA: pp. 410–446.

JORDAN, C. F., F. GOLLEY, J. D. HALL, AND J. HALL. 1979. Nutrient scavenging of rainfall by the canopy of an Amazonian rain forest. *Biotropica* 12:61–66.

JORDAN, C. F., J. R. KLINE, AND D. S. SASSIER. 1972. Relative stability of mineral cycles in forest ecosystems. *Am. Nat.* 106:237–253.

JORDAN, W. R., III, M. E. GILPIN, AND J. O. ABER. 1987. *Restoration Ecology: A Synthetic Approach to Ecological Research.* Cambridge University Press, Cambridge, England.

JORDEN, T. E., AND I. VALIELA. 1982. The nitrogen cycle of the ribbed mussel *Geukinsea demissa* and its significance in nitrogen flow in a New England salt marsh. *Limno. Oceanogr.* 27:75–90.

JORGENSEN, J. R., AND C. G. WELLS. 1986. *Forester's Primer in Nutrient Cycling.* U.S.D.A. For. Serv. Gen. Tech. Rept. SE-37.E.

JUDAY, C. 1940. The annual energy budget of an inland lake. *Ecology* 21:438–450.

KABAT, C., AND D. R. THOMPSON. 1963. *Wisconsin Quail, 1834–1962: Population Dynamics and Habitat Management.* Tech. Bull. No. 30. Wisconsin Conserv. Dept., Madison.

KAJAH, Z., G. BRETSCHKO, F. SCHIEMER, AND C. LEVEGUE. 1980. Zoobenthos. In E. D. LeCren and R. H. Lowe-McConnell

(eds.), *The Functioning of Freshwater Ecosystems* (International Biological Programme No. 22). Cambridge University Press, Cambridge, England, pp. 285–307.

KAKENOSKY, G. 1972. Wolf predation on wintering deer in east central Ontario. *J. Wildl. Manage.,* 36:357–367.

KALININ, G. R., AND V. D. BYKOV. 1969. The world's water resources, present and future. *Impact of Science on Society,* 19:135–150.

KALLIO, S., AND P. KALLIO. 1975. Nitrogen fixation in lichens at Kevo, North Finland. In F. E. Wiegolaski (ed.), *Fennoscandian Tundra Ecosystems: Part 1, Plants and Microorganisms.* Springer Verlag, New York, pp. 292–304.

KAMIL, A. C., J. R. KREBS, AND H. R. PULLIAN (EDS.). 1987. *Foraging Behavior.* Plenum, New York.

KAMIL, A. C., AND T. D. SARGENT (EDS.). 1981. *Foraging Behavior: Ecological, Ethological, and Psychological Approaches.* Garland STMP Press, New York.

KAMMER, A. E. 1981. Physiological mechanisms of thermoregulation. In O. L. Lange, P. S. Nobel, C. B. Osmond, and H. Zeigler (eds.), *Physiological Plant Ecology. I. Responses to the Physical Environment.* Springer-Verlag, Berlin, pp. 115–168.

KANWISHER, J. W. 1955. Freezing in intertidal animals. *Biol. Bull. Woods Hole* 109:56–63.

———. 1959. Histological metabolism of frozen intertidal animals. *Biol. Bull. Woods Hole* 116:285–264.

KAPPER, L. 1981. Ecological significance of resistance to high temperature. In O. Lange, P. S. Nobel, C. B. Osmond, and H. Zeigler (eds.), *Physiological Plant Ecology. I. Responses to the Environment.* Springer-Verlag, New York, pp. 439–474

KAREIVA, P. M., J. G. KINGSOLVER, AND R. B. HUEY. 1993. Introduction. In P. M. Kareiva, J. G. Kingsolver, and R. B. Huey (eds.), *Biotic Interactions and Global Change.* Sinauer Associates, Sunderland MA, pp. 1–6.

KARR, J. R. 1975. Production and energy pathways, and community diversity in forest birds. In F. Golley and E. Medina (eds.), *Tropical Ecological Systems: Trends in Terrestial and Aquatic Research.* Springer-Verlag, New York, pp. 161–176.

KARR, J. R., AND R. R. ROTH. 1971. Vegetation structure and avian diversity in several new world areas. *Am. Nat.* 105:423–435.

KASTINGS, J. F. 1993. Earth's early atmosphere. *Science* 259:921–926.

KASTINGS, J. F., AND T. P. ACKERMAN. 1986. Climatic consequences of very high carbon dioxide levels in the earth's early atmosphere. *Science* 234:1383–1385.

KAYS, S., AND J. L. HARPER. 1974. The regulation of plant and tiller density in a grass sward. *J. Ecol.* 62:97–105.

KEELER, K. 1981. A model of selection for facultative mutualism. *Am. Nat.* 18:488–498.

KEELING, C. D. 1968. Carbon dioxide in surface ocean waters. *J. Geophys. Res.* 73:4543.

KEELING, C. D. 1991. $CO_2$ global emissions. In T.A. Boden, R. J. Sepanski, and F. W. Stoss (eds.), *Trend 91. A Compendium of Data on Global Change.* Oak Ridge National Laboratory, Oak Ridge, TN, pp. 382–383.

KEELING, C. D. AND T. P. WHORF. 1991. Mauna Loa. In T. A. Boden, R. J. Sepanski, and F. W. Stoss (eds.), *Trend 91. A Compendium of Data on Global Change.* Oak Ridge National Laboratory, Oak Ridge, TN, pp. 12–13.

KEITH, L. B. 1974. Some features of population dynamics in mammals. *Proc. Inter. Congr. Game Biol. Stockholm,* 11:17–58.

———. 1983. Role of food in hare population cycles. *Oikos* 40:385–395.

KEITH, L. B, J. R. CARY, O. J. RONGSTAD, AND M. C. BRITTINGHAM. 1984. Demography and ecology of declining snowshoe hare population. *J. Wildl. Manage.* 90:1–43.

KEITH, L. B., AND L. A. WINDBERG. 1978. A demographic analysis of the snowshoe hare cycle. *Wildl. Monogr.* 58.

KELEY, J. R., JR., M. K. LAUBHAN, F. A. REID, J. S. WORTHAM, AND L. H. FREDRICKSON. 1994. *Options for water-level control in developed wetlands.* U.S. Fish and Wildlife Service Fish and Wildlife Leaflet 13.4.8. 7 pp.

KENDEIGH, S. C. 1961. Energy of birds conserved by roosting in cavities. *Wilson Bull.* 73:140–147.

KERBES, R. H., P. M. KOTANEN, AND R. L. JEFFRIES. 1990. Destruction of wetland habitats by lesser snow geese: A keystone species on the west coast of Hudson Bay. *J. Appl. Ecol.* 27:242–258.

KERNER, A. 1904. *The Natural History of Plants, Their Forms, Growth, Reproduction, and Distribution.* Greshma, London.

KERR, R. A. 1987. Halocarbons linked to ozone hole. *Science* 236:1182–1183.

———. 1992a. Pollution haze cools the greenhouse. *Science* 255:682–683.

———. 1992b. New assaults seen on Earth's ozone shield. *Science* 255:797–798.

KESSEL, S. R., AND W. C. FISCHER. 1981. *Predicting post-fire plant succession for fire management planning.* USDA For. Serv., Gen. Tech. Rept. Int-94.

KETCHUM, B. H. (ED). 1983. *Estuaries and Enclosed Seas* (*Ecosystems of the World,* Vol. 26). Elsevier, Amsterdam.

KETTLEWELL, H. B. D. 1961. The phenomenon of industrial melanism in Lepidoptera. *Ann. Rev. Entomol.* 6:245–262.

———. 1965. Insect survival and selection for pattern. *Science* 148:1290–1296.

KHALIL, M. A. K., AND R. A. RASMUSSEN. 1984. Carbon monoxide in the earth's atmosphere: Increasing trend. *Science* 224:54–56.

KIESTER, A. R. 1971. Species density of North American amphibians and reptiles. *Syst. Zool.* 20:127–137.

KILGORE, B. M. 1973. Impact of prescribed burning on a sequoia–mixed conifer forest. *Proc. Tall Timbers Fire Ecol. Conf.* 12:345–375.

———. 1976. Fire management in the National Parks: An overview. *Proc. Tall Timbers Fire Ecol. Conf.* 14:45–57.

KILRICH, M. I., J. A. MACMAHON, R. R. PARAMETER, AND D. V. SISSON. 1986. Native seed preferences of shrub-steppe rodents, birds, and ants: The relationship of seed attributes and seed use. *Oecologica* 68:327–337.

KINGSLAND, S. 1985. *Modelling Nature.* University of Chicago Press, Chicago.

KINGSTON, T. J., AND M. J. COE. 1977. The biology of the giant dung beetle (*Heliocopris dilloni*) (Coleoptera: Scarabaeidae). *J. Zool.* (London) 181:243–263.

KIRA, K., H. OGAWA, AND K. SHINOZOKI. 1953. Intraspecific competition among higher plants. I. Competition and density yield interrelations in regularly dispersed populations. *J. Polytechnic Institute,* Osaka City University 4 (4):1–16.

KIRK, J. T. O. 1983. *Light and Photosynthesis in Aquatic Systems.* Cambridge University Press, New York.

KIRKPATRICK, M. 1986. The handicap mechanism does not work. *Am. Nat.* 127:222–240.

———. 1987. Sexual selection by female choice in polygynous animals. *Ann. Rev. Ecol. Syst.* 18:43–70.

KIRTPATRICK, M. 1981. Spatial and age dependent patterns of growth in New England black birch. *Am. J. Bot.* 68:535–543.

KITCHELL, J. F., R. V. O'NEILL. D. WEBB, G. W. GALLEPP, S. M. BARTELL, J. F. KOONCE, AND B. S. AUSMUS. 1979. Consumer regulation of nutrient cycling. *Bioscience* 29:28–34.

KITTREDGE, J. 1948. *Forest Influences.* McGraw-Hill, New York.

KLEMOW, K. M. AND D. J. RAYNAL. 1983. Population biology of an annual plant in a temporally variable habitat. *J. Ecol.* 71:691–703.

———. 1985. Demography of two facultative biennial plant species in an unproductive habitat. *J. Ecology* 73:147–167.

KLOPFER, P. 1963. Behavioral aspects of habitat selection: The role of early experience. *Wilson Bull.* 75:15–22.

KLOPFER, P. H., AND J. U. GANZHORN. 1985. Habitat selection: Behavioral aspects. In M. Cody (ed.), *Habitat Selection in Birds.* Academic, Orlando FL.

KLUGE, M. 1982. Crassulacean acid metabolism (CAM). In Govindjee (ed.), *Development, Carbon Metabolism, and Plant Productivity.* Academic, New York, pp. 232–264.

KNAPP, A. K., AND T. R. SEASTEDT. 1986. Detritus accumulation limits productivity of tall grass prairie. *Bioscience* 36:662–668.

KNAPP, R. A., AND R. C. SARGENT. 1989. Egg-mimicry as a mating strategy in the fantail darter *Etherostoma flabellare:* Females prefer males with eggs. *Behav. Ecol. Sociobiol.* 25:321–326.

KNAPTON, R. W., AND J. R. KREBS. 1976. Settlement patterns, territory size, and breeding density in the song sparrow *(Melospiza melodia). Can. J. Zool.* 52:1413–1420.

KNIGHT, R. R., AND L. L. EBERHARDT. 1985. Population dynamics of Yellowstone grizzly bears. *Ecology* 66:323–334.

KNOLL, A. H., J. M. HAYES, A. J. KAUFMAN, K. SWETT, AND J. B. LAMBERT. 1986. Secular variation in carbon isotope ratios from Upper Protero successions of Svalbard and East Greenland. *Nature* 321:832–838.

KNUTSON, R. M. 1974. Heat production and temperature regulation in eastern skunk cabbage. *Science* 186:745–747.

KNOX, E. G. 1952. *Jefferson County (NY) Soils and Soil Map.* NY State College Agr., Cornell University, Ithaca, NY.

KOBLENTZ-MISHKE, J. J., V. V. VOLKOVINSKY, AND J. G. KABANOVA. 1970. Plankton primary production of the world's oceans. In W. S. Wooster (ed.), *Scientific Exploration of the South Pacific.* National Academy of Science, Washington, DC, pp. 183–193.

KODRICH-BROWN, A. K., AND J. H. BROWN. 1984. Truth in advertising: The kinds of traits favored by sexual selection. *Am. Nat.* 124:309–3223.

KONISHI, M. 1973. How the owl tracks its prey. *Am. Sci.* 61:414–424.

KOK, B. 1965. Photosynthesis: The pathway of energy. In J. Bonner and J. E. Varner (eds.), *Plant Biogeochemistry.* Academic Press, New York, pp. 904–960.

KOPEC, R. J. 1970. Further observations of the urban heat island of a small city. *Bull. Am. Meteorol. Soc.* 51:602–606.

KORCHAGIN, A. A., AND V. G. KARPOV. 1974. Fluctuations in coniferous taiga communities. In R. Knapp (ed.), *Vegetation Dynamics.* Part 8. *Handbook of Vegetation Science.* Junk, The Hague, Netherlands, pp. 225–231.

KORPIMAKI, E., AND K. NORRDAHL. 1991. Numerical and functional responses of kestrels, short-eared owls, and long-eared owls to vole densities. *Ecology* 72:814–826.

KORSCHGEN, C. E., AND R. B. DAHLGREN. 1992. *Human disturbances of waterfowl: Causes, effects, and management.* U.S. Fish and Wildlife Service Fish and Wildlife Leaflet 13.2.15. 8 pp.

KOWAL, N. E. 1966. Shifting agriculture, fire, and pine forests in the Cordillera, Central Luzon, Philippines. *Ecol. Monogr.* 56:389–419.

KOZLOWSKI, T. T. 1984. Response of woody plants to flooding. In T. T. Kozlowski (ed.), *Flooding and Plant Growth.* Academic, Orlando, FL., pp. 129–163.

———. 1984. Plant responses to flooding of soil. *Bioscience* 34:162–168.

KOZLOWSKI, T. T., AND S. G. PALLARDY. 1984. Effects of flooding on water, carbohydrate, and mineral relations. In T. T. Kozlowski (ed.), *Flooding and Plant Growth.* Academic, Orlando, FL, pp. 165–194.

KRAMER, P. J. 1981. Carbon dioxide concentration, photosynthesis, and dry matter production. *Bioscience* 31:29–33.

———. 1983. *Water Relations of Plants.* Academic, Orlando, FL.

KRATZ, T. K., AND C. D. DEWITT. 1986. Internal factors controlling peatland-lake ecosystem development. *Ecology* 67:100–107.

KREBS, C. J. 1978. A review of the Chitty hypothesis of population regulation. *Can. J. Zool.* 56:2463–2480.

———. 1985. *Ecology: The Experimental Analysis of Distribution and Abundance,* 3rd ed. Harper & Row, New York.

———. 1989. *Ecological Methodology.* Harper & Row, New York.

KREBS, C. J. B. L. KELLER, AND R. H. TAMARIN. 1969. *Microtus* population biology: Demographic changes in fluctuating populations of *M. ochrogaster* and *M. pennsylvanicus* in southern Indiana. *Ecology* 75:214–223.

KREBS, C. J., I. WINGATE, J. LEDUC, J. A. REDFIELD, M. TAITT, AND R. HILBORN. 1976. *Microtus* population biology: Dispersal in fluctuating populations of *M. townsendii. Can. J. Zool.* 54:79–95.

KREBS, J. R. 1971. Territory and breeding density in the great tit *Parus major. Ecology* 52:2–22.

KREBS, J. R., AND N. B. DAVIES (EDS.). 1978. *Behavioral Ecology: An Evolutionary Approach.* Blackwell, Oxford, England.

———. 1984. *Behavioral Ecology: An Evolutionary Approach,* 2nd ed. Blackwell, Oxford, England.

KREBS, J. R., A. KACELNIK, AND P. TAYLOR. 1978. Optimal sampling by foraging birds: An experiment with great tits *(Parus major). Nature* 275:27–31.

KREMER, J. N., AND S. W. NIXON. 1978. *A Coastal Marine Ecosystem.* Springer-Verlag, New York.

KRUCKEBERG, A. R. 1954. The ecology of serpentine soils. III. Plant species in relation to serpentine soils. *Ecology* 35:267–274.

KRUG E. C. AND C. R. FRINK. 1983. Acid rain on acid soil: A new perspective. *Science* 221:520–525.

KRUGER, F. J. 1979. South African heathlands. In R. L. Specht (ed.), *Heathlands and Related Shrublands Descriptive Studies (Ecosystems of the World,* Vol. 9A), Elsevier, Amsterdam.

KRUUK, H. 1972. *The Spotted Hyena.* University of Chicago Press, Chicago.

KUCERA, C. L., R. C. DAHLMAN, AND M. R. KOELLING. 1967. Total net productivity and turnover on an energy basis for tallgrass prairie. *Ecology* 48:536–541.

KUCHLER, A. W. 1964. *Potential Natural Vegetation of the United*

*States, U. S. Geological Survey National Atlas.* U.S. Geological Survey, Washington, DC.

KUENZLER, E. J. 1958. Niche relations of three species of Lycosid spiders. *Ecology* 39:494–500.

LACK, D. L., AND L. S. V. VENABLES. 1939. The habitat distribution of British woodland birds. *J. Anim. Ecol.* 8:39–71.

LAMOTTE, M. 1975. The structure and function of a tropical savannah ecosystem. In F. B. Holley and E. Medina (eds.), *Tropical Ecological Systems: Trends in Terrestrial and Aquatic Research.* Springer-Verlag, New York, pp. 179–222.

LAMPSON, N. I. 1987. *D.b.h./crown diameter relationships in mixed Appalachian hardwood stands.* Res. Pap. NE-610, USDA Forest Service, Northeastern Forest Experiment Station.

LANCINANI, C. A. 1975. Parasite-induced alternations in host reproduction and survival. *Ecology* 56:689–695.

LANDAHL, J., AND R. B. ROOT. 1969. Differences in the life tables of tropical and temperate milkweed bugs, genus *Oncopeltus* (Hemiptera: Lygaeidae). *Ecology* 50:734–737.

LANDE, R., AND G. F. BARROWCLOUGH. 1987. Effective population size, genetic variation, and their use in population management. In M. E. Soule and B. A. Wilcox (eds.), *Conservation Biology: An Evolutionary-Ecological Approach.* Sinauer Associates, Sunderland, MA, pp. 135–149.

LANDSBERG, H. E. 1970. Man-made climatic changes. *Science* 170:1265–1274.

LANE, P. A. 1985. A food web approach to mutualism in lake communities. In D. H. Boucher (ed.), *The Biology of Mutualism.* Oxford University Press, New York, pp. 344–374.

LANG, G. E. 1985. Forest turnover and the dynamics of bole wood litter in subalpine balsam fir forest. *Can. J. For. Res.* 15:262–288.

LANG G. E., AND R. T. FORMAN. 1978. Detrital dynamics in a mature forest: Hutchinson Memorial Forest. *Ecology* 59:580–595.

LANG, G. E., W. A. REINERS, AND R. R. HEIER. 1976. Potential alterations of precipitation chemistry by epiphytic lichens. *Oecologica* 25:229–241.

LARCHER, W. 1980. *Physiological Plant Ecology.* Springer-Verlag, New York.

———., A. CERNUSA, L. SCHMIDT, G. GRABHERR, E. NÖTZEL AND N. SMEETS. 1975. Mt. Patscherfofel, Austria. In T. Rosswall and O. W. Heal (eds.), *Structure and Function of Tundra Ecosystems.* Swedish Natural Science Research Council, Stockholm, pp. 125–139.

———., AND H. BAUER. 1981. Ecological significance of resistance to low temperature. In O. L. Lange, P. S. Nobel, C. B. Osmond, and H. Zeigler (eds.), *Physiological Plant Ecology. I. Responses to the Physical Environment.* Springer-Verlag, Berlin, pp. 403–437.

LARSEN, K. W., AND S. BOUTIN. 1994. Movements, survival, and settlement of red squirrel (*Tamiasciurus hudsonicus*) offspring. *Ecology* 75:214–223.

LASSOIE, J. P., T. M. HINCKLEY, AND C. C. GRIER. 1985. Coniferous forests of the Pacific Northwest. In B. F. Chabot and H. A. Mooney (eds.), *Physiological Ecology of North American Plant Communities.* Chapman & Hall, New York, pp. 127–161.

LAUBHAN, M. 1992. A technique for estimating seed production of common moist-soil plants. U.S. Fish and Wildlife Service Fish and Wildlife Leaflet 13.4.5.8 pp.

LAUENROTH, W. K. 1979. Grassland primary production: North American grasslands in perspective. In N.F. French (ed.),

*Perspectives in Grassland Ecology.* Springer-Verlag, New York, pp. 3–24.

LAUFF, G. (ED.). 1967. *Estuaries.* American Association for the Advancement of Science, Washington, DC.

LAURANCE, W. F. 1991. Edge effects in tropical forest fragments: Application of a model design of nature reserves. *Biol. Cons.* 57:203–219.

LAURANCE, W. F., AND E. YENSEN. 1991. Predicting the impacts of edge effects in fragmented habitats. *Biol. Cons.* 55:77–92.

LAW, R., A. D. BRADSHAW, AND P. W. PUTWAIN. 1977. Life history variation in *Poa annua. Evolution* 3:233–246.

LAWLOR, L. R. 1976. Molting, growth and reproductive strategies in the terrestrial isopod *Armadillidum vulgare. Ecology* 57:1179–1194.

LAWRENCE, D. B. 1958. Glaciers and vegetation in southeastern Alaska. *Am. Sci.* 46:89–122.

LAWTON, J. H., M. P. HASSELL, AND J. R. BEDDINGTON. 1975. Prey death rates and rate of increase of arthropod predator populations. *Nature* 255:60–62.

LAWTON, J. H., AND D. R. STRONG. 1981. Community pattern and composition in folivorous insects. *Am. Nat.* 118:317–338.

LEAN, D. R. S. 1973a. Movements of phosphorus between its biologically important forms in lake water. *J. Fish. Res. Board Can.* 30:1525–1536.

———. 1973b. Phosphorus dynamics in lake water. *Science* 179:678–680.

LEES, D. R., AND E. R. CREED. 1975. Industrial melanism in *Biston betularia:* The role of selective predation. *J. Anim. Ecol.* 44:67–83.

LEIBIG, J. 1840. *Organic Chemistry and Its Application to Vegetable Physiology and Agriculture.* T. B. Peterson, Philadelphia.

LEITH, H. (ED.). 1974. *Phenology and Seasonality Modeling.* Springer-Verlag, New York.

LEMON, E. R. 1960. Photosynthesis under field conditions. II. An aerodynamic method for determining the turbulent carbon dioxide exchange between the atmosphere and a cornfield. *Agron. J.* 52:697–703.

LEMON, E. R. (ED.). 1983. $CO_2$ *and Plants: The Response of Plants to Rising Levels of Atmospheric Carbon Dioxide.* Westview Press, Boulder, CO.

LENINGTON, S. 1980. Female choice and polygyny in redwinged blackbirds. *Behaviour* 28:347–361.

LESLIE, P. H. 1959. The properties of a certain time lag of population growth and the influence of an external random factor on a number of such populations. *Physiol. Zool.* 32:151–159.

LESSELS, C. M. 1986. Brood size in Canada geese: A manipulation experiment. *J. Anim. Ecol.* 55:669–689.

———. 1991. The evolution of life histories. In J. R. Krebs and N. B. Davies (eds.), *Behavioural Ecology,* 3rd ed. Blackwell, London, pp. 32–68.

LETT, P. F., R. K. MOHN, AND D. F. GRAY. 1981. Density-dependent processes and management strategy for the northwest Atlantic harp seal populations. In C. W. Fowler and T. D. Smith (eds.), *Dynamics of Large Mammal Populations.* Wiley, New York, pp. 135–158.

LEVENSON, J. B. 1981. Woodlots as biogeographic islands in southeastern Wisconsin. In R. L. Burgess and D. M. Sharpe (eds.), *Forest Island Dynamics in Man-Dominated Landscapes.* Springer-Verlag, New York, pp. 13–39.

LEVERLICH, W. J., AND D. A. LEVIN. 1979. Age specific survivorship and reproduction in *Phlox drummondii*. *Am. Nat.* 113:881–903.

LEVIN, D. A. 1976. The chemical defenses of plants to pathogens and herbivores. *Ann. Rev. Ecol. Syst.* 7:121–159.

LEWIS, J. K. 1971. The grassland biome: A synthesis of structure and function, 1970. In N. R. French (ed.), *Preliminary Analysis of Structure and Function in Grasslands*. Range Sci. Dept. Sci. Ser. No. 10. Colorado State University, Fort Collins, pp. 317–387.

LEWONTIN, R. C. 1970. The units of selection. *Ann. Rev. Ecol. Syst.* 1:1–18.

LIDICKER, W. Z. 1975. The role of dispersal in the demography of small mammals. In F. B. Golley, E. Petrusewicz, and L. Ryszkowski (eds.), *Small Mammals, Their Productivity and Population Dynamics*. Cambridge University Press, Cambridge, England, pp. 103–128.

———. 1985. Dispersal. In R. H. Tamarin (ed.), *Biology of New World* Microtus, Sp. Pub. 8, American Society of Mammalogists.

LIGNON, J. D. 1981. Demographic patterns and communal breeding in the green hoopoo, *Phoeniculus purpureus*. In R. D. Alexander and D. W. Tinkle (eds.), *Natural Selection and Social Behavior: Recent Research and New Theories*. Chiron, New York, pp. 231–246.

LIKENS, G. E., AND F. H. BORMANN. 1974. Acid rain: A serious regional environmental problem. *Science* 184:1176–1179.

———. 1974. Linkages between terrestrial and aquatic ecosystems. *Bioscience* 24(8):447–456.

———. 1975. Nutrient-hydrologic interactions (eastern United States). In A. D. Hasler (ed.), *Coupling of Land and Water Systems*. Springer-Verlag, New York, pp. 1–5.

LIKENS, G. E., F. H. BORMANN, M. N. JOHNSON, AND R. S. PIERCE. 1967. The calcium, magnesium, potassium, and sodium budgets for a small forested ecosystem. *Ecology* 38:46–49.

LIKENS, G. E., F. H. BORMANN, R. S. PIERCE, AND D. W. FISHER. 1971. Nutrient hydrologic cycle interaction in small forested watershed ecosystems. In P. Duvigneaud (ed.), *Productivity of Forest Ecosystems*. UNESCO, Paris, pp. 553–563.

LILLYWHITE, H. B. 1970. Behavioral temperature regulation in the bullfrog, *Rana catesbeiana*. *Copeia* 1970:158–168.

LIMA, S. L., AND L. M. DILL. 1990. Behavioural decisions make under the risk of predation: A review and prospectus. Can. J. Zool. 68:619–640.

LINDBERG, S. E., R. C. HARRISS, AND R. R. TURNER. 1982. Atmospheric deposition of metals to forest vegetation. *Science* 215:1069–1611.

LINDEMAN, R. 1942. Trophic-dynamic aspects of ecology. *Ecology* 23:399–418.

LINDQUIST, B. 1942. Experimentelle Untersuchingen uber die Bedeutung einiger Landmollusken fur die zersetgung der Waldstreu. *Kgl. Fysiograf. Sallskap. Lund. Forh.* 11:144–156.

LINDSAY, J. H. 1971. Annual cycle of leaf water potential in *Picea engelmannii* and *Abies lasiocarpa* at timberline in Wyoming. *Arctic and Alpine Research* 3:131–138.

———. 1985. A food web approach to mutualism in lake communities. In D. H. Boucher (ed.), *The Biology of Mutualism*. Oxford University, Oxford, England, pp. 344–374.

LINDSTROM, E. R., H. ANDREN, P. ANGELSTAM, G. CEDERLUND. 1994. Disease reveals the predator: Sarcoptic mange, red fox predation, and prey population. *Ecology* 75:1042–1049.

LITTLE, S., AND H. A. SOMES. 1965. *Atlantic white cedar being eliminated by excessive animal damage in south Jersey*. USDA For. Serv. Res. Note NE33.

LITVIATIS, J. A., J. A. SHERBURNE, AND J. A. BISSONETTE. 1985. Influence of understory characteristics on a snowshoe hare habitat use and density. *J. Wildl. Manage.* 49:866–873.

LLOYD, D. G. 1987. Selection of offspring size at independence and other size vs. number strategies. *Am. Nat.* 129:800–817.

LLOYD, J. E. 1951. Mimicry in the sexual signal of fireflies. *Sci. Amer.* 245:110–117.

LLOYD, P. H., AND O. A. E. RASA. 1989. Status, reproductive success and fitness in the Cape Mountain zebra (*Equus zebra zebra*). *Behav. Ecol. Sociobiol.* 25:411–420.

LOACH, K. 1967. Shade tolerance in tree seedlings. 1. Leaf photosynthesis and respiration in plants raised under artificial shade. *New Phytol.* 66:607–621.

———. 1970. Shade tolerance in tree seedlings. 2. Growth analysis of plants raised under artificial shade. *New Phytol.* 69:273–286.

LOCKE, M., AND H. NICHOL. 1992. Iron economy in insects: Transport, metabolism and storage. *Ann. Rev. Entomol.* 37:195–215.

LODGE, D. J., F. N. SCATENA, C. E. ASBURY, AND M. J. SANCHEZ. 1991. Fine litterfall and related nutrient inputs resulting from Hurricane Hugo in subtropical wet and lower montane rain forests of Puerto Rico. *Biotropica* 23:36–342.

LOEHLE, C., AND J. H. K. PECKMANN. 1988. Evolution: The missing ingredient in systems ecology. *Am. Nat.* 132:884–899.

LOERY, G., AND J. D. NICHOLS. 1985. Dynamics of a black-capped chickadee population, 1958–1983. *Ecology* 66:1195–1203.

LOKEMOEN, J. T. 1993. *Increasing waterfowl nesting success on islands and peninsulas*. U.S. Fish and Wildlife Service Fish and Wildlife Leaflet 13.2.11. 7 pp.

LOMBARDI, J. R., AND J. B. VANDENBERGH. 1977. Pheromonially induced sexual maturation in females: Regulation by the social environment of the male. *Science* 196:545–546.

LONG, G. 1982. Productivity of western coniferous forests. In R. L. Edmonds (ed.), *Analysis of Coniferous Forest Ecosystems in Western United States*. Hutchinson, Ross, Stroudsburg, PA, pp. 89–125.

LONGCORE, J. R., AND G. F. SEPIK (EDS.). 1994. *Proceedings of the Eighth American Woodcock Symposium*. U.S. Fish and Wildlife Service Biological Report 16. 139 pp.

LONSDALE, W. M., AND A. R. WATKINSON. 1982. Light and self-thinning. *New Phytol.* 90:431–435.

LORD, R. D. 1960. Litter size and latitude in North American mammals. *Am. Midl. Nat.* 64:488–499.

LORIMER, C. G. 1989. Relative effects of small and large disturbances on temperate hardwood forest structure. *Ecology* 70:565–567.

LOSSAINT, P. 1973. Soil-vegetation relationships in Mediterranean ecosystems of southern France. In F. di Castri and H. A. Mooney (eds.), *Mediterranean Type Ecosystems: Origin and Structure*. Springer-Verlag, New York, pp. 199–210.

LOTKA, A. J. 1925. *Elements of Physical Biology*. Williams and Wilkens, Baltimore.

LOVEJOY, T. E., R. O. BIERREGAARD, JR., H. B. RYLANDS, J. R. MALCOLM, C. E. QUINTELA, L. H. HARPER, K. S. BROWN, JR., A. H. POWELL, G. V. N. POWELL, H. O. R. SCHUBERT, AND M. B. HAYS. 1986. Edge and other effects of isolation on Amazonian

forest fragments. In M. Soule (ed.), *Conservation Biology.* Sinauer Associates, Sunderland, MA, pp. 257–285.

LOVETT, G. M., W. A. REINERS, AND R. K. OLSON. 1982. Cloud droplet deposition in subalpine balsam fir forests: Hydrological and chemical input. *Science* 218:1303–1304.

LOVETT DOUST, J., AND P. B. CAVERS. 1982. Sex and gender dynamics in jack-in-the-pulpit *Arisaema triphyllum* (Araceae). *Ecology* 63:797–808.

LOWE, V. P. Q. 1969. Population dynamics of red deer (*Cervus elaphus L.*) on Rhum. *J. Anim. Ecol.* 38:425–457.

LOWE-MCCONNELL, R. H. 1969. Speciation in tropical freshwater fishes. *Biol. J. Linn. Soc.* 1:51–75.

LUBCHENCO, J. 1978. Plant species diversity in a marine intertidal community: Importance of herbivore food preferences and algal competitive abilities. *Am. Nat.* 112:23–29.

———. 1980. Algal zonation in the New England rocky intertidal community: An experimental analysis. *Ecology* 61:333–344.

———. 1983. *Littorina* and *Fucus:* Effects of herbivores, substratum heterogeneity, and plant escapes during succession. *Ecology* 64:1116–1123.

———. 1986. Relative importance of competition and predation: Early colonization by seaweeds in New England. In J. Diamond and T. Case (eds.), *Community Ecology.* Harper & Row, New York, pp. 537–555.

LUBCHENCO, J. A., M. OLSON, L. B. BRUBAKER, ET AL. 1992. The sustainable biosphere initiative: An ecological research agenda. *Ecology* 72:371–412.

LUCUS, W., AND J. A. BERRY. 1985. *Inorganic Carbon Uptake by Aquatic Photosynthetic Organisms.* American Society of Plant Physiology, Washington, DC.

LUDWIG, J. A., AND J. F. REYNOLDS. 1988. *Statistical Ecology.* Wiley, New York.

LUGO, A., AND S. BROWN. 1992. Tropical forests as sinks for organic carbon. *For. Ecol. and Manage.* 54:239–255.

LUGO, A. E. 1980. Mangrove ecosystems: Successional or steady state? *Biotropica* 12:65–72.

———. 1992. The search for carbon sinks in the tropics. *Water, Air and Soil Pollution* 64:3–9.

LUGO, A. E. (ED.). 1990. *The Forested Wetland.* Elsevier, Amsterdam.

LUGO, A. E., AND S. BROWN. 1986. Steady state terrestrial ecosystem and the global carbon cycle. *Vegetatio* 68:83–90.

LUGO, A. E., AND S. C. SNEDAKER. 1974. The ecology of mangroves. *Ann. Rev. Ecol. Syst.* 5:39–64.

LULL, H. W., AND W. E. SOPPER. 1969. *Hydrologic effects from urbanization of forested watersheds in the northeast.* USDA Forest Serv. Res. Paper, NE-146.

LUNDBERG, P. 1988. Functional response of a small mammalian herbivore: The disk equation revisited. *J. Anim. Ecol.* 57:999–1006.

LUSSENHOP, J. 1992. Mechanism of microarthropod-microbial interactions in soil. *Adv. Ecol. Res.* 23:1–33.

LUTZ, H., AND R. F. CHANDLER. 1954. *Forest Soils.* Wiley, New York.

LUTZ, H. J. 1956. *Ecological effects of forest fires in the interior of Alaska.* USDA Tech. Bull. No. 1133.

LUXTON, M. 1982. Quantitative utilization of energy by soil fauna. *Oikos* 39:342–354.

LYMAN, C. P., J. S. WILLIS, A. MALAN, AND L. C. H. WANG. 1982.

*Hibernation and Torpor in Mammals and Birds.* Academic Press, New York.

LYNCH, J. F., AND N. K. JOHNSON. 1974. Turnover and equilibria in insular avifaunas with special reference to California Channel Islands. *Condor* 76:370–384.

LYNCH, J. F., AND R. F. WHITCOMB. 1977. Effects of insularization of the eastern deciduous forest and avifaunal diversity and turnover. In *Classification, Inventory, and Analysis of Fish and Wildlife Habitat.* FWS/OBS-78/76, U. S. Fish and Wildlife Service, Washington, DC, pp. 461–489.

MABBERLEY, D. J. 1983. *Tropical Rain Forest Ecology.* Blackie, London.

MCALLEN, B. M., AND C. V. R. DICKMAN. 1986. The role of photoperiod in the timing of reproduction in the Dasyurid marsupial *Antechinus stuartii. Oecologica* 68:259–264.

MCARDLE, R. E., W. H. MEYER, AND D. BRUCE. 1949. *The yield of Douglas-fir in the Pacific Northwest.* USDA Tech. Bull. No. 201 (rev.).

MACARTHUR, R. H. 1955. Fluctuations of animal populations and a measure of community stability. *Ecology* 26:533–536.

———. 1958. Population ecology of some warblers of northeastern coniferous forests. *Ecology* 39:599–619.

———. 1960. On the relative abundance of species. *Am. Nat.* 94:25–36.

———. 1972. *Geographical Ecology.* Harper & Row, New York.

MACARTHUR, R. H., AND R. LEVINS. 1967. The limiting similarity, convergence, and divergence of coexisting species. *Am. Nat.* 101:377–385.

MACARTHUR, R. H., AND J. W. MACARTHUR. 1961. On bird species diversity. *Ecology* 42:594–598.

MACARTHUR, R. H., AND E. R. PIANKA. 1966. On optimal use of a patchy environment. *Am. Nat.* 100:603–609.

MACARTHUR, R. H., AND E. O. WILSON. 1963. An equilibrium theory of insular zoogeography. *Evolution* 17:373–387.

———. 1967. *The Theory of Island Biogeography.* Princeton University Press, Princeton, NJ.

MCAULIFFE, J. R. 1984. Competition for space, disturbance, and the structure of a benthic stream community. *Ecology* 65:894–908.

MCBEE, R. H. 1971. Significance of intestinal microflora in herbivory. *Ann. Rev. Ecol. Syst.* 2:165–176.

MCBRAYER, J. F. 1977. Contributions of cryptozoa to forest nutrient cycles. In W. J. Mattson (ed.), *The Role of Arthropods in Forest Ecosystems.* Springer-Verlag, New York, pp. 70–77.

MCBRIDE, G. I., P. PARE, AND F. FOENANDER. 1969. The social organization and behavior of the feral domestic fowl. *Anim. Behav. Monogr.* 2:127–181.

MCCALLA, T. M. 1943. Microbiological studies of the effects of straw used as mulch. *Trans. Kansas Acad. Sci.* 43:52–56.

MCCULLOUGH, D. R. 1979. *The George Reserve Deer Herd: Population Ecology of a K-selected Species.* Univ. Michigan Press, Ann Arbor.

———. 1981. Population dynamics of the Yellowstone grizzly. In C. W. Fowler and T. D. Smith (eds.), *Dynamics of Large Mammal Populations.* Wiley, New York, pp. 173–196.

———. 1982. Antler characteristics of George Reserve white-tailed deer. *J. Wildl. Manage.* 46:823–826.

MCCUNE, B., AND G. COTTAM. 1985. The successional status of a southern Wisconsin woods. *Ecology* 66:1270–1278.

MacDonald, G. J., and L. Sertario (eds.). *Global Climate and Ecosystem Change.* Plenum, New York.

McDowell, D. M., and R. J. Maiman. 1986. Structure and function of a benthic invertebrate stream community as influenced by beaver (*Castor canadensis*). *Oecologica* 68:481–489.

McElveen, J. 1977. The edge effect on a forest bird community in North Florida. *Proc. Southeastern Game and Fish Comm. Conf.* 31:212–215.

McElroy, M. B., and R. J. Salawith. 1989. Changing composition of the global stratosphere. *Science* 243:763–770.

McGee, C. E. 1984. *Heavy mortality and succession in a virgin mixed mesophytic forest.* USDA Southern For. Exp. Stat. Res. Paper SO-209.

McGinnes, W. G. 1972. North America In C. M. McKella, J. P. Blaisdell, and J. R. Goodwin (eds.), *Wildland Shrubs: Their Biology and Utilization.* USDA For. Serv. Gen. Tech. Rept. INT-1, pp. 55–66.

McGraw, J. B. 1985a. Experimental ecology of *Dryas actopetela* ecotypes: Relative response to competition. *New Phytologist* 100:23–241.

———. 1985b. Experimental ecology of *Dryas actopetela* ecotypes. III. Environmental factors and plant growth. *Arctic and Alpine Res.* 17:229–239.

———. 1989. Effects of age and size on life histories and population growth of *Rhododendron maximum* shoots. *Am. J. Bot.* 76:113–123.

McGraw, J. B., and J. Antonovics. 1983. Experimental ecology of *Dryas actopetela* ecotypes. I. Ecotypic differentiation and life cycles stages of selection. *J. Ecol.* 71:879–897.

McGraw, J. B., and F. S. Chapin. 1989. Competitive ability and adaptation to fertile and infertile soils in two *Eriophorum* species. *Ecology* 70:736–749.

McIlroy, R. J. 1972. *An Introduction of Tropical Grassland Husbandry,* 2nd ed. Oxford, London.

McIntosh, R. P. 1976. Ecology since 1900. In B. J. and T. J. White (eds.), *Issues and Ideas in America.* University of Oklahoma Press, Norman, OK.

———. 1980. The background of some current problems of theoretical ecology. *Synthese* 43:195–255.

———. 1985. *The Background of Ecology: Concept and Theory.* Cambridge University Press, Cambridge, England.

———. 1987. Pluralism in ecology. *Ann. Rev. Ecol. Syst.* 18:2321–341.

MacKenzie, J. J., and M. T. El-asbry (eds.). 1989. *Air Pollution's Toll on Forests and Crops.* Yale University Press, New Haven.

McKell, C. M., J. P. Blaisdell, and J. R. Goodwin (eds). 1972. *Wildland Shrubs: Their Biology and Utilization.* USDA For. Serv. Gen. Tech. Rept. INT-1.

McLaughlin, S. B., T. J. Blasing, L. K. Mann, and D. N. Duvick. 1983. Effects of acid rain and gaseous pollutants on forest productivity: A regional scale approach. *J. Air Pollution Control Assoc.* 33:1042–1045.

MacLean, S. F., Jr. 1980. The detritus-based trophic system. In J. Brown, P. C. Miller, L. L. Tieszen, and F. L. Bunnell (eds.), *An Arctic Ecosystem: The Coastal Tundra at Barrow, Alaska.* Dowden, Hutchinson and Ross, Stroudsburg, PA, pp. 411–457.

McLellan, C. H., A. D. Dobson, D. S. Wilcove, and J. F. Lynch. 1986. Effects of forest fragmentation on new- and old-world bird communities: Empiricial observations and theoretical im-

plications. In J. Verner, M. L.. Morrison, and C. J. Ralph (eds.), *Wildlife 2000: Modeling Habitat Relationships of Terrestrial Vertebrates.* University of Wisconsin Press, Madison, pp. 305–313.

MacLintock, L., R. F. Whitcomb, and B. L. Whitcomb. 1977. Island biogeography and the "habitat islands" of eastern forest. II. Evidence for the value of corridors and minimization of isolation in preservation of biotic diversity. *Am. Birds* 31:6–12.

MacLulich, D. A. 1937. Fluctuations in the numbers of varying hare (*Lepus americanus*). Univ. Toronto Biol. Ser. No. 43.

MacMahon, J. A., and F. H. Wagner. 1985. The Mojave, Sonoran, and Chihuahuan deserts of North America. In M. Evenardi, I. Noy-Meir, and D. W. Goodall (eds.), *Hot Deserts and Arid Shrublands (Ecosystems of the World,* 12A), Elsevier, Amsterdam, pp. 105–202.

MacMillan, P. C. 1981. Log decomposition in Donaldson's woods, Spring Mill State Park, Indiana. *Am. Mid. Nat.* 106:335–344.

McNab, B. K. 1963. Bioenergetics and the determination of home range size. *Am. Nat.* 97:133–140.

———. 1973. Energetics and the distribution of vampires. *J. Mammal.* 54:131–144.

———. 1978. The evolution of endothermy in the phylogeny of mammals. *Am. Nat.* 112:1–21.

———. 1980. Food habits, energetics, and the population biology of mammals. *Am. Nat.* 116:106–124.

———. 1982. Evolutionary alternatives in the physiological ecology of bats. In T. Kunz (ed.), *Ecology of Bats.* Plenum Press, New York, pp. 151–200.

McNaughton, S. J. 1975. *r* and *K* selection in *Typhya. Am. Nat.* 109:215–261.

———. 1979. Grazing as an optimization process: Grass-ungulate relationships in the Serengeti. *Am. Nat.* 113:691–703.

———. 1983. Serengeti grassland ecology: The role of composite environmental factors and contingency in community organization. *Ecol. Monogr.* 53:291–320.

———. 1984. Grazing lawns: Animals in herds, plant forms, and coevolution. *Am. Nat.* 124:863–886.

———. 1985. Ecology of a grazing ecosystem: The Serengeti. *Ecol. Monogr.* 55:259–294.

McNaughton, S. J., and N. J. Georgiadis. 1988. Ecology of African grazing and browsing mammals. *Am. Rev. Ecol. Syst.* 17:39–65.

McNaughton, S. J., M. Osterheld, D. A. Frank, and K. J. Williams. 1989. Ecosystem-level patterns of primary productivity and herbivory in terrestrial habitats. *Nature* 341:142–144.

McNaughton, S. J., R. W. Ruess, and S. W. Seagle. 1988. Large mammals and process dynamics in African ecosystems. *Bioscience* 38:794–800.

McNaughton, S. J., J. L. Tarrants. M. M. McNaughton, and R. T. H. Davis. 1985. Silica as a defense against herbivory and a growth promoter in African grasses. *Ecology* 66:528–535.

McNaughton, S. J., and L. L. Wolf. 1979. *General Ecology.* Holt, Rinehart and Winston, New York.

McPherson, J. K. And C. H. Muller. 1969. Allelopathic effects of *Adenostoma fasciculatum* "chamise" in the California chaparral. *Ecol. Monogr.* 39:177–179.

Mader, H. J. 1984. Animal isolation by roads and agricultural fields. *Biol. Cons.* 29:81–96.

MADGWICK, H. A. I., AND J. D. OVINGTON. 1959. The chemical composition of precipitation in adjacent forest and open plots. *Forestry* 32:14–22.

MAGEE, P. A. 1993. *Detrital accumulation and processing in wetlands.* U.S. Fish and Wildlife Service Fish and Wildlife Leaflet 13.3.14. 7 pp.

MAGUIRE, D. A., AND R. T. T. FORMAN. 1983. Herb cover effects on tree seedling patterns in a mature hemlock-hardwood forest. *Ecology* 64:1367–1380.

MAHALL, B. E., AND R. B. PORK. 1976. The ecotone between *Spartina foliosa.* Trin. and *Salicornia virginica L.* in salt marshes of northern San Francisco Bay. 1. Biomass and production. *J. Ecol.* 64:421–433.

MAIN, A. R. 1981. Fire tolerance of heathland animals. In R. Specht (ed.), *Heathlands and Related Shrublands (Ecosystems of the World,* Vol. 9b), Elsevier, Amsterdam, pp. 85–90.

MALCOLM, S. B. 1990. Mimicry: Status of a classical evolutionary paradigm. *TREE* 5:57–62.

MALTHUS, T. R. 1798. *An Essay on Principles of Population.* Johnson, London (numerous reprints).

MANABE, S., AND R. T. WETHERALD. 1986. Reduction in summer soil wetness induced by an increase in atmospheric carbon dioxide. *Science* 232:626–628.

MANUAT, J. 1983. The vegetation of African savannas. In F. Bourliere (ed.), *Tropical Savannas (Ecosystems of the World,* Vol. 13). Elsevier, Amsterdam, pp. 109–149.

MARGALEF, R. 1963. On certain unifying principles in ecology. *Am. Nat.* 47:357–374.

———. 1968. *Perspectives in Ecological Theory.* University of Chicago Press, Chicago.

MARGULES, C., AND M. B. USHER. 1981. Criteria used in assessing wildlife conservation potential: A review. *Biol. Cons.* 22:217–227.

MARKS, M. E. 1992. *Salmonid whirling disease.* U.S. Fish and Wildlife Service Fish and Wildlife Leaflet 17. 11 pp.

MARKS, P. L. 1974. The role of pin cherry (*Prunus pensylvanica L.*) in the maintenance of stability in northern hardwood ecosystems. *Ecol. Monogr.* 44:73–88.

MARQUIS, D. A. 1974. *The impact of deer browsing on Allegheny hardwood regeneration.* USDA. Forest Serv. Res. Paper NE-308.

———. 1981. *Effect of deer browsing on timber production in Allegheny hardwood forests of northwestern Pennsylvania.* USDA For. Ser. Res. Paper NE-475.

MARQUIS, D. A., AND T. J. GRISEZ. 1978. *The effect of deer exclosures on the recovery of vegetation in failed clearcuts on the Allegheny plateau.* USDA For. Res. Note NE-270.

MARTIN, C. W., R. S. PIERCE, G. E. LIKENS, AND F. H. BORMANN. 1986. *Clearcutting affects stream chemistry in the White Mountains of New Hampshire.* USDA For. Serv. Res. Paper NE-579.

MARTIN, L. F., AND E. H. BUCHER. 1993. Natal dispersal and first breeding age in monk parakeets. *Auk* 110:930–933.

MARTIN, M. M. 1970. The biochemical basis of the fungus-attine ant symbiosis. *Science* 169:16–20.

MARTINAT, P. J. 1987. The role of climatic variation and weather in forest insect outbreaks. In P. Barbosa and J. C. Schultz (eds.), *Insect Outbreaks.* Academic Press, San Diego, pp. 241–268.

MARX, D. H. 1971. Ectomycorrhizae as biological deterrents to pathogenic root infections. In E. Hacskaylo (ed.), *Mycorrhizae.* USDA Misc. Pub. 1189, pp. 81–96.

———. 1973. Mycorrhiza and feeder root disease. Pp. 351–382 in A. D. Monks and T. T. Kozlowski (eds.), Ectomycorrhizae. New York: Academic Press.

MASER, C., R., G. ANDERSON, K. CROMAC, J. T. WILLIAMS, AND R. E. MARTIN. 1979. Dead and down woody material. In J. W. Thomas (ed.), *Wildlife Habitats in Managed Forests: The Blue Mountains of Washington and Oregon.* Agr. Handbook 533, U. S. Department of Agriculture, Washington, DC, pp. 78–95.

MASER, C., AND J. M. TRAPPE (EDS.). 1984. *The Seen and the Unseen World of the Fallen Tree.* USDA For. Serv. Gen. Tech. Rept. PNW-164.

MASER, C., J. M. TRAPPE, AND R. A. NUSSBAUM. 1978. Fungal-small mammal interrelationship with emphasis on Oregon coniferous forests. *Ecology* 59:799–809.

MASSEY, A., AND J. D. VANDENBERG. 1980. Puberty delay by a urinary cue from female house mice in feral populations. *Science* 209:821–822.

MATLACK, G. 1994. Plant demography, land use history, and the commercial use of forests. *Cons. Biol.* 8:298–299.

MATTHIESSEN, P. 1985. Contamination of wildlife with DDT insecticides in relation to tsetse fly control operations. *Env. Pollution (B)* 10:189–211.

MATTSON, W. J., JR. 1980. Herbivory in relation to plant nitrogen content. *Ann. Rev. Ecol. Syst.* 11:119–161.

MATTSON, W. J., AND N. D. ADDY. 1975. Phytophagous insects as regulators of forest primary productivity. *Science* 190:515–521.

MATTSON, W. J., AND R. H. HOACH. 1987. The role of drought stress in provoking outbreaks of phytophagous insects. In P. Barbosa and J. Schultz (eds.), *Insect Outbreaks: Ecological and Evolutionary Perspectives.* Academic Press, Orlando, FL.

———. 1987. The role of drought in outbreaks of plant-eating insects. *Bioscience* 37:110–118.

MAY, M. L. 1976. Thermoregulation and adaptation to temperature in dragonflies (Odonata: Anisoptera). *Ecol. Monogr.* 46:1–32.

———. 1979. Insect thermoregulation. *Ann. Rev. Entomol.* 24:313–350.

MAY, R. 1973. *Stability and Complexity in Model Ecosystems.* Princeton University Press, Princeton, NJ.

MAY, R. M. 1976. Models for single populations. In R. M. May (ed.), *Theoretical Ecology: Principles and Applications.* Saunders, Philadelphia, pp. 4–29.

———. 1981. Models for two interacting populations. In R. M. May (ed.), *Theoretical Ecology,* 2nd ed. Sinauer Associates, Sunderland, MA, pp. 78–104.

———. 1983. Parasitic infections as regulators of animal populations. *Am. Sci.* 71:36–45.

MAY, R. M., AND R. M. ANDERSON. 1978. Regulation and stability of host-parasite population interactions. II. Destabilizing processes. *J. Anim. Ecol.* 47:249–267.

———. 1979. Population biology of infectious diseases: Part II. *Nature* 280:455–461.

———. 1983. Parasite-host coevolution. In D. J. Futuyma and M. Slatkin (ed.), *Coevolution.* Sinauer Associates, Sunderland, MA, pp. 186–206.

MAY, R. M., AND D. I. RUBENSTEIN. 1985. Mammalian reproductive strategies. In C. R. Austin and R. V. Short (eds.), *Reproduction in Mammals,* 4. *Reproductive Fitness,* 2nd ed. Cambridge University Press, Cambridge, England, pp. 1–23.

MAYER, H. M. 1969. *The Spatial Expression of Urban Growth.* Commission on College Geography Resource Paper No. 7. Association of American Geographers, Washington, DC.

MAYFIELD, H. R. 1960. The Kirtland's warbler. *Cranbrook Inst. Sci. Bull.* No. 40.

MAYNARD SMITH, J. 1956. Fertility, mating behavior, and sexual selection in *Drosophila subobscura. J. Genet.* 54:261–279.

MAYNARD SMITH, J. 1971. The origin and maintenance of sex. In G. C. Williams (ed.), *Group Selection.* Aldine, Chicago, pp. 163–175.

———. 1976. A comment on the Red Queen. *Am. Nat.* 110:325–330.

———. 1991. Theories of sexual selection. *TREE* 6:146–151.

MAYO, J. M., A. P. HARTGERING, D. E. DESPAIN, R. G. THOMPSON, E. M. B. VAN ZINDERIN BAKER, AND S. D. NELSON. 1977. Gas exchange studies of *Carex* and *Dryas* Truelove lowland. In L. C. Bliss (ed.), *Truelove Lowland, Devon Island, Canada: A High Arctic Ecosystem.* University of Alberta Press, Edmonton, pp. 265–280.

MAYR, E. 1963. *Animal Species and Evolution.* Harvard University Press, Cambridge, MA.

MECH, L. D. 1970. *The Wolf: The Ecology and Behavior of an Endangered Species.* Doubleday, Garden City, NY.

MECH, L. D., R. E. MCROBERTS, R. O. PETERSON, AND R. E. PAGE. 1987. Relationships of deer and moose populations to previous winter's snow. *J. Anim. Ecol.* 56:615–627.

MEDER, H. J. 1984. Animal isolation by roads and agricultural fields. *Biol. Cons.* 29:81–96.

MEETINTEMEYER, V., E. O. BOX, AND R. THOMPSON. 1982. World patterns and amounts of terrestrial plant litter production. *Bioscience* 32:108–113.

MEEUSE, B. J. 1975. Thermogenic respiration in aroids. *Ann. Rev. Plant Physiol.* 25:117–126.

MEFFEE, G. K., AND M. L. CRUMP. 1987. Possible growth and reproductive benefits of cannibalism in the mosquito fish. *Am. Nat.* 129:203–212.

MENDELSSOHN, I. A. 1979. Nitrogen metabolism in the height form of *Spartina alterniflora* in North Carolina. *Ecology* 60:514–584.

MENGE, B. A., AND J. P. SUNDERLAND. 1976. Species diversity gradients: Synthesis of roles of predation, competition, and temporal heterogeneity. *Am. Nat.* 110:351–369.

MENZIE, C. M. 1969. *Metabolism of pesticides.* U.S. Fish and Wildl. Serv. Sp. Sci. Rept. Wildl. No. 127.

MERENDINO, M. T., C. O. ANKNEY AND D. G. DENNIS. 1993. Increasing mallards, decreasing black ducks: More evidence for cause and effect. *J. Wildl. Manage.* 57:199–208.

MESLOW, E. C., AND L. B. KEITH. 1968. Demographic parameters of a snowshoe hare population. *J. Wildl. Manage.* 32:812–835.

METTLER, L. E., AND T. G. GREGG. 1969. *Population Genetics and Evolution.* Prentice-Hall, Englewood Cliffs, NJ.

MICHOD, R. E. 1982. The theory of kin selection. *Ann. Rev. Ecol. Syst.* 13:23–56.

MILLER, A. H. 1942. Habitat selection among higher vertebrates and its relation to intraspecific variation. *Am. Nat.* 76:25–35.

MILLER, G. R., AND A. WATSON. 1978a. Territories and the food plants of individual red grouse: 1. Territory size, number of mates, and brood size compared with the abundance; production, and diversity of heather. *J. Anim. Ecol.* 47:293–305.

———1978b. Heather productivity and its relevance to the regula-

tion of red grouse populations. In O. W. Heal and D. F. Perkins (eds.), *Production Ecology of British Moors and Grasslands.* Springer-Verlag, New York, pp. 277–285.

MILNE, A. 1957. Theories of natural control of insect populations. *Cold Spring Harbor Symp. Quant. Biol.* 22:253–271.

MINCHELLA, D. J., AND M. E. SCOTT. 1991. Parasitism: A cryptic determinant of animal community structure. *TREE* 6:250–254.

MINCHELLA, D. J., B. K. LEATHERS, K. M. BROWN, AND J. N. MCNAIR. 1985. Host and parasite counteradaptation: An example from a freshwater snail. *Am. Nat.* 126:843–854.

MINSHALL, G. W., K. W. CUMMINS, R. C. PETERSEN, C. E. CUSHING, D. A. BRUNS, J. R. SEDELL, AND R. L. VANNOTE. 1985. Developments in stream ecology. *Can. J. Fish. Aquat. Sci.* 42:1045–1055.

MINSHALL, G. W., R. C. PETERSEN, K. W. CUMMINS, T. L. BOTT, J. R. SEDELL, C. E. CUSHING, AND R. L. VANNOTE. 1983. Interbiome comparison of stream ecosystem dynamics. *Ecol. Monogr.* 53:1–25.

MOBIUS, K. 1877. An oyster bank is a bioconose, or a social community. Transl. by H. J. Rice from Die Auster und die Austerwirtschaft. Wiegundt, Hemfel, and Parey, Berlin. In *Report of U.S. Commission of Fisheries 1889,* pp. 683–675.

MOCK, D. W., AND M. FUJIOKA. 1990. Monogamy and long-term pair bonding in vertebrates. *TREE.* 5:39–43.

MOEHLMAN, P. D. 1979. Jackal helpers and pup survival. *Nature* 277:382–383.

———. 1983. Socioecology of silverbacked and golden jackals, *Canis mesomelas* and *C. aureus.* In J. F. Eisenberg and D. G. Kleinman (eds.), *Recent Advances in the Study of Mammalian Behavior.* Sp. Pub. 7, American Society of Mammalogists, pp. 423–453.

MOHLER, C. I., P. I. MARKS, AND D. G. SPRUGEL. 1978. Stand structure and allometry of trees during self-thinning of pure stands. *J. Ecol.* 66:599–614.

MØLLER, A. 1991. Parasites, sexual ornamentation, and male choice in barn swallows. In J. E. Loye and M. Zuk (eds.), *Bird-Parasite Interactions.* Oxford University Press, Oxford, pp. 328–343.

MØLLER, A. P. 1991. Parasite load reduces song output in a passerine bird. *Anim. Behav.* 41:723–730.

MONRO, J. 1967. The exploitation and conservation of resources by populations of insects. *J. Anim. Ecol.* 36:531–547.

MONSI, M. 1968. Mathematical models of plant communities. In F. E. Eckardt (ed.), *Functioning of Terrestrial Ecosystems at the Primary Production Level.* UNESCO, Paris, pp. 131–149.

MONTEITH, J. L., AND G. SZEICZ. 1960. The carbon dioxide flux over a field of sugar beet. *Quart. J. Roy. Meteor. Soc.* 86:205–213.

MONSON, R. K., G. E. EDWARDS, AND M. S. B. KU. 1984. $C_3$-$C_4$ intermediate photosynthesis in plants. *Bioscience* 34:563–574.

MOOK, L. J. 1963. Birds and spruce budworm. In R. Morris (ed.), *Entomol. Soc. Can. Mem.* 31, pp. 244–248.

MOONEY, H. A. 1981. Primary production in mediterranean-type shrubland. In F. di Castri, D. Goodall, and R. Specht (eds.), *Mediterranean-type Shrublands (Ecosystems of the World,* Vol. 11). Elsevier, Amsterdam, pp. 249–256.

MOONEY, H. A. (ED.). 1977. *Convergent Evolution in Chile and California Mediterranean Climate Ecosystems.* Academic Press, New York.

MOONEY, H. A., AND W. D. BILLINGS. 1961. Comparative physio-

logical ecology of arctic and alpine populations of *Oxyria digyna*. *Ecol. Monogr.* 31:1–29.

MOONEY, H. A., AND C. E. CONRAD. 1977. *Proc. Symp. Environmental Consequences of Fire and Fuel Management in Mediterranean Ecosystems.* USDA For. Serv. Gen. Tech. Rept. WO-3, U.S. Department of Agriculture, Washington, DC.

MOONEY, H. A., AND E. L. DUNN. 1970. Convergent evolution of Mediterranean climate evergreen sclerophyll shrubs. *Evolution* 24:292–303.

MOONEY, H. A., J. EHLERINGER, AND O. BJORKMAN. 1977. The energy balance of leaves of the evergreen desert shrub *Atriplex hymenelytra*. *Oecologica* 29:301–310.

MOONEY, H. A., AND P. C. MILLER. 1985. Chaparral. In B. F. Chabot and H. A. Mooney (eds.), *Physiological Ecology of North American Plant Communities.* Chapman and Hall, New York, pp. 213–231.

MOONEY, H. A., AND D. J. PARSONS. 1973. Structure and function of the California chaparral: An example from San Dimas. In F. di Castri and H. A. Mooney (eds.), *Mediterranean-type Ecosystems: Origin and Structure.* Springer-Verlag, New York, pp. 83–112.

MOORE, H. B. 1958. *Marine Ecology.* Wiley, New York.

MOORE, J. J., P. DOUDING, AND B. HEALY. 1975. Glenamoy, Ireland. In T. Rosswall and O. W. Heal (eds.), *Structure and Function of Tundra Ecosystems.* Swedish Natural Science Research Council, Stockholm, pp. 321–343.

MOORE, N. W., AND M. D. HOOPER. 1975. On the number of bird species in British woods. *Biol. Cons.* 8:239–250.

MOORE, P. D., AND D. J. BELLAMY. 1974. *Peatlands.* Springer-Verlag, New York.

MOORE, W. S., AND R. A. DOLBEER. 1989. The use of banding recovery data to estimate dispersal rates and gene flow in avian species: Case studies in the red-winged blackbird and common grackle. *Condor* 91:242–253.

MORAN, P. A. 1949. The statistical analysis of the Canadian lynx cycle. *Aus. J. Zoo.* 1:291–298.

MORRIS, R. F., W. F. CHESHIRE, C. A. MILLER, AND D. G. MOTT. 1958. The numerical response of avian and mammalian predators during a gradation of the spruce budworm. *Ecology* 39:487–494.

MOSSE, B., D. P. STRIBLEY, AND F. LETACON. 1981. Ecology of mycorrhizae and mycorrhizal fungi. *Adv. Microb. Ecol.* 5:137–210.

MOULDER, B. C., AND D. E. REICHLE. 1974. Significance of spider predation in the energy dynamics of forest floor arthropod communities. *Ecol. Monogr.* 42:473–498.

MOULDER, B. C., D. E. REICHLE, AND S. I. AUERBACH. 1970. *Significance of spider predation in the energy dynamics of forest floor arthropod communities.* Oak Ridge National Laboratory Report ORNL 4452.

MOULTON, M. P., AND S. L. PIMM. 1986. The extent of competition in shaping an introduced avifauna. In J. Diamond and T. J. Case (eds.), *Community Ecology.* Harper & Row, New York, pp. 80–97.

MUELLER-DOMBOIS, D. 1987. Natural dieback in forests. *Bioscience* 37:575–585.

MUELLER-DOMBOIS, D., AND H. ELLENBERG. 1974. *Aims and Methods of Vegetation Ecology.* Wiley, New York.

MUGAAS, J. N., J. SEIDENSTICKER, AND K. D. MAHLKE-JOHNSON. 1993. Metabolic adaptation to climate and distribution of the raccoon *Procyon lotor* and other Procyonidae. *Smithsonian Contr. Zool.* 542:1–34.

MUKERJI, M. K., AND S. H. GAGE. 1978. A model for estimating hatch and mortality of grasshopper egg population based on soil moisture and heat. *Ann. Entomol. Soc. Am.* 71:183–190.

MULLER, C. H., R. B. HANAWALT, AND J. K. McPHERSON. 1968. Allelopathic control of herb growth in the fire cycle of California chaparral. *Bull. Torrey Bot. Club* 95:225–231.

MULROY, T. W., AND P. W. RUNDEL. 1977. Annual plants: Adaptations to desert environments. *Bioscience* 27:109–114.

MURDOCH, W. W. 1969. Switching in general predators: Experiments on predator specificity and stability of prey populations. *Ecol. Monogr.* 39:335–354.

MURDOCH, W. W., AND A. OATEN. 1975. Predation and population stability. *Adv. Ecol. Res.* 9:1–131.

MURPHY, G. I. 1966. Population biology on the Pacific sardine. *Proc. Calif. Acad. Sci.* 4th Series, 34:1–84.

———. 1967. Vital statistics of the Pacific sardine and the population consequences. *Ecology* 48:731–736.

MURPHY, P. G., AND A. E. LUGO. 1986. Ecology of tropical dry forests. *Ann. Rev. Ecol. Syst.* 17:67–85.

MURRAY, B. G., JR. 1967. Dispersal in vertebrates. *Ecology* 48:975–978.

MUSCATINE, L., AND J. W. PORTER. 1977. Reef corals: Mutualistic symbioses adapted to nutrient-poor environments. *Bioscience* 27:454–460.

MUTCH, R. W. 1970. Wildland fires and ecosystems—a hypothesis. *Ecology* 51:1046–1051.

MUUL, I. 1969. Photoperiod and reproduction in flying squirrels, *Glaucomys volans*. *J. Mammal.* 50:542–549.

MYERS, J. H. 1988. Can a general hypothesis explain population cycles of forest Lepidoptera? *Adv. Ecol. Res.* 18:179–284.

———. 1990. Population cycles of western tent caterpillars: Experimental introduction and synchrony of fluctuations. *Ecology* 71:986–995.

———. 1993. Population outbreaks in forest Lepidoptera. *Am. Sci.* 81:240–251.

MYERS, J. H., AND C. J. KREBS. 1971. Genetic, behavioral, and reproductive attributes of dispersing field voles *Microtus pennsylvanicus* and *Microtus ochrogaster*. *Ecol. Monogr.* 41:53–78.

MYERS, K., C. S. HALE, R. MYKYTOWYCZ, AND R. L. HUGHS. 1971. The effects of varying density and space on sociality and health in animals. In A. H. Esser, *Behavior and Environment: The Use of Space by Animals and Men.* Plenum, New York, pp. 148–187.

MYERS, N. 1983. *A Wealth of Wild Species.* Westview, Boulder, CO.

NACE, R. L. 1969. Human uses of ground water. In R. J. Chorley (ed.), *Water, Earth, and Man.* Methuen, London, pp. 285–294.

NADELHOFFER, K., J. D. ABER, AND J. M. MELILLO. 1985. Fine roots, net primary production, and soil nitrogen availability: A new hypothesis. *Ecology* 66:1377–1390.

NAIMAN, R. J., J. M. MELILLO, AND J. E. HOBBIE. 1986. Ecosystem alteration of boreal forest streams by beaver (*Castor canadensis*). *Ecology* 67:1254–1269.

NATIONAL RESEARCH COUNCIL. 1983. *Acid Deposition: Atmospheric Processes in Eastern North America.* National Academy Press, Washington, DC.

NATIONAL RESEARCH COUNCIL OF CANADA. 1981. Acidification in

the Canadian aquatic environment: Scientific criteria for assessing the effects of acid deposition on aquatic ecosystems. NRCC No. 18475. National Research Council of Canada, Ottawa.

NEDWELL, D. B. 1984. The input and mineralization of organic carbon in anaerobic aquatic sediments. *Adv. Microb. Ecol.* 1:93–131.

NEGUS, N. C., P. J. BERGER, AND L. G. FORSLUND. 1977. Reproductive strategy of *Microtus montanus*. *J. Mammal.* 58:347–353.

NELSON, D. J., AND C. C. SCOTT. 1962. Role of detritus in the productivity of a rock-outcrop community in a Piedmont stream. *Limnol. Oceanogr.* 3:396–413.

NELSON, M. E., AND L. D. MECH. 1981. Deer social organization and wolf predation in northeastern Minnesota. *Wildl. Monogr.* 77, 53pp.

NELSON, R. A. 1980. Protein and fat metabolism in hibernating bears. *Fed. Proc.* 39:2955–2958.

NELSON, R. A., AND T. D. I. BECK. 1984. Hibernation adaptation in the black bear: Implications for management. *Proc. East. Workshop Black Bear Manage. Res.* 7:48–53.

NELSON, R. A., T. D. I. BECK, AND D. L. STENGER. 1983. Ratio of serum urea to serum creatinine in wild black bears. *Science* 226:841–842.

NELSON, R. A., G. E. FOLK, JR., E. W. PFEIFFER, J. J. CRAIGHEAD, C. J. JONKEL, AND D. L. STEIGER. 1983. Behavior, biochemistry, and hibernation in black, grizzly, and polar bears. *Int. Conf. Bear Res. Manage.* 5:284–290.

NEUSOME, A. E., AND P. C. CATLING. 1979. Habitat preferences of mammals inhabiting heathlands of warm temperate coastal south-eastern Queensland. In R. L. Specht, (ed.), *Heathlands and Related Shrubland (Ecosystems of the World,* Vol. 9A). Elsevier, Amsterdam, pp. 301–316.

NEWBOLD, J. D., J. W. ELWOOD, R. V. O'NEILL, AND A. L. SHELDON. 1983. Phosphorus dynamics in a woodland stream: A study of nutrient spiraling. *Ecology* 65:1249–1265.

———. 1984. Phosphorus dynamics in a woodland stream ecosystem. *Bioscience* 34:43–44.

NEWBOLD, J. D., R. V. O'NEILL, J. W. ELWOOD, AND W. VAN WINKLE. 1982. Nutrient spiraling in streams: Implications for nutrient and invertebrate activity. *Am. Nat.* 20:628–652.

NEWBOULD, P. J. 1967. *Methods in Estimating the Primary Productivity of Forests.* IBP Handbook No. 2. Blackwell, Oxford.

NEWELL, R. C. 1965. The role of detritus on the nutrition of two marine deposit feeders, the prosobranch *Hydobia ulvae* and the bivalve *Macoma balthica. Proc. Zool. Soc. London* 144:25–45.

NEWELL, S. Y., R. D. FALLON, R. M. C. RODRIGUEZ, AND L. C. GROENE. 1985. Influence of rain, tidal wetting, and relative humidity on release of carbon dioxide by standing-dead salt marsh plants. *Oecologica* 68:73–79.

NEWMAN, E. I. 1988. Mycorrhizal links between plants: Their functioning and significance. *Adv. Ecol. Res.* 18:243–270.

NEWMAN, J. A., AND M. A. ELGAR. 1991. Sexual cannibalism in orb-weaving spiders: An economic model. *Am. Nat.* 138:1372–1395.

NEWSOME, A. E. 1990. The control of vertebrate pests by vertebrate predators. *TREE* 5:187–191.

NEWSOME, A. E., I. PARER, AND P. C. COTLING. 1989. Prolonged

prey suppression by carnivores: Predator removal experiments. *Oecologica* 78:458–467.

NICHOLLS, A. G. 1933. On the biology of *Calanus finmarchieus.* III. Vertical distribution and diurnal migration in the Clyde Sea area. *J. Marine Biol. Assoc. U.K.* 19:139–164.

NICE, M. M. 1943. Studies in the life history of the song sparrow: 2. *Trans. Linn. Soc. New York* 6:1–329.

———. 1962. Development of behavior in precocial birds. *Trans. Linn. Soc. New York* 8.

NICHOLSON, A. J. 1954. An outline of the dynamics of animal populations. *Aust. J. Zool.* 2:9–65.

———. 1957. The self-adjustment of populations to change. *Cold Spring Harbor Symp. Quant. Biol.* 22:153–173.

NICHOLSON, A. J., AND V. A. BAILEY. 1935. The balance of animal populations: Part 1. *Proc. Zool. Soc. London,* 3:551–598.

NIERING, W. A., AND F. E. EGLER. 1955. A shrub community of *Viburnum lentago* stable for twenty-five years. *Ecology* 36:356–360.

NOBLE, J. C., A. D. BELL, AND J. L. HARPER. 1979. The population biology of plants with clonal growth. I. Morphology and structural demography of *Carex Arenaria. J. Ecol.* 67:983–1008.

NOBLE, J. R., AND R. O. SLATYER. 1980. The use of vital attributes to predict successional changes in plant communities subject to recurrent disturbances. *Vegetatio* 43:5–21.

NOBEL, P. S. 1978. Surface temperature of cacti: Influence of environmental and morphological factors. *Ecology* 59:986–996.

———. 1985. Desert succulents. In B. F. Chabot and H. A. Mooney (eds.), *Physiological Ecology of North American Plant Communities.* Chapman and Hall, New York, pp. 131–197.

NOY-MEIR, I. 1973. Desert ecosystems: Environment and producers. *Ann. Rev. Ecol. Syst.* 4:25–51.

———. 1974. Desert ecosystems: Higher trophic levels. *Ann. Rev. Ecol. Syst.* 5:195–214.

———. 1975. Stability of grazing systems: An application of predator-prey graphs. *J. Ecol.* 63:459–481.

———. 1985. Desert ecosystem structure and function. In M. I. Evenardi, I. Noy-Meir, and D. W. Goodall (eds.), *Hot Deserts and Arid Shrublands,* Vol. 12A Elsevier, Amsterdam, pp. 93–103.

NYBAKKEN, J. W. 1988. *Marine Biology: An Ecological Approach,* 2nd ed. Harper & Row, New York.

OBERNDORFER, R. Y., J. V. MCARTHUR, J. R. BARNES, AND J. DIXON. 1984. The effects of invertebrate predators of leaf litter processing in an alpine stream. *Ecology* 65:1325–1331.

O'BRIEN, S. J., AND J. F. EVERMANN. 1988. Interactive influence of infectious disease and genetic diversity in natural populations. *TREE* 3:254–259.

O'BRIEN, S. J., AND D. E. WILDT, D. GOLDMAN, D. R. MERREL, AND M. BASH. 1983. The cheetah is depauperate in genetic variation. *Science* 221:459–462.

O'DOWD, D. J., AND M. E. HAY. 1980. Mutualism between harvester ants and a desert ephemeral: Seed escape from rodents. *Ecology* 61:531–540.

ODUM, E. P. 1964. The new ecology. *Bioscience* 14:14–16.

———. 1971. *Fundamentals of Ecology,* 3rd ed. Saunders, Philadelphia.

———. 1983. *Basic Ecology.* Saunders, Philadelphia.

ODUM, E. P., AND L. J. BIEVER. 1984. Resource policy, mutualism, and energy partitioning in food chains. *Am. Nat.* 100:65–75.

ODUM, H. T. 1970. Summary: An emerging view of the ecological system at El Verde. In H. T. Odum and R. F. Pigeon (eds.), *A Tropical Rain Forest,* US Atomic Energy Commission, Washington, DC. pp. I191–I218.

———. 1983. *Systems Ecology: An Introduction.* Wiley, New York.

ODUM, W. E., AND M. A. HEYWOOD. 1978. Decomposition of intertidal freshwater marsh plants. In R. E. Good, D. F. Whigham, and R. L. Simpson (eds.), *Freshwater Wetlands.* Academic Press, New York, pp. 89–97.

ODUM, W. E., T. J. SMITH III, J. K. HOOVER, AND C. C. McIVOR. 1984. *The Ecology of Tidal Freshwater Marshes of the United East Coast: A Community Profile.* U. S. Fish and Wildlife Service FWS/OBS-87/17, Washington, DC, 177 pp.

OECHEL, W. C., AND W. T. LAWRENCE. 1985. Taiga. In B. F. Chabot and H. A. Mooney (eds.), *Physiological Ecology of North American Plant Communities.* Chapman and Hall, New York, pp. 66–94.

OFFICER, C. B. 1983. Physics of estuarine circulation. In B. Ketchum (ed.), *Estuaries and Enclosed Seas (Ecosystems of the World,* Vol. 26). Elsevier, Amsterdam, pp. 15–42.

OGREN, W. L., AND R. CHOLLET. 1982. Photorespiration. In Govindjee (ed.), *Photosynthesis, Development, Carbon Metabolism, and Plant Productivity.* Academic Press, New York, pp. 191–230.

OHMART, C. P., G. G. STEWART, AND J. R. THOMAS. 1983. Leaf consumption by insects in three *Eucalyptus* forest types in southeastern Australia and their role in short-term nutrient cycling. *Oecologica* 59:322–330.

OJASTI, J. 1983. Ungulates and large rodents of South America. In F. Bourliere (ed.), *Tropical Savannas (Ecosystems of the World,* Vol. 9). Elsevier, Amsterdam, pp. 427–440.

OKE, T. R., AND C. EAST. 1971. The urban boundary layer in Montreal. *Boundary-layer Meteorol.* 1:411.

OLDEMEYER, J. L., D. E. BIGGINS, B. J. MILLER, AND R. CRETE. 1994. *Proc. symposium on the management of prairie dog complexes for the reintroduction of the black-footed ferret.* U.S. Fish and Wildlife Service Biological Report 13. 96 pp.

OLSON, J. S. 1958. Rates of succession and soil changes on southern Lake Michigan sand dunes. *Bot. Gazette* 119:125–170.

———. 1970. Carbon cycles and temperate woodlands. In D. E. Reichle (ed.), *Analysis of Temperate Forest Ecosystems.* Springer-Verlag, New York, pp. 226–241.

O'NEILL, R. V. 1976. Ecosystem persistence and heterotrophic regulation. *Ecology* 57:1244–1253.

O'NEILL, R. V., W. F. HARRIS, B. S. AUSMUS, AND D. E. REICHLE. 1975. A theoretical basis for ecosystem analysis with particular reference to element cycling. In F. G. Howell, J. B. Gentry, and M. H. Smith (eds.), *Mineral Cycling in Southeastern Ecosystems.* ERDA Symposium Series, National Technical Information Service, U. S. Department of Commerce, pp. 28–40.

ORIANS, G. H. 1969. On the evolution of mating systems in birds and mammals. *Am. Nat.* 103:589–603.

ORIANS, G. H., AND J. F. WHITTENBERGER. 1991. Spatial and temporal scales in habitat selection. *Am. Nat.* 137:S29–S49.

ORITZ, C. L., B. J. LEBOEUF, AND D. P. COSTA. 1984. Milk intake of elephant seal pups: An index of parental investment. *Am. Nat.* 124:416–422.

OSBORN, F. 1949. *Our Plundered Planet.* Little, Brown, Boston.

OSMUND, C. B. 1978. Crassulacean acid metabolism: A curiosity in context. *Ann Rev. Plant Physiol.* 35:415–442.

OVERGAARD, C. 1949. Studies on the soil microfauna: II. The soil-inhabiting nematodes. *Nat. Jutland* 2:131–150.

OWEN, D. F. 1980. How plants may benefit from animals that eat them. *Oikos* 35:230–235.

OWEN, D. F., AND R. G. WIEGERT. 1981. Mutualism between grasses and grazers: An evolutionary hypothesis. *Oikos* 36:376–378.

OZOGA, J. J., AND W. W. GYSEL. 1972. Response of white-tailed deer to winter weather. *J. Wildl. Manage.* 36:892–896.

OZOGA, J. J., AND L. J. VERME. 1986. Initial and maternal success of white-tailed deer. *J. Wildl. Manage.* 50:122–124.

OZOGA, J. J., L. J. VERME, AND C. S. BIENG. 1982. Parturition behavior and territoriality in deer: Impact on neonatal mortality. *J. Wildl. Manage.* 446:1–11.

PACKER, C. L. HERBST, A. E. PUSEY, J. P. BYGOTT, J. P. HANBY, S. J. CAIRNS, AND M. B. MULDER. 1988. Reproductive success in lions. In T. H. Clutton-Brock (ed.), *Reproductive Success: Studies of Individual Variation in Contrasting Breeding Systems,* University of Chicago Press, Chicago, pp. 363–383.

PACKER, C., AND L. RUTTAN. 1988. The evolution of cooperative hunting. *Am. Nat.* 132:159–198.

PAGE, C. N. 1982. The history and spread of bracken in Britain. *Proc. Royal Soc. Edinburgh* 81B:3–10.

PAGE, K. N. AND T. G. WHITHAM. 1987. Overcompensation in response to mammalian herbivory: The advantage of being eaten. *Am. Nat.* 129:407–416.

PAINE, R. T. 1966. Food web complexity and species diversity. *Am. Nat.* 100:65–75.

———. 1969. The *Pisaster-Tegula* interaction: Prey patches, predator food preference and intertidal community structure. *Ecology* 50:950–961.

PALACA, S., AND J. ROUGHGARDEN. 1984. Control of arthropod abundance by *Anolis* lizards on St. Eustatius (Nethl. Antilles). *Oecologica* 64:160–162.

PALMER, H. E., W. C. HANSON, B. I. GRIFFIN, AND W. C. ROESCH. 1963. Cesium-137 in Alaskan Eskimos. *Science* 142(3588):64–65.

PALMER, J. D. 1976. *An Introduction to Biological Rhythms.* Academic Press, New York.

———. 1990. The rhythmic lives of crabs. *Bioscience* 40:352–358.

PALMGREN, P. 1949. Some remarks on the short-term fluctuations in the numbers of northern birds and mammals. *Oikos* 1:114–121.

PALO, R. T., A. PEHRSON, AND P. KNUTSSON. 1983. Can birch phenolics be of importance in the defense against browsing vertebrates? *Finn. Game Res.* 41:75–80.

PARIS, O. H. 1969. The function of soil fauna in grassland ecosystems. In R. L. Dix and R. G. Beidleman (eds.), *The Grassland Ecosystem: A Preliminary Synthesis.* Range Sci. Dept. Sci. Ser. No. 2, Colorado State University, Fort Collins, pp. 331–360.

PARK, T. 1954. Experimental studies of interspecies competition: 2. Temperature, humidity and competition in two species of *Trilobium. Physiol. Zool.* 27:177–238.

PARKER, G. A., AND R. A. STUART. 1976. Animal behavior as a strategy optimizer: Evolution of resource assessment strata and optimal emigration thresholds. *Am. Nat.* 110:1055–1076.

PARRY, M. 1992. The potential effect of climate changes in agriculture and land use. *Adv. Ecol. Res.* 22:63–91.

PARSONS, J. 1971. Cannibalism in herring gulls. *Brit. Birds* 64:528–537.

PASTEUR, G. 1982. A classificatory review of mimicry systems. *Ann. Rev. Ecol. Syst.* 13:169–199.

PASTOR, J., R. J. NAIMAN, B. DEWEY, AND P. MCINNES. 1988. Moose, microbes, and boreal forest. *Bioscience* 88:770–777.

PATE, V. S. L. 1933. *Studies on fish food in selected areas: A biological survey of Raquette Watershed,* N.Y. State Conserv. Dept. Biol. Survey No. 8, pp. 136–157.

PATTERSON, D. T. 1975. Nutrient return in stemflow and throughfall of individual trees in the Piedmont deciduous forest. In F. G. Howell et al. (eds.), *Mineral Cycling in Southeastern Ecosystems.* National Technical Information Service, U.S. Dept. Commerce, pp. 800–812.

PAYETTE, S. 1988. Late-Holocene development of subarctic ombrotrophic peatlands: Allogenic and autogenic succession. *Ecology* 62:516–531.

PAYNE, R. 1968. Among wild whales. *N.Y. Zool. Soc. Newsletter,* November 1968.

PEAKALL, D. B. 1970. Pesticides and the reproduction of birds. *Sci. Am.* 222:72–78.

PEARCY, R. W. 1976. Temperature effects on growth and $CO_2$ exchange rates of *Atriplex leniformis. Oecologica* 26:245–255.

———. 1977. Acclimation of photosynthetic and respiratory $CO_2$ to growth temperature in *Atriplex lentiformis* (Torr.) Wats. *Plant Physiol.* 61:484–486.

PEARL, R. 1927. The growth of populations. *Quart. Rev. Biol.* 2:532–548.

PEARL, R., AND L. J. REED. 1920. On the rate of growth of the population of the United States since 1790 and its mathematical representation. *Proc. Nat. Acad. Sci.* 6:275–288.

PEARMAN, G. I., AND P. HYSON. 1981. The annual variation of atmospheric $CO_2$ concentration observed in the Northern Hemisphere. *J. Geophysical Res.* 86:9839.

PEARSON, D. L. 1971. Vertical stratification of birds in a tropical dry forest. *Condor* 73:46–55.

PEARSON, W. D., AND B. J. KRAMER. 1972. Drift and production of two aquatic insects in a mountain stream. *Ecol. Monogr.* 42:365–385.

PEASE, J. L., R. H. VOWLES, AND L. B. KEITH. 1979. Interaction of snowshoe hares and woody vegetation. *J. Wildl. Manage.* 43:43–60.

PECKARSKY, B. L. 1982. Aquatic insect predator-prey relations. *Bioscience* 32:261–266.

PEET, R. K. 1974. The measurement of species diversity. *Ann. Rev. Ecol. Syst.* 5:285–307.

———. 1981. Changes in biomass and production during secondary forest succession. In D. C. West, H. H. Shugart, and D. B. Botkin (eds.), *Forest Succession: Concepts and Applications.* Springer-Verlag, New York, pp. 324–338.

PEET, R. K., AND N. L. CHRISTENSEN. 1980. Succession: A population process. *Vegetatio* 43:131–140.

PEFRANKA, J. W., AND A. SIH. 1986. Environmental instability, competition, and density-dependent growth and survivorship of a stream-dwelling salamander. *Ecology* 67:729–736.

PENGELLEY, E. T., AND S. J. ASMUNDSON. 1974. Circannual rhythmicity in hibernating mammals. In E. T. Pengelley (ed.), *Circannual Clocks: Annual Biological Rhythms.* Academic Press, New York, pp. 95–106.

PERRY, O. A., H. MARGOLIS, C. CHOQUETTE, R. MOLINA, AND J. M.

TRAPPE. 1989. Ectomycorrhizal mediation of competition between coniferous tree species. *New Phytologist* 111:501–511.

PETERMAN R. M. 1990. The importance of reporting statistical power: Forest decline and acidic deposition. *Ecology* 71:2024–2027.

PETERS, R. L., AND J. D. S. DARLING. 1985. The greenhouse effect and nature reserves. *Bioscience* 3:707–717.

PETERS, R. P., AND L. D. MECH. 1975. Scent-marking in wolves. *Am. Sci.* 63:628–637.

PETERSON, D. L., AND F. A. BAZZAZ. 1978. Life cycle characteristics of *Aster pilosus* in early successional habitats. *Ecology* 59:1005–1013.

PETERSON, H. 1982. Structure and size of soil animal populations. *Oikos* 39:306–329.

PETERSON, H., AND M. LUXTON. 1982. A comparative analysis of soil fauna populations and their role in the decomposition process. *Oikos* 39:287–388.

PETERSON, R. O. 1977. *Wolf ecology and prey relationships in Isle Royale.* U. S. Nat. Park Serv. Sci. Monogr. Ser. 11. 210 pp.

PETRIE, M. AND A. P. MOLLER. 1991. Laying eggs in others' nests: Intraspecific brood parasitism in birds. TREE 6:315–320.

PETRINVICH, L., AND T. L. PATTERSON. 1982. The white-crowned sparrow: Stability, recruitment, and population structure in the Nuttal subspecies (1975–1980). *Auk* 99:1–14.

PFEIFFER, W. 1962. The fright reaction of fish. *Biol. Rev.* 37:495–511.

PFISTER, R. D., B. L. KOVALCHIK, S. E. ARNO, AND P. C. PRESBY. 1977. *Forest Habitat Types of Montana.* USDA. For. Serv. Gen. Tech. Rept. INT-34.

PHILANDER, S. G. H. 1983. El Nino southern oscillation phenomenon. *Nature* 302:295–301.

PHILPOT, C. W. 1977. Vegetation features as determinants of fire frequency and intensity. In H. A. Mooney and C. E. Conrad (eds.), *Environmental Consequences of Fire and Fuel Management in Mediterranean Ecosystems,* USDA For. Ser. Gen. Tech. Rept. WO-26, Washington, DC, pp. 12–16.

PIANKA, E. 1967. On lizard species diversity, North American flatlands desert. *Ecology* 48:333–351.

———. 1972. *r* and *k* selection or *b* and *d* selection? *Am. Nat.* 100:65–75.

———. 1975. Niche relations of desert lizards. In M. Cody and J. Diamond (eds.), *Ecology and Evolution of Communities.* Harvard University Press, Cambridge, pp. 292–314.

———. 1978. *Evolutionary Ecology,* 3rd ed. Harper & Row, New York.

———. 1980. On *r* and *K* selection. *Am. Nat.* 102:592–597.

———. 1981. Competition and niche theory. In R. M. May (ed.), *Theoretical Ecology: Principles and Application.* Blackwell, Oxford, pp. 167–196.

———. 1994. *Evolutionary Ecology,* 5th ed. HarperCollins, New York.

PICKETT, S. T. A., AND P. WHITE (EDS.). 1985. *The Ecology of Natural Disturbances and Patch Dynamics.* Academic Press, Orlando, FL.

PIEHLER, K. G. 1987. Habitat relationships of three grassland sparrow species on reclaimed surface mines in Pennsylvania. Unpublished MS thesis, West Virginia University, Morgantown.

PIELOU, E. 1994. *After the Ice Age: The Return of Life to Glaciated North America.* University of Chicago Press, Chicago.

PIELOU, E. C. 1972. Niche width and niche overlap: A method for measuring them. *Ecology* 53:687–692.

———. 1974. *Population and Community Ecology.* Gordon and Breach, New York.

———. 1975. *Ecological Diversity.* Wiley, New York.

———. 1981. The usefulness of ecological models: A stock-taking. *Quart. Rev. Biol.* 56:1423–1437.

PIERCE, B. A. 1985. Acid tolerance in amphibians. *Bioscience* 35:239–243.

PIJL, L. VAN DER, AND C. L. DOTSON. 1966. *Orchid Flowers: Their Pollination and Evolution.* University of Miami Press, Miami, FL.

PILCHER, J. R., AND B. GRAY. 1982. The relationship between oak tree growth and climate in Butran. *J. Ecol.* 70:297–304.

PILSON, D., AND M. D. RAUSHER. 1988. Clutch size adjustment by a swallowtail butterfly. *Nature* 333:361–363.

PIMENTEL, D., J. E. DEWEY, AND H. H. SCHWARDT. 1951. Space-time structure of the environment and the survival of the parasite-host system. *Am. Nat.* 97:141–167.

PIMM, S. L. 1980. Food web design and the effect of species deletion. *Oikos* 35:139–149.

———. 1982. *Food Webs.* Chapman and Hall, London.

———. 1987. Determining the effects of introduced species. *TREE* 2:106–107.

———. 1991. *The Balance of Nature: Ecological Issues in the Conservation of Species and Communities.* University of Chicago Press, Chicago.

PIMM, S. L., H. L. JONES, AND J. DIAMOND. 1988. On the risk of extinction. *Am Nat.* 132:757–785.

PIMM, S. L., AND R. L. KITCHING. 1987. The determinants of food chain lengths. *Oikos* 50:302–307.

PIMM, S. L., AND J. H. LAWTON. 1977. The number of trophic levels in ecological communities. *Nature* 268:329–331.

PIMM, S. L., AND J. W. PIMM. 1982. Resource use, competition, and resource availability in Hawaiian honeycreepers. *Ecology* 63:1468–1486.

PISTOLE, D. H., AND J. A. CRANFORD. 1982. Photoperiodic effects of growth in *Microtus pennsylvanicus. J. Mammal.* 63:547–553.

PITELKA, F. A. 1973. Cyclic patterns in lemming populations near Barrow, Alaska. In M. E. Butler (ed), *Alaskan Arctic Tundra.* Arctic Institute of America Tech. Paper No. 25, pp. 199–216.

PITTENDRIDGH, C. S. 1966. The circadian oscillation in *Drosophila pseudoobscura* pupae: A model of the photoperiodic clock. *Z. Pflanzenphysiol* 54:275–307.

PIVNICK, K. A., AND J. N. MCNEIL. 1986. Sexual differences in the thermoregulation of *Thymelicus lineola* adults (Lepidoptera: Hesperiidae). *Ecology* 67:1024–1035.

PLATTS, W. J., AND D. R. STRONG. 1989. Tree fall gaps and forest dynamics. *Ecology* 70:535–576.

PLESCZYNSKA, W. K., AND R. HANSELL. 1980. Polygyny and decision theory: Testing of a model in lark buntings (*Calamospiza melanocorys*). *Am. Nat.* 116:821–830.

PLOWRIGHT, W. 1982. The effects of rinderpest and rinderpest control on wildlife in Africa. *Symp. Zool. Soc. Lond.* 50:1–28.

POLICANSKY, D. 1982. Sex change in plants and animals. *Ann. Rev. Ecol. Syst.* 13:471–495.

POLIS, G. 1981. The evolution of intraspecific predation. *Ann. Rev. Ecol. Syst.* 12:225–251.

POLIS, G. AND R. D. HOLT. 1992. Intraguild predation: The dynamic of complex trophic interactions. *TREE* 7:151–154.

POLIS, G. A., AND R. D. FARLEY. 1980. Population biology of a desert scorpion: Survivorship, microhabitat, and the evolution of life history strategy. *Ecology* 61:620–629.

POLIS, G. A., C. A. MYERS, AND R. D. HOLT. 1989. The ecology and evolution of intraguild predation: Potential competitors that eat each other. *Ann. Rev. Ecol. Syst.* 20:297–330.

POLUNIN, N. 1955. Aspects of arctic botany. *Am. Sci.* 43:307–322.

POMEROY, L. R., AND E. J. KUENZLER. 1969. Phosphorus turnover by coral reef animals. In D. J. Nelson and F. E. Evans (eds.), *Symposium on Radioecology, Conf. 670503.* National Technical Information Services, Springfield, VA, pp. 478–483.

POMEROY, L. R., H. M. MATHEWS, AND H. SHIKMIN. 1963. Excretion of phosphate and soluble organic phosphorus compounds by zooplankton. *Limnol. Oceanogr.* 4:50–55.

POMEROY, L. R., AND R. G. WIEGERT (EDS.). 1981. *The Ecology of a Salt Marsh.* Springer-Verlag, New York.

POOLE, R. W. 1974. *An Introduction to Quantitative Ecology.* McGraw-Hill, New York.

POORE, M. E. D. 1968. Studies in Malaysian rain forests. 1. The forest on Triassic sediments in the Jenka forest reserve. *J. Ecol.* 56:143–196.

POST, W. M., C. C. TRAVIS, AND D. L. DEANGELIS. 1980. Evolution of mutualism between species. In C. L. Cooke and S. Brisenberg (eds.), *Differential Equations and Applications in Ecology, Epidemics, and Population Problems.* Academic Press, New York, pp. 183–201.

———. 1985. Mutualism, limited competition, and positive feedback. In D. H. Boucher (ed.), *The Biology of Mutualism.* Oxford University Press, New York, pp. 305–325.

POSTEL, S. 1984. Air pollution, acid rain, and the future of forests. *Worldwatch Paper* 58:1–22, 44–49.

POTTER, D. A., AND T. W. KIMMERER. 1989. Inhibition of herbivory on young holly leaves: Evidence for the role of saponins. *Oecologica* 78:322–329.

POTTS, G. R., S. C. TAPPE, AND P. J. HUDSON. 1984. Population fluctuations in red grouse: Analysis of bog records and a simulation model. *J. Anim. Ecol.* 53:21–36.

POUGH, F. H. 1988. Mimicry of vertebrates: Are the rules different? *Am. Nat.* 131:S67–S192.

POWELL, M. A., AND G. N. SOMERO. 1983. Blood components prevent sulfide poisoning of respiration of the hydrothermal vent tube worm *Riftia pachyptila. Science* 219:297–299.

PRENTKI, R. T., T. D. GUFASON, AND M. S. ADAMS. 1978. Nutrient movements in lakeside marshes. In R. E. Good, D. F. Whigham, and R. L. Simpson (eds.), *Freshwater Wetlands: Ecological Processes and Management Potential.* Academic Press, New York, pp. 169–194.

PRESTON, F. W. 1960. Time and space and the variation of species. *Ecology* 41:611–627.

———. 1962. The canonical distribution of commonness and rarity: Parts 1 and 2. *Ecology* 43:185–215, 410–432.

PRICE, P. W. 1975. Reproductive strategies of parasitoids. In P. W. Price (ed.), *Evolutionary Strategies of Parasitic Insects and Mites.* Plenum, New York, pp. 87–111.

———. 1980. *Evolutionary Ecology of Parasites.* Princeton University Press, Princeton, NJ.

PRICE, P. W., C. E. BOUTON, P. GROSS, B. A. MCPHERSON, J. N. THOMPSON, AND E. E. WEIS. 1980. Interactions among three trophic levels: Influence of plants on interactions between insect herbivores and natural enemies. *Ann. Rev. Ecol. Syst.* 11:41–65.

PRICE, P. W., M. WESTOBY, B. RICE, P. R. ATSATT, R. S. FRITZ, J. N. THOMPSON, AND K. MOBLEY. 1986. Parasite mediation in ecological interactions. *Ann. Rev. Ecol. Syst.* 17:487–505.

PRIMACK, R. B. 1979. Reproductive effort in annual and perennial species of *Plantago* (Plantaginaceae). *Am. Nat.* 114:51–62.

PROCTOR, J., AND S. R. J. WOODWELL. 1975. The ecology of serpentine soils. *Adv. Ecol. Res.* 9:256–366.

PRUITT, W. O., JR. 1970. Some aspects of interrelationships of permafrost and tundra biotic communities. In *Productivity and Conservation in Northern Circumpolar Lands,* IUCN Publ. 10:33–41.

PUCKETT, L. J. 1982. Acid rain, air pollution, and tree growth in southeastern New York. *J. Environ. Qual.* 11:376–381.

PULLIAM, H. R. 1988. Sources, sinks, and population regulation. *Am. Nat.* 132:652–661.

PULLIAM, H. R., AND B. J. DANIELSON. 1991. Sources, sinks, and habitat selections: A landscape perspective on population dynamics. *Am. Nat.* 137:S50–S66.

PUSEY, A. E., AND C. PACKER. 1986. The evolution of sex-biased dispersal in lions. *Behaviour* 101:275–310.

PUTMAN, R. J. 1978a. Patterns of carbon dioxide evolution from decaying carrion: Decomposition of small mammal carrion in temperate systems. *Oikos* 31:49–57.

———. 1978b. Flow of energy and organic matter from a carcass during decomposition: Decomposition of small mammal carrion in temperate systems. 2. *Oikos* 31:58–68.

———. 1983. *Carrion and Dung: The Decomposition of Animal Wastes.* Edward Arnold, London.

PUTMAN, R. J., AND S. D. WRATTEN. 1984. *Principles of Ecology.* University of California Press, Berkeley.

PUTWAIN, P. D., AND J. L. HARPER. 1970. Studies of dynamics of plant populations: 3. The influence of associated species on populations of *Rumex acetosa* L. and *R. acetosella* L. in grassland. *J. Ecol.* 58:251–264.

QUAY, P. D., B. TILBROOK, AND C. S. WONG. 1992. Oceanic uptake of fossil fuel $CO_2$: Carbon-13 evidence. *Science* 256:74–79.

RALLS, K, P. H. HARVEY, AND M. A. LYLES. 1986. Inbreeding in natural populations of birds and mammals. In M. Soule (ed.), *Conservation Biology: The Science of Diversity,* Sinauer Associates, Sunderland, MA, pp. 35–56.

RAMUS, J. 1983. A physiological test of the theory of complementary chromatic adaptation. II. Brown, green, and red seaweeds. *J. Phycol.* 19:173–178.

RANDOLPH, S. E. 1975. Patterns of distribution of the tick *Ixodes trianguliceps* Birula on its host. *J. Anim. Ecol.* 44:451–474.

RANNEY, J. W. 1977. *Forest island edges: Their structure, development, and importance to regional forest ecosystem dynamics.* EDFB/IBP Cont. No. 77/1, Oak Ridge National Laboratory, Oak Ridge, TN.

RANNEY, J. W., M. C. BRUNNER, AND J. B. LEVENSON. 1981. The importance of edge in the structure and dynamics of forest islands. In R. L. Burgess and D. M. Sharpe (eds.), *Forest Island Dynamics in Man-dominated Landscapes* (Ecological Studies No. 41). Springer-Verlag, New York, pp. 67–95.

RANWELL, D. S. 1961. *Spartina* salt marshes in southern England: 1. The effects of sheep grazing at the upper limits of *Spartina* marsh in Bridgewater Bay. *J. Ecol.* 49:325–340.

RAO, S. S., A. A. JURKOVIC, AND O. NRIAGU. 1984. Bacterial activity in sediments of lakes receiving acid precipitation. *Environ. Pollution (A)* 36:195–205.

RAPP, M., AND P. LOSSAINT. 1981. Some aspects of mineral cycling in the garrigue of southern France. In F. di Castri, D. Goodall, and R. L. Specht (eds.), *Mediterranean-type Shrublands (Ecosystems of the World,* Vol. 11). Elsevier, Amsterdam, pp. 289–302.

RASMUSSEN, R. A., AND M. A. K. KAHLIL. 1986. Atmospheric trace gases: Trends and distribution over the last decade. *Science* 232:1623–1624.

RATLIFF, R. D. 1982. A correction of Cole's $C_7$ and Hurlbert's $C_8$ coefficients of interspecific association. *Ecology* 63:1605–1606.

RATTI, J. T., AND K. P. REESE. 1988. Preliminary test of the ecological trap hypothesis. *J. Wildl. Manage.* 52:484–491.

RATTI, J. T., AND K. P. REESE. 1988. Edge effect: A concept under scrutiny. *Trans. N. A. Wildl. & Nat. Res. Conf.* 53:127–136.

RAVEN, P. H. 1973. The evolution of Mediterranean flora. In F. di Castri and H. A. Mooney (eds.), *Mediterranean-type Ecosystems: Origin and Structure.* Springer-Verlag, New York, pp. 213–224.

RAWLINS, J. E. 1980. Thermoregulation by the black swallowtail butterfly *Papilio polypenes. Ecology* 61:345–357.

RAYMOND, D., D. J. JOUREL, J. M. BARNOLA, J. CAPPELLAZ, R. J. DELMAS, AND C. LORIUS. 1993. The ice record of greenhouse gases. *Science* 259:934–941.

RAYNER, A. D. M., AND C. BIDDY. 1988. Fungal communities in the decay of wood. *Adv. Microb. Ecol.* 10:115–166.

RAYNOL, D. J. 1983. Atmospheric deposition and ionic input in Adirondack forests. *J. Air Pollution Control Assoc.* 33:1032–1036.

READ, E. A. 1994. The importance of community classification to mitigation and restoration of coastal sagebrush. *Restoration Ecol.* 2:80–86.

REAL, L. 1977. The kinetics of functional response. *Am. Nat.* 111:289–300.

———. 1983. *Pollination Ecology.* Academic Press, Orlando FL.

REAL, L., AND T. CARACO. 1986. Risk and foraging in stochastic environments. *Ann. Rev. Ecol. Syst.* 17:371–390.

REED, D. C., AND M. S. FOSTER. 1984. The effects of canopy shading on algal recruitment and growth in a giant kelp forest. *Ecology* 65:937–948.

REEKIE, E. G., AND F. A. BAZZAZ. 1987. Reproductive efforts in plants. 3. Effect of reproduction on vegetative activity. *Am. Nat.* 29:907–919.

REEM, C. H. 1976. Loon productivity, human disturbance, and pesticide residues in northern Minnesota. *Wilson Bull.* 88:427–431.

REGEHR, D. L., AND F. A. BAZZAZ. 1976. Low temperature photosynthesis in successional winter annuals. *Ecology* 57:1297–1303.

REGIER, H. A., AND K. H. LOFTUS. 1972. Effects of fisheries exploitation on salmonid communities in oligotrophic lakes. *J. Fish. Res. Board Can.* 29:959–968.

REICE, S. R. 1994. Nonequilibrium determinants of biological community structure. *Amer. Sci.* 82:424–435.

REICHARDT, P. B., J. P. BRYANT, T. P. CLAUSEN, AND G. WIELAND. 1984. Defense of winter-dormant Alaska paper birch against snowshoe hare. *Oecologica* 68:58–59.

REICHLE, D. E. 1971. Energy and nutrient metabolism of soil and litter invertebrates. In P. Duvigneaud (ed.), *Productivity of Forest Ecosystems.* UNESCO, Paris, pp. 465–477.

REICHLE, D. E. (ED.). 1981. *Dynamic Properties of Forest Ecosystems.* Cambridge University Press, Cambridge, England.

REICHLE, D. E., B. E., DINGER, N. T. EDWARDS, W. F. HARRIS, AND P. SOLLINS ET AL. 1973. Carbon flow and storage in a forest ecosystem. In G. M. Woodwell and E. V. Pecan (eds.), *Carbon and the Biosphere,* National Technical Information Service, Springfield, VA, pp. 345–365.

REIFSNYDER, W. E., AND H. W. LULL. 1965. Radiant energy in relation to forests. USDA Tech. Bull. No. 1344.

REINECKE, K. J., AND D. DELNICKI. 1992. *DUCKDATA: A bibliographic data base for North American waterfowl (Anatidae) and their wetland habitats.* U.S. Fish and Wildlife Service Resource Publication 186.7 pp.

REITER, R. J. 1981. The mammalian pineal gland: Structure and function. *Am. J. Anat.* 162:287–313.

REX, M. A. 1981. Community structure in the deep-sea benthos. *Ann. Rev. Ecol. Syst.* 12:331–354.

REY, J. R. 1981. Ecological biogeography of arthropods on *Spartina* islands in northwest Florida. *Ecol. Monogr.* 51:237–265.

REYNOLDS, J. C. 1985. Details of the geographic replacement of the red squirrel *Sciurus vulgaris* by the gray squirrel *Sciurus carolinensis* in eastern England. *J. Anim. Ecol.* 54:149.

REYNOLDS, D. N. 1984. Alpine annual plants: Phenology, germination, photosynthesis, and growth of three Rocky Mountain species. Ecology, 65:759–766.

RHOADES, D. F., AND R. G. CATES. 1976. A general theory of plant antiherbivore chemistry. *Recent Adv. Phytochemistry* 10:168–213.

RICE, W. R. 1982. Acoustical location of prey by the marsh hawk: Adaptation to concealed prey. *Auk* 99:403–413.

RICE, E. L. 1972. Allelopathic effects of *Andropogon virginicus* and its persistence in old fields. *Am. J. Bot.* 59:752–755.

RICKLEFS, R. 1979. *Ecology,* 2nd ed. Chiron Press, New York.

———. 1980. Geographical variation in clutch size in passerine birds: Ashmoles's hypothesis. *Auk* 97:38–49.

———. 1987. Community diversity: Relative roles of local and regional processes. *Science* 235:167–171.

RIEBESELL, J. F. 1981. Photosynthetic adaptations in bog and alpine populations of *Ledum groenlandica. Ecology* 62:579–586.

RIECHERT, S. E. 1981. The consequences of being territorial: Spiders, a case study. *Am. Nat.* 117:871–892.

RICH, P. H., AND R. G. WETZEL. 1978. Detritus in the lake ecosystem. *Am. Nat.* 112:57–71.

RIEDMAN, M. L. 1982. The evolution of alloparental care and adoption in mammals and birds. *Q. Rev. Biol.* 57:405–435.

RIGBY, C., AND J. H. LAWTON. 1981. Species-area relationships of arthropods on host plants: Herbivores on bracken. *J. Biogeog.* 8:125–133.

RILEY, G. A. 1973. Particulate and dissolved organic carbon in the oceans. In G. M. Woodwell and E. V. Pecan (eds.), *Carbon and the Biosphere* National Technical Information Service, Springfield, VA, pp. 204–220.

RINGELMAN, J. K. 1992a. *Ecology of montane wetlands.* U.S. Fish and Wildlife Service Fish and Wildlife Leaflet 13.3.6.7 pp.

———. 1992b. *Identifying the factors that limit duck production.* U.S. Fish and Wildlife Service Fish and Wildlife Leaflet 13.2.7.8 pp.

RINGLER, N. 1979. Selective predation by drift-feeding brown trout (*Salmo trutta*). *J. Fish. Res. Bd. Can.* 26:392–403.

RISEBROUGH, R. W., W. WALKER, T. T. SCHMIDT, B. W. DELAPPE, AND C. W. CONNERS. 1976. Transfer of chlorinated biphenyls to Antarctica. *Nature* (London) 264:738–739.

RISSER, P. G. 1985. Grasslands. In B. F. Chabot and H. A. Mooney (eds.), *Physiological Ecology of North American Plant Communities.* Chapman and Hall, New York, pp. 236–256.

RISSER, P. G., E. C. BIRNEY, H. D. BLOCKER, S. W. MAY, W. J. PARTON, AND J. A. WIENS. 1981. *The True Prairie Ecosystem* (US/IBP Synthesis Series 16). Hutchinson, Ross, Stroudsburg, PA.

ROBBINS, C. S., D. K. DAWSON, AND B. A. DOWELL. 1989. Habitat requirements of breeding forest birds of the Middle Atlantic States. *Wildl. Monogr.* 103. The Wildlife Society.

ROBBINS, C. T., S. MOLE, A. E. HAGERMAN, AND T. A. HANLEY. 1987. Role of tannins in defending plants against ruminants: reduction dry matter digestion? *Ecology* 68:1606–1615.

ROBERTSON, G. J., M. D. WATSON, AND F. COOK. 1992. Frequency, timing, and costs of interspecific nest parasitism in the common eider. *Condor* 94:871–879.

ROBINSON, S. K., AND R. T. HOLMES. 1984. Effects of plant species and foliage structure on the foraging behavior of forest birds. *Auk,* 101:672–684.

RODIN, L. Y., AND N. I. BAZILEVIC. 1967. *Production and Mineral Cycling in Terrestrial Vegetation.* (Translated from Russian by Scripta Technica). G.E. Fogg (ed.) Oliver and Boyd, Edinburgh.

———. 1968. World distribution of biomass. In F. E. Eckhard (ed.), *Functioning of Terrestrial Ecosystems at the Primary Production Level.* UNESCO, Paris, pp. 45–52.

ROGERS, L. L. 1987. Effects of food supply and kinship on social behavior, movements, and population dynamics of black bears in northeastern Minnesota. *Wildl. Monogr.* 97. The Wildlife Society Washington DC.

———. 1987. Factors influencing dispersal in black bears. In B. D. Chepko-Sade and Z. T. Halpin (eds.), *Mammalian Dispersal Patterns.* University of Chicago Press, Chicago, pp. 75–84.

ROHWER, F. C., AND S. FREEMAN. 1989. The distribution of conspecific nest parasitism in birds. *Can. J. Zool.* 67:239–253.

ROMESBURG, H. C. 1981. Wildlife science: Gaining reliable knowledge. *J. Wildl. Manage.* 45:293–313.

ROMME, W. H., AND D. H. KNIGHT. 1982. Landscape diversity: The concept applied to Yellowstone Park. *Bioscience* 32:664–670.

ROOT, R. B. 1967. The niche exploitation pattern of the blue-gray gnatcatcher. *Ecol. Monogr.* 37:317–350.

ROSEBERRY, J. L., AND W. D. KLIMSTRA. 1984. *Population Ecology of the Bobwhite.* Southern Illinois University Press, Carbondale.

ROSENZWEIG, M. L., AND R. H. MACARTHUR. 1963. Graphical representation and stability conditions of predator-prey interactions. *Am. Nat.* 97:209–223.

ROSS, J. 1982. Myxomatosis: The natural evolution of disease. *Symp. Zool. Soc. Lond.* 50:77–95.

ROSSWALL, T., J. G. K. FLOWER-ELLIS, L. G. JOHANSSON, S. JOHANSSON, B. E. RYDÉN AND M. SONESSON. 1975. Stordalen (Abisko), Sweden. In T. Rosswall and O. W. Heal (eds.), *Structure and Function of Tundra Ecosystems.* Swedish Natural Science Research Council, Stockholm, pp. 265–294.

ROSSWALL, T., AND U. GRANHALL. 1980. Nitrogen cycling in a subarctic ombrotrophic mire. In M. Sonesson (ed.), *Ecology of a Subarctic Mire. Ecol. Bull.* 30. Swedish Natural Science Research Council, Stockholm.

ROTENBERRY, J. T., AND J. A. WEINS. 1985. Statistical power analysis and community-wide patterns. *Am. Nat.* 125:164–168.

ROUGHGARDEN, J. 1974. Species packing and the competition function with illustrations from coral reef fish. *Theor. Pop. Biol.* 5:163–186.

———. 1986. A comparison of food-limited and space-limited animal competition communities. In J. Diamond and T. J. Case (eds.), *Community Ecology.* Harper & Row, New York, pp. 492–516.

ROYAMA, T. 1970. Factors governing the hunting behavior and selection of food by the great tit. *J. Anim. Ecol.* 39:619–668.

RUINEN, J. 1962. The phyllosphere: An ecologically neglected region. *Plant Soil* 15:81–109.

RUNDEL, P. W. 1980. The ecological distribution of $C_3$ and $C_4$ grasses in the Hawaiian Islands. *Oecologica* 45:354–359.

———. 1981. The mattoral zone of central Chile. In F. di Castri, D. Goodall, and R. Specht (eds.), *Mediterranean-type Ecosystems.* Elsevier, Amsterdam, pp. 175–210.

RUNKLE, J. R. 1981. Gap regeneration in some old-growth forests of eastern United States. *Ecology* 62:1041–1051.

———. 1984. Development of woody vegetation in treefall gaps in a beech-maple forest. *Holarctic Ecol.* 7:157–164.

———. 1985. Disturbance regimes in temperate forests. In S. T. A. Pickett and P. S. White (eds.), *Ecology of Natural Disturbance and Patch Dynamics.* Academic Press, Orlando, Fl, pp. 17–34.

RUNKLE, J. R., AND T. C. YETTER. 1987. Treefalls revisited: Gap dynamics in the southern Appalachians. *Ecology* 68:417–424.

RUSCH, D. H., E. C. MESLOW, P. D. DOERR, AND L. B. KEITH. 1972. Response of great horned owl populations to changing prey densities. *J. Wildl. Manage.* 36:282–296.

RUTHERFORD, M. C. 1978. Primary production ecology in southern Africa. In M. J. Wegner (ed.), *Biogeography and Ecology of Southern Africa.* Junk, The Hague, Netherlands, pp. 621–659.

RYAN, D. F., AND F. H. BORMAN. 1982. Nutrient resorption in northern hardwoods forests. *Bioscience* 32:29–32.

RYAN, M. G. AND R. H. WARING. 1992. Maintenance respiration and stand development in a subalpine lodgepole pine forest. *Ecology* 73:2100–2108.

RYTHER, J. H., AND C. S. YENTSCH. 1957. The estimation of phytoplankton production in the ocean from the chlorophyll and light data. *Limno. Oceanogr.* 2:381–386.

SADLIER, R. M. F. 1969. *The Ecology of Reproduction in Wild and Domestic Mammals.* Methuen, London.

SAFFO, M. B. 1987. New light on seaweeds. *Bioscience.* 37:654–664.

SAKAI, A. K., AND T. A. BURRIS. 1985. Growth in male and female aspen clones: A twenty-five year longitudinal study. *Ecology* 66:1921–1927.

SALE, P. F. 1980. The ecology of fishes on coral reefs. *Oceanogr. Mar. Biol. Ann. Rev.* 18:367–421.

SALISBURY, E. J. 1929. The biological equipment of species in relation to competition. *J. Ecol.* 17:197–222.

SANDERS, H. L. 1968. Marine benthic diversity: A comparative study. *Am. Nat.* 102:243–282.

SANDON, H. 1927. *The Composition and Distribution of Protozoan Fauna of the Soil.* Oliver and Boyd, London.

SANTOS, P. F., N. Z. ELKINS, Y. STEINBERGER, AND W. G. WHITFORD. 1984. A comparison of surface and buried *Larrea tridentata* leaf litter decomposition in North American hot deserts. *Ecology* 65:278–284.

SANTOS, P. F., J. PHILLIPS, AND W. G. WHITFORD. 1981. The role of mites and nematodes in early stages of buried litter decomposition in a desert. *Ecology* 62:664–669.

SARGEANT, A. B., R. J. GREENWOOD, M. A. SOVADA, AND T. L. SHAFFER. 1993. *Distribution and abundance of predators that affect duck production in the prairie pothole region.* U.S. Fish and Wildlife Service Res Pub 194. 96 pp.

SARGENT, R. C. 1989. Allopaternal care in the fathead minnow, *Pimiphales promelas:* Stepfathers discriminate against their adopted eggs. *Behav. Ecol. Sociobiol.* 25:379–385.

SARGENT, R. C., AND M. R. GROSS. 1985. Parental investment decision rules and the Concorde fallacy. *Behav. Ecol. Sociobiol.* 17:43–45.

SARGENT, R. C., P. D. TAYLOR, AND M. P. GROSS. 1987. Parental care and the evolution of egg size in fish. *Am. Nat.* 129:32–46.

SARMIENTO, G. 1984. *The Ecology of Neotropical Savannas.* Harvard University Press, Cambridge, MA.

SARMIENTO, G., AND M. MONASTERRIO. 1975. A critical consideration of environmental conditions associated with the occurrence of savanna ecosystems in tropical America. In F. B. Golly and E. Medina (eds.), *Tropical Ecological Systems.* Springer-Verlag, Berlin, pp. 223–250.

SARUKHAN, J., M. MARTINEZ-RAMOS, AND D. PINERO. 1984. The analysis of demographic variability at the individual level and its population consequences. In R. Drizo and J. Sarukhan (eds), *Perspectives in Plant Population Ecology.* Sinauer Associates, Sunderland, MA, pp. 83–106.

SASEKUMAR, A. AND C. V. CHING. 1987. Mangroves and praws: Further perspectives. *Proc. 10th Annual Seminar of the Malaysian Society of Marine Sciences,* pp. 10–22.

SAUGER, E. A., AND B. RIPLEY. 1974. Microclimate and production of a native grassland: A micrometeorological study. *Oecologica Plantarum* 9:33–363.

SAUNDERS, D. S. 1982. *Insect Clocks,* 2nd ed. Pergamon Press, Oxford.

SAUNDERS, G. W. 1976. Decomposition in fresh water. In J. M. Anderson and A. Macfadyin (eds.), *The Role of Terrestrial and Aquatic Organisms in the Decomposition Process.* Blackwell, Oxford, pp. 341–373.

SCHAEFER, R. 1973. Microbial activity under seasonal conditions of drought in Mediterranean climates. In F. di Castri and H. A. Mooney (eds.), *Mediterranean-type Ecosystems: Origin and Structure.* Springer-Verlag, New York, pp. 191–198.

SCHAFFER, W. M. 1974. Optimal reproductive effort in fluctuating environments. *Am. Nat.* 108:783–790.

———. 1981. Ecological abstraction: The consequences of reduced dimensionality in ecological models. *Ecol. Monogr.* 51:383–401.

———. 1985. Order and chaos in ecological systems. *Ecology* 66:93–106.

SCHAFFER, W. M., AND M. KOT. 1985. Differential systems in ecology and epidemiology. In A. V. Holder (ed.), *Chaos.* Manchester University Press, Manchester.

SCHALL, B. A. 1978. Age structure in *Liatris acidota. Oecologica* 32:93–100.

———. 1984. Life history variation, natural selection, and maternal effects in plant populations. In R. Drizo and J. Sarukhan (eds.), *Perspectives in Plant Population Ecology.* Sinauer, Sunderland, MA, pp. 188–211.

SCHALLER, G. B. 1972. *Serengeti: A Kingdom of Predators.* Knopf, New York.

SCHEEL, D., AND C. PACKER. 1991. Group hunting behavior of lions: A search for cooperation. *Anim Behav.* 41:697–709.

SCHEFFER, V. C. 1951. The rise and fall of a reindeer herd. *Sci. Month.* 73:356–362.

SCHEINER, S. M. 1993. Genetics and the evolution of phenotypic plasticity. *Ann Rev. Ecol. Syst.* 24:35–68.

SCHEMISKE, D. W., AND T. BROKAW. 1981. Treefalls and the distribution of understory birds in a tropical forest. *Ecology* 62:938–945.

SCHLESINGER, W. H. 1977. Carbon balance in terrestrial detritus. *Ann. Rev. Ecol. Syst.* 8:51–81.

SCHLICHTER, L. C. 1981. Low pH affects the fertilization and development of *Rana pipiens* eggs. *Can. J. Zool.* 59:1693–1699.

SCHMID, A. 1982. Survival of frogs in low temperature. *Science* 215:312–315.

SCHMIDT-NIELSON, K. 1960. The salt secreting gland of marine birds. *Circulation* 21:955–967.

SCHMITT, C. J., A. D. LEMLY, AND P. V. WINGER. 1993. *Habitat suitability index model for brook trout in streams of the southern Blue Ridge Province: Surrogate variables, model evaluation, and suggested improvements.* U.S. Fish and Wildlife Service Biological Report 18. 43 pp.

SCHODDE, R. 1981. Bird communities of the Australian mallee: Composition, derivation, distribution, structure, seasonal cycles. In F. di Castri, D. Goodall, and R. Specht (eds.), *Mediterranean-type Shrublands* (*Ecosystems of the World,* Vol. 11). Elsevier, Amsterdam, pp. 387–416.

SCHOENER, T. 1982. The controversy over interspecific competition. *Am. Sci.* 70:586–595.

———. 1983. Field experiments on interspecific competition. *Am. Nat.* 122:240–285.

———. 1983. Simple models of optimal feeding-territory size: A reconciliation. *Am. Nat.* 121:608–629.

SCHOENER, T. W., AND D. A. SPILLER. 1987. Effect of lizards on spider populations: Manipulative reconstruction of a natural experiment. *Science* 236:949–953.

SCHOENER, T. W., AND C. A. TOFT. 1983. Spider populations: Extraordinarily high densities on islands without top predators. *Science* 21:1353–1355.

SCHOFIELD, C. L., D. JOSEPHSON, C. KELEHER, AND S. P. GLOSS. 1993. *Thermal stratification of dilute lakes—evaluation of regulatory processes and biological effects before and after base addition: Effects on brook trout habitat and growth.* U.S. Fish and Wildlife Service Biological Report 9. 36 pp.

SCHOFIELD, C. L., AND J. R. TROJNAR. 1980. Aluminum toxicity to brook trout (*Salvelinus fontinali*) in acidified waters. In T. Y. Toribara, M. W. Miller, and P. E. Morrow (eds.), *Polluted Rain.* Plenum, New York, pp. 341–365.

SCHOLANDER, P. F., R. HOCK, V. WALTHERS, F. JOHNSON, AND L. IRVING. 1950. Heat regulation in some arctic and tropical birds and mammals. *Biol. Bull.* 99:237–258.

SCHOLANDER, P. F., V. WALTERS, R. HOCK, L. IRVING, AND F. JOHNSON. 1950. Body insulation of some arctic and tropical mammals and birds. *Biol. Bull.* 99:225–236.

SCHONEWALD-COX, C., AND M. BUECHNER. 1992. Park protection and public roads. In P. L. Feidler and S. K. Jain (eds.), *Conservation Biology: The Theory and Practice of Nature Conservation, Preservation, and Management.* Chapman and Hall, New York, pp. 373–395.

SCHROEDER, M. J., AND C. C. BUCK. 1970. *Fire Weather.* USDA Ag. Handbook 360, USDA Forest Service, Washington, DC.

SCHROEDER, R. L., AND A. W. ALLEN. 1992. *Assessment of habitat of wildlife communities on the Snake River, Jackson, Wyoming.* U.S. Fish and Wildlife Service Resource Publication 190. 21 pp.

SCHROEDER, R. L., AND S. L. HAIRE. 1993. *Guidelines for the development of community-level habitat evaluation models.* U.S. Fish and Wildlife Service Biological Report 8. 8 pp.

SCHULTZ, J. C., AND I. T. BALDWIN. 1982. Oak leaf quality declines in response to defoliation by gypsy moth larvae. *Science* 217:149–151.

SCHULZE, E. D., R. H. ROBICHAUX, J. GRACE, P. W. RUNDEL, AND J. R. EHLERINGER. 1987. Plant water balance. *Bioscience* 37:30–37.

SCHWARTZ, S. E. 1989. Acid deposition: Unraveling a regional phenomenon. *Science* 243:753–763.

SCRIBER, J. M. A. AND F. SLANSKY, JR. 1981. The nutritional ecology of immature insects. *Ann. Rev. Entomol.* 26:182–211.

SEARCY, W. A. 1979. Male characteristics and pairing success in red-winged blackbird. *Auk* 96:353–363.

SEARCY, W. A., AND M. ANDERSSON. 1986. Sexual selection and the evolution of song. *Ann. Rev. Ecol. Syst.* 17:507–534.

SEARCY, W. A., P. D. MCARTHUR, AND K. YASUKAWA. 1985. Song repertoire size and male quality in song sparrows. *Condor* 87:222–228.

SEARS, P. B. 1935. *Deserts on the March.* University of Oklahoma Press, Normal, OK.

SEASTEDT, T. R. 1984. The role of microarthropods in the decomposition and mineralization process. *Ann. Rev. Entomol.* 29:25–46.

SEGERSTRALE, S. G. 1947. New observations on the distribution and morphology of the amphipod *Gammarus zaddachi* Sexon, with notes on related species. *J. Marine Biol. Assoc. U.K.* 27:219–244.

SEMIL, B., AND P. W. SHERMAN. 1986. Dynamics of nest parasitism in wood ducks. *Auk* 103:813–816.

SEMLER, D. E. 1971. Some aspects of adaptation in a polymorphism for breeding in the three-spined stickleback (*Gasterostermus aculeatus*). *J. Zool.* (Lond.) 165:291–302.

SHAPIRO, D. Y. 1979. Social behavior, group structure, and the control of sex reversal in hermaphroditic fish. *Adv. Study Behav.* 10:43–102.

———. 1980. Serial female sex changes after simultaneous removal of males from social groups of a coral reef fish. *Science* 209:1136–1137.

———. 1987. Differentiation and evolution of sex change in fishes. *Bioscience* 37:490–497.

SHAW, S. P., AND C. G. FREDINE. 1956. *Wetlands of the United States.* U.S. Fish and Wildl. Circ. 39.

SHEA, M. L., R. S. WARREN, AND W. A. NEIRING. 1975. Biochemical and transplantation studies of the growth form of *Spartina alterniflora* on Connecticut salt marshes. *Ecology* 56:461–466.

SHELDON, W. G. 1967. *The Book of the American Woodcock.* University of Massachusetts Press, Amherst.

SHELFORD, V. E. 1911. Physiological animal geography. *J. Morph.* 22:551–618.

———. 1913. Animal communities in temperate America. *Bull. Geog. Soc. Chicago* 5:1–368.

SHERMAN, P. W. 1977. Nepotism and the evolution of alarm calls. *Science* 197:1246–1253.

———. 1981. Kinship, demography and Belding's ground squirrel nepotism. *Behav. Ecol. Sociobiol.* 8:251–259.

SHERMAN, P. W., J. V. M. JARVIS, AND R. D. ALEXANDER. 1991. *The Biology of the Naked Mole Rat.* Princeton University Press, Princeton, NJ.

SHERR, E. B., AND B. F. SHERR. 1991. Planktonic microbes:Tiny cells at the base of the ocean's food webs. *Tree* 6:50–54.

SHIELDS, W. M. 1987. Dispersal and mating systems: Investigating their causal connections. In B. D. Chepho-Sade and Z. T. Halpi (eds.), *Mammalian Dispersal Patterns.* University of Chicago Press, Chicago, pp. 3–24.

SHIPLEY, L. A., J. E. GROSS, D. E. SPALINGER, N. T. HOBBS, AND B. A. WUNDER. 1994. The scaling intake rate in mammalian herbivores. *Am. Nat.* 143:1055–1082.

SHUGART, H. H. 1984. *A Theory of Forest Dynamics: The Ecological Implications of Forest Succession.* Springer-Verlag, New York.

SHUGART, H. H., AND D. C. WEST. 1977. Development of an Appalachian deciduous forest model and its application to assessment of the impact of the chestnut blight. *J. Environ. Manage.* 5:161–169.

SHURE, D. J., AND H. S. RAGSDALE. 1977. Patterns of primary succession on granite outcrop surfaces. *Ecology* 58:993–1006.

SIEBURTH, J. M., AND A. JENSEN. 1970. Production and transformation of extracellular organic matter from marine littoral algae: A resume. In D. E. Hood (ed.), *Organic Matter in Natural Waters.* University of Alaska, Fairbanks, pp. 203–223.

SIH, A., P. C. ROWLEY, M. MCPEEK, J. PETRANKA, AND K. STROHMEIER. 1985. Predation, competition, and prey communities: A review of field experiments. *Ann Rev. Ecol. Syst.* 16:269–311.

SILK, J. B. 1983. Local resource competition and facultative adjustment of sex ratio in relation to competitive abilities. *Am. Nat.* 121:56–66.

SIMBERLOFF, D., AND N. GOTELLI. 1984. Effects of insularization on plant species richness in the prairie-forest ecotone. *Biol. Cons.* 29:27–46.

SIMBERLOFF, D. S. 1974. Equilibrium theory of island biogeography and ecology. *Ann. Rev. Ecol. Syst.* 5:161–182.

———. 1978. Using biogeographic distributions to determine if colonization is stochastic. *Am. Nat.* 112:723–726.

———. 1983. Competition theory, hypothesis testing, and other community ecological buzzwords. *Am. Nat.* 122:626–635.

SIMBERLOFF, D. S., AND L. G. ABELE. 1976. Island biogeographic theory and conservation practice. *Science* 191:285–286.

———. 1982. Refuge design and island biogeographic theory: Effects of fragmentation. *Am. Nat.* 120:41–50.

———. 1984. Conservation and obfuscation: Subdivision of reserves. *Oikos* 42:399–401.

SIMBERLOFF, D. S., AND E. O. WILSON. 1969. Experimental zoogeography of islands: The colonization of empty islands. *Ecology* 50:278–296.

———. 1970. Experimental zoogeography of islands: A two-year record of colonization. *Ecology* 50:278–296.

SIMPSON, G. G. 1964. Species density of North American recent mammals. *Syst. Zool.* 13:57–73.

SIMS, P. L. 1988. Grasslands. In M. G. Barbour and W. D. Billings (eds.), *North American Terrestrial Vegetation.* Cambridge University Press, New York, pp. 266–286.

SIMS, P. L., AND J. S. SINGH. 1971. Herbage dynamics and net primary production in certain grazed and ungrazed grasslands in North America. In N. R. French (ed.), *Preliminary Analysis of Structure and Function in Grasslands.* Range Sci. Dept. Sci. Ser. No. 10. Colorado State University, Fort Collins, pp. 59–124.

SINCLAIR, A. R. E. 1977. *The African Buffalo: A Study of Resource Limitation of Populations.* University of Chicago Press, Chicago.

———. 1979. Dynamics of the Serengeti Ecosystem. In A. R. E. Sinclair and M. Norton-Griffiths (eds.), *Serengeti: Dynamics of an Ecosystem.* University of Chicago Press, Chicago, pp. 1–30.

SINCLAIR, A. R. E., J. M. GISLINE, G. HOLDSWORTH, C. T. KREBS, S. BOUTIN, J. N. M. SMITH, R. BOONSTRA, AND M. DALE. 1993. Can the solar cycle and climate synchronize the snowshoe hare cycle in Canada? Evidence of tree rings and ice cores. *Am. Nat.* 141:173–198.

SINCLAIR, A. R. E., C. J. KREBS, J. N. M. SMITH, AND S. BOUTIN. 1988. Population biology of snowshoe hares. III. Nutrition, plant secondary compounds, and food limitation. *J. Anim. Ecol.* 57:787–806.

SINCLAIR, A. R. E., AND J. N. M. SMITH. 1984. Do secondary compounds determine feeding preferences of snowshoe hares? *Oecologica* 61:403–410.

SKUTCH, A. 1986. *Helpers at Birds' Nests: A World-wide Survey of Cooperative Breeding and Related Behavior.* University of Iowa Press, Iowa City.

SLATKIN, M. 1987. Gene flow and the geographic structure of natural populations. *Science* 236:787–792.

SLOBODKIN, L. B. 1962. *Growth and Regulation of Animal Population.* Holt, Rinehart and Winston, New York.

SMALLEY, A. E. 1960. Energy flow of a salt marsh grasshopper population. *Ecology* 41:672–677.

SMAYDA, T. J. 1983. The phytoplankon of estuaries. In B. Ketchum (ed.), *Estuaries and Enclosed Seas* (*Ecosystems of the World,* Vol. 26). Elsevier, Amsterdam.

SMITH, A. D. M. 1985. A continuous time model: A deterministic model of temporal rabies. In P. J. Bacon (ed.), *Population Dynamics of Rabies in Wildlife.* Academic Press, London, pp. 131–146.

SMITH, A. P., AND T. P. YOUNG. 1987. Tropical alpine plant ecology. *Ann. Rev. Ecol. Syst.* 18:137–158.

SMITH, B. D. 1978. *Mississippi Settlement Patterns.* Academic Press, New York.

SMITH, C. C., AND S. D. FRETWELL. 1974. The optimal balance between size and number of offspring. *Am. Nat.* 108:499–506.

SMITH, D. W. 1986. *Principles of Silviculture.* McGraw-Hill, New York.

SMITH, H. J. 1978. Parasites of red foxes in New Brunswick and Nova Scotia. *J. Wildl. Dis.* 14:366–370.

SMITH, M. H., R. K. CHESSER, E. C. COTHRAN, AND P. E. JOHNS. 1983. Genetic variability and antler growth in a natural population of white-tailed deer. In R. D. Brown (ed.), *Antler Development in Cervidae.* Caesar Kleberg Wildl. Res. Inst., Kingsville, TX.

SMITH, R. A. H., AND A. D. BRADSHAW. 1979. The use of heavy metal tolerant plant populations for the reclamation of metalliferous wastes. *J. App. Ecol.* 16:595–612.

SMITH, R. E., AND B. A. HORWITZ. 1969. Brown fat and thermogenesis. *Physiol. Rev.* 49:330–425.

SMITH, R. L. 1956. An evaluation of conifer plantations as wildlife habitat. PhD Dissertation, Cornell University, Ithaca, NY.

———. 1959. Conifer plantations as wildlife habitat. *N.Y. Fish Game J.* 5:101–132.

———. 1962. *Acorn consumption by white-footed mice* (Peromyscus leucopus). Bull. 482T, WV Univ. Agr. Expt. Sta.

———. 1963. Some ecological notes on the grasshopper sparrow. *Wilson Bull.* 75:159–165.

———. 1976. Socio-ecological evolution in the hill country of southwestern West Virginia. In J. Luchok et al. (eds.), *Hill Lands.* Proc. International Symposium, West Virginia University Office of Publications, Morgantown WV pp. 198–202.

SMITH, S. M. 1978. The "underworld" in a territorial adaptive strategy for floaters. *Am. Nat.* 112:570–582.

SMITH, T. M., AND M. HUSTON. 1987. A theory of spatial and temporal dynamic of plant communities. *Vegetatio* 83:49–69.

SMITH, T. M., AND P. GOODMAN. 1987. The effect of competition on the structure and dynamics of *Acacia* savannas in southern Africa. *J. Ecol.* 75:1013–1044.

SMITH, T. M., AND K. GRANT. 1986. The role of competition in the spacing of trees in a *Burkea africana-Terminalia sericea* savanna. *Biotropica* 18:219–223.

SMITH, T. M., AND H. H. SHUGART. 1987. Territory size variation in the ovenbird: The role of habitat structure. *Ecology* 68:695–704.

SMITH, T. M., H. H. SHUGART, G. B. BONAN, AND H. B. SMITH. 1992. Modeling the potential response of vegetation to global climate change. *Adv. Ecol. Res.* 22:93–116.

SMITH, T. M., AND B. H. WALKER. 1983. The role of competition in the spacing of savanna trees. *Proc. Grassland Soc. S. Africa* 18:159–164.

SMITH, W. 1976. Lead contamination of the roadside ecosystem. *J. Air Pollution Control Assoc.* 26:753–766.

SMITH, W. H. 1981. *Air Pollution and Forests: Interactions between Air Contamination and Forest Ecosystems.* Springer-Verlag, New York.

SMITH, W. P. 1991. Ontogeny and adaptiveness of tail-flagging behavior in white-tailed deer. *Am. Nat.* 138:190–200.

SMOCK, L. A., AND K. L. HARLOWE. 1983. Utilization and processing of freshwater wetland macrophytes by the detritivore *Asellus forbesi. Ecology* 64:1156–1565.

SOJDA, R. S., AND K. L. SOLBERG. 1993. *Management and control of cattails.* U.S. Fish and Wildlife Leaflet 13.4.13.8 pp.

SOKAL, R. R., AND F. J. ROHLF. 1981. *Biostatistics,* 2nd ed. W.H. Freeman, San Francisco.

SOLLINS, P. 1982. Input and decay of coarse woody debris in coniferous forest stands in western Oregon and Washington. *Can. J. For.* 12:18–28.

SOLLINS, P., C. C. GRIER, F. M. MCCORSIN, K. CROMACK, JR., R. FOGEL, AND R. L. FREDRIKSEN. 1980. The internal element cycles of an old-growth Douglas-fir ecosystem in western Washington. *Ecol. Monogr.* 50:275–282.

SOLOMON, A. M. 1986. Transient response of forests to CO$_2$ induced climate change: Simulation modeling experiments in eastern North America. *Oecologica* 68:567–579.

SOLOMON, M. E. 1949. The natural control of animal populations. *J. Anim. Ecol.* 18:1–32.

———. 1957. Dynamics of insect populations. *Ann. Rev. Entomol.* 2:121–142.

SOLOMON, P. M., R. DE ZAFRA, A. PARRISH, AND J. W. BARRET. 1984. Diurnal variation of stratospheric chlorine monoxide: Critical test of chlorine chemistry in the ozone layer. *Science* 224:1210–1214.

SORENSON, M. D. 1991. The functional significance of parasitic egg laying and typical nesting in redhead ducks: An analysis of individual behavior. *Anim. Behav.* 42:771–796.

———. 1993. Parasitic egg laying in canvasbacks: Frequency, success, and individual behavior. *Auk* 110:57–69.

SORIANO, A. 1972. South America. In C. M. McKell et al. (eds.), *Wildland Shrubs: Their Biology and Utilization.* USDA Forest Serv. Gen. Tech. Rept. INT-1, pp. 31–54.

SOULE, M. E. 1986. *Conservation Biology: The Science of Scarcity and Diversity.* Sinauer Associates, Sunderland, MA.

SOUSA, W. P. 1979. Disturbance in marine intertidal boulder fields: The nonequilibrium maintenance of species diversity. *Ecology* 60:1225–1239.

———. 1984. Intertidal mosaics: Patch size, propagule availability, and spatially variable patterns of succession. *Ecology* 65:1918–1935.

SOWLS, L. K. 1960. Results of a banding study of Gambel's quail in southern Arizona. *J. Wildl. Manage.* 24:185–190.

SPAETH, J. N, AND C. H. DIEBOLD. 1938. Some interrelations between soil characteristics, water tables, soil temperature, and snow cover in the forest and adjacent open areas in south central New York. *Cornell Univ. Agr. Exp. Stat. Mem.* 213.

SPALINGER, D. E., AND N. T. HOBBS. 1992. Mechanisms of foraging in mammalian herbivores: New models of functional response. *Am. Nat.* 140:325–348.

SPECHT, R. L. 1979. Heathlands and related shrublands of the world. In R. L. Specht (ed.), *Heathlands and Related Shrublands: Descriptive Studies (Ecosystems of the World,* Vol. 9A). Elsevier, Amsterdam, pp. 1–18.

———. 1981. Mallee ecosystems in southern Australia. In F. di Castri, D. W. Goodall, and R. L. Specht (eds.), *Mediterranean-type Shrublands (Ecosystems of the World,* Vol. 11). Elsevier, Amsterdam, pp. 203–231.

SPENSER, C. N., B. MCCLELLAND, AND J. A. STANFORD. 1991. Shrimp stocking, salmon collapse, and eagle displacement. *Bioscience* 41:14–21.

SPICER, R. A., AND J. L. CHAPMAN. 1990. Climate change and the evolution of high latitude terrestrial vegetation. *TREE* 5:279–284.

SPOONER, G. M. 1947. The distribution of *Gammarus* species in estuaries: Part I. *J. Marine Biol. Assoc. U. K.* 27:1–52.

SPRUGEL, D. G. 1976. Dynamic structure of wave generated *Abies balsamea* forests in northeastern United States. *J. Ecol.* 64:889–911.

SPURR, S. H. 1957. Local climate in the Harvard Forest. *Ecology* 38:37–56.

STACEY, P. B., AND M. TAPER. 1992. Environmental variation and the persistence of small populations. *Ecol. Appl.* 21:18–29.

STAMPS, J. A., AND V. V. KRISHNAN. 1990. The effect of settlement tactics on territory sizes. *Am. Nat.* 135:527–546.

STANTON, N. L. 1988. The underground in grasslands. *Ann. Rev. Ecol. Syst.* 19:573–589.

STARK, N. 1973. *Nutrient Cycling in a Jeffrey Pine Forest Ecosystem.* Montana Forest and Conservation Experiment Station, Missoula.

———. 1976. Fuel reduction and nutrient status and cycling relationships associated with understory burning in larch–Douglas-fir stands. In *Proc. Tall Timbers Fire Ecology Conf.* 143:573–596.

STAUFFER, B., A. NEFTEL, G. FISCHER, AND H. OESCHGER. 1991.

Siple Station (Atmospheric H₄—Historical record from ice cores). In T. A. Boden, R. J. Sepanski, and F. W. Stoss (eds.), *Trend 91: A Compendium of Data on Global Change.* Oak Ridge National Laboratory, Oak Ridge, TN, pp. 220–221.

STEARNS, S. C. 1976. Life history tactics: A review of ideas. *Quart. Rev. Biol.* 51:3–47.

STEBBINS, G. L. 1972. Evolution and diversity of arid-land shrubs. In C. M. McKell et al. (eds.), *Wildland Shrubs: Their Biology and Utilization.* USDA Forest Serv. Gen. Tech. Rept. INT

STECK, F. 1982. Rabies in wildlife. *Symp. Zool. Soc. Lond.* 50:57–75.

STEINBERG, P. D. 1984. Algal˙ defense against herbivores: Allocation of phenolic compounds in the kelp *Alaria marginala. Science* 223:405–407.

STENGER, J., AND J. B. FALLS. 1959. The utilized territory of the ovenbird. *Wilson Bull.* 71:125–140.

STENSETH, N. C. 1983. Causes and consequences of dispersal in small mammals. In I. R. Swingland and P. J. Greenwood (eds.), *The Ecology of Animal Movements.* Oxford, Oxford, pp. 63–101.

STEPHENS, D. W. 1981. The logic of risk-sensitive foraging preferences. *Anim. Behav.* 29:628–629.

STEPHENSON, T. A., AND A. STEPHENSON. 1954. Life between the tide-marks in North America: 3A. Nova Scotia and Prince Edward Island: The geographical features of the region. *J. Ecol.* 42:14–45, 46–70.

STEPONKUS, P. L. 1981. Responses to extreme temperatures: Cellular and subcellular bases. In O. Lange, P. S. Nobel, C. B. Osmund, and H. Zeigler (eds.), *Physiological Plant Ecology. I. Vol 12A. Encyclopedia of Plant Physiology.* Springer Verlag, New York. pp. 371–402.

STERN, J. E., T. HOM, E. CASILLAS, A. FRIEDMAN, AND U. VARANASI. 1987. Simultaneous exposure of English sole (*Parophrys vetulus*) to sediment-associated xenobiotics. Part 2: Chronic exposure to an urban estuarine sediment with added ³H-benzo(a)pyrene and ¹⁴C-Polychlorinated biphenyls. *Marine Environ. Res.* 22:123–151.

STERN, W. L., AND M. F. BUELL. 1951. Life-form spectra in a New Jersey pine barren forest and Minnesota jack pine forest. *Bull. Torrey Bot. Club* 78:61–65.

STEVENS, D. W., AND J. R. KREBS. 1986. *Foraging Theory.* Princeton, NJ: Princeton University Press.

STEVENS, G. C. 1989. The latitudinal gradient in geographical range: How so many species can coexist in the tropics. *Am. Nat.* 133:240–256.

STEVENS, G. C., AND J. F. FOX. 1991. The causes of treelines. *Ann. Rev. Ecol. Syst.* 22:177–192.

STEVENSON, J. R., AND E. F. STOERMER. 1982. Luxury consumption of phosphorus by benthic algae. *Bioscience* 31:682–683.

STEVENSON, R. D. 1985. Body size and limits to the daily range of body temperature in terrestrial ectotherms. *Am. Nat.* 125:102–117.

STEWART, B. A., L. K. PORTER, AND F. G. VIETS. 1966a. Effects of sulfur content of straws on rates of decomposition and plant growth. *Soil Sci. Soc. Am. Proc.* 30:355–358.

———. 1966b. Sulfur requirements for decomposition of cellulose and glucose in soil. *Soil Sci. Soc. Am. Proc.* 30:453–456.

STILES, E. W. 1980. Patterns of fruit presentation and seed dispersal in bird disseminated woody plants in the eastern deciduous forest. *Am. Nat.* 116:670–688.

———. 1982. Fruit flags: Two hypotheses. *Am. Nat.* 120:500–509.

STILES, F. G. 1975. Ecology, flowering phenology, and hummingbird pollination of some Costa Rican *Heliconia* species. *Ecology* 56:285–301.

STODDARD, H. 1932. *The Bobwhite Quail: Its Habits, Preservation, and Increase.* Charles Scribner, New York.

STOWE, L. G., AND J. A. TEERI. 1978. The geographic distribution of C₄ species of the Dicotyledonae in relation to climate. *Am. Nat.* 112:609–623.

STRAHLER, A. 1971. *The Earth Sciences.* Harper & Row, New York.

STRONG, D. R., J. H. LAWTON, AND R. SOUTHWOOD. 1984. *Insects and Plants: Community Patterns and Mechanisms.* Harvard University Press, Cambridge, MA.

STRONG, D. R., JR., AND J. R. REY. 1982. Testing for MacArthur-Wilson equilibrium with the arthropods of the miniature *Spartina archipelago* at Oyster Bay, Florida. *Am. Zool.* 22:350–360.

STRONG, D. R., D. SIMBERLOFF, L. G. ABELE, AND A. B. THISTLE (EDS.). 1984. *Ecological Communities: Conceptual Issues and the Evidence.* Princeton University Press, Princeton, NJ.

STUIVER, M. 1979. Atmospheric carbon dioxide in the nineteenth century. *Science* 202:1109.

SUBRAMANIAN, A. N., S. TANABE, H. TANEKA, H. HIDAKA, AND R. TATSUKAWA. 1987. Gain and loss rates and biological half-life of PCBs and DDT in the bodies of Adelie penguins. *Environ. Pollution* 43:39–46.

SUHONEN, J. 1993. Predation risk influences the use of foraging sites by tits. *Ecology* 74:1197–1203.

SUNDQUIST, E. T. 1993. The global carbon budget. *Science* 259:934–941.

SWANK, W. T., J. W. FITZGERALD, AND J. T. ASH. 1983. Microbial transformation of sulfate in forest soils. *Science* 223:182–184.

SWANK, W. T., AND J. B. WAIDE. 1979. Interpretation of nutrient cycling research in a management context: Evaluating potential effects of alternative management strategies on site productivity. In R. Warming (ed.), *Forest: Fresh Perspectives from Ecosystem Analysis.* Proc. Ann. Biol. Coll. Oregon State University Press, Corvallis, pp. 137–158.

SWANK, W. T., J. B. WAIDE, D. A. CROSSLEY, AND R. L. TODD. 1981. Insect defoliation enhances nitrate export from forest ecosystems. *Oecologica* 51:297–299.

SWIFT, M. J., O. W. HEAL, AND J. M. ANDERSON. 1979. *Decomposition in Terrestrial Ecosystems.* Oxford, Blackwell.

SYDEMAN, W. J., AND S. D. EMSLIE. 1992. Effect of parental age on hatching asynchrony, egg size, and third chick disadvantage in western gulls. *Auk* 109:242–248.

TABER, R. D., AND R. F. DASMANN. 1958. The black-tailed deer of the chaparral. Calif. Dept. Fish Game, *Game Bull.* No. 8.

TAHVANAINEN, J., E. HELLE, R. JULKUNEN-TITTO, AND A. LAVOLA. 1985. Phenolic compounds of willow bark as deterrents against feeding by mountain hare. *Oecologica* 65:319–323.

TAIT, D. E. N. 1980. Abandonment as a reproductive tactic: The example of grizzly bears. *Am. Nat.* 115:800–808.

TAIT, R. V. 1968. *Elements of Marine Ecology.* Plenum, New York.

TAJCHMAN, S. J., M. H. HARRIS, AND E. C. TOWNSEND. 1988. Variability of the radiative index of dryness in an Appalachian watershed. *Agric. For. Meteorol.* 42:199–207.

TALLAMY, D. W. 1984. Insect parental care. *Bioscience* 34:20–24.

TAMARIN, R. H. 1978. Dispersal, population regulation, and K-selection in field mice. *Am. Nat.* 112:545–555.

TAMM, C. O. 1951. Removal of plant nutrients from tree crowns by rain. *Physiol. Plant* 4:184–188.

TANNER, J. T. 1975. The stability and intrinsic growth rates of prey and predator populations. *Ecology* 56:855–867.

TANSLEY, A. G. 1935. The use and abuse of vegetational concepts and terms. *Ecology* 16:284–307.

TANSLEY, A. G., AND R. S. ADAMSON. 1925. Studies on the vegetation of English chalk. III. The chalk grasslands of the Hampshire-Sussex border. *J. Ecol.* 13:177–223.

TASKEY, R. D., C. L. CURTIS, AND J. STONE. 1989. Wildfire rye grass seeding and watershed rehabilitation. In N. H. Berg (tech. coordinator), *Proc. Symp. Fire and Watershed Manage. Gen. Tech. Rept.* PSW-109, USDA Forest Service Pacific Southwest Forest and Range Exp. Stat., Berkeley, CA, pp. 115–124.

TATE, R. L., III. 1987. *Soil Organic Matter: Biological and Ecological Effects.* Wiley, New York.

TAYLOR, C. R. 1969. The eland and the oryx. *Sci. Am.* 220(1):88–95.

TAYLOR, C. R., AND C. P. LYMAN. 1972. Heat storage in running antelopes: Independence of brain and body temperatures. *Am. J. Physiol.* 222:114–117.

TAYLOR, R. J. 1984. *Predation.* Chapman and Hall, New York.

TAYLOR, W. R. 1961. Distribution in depth of marine algae in the Caribbean and adjacent seas. In *Recent Advances in Botany,* University of Toronto Press, Toronto, pp. 193–197.

TEAL, J. M. 1962. Energy flow in the salt marsh ecosystem of Georgia. *Ecology* 43:614–624.

TEMPLE, S. A. 1977. The dodo and the tambalacoque tree. *Science* 203:1364.

———. 1986. Predicting impacts of habitat fragmentation on forest birds: A comparison of two models. In J. Verner, M. L. Morrison, and C. T. Ralph (eds.), *Wildlife 2000: Modeling Habitat Relations of Terrestrial Vertebrates.* University of Wisconsin Press, Madison, pp. 301–304.

TEMPLETON, J. W., R. M. SHARP, J. WILLIAMS, D. DAVIS, D. HARMEL, B. ARMSTRONG, AND S. WARDROUP. 1983. Single dominant gene effect on the expression of antler point numbers in the white-tailed deer. In R. D. Brown (ed.), *Antler Development in Cervidae.* Caesar Kleberg Wildlife Research Institute, Kingsville, TX, pp. 469–470.

TERAMURA, A. H. 1990. Implication of stratospheric ozone depletion upon plant production. *Hort. Sci.* 25:1557–1559.

TERRI, J. A. 1979. The climatology of the $C_4$ photosynthetic pathway. In O. T. Solbrig, S. Jain, G. B. Johnson, and P. H. Raven (eds.), *Topics in Plant Population Biology.* Columbia University Press, New York, pp. 356–374.

TERRI, J. A., AND L. STOWE. 1976. Climate patterns and distribution of $C_4$ grasses in North America. *Oecologia* 23:1–12.

TESTER, J. R., AND W. H. MARSHALL. 1961. A study of certain plant and animal interrelations on a native prairie in Northwestern Minnesota. *Minn. Mus. Nat. Hist.* Occasional Paper No. 8.

TEVINI, M., AND A. H. TERAMURA. 1989. UV-B effects on terrestrial plants. *Photochem. Photobiol.* 50:479–487.

THAYER, G. W., W. J. KENWORTHY, AND M. S. FONSECA. 1984. *The Ecology of Eelgrass Meadows of the Atlantic Coast.* U.S. Fish and Wildlife Service FWS/OBS-84/02.

THIENEMANN, A. 1927. Der Bau des Seebeckens in seiner Bedeutung fur den Ablouf des Lebens im See. *Zool. Bot Ges. Vienna Verhandl.* 77:87–91.

THOMAS, J. W., R. G. ANDERSON, C. MASER, AND E. L. BULL. 1979.

Snags. In J. W. Thomas (ed.)., *Wildlife Habitats in Managed Forests (The Blue Mountains of Oregon and Washington).* USDA Forest Serv. Ag. Handb. No. 553, pp. 60–77.

THOMAS, R. D. K., AND E. C. OLSON (EDS.). 1980. *A Cold Look at the Warm-blooded Dinosaurs,* AAAS Selected Symposia 28. Westview Press, Boulder, CO.

THOREAU, H. D. 1860. Succession of forest trees. Mass. Board Agric. Rept. VIII.

THORNTHWAITE, C. W. 1948. An approach to a rational classification of climate. *Geog. Rev.* 38:55–94.

THORNTHWAITE, C. W. 1931. Climates of North America according to a new classification. *Geog. Rev.* 21:633–655.

THURMAN, N. C., AND J. C. SENCINDIVER. 1986. Properties and classification of mine soils at two sites in West Virginia. *Soil. Sci. Soc. Amer. J.* 36:181–185.

THURSTON, J. M. 1969. The effect of liming and fertilizers on the botanical composition of permanent grassland and on the yield of hay. In I. H. Rorison (ed.), *Ecological Aspects of Mineral Nutrition of Plants.* Blackwell, Oxford, pp. 3–10.

TIEDEMANN, A. R., E. D. MCARTHUR, H. C. STUTZ, R. STEVENS, AND K. L. JOHNSON (COMPILERS). 1984. *Proceedings Symposia on the Biology of* Atriplex *and Related Chenopods.* USDA For. Serv. Gen. Tech. Rep. INT-172.

TIESZEN, L. L. 1978. Photosynthesis in the principal Barrow, Alaska species: A summary of field and laboratory responses. In *Vegetation and Production Ecology of an Alaska Arctic Tundra.* Springer-Verlag, New York, pp. 241–268.

TIESZEN, L. L., P. C. MILLER, AND W. C. OECHEL. 1980. Photosynthesis. In J. Brown, P. C. Miller, L. L. Tieszen, and F. L. Bunnell (eds.), *The Arctic Ecosystem: The Coastal Tundra at Barrow, Alaska,* (US/IBP Synthesis Ser. No. 12). Dowden, Hutchinson & Ross, Stroudsburg, PA, pp. 102–139.

TIESZEN, L. L., M. M. SENYIMBA, S. K. IMBAMBA, AND J. H. TROUGHTON. 1979. The distribution of $C_3$ and $C_4$ grasses and carbon isotope discrimination along an altitudinal and moisture gradient in Kenya. *Oecologica* 37:337–350.

TILMAN, D. 1980. Resources: A graphical-mechanistic approach to competition and predation. *Am. Nat.* 116:362–393.

———. 1982. *Resources: Competition and Community Structure.* Princeton University Press, Princeton, NJ.

———. 1985. The resource ratio hypothesis of succession. *Am. Nat.* 125:827–852.

———. 1986. Evolution and differentiation in terrestrial plant communities: The importance of the soil resource-light gradient. In J. Diamond and T. Case (eds), *Community Ecology.* Harper & Row, New York, pp. 359–380.

———. 1987. The importance of the mechanisms of the interspecific interaction. *Amer. Nat.* 129:769–774.

———. 1988. *Plant Strategies and the Dynamics and Structure of Plant Communities.* Princeton University Press, Princeton, NJ.

TILMAN, D. M. MATTSON, AND S. LANGER. 1981. Competition and nutrient kinetics along a temperature gradient: An experimental test of a mechanistic approach to niche theory. *Limno. Oceanogr.* 26:1020–1033.

TINBERGEN, N. 1951. *A Study of Instinct.* Oxford University Press, Oxford.

———. 1953. *The Herring Gull's World.* London: Collins.

———. 1960. The natural control of insects in pinewoods: 1. Factors influencing the intensity of predation by songbirds. *Arch. Neerl. Zool.* 13:265–343.

TINER, R. W. 1991. The concept of a hydrophyte for wetland identification. *Bioscience* 41:236–237.

TINKLE, D. W. 1969. The concept of reproductive effort and its relation to the evolution of life histories of lizards. *Am. Nat.* 103:501–516.

TINKLE, D. W., AND R. E. BALLINGER. 1972. *Sceloporus undulatus,* a study of the intraspecific comparative demography of a lizard. *Ecology* 53:570–585.

TINKLE, D. W., J. W. CONDON, AND P. C. ROSEN. 1981. Nesting season and success: Implications for the demography of painted turtles. *Ecology* 62:1426–1432.

TINLEY, K. L. 1982. The influence of soil moisture balance on ecosystem patterns in southern Africa. In B. J. Huntley and B. H. Walker (eds.), *Ecology of Tropical Savannas.* Springer-Verlag, New York, pp. 175–192.

TOBIN, M. E., AND M. E. RICHMOND. 1993. *Vole management in fruit orchards.* U.S. Fish and Wildlife Service Biological Report 5. 18 pp.

TOFT, C. A., AND P. J. SHEA. 1983. Detecting community-wide patterns: Estimating power strengthens statistical inference. *Am. Nat.* 122:618–625.

TOMASELLE, R. 1981a. Main physiognomic types and geographic distribution of shrub systems related to Mediterranean climates. In F. di Castri, D. W. Goodall, and R. L. Specht (eds.), *Mediterranean-type Shrublands (Ecosystems of the World,* Vol. 11). Elsevier, Amsterdam, pp. 95–106.

———. 1981b. Relations with other ecosystems: Temperate evergreen forests, coniferous forests, savannas, steppes, and desert shrubland. In F. di Castri, D. W. Goodall, and R. L. Specht (eds.), *Mediterranean-type Shrublands (Ecosystems of the World,* Vol. 11). Elsevier, Amsterdam., pp. 123–136.

TOMBACH, D. F. 1982. Dispersal of whitebark pine seeds by Clark's nutcracker: A mutualism hypothesis. *J. Anim. Ecol.* 51:451–467.

TOMLINSON, P. B. 1983. Structural elements of the rain forest. In F. B. Golley (ed.), *Tropical Rain Forest Ecosystems: Structure and Function (Ecosystems of the World,* Vol. 14A). Elsevier, Amsterdam, pp. 9–28.

TRABAUD, D. L. 1981. Man and fire: Impacts on Mediterranean vegetation. In F. di Castri, D. W. Goodall, and R. L. Specht (eds.), *Mediterranean-type Shrublands (Ecosystems of the World,* Vol. 11). Elsevier, Amsterdam, pp. 523–538.

TRACY, C. R. 1976. A model of the dynamic exchanges of water and energy between a terrestrial amphibian and its environment. *Ecol. Monogr.* 46:293–326.

TRANSEAU, E. N. 1926. The accumulation of energy by plants. *Ohio J. Sci.* 26:1–10.

TRIMBLE, G., JR. 1973. *The regeneration of central Appalachian hardwoods with emphasis on the effects of site quality and harvesting.* USDA For. Serv. Res. Paper NE 282.

TRIMBLE, G., JR., AND E. H. TYRON. 1966. Crown encroachment into openings cut into Appalachian hardwood stands. *J. For.* 64:104–108.

TRISKA, F. J., AND K. CROMACH, JR. 1980. The role of wood debris in forests and streams. In R. H. Waring (ed.), *Forests: Fresh Perspectives from Ecosystem Analysis.* Oregon State University, Corvallis, pp. 171–190.

TRISKA, F. J., J. R. SEDELL, K. CROMACH, JR., S. V. GREGORY, AND F. M. McCOUSON. 1984. Nitrogen budget for a small coniferous forest stream. *Ecol. Monogr.* 54:119–140.

TRISKA, F. J., J. R. SEDELL, S. V. GREGORY. 1982. Coniferous forest streams. In R. L. Edmonds (ed.), *Analysis of Coniferous Forest Ecosystems in Western United States,* US/IBP Synthesis Ser. No. 14. Dowden, Hutchinson & Ross, Stroudsburg, PA, pp. 292–332.

TRIVERS, R. L., AND D. E. WILLARD. 1973. Natural selection of parental ability to vary the sex ratio of offspring. *Science* 179:90–92.

TRUMBLE, J. T., D. M. KOLODNY-HIRSCH, AND I. P. TING. 1989. Plant compensation for arthropod herbivory. *Ann Rev. Entomol.* 38:93–119.

TRYON, E. H., AND G. R. TRIMBLE, JR. 1969. Effect of distance from stand border on height of hardwood reproduction in openings. *WV Acad. Sci. Proc.* 41:125–132.

TULLAR, B. F. 1979. The management of foxes in New York State. *Conservationist* 34(3):33–36.

TUNNICLIFFE, V. 1992. Hydrothermal vent communities of the deep sea. *Amer. Sci.* 80:336–349.

TURKINGTON, R. 1983. Leaf and flower demography of *Trifolium repens* L. growth in a mixture with grasses. *New Phytol.* 93:599–616.

TURNER, G. F., AND T. J. PITCHER. 1986. Attack abatement: A model for group protection by combined avoidance and dilution. *Am. Nat.* 128:228–240.

TYRTIKOV, A. P. 1959. Perennially frozen ground. In *Principles of Geocryology: Part I, General Geocryology* (trans. from Russian by R. E. Brown). *Nat. Res. Cun. Canada Tech. Trans.* 1163 (1964):399–421.

UDVARDY, M. D. F. 1958. Ecological and distributional analysis of North American birds. *Condor* 60:50–66.

UHL, C., AND C. F. JORDAN. 1984. Succession and nutrient dynamics following forest cutting and burning in Amazonia. *Ecology* 65:1467–1492.

UNDERWOOD, A. J. 1986. The analysis of competition by field experiments. In J. Kikkawa and D. J. Anderson (eds.), *Community Ecology: Pattern and Process.* Blackwell, Melbourne, pp. 240–268.

UNDERWOOD, A. J., AND E. J. DENLEY. 1984. Paradigms, explanations, and generalizations in models for the structure of intertidal communities on rocky shores. In D. R. Strong, Jr., D. Simberloff, L. G. Abele, and A. B. Thistle (eds.), *Ecological Communities: Conceptual Issues and the Evidence.* Princeton University Press, Princeton, NJ, pp. 151–180.

UNDERWOOD, A. J., E. J. DENLEY, AND M. J. MORAN. 1983. Experimental analyses of the structure and dynamics of midshore rocky intertidal communities in New South Wales. *Oecologica* 56:202–219.

USDA SOIL CONSERVATION SERVICE SOIL SURVEY STAFF. 1975. *Soil Taxonomy: A Basic System of Soil Classification for Making and Interpreting Soil Surveys.* Agr. Handbook 436. USDA, Washington, DC.

UYENOYAMA, M. K., AND M. W. FELDMAN. 1980. Theories of kin and group selection in large and small populations in fluctuating environments. *Theor. Popul. Biol.* 17:380–414.

VADAS, R. L. 1977. Preferential feeding: An optimization strategy in sea urchins. *Ecol. Monogr.* 47:337–371.

VALIELA, I., AND J. M. TEAL. 1979. The nitrogen budget of a salt marsh ecosystem. *Nature* 20:652–656.

VALIELA, I., J. M. TEAL, AND W. G. DENSER. 1978. The nature of growth forms in salt marsh grass *Spartina alterniflora. Am. Nat.* 112:461–470.

VALLENTYNE, J. R. 1974. *The Algal Bowl-Lakes and Man.* Misc. Spec. Publ. 22. Department of Environment, Ottawa, Canada.

VAN DER HAMMER, T., T. A. WIJMSTRA, AND W. H. ZAGWIGN. 1971. The floral record of late Cenozoic of Europe. In K. K. Turekian (ed.), *The Late Cenozoic Glacial Ages.* Yale University, New Haven, CT, pp. 391–424.

VANDERMEER, J. H. 1980. Indirect mutualism: Variations on a theme by Stephen Levine. *Am. Nat.* 116:441–448.

VANDERMEER, J. H., AND D. H. BOUCHER. 1978. Varieties of mutualistic interaction in population models. *J. Theor. Biol.* 74:594–558.

VAN BALLENBERGHE, V. 1984. Wolf predation on caribou: The Netchina herd case history, *J. Wildl. Manage.* 48:71–720.1

VAN DER VALK, A. G. 1981. Succession in wetlands: A Gleasonian approach. *Ecology* 62:68–696.

VAN DER VALK, A. G., AND C. B. DAVIS. 1978. The role of seed banks in the vegetation dynamics of prairie glacial marshes. *Ecology* 59:322–335.

VAN DOVER, C. L., C. J. BERG, AND R. D. TURNER. 1988. Recruitment of marine invertebrates to hard substrates at deep-sea hydrothermal vents on the East Pacific Rise and Galapagos spreading center. *Deep-Sea Res.* 35:1833–1849.

VAN HOOKE, R. I. 1971. Energy and nutrient dynamics of spider and orthopteran populations in a grassland ecosystem. *Ecol. Monogr.* 41:1–26.

VANNOTE, R. L., G. W. MINSHALL, K. W. CUMMINS, J. R. SCHELL, AND C. E. CUSHING. 1980. The river continuum concept. *Can. J. Fish. Aq. Sci.* 37:130–137.

VASEK, F. C., AND L. J. LUND. 1980. Soil characteristics associated with a primary plant succession on a Mojave Desert dry lake. *Ecology* 61:1013–1018.

VAUGHAN, T. A. 1978. *Mammalogy,* 2nd ed. Saunders, Philadelphia.

VAUGHN, M. R., AND L. B. KEITH. 1981. Demographic response of experimental snowshoe hare populations to overwinter food shortage. *J. Wildl. Manage.* 45:354–380.

VEALE, P. T., AND H. L. WASCHER. 1956. *Henderson County soils.* Illinois Univ. Agr. Expt. Sta. Soil Rept. No. 77.

VERME, L. J., AND D. I. 1984. Physiology and nutrition. In L. Halls (ed.), *White-tailed Deer Ecology and Management.* Stackpole, Harrisburg, pp. 91–118.

VERNER, J., M. L. MORRISON, AND C. J. RALPH. 1986. *Wildlife 2000: Modeling Habitat Relationships of Terrestrial Vertebrates.* University of Wisconsin Press, Madison.

VEZINA, P., AND D. W. K. BOULTER. 1966. The spectral composition of near ultraviolet and visible radiation beneath forest canopies. *Can. J. Bot.* 44:1267–1283.

VITOUSAK, P. M., AND J. M. MELILLO. 1979. Nutrient losses from disturbed forests: Patterns and mechanisms. *Forest Sci.* 25:605–619.

VITOUSAK, P. M., AND R. L. SANFORD, JR. 1986. Nutrient cycling in moist tropical forests. *Ann. Rev. Ecol. Syst.* 17:131–167.

VITT, L. J. 1992. Lizard mimics millipede. *Nat. Geog. Res. and Explor.* 8:76–95.

VOGEL, S. 1969. Flowers offering fatty oil instead of nectar. *XI Proc. Intl. Bot. Congress,* Seattle, WA. p. 229.

VOIGHT, W. 1948. *The Road to Survival.* William Sloan, New York.

VOIGT, G. K. 1971. Mycorrhizae and nutrient mobilization. In E. Hacskaylo (ed.), *Mycorrhizae.* USDA For. Serv. Misc. Pub. No. 1189, pp. 122–131.

VOLTERRA, V. 1926. Variation and fluctuations of the numbers of individuals in animal species living together. Reprinted in R. M. Chapman (1931), *Animal Ecology.* McGraw-Hill, New York, pp. 409–448.

WAIDE, J. B., W. H. CASHEY, R. I. TODD, AND L. R. BORING. 1988. Changes in soil nitrogen pools and transformations following forest clear-cutting. In W. T. Swank and D. A. Crossley (eds.), *Forest Hydrology and Ecology at Coweeta.* Springer-Verlag, New York.

WALBOT, V., AND C. A. CULLIS. 1985. Rapid genomic changes in higher plants. *Ann. Rev. Plant Physiol.* 36:367–396.

WALKER, B. 1989. Diversity and stability in ecosystem conservation. In D. Western and M. C. Pearl (eds.), *Conservation for the Twenty-first Century.* Oxford, New York, pp. 125–130.

WALDMAN, J. M., J. W. MUNGER, D. J. JACOB, R. C. GLAGAN, J. J. MORGAN, AND M. R. HOFFMAN. 1982. Chemical composition of acid fog. *Science* 218:677–680.

WALKER, J. C. 1980. The oxygen cycle. In O. Hutzinger (ed.), *The Natural Environment and the Biogeochemical Cycles.* Springer-Verlag, New York, pp. 87–104.

———. 1984. How life affects the atmosphere. *Bioscience* 43:486–491.

WALKER, J. M., A. GARGER, R. H. J. BERGER, AND H. C. HELLER. 1979. Sleep and aestivation (shallow torpor): Continuous process of energy conservation. *Science* 204:1098–1100.

WALKER, L. R., J. C. ZASADA, AND F. S. CHAPIN, III. 1986. The role of life history processes in primary succession on Alaskan floodplain. *Ecology* 67:1243–1253.

WALKER, P. C., AND R. T. HARTMAN. 1960. Forest sequence of the Hartstown bog area in western Pennsylvania. *Ecology* 41:461–474.

WALKER, R. B. 1954. Ecology of serpentine soils. II. Factors affecting plant growth on serpentine soils. *Ecology.* 35:259–274.

WALKINGSHAW, L. 1983. *Kirtland's Warbler.* Cranbrook Institute of Science, Bloomfield Hills, MI.

WALLACE, B. 1968. *Topics in Population Genetics.* Norton, New York.

WALLACE, J. B., J. R. WEBSTER, AND T. F. COFFNEY. 1982. Stream detritus dynamics regulation by invertebrate consumers. *Oecologica* 53:197–200.

WALLER, D. M. 1982. Jewelweed's sexual skills. *Nat. History* 91(5):32–39.

WALTER, H. 1977. *Ecology of Tropical and Subtropical Vegetation.* Oliver and Boyd, Edinburgh.

———. 1979. *Vegetation of the Earth and Ecological Systems of the Geosphere,* 2nd ed. Springer-Verlag, New York.

WALTERS, C. J. 1986. *Adaptive Management of Natural Resources.* Macmillan, New York.

WARNER, R. E. 1968. The role of introduced diseases in the extinction of the endemic Hawaiian avifauna. *Condor* 70:101–120.

WASER, N. M., AND L. A. REAL. 1979. Effective mutualism between sequentially flowering plant species. *Nature* 281:670–672.

WASER, P. M. 1985. Does competition drive dispersal? *Ecology* 66:1170–1175.

WASSINK, E. C. 1968. Light energy conversion in photosynthesis and growth of plants. In F. E. Eckhardt (ed.), *Functioning of Terrestrial Ecosystems of the Primary Production Level.* UNESCO, Paris, pp. 53–66.

WATERHOUSE, F. L. 1955. Microclimatological profiles in grass cover in relation to biological problems. *Quart. J. Roy. Meteorol. Soc.* 81:63–71.

WATERS, T. F. 1972. The drift of stream insects. *Ann. Rev. Entomol.* 17:253–212.

WATT, A. S. 1947. Pattern and process in the plant community. *J. Ecol.* 35:1–22.

———. 1955. Bracken versus heather: A study in plant sociology. *J. Ecol.* 43:490–406.

WATT, R. E. F. 1968. *Ecology and Resource Management: A Quantitative Approach.* McGraw-Hill, New York.

WAUTERS, L., AND A. A. DOHONDT. 1989. Body weight, longevity, and reproductive success in red squirrels (*Sciurus vulgaris*). *J. Anim. Ecol.* 58:637–651.

WEAVER, J. E. 1954. *North American Prairie.* Johnson, Lincoln, NE.

WEAVER, J. E., AND F. W. ALBERTSON. 1956. *Grasslands of the Great Plains: Their Nature and Use.* Johnson, Lincoln, NE.

WEAVER, M., AND M. KELLMAN. 1981. The effects of forest fragmentation on woodlot tree biotas in Southern Ontario. *J. Biogeogr.* 8:199–210.

WEAVER, P. L. 1989. Forest changes influence hurricanes in Puerto Rico's Liquillo Mountains. *Interciencia* 14:181–191.

WEBBER, P. J., P. C. MILLER, F. S. CHAPIN, III, AND B. H. McCOWN. 1980. The vegetations: Pattern and succession. In J. Brown, P. C. Miller, L. L. Trieszen, F. L. Bunnell (eds.), *An Arctic Ecosystem: The Coastal Tundra at Barrow, Alaska.* Dowden, Hutchinson, Ross, Stroudsburg, PA, pp. 186–218.

WEBER, C. A. 1907. Aufbau und Vegetation der Moore Norddeutschlands. *Bot. Jahrb.* 90:19–34.

WEBSTER, J. 1956–1957. Succession of fungi on decaying cocksfoot culms: Parts 1 and 2. *J. Ecol.* 44:517–544; 45:1–30.

WEBSTER, J. R., D. J. D'ANGELO, AND G. T. PETERS. 1991. Nitrate and phosphate uptake in streams at Coweeta Hydrological Laboratory. *Vehn. Internat. Verein Limnol.* 24:1681–1686.

WEBSTER, J. R., J. B. WAIDE, AND B. C. PATTEN. 1975. Nutrient recycling and the stability of ecosystems. In F. G. Howell, J. B. Gentry, and M. H. Smith (eds.), *Mineral Cycling in Southeastern Ecosystems.* National Technical Information Service, U.S. Dept. Commerce, pp. 1–27.

WECKER, S. C. 1963. The role of early experience in habitat selection by the prairie deer mouse, *Peromyscus maniculatus bairdi. Ecol. Monogr.* 33:307–325.

WEEKS, H. P., JR., AND C. M. KIRKPATRICK. 1976. Adaptations of white-tailed deer to naturally occurring sodium deficiencies. *J. Wildl. Manage.* 40:610–625.

———. 1978. Salt preferences and sodium drive phenology in fox squirrels and woodchuck. *J. Mamm.* 59:531–542.

WEIR, J. S. 1972. Spatial distribution of elephants in an African national park in relation to environmental sodium. *Oikos* 23:1–13.

WEISE, C. M. 1974. Seasonality in birds. In H. Leith (ed.), *Phenology and Seasonality Modeling.* Springer-Verlag, New York, pp. 139–147.

WELLER, D. E. 1987a. A revaluation of the −3/2 power rule of plant self-thinning. *Ecol. Monogr.* 57:23–43.

———. 1987b. Self-thinning exponent correlated with allometric measures of plant geometry. *Ecology* 68:813–821.

———. 1989. The interspecific size-density relationship among crowded plant stands and its implications for the −3/2 power rule of self-thinning. *Am. Nat.* 133:20–41.

WELLER, M. W. 1981. *Freshwater Marshes.* University of Minnesota Press, Minneapolis.

WELLINGTON, G. W. 1982a. An experimental analysis of the effects of light and zooplankton on coral zonation. *Oecologia* 52:311–320.

———. 1982b. Depth zonation of corals in the Gulf of Panama: Control and facilitation by resident reef fishes. *Ecol. Monogr.* 52:223–241.

WELLS, K. B. 1977. The social behavior of anuran amphibians. *Anim. Behav.* 25:666–693.

WENT, F. W., AND N. STARK. 1968. Mycorrhiza. *Bioscience* 18(11):1035–1039.

WERNER, E. E., AND D. J. HALL. 1976. Niche shift in sunfishes: Experimental evidence and significance. *Science* 191:404–406.

———. 1977. Competition and habitat shift in two sunfishes (Centrarchidae). *Ecology* 58:867–876.

———. 1979. Foraging efficiency and habitat switching in sunfishes. *Ecology* 60:256–264.

WEST, N. E. 1979. Formation, distribution, and function of plant litter in desert ecosystems. In D. W. Goodall and R. A. Perry (eds.), *Arid-Land Ecosystems: Structure, Functioning and Management,* Vol. 1. Cambridge University Press, London, pp. 647–659.

———. 1981. Nutrient cycling in desert ecosystems. In D. W. Goodall, R. A. Perry, and K. M. W. Howes (eds.), *Arid-Land Ecosystems: Structure, Functioning and Management,* Vol. 2. Cambridge University Press, Cambridge, pp. 301–324.

WEST, N. E., AND J. J. SKUJINS (EDS.). 1978. *Nitrogen in Desert Ecosystems,* US/IBP Synthesis Series 9. Dowden, Hutchinson & Ross, Stroudsburg, PA.

WEST, N. E., AND P. T. TUELLER. 1972. Special approaches to studies of competition and succession in shrub communities. In C. M. McKell, J. P. Blaisdell, and J. R. Goodin, *Wildland Shrubs; Their Biology and Utilization.* USDA Forest Service Gen. Tech. Rept INT-1, pp. 172–181.

WESTMORELAND, D., AND L. B. BEST. 1987. What limits mourning dove nests to a clutch of two? *Condor* 89:489–493.

WETZEL, R. 1975. *Limnology.* Saunders, Philadelphia.

WETZEL, R. G., AND H. L. ALLEN. 1970. Function and interactions of dissolved organic matter and the littoral zone in lake metabolism and eutrophication. In Z. Kabap and A. Hillbricht-Ilkowska (eds.), *Productivity Problems of Freshwater.* PWN Polish Sci. Publ., Warsaw, pp. 333–347.

WETZEL, R. G., P. H. RICH, M. C. MILLER, AND H. L. ALLEN. 1972. Metabolism of dissolved and particulate detrital carbon in a temperate hardwater lake. *Mem. 1st. Ital. Idrobiol.* 29(Supp.), pp. 185–243.

WHICKER, A. D., AND J. K. DETLING. 1988. Ecological consequences of prairie dog disturbances. *Bioscience* 38:778–785.

WHIGHAM, D. F., J. McCORMICK, R. E. GOOD, AND R. L. SIMPSON. 1978. Biomass and primary production of freshwater tidal marshes. In R. E. Good, D. F. Whigham, and R. L. Simpson (eds.), *Freshwater Wetlands: Ecological Processes and Management Potentials.* Academic, New York, pp. 243–257.

WHITCOMB, R. F., J. F. LYNCH, P. A. OPLER, AND C. S. ROBBINS. 1976. Island biogeography and conservation: Strategy and limitations. *Science* 193:1030–1032.

WHITE, J. 1979. The plant as a metapopulation. *Ann. Rev. Ecol. Syst.* 10:109–145.

WHITFORD, W. G., D. W. FRECHMAN, N. Z. ELKINS, L. W. PARKER, R. PARMALEO, J. PHILLIPS, AND S. TUCKER. 1981. Diurnal migration and response to simulated rainfall in desert soil microarthropods and nematodes. *Soil Biol. and Biogeochem.* 13:17–425.

WHITHAM, T. G. 1980. The theory of habitat selection: Examined and extended using *Pemphigus* aphids. *Am. Nat.* 115:449–466.

WHITMORE, T. C. 1984. *Tropical Rain Forest of the Far East,* 2nd ed. Oxford, London.

WHITTAKER, R. H. 1953. A consideration of the climax theory: The climax as a population and pattern. *Ecol. Monogr.* 23:41–78.

———. 1954. The ecology of serpentine soils. IV. The vegetational response to serpentine soils. *Ecology* 35:275–288.

———. 1956. Vegetation of the Great Smoky Mountains. *Ecol. Monogr.* 26:1–80.

———. 1960. Vegetation of the Siskiyou Mountains, Oregon and California. *Ecol. Monogr.* 30:279–338.

———. 1962. Classification of natural communities. *Bot. Rev.* 28:1–239.

———. 1963. Net production of heath balds and forest heaths in the Great Smoky Mountains. *Ecology* 44:176–182.

———. 1965. Dominance and diversity in land plant communities. *Science* 147:250–260.

———. 1967. Gradient analysis of vegetation. *Biol. Rev.* 42:207–264.

———. 1972. Evolution and the measurement of species diversity. *Taxon.* 21:213–251.

———. 1974. Climax concepts and recognition. In R. Knapp (ed.), *Vegetation Dynamics.* Junk, The Hague, pp. 137–154.

———. 1975. *Communities and Ecosystems,* 2nd ed. Macmillan, New York.

WHITTAKER, R. H., F. H. BORMANN, G. E. LIKENS, AND T. G. SICCAMA. 1974. The Hubbard Brook ecosystem study: Forest biomass and production. *Ecol. Monogr.* 44:233–252.

WHITTAKER, R. H., AND P. R. FEENEY. 1971. Allelochemics: Chemical interactions between species. *Science* 171:757–770.

WHITTAKER, R. H., S. A. LEVIN, AND R. B. ROOT. 1973. Niche, habitat, and ecotope. *Am. Nat.* 107:321–338.

WHITTAKER, R. H., AND G. E. LIKENS. 1973. Carbon in the biota. In G. M. Woodwell and E. V. Pecan (eds.), *Carbon and the Biosphere Conf. 72501.* National Technical Information Service, Springfield, VA, pp. 281–300.

WHITTAKER, R. H., AND P. L. MARKS. 1975. Methods of assessing terrestrial productivity. In H. Leith and R. H. Whittaker (eds.), *Primary Productivity of the Biosphere.* Springer Verlag, New York, pp. 55–118.

WHITTAKER, R. H., AND G. M. WOODWELL. 1968. Dimension and production relations of trees and shrubs in the Brookhaven forest, New York. *Ecology* 56:1–25.

———. 1969. Structure, production, and diversity of the oak-pine forest at Brookhaven, New York. *J. Ecol.* 57:155–174.

WIEBES, J. T. 1979. Coevolution of figs and their insect pollinators. *Ann. Rev. Ecol. Syst.* 10:1–12.

WIEGANDT, E. 1976. Past and present in the Swiss Alps. In J. Luchok, J. D. Cawthon, and M. J. Breslen (eds.), *Hill Lands Proc. Int. Symp.* West Virginia University, Morgantown, pp. 203–208.

WIELAND, N. K., AND F. A. BAZZAZ. 1975. Physiological ecology of three codominant successional annuals. *Ecology* 56:681–688.

WIELGOLASKI, F. E. 1975. Productivity of tundra ecosystems. In National Academy of Science, *Fennoscandian Tundra Ecosystems. Part 1, Plants and Microorganisms.* Springer-Verlag, New York, pp. 1–12.

WIELGOLASKI, F. E., L. C. BLISS, J. SUOBODA, AND G. DOYLE. 1981. Primary production of tundra. In L. C. Bliss, O. W. Heal, and J. J. Moore (eds.), *Tundra Ecosystems: A Comparative Analysis.* Cambridge University Press, Cambridge, pp. 187–225.

WIELGOLASKI, F. E., AND S. KJELVIK. 1975. Energy content and use of solar radiation of Fennoscandian tundra plants. In F. E. Wielgolaski (ed.), *Fennoscandian Tundra Ecosystems. Part 1, Plants and Microorganisms.* Springer-Verlag, New York, pp. 201–207.

WIENS, J. A. 1973. Pattern and process in grassland bird communities. *Ecol. Monogr.* 43:237–270.

———. 1975. Avian communities, energetics and functions in coniferous forest habitats. In *Proc. Symp. Manage. Forest Range Habitats for Nongame Birds.* USDA. Forest Serv. Gen. Tech. Rept. WO-I.

———. 1976. Population responses to patchy environments. *Ann. Rev. Ecol. Syst.* 7:81–120.

———. 1977. On competition and variable environments. *Am. Sci.* 65:590–597.

———. 1985. Habitat selection in variable environments: Shrub-steppe birds. In M. Cody (ed.), *Habitat Selection in Birds.* Academic Press, Orlando, FL, pp. 227–251.

WIENS, J. A., J. F. ADDICOTT, T. J. CASE, AND J. DIAMOND. 1986. Overview of the importance of spatial and temporal scale in ecological investigations. In J. Diamond and T. Case (eds.), *Community Ecology.* Harper & Row, New York, pp. 229–239.

WIENS, J. A., AND J. T. ROTENBERRY. 1981. Habitat associations and community structure of birds in shrub-steppe environments. *Ecol. Monogr.* 51:21–41.

WILBUR, H. M., AND R. A. ALFORD. 1985. Priority effect in experimental pond communities: Response of *Hyla* to *Bufo* and *Rana*. *Ecology* 66:1106–1114.

WILBUR, H. M., AND J. E. FAUTH. 1990. Experimental aquatic food webs: Interactions between two predators and two prey. *Am. Nat.* 135:176–204.

WILCOVE, D. S. 1986. Nest predation in forest tracts and the decline of migratory songbirds. *Ecology* 66:1211–1214.

WILCOX, B. A. 1980. Insular ecology and conservation. In M. E. Soule and B. A. Wilcox (eds.), *Conservation Biology: An Evolutionary Ecological Perspective.* Sinauer Associates, Sunderland, MA, pp. 95–117.

WILLIAMS, C. B. 1964. *Patterns in the Balance of Nature.* Academic Press, New York.

———. 1966. *Adaptation and Natural Selection.* Princeton University Press, Princeton, NJ.

WILLIAMS, C. E. 1965. Soil fertility and cottontail body weights: A reexamination. *J. Wildl. Manage.* 28:329–337.

WILLIAMS, C. E., AND A. I. CASHEY. 1965. Soil fertility and cottontail fecundity in southeastern Missouri. *Am. Midl. Nat.* 74:211–224.

WILLIAMS, G. C. 1966. *Adaptation and Natural Selection.* Princeton University Press, Princeton, NJ.

———. 1975. *Sex and Evolution.* Princeton University Press, Princeton, NJ.

WILLIAMS, K., K. G. SMITH, AND F. M. STEVEN. 1993. Emergence of 13-year periodical cicadas (Cicadidae: Magicicada): Phenology, mortality, and predator satiation. *Ecology* 74:1143–1152.

WILLIAMS, K. S., AND L. E. GILBERT. 1981. Insects as selective agents on plant vegetative morphology: Egg mimicry reduces egg laying by butterflies. *Science* 212:467–469.

WILLIS, A. J. 1963. Braunton burrows: The effects on vegetation of the addition of mineral nutrients to the dune soils. *J. Ecol.* 51:353–374.

WILLIS, E. O. 1963. Is the zone-tailed hawk a mimic of the turkey vulture? *Condor* 65:313–317.

———. 1984. Conservation, subdivision of reserves, and antidismemberment hypothesis. *Oikos* 42:396–398.

WILLOUGHBY, L. C. 1974. Decomposition of litter in freshwater. In C. H. Dickinson and G. J. F. Pugh (eds.), *Biology of Plant Litter Decomposition,* Vol. II. Academic, London, pp. 659–661.

WILSON, D. 1975. A theory of group selection. *Proc. Nat. Acad. Sci.* 72:143–146.

———. 1977. Structured demes and the evolution of group-advantageous traits. *Am. Nat.* 111:157–185.

———. 1979. Structured demes and trait-group variation. *Am. Nat.* 113:606–610.

———. 1980. *The Natural Selection of Populations and Communities.* Benjamin Cummings, Menlo Park, CA.

———. 1983. The group selection controversy: History and current status. *Ann. Rev. Ecol. Syst.* 14:159–187.

WILSON, D. E., AND G. L. GRAHAM (EDS.). 1992. *Pacific Island flying foxes: Proceedings of an international conservation conference.* U.S. Fish and Wildlife Service Biological Report 90(23). 176 pp.

WILSON, E. O. 1971. Competitive and aggressive behavior. In J. Eisenberg and W. Dillim (eds.), *Man and Beast: Comparative Social Behavior.* Smithsonian Institute Press, Washington, DC, pp. 522–533.

———. 1975. *Sociobiology: The New Synthesis.* Harvard University Press, Cambridge, MA.

WILSON, E. O., AND W. H. BOSSERT. 1971. *A Primer of Population Biology.* Sinauer Associates, Sunderland, MA.

WILSON, E. O., AND D. S. SIMBERLOFF. 1969. Experimental zoogeography of islands: Defaunation monitoring techniques. *Ecology* 50:267–278.

WILSON, E. O., AND E. O. WILLIS. 1975. Applied biogeography. In M. L. Cody and J. Diamond (eds.), *Ecology and Evolution of Communities.* Cambridge University Press, Cambridge, England, pp. 522–533.

WING, L. D., AND I. D. BUSS. 1970. Elephants and forests. *Wildlife Monogr.* 19. The Wildlife Society, Washington, DC.

WITHERSPOON, JR., J. P. 1964. Cycling of cesium-134 in white oak trees. *Ecol. Monogr.* 34:403–420.

WITHERSPOON, J. P., S. I. AVERBACH, AND J. S. OLSON. 1962. Cycling of cesium-134 in white oak trees on sites of contrasting soil type and moisture. *Oak Ridge Nat. Lab.* 3328:1–143.

WITKAMP, M., AND D. A. CROSSLEY. 1966. The role of arthropods and microflora on the breakdown of white oak litter. *Pedobiologia* 6:293–303.

WOLF, F. M. 1986. *Meta-analysis: Quantitative methods for research synthesis.* Sage, Beverly Hills, CA.

WOLFERT, R. R., AND M. T. BUR. 1992. *Selection of prey by walleyes in the Ohio waters of the central basin of Lake Erie, 1985–1987.* U.S. Fish and Wildlife Service Resource Publication 182. 14pp.

WOLFF, J. O. 1980. The role of habitat patchiness in the population dynamics of snowshoe hares. *Ecol. Monogr.* 50:111–130.

———. 1988. Maternal investment and sex ratio adjustment in American bison calves. *Behav. Ecol. Sociobiol.* 23:127–133.

WOLIN, M. J. 1979. The rumen fermentation: A model for microbial interactions in anaerobic systems. *Adv. Microbial Ecol.* 3:49–77.

WOODWARD, I. 1987. *Climate and Plant Distribution.* Cambridge University Press, Cambridge, England.

WOODWELL, G. M., AND D. B. BOTKIN. 1970. Metabolism of terrestrial ecosystems by gas exchange techniques: The Brookhaven approach. In D. Reichle (ed.), *Analysis of Temperate Forest Ecosystems. Studies in Ecology* I. New York, pp. 73–85.

WOODWELL, G. M., R. A. HOUGHTON, C. A. S. HALL, D. E. WHITNEY, R. A. MOLL, AND D. W. JUERS. 1979. The Flax Pond ecosystem study: The annual metabolism and nutrient budgets of a salt marsh. In R. L. Jeffries and A. J. Davies (eds.), *Ecological Processes in Coastal Environments.* Blackwell, Oxford, pp. 491–511.

WOODWELL, G. M., R. H. WHITTAKER, W. A. REINERS, G. E. LIKENS, C. C. DELWICHE, AND D. B. BOTKIN. 1978. The biota and the world carbon cycle. *Science* 199:141–145.

WOOLFENDEN, G. E. 1975. Florida scrub jay helpers at the nest. *Auk* 92:1–15.

WOOLFENDEN, G. E., AND J. W. FITZPATRICK. 1984. *The Florida Scrub Jay: Demography of a Cooperatively Breeding Bird.* Princeton University, Princeton, NJ.

WRIGHT, H. A., AND R. W. BAILEY. 1982. *Fire Ecology: United States and Southern Canada.* Wiley, New York.

WRIGHT, W. G. 1988. Sex change in Mollusca. *TREE* 3:137–140.

WUNDERLE, J. M., JR., AND Z. COTTO-NAVARRO. 1988. Constant vs. variable risk-aversion in foraging bananaquits. *Ecology* 69:1434–1438.

YAHNER, R. H. 1984. Effects of habitat patchiness created by a ruffed grouse management plan on breeding bird communities. *Am. Mid. Nat.* 96:179–194.

———. 1986. Structure, seasonal dynamics, and habitat relationships of avian communities in small, even-aged forest stands. *Wilson Bull.* 98:61–82.

———. 1988. Changes in wildlife communities near edges. *Cons. Biol.* 2:333–339.

YARIE, J. 1981. Forest fire cycles and life tables: A case study from interior Alaska. *Can. J. For. Res.* 11:554–562.

YEATON, R. I. 1978. A cyclic relationship between *Larrea tridentata* and *Opuntia lepticaulis* in the northern Chihuahuan desert. *J. Ecol.* 66:651–656.

YODA, K., T. KIRA, H. OGAWA, AND K. HOZUMI. 1963. Self-thinning in overcrowded pure stands under cultivated and natural conditions. *J. Biol. Osaka Univ.* 14:107–129.

YODZIS, P. 1981a. The connectance of real ecosystems. *Nature* 284:544–545.

———. 1981b. The structure of assembled communities. *J. Theor. Biol.* 92:103–117.

———. 1982. The compartmentation of real and assembled communities. *Am. Nat.* 120:551–570.

———. 1983. Community assembly, energy flow, and food web structure. In D. L. DeAngelis, W. M. Post, and G. Sugihara (eds.), *Current Trends in Food Web Theory.* ORNL 5983, Oak Ridge National Laboratory, Oak Ridge, TN, pp. 41–44.

———. 1988. *An Introduction to Theoretical Ecology.* Harper & Row, New York.

YOM-TOV, Y. 1980. Intraspecific nest parasitism in birds. *Biol. Rev. Cont. Phil. Soc.* 55:93–108.

YONGE, C. M. 1963. The biology of coral reefs. *Adv. Mar. Biol.* 1:209–260.

ZACK, R., AND J. B. FALLS. 1976a. Ovenbird hunting behavior in a patchy environment: An experimental study. *Can. J. Zool.* 54:1863–1879.

———. 1976b. Foraging behavior, learning, and exploration by captive ovenbirds. *Can. J. Zool.* 54:1880–1893.

———. 1976c. Do ovenbirds hunt by expectation? *Can. J. Zool.* 54:1894–1903.

ZACK, R., AND J. N. M. SMITH. 1981. Optimal foraging in wild birds. In A. R. Kamil and T. D. Sargent (eds.), *Foraging Behavior: Ecological, Ethological and Psychological Approaches.* Garland STMP Press, New York, pp. 95–109.

ZAHAVI, A. 1975. Mate selection for a handicap. *J. Theor. Biol.* 53:205–214.

———. 1977. The cost of honesty (Further remarks on the handicap principle). *J. Theor. Biol.* 67:603–605.

ZAMMUTO, R. M., AND J. S. MILLAR. 1985a. Environmental predictability, variability, and *Spermophilus columbianus* life history over an environmental gradient. *Ecology* 66:1784–1794.

———. 1985b. A consideration of bet hedging in *Spermophilus columbianus. J. Mamm.* 66:652–660.

ZEDLER, P. H., C. R. GAUTIER, AND G. S. MCMASTER. 1983. Vegetation change in response to extreme events: The effect of a short interval between fires in California chaparral and coastal shrub. *Ecology* 64:809–818.

ZEEVALKING, H. S., AND L. F. M. FRESCO. 1977. Rabbit grazing and diversity in a dune area. *Vegetatio,* 35:193–196.

ZELLER, D. 1961. Certain mulch and soil characteristics of major range sites in western North Dakota as related to range conditions. MA thesis, North Dakota State University, Fargo.

ZIEMAN, J. C. 1982. *The ecology of the seagrasses of South Florida: A community profile.* U.S. Fish and Wildlife Service program FWS/OBS-82/25.

ZIMEN, E. 1978. *The Wolf: A Species in Danger* (1981 translation). Dell, New York.

ZIMMERMAN, J. L. 1971. The territory and its density dependent effect in *Spiza americana.* Auk 88:591–612.

ZOELLICK, B. W., AND N. S. SMITH. 1992. Size and spatial organization of home ranges of kit foxes in Arizona. *J. Mamm.* 73:83–88.

ZUK, M. 1991. Parasites and bright birds: New data and new predictions. In J. E. Loge and M. Zuk (eds.), *Bird-Parasite Interactions.* Oxford University Press, Oxford, pp. 317–327.

## Addendum to Bibliography

GASAWAY, W. C., R. O. STEVENSON, J. L. DAVIS, P. E. K. SHEPHERD, AND O. E. BURRIS. 1983. Interrelationships of wolves, prey, and man in interior Alaska. *Wildl. Monogr.* 84.

GOLTERMAN, H. L., AND F. A. KOUWE. 1980. Chemical budgets and nutrient pathways. In E. D. Cren and R. H. Lowe-McConnell (eds.), *The Functioning of Freshwater Ecosystems* (International Biological Programmes No. 22). Cambridge University Press, Cambridge, England. pp. 88-140.

PETERSON, R. O., AND R. E. PAGE. 1983. Wolf-moose fluctuations at Isle Royale National Park, Michigan, U.S.A. *Acta Zool. Fenn.* 174:251-253.

PETERSON, R. O., R. E. PAGE, AND K. M. DODGE. 1984. Wolves, moose, and the allometry of population cycles. *Science* 244:1350-1352.

# Acknowledgments

## PHOTO ACKNOWLEDGMENTS

Unless otherwise acknowledged, all photographs are the property of Scott, Foresman and Company. Page abbreviations are as follows: (T) top, (C) center, (B) bottom, (L) left, (R) right, (INS) inset.

TITLE PAGE PHOTO: Richard Mariscal/Bruce Coleman Inc.

PART-OPENING PHOTOS: Page 1: © Sharon Gerig/Tom Stack & Associates / Page 3: Gerry Ellis Nature Photography/ Page 149: NASA / Page 223: David Muench Photography / Page 359: Wendy Shattil/Bob Rozinski/Tom Stack & Associates ©1989 / Page 595: David Madison/Bruce Coleman Inc.

Page 3: Michio Hoshino/Minden Pictures / Page 5L: Culver Pictures / Page 5C&R: Courtesy Hunt Inst. for Botanical Documentation/Carnegie Mellon University, Pittsburgh, PA / Page 7: Courtesy Mr. G. P. Darwin/By permission of the Darwin Museum, Down House / Page 11: Erich Hartmann/Magnum Photos / Page 15: Yoav Levy/Phototake / Page 17L: R. L. Smith / Page 17R: U.S.Forestry Service / Page 29: Zig Leszczynski/ANIMALS ANIMALS / Page 35: Jim Brandenburg/Minden Pictures / Page 63: Brian Parker/Tom Stack & Associates / Page 73L: Patti Murray/Earth Scenes / Page 73R: Larry Ulrich/DRK Photo / Page 74T: Walter Chandoha / Page 74B: John Eastcott/YVA Momatiuk/DRK Photo / Page 79: John Gerlach/DRK Photo / Page 99: Scott Blackman/Tom Stack & Associates / Page 104: R.

L. Smith / Page 115: Jim Brandenburg/Minden Pictures / Page 122L: R. F. Head/Earth Scenes / Page 122R: R. L. Smith / Page 123: Mickey Gibson/Earth Scenes / Pages 125, 126: R. L. Smith / Page 129: Rod Planck/Tom Stack & Associates / Page 146: Courtesy of Soil Conservation Service / Page 147: Mike Andrews/Earth Scenes / Page 152: John Eastcott/YVA Momatiuk/Earth Scenes / Page 156ALL: R. L. Smith / Page 160TL: Michael Fogden/Earth Scenes / Page 160TR: Kevin Byron/Bruce Coleman Inc. / Page 160C: Oxford Scientific Films/ANIMALS ANIMALS / Page 160BL: Donald Specker/ANIMALS ANIMALS / Page 160BR: Robert Maier/ANIMALS ANIMALS / Page 164: William E. Ferguson Photography / Page 167: Michio Hoshino/Minden Pictures / Page 171TL: Stephen J. Krasemann/DRK Photo / Page 171TR: R. L. Smith / Page 171B: Joseph R. Pearce/DRK Photo / Page 195: Frans Lanting/Minden Pictures / Page 207L: Breck P. Kent/Earth Scenes / Page 207R: E. R. Degginger/Earth Scenes / Page 211: R. L. Smith / Page 217: Cub Kahn/f/Stop Pictures, Inc. / Page 225: Diana L. Stratton/Tom Stack & Associates / Page 228L: Michael Black/Bruce Coleman Inc. / Page 228R: Breck P. Kent/Earth Scenes / Page 229TL: Tom Bean/DRK Photo / Page 229TR: Kevin Mageel/Tom Stack & Associates / Page 229B: Stephen J. Krasemann/DRK Photo / Page 230L: Walt Anderson / Page 230R: Neil Cooper/Panos Pictures / Page 232: Dave Watts/Tom Stack & Associates / Page 238TL: Jim Holmes/Panos Pictures / Page 238TR: Bill Ruth/Bruce Coleman Inc. / Page 238BL: Dale & Marian Zimmerman/Bruce Coleman Inc. / Page 238BR: Ell A. Mittermeier/Bruce Coleman Inc. / Page 239: Stephen J. Krasemann/DRK Photo / Page 241: R. L. Smith / Page 242: Dan Suzio / Page 243B: Matthew Kneale/Panos Pictures / Page 244:

The Photo Library-Sydney/Michael James / *Page 245:* Michael P. Gradomski/Bruce Coleman Inc. / *Page 248T:* Brian Parker/Tom Stack & Associates / *Page 248B:* Andris Apse/Bruce Coleman Inc. / *Page 249:* Stephen J. Krasemann/DRK Photo / *Page 250:* Des Bartlett/Bruce Coleman Inc. / *Page 251:* John Eastcott/YVA Momatiuk/DRK Photo / *Page 253T:* Dan Suzio / *Page 253:* R. L. Smith / *Page 254TL:* Steve McCutcheon / *Pages 254TR, 254BL, 254BR:* R. L. Smith / *Page 256:* Fred Bruemmer/DRK Photo / *Page 257TL:* Steve McCutcheon / *Page 257TR:* Robin Smith/The Photo Library-Sydney / *Page 257B:* Walt Anderson / *Page 263:* Tom Bean/DRK Photo / *Page 265:* Rich Buzzeli/Tom Stack & Associates / *Page 266TL:* Patti Murray/Earth Scenes / *Page 266TR:* Tom Bean/DRK Photo / *Page 266B:* Larry Ulrich/DRK Photo / *Page 267TL:* Heather Angel/Biophoto Associates / *Page 267TR:* Hans Reinhard/Bruce Coleman Inc. / *Page 267B:* Tom Bean/DRK Photo / *Page 268:* Ken Lewis/Earth Scenes / *Page 274:* Jeremy Hartley/Panos Pictures / *Page 275TL:* Larry Ulrich/DRK Photo / *Page 275TR:* Tom Coker/Yesteryear Photography / *Page 275B:* Jack Wilburn/Earth Scenes / *Page 276L:* David C. Fritts/Earth Scenes / *Page 276R:* Doug Wechsler/Earth Scenes / *Page 277:* R. L. Smith / *Page 281TL:* Dr. Nigel Smith/Earth Scenes / *Page 281TR:* James P. Rowan/DRK Photo / *Page 281BL:* Belinda Wright/DRK Photo / *Page 281BR:* Peter French/Bruce Coleman Inc. / *Page 282:* Michael Fogden/Bruce Coleman Inc. / *Page 283:* N. H.Cheatham/DRK Photo / *Page 285:* Michael Fogden/DRK Photo / *Page 287:* Dr. Nigel Smith/Earth Scenes / *Page 289:* Kirkendall/Spring Photography, Edmonds, WA / *Pages 291T, 291CL:* Bob and Ira Spring / *Page 291CR:* Alex Maclean/Landslides / *Page 291BL:* Landslides / *Page 291BR:* Rich Buzzelli/Tom Stack & Associates / *Pages 295TL, 295 TR:* Gary Braasch / *Page 295BL:* Brian Parker/Tom Stack & Associates / *Page 295BR:* David Wrobel/Biological Photo Service / *Page 299T:* Gary Braasch / *Page 299B:* Brian Parker/Tom Stack & Associates / *Page 302T:* Dale Jorgenson/Tom Stack & Associates / *Page 302B:* TSA/Tom Stack & Associates / *Page 303:* B. "Moose" Peterson/Wildlife Research Photography / *Page 305:* Bob and Ira Spring / *Page 307:* Kirkendall/Spring Photography, Edmonds, WA / *Pages 313, 317TL, 317TR:* Gary Braasch / *Page 317B:* B. "Moose" Peterson/Wildlife Research Photography / *Page 318L:* John Gerlach/DRK Photo / *Page 318R:* Mark Rollo/Photo Researchers / *Page 325:* T.A.Wiewandt/DRK Photo / *Page 331T:* Runk/Schoenberger/Grant Heilman Photography / *Page 331B:* Bill Curtsinger / *Page 337:* Stephen J. Krasemann/AllStock Inc./Tony Stone Images / *Page 340:* Tom Bean/AllStock Inc./Tony Stone Images / *Page 341:* Marty Snyderman / *Page 342:* R. L. Smith / *Page 345T, 345B:* Marty Snyderman / *Pages 350, 351ALL, 352, 353L:* R. L. Smith / *Page 353R:* Bill Curtsinger / *Page 356L:* Gary Braasch / *Page 356R:* Gerry Ellis Nature Photography / *Page 361:* Darrell Gulin/AllStock Inc./Tony Stone Images / *Pages 389, 411:* Frans Lanting/Minden Pictures / *Page 427:* Jim Brandenburg/Minden Pictures / *Page 431:* C. C. Lockwood/DRK Photo / *Page 433:* M. Andersson/VIREO / *Page 435:* Tom Bledsoe/AllStock Inc./Tony Stone Images / *Page 436:* Daniel J. Cox/AllStock Inc./Tony Stone Images / *Page 438:* James Randkler/AllStock Inc./Tony Stone Images / *Page 446:* Doug Wechsler/VIREO / *Page 455:* Art Wolfe/AllStock Inc./Tony Stone Images / *Page 466T:* A. & E. Morris/VIREO / *Page 466B:* Nigel Dennis/Photo Researchers / *Page 479:* Beverly Joubert/Copyright, National Geographic Society / *Page 488B:* Leonard Rue Enterprises / *Page 489:* Ken Brate/Photo Researchers / *Page 501:* Merlin D. Tuttle/AllStock Inc./Tony Stone Images / *Page 521:* Rod Planck / *Page 522:* John Bora/Photo Researchers / *Page 523:* S. J. Krasemann/Peter Arnold, Inc. / *Page 525:* John Serrao/Photo Researchers / *Page 527:* R. L. Smith / *Page 528:* Courtesy D. M. Ball/Auburn University / *Page 530:* D. Cavagnaro/DRK Photo / *Page 535:* Frans Lanting/Minden Pictures / *Page 537T:* Dr. Fred A. Urquhart, Insect Migration Studies, University of Toronto / *Page 537B:* John Cancalosi/Peter Arnold, Inc. / *Page 538:* Gunter Ziesler/Peter Arnold, Inc. / *Page 539:* S. J. Krasemann/Peter Arnold, Inc. / *Page 557:* Runk/Schoenberger/Grant Heilman Photography / *Page 558:* Robert & Linda Mitchell / *Page 559L:* J. D. Eisenbach/Photo Researchers / *Page 559R:* James Webb/Phototake / *Page 561ALL:* R. L. Smith / *Page 566:* Gary R. Zahm/DRK Photo / *Page 568:* S. J. Krasemann/Peter Arnold, Inc. / *Page 574:* Richard R. Hansen/Photo Researchers / *Page 576:* Robert & Linda Mitchell / *Page 581:* Harvey Lloyd/Peter Arnold, Inc. / *Page 584TL:* Gary Braasch / *Page 584TR:* Bruce Iverson / *Page 584B:* Bruce Coleman Inc. / *Page 588TL:* Darrell Gulin/AllStock Inc./Tony Stone Images / *Page 588TR:* Robert & Linda Mitchell / *Page 588B:* Dr. Morley Read/SPL/Photo Researchers / *Page 589:* Kim Taylor/Bruce Coleman Inc. / *Pages 591, 592:* R. L. Smith / *Page 593T:* Gregory K. Scott/Photo Researchers / *Page 593BL:* Dr. Paul A. Zahl/Photo Researchers / *Page 593BR:* Norbert Wu/AllStock Inc./Tony Stone Images / *Page 597:* Gary Braasch / *Page 633:* Jim Brandenburg/Minden Pictures / *Page 634L:* W. H. Culver / *Page 634R:* W. J. Reich/Montana Historical Society, Helena / *Page 639ALL:* Photos Courtesy NIFC/Photography by Tianna Glenn / *Page 640:* Tony Arruza/Bruce Coleman Inc. / *Page 642T:* Walt Anderson / *Page 642B:* R. L. Smith / *Page 643:* Courtesy Yellowstone National Park / *Page 644T:* R. L. Smith / *Page 644B:* Jay Brown/Courtesy Robert Clark Photography / *Page 645:* Stephen J. Krasemann/Photo Researchers / *Page 646:* Peter D. Pickford/DRK Photo / *Page 647L:* Tom Bean/DRK Photo / *Page 647R:* T. A. Wiewandt/DRK Photo / *Page 650:* R. L. Smith / *Page 651:* Academy of Natural Sciences Philadelphia/VIREO / *Page 655:* Alan Pitcairn/Grant Heilman Photography / *Pages 657, 658, 676:* R. L. Smith.

# ILLUSTRATION ACKNOWLEDGMENTS

The author wishes to acknowledge the following for kind permission to adapt, reprint, or redraw from their publications the figures listed below. Full information on source and citation can be found in the captions and bibliography.

*Academic Press, Inc.*
Figures 10.3 © 1973, 15.30 © 1978, 15.31 © 1978, 15.32 © 1978.

*American Association for the Advancement of Science*
Figures 28.7 © 1965, 28.26 © 1965.

*American Institute of Biological Sciences*
Figures 11.10 © 1967, 15.18 © 1974, 16.26 © 1978.

*American Ornithological Union*
Figure 19.1a © 1993

*American Society of Mammalogists*
Figures 6.18 © 1973, 19.12 © 1992.

*Annual Reviews, Inc.*
Figures 4.9 © 1966, 21.9 © 1983.

*Australian Academy of Sciences*
Figure 13.15 © 1985.

*Board of Regents of the Wisconsin System (University of Wisconsin Press)*
Figure 28.20 © 1986.

*Cambridge University Press*
Figures 6.14 © 1979, 13.11 © 1980. 13.23 © 1973, 13.26 © 1981, 21.17 © 1981.

*The Company of Biologists, Cambridge, England*
Figure 6.16 © 1976.

*Crane Russak*
Figure 15.21 © 1971.

*Ecological Society of America*
Figures 6.1 © 1976, 11.12 © 1972, 11.15 © 1961, 16.35 © 1962, 17.8 © 1973, 17.20 © 1992, 17.24 © 1986, 18.9 © 1980, 18.13 © 1981, 19.7 © 1986, 19.13 © 1979, 21.1 © 1962, 23.11 © 1974, 25.14 © 1994, 28.14 © 1962, 29.15 © 1974, 30.6 © 1978, 30.8 © 1975, 30.15 © 1960.

*Elsevier Science Publishers*
Figures 2.8 © 1984, 20.4 © 1987, 25.9 © 1992, 28.21 © 1987, 28.22 © 1990.

*HarperCollins Publishers (Harper & Row)*
Figures 4.1 © 1963, 6.6 © 1989, 6.12 © 1989, 22.18 © 1988, 22.20 © 1988, 22.23 © 1978.

*Hutchinson and Ross*
Figure 14.10 © 1982.

*International Institute of Applied Systems Analysis (IIASA)*
Figure 25.9 © 1982.

*Klewer Academic Publishers*
Figures 28.25 © 1987, 28.26 © 1987.

*Macmillan Publications, Ltd.*
Figures 20.6 © 1982, 21.12 © 1979, 24.5 © 1989, 28.5 © 1987.

*Macmillan Publishing Company*
Figures 4.29 © 1970, 17.26 © 1982, 18.14 ©1982, 28.24 © 1975.

*Martin Nijhoff Publishers, The Hague, The Netherlands*
Figure 19.9 © 1969.

*Oliver and Boyd*
Figures 4.5 © 1971, 4.27 © 1971.

*Oregon State University Press*
Figure 6.3 © 1968.

*Oxford University Press*
Figure 28.15 © 1981.

*Plenum Press*
Figure 30.20 © 1984.

*Prentice Hall*
Figure 21.5 © 1969.

*Regents of the University of California*
Figure 6.7 © 1965.

*Regents of the University of Southern Illinois*
Figure 18.12 © 1984.

*Scientific American*
Figure 4.1 © 1987. All rights reserved.

*Sinauer Associates*
Figure 21.3 © 1971.

*Society of Limnology and Oceanography*
Figure 22.4 © 1981.

*Springer-Verlag, Inc.*
Figures 6.21 © 1976, 7.13 © 1974, 7.15 © 1974, 9.4 © 1980, 13.9 © 1979, 13.10 © 1979, 15.7 © 1974, 28.10 © 1981, 28.16 © 1981, 30.10 © 1979.

*University of Chicago Press*
Figures 15.8 © 1978, 15.19 © 1982, 17.18 © 1979, 17.29 © 1984, 19.10 © 1978, 20.13 © 1984, 20.3 © 1986, 20.14 © 1976, 20.21 © 1977, 25.13 © 1992, 27.14 © 1988, 30.4 © 1987, 30.9 © 1971.

*University of Georgia Press*
Figure 14.25 © 1982.

*University of North Carolina Press*
Figure 4.3 © 1947.

*The Wildlife Society*
Figures 2.6 © 1981, 8.7 © 1985, 28.19 © 1989.

*John Wiley and Sons, Inc.*
Figures 11.6 © 1975, 18.10 © 1981, 18.11 ©1981, 18.14 © 1981, 30.17 © 1971, 30.19 © 1981.

*Yale University Press*
Figure 30.18 © 1974.

*The Zoological Society of London*
Figures 26.4 © 1982, 26.10 © 1982.

# Index

# Ecoregions of the Continents

**Polar Domain**

Icecap Division

Icecap Regime Mountains

Tundra Division

Tundra Regime Mountains

Subarctic Division

Subarctic Regime Mountains

**Humid Temperate Domain**

Warm Continental Division

Warm Continental Regime Mountains

Hot Continental Division

Hot Continental Regime Mountains

Subtropical Division

Subtropical Regime Mountains

Marine Division

Marine Regime Mountains

Prairie Division

Prairie Regime Mountains

Mediterranean Division

Mediterranean Regime Mountains

**Dry Domain**

Tropical / Subtropical Steppe Division

Tropical / Subtropical Steppe Regime Mountains

Tropical / Subtropical Desert Division

Tropical / Subtropical Desert Regime Mountains

Temperate Steppe Division

Temperate Steppe Regime Mountains

Temperate Desert Division

Temperate Desert Regime Mountains

**Humid Tropical Domain**

Savanna Division

Savanna Regime Mountains

Rainforest Division

Rainforest Regime Mountains

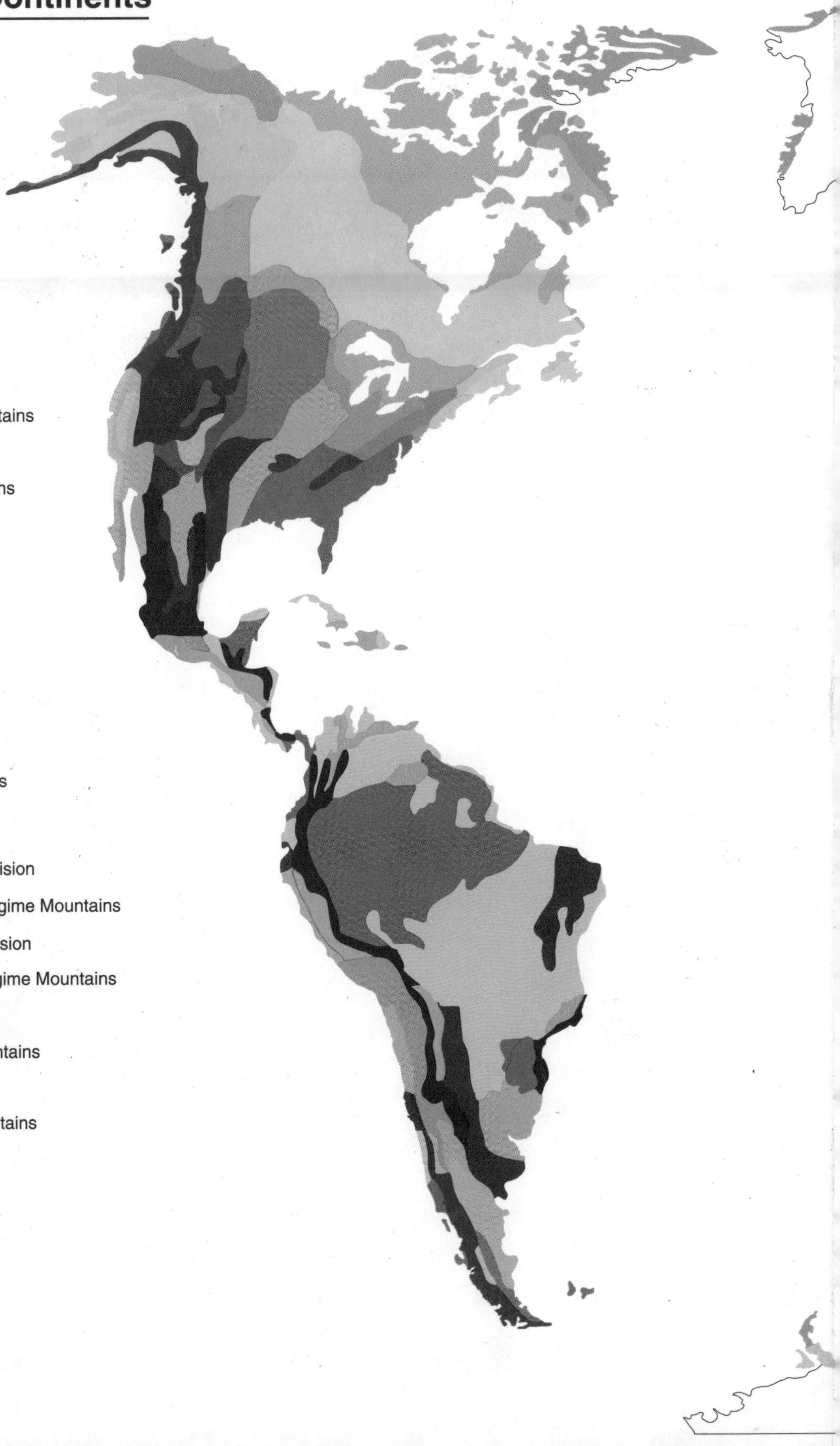

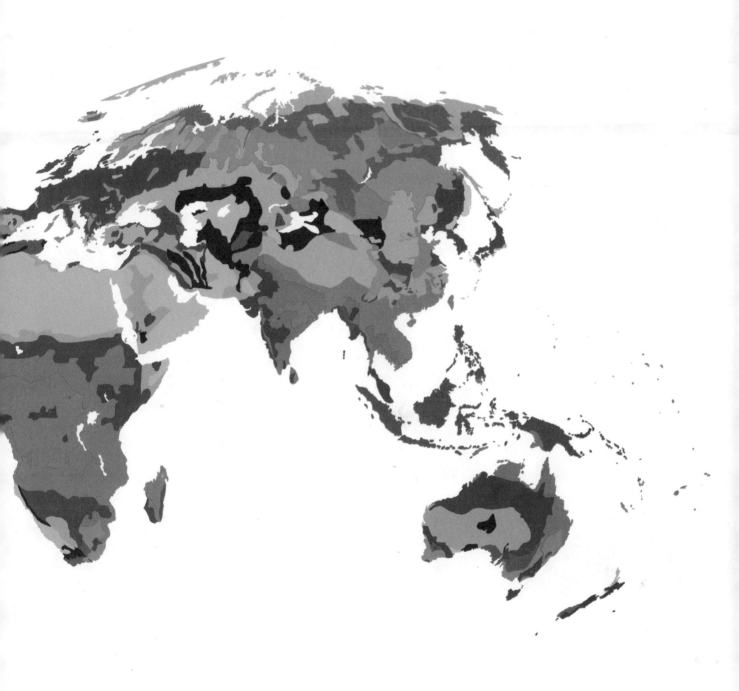